Biosensing and Diagnosis

Biosensing and Diagnosis

Editors

Yan Zhang
Sai Bi
Qin Xu
Yingju Liu

MDPI

Basel • Beijing • Wuhan • Barcelona • Belgrade • Novi Sad • Cluj • Manchester

Editors
Yan Zhang
University of Jinan
Jinan
China

Sai Bi
Qingdao University
Qingdao
China

Qin Xu
Yangzhou University
Yangzhou
China

Yingju Liu
South China Agricultural
University
Guangzhou
China

Editorial Office
MDPI AG
Grosspeteranlage 5
4052 Basel, Switzerland

This is a reprint of articles from the Special Issue published online in the open access journal *Biosensors* (ISSN 2079-6374) (available at: https://www.mdpi.com/journal/biosensors/special_issues/X28X4MH3O0).

For citation purposes, cite each article independently as indicated on the article page online and as indicated below:

Lastname, A.A.; Lastname, B.B. Article Title. *Journal Name* **Year**, *Volume Number*, Page Range.

ISBN 978-3-7258-2601-8 (Hbk)
ISBN 978-3-7258-2602-5 (PDF)
doi.org/10.3390/books978-3-7258-2602-5

Contents

biosensors

MDPI

Editorial

Trends and Perspectives in Biosensing and Diagnosis

Yan Zhang [1,*], Sai Bi [2,*], Qin Xu [3,*] and Yingju Liu [4,*]

[1] School of Chemistry and Chemical Engineering, University of Jinan, Jinan 250022, China
[2] College of Chemistry and Chemical Engineering, Key Laboratory of Shandong Provincial Universities for Functional Molecules and Materials, Qingdao University, Qingdao 266071, China
[3] Institute of Innovation Materials and Energy, School of Chemistry and Chemical Engineering, Yangzhou University, Yangzhou 225002, China
[4] College of Materials and Energy, South China Agricultural University, Guangzhou 510642, China
* Correspondence: chm_zhangyan@hotmail.com (Y.Z.); bisai11@126.com (S.B.); xuqin@yzu.edu.cn (Q.X.); liuyingju@hotmail.com (Y.L.)

Biosensors are attractive tools for detecting molecules and small particles, as they can produce rapid, sensitive, and specific signals [1,2]. Recently, innovative research on nanomaterials, advanced manufacturing technologies, and interdisciplinary collaboration has significantly facilitated the development of simple, cost-effective diagnostic tools [3–5], allowing for immediate analysis and interpretation of complex samples. As a consequence, biosensing devices have been fabricated to provide rich information for applications in home healthcare [6–8], food quality control [9–11], environmental monitoring [12–14], and emergency security [15,16].

Generally, biosensors comprise three key components, including the target recognition element, transducer, and signal processing element. To obtain better output signals, various strategies have been introduced into the field of biosensing to improve the performance of different components [17,18]. For instance, nanozymes with intrinsic enzyme-like properties and better stability under a harsh environment have been widely employed as ideal recognition elements in the biosensing field, effectively overcoming the inactivation of natural enzymes [19–21]. In addition, metal–organic frameworks and their derivatives have been utilized as ideal carriers for the fabrication of biosensing interfaces by dispersing electrochemical or photochemical active substances [22,23]. Moreover, the ingenious and rational integration of signal amplification strategies can remarkably enhance the sensitivity of biosensing systems, allowing for rapid and accurate detection results [24,25]. As for the signal acquisition process, by integrating mobile devices into biosensing platforms [26,27], real-time monitoring of target conditions has been realized, which lays a good foundation for the point-of-care of diagnosis. It is foreseeable that the rapid disease detection based on biosensors can be realized soon, allowing individuals to take control of their own health while providing healthcare professionals with valuable data for decision-making [28,29].

Despite significant advances in biosensing and diagnosis, integrating biosensors into practice still requires robust validation studies to determine their effectiveness and reliability. It is essential to strengthen collaboration between academia, industry, and regulatory bodies to facilitate the translation of laboratory innovations into real-world applications [30–32]. Moreover, the convergence of machine learning, big data analytics, and the Internet of Things is poised to revolutionize the field of biosensing and diagnosis [33,34]. For example, machine learning algorithms can analyze the vast datasets generated by biosensors, identifying patterns and correlations that may not be obvious through traditional statistical approaches [35]. This capability enhances the accuracy of diagnostics, enabling personalized treatment plans tailored to the needs of individual patients. Following the evolving consumer demand and emerging technologies, it is believed that significant advances and major breakthroughs will predictably occur in this attractive field.

Citation: Zhang, Y.; Bi, S.; Xu, Q.; Liu, Y. Trends and Perspectives in Biosensing and Diagnosis. *Biosensors* **2024**, *14*, 499. https://doi.org/10.3390/bios14100499

Received: 4 October 2024
Accepted: 11 October 2024
Published: 12 October 2024

With the purpose of witnessing recent exciting achievements in such an attractive field, this Special Issue reports the innovations in a variety of detection techniques in this area, including electrochemistry [36], photoelectrochemistry [37], fluorescence [38], etc. It consists of 28 research articles, 3 communications, and 4 reviews, covering almost all the applications of biosensing and diagnosis mentioned above. It is believed that this Special Issue will be of great interest to early career researchers who are working in the domains of analytical chemistry, biotechnology, bioelectronics, and nanomedicine, and promote further high-quality research in such emerging fields.

Funding: This work was financially supported by the National Natural Science Foundation of China (22474067, 22474123, 22274059, 22076161), the Special Funds of the Taishan Scholar Program of Shandong Province (tstp20230623, tsqn202103082), the Excellent Youth Innovation Team in Universities of Shandong (2021KJ021), China Postdoctoral Science Foundation (2023M731302), Science and Technology Program of University of Jinan (XKY2203). Yan Zhang was sponsored by the Fund for Overseas Study from the Education Department of Shandong Province.

Acknowledgments: As Guest Editors of the Special Issue "Biosensing and Diagnosis", we would like to express our deep appreciation to all authors whose valuable work was published under this issue, as well as all the reviewers for their constructive comments on the manuscripts, which made this Special Issue possible.

Conflicts of Interest: The authors declare no conflicts of interest.

References

1. Hamidi, S.V.; Jahromi, A.K.; Hosseini, I.I.; Moakhar, R.S.; Collazos, C.; Pan, Q.; Liang, C.; Mahshid, S. Surface-Based Multimeric Aptamer Generation and Bio-Functionalization for Electrochemical Biosensing Applications. *Angew. Chem. Int. Ed.* **2024**, *136*, e202402808. [CrossRef]
2. Kim, K.R.; Yeo, W.-H. Advances in Sensor Developments for Cell Culture Monitoring. *BMEMat* **2023**, *1*, e12047. [CrossRef]
3. Li, L.; Fang, S.; Chen, W.; Li, Y.; Vafadar, M.F.; Wang, D.; Kang, Y.; Liu, X.; Luo, Y.; Liang, K.; et al. Facile Semiconductor p-n Homojunction Nanowires with Strategic p-Type Doping Engineering Combined with Surface Reconstruction for Biosensing Applications. *Nano-Micro Lett.* **2024**, *16*, 192. [CrossRef] [PubMed]
4. Veenuttranon, K.; Kaewpradub, K.; Jeerapan, I. Screen-Printable Functional Nanomaterials for Flexible and Wearable Single-Enzyme-Based Energy-Harvesting and Self-Powered Biosensing Devices. *Nano-Micro Lett.* **2023**, *15*, 85. [CrossRef]
5. Xue, H.; Peng, Y.; Jing, Q.; Zhou, J.; Han, G.; Fu, W. Sensing with Extended Gate Negative Capacitance Ferroelectric Field-Effect Transistors. *Chip* **2024**, *3*, 100074. [CrossRef]
6. Zhu, X.Y.; Li, T.; Hai, X.; Bi, S. A Nanozyme-Based Colorimetric Sensor Array as Electronic Tongue for Thiols Discrimination and Disease Identification. *Biosens. Bioelectron.* **2022**, *213*, 114438. [CrossRef]
7. Ye, Q.; Dai, T.; Shen, J.; Xu, Q.; Hu, X.; Shu, Y. Incorporation of Fluorescent Carbon Quantum Dots into Metal–Organic Frameworks with Peroxidase-Mimicking Activity for High-Performance Ratiometric Fluorescent Biosensing. *J. Anal. Test.* **2023**, *7*, 16–24. [CrossRef]
8. Duan, Y.; Xu, L.; Ma, F. Label-Free Electrochemiluminescence Biosensor for Quantization of Matrix Metalloproteinase-14. *J. Liaocheng Univ. (Nat. Sci. Ed.)* **2023**, *36*, 66–72.
9. Li, P.; Zhang, S.; Xu, C.; Zhang, L.; Liu, Q.; Chu, S.; Li, S.; Mao, G.; Wang, H. Coating Fe_3O_4 Quantum Dots with Sodium Alginate Showing Enhanced Catalysis for Capillary Array-based Rapid Analysis of H_2O_2 in Milk. *Food Chem.* **2022**, *380*, 132188. [CrossRef]
10. Lakshmipathy, K.; Sindhu, S.; Singh, A.; Chikkaballapur Krishnappa, S.; Duggonahally Veeresh, C. A Review on Pesticides Degradation by Using Ultraviolet Light Treatment in Agricultural Commodities. *eFood* **2024**, *5*, e129. [CrossRef]
11. Lv, C.-C.; Zhang, L.; Hou, Y.; Liu, W. Paper-based Chip for Detection of Melamine in Milk. *J. Liaocheng Univ. (Nat. Sci. Ed.)* **2020**, *33*, 45–50.
12. Wang, Y.; Rong, Y.; Ma, T.; Li, L.; Li, X.; Zhu, P.; Zhou, S.; Yu, J.; Zhang, Y. Photoelectrochemical Sensors Based on Paper and Their Emerging Applications in Point-of-Care Testing. *Biosens. Bioelectron.* **2023**, *236*, 115400. [CrossRef]
13. Zhao, C.; Xie, X.; Wan, S.; Jin, C.; Zhao, L.; Yan, B. Covalent Organic Framework Modified Polyaniline Electrospun Nanofiber Membrane for Electrochemical Detecting of Pb^{2+}. *J. Liaocheng Univ. (Nat. Sci. Ed.)* **2023**, *36*, 74–81.
14. Fan, P.-F.; Liu, C.; Li, Q.-J.; Hu, C.-C.; Wu, X.-W.; Zhang, X.-H.; Liang, H.; Yang, S.-Y. Microwave-assisted Rapid Synthesis of Ovalbumin-Stabilized Gold Nanoclusters for Picric Acid Determination. *J. Cent. South Univ.* **2023**, *30*, 74–84. [CrossRef]
15. Wang, S.; Shu, J.; Lyu, A.; Huang, X.; Zeng, W.; Jin, T.; Cui, H. Label-Free Immunoassay for Sensitive and Rapid Detection of the SARS-CoV-2 Antigen Based on Functionalized Magnetic Nanobeads with Chemiluminescence and Immunoactivity. *Anal. Chem.* **2021**, *93*, 14238–14246. [CrossRef]

16. Feng, L.; Zhang, L.; Chen, X.; Zhang, C.; Mao, G.; Wang, H. A Visible Light-driven Photoelectrochemical Sensor for Mercury (II) with "Turn-on" Signal Output through In-Situ Formation of Double Type-II Heterostructure Using CdS Nanowires and ZnS Quantum Dots. *Chem. Eng. J.* **2022**, *441*, 136073. [CrossRef]
17. Liu, Q.; Xu, C.; Chu, S.; Li, S.; Wang, F.; Si, Y.; Mao, G.; Wu, C.; Wang, H. Covalent Organic Framework-Loaded Silver Nanoparticles as Robust Mimetic Oxidase for Highly Sensitive and Selective Colorimetric Detection of Mercury in Blood. *J. Mater. Chem. B* **2022**, *10*, 10075–10082. [CrossRef]
18. Luo, Y.; Ye, Q.; Xie, T.; Xie, J.; Mao, K.; Zou, H.; Li, Y.; Huang, C.; Zhen, S. A Novel Molecular Imprinted Polymers-Based Lateral Flow Strip for Sensitive Detection of Thiodiglycol. *J. Anal. Test.* **2023**, *7*, 110–117. [CrossRef]
19. Chen, M.T.; Qileng, A.; Chen, S.Z.; Huang, H.L.; Xu, Z.L.; Liu, W.P.; Liu, Y.J. Advanced Enzyme Mimicking Engineering: 3D Biomimetic Pt Single-Atom Nanozymes Initiating Pressure-Driven Device. *Adv. Funct. Mater.* **2024**, *34*, 2402552. [CrossRef]
20. Zhang, Y.; Xu, J.; Zhou, S.; Zhu, L.; Lv, X.; Zhang, J.; Zhang, L.; Zhu, P.; Yu, J. DNAzyme-Triggered Visual and Ratiometric Electrochemiluminescence Dual-Readout Assay for Pb(II) Based on an Assembled Paper Device. *Anal. Chem.* **2020**, *92*, 3874–3881. [CrossRef]
21. Hu, L.; Li, J.; Zhai, Y. Fe-N-C Single-Atom Nanozymes for Colorimetric and Electrochemical Detection of H_2O_2. *J. Liaocheng Univ. (Nat. Sci. Ed.)* **2024**, *37*, 62–68.
22. Qi, H.; Wang, Z.; Li, H.; Li, F. Directionally In Situ Self-Assembled Iridium(III)-Polyimine Complex-Encapsulated Metal-Organic Framework Two-Dimensional Nanosheet Electrode To Boost Electrochemiluminescence Sensing. *Anal. Chem.* **2023**, *95*, 12024–12031. [CrossRef] [PubMed]
23. Liu, X.; Gao, X.; Yang, L.; Zhao, Y.; Li, F. Metal-Organic Framework-Functionalized Paper-Based Electrochemical Biosensor for Ultrasensitive Exosome Assay. *Anal. Chem.* **2021**, *93*, 11792–11799. [CrossRef] [PubMed]
24. Klemm, B.; Roshanasan, A.; Piergentili, I.; van Esch, J.H.; Eelkema, R. Naked-Eye Thiol Analyte Detection via Self-Propagating, Amplified Reaction Cycle. *J. Am. Chem. Soc.* **2023**, *145*, 21222–21230. [CrossRef]
25. Guo, M.; Chen, X.; Yuan, Z.; Lu, M.; Xie, X. Upconversion Nanoparticle-Organic Dye Nanocomposites for Chemo- and Biosensing. *J. Anal. Test.* **2023**, *7*, 345–368. [CrossRef]
26. Lin, Q.; Yu, Z.; Lu, L.; Huang, X.; Wei, Q.; Tang, D. Smartphone-Based Photoelectrochemical Immunoassay of Prostate-Specific Antigen Based on Co-doped Bi_2O_2S Nanosheets. *Biosens. Bioelectron.* **2023**, *230*, 115260. [CrossRef]
27. Geng, C.; Huang, Y.; Li, B.; Wang, Y.; Zhu, L.; Xu, Y.; Gao, K.; Mu, Y.; Su, Y.; Deng, S.; et al. Point-of-Care Testing of Chloramphenicol in Food Production Using Smartphone-Based Electrochemical Detector. *J. Anal. Test.* **2023**, *7*, 33–39. [CrossRef]
28. Zhou, Y.; Yan, S.; Dong, W.; Wu, C.; Zhao, Z.; Wang, R.; Duo, Y.; Huang, Y.; Xu, D.; Jiang, C. Biosensing Strategies for Amyloid-Like Protein Aggregates. *BMEMat* **2024**, *2*, e12053. [CrossRef]
29. Rink, S.; Baeumner, A.J. Progression of Paper-Based Point-of-Care Testing toward Being an Indispensable Diagnostic Tool in Future Healthcare. *Anal. Chem.* **2023**, *95*, 1785–1793. [CrossRef]
30. Wang, Q.; Han, N.; Shen, Z.; Li, X.; Chen, Z.; Cao, Y.; Si, W.; Wang, F.; Ni, B.-J.; Thakur, V.K. MXene-Based Electrochemical (Bio) Sensors for Sustainable Applications: Roadmap for Future Advanced Materials. *Nano Mater. Sci.* **2023**, *5*, 39–52. [CrossRef]
31. Tian, Y.; Zhang, L.; Zhang, C.; Bao, B.; Li, Q.; Wang, L.; Song, Z.; Li, D. Deep-Learning Enabled Smart Insole System Aiming for Multifunctional Foot-Healthcare Applications. *Exploration* **2024**, *4*, 20230109. [CrossRef] [PubMed]
32. Zhu, L.; Lv, X.; Yu, H.; Tan, X.; Rong, Y.; Feng, W.; Zhang, L.; Yu, J.; Zhang, Y. Paper-Based Bipolar Electrode Electrochemiluminescence Platform Combined with Pencil-Drawing Trace for the Detection of M.SssI Methyltransferase. *Anal. Chem.* **2022**, *94*, 8327–8334. [CrossRef] [PubMed]
33. Lu, D.; Chen, Y.; Ke, L.; Wu, W.; Yuan, L.; Feng, S.; Huang, Z.; Lu, Y.; Wang, J. Machine Learning-Assisted Global DNA Methylation Fingerprint Analysis for Differentiating Early-Stage Lung Cancer from Benign Lung Diseases. *Biosens. Bioelectron.* **2023**, *235*, 115235. [CrossRef] [PubMed]
34. Sun, X.; Shan, Y.; Jian, M.; Wang, Z. A Multichannel Fluorescence Isothermal Amplification Device with Integrated Internet of Medical Things for Rapid Sensing of Pathogens through Deep Learning. *Anal. Chem.* **2023**, *95*, 15146–15152. [CrossRef] [PubMed]
35. Liu, D.-Y.; Xu, L.-M.; Lin, X.-M.; Wei, X.; Yu, W.-J.; Wang, Y.; Wei, Z.-M. Machine Learning for Semiconductors. *Chip* **2022**, *1*, 100033. [CrossRef]
36. Xu, C.-Y.; Ning, K.-P.; Wang, Z.; Yao, Y.; Xu, Q.; Hu, X.-Y. Flexible Electrochemical Platform Coupled with In Situ Prepared Synthetic Receptors for Sensitive Detection of Bisphenol A. *Biosensors* **2022**, *12*, 1076. [CrossRef]
37. Shi, H.; Che, Y.; Rong, Y.; Wang, J.; Wang, Y.; Yu, J.; Zhang, Y. Visual/Photoelectrochemical Off-On Sensor Based on Cu/Mn Double-Doped CeO_2 and Branched Sheet Embedded Cu_2O/CuO Nanocubes. *Biosensors* **2023**, *13*, 227. [CrossRef]
38. Wang, C.; Zhu, K.; Yu, J.; Shi, P. Complementary DNA Significantly Enhancing Signal Response and Sensitivity of a Molecular Beacon Probe to Aflatoxin B1. *Biosensors* **2023**, *13*, 195. [CrossRef]

biosensors

MDPI

Article

Electrochemical Determination of 4-Bromophenoxyacetic Acid Based on CeO_2/eGr Composite

Haijun Du [1,†], Yan Zhang [1,†], Xin Wang [1,2], Huali Hu [1], Jixing Ai [1], Huanxi Zhou [2,3], Xia Yan [3], Yang Yang [2,*] and Zhiwei Lu [4,*]

1 School of Chemical Engineering, Guizhou Minzu University, Guiyang 550025, China
2 School of Materials Science and Engineering, Guizhou Minzu University, Guiyang 550025, China
3 School of Mechatronics Engineering, Guizhou Minzu University, Guiyang 550025, China
4 College of Science, Sichuan Agricultural University, Ya'an 625014, China
* Correspondence: y.yang@gzmu.edu.cn (Y.Y.); zhiweilu@sicau.edu.cn (Z.L.)
† These authors contributed equally to this work.

Abstract: The determination of plant growth regulators is of great importance for the quality monitoring of crops. In this work, 4-bromophenoxyacetic acid (4-BPA), one of the phenoxyacetic acids, was detected via the electrochemical method for the first time. A CeO_2-decorated electrochemical exfoliated graphene (eGr) composite (CeO_2/eGr) was constructed as the sensor for sensitive detection of 4-BPA due to the synergistic effect of the excellent catalytic active sites of CeO_2 and good electron transference of the eGr. The developed CeO_2/eGr sensor displayed a good linearity in a wide range from 0.3 to 150 μmol/L and the lowest detection limit of 0.06 μmol/L for 4-BPA detection. Electrochemical oxidation of 4-BPA follows a mix-controlled process on the CeO_2/eGr electrode, which involves 2e in the transference process. This developed CeO_2/eGr sensor has excellent repeatability with a relative standard deviation (RSD) of 2.35% in 10 continuous measurements. Moreover, the practical application of the sensor for 4-BPA detection in apple juice has recoveries in the range of 90–108%. This proposed CeO_2/eGr sensor has great potential for detecting plant growth regulators in the agricultural industry.

Keywords: plant growth regulators; 4-bromophenoxyacetic acid; electrochemical determination; lowest detection limit; CeO_2/eGr composite

Citation: Du, H.; Zhang, Y.; Wang, X.; Hu, H.; Ai, J.; Zhou, H.; Yan, X.; Yang, Y.; Lu, Z. Electrochemical Determination of 4-Bromophenoxyacetic Acid Based on CeO_2/eGr Composite. *Biosensors* **2022**, *12*, 760. https://doi.org/10.3390/bios12090760

Received: 30 August 2022
Accepted: 13 September 2022
Published: 15 September 2022

1. Introduction

Plant growth regulators (PGRs) are widely used to promote a crop's productivity and quality [1,2], control crop type [3], resist biotic and abiotic stress [4], regulate differentiation of cells, control weeds [5], and for phytoremediation [6]. 4-bromophenoxyacetic acid (4-BPA) is a PGR that can control weeds, accelerate plant growth, and enhance the fruit setting rate. However, inappropriate usage of 4-BPA will cause malformations and affect the quality of the crop. Furthermore, accumulation in the crop can be a detriment to other plants, animals, and to human health. Therefore, it is of great necessity to develop convenient, sensitive, and reliable analytical methods for 4-BPA determination. Unfortunately, to the authors' knowledge, only Sutcharitchan [7] has developed a liquid chromatography-tandem mass spectrometry (LC-MS) method for 4-BPA determination in Chinese herbs.

4-BPA is a phenoxyacetic acid, and a variety of analytical methods for phenoxyacetic acid detection have been developed, including capillary electrophoresis with laser-induced fluorescence [8], ultra-high liquid chromatography-mass spectrometry [9], headspace gas chromatography high-performance liquid chromatography [10] and electrochemical methods [11–15]. Compared to these methods, electrochemical methods have the advantage of high sensitivity, low cost, portability, and simple operation [16,17]. The direct interaction between the electrode surface and analyte can reduce the lowest limitation and improve the detection range, which has the potential to establish a rapid on-site inspection method. In addition, the advantages of the electrochemical sensors include fast response, on-site deployment, effective detection without sample pretreatment and the modified nanocomposite exhibits high electrocatalytic activity for an analyte [18]. Because of its electrochemical activity, 4-BPA is detectable via electrochemical methods.

Ceria (CeO_2) is an n-type semiconductor that has the potential to be used as a sensing material, since it has excellent redox characteristics and is highly catalytic, biocompatible, and non-toxic [19,20]. Moreover, CeO_2 can selectively bind with organic molecules due to its characteristics of oxygen vacancies, free electrons, and high chemical stability [21]. However, it is hard to directly use as modifier material in electrochemical sensors as CeO_2 suffers from poor conductivity and easy aggregation. Therefore, it is essential to enhance its sensing activities by introducing conductive support.

Graphene (Gr) is one of the best candidates for the construction of electrochemical sensors, as it has unique and extraordinary physical and chemical properties, such as high specific surface area, low charge-transfer resistance, excellent electrochemical activity and high electrical conductivity [22,23], which improve the value of heterogeneous electron transfer (HEF) and the output signal intensity [24,25]. Over the past few decades, different protocols, including top-down approach (e.g., chemical vapor deposition) and bottom-up methods (e.g., electrochemical exfoliation) have been developed [26,27]. Among these methods, electrochemical exfoliation of graphite has the advantages of being low-cost and environmentally friendly, easily made and scalable, and the electrochemical conditions are controllable. Moreover, the exfoliated graphene is a zero-gap semiconductor; it can be doped with p-block elements (N, S, P, B, and metal oxides) and d- block elements (inherent impurities) [26,28] to improve the HEF to promote the interaction between target molecules and the electrode. In the study of Liu [29], gold-palladium nanoparticles were cast on the graphene nano-platelets, and the nanocomposites showed high electrocatalytic ability towards the oxidation of hydrazine. Li [30] synthesized three kinds of CeO_2 nanostructures and then loaded them on the graphene nanoplatelets to detect phenolic pollutants, which exhibited excellent electrochemical activity. Previously, our group has developed an electrochemical exfoliation graphene (eGr) sensor, which displayed the lowest detection limit (LOD) of 0.15 μM for electrochemical determination of Kinetin [31]. Therefore, CeO_2 was decorated on the surface of eGr to form a nanocomposite that will have a synergetic effect between eGr and CeO_2, thus forming a sensitive, selective and promising electrode system for 4-BPA detection. In the electrochemical determination process, several types of electrodes are suitable for electroanalytical applications, such as glassy carbon electrodes (GCE), gold electrodes (GE), carbon paste electrodes (CPE) and pencil electrodes. Among these electrodes, GCE shows attractive electrochemical reactivity, negligible porosity, good mechanical rigidity and good repeatability and reproducibility [32].

Herein, CeO_2 nanocubes were synthesized via the hydrothermal method and electrochemical exfoliated graphene was prepared through the previous method [25]. A CeO_2 decorated eGr composite was constructed and employed to detect 4-BPA for the first time. This developed eGr/CeO_2 sensor has a linear range from 0.3 to 150 μM and the LOD of 0.06 μM for 4-BPA determination, which shows great potential for the detection of plant regulators in the agricultural industry.

2. Materials and Methods

2.1. Reagents and Materials

All reagents used in the experiments are analytical reagent grade and without any treatment. Graphite sheets were purchased from the local electronic market. $Ce(NO_3)_3 \cdot 6H_2O$, 4-BPA, indole 3-acetic acid, naphthalene acetic acid and 6-benzylaminopurine were purchased from Macklin biochemical Technology Co., Ltd. (Shanghai, China). A solution of 0.1 M 4-BPA was prepared by dissolving a suitable amount of 4-BPA in alcohol and diluting the mixture to 10 mL, then the solution was stored in a refrigerator at 4 °C. Phosphate buffer solution (PBS) was used as a supporting electrolyte by combining a stock solution of 0.1 M KH_2PO_4 (Aladdin Reagent Co., Ltd., Shanghai, China) and 0.1 M NaH_2PO_4 (Aladdin

Reagent Co., Ltd., Shanghai, China), then 0.1 M H_3PO_4 (Aladdin Reagent Co., Ltd., Shanghai, China) and 0.1 M NaOH (Sinopharm Group Chemical Reagent Co., Ltd., Shanghai, China) were, respectively, used to adjust the pH to the desired value.

2.2. Preparation of CeO_2, eGr, and eGr/CeO_2 Composites

CeO_2 nanocubes were prepared using the hydrothermal method. Firstly, 0.6948 g $Ce(NO_3)_3 \cdot 6H_2O$ and 0.0224 g hexamethylenetetramine (HMT) was dissolved in 40 mL distilled water and 40 mL ethanol. The resulting solution was vigorously stirred for 20 min at room temperature, then it was transferred into a 100 mL Teflon-lined stainless-steel autoclave, and hydrothermally heated at 180 °C for 20 h. After that, the product was collected by centrifuging, and alternatively washed with distilled water and ethanol to a neutral pH, and then dried in an oven at 80 °C for 12 h. Finally, the obtained yellow powder was calcined at 400 °C for 5 h.

For eGr preparation, a two-electrode cell was used including graphite foil as an anode, a platinum net as a cathode, and 0.1 M $(NH_4)_2SO_4$ as a supporting electrolyte, whereas the eGr was obtained with the aid of SO_4^{2-} intercalation and oxidization to produce sulfur dioxide and oxygen gases, then the product was centrifuged, filtrated and dried overnight.

Next, 9 mg CeO_2 and/or 9 mg eGr were dispersed in 9 mL *N, N'*-dimethylformamide (DMF) solution with vigorous stirring and then ultra-sonicating for 2 h to obtain the CeO_2, eGr, and CeO_2/eGr suspensions. The glassy carbon electrodes (GCEs) were polished to a mirror-like surface with 0.3 and 0.5 μm Al_2O_3 slurries on chamois leather, then alternately rinsing with distilled water/ethanol (1:1, *v*/*v*) solution and double distilled water for 3 min. Finally, CeO_2, eGr, and CeO_2/eGr suspensions were, respectively, drop-coated on the mirror-like surface of GCEs to obtain CeO_2/GCE, eGr/GCE, and CeO_2/eGr/GCE electrodes.

2.3. Characterization

Scanning electron microscope (SEM, Thermo scientific Apreo 2C, Waltham, MA, USA) and transmission electron microscope (TEM, FEI Tecnai F20, Hillsboro, OR, USA) and High-resolution transmission electron microscope (HRTEM, FEI Tecnai F20, Hillsboro, OR, USA) were used to analyze the surface morphologies of CeO_2, eGr and CeO_2/eGr nanocomposite. Raman spectra were collected with a 532 nm diode laser by Thermo Fisher Dxr2xi (Waltham, MA, USA). X-ray diffraction (XRD) patterns were recorded by the PANayltical Empyrean system (Almelo, The Netherlands). X-ray photoelectron spectra (XPS) were recorded by Thermo Scientific K-Alpha (Waltham, MA, USA).

Cyclic voltammetry (CV), linear sweep voltammetry (LSV), and differential pulse voltammetry (DPV) measurements were performed on a CHI 660E electrochemical workstation (CH Instruments ins., Shanghai, China) and a classic three-electrode system. CVs were carried out with the potential range from 0.8 to 1.5 V at a scan rate of 0.1 V s^{-1}. LSVs were carried out with the potential range from 0.9 to 1.6 V at a scan rate of 0.1 V s^{-1}. DPVs were carried out from 0.8 to 1.5 V with parameters of 0.05 V amplitude, 0.06 s pulse width, 0.02 s sampling width, 0.5 s pulse period and 30 s rest time. A glassy carbon electrode (GCE, diameter 3 mm) or modified GCE as the working electrode, a platinum wire served as a counter electrode and Ag/AgCl was used as the reference electrode.

2.4. Sample Pretreatment

The purchased apple was squeezed into juice (120 g) and the fresh apple juice was mixed with distilled water and sonicated for 20 min (at room temperature), then centrifuged for 5 min with 10,000 r/min to obtain the supernatants. Supernatants were collected for further quantification of 4-BPA.

3. Results and Discussion

3.1. Morphology and Phase Structure Characterization

The morphology of the synthesized CeO_2 was observed by SEM, which presents a nano-cubic structure, and agglomerated together due to the nanosize effect (Figure 1a). The diffraction rings in the selected area electron diffraction (SAED) suggest the as-synthesized CeO_2 is polycrystalline and mainly exists in (111), (200), (220), and (311) crystallite planes (Figure 1b). From HRTEM (Figure 1c), there is a main lattice space distance of 0.314 nm which belongs to (111) crystallite plane of CeO_2. For the eGr sample, the TEM image indicates the prepared eGr displays a layered structure (Figure 1d), and the SAED indicates that the eGr mainly presents (002), and (004) crystallite planes (Figure 1e), the existed lattice space distance of 0.34 nm matches well with the theoretical value of graphene (002) crystallite plane (Figure 1f); it verifies the high quality of the prepared eGr. The CeO_2/eGr nanocomposite used here was 1:1 in a weight ratio (1:1 wt.). Figure 1g shows that CeO_2 is uniformly loaded on the surface of eGr and forms on the selected area (Figure 1h); the corresponding elemental mapping illustrates that the elements of C, O and Ce exist in the CeO_2/eGr nanocomposite (Figure 1i–k).

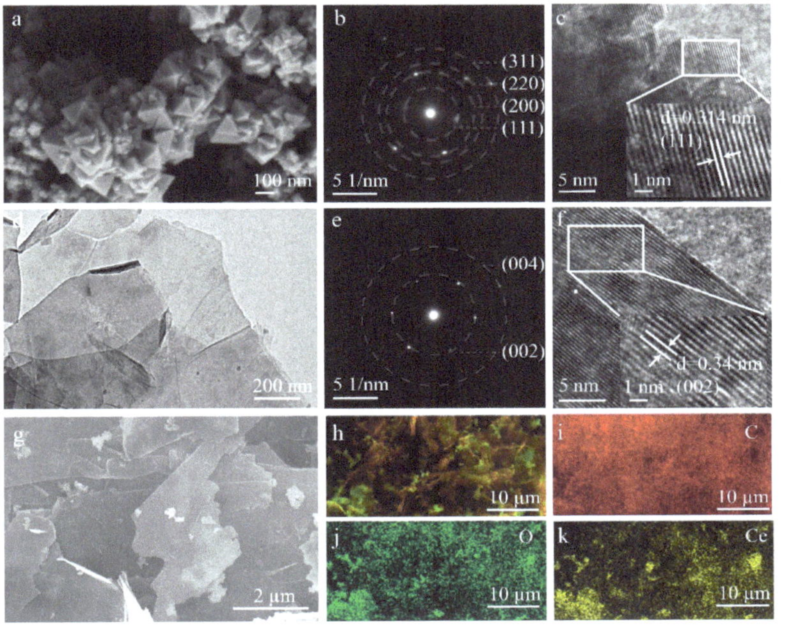

Figure 1. SEM image (**a**), SEAD (**b**) and HRTEM (**c**) of CeO_2, TEM (**d**), SEAD (**e**) and HRTEM (**f**) of eGr, SEM image (**g**) and corresponding elemental mapping of the selected area (**h**), C (**i**), O (**j**) and Ce (**k**) of CeO_2/eGr.

The crystallite structure and composition of CeO_2, eGr and CeO_2/eGr composite (1:1 wt.) were evaluated by Raman spectra, XRD and XPS. In Raman spectra of Figure 2a, it shows a characteristic peak located at 461 cm^{-1} for the CeO_2 sample, which stems from the symmetrical stretching of Ce-O vibrational and originates from the F2g vibrational mode of the CeO_2 phase [33]. For eGr, the peaks at 1356, 1580 and 2710 cm^{-1} were assigned to D, G and 2G bands of graphene, respectively [34]. The D band at ~1356 cm^{-1} is derived from the defects and structural disorder in the sp^2-carbon nanomaterials. The G band at 1580 cm^{-1} is related to the in-plane vibrations of the 2D hexagonal graphene lattice. The CeO_2/eGr composite sample possesses both Raman characteristics of CeO_2 and eGr. XRD was used to analyze the structure of the prepared materials. In Figure 2b, the strong and sharp diffraction peaks indicate all the samples are in good crystallinity. For eGr, two diffraction peaks at 26.4°and 54.5° are observed, which are related to (002) and (004) planes of graphene, this is in accordance with the SEAD result (Figure 1e). For CeO_2, the diffraction peaks located at 28.5°, 33.1°, 47.5°, 56.3°, 59.1°, 69.4°, 76.7°, 79.1°and 88.4° correspond to

(111), (200), (220), (311), (222), (400), (331), (420) and (422) planes (JCPDS 81-0792). The CeO_2/eGr composite also contains all the characteristic peaks of CeO_2 and eGr. Based on the CeO_2/eGr composite containing all the features of CeO_2 and eGr, the chemical composition of the CeO_2/eGr composite was further characterized by XPS. The XPS survey spectrum (Figure 2c) reveals the existence of Ce, O, and C elements in the CeO_2/eGr composite. The Ce3d electron core line was analyzed and is depicted in Figure 2d; it can be deconvoluted into eight peaks and labeled as v_0, v_1, v_2, v_3 ($3d_{3/2}$ region), and u_0, u_1, u_2, u_3 ($3d_{5/2}$ region). Peaks v_0, v_2, v_3 and u_0, u_2, u_3 are characteristics of Ce(IV) 3D final states, while, v_1 and u_1 are Ce(III) 3D final states [35]. Therefore, the as-prepared CeO_2 contains part of Ce(III), and the percentage of Ce(III) was calculated by Equation (1), which is based on the fitted areas of the corresponding peaks of Ce(III) and Ce(IV) [36].

$$\left(Ce^{3+}\right)_{surf} = \frac{Ce(III)}{Ce(III) + Ce(IV)} \quad (1)$$

The calculated percentage of Ce(III) is ~20%, which is similar to the previously reported CeO_2 nanomaterials [36]. The presence of Ce(III) indicates the formation of oxygen vacancies, which can provide catalytically active sites for the sensor. The O1s spectrum can be separated into three peaks as illustrated in Figure 2e, the peak located at ~529.9 eV corresponds to the crystal lattice oxygen in CeO_2. The peak located at 532.2 eV and 533.4 eV could be, respectively, related to the oxygen vacancies and the adsorbed oxygen on the composite [37,38]. The C1s spectrum can be separated into three peaks (Figure 2f). The peak placed at 284.6 eV corresponds to the sp2 carbon atoms or can be attributed to C=C [39]. The other small peaks at 286.1 and 287.9 eV correspond to C–O and C=O on the surface of the composite, respectively.

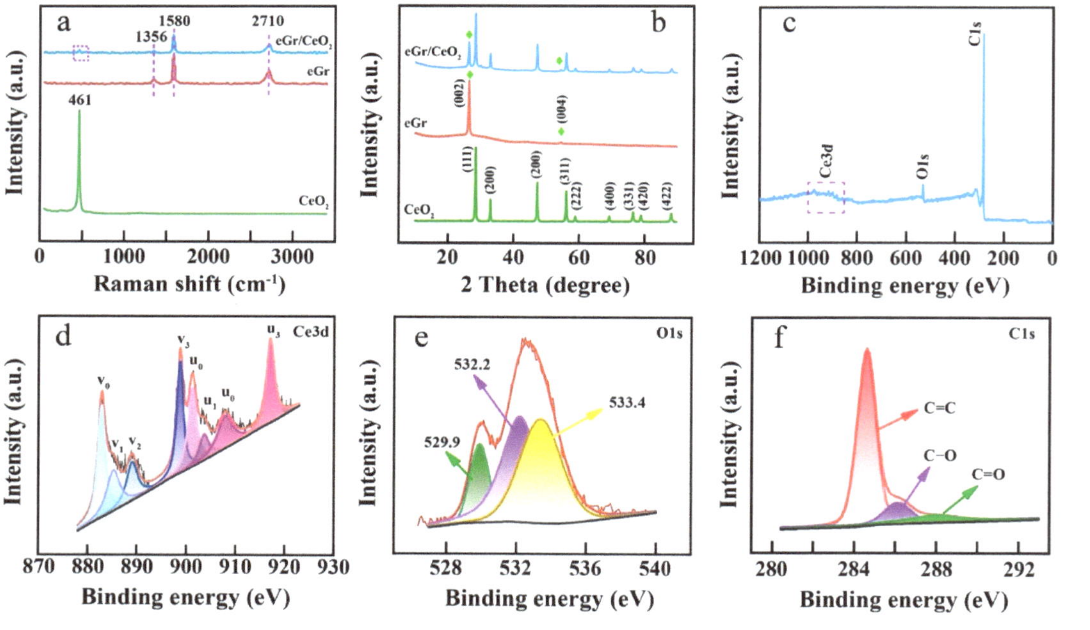

Figure 2. (**a**) Raman spectra, (**b**) XRD patterns of eGr, CeO_2, and CeO_2/eGr, (**c**) XPS survey spectrum, (**d**) Ce3d, (**e**) O1s and (**f**) C1s spectra of CeO_2/eGr composite, respectively.

3.2. The Electrochemical Characteristic of the Prepared Electrode

The prepared eGr, CeO_2, and CeO_2/eGr composites (1:1 wt.) were, respectively, cast on a glassy carbon electrode (eGr/GCE, CeO_2/GCE and CeO_2/eGr/GCE), and their electrochemical performances were firstly estimated by Cyclic voltammetry (CV) at 50 mV s^{-1} in the solution of 5 mM $[Fe(CN)_6]^{3-/4-}$ and 0.1 M KCl. The bare GCE electrode was conducted as the control sample. From Figure 3a, all electrodes show different levels of electrochemical activity, after evaluating the redox peak current densities and CV curve area, the electrochemically active follows the order of CeO_2/eGr/GCE > eGr/GCE > CeO_2/GCE > GCE. This suggests that the CeO_2/eGr/GCE has the largest specific surface area, and the best electrochemically active and kinetic, which could arise from the synergistic effects of excellent catalytic active sites of CeO_2 and good electron transference of eGr. In addition, the standard heterogeneous rate constant (k^0) for bare GCE, CeO_2/eGr/GCE, eGr/GCE, CeO_2/GCE were calculated by Nicholson's equation [40] and the values are, respectively, 0.0041 cm·s^{-1}, 0.0077 cm·s^{-1}, 0.0045 cm·s^{-1}, 0.0049 cm·s^{-1}. The CeO_2/eGr/GCE has the highest value of 0.0077 cm s^{-1} that verifies CeO_2/eGr composite provides the best conditions for electron transfer.

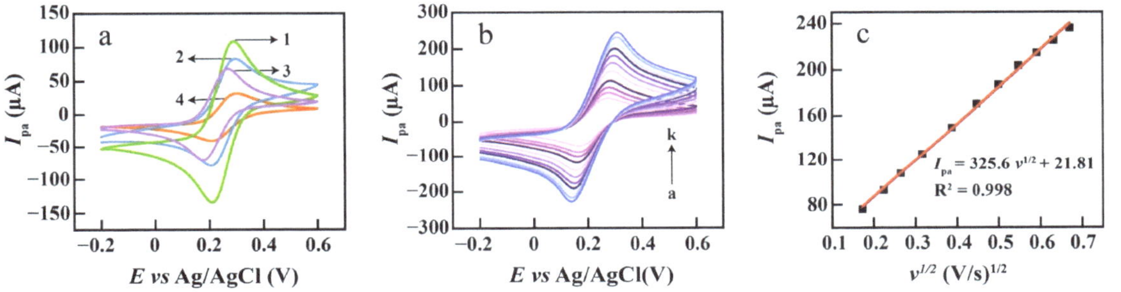

Figure 3. (**a**) CVs of CeO_2/eGr/GCE in the presence of 5 mM $[Fe(CN)6]^{3-/4-}$ solution in aqueous 0.1 M KCl. CeO_2/eGr/GCE (1), eGr/GCE (2), CeO_2/GCE (3), bare GCE (4); (**b**) CVs of CeO_2/eGr/GCE in the presence of 5 mM $[Fe(CN)6]^{3-/4-}$ solution in aqueous 0.1 M KCl at various scan rate (from a to k): 0.03, 0.05, 0.07, 0.1, 0.15, 0.2, 0.25, 0.3, 0.35, 0.4, 0.45 V s^{-1}. (**c**) The plot of peak currents vs. $v^{1/2}$.

The electroactive surface area is a critical factor for the electrochemical sensor, which was estimated in $[Fe(CN)_6]^{3-/4-}$ solution with scan rates ranging from 0.03 to 0.45 V s^{-1} via the Randles-Sevcik equation (Equation (2)) [41].

$$I_{pa} = (2.69 \times 10^5) n^{3/2} A D^{1/2} C v^{1/2} \quad (2)$$

where n refers to electron transfer number, A is the active surface area, C is the concertation of $[Fe(CN)_6]^{3-/4-}$, v is the scan rate and D is the diffusion coefficient. Here, n = 1, D = 6.6 × 10^{-6} cm^2 s^{-1} [34] for 5 mM $K_3[Fe(CN)_6]$ solution containing 0.1 M KCl. For CeO_2/eGr/GCE (Figure 3b), the active surface area of CeO_2/eGr/GCE was calculated to be 0.097 cm^2 (Figure 3c), which is higher than that of the electroactive surface areas of GCE (0.04 cm^2), eGr/GCE (0.08 cm^2), and CeO_2/GCE (0.045 cm^2), which is displayed in the Supporting Information, Figure S1. This result agrees well with the electrochemical activity order of the prepared sensors. The roughness factor (f_r) of the electrochemical sensors was calculated to evaluate the actual active surface area by comparing the oxidation peak current (I_{pa}) of the prepared sensor to bare GCE for $[Fe(CN)_6]^{3-/4-}$ reaction [42]. Peaks ratio is equal to areas ratio according to the proposed Equation (3) [42].

$$f_r = \frac{I_{P2}}{I_{P1}} = \frac{A_2}{A_1} \quad (3)$$

The f_r determined by electrochemical methods depends not only on the size of the electrode (the actual surface), but also on the number of redox centers that can be reached on the surface. Therefore, the f_r was calculated to be 2.425, 2, and 1.625 for CeO_2/eGr/GCE, eGr/GCE, and CeO_2/GCE, respectively.

3.3. The Electrochemical Performance of the Prepared Electrode for 4-BPA Detection

The GCE, eGr/GCE, CeO_2/GCE and CeO_2/eGr/GCE (1:1 wt.) for 4-BPA detection were characterized by CV in the electrolyte with and without 50 μmol L^{-1} 4-BPA in 0.1 mol L^{-1} phosphate buffer (pH = 3). As displayed in Figure 4a, when the presence of 50 μmol L^{-1} 4-BPA, all electrodes present one oxidation peak, which indicates the 4-BPA is electrochemically detectable and the reaction of 4-BPA is irreversible. The CeO_2/eGr/GCE (1:1 wt.) shows the highest oxidation peak current (I_{pa}) and the lowest onset potential; this verified that the CeO_2/eGr composite has the best sensitivity for electrochemical detection of 4-BPA, which should be attributed to the synergetic effect of the catalytic properties of CeO_2 and the fast electron transference of eGr. The ratios between CeO_2 and eGr have further been measured and shown in Figure 4b. With the CeO_2:eGr ratio increasing from 0:4 to 1:1, the oxidation peak current of 4-BPA increases and reaches the maximum at the ratio of 1:1, then the peak current drops with the further increase in the CeO_2 content. The reason could be that CeO_2 is a semiconductor, and it provides electrocatalytic activity sites. When the CeO_2 content is too low, it will not create enough activity sites. While the content is higher than 1:1, the conductivity and electron transference of the electrode will decrease. Therefore, the optimum ratio was 1:1 for 4-BPA detection and selected in the following study.

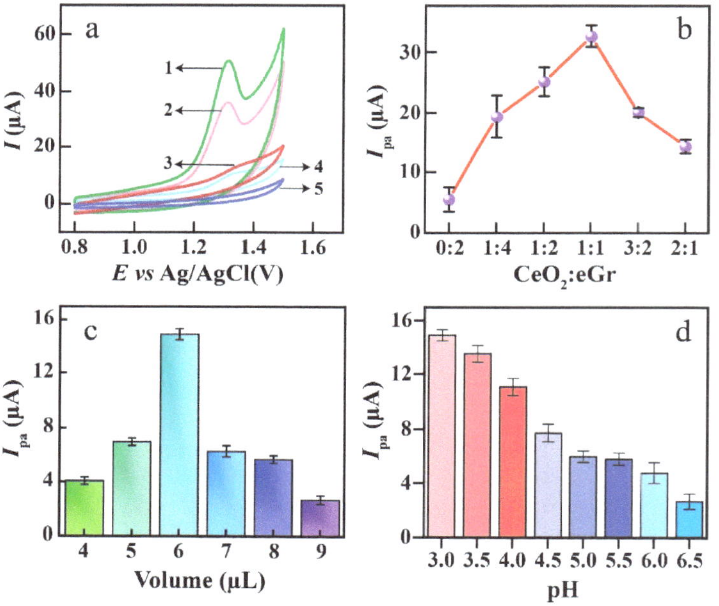

Figure 4. (**a**) CVs of CeO_2/eGr/GCE (1), eGr/GCE (2), bare GCE (3) and CeO_2/GCE (4) in 0.1 M PBS solution containing 50 μmol L^{-1} 4-BPA at the scan rate of 100 mV s^{-1}, CeO_2/eGr/GCE (5) in PBS solution without 4-BPA; (**b**) Influence of the ratio between CeO_2 and eGr on the response of 4-BPA based on LSVs. (**c**) the response of 10 μmol L^{-1} 4-BPA to different volumes of composite nanomaterial in PBS solution (pH = 3). (**d**) Influence of pH on the peak current of 4-BPA in 0.1 M PBS solution.

The different loading amounts of CeO_2/eGr composite on GCE were measured with 10 μmol L^{-1} 4-BPA (Supporting information, Figure S2). It was found that the oxidation peak current of 4-BPA increases with the loading volume, increasing to 6 μL, while the response decreased when the loading amount further increased (Figure 4c). This could be due to the lack of cover on the electrode surface when the loading composites were below

6 μL, while too much loading amount would have hindered the activity sites that cause the decrease in the response peak current [43].

The PBS, Britton–Robison (B-R), and acetic acid-sodium acetate buffer solutions were evaluated as supporting electrolytes for 4-BPA detection (Supporting information, Figure S3). Among these supporting electrolytes, the PBS buffer solution shows more sensitivity for 4-BPA detection. Therefore, PBS was selected as a supporting electrolyte. Moreover, pH is another key impact factor for electrochemical analysis. The pH values ranging from 3 to 6.5 were evaluated in 0.1 M PBS buffer solution (Supporting information, Figure S4). It can be seen that the optimum response pH for 4-BPA was 3, and the response gradually decreased as the pH increased (Figure 4d). This phenomenon could be due to the conductivity loss and the presence of carboxyl groups with the increase in pH [44].

The oxidation process of 4-BPA on CeO_2/eGr/GCE was further studied by linear sweep voltammogram (LSV); different scan rates (50 mV s^{-1} to 450 mV s^{-1}) were conducted and 20 μmol L^{-1} 4-BPA was used. As exhibited in Figure 5a, the I_{pa} increased when the scan rates increased. Moreover, the oxidation peaks positively shifted. More importantly, the oxidation peak current increased linearly with the square root of scan rates (Figure 5b), the linear regression is $I_{pa} = 3.936v^{1/2} - 9.318$, $R^2 = 0.999$ and the linear relationship for $\ln(I)$ versus $\ln(v)$ was established and the slope was found to be 0.82 (Supporting information, Figure S5), which indicates that the electrochemical oxidation of 4-BPA on CeO_2/eGr/GCE was controlled by a mixed process [45]. The relationship between E_{pa} and $\ln v$ is presented by Laviron's theory [46]:

$$E_{pa} = E^0 + \left(\frac{RT}{\alpha nF}\right) \ln\left(\frac{RTk^0}{\alpha nF}\right) + \left(\frac{RT}{\alpha nF}\right) \ln v \tag{4}$$

where α is the charge transfer coefficient, E^0 is the apparent potential, n is the number of the electron, v is the scan rate, the values of R, T and F are 8.314 J K^{-1} mol^{-1}, 298 K and 96485 C mol^{-1}, respectively. Therefore, the number of electrons can be calculated via the linear equations of $E_{pa} - \ln v$ (Supporting information, Figure S6). Generally, for an irreversible electrode process, the value of α is assumed to be 0.5. Hence, the value of n is calculated to be 2. Therefore, the electrocatalytic oxidation mechanism of 4-BPA is proposed in Figure 5c, where 4-BPA will firstly be degraded to 4-bromophenol. Then, the 4-bromophenol will be electrochemically oxidized to enzoquinone [47]. The whole process has two electrons involved, which is in accordance with the calculated results from Laviron's theory.

DPV shows the sensitive response to low concentrations as compared to LSV. Therefore, DPV was used to detect 4-BPA in PBS solution with different concentrations. As illustrated in Figure 5d, the peak current increases linearly with the concentrations of 4-BPA varying from 0.3 to 150 μM. However, there are two linear relationships obtained. From Figure 5e, in the range of 0.3 to 20 μM, the linear regression equation is $I_{pa} = 0.75c + 0.08$, ($R^2 = 0.991$), and from 20 to 150 μM, the linear relationship is $I_{pa} = 0.199c + 11.24$, ($R^2 = 0.993$). Moreover, the lowest detection limit (LOD) was calculated to be 0.06 μmol L^{-1} according to the following equation of 3 s/m, where m is the slope of the regression equation and s is the standard deviation of the response.

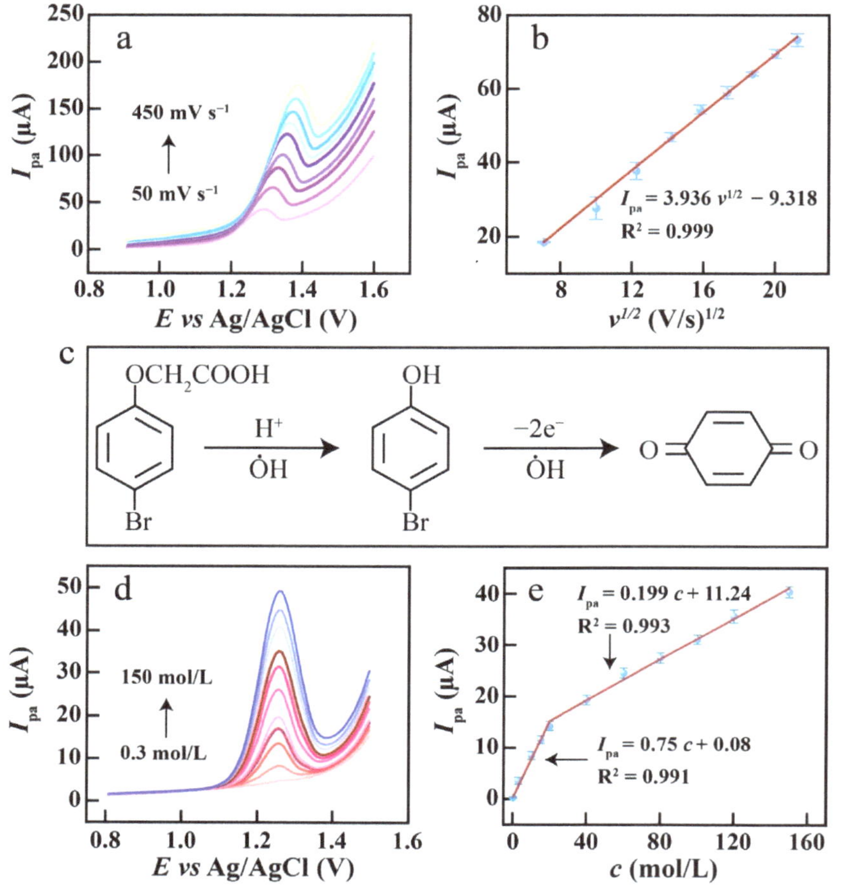

Figure 5. (**a**) LSVs of CeO_2/eGr/GCE in 0.1 M PBS solution containing 20 μmol L^{-1} 4-BPA at 50, 100, 150, 200, 250, 300, 350, 400, 450 mV s^{-1}, (**b**) the plot of I_{pa} vs. $v^{1/2}$, (**c**) the proposed electrochemical reaction process of 4-BPA, (**d**) DPVs of CeO_2/eGr/GCE in 0.1 M PBS solution containing 0.3, 3, 10, 15, 20, 40, 60, 80, 100, 120, 150 μmol L^{-1} 4-BPA, (**e**) the plot of the oxidation peak current vs. the concentration of 4-BPA.

3.4. The Repeatability and the Anti-Interference Ability of the CeO_2/eGr/GCE

The repeatability of the CeO_2/eGr/GCE was carried out with 10 μmol L^{-1} 4-BPA by means of LSV (Supporting information, Figure S7). After 10 continuous measurements (Figure 6a), the relative standard deviation (RSD) of the oxidation peak currents was found to be 2.35% for 4-BPA. After storing the electrode at 4 °C for 15 days, the electroactive oxidation currents of 4-BPA reduced 3.21% compared to the original value. These results indicate that the proposed sensor has good stability and repeatability.

To estimate the anti-interference ability of the CeO_2/eGr/GCE, some regular interfering species were tested. From Figure 6b, no considerable interferences were observed in the presence of fifty-fold excess K^+, Na^+, Mg^{2+}, glucose, sucrose, and ten-fold rutin, quercetin, fenitrothion, imidacloprid, clothianidin, IAA and SA (peak current change < 6%).

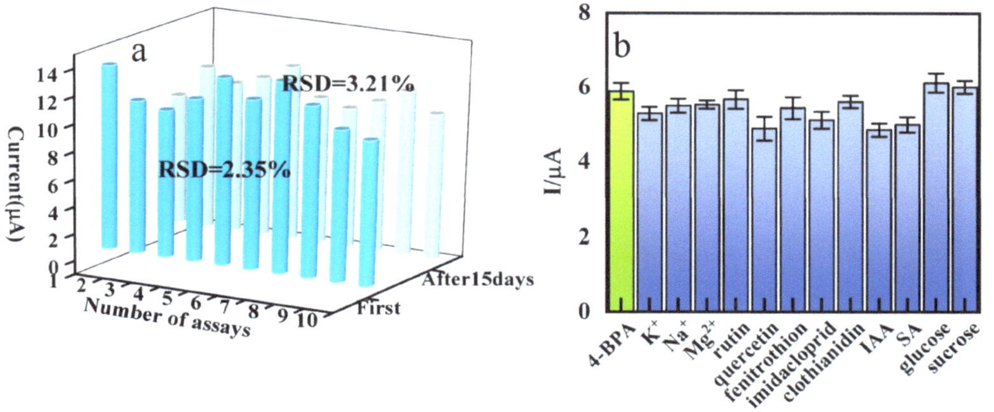

Figure 6. (**a**) The repeatability of current response for the CeO_2/eGr/GCE in 0.1 M PBS (pH = 3) with 10 successive assays at first day and after 15 days. (**b**) The influence of other compounds on the detection of 4-BPA on CeO_2/eGr/GCE.

3.5. The Practical Application of the CeO_2/eGr/GCE

To evaluate the practicability of CeO_2/eGr/GCE, the sensor was used to detect 4-BPA in real apple samples; the analytical results are listed in Table 1. No response of 4-BPA was found in the apple sample, and the recoveries were evaluated by the standard addition method. The recoveries are in the range of 90–108%. This proposed CeO_2/eGr sensor shows great potential for the detection of plant growth regulators in the agricultural industry.

Table 1. Results of the recovery analysis of 4-BPA in apple sample (n = 3).

Sample	Added Value (μM)	Determined Value (μM)	Recovery (%)
1	0	Not detected (a, b)	-
2	0.5	0.45 ± 0.01	90%
3	1	1.08 ± 0.04	108%
4	3	2.78 ± 0.03	93%
5	5	4.86 ± 0.04	97%

a The 4-BPA level determined by the proposed CeO_2/eGr/GCE. b The 4-BPA level determined by the HPLC system.

4. Conclusions

In this work, we used an eco-friendly method of electrochemical exfoliation to prepare eGr and the hydrothermal method to prepare CeO_2. Then, we constructed a selective, sensitive electrochemical method based on a eGr/CeO_2 composite and modified GCE to electrochemically detect 4-BPA. The prepared CeO_2/eGr sensor exhibited excellent electrocatalytic activity due to the synergistic effect of the excellent catalytic active sites of CeO_2 and good electron transference of the eGr. The developed CeO_2/eGr sensor has an active surface area of 0.097 cm^2 and a roughness factor of 2.425. The optimized ratio of CeO_2:eGr is 1:1 for 4-BPA determination. The CeO_2/eGr sensor exhibited good linearity in a wide range from 0.3 to 150 μmol/L and the lowest detection limit of 0.06 μmol/L for 4-BPA detection. Electrochemical oxidation of 4-BPA followed a mix-controlled process, which involves 2e in the transference processes. In addition, there were no significant interfering substances among K^+, Na^+, Mg^{2+}, rutin, quercetin, fenitrothion, imidacloprid, clothianidin, IAA, SA, glucose, and sucrose. The proposed electrochemical sensor showed excellent repeatability with the RSD of 2.35% for 10 measurements. In addition, the recoveries of the proposed CeO_2/eGr sensor were evaluated by the standard addition method, and are in the range of 90–108%. The low cost and easily-made sensor has great potential for detecting other plant growth regulators.

Supplementary Materials: The following supporting information can be downloaded at: https://www.mdpi.com/article/10.3390/bios12090760/s1, Figure S1: the plot of peak currents vs. $v^{1/2}$ for (a) CeO_2/GCE, (b) GCE, (C) eGr/GCE in the presence of 5 mM $[Fe(CN)_6]^{3-/4-}$ solution in aqueous of 0.1 M KCl. Figure S2: LSVs response of 10 μM 4-BAP to different volumes of composite in PBS solution (pH = 3). Figure S3: LSVs for 4-BPA detection in different buffer solution. Figure S4: Influence of pH on 4-BPA detection in 0.1 M PBS solution. Figure S5: the plots of ln I vs. ln v for 20 μM 4-BPA at 0.05, 0.1, 0.15,0.2, 0.25, 0.3, 0.35, 0.4, 0.45 V s^{-1}. Figure S6: the plots of E_{pa} vs. ln v for 20 μM 4-BPA at 0.05, 0.1, 0.15,0.2, 0.25, 0.3, 0.35, 0.4, 0.45 V s^{-1}. Figure S7: (a) The LSVs of the CeO_2/eGr/GCE in 0.1 M PBS (pH = 3) at first day and after 15 days. (b) 10 continuous measurements of 4-BPA on CeO_2/eGr/GCE.

Author Contributions: Conceptualization, project administration, and funding acquisition, H.D., Y.Y. and Z.L.; methodology and data curation, Y.Z.; validation and formal analysis, X.W. and H.H.; investigation, J.A., H.Z. and X.Y.; writing—original draft preparation, H.D. and Y.Z.; writing—review and editing, Y.Y. and Z.L. All authors have read and agreed to the published version of the manuscript.

Funding: This research was funded by the National Natural Science Foundation of China (No. 51801039, 81860701, 82060714, 22104102), Guizhou Provincial Science and Technology Projects (NO. ZK [2021]242), The Innovation Team Project of Guizhou Higher Education ([2022]013), The open project of Laboratory of Xinjiang Native Medicinal and Edible Plant Resources Chemistry (NO. KSUZDSYS202104). Sichuan Science and Technology Program (Grant No. 2021YJ0316).

Institutional Review Board Statement: Not applicable.

Informed Consent Statement: Not applicable.

Data Availability Statement: Not applicable.

Conflicts of Interest: The authors declare no conflict of interest.

References

1. Bons, H.K.; Kaur, M. Role of plant growth regulators in improving fruit set, quality and yield of fruit crops: A review. *J. Hortic. Sci. Biotechnol.* **2019**, *95*, 137–146. [CrossRef]
2. Kahlina, K.V.; Jurić, S.; Marijan, M.; Mutaliyeva, B.; Khalus, S.V.; Prosyanik, A.V.; Vinceković, M. Synthesis, Characterization, and Encapsulation of Novel Plant Growth Regulators (PGRs) in Biopolymer Matrices. *Int. J. Mol. Sci.* **2021**, *22*, 1847. [CrossRef] [PubMed]
3. Wei, Y.Z.; Zhang, H.N.; Li, W.C.; Xie, J.H.; Wang, Y.C.; Liu, L.Q.; Shi, S.Y. Phenological growth stages of lychee (*Litchi chinensis* Sonn.) using the extended BBCH-scale. *Sci. Hortic.* **2013**, *161*, 273–277. [CrossRef]
4. Fan, Y.X.; Yang, W.; Yan, Q.Q.; Chen, C.R.; Li, J.H. Genome-Wide Identification and Expression Analysis of the Protease Inhibitor Gene Families in Tomato. *Genes* **2020**, *11*, 1. [CrossRef] [PubMed]
5. Brondi, S.H.G.; Lanças, F.M. Development and validation of a multi-residue analytical methodology to determine the presence of selected pesticides in water through liquid chromatography. *J. Braz. Chem. Soc.* **2005**, *16*, 650–654. [CrossRef]
6. Rostami, S.; Azhdarpoor, A. The application of plant growth regulators to improve phytoremediation of contaminated soils: A review. *Chemosphere* **2019**, *220*, 818–827. [CrossRef]
7. Sutcharitchan, C.; Miao, S.; Li, W.T.; Liu, J.M.; Zhou, H.; Ma, Y.X.; Ji, S.; Cui, Y.J. High performance liquid chromatography-tandem mass spectrometry method for residue determination of 39 plant growth regulators in root and rhizome Chinese herbs. *Food Chem.* **2020**, *322*, 126766. [CrossRef]
8. Chen, H.; Feng, X.; Guo, H.S.; Zhang, H.W. Simultaneous determination of phytohormones containing carboxyl in crude extracts of fruit samples based on chemical derivatization by capillary electrophoresis with laser-induced fluorescence detection. *J. Chromatogr. B* **2011**, *879*, 1802–1808. [CrossRef]
9. Jiang, C.L.; Dai, J.X.; Han, H.L.; Wang, C.; Li, Z.; Lu, C.Y.; Chen, H.P. Determination of thirteen acidic phytohormones and their analogues in tea (*Camellia sinensis*) leaves using ultra high performance liquid chromatography tandem mass spectrometry. *J. Chromatogr. B* **2020**, *1149*, 122144. [CrossRef]
10. Qin, G.F.; Zou, K.T.; Tian, L.; Li, Y.B. Determination of Five Plant Growth Regulator Containing Carboxyl in Bean Sprouts Based on Chemical Derivatization by GC-MS. *Food Anal. Methods* **2018**, *11*, 2628–2635. [CrossRef]
11. Arduini, F.; Cinti, S.; Caratelli, V.; Amendola, L.; Palleschi, G.; Moscone, D. Origami multiple paper-based electrochemical biosensors for pesticide detection. *Biosens. Bioelectron.* **2019**, *126*, 346–354. [CrossRef]
12. Fusco, G.; Gallo, F.; Tortolini, C.; Bollella, P.; Ietto, F.; Mico, A.D.; D'Annibale, A.; Antiochia, R.; Favero, G.; Mazzei, F. AuNPs-functionalized PANABA-MWCNTs nanocomposite-based impedimetric immunosensor for 2,4-dichlorophenoxy acetic acid detection. *Biosens. Bioelectron.* **2017**, *93*, 52–56. [CrossRef]

13. Liu, F.P.; Qin, X.W.; Huang, Z.F.; Zhang, G.; Xiang, C.Z.; Zhang, C.Y.; Liang, H.L.; Peng, J.Y. Green synthesis of porous graphene and its application for sensitive detection of hydrogen peroxide and 2,4-dichlorophenoxyacetic acid. *Electrochim. Acta.* **2019**, *295*, 615–623. [CrossRef]
14. Skrzypczyńska, K.; Kuśmierek, K.; Świątkowski, A. Carbon paste electrodes modified with various carbonaceous materials for the determination of 2,4-dichlorophenoxyacetic acid by differential pulse voltammetry. *J. Electroanal. Chem.* **2016**, *766*, 8–15. [CrossRef]
15. Wang, H.M.; Xu, Q.; Wang, J.; Du, W.; Liu, F.P.; Hu, X.Y. Dendrimer-like amino-functionalized hierarchical porous silica nanoparticle: A host material for 2,4-dichlorophenoxyacetic acid imprinting and sensing. *Biosens. Bioelectron.* **2018**, *100*, 105–114. [CrossRef]
16. Arvand, M.; Gaskarmahalleh, A.A.; Hemmati, S. Enhanced-Oxidation and Highly Sensitive Detection of Tartrazine in Foodstuffs via New Platform Based on Poly(5-Sulfosalicylic Acid)/Cu(OH)2 Nanoparticles. *Food Anal. Methods* **2017**, *10*, 2241–2251. [CrossRef]
17. Su, D.D.; Zhang, Y.Y.; Wang, Z.J.; Wan, Q.J.; Yang, N.J. Decoration of graphene nano platelets with gold nanoparticles for voltammetry of 4-nonylphenol. *Carbon* **2017**, *117*, 313–321. [CrossRef]
18. Arumugasamy, S.K.; Govindaraju, S.; Yun, K. Electrochemical sensor for detecting dopamine using graphene quantum dots incorporated with multiwall carbon nanotubes. *Appl. Surf. Sci.* **2020**, *508*, 145294. [CrossRef]
19. Kim, C.K.; Kim, T.; Choi, I.Y.; Soh, M.; Kim, D.H.; Kim, Y.J.; Jang, H.D.; Yang, H.S.; Kim, J.Y.; Park, H.K.; et al. Rücktitelbild: Ceria Nanoparticles that can Protect against Ischemic Stroke. *Chem. Int. Ed. Eng.* **2012**, *51*, 11039–11043. [CrossRef]
20. Wang, Z.L.; Li, G.R.; Ou, Y.N.; Feng, Z.P.; Qu, D.L.; Tong, Y.X. Electrochemical deposition of Eu3+-doped CeO_2 nanobelts with enhanced optical properties. *J. Phys. Chem. C.* **2011**, *115*, 351–356. [CrossRef]
21. Ansari, A.A.; Alam, M. Nickle-ion-substituted ceria nanoparticles-based electrochemical sensor for sensitive detection of thiourea. *J. Mater. Sci.* **2021**, *32*, 23266–23274. [CrossRef]
22. Chen, C.H.; Yang, S.W.; Chuang, M.C.; Woon, W.Y.; Su, C.Y. Towards the continuous production of high crystallinity graphene via electrochemical exfoliation with molecular in situ encapsulation. *Nanoscale* **2015**, *7*, 15362–15373. [CrossRef]
23. Ambrosi, A.; Chua, C.K.; Latiff, N.M.; Loo, A.H.; Wong, C.H.A.; Eng, A.Y.S.; Bonanni, A.; Pumera, M. Graphene and its electrochemistry–an update. *Chem. Soc. Rev.* **2010**, *39*, 4146–4157. [CrossRef]
24. Pumera, M. Electrochemistry of graphene, graphene oxide and other graphenoids: Review. *Electrochem. Commun.* **2013**, *36*, 14–18. [CrossRef]
25. Ambrosi, A.; Chua, C.K.; Bonanni, A.; Pumera, M. Electrochemistry of Graphene and Related Materials. *Chem. Rev.* **2014**, *114*, 7150–7188. [CrossRef]
26. Brownson, D.A.C.; Banks, C.E. *The Handbook of Graphene Electrochemistry*, 1st ed; Springer: London, UK, 2014; pp. 6–20. [CrossRef]
27. Jayaprakash, G.K. Pre-post redox electron transfer regioselectivity at the alanine modified nano graphene electrode interface. *Chem. Phys. Lett.* **2022**, *789*, 139295. [CrossRef]
28. Sun, J.J.; Xie, X.H.; Xie, K.; Xu, S.C.; Jiang, S.Z.; Ren, J.F.; Zhao, Y.F.; Xu, H.Q.; Wang, J.J.; Yue, W.W. Magnetic Graphene Field-Effect Transistor Biosensor for Single-Strand DNA Detection. *Nanoscale Res. Lett.* **2019**, *14*, 248. [CrossRef]
29. Liu, Y.; Qiu, Z.P.; Wan, Q.J.; Wang, Z.H.; Wu, K.B.; Yang, N.J. High-Performance Hydrazine Sensor Based on Graphene Nano Platelets Supported Metal Nanoparticles. *Electroanalysis* **2016**, *28*, 126–132. [CrossRef]
30. Li, C.; Zhang, Y.Y.; Zeng, T.; Chen, X.Y.; Wang, W.Q.; Wan, Q.J.; Yang, N.J. Graphene nanoplatelet supported CeO_2 nanocomposites towards electrocatalytic oxidation of multiple phenolic pollutan. *Anal. Chim. Acta* **2019**, *1088*, 45–53. [CrossRef]
31. Zhang, Y.; Ai, J.X.; Hu, H.L.; Wang, X.; Zhou, H.X.; Du, K.Z.; Du, H.J.; Yang, Y. Highly Sensitive Detection of Kinetin with Electrochemical Exfoliation of Graphene Nanosheets. *Appl. Phys. A* **2022**, *128*, 350. [CrossRef]
32. Kumar, A.A.; Swamy, B.E.K.; Rani, T.S.; Ganesh, P.S.; Raj, Y.P. Voltammetric determination of catechol and hydroquinone at poly(murexide) modified glassy carbon electrode. *Mater. Sci. Eng. C.* **2019**, *98*, 746–752. [CrossRef] [PubMed]
33. Tan, H.Y.; Wang, J.; Yu, S.Z.; Zhou, K.B. Support Morphology-Dependent Catalytic Activity of Pd/CeO_2 for Formaldehyde Oxidation. *Sci. Technol.* **2015**, *49*, 8675–8682. [CrossRef] [PubMed]
34. Yu, J.; Zhang, Y.Y.; Li, H.; Wan, Q.J.; Li, Y.W.; Yang, N.J. Electrochemical properties and sensing applications of nanocarbons: A comparative study. *Carbon* **2018**, *129*, 301–309. [CrossRef]
35. Mullins, D.R.; Overbury, S.H.; Huntley, D.R. Electron spectroscopy of single crystal and polycrystalline cerium oxide surfaces. *Surf. Sci.* **1998**, *409*, 307–319. [CrossRef]
36. Lyu, L.; Xie, Q.; Yang, Y.Y.; Wang, R.R.; Cen, W.F.; Luo, S.Y.; Yang, W.S.; Gao, Y.; Xiao, Q.Q.; Zou, P.; et al. A Novel CeO_2 Hollow-Shell Sensor Constructed for High Sensitivity of Acetone Gas Detection. *Appl. Surf. Sci.* **2022**, *571*, 151337. [CrossRef]
37. Meng, F.M.; Wang, L.N.; Cui, J.B. Controllable synthesis and optical properties of nano-CeO_2 via a facile hydrothermal route. *J. Alloys Compd.* **2013**, *556*, 102–108. [CrossRef]
38. Yang, Y.; Yang, Y.M.; Fu, T.W.; Zhu, J.; Fan, J.P.; Zhang, Z.; Zhang, J.Q. Influence of ethanol content in the precursor solution on anodic electrodeposited CeO_2 thin films. *Thin Solid Film.* **2014**, *556*, 128–136. [CrossRef]
39. Parvez, K.; Wu, Z.S.; Li, R.; Liu, X.; Graf, R.; Feng, X.; Müllen, K. Exfoliation of Graphite into Graphene in Aqueous Solutions of Inorganic Salts. *J. Am. Chem. Soc.* **2014**, *136*, 6083–6091. [CrossRef]
40. Nicholson, R.S. Theory and Application of Cyclic Voltammetry for Measurement of Electrode Reaction Kinetics. *Anal. Chem.* **1965**, *37*, 1351–1355. [CrossRef]

41. Sha, T.; Liu, J.; Sun, M.; Li, L.; Bai, J.; Hu, Z.; Zhou, M. Green and low-cost synthesis of nitrogen-doped graphene-like mesoporous nanosheets from the biomass waste of okara for the amperometric detection of vitamin C in real samples. *Talanta* **2019**, *200*, 300–306. [CrossRef]
42. Krzyczmonik, P.; Socha, E.; Skrzypek, S.; Soliwoda, K.; Celichowski, G.; Grobelny, J. Honeycomb-structured porous poly(3,4-ethylenedioxythiophene) composite layers on a gold electrode. *Thin Solid Films* **2014**, *565*, 54–61. [CrossRef]
43. Zamiri, R.; Salehizadeh, S.A.; Ahangar, H.A.; Shabani, M.; Rebelo, A.; Ferreira, J.M.F. Dielectric and optical properties of Ni-and Fe-doped CeO_2 Nanoparticles. *Appl. Phys. A* **2019**, *125*, 1–7. [CrossRef]
44. Chen, Y.N.; Chen, Q.; Zhao, H.; Dang, J.Q.; Jin, R.; Zhao, W.; Li, Y.H. Wheat Straws and Corn Straws as Adsorbents for the Removal of Cr(VI) and Cr(III) from Aqueous Solution: Kinetics, Isotherm, and Mechanism. *ACS Omega* **2020**, *5*, 6003–6009. [CrossRef]
45. Huang, Q.; Lin, X.; Tong, L.; Tong, Q.X. Graphene Quantum Dots/Multiwalled Carbon Nanotubes Composite-Based Electrochemical Sensor for Detecting Dopamine Release from Living Cells. ACS Sustain. *Chem. Eng.* **2020**, *8*, 1644–1650. [CrossRef]
46. Laviron, E. General expression of the linear potential sweep voltammogram in the case of diffusionless electrochemical systems. *J. Electroanal. Chem.* **1979**, *101*, 19–28. [CrossRef]
47. Xu, D.D.; Song, X.Z.; Qi, W.Z.; Wang, H.; Bian, Z.Y. Degradation mechanism, kinetics, and toxicity investigation of 4-bromophenol by electrochemical reduction and oxidation with Pd-Fe/graphene catalytic cathodes. *Chem. Eng. J.* **2018**, *333*, 477–485. [CrossRef]

biosensors

MDPI

Article

A Core–Shell Au@TiO_2 and Multi-Walled Carbon Nanotube-Based Sensor for the Electroanalytical Determination of H_2O_2 in Human Blood Serum and Saliva

Ayman Ali Saeed [1], Mohammed Nooredeen Abbas [1], Waheed Fathi El-Hawary [2], Yousry Moustafa Issa [2] and Baljit Singh [3,*]

1 Applied Organic Chemistry Department, Chemical Industries Research Institute, National Research Centre (NRC), Dokki, Giza 12622, Egypt
2 Chemistry Department, Faculty of Science, Cairo University, Giza 12613, Egypt
3 MiCRA Biodiagnostics Technology Gateway & Centre of Applied Science for Health, Technological University Dublin (TU Dublin), D24 FKT9 Dublin 24, Ireland
* Correspondence: baljit.singh@tudublin.ie; Tel.: +353-12-207-863

Abstract: A hydrogen peroxide (H_2O_2) sensor was developed based on core–shell gold@titanium dioxide nanoparticles and multi-walled carbon nanotubes modified glassy carbon electrode (Au@TiO_2/MWCNTs/GCE). Core–shell Au@TiO_2 material was prepared and characterized using a scanning electron microscopy and energy dispersive X-ray analysis (SEM/EDX), transmission electron microscopy (TEM), atomic force microscopy (AFM), Raman spectroscopy, X-ray diffraction (XRD) and Zeta-potential analyzer. The proposed sensor (Au@TiO_2/MWCNTs/GCE) was investigated electrochemically using cyclic voltammetry (CV) and electrochemical impedance spectroscopy (EIS). The analytical performance of the sensor was evaluated towards H_2O_2 using differential pulse voltammetry (DPV). The proposed sensor exhibited excellent stability and sensitivity with a linear concentration range from 5 to 200 μM (R^2 = 0.9973) and 200 to 6000 μM (R^2 = 0.9994), and a limit of detection (LOD) of 1.4 μM achieved under physiological pH conditions. The practicality of the proposed sensor was further tested by measuring H_2O_2 in human serum and saliva samples. The observed response and recovery results demonstrate its potential for real-world H_2O_2 monitoring. Additionally, the proposed sensor and detection strategy can offer potential prospects in electrochemical sensors development, indicative oxidative stress monitoring, clinical diagnostics, general cancer biomarker measurements, paper bleaching, etc.

Keywords: electrochemical sensor; core–shell; titanium dioxide; gold nanoparticles; carbon nanotubes; hydrogen peroxide

Citation: Saeed, A.A.; Abbas, M.N.; El-Hawary, W.F.; Issa, Y.M.; Singh, B. A Core–Shell Au@TiO_2 and Multi-Walled Carbon Nanotube-Based Sensor for the Electroanalytical Determination of H_2O_2 in Human Blood Serum and Saliva. *Biosensors* **2022**, *12*, 778. https://doi.org/10.3390/bios12100778

Received: 19 August 2022
Accepted: 15 September 2022
Published: 20 September 2022

1. Introduction

Carbon nanomaterials, including single-walled carbon nanotubes (SWCNTs), multi-walled carbon nanotubes (MWCNTs), nanofibers, activated carbon and graphene, have unique properties such as chemical stability and durability, high electrical conductivity, mechanical strength and a high surface-to-volume ratio which collectively makes them an excellent choice for electrochemical sensors and biosensors development [1–10]. Carbon nanotubes (CNTs) are an excellent and promising option in fabricating electrochemical sensors owing to their remarkable electrical and thermal conductivity, high surface area, chemical stability and mechanical properties [11–14]. Metal nanoparticles, due to their excellent conductivity, surface area and remarkable electrocatalytic properties, are considered ideal candidates for the electrochemical detection of hydrogen peroxide (H_2O_2). They are capable of promoting electron transfer processes and offer abundant catalytic active sites during the H_2O_2 redox reaction. Carbon-based nanomaterials can be mixed with metal nanoparticles to create composites that have excellent synergistic features, which can improve the sensitivity and overall performance of the modified electrodes.

Core–shell nanomaterials (CSNs) are composite materials with a core–shell structure formed by an inner layer of one material (core) and an exterior layer of another material (shell) [15]. The shell materials are frequently chosen based on their nature and the targeted application. Many advantages can be achieved by carefully selecting shell material, including improved optical/electrical/magnetic properties, multifunctional capability, thermal stability or dispersibility of the materials, reduced precious material content and efficient use. In the core–shell structure, the size and kind of core are also crucial factors. The main advantage of CSNs is that the distinct properties of the core and shell can be combined in a single material, resulting in improved electrocatalytic activity and new physical and chemical properties that are inaccessible or unavailable from the individual components due to lack of this synergistic effect. In order to maintain the nanoparticles' stability and chemical activity, the core material is also safeguarded against migration and aggregation. CSNs have been studied extensively for a variety of biological applications, including drug administration, cancer treatment, bioimaging, cell labelling, genetic engineering, methanol electrooxidation and so on [16,17]. Due to their higher surface area, superior catalytic activities and biocompatibility, CSNs have been exploited as signal amplifiers and considered promising electron modifiers to create novel sensing platforms, including electrochemical sensors and biosensors. The excellent features of CSNs make them ideal candidates for developing sensitive and novel electrochemical sensors [18–26].

Accurate, sensitive and reliable detection of H_2O_2 has received substantial attention in analytical applications due to the importance of H_2O_2 in various areas. Hydrogen peroxide is a common peroxide that is used in biological systems, medical diagnosis, environmental analysis, food and many other applications [27–29]. Furthermore, H_2O_2 as a member of reactive oxygen species (ROS), is considered one of the crucial oxidative stress biomarkers. The excessive level of H_2O_2 can cause a series of diseases such as Alzheimer's, Parkinson's, myocardial infarction, inflammatory lung diseases, cancer, etc. [30,31]. Therefore, it is essential to develop an effective method for rapid and reliable monitoring of H_2O_2 [32,33].

Several analytical methods have been developed for H_2O_2 determination, including spectrophotometry [34,35], chemiluminescence [36–38], fluorescence [39–41], chromatography [42,43] and electrochemical sensors [44–65]. However, due to the technical drawbacks of the traditional methods (low sensitivity and selectivity, laborious, time-consuming and complicated instrumentation), electrochemical sensors have received more attention due to their associated practical advantages, including high sensitivity, portability, simplicity, cost-effectiveness, rapid response time and ease of fabrication and operation. Electrochemical sensors for H_2O_2 determination are mainly based on enzymatic and non-enzymatic approaches, but due to drawbacks of enzymatic sensors such as instability (temperature, pH-related challenges), shelf-life and immobilization procedures, non-enzymatic sensors have received greater attention [44–61].

In this paper, we describe the fabrication of a hydrogen peroxide sensor based on modifying the surface of a glassy carbon electrode (GCE) with a gold@titanium dioxide (Au@TiO_2) core–shell nanoparticle and multi-walled carbon nanotubes. The characterization of the core–shell material and fabricated electrode was investigated thoroughly, and the electroanalytical performance of the sensor was studied and discussed in detail. The proposed sensor exhibited excellent electroanalytical performance and electrocatalytic activity toward H_2O_2 reduction. The practicality of the proposed sensor was tested by measuring H_2O_2 in human serum and saliva samples which demonstrates its potential for H_2O_2 monitoring in real-world samples. Additionally, the proposed sensor and detection strategy may offer potential prospects in electrocatalysts and electrochemical sensors development, as well as in other applications, including indicative oxidative stress monitoring, clinical diagnostics, general cancer biomarker measurements, food processing, paper bleaching and environmental analysis.

2. Materials and Methods

2.1. Materials

Gold(III) chloride hydrate ($HAuCl_4.H_2O$), trisodium citrate and multi-walled carbon nanotubes (MWCNTs) were purchased from Sigma-Aldrich. Titanium(IV) tetraisopropoxide (TTIP, 98%) and sodium dodecyl sulphate (SDS, 85%) were purchased from ACROS ORGANICS. H_2O_2 (30%) was purchased from Advent Chembio Pvt. Ltd. Potassium ferricyanide was purchased from VEB Laborchemie Apolda, while potassium ferrocyanide was purchased from BDH.

2.2. Apparatus and Measurements

The cyclic voltammetry (CV), electrochemical impedance spectroscopy (EIS) and differential pulse voltammetry (DPV) measurements were performed using CH Instruments Inc. (CHI 660D). A glassy carbon electrode (GCE, 3 mm diameter) was used as a working electrode, while a platinum wire and an Ag/AgCl reference electrode (3 M KCl) were used as counter and reference electrodes, respectively. The experiments of electrochemical characterization were carried out in 5 mM $[Fe(CN)_6]^{3-/4-}$ solution as a redox probe from +0.7 V to −0.4 V, while the different concentrations of H_2O_2 (5 to 6000 μM) were prepared in 0.1 M PBS saturated with N_2.

Raman spectroscopy was carried out using WITec alpha300 R at 532 nm. Zeta potential was measured using Malvern Zetasizer zs. Transmission electron microscopy (TEM) images were recorded using a JEOL JEM-1230. Surface morphology characterization and identification were performed using scanning electron microscopy (SEM, TESCAN VEGA3) coupled with energy dispersive X-ray analysis (EDX, BRUKER) and an atomic force microscope (5600LS, Agilent, Santa Clara, CA, USA). X-ray diffraction (XRD) was performed using an X-ray diffractometer (BRUKER, D8 DISCOVER).

2.3. Synthesis of Gold Nanoparticles (AuNPs)

Gold(III) chloride solution (147 mM, 100 μL) was added to distilled water (50 mL) to prepare 0.01% (*w/v*) solution and heated until boiling. In total, 1% (*w/v*, 2 mL) trisodium citrate was added under vigorous stirring, and the color of the solution changed from pale yellow to blue and finally red. The solution was cooled to room temperature and stored at 4 °C. The final concentration of the as-prepared gold nanoparticles was 58 ppm.

2.4. Synthesis of Core–Shell AuNPs@TiO$_2$ and AuNPs@TiO$_2$/MWCNTs

Au@TiO_2 was prepared by modifying the preparation methods mentioned in previous reports [66–68]. Au@TiO_2 with different wt% of core Au (1.09, 2.17, 4.24, 8.14 and 11.74%) was prepared. For Au@TiO_2 (8.14% Au), 20 mL of the prepared AuNPs stock solution was added dropwise to 20 mL of 0.1 mol/L SDS solution under vigorous stirring for 15 min. The mixture was centrifuged at 12,000 rpm for 10 min. The SDS-capped AuNPs were settled at the bottom of the centrifuge tube while the supernatant was removed. The SDS-capped AuNPs were washed three times with distilled water to remove the excess free SDS. In total, 50 μL of TTIP were mixed with 500 μL of isopropanol, and the mixture was added dropwise to the washed SDS-capped AuNPs under vigorous stirring. The formed precipitate was washed with ethanol and allowed to dry. The dried product was calcinated at 500 °C for 4 hrs. Then, 1 mg of AuNPs@TiO_2 was suspended in 980 μL distilled water, and 20 μL of 1 mg/mL MWCNTs were added. Finally, the mixture was sonicated for 30 min.

2.5. Fabrication of H_2O_2 Electrode (Au@TiO$_2$/MWCNTs/GCE)

Before making the electrochemical measurements, glassy carbon working electrodes were polished using alumina powders (1.0, 0.3 and 0.05 μm size), followed by washing with acetone and finally with deionized water and dried at room temperature. The glassy carbon electrode was firstly activated by recording CV in 0.5 M H_2SO_4 from 1.4 to −0.2 V for 20 cycles at a scan rate of 0.1 V/s. The electrode was then washed with distilled water and allowed to dry. Then, 5 μL of the Au@TiO_2/MWCNTs were drop-casted onto the electrode surface and allowed to dry at room temperature.

2.6. Real Sample Analysis

The proposed H_2O_2 sensor (Au@TiO_2/MWCNTs/GCE) was tested for the determination of H_2O_2 in human serum and saliva samples. Three volunteers participated in this study, two males and one female, with different ages varying from 24 to 40 years. Saliva samples were collected in the morning, volunteers were asked to rinse their mouths with water for 1 min, and then samples were collected after 3 min. Collected saliva samples were centrifuged for three minutes at 4000 rpm in order to settle down any food residues, and clear saliva was diluted 10 times with 0.1 M PBS (pH 7.4) for analysis. For the serum analysis, a 10 mL sample was collected, and the blood was allowed to coagulate for 1 h, then centrifuged at 5000 rpm for 5 min, and finally, the upper serum layer was collected and diluted 20 times with PBS for analysis.

3. Results and Discussion

3.1. UV-Visible Spectroscopy

In order to ensure the success of Au@TiO_2 preparation, UV-Visible absorption spectra were recorded for AuNPs, TiO_2 and Au@TiO_2. As shown in Figure 1, in the case of AuNPs, the spectrum showed the characteristic absorption peak at 524 nm. TiO_2 nanoparticles are commonly used as an additive in most sunscreen formulations due to their broad absorption within the UVB and UVA region, with a maximum of 327.5 nm. The successful preparation of Au@TiO_2 was accompanied and confirmed by the disappearance of the AuNPs peak, which is attributed to the complete coverage of the core (Au) by the TiO_2 shell. Moreover, a slight blue shift was observed from 327.5 nm to 323.5 nm.

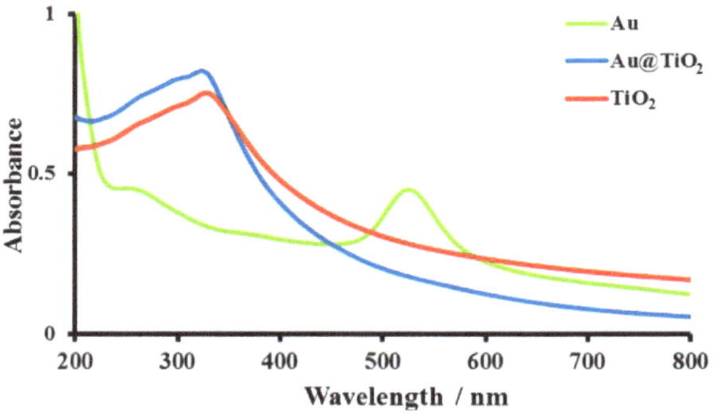

Figure 1. UV-Visible absorption spectra of AuNPs, TiO_2 and Au@TiO_2 materials.

3.2. Raman Spectroscopy

The TiO_2 and Au@TiO_2 materials were characterized by Raman spectroscopy, and the associated spectra are shown in Figure 2. The existence of bands at 144 (E_g), 198 (E_g), 396 (B_{1g}), 515 (A_{1g}) and 638 (E_g) cm^{-1} are unquestionably linked to the TiO_2 anatase phase [69–71]. In the case of Au@TiO_2, the typical E_g peak of TiO_2 at 144 cm^{-1} shifted to a higher wavenumber (146 cm^{-1}), indicating more crystalline disorders in the anatase TiO_2. These crystalline disorders, which developed at the point where Au and TiO_2 came into contact, impact the vibrational frequency of anatase TiO_2 and serve as traps for the produced photoelectrons.

3.3. X-ray Diffraction (XRD)

The crystalline nature of the core–shell Au@TiO_2 material was characterized by X-ray diffraction (XRD), as shown in Figure 3. The samples show two series of peaks, which can be attributed to the anatase-TiO_2 and the face-centered-cubic Au. Planes (101), (200), (105) and (211) of anatase-TiO_2 are responsible for the peaks at 2θ of 25.7°, 44.4°, 54.0° and 54.3°, while planes (111), (200), (220) and (311) of face-centered-cubic Au are associated with the remaining peaks at 2θ of 38.2°, 44.4°, 64.7° and 77.7°, in agreement with the literature [72].

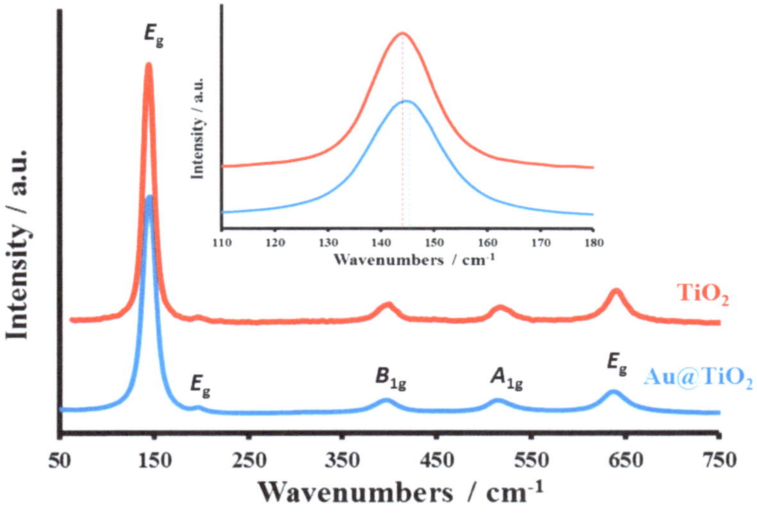

Figure 2. Raman spectra of the prepared TiO_2 and Au@TiO_2 (Au is 8.14%) materials.

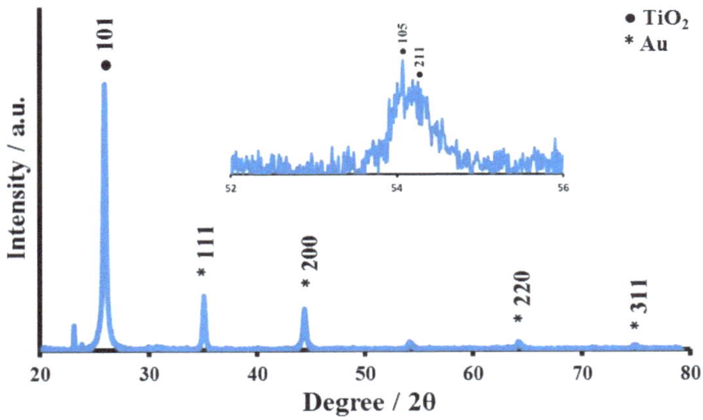

Figure 3. X-ray diffraction pattern for the prepared core–shell Au@TiO_2 material (Au = 8.14%).

3.4. Zeta-Potential Analyzer and Size Distribution

Higher zeta potential values indicate that the periphery surface charge of the nanoparticles is higher, which encourages repulsion and prevents the formation of aggregates, a sign of the stability of the core–shell nanoparticles. Au@TiO_2 core–shell with different wt% of Au (1.09, 2.17, 4.24, 8.14 and 11.74%) were prepared, and their Zeta potentials were measured to evaluate the effect of the weight percent of the core gold on the surface charge and stability of the whole core-shell material. As seen in Figure 4, the increase in the weight percent of core gold is accompanied by a positive increase in the value of Zeta potential until it reaches nearly stable values between 4.24% and 8.14%. Therefore, we decided to use 8.14% of Au content in further studies.

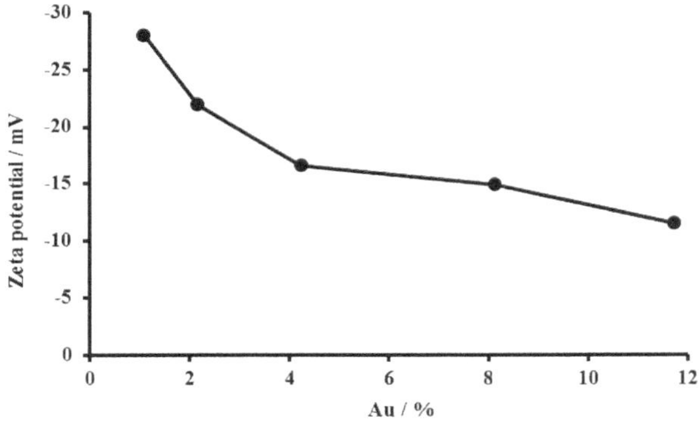

Figure 4. Zeta potential of the prepared core–shell Au@TiO_2 material with varying Au content (1.09, 2.17, 4.24, 8.14 and 11.74%).

3.5. *Transmission Electron Microscopy (TEM)*

TEM analysis was performed to directly measure the size, size distribution and morphology of Au@TiO_2 and Au@TiO_2/MWCNTs, which are shown in Figure 5a–d. The spherical gold nanoparticles were observed as a dark core, with about 13–15 nm diameter, in the center and totally encapsulated by a brighter shell of TiO_2. It can be observed that gold nanoparticles are well-dispersed and uniformly incorporated in the TiO_2 matrix without any significant agglomeration. Figure 5c shows the selected area electron diffraction pattern, which agrees with the XRD data and reveals that the compositions of core–shell (Au and TiO_2) reflect their signatures of crystal planes in the hybrid or composite form. The combination of MWCNTs with core–shell Au@TiO_2 is confirmed and presented in Figure 5d.

Figure 5. TEM images of (**a**,**b**) Au@TiO_2 with 8.14% Au; (**c**) selected area electron diffraction pattern; (**d**) TEM image of Au@TiO_2/MWCNTs composite material.

3.6. *Scanning Electron Microscopy and Energy Dispersive X-ray Analysis (SEM and EDX)*

SEM images were recorded in order to describe the surface morphology and distribution of the Au@TiO_2 material. As shown in Figure 6a,b, the presence of spherical and uniformly distributed particles was observed with a size of approximately 30.29 nm (diameter). The modification of the electrode surface with such core–shell material promises a high surface area and better electrocatalytic activity. The elemental analysis for Au@TiO_2 and

Au@TiO_2/MWCNT materials was performed, and EDX profiles are shown in Figure 6c,d. The EDX profile for Au@TiO_2/MWCNT (Figure 6d) confirms the presence of Ti, O and Au and C with a weight percent of 39.88%, 46.92%, 7.76% and 5.45%, respectively. The Au amount observed from the EDX measurements is in approximate agreement with the theoretical value of 8.14%.

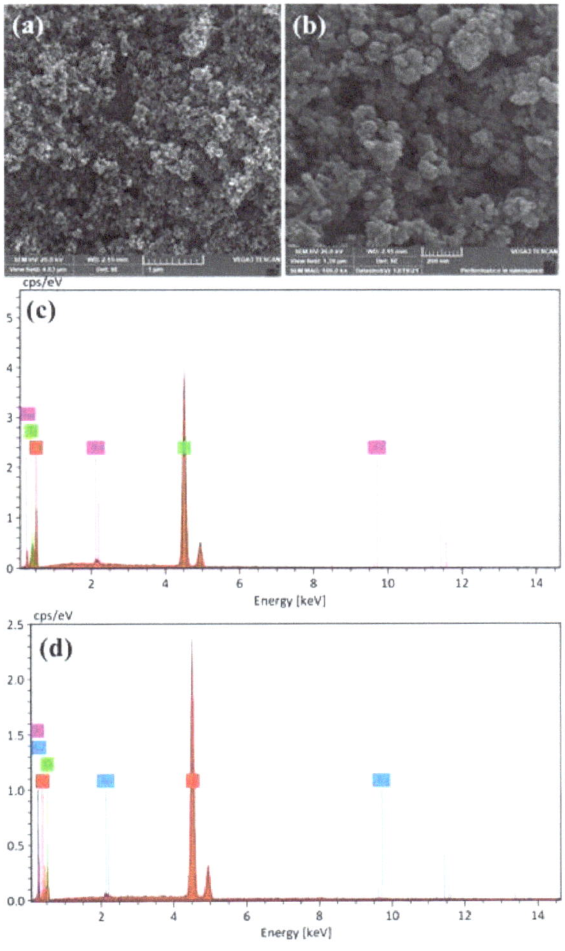

Figure 6. SEM images of Au@TiO_2 material (Au = 8.14%) at different magnifications; (**a**) 30 kx and (**b**) 100 kx. (**c**,**d**) EDX profiles for Au@TiO_2 and Au@TiO_2/MWCNT materials confirming the presence of elements as expected.

3.7. Atomic Force Microscopy (AFM)

AFM analysis was performed to characterize the surface morphology of the Au@TiO_2-modified surface, and the images are shown in Figure 7. It is obvious that the topography of the core–shell-modified surface has a high surface roughness with a root mean square height (Sq) of 35.4 nm. As shown in Figure 7a,c, the 3D images confirm the success of the core–shell preparation process. Protrusions were observed in the TiO_2 shell as the embedded gold nanoparticles in the core push the TiO_2 shell out.

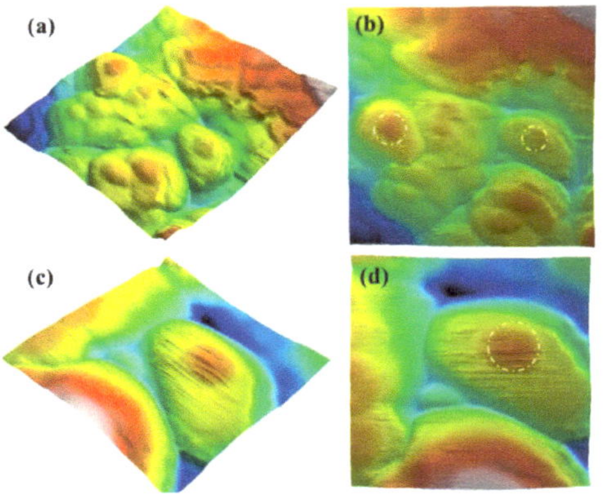

Figure 7. AFM images of the core–shell Au@TiO_2 material; (**a**,**c**) 3D images and (**b**,**d**) plane images.

3.8. Cyclic Voltammetry

Cyclic voltammetry characterization (Figure 8 and Table 1) was performed by recording the voltammograms for various stages of electrode modification in 5 mM $[Fe(CN)_6]^{3-/4-}$ (1:1) in 0.1 M KCl at a scan rate of 100 mV/s (potential window 0.7 to −0.4 V). The current response for the MWCNTs modified electrode (I_{Oxi} = 60.71 µA and I_{Red} = −61.61 µA, ΔE = 129 mV) was improved very significantly compared to the bare GCE (I_{Oxi} = 45.83 µA, I_{Red} = −44.61 µA, ΔE = 220 mV). This is attributed to the high conductivity of MWCNTs, which improve the electron transfer kinetics between the electrode surface and redox couple.

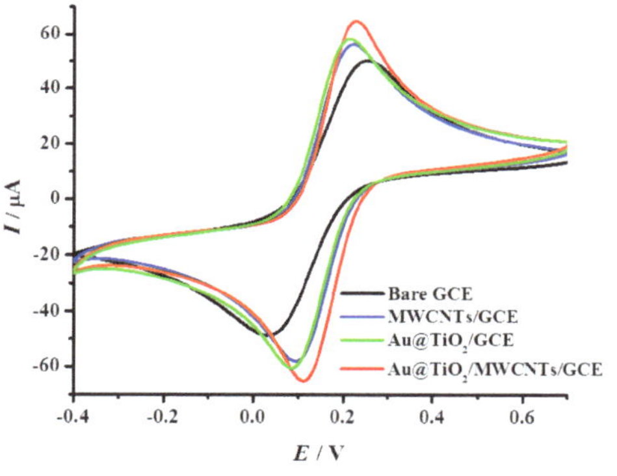

Figure 8. Cyclic voltammetric responses for bare GCE, MWCNTs/GCE, Au@TiO_2/GCE and Au@TiO_2/MWCNTs/GCE in 5 mM $[Fe(CN)_6]^{3-/4-}$/0.1 M KCl at a scan rate of 100 mV/s.

Table 1. CV and EIS performance data for modified and bare electrodes in $[Fe(CN)_6]^{3-/4-}$.

Electrode	I_{Oxi} (µA)	E (mV)	I_{Red} (µA)	E (mV)	ΔE (mV)	R_s (Ω)	R_{ct} (Ω)	C (µF)	W (mΩ)
Bare GCE	45.83	252	−44.61	32	220	128.4	2610	0.49	0.143
MWCNTs/GCE	60.71	223	−61.61	94	129	113.9	634	0.91	0.167
Au@TiO_2/GCE	61.81	213	−63.04	87	126	118.4	529	1.13	0.149
Au@TiO_2/MWCNTs/GCE	69.77	228	−69.56	113	115	124.8	281	1.00	0.266

The drop-casting of the as-prepared core–shell Au@TiO_2 material onto GCE also showed an increase in the current response (I_{Oxi} = 61.81 µA, I_{Red} = −63.04 µA) with a decrease in ΔE value (126 mV) and this is attributed to the combined catalytic activity of TiO_2 and Au.

Au@TiO_2/MWCNTs/GCE showed a much better current response compared to bare GCE, MWCNTs/GCE and Au@TiO_2/GCE with I_{Oxi} = 69.77 μA, I_{Red} = −69.56 μA and the lowest ΔE (115 mV) was observed. This improvement is attributed to the synergistic effect of the core–shell Au@TiO_2 structure and the high electrical conductivity of MWCNTs.

3.9. Electrochemical Impedance Spectroscopy

EIS analysis was performed and recorded in 5 mM $[Fe(CN)_6]^{3-/4-}$ solution at 0.22 V with a frequency range from 0.1 Hz to 10^5 Hz (Figure 9 and Table 1) to support the cyclic voltammetry results and to confirm the R_{ct} values which agree well with the voltammetry findings. The bare GCE showed an R_{ct} of 2610 Ω, which was decreased very significantly (634 Ω) after modifying the GCE surface with MWCNTs. This is due to the reason that the electron process kinetics is faster due to the high electrical conductivity of MWCNTs, which ultimately facilitates the electron transfer and decreases the resistance. Au@TiO_2-modified GCE also showed a decrease in the R_{ct} value (529 Ω), but the proposed electrode material Au@TiO_2/MWCNTs/GCE exhibited the best and lowest charge transfer resistance (281 Ω) among all the tested materials and is attributed to the synergistic effect of the core–shell Au@TiO_2 nanostructure and high conductivity of MWCNTs.

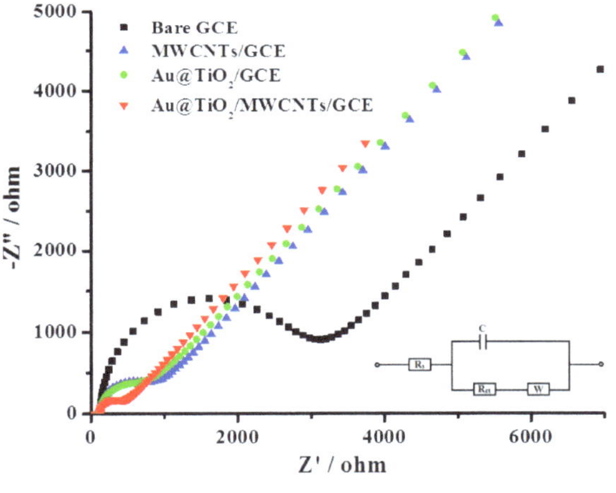

Figure 9. Nyquist plot for modified (labeled) and bare glassy carbon electrodes in 5 mM $[Fe(CN)_6]^{3-/4}$/0.1 M KCl. Insert shows the Randel circuit used for data fitting.

3.10. Analytical Performance (H_2O_2 Sensing)

In order to confirm the synergistic effect on H_2O_2 sensing, DPVs were recorded in N_2-saturated PBS for the different stages of electrode modification. As shown in Figure 10, the increase in the current response was 1.5 and 1.9 times in the case of MWCNTs/GCE and Au@TiO_2/GCE, respectively, while an increase of 4.6 times was observed in the case of Au@TiO_2/MWCNTs/GCE compared to the bare glassy carbon electrode.

The DPV response of the proposed electrode (Au@TiO_2/MWCNTs/GCE) towards different H_2O_2 concentrations (PBS, pH 7.4) was recorded and shown in Figure 11. The reduction current of the hydrogen peroxide (at −0.62 V) increased with H_2O_2 concentration and showed a linear response from 5 to 200 μM (y = −0.0046x − 0.015) and from 200 to 6000 μM (y = −0.0074x + 0.7217) with coefficient of determination (R^2) of 0.9973 and 0.9994, respectively. The calculated limit of detection (LOD) was 1.4 μM based on 3σ calculations. The analytical performance of the proposed sensor was compared to other relevant and reported results and is summarized in Table 2. It is clearly evident that the proposed sensor exhibited an excellent response in terms of linear range and detection limit.

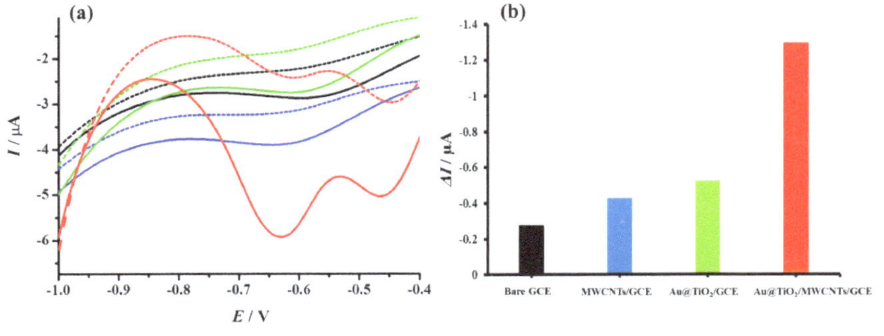

Figure 10. (**a**) Differential pulse voltammograms recorded in PBS (dashed lines) and 300 μM H_2O_2 (solid lines) for various stages of electrode modification, Bare GCE (black); MWCNTs/GCE (blue); Au@TiO_2/GCE (green); Au@TiO_2/MWCNTs/GCE (red). (**b**) Histogram represents the corresponding change in current values (ΔI).

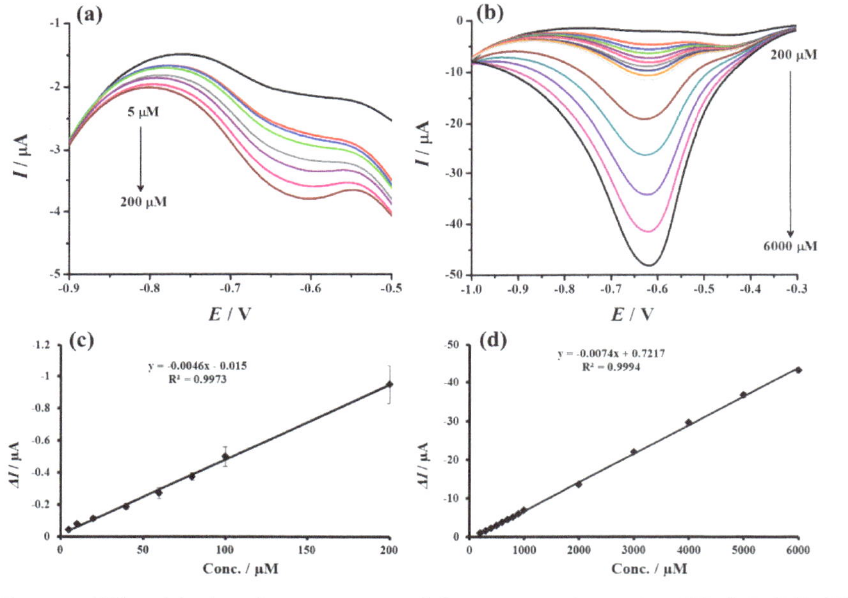

Figure 11. Differential pulse voltammograms recorded using proposed sensor (Au@TiO_2/MWCNTs/GCE towards different concentrations of H_2O_2 (**a**) 5 to 200 μM and (**b**) 200 to 6000 μM) in N_2-saturated PBS (pH 7.4)). (**c**,**d**) Corresponding calibration curves from the current responses vs. H_2O_2 concentrations plot.

3.10.1. Stability and Reproducibility of H_2O_2 Sensor

Stability is an important parameter to describe the performance of an electrochemical sensor. The stability of the Au@TiO_2/MWCNTs/GCE sensor was investigated by measuring the current responses for different concentrations of H_2O_2 (20–1000 μM, Figure 12) after 50 days and compared with the initial response recorded. The current response for H_2O_2 after 50 days was found to be 97.1% of its initial response, and this reveals the acceptable stability and sufficient lifetime of the proposed sensor. The reproducibility of three freshly prepared Au@TiO_2/MWCNTs/GCEs was tested by measuring the current responses of 200 μM H_2O_2, and the calculated %RSD was 4.63%.

Table 2. Comparison of the proposed sensor with the recently reported H_2O_2 sensors.

Electrode Material	Linear Range	LOD (μM)	Ref.
GO-Fe_3O_4-PAMAM-Pd/GCE	0.05–160 μM	0.01	[48]
Pd/TNM@rGO	up to 12 mM	0.0025	[49]
GQDs-CS/MB/GCE	1.0 μM–2.9 mM 2.9–11.78 mM	0.7	[50]
Paper/CNTs/AgNPs	1 μM–700 μM	-	[51]
CuNPs-rGO	up to 18 mM	601	[52]
LSG-Ag	0.1–10 mM	7.9	[53]
α-MoO_3/GO/GCE	0.92 μM–2.46 mM	0.31	[54]
PtNPs@SPCEs	0–215 μM	1.9	[55]
MPS electrode	10 and 5000 μM	4.35	[56]
Pt-Pd/CFME	5–3920 μM	0.42	[57]
$NiCoSe_2$/GCE	0.05 to 402 μM	0.03	[58]
Cu@Pt/C	0.50 μM–32.56 mM	0.15	[59]
Fe_3O_4@PEI@AuNPs-GRE	0.2–500.0 μM	0.07	[60]
Cu/Cu_2O/FTO	0.2–2000 μM	0.04	[61]
AuNPs/*n*-GaN	40 μM–1 mM	10	[62]
Pd/AuNPs	0.5–6 mM	-	[63]
Ni–Bi/CC	0.1 μM–0.5 mM	0.00085	[64]
PtNP/rGO–CNT/PtNP/SPCE	25–1000 μM	4.3	[65]
Au@TiO_2/MWCNTs/GCE	5–200 μM and 200 μM–6 mM	1.4	This work

PAMAM: Poly(amidoamine) dendrimer; TNM: *tert*-Nonyl mercaptan; GQDs: Graphene quantum dots; CS: Chitosan; MB: Methylene blue; LSG; Laser scribed graphene; SPCEs: Screen-printed carbon electrodes; MPS: Macroporous Silicon; CFME: Carbon fiber microelectrode; PEI: Polyethyleneimine; GRE: Graphite rod electrode; FTO: Fluorine doped tin oxide; CC: Carbon cloth.

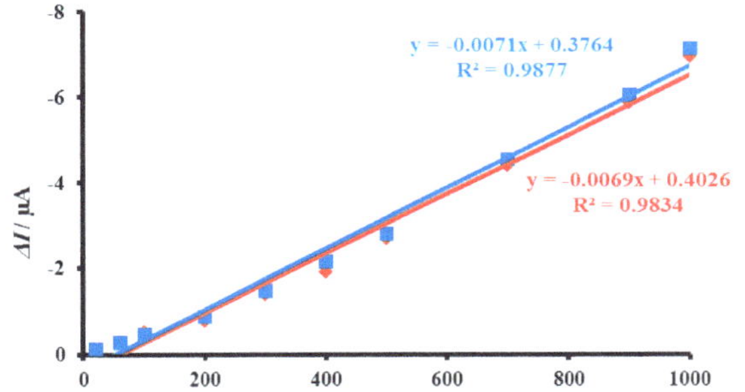

Figure 12. Stability study for the determination of H_2O_2 (20–1000 μM) after 50 days (red line) using the proposed sensor (Au@TiO_2/MWCNTs/GCE) compared to the response observed initially (blue line).

3.10.2. Selectivity of H_2O_2 Sensor

Possible interferences which may occur during the determination of H_2O_2 in the biological samples were examined using the proposed sensor (Au@TiO_2/MWCNTs/GCE). The interference study was performed by measuring the current response of 200 μM H_2O_2, and then the change in the current response was recorded in the presence of the common interferents (ascorbic acid, glucose, methionine, cysteine and uric acid) with 10-fold concentration (2 mM) compared to H_2O_2. As shown in Table 3, the effect of these interferents on H_2O_2 determination is expressed as a percentage recovery from the initial H_2O_2 current response (before the interferents were added). Even with such high concentrations of interferents (10-fold), the results reveal that the proposed sensor is quite selective towards H_2O_2.

Table 3. The H_2O_2 (200 µM) recovery data for the proposed electrode in the presence of common interferents (10-fold concentration, 2 mM).

Interferent (10-Folds)	Recovery (%) (n = 3)
Ascorbic acid	89.11 ± 0.50
Glucose	102.45 ± 0.87
Methionine	96.21 ± 1.23
Cysteine	105.65 ± 0.68
Uric acid	97.31 ± 0.73

3.11. Real Sample Analysis

The diluted real samples (human saliva and serum samples) were considered, and analysis was performed in triplicate for different concentrations of H_2O_2 by the standard addition method. The recovery values of the proposed sensor are shown in Table 4, which vary from 85.3 to 117.9%. The obtained results reveal that the proposed sensor exhibits a reliable response and excellent practicality for the electroanalytical determination of H_2O_2 in real biological samples.

Table 4. Application of the proposed sensor Au@TiO_2/MWCNTs/GCE for real sample analysis (human serum and saliva samples).

Sample	Added (µM)	Found in Serum (µM) (n = 3)	Recovery (%)	Found in Saliva (µM) (n = 3)	Recovery (%)
1	20	18.02 ± 0.94	90.09	18.33 ± 0.71	91.63
	40	38.10 ± 0.89	95.24	38.28 ± 0.60	95.71
	60	51.17 ± 0.70	85.29	70.74 ± 1.43	117.90
2	20	20.70 ± 1.20	103.48	21.30 ± 0.87	106.52
	40	37.41 ± 0.64	93.53	43.98 ± 0.69	109.95
	60	57.39 ± 0.56	95.65	63.85 ± 0.45	106.41
3	20	22.11 ± 0.88	110.54	22.76 ± 1.02	113.80
	40	38.10 ± 1.09	95.24	39.63 ± 0.70	99.08
	60	51.17 ± 0.79	85.29	68.30 ± 0.83	113.84

4. Conclusions

An electrochemical non-enzymatic biosensor based on Au@TiO_2 core–shell nanoparticles and muti-walled carbon nanotubes was developed and investigated for H_2O_2 determination. The success of the Au@TiO_2 preparation process was confirmed and characterized by various techniques. The different electrode modification stages were electrochemically characterized using cyclic voltammetry and electrochemical impedance spectroscopy in 5 mM $[Fe(CN)_6]^{3-/4-}$ in order to confirm the synergistic effect of Au@TiO_2 and MWCNTs. The synergistic effect was further confirmed during differential pulse voltammetry analysis of H_2O_2 (300 µM) in 0.1 M N_2-saturated PBS (pH 7.4). The analytical performance of the modified electrode (Au@TiO_2/MWCNTs/GCE) was investigated by plotting the calibration curves of different concentrations of H_2O_2 varying from 5 to 6000 µM. The results showed that by increasing the H_2O_2 concentration, the reduction current of the hydrogen peroxide at −0.62 V increases, and the current values showed a linear response from 5 to 200 µM (R^2 = 0.9973) and 200 to 6000 µM (R^2 = 0.9994). The calculated limit of detection was 1.4 µM and attributed to the synergistic effect of Au@TiO_2 and MWCNTs.

The proposed sensor exhibited good selectivity over the possible interferents (ascorbic acid, glucose, methionine, cysteine and uric acid at a 10-fold concentration level compared to H_2O_2). The lifetime of the sensor/electrode reached 50 days with a decrease of only 2.9% of the original response (current). The developed sensor was further tested for the determination of H_2O_2 in real biological samples (human serum and saliva) using the standard addition method. The observed response and recovery results support the potential of the proposed sensor for H_2O_2 monitoring in future real-world sample analysis.

Author Contributions: Conceptualization, M.N.A. and B.S.; methodology, A.A.S., M.N.A. and B.S.; software, A.A.S. and B.S.; validation, A.A.S., M.N.A. and B.S.; formal analysis, A.A.S.; investigation, A.A.S., M.N.A. and B.S.; resources, A.A.S., M.N.A., W.F.E.-H., Y.M.I. and B.S.; data curation, A.A.S., M.N.A. and B.S.; writing—original draft preparation, A.A.S., M.N.A. and B.S.; writing—review and editing, A.A.S., M.N.A., B.S., W.F.E.-H. and Y.M.I.; visualization, A.A.S. and B.S.; supervision, M.N.A., B.S., W.F.E.-H. and Y.M.I.; project administration, M.N.A.; funding acquisition, M.N.A. and A.A.S. All authors have read and agreed to the published version of the manuscript.

Funding: This research was partially funded by the National Research Centre (NRC, Egypt) as a PhD research support.

Informed Consent Statement: Informed consent was obtained from all subjects involved in the study.

Data Availability Statement: The data presented in this study are available on request from the corresponding author.

Acknowledgments: This paper is dedicated to the memory of Mohammed Nooredeen Abbas. We (authors) dedicate this paper to the memory of our friend 'Mohammed Nooredeen Abbas' (who passed away recently) and are grateful to him for his collaborative activities and contributions to the scientific community.

Conflicts of Interest: The authors declare no conflict of interest.

References

1. Lan, L.; Yao, Y.; Ping, J.; Ying, Y. Recent advances in nanomaterial-based biosensors for antibiotics detection. *Biosens. Bioelectron.* **2017**, *91*, 504–514. [CrossRef]
2. Teradal, N.L.; Jelinek, R. Carbon nanomaterials in biological studies and biomedicine. *Adv. Healthc. Mater.* **2017**, *6*, 1700574. [CrossRef]
3. Porto, L.S.; Silva, D.N.; de Oliveira, A.E.F.; Pereira, A.C.; Borges, K.B. Carbon nanomaterials: Synthesis and applications to development of electrochemical sensors in determination of drugs and compounds of clinical interest. *Rev. Anal. Chem.* **2019**, *38*, 20190017. [CrossRef]
4. Bobrinetskiy, I.I.; Knezevic, N.Z. Graphene-based biosensors for on-site detection of contaminants in food. *Anal. Methods* **2018**, *10*, 5061–5070. [CrossRef]
5. Bounegru, A.V.; Apetrei, C. Carbonaceous nanomaterials employed in the development of electrochemical sensors based on screen-printing technique—A review. *Catalysts* **2020**, *10*, 680. [CrossRef]
6. Kirchner, E.-M.; Hirsch, T. Recent developments in carbon-based two-dimensional materials: Synthesis and modification aspects for electrochemical sensors. *Microchim. Acta* **2020**, *187*, 441. [CrossRef]
7. Kour, R.; Arya, S.; Young, S.-J.; Gupta, V.; Bandhoria, P.; Khosla, A. Recent advances in carbon nanomaterials as electrochemical biosensors. *J. Electrochem. Soc.* **2020**, *167*, 037555. [CrossRef]
8. Pandey, H.; Khare, P.; Singh, S.; Singh, S.P. Carbon nanomaterials integrated molecularly imprinted polymers for biological sample analysis: A critical review. *Mater. Chem. Phys.* **2020**, *239*, 121966. [CrossRef]
9. Wang, J. Carbon-nanotube based electrochemical biosensors: A review. *Electroanal. Int. J. Devoted Fundam. Pract. Asp. Electroanal.* **2005**, *17*, 7–14. [CrossRef]
10. Nooredeen, N.M.; El-Ghaffar, A.; Darwish, W.; Elshereafy, E.; Radwan, A.; Abbas, M. Graphene oxide with covalently attached zinc monoamino-phthalocyanine coated graphite electrode as a potentiometric platform for citrate sensing in pharmaceutical preparations. *J. Solid State Electrochem.* **2015**, *19*, 2141–2154. [CrossRef]
11. Zaporotskova, I.V.; Boroznina, N.P.; Parkhomenko, Y.N.; Kozhitov, L.V. Carbon nanotubes: Sensor properties. A review. *Mod. Electron. Mater.* **2016**, *2*, 95–105. [CrossRef]
12. Norizan, M.N.; Moklis, M.H.; Demon, S.Z.N.; Halim, N.A.; Samsuri, A.; Mohamad, I.S.; Knight, V.F.; Abdullah, N. Carbon nanotubes: Functionalisation and their application in chemical sensors. *RSC Adv.* **2020**, *10*, 43704–43732. [CrossRef] [PubMed]
13. Arlyapov, V.; Kharkova, A.; Kurbanaliyeva, S.; Kuznetsova, L.; Machulin, A.; Tarasov, S.; Melnikov, P.; Ponamoreva, O.; Alferov, V.; Reshetilov, A. Use of biocompatible redox-active polymers based on carbon nanotubes and modified organic matrices for development of a highly sensitive BOD biosensor. *Enzym. Microb. Technol.* **2021**, *143*, 109706. [CrossRef]
14. Schroeder, V.; Savagatrup, S.; He, M.; Lin, S.; Swager, T.M. Carbon nanotube chemical sensors. *Chem. Rev.* **2018**, *119*, 599–663. [CrossRef] [PubMed]
15. Wang, H.; Chen, L.; Feng, Y.; Chen, H. Exploiting core–shell synergy for nanosynthesis and mechanistic investigation. *Acc. Chem. Res.* **2013**, *46*, 1636–1646. [CrossRef]

16. Core, R.G.C. Classes, Properties, Synthesis Mechanisms, Characterization, and Applications/Rajib Ghosh Chaudhuri, Santanu Paria. *Chem. Rev.* **2012**, *112*, 2373.
17. Singh, B.; Seddon, B.; Dempsey, E.; Redington, W.; Dickinson, C. Porous Core-Shell Platinum-Silver Nanocatalyst for the Electrooxidation of Methanol. *Electroanalysis* **2015**, *27*, 135–143. [CrossRef]
18. Krishnan, S.K.; Prokhorov, E.; Bahena, D.; Esparza, R.; Meyyappan, M. Chitosan-covered Pd@Pt core–shell nanocubes for direct electron transfer in electrochemical enzymatic glucose biosensor. *ACS Omega* **2017**, *2*, 1896–1904. [CrossRef]
19. Li, M.; Wang, P.; Li, F.; Chu, Q.; Li, Y.; Dong, Y. An ultrasensitive sandwich-type electrochemical immunosensor based on the signal amplification strategy of mesoporous core–shell Pd@Pt nanoparticles/amino group functionalized graphene nanocomposite. *Biosens. Bioelectron.* **2017**, *87*, 752–759. [CrossRef]
20. Wang, R.; Liu, W.-D.; Wang, A.-J.; Xue, Y.; Wu, L.; Feng, J.-J. A new label-free electrochemical immunosensor based on dendritic core-shell AuPd@ Au nanocrystals for highly sensitive detection of prostate specific antigen. *Biosens. Bioelectron.* **2018**, *99*, 458–463. [CrossRef]
21. Zhang, X.; Li, Y.; Lv, H.; Feng, J.; Gao, Z.; Wang, P.; Dong, Y.; Liu, Q.; Zhao, Z. Sandwich-type electrochemical immunosensor based on Au@Ag supported on functionalized phenolic resin microporous carbon spheres for ultrasensitive analysis of α-fetoprotein. *Biosens. Bioelectron.* **2018**, *106*, 142–148. [CrossRef] [PubMed]
22. Yang, F.; Yang, Z.; Zhuo, Y.; Chai, Y.; Yuan, R. Ultrasensitive electrochemical immunosensor for carbohydrate antigen 19-9 using Au/porous graphene nanocomposites as platform and Au@Pd core/shell bimetallic functionalized graphene nanocomposites as signal enhancers. *Biosens. Bioelectron.* **2015**, *66*, 356–362. [CrossRef] [PubMed]
23. Xu, W.; Yi, H.; Yuan, Y.; Jing, P.; Chai, Y.; Yuan, R.; Wilson, G.S. An electrochemical aptasensor for thrombin using synergetic catalysis of enzyme and porous Au@Pd core–shell nanostructures for signal amplification. *Biosens. Bioelectron.* **2015**, *64*, 423–428. [CrossRef] [PubMed]
24. Tabrizi, M.A.; Shamsipur, M.; Saber, R.; Sarkar, S.; Sherkatkhameneh, N. Flow injection amperometric sandwich-type electrochemical aptasensor for the determination of adenocarcinoma gastric cancer cell using aptamer-Au@Ag nanoparticles as labeled aptamer. *Electrochim. Acta* **2017**, *246*, 1147–1154. [CrossRef]
25. Mazloum-Ardakani, M.; Hosseinzadeh, L.; Taleat, Z. Synthesis and electrocatalytic effect of Ag@Pt core–shell nanoparticles supported on reduced graphene oxide for sensitive and simple label-free electrochemical aptasensor. *Biosens. Bioelectron.* **2015**, *74*, 30–36. [CrossRef] [PubMed]
26. Wang, M.; Hu, B.; Ji, H.; Song, Y.; Liu, J.; Peng, D.; He, L.; Zhang, Z. Aptasensor based on hierarchical core–shell nanocomposites of zirconium hexacyanoferrate nanoparticles and mesoporous mFe_3O_4@ mC: Electrochemical quantitation of epithelial tumor marker Mucin-1. *ACS Omega* **2017**, *2*, 6809–6818. [CrossRef]
27. Bai, J.; Jiang, X. A facile one-pot synthesis of copper sulfide-decorated reduced graphene oxide composites for enhanced detecting of H_2O_2 in biological environments. *Anal. Chem.* **2013**, *85*, 8095–8101. [CrossRef]
28. Rhee, S.G. H_2O_2, a necessary evil for cell signaling. *Science* **2006**, *312*, 1882–1883. [CrossRef]
29. Wu, W.; Li, J.; Chen, L.; Ma, Z.; Zhang, W.; Liu, Z.; Cheng, Y.; Du, L.; Li, M. Bioluminescent probe for hydrogen peroxide imaging in vitro and in vivo. *Anal. Chem.* **2014**, *86*, 9800–9806. [CrossRef]
30. Hang, T.; Xiao, S.; Yang, C.; Li, X.; Guo, C.; He, G.; Li, B.; Yang, C.; Chen, H.-j.; Liu, F.; et al. Hierarchical graphene/nanorods-based H_2O_2 electrochemical sensor with self-cleaning and anti-biofouling properties. *Sens. Actuators B Chem.* **2019**, *289*, 15–23. [CrossRef]
31. Dong, Y.; Zheng, J. Environmentally friendly synthesis of Co-based zeolitic imidazolate framework and its application as H_2O_2 sensor. *Chem. Eng. J.* **2020**, *392*, 123690. [CrossRef]
32. Shang, L.; Zeng, B.; Zhao, F. Fabrication of novel nitrogen-doped graphene–hollow AuPd nanoparticle hybrid films for the highly efficient electrocatalytic reduction of H_2O_2. *ACS Appl. Mater. Interfaces* **2015**, *7*, 122–128. [CrossRef] [PubMed]
33. Yang, K.; Zhong, H.; Cheng, Z.; Li, X.; Zhang, A.; Li, T.; Zhang, Y.; Liu, G.; Qian, H. Magnetic Fe_3O_4 stacked sphere-like nanocomposite and its application as platform for H_2O_2 sensing. *J. Electroanal. Chem.* **2018**, *814*, 1–6. [CrossRef]
34. Nagaraja, P.; Prakash, J.; Asha, S.; Bhaskara, B.; Kumar, S.A. Dibenzazepin hydrochloride as a new spectrophotometric reagent for determination of hydrogen peroxide in plant extracts. *Environ. Monit. Assess.* **2012**, *184*, 5983–5988. [CrossRef]
35. Hoshino, M.; Kamino, S.; Doi, M.; Takada, S.; Mitani, S.; Yanagihara, R.; Asano, M.; Yamaguchi, T.; Fujita, Y. Spectrophotometric determination of hydrogen peroxide with osmium(VIII) and m-carboxyphenylfluorone. *Spectrochim. Acta Part A Mol. Biomol. Spectrosc.* **2014**, *117*, 814–816. [CrossRef]
36. Jamil, L.A.; Faizullah, A.T.; Saleem, P.H. Flow Injection Analysis of Hydrogen Peroxide with Peroxyoxalate Chemiluminescence Detection. *Sci. J. Univ. Zakho* **2017**, *5*, 88–92. [CrossRef]
37. Yu, D.; Wang, P.; Zhao, Y.; Fan, A. Iodophenol blue-enhanced luminol chemiluminescence and its application to hydrogen peroxide and glucose detection. *Talanta* **2016**, *146*, 655–661. [CrossRef]
38. Wang, K.; Liu, Q.; Wu, X.-Y.; Guan, Q.-M.; Li, H.-N. Graphene enhanced electrochemiluminescence of CdS nanocrystal for H_2O_2 sensing. *Talanta* **2010**, *82*, 372–376. [CrossRef]
39. Sasakura, K.; Hanaoka, K.; Shibuya, N.; Mikami, Y.; Kimura, Y.; Komatsu, T.; Ueno, T.; Terai, T.; Kimura, H.; Nagano, T. Development of a highly selective fluorescence probe for hydrogen sulfide. *J. Am. Chem. Soc.* **2011**, *133*, 18003–18005. [CrossRef]
40. Xu, M.; Han, J.-M.; Wang, C.; Yang, X.; Pei, J.; Zang, L. Fluorescence ratiometric sensor for trace vapor detection of hydrogen peroxide. *ACS Appl. Mater. Interfaces* **2014**, *6*, 8708–8714. [CrossRef]

41. Chang, J.; Li, H.; Hou, T.; Duan, W.; Li, F. Paper-based fluorescent sensor via aggregation induced emission fluorogen for facile and sensitive visual detection of hydrogen peroxide and glucose. *Biosens. Bioelectron.* **2018**, *104*, 152–157. [CrossRef] [PubMed]
42. Tarvin, M.; McCord, B.; Mount, K.; Sherlach, K.; Miller, M.L. Optimization of two methods for the analysis of hydrogen peroxide: High performance liquid chromatography with fluorescence detection and high performance liquid chromatography with electrochemical detection in direct current mode. *J. Chromatogr. A* **2010**, *1217*, 7564–7572. [CrossRef] [PubMed]
43. Hu, H.-C.; Jin, H.-J.; Chai, X.-S. Rapid determination of hydrogen peroxide in pulp bleaching effluents by headspace gas chromatography. *J. Chromatogr. A* **2012**, *1235*, 182–184. [CrossRef] [PubMed]
44. Yu, G.; Wu, W.; Pan, X.; Zhao, Q.; Wei, X.; Lu, Q. High Sensitive and Selective Sensing of Hydrogen Peroxide Released from Pheochromocytoma Cells Based on Pt-Au Bimetallic Nanoparticles Electrodeposited on Reduced Graphene Sheets. *Sensors* **2015**, *15*, 2709–2722. [CrossRef]
45. He, G.; Gao, F.; Li, W.; Li, P.; Zhang, X.; Yin, H.; Yang, B.; Liu, Y.; Zhang, S. Electrochemical sensing of H 2 O 2 released from living cells based on AuPd alloy-modified PDA nanotubes. *Anal. Methods* **2019**, *11*, 1651–1656. [CrossRef]
46. Dai, H.; Lü, W.; Zuo, X.; Zhu, Q.; Pan, C.; Niu, X.; Liu, J.; Chen, H.; Chen, X. A novel biosensor based on boronic acid functionalized metal-organic frameworks for the determination of hydrogen peroxide released from living cells. *Biosens. Bioelectron.* **2017**, *95*, 131–137. [CrossRef]
47. Xi, J.; Xie, C.; Zhang, Y.; Wang, L.; Xiao, J.; Duan, X.; Ren, J.; Xiao, F.; Wang, S. Pd nanoparticles decorated N-doped graphene quantum dots@N-doped carbon hollow nanospheres with high electrochemical sensing performance in cancer detection. *ACS Appl. Mater. Interfaces* **2016**, *8*, 22563–22573. [CrossRef]
48. Baghayeri, M.; Alinezhad, H.; Tarahomi, M.; Fayazi, M.; Ghanei-Motlagh, M.; Maleki, B. A non-enzymatic hydrogen peroxide sensor based on dendrimer functionalized magnetic graphene oxide decorated with palladium nanoparticles. *Appl. Surf. Sci.* **2019**, *478*, 87–93. [CrossRef]
49. Bozkurt, S.; Tosun, B.; Sen, B.; Akocak, S.; Savk, A.; Ebeoğlugil, M.F.; Sen, F. A hydrogen peroxide sensor based on TNM functionalized reduced graphene oxide grafted with highly monodisperse Pd nanoparticles. *Anal. Chim. Acta* **2017**, *989*, 88–94. [CrossRef]
50. Mollarasouli, F.; Asadpour-Zeynali, K.; Campuzano, S.; Yáñez-Sedeño, P.; Pingarrón, J.M. Non-enzymatic hydrogen peroxide sensor based on graphene quantum dots-chitosan/methylene blue hybrid nanostructures. *Electrochim. Acta* **2017**, *246*, 303–314. [CrossRef]
51. Shamkhalichenar, H.; Choi, J.-W. An inkjet-printed non-enzymatic hydrogen peroxide sensor on paper. *J. Electrochem. Soc.* **2017**, *164*, B3101–B3106. [CrossRef]
52. Nia, P.M.; Woi, P.M.; Alias, Y. Facile one-step electrochemical deposition of copper nanoparticles and reduced graphene oxide as nonenzymatic hydrogen peroxide sensor. *Appl. Surf. Sci.* **2017**, *413*, 56–65.
53. Aparicio-Martínez, E.; Ibarra, A.; Estrada-Moreno, I.A.; Osuna, V.; Dominguez, R.B. Flexible electrochemical sensor based on laser scribed Graphene/Ag nanoparticles for non-enzymatic hydrogen peroxide detection. *Sens. Actuators B Chem.* **2019**, *301*, 127101. [CrossRef]
54. Li, B.; Song, H.-Y.; Deng, Z.-P.; Huo, L.-H.; Gao, S. Novel sensitive amperometric hydrogen peroxide sensor using layered hierarchical porous α-MoO3 and GO modified glass carbon electrode. *Sens. Actuators B Chem.* **2019**, *288*, 641–648. [CrossRef]
55. Agrisuelas, J.; González-Sánchez, M.-I.; Valero, E. Hydrogen peroxide sensor based on in situ grown Pt nanoparticles from waste screen-printed electrodes. *Sens. Actuators B Chem.* **2017**, *249*, 499–505. [CrossRef]
56. Al-Hardan, N.H.; Abdul Hamid, M.A.; Shamsudin, R.; Al-Khalqi, E.M.; Kar Keng, L.; Ahmed, N.M. Electrochemical hydrogen peroxide sensor based on macroporous silicon. *Sensors* **2018**, *18*, 716. [CrossRef]
57. Li, H.; Zhao, H.; He, H.; Shi, L.; Cai, X.; Lan, M. Pt-Pd bimetallic nanocoral modified carbon fiber microelectrode as a sensitive hydrogen peroxide sensor for cellular detection. *Sens. Actuators B Chem.* **2018**, *260*, 174–182. [CrossRef]
58. Sakthivel, M.; Sukanya, R.; Chen, S.-M.; Pandi, K.; Ho, K.-C. Synthesis and characterization of bimetallic nickel-cobalt chalcogenides (NiCoSe2, NiCo2S4, and NiCo2O4) for non-enzymatic hydrogen peroxide sensor and energy storage: Electrochemical properties dependence on the metal-to-chalcogen composition. *Renew. Energy* **2019**, *138*, 139–151. [CrossRef]
59. Zhao, W.; Jin, J.; Wu, H.; Wang, S.; Fneg, C.; Yang, S.; Ding, Y. Electrochemical hydrogen peroxide sensor based on carbon supported Cu@Pt core-shell nanoparticles. *Mater. Sci. Eng. C* **2017**, *78*, 185–190. [CrossRef]
60. Ghanei-Motlagh, M.; Hosseinifar, A. A novel amperometric hydrogen peroxide sensor based on gold nanoparticles supported on Fe_3O_4@polyethyleneimine. *Int. J. Environ. Anal. Chem.* **2020**, *100*, 591–601. [CrossRef]
61. Han, L.; Tang, L.; Deng, D.; He, H.; Zhou, M.; Luo, L. A novel hydrogen peroxide sensor based on electrodeposited copper/cuprous oxide nanocomposites. *Analyst* **2019**, *144*, 685–690. [CrossRef] [PubMed]
62. Ma, C.; Yang, C.; Zhang, M. A novel electrochemical hydrogen peroxide sensor based on AuNPs/n-type GaN electrode. *Chem. Lett.* **2020**, *49*, 656–658. [CrossRef]
63. Banerjee, S.; Hossain, M.F.; Slaughter, G. A Highly Sensitive Non-Enzymatic Hydrogen Peroxide Sensor based on Palladium-Gold Nanoparticles. In Proceedings of the 2020 IEEE 15th International Conference on Nano/Micro Engineered and Molecular System (NEMS), San Diego, CA, USA, 27–30 September 2020; pp. 286–289.
64. Wang, Z.; Xie, F.; Liu, Z.; Du, G.; Asiri, A.M.; Sun, X. High-performance non-enzyme hydrogen peroxide detection in neutral solution: Using a nickel borate nanoarray as a 3D electrochemical sensor. *Chem. Eur. J.* **2017**, *23*, 16179–16183. [CrossRef] [PubMed]

65. Lee, S.; Lee, Y.J.; Kim, J.H.; Lee, G.-J. Electrochemical Detection of H_2O_2 Released from Prostate Cancer Cells Using Pt Nanoparticle-Decorated rGO–CNT Nanocomposite-Modified Screen-Printed Carbon Electrodes. *Chemosensors* **2020**, *8*, 63. [CrossRef]
66. Oh, J.T.; Chowdhury, S.R.; Lee, T.L.; Misra, M. Synergetic influence of Au/Cu_2O core-shells nanoparticle on optical, photo-electrochemical, and catalytic activities of Au/Cu_2O/TiO_2 nanocomposite. *Dye. Pigment.* **2019**, *160*, 936–943. [CrossRef]
67. Chen, T.-M.; Xu, G.-Y.; Ren, H.; Zhang, H.; Tian, Z.-Q.; Li, J.-F. Synthesis of Au@TiO_2 core–shell nanoparticles with tunable structures for plasmon-enhanced photocatalysis. *Nanoscale Adv.* **2019**, *1*, 4522–4528. [CrossRef]
68. Goebl, J.; Joo, J.B.; Dahl, M.; Yin, Y. Synthesis of tailored Au@TiO_2 core–shell nanoparticles for photocatalytic reforming of ethanol. *Catal. Today* **2014**, *225*, 90–95. [CrossRef]
69. Sun, H.; He, Q.; Zeng, S.; She, P.; Zhang, X.; Li, J.; Liu, Z. Controllable growth of Au@TiO_2 yolk–shell nanoparticles and their geometry parameter effects on photocatalytic activity. *New J. Chem.* **2017**, *41*, 7244–7252. [CrossRef]
70. Wang, Y.; Yang, C.; Chen, A.; Pu, W.; Gong, J. Influence of yolk-shell Au@TiO_2 structure induced photocatalytic activity towards gaseous pollutant degradation under visible light. *Appl. Catal. B Environ.* **2019**, *251*, 57–65. [CrossRef]
71. Baba, K.; Bulou, S.; Quesada-Gonzalez, M.; Bonot, S.; Collard, D.; Boscher, N.D.; Choquet, P. Significance of a noble metal nanolayer on the UV and visible light photocatalytic activity of anatase TiO_2 thin films grown from a scalable PECVD/PVD approach. *ACS Appl. Mater. Interfaces* **2017**, *9*, 41200–41209. [CrossRef]
72. Sun, H.; He, Q.; She, P.; Zeng, S.; Xu, K.; Li, J.; Liang, S.; Liu, Z. One-pot synthesis of Au@TiO_2 yolk-shell nanoparticles with enhanced photocatalytic activity under visible light. *J. Colloid Interface Sci.* **2017**, *505*, 884–891. [CrossRef] [PubMed]

biosensors

MDPI

Review

Biosensors and Drug Delivery in Oncotheranostics Using Inorganic Synthetic and Biogenic Magnetic Nanoparticles

Tatiana M. Zimina [1,2,*,†], Nikita O. Sitkov [1,2,†], Kamil G. Gareev [1,2], Viacheslav Fedorov [2], Denis Grouzdev [3], Veronika Koziaeva [2,4], Huile Gao [5], Stephanie E. Combs [6] and Maxim Shevtsov [2,6,7,*]

1 Department of Micro and Nanoelectronics, Saint Petersburg Electrotechnical University "LETI", 197022 Saint Petersburg, Russia
2 Laboratory of Biomedical Nanotechnologies, Institute of Cytology of the Russian Academy of Sciences, 194064 Saint Petersburg, Russia
3 SciBear OU, Tartu mnt 67/1-13b, Kesklinna Linnaosa, 10115 Tallinn, Estonia
4 Research Center of Biotechnology of the Russian Academy of Sciences, Institute of Bioengineering, 119071 Moscow, Russia
5 Key Laboratory of Drug-Targeting and Drug Delivery System of the Education Ministry, West China School of Pharmacy, Sichuan University, Chengdu 610041, China
6 Department of Radiation Oncology, Klinikum Rechts der Isar, Technical University of Munich, 81675 Munich, Germany
7 National Center for Neurosurgery, Nur-Sultan 010000, Kazakhstan
* Correspondence: tmzimina@gmail.com (T.M.Z.); maxim.shevtsov@tum.de (M.S.); Tel.: +7-9219413916 (T.M.Z.); +49-1731488882 (M.S.)
† These authors contributed equally to this work.

Citation: Zimina, T.M.; Sitkov, N.O.; Gareev, K.G.; Fedorov, V.; Grouzdev, D.; Koziaeva, V.; Gao, H.; Combs, S.E.; Shevtsov, M. Biosensors and Drug Delivery in Oncotheranostics Using Inorganic Synthetic and Biogenic Magnetic Nanoparticles. *Biosensors* **2022**, *12*, 789. https://doi.org/10.3390/bios12100789

Received: 25 August 2022
Accepted: 18 September 2022
Published: 25 September 2022

Abstract: Magnetic nanocarriers have attracted attention in translational oncology due to their ability to be employed both for tumor diagnostics and therapy. This review summarizes data on applications of synthetic and biogenic magnetic nanoparticles (MNPs) in oncological theranostics and related areas. The basics of both types of MNPs including synthesis approaches, structure, and physicochemical properties are discussed. The properties of synthetic MNPs and biogenic MNPs are compared with regard to their antitumor therapeutic efficiency, diagnostic potential, biocompatibility, and cellular toxicity. The comparative analysis demonstrates that both synthetic and biogenic MNPs could be efficiently used for cancer theranostics, including biosensorics and drug delivery. At the same time, reduced toxicity of biogenic particles was noted, which makes them advantageous for in vivo applications, such as drug delivery, or MRI imaging of tumors. Adaptability to surface modification based on natural biochemical processes is also noted, as well as good compatibility with tumor cells and proliferation in them. Advances in the bionanotechnology field should lead to the implementation of MNPs in clinical trials.

Keywords: magnetic nanoparticles; biogenic magnetic nanoparticles; magnetotactic bacteria; magnetosomes; biosensors; drug delivery; oncotheranostics

1. Introduction

Screening and early diagnostics of oncological diseases are important factors of successful treatment. At the same time, for timely diagnosis, it is necessary to detect minor amounts of markers of such diseases [1]. The modern development of nanotechnologies opens up new opportunities for creating highly sensitive means of diagnostics and treatment of cancer. Currently, such problems are solved by developing new biosensor systems based on various materials and detection principles.

According to the IUPAC definition, a biosensor is "an autonomous complex device capable of obtaining quantitative or semi-quantitative data using a biorecognizing element (bioreceptor) that is in direct spatial contact with a transforming element (transducer)" [2]. In a broad sense, a biosensor is a device that converts a physical or chemical effect on biological objects into a measurable signal. Three main parts can be distinguished in the biosensor structure as follows: (*i*) *biorecognizing element* (e.g., oligonucleotide or peptide aptamers, antibodies, enzymes, proteins, microorganisms, organelles, cell receptors, etc.), which is a material or biomimetic component with a high degree of selectivity to the target analyte; (*ii*) *transducer* (e.g., optical, electrochemical, acoustic, etc.) that converts the signal of interaction between the biorecognizing element and the analyte into a measurable and quantifiable signal (in most cases, an electrical signal); and (*iii*) *data processing system* that analyzes the received signals and visualizes the measurement results conveniently for the operator.

Modern biosensor technology is an example of the convergence of various scientific and technical areas. The use of a variety of nanomaterials (e.g., nanorods, carbon nanotubes, graphene, quantum dots, etc.) is one of the main means of increasing the sensitivity and selectivity of biosensors [3].

One rapidly developing direction in the field of theranostics is the utilization of magnetic nanoparticles as transport elements both in sensorics and in therapy by addressing drug delivery. Two main domains of MNPs are used for these purposes: chemically synthesized nanoparticles (which can be synthesized by various methods, such as sol–gel, chemical reduction, co-precipitation, hydrothermal synthesis, and pulsed laser ablation in dimethylformamide and by green methods using plant extracts) [4–8], and biogenic nanoparticles [9–11]. There are also other important applications that make magnetic nanoparticles suitable for theranostics (Figure 1).

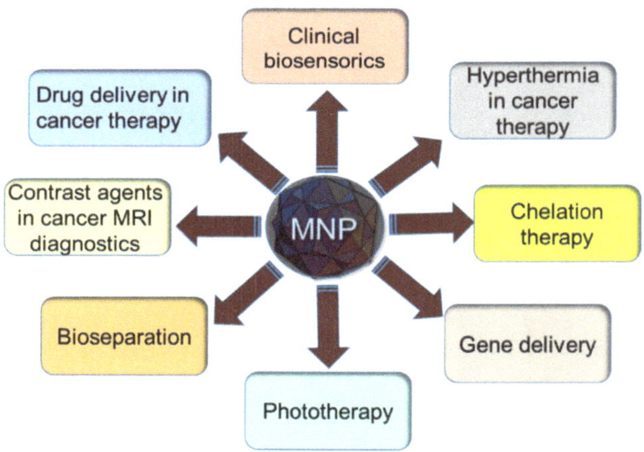

Figure 1. Main clinical applications of magnetic nanoparticles (MNPs).

In recent years, a particular interest is attracted to biogenic magnetic nanoparticles. Since "bacteria with motility directed by the local geomagnetic field have been observed in marine sediments" by Richard Blakemore in 1975 [12], a large number of studies have been carried out in order to determine the properties of such particles, or, so-called magnetosomes, which are the intracytoplasmic membrane vesicles, containing magnetic nanocrystal covered in a lipid bilayer with proteins [13–20]. The most important applications of such biogenic nanoparticles include the biosensors for medical diagnostics, and cancer diagnostics, in particular, vehicles for target delivery of anticancer agents [21], hyperthermia treatment employing alternating electromagnetic field [22], and tumor diagnostics using magnetic resonance imaging (MRI) [23].

In this review, synthetic inorganic and biogenic magnetic nanoparticles will be considered particularly with regard to their applications in biosensors and drug delivery.

2. Synthetic Magnetic Nanoparticles

2.1. Characterization of Synthetic Magnetic Nanoparticles

Magnetic nanoparticles are biocompatible and lend themselves well to modification by various biorecognition ligands [24]. The main characteristics of MNPs are their subcellular size, ranging from a few nanometers to tens of nanometers, allowing them to interact with nano-molecular-sized biomolecules [25]. Due to their unique properties, magnetic nanoparticles can be used in various biomedical applications including diagnostics [26–28], drug delivery [29], hyperthermia treatment [30], tumor cell isolation [31], and precise reagent manipulation [32].

Ferromagnetic materials are composed of electrically charged particles connected in groups that create magnetic dipoles, which are called magnetons. The volume of a magnetic material is a domain structure in which all magnetons are aligned in a single direction by the exchange forces. This domain structure distinguishes ferromagnetics from paramagnetics [33]. The domain structure of a material is capable of rearrangement as its particle size decreases. This property is one of the main factors determining its ferromagnetic behavior. Figure 2 demonstrates the dependence of the coercive force on the particle size. Reducing the particle size entails an increase in the coercive force up to a certain point—the transition between the single-domain and multi-domain states. The formation of two separate domains below this critical size will be energetically unprofitable. A further decrease in particle size leads to the fact that the coercive force decreases sharply and reaches a minimum in the superparamagnetic state. The transition to the superparamagnetic state induced by thermal effects is not a direct phase transition [34]. MNPs are strongly influenced by an external magnetic field due to the magnetic moment of the network element and the field tension. Therefore, when the external magnetic field disappears, they behave as inactive particles [29]. Synthetic MNPs can take various structural-hierarchical forms depending on the applied magnetic field, such as nanochains, nanorings, nanosheets, and large cuboids [35].

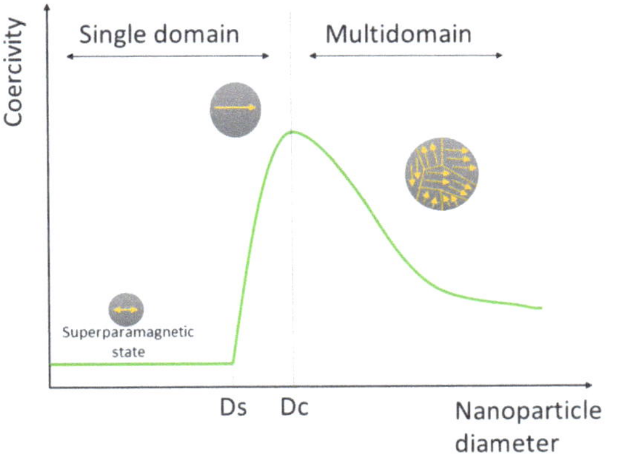

Figure 2. Dependence of coercivity on particle size: Ds and Dc are the thresholds of superparamagnetic and critical size, respectively.

Magnetic nanoparticles based on iron oxides (IONPs), such as maghemite (γ-Fe_2O_3), and magnetite (Fe_3O_4), are the most common for biomedical applications. Such particles are able to decompose into oxygen and iron and are easily excreted from the organism. When nanoparticles are fabricated of a size of approximately 10 nm in diameter, IONPs exhibit superparamagnetic behavior (superparamagnetic iron oxide nanoparticles, SPIONs), which is due to their better dispersibility without a magnetic field [36]. They accumulate in the target region in the presence of a magnetic field, which is of great importance for use in theranostics. IONPs can be doped with elements with high magnetic susceptibility (e.g.,

$CoFe_2O_4$, $MnFe_2O_4$, etc.) and metal alloys (e.g., FePt, FeCo, etc.), but their use in living organisms is difficult due to their rapid oxidation and potential cytotoxicity. Multiferroic magnetoelectric (ME) materials such as cobalt ferrite ($CoFe_2O_4$)/barium titanate ($BaTiO_3$) nanoparticles have been used for numerous applications, such as biosensors and drug delivery [37–40]. Their important feature is the presence of a cross-coupling between the electrical and magnetic phases to combine the properties of the individual phases. In ME composites, the application of an external electric field affects the magnetization and vice versa. Crosslinking in composite materials depends on the electrical and magnetic phases, the interface between them, the size, the nature of the connection between the phases, and deformation [41]. One of the most types of common magnetic nanoparticles is «core–shell» structures [42], where the core is a magnetic nanoparticle and its shell is a coating of biocompatible or bioselective material, which allows such structures to be used for drug delivery and in biosensor technology. Biocompatibility and cytotoxicity are key factors to consider when using core–shell MNPs in biomedical applications. Such particles can easily pass through biological membranes and move through the bloodstream, which can lead to disruption of the normal functioning of the systems of a living organism [43]. The presence of a shell on the surface of MNPs is necessary for their stabilization in a colloid system. This will ensure the safety of their interaction with living systems [44]. Hybrid core–shell structures using noble metals such as gold, silver, and platinum are also known [45–47]. The optical properties of such MNPs can be precisely tuned by changing the core size, and shell thickness as well as the core and shell shapes [48]. For example, iron oxide–Au MNPs generally exhibit the same magnetic properties as the cores with reduced saturation magnetization due to the mass contribution of the diamagnetic Au and iron oxide–Au core–shell nanostars exhibit multiple plasmonic resonances due to the coupling of the core and tip plasmons [48].

There is a variety of technologies for producing synthetic MNPs that control their size, shape, surface coating, colloidal stability, and other properties, which is especially important for biomedical applications. Magnetic nanoparticles can be manufactured either in a «top-down» or «bottom-up» approach. The «top-down» method involves a high-energy ball milling process of a magnetic sample until the desired nanoscale size is achieved. The advantage of the «top-down» method is that a large number of particles can be produced within a single batch, while the disadvantage is that the control over particle shape and size which is important in biomedical applications, is compromised. The «bottom-up» method starts with a salt of ferrous (Fe^{2+}) or ferric (Fe^{3+}) ions which then undergo a chemical process to nucleate and induce seeded growth to obtain particles of the desired hydrodynamic diameter [49]. The «bottom-up» approach includes such techniques as hydrothermal, solvothermal, sol–gel, co-precipitation, flow injection syntheses, electrochemical, and laser pyrolysis [50]. Microfluidic methods of synthesizing magnetic nanoparticles, which are based on "lab-on-a-chip" methods and approaches, were also described [51,52]. Such techniques include continuous-flow microreactors and droplet-based microreactors, based on cross-flow, co-flow, and flow-focusing methods, etc. [53,54].

2.2. Applications of Biosensors for Cancer Diagnostics

Magnetic nanoparticles (MNPs) are attractive for use in biosensors, since most biological samples have an insignificant magnetic susceptibility, and therefore the background against which measurements are performed is extremely low [55]. MNPs are biocompatible and can be easily modified with various biorecognizing ligands [24].

In biosensorics, the most widely used detection methods, in which MNPs are applied, are optical and electrochemical. Specialized magnetic detection techniques and other techniques are also known. Optical detection methods such as luminescent, fluorescent, colorimetric, etc., are very sensitive and specific. Their principles of operation are based on a change in the phase, amplitude, polarization, or frequency of incoming light in response to biorecognition processes. In such methods, to increase sensitivity, the target molecule or biorecognition element is labeled with a chromogenic or fluorescent label, such as a dye.

A change in color/fluorescence intensity indicates the presence of target molecules, which provides high sensitivity with a detection limit of up to one molecule [56]. Electrochemical biosensors use electrodes with recognition elements immobilized on their surface, capable of selectively binding to target molecules. Detection of a target binding to a recognizer in solution is based on the detecting of changes in currents and/or voltages. Electrochemical biosensors (potentiometric, amperometric, and impedimetric) represent detection systems that convert a chemical reaction into a measurable electrical signal. Due to their low cost, low power, and ease of miniaturization, electrochemical biosensors hold great promise for various biomedical applications, especially for Point-of-care Testing (PocT) devices. However, the functioning of these sensors can be influenced by various effects on the electrode surfaces related to pH, ionic strength, and the chemical composition of biological fluids [57]. Magnetic detection methods include various sensing techniques based on giant magnetoresistance, tunneling magnetoresistance, planar Hall effect, etc. These methods are used to measure the magnetic response in the form of susceptibility, relaxation, residual magnetization, and even frequency mixing [58].

Successful cancer treatment remains a challenge where the importance of early tumor diagnostics cannot be overestimated. Therefore, a major goal in modern biosensorics is to increase the sensitivity of detection methods and decrease the analysis time. Magnetic nanoparticles can be easily integrated onto the surface of the transducer or used as a sample preparation component with further concentration of the biological sample in the active region of the biosensor. Due to the growing interest in increasing sensitivity and selectivity, the optimization of MNPs for specific applications and the choice of optimal detection methods are important challenges for modern nanoscience [58]. MNPs can greatly increase the sensitivity of biosensor devices. MNPs exhibit different magnetic properties compared to the bulk material due to the reduced number of magnetic domains, which leads to the appearance of a superparamagnetic state. In this state, the magnetization can randomly change direction in a very short time. This superparamagnetic behavior prevents attractive or repulsive forces between magnetic nanoparticles until an external magnetic field is applied [59].

To diagnose oncological diseases, biosensors based on magnetic nanoparticles use various techniques for modifying them in order to bind highly specific biorecognizing agents, which should capture extremely low analyte concentrations. The variety of detection techniques and biorecognition interfaces used for the recognition of cancer cells are presented in Table 1. Some examples of applications of biosensor devices based on various detection methods using magnetic nanoparticles for cancer diagnosis will be discussed below.

Table 1. Applications of synthetic magnetic nanoparticles in biosensors for cancer diagnostics.

Detection Principle		Biorecognition Interface	Target	Detection Limit	Refs.
Electrochemical	Square wave voltammetry (SWV)	DNA-modified gold-coated magnetic nanoparticles (DNA-Au@MNPs)	DNA methylation for ovarian cancer diagnosis	2 aM	[60]
	SWV	DNA-Au@MNPs	Circulating tumor DNA (ctDNA)	5 fM	[61]
	Differential pulse voltammetry (DPV)	MWCNT/Fe_3O_4 modified with anti-PSA antibodies	Prostate-specific antigen (PSA)	0.39 pg·mL^{-1}	[62]
	DPV	Apt-GMNPs	Human T-cell acute lymphocytic leukemia cells (CCRF-CEM)	10 cells·mL^{-1}	[63]
	Amperometry	Fe_3O_4@GO modified with anti-PSA antibodies	PSA and prostate-specific membrane antigen (PSMA)	15 fg·mL^{-1} and 4.8 fg·mL^{-1}, respectively	[64]

Table 1. *Cont.*

	Detection Principle	Biorecognition Interface	Target	Detection Limit	Refs.
	Amperometry	Sox/Pt–Fe_3O_4@C/GCE	Sarcosine (prostate cancer biomarker)	0.43 μM	[65]
	Electrochemical impedance spectroscopy (EIS)	MBCPE/Fe_3O_4-RGO/PANHS/ssDNA	Breast cancer mutation *BRCA1 5382 insC*	2.8×10^{-19} mol·L^{-1}	[66]
	EIS	MNPs + antibodies	EpCAM, MUC-1, and HER-2	0.5 μg, 1.0 μg and 0.125 μg per 10^6 cells, respectively	[67]
	Chronoamperometry	γ-Fe_2O_3 /Cr^{VI}/Amine Oxidase	Polyamine in tumor tissue	2.47 μM	[68]
	Potentiometry	Anti-AFP with the nanogold/MPS–$CoFe_2O_4$ particles	AFP (α-1-fetoprotein)	0.3 ng·mL^{-1}	[69]
Optical	Surface-enhanced Raman spectroscopy (SERS)	Magnetic nanoparticle–antibody–CEA–antibody–gold nanoparticle–Raman reporter	Carcinoembryonic antigen (CEA)	10^{-12} M	[70]
	SERS	Raman tags-DNA probes modified Fe_3O_4@Ag NPs	MicroRNA in cancer cells (HeLa, MCF-7, A549)	0.3 fM	[71]
	SERS	Magnetic molecularly imprinted polymers (MMIPs) with anti-PSA@DTNB@Au nanoparticles	Prostate-specific antigen (PSA)	0.9 pg·mL^{-1}	[72]
	Surface plasmon resonance and MPQ cytometry	Magnetite nanoparticles modified by phytolectins (SBA, WGA, ConA)	Epidermoid carcinoma cells	up to 4.2 ± 0.1 pg·cell^{-1} 2.2 ± 0.5 pg·cell^{-1} and 0.45 ± 0.07 pg·cell^{-1}, respectively	[73]
	Surface plasmon resonance	Erlotinib conjugated MNP (erlotinib-MNP)	Human lung cancer cells (A549 cells)	5 μg·mL^{-1}	[74]
	UV–vis spectrometry	Au nanoparticles/DNA/magnetic beads	Anterior gradient homolog 2 (AGR2)	6.6 pM	[75]
	Fluorescent detection	DNA/dextran/PAA/Fe_3O_4 NPs	p53 protein	8 pM	[76]
	Magnetofluoro-immunosensing (MFI) system	Ag/iron oxide NP-decorated graphene	Prostate-cancer-cell-derived exosome	134.32 NPs·mL^{-1}	[77]
	Colorimetry	superparamagnetic iron oxide nanoparticles (SPIONs)/NanoZyme/Transferrine	Transferrin receptors in human U87MG glioblastoma cells	50 cells	[78]
	Colorimetry	Nanocomposite MNP and Pt NP in ordered mesoporous carbon	Human epidermal growth factor receptor 2 (HER2)	1.5 ng·mL^{-1}	[79]
Other principles	Loop-mediated isothermal amplification (LAMP) and lateral flow device (LFD) with magnetometric detection	Biotin-labeled inner primer and digoxigenin-labeled dUTP and gold magnetic nanoparticle (GMNP) as a signal generator	DNA methylation pattern of miR-34a	-	[80]

Table 1. *Cont.*

Detection Principle	Biorecognition Interface	Target	Detection Limit	Refs.
Methylation-specific lateral flow assay (MS-LFA) with magnetometric detection	Amplicon recognizing and capture by gold magnetic nanoparticles (GMNPs)	DNA methylation pattern of miR-34a	0.01 pg	[81]
Magnetic flow cytometry	Magnetic nanoparticles with aptamers	Pancreatic cancer cells	-	[82]
Magnetoresistance	Fe_3O_4 NPs/Ab in InSb-based semiconductor channel	Liver cancer antigen	0.14 pg·mL^{-1}	[83]
Nanoprobe-based nuclear magnetic resonance (NMR) spectroscopy	Core-shell $CoFe_2O_4$@$BaTiO_3$ magnetoelectric (ME) nanoparticles (MENs)	Ovarian carcinoma cells Skov3, glioblastoma cells U87-MG, and breast adenocarcinoma cells MCF-7	-	[84]
Giant magnetoresistance detection	MoS_2–Fe_3O_4-Aptamer	Exosomes derived from human A431 epidermoid carcinoma cells	100 exosomes	[85]

Electrochemical biosensors have become widespread due to a wide range of detection techniques, ease of technological implementation, availability of measuring equipment, and wide potential for miniaturization and integration into lab-on-a-chip systems. The current level of micro- and nanotechnologies has made it possible to create a variety of components of biosensor systems using magnetic nanoparticles, thus providing increasingly lower detectable concentrations of target substances, which is extremely important in cancer diagnostics.

The use of noble metal coatings has become a common technique for creating biorecognizing structures based on magnetic nanoparticles for electrochemical biosensors [60]. Thus, Chen et al. reported the development of an electrochemical biosensor for the detection of DNA methylation in blood-based on square wave voltammetry [60]. This process involves hybridization in a probe network of DNA-modified gold-coated magnetic nanoparticles (DNA-Au@MNP) complementary to the target DNA and subsequent enzymatic digestion to differentiate between methylated and unmethylated DNA strands. The detection limit for the developed biosensor was 2 aM. Another electrochemical sensing assay in combination with the DNA-Au@MNPs was used for the direct detection of the levels of circulating tumor DNA from whole blood [61]. This biosensor can selectively detect short- and long-strand DNA targets with a dynamic range from 2 aM to 20 nM for 22 nucleotide targets and from 200 aM to 20 nM for 101 nucleotide targets, respectively. In another study biosensor based on hollow hybrid magnetic Pt–Fe_3O_4@C nanospheres for sarcosine detection was presented by Yang et al. [65]. In order to achieve excellent electron transfer, polyaniline was used as a coating of Pt-Fe_3O_4 nanoparticles, which were then pyrolyzed to carbon.

A promising approach for the implementation of high-sensitivity electrochemical biosensors is the use of magnetic nanoparticles combined with carbon nanomaterials. The biosensor for prostate-specific antigen (PSA) detection was developed based on modification of the glassy carbon electrode (GCE) surface with a nanocomposite of carboxyl functionalized multi-walled carbon nanotubes (MWCNTs), and Fe_3O_4 nanoparticles for signal amplification and facilitating electron transfer [62]. The biosensor demonstrated the detection limit of 0.39 pg·mL^{-1} to measure PSA with a linear range from 2.5 pg·mL^{-1} to 100 ng·mL^{-1}. Khoshfetrat and Mehrgardi [63] reported an aptamer-based electrochemical biosensor with carbon-modified electrodes for quantitative detection of leukemia cells using MNPs. A composition of Fe_3O_4 nanoparticles coated with gold nanoparticles was used to immobilize the thiolated sgc8c aptamer on the surface. The binding of tumor cells to the aptamer leads to the destruction of its hairpin structure. As a result, intercalator molecules are released (ethidium bromide was used for this purpose), which leads to a decrease in the electrochemical signal. Amplification of the signal of the electrochemical platform was provided by the immobilization of nitrogen-doped graphene nanosheets on the electrode surface. Jahanbani et al. [66] designed a label-free DNA biosensor for breast cancer mutation detection (BRCA1 5382 insC) based on a magnetic bar carbon paste electrode (MBCPE) modified with Fe_3O_4/reduced graphene oxide (Fe_3O_4NP-RGO) as a composite and 1-pyrenebutyric acid-N-hydroxysuccinimide ester (PANHS) as a linker for DNA sequences detection (Figure 3). The MBCPE/Fe_3O_4-RGO/PANHS electrode was modified with probe strands for the exact incubation time. This biosensor showed a detection limit of 2.8×10^{-19} mol·L^{-1} in a linear range from 1.0×10^{-18} mol·L^{-1} to 1.0×10^{-8} mol·L^{-1}.

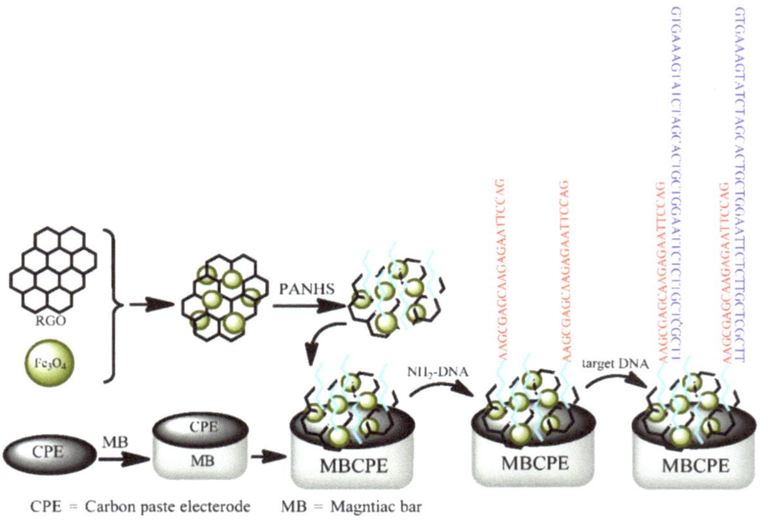

Figure 3. Schematic representation of the modified electrochemical biosensor based on MBCPE/Fe_3O_4-RGO/PANHS platform. Reprinted from [66] with permission of Elsevier provided by Copyright Clearance Center.

MNPs with a "core–shell" structure are among the most frequently used in biosensor applications. At the same time, new variants of shells are being created, which make it possible to detect various cancer cells with high sensitivity. Thus, surface-active maghemite nanoparticles (SAMNs) with a self-assembled coating of Cr^{VI} were used to make a nanostructure (SAMN@Cr^{VI}) with immobilized bovine serum amine oxidase (BSAO) [70]. The use of chromate made it possible to bind BSAO, which does not spontaneously bind to the SAMN surface, and the electrochemical signal of the SAMNs was radically changed on the formation of the self-assembled Cr^{VI} shell. The obtained nanoconjugate was employed for interference-free polyamine determination in liver cancer tissues. Wang et al. provided an electrochemical immunoassay method for the detection of α-1-fetoprotein (AFP) based on core–shell–shell nanoparticles functionalized with antibodies [69]. The basic $CoFe_2O_4$/(3-mercaptopropyl) trimethoxysilane nanostructure ($CoFe_2O_4$–MPS) was synthesized by covalent conjugation. Then, gold nanoparticles were sorbed onto the surface of this nanostructure using the Au–S bond, and then anti-AFP antibodies were immobilized. MNPs doped with biomolecules were attached to the electrode surface by applying an external magnetic field. The voltammetric biosensor was performed in a linear range from 0.8 to 120 ng·mL^{-1} AFP concentration range with a detection limit of 0.3 ng·mL^{-1}.

Optical detection systems are also widely used in biosensors due to their high sensitivity. Nevertheless, in this group of methods, it is also important to provide signal amplification for the detection of ultra-low analyte concentrations. Therefore, magnetic nanoparticles are also used in this group of methods.

A common optical detection method is surface-enhanced Raman scattering (SERS). This method offers the creation of quite low-cost multisensor systems, which is especially important for the development of Point-of-care Testing devices. Pang et al. [71] reported a surface-enhanced Raman scattering (SERS) detection system of the total RNA extracted from tumor cells using a hybrid Fe_3O_4@Ag MNPs biosensor functionalized with DNA probes. A single target miRNA molecule can rehybridize thousands of DNA probes to trigger signal-enhancing recycling in the presence of an endonuclease duplex-specific nuclease (DSN). The superparamagnetic properties of Fe_3O_4@Ag hybrid MNPs allowed capturing, concentration, and direct quantification of target miRNA let-7b without any PCR preamplification treatment. Additionally, MNPs may be used as a structural component of molecularly imprinted polymers (MMIPs) for SERS biosensors [72]. This sandwich structure served as an antibody-free probe and was labeled with gold nanoparticles that were modified with anti-PSA and a Raman reporter. This allowed to create a plasmonic structure between the MMIP and the SERS label. The detection and quantification limits of the developed sensor were 0.9 pg·mL^{-1} and 3.2 pg·mL^{-1}, respectively (Figure 4).

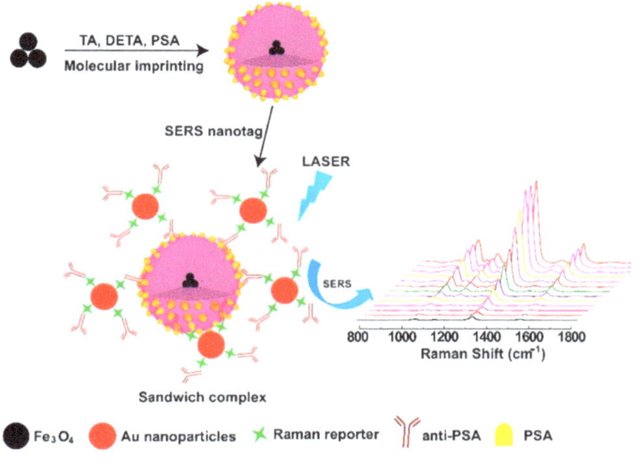

Figure 4. The schematic of plasmonic biosensor for prostate-specific antigen by combining magnetic molecularly imprinted polymer and surface-enhanced Raman spectroscopy. Reprinted from [72] with permission of Elsevier provided by Copyright Clearance Center.

Multiparametric surface plasmon resonance can be effectively used as a label-free technique for studying the process of dynamic mass transfer in the nanoparticle/cell system in the fluid cell. Shipunova et al. [73] obtained a spectrum of colloidally stable MNPs modified with phytolectins (SBA, WGA, ConA) of different specificity for monosaccharides (GalNAc, GlcNAc, and Man, respectively) and studied the interaction of these conjugates with A431 human epidermoid carcinoma cells. The authors showed that not only the angle of the minimum peak in the full angular spectrum but also the intensity of this peak can be used to study the binding of target MNPs to living cells in dynamics. This is explained by the contribution of metal nanoparticles to the absorption of incident electromagnetic radiation by free electrons by the resonance mechanism at the interface between media with different refractive indices. The combination of label-free SPR and magnetometric MPQ cytometry techniques allowed to establish that MNPs modified with soybean agglutinin bind to epidermoid carcinoma cells reaching saturation in 12 min to 4.2 ± 0.1 pg·cell^{-1} [73].

Spectrometric techniques can also be used for high-sensitive cancer biomarkers detection. One example of such a system is the work [75]. Here, the principle of UV–visible spectrometry was used to develop a sandwich-type aptasensor based on gold nanoparticles/DNA/magnetic beads to detect the cancer biomarker protein AGR2. The obtained structure (Figure 5), built on the high affinity between the aptamer and the target protein and providing a picomolar detection limit, made it possible to determine the target analyte with a sample volume of up to 20 μL. It was also noted that the sensitivity and selectivity of the developed sensor can be improved using magnetic separation.

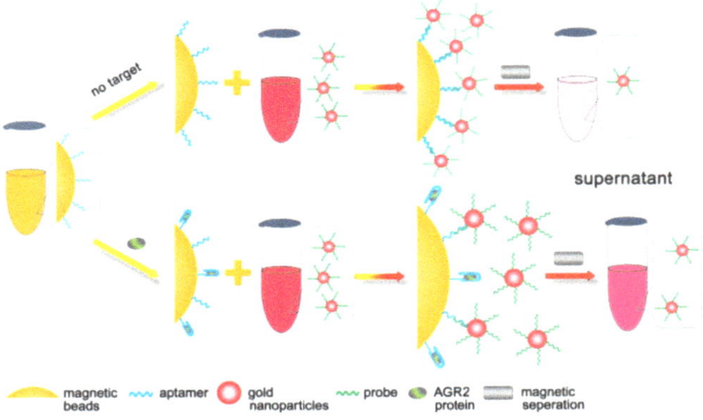

Figure 5. Schematic representation of AGR2 protein detection procedure. Reprinted from [75] with permission of Elsevier provided by Copyright Clearance Center.

Another example of spectrometric detection in biosensor systems is fluorescence registration. A fluorescent biosensor for p53 protein quantification was explored using DNA/dextran/PAA/Fe_3O_4 nanocarriers by Xu et al. [76]. Dextran-aminated MNPs were used to functionalize the consensus DNA that can selectively bind wild-type p53 protein (Figure 6). Dextran coating reduced nonspecific protein absorption and the sensitivity for p53 protein was achieved due to the facile magnetic separation from the complex condition. Inhibition of the process of DNA replacement by the captured p53 protein on the DNA consensus domain provided a decrease in fluorescent emission. Another promising approach for cancer detection war presented by Lee et al. [77], who developed cancer-cell-derived exosomes biosensor via the magnetofluoro-immunosensing (MFI) system using hybrid Ag/iron oxide NP-decorated graphene (Ag/IO-GRP) without purification and concentration processes. The authors successfully detected a prostate-cancer-cell-derived exosome from non-purified exosomes in a culture media sample in a concentration range from 10^2 NPs·mL^{-1} to 10^6 NPs·mL^{-1}.

Colorimetric methods of registration are based on measuring the optical density at a given wavelength, thus making it possible to determine the concentration of the substance in question. The result of the analysis can be recorded quite easily with a mobile device. A fast colorimetric immunosensor was developed on the basis of a nanocomposite of platinum and magnetic nanoparticles incorporated into mesoporous carbon [79]. This system allowed highly specific detection of human epidermal growth factor receptor 2 (HER2) at room temperature within 3 min. The technology exemplifies a strategy, in which nanocomposites are utilized for rapid, robust, and convenient identification of target pathogens. Consequently, the approach has potential applications for point-of-care (PoC) detection in clinical diagnostics.

The combination of Immunochromatographic and magnetometric techniques is quite promising for use in biosensors. The combination of loop-mediated isothermal amplification (LAMP) and lateral flow device (LFD) was used to identify traces of DNA methylation from highly heterogeneous cancerous specimens [80]. Gold magnetic nanoparticle (GMNP), working as a signal generator in this biosensor, enabled to interpret DNA methylation patterns through both visual and magnetic representation. The result can be obtained through both visual and magnetic detection. The performance of this biosensor was verified with real-world samples in the determination of the DNA methylation pattern of miR-34a promoter (Figure 7). Another site-specific biosensor based on lateral flow assay was established for both visual and magnetic DNA methylation determination [81]. The introduction of primers labeled with digoxin and biotin into PCR made it possible to recognize amplicons that can be recognized and captured by gold magnetic nanoparticles (GMNPs) using the developed biosensor device. The optical properties of GMNPs make it possible to use them as a signal probe and interpret amplicons even with the naked eye. The magnetic properties of these particles make it possible to register a signal using a magnetometer. The combination of such detection techniques is promising for use in clinical practice.

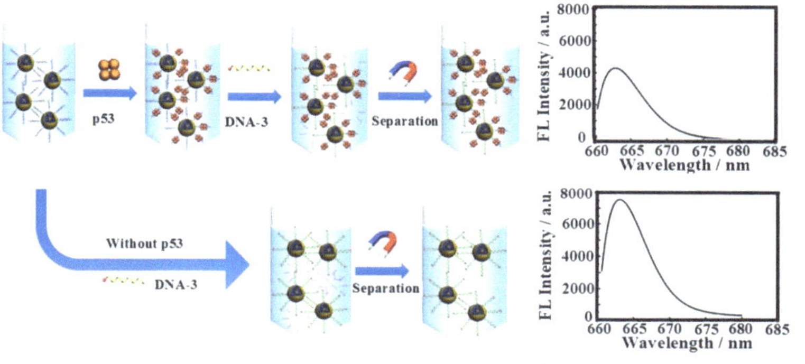

Figure 6. A fluorescent sensor for p53 protein expression was developed by combination of functional consensus DNA and magnetic nanoparticles. The sensor can realize ultrasensitive detection of p53 protein in real cell lysate. Reprinted from [76] with permission of Elsevier provided by Copyright Clearance Center.

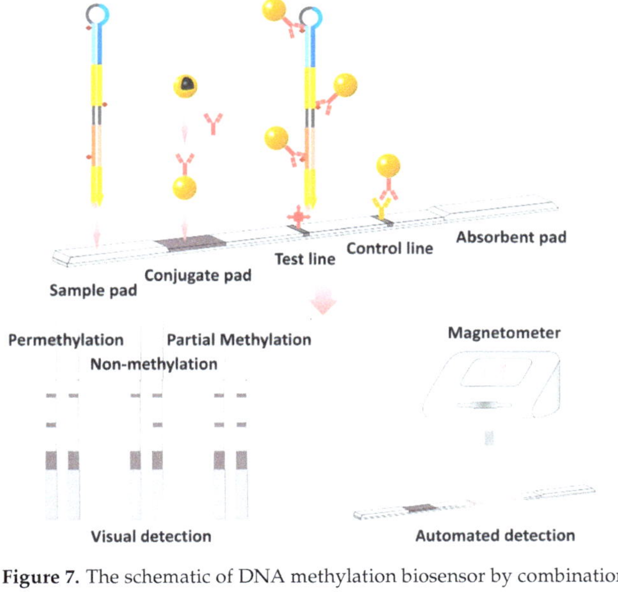

Figure 7. The schematic of DNA methylation biosensor by combination of isothermal amplification and lateral flow device. Reprinted from [80] with permission of Elsevier provided by Copyright Clearance Center.

The unique magnetic properties have also found their application in biosensors for cancer diagnostics. One of these properties is giant magnetoresistance, devices based on which can be integrated into miniature analytical systems. Magnetic flow cytometers are a possible solution for rapid cancer cellular detection in PoC testing devices. Thus, Huang et al. [82] reported an array of microfluidic biosensors based on a giant magnetoresistive spin valve with multiband sensor geometry and matched filtering to improve detection accuracy without sacrificing throughput (Figure 8). When cells labeled with MNPs pass by, the sensor generates a characteristic signal, which allows measurements to be taken in a multiparametric mode. The throughput of the developed device for multiparametric measurements was 37–2730 cells/min.

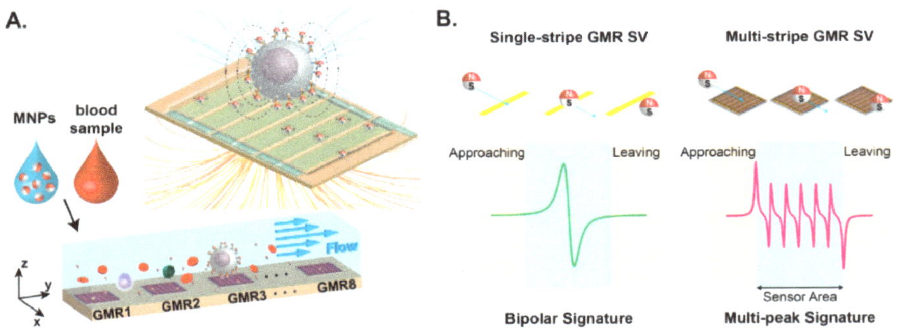

Figure 8. Magnetic flow cytometer (MFC) concept: (**A**). Operation of a GMR SV-based MFC where MNP decorated cells flow over GMR SV sensors. (**B**). Signature from conventional single-stripe sensors with a simple bipolar peak that increases the false detection events and the proposed multi-stripe configuration that enhances the signal differentiation by creating a unique magnetic signature. Reprinted from [82] with permission of Elsevier provided by Copyright Clearance Center.

Another magnetoresistive biosensor based on an InSb channel was demonstrated by Kim et al. [83]. The Fe_3O_4 nanoparticles bound to the target antigen created a stray magnetic field, which induced a change in semiconductor channel resistance due to the Lorentz force. The antigen concentration was proportional to the number of MNPs attached to the sensor surface and, therefore, could be determined by measuring the magnetoresistance of the sensor channel. Zhu et al. presented a GMR biosensor for exosome detection based on aptamer-modified 2D MoS_2-Fe_3O_4 nanostructures providing both multidentate targeting and signal amplification [85]. Unlike pure MNPs, layered MoS_2 nanostructure acts as a recruitment matrix for high-density MNPs as magnetic probes. The developed GMR sensor using 2D magnetic nanocomposites demonstrated reproducibility and selectivity with a detection limit of 100 exosomes.

Multiferroic magnetoelectric nanoparticles (MENs) are an attractive tool for the development of new magnetic tools for cancer diagnostics. For this, Nagecetti et al. [84] used such 30-nm core–shell particles as probes synthesized by the solvothermal method. In such a system, the electric and magnetic fields are inextricably linked. Due to the clear association with cells and the magnetoelectric effect, the NMR absorption spectra for cells incubated with MENs differed significantly from cells without such particles. Ordinary MNPs caused only minor changes in the adsorption spectra or did not cause them at all. The authors concluded that the minimization of the magnetoelectric energy upon binding of nanoparticles to cells is responsible for the change in the NMR adsorption spectrum in the case of MEN.

Thus, the variety of implementations of biosensor systems using MNPs for cancer diagnostics clearly indicates broad prospects for the introduction of these systems into clinical practice. Undoubtedly, the key task in creating such systems is to increase their sensitivity, stability, reproducibility, reliability, as well as the economic availability of testing.

Nevertheless, the creation of multiparametric biosensor systems for cancer diagnostics remains an important task, since large-format POCT screening makes it possible to form a map of human health and exclude many causes of diseases at once. This will allow us to formulate the optimal strategy for the treatment or preventive correction of the patient's condition.

2.3. Synthetic Magnetic Nanoparticles for Drug Delivery

Following diagnostics, modern technologies should provide conditions for the successful treatment of oncological diseases. Most anticancer drugs (e.g., doxorubicin, paclitaxel, curcumin, etc.) are administered intravenously, which often leads to a significant number of side effects [86–91]. The major limitation of present chemotherapeutic agents is the poor selectivity and the resultant toxicity [92]. Therefore, to minimize the negative effects of anticancer drugs, various systems are being developed for their targeted delivery directly to tumor cells [93].

Magnetic nanoparticles have become one of the most developing means of targeted drug delivery against cancer due to the possibility of their modification with various shells that improve their biocompatibility, allowing them to attach more active substances to them and avoid aggregation in the blood vessels. The size of MNPs is an extremely important factor for their therapeutic use since particles with a diameter of less than 10 nm are quite easily excreted through the renal clearance, while those larger than 200 nm are absorbed by the spleen [94]. To bring MNPs with an anticancer drug to the site of the tumor, high-gradient rare-earth metal magnets are mainly used, which make it possible to focus the magnetic field in the desired area. However, as the distance of the tumor from the body surface increases, the effectiveness of such targeting decreases, since the strength of the applied magnetic field decreases [95]. Figure 9 shows a scheme for targeted delivery of anticancer drugs using MNPs.

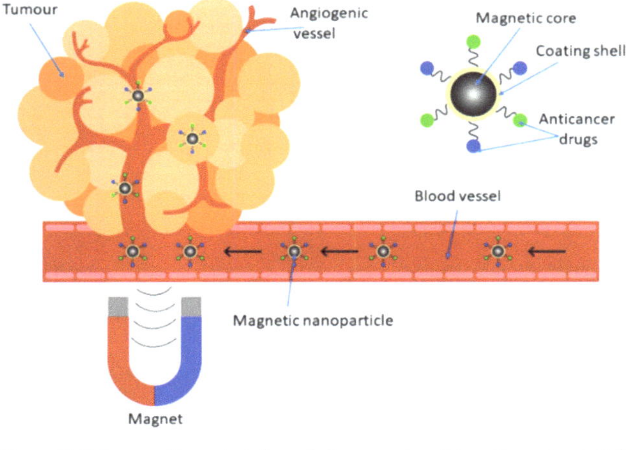

Figure 9. Scheme of targeted drug delivery using magnetic nanoparticles.

Since toxicity remains the main challenge in the use of magnetic nanoparticles in anticancer drug delivery, to overcome this, new methods of modifying MNPs into biocompatible shells with chemotherapeutic agents attached to them are being developed. In addition, it is necessary to overcome the agglomeration and aggregation of MNPs, which can also cause negative consequences for the body and complicate the complete release of drugs in the tumor site.

The variety of anticancer drugs necessitates the creation of an optimal delivery system for each of them to the tumor site in order to provide the most effective treatment. Oncotheranostics and nanomedicine are currently developing quite intensively and are now becoming one of the key areas in the creation of new methods of cancer treatment. Therefore, we propose to consider in this part of the review only fairly recent examples of the use of magnetic nanoparticles for targeted drug delivery. Table 2 shows the structures of systems for targeted delivery of various anticancer drugs based on the work of the last 5 years.

Table 2. The diversity of magnetic nanoparticle cores and shells modification agents for targeted drug delivery in oncology.

Anticancer Drug	Type of MNPs	Coating Agents	Target Cell	Refs.
Adriamycin	Fe_3O_4	Homogenous gelatin microspheres	Hepatocellular carcinoma (HCC)	[96]
Bufalin	Fe_3O_4	Liposomes	4T1 breast cancer cells	[97]
Camptothecin (CPT)	Fe_3O_4	Dextran + folate	Prostate cancer cells	[98]
Cisplatin	Fe_3O_4	Amphiphilic polymer + near-infrared dye-labeled HER2 affibody	HER2-expressing tumor cells	[99]
Curcumin (Cur)	Fe_3O_4	Bovine serum albumin	MCF7 cells	[90]
Cur	$ZnFe_2O_4$	L-cysteine (L-Cys) + oxygen-containing functional groups and nitrogen-rich mesoporous graphite-phase carbon nitride (Ox, N-rich mpg-C_3N_4)	Human lung adenocarcinoma A549 cells	[100]
Cur	Fe_3O_4	Hyperbranched polyglycerol (HPG) and folic acid (FA)	HeLa cells	[101]
Doxorubicin (DOX)	Fe_3O_4	Polyethylene Glycol (PEG) + polyarabic acid	Human breast cancer cell line MDA-MB-231	[102]
DOX	Superparamagnetic iron oxide nanoparticles (SPIONs)	Poly(ethylene glycol)-poly(aspartic acid) [PEG-P(Asp)] copolymer	Colon carcinoma and fibroblast cell lines	[103]
DOX	mesoporous haematite Fe_2O_3	-	Human breast cancer, MCF-7	[104]
DOX	$CoFe_2O_4$	Leucine (Leu)	HeLa cells	[105]
DOX	Fe_3O_4	Magnetic molybdenum disulfide (mMoS_2) + Liposomes	Human breast cancer, MCF-7	[106]
DOX	Ag-Fe_3O_4	Dextrin + cell-penetrating peptide (Tat)	MCF-7 cells	[107]
DOX and methotrexate	$CoFe_2O_4$@$BaTiO_3$	-	Human hepatocellular carcinoma (HepG2) and human malignant melanoma (HT144)	[108]
Erlotinib (ERL)	SPIONs	Poly N-isopropyl acrylamide (PNIPAM) with aptamer AS1411	Prostate cancer cells	[109]
Growth hormone-releasing hormone antagonist of the MIA class (MIA690)	$CoFe_2O_4$@$BaTiO_3$	-	Human glioblastoma cells (U-87MG)	[110]
Hydrophobic anticancer agent ASC-J9	Fe_3O_4	Silk fibroin + cationic amphiphilic anticancer peptide, G(IIKK)$_3$I-NH_2 (G3)	Colorectal cancer cells HCT 116	[111]
Methotrexate	Fe_3O_4	Arginine	MCF-7, 4T1, and HFF-2 cell lines	[90]

Table 2. *Cont.*

Anticancer Drug	Type of MNPs	Coating Agents	Target Cell	Refs.
Oxaliplatin (OXA), and irinotecan (IRI)	Fe_3O_4	Chitosan (CS)	CT-26 cancer cells	[112]
Paclitaxel (PTX)	SPIONs	FA-conjugated Polyethylene glycol (PEG)/ polyethyleneimine (PEI)-SPIONs SPTX-loaded nanoparticles (SPTX@FA@PEG/PEI-SPIONs)	Nasopharyngeal carcinoma	[88]
siRNA	Fe_3O_4	Polyethyleneimine (PEI)	B-cell lymphoma-2 (BCL2), Ca9-22 oral cancer cells	[113]
Sorafenib	Fe_3O_4	Mesoporous organosilica + MnO_2 + hyaluronic acid	Human lung adenocarcinoma A549 cells	[114]
Quercetin 5-fluorouracil	SPIONs	Zeolitic imidazolate frameworks (ZIF) + FA	Breast cancer MDA-MB-231 cells	[115]
Quercetin	$MnFe_2O_4$	Mesoporous hydroxyapatite (HA)	Human breast cancer MCF-7 cells	[116]
Ursolic acid (UA)	Fe_3O_4	β-cyclodextrin, folate	Human breast cancer MCF-7 cells	[117]
Violacein	Fe_3O_4	Polylactic acid	Glioblastoma and melanoma cancer cell lines	[118]
Zidovudine	$NiFe_2O_4$	Poly(vinyl alcohol)/stearic acid with poly(ethylene glycol) PEG	Human SK-BR-3 breast cancer cell lines	[119]
5-fluorouracil (FLU)	Fe_3O_4	(3-aminopropyl) triethoxysilane + tryptophan (TRP)	Human breast cancer MCF-7 cells	[120]
FLU	Fe_3O_4-Pt	FLU@PEG nanospheres	4T1 cells	[121]

As can be seen from Table 2, a very wide range of coatings for magnetic nanoparticles is used to deliver various anticancer drugs. The variety of coatings used to create targeted drug delivery systems based on magnetic nanoparticles suggests that the optimal composition of such a coating has not yet been found, which would minimize the toxicity of the system and make it fully biocompatible. Below we consider some examples of such structures with different coatings.

One of the most common coatings for drug delivery using MNPs is liposomes. A nanocomposite based on Fe_3O_4 nanoparticles coated with liposomes loaded with bufalin was presented as a drug for inhibiting lymphatic metastasis in breast cancer [97]. MNPs performed a targeting and photothermal function, accumulating in the sentinel lymph nodes of laboratory mice. The proposed technique allowed to reduce the incidence of lung metastases by 81% and achieve 91% tumor inhibition in the sentinel lymph nodes of mice. Lee et al. reported on doxorubicin nanocarriers based on liposome-coated magnetic molybdenum disulfide ($mMoS_2$) used for combined photochemotherapy [106]. The nanocarriers demonstrated a rather low rate of nonspecific protein adsorption and a low degree of aggregation in physiological saline. A reasonably successful cellular uptake profile of MCF-7 cells without significant cytotoxicity was obtained from in vitro studies. While in vivo studies (Figure 10) demonstrated that a drug delivery system based on $mMoS_2$ and liposomes provides tumor inhibition in mice with fewer negative effects.

Amino acids are also one of the widely used coatings for magnetic nanoparticles in oncotheranostics. Thus, L-cysteine-encapsulated $ZnFe_2O_4$ nanoparticles in combination with oxygen-containing functional groups and a nitrogen-rich mesoporous graphite phase with carbon nitride were used as a biodegradable target sonodynamic chemotherapeutic agent for tumor eradication [100]. The developed nanocomposite served as a carrier of the anticancer drug curcumin with a pH and ultrasound trigger, as well as to perform a semi-enzymatic sonocatalytic function. In another study, Attari et al. [90] proposed a method for the preparation of iron oxide MNPs coated with arginine using the in situ co-precipitation method and the one-pot method. The obtained nanoparticles were covalently bound to methotrexate and can target most cancer cells whose surface is overexpressed with folate receptors. Due to the functionalization of nanoparticles with arginine, an amide bond appeared on their surface between the amino groups and terminal carboxylic acid grappa on methotrexate, which was released from the nanoparticles in the presence of protease-like lysosomal conditions. Experiments on cell lines MCF-7, 4T1, and HFF-2 demonstrated the absence of cytotoxicity, which makes the developed system promising for use in clinical practice. In [105], solvothermally synthesized nanoparticles of cobalt ferrite ($CoFe_2O_4$) coated with leucine were used as a doxorubicin delivery system. The developed nanocarriers not only showed the ability to effectively inhibit the proliferation of HeLa cells, exerting an obvious cytotoxic effect on them but also demonstrated high sensitivity to a magnetic field in comparison with $CoFe_2O_4$ nanoparticles without leucine coating (Figure 11).

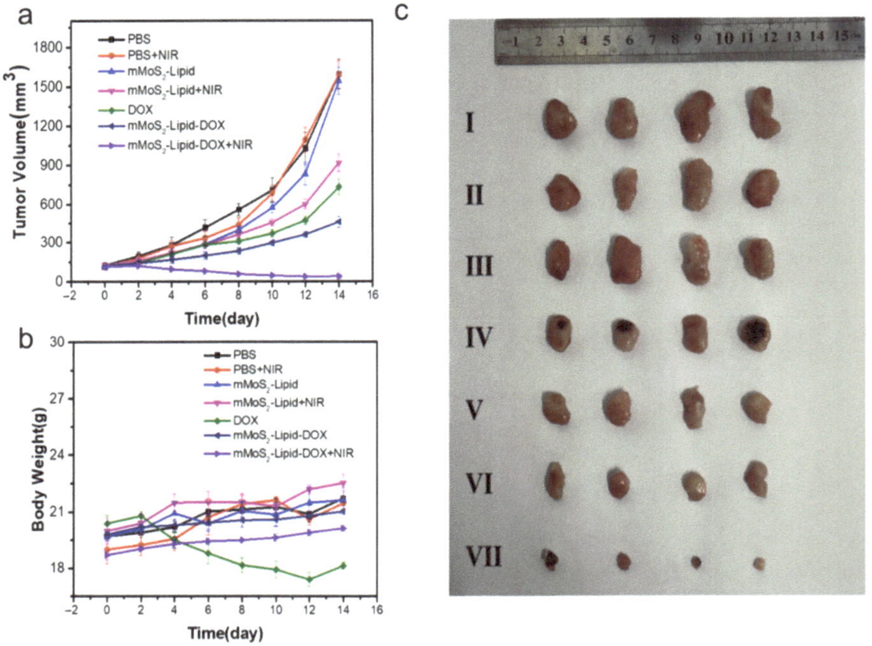

Figure 10. The changing curve of tumor volume from the beginning to the end of treatment (mean ± SD, $n = 4$) (**a**). The weight change curve of mice in each group (mean ± SD, $n = 4$) (**b**). The tumor image of different groups of mice on the 14th day following the treatment (mean ± SD, $n = 4$) (**c**). Reprinted from [106] with permission of Elsevier provided by Copyright Clearance Center.

Another promising direction in the development of magnetotherapeutic preparations is the use of polymer compositions as coatings for magnetic nanoparticles. Jin et al. [113] proposed polyethyleneimine-coated Fe_3O_4 MNPs for delivery of therapeutic siRNAs targeting B-cell lymphoma-2 and Baculoviral IAP repeat-containing 5 into Ca9-22 oral cancer cells. The study demonstrated significant inhibition of Ca9-22 cell viability and migration as a result of the use of nanoparticle-delivered siRNAs. Noh et al. [103] obtained a nanocomplex responsive to the tumor intracellular microenvironment, co-assembled from a copolymer of polyethylene glycol and poly (aspartic acid), superparamagnetic iron oxide nanoparticles (SPIONs), and doxorubicin. The performance of this therapeutic nanocomplex was studied on cell lines of colon carcinoma and fibroblasts. Moreover, NPs showed enzymatic degradation in the presence of protease, as well as a contrast effect on magnetic resonance imaging. Gui et al. presented a composite nanosystem based on folic acid (FA)-loaded SPIONs designed to reduce adverse reactions to water-insoluble parcitaxel (PTX) [88]. An increase in the hydrophilicity of PTX was achieved by modifying it with succinic anhydride, thus obtaining'-succinate parcitaxel (SPTX). SPTX-loaded FA-conjugated polyethylene glycol (PEG)/polyethyleneimine (PEI)-SPION (SPTX@FA@PEG/PEI-SPION) was prepared by solvent volatilization and hydrogen bond adsorption. Pharmacokinetic studies in rats in vivo on nasopharyngeal carcinoma cells (Figure 12) showed that SPTX@FA@PEG/PEI-SPIONs particles had a longer duration of action ($t_{1/2}$ = 3.41 h) than free SPTX or PTX ($t_{1/2}$ = 1.67 h).

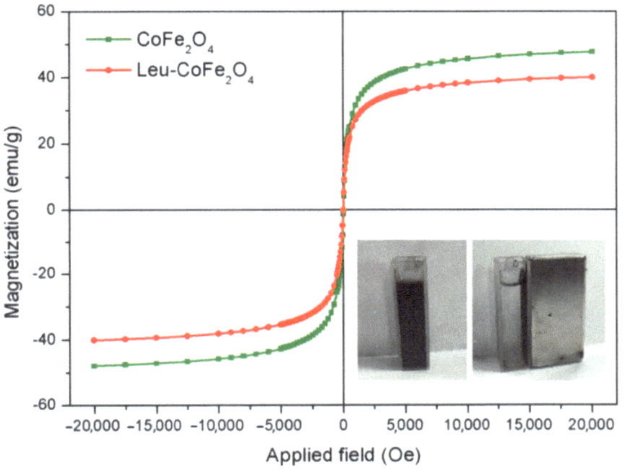

Figure 11. Magnetization curves of bare $CoFe_2O_4$ and Leu-coated $CoFe_2O_4$ nanoparticles. The inset shows the process of dispersion and magnetic separation. Reprinted from [105] with permission of Elsevier provided by Copyright Clearance Center.

To reduce toxicity and increase biocompatibility, magnetic nanoparticles can be coated with proteins. Chen et al. used high-voltage electrospray technology to develop microspheres based on Fe_3O_4 nanoparticles with a gelatin shell for loading them with adriamycin [96]. In addition to a good antitumor efficacy, the obtained nanocomplex activated ferroptosis in tumor cells (the ferroptosis marker GPX4 was significantly decreased, and ACSL4 was significantly increased) together with exposure to microwave hyperthermia, and also showed excellent properties for magnetic resonance imaging. Another interesting application of protein coatings is work [90], where bovine serum albumin-coated MNPs were used as curcumin carriers. Nanospheres prepared through desolvation and chemical co-precipitation process demonstrated cytotoxic activity on the MCF-7 cell line and sustained release of curcumin at 37 °C in different buffer solutions. Tomeh et al. [111] developed a microfluidic method for the production of peptide-functionalized magnetic silk nanoparticles based on silk fibroin for targeted delivery of the hydrophobic anticancer agent ASC-J9 (Figure 13). A swirl mixer integrated into a microfluidic chip allowed to achieve the required shape and size for the synthesized MNPs. Their surface was functionalized with a cationic amphiphilic antitumor peptide G $(IIKK)_3I\text{-}NH_2$ (G3) in order to increase the selectivity to cancer cells. The resulting complex increased the anticancer activity and cellular uptake of the G3 peptide in HCT 116 colorectal cancer cells as compared to the free G3 peptide.

The use of polysaccharides as coatings for MNPs is another common option for creating nanosystems for targeted drug delivery based on MNPs. Dextran is widely used for this purpose. In the study [98], SPIONs coated with dextran and conjugated with folic acid were synthesized by co-precipitation to deliver camptothecin to prostate cancer cells. The nanocarriers, which were spherical with an average diameter of 63.31 nm, demonstrated antitumor activity in AT3B-1 cancer cells by actively releasing and delivering camptothecin at 37 °C in phosphate and citrate buffers. The MNPs presented in [102] coated with polyarabic acid (a water-soluble polysaccharide molecule) and loaded with doxorubicin demonstrated effective penetration through cell membranes and internalization into breast cancer cells in a mouse model. The developed nanomaterials have demonstrated good biocompatibility, low cytotoxicity in vitro and in vivo, as well as the possibility of using them as a contrast agent in MRI.

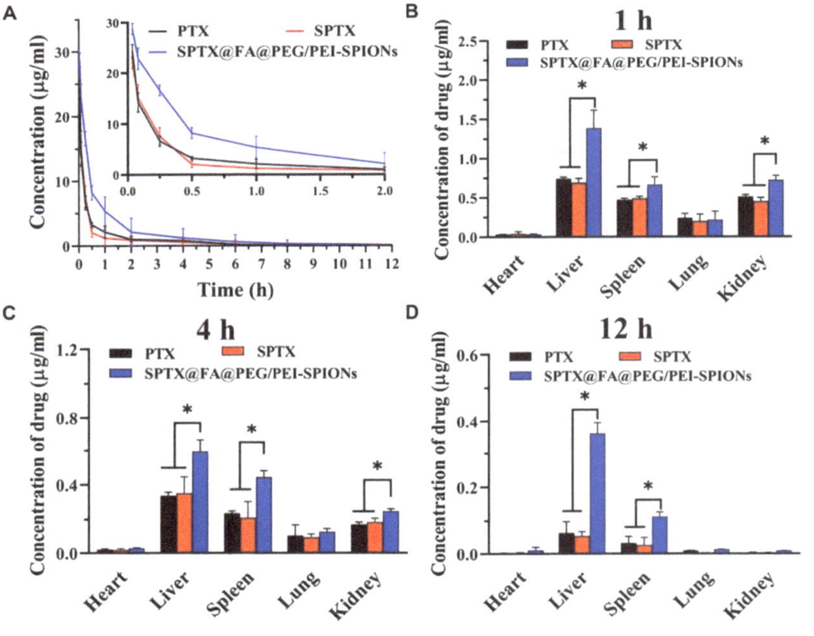

Figure 12. Plasma concentration–time curves of free PTX, SPTX, and SPTX@FA@PEG/PEI-SPIONs in vivo, data are expressed as the mean ± SD (n = 4) (**A**). Tissue distributions of free PTX, SPTX and SPTX@FA@PEG/PEI-SPIONs at 1 h (**B**), 4 h (**C**) and 12 h (**D**) post-intravenous injection (n = 4), * $p < 0.05$. Reprinted from [88], license CC BY 3.0.

It is also of interest to form non-toxic coatings on magnetic nanoparticles based on mesoporous materials. Recently, Abolhasani Zadeh et al. [104] proposed mesoporous hematite MNPs loaded with doxorubicin as a multifunctional theranostic agent exhibiting therapeutic activity against human MCF-7 breast cancer cells. These biomimetic mesoporous MNPs have over 71% doxorubicin loading efficiency resulting in a 50% reduction in cancer cells at a concentration of 0.5 $\mu g \cdot ml^{-1}$. The obtained MNPs, having a polygonal structure with an increased surface area and high porosity, became suitable nanocontainers for a high loading of doxorubicin. Another promising method for effective tumor cell killing is ferroptosis, which bypasses apoptosis and overcomes tumor drug resistance. Thus, acid- and redox-sensitive MNPs loaded with sorafenib developed by Chi et al. effectively stimulated tumor ferroptosis and inhibited tumor growth in vivo [114] (Figure 14). Mesoporous organosilicon nanoparticles (MONs) were coated on the outside with Fe_3O_4 MNPs, which provided sufficient iron ions for ferroptosis and magnetic targeting. As a result, a core–shell nanostructure was formed, which contained a disulfide bond with a redox reaction. MnO_2 was dropped onto the surface of the MON as a pylorus, which degraded to O_2 at low pH to promote sorafenib release. Hyaluronic acid acted as a protector of the nanoparticles from removal by the immune system and promoted active targeting of cancer cells. Another interesting study demonstrated the use of mesoporous magnetic $MnFe_2O_4$ core–shell nanocomposite particles for poorly water-soluble quercetin delivery [116]. $MnFe_2O_4$ nanoparticles as a core of nanostructure were made with the co-precipitation method. Then, the synthesized MNPs were coated with mesoporous hydroxyapatite (HA) shell as a new perspective for drug loading. The magnetic mesoporous nanostructure had a specific surface area and mean pore size of 165.44 m^2/g and 11.561 nm, respectively, which provided the possibility to efficiently load QC into the MNPs' pores with the subsequent pH-dependent release of the agent.

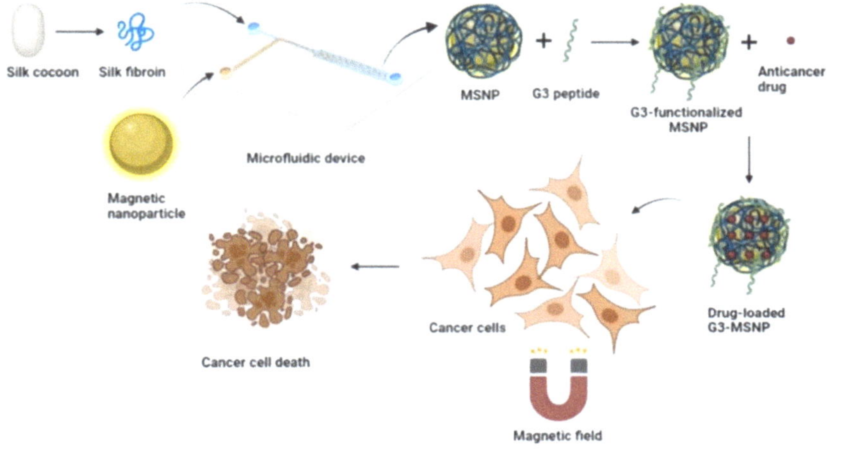

Figure 13. A schematic of peptide-functionalized magnetic silk nanoparticles produced by a swirl mixer for enhanced anticancer activity of ASC-J9. Reprinted from [111] with permission of Elsevier provided by Copyright Clearance Center.

Multiferroic magnetoelectric nanoparticles are also used in the targeted delivery of anticancer drugs. Stewart et al. [110] studied the externally controlled anticancer effects of binding synthetic tumor growth-inhibiting peptides to CoFe2O4@BaTiO3 magnetoelectric nanoparticles (MENs) in the treatment of glioblastomas. MIA-class growth hormone-releasing hormone antagonist molecules (MIA690) were chemically linked to these particles and then tested in vitro on human glioblastoma (U-87MG) cells. Studies have demonstrated externally controlled, highly efficient binding of MIA690 to MEN, specificity for glioblastoma cells, and on-demand release of the peptide using d.c. and a.c. magnetic fields, respectively. The work [108] presents colloidally stable MENs of cobalt ferrite @ barium titanate ($CoFe_2O_4@BaTiO_3$) synthesized by the sonochemical method and further functionalized with doxorubicin and methotrexate. In vitro cytotoxicity studies performed on hepatocellular carcinoma (HepG2) and human malignant melanoma (HT144) cells confirmed the magnetoelectric properties of $CoFe_2O_4@BaTiO_3$ NPs in the presence of an external magnetic field (5 mT) with significantly increased cytotoxicity compared to free preparations and without field replicates.

Metal–organic frameworks (MOFs) are becoming a promising tool for drug delivery applications. Pandit et al. [115] reported the development and fabrication of a dual MOF composite with encapsulated iron oxide (IO) nanoparticles coated with folic acid (FA) as the targeting agent and quercetin (Q) as the drug agent. Due to the presence of SPION, composites inherently show potential use for MRI. The integration of dual zeolite imidazolate frameworks (ZIF-8/ZIF-67) with targeting agents and drugs demonstrated the effective anticancer activity of the obtained nanocomposites (IO/Z8-Z67/FA/Q) in an FA receptor-positive breast cancer cell culture model (MDA-MB-231). The resulting nanocomposite enhanced apoptosis and cytotoxicity in the MDA-MB-231 cell line (expressing folate receptors) compared to the MCF-7 cell line, in which folate receptors were absent. Mechanically, the folic acid receptor targeting the delivery of the IO/Z8-Z67/FA/Q nanocomposite to MDA-MB-231 cells caused high ROS generation and nuclear fragmentation, which led to cell death. The proposed nanocomposite was also used for 5-fluorouracil loading, and the results of cytotoxicity suggest that it is a versatile nanocarrier for targeted drug delivery.

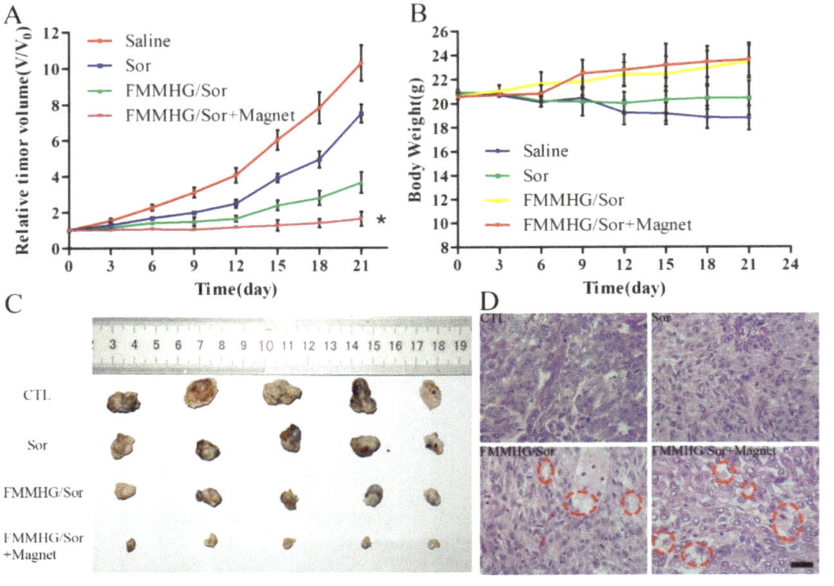

Figure 14. In vivo antitumor therapeutic effect of FMMHG/Sor. (**A**) The tumor volume of mice after being treated with saline, free Sor, FMMHG/Sor, and FMMHG/Sor + magnet for 21 days. * $p < 0.05$. (**B**) The body weight of mice in each group (saline, free Sor, FMMHG, and FMMHG + magnet) with treatment for 21 days. (**C**) Representative photos of tumor tissues obtained in different groups (saline, free Sor, FMMHG, and FMMHG + magnet) after 21 days of treatment. (**D**) Histological section of tumor tissues with H&E staining of different groups (saline, free Sor, FMMHG, and FMMHG + magnet) after 21 days of treatment. The fat vacuoles were marked with a red circle, which was the sign of ferroptosis. Scale bars, 100 µm. Reprinted from [114] with permission of Elsevier provided by Copyright Clearance Center.

Hybrid structures containing nanoparticles of iron oxide and noble metals have also found application in the targeted delivery of anticancer drugs. The system prepared by Liu et al. [107], was based on dextrin-coated silver NPs, which were then cross-linked with iron oxide NPs and a cell-penetrating peptide (Tat), resulting in dual-functional Tat-FeAgNPs with both superparamagnetic and cell-penetrating properties. The resulting nanocomplex can first overcome the blood flow shear force and reach the target organ under the action of an external magnetic field, and then the surface-modified Tat can further promote tissue penetration, which can effectively improve the efficiency of targeted drug delivery. The results showed that the obtained nanocomplex can promote cellular uptake and cytotoxicity of nanoparticles loaded with doxorubicin, while the IC_{50} of Tat-FeAgNP-Dox was 0.63 $\mu mol \cdot L^{-1}$. Nie et al. [121] fabricated platinum (Pt) nanozymes dispersed on the surface of iron oxide (Fe_3O_4) nanospheres loaded with 5-fluorouracil (FLU), which, in addition to enhancing peroxidase-mimic activity and catalsimic activity, led to the formation of a pH-sensitive nanoplatform for drug delivery for breast cancer treatment. Cytotoxicity tests showed that the obtained Fe_3O_4/Pt-FLU@PEG nanospheres moderate the proliferation of 4T1 cancer cells mediated by apoptosis and intracellular production of reactive oxygen species. In vivo assays have shown a significant reduction in tumor size and overcoming tumor hypoxia.

Since MNPs are foreign objects to the body, therefore, the immune system rejects them and various toxic reactions are caused. Therefore, the targeted delivery of drugs to the human body requires the control of such parameters of MNPs as shape, size, homogeneity, and coating composition. The coating plays an essential role in various applications: being highly biocompatible and stable, it would allow the adhered biomolecules to remain active for longer within the body and satisfactorily control their release or remaining attached for long periods to be stored and used for diagnosis [50]. MNPs synthesis and formulations face critical biological barriers, such as localization at the target site, the effective delivery of the drug to the target site, cross-physiological talk, and the other technical obstacles specific to cancer [122]. The development of precision drug delivery systems based on magnetic nanoparticles will make it possible to implement highly effective oncotheranostic techniques and improve the quality of life of patients.

3. Biogenic Nanoparticles

3.1. Biogenic Synthesis and Diversity of Magnetic Nanoparticles

Biogenic synthesis of nanoparticles can be carried out by organisms such as bacteria [123–126], fungi [127], lichens [128], and algae [129]. The production of biogenic nanoparticles is environmentally friendly since the synthesis process takes place at ambient temperature and pressure, and no toxic chemicals are used [130]. Hence, many researchers are focusing on synthesizing biogenic nanoparticles over chemically or physically synthesized ones to produce inexpensive, energy-efficient, and non-toxic metal nanoparticles [131]. Various types of naturally synthesized metallic nanoparticles consist mainly of Ce, Ag, Au, Pt, Pd, Cu, Ni, Se, Fe, or their oxides [132]. Among them, a particular interest is attracted to magnetic nanoparticles [133].

Magnetotactic bacteria (MTB) synthesize magnetosome magnetic particles with a well-controlled size and morphology, covered with an organic membrane 3–4 nm thick, which provides high and uniform, compared to artificial magnetite, dispersion in aqueous media, making them ideal biotechnological materials [134] (Figure 15).

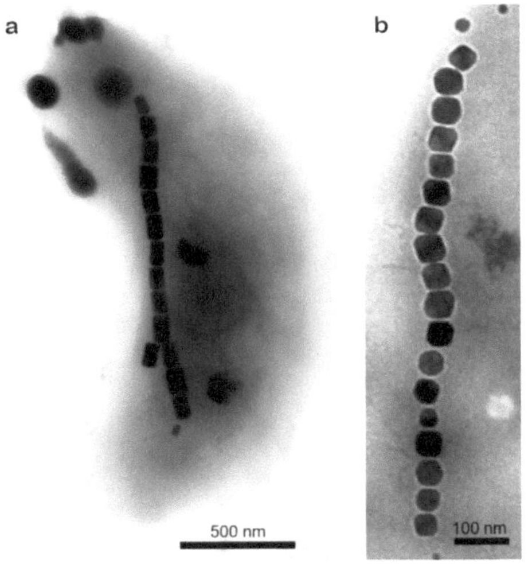

Figure 15. TEM image of (**a**) MTB with a chain of prismatic magnetosomes, (**b**) a chain of magnetosomes with a visible membrane.

Based on the results obtained in a detailed study of *Magnetospirillum gryphiswaldense* MSR-1 and *M. magneticum* AMB-1 strains, the mechanism of magnetosome biomineralization was suggested, with *mam* genes being mainly involved. Magnetosome formation is a complex process that has been divided into 4 steps, each of which involves certain

Mam proteins [135–138]. The magnetosome membrane is the result of the invagination of the cytoplasmic membrane [139,140]. The initiation point of the invagination process is apparently not determined by the specific composition of lipids in the membrane but rather is triggered by the presence of certain Mam proteins on it [137]. The minimal protein complex MamLQBIEMO enables proper invagination, whereas magnetite biomineralization requires additional proteins [119]. Among them, the membrane protein MamB is probably the most crucial [135,136]. The sorting stage inlolves the addressing of magnetosome proteins to the forming vesicle. Presumably, MamA protein plays an important role in this process, since it is present on the magnetosome membrane surface in large quantities [141,142] (Figure 16). MamA contains a repetitive protein–protein interaction site, which provides its oligomerization and the ability to bind other proteins [143]. Since MamA completely covers the magnetosome membrane, it can serve as a receptor for other magnetosome proteins [141,144]. For example, MamC, which is one of the most abundant proteins on magnetosome membranes in wild-type cells, was found to be mislocalized in *Magnetospirillum gryphiswaldense* MSR-1 mutants with MamA deletion [145].

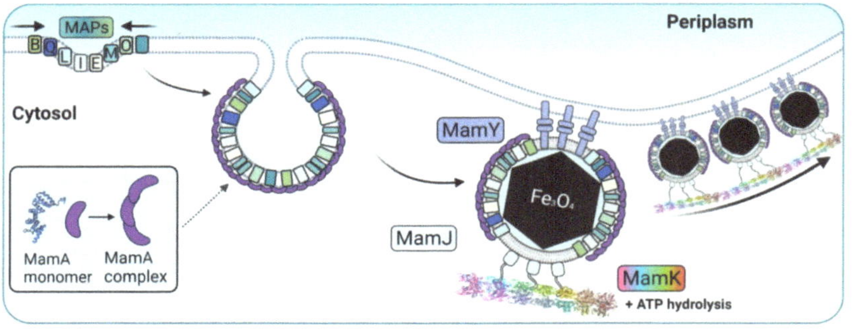

Figure 16. Suggested model of protein sorting, membrane invagination, and magnetosome assembly into an organized chain. Proteins solved structures are in ribbon representation. Reprinted from [137], license CC BY-NC-ND 4.0.

After the creation of the magnetosome vesicle, the transport of the corresponding ions in and out of the vesicle takes place to synthesize magnetite or greigite in the magnetosomes [146]. Such proteins as MamB and MamM, two carriers of divalent iron cations from the cell into the magnetosome, are involved [147]. When optimal physicochemical conditions are achieved in the magnetosome vesicle, one magnetite crystal per magnetosome is synthesized, which has a species-specific morphology [148]. Alignment into the chains occurs simultaneously with nucleation and crystal growth [137]. MamK, MamJ, and MamY are taking part in the chain organization [139,149].

The ability to form magnetosomes has been found in many bacteria from more than 10 prokaryotic phyla with different physiology [150–153]. However, all isolated in axenic culture MTB or those that have morphological descriptions and genomic sequences belong to the phyla *Pseudomonadota (classes Alphaproteobacteria, Gammaproteobacteria*, and *Magnetococcia), Thermodesulfobacteriota Nitrospirota*, and *Omnitrophota* (Table 3). Most MTB have magnetosomes organized into one or more chains. On average, MTB contains several tens of magnetic particles, and some species, such as *Candidatus* Magnetobacterium bavaricum, contain up to 1000 magnetosomes per cell [154]. The magnetic core crystals of magnetotactic bacteria are of different shapes which depends on MTB species. The majority of MTB can synthesize only one type of crystal, either magnetite or greigite. BW-1, however, was shown to be able to synthesize both types of crystals, depending on the sulfide concentrations in the medium [155]. The crystal size, crystallographic orientation, and arrangement of the magnetosomes in the MTB are crucial for the magnetic properties of the cell [156].

Table 3. Characteristics of magnetosomes from phylogenetically and morphologically identified MTB.

Name of Organism	Crystal Composition	Crystal Shape	Magnetosome			Ref.
			Number	Length (nm)	Width (nm)	
Alphaproteobacteria						
Magnetospirillum caucaseum SO-1.	Fe_3O_4	cuboctahedral	~25	40–50	40–50	[157–160]
Magnetospirillum gryphiswaldense MSR-1	Fe_3O_4	cuboctahedral	~30	32–45	32–45	[161–163]
Magnetospirillum kuznetsovii LBB-42	Fe_3O_4	cuboctahedral	~25	40–50	40–50	[164]
Magnetospirillum magneticum AMB-1	Fe_3O_4	cuboctahedral	~20	~45	~40	[165–167]
Magnetospirillum magnetotacticum MS-1	Fe_3O_4	cuboctahedral	~25	40–50	40–50	[167–169]
Magnetospirillum marisnigri SP-1	Fe_3O_4	cuboctahedral	~25	40–50	40–50	[157,170]
Magnetospirillum moscoviense BB-1	Fe_3O_4	cuboctahedral	~25	40–50	40–50	[157,171]
Ca. Magneticavibrio boulderlitore LM-1	Fe_3O_4	prismatic	~15	~50	~40	[172,173]
Magnetovibrio blakemorei MV-1	Fe_3O_4	prismatic	~10	~55	~35	[174–176]
Ca. Terasakiella magnetica PR-1	Fe_3O_4	prismatic	~15	~45	~35	[177]
Magnetospira sp. QH-2	Fe_3O_4	prismatic	~15	~80	~60	[178]
Gammaproteobacteria						
BW-2	Fe_3O_4	octahedral	~30	~65	~60	[179,180]
GRS-1	Fe_3O_4	octahedral	~300	~65	~55	[181]
FZSR-1	Fe_3O_4	prismatic	~20	~80	~55	[182]
FZSR-2	Fe_3O_4	prismatic	~20	~80	~55	[182]
NS-1	Fe_3O_4	prismatic	~10	~70	~60	[183]
SHHR-1	Fe_3O_4	prismatic	~15	~75	~55	[184]
SS-5	Fe_3O_4	prismatic	~20	~85	~65	[180,185]
Magnetococcia						
Magnetococcus marinus MC-1	Fe_3O_4	prismatic	~15	~80	~70	[186–188]
Ca. Magnetaquicoccus inordinatus UR-1	Fe_3O_4	prismatic	~30	~75	~45	[189]
Ca. Magnetococcus massalia MO-1	Fe_3O_4	cuboctahedral	~20	~65	~55	[190,191]
Magnetofaba australis IT-1	Fe_3O_4	cuboctahedral	~10	~85	~75	[192,193]
Thermodesulfobacteriota						
Ca. Belliniella magnetica LBB04	Fe_3O_4	bullet	~35	~100	~35	[194,195]
Desulfamplus magnetovallimortis BW-1	$Fe_3O_4Fe_3S_4$	bulletpleomorphic	NDND	~55~33	~35~32	[155,156, 196]
Desulfovibrio magneticus RS-1	Fe_3O_4	irregular/bullet	~10	~40	~20	[197–199]
Ca. Magnetananas rongchenensis RPA	Fe_3O_4	bullet	~70	~115	~40	[200,201]
Ca. Magnetoglobus multicellularis	Fe_3S_4	pleomorphic	60–100	~90	~70	[202–204]
Nitrospirota						
Ca. Magnetobacterium bavaricum	Fe_3O_4	bullet	~1000	~130	~40	[154,205]
Ca. Magnetobacterium casensis MYR-1	Fe_3O_4	bullet	~1000	~105	~40	[206,207]
Ca. Magnetobacterium cryptolimnobacter XYR	Fe_3O_4	bullet	~150	~130	~30	[208]
Ca. Magnetomicrobium cryptolimnococcus XYC	Fe_3O_4	bullet	~100	~135	~45	[208]
Ca. Magnetominusculus linsii LBB02	Fe_3O_4	bullet	~40	~120	~40	[194,195]
Ca. Magnetomonas plexicatena LBB01	Fe_3O_4	bullet	~35	~110	~45	[194,195]
Omnitrophota						
Ca. *Omnitrophus magneticus* SKK-01	Fe_3O_4	bullet	~175	~110	~35	[205,209]

The magnetosomes extracted from MTB meet all requirements in terms of size, morphology, biocompatibility, and magnetization capability [210–214]. However, despite the wide diversity of MTB and the remarkable properties of magnetosomes, only a few species from the phylum *Pseudomonadota* are cultivated and used to study the mechanisms of magnetosome formation and their applications [215]. Considerable efforts have been devoted to the production and purification of magnetosomes to obtain large yields of stable magnetic nanoparticles [216]. For example, studies have been carried out to optimize growth conditions [217,218] and transfer magnetosome genes into fast-growing non-magnetotactic strains [219].

Separation of MTB with different magnetic contents with subsequent isolation and enrichment can be performed using microfluidic devices based on their magnetic contents [220]. The motility of MTB can overcome magnetic forces, causing false positives (reduced purity) and false negatives (reduced yield). To overcome the movement of bacteria, MTB strains were treated with a cold/alkaline medium (10 °C, pH 8.5). Magnetosome production and growth were unaffected by this treatment. Thus, high-throughput separation of *Magnetospirillum gryphiswaldense* MSR-1 (1.000 cells/μL × 25 μL/min = 25,000 cells/min) was achieved with up to 80% sensitivity and 95% isolation purity. This demonstrates that microfluidic technology can greatly facilitate the separation of MTB cells with the required magnetic properties.

Isolating intact magnetosome organelles is an essential technique used in biotechnological applications. Magnetosomes from disrupted cells can be purified by means of their magnetic attraction with a permanent magnet and further ultracentrifugation in a sucrose density gradient [142,221,222]. However, despite extensive washing magnetically enriched magnetosomes still contained numerous contaminating proteins from other cellular fractions [223]. The use of nano and microfluidic technologies can be adapted for the study and isolation of magnetosomes and MTB [220,223,224].

3.2. *Applications of Magnetosomes in Cancer Theranostics*

3.2.1. Biosensors on the Basis of Magnetosomes

Magnetosomes (MS), due to uniform size and morphology (Figure 15), highly ordered organic membrane (Figure 16), and ability to form homogeneous dispersions offer a perspective substrate in biotechnological applications, particularly biosensors [134]. Thus, their advantages over inorganic magnetite nanoparticles, include, first of all, a stable single-domain form of the magnetic core which provides a permanent magnetic state at ambient temperature, then—a high chemical purity, a narrow size distribution, etc. These properties enable MS to be applied for cell identification and isolation on-a-chip, antigen detection and recovery, enzyme immobilization and capture of target proteins, and contrast enhancement in magnetic resonance imaging (MRI) [224–226].

In [227], the authors described an original method of toxicity detection, in which the magnetic properties of magnetosomes within the magnetotactic bacteria are combined with bioluminescence ability. These features are achieved by genetic engineering on the basis of magnetotactic bacteria *Magnetospirillum gryphiswaldense,* strain MSR-1. The approach made it possible to obtain a hybrid organism (BL-MTB) that combines the magnetic properties of navigation with the ability to emit a red glow of click beetle luciferase, and the latter property turned out to be proportional to the viability of the bacterium [227]. The magnetic navigation ability of bacterium served as a "natural actuato" to provide transport of bacteria within a microfluidic chip from the reactor to the detection volume. As a result, a cost-effective biosensor for toxicity monitoring was developed using microfluidic technology implemented on a polydimethylsiloxane (PDMS) molded chip. This analytical technology is quite express, since BL-MTB are incubated for 30 min with the sample, moved by microfluidics, trapped, and concentrated in detection chambers by an array of neodymium–iron–boron magnets [227].

A biosensor for white spot syndrome virus (WSSV) detection was presented in [228]. The biosensor is implemented on the basis of antigen-antibody reaction of VP28-specific antibodies conjugated with magnetosomes at concentrations of 1 and 2 $mg \cdot mL^{-1}$ and VP28 antigen at concentrations of 0.025~10 $ng \cdot \mu L^{-1}$. The complex was transported to carbon planar electrodes in a magnetic field applied externally and the antigen concentration was determined using an electrical impedance measurement principle. The assay was applied in monitoring seafood samples contaminated with WSSV and VP28 antigen of concentration as low as 0.01 $ng \cdot \mu L^{-1}$ was detected. Thus, magnetosomes were successfully applied in biosensors for detecting viruses, due to the possibility of biorecognition ligands conjugation to their native surface and the capability of addressing positioning and concentration of target particles at detection areas due to magnetic carriers [228].

A similar detection principle was implemented in biosensors for the determination of pathogenic bacteria and their toxins [229]. The authors presented a biosensor of *Salmonella typhimurium* employing a magnetosome with immobilized antibodies for the "O" antigen of *Salmonella* lipopolysaccharide (Figure 17). The optimal MS-Ab complex concentrations for detection of lipopolysaccharide concentration of 1 ng·mL^{-1}, were in the range of 2 mg·mL^{-1}~0.8 μg·mL^{-1}. An external magnet was used for the concentration of the probe at the area of the electrode. The reaction was detected using the electrical impedance principle. In real samples, the biosensor demonstrated high sensitivity with the bacteria detection limit of 101 CFU·mL^{-1} [229]. Figure 17 shows schematically the advantages of the developed biosensor, which integrates on a single chip a number of operations applied in conventional analysis of pathogenic organisms.

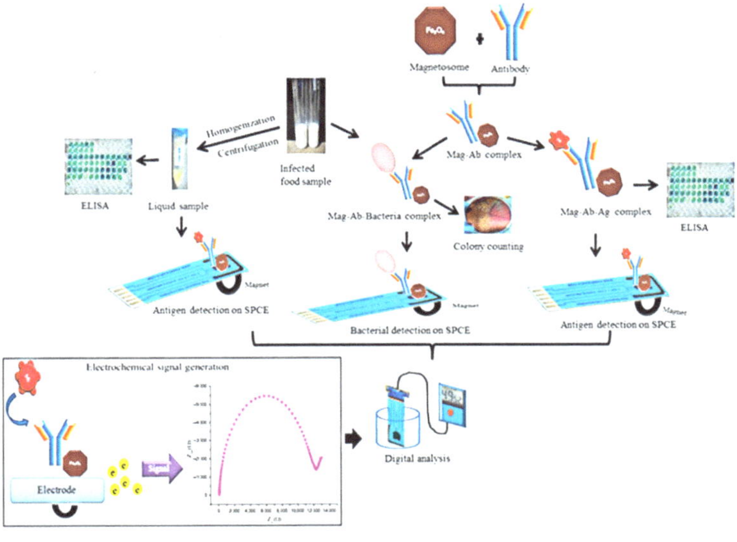

Figure 17. Schematic illustration of lipopolysaccharide and *Salmonella typhimurium* detection using magnetosome-anti-Salmonella antibody complex. Reprinted from [229] with permission of Elsevier provided by Copyright Clearance Center.

An important issue in the applications of magnetosomes is the efficiency of biogenic particles in comparison with synthetic magnetic nanoparticles. In [230], the functional applicability of genetically engineered magnetosomes was evaluated and compared with that of commercial immunomagnetic beads. The engineered magnetosomes were fused to protein A and then bonded to antibodies. A previously constructed recombinant MTB strain, *Magnetospirillum gryphiswaldense* ΔF-FA, appeared capable of forming an engineered BMP with protein A on its surface. It has been demonstrated that magnetosomes are characterized by ordered arrangements of bonded antibodies on the surface with fused protein A with a linkage rate of 962 μg Ab per mg of magnetosomes [230]. The complex was used for the detection of *V. parahaemolyticus* surface antigen and hapten, whereas the maximal capture rate was 90% and detection sensitivity was 5 CFU·mL^{-1}. Thus, a new engineered BMP fused with protein A (ΔF-BMP-FA), coupled with an antibody demonstrated a higher capacity for adsorption of antigen and gentamicin as compared with those of commercial immunomagnetic beads. It has been shown that such particles are inexpensive, eco-friendly, and show a strong potential of applicability as alternatives to commercial immunomagnetic beads, having high Ab-conjugation and antigen-adsorption capacity [230].

A number of authors note that magnetosomes are perspective also as mediators for magnetic fluid hyperthermia and as contrast agents for magnetic resonance imaging, both in vitro and in vivo. Using magnetosomes produced by the magnetotactic bacteria *Magnetospirillum gryphiswaldence* authors of [231] demonstrated that the phospholipid membrane of magnetosomes provides good protection against oxidation and particles are stable over a period of several months. The temperature kinetic relationships obtained for magnetosomes dispersed in an agarose gel under an alternate magnetic field of 17 $kA \cdot m^{-1}$ at 183 kHz frequency demonstrated a rate of temperature rise of 1 $°C \cdot min^{-1}$, which corresponds to a high specific absorption rate (SAR) of 482.7 ± 50.8 $W \cdot g^{-1}$ per mass of iron. Further in [231] the MRI contrast efficiency was also evaluated by means of the acquisition of NMRD profiles for magnetosomes dispersed in agarose gel and in water, showing good results as a negative MRI contrast agent. The MRI experiments on an animal model were carried out with the human glioblastoma–astrocytoma (U87MG) cells inoculated into mice and their presence was detected by magnetic resonance images two weeks after the injection of magnetosomes into the tumor mass thus proving the diagnostic potential of this approach. The high values of relaxivity r2 and the r2/r1 ratio presented in the article [231] show that magnetosomes are efficient superparamagnetic contrast agents for MRI. Further progress in increasing MRI sensitivity and contrast is achieved in [215], where authors developed genetically engineered magnetosomes showing an extremely high relaxivity value of 599.74 $mM^{-1} \cdot s^{-1}$. The magnetosomes were extracted from *Magnetospirillum gryphiswaldense* MSR-1 and genetically engineered protein structures of anti-HER2 with the ability to target HER2 were conjugated to the surface layer of the magnetosomes via the anchor protein MamC. This allowed the magnetosomes to target tumors in vitro and in vivo. The magnetosomes did not cause any notable pathogenic effect in the animals, which will greatly advance the development of biogenic magnetic nanoparticles for noninvasive cancer imaging [232].

An effect of increased transverse relaxivity r2 in biogenic MNPs was also noted in earlier works [233]. Studies were carried out with magnetosomes isolated from *Magnetovibrio blakemorei* strain MV-1 and *Magnetospirillum magneticum* AMB-1 which were compared with commercial ferumoxide. The dispersions were studied in vitro and in vivo. Thus, relaxometry measurements at 17.2 T and 20 °C were carried out with phantoms containing agar. The estimated transverse relaxivities r2 for ferumoxide, cuboctahedral magnetosomes from *Magnetospirillum magneticum* AMB-1, and elongated-prismatic magnetosomes from *Magnetovibrio blakemorei* MV-1 were 17.3 ± 15 $mM^{-1} \cdot s^{-1}$, 489 ± 26 $mM^{-1} \cdot s^{-1}$, 728 ± 35 $mM^{-1} \cdot s^{-1}$, correspondingly. Aqueous dispersions were tested in the mouse model and the gain in sensitivity by T2*-weighted imaging at 17.2 T of the mouse brain vasculature was observed after injection of magnetosomes at low concentrations of iron (20 μmol iron kg^{-1}). Commercial ferumoxide with the same level of iron did not allow such a phenomenon to be observed [233].

An important issue in theranostic approaches on the basis of biogenic magnetic nanoparticles is the modification of the surface. It is noted that such particles are easier to modify due to the specific properties of their surface. In [234], the authors proposed a peptide, for modification of magnetosome surface, showing complementarity to human epithelial growth factor receptor (EGFR) with nanomolar affinity and to epithelial growth factor receptor-2 (HER2) with a lower affinity but comparable to other reported peptides [234]. EGFR is known for being overexpressed in many human epithelial cancers, and thus, could serve as a target for cancer diagnosis and therapy. Authors developed the peptide by screening a computational-aided design of one-bead-one-compound (OBOC) peptide library followed by in situ single-bead sequencing microarray. Two peptides, P75 and P19, were selected to be used as probes for breast cancer cell imaging. The specificity of peptides was tested by confocal fluorescence imaging of the culture with FITC-labeled peptides, incubated for 20 min, and washed. Co-localization analysis was also performed for which AlexaFluor555 conjugated anti-human EGFR antibody was used (Figure 18). Magnetosomes were isolated from *Magnetospirillum gryphiswaldense* (MSR-1) and coupled to targeting peptide P75 by the one-step condensation reaction of amino and carboxyl groups. The transmission electron microscopy (TEM) images of the intact bacteria, magnetosomes, and Mag-P75 with fluorophore indicated that peptide P75 functionalized magnetosomes are well dispersed and have a narrow size distribution. The modified-with-P-75 magnetosome nanoparticles were used for targeted magnetic resonance imaging on the mouse model. The results demonstrated the potential of this peptide for EGFR and HER2-positive tumor theranostics [234] (Figure 18).

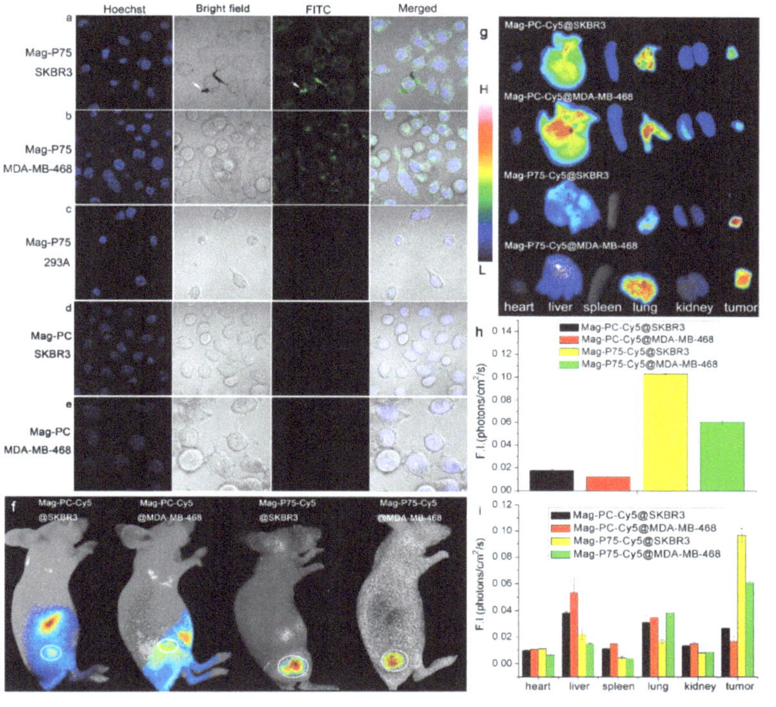

Figure 18. Validation of targeting ability and specificity of Mag-P75 in vivo and in vitro by fluorescence imaging. Confocal analysis of Mag-P75 (labeled with FITC, green) NPs in SKBR3 (**a**), MDA-MB-468 (**b**), and 293A (**c**) cell lines, and Mag-PC (labeled with FITC, green) NPs in SKBR3 (**d**) and MDA-MB-468 (**e**) cell lines, the white arrows indicate the targeting peptides were successfully coupled onto magnetosomes; (**f**) in vivo fluorescence imaging of Mag-PC and Mag-P75 NPs to SKBR3 and MDA-MB-468 tumor-bearing mice; (**g**) ex vivo fluorescence imaging of Mag-PC and Mag-P75 NPs accumulation in tumors and normal organs, and (**h**,**i**) quantification of the fluorescence signals of tumors in vivo and tumors vs. normal organs ex vivo, respectively. Fluorescence intensity was measured in terms of counts/energy/area and is presented with the average value (n = 3). Reprinted from [234] with permission of Elsevier provided by Copyright Clearance Center.

3.2.2. Drug Delivery in Cancer Theranostics Using Magnetosomes

The application of nanoscale vesicles for drug delivery in biomedicine has accelerated in recent years and they are now extensively used in patient treatment [235,236]. Magnetosomes isolated from magnetotactic bacteria can be used as carriers of anticancer drugs embedded in their membranes. In [237], the authors bound cytosine arabinoside (Ara-C), in order to reduce its toxic effect, on magnetosome membrane through crosslinking stimulated by the natural biological agent—genipin (GP). The magnetosomes were isolated from *Magnetospirillum magneticum* AMB-1. The complex showed a strongly enhanced controlled drug release effect relieving thus the severe side effects of the drug.

There are many methods of synthesizing MNPs of different sizes and coating, which can influence their performance as drug delivery vesicles, but magnetosomes are naturally occurring, lipid-coated MNPs that exhibit good stability, homogeneity, biocompatibility, etc. However, their properties, such as cancer cell uptake, toxicity, etc., need further studies. In [238], magnetosomes, synthetic MNP of different sizes, and coated biomimetic MNP were studied for their uptake by MDA-MB-231 cells (estrogen, progesterone, and Her-2 receptor-negative cell line, which serves as a good model of late-stage triple-negative breast cancers). The magnetosome mimics are MNP coated with the oleic acid (OA@MNP) and with silica (Si@MNP) (Figure 19), the latter are of two sizes.

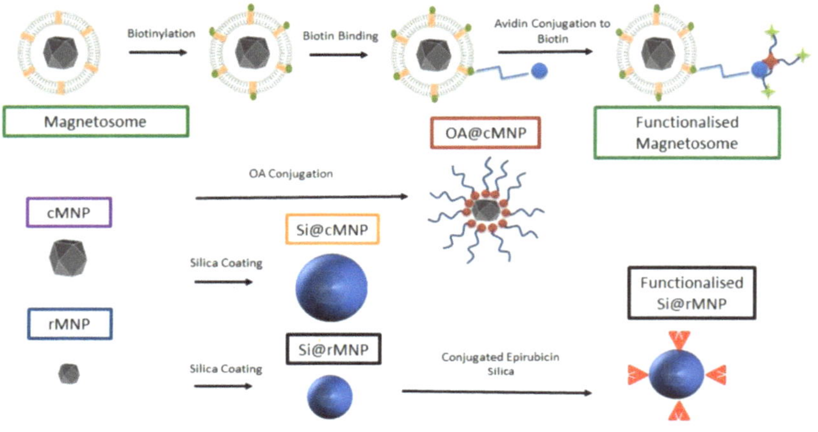

Figure 19. A schematic of the experimental design and samples produced. The top labeled in green depicts the magnetosome and how it is biotinylated for functionalization. Center-left shows the control cMNP labeled in purple with surface coatings of oleic acid (red) and silica (orange label). The bottom-left shows smaller control rMNP (blue label) coated with silica (black label) and with conjugated epirubincin (bottom right). Sample color-code used throughout in figures. Reprinted from [239], license CC BY 4.0.

The experiments demonstrated the uptake of particles by MDA-MB-231 cells through inclusion bodies and those particles were located intracellularly. The authors consider that due to the size of the particles, intracellular uptake most likely occurred via pinocytosis with the inclusion bodies being pinosomes or lysosomes, although other processes of MNP internalization including clathrin-mediated endocytosis, etc., are also possible [239]. The particle size was shown to have a negligible effect on overall iron uptake by the MDA-MB-231 cell line. The observed effects of internalization offer the ability to deliver therapeutic compounds directly into the cell and the use of their magnetism to steer the MNP within the body [238].

The magnetic properties of magnetosomes significantly improve the targeting potential for drug delivery in the presence of a magnetic field. Although the application of such processes obviously demands high levels of monodispersity and reproducibility of size and physical-chemical properties of magnetosomes which in turn ensures a reliable and consistent magnetic response and precise positioning at target tumors. In [240], the controlled navigation capabilities of *Magnetospirillum magneticum* strain AMB-1 (AMB-1) in a magnetic field to target a group of mammalian cells using an in vitro monolayer of Chinese hamster ovary (CHO) cells, including both healthy and tumor cells, was implemented and studied. The motility of MTB cells was controlled by a locally generated magnetic field using ~3-mm-sized solenoid coils forming a network of tracks (Figure 20). At initial time, AMB-1 cells interact with the neighboring CHO cells. When the next coil is charged CHO cells integrated with AMB-1 bacteria are observed moving towards the charged coil (marked with the dashed circle in Figure 20). The direction of the cells' movement is reversed by switching the order of the coils in the other direction. The authors also studied the interaction of AMB-1 and CHO via computer simulation, by selecting the surface proteins MSP-1 and flagellin of AMB-1 and about 14 potential candidates for CHO. The authors concluded that the mammalian cell surface proteins, which are predominantly responsible for cell signaling, are the primary targets of the AMB-1 cells that are dissuaded by flagellin of AMB-1 (Figure 21). On similar lines, the plasma membrane proteins, whose primary function is to maintain the mammalian cell structure and function, are the targets of the AMB-1 cell surface protein, MSP-1.

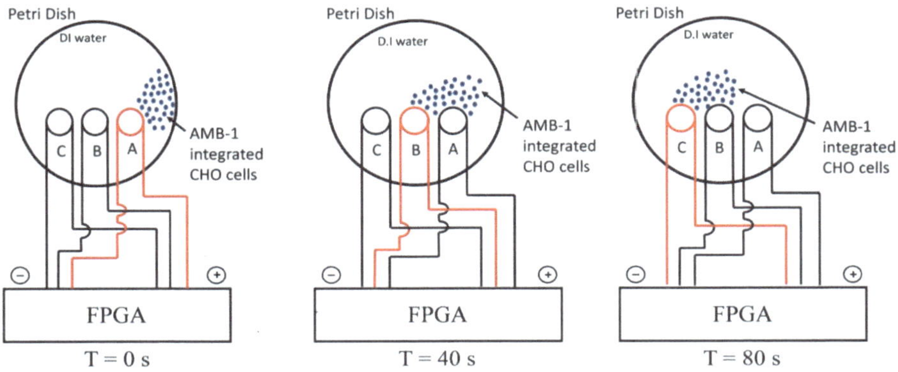

Figure 20. Schematic illustrating the experimental setup for directional control displaying the movement of AMB-1-integrated CHO cells from coil A to coil C. The charged coil is shown in red, and the AMB-1-integrated CHO cells are displayed as blue dots. Reprinted from [240], license CC BY 4.0.

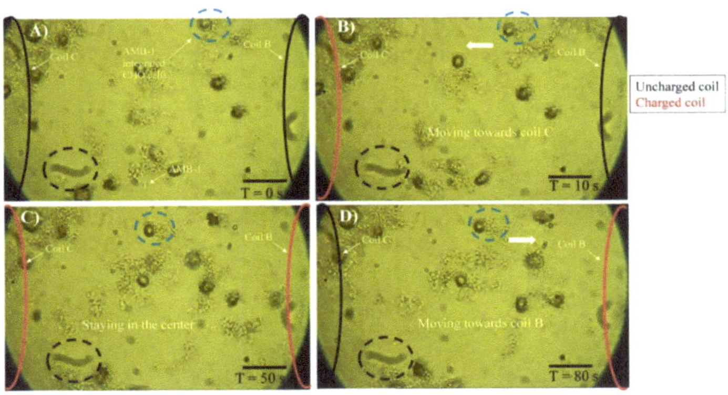

Figure 21. Motility of AMB-1-integrated CHO cells with time. (**A**) At T = 0 s, AMB-1-integrated CHO cells are moving randomly. (**B**) At T = 10 s, coil C on the left is charged and an AMB-1-integrated CHO cell, highlighted by a blue dotted circle, starts to navigate toward the charged coil C. (**C**) At T = 50 s, coils B (right) and C (left) are charged and the AMB-1-integrated CHO cell in the blue dotted circle stays in the center. (**D**) At T = 80 s, only coil C (left) is discharged and the AMB-1-integrated CHO cells can be seen moving toward coil B to the right. A stationary black dotted circle is added at the bottom of every frame to serve as a reference point. The reversal was achieved in the same manner (scale bar = 50 μm). Reprinted from [240], license CC BY 4.0.

Figure 21 presents images obtained with a phase contrast microscope at a magnification of 40×, showing experiments with controlled movement of AMB-1 cells, integrated with

CHO cells in a magnetic field. The direction of magnetic field lines is defined by switching the particular coil (C, B). At time T = 0 s, AMB-1 cells exhibit Brownian motion and interact with the neighboring CHO cells. When the coil B is charged at T = 10 s, the CHO cells integrated with AMB-1 bacteria move toward coil B (marked with a dashed blue circle), when coils C and B are charged the cell complex does not move, and when coil B is charged, and magnetic field changes polarity to the opposite, the complex of cells moved in the reverse direction [240].

The investigations in magnetosome applications in drug delivery for the treatment of oncological diseases are aimed mainly at achieving the chemical stability of preparations after administration as well as their precise delivery to target tumors and their nearest environment, keeping healthy tissues intact [241]. Biogenic magnetic particles—magnetosomes, perform this task better, due to the stability of their surface layer and their ability to bind specific medications and provide their gradual release. Furthermore, by exploiting their natural magnetotaxis they can be controlled with an externally generated magnetic field. This gives a prospect of guiding magnetosomes in the human body towards the target locations. [241]. Thus, by altering the magnetic field, as shown in [240], it may be possible to control the drug delivery process and move them to the tumor. Some authors [242–244] within the framework of this concept call these bacteria specialized nanorobots (Figure 22).

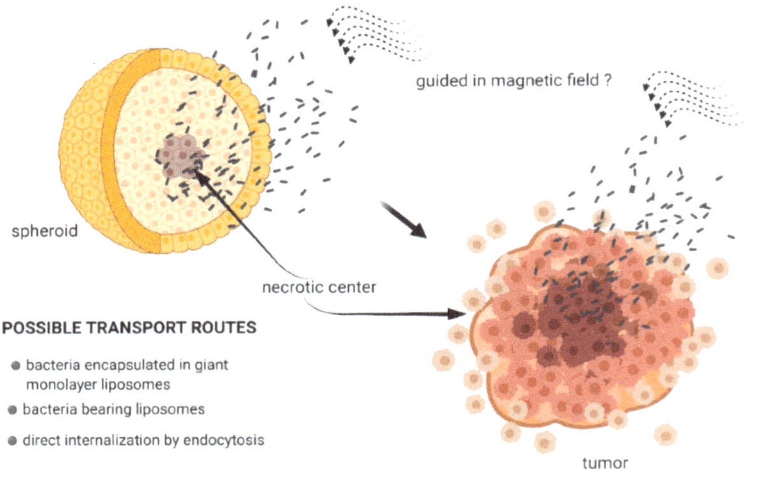

Figure 22. Magnetotactic bacteria as potential drug carriers capable of penetrating the tumor. Reprinted from [242], license CC BY 4.0.

In [245], a method for addressing targeting antitumor preparation using magnetosomes was developed on the basis of *Magnetospirillum gryphiswaldense* strain MSR-1, loaded with doxorubicin (DOX) and transferrin (Tf) towards human hepatoma cell line HepG2 and human normal hepatic cell line HL-7702. The simultaneous loading of DOX and Tf on the magnetosomes (Tf-BMs-DOX) enabled an address delivery and binding of complex to the target tumor cells via transferrin receptors (TfR), which are represented on the tumor cells in concentrations of about 100 times higher than in normal cells (Figure 23). The comparative studies with cancer and normal cells demonstrated that the complex Tf-BMs-DOX recognized HepG2 cells more specifically in comparison with HL-7702 because of the high expression of TfR on the surface of HepG2 cells. Data on drug release showed that magnetosomes loaded with DOX were capable of sustained drug release. This means that the frequency of administration and doses could be reduced and the therapeutic effect enhanced. Furthermore, it was observed that the complex Tf-BMs-DOX shows increased tumor cytotoxicity than free DOX or BMs-DOX. The tumor suppression rate was 56.78%, while in free DOX—31.26%. The results obtained in [245] show that magnetosomes modified with DOX and Tf are able to actively target the tumor via intravenous injection (Figure 23).

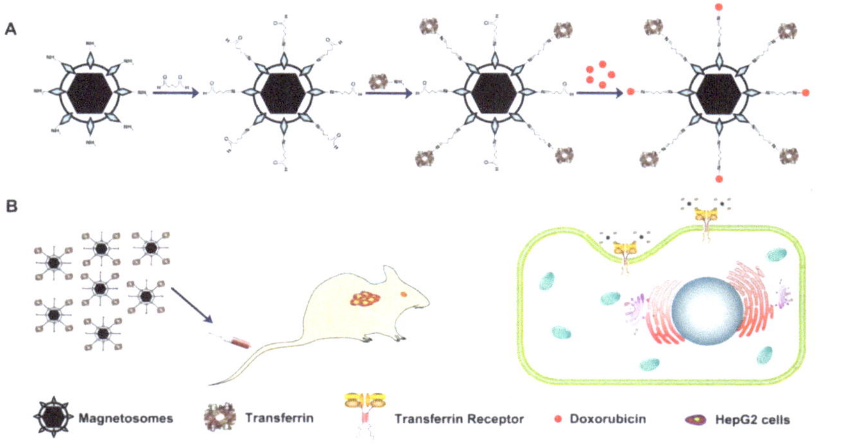

Figure 23. Schematic depiction of formulation of Tf-BMs-DOX (**A**). The antitumor of Tf-BMs-DOX in vivo (**B**). Reprinted from [245], license CC BY 4.0.

4. Comparative Analysis of the Relevance of Synthesized and Biogenic Particles in Biosensors and Drug Delivery for Cancer Theranostics

Biogenic nanoparticles (BNPs) have been evaluated in a number of studies as eco-friendly and a cost-effective alternative to the chemical synthesis processes [245,246]. The authors of [245] suggest that the advantages of biogenic nanoparticles are due to the natural thermodynamic stability of an organic layer surrounding the magnetic core in these structures. The stability of the structure of biogenic particles could be explained by the presence of different biological macromolecules, such as proteins, lipids, DNA, and polysaccharides, as well as low molecular weight metabolites, such as flavonoids, terpenoids, glycosides, organic acids, and alkaloids—all naturally produced by organisms [246]. Generally, all nanostructures are thermodynamically not stable due to high values of specific surface areas and energy, which leads to the necessity to stabilize them via adding components providing electrostatic, steric, dielectric, etc., stabilizing effects on the dispersion of nanoparticles. The diversity of biological molecules enables various stabilizing effects to be implemented in the nanosystems, including electrostatic repulsion, steric hindrance, van der Waals interaction, etc., which lead to a high degree of stability [246] (Figure 24).

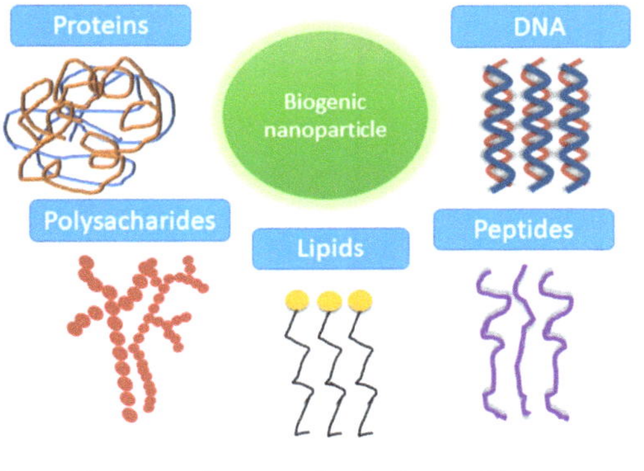

Figure 24. Classes of macromolecules involved in thermodynamic stabilization of biogenic nanoparticles.

Thus, a key feature of bacterial magnetosomes is the presence of a biological membrane with a defined biochemical composition [247]. This particular coating ensures high quality and homogeneity of dispersions and provides thermodynamic advantages for surface modification [247]. The modification of the surface layer is possible either chemically, or genetically. The second approach provides many advantages since it enables to implement a number of functions at the stage of magnetosome biomineralization [247].

Biosensors are mostly used for in vitro analysis, for which the toxicity issue is not of the primary importance, as that for the drug delivery procedures in vivo. (Although the direction of biosensors for in vivo monitoring begins to develop, it has not yet become a well-established approach [248].) The important features of magnetic nanoparticles for biosensors are: spatial order and stability of the surface layer, capability of chemical modification and bonding of ligands, uniformity of size, homogeneity of magnetic properties, high magnetic relaxivity value [215,231,233,249], and some others. The stability of the organic surface layer in biogenic particles is higher and the structure of the layer is more ordered, respectively the uniformity of the ligands layer is higher in biogenic particles, as noted by a number of authors [241,246]. At the same time, there is a vast diversity of chemically synthesized MNPs, such as "core–shell" type structures (Table 1), and a considerable amount of variants of shells are created, which make it possible to detect various cancer cells with high sensitivity, as well as the availability of high yield processes for the preparation of such MNPs which convinces us of the rationality of application of such particles in mass analysis, PoCT systems and such like applications [69,70]. A unique feature of biogenic nanoparticles is the possibility of application of genetic engineering approaches for their chemical modification which makes it possible to design unique analytical protocols [247]. Thus, it is possible to conclude that in biosensoric applications in vitro, magnetic nanoparticles of biological and inorganic origin, demonstrate close performance with some advantages of biogenic particles. At the same time, the area of MNP synthesis and applications developed a vast diversity of organic and inorganic coatings for MNPs, which is of great value and an opportunity to develop analytical methods for mass biomedical monitoring.

Mobility and targeted delivery problems for magnetic nanoparticles are becoming an actual and important direction of research. Since the formulation of the "magic bullet" concept by Paul Erlich [250], this is the closest perspective of its realization [244]. These bacterial microrobots can be remotely controlled using magnetic fields due to their internal chain of iron oxide nanoparticles acting like a compass needle, as well as, which seems to have more perspective—using magnetosomes as unidomain particles. A comparison of three-varying magnetic field sequences generated by three orthogonal pairs of electromagnets able to generate controllable 3D aggregations of MTB gives a prototype of nanorobots for targeted drug delivery [244]. In cancer therapy, the problem of low internalization of medications in tumor cells and the problem of low internalization of anticancer drugs remains very acute. At the same time, many cancer drugs are expensive and not readily available. These problems make the use of nanocarriers an efficient solution, which improves its therapeutic index via elevating tumor cell internalization and reducing the dose of medication [101]. Another problem is the poor selectivity of anticancer drugs and as a result high toxicity. The address delivery could resolve the toxicity issue [101]. At the same time, many problems with synthetic and biogenic nanoparticles are yet to be solved. Interactions of polymers, including proteins, resulting surface charges, geometry, and energy could have some advantages and disadvantages regarding drug delivery applications [247]. Surface charges can cause aggregation, as well as repulsion, as well as increasing or decreasing the adsorption of the drug onto the surfaces depending on the charges of the drug used.

In [251], the authors compared two different nanoparticles: bacterial magnetosome and HSA-coated iron oxide nanoparticles for targeting breast cancer. Both magnetosomes and HSA-coated iron oxide nanoparticles were chemically conjugated to fluorescent-labeled anti-EGFR antibodies. In vivo MR imaging in a mouse breast cancer model shows the effective intratumoral distribution of both nanoparticles in the tumor tissue. Magnetosomes demonstrated higher distribution than HSA-coated iron oxide nanoparticles according to fluorescence microscopy evaluation. According to the results of in vitro and in vivo study results, magnetosomes are promising for targeting and therapy applications of breast cancer cells [251].

5. Conclusions

In recent years, studies demonstrated the advantages of using biogenic nanoparticles in cancer therapy, as well as in vivo visualization of tumors and other pathological neoplasms. These advantages are due to low toxicity, high stability, and spatial order of the organic surface layers. An important property of magnetosomes is the ability to modify their biochemical properties by genetic engineering, which makes it possible to implement unique analytical protocols on biosensor platforms. At the same time, as regards the biosensors and various in vitro applications, the analysis of scientific articles shows that synthetic nanoparticles, as well as biogenic ones, are equally perspective, and further developments as regard shape, surface modification, and analytical protocols are actual and important. In a number of application types, such as mass monitoring of the population, or PoCT, the chemically synthesized MNPs are even preferable, due to relatively low production cost, considerably high yield of the manufacturing processes, sufficient control on the particles size and size distribution, as well as magnetic properties and chemical modification of the surface.

Author Contributions: T.M.Z. and N.O.S., Writing—Original Draft, Review and Editing; D.G., V.K., K.G.G., V.F., H.G., S.E.C. and M.S., Writing—Original draft, Review and Editing; M.S. and H.G., Funding Acquisition; M.S., Project Administration. All authors have read and agreed to the published version of the manuscript.

Funding: This research was funded by the Russian Science Foundation (Grant Number 22-15-00240). Huile Gao was supported by the National Natural Science Foundation of China (81961138009).

Institutional Review Board Statement: Not applicable.

Informed Consent Statement: Not applicable.

Data Availability Statement: Not applicable.

Conflicts of Interest: The authors declare no conflict of interest.

References

1. Schiffman, J.D.; Fisher, P.G.; Gibbs, P. Early Detection of Cancer: Past, Present, and Future. *Am. Soc. Clin. Oncol. Educ. Book* **2015**, *35*, 57–65. [CrossRef]
2. Thévenot, D.R.; Toth, K.; Durst, R.A.; Wilson, G.S. Electrochemical biosensors: Recommended definitions and classification. *Biosens. Bioelectron.* **2001**, *16*, 121–131. [CrossRef]
3. Solaimuthu, A.; Vijayan, A.N.; Murali, P.; Korrapati, P.S. Nano-biosensors and their relevance in tissue engineering. *Curr. Opin. Biomed. Eng.* **2020**, *13*, 84–93. [CrossRef]
4. Singh, A.; Sahoo, S.K. Magnetic nanoparticles: A novel platform for cancer theranostics. *Drug Discov. Today* **2013**, *19*, 474–481. [CrossRef] [PubMed]
5. Veiseh, O.; Gunn, J.W.; Zhang, M. Design and fabrication of magnetic nanoparticles for targeted drug delivery and imaging. *Adv. Drug Deliv. Rev.* **2010**, *62*, 284–304. [CrossRef]
6. Sun, C.; Lee, J.S.H.; Zhang, M. Magnetic nanoparticles in MR imaging and drug delivery. *Adv. Drug Deliv. Rev.* **2008**, *60*, 1252–1265. [CrossRef]
7. Shubayev, V.I.; Pisanic, T.R., II; Jin, S. Magnetic nanoparticles for theragnostics. *Adv. Drug Deliv. Rev.* **2009**, *61*, 467–477. [CrossRef]
8. Al-Karagoly, H.; Rhyaf, A.; Naji, H.; Albukhaty, S.; AlMalki, F.A.; Alyamani, A.A.; Albaqami, J.; Aloufi, S. Green synthesis, characterization, cytotoxicity, and antimicrobial activity of iron oxide nanoparticles using *Nigella sativa* seed extract. *Green Process. Synth.* **2022**, *11*, 254–265. [CrossRef]
9. Kuchma, E.; Kubrin, S.; Soldatov, A. The Local Atomic Structure of Colloidal Superparamagnetic Iron Oxide Nanoparticles for Theranostics in Oncology. *Biomedicines* **2018**, *6*, 78. [CrossRef]
10. Mughal, B.; Zaidi, S.Z.J.; Zhang, X.; Hassan, S.U. Biogenic Nanoparticles: Synthesis, Characterisation and Applications. *Appl. Sci.* **2021**, *11*, 2598. [CrossRef]

11. Šafařík, I.; Šafaříková, M. Magnetic Nanoparticles and Biosciences. In *Nanostructured Materials*; Hofmann, H., Rahman, Z., Schubert, U., Eds.; Springer: Vienna, Austria, 2002. [CrossRef]
12. Blakemore, R. Magnetotactic Bacteria. *Science* **1975**, *190*, 377–379. [CrossRef] [PubMed]
13. Markande, A.R.; Mistry, K.; Undaviya, S.; Jha, A. Magnetic Nanoparticles from Bacteria. In *Biobased Nanotechnology for Green Applications. Nanotechnology in the Life Sciences*; Sarma, H., Joshi, S.J., Prasad, R., Jampilek, J., Eds.; Springer: Cham, Switzerland, 2021. [CrossRef]
14. Tay, A.; McCausland, H.; Komeili, A.; Di Carlo, D. Nano and Microtechnologies for the Study of Magnetotactic Bacteria. *Adv. Funct. Mater.* **2019**, *29*, 1904178. [CrossRef]
15. Rajalakshmi, A.; Anjukam, E.; Ramesh, M.; Kavitha, K.; Puvanakrishnan, R.; Ramesh, B. A novel colorimetric technique for estimating iron in magnetosomes of magnetotactic bacteria based on linear regression. *Arch. Microbiol.* **2022**, *204*, 282. [CrossRef] [PubMed]
16. Bhat, B.; Pai, S.; Panda, M.; Aadrika; Anjali, K.; Ram, H.N.A.; Pai, A.; Venkatesh, K.B. Isolation of microbes possessing magnetosomes and their potential role in drug delivery. *Res. J. Pharm. Technol.* **2020**, *13*, 5042. [CrossRef]
17. Wang, X.; Li, Y.; Zhao, J.; Yao, H.; Chu, S.; Song, Z.; He, Z.; Zhang, W. Magnetotactic bacteria: Characteristics and environmental applications. *Front. Environ. Sci. Eng.* **2020**, *14*, 56. [CrossRef]
18. Ying, G.; Zhang, G.; Yang, J.; Hao, Z.; Xing, W.; Lu, D.; Zhang, S.; Yan, L. Biomineralization and biotechnological applications of bacterial magnetosomes. *Colloids Surfaces. B Biointerfaces* **2022**, *216*, 112556. [CrossRef]
19. Yan, L.; Xing, W. Methods to Study Magnetotactic Bacteria and Magnetosomes. *Methods Microbiol.* **2018**, *45*, 357–386. [CrossRef]
20. Lin, W.; Pan, Y.; Bazylinski, D.A. Diversity and ecology of and biomineralization by magnetotactic bacteria. *Environ. Microbiol. Rep.* **2017**, *9*, 345–356. [CrossRef]
21. Lins, U.; Farina, M. Magnetosome chain arrangement and stability in magnetotactic cocci. Antonie van Leeuwenhoek. *Int. J. Gen. Mol. Microbiol.* **2004**, *85*, 335–341. [CrossRef]
22. Alphandery, E.; Guyot, F.; Chebbi, I. Preparation of chains of magnetosomes, isolated from Magnetospirillum magneticum strain AMB-1 magnetotactic bacteria, yielding efficient treatment of tumors using magnetic hyperthermia. *Int. J. Pharm.* **2012**, *434*, 444–452. [CrossRef]
23. Benoit, M.R.; Mayer, D.; Barak, Y.; Chen, I.Y.; Hu, W.; Cheng, Z.; Wang, S.X.; Spielman, D.M.; Gambhir, S.S.; Matin, A. Visualizing Implanted Tumors in Mice with Magnetic Resonance Imaging Using Magnetotactic Bacteria. *Clin. Cancer Res.* **2009**, *15*, 5170–5177. [CrossRef] [PubMed]
24. Yoon, T.-J.; Lee, H.; Shao, H.; Hilderbrand, S.A.; Weissleder, R. Multicore Assemblies Potentiate Magnetic Properties of Biomagnetic Nanoparticles. *Adv. Mater.* **2011**, *23*, 4793–4797. [CrossRef] [PubMed]
25. Nuñez-Magos, L.; Lira-Escobedo, J.; Rodríguez-López, R.; Muñoz-Navia, M.; Castillo-Rivera, F.; Viveros-Méndez, P.X.; Araujo, E.; Encinas, A.; Saucedo-Anaya, S.A.; Aranda-Espinoza, S. Effects of DC Magnetic Fields on Magnetoliposomes. *Front. Mol. Biosci.* **2021**, *8*, 703417. [CrossRef] [PubMed]
26. Dadfar, S.M.; Roemhild, K.; Drude, N.I.; von Stillfried, S.; Knüchel, R.; Kiessling, F.; Lammers, T. Iron oxide nanoparticles: Diagnostic, therapeutic and theranostic applications. *Adv. Drug Deliv. Rev.* **2019**, *138*, 302–325. [CrossRef] [PubMed]
27. Wu, K.; Su, D.; Liu, J.; Saha, R.; Wang, J.-P. Magnetic nanoparticles in nanomedicine: A review of recent advances. *Nanotechnology* **2019**, *30*, 502003. [CrossRef] [PubMed]
28. Ryvolova, M.; Chomoucka, J.; Drbohlavova, J.; Kopel, P.; Babula, P.; Hynek, D.; Adam, V.; Eckschlager, T.; Hubalek, J.; Stiborova, M.; et al. Modern Micro and Nanoparticle-Based Imaging Techniques. *Sensors* **2012**, *12*, 14792–14820. [CrossRef] [PubMed]
29. Kianfar, E. Magnetic Nanoparticles in Targeted Drug Delivery: A Review. *J. Supercond. Nov. Magn.* **2021**, *34*, 1709–1735. [CrossRef]
30. Khot, V.; Pawar, S. Magnetic Hyperthermia with Magnetic Nanoparticles: A Status Review. *Curr. Top. Med. Chem.* **2014**, *14*, 572–594. [CrossRef]
31. Elingarami, S.; Zeng, X. A Short Review on Current Use of Magnetic Nanoparticles for Bio-Separation, Sequencing, Diagnosis and Drug Delivery. *Adv. Sci. Lett.* **2011**, *4*, 3295–3300. [CrossRef]
32. Liu, Y.; Gao, Y.; Xu, C. Using magnetic nanoparticles to manipulate biological objects. *Chin. Phys. B* **2013**, *22*, 097503. [CrossRef]
33. Akbarzadeh, A.; Samiei, M.; Davaran, S. Magnetic nanoparticles: Preparation, physical properties, and applications in biomedicine. *Nanoscale Res. Lett.* **2012**, *7*, 144. [CrossRef] [PubMed]
34. Witte, K.; Grüttner, C.; Bodnar, W.; Burkel, E. Magnetic Nanoparticles for Biomedical Applications. In *Encyclopedia of Nanotechnology*; Bhushan, B., Ed.; Springer: Dordrecht, The Netherlands, 2016. [CrossRef]
35. Li, K.; Xu, J.; Li, P.; Fan, Y. A review of magnetic ordered materials in biomedical field: Constructions, applications and prospects. *Compos. Part B Eng.* **2022**, *228*, 109401. [CrossRef]
36. Khizar, S.; Ahmad, N.M.; Zine, N.; Jaffrezic-Renault, N.; Errachid-El-Salhi, A.; Elaissari, A. Magnetic Nanoparticles: From Synthesis to Theranostic Applications. *ACS Appl. Nano Mater.* **2021**, *4*, 4284–4306. [CrossRef]
37. Kargol, A.; Malkinski, L.; Caruntu, G. Biomedical Applications of Multiferroic Nanoparticles. In *Advanced Magnetic Materials*; IntechOpen: London, UK, 2012. [CrossRef]
38. Kopyl, S.; Surmenev, R.; Surmeneva, M.; Fetisov, Y.; Kholkin, A. Magnetoelectric effect: Principles and applications in biology and medicine—A review. *Mater. Today Bio* **2021**, *12*, 100149. [CrossRef] [PubMed]
39. Bedoya-Hincapié, C.M.; Restrepo-Parra, E.; López-Carreño, L.D. Applications of magnetic and multiferroic core/shell nanostructures and their physical properties. *DYNA* **2018**, *85*, 29–35. [CrossRef]
40. Rao, B.N.; Kaviraj, P.; Vaibavi, S.R.; Kumar, A.; Bajpai, S.K.; Arockiarajan, A. Investigation of magnetoelectric properties and biocompatibility of $CoFe_2O_4$-$BaTiO_3$ core-shell nanoparticles for biomedical applications. *J. Appl. Phys.* **2017**, *122*, 164102. [CrossRef]

41. Praveena, M.G.; Thoufeeq, S.; Manikanta, B.; Rahul, M.T.; Bhowmik, R.N.; Nair, S.S.; Kalarikkal, N.; Mohammed, E.M.; Kala, M.S.; Anantharaman, M.R. A magnetoelectric nanocomposite based on two dimensional Cr2O3 and CoFe2O4. *Solid State Commun.* **2022**, *354*, 114865. [CrossRef]
42. Adamiano, A.; Iafisco, M.; Tampieri, A. Magnetic core-shell nanoparticles: Remote driving, hyperthermia, and controlled drug release. In *Core-Shell Nanostructures for Drug Delivery and Theranostics*; Woodhead Publishing: Sawston, UK, 2018; pp. 259–296. [CrossRef]
43. Jenjob, R.; Phakkeeree, T.; Crespy, D. Core–shell particles for drug-delivery, bioimaging, sensing, and tissue engineering. *Biomater. Sci.* **2020**, *8*, 2756–2770. [CrossRef]
44. Mosayebi, J.; Kiyasatfar, M.; Laurent, S. Synthesis, Functionalization, and Design of Magnetic Nanoparticles for Theranostic Applications. *Adv. Heal. Mater.* **2017**, *6*, 23–29. [CrossRef]
45. Chen, Y.; Xianyu, Y.; Dong, M.; Zhang, J.; Zheng, W.; Qian, Z.; Jiang, X. Cascade Reaction-Mediated Assembly of Magnetic/Silver Nanoparticles for Amplified Magnetic Biosensing. *Anal. Chem.* **2018**, *90*, 6906–6912. [CrossRef]
46. Chen, D.; Wu, Y. Rapid and Ultrasensitive Electrochemical Detection of TP53 Gene Mutation in Blood: Hybridization with a DNA/Gold-Coated Magnetic Nanoparticle Network. *Anal. Sens.* **2022**, *2*, e202200032. [CrossRef]
47. Yu, H.; Yu, J.; Li, L.; Zhang, Y.; Xin, S.; Ni, X.; Sun, Y.; Song, K. Recent Progress of the Practical Applications of the Platinum Nanoparticle-Based Electrochemistry Biosensors. *Front. Chem.* **2021**, *9*, 677876. [CrossRef] [PubMed]
48. Kwizera, E.A.; Chaffin, E.; Wang, Y.; Huang, X. Synthesis and properties of magnetic-optical core–shell nanoparticles. *RSC Adv.* **2017**, *7*, 17137–17153. [CrossRef]
49. Anik, M.I.; Hossain, M.K.; Hossain, I.; Mahfuz, A.M.U.B.; Rahman, M.T.; Ahmed, I. Recent progress of magnetic nanoparticles in biomedical applications: A review. *Nano Sel.* **2021**, *2*, 1146–1186. [CrossRef]
50. Materón, E.M.; Miyazaki, C.M.; Carr, O.; Joshi, N.; Picciani, P.H.; Dalmaschio, C.J.; Davis, F.; Shimizu, F.M. Magnetic nanoparticles in biomedical applications: A review. *Appl. Surf. Sci. Adv.* **2021**, *6*, 100163. [CrossRef]
51. Chen, Z.; Song, S.; Ma, J.; Da Ling, S.; Wang, Y.D.; Kong, T.T.; Xu, J.H. Fabrication of magnetic core/shell hydrogels via microfluidics for controlled drug delivery. *Chem. Eng. Sci.* **2022**, *248*, 117216. [CrossRef]
52. Zou, L.; Huang, B.; Zheng, X.; Pan, H.; Zhang, Q.; Xie, W.; Zhao, Z.; Li, X. Microfluidic synthesis of magnetic nanoparticles in droplet-based microreactors. *Mater. Chem. Phys.* **2022**, *276*, 125384. [CrossRef]
53. Abedini-Nassab, R.; Pouryosef Miandoab, M.P.; Şaşmaz, M. Microfluidic Synthesis, Control, and Sensing of Magnetic Nanoparticles: A Review. *Micromachines* **2021**, *12*, 768. [CrossRef] [PubMed]
54. Fattahi, Z.; Hasanzadeh, M. Nanotechnology-assisted microfluidic systems for chemical sensing, biosensing, and bioanalysis. *TrAC Trends Anal. Chem.* **2022**, *152*, 116637. [CrossRef]
55. Haun, J.B.; Yoon, T.-J.; Lee, H.; Weissleder, R. Molecular Detection of Biomarkers and Cells Using Magnetic Nanoparticles and Diagnostic Magnetic Resonance. *Methods Mol. Biol.* **2011**, *726*, 33–49. [CrossRef]
56. Spitzberg, J.D.; Zrehen, A.; Van Kooten, X.F.; Meller, A. Plasmonic-Nanopore Biosensors for Superior Single-Molecule Detection. *Adv. Mater.* **2019**, *31*, e1900422. [CrossRef] [PubMed]
57. Qureshi, A.; Gurbuz, Y.; Niazi, J.H. Biosensors for cardiac biomarkers detection: A review. *Sens. Actuators B Chem.* **2012**, *171–172*, 62–76. [CrossRef]
58. Chen, Y.-T.; Kolhatkar, A.G.; Zenasni, O.; Xu, S.; Lee, T.R. Biosensing Using Magnetic Particle Detection Techniques. *Sensors* **2017**, *17*, 2300. [CrossRef] [PubMed]
59. Holzinger, M.; Le Goff, A.; Cosnier, S. Nanomaterials for biosensing applications: A review. *Front. Chem.* **2014**, *2*, 63. [CrossRef]
60. Chen, D.; Wu, Y.; Tilley, R.D.; Gooding, J.J. Rapid and ultrasensitive electrochemical detection of DNA methylation for ovarian cancer diagnosis. *Biosens. Bioelectron.* **2022**, *206*, 114126. [CrossRef]
61. Chen, D.; Wu, Y.; Hoque, S.; Tilley, R.D.; Gooding, J.J. Rapid and ultrasensitive electrochemical detection of circulating tumor DNA by hybridization on the network of gold-coated magnetic nanoparticles. *Chem. Sci.* **2021**, *12*, 5196–5201. [CrossRef]
62. Shamsazar, A.; Asadi, A.; Seifzadeh, D.; Mahdavi, M. A novel and highly sensitive sandwich-type immunosensor for prostate-specific antigen detection based on MWCNTs-Fe3O4 nanocomposite. *Sens. Actuators B Chem.* **2021**, *346*, 130459. [CrossRef]
63. Khoshfetrat, S.M.; Mehrgardi, M.A. Amplified detection of leukemia cancer cells using an aptamer-conjugated gold-coated magnetic nanoparticles on a nitrogen-doped graphene modified electrode. *Bioelectrochemistry* **2017**, *114*, 24–32. [CrossRef]
64. Sharafeldin, M.; Bishop, G.W.; Bhakta, S.; El-Sawy, A.; Suib, S.L.; Rusling, J.F. Fe3O4 nanoparticles on graphene oxide sheets for isolation and ultrasensitive amperometric detection of cancer biomarker proteins. *Biosens. Bioelectron.* **2016**, *91*, 359–366. [CrossRef]
65. Yang, Q.; Li, N.; Li, Q.; Chen, S.; Wang, H.-L.; Yang, H. Amperometric sarcosine biosensor based on hollow magnetic Pt–Fe3O4@C nanospheres. *Anal. Chim. Acta* **2019**, *1078*, 161–167. [CrossRef]
66. Jahanbani, S.; Benvidi, A. A novel electrochemical DNA biosensor based on a modified magnetic bar carbon paste electrode with Fe3O4NPs-reduced graphene oxide/PANHS nanocomposite. *Mater. Sci. Eng. C* **2016**, *68*, 1–8. [CrossRef] [PubMed]
67. Huerta-Nuñez, L.F.E.; Gutierrez-Iglesias, G.; Martínez-Cuazitl, A.; Mata-Miranda, M.M.; Alvarez-Jiménez, V.D.; Sánchez-Monroy, V.; Golberg, A.; González-Díaz, C.A. A biosensor capable of identifying low quantities of breast cancer cells by electrical impedance spectroscopy. *Sci. Rep.* **2019**, *9*, 6419. [CrossRef] [PubMed]
68. Bonaiuto, E.; Magro, M.; Baratella, D.; Jakubec, P.; Sconcerle, E.; Terzo, M.; Miotto, G.; Macone, A.; Agostinelli, E.; Fasolato, S.; et al. Ternary Hybrid γ-Fe_2O_3/Cr^{VI}/Amine Oxidase Nanostructure for Electrochemical Sensing: Application for Polyamine Detection in Tumor Tissue. *Chem.–A Eur. J.* **2016**, *22*, 6846–6852. [CrossRef] [PubMed]

69. Wang, L.; Gan, X. Antibody-functionalized magnetic nanoparticles for electrochemical immunoassay of α-1-fetoprotein in human serum. *Mikrochim. Acta* **2009**, *164*, 231–237. [CrossRef]
70. Zou, K.; Gao, Z.; Deng, Q.; Luo, Y.; Zou, L.; Lu, Y.; Zhao, W.; Lin, B. Picomolar detection of carcinoembryonic antigen in whole blood using microfluidics and surface-enhanced Raman spectroscopy. *Electrophoresis* **2016**, *37*, 786–789. [CrossRef] [PubMed]
71. Pang, Y.; Wang, C.; Wang, J.; Sun, Z.; Xiao, R.; Wang, S. Fe_3O_4@Ag magnetic nanoparticles for microRNA capture and duplex-specific nuclease signal amplification based SERS detection in cancer cells. *Biosens. Bioelectron.* **2016**, *79*, 574–580. [CrossRef]
72. Turan, E.; Zengin, A.; Suludere, Z.; Kalkan, N.; Tamer, U. Construction of a sensitive and selective plasmonic biosensor for prostate specific antigen by combining magnetic molecularly-imprinted polymer and surface-enhanced Raman spectroscopy. *Talanta* **2022**, *237*, 122926. [CrossRef]
73. Shipunova, V.O.; Nikitin, M.P.; Belova, M.M.; Deyev, S.M. Label-free methods of multiparametric surface plasmon resonance and MPQ-cytometry for quantitative real-time measurements of targeted magnetic nanoparticles complexation with living cancer cells. *Mater. Today Commun.* **2021**, *29*, 102978. [CrossRef]
74. Mohammadzadeh-Asl, S.; Aghanejad, A.; de la Guardia, M.; Dolatabadi, J.E.N.; Keshtkar, A. Surface plasmon resonance signal enhancement based on erlotinib loaded magnetic nanoparticles for evaluation of its interaction with human lung cancer cells. *Opt. Laser Technol.* **2021**, *133*, 106521. [CrossRef]
75. Hu, Y.; Li, L.; Guo, L. The sandwich-type aptasensor based on gold nanoparticles/DNA/magnetic beads for detection of cancer biomarker protein AGR2. *Sens. Actuators B Chem.* **2015**, *209*, 846–852. [CrossRef]
76. Xu, Q.; Liang, K.; Liu, R.-Y.; Deng, L.; Zhang, M.; Shen, L.; Liu, Y.-N. Highly sensitive fluorescent detection of p53 protein based on DNA functionalized Fe3O4 nanoparticles. *Talanta* **2018**, *187*, 142–147. [CrossRef] [PubMed]
77. Lee, J.; Lee, J.-H.; Mondal, J.; Hwang, J.; Kim, H.S.; Kumar, V.; Raj, A.; Hwang, S.R.; Lee, Y.-K. Magnetofluoro-Immunosensing Platform Based on Binary Nanoparticle-Decorated Graphene for Detection of Cancer Cell-Derived Exosomes. *Int. J. Mol. Sci.* **2022**, *23*, 9619. [CrossRef] [PubMed]
78. Weerathunge, P.; Pooja, D.; Singh, M.; Kulhari, H.; Mayes, E.L.; Bansal, V.; Ramanathan, R. Transferrin-conjugated quasi-cubic SPIONs for cellular receptor profiling and detection of brain cancer. *Sens. Actuators B Chem.* **2019**, *297*, 126737. [CrossRef]
79. Kim, M.I.; Ye, Y.; Woo, M.-A.; Lee, J.; Park, H.G. A Highly Efficient Colorimetric Immunoassay Using a Nanocomposite Entrapping Magnetic and Platinum Nanoparticles in Ordered Mesoporous Carbon. *Adv. Healthc. Mater.* **2014**, *3*, 36–41. [CrossRef]
80. Liu, X.; Zhang, J.; Cai, Y.; Zhang, S.; Ma, K.; Hua, K.; Cui, Y. A novel DNA methylation biosensor by combination of isothermal amplification and lateral flow device. *Sens. Actuators B Chem.* **2021**, *333*, 129624. [CrossRef]
81. Zhang, J.; Liu, X.; Zhang, S.; Cai, Y.; Ma, K.; Hua, K.; Cui, Y. A site-specific DNA methylation biosensor for both visual and magnetic determination based on lateral flow assay. *Analyst* **2021**, *146*, 2248–2254. [CrossRef]
82. Huang, C.-C.; Ray, P.; Chan, M.; Zhou, X.; Hall, D.A. An aptamer-based magnetic flow cytometer using matched filtering. *Biosens. Bioelectron.* **2020**, *169*, 112362. [CrossRef]
83. Kim, S.J.; Lee, S.-W.; Song, J.D.; Kwon, Y.-W.; Lee, K.-J.; Koo, H.C. An InSb-based magnetoresistive biosensor using Fe3O4 nanoparticles. *Sens. Actuators B Chem.* **2018**, *255*, 2894–2899. [CrossRef]
84. Nagesetti, A.; Rodzinski, A.; Stimphil, E.; Stewart, T.; Khanal, C.; Wang, P.; Guduru, R.; Liang, P.; Agoulnik, I.; Horstmyer, J.; et al. Multiferroic coreshell magnetoelectric nanoparticles as NMR sensitive nanoprobes for cancer cell detection. *Sci. Rep.* **2017**, *7*, 1610. [CrossRef]
85. Zhu, F.; Li, D.; Ding, Q.; Lei, C.; Ren, L.; Ding, X.; Sun, X. 2D magnetic MoS2–Fe3O4 hybrid nanostructures for ultrasensitive exosome detection in GMR sensor. *Biosens. Bioelectron.* **2020**, *147*, 111787. [CrossRef]
86. Javed, K.R.; Ahmad, M.; Ali, S.; Butt, M.Z.; Nafees, M.; Butt, A.; Nadeem, M.; Shahid, A. Comparison of Doxorubicin Anticancer Drug Loading on Different Metal Oxide Nanoparticles. *Medicine* **2015**, *94*, e617. [CrossRef]
87. Li, H.; Xu, W.; Li, F.; Zeng, R.; Zhang, X.; Wang, X.; Zhao, S.; Weng, J.; Li, Z.; Sun, L. Amplification of anticancer efficacy by co-delivery of doxorubicin and lonidamine with extracellular vesicles. *Drug Deliv.* **2022**, *29*, 192–202. [CrossRef] [PubMed]
88. Gui, G.; Fan, Z.; Ning, Y.; Yuan, C.; Zhang, B.; Xu, Q. Optimization, Characterization and in vivo Evaluation of Paclitaxel-Loaded Folate-Conjugated Superparamagnetic Iron Oxide Nanoparticles. *Int. J. Nanomed.* **2021**, *16*, 2283–2295. [CrossRef]
89. Pérez-Herrero, E.; Fernández-Medarde, A. Advanced targeted therapies in cancer: Drug nanocarriers, the future of chemotherapy. *Eur. J. Pharm. Biopharm.* **2015**, *93*, 52–79. [CrossRef] [PubMed]
90. Nosrati, H.; Sefidi, N.; Sharafi, A.; Danafar, H.; Manjili, H.K. Bovine Serum Albumin (BSA) coated iron oxide magnetic nanoparticles as biocompatible carriers for curcumin-anticancer drug. *Bioorg. Chem.* **2018**, *76*, 501–509. [CrossRef]
91. Attari, E.; Nosrati, H.; Danafar, H.; Manjili, H.K. Methotrexate anticancer drug delivery to breast cancer cell lines by iron oxide magnetic based nanocarrier. *J. Biomed. Mater. Res. Part A* **2019**, *107*, 2492–2500. [CrossRef] [PubMed]
92. Ayyanaar, S.; Bhaskar, R.; Esthar, S.; Vadivel, M.; Rajesh, J.; Rajagopal, G. Design and development of 5-fluorouracil loaded biodegradable magnetic microspheres as site-specific drug delivery vehicle for cancer therapy. *J. Magn. Magn. Mater.* **2022**, *546*, 168853. [CrossRef]
93. Zhu, Y.; Liao, L. Applications of Nanoparticles for Anticancer Drug Delivery: A Review. *J. Nanosci. Nanotechnol.* **2015**, *15*, 4753–4773. [CrossRef] [PubMed]
94. Mandal, A.; Bisht, R.; Pal, D.; Mitra, A.K. Diagnosis and Drug Delivery to the Brain: Novel strategies. In *Emerging Nanotechnologies for Diagnostics, Drug Delivery and Medical Devices*; Elsevier: Amsterdam, The Netherlands, 2017; pp. 59–83. [CrossRef]
95. McBain, S.C.; Yiu, H.H.; Dobson, J. Magnetic nanoparticles for gene and drug delivery. *Int. J. Nanomed.* **2008**, *3*, 169–180. [CrossRef]

96. Chen, M.; Li, J.; Shu, G.; Shen, L.; Qiao, E.; Zhang, N.; Fang, S.; Chen, X.; Zhao, Z.; Tu, J.; et al. Homogenous multifunctional microspheres induce ferroptosis to promote the anti-hepatocarcinoma effect of chemoembolization. *J. Nanobiotechnol.* **2022**, *20*, 179. [CrossRef]
97. Hu, W.; Qi, Q.; Hu, H.; Wang, C.; Zhang, Q.; Zhang, Z.; Zhao, Y.; Yu, X.; Guo, M.; Du, S.; et al. Fe3O4 liposome for photothermal/chemo-synergistic inhibition of metastatic breast tumor. *Colloids Surfaces A Physicochem. Eng. Asp.* **2022**, *634*, 127921. [CrossRef]
98. Al-Musawi, S.; Albukhaty, S.; Al-Karagoly, H.; Sulaiman, G.M.; Jabir, M.S.; Naderi-Manesh, H. Dextran-coated superparamagnetic nanoparticles modified with folate for targeted drug delivery of camptothecin. *Adv. Nat. Sci. Nanosci. Nanotechnol.* **2020**, *11*, 045009. [CrossRef]
99. Satpathy, M.; Wang, L.; Zielinski, R.J.; Qian, W.; Wang, Y.A.; Mohs, A.; Kairdolf, B.A.; Ji, X.; Capala, J.; Lipowska, M.; et al. Targeted Drug Delivery and Image-Guided Therapy of Heterogeneous Ovarian Cancer Using HER2-Targeted Theranostic Nanoparticles. *Theranostics* **2019**, *9*, 778–795. [CrossRef] [PubMed]
100. Rostami, M.; Badiei, A.; Sorouri, A.M.; Fasihi-Ramandi, M.; Ganjali, M.R.; Rahimi-Nasrabadi, M.; Ahmadi, F. Cur-loaded magnetic ZnFe2O4@L-cysteine—Ox, N-rich mesoporous -gC3N4 nanocarriers as a targeted sonodynamic chemotherapeutic agent for enhanced tumor eradication. *Surf. Interfaces* **2022**, *30*, 101900. [CrossRef]
101. Ramezani Farani, M.; Azarian, M.; Heydari Sheikh Hossein, H.; Abdolvahabi, Z.; Mohammadi Abgarmi, Z.; Moradi, A.; Mousavi, S.M.; Ashrafizadeh, M.; Makvandi, P.; Saeb, M.R.; et al. Folic Acid-Adorned Curcumin-Loaded Iron Oxide Nanoparticles for Cervical Cancer. *ACS Appl. Bio Mater.* **2022**, *5*, 1305–1318. [CrossRef]
102. Patitsa, M.; Karathanou, K.; Kanaki, Z.; Tzioga, L.; Pippa, N.; Demetzos, C.; Verganelakis, D.A.; Cournia, Z.; Klinakis, A. Magnetic nanoparticles coated with polyarabic acid demonstrate enhanced drug delivery and imaging properties for cancer theranostic applications. *Sci. Rep.* **2017**, *7*, 775. [CrossRef]
103. Noh, K.; Uthaman, S.; Lee, C.-S.; Kim, Y.; Pillarisetti, S.; Hwang, H.S.; Park, I.-K.; Huh, K.M. Tumor intracellular microenvironment-responsive nanoparticles for magnetically targeted chemotherapy. *J. Ind. Eng. Chem.* **2022**, *111*, 121–128. [CrossRef]
104. Abolhasani Zadeh, F.; Abdalkareem Jasim, S.; Atakhanova, N.E.; Majdi, H.S.; Abed Jawad, M.; Khudair Hasan, M.; Borhani, F.; Khatami, M. Drug delivery and anticancer activity of biosynthesised mesoporous Fe_2O_3 nanoparticles. *IET Nanobiotechnol.* **2022**, *16*, 85–91. [CrossRef]
105. Zhang, H.; Wang, J.; Zeng, Y.; Wang, G.; Han, S.; Yang, Z.; Li, B.; Wang, X.; Gao, J.; Zheng, L.; et al. Leucine-coated cobalt ferrite nanoparticles: Synthesis, characterization and potential biomedical applications for drug delivery. *Phys. Lett. A* **2020**, *384*, 126600. [CrossRef]
106. Li, J.; Yang, N.; Yang, M.; Lu, C.; Xie, M. Development of a magnetic MoS2 system camouflaged by lipid for chemo/phototherapy of cancer. *Colloids Surf. B Biointerfaces* **2022**, *213*, 112389. [CrossRef]
107. Liu, E.; Zhang, M.; Cui, H.; Gong, J.; Huang, Y.; Wang, J.; Cui, Y.; Dong, W.; Sun, L.; He, H.; et al. Tat-functionalized Ag-Fe3O4 nano-composites as tissue-penetrating vehicles for tumor magnetic targeting and drug delivery. *Acta Pharm. Sin. B* **2018**, *8*, 956–968. [CrossRef] [PubMed]
108. Shahzada, K.; Mushtaqbc, S.; Rizwana, M.; Khalida, W.; Atifa, M.; Din, F.U.; Ahmadc, N.; Abbasic, R.; Alia, Z. Field-controlled magnetoelectric core-shell CoFe2O4@BaTiO3 nanoparticles as effective drug carriers and drug release in vitro. *Mater. Sci. Eng. C* **2021**, *119*, 111444. [CrossRef] [PubMed]
109. Foroughi-Nia, B.; Aghanejad, A.; Kadkhoda, J.; Barar, J.; Nosrati, H.; Davaran, S. AS1411 conjugated magnetic-based poly *N* -isopropyl acrylamide nanoparticles for delivery of erlotinib to prostate cancer cells. *Appl. Organomet. Chem.* **2022**, *36*, e6691. [CrossRef]
110. Stewart, T.S.; Nagesetti, A.; Guduru, R.; Liang, P.; Stimphil, E.; Hadjikhani, A.; Salgueiro, L.; Horstmyer, J.; Cai, R.; Schally, A.; et al. Magnetoelectric nanoparticles for delivery of antitumor peptides into glioblastoma cells by magnetic fields. *Nanomedicine* **2018**, *13*, 423–438. [CrossRef]
111. Tomeh, M.A.; Hadianamrei, R.; Xu, D.; Brown, S.; Zhao, X. Peptide-functionalised magnetic silk nanoparticles produced by a swirl mixer for enhanced anticancer activity of ASC-J9. *Colloids Surf. B Biointerfaces* **2022**, *216*, 112549. [CrossRef]
112. Farmanbar, N.; Mohseni, S.; Darroudi, M. Green synthesis of chitosan-coated magnetic nanoparticles for drug delivery of oxaliplatin and irinotecan against colorectal cancer cells. *Polym. Bull.* **2022**, 1–19. [CrossRef]
113. Jin, L.; Wang, Q.; Chen, J.; Wang, Z.; Xin, H.; Zhang, D. Efficient Delivery of Therapeutic siRNA by Fe3O4 Magnetic Nanoparticles into Oral Cancer Cells. *Pharmaceutics* **2019**, *11*, 615. [CrossRef]
114. Chi, H.; Zhu, G.; Yin, Y.; Diao, H.; Liu, Z.; Sun, S.; Guo, Z.; Xu, W.; Xu, J.; Cui, C.; et al. Dual-Responsive multifunctional "core-shell" magnetic nanoparticles promoting Fenton reaction for tumor ferroptosis therapy. *Int. J. Pharm.* **2022**, *622*, 121898. [CrossRef]
115. Pandit, P.; Bhagat, S.; Rananaware, P.; Mohanta, Z.; Kumar, M.; Tiwari, V.; Singh, S.; Brahmkhatri, V.P. Iron oxide nanoparticle encapsulated; folic acid tethered dual metal organic framework-based nanocomposite for MRI and selective targeting of folate receptor expressing breast cancer cells. *Microporous Mesoporous Mater.* **2022**, *340*, 112008. [CrossRef]
116. Arabzadeh, A.; Akhlaghi, N.; Najafpour-Darzi, G. Quercetin loading on mesoporous magnetic MnFe2O4@ hydroxyapatite core-shell nanoparticles for treating cancer cells. *Adv. Powder Technol.* **2022**, *33*, 103609. [CrossRef]
117. Ghasemzadeh, F.; Mohammadi, M.; Najafpour, G.D.; Moghadamnia, A.A. Ursolic acid loaded β-cyclodextrin/folic acid/Fe3O4 nanocomplex for drug delivery to tumor cells. *J. Drug Deliv. Sci. Technol.* **2022**, *72*, 103412. [CrossRef]
118. Kanelli, M.; Saleh, B.; Webster, T.J.; Vouyiouka, S.; Topakas, E. Co-Encapsulation of Violacein and Iron Oxide in Poly(lactic acid) Nanoparticles for Simultaneous Antibacterial and Anticancer Applications. *J. Biomed. Nanotechnol.* **2022**, *18*, 729–739. [CrossRef] [PubMed]

119. Joshy, K.S.; Augustine, R.; Mayeen, A.; Alex, S.M.; Hasan, A.; Thomas, S.; Chi, H. $NiFe_2O_4$/poly(ethylene glycol)/lipid–polymer hybrid nanoparticles for anti-cancer drug delivery. *New J. Chem.* **2020**, *44*, 18162–18172. [CrossRef]
120. Tokmedash, M.A.; Zadeh, E.S.; Balouchi, E.N.; Salehi, Z.; Ardestani, M.S. Synthesis of smart carriers based on tryptophan-functionalized magnetic nanoparticles and its application in 5-fluorouracil delivery. *Biomed. Mater.* **2022**, *17*, 045026. [CrossRef]
121. Nie, Z.; Vahdani, Y.; Cho, W.C.; Bloukh, S.H.; Edis, Z.; Haghighat, S.; Falahati, M.; Kheradmandi, R.; Jaragh-Alhadad, L.A.; Sharifi, M. 5-Fluorouracil-containing inorganic iron oxide/platinum nanozymes with dual drug delivery and enzyme-like activity for the treatment of breast cancer. *Arab. J. Chem.* **2022**, *15*, 103966. [CrossRef]
122. Ali, A.; Shah, T.; Ullah, R.; Zhou, P.; Guo, M.; Ovais, M.; Tan, Z.; Rui, Y. Review on Recent Progress in Magnetic Nanoparticles: Synthesis, Characterization, and Diverse Applications. *Front. Chem.* **2021**, *9*, 629054. [CrossRef]
123. Husseiny, M.I.; El-Aziz, M.A.; Badr, Y.; Mahmoud, M.A. Biosynthesis of gold nanoparticles using Pseudomonas aeruginosa. *Spectrochim. Acta Part A Mol. Biomol. Spectrosc.* **2007**, *67*, 1003–1006. [CrossRef]
124. Khan, A.A.; Khan, S.; Khan, S.; Rentschler, S.; Laufer, S.; Deigner, H.-P. Biosynthesis of iron oxide magnetic nanoparticles using clinically isolated Pseudomonas aeruginosa. *Sci. Rep.* **2021**, *11*, 20503. [CrossRef]
125. Rai, M.; Gade, A.; Yadav, A. Biogenic Nanoparticles: An Introduction to What They Are, How They Are Synthesized and Their Applications. In *Metal Nanoparticles in Microbiology*; Rai, M., Duran, N., Eds.; Springer: Berlin/Heidelberg, Germany, 2011; pp. 1–14. ISBN 978-3-642-18311-9.
126. Golinska, P.; Wypij, M.; Ingle, A.P.; Gupta, I.; Dahm, H.; Rai, M. Biogenic synthesis of metal nanoparticles from actinomycetes: Biomedical applications and cytotoxicity. *Appl. Microbiol. Biotechnol.* **2014**, *98*, 8083–8097. [CrossRef]
127. Govender, Y.; Riddin, T.; Gericke, M.; Whiteley, C.G. Bioreduction of platinum salts into nanoparticles: A mechanistic perspective. *Biotechnol. Lett.* **2009**, *31*, 95–100. [CrossRef]
128. Siddiqi, K.S.; Rashid, M.; Rahman, A.; Tajuddin; Husen, A.; Rehman, S. Biogenic fabrication and characterization of silver nanoparticles using aqueous-ethanolic extract of lichen (Usnea longissima) and their antimicrobial activity. *Biomater. Res.* **2018**, *22*, 23. [CrossRef] [PubMed]
129. Chakraborty, N.; Banerjee, A.; Lahiri, S.; Panda, A.; Ghosh, A.N.; Pal, R. Biorecovery of gold using cyanobacteria and an eukaryotic alga with special reference to nanogold formation—A novel phenomenon. *J. Appl. Phycol.* **2009**, *21*, 145–152. [CrossRef]
130. Gade, A.K.; Bonde, P.; Ingle, A.P.; Marcato, P.D.; Durán, N.; Rai, M.K. Exploitation of Aspergillus niger for Synthesis of Silver Nanoparticles. *J. Biobased Mater. Bioenergy* **2008**, *2*, 243–247. [CrossRef]
131. Thakkar, K.N.; Mhatre, S.S.; Parikh, R.Y. Biological synthesis of metallic nanoparticles. *Nanomed. Nanotechnol. Biol. Med.* **2010**, *6*, 257–262. [CrossRef] [PubMed]
132. Alphandéry, E. Natural Metallic Nanoparticles for Application in Nano-Oncology. *Int. J. Mol. Sci.* **2020**, *21*, 4412. [CrossRef] [PubMed]
133. Assa, F.; Jafarizadeh-Malmiri, H.; Ajamein, H.; Anarjan, N.; Vaghari, H.; Sayyar, Z.; Berenjian, A. A biotechnological perspective on the application of iron oxide nanoparticles. *Nano Res.* **2016**, *9*, 2203–2225. [CrossRef]
134. Arakaki, A.; Nakazawa, H.; Nemoto, M.; Mori, T.; Matsunaga, T. Formation of magnetite by bacteria and its application. *J. R. Soc. Interface* **2008**, *5*, 977–999. [CrossRef]
135. Uebe, R.; Schüler, D. Magnetosome biogenesis in magnetotactic bacteria. *Nat. Rev. Microbiol.* **2016**, *14*, 621–637. [CrossRef]
136. Barber-Zucker, S.; Keren-Khadmy, N.; Zarivach, R. From invagination to navigation: The story of magnetosome-associated proteins in magnetotactic bacteria. *Protein Sci.* **2016**, *25*, 338–351. [CrossRef]
137. Ben-Shimon, S.; Stein, D.; Zarivach, R. Current view of iron biomineralization in magnetotactic bacteria. *J. Struct. Biol. X* **2021**, *5*, 100052. [CrossRef]
138. Komeili, A. Molecular mechanisms of compartmentalization and biomineralization in magnetotactic bacteria. *FEMS Microbiol. Rev.* **2012**, *36*, 232–255. [CrossRef]
139. Komeili, A.; Li, Z.; Newman, D.K.; Jensen, G.J. Magnetosomes Are Cell Membrane Invaginations Organized by the Actin-Like Protein MamK. *Science* **2006**, *311*, 242–245. [CrossRef] [PubMed]
140. Tanaka, M.; Okamura, Y.; Arakaki, A.; Tanaka, T.; Takeyama, H.; Matsunaga, T. Origin of magnetosome membrane: Proteomic analysis of magnetosome membrane and comparison with cytoplasmic membrane. *Proteomics* **2006**, *6*, 5234–5247. [CrossRef]
141. Zeytuni, N.; Ozyamak, E.; Ben-Harush, K.; Davidov, G.; Levin, M.; Gat, Y.; Moyal, T.; Brik, A.; Komeili, A.; Zarivach, R. Self-recognition mechanism of MamA, a magnetosome-associated TPR-containing protein, promotes complex assembly. *Proc. Natl. Acad. Sci. USA* **2011**, *108*, E480–E487. [CrossRef] [PubMed]
142. Raschdorf, O.; Bonn, F.; Zeytuni, N.; Zarivach, R.; Becher, D.; Schüler, D. A quantitative assessment of the membrane-integral sub-proteome of a bacterial magnetic organelle. *J. Proteom.* **2018**, *172*, 89–99. [CrossRef] [PubMed]
143. Vargas, G.; Cypriano, J.; Correa, T.; Leão, P.; Bazylinski, D.A.; Abreu, F. Applications of Magnetotactic Bacteria, Magnetosomes and Magnetosome Crystals in Biotechnology and Nanotechnology: Mini-Review. *Molecules* **2018**, *23*, 2438. [CrossRef] [PubMed]
144. Yamamoto, D.; Taoka, A.; Uchihashi, T.; Sasaki, H.; Watanabe, H.; Ando, T.; Fukumori, Y. Visualization and structural analysis of the bacterial magnetic organelle magnetosome using atomic force microscopy. *Proc. Natl. Acad. Sci. USA* **2010**, *107*, 9382–9387. [CrossRef] [PubMed]
145. Lohße, A.; Borg, S.; Raschdorf, O.; Kolinko, I.; Tompa, E.; Pósfai, M.; Faivre, D.; Baumgartner, J.; Schüler, D. Genetic Dissection of the *mamAB* and *mms6* Operons Reveals a Gene Set Essential for Magnetosome Biogenesis in Magnetospirillum gryphiswaldense. *J. Bacteriol.* **2014**, *196*, 2658–2669. [CrossRef] [PubMed]

146. Schüler, D. Genetics and cell biology of magnetosome formation in magnetotactic bacteria. *FEMS Microbiol. Rev.* **2008**, *32*, 654–672. [CrossRef]
147. Uebe, R.; Junge, K.; Henn, V.; Poxleitner, G.; Katzmann, E.; Plitzko, J.M.; Zarivach, R.; Kasama, T.; Wanner, G.; Pósfai, M.; et al. The cation diffusion facilitator proteins MamB and MamM of Magnetospirillum gryphiswaldense have distinct and complex functions, and are involved in magnetite biomineralization and magnetosome membrane assembly. *Mol. Microbiol.* **2011**, *82*, 818–835. [CrossRef]
148. Arakaki, A.; Yamagishi, A.; Fukuyo, A.; Tanaka, M.; Matsunaga, T. Co-ordinated functions of Mms proteins define the surface structure of cubo-octahedral magnetite crystals in magnetotactic bacteria. *Mol. Microbiol.* **2014**, *93*, 554–567. [CrossRef] [PubMed]
149. Toro-Nahuelpan, M.; Giacomelli, G.; Raschdorf, O.; Borg, S.; Plitzko, J.M.; Bramkamp, M.; Schüler, D.; Müller, F.-D. MamY is a membrane-bound protein that aligns magnetosomes and the motility axis of helical magnetotactic bacteria. *Nat. Microbiol.* **2019**, *4*, 1978–1989. [CrossRef]
150. Gareev, K.G.; Grouzdev, D.S.; Kharitonskii, P.V.; Kosterov, A.; Koziaeva, V.V.; Sergienko, E.S.; Shevtsov, M.A. Magnetotactic Bacteria and Magnetosomes: Basic Properties and Applications. *Magnetochemistry* **2021**, *7*, 86. [CrossRef]
151. Uzun, M.; Koziaeva, V.; Dziuba, M.; Alekseeva, L.; Krutkina, M.; Sukhacheva, M.; Baslerov, R.; Grouzdev, D. Looking for a Needle in a Haystack: Magnetotactic Bacteria Help in "Rare Biosphere" Investigations. *bioRxiv* **2022**. [CrossRef]
152. Uzun, M.; Alekseeva, L.; Krutkina, M.; Koziaeva, V.; Grouzdev, D. Unravelling the diversity of magnetotactic bacteria through analysis of open genomic databases. *Sci. Data* **2020**, *7*, 252. [CrossRef]
153. Lin, W.; Zhang, W.; Paterson, G.A.; Zhu, Q.; Zhao, X.; Knight, R.; Bazylinski, D.A.; Roberts, A.P.; Pan, Y. Expanding magnetic organelle biogenesis in the domain Bacteria. *Microbiome* **2020**, *8*, 152. [CrossRef]
154. Spring, S.; Amann, R.; Ludwig, W.; Schleifer, K.-H.; van Gemerden, H.; Petersen, N. Dominating Role of an Unusual Magnetotactic Bacterium in the Microaerobic Zone of a Freshwater Sediment. *Appl. Environ. Microbiol.* **1993**, *59*, 2397–2403. [CrossRef]
155. Descamps, E.C.T.; Monteil, C.L.; Menguy, N.; Ginet, N.; Pignol, D.; Bazylinski, D.A.; Lefèvre, C.T. Desulfamplus magnetovallimortis gen. nov., sp. nov., a magnetotactic bacterium from a brackish desert spring able to biomineralize greigite and magnetite, that represents a novel lineage in the Desulfobacteraceae. *Syst. Appl. Microbiol.* **2017**, *40*, 280–289. [CrossRef]
156. Epósfai, M.; Lefèvre, C.T.; Etrubitsyn, D.; Bazylinski, D.A.; Frankel, R.B. Phylogenetic significance of composition and crystal morphology of magnetosome minerals. *Front. Microbiol.* **2013**, *4*, 344. [CrossRef]
157. Dziuba, M.; Koziaeva, V.; Grouzdev, D.; Burganskaya, E.; Baslerov, R.; Kolganova, T.; Chernyadyev, A.; Osipov, G.; Andrianova, E.; Gorlenko, V.; et al. Magnetospirillum caucaseum sp. nov., Magnetospirillum marisnigri sp. nov. and Magnetospirillum moscoviense sp. nov., freshwater magnetotactic bacteria isolated from three distinct geographical locations in European Russia. *Int. J. Syst. Evol. Microbiol.* **2016**, *66*, 2069–2077. [CrossRef]
158. Gareev, K.G.; Grouzdev, D.S.; Kharitonskii, P.V.; Kirilenko, D.A.; Kosterov, A.; Koziaeva, V.V.; Levitskii, V.S.; Multhoff, G.; Nepomnyashchaya, E.K.; Nikitin, A.V.; et al. Magnetic Properties of Bacterial Magnetosomes Produced by *Magnetospirillum caucaseum* SO-1. *Microorganisms* **2021**, *9*, 1854. [CrossRef] [PubMed]
159. Grouzdev, D.S.; Dziuba, M.V.; Sukhacheva, M.S.; Mardanov, A.V.; Beletskiy, A.V.; Kuznetsov, B.B.; Skryabin, K.G. Draft Genome Sequence of Magnetospirillum sp. Strain SO-1, a Freshwater Magnetotactic Bacterium Isolated from the Ol'khovka River, Russia. *Genome Announc.* **2014**, *2*, e00235-14. [CrossRef] [PubMed]
160. Grouzdev, D.S.; Dziuba, M.V.; Kurek, D.V.; Ovchinnikov, A.I.; Zhigalova, N.A.; Kuznetsov, B.B.; Skryabin, K.G. Optimized Method for Preparation of IgG-Binding Bacterial Magnetic Nanoparticles. *PLoS ONE* **2014**, *9*, e109914. [CrossRef]
161. Faivre, D.; Menguy, N.; Pósfai, M.; Schüler, D. Environmental parameters affect the physical properties of fast-growing magnetosomes. *Am. Miner.* **2008**, *93*, 463–469. [CrossRef]
162. Schleifer, K.H.; Schüler, D.; Spring, S.; Weizenegger, M.; Amann, R.; Ludwig, W.; Köhler, M. The Genus Magnetospirillum gen. nov. Description of Magnetospirillum gryphiswaldense sp. nov. and Transfer of Aquaspirillum magnetotacticum to Magnetospirillum magnetotacticum comb. nov. *Syst. Appl. Microbiol.* **1991**, *14*, 379–385. [CrossRef]
163. Fdez-Gubieda, M.L.; Muela, A.; Alonso, J.; García-Prieto, A.; Olivi, L.; Fernández-Pacheco, R.; Barandiarán, J.M. Magnetite Biomineralization in *Magnetospirillum gryphiswaldense*: Time-Resolved Magnetic and Structural Studies. *ACS Nano* **2013**, *7*, 3297–3305. [CrossRef]
164. Koziaeva, V.V.; Rusakova, S.A.; Slobodova, N.V.; Uzun, M.; Kolganova, T.V.; Skryabin, K.G.; Grouzdev, D.S. Magnetospirillum kuznetsovii sp. nov., a novel magnetotactic bacterium isolated from a lake in the Moscow region. *Int. J. Syst. Evol. Microbiol.* **2019**, *69*, 1953–1959. [CrossRef]
165. Li, J.; Pan, Y.; Chen, G.; Liu, Q.; Tian, L.; Lin, W. Magnetite magnetosome and fragmental chain formation of *Magnetospirillum magneticum* AMB-1: Transmission electron microscopy and magnetic observations. *Geophys. J. Int.* **2009**, *177*, 33–42. [CrossRef]
166. Chen, H.; Zhang, S.-D.; Chen, L.; Cai, Y.; Zhang, W.-J.; Song, T.; Wu, L.-F. Efficient Genome Editing of Magnetospirillum magneticum AMB-1 by CRISPR-Cas9 System for Analyzing Magnetotactic Behavior. *Front. Microbiol.* **2018**, *9*, 1569. [CrossRef]
167. Matsunaga, T.; Okamura, Y.; Fukuda, Y.; Wahyudi, A.T.; Murase, Y.; Takeyama, H. Complete Genome Sequence of the Facultative Anaerobic Magnetotactic Bacterium Magnetospirillum sp. strain AMB-1. *DNA Res.* **2005**, *12*, 157–166. [CrossRef]
168. Devouard, B.; Posfai, M.; Hua, X.; Bazylinski, D.A.; Frankel, R.B.; Buseck, P.R. Magnetite from magnetotactic bacteria; size distributions and twinning. *Am. Miner.* **1998**, *83*, 1387–1398. [CrossRef]
169. Maratea, D.; Blakemore, R.P. Aquaspirillum magnetotacticum sp. nov., a Magnetic Spirillum. *Int. J. Syst. Bacteriol.* **1981**, *31*, 452–455. [CrossRef]

170. Kozyaeva, V.V.; Grouzdev, D.S.; Dziuba, M.V.; Kolganova, T.V.; Kuznetsov, B.B. Diversity of magnetotactic bacteria of the Moskva River. *Microbiology* **2017**, *86*, 106–112. [CrossRef]
171. Koziaeva, V.V.; Dziuba, M.V.; Ivanov, T.M.; Kuznetsov, B.B.; Skryabin, K.G.; Grouzdev, D.S. Draft Genome Sequences of Two Magnetotactic Bacteria, *Magnetospirillum moscoviense* BB-1 and *Magnetospirillum marisnigri* SP-1. *Genome Announc.* **2016**, *4*, e00814-16. [CrossRef] [PubMed]
172. Monteil, C.; Grouzdev, D.S.; Perrière, G.; Alonso, B.; Rouy, Z.; Cruveiller, S.; Ginet, N.; Pignol, D.; Lefevre, C.T. Repeated horizontal gene transfers triggered parallel evolution of magnetotaxis in two evolutionary divergent lineages of magnetotactic bacteria. *ISME J.* **2020**, *14*, 1783–1794. [CrossRef] [PubMed]
173. Lefèvre, C.T.; Schmidt, M.L.; Viloria, N.; Trubitsyn, D.; Schüler, D.; Bazylinski, D.A. Insight into the Evolution of Magnetotaxis in Magnetospirillum spp., Based on *mam* Gene Phylogeny. *Appl. Environ. Microbiol.* **2012**, *78*, 7238–7248. [CrossRef] [PubMed]
174. Bazylinski, D.A.; Williams, T.J.; Lefèvre, C.T.; Trubitsyn, D.; Fang, J.; Beveridge, T.J.; Moskowitz, B.M.; Ward, B.; Schübbe, S.; Dubbels, B.L.; et al. Magnetovibrio blakemorei gen. nov., sp. nov., a magnetotactic bacterium (Alphaproteobacteria: Rhodospirillaceae) isolated from a salt marsh. *Int. J. Syst. Evol. Microbiol.* **2013**, *63*, 1824–1833. [CrossRef]
175. Bazylinski, D.A.; Frankel, R.B.; Jannasch, H.W. Anaerobic magnetite production by a marine, magnetotactic bacterium. *Nature* **1988**, *334*, 518–519. [CrossRef]
176. Clemett, S.J.; Thomas-Keprta, K.L.; Shimmin, J.; Morphew, M.; McIntosh, J.R.; Bazylinski, D.A.; Kirschvink, J.L.; Wentworth, S.J.; McKay, D.S.; Vali, H.; et al. Crystal morphology of MV-1 magnetite. *Am. Miner.* **2002**, *87*, 1727–1730. [CrossRef]
177. Monteil, C.L.; Perrière, G.; Menguy, N.; Ginet, N.; Alonso, B.; Waisbord, N.; Cruveiller, S.; Pignol, D.; Lefèvre, C.T. Genomic study of a novel magnetotactic *Alphaproteobacteria* uncovers the multiple ancestry of magnetotaxis. *Environ. Microbiol.* **2018**, *20*, 4415–4430. [CrossRef]
178. Zhu, K.; Pan, H.; Li, J.; Yu-Zhang, K.; Zhang, S.-D.; Zhang, W.-Y.; Zhou, K.; Yue, H.; Pan, Y.; Xiao, T.; et al. Isolation and characterization of a marine magnetotactic spirillum axenic culture QH-2 from an intertidal zone of the China Sea. *Res. Microbiol.* **2010**, *161*, 276–283. [CrossRef]
179. Geurink, C.; Lefevre, C.T.; Monteil, C.L.; Morillo-Lopez, V.; Abreu, F.; Bazylinski, D.A.; Trubitsyn, D. Complete Genome Sequence of Strain BW-2, a Magnetotactic Gammaproteobacterium in the Family *Ectothiorhodospiraceae*, Isolated from a Brackish Spring in Death Valley, California. *Microbiol. Resour. Announc.* **2020**, *9*, e01144-19. [CrossRef]
180. Lefèvre, C.T.; Viloria, N.; Schmidt, M.L.; Pósfai, M.; Frankel, R.B.; Bazylinski, D.A. Novel magnetite-producing magnetotactic bacteria belonging to the Gammaproteobacteria. *ISME J.* **2012**, *6*, 440–450. [CrossRef] [PubMed]
181. Taoka, A.; Kondo, J.; Oestreicher, Z.; Fukumori, Y. Characterization of uncultured giant rod-shaped magnetotactic Gammaproteobacteria from a freshwater pond in Kanazawa, Japan. *Microbiology* **2014**, *160*, 2226–2234. [CrossRef]
182. Liu, P.; Tamaxia, A.; Liu, Y.; Qiu, H.; Pan, J.; Jin, Z.; Zhao, X.; Roberts, A.P.; Pan, Y.; Li, J. Identification and characterization of magnetotactic Gammaproteobacteria from a salt evaporation pool, Bohai Bay, China. *Environ. Microbiol.* **2021**, *24*, 938–950. [CrossRef]
183. Leão, P.; Teixeira, L.C.R.S.; Cypriano, J.; Farina, M.; Abreu, F.; Bazylinski, D.A.; Lins, U. North-Seeking Magnetotactic Gammaproteobacteria in the Southern Hemisphere. *Appl. Environ. Microbiol.* **2016**, *82*, 5595–5602. [CrossRef] [PubMed]
184. Li, J.; Zhang, H.; Menguy, N.; Benzerara, K.; Wang, F.; Lin, X.; Chen, Z.; Pan, Y. Single-Cell Resolution of Uncultured Magnetotactic Bacteria via Fluorescence-Coupled Electron Microscopy. *Appl. Environ. Microbiol.* **2017**, *83*, e00409-17. [CrossRef] [PubMed]
185. Trubitsyn, D.; Monteil, C.L.; Geurink, C.; Morillo-Lopez, V.; de Almeida, L.G.P.; de Vasconcelos, A.T.R.; Abreu, F.; Bazylinski, D.A.; Lefevre, C.T. Complete Genome Sequence of Strain SS-5, a Magnetotactic Gammaproteobacterium Isolated from the Salton Sea, a Shallow, Saline, Endorheic Rift Lake Located on the San Andreas Fault in California. *Microbiol. Resour. Announc.* **2021**, *10*, e00928-20. [CrossRef]
186. Meldrum, F.C.; Mann, S.; Heywood, B.R.; Frankel, R.B.; Bazylinski, D.A. Electron microscopy study of magnetosomes in a cultured coccoid magnetotactic bacterium. *Proc. R. Soc. B Boil. Sci.* **1993**, *251*, 231–236. [CrossRef]
187. Schübbe, S.; Williams, T.J.; Xie, G.; Kiss, H.E.; Brettin, T.S.; Martinez, D.; Ross, C.A.; Schüler, D.; Cox, B.L.; Nealson, K.H.; et al. Complete Genome Sequence of the Chemolithoautotrophic Marine Magnetotactic Coccus Strain MC-1. *Appl. Environ. Microbiol.* **2009**, *75*, 4835–4852. [CrossRef]
188. Bazylinski, D.A.; Williams, T.J.; Lefevre, C.T.; Berg, R.J.; Zhang, C.L.; Bowser, S.S.; Dean, A.J.; Beveridge, T.J. Magnetococcus marinus gen. nov., sp. nov., a marine, magnetotactic bacterium that represents a novel lineage (Magnetococcaceae fam. nov., Magnetococcales ord. nov.) at the base of the Alphaproteobacteria. *Int. J. Syst. Evol. Microbiol.* **2013**, *63*, 801–808. [CrossRef]
189. Koziaeva, V.; Dziuba, M.; Leão, P.; Uzun, M.; Krutkina, M.; Grouzdev, D. Genome-Based Metabolic Reconstruction of a Novel Uncultivated Freshwater Magnetotactic coccus "Ca. Magnetaquicoccus inordinatus" UR-1, and Proposal of a Candidate Family "Ca. Magnetaquicoccaceae". *Front. Microbiol.* **2019**, *10*, 2290. [CrossRef]
190. Lefevre, C.T.; Bernadac, A.; Yu-Zhang, K.; Pradel, N.; Wu, L.-F. Isolation and characterization of a magnetotactic bacterial culture from the Mediterranean Sea. *Environ. Microbiol.* **2009**, *11*, 1646–1657. [CrossRef]
191. Ji, B.; Zhang, S.-D.; Zhang, W.-J.; Rouy, Z.; Alberto, F.; Santini, C.-L.; Mangenot, S.; Gagnot, S.; Philippe, N.; Pradel, N.; et al. The chimeric nature of the genomes of marine magnetotactic coccoid-ovoid bacteria defines a novel group of *Proteobacteria*. *Environ. Microbiol.* **2017**, *19*, 1103–1119. [CrossRef]
192. Werckmann, J.; Cypriano, J.; Lefèvre, C.T.; Dembelé, K.; Ersen, O.; Bazylinski, D.A.; Lins, U.; Farina, M. Localized iron accumulation precedes nucleation and growth of magnetite crystals in magnetotactic bacteria. *Sci. Rep.* **2017**, *7*, 8291. [CrossRef]

193. Morillo, V.; Abreu, F.; Araujo, A.C.; De Almeida, L.G.P.; Enrich-Prast, A.; Farina, M.; De Vasconcelos, A.T.R.; Bazylinski, D.A.; Lins, U. Isolation, cultivation and genomic analysis of magnetosome biomineralization genes of a new genus of South-seeking magnetotactic cocci within the Alphaproteobacteria. *Front. Microbiol.* **2014**, *5*, 72. [CrossRef] [PubMed]
194. Uzun, M.; Koziaeva, V.; Dziuba, M.; Leão, P.; Krutkina, M.; Grouzdev, D. Detection of interphylum transfers of the magnetosome gene cluster in magnetotactic bacteria. *Front. Microbiol.* **2022**, *13*, 945734. [CrossRef]
195. Koziaeva, V.V.; Alekseeva, L.M.; Uzun, M.M.; Leão, P.; Sukhacheva, M.V.; Patutina, E.O.; Kolganova, T.V.; Grouzdev, D.S. Biodiversity of Magnetotactic Bacteria in the Freshwater Lake Beloe Bordukovskoe, Russia. *Microbiology* **2020**, *89*, 348–358. [CrossRef]
196. Lefèvre, C.T.; Menguy, N.; Abreu, F.; Lins, U.; Pósfai, M.; Prozorov, T.; Pignol, D.; Frankel, R.B.; Bazylinski, D.A. A Cultured Greigite-Producing Magnetotactic Bacterium in a Novel Group of Sulfate-Reducing Bacteria. *Science* **2011**, *334*, 1720–1723. [CrossRef] [PubMed]
197. Nakazawa, H.; Arakaki, A.; Narita-Yamada, S.; Yashiro, I.; Jinno, K.; Aoki, N.; Tsuruyama, A.; Okamura, Y.; Tanikawa, S.; Fujita, N.; et al. Whole genome sequence of *Desulfovibrio magneticus* strain RS-1 revealed common gene clusters in magnetotactic bacteria. *Genome Res.* **2009**, *19*, 1801–1808. [CrossRef] [PubMed]
198. Pósfai, M.; Moskowitz, B.M.; Arató, B.; Schüler, D.; Flies, C.; Bazylinski, D.A.; Frankel, R.B. Properties of intracellular magnetite crystals produced by Desulfovibrio magneticus strain RS-1. *Earth Planet. Sci. Lett.* **2006**, *249*, 444–455. [CrossRef]
199. Sakaguchi, T.; Burgess, J.G.; Matsunaga, T. Magnetite formation by a sulphate-reducing bacterium. *Nature* **1993**, *365*, 47–49. [CrossRef]
200. Chen, Y.-R.; Zhang, R.; Du, H.-J.; Pan, H.-M.; Zhang, W.-Y.; Zhou, K.; Li, J.-H.; Xiao, T.; Wu, L.-F. A novel species of ellipsoidal multicellular magnetotactic prokaryotes from Lake Yuehu in China: The Ellipsoidal MMPs from the Lake Yuehu. *Environ. Microbiol.* **2014**, *17*, 637–647. [CrossRef] [PubMed]
201. Leão, P.; Chen, Y.-R.; Abreu, F.; Wang, M.; Zhang, W.-J.; Zhou, K.; Xiao, T.; Wu, L.-F.; Lins, U. Ultrastructure of ellipsoidal magnetotactic multicellular prokaryotes depicts their complex assemblage and cellular polarity in the context of magnetotaxis: Ultrastructure and Magnetotaxis in Ellipsoidal MMP. *Environ. Microbiol.* **2017**, *19*, 2151–2163. [CrossRef]
202. Abreu, F.; Cantão, M.E.; Nicolás, M.F.; Barcellos, F.G.; Morillo, V.; Almeida, L.G.; do Nascimento, F.F.; Lefèvre, C.T.; Bazylinski, D.A.; de Vasconcelos, A.T.R.; et al. Common ancestry of iron oxide- and iron-sulfide-based biomineralization in magnetotactic bacteria. *ISME J.* **2011**, *5*, 1634–1640. [CrossRef] [PubMed]
203. Abreu, F.; Martins, J.L.; Silveira, T.S.; Keim, C.N.; De Barros, H.G.P.L.; Filho, F.J.G.; Lins, U. 'Candidatus Magnetoglobus multicellularis', a multicellular, magnetotactic prokaryote from a hypersaline environment. *Int. J. Syst. Evol. Microbiol.* **2007**, *57*, 1318–1322. [CrossRef]
204. Abreu, F.; Silva, K. Greigite magnetosome membrane ultrastructure in 'Candidatus Magnetoglobus multicellularis'. *Int. Microbiol.* **2008**, *1*, 75–80. [CrossRef]
205. Kolinko, S.; Richter, M.; Glöckner, F.-O.; Brachmann, A.; Schüler, D. Single-cell genomics of uncultivated deep-branching magnetotactic bacteria reveals a conserved set of magnetosome genes. *Environ. Microbiol.* **2016**, *18*, 21–37. [CrossRef] [PubMed]
206. Lin, W.; Deng, A.; Wang, Z.; Li, Y.; Wen, T.; Wu, L.-F.; Wu, M.; Pan, Y. Genomic insights into the uncultured genus 'Candidatus Magnetobacterium' in the phylum Nitrospirae. *ISME J.* **2014**, *8*, 2463–2477. [CrossRef]
207. Li, J.; Menguy, N.; Gatel, C.; Boureau, V.; Snoeck, E.; Patriarche, G.; Leroy, E.; Pan, Y. Crystal growth of bullet-shaped magnetite in magnetotactic bacteria of the *Nitrospirae* phylum. *J. R. Soc. Interface* **2015**, *12*, 20141288. [CrossRef] [PubMed]
208. Zhang, W.; Wang, Y.; Liu, L.; Pan, Y.; Lin, W. Identification and Genomic Characterization of Two Previously Unknown Magnetotactic Nitrospirae. *Front. Microbiol.* **2021**, *12*, 690052. [CrossRef]
209. Kolinko, S.; Jogler, C.; Katzmann, E.; Wanner, G.; Peplies, J.; Schüler, D. Single-cell analysis reveals a novel uncultivated magnetotactic bacterium within the candidate division OP3. *Environ. Microbiol.* **2012**, *14*, 1709–1721. [CrossRef]
210. Jacob, J.J.; Suthindhiran, K. Magnetotactic bacteria and magnetosomes—Scope and challenges. *Mater. Sci. Eng. C* **2016**, *68*, 919–928. [CrossRef]
211. Han, L.; Li, S.; Yang, Y.; Zhao, F.; Huang, J.; Chang, J. Comparison of magnetite nanocrystal formed by biomineralization and chemosynthesis. *J. Magn. Magn. Mater.* **2007**, *313*, 236–242. [CrossRef]
212. Sun, J.; Li, Y.; Liang, X.-J.; Wang, P.C. Bacterial Magnetosome: A Novel Biogenetic Magnetic Targeted Drug Carrier with Potential Multifunctions. *J. Nanomater.* **2011**, *2011*, 469031. [CrossRef]
213. Alphandéry, E. Applications of Magnetosomes Synthesized by Magnetotactic Bacteria in Medicine. *Front. Bioeng. Biotechnol.* **2014**, *2*, 5. [CrossRef]
214. Lower, B.H.; Bazylinski, D.A. The Bacterial Magnetosome: A Unique Prokaryotic Organelle. *Microb. Physiol.* **2013**, *23*, 63–80. [CrossRef]
215. Basit, A.; Wang, J.; Guo, F.; Niu, W.; Jiang, W. Improved methods for mass production of magnetosomes and applications: A review. *Microb. Cell Fact.* **2020**, *19*, 197. [CrossRef] [PubMed]
216. Alphandéry, E. Applications of magnetotactic bacteria and magnetosome for cancer treatment: A review emphasizing on practical and mechanistic aspects. *Drug Discov. Today* **2020**, *25*, 1444–1452. [CrossRef] [PubMed]
217. Liu, Y.; Li, G.R.; Guo, F.F.; Jiang, W.; Li, Y.; Li, L.J. Large-scale production of magnetosomes by chemostat culture of Magnetospirillum gryphiswaldense at high cell density. *Microb. Cell Fact.* **2010**, *9*, 99. [CrossRef] [PubMed]
218. Zhang, Y.; Zhang, X.; Jiang, W.; Li, Y.; Li, J. Semicontinuous Culture of Magnetospirillum gryphiswaldense MSR-1 Cells in an Autofermentor by Nutrient-Balanced and Isosmotic Feeding Strategies. *Appl. Environ. Microbiol.* **2011**, *77*, 5851–5856. [CrossRef]

219. Kolinko, I.; Lohße, A.; Borg, S.; Raschdorf, O.; Jogler, C.; Tu, Q.; Pósfai, M.; Tompa, É.; Plitzko, J.M.; Brachmann, A.; et al. Biosynthesis of magnetic nanostructures in a foreign organism by transfer of bacterial magnetosome gene clusters. *Nat. Nanotechnol.* **2014**, *9*, 193–197. [CrossRef]
220. Tay, A.; Pfeiffer, D.; Rowe, K.; Tannenbaum, A.; Popp, F.; Strangeway, R.; Schüler, D.; Di Carlo, D. High-Throughput Microfluidic Sorting of Live Magnetotactic Bacteria. *Appl. Environ. Microbiol.* **2018**, *84*, e01308-18. [CrossRef]
221. Gorby, Y.A.; Beveridge, T.J.; Blakemore, R.P. Characterization of the bacterial magnetosome membrane. *J. Bacteriol.* **1988**, *170*, 834–841. [CrossRef]
222. Grünberg, K.; Wawer, C.; Tebo, B.M.; Schüler, D. A Large Gene Cluster Encoding Several Magnetosome Proteins Is Conserved in Different Species of Magnetotactic Bacteria. *Appl. Environ. Microbiol.* **2001**, *67*, 4573–4582. [CrossRef] [PubMed]
223. Salam, M.A. Isolation of magnetotactic bacteria from environmental samples and optimization and characterization of extracted magnetosomes. *Appl. Ecol. Environ. Res.* **2019**, *17*, 5355–5367. [CrossRef]
224. Baki, A.; Wiekhorst, F.; Bleul, R. Advances in Magnetic Nanoparticles Engineering for Biomedical Applications—A Review. *Bioengineering* **2021**, *8*, 134. [CrossRef] [PubMed]
225. Marcuello, C.; Chambel, L.; Rodrigues, M.S.; Ferreira, L.P.; Cruz, M.M. Magnetotactic Bacteria: Magnetism Beyond Magnetosomes. *IEEE Trans. NanoBiosci.* **2018**, *17*, 555–559. [CrossRef]
226. Mathuriya, A.S. Magnetotactic bacteria: Nanodrivers of the future. *Crit. Rev. Biotechnol.* **2016**, *36*, 788–802. [CrossRef]
227. Roda, A.; Cevenini, L.; Borg, S.; Michelini, E.; Calabretta, M.M.; Schüler, D. Bioengineered bioluminescent magnetotactic bacteria as a powerful tool for chip-based whole-cell biosensors. *Lab Chip* **2013**, *13*, 4881–4889. [CrossRef] [PubMed]
228. Sannigrahi, S.; Arumugasamy, S.K.; Mathiyarasu, J.; Sudhakaran, R.; Suthindhiran, K. Detection of white spot syndrome virus in seafood samples using a magnetosome-based impedimetric biosensor. *Arch. Virol.* **2021**, *166*, 2763–2778. [CrossRef] [PubMed]
229. Sannigrahi, S.; Arumugasamy, S.K.; Mathiyarasu, J.; Suthindhiran, K. Magnetosome-anti-Salmonella antibody complex based biosensor for the detection of Salmonella typhimurium. *Mater. Sci. Eng. C* **2020**, *114*, 111071. [CrossRef] [PubMed]
230. Xu, J.; Liu, L.; He, J.; Ma, S.; Li, S.; Wang, Z.; Xu, T.; Jiang, W.; Wen, Y.; Li, Y.; et al. Engineered magnetosomes fused to functional molecule (protein A) provide a highly effective alternative to commercial immunomagnetic beads. *J. Nanobiotechnol.* **2019**, *17*, 37. [CrossRef] [PubMed]
231. Orlando, T.; Mannucci, S.; Fantechi, E.; Conti, G.; Tambalo, S.; Busato, A.; Innocenti, C.; Ghin, L.; Bassi, R.; Arosio, P.; et al. Characterization of magnetic nanoparticles from *Magnetospirillum Gryphiswaldense* as potential theranostics tools. *Contrast Media Mol. Imaging* **2016**, *11*, 139–145. [CrossRef] [PubMed]
232. Zhang, Y.; Ni, Q.; Xu, C.; Wan, B.; Geng, Y.; Zheng, G.; Yang, Z.; Tao, J.; Zhao, Y.; Wen, J.; et al. Smart Bacterial Magnetic Nanoparticles for Tumor-Targeting Magnetic Resonance Imaging of HER2-Positive Breast Cancers. *ACS Appl. Mater. Interfaces* **2019**, *11*, 3654–3665. [CrossRef] [PubMed]
233. Mériaux, S.; Boucher, M.; Marty, B.; Lalatonne, Y.; Prévéral, S.; Motte, L.; Lefèvre, C.T.; Geffroy, F.; Lethimonnier, F.; Péan, M.; et al. Magnetosomes, Biogenic Magnetic Nanomaterials for Brain Molecular Imaging with 17.2 T MRI Scanner. *Adv. Healthc. Mater.* **2015**, *4*, 1076–1083. [CrossRef]
234. Xiang, Z.; Yang, X.; Xu, J.; Lai, W.; Wang, Z.; Hu, Z.; Tian, J.; Geng, L.; Fang, Q. Tumor detection using magnetosome nanoparticles functionalized with a newly screened EGFR/HER2 targeting peptide. *Biomaterials* **2017**, *115*, 53–64. [CrossRef]
235. Ertas, Y.N.; Abedi Dorcheh, K.; Akbari, A.; Jabbari, E. Nanoparticles for Targeted Drug Delivery to Cancer Stem Cells: A Review of Recent Advances. *Nanomaterials* **2021**, *11*, 1755. [CrossRef]
236. Zhao, N.; Honert, J.; Schmid, B.; Klas, M.; Isoya, J.; Markham, M.; Twitchen, D.; Jelezko, F.; Liu, R.-B.; Fedder, H.; et al. Sensing single remote nuclear spins. *Nat. Nanotechnol.* **2012**, *7*, 657–662. [CrossRef]
237. Ganta, S.; Devalapally, H.; Shahiwala, A.; Amiji, M. A review of stimuli-responsive nanocarriers for drug and gene delivery. *J. Control. Release* **2008**, *126*, 187–204. [CrossRef]
238. Tanaka, M.; Mazuyama, E.; Arakaki, A.; Matsunaga, T. MMS6 Protein Regulates Crystal Morphology during Nano-sized Magnetite Biomineralization in Vivo. *J. Biol. Chem.* **2011**, *286*, 6386–6392. [CrossRef]
239. Taher, Z.; Legge, C.; Winder, N.; Lysyganicz, P.; Rawlings, A.; Bryant, H.; Muthana, M.; Staniland, S. Magnetosomes and Magnetosome Mimics: Preparation, Cancer Cell Uptake and Functionalization for Future Cancer Therapies. *Pharmaceutics* **2021**, *13*, 367. [CrossRef] [PubMed]
240. Xavierselvan, M.; Divecha, H.R.; Hajra, M.; Silwal, S.; Macwan, I. Towards Tumor Targeting via Invasive Assay Using Magnetospirillum magneticum. *Front. Microbiol.* **2021**, *12*, 697132. [CrossRef] [PubMed]
241. Martel, S.; Mohammadi, M.; Felfoul, O.; Lu, Z.; Pouponneau, P. Flagellated Magnetotactic Bacteria as Controlled MRI-trackable Propulsion and Steering Systems for Medical Nanorobots Operating in the Human Microvasculature. *Int. J. Robot. Res.* **2009**, *28*, 571–582. [CrossRef] [PubMed]
242. Kuzajewska, D.; Wszołek, A.; Żwierełło, W.; Kirczuk, L.; Maruszewska, A. Magnetotactic Bacteria and Magnetosomes as Smart Drug Delivery Systems: A New Weapon on the Battlefield with Cancer? *Biology* **2020**, *9*, 102. [CrossRef] [PubMed]
243. Majedi, Y.; Loghin, D.; Mohammadi, M.; Martel, S. Characterizations of magnetotactic bacteria conjugated versus unconjugated with carboxylate-Functionalized superparamagnetic iron oxide nanoparticles for tumor targeting purposes. In Proceedings of the 2017 International Conference on Manipulation, Automation and Robotics at Small Scales (MARSS), Montreal, QC, Canada, 17–21 July 2017; pp. 1–6. [CrossRef]

244. De Lanauze, D.; Felfoul, O.; Turcot, J.-P.; Mohammadi, M.; Martel, S. Three-dimensional remote aggregation and steering of magnetotactic bacteria microrobots for drug delivery applications. *Int. J. Robot. Res.* **2013**, *33*, 359–374. [CrossRef]
245. Wang, J.; Geng, Y.; Zhang, Y.; Wang, X.; Liu, J.; Basit, A.; Miao, T.; Liu, W.; Jiang, W. Bacterial magnetosomes loaded with doxorubicin and transferrin improve targeted therapy of hepatocellular carcinoma. *Nanotheranostics* **2019**, *3*, 284–298. [CrossRef]
246. Piacenza, E.; Presentato, A.; Turner, R.J. Stability of biogenic metal(loid) nanomaterials related to the colloidal stabilization theory of chemical nanostructures. *Crit. Rev. Biotechnol.* **2018**, *38*, 1137–1156. [CrossRef]
247. Lang, C.; Schüler, D.; Faivre, D. Synthesis of Magnetite Nanoparticles for Bio- and Nanotechnology: Genetic Engineering and Biomimetics of Bacterial Magnetosomes. *Macromol. Biosci.* **2007**, *7*, 144–151. [CrossRef]
248. Rong, G.; Corrie, S.R.; Clark, H.A. In Vivo Biosensing: Progress and Perspectives. *ACS Sens.* **2017**, *2*, 327–338. [CrossRef]
249. Ozbakir, H.F.; Miller, A.D.C.; Fishman, K.B.; Martins, A.F.; Kippin, T.E.; Mukherjee, A. A Protein-Based Biosensor for Detecting Calcium by Magnetic Resonance Imaging. *ACS Sens.* **2021**, *6*, 3163–3169. [CrossRef] [PubMed]
250. Finsen, N.R. *Nobel Lectures, Physiology or Medicine 1901–1921*; Elsevier Publishing Company: Amsterdam, The Netherlands, 1967.
251. Erdal, E.; Demirbilek, M.; Yeh, Y.; Akbal, Ö.; Ruff, L.; Bozkurt, D.; Cabuk, A.; Senel, Y.; Gumuskaya, B.; Algın, O.; et al. A Comparative Study of Receptor-Targeted Magnetosome and HSA-Coated Iron Oxide Nanoparticles as MRI Contrast-Enhancing Agent in Animal Cancer Model. *Appl. Biochem. Biotechnol.* **2018**, *185*, 91–113. [CrossRef] [PubMed]

biosensors

MDPI

Article

High-Density Gold Nanoparticles Implanted on Mg/Fe LDH Nanoflowers Assisted Lateral Flow Immuno-Dipstick Assay for Visual Detection of Human Epididymal Protein 4

Hao Liu [1], Mei-Xia Wu [2] and Shou-Nian Ding [1,*]

[1] Jiangsu Province Hi-Tech Key Laboratory for Bio-Medical Research, School of Chemistry and Chemical Engineering, Southeast University, Nanjing 211189, China
[2] Lianshui Peoples Hospital, Huaian 223400, China
* Correspondence: snding@seu.edu.cn

Abstract: The timelier and more accurate the diagnosis of the disease, the higher the patient's survival rate. Human epididymal protein 4 (HE4) has great significance as a biomarker of concern for reflecting ovarian cancer. Herein, we prepared a novel optical label that can be used in lateral-flow immuno-dipstick assay (LFIA) for sensitive visual detection of HE4 by implanting hydrophobic gold nanoparticles (Au NPs) at high density in Mg/Fe LDH nanoflowers (MF NFs). MF NFs with large specific surface area, high porosity, abundant active binding sites, and stable structure were employed for the first time as templates to directly anchor Au NPs in the organic phase. After simple modification with an optimized amount of branched polyethyleneimine, not only could MF@Au NFs be dispersed in the aqueous phase, but also amino functional groups were introduced on its surface to facilitate subsequent antibody coupling steps. The limit of detection reaches 50 pM with a detection range of 50 to 1000 pM. This work initially explored how MF NFs can be used to load signal labels with ideal stability and signal amplification capabilities, which greatly improves the practicability of LFIA and highlights its important role in the field of rapid diagnostics.

Keywords: layered double hydroxide nanoflowers; gold nanoparticles; human epididymal protein 4; lateral flow immuno-dipstick assay

Citation: Liu, H.; Wu, M.-X.; Ding, S.-N. High-Density Gold Nanoparticles Implanted on Mg/Fe LDH Nanoflowers Assisted Lateral Flow Immuno-Dipstick Assay for Visual Detection of Human Epididymal Protein 4. *Biosensors* **2022**, *12*, 797. https://doi.org/10.3390/bios12100797

Received: 31 August 2022
Accepted: 25 September 2022
Published: 27 September 2022

Publisher's Note: MDPI stays neutral with regard to jurisdictional claims in published maps and institutional affiliations.

1. Introduction

Ovarian cancer is one of the three most common malignant tumors in the female reproductive system [1,2]. Most ovarian cancer patients are in the middle and late stages when they are first diagnosed and thus lose the best time for treatment [3–5]. Therefore, early diagnosis has important clinical significance for the treatment and prognosis of ovarian cancer. As a new tumor marker discovered in recent years, human epididymal protein 4 (HE4) has been studied by many researchers in various respects [6–12]. HE4 is not expressed in normal ovarian tissue but is abundantly expressed in ovarian cancer tissue. Ingegerd Hellström et al. found that HE4 was overexpressed in epithelial ovarian cancer in 2003, demonstrating for the first time that HE4 protein has potential as a biomarker for ovarian cancer [13]. As we all know, early detection and early treatment is the mainstream idea of current diagnosis, which will help reduce the morbidity and mortality of the disease and ultimately improve the quality of life of patients. Although traditional methods such as enzyme-linked immunosorbent assay and quantitative real-time PCR can help solve the early diagnosis of diseases, they are limited by expensive and complicated laboratory equipment and well-trained operators [14].

Lateral flow immuno-dipstick assay (LFIA), as a star product in point-of-care devices, has already played an important role in medical diagnosis, food analysis, and environmental monitoring [15–17]. Gold nanoparticles (Au NPs) and latex microbeads are the main tags currently used to construct LFIA. Among them, Au NPs are favored by researchers in photonics, catalysis, and bio-nanotechnology due to their inherently superior physicochemical properties and easy functionalization [18–21]. Notably, since the localized surface plasmon resonance effect is very sensitive to particle size, structure, composition, and interparticle distance, Au NPs have been further utilized in optical detection platforms. This colorimetric assay based on a single gold nanoparticle as a signal reporter molecule usually suffers from low sensitivity in low-abundance biomarker systems, which seriously hinders its application in early diagnosis [22]. In addition, the currently widely used optical labels are usually water-phase synthesized Au NPs, which have inherent disadvantages such as wide size distribution and insufficient stability. These shortcomings can be addressed by the oil-phase synthesized Au NPs because these nanocrystals have the advantages of controllable particle size and uniform morphology, and the surface of the nanoparticles is capped by alkyl ligands with stable surface chemical properties [23,24]. Several studies have demonstrated that template-based loading is an effective strategy to obtain composites with good size, shape, and configuration [25–27]. The three-dimensional incorporation of gold units into the template is highly desirable, which allows for a uniform distribution of Au NPs and is beneficial for preserving the plasmonic properties of individual Au NPs.

Layered double hydroxides (LDHs) in the form of anionic clays have attracted increasing attention from researchers due to their layered structure, high surface area, porous structure, and interlayer ion exchange [28]. Among them, spherical LDH nanoparticles with porous structure have attracted much attention due to their structural stability and high surface area. Mahfuza Mubarak et al. synthesized Mg/Fe-LDH hollow nanospheres with high specific surface areas by a simple ethylene glycol-mediated thermal method using only two metal precursors, Mg^{2+} and Fe^{3+} [29]. In this work, Mg/Fe LDH nanoflowers (MF NFs) were used for the first time as templates for loading hydrophobic Au NPs due to their advantages of large surface area, high porosity, abundant active binding sites, and stable structure [30–32]. As illustrated in Figure 1, a large amount of hydrophobic Au NPs were loaded into MF NFs through thiol-metal covalent bonds. The outer layer of MF@Au nanocomposites consists of branched polyethyleneimine (PEI), which can not only serve as a hydrophilic modification layer but also introduce amino functional groups to facilitate subsequent functional derivatization. Finally, a new type of high-density Au NP-incorporated MF NFs were successfully prepared. In the MF@Au@PEI−LFIA, MF@Au@PEI NFs containing amino functional groups on the surface were carboxylated by succinic anhydride and combined with anti-HE4 labeled antibodies to prepare the bioconjugates. Subsequently, different concentrations of HE4 antigen are combined with bioconjugates and applied to the LFIA strip. Qualitative results can be measured by using the naked eye within 15 min. MF@Au@PEI−LFIA exhibits higher sensitivity and a broader linear region compared with those of colloidal gold test strips.

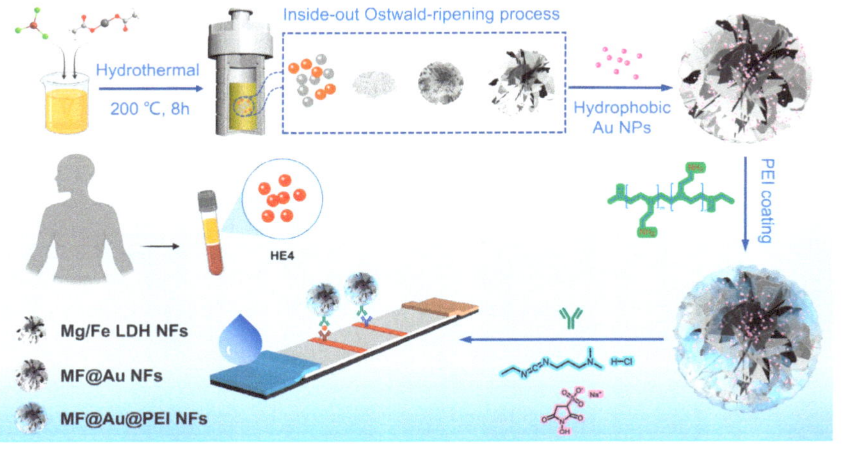

Figure 1. Schematic diagram of the preparation of high-density Au NPs-implanted Mg/Fe-LDH nanoflowers and detection process for MF@Au@PEI−LFIA.

2. Materials and Methods

2.1. Chemicals and Biological Reagents

Iron (III) chloride hexahydrate, magnesium acetate tetrahydrate, tetrachloroauric (III) acid tetrahydrate, glutaraldehyde, ethylene glycol, anhydrous ethanol, oleylamine (OLM), and Triton X-100 were purchased from Sinopharm Chemical Reagent Co., Ltd. Branched polyethyleneimine (PEI, MW 25000) was purchased from Sigma-Aldrich (Burlington, MA, USA). Bovine serum albumin (BSA), trimethoxysilylpropanethiol (3-MPTMS), 1-ethyl-3-[3-dimethylaminopropyl] carbodiimide hydrochloride (EDC·HCl), and sodium 1-hydroxy-2,5-dioxopyrrolidine-3-sulfonate (Sulfo-NHS) were purchased from Energy Chemical Co., Ltd. (Bingham, Nottinghamshire, UK). All chemicals were used as received without purification. Phosphate buffer solution (PBS, 0.01 M, pH 7.4), MES buffer solution (0.01 M, pH = 6.0), and HEPES buffer solution (0.01 M, pH = 7.4) were freshly prepared before use. The Anti-HE4 monoclonal antibody (Ab_1&Ab_2), AFP, CA199, and CEA were purchased from Shanghai Linc-Bio Science Co., LTD (Shanghai, China). Goat anti-mouse IgG antibody, sample pads, polyvinyl chloride (PVC) substrate, nitrocellulose (NC) membranes, absorbent pads, and plastic adhesive cards were purchased from Shanghai Joey Biotechnology Co. Ltd. (Shanghai, China).

2.2. Characterization

^{13}C CP/MAS NMR spectra were obtained using a Bruker ASCEND^{TM} 400WB spectrometer at 400 MHz and reported as parts per million (ppm). UV-vis absorbance spectra were obtained on a Shimadzu UV 2600 spectropolarimeter (Japan). Scanning electron microscopy (SEM) measurements were carried out under field emission scanning electron microscopy (Thermo Scientific, Waltham, MA, USA) operated at 20 kV. Transmission electron microscopy (TEM) measurements were carried out under a field-emission high-resolution transmission electron microscopy Talos F200X (Thermo Scientific, Waltham, MA, USA). Fourier transform infrared (FT-IR) spectrum was collected from a Nicolet 5700 (Thermo Nicolet Corporation, Waltham, MA, USA) IR spectrometer in the range of 4000–400 cm^{-1}. Powder X-ray diffraction (XRD) analysis was carried out using Rigaku Ultima IV multifunctional horizontal X-ray diffractometer. X-ray photoelectron spectroscopy (XPS) studies were performed using the XPS-2 (PreVac, Poland) system. Dynamic light scattering (DLS) and ζ-potential data were collected from NanoBrook 90 Plus PALS.

2.3. Preparation of Mg/Fe LDH Nanoflowers

Mg/Fe LDH nanoflowers (MF NFs) were synthesized according to a previous study with some minor modifications [29]. Briefly, Mg $(OAc)_2{\cdot}4H_2O$ (3.0 g, 14.1 mmol) was put into a three-necked flask containing 120 mL of ethylene glycol and ultrasonically dissolved quickly. Then, $FeCl_3{\cdot}6H_2O$ (285 mg, 1.05 mmol) dissolved in 30 mL of ethylene glycol was added to the flask through a dropping funnel, and the mixture was allowed to form a homogeneous yellow clear solution with magnetic stirring for 4 h. Afterwards, the solution was transferred to a Teflon-lined autoclave and heated at 200 °C for 8 h. The final product was cooled to room temperature and then washed three times with ethanol and water. The yellowish MF NFs were obtained after vacuum drying at 60 °C overnight.

2.4. Synthesis of OLM Capped Gold Nanoparticles (Au NPs)

Au NPs were synthesized by rapidly injecting gold precursor into a pre-heated surfactant solution [24]. Briefly, 10 mL of OLM was injected into a 25 mL three-neck flask and refluxed at 150 °C under an N_2 atmosphere. Then, $HAuCl_4{\cdot}4H_2O$ (0.2472 g, 0.6 mmol) was dissolved in OLM (2 mL), and it was quickly injected into the flask to continue the heating reaction for 2 h to obtain monodisperse Au NPs.

2.5. Preparation of MF@Au@PEI Nanoflowers

Firstly, MF NFs needed to be modified with sulfhydryl functional groups. MF NFs (50 mg) were dispersed in absolute ethanol by ultrasonication, and then $NH_3{\cdot}H_2O$ (500 μL) and MPTMS (500 μL) were added, followed by vigorous stirring at 25 °C for 6 h. The sulfhydryl-terminated MF NFs were harvested by centrifugation, washed thoroughly with ethanol and H_2O, and dispersed in 50 mL of chloroform. Secondly, excess Au NPs chloroform dispersion was added to the above sulfhydryl-terminated MF NFs chloroform dispersion, and a clear and transparent solution was obtained via ultrasonically treatment for 20 min. The complexes were washed twice with chloroform to remove unlinked AuNPs. The MF@Au NFs were properly dried in airflow and then dispersed in a 20 mL chloroform solution containing PEI (0.2 mg/mL) under ultrasonication conditions for 20 min. After centrifugation and washing, the MF@Au@PEI NFs can be directly dispersed in PBS solution and stored at 4 °C for later use.

2.6. Preparation of Ab_2-MF@Au@PEI

The specific preparation process is roughly the same as that reported in the previous literature [33]. Briefly, MF@Au@PEI NFs with amino functional groups on the surface were dispersed in water at a concentration of 1 mg/mL, and succinic anhydride solution (5 mg/mL) was added and stirred for 4 h to modify MF@Au@PEI with carboxyl groups. After that, it was washed several times with ethanol and water and dispersed in MES buffer solution (0.01 M, pH = 6.0) for further use. One milligram of carboxyl-modified MF@Au@PEI NFs were ultrasonically dispersed into 0.5 mL of MES buffer solution, 4 mg of EDC and 6 mg of Sulfo-NHS were added in sequence, and the reaction was shaken for 15 min. The activated MF@Au@PEI NFs were washed with water twice and redispersed in 0.5 mL of HEPES buffer (0.01 M, pH = 7.4). Seventy-five micrograms of Ab_2 (0.5 mL, 0.15 mg/mL) was added to the above solution and incubated for 2 h at room temperature with shaking. Finally, Ab_2-MF@Au@PEI bioconjugates were collected by centrifugation, washed with HEPES buffer, and redispersed in 1 mL of HEPES buffer solution (0.01 M, pH = 7.4, containing 1% BSA) to form a storage dispersion for later use.

2.7. Fabrication of the MF@Au@PEI−LFIA Test Strips

LFIA test strips were prepared by our previous reported literature [33,34]. The MF@Au@PEI−LFIA consists of six parts, including polyvinyl chloride (PVC) substrate, nitrocellulose (NC) membrane, absorbent paper, sample pad, HE4 test line, and control line. The sample pad was pretreated with PBS (0.01 M, pH = 7.4) containing 1% Triton X-100 and 2% NaCl solution and dried overnight at room temperature. The capture antibody (Ab_1, 2 mg/mL) and goat anti-mouse IgG antibody (2 mg/mL) were dispersed in the test line and control line on the NC membrane, respectively. After that, the absorbent-pad-modified NC membrane and the pretreated sample pad were assembled in sequence on the PVC adhesive backing. Then, the assembled strips were cut into 4 mm wide pieces. Finally, the prepared test strips were sealed and stored in a light-proof box.

2.8. Detection of HE4 with the MF@Au@PEI−LFIA

Different concentrations of 60 μL HE4 standard target solutions (0, 5, 10, 20, 50, 100, 200, 400, 800, and 1000 pM) were premixed with Ab_2-MF@Au@PEI NFs (40 μL), respectively. Afterward, the mixtures were applied to the sample pad. Each concentration was detected three times. After 15 min, qualitative results could be obtained by observing the red bands on the strips. An Apple mobile phone camera was used to take pictures for qualitative measurements, and ImageJ software was used to digitally process the images for T-line and C-line intensities. To compensate for potential intensity variations caused by acquisition conditions such as lighting and camera settings, background subtraction was required, and this work used the ratio of T-line intensity to C-line intensity (T/C) to quantify the signal [35,36].

3. Results and Discussion

3.1. Synthesis and Characterization of the MF@Au@PEI NFs

Mg/Fe LDH NFs were synthesized by using magnesium acetate and ferric chloride as precursors and ethylene glycol as a solvent in an autoclave. Figure 2a,b and Figure S3a are typical images of the as-prepared MF NFs. These flower-like nanospheres consist of many nanosheets with a thickness of about 10 nm interconnected to form highly open structures with dimensions of 380–400 nm (Figure S5a in Supplementary Materials). The growth mechanism of MF can be explained by the so-called inside-out Ostwald maturation process proposed by Song et al. [37–39]. Firstly, magnesium ions and iron ions coagulate and self-assemble into solid flower-like spheres. Then, nanosheets on the spherical shell begin to grow, while the inner core begins to gradually form voids, and finally, 3D hierarchical flower-like hollow nanospheres are obtained. After the sulfhydryl functional group modification of MF, OLM-capped AuNPs with uniform size (10 nm, Figures 2c, S1 and S5d) prepared by reducing chloroauric acid in the organic phase were reacted with SH-MF ultrasonically for 20 min in chloroform solution to obtain MF@Au NFs. It can be seen from Figures 2d and S3b that there are many prominent Au NPs on the surface of MF NFs, which proves that the Au NPs are abundantly loaded into the MF NFs. Likewise, the high density of Au NPs bound to MF can be more clearly seen by TEM (Figures 2f and S3c,d). It can be seen from Figure 2e that after the MF@Au NFs are coated with PEI hydrophilic modification, the Au NPs on its surface are wrapped. Energy-dispersive X-ray spectroscopy (EDS) elemental mapping, nuclear magnetic resonance (NMR), and Fourier transform infrared (FT-IR) techniques were employed to investigate the composition of MF@Au@PEI NFs. As shown in Figure 2g–l and Figure S4, MF@Au@PEI NFs are composed of Mg, Fe, Au, C, and N elements, which can demonstrate the successful preparation of the MF@Au@PEI composite structure. In the FT-IR spectra (Figure 3a), the MF NFs prepared in this work are in good agreement with the characteristics of LDH-type materials reported in the literature [29,40]. The characteristic peaks located at 2918 cm^{-1} (stretching vibration of -CH_2), 2850 cm^{-1} (stretching vibration of -CH_2), 1460 cm^{-1} (stretching vibration of -CH_2), and 1380 cm^{-1} (stretching vibration of -CH_2) indicate the successful preparation of OLM-capped Au NPs. The curves of MF@Au are in good agreement with OLM-capped

Au NPs, which indicates the formation of MF@Au composites. After PEI coating, the characteristic peaks at 3460 cm^{-1}, and 1630 cm^{-1} indicate the presence of amino functional groups. Not only did the MF@Au@PEI NFs show good hydrophilicity, but their surface was also modified with amino groups, which facilitated the subsequent antibody conjugation procedures. Figure 3b shows the UV-vis absorption data that the absorption peaks of the three are all around 520 nm. The X-ray powder diffraction (XRD) patterns of the MF NFs and MF@Au are shown in Figure 3c. The MF NFs prepared in this work have five main peaks at 9.26°, 22.12°, 34.1°, 42.3°, and 59.7°, corresponding to the (003), (006), (012), (015) and (110) planes, respectively [29]. The Au diffraction peaks in the MF@Au composite are well preserved, and the peak shape corresponds to the standard card one-to-one (JCPDS#04–0784), showing the cubic structure of Au NPs. These indicate that the crystal structure of the oil-phase AuNPs prepared in this work were not affected during the entire assembly process. In the ^{13}C CP/MAS NMR spectrum of SH-MF NFs (Figure 3d), three peaks at 17.30, 20.48, and 32.92 ppm can be assigned to methylene carbons originated from the MPTMS, which indicates that thiol functional groups have been modified onto the MF NFs. The ζ-potential results of MF NFs, hydrophobic Au NPs, MF@Au, and MF@Au@PEI are shown in Figure S5f. The ζ-potential of MF NFs in the water phase is negative (−54.92 mV) due to the large number of hydroxyl functional groups. The change in the ζ-potential value of MF@Au indicates that Au NPs are loaded in MF NFs, which turns the ζ-potential into a positive charge (53.26 mV). The surface composition and chemical state of the MF@Au NFs were investigated by X-ray photoelectron spectroscopy (XPS). The MF NFs are mainly composed of C, O, Mg, and Fe elements (Figure S6), which is consistent with the XPS data reported in the literature [29]. As shown in Figure 4a, MF@Au NFs are mainly composed of C, O, Mg, and Au elements. High-resolution XPS spectra were recorded for C 1s, Mg 2p, and Au 4f, as shown in Figure 4b–d. The C 1s spectrum consists of two components with binding energies around 284.6 eV and 285.5 eV, corresponding to C–C and C–OH bonds, respectively [29,41]. The Mg 2p spectrum has two peaks at about 49.0 and 49.9 eV, which can be ascribed to Mg–OH and Mg–O bonds, respectively [29,42]. Figure 4d shows the XPS spectrum of Au 4f, and the curve fitting analysis shows that Au $4f_{7/2}$ and Au $4f_{5/2}$ have two peaks at 83.2 eV and 87.0 eV, respectively. The Au 4f doublet of the sample is split to 3.8 eV, indicating that Au exists only in the metallic state [43]. The above characterization results all prove the successful preparation of MF@Au@PEI NFs.

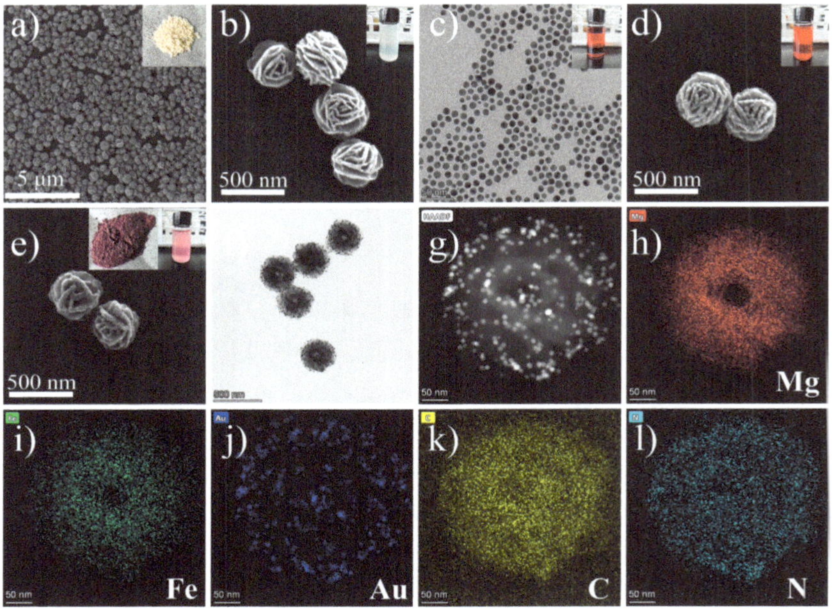

Figure 2. (**a**,**b**) SEM and TEM images of Mg/Fe LDH NFs. (The two insets are MF NFs solid powder and the state dispersed in H_2O.) (**c**) OLM capped Au NPs. (The inset is the state dispersed in $CHCl_3$.) (**d**) SEM images of MF@Au NFs. (The inset is the state dispersed in $CHCl_3$.) (**e**,**f**) SEM and TEM images of MF@Au@PEI NFs. (The two insets are the MF@Au@PEI solid powder and the state dispersed in H_2O.) (**g–l**) HAADF STEM image and EDS mapping images of MF@Au@PEI NFs.

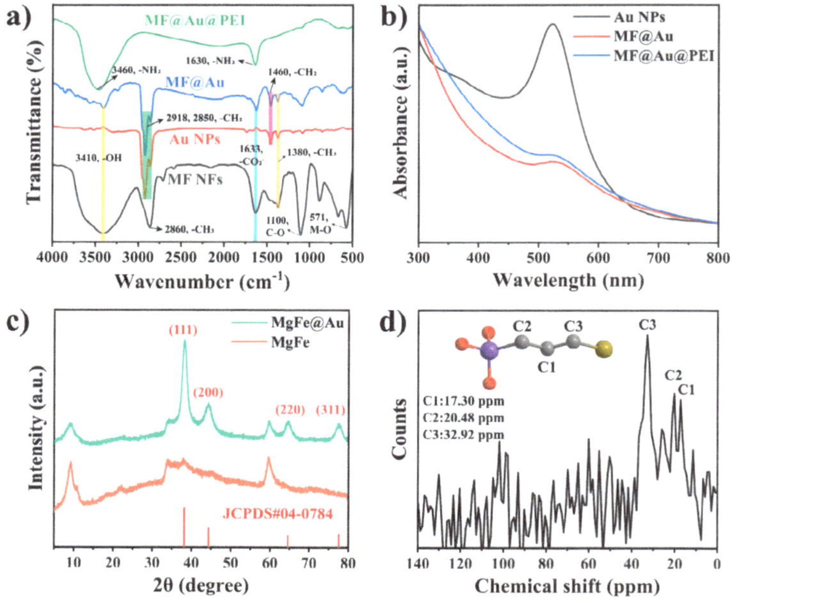

Figure 3. (**a**) FT-IR spectra of MF NFs, Au NPs, MF@Au NFs, and MF@Au@PEI NFs. (**b**) UV-vis absorption spectra of Au NPs, MF@Au NFs, and MF@Au@PEI NFs. (**c**) The XRD patterns of MF and

MF@Au NFs. (The vertical red solid lines stand for diffractions according to JCPDS#04-0784.s) (**d**) Solid-state ^{13}C CP/MAS NMR spectrum of SH-MF NFs. (The inset is the schematic diagrams of 3-MPTMS.)

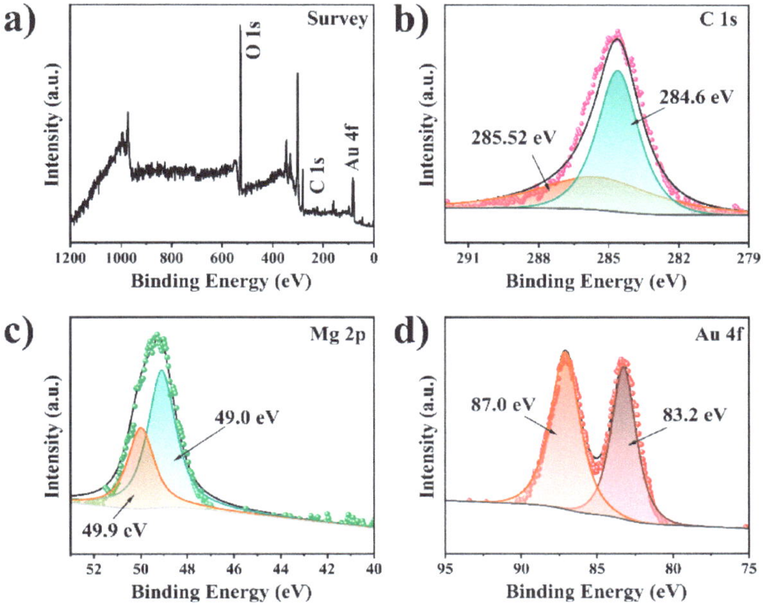

Figure 4. (**a**) XPS survey spectra, and high resolution (**b**) C 1s, (**c**) Mg 2p, and (**d**) Au 4f spectra of MF@Au NFs.

3.2. Performance of MF@Au@PEI−LFIA for HE4 Detection

Before implementing the MF@Au@PEI−LFIA, the optimization of the concentrations of PEI is very critical to achieve the best detection effect. Four different concentrations of PEI (5, 1, 0.5, 0.2 mg/mL) were used to coat MF@Au NFs. As shown in Figure S7, when the concentration exceeds 0.2 mg/mL, the MF@Au NFs experience different degrees of coagulation. We believe that the excessive PEI coating may lead to the mutual adhesion of the NFs, and the sedimentation occurs under the action of gravity. Therefore, the PEI concentration used in this work was 0.2 mg/mL. Various concentrations of HE4 antigen standard solutions were applied to analyze the sensitivity of the MF@Au@PEI−LFIA. The antigen standard solutions of different concentrations were premixed with Ab_2-MF@Au@PEI NFs and then added to the sample pads. The qualitative model of MF@Au@PEI−LFIA is based on visual observation of the line color. The test result images of the test strip after 15 min are shown in Figure 5a. All test strips showed a control line, which proved that the test results are valid. The results indicated that the limit of detection using the naked eye observation was approximately 50 pM for HE4. The signal intensity values of the test line and the control line were obtained by taking advantage of ImageJ software. As shown in Figure 5b, the linear relationship of HE4 between the T line intensity value/C line intensity value and the antigen concentrations is $y = 0.013x - 0.024$ ($R^2 = 0.983$). Clinical medicine considers the concentration of HE4 to be positive for ovarian cancer when it exceeds 70 pM [44]. As a comparison, we detected HE4 standard samples at various concentrations using traditional citrate-capped Au NPs as optical labels. As shown in Figure S8, when the concentration of HE4 was 800 pM, the T line still did not show a red band, indicating that the traditional citrate-capped Au NPs as optical labels had low detection sensitivity for HE4. Meanwhile, this result verifies the optical signal amplification of MF@Au@PEI as an optical label in antigen immune response. In addition, the assessment of three biomarkers (AFP, CEA, and CA199) at a concentration of 1 μg/mL verified the specificity of MF@Au@PEI−LFIA. The experimental results are shown in Figure 5c, and none of the other three biomarkers can produce a red band on the T line, except for HE4, which proved that MF@Au@PEI−LFIA has good selectivity.

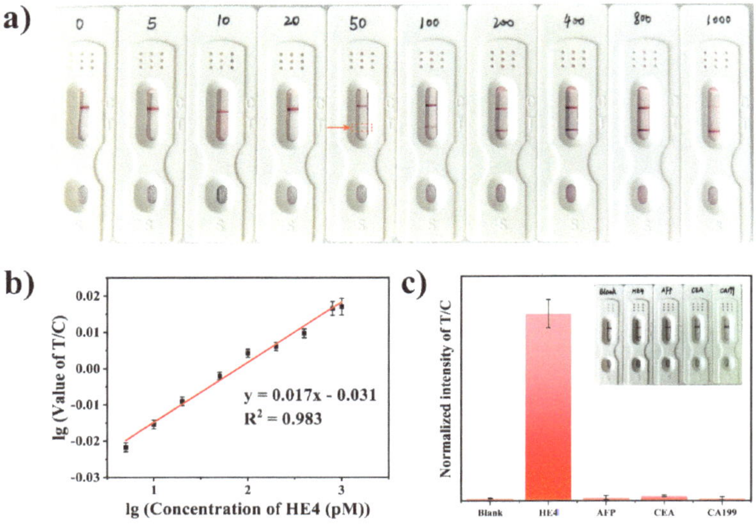

Figure 5. (**a**) Photograph of the MF@Au@PEI−LFIA when detecting different concentrations of HE4 (0, 5, 10, 20, 50, 100, 200, 400, 800, and 1000 pM). (**b**) Linear response of MF@Au@PEI−LFIA for detection of HE4 in the concentration range of 50−1000 pM. (**c**) Specificity of MF@Au@PEI−LFIA for different biomarkers (AFP, CEA, CA199).

4. Conclusions

In summary, we prepared a novel optical label by efficiently implanting hydrophobic Au NPs into Mg/Fe LDH nanoflowers. Compared with traditional colloidal gold nanotags, the incorporation of high-density Au NPs greatly enhanced the optical signal intensity of each carrier, and the detection sensitivity of HE4 was greatly improved. Under optimal experimental conditions, the limit of detection for HE4 was 50 pM with a detection range of 50 to 1000 pM. The study found that MF@Au@PEI−LFIA is highly selective for HE4 and has little mutual interference with other biomarkers (AFP, CA199, CEA). This work initially explores how Mg/Fe LDH NFs can be used to load signal labels with desirable stability and signal amplification capabilities, which greatly broadens the practicability of LFIA and highlights its important role in rapid diagnosis.

Supplementary Materials: The following supporting information can be downloaded at: https://www.mdpi.com/article/10.3390/bios12100797/s1, Figure S1. TEM images of OLM capped Au NPs. Figure S2. TEM images of Citrate capped Au NPs. Figure S3. (a). TEM image of MF NFs. (b). SEM image of MF@Au NFs. (c–d). TEM image of MF@Au@PEI NFs. Figure S4. The EDS spectrum of MF@Au@PEI NFs. Figure S5. (a–e) DLS data of hydrophobic Au NPs, MF NFs, MF@Au NFs, MF@Au@PEI NFs and citrate capped Au NPs. (f) ζ-potential of MF NFs, hydrophobic Au NPs, MF@Au NFs and MF@Au@PEI NFs. Figure S6. High resolution (a) C 1s, (b) Mg 2p, (c) Fe 2p, and (d) O 1s spectra of MF NFs. Figure S7. MF@Au NFs coated with different concentrations of PEI (concentrations of (a–d) are 5, 1, 0.5, 0.2 mg/mL in order, the scale bar is 2 μm). Figure S8. Photographs of the detection results of different concentrations of HE4 (0, 10, 50, 100, 200, 400, and 800 pM) using citrate capped Au NPs as optical labels [45–47].

Author Contributions: H.L.: Methodology, Investigation, Writing—original draft, Writing—review and editing, Visualization. M.-X.W.: Validation, Resources. S.-N.D.: Conceptualization, Methodology, Resources, Funding acquisition, Writing—review and editing, Supervision. All authors have read and agreed to the published version of the manuscript.

Funding: This work was supported by the National Key Research and Development Program of China (2017YFA0700404), and the National Natural Science Foundation of China (22174015).

Institutional Review Board Statement: Not applicable.

Informed Consent Statement: Not applicable.

Data Availability Statement: Not applicable.

Conflicts of Interest: The authors declare no conflict of interest.

References

1. Morice, P.; Gouy, S.; Leary, A. Mucinous Ovarian Carcinoma. *N. Engl. J. Med.* **2019**, *380*, 1256–1266. [CrossRef] [PubMed]
2. Lheureux, S.; Braunstein, M.; Oza, A.M. Epithelial ovarian cancer: Evolution of management in the era of precision medicine. *CA Cancer J. Clin.* **2019**, *69*, 280–304. [CrossRef] [PubMed]
3. Anp, P. The American Cancer Society's Facts & Figures: 2020 Edition. *J. Adv. Pract. Oncol.* **2020**, *11*, 135. [CrossRef]
4. Zheng, L.; Cui, C.; Shi, O.; Lu, X.; Li, Y.-K.; Wang, W.; Li, Y.; Wang, Q. Incidence and mortality of ovarian cancer at the global, regional, and national levels, 1990–2017. *Gynecol. Oncol.* **2020**, *159*, 239–247. [CrossRef] [PubMed]
5. Bray, F.; Ferlay, J.; Soerjomataram, I.; Siegel, R.L.; Torre, L.A.; Jemal, A. Global cancer statistics 2018: GLOBOCAN estimates of incidence and mortality worldwide for 36 cancers in 185 countries. *CA Cancer J. Clin.* **2018**, *68*, 394–424. [CrossRef] [PubMed]
6. Li, J.; Dowdy, S.; Tipton, T.; Podratz, K.; Lu, W.-G.; Xie, X.; Jiang, S.-W. HE4 as a biomarker for ovarian and endometrial cancer management. *Expert Rev. Mol. Diagn.* **2009**, *9*, 555–566. [CrossRef] [PubMed]
7. Montagnana, M.; Danese, E.; Giudici, S.; Franchi, M.; Guidi, G.C.; Plebani, M.; Lippi, G. HE4 in ovarian cancer: From discovery to clinical application. In *Advances in Clinical Chemistry*; Makowski, G.S., Ed.; Elsevier: Amsterdam, The Netherlands, 2011; Volume 55, pp. 1–20.
8. Zheng, L.-E.; Qu, J.-Y.; He, F. The diagnosis and pathological value of combined detection of HE4 and CA125 for patients with ovarian cancer. *Open Med.* **2016**, *11*, 125–132. [CrossRef]
9. Scaletta, G.; Plotti, F.; Luvero, D.; Capriglione, S.; Montera, R.; Miranda, A.; Lopez, S.; Terranova, C.; De Cicco Nardone, C.; Angioli, R. The role of novel biomarker HE4 in the diagnosis, prognosis and follow-up of ovarian cancer: A systematic review. *Expert Rev. Anticancer Ther.* **2017**, *17*, 827–839. [CrossRef]
10. Furrer, D.; Grégoire, J.; Turcotte, S.; Plante, M.; Bachvarov, D.; Trudel, D.; Têtu, B.; Douville, P.; Bairati, I. Performance of preoperative plasma tumor markers HE4 and CA125 in predicting ovarian cancer mortality in women with epithelial ovarian cancer. *PLoS ONE* **2019**, *14*, e0218621. [CrossRef]
11. Liu, D.; Kong, D.; Li, J.; Gao, L.; Wu, D.; Liu, Y.; Yang, W.; Zhang, L.; Zhu, J.; Jin, X. HE4 level in ascites may assess the ovarian cancer chemotherapeutic effect. *J. Ovarian Res.* **2018**, *11*, 47. [CrossRef]
12. Liu, J.; Han, L.; Jiao, Z. The diagnostic value of human epididymis protein 4 for endometrial cancer is moderate. *Sci. Rep.* **2021**, *11*, 575. [CrossRef] [PubMed]
13. Hellström, I.; Raycraft, J.; Hayden-Ledbetter, M.; Ledbetter, J.A.; Schummer, M.; McIntosh, M.; Drescher, C.; Urban, N.; Hellström, K.E. The HE4 (WFDC2) protein is a biomarker for ovarian carcinoma. *Cancer Res.* **2003**, *63*, 3695–3700. [PubMed]
14. Wender, R.C.; Brawley, O.W.; Fedewa, S.A.; Gansler, T.; Smith, R.A. A blueprint for cancer screening and early detection: Advancing screening's contribution to cancer control. *CA Cancer J. Clin.* **2019**, *69*, 50–79. [CrossRef] [PubMed]
15. Ghorbanizamani, F.; Tok, K.; Moulahoum, H.; Harmanci, D.; Hanoglu, S.B.; Durmus, C.; Zihnioglu, F.; Evran, S.; Cicek, C.; Sertoz, R.; et al. Dye-Loaded Polymersome-Based Lateral Flow Assay: Rational Design of a COVID-19 Testing Platform by Repurposing SARS-CoV-2 Antibody Cocktail and Antigens Obtained from Positive Human Samples. *ACS Sens.* **2021**, *6*, 2988–2997. [CrossRef]
16. Ren, Y.; Wei, J.; He, Y.; Wang, Y.; Bai, M.; Zhang, C.; Luo, L.; Wang, J.; Wang, Y. Ultrasensitive label-free immunochromatographic strip sensor for Salmonella determination based on salt-induced aggregated gold nanoparticles. *Food Chem.* **2021**, *343*, 128518. [CrossRef]
17. Zhu, C.; Zhang, G.; Huang, Y.; Yang, S.; Ren, S.; Gao, Z.; Chen, A. Dual-competitive lateral flow aptasensor for detection of aflatoxin B1 in food and feedstuffs. *J. Hazard. Mater.* **2018**, *344*, 249–257. [CrossRef]
18. Zhao, J.; Xue, S.; Ji, R.; Li, B.; Li, J. Localized surface plasmon resonance for enhanced electrocatalysis. *Chem. Soc. Rev.* **2021**, *50*, 12070–12097. [CrossRef]
19. Montes-García, V.; Squillaci, M.A.; Diez-Castellnou, M.; Ong, Q.K.; Stellacci, F.; Samorì, P. Chemical sensing with Au and Ag nanoparticles. *Chem. Soc. Rev.* **2021**, *50*, 1269–1304. [CrossRef]
20. Kang, X.; Li, Y.; Zhu, M.; Jin, R. Atomically precise alloy nanoclusters: Syntheses, structures, and properties. *Chem. Soc. Rev.* **2020**, *49*, 6443–6514. [CrossRef]

21. Yeh, C.-M.; Chen, M.-C.; Wu, T.-C.; Chen, J.-W.; Lai, C.-H. Lectin-Triggered Aggregation of Glyco-Gold Nanoprobes for Activity-based Sensing of Hydrogen Peroxide by the Naked Eye. *Chem. Asian J.* **2021**, *16*, 3462–3468. [CrossRef]
22. Zhang, D.; Huang, L.; Liu, B.; Ni, H.; Sun, L.; Su, E.; Chen, H.; Gu, Z.; Zhao, X. Quantitative and ultrasensitive detection of multiplex cardiac biomarkers in lateral flow assay with core-shell SERS nanotags. *Biosens. Bioelectron.* **2018**, *106*, 204–211. [CrossRef] [PubMed]
23. Park, J.; Joo, J.; Kwon, S.G.; Jang, Y.; Hyeon, T. Synthesis of Monodisperse Spherical Nanocrystals. *ChemInform* **2007**, *38*, 4630–4660. [CrossRef]
24. Liu, S.; Chen, G.; Prasad, P.N.; Swihart, M.T. Synthesis of Monodisperse Au, Ag, and Au–Ag Alloy Nanoparticles with Tunable Size and Surface Plasmon Resonance Frequency. *Chem. Mater.* **2011**, *23*, 4098–4101. [CrossRef]
25. Yoon, M.; Kim, Y.; Cho, J. Multifunctional Colloids with Optical, Magnetic, and Superhydrophobic Properties Derived from Nucleophilic Substitution-Induced Layer-by-Layer Assembly in Organic Media. *ACS Nano* **2011**, *5*, 5417–5426. [CrossRef]
26. Lee, B.; Kim, Y.; Lee, S.; Kim, Y.S.; Wang, D.; Cho, J. Layer-by-Layer Growth of Polymer/Quantum Dot Composite Multilayers by Nucleophilic Substitution in Organic Media. *Angew. Chem. Int. Ed.* **2010**, *49*, 359–363. [CrossRef]
27. Kim, J.; Lee, J.E.; Lee, J.; Jang, Y.; Kim, S.-W.; An, K.; Yu, J.H.; Hyeon, T. Generalized Fabrication of Multifunctional Nanoparticle Assemblies on Silica Spheres. *Angew. Chem. Int. Ed.* **2006**, *45*, 4789–4793. [CrossRef]
28. Abellán, G.; Martí-Gastaldo, C.; Ribera, A.; Coronado, E. Hybrid Materials Based on Magnetic Layered Double Hydroxides: A Molecular Perspective. *Acc. Chem. Res.* **2015**, *48*, 1601–1611. [CrossRef]
29. Mubarak, M.; Jeon, H.; Islam, M.S.; Yoon, C.; Bae, J.-S.; Hwang, S.-J.; Choi, W.S.; Lee, H.-J. One-pot synthesis of layered double hydroxide hollow nanospheres with ultrafast removal efficiency for heavy metal ions and organic contaminants. *Chemosphere* **2018**, *201*, 676–686. [CrossRef]
30. Zong, Y.; Li, K.; Tian, R.; Lin, Y.; Lu, C. Highly dispersed layered double oxide hollow spheres with sufficient active sites for adsorption of methyl blue. *Nanoscale* **2018**, *10*, 23191–23197. [CrossRef]
31. Lin, Y.; Zeng, Z.; Zhu, J.; Wei, Y.; Chen, S.; Yuan, X.; Liu, L. Facile synthesis of ZnAl-layered double hydroxide microspheres with core–shell structure and their enhanced adsorption capability. *Mater. Lett.* **2015**, *156*, 169–172. [CrossRef]
32. Shan, R.-R.; Yan, L.-G.; Yang, Y.-M.; Yang, K.; Yu, S.-J.; Yu, H.-Q.; Zhu, B.-C.; Du, B. Highly efficient removal of three red dyes by adsorption onto Mg–Al-layered double hydroxide. *J. Ind. Eng. Chem.* **2015**, *21*, 561–568. [CrossRef]
33. Xu, L.-D.; Zhang, Q.; Ding, S.-N.; Xu, J.-J.; Chen, H.-Y. Ultrasensitive Detection of Severe Fever with Thrombocytopenia Syndrome Virus Based on Immunofluorescent Carbon Dots/SiO_2 Nanosphere-Based Lateral Flow Assay. *ACS Omega* **2019**, *4*, 21431–21438. [CrossRef] [PubMed]
34. Xu, L.D.; Zhu, J.; Ding, S.N. Immunoassay of SARS-CoV-2 nucleocapsid proteins using novel red emission-enhanced carbon dot-based silica spheres. *Analyst* **2021**, *146*, 5055–5060. [CrossRef]
35. Kim, M.; Kim, M.S.; Kweon, S.H.; Jeong, S.; Kang, M.H.; Kim, M.I.; Lee, J.; Doh, J. Simple and Sensitive Point-of-Care Bioassay System Based on Hierarchically Structured Enzyme-Mimetic Nanoparticles. *Adv. Healthc. Mater.* **2015**, *4*, 1311–1316. [CrossRef] [PubMed]
36. Kim, M.S.; Kweon, S.H.; Cho, S.; An, S.S.A.; Kim, M.I.; Doh, J.; Lee, J. Pt-Decorated Magnetic Nanozymes for Facile and Sensitive Point-of-Care Bioassay. *ACS Appl. Mater. Interfaces* **2017**, *9*, 35133–35140. [CrossRef]
37. Huang, P.; Liu, J.; Wei, F.; Zhu, Y.; Wang, X.; Cao, C.; Song, W. Size-selective adsorption of anionic dyes induced by the layer space in layered double hydroxide hollow microspheres. *Mater. Chem. Front.* **2017**, *1*, 1550–1555. [CrossRef]
38. Géraud, E.; Rafqah, S.; Sarakha, M.; Forano, C.; Prevot, V.; Leroux, F. Three Dimensionally Ordered Macroporous Layered Double Hydroxides: Preparation by Templated Impregnation/Coprecipitation and Pattern Stability upon Calcination. *Chem. Mater.* **2008**, *20*, 1116–1125. [CrossRef]
39. Wang, B.; Wu, H.; Yu, L.; Xu, R.; Lim, T.-T.; Lou, X.W. Template-free Formation of Uniform Urchin-like α-FeOOH Hollow Spheres with Superior Capability for Water Treatment. *Adv. Mater.* **2012**, *24*, 1111–1116. [CrossRef]
40. Mubarak, M.; Islam, M.S.; Yoon, D.-Y.; Lee, J.H.; Park, H.J.; Bae, J.-S.; Lee, H.-J. Flower-like Mg/Fe-layered double oxide nanospheres with ultrahigh adsorption efficiency for anionic organic dyes. *Colloids Surf. A Physicochem. Eng. Asp.* **2021**, *618*, 126446. [CrossRef]
41. Cano, E.; Bastidas, J.M.; Polo, J.L.; Mora, N. Study of the Effect of Acetic Acid Vapor on Copper Corrosion at 40 and 80% Relative Humidity. *J. Electrochem. Soc.* **2001**, *148*, B431–B437. [CrossRef]
42. Ardizzone, S.; Bianchi, C.L.; Fadoni, M.; Vercelli, B. Magnesium salts and oxide: An XPS overview. *Appl. Surf. Sci.* **1997**, *119*, 253–259. [CrossRef]
43. Daware, K.; Kasture, M.; Kalubarme, R.; Shinde, R.; Patil, K.; Suzuki, N.; Terashima, C.; Gosavi, S.; Fujishima, A. Detection of toxic metal ions Pb2+ in water using SiO_2@Au core-shell nanostructures: A simple technique for water quality monitoring. *Chem. Phys. Lett.* **2019**, *732*, 136635. [CrossRef]
44. Nicholson, B.D.; Lee, M.-M.; Wijeratne, D.; James, T.; Shine, B.; Oke, J.L. Trends in Cancer Antigen 125 testing 2003–2014: A primary care population-based cohort study using laboratory data. *Eur. J. Cancer Care* **2019**, *28*, e12914. [CrossRef] [PubMed]
45. Zhou, Y.; Pan, F.-G.; Li, Y.-S.; Zhang, Y.-Y.; Zhang, J.-H.; Lu, S.-Y.; Ren, H.-L.; Liu, Z.-S. Colloidal gold probe-based immunochromatographic assay for the rapid detection of brevetoxins in fishery product samples. *Biosens. Bioelectron.* **2009**, *24*, 2744–2747. [CrossRef]

46. Frens, G. Controlled Nucleation for the Regulation of the Particle Size in Monodisperse Gold Suspensions. *Nat. Phys. Sci.* **1973**, *241*, 20–22. [CrossRef]
47. Zuo, J.-Y.; Jiao, Y.-J.; Zhu, J.; Ding, S.-N. Rapid Detection of Severe Fever with Thrombocytopenia Syndrome Virus via Colloidal Gold Immunochromatography Assay. *ACS Omega* **2018**, *3*, 15399–15406. [CrossRef]

biosensors

Article

Defect Surface Engineering of Hollow $NiCo_2S_4$ Nanoprisms towards Performance-Enhanced Non-Enzymatic Glucose Oxidation

Xiaomin Lang [1,†], Dandan Chu [1,†], Yan Wang [1], Danhua Ge [1,*] and Xiaojun Chen [1,2,*]

1 College of Chemistry and Molecular Engineering, Nanjing Tech University, Nanjing 211800, China
2 Jiangsu Key Laboratory of Molecular Biology for Skin Diseases and STIs, Nanjing 210042, China
* Correspondence: gedanhua@njtech.edu.cn (D.G.); chenxj@njtech.edu.cn (X.C.)
† These authors contributed equally to this work.

Abstract: Transition metal sulfides have been explored as electrode materials for non-enzymatic detection. In this work, we investigated the effects of phosphorus doping on the electrochemical performances of $NiCo_2S_4$ electrodes (P-$NiCo_2S_4$) towards glucose oxidation. The fabricated non-enzymatic biosensor displayed better sensing performances than pristine $NiCo_2S_4$, with a good sensitivity of 250 μA mM^{-1} cm^{-2}, a low detection limit (LOD) of 0.46 μM (S/N = 3), a wide linear range of 0.001 to 5.2 mM, and high selectivity. Moreover, P-$NiCo_2S_4$ demonstrated its feasibility for glucose determination for practical sample testing. This is due to the fact that the synergetic effects between Ni and Co species, and the partial substitution of S vacancies with P can help to increase electronic conductivity, enrich binary electroactive sites, and facilitate surface electroactivity. Thus, it is found that the incorporation of dopants into $NiCo_2S_4$ is an effective strategy to improve the electrochemical activity of host materials.

Keywords: phosphorus doping; $NiCo_2S_4$; non-enzymatic sensor; electrochemistry

Citation: Lang, X.; Chu, D.; Wang, Y.; Ge, D.; Chen, X. Defect Surface Engineering of Hollow $NiCo_2S_4$ Nanoprisms towards Performance-Enhanced Non-Enzymatic Glucose Oxidation. *Biosensors* **2022**, *12*, 823. https://doi.org/10.3390/bios12100823

Received: 6 September 2022
Accepted: 3 October 2022
Published: 4 October 2022

1. Introduction

The accurate measurement of glucose concentration from blood or other sources is of great significance to diagnose diabetes and monitor food quality in the field of pharmaceuticals and foods [1]. Recently, electrochemical sensors provide such simple operation, rapid response, and high sensitivity that they have aroused wide attention for use in glucose detection. Commercial glucose biosensors are generally based on the glucose oxidase enzyme (GO_x), where glucose is converted into gluconolactone in the presence of saturated oxygen. The enzymatic sensors exhibit outstanding sensitivity and selectivity but suffer from several limitations caused by environmental effects, high cost, and tedious enzyme immobilization process [2]. In order to break the above-mentioned drawbacks, it is essential to develop and explore the non-enzymatic sensors for direct electrooxidation of glucose. Over the past decades, thanks to their highly active area, excellent stability, and relatively cheap price, various transition metal chalcogenides, including oxides, sulfides, and selenides, have been of great promise in electrochemical sensors. Among these, transition metal sulfides with high electrochemical activity have drawn extensive attention [3–6]. Compared with oxides, sulfides have a more striking electrochemical performance with outstanding electronic conductivity and abundant redox chemical properties, because of their more flexible structure with an elongation of chemical bonds constructed by replacing O with S, which benefits electron transport [7,8]. In particular, $NiCo_2S_4$ usually exhibits high electrochemical activities as a kind of single-phase binary metal sulfide, rather than the simple mixture of NiS_x and CoS_x, due to rich chemical redox, low cost, complex chemical compositions, abundant resources, environmental friendliness, and the synergetic effect of both individual components [9]. In spite of promising results, $NiCo_2S_4$ still suffers from a reduced density of electrochemically active sites and severe polarization on account of lower conductivity and volume changes during electrochemical reactions. It still cannot meet the demands of high capacitance, which brings a challenge to its practical utilization [10,11]. Therefore, enormous research efforts have been devoted to enhancing the electrochemical performance of $NiCo_2S_4$ through regulating nanostructures or combinations with conductive materials.

Typical nanostructures of nanowires, nanotubes, or nanorods could remarkably enlarge the accessible areas between the electrolyte and electrode and shorten the ionic diffusion distance, while introducing conductive materials may optimize the electrical conductivity and extend the electroactive surface [3,12,13]. For example, Huang's group fabricated a non-enzymatic glucose sensor based on the 3D flower-like $NiCo_2S_4$ deposited Ni-modified cellulose filter paper, exhibiting a wide linear range of 0.5 μM–6 mM, high sensitivity of 283 $\mu A\ mM^{-1}\ cm^{-2}$, and a low detection limit (LOD) of 50 nM [14]. Guo et al. reported a $NiCo_2S_4$ nanowire array with a unique core-shell structure grown on electrospun graphitic nanofiber film, endowed with a wide linear range and a low LOD for glucose sensing [15]. Although $NiCo_2S_4$ could achieve outstanding electrochemical properties through the above two methods, it is not always feasible to modulate the inherent electronic properties to ideal levels. Recent reports have demonstrated that an additional surface-modified coating can change surface charge, microenvironment, and active site exposure, to achieve the purpose of adjusting activity [16]. Especially, anionic doping is verified as an effective method to introduce the surface defects for altering electron density, thereby improving the redox reactivity of the $NiCo_2S_4$ complex [13,17,18]. Among them, the phosphorus (P) element with fully vacant 3D orbitals and lone-pair electrons has drawn enormous attention for promoting the surface electroactivity of host materials. Thereby, the surface P-anion doping of $NiCo_2S_4$ might accommodate the surface charge state by tuning the partial charge density of neighboring bonded Ni and Co atoms. This method would enhance the redox activity of $NiCo_2S_4$ by optimizing the adsorption energy of electrochemical analytes and reducing the strain during redox reactions [13,19–21].

Inspired by these analyses, P-doping $NiCo_2S_4$ with a hollow nanoprism-like structure (P-$NiCo_2S_4$ HNPs) is successfully synthesized in this work through a simple phosphatization reaction. Benefiting from the unique hollow structure and surface engineering of $NiCo_2S_4$, the P-$NiCo_2S_4$ HNPs have offered a better electrochemical sensing performance than that of pristine $NiCo_2S_4$, such as higher sensitivity, better selectivity, and reproducibility, as well as excellent feasibility toward glucose determination in practical samples.

2. Materials and Methods

2.1. Chemicals and Materials

All reagents are of analytical grade and used without further purification. Double distilled water was used throughout the experiments. Cobalt (II) acetate tetrahydrate ($Co(OAc)_2{\cdot}4H_2O$), nickel (II) acetate tetrahydrate ($Ni(Oac)_2{\cdot}4H_2O$), potassium hexacyanoferrate (II) ($K_4[Fe(CN)_6]$), potassium ferricyanide ($K_3[Fe(CN)_6]$), D-(+)-glucose, thioacetamide (TAA), uric acid (UA), and sodium hypophosphite (NaH_2PO_2) were purchased from Sinopharm Chemical Reagent Co., Ltd. (Shanghai, China). Ascorbic acid (AA), ethanol (EtOH), potassium chloride (KCl), and 4-acetamidophenol (AP) were ordered from Aladdin Industrial Co., Ltd. (Shanghai, China), Yasheng Chemical Co., Ltd. (Jiangsu, China), Lingfeng Chemical Reagent Co., Ltd. (Shanghai, China), and Shanghai Macklin Biochemical Co., Ltd. (Shanghai, China), respectively. Urea and sodium chloride (NaCl) were acquired from Xilong Chemical Co., Ltd. (Guangdong, China), while L-cysteine (Lcy) and D-fructose (Fru) were obtained from Huixing Biochemical Reagent Co., Ltd. (Shanghai, China).

2.2. Apparatus

The morphologies and structures of the as-prepared samples were characterized by using scanning electron microscopy (SEM, Zelss Sigma300) at 3 kV and transmission electron microscopy (TEM, FEI Talos-F200s) at a voltage of 200 kV. The crystalline structure and chemical states of the samples were verified by powder X-ray diffraction (XRD, D/max-2500 diffractometer, Rigaku, Japan), energy-dispersive spectroscopy (EDS), and X-ray photoelectron spectroscopy (XPS, Kratos Axis Ultra DLD spectrometer). Surface area allowing Brunauer-Emmett-Teller (BET) isotherms was carried out by monitoring N_2 adsorption/desorption using a NOVA 2000 surface area analyzer (Quantachrome) at 77 K.

2.3. Fabrication of the P-$NiCo_2S_4$ HNPs

According to the previous literature, $NiCo_2S_4$ NPs were firstly synthesized [22]. Briefly, $Co(OAc)_2{\cdot}4H_2O$ (0.25 g), $Ni(OAc)_2{\cdot}4H_2O$ (0.12 g), and urea (0.56 g) were successively added into EtOH (80 mL) under stirring. Then, the resulting mixture was heated to 65 °C for 4 h, and Co/Ni precursors were obtained after processing. Then, the Co/Ni precursors were re-dispersed into EtOH, followed by the addition of TAA. Subsequently, the resulting solution was transferred into a Teflon-lined stainless-steel autoclave and heated to 160 °C for 6 h. $NiCo_2S_4$ NPs were obtained after processing and drying in the air. Finally, $NiCo_2S_4$ (0.05 g) and NaH_2PO_2 (0.2 g) were placed in tandem and annealed at 350 °C for 2 h under a flowing N_2 atmosphere to produce the P-$NiCo_2S_4$ HNPs.

2.4. Electrochemical Measurements

Before preparing modified electrodes, the indium tin oxide (ITO, diameter of 3 mm) electrodes were successively sonicated and cleaned with acetone, EtOH, and water for use. After sonication for an adequate time, 6 μL of a uniform dispersion of P-$NiCo_2S_4$ HNPs in EtOH (1 mg mL^{-1}) was drop-casted onto the prepared ITO and dried at room temperature for next use. Cyclic voltammogram (CV) and chronoamperometry were performed by a CHI 660D instrument (Chenhua Instrument Co., Shanghai, China) to evaluate the electrochemical performance of the samples. In addition, the glucose detection was carried out in 0.2 M NaOH solution.

3. Results and Discussion

As illustrated in Figure 1A, the $NiCo_2S_4$ was synthesized using our previous method [22], and the synthetic procedure for P-$NiCo_2S_4$ was described through a facile P-doping. Utilizing $CO_3{}^{2-}$ and OH^- derived from the hydrolysis of urea in the presence of Co^{2+} and Ni^{2+}, the Co/Ni precursors were obtained by a solution reaction and then employed as the self-engaged templates. The XRD pattern confirmed that the crystalline Co/Ni precursors possessed a tetragonal cobalt/nickel acetate hydroxide phase (Figure S1, Supplementary Materials) [22]. Moreover, The EDS spectrum substantiated the successful synthesis of Co/Ni precursors with the mole ratio of Co to Ni of 2:1 (Figure S2). In addition, the morphologies and structures were further characterized by SEM. As shown in Figure 1B,C, the uniform prism-like Co/Ni precursors possessed a length of ~1.3 μm and a width of ~280 nm with a smooth surface. After sulfidation and the follow-up phosphating process, the length and width of the resulting P-$NiCo_2S_4$ were ~1.5 μm and ~250 nm, respectively (Figure 1D,E). It could be seen that the P-$NiCo_2S_4$ HNPs still maintained the original prism-like appearance with a hollow structure and rough surface, which was of benefit to enhancing the electron transfer between interfaces. As shown in Figure 1F of the TEM image of the P-$NiCo_2S_4$ HNPs, this further demonstrated the detailed geometrical morphology and the hollow interior structure. The XRD pattern of pristine P-$NiCo_2S_4$ was presented in Figure 2A. Obviously, all the diffraction peaks can well correspond to the cubic $NiCo_2S_4$ phase (JCPDS Card. No. 20-0782) [13,22,23]. The result indicated that the $NiCo_2S_4$ crystal structure was not distorted significantly by the introduction of P [13,23]. In accordance with the XRD analysis, the high-resolution TEM (HRTEM) depicted that the inter-planar distances of P-$NiCo_2S_4$ crystals were ~0.28 nm and ~0.56 nm, corresponding to the (311) and (111) planes of P-$NiCo_2S_4$, respectively (Figure 1G). Additionally, the corresponding EDS mapping further confirmed the uniform distribution of Co, Ni, S, and P elements within the hollow prisms (Figure 1H). Such results stayed in step with the EDS spectrum (Figure S3).

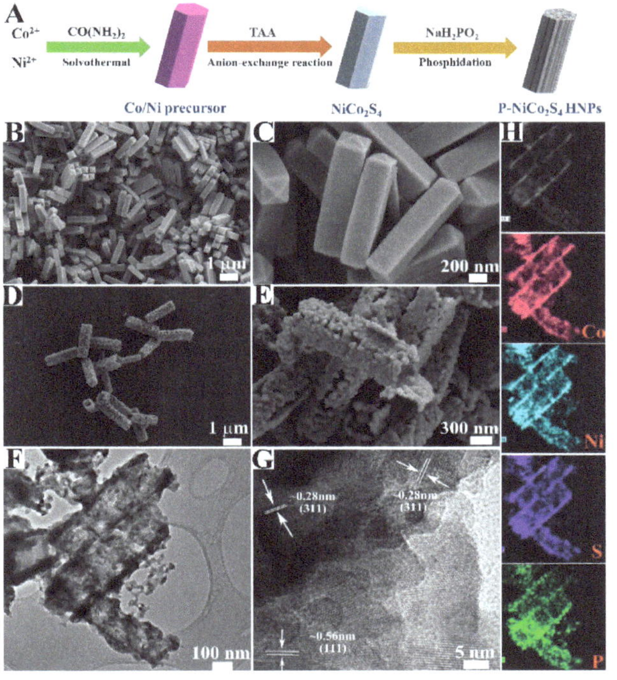

Figure 1. (**A**) Schematic illustration of P-$NiCo_2S_4$ HNPs preparation. SEM images of: (**B**,**C**) Co/Ni precursor, and (**D**,**E**) P-$NiCo_2S_4$. (**F**) TEM, (**G**) HRTEM image, and (**H**) EDS mapping of P-$NiCo_2S_4$ HNPs.

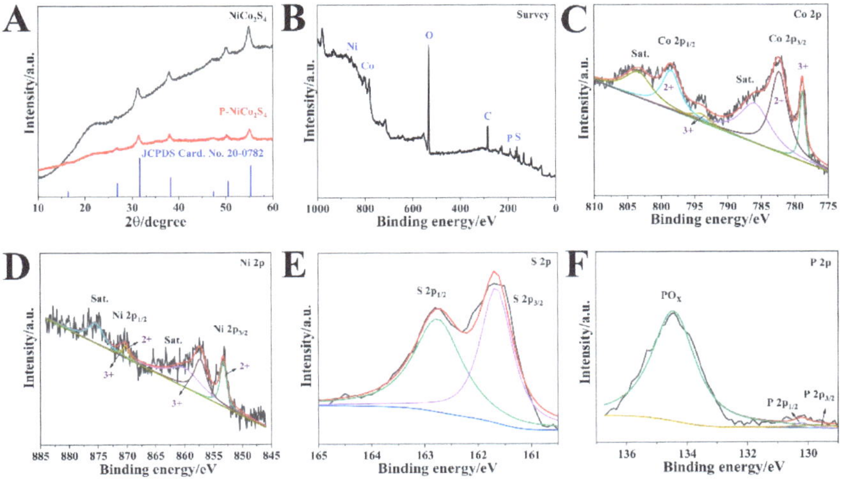

Figure 2. (**A**) XRD patterns of P-$NiCo_2S_4$ and $NiCo_2S_4$. (**B**) Full range XPS survey spectrum. The 2p spectrum of: (**C**) Co, (**D**) Ni, (**E**) S, and (**F**) P elements in P-$NiCo_2S_4$ HNPs.

As shown in Figure 2, XPS was performed to further identify the chemical environment and surface element state of the as-prepared P-$NiCo_2S_4$ HNPs. The XPS survey spectrum confirmed the presence of P, S, O, Co, and Ni elements on the surface of P-$NiCo_2S_4$ HNPs (Figure 2B). For the 2p spectrum of Co (Figure 2C), it could be noticed that two main peaks at

782.4 and 798.7 eV were consistent with Co $2p_{3/2}$ and Co $2p_{1/2}$ of Co^{2+}. Two other peaks at 778.8 and 793.9 eV could be indexed to Co $2p_{3/2}$ and Co $2p_{1/2}$ of Co^{3+}, respectively, together with two satellite peaks at 786.2 and 803.5 eV [5,24]. The Ni 2p spectrum exhibited similar characteristics, as shown in Figure 2D. Two peaks at 853.3 and 870.2 eV were attributed to Ni $2p_{3/2}$ and Ni $2p_{1/2}$ of Ni^{2+}, while two additional peaks at 857.4 and 871.3 eV resulted from Ni $2p_{3/2}$ and Ni $2p_{1/2}$ of Ni^{3+}, respectively [25]. For the 2p spectrum of the S element (Figure 2E), the peak at 161.7 eV (S $2p_{3/2}$) was ascribed to the metal-sulfur bonds, while the binding energy of 162.8 eV (S $2p_{1/2}$) was linked to S^{2-} with low coordination at the surface. In addition, the two signals at 129.6 and 130.2 eV were assigned to P $2p_{3/2}$ and P $2p_{1/2}$, relating to metal phosphides. Moreover, the peak at 133.4 eV belonged to the oxidized phosphorus species (PO_x), which originated from the exposure of its surface to air for ages [26]. Finally, the XPS result suggested a P content of about 3.8 at%. Therefore, the XPS results demonstrated the doping of P in $NiCo_2S_4$, which would increase the number of active sites, improve its conductivity, and thus enhance its electrochemical performance.

The CV was conducted to investigate the mechanism and electrochemical sensor activity of the P-$NiCo_2S_4$ HNPs modified electrode for glucose oxidation via a standard three-electrode system in a 0.2 M NaOH solution at a scan rate of 50 mV s^{-1}. As shown in Figure 3A, although the CVs of bare ITO in the absence (black line) and presence (red line) of 1 mM glucose in an applied potential range of 0–0.6 V were active for glucose electrooxidation, the response was very weak. Figure 3B displays the CV curves of P-$NiCo_2S_4$/ITO with cathodic and anodic peaks at about 0.4 and 0.45 V, respectively. It was noticeable that a higher glucose oxidation peak was observed when adding 1 mM glucose, suggesting the excellent electrocatalytic activity of P-$NiCo_2S_4$/ITO towards glucose oxidation. The corresponding reaction mechanism for the glucose oxidation on the P-$NiCo_2S_4$ HNPs modified electrode can be briefly described as follows:

$$CoS + OH^- \leftrightarrow CoSOH + e^- \quad (1)$$

$$CoSOH + OH^- \leftrightarrow CoSO + H_2O + e^- \quad (2)$$

$$NiS + OH^- \leftrightarrow NiSOH + e^- \quad (3)$$

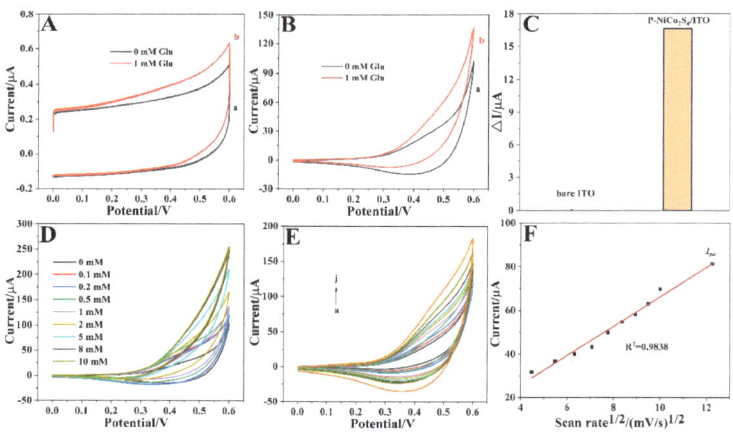

Figure 3. CVs of: (**A**) bare ITO, and (**B**) P-$NiCo_2S_4$/ITO in 0.2 M NaOH without (a) and with (b) 1 mM glucose (scan rate: 50 mV s^{-1}). (**C**) A bar chart for comparison of current increments (ΔI) for bare ITO and P-$NiCo_2S_4$/ITO. (**D**) CVs of P-$NiCo_2S_4$/ITO in 0.2 M NaOH in the presence of different glucose concentrations at a scan rate of 50 mV s^{-1}. (**E**) CVs of P-$NiCo_2S_4$/ITO in 0.2 M NaOH with 1 mM glucose at different scan rates (a to j: 20, 30, 40, 50, 60, 70, 80, 90, 100, and 150 mV s^{-1}). (**F**) The corresponding plot of anodic current density (I_{pa}) vs. the square root of the scan rate.

In addition, it can be easily seen from a comparison of current increments (ΔI) for bare ITO and P-$NiCo_2S_4$/ITO in Figure 3C that the electrochemical performance of the electrode modified with P-$NiCo_2S_4$ was greatly improved. Figure 3D presents the CVs of P-$NiCo_2S_4$/ITO with different glucose concentrations to evaluate the feasibility of glucose sensing, and the anodic peak current significantly increased with increasing concentration from 0 to 10 mM, representing typical catalytic oxidation of glucose. As shown in Figure 3E, the CVs of P-$NiCo_2S_4$/ITO at a series of scan rates with 1 mM glucose suggested that increasing the scan rate will lead to a rise in currents and a shift in potentials, which mainly comes from the increasing internal diffusion resistance within P-$NiCo_2S_4$ HNPs as the scan rate increases [2,27]. Furthermore, Figure 3F exhibits a good linear dependence relation between the anodic peak current and the square root of the scan rate with a linear regression coefficient (R^2) of 0.9838, implying that the glucose oxidation on the P-$NiCo_2S_4$/ITO is a reversible and typical diffusion-controlled electrochemical process [28].

The identification of the optimal concentration of NaOH solution for an effective non-enzymatic glucose sensor was monitored at P-$NiCo_2S_4$/ITO for a range of 0.01~0.5 M (Figure 4A). Upon the increment in concentration from 0.01 to 0.2 M, the electrooxidation kinetics of glucose was dramatically enhanced, and the utmost response towards glucose was observed in 0.2 M NaOH solution. High-alkaline conditions will generate higher oxidation state species (Ni^{2+}/Ni^{3+} and Co^{3+}/Co^{4+}), which provide the maximal glucose oxidation responses at P-$NiCo_2S_4$/ITO. However, stronger alkaline conditions tend to corrode electrodes resulting in limiting the electrochemical reaction [14]. Next, the optimal value of the applied potential was achieved in order to investigate the effect of detection potential on the amperometric response of P-$NiCo_2S_4$/ITO to glucose. As shown in Figure 4B, the oxidation current increased with increasing the detection potential and reached the highest value at 0.4 V. Therefore, 0.4 V was chosen as the optimal working potential for the follow-up study. Subsequently, for a systematic comparison, the P-$NiCo_2S_4$ nanoprisms with different amounts of NaH_2PO_2 were also prepared under the same synthetic conditions. In addition, the amperometric current responses of P-$NiCo_2S_4$/ITO with 0.1 g, 0.2 g, and 0.4 g NaH_2PO_2 at 0.4 V were represented in Figure 4C, which exhibited an almost obvious current difference. It was found that 0.2 g NaH_2PO_2 was the optimal admixing quantity in the following experiments. To analyze the interfacial behaviors of P-$NiCo_2S_4$ with 0.1 g, 0.2 g, and 0.4 g NaH_2PO_2, the electrochemical impedance spectroscopy (EIS) was performed in 0.1 M KCl solution containing 5 mM $K_3[Fe(CN)_6]/K_4[Fe(CN)_6]$. Obviously, the P-$NiCo_2S_4$ with 0.2 g NaH_2PO_2 showed a lower charge-transfer resistance (diameter of the semicircle) value than those of the other two, revealing the highest electrical conductivity of P-$NiCo_2S_4$ with 0.2 g NaH_2PO_2, and the result was consistent with Figure 4C. P-doping content has an effect on the electrical conductivity and electrocatalytic activity of materials. The high content of P-doping can significantly improve the fixation site and catalytic site of polysulfide, and thus promote the catalytic activity of materials. However, excessive P-doping will seriously affect the structural integrity of the materials, which significantly reduces the electrical conductivity [19,29,30]. The appropriate amount of P-doping can accelerate electron transport, and thus it would be a promising electrode to fabricate sensitive sensors. As shown in Figure S4, the BET surface areas of P-$NiCo_2S_4$ with 0.2 g and 0.4 g NaH_2PO_2 were 53.679 and 56.725 $m^2\ g^{-1}$, respectively, which were slightly higher than that of P-$NiCo_2S_4$ with 0.1 g NaH_2PO_2 (43.558 $m^2\ g^{-1}$). However, compared with pure $NiCo_2S_4$, P-doping can improve the specific surface area and electrochemical activity. Besides, the effective surface area of working electrodes will directly influence their sensitivity, thus playing a large role in developing electrochemical sensors. The CVs of P-$NiCo_2S_4$/ITO were measured in 5 mM $K_3Fe(CN)_6$ containing 0.1 M KCl solution, shown in Figure S5A. Additionally, Figure S5B indicates that the I_p values of P-$NiCo_2S_4$/ITO increase linearly with the square root of scan rates. Therefore, the effective surface area of the P-$NiCo_2S_4$ was assessed by the Randles–Sevcik equation: $I_p = (2.69 \times 10^5)n^{3/2}AD^{1/2}v^{1/2}C$, where n was the number of electrons transferred, D was the diffusion coefficient of $Fe(CN)_6{}^{3-}$ (7.6×10^{-6} $cm^2\ s^{-1}$), C was the reactant concentration (mol cm^{-3}), v is the scan rate (V s^{-1}), A was the effective area of the electrode, and I_p was the peak current [31,32]. The slope was the ratio of I_p to $v^{1/2}$ (Figure S5B); thus, the effective surface area of the $P\text{-}NiCo_2S_4$ was calculated as $0.134\ cm^2$.

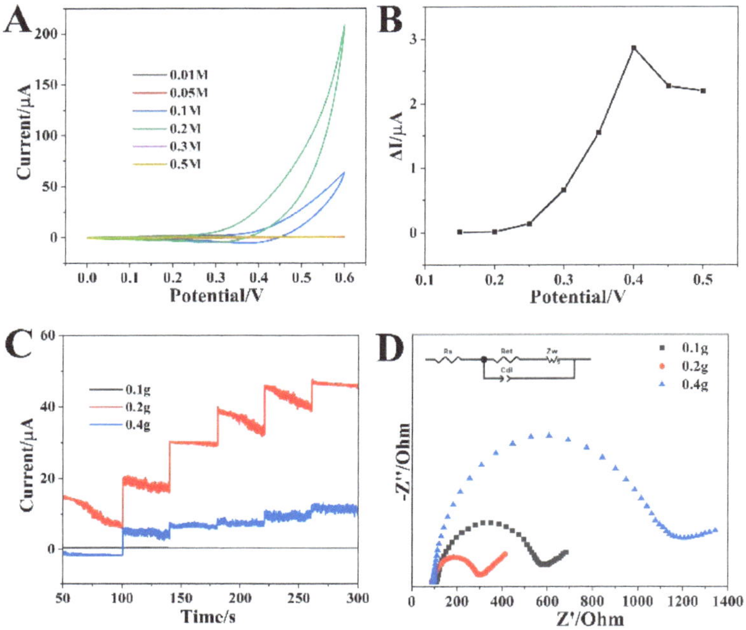

Figure 4. (**A**) CVs of $P\text{-}NiCo_2S_4/ITO$ in different concentrations of NaOH solution with 1 mM glucose (scan rate: 50 mV s^{-1}). (**B**) Current responses of $P\text{-}NiCo_2S_4/ITO$ at different detection potentials with 1 mM glucose in 0.2 M NaOH. (**C**) Amperometric responses of $P\text{-}NiCo_2S_4/ITO$ with different amounts of NaH_2PO_2 at 0.40 V with continuous addition of 1 mM glucose in 0.2 M NaOH. (**D**) Nyquist plots for $P\text{-}NiCo_2S_4$ electrodes with different amounts of NaH_2PO_2.

The amperometric analysis was used as an extremely attractive electrochemical technique to study the sensitivity, selectivity, and detection limit of the proposed glucose sensor. Under the above-optimized conditions, Figure 5A depicts a typical current-time plot of the $P\text{-}NiCo_2S_4/ITO$ on the successive step-wise addition of glucose. The inset in Figure 5A represents the magnification of glucose concentration ranging from 1 to 4 μM. A wide linear response of $P\text{-}NiCo_2S_4/ITO$ for glucose sensing from 0.001 to 5.2 mM was shown in Figure 5B. The linear regression equation for $P\text{-}NiCo_2S_4/ITO$ was $I/\mu A = 17.66C/mM + 1.72$ ($R^2 = 0.9903$) with a high sensitivity of 250.0 μA mM^{-1} cm^{-2}. As a consequence, the LOD of the sensor was calculated to be 0.46 μM (S/N = 3). In addition, after adding glucose to the NaOH solution, the $P\text{-}NiCo_2S_4$ sensor produced steady-state signals less than 0.1s (Figure 5C). Table 1 lists the detection performance comparison of $P\text{-}NiCo_2S_4/ITO$ with some similar non-enzymatic glucose sensors. It is clear that the performance of $P\text{-}NiCo_2S_4/ITO$ is comparable or superior to some reports, for example, $Co_3O_4/NiCo_2O_4$, $NiCo_2S_4/Pt$, and $CoNi_2S_4$@NCF [33–35]. The remarkable electrochemical performance of $P\text{-}NiCo_2S_4/ITO$ could stem from the hollow structures, synergetic effects between Ni and Co species, and partial substitution of S vacancies with P. This can provide more electrochemical active sites and enhance electrical conductivity, implying the potential value of $P\text{-}NiCo_2S_4$ in many systems for glucose detection.

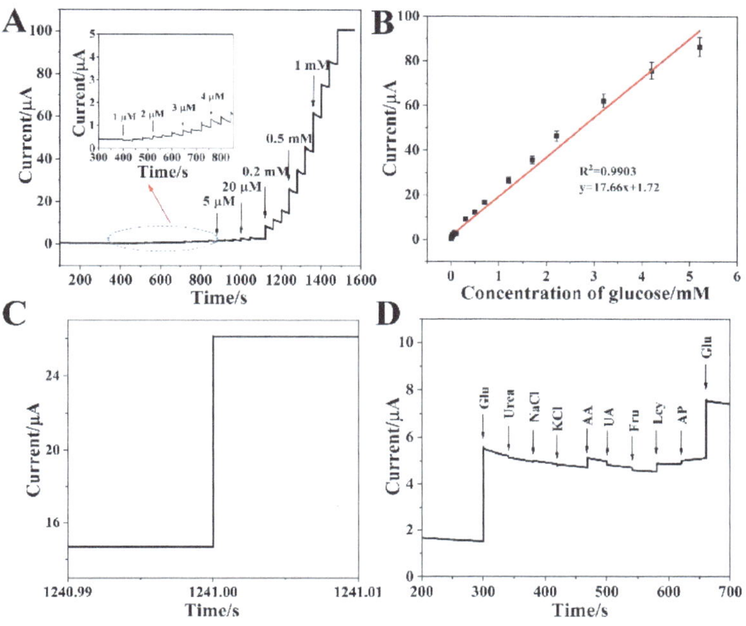

Figure 5. (**A**) Amperometric response of the P-$NiCo_2S_4$/ITO with successive addition of glucose at 0.4 V in 0.2 M NaOH (inset: the current response of electrode towards adding glucose from 1 to 4 μM). (**B**) The corresponding calibration curve of the P-$NiCo_2S_4$/ITO electrode to successive additions of glucose ranging from 1 μM to 5.2 mM. (**C**) The response time of the P-$NiCo_2S_4$/ITO electrode. (**D**) Amperometric response of the P-$NiCo_2S_4$/ITO with successive addition of glucose (1 mM), interfering species (0.1 mM), and the second addition of glucose (1 mM) in 0.2 M NaOH.

Table 1. Performance comparison of P-$NiCo_2S_4$/ITO sensor toward glucose oxidation with some similar electrodes.

Electrode Materials	Sensitivity (μA mM^{-1} cm^{-2})	Linear Range (mM)	LOD (μM)	Ref.
Co_3O_4/$NiCo_2O_4$	304	0.01–3.52	0.384	[33]
$NiCo_2S_4$/Pt	5.14	0.001–0.664	1.2	[34]
$CoNi_2S_4$@NCF	6.675	0.5–12.5	NR	[35]
	54.82	12.5–30		
Ni_5P_4	149.6	0.002–5.3	0.7	[36]
NiS-rGO	NR	0.05–1.7	10	[37]
CoP/GCE	116.8	0.5–5.5	9	[38]
P-$NiCo_2S_4$/ITO	**250**	**0.001–5.2**	**0.46**	**This work**

Selectivity and stability are the major factors to evaluate the performance of the non-enzymatic glucose sensor. It is clear that the P-$NiCo_2S_4$/ITO displays an anti-interference advantage after the addition of glucose (1 mM) and a series of interfering species (0.1 mM each) at a working potential of 0.4 V, shown in Figure 5D, indicating the good ability of anti-interference of P-$NiCo_2S_4$ with negligible interference from urea, NaCl, KCl, AA, UA, Fru, Lcy, and AP. The long-term stability of the P-$NiCo_2S_4$/ITO electrode was evaluated by measuring the current peak towards 1 mM glucose over a week at room temperature. The electrode basically kept the same value as the original current after 7 days, showing good stability of the P-$NiCo_2S_4$ electrode at room temperature (Figure 6). The enhanced electrochemical behavior may be attributed to: (i) more redox reactions from multiple oxide states of the Ni and Co species, (ii) a much lower optical band gap energy and higher electric conductivity of binary metal sulfides, (iii) its unique hollow structure that enlarges surface area and shortens electronic transmission, and (iv) the anionic phosphorus doping that provides some new active sites and introduces surface defects.

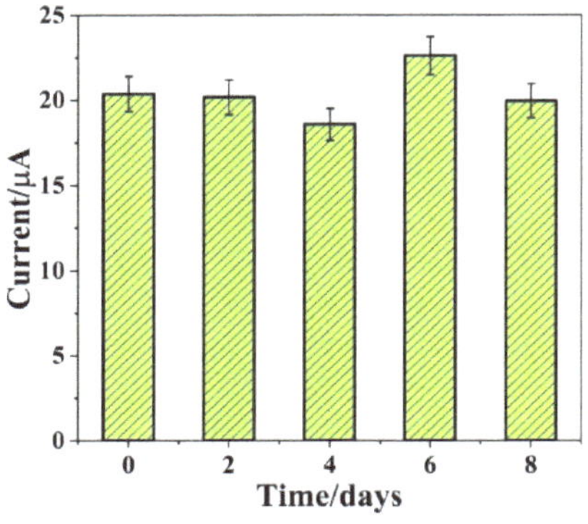

Figure 6. The stability of P-$NiCo_2S_4$/ITO stored at room temperature over 7 days in the presence of 1 mM glucose.

In addition, the feasibility of the proposed electrode was evaluated by monitoring glucose in human serum. The serum samples were donated by Jiangsu Province Hospital. The serum sample was firstly separated by centrifugation at 8000 rpm for 15 min, then the collected supernatant was obtained for further electrochemical tests. The standard addition method was employed to detect the glucose concentration in serum under optimal experimental conditions. The successive addition of glucose was continuously added into the mixture of 9.5 mL of 0.2 M NaOH and 0.5 mL of the treated serum supernatant (Figure S6A). The corresponding calibration curve of glucose ranging from 0.005 to 5.2 mM on the P-$NiCo_2S_4$/ITO was illustrated in Figure S6B, and the linear regression equation was $\Delta I/\mu A = 4.62C/mM + 36.51$ ($R^2 = 0.9891$), disclosing good stability and anti-interference of the proposed glucose sensor for real blood samples. Moreover, we also measured the recoveries in serum samples by adding glucose solution with different concentrations. As presented in Table 2, the recoveries were calculated to be in the range of 96.4–101.8%. The above results demonstrated that the P-$NiCo_2S_4$/ITO had favorable feasibility of glucose detection in real serum samples.

Table 2. The recovery of glucose by standard addition method in the serum sample.

Sample	Added Concentration/mM	Measured Concentration/mM	Recovery/%
Serum	0.055	0.056	101.8
	0.275	0.271	98.6
	0.475	0.458	96.4

4. Conclusions

In summary, we have rationally synthesized hollow $NiCo_2S_4$ nanoprisms in situ using a urea precursor as a sacrifice template by an anion exchange reaction, followed by constructing P-$NiCo_2S_4$ through simple P element doping for enhancing the electrochemical performance towards glucose oxidation. Compared with pristine $NiCo_2S_4$, a non-enzymatic sensor with good sensitivity, low detection limit, wide linear range, and high selectivity was established. It is confirmed that the incorporation of P can induce the surface defect at the atomic level to improve the electrochemical activity of host materials, which often suffer from slow ion diffusion and charge transfer kinetics, and the deficiency of electrochemically active sites. A series of electrodes with different amounts of P suggest that an appropriate amount of P-doping can accelerate electron transport. Moreover, P-$NiCo_2S_4$ demonstrates its feasibility for glucose determination for practical sample testing. These results reveal that the synergetic effects between Ni and Co species and the partial substitution of S vacancies with P can help to increase electronic conductivity, enrich binary electroactive sites, and boost surface electroactivity. Thus, our proposed non-enzymatic sensor has great prospects in the application of detecting glucose rapidly and effectively. In addition, the reported nanomaterials-based electrodes have been mostly applied in strongly alkaline conditions for non-enzymatic glucose sensors so far. However, the electrochemical characteristics in weakly alkaline or even neutral environments close to human body fluid are rarely reported. It is possible that, before the test, a negative potential is first employed to reduce the water in situ, and the hydrogen gets released resulting in the formation of OH^-; thus, it produces an alkaline microenvironment, so as to realize glucose detection in a neutral environment. This will be further studied in the future.

Supplementary Materials: The following supporting information can be downloaded at: https://www.mdpi.com/article/10.3390/bios12100823/s1, Figure S1: XRD pattern of Co/Ni precursors; Figure S2: EDS spectrum of Co/Ni precursors; Figure S3: EDS spectrum of the P-$NiCo_2S_4$ HNPs; Figure S4: N_2 adsorption/desorption isotherms of P-$NiCo_2S_4$/ITO with different amounts of NaH_2PO_2; Figure S5: (A) CVs of P-$NiCo_2S_4$/ITO in 5 mM $K_3Fe(CN)_6$ solution containing 0.1 M KCl at different scan rates. (B) The linear plot of I_p vs. $v^{1/2}$ of P-$NiCo_2S_4$/ITO; Figure S6: (A) Amperometric response of P-$NiCo_2S_4$/ITO with successive addition of glucose at 0.4 V in 0.2 M NaOH in healthy human serum samples. (B) The corresponding calibration curve of the glucose sensor.

Author Contributions: Conceptualization, X.L. and D.G.; methodology, D.C.; software, Y.W.; validation, X.L. and D.C.; formal analysis, X.L., X.L., and D.C.; investigation, Y.W.; resources, X.C.; data curation, X.L. and D.C.; writing-original draft preparation, D.G.; writing-review and editing, D.G. and X.C.; supervision, D.G. and X.C.; project administration, X.C.; funding acquisition, X.C. All authors have read and agreed to the published version of the manuscript.

Funding: This research was funded by the National Natural Science Foundation of China (21575064) and the Jiangsu Graduate Scientific Research Innovation Program (KYCX21_1075).

Institutional Review Board Statement: Not applicable.

Informed Consent Statement: Not applicable.

Data Availability Statement: Not applicable.

Acknowledgments: This work was financially supported by the project fund of Jiangsu Key Laboratory of Molecular Biology for Skin Diseases and STIs (2021).

Conflicts of Interest: The authors declare no conflict of interest.

References

1. Chen, D.; Wang, H.; Yang, M. A Novel Ball-in-Ball Hollow $NiCo_2S_4$ Sphere based Sensitive and Selective Nonenzymatic Glucose Sensor. *Anal. Methods* **2017**, *9*, 4718–4725. [CrossRef]
2. Zhou, Y.; Hu, Q.; Yu, F.; Ran, G.-Y.; Wang, H.-Y.; Shepherd, N.D.; D'Alessandro, D.M.; Kurmoo, M.; Zuo, J.-L. A Metal-Organic Framework based on a Nickel Bis(Dithiolene) Connector: Synthesis, Crystal Structure, and Application as an Electrochemical Glucose Sensor. *J. Am. Chem. Soc.* **2020**, *142*, 20313–20317. [CrossRef]
3. Wang, M.; Ma, J.; Guan, X.; Peng, W.; Fan, X.; Zhang, G.; Zhang, F.; Li, Y. A Novel H_2O_2 Electrochemical Sensor based on $NiCo_2S_4$ Functionalized Reduced Graphene Oxide. *J. Alloy Compd.* **2019**, *784*, 827–833. [CrossRef]
4. Ahmad, R.; Vaseem, M.; Tripathy, N.; Hahn, Y.B. Wide Linear-Range Detecting Nonenzymatic Glucose Biosensor based on CuO Nanoparticles Inkjet-Printed on Electrodes. *Anal. Chem.* **2013**, *85*, 10448–10454. [CrossRef] [PubMed]
5. Kim, W.B.; Lee, S.H.; Cho, M.; Lee, Y. Facile and Cost-Effective CuS Dendrite Electrode for Non-Enzymatic Glucose Sensor. *Sens. Actuators B Chem.* **2017**, *249*, 161–167. [CrossRef]
6. Warkhade, S.K.; Singh, R.P.; Das, R.S.; Gaikwad, G.S.; Zodape, S.P.; Pratap, U.R.; Maldhure, A.; Wankhade, A.V. $CoSe_2$ Nanoflakes: An Artificial Nanoenzyme with Excellent Peroxidase like Activity. *Inorg. Chem. Commun.* **2021**, *126*, 108461. [CrossRef]
7. Gao, Y.-P.; Huang, K.-J. $NiCo_2S_4$ Materials for Supercapacitor Applications. *Chem. Asian J.* **2017**, *12*, 1969–1984. [CrossRef]
8. Wang, B.; Cheng, Y.; Su, H.; Cheng, M.; Li, Y.; Geng, H.; Dai, Z. Boosting Transport Kinetics of Cobalt Sulfides Yolk-Shell Spheres by Anion Doping for Advanced Lithium and Sodium Storage. *ChemSusChem* **2020**, *13*, 4078–4085. [CrossRef]
9. Yin, H.; Zhan, T.; Qin, D.; He, X.; Nie, Q.; Gong, J. Self-Assembly of Dandelion-Like $NiCo_2O_4$ Hierarchical Microspheres for Non-Enzymatic Glucose Sensor. *Inorg. Nano-Met. Chem.* **2017**, *47*, 1560–1567. [CrossRef]

10. Wang, X.; Zhou, R.; Zhang, C.; Xi, S.; Jones, M.W.M.; Tesfamichael, T.; Du, A.; Gui, K.; Ostrikov, K.; Wang, H. Plasma-Induced On-Surface Sulfur Vacancies in $NiCo_2S_4$ Enhance the Energy Storage Performance of Supercapatteries. *J. Mater. Chem. A* **2020**, *8*, 9278–9291. [CrossRef]
11. Li, Y.; Kamdem, P.; Jin, X.-J. In situ Growth of Chrysanthemum-like $NiCo_2S_4$ on MXene for High-Performance Supercapacitors and Non-Enzymatic H_2O_2 Sensor. *Dalton Trans.* **2020**, *49*, 7807–7819. [CrossRef] [PubMed]
12. Wan, Z.; Zhang, R.; Yan, X.; Fan, K. Structure and Activity of Nanozymes: Inspirations for De Novo Design of Nanozymes. *Mater. Today* **2020**, *41*, 81–119.
13. Kang, L.; Zhang, M.; Zhang, J.; Liu, S.; Zhang, N.; Yao, W.; Ye, Y.; Luo, C.; Gong, Z.; Wang, C.; et al. Dual-Defect Surface Engineering of Bimetallic Sulfide Nanotubes towards Flexible Asymmetric Solid-State Supercapacitors. *J. Mater. Chem. A* **2020**, *8*, 24053–24064. [CrossRef]
14. Babu, K.J.; Kumar, T.R.; Yoo, D.J.; Phang, S.-M.; Kumar, G.G. Electrodeposited Nickel Cobalt Sulfide Flowerlike Architectures on Disposable Cellulose Filter Paper for Enzyme-Free Glucose Sensor Applications. *ACS Sustain. Chem. Eng.* **2018**, *6*, 16982–16989. [CrossRef]
15. Guo, Q.; Wu, T.; Liu, L.; He, Y.; Liu, D.; You, T. Hierarchically Porous $NiCo_2S_4$ Nanowires Anchored on Flexible Electrospun Graphitic Nanofiber for High-Performance Glucose Biosensing. *J. Alloys Compd.* **2020**, *819*, 153376. [CrossRef]
16. Wu, J.; Wang, X.; Wang, Q.; Lou, Z.; Li, S.; Zhu, Y.; Qin, L.; Wei, H. Nanomaterials with Enzyme-like Characteristics (Nanozymes): Next-Generation Artificial Enzymes (II). *Chem. Soc. Rev.* **2019**, *48*, 1004–1076. [CrossRef]
17. Chen, M.; Xiao, J.; Hua, W.; Hu, Z.; Wang, W.; Gu, Q.; Tang, Y.; Liu, S.C.H.; Dou, S. A Cation and Anion Dual Doping Strategy for the Elevation of Titanium Redox Potential for High-Power Sodium-Ion Batteries. *Angew. Chem. Int. Ed.* **2020**, *59*, 12076–12083. [CrossRef]
18. Liu, S.; Kang, L.; Zhang, J.; Jung, E.; Lee, S.; Jun, S.C. Structural Engineering and Surface Modification of MOF-Derived Cobalt-based Hybrid Nanosheets for Flexible Solid-State Supercapacitors. *Energy Stor. Mater.* **2020**, *32*, 167–177. [CrossRef]
19. Lin, J.; Wang, Y.; Zheng, X.; Liang, H.; Jia, H.; Qi, J.; Cao, J.; Tu, J.; Fei, W.; Feng, J. P-Doped $NiCo_2S_4$ Nanotubes as Battery-Type Electrodes for High-Performance Asymmetric Supercapacitors. *Dalton Trans.* **2018**, *47*, 8771–8778. [CrossRef]
20. Li, B.; Gu, P.; Feng, Y.; Zhang, G.; Huang, K.; Xue, H.; Pang, H. Ultrathin Nickel-Cobalt Phosphate 2D Nanosheets for Electrochemical Energy Storage under Aqueous/Solid-State Electrolyte. *Adv. Funct. Mater.* **2017**, *27*, 1605784. [CrossRef]
21. Zhang, J.; Liu, Y.; Xia, B.; Sun, C.; Liu, Y.; Liu, P.; Gao, D. Facile One-Step Synthesis of Phosphorus-Doped CoS_2 as Efficient Electrocatalyst for Hydrogen Evolution Reaction. *Electrochim. Acta* **2018**, *259*, 955–961. [CrossRef]
22. Chu, D.; Yan, L.; Chen, Q.; Chu, X.-Q.; Ge, D.; Chen, X. Efficient Improvement in Non-Enzymatic Glucose Detection Induced by the Hollow Prism-Like $NiCo_2S_4$ Electrocatalyst. *Dalton Trans.* **2021**, *50*, 15162–15169. [CrossRef] [PubMed]
23. Gu, H.; Fan, W.; Liu, T. Phosphorus-Doped $NiCo_2S_4$ Nanocrystals Grown on Electrospun Carbon Nanofibers as Ultra-Efficient Electrocatalysts for the Hydrogen Evolution Reaction. *Nanoscale Horiz.* **2017**, *2*, 277–283. [CrossRef]
24. Ding, J.; Zhong, L.; Wang, X.; Chai, L.; Wang, Y.; Jiang, M.; Li, T.-T.; Hu, Y.; Qian, J.; Huang, S. General Approach to MOF-Derived Core-Shell Bimetallic Oxide Nanowires for Fast Response to Glucose Oxidation. *Sens. Actuators B Chem.* **2020**, *306*, 127551. [CrossRef]
25. Xu, Z.; Wang, Q.; Zhangsun, H.; Zhao, S.; Zhao, Y.; Wang, L. Carbon Cloth-Supported Nanorod-like Conductive Ni/Co Bimetal MOF: A Stable and High-Performance Enzyme-Free Electrochemical Sensor for Determination of Glucose in Serum and Beverage. *Food Chem.* **2021**, *349*, 129202. [CrossRef]
26. Yang, J.; Guo, D.; Zhao, S.; Lin, Y.; Yang, R.; Xu, D.; Shi, N.; Zhang, X.; Lu, L.; Lan, Y.-Q.; et al. Cobalt Phosphides Nanocrystals Encapsulated by P-Doped Carbon and Married with P-Doped Graphene for Overall Water Splitting. *Small* **2019**, *15*, 1804546. [CrossRef] [PubMed]
27. Li, L.; Xu, J.; Lei, J.; Zhang, J.; McLarnon, F.; Wei, Z.; Li, N.; Pan, F. A One-Step, Cost-Effective Green Method to in Situ Fabricate $Ni(OH)_2$ Hexagonal Platelets on Ni Foam as Binder-Free Supercapacitor Electrode Materials. *J. Mater. Chem. A* **2015**, *3*, 1953–1960. [CrossRef]
28. Cao, X.; Wang, K.; Du, G.; Asiri, A.M.; Ma, Y.; Lu, Q.; Sun, X. One-Step Electrodeposition of a Nickel Cobalt Sulfide Nanosheet Film as a Highly Sensitive Nonenzymatic Glucose Sensor. *J. Mater. Chem. B* **2016**, *4*, 7540–7544. [CrossRef]
29. Wang, J.; Han, W.-Q. A Review of Heteroatom Doped Materials for Advanced Lithium-Sulfur Batteries. *Adv. Funct. Mater.* **2022**, *32*, 2107166. [CrossRef]
30. Paraknowitsch, J.P.; Thomas, A. Doping Carbons Beyond Nitrogen: An Overview of Advanced Heteroatom Doped Carbons with Boron, Sulphur and Phosphorus for Energy Applications. *Energy Environ. Sci.* **2013**, *6*, 2839–2855. [CrossRef]
31. Xiao, X.; Zhang, X.; Zhang, Z.; You, J.; Liu, S.; Wang, Y. Macro-/Meso-Porous $NiCo_2O_4$ Synthesized by Template-Free Solution Combustion to Enhance the Performance of a Nonenzymatic Amperometric Glucose Sensor. *Microchim. Acta* **2020**, *187*, 64. [CrossRef] [PubMed]
32. Zhou, Y.; Ni, X.; Ren, Z.; Ma, J.; Xu, J.; Chen, X. A Flower-Like NiO-SnO_2 Nanocomposite and Its Non-Enzymatic Catalysis of Glucose. *RSC Adv.* **2017**, *7*, 45177–45184. [CrossRef]
33. Xue, B.; Li, K.; Feng, L.; Lu, J.; Zhang, L. Graphene Wrapped Porous $Co_3O_4/NiCo_2O_4$ Double-Shelled Nanocages with Enhanced Electrocatalytic Performance for Glucose Sensor. *Electrochim. Acta* **2017**, *239*, 36–44. [CrossRef]
34. Kannan, P.K.; Hu, C.; Morgan, H.; Rout, C.S. One-Step Electrodeposition of $NiCo_2S_4$ Nanosheets on Patterned Platinum Electrodes for Non-Enzymatic Glucose Sensing. *Chem. Asian J.* **2016**, *11*, 1837–1841. [CrossRef]
35. Dong, M.; Hu, H.; Ding, S.; Wang, C.; Li, L. Flexible Non-Enzymatic Glucose Biosensor Based on $CoNi_2S_4$ Nanosheets Grown on Nitrogen-Doped Carbon Foam Substrate. *J. Alloys Compd.* **2021**, *883*, 160830. [CrossRef]
36. Xin, X.; Wang, Y.; Han, C.; Cui, Y.; Xu, Y.; Tao, Y.; Zhang, D.; Xu, X. Porous Flower-Like Ni_5P_4 for Non-Enzymatic Electrochemical Detection of Glucose. *Mater. Chem. Phys.* **2020**, *240*, 122202.

37. Radhakrishnan, S.; Kim, S.J. Facile Fabrication of NiS and a Reduced Graphene Oxide Hybrid Film for Nonenzymatic Detection of Glucose. *RSC Adv.* **2015**, *5*, 44346–44352. [CrossRef]
38. Sun, Q.-Q.; Wang, M.; Bao, S.-J.; Wang, Y.C.; Gu, S. Analysis of Cobalt Phosphide (CoP) Nanorods Designed for Non-Enzyme Glucose Detection. *Analyst* **2016**, *141*, 256–260. [CrossRef]

biosensors

MDPI

Article

Polydopamine-Functionalized Copper Peroxide/ZIF-8 Nanoparticle-Based Fluorescence-Linked Immunosorbent Assay for the Sensitive Determination of Carcinoembryonic Antigen by Self-Supplied H_2O_2 Generation

Juanjuan Huang [1,†], Yiyun Yao [1,†], Yanling Chen [1], Tianran Lin [1], Li Hou [1,*] and Dianping Tang [2,*]

1 School of Chemistry and Pharmaceutical Sciences, State Key Laboratory for Chemistry and Molecular Engineering of Medicinal Resources, Guangxi Normal University, Guilin 541004, China
2 Key Laboratory of Analytical Science for Food Safety and Biology (MOE & Fujian Province), Department of Chemistry, Fuzhou University, Fuzhou 350108, China
* Correspondence: houli@mailbox.gxnu.edu.cn (L.H.); dianping.tang@fzu.edu.cn (D.T.)
† These authors contributed equally to this paper.

Abstract: Copper peroxide/zeolitic imidazolate framework/polydopamine nanoparticles (CP/ZIF-8/PDA)-based fluorescence-linked immunosorbent assay (FLISA) was designed for the sensitive and high-throughput determination of carcinoembryonic antigen (CEA) by self-supplied H_2O_2 generation. Specifically, the CEA aptamer was modified on the surface of CP/ZIF-8/PDA to form an immunoprobe. The structures of CP and ZIF-8 could be broken under acidic conditions, and produced the Cu^{2+} and H_2O_2 due to the dissociation the CP. A subsequent Fenton-type reaction of Cu^{2+} and H_2O_2 generated hydroxyl radical (·OH). o-phenylenediamine (OPD) was oxidized by the ·OH to form 2, 3-diaminophenazine (DPA) with a significant fluorescence signal. CP/ZIF-8/PDA could be used as an efficient Fenton-type reactant to generate a large amount of ·OH to promote OPD oxidation. The sensitive detection of CEA could be realized. Under optimal conditions, the FLISA platform displayed a linear detection range from 0.01 to 20 ng mL^{-1} with a detection limit of 7.6 pg mL^{-1} for CEA. This strategy has great application potential for sensitive and high-throughput determination for other biomarkers in the field of biomedicine.

Keywords: fluorescence-linked immunosorbent assay; Fenton-type reaction; carcinoembryonic antigen

Citation: Huang, J.; Yao, Y.; Chen, Y.; Lin, T.; Hou, L.; Tang, D. Polydopamine-Functionalized Copper Peroxide/ZIF-8 Nanoparticle-Based Fluorescence-Linked Immunosorbent Assay for the Sensitive Determination of Carcinoembryonic Antigen by Self-Supplied H_2O_2 Generation. *Biosensors* **2022**, *12*, 830. https://doi.org/10.3390/bios12100830

Received: 3 September 2022
Accepted: 1 October 2022
Published: 6 October 2022

1. Introduction

The enzyme-linked immunosorbent assay (ELISA) is the gold standard for the detection of a large number of routine biological analytes [1]. Nevertheless, conventional ELISA still suffers from high background, low sensitivity, false-positive results, it is expensive and the labeling enzyme is highly susceptible to inactivation [2,3], so the practical applications of the ELISA technique is restricted. Normally, enzyme-labeled secondary antibodies are need for ELISA to determine the absorbance of the substrate [4]. Distinct from the ELISA, fluorescence-linked immunosorbent assay (FLISA) employs a fluorescent substance as a label to measure the fluorescence intensity [5]. Both ELISA and FLISA are efficient high-throughput detection methods. The principles of their operation and detection are similar. The rapid development of nanotechnology has made tremendous progress in the development of the FLISA [6]. In recent years, many fluorescent materials have been developed for the FLISA method, including graphene oxide (GO) [7] and metal-organic frameworks [8]. These materials generally use the fluorescence resonance energy transfer (FRET) strategy to achieve a sensitive signal output, but the donor and acceptor of FRET are susceptible to distance and depend on the degree of overlap between the emission spectrum of the donor and the excitation spectrum of the acceptor. In addition, many signal amplification methods of enhancing the sensitivity of FLISA have been investigated, including carbon dots (o-CDs) aggregation fluorescence quenching [9], the simple enhancement of fluorescence by Triton-X 100 micelles [10] and signal amplification by fluorescein-labeled DNA strand hybridization reaction [11]. However, these methods have limitations for analytical applications, such as their susceptibility to interference by metal ions and the high cost of reagent. Therefore, the exploration of the fluorescence output strategy and appropriate nanomaterial in FLISA are still very meaningful.

Zeolitic imidazolate frameworks (ZIFs) are typical materials for metal organic frameworks (MOFs), which have been widely used in sensors and drug delivery due to their large specific surface area, good porosity and biocompatibility [12,13]. Because of the high affinity of ZIFs for enzymes, Guo et al. [14] designed a sensor based on the enzymatic cascade reaction of metal-organic framework composites (AuNCs/β-Gal/GOx@ZIF-8) for the detection of lactose. The sensitivity of this sensor was effectively enhanced by catalytic signal amplification. However, the biological enzymes are expensive and easily lose activity since they are susceptible to external conditions, including temperature and pH changes, as well as a complicated modification process. Therefore, enzyme-mediated catalytic signal amplification by assembling enzyme-ZIFs nanocomposites is not an ideal signal output strategy. Wang et al. [15] synthesized a copper peroxide/zeolitic imidazolate framework/polydopamine (CP/ZIF-8/PDA) composite, which was combined with a 3-aminobenzeneboronic acid (ABA)/polyvinyl alcohol (PVA) film-modified electrode and self-supplied H_2O_2 generation for the electrochemical immunoassay of the glycoantigen CA19-9. This method used HCl-triggered cascade reaction without introducing enzymes and metal ions, which had good sensitivity, selectivity and stability, and provided a new idea for designing simple, fast and effective immunosensors. Above all, the known ZIFs-based FLISA for the sensitive and high-throughput determination of CEA by self-supplied H_2O_2 generation has not reported so far.

Inspired by the above work, a FLISA platform for the sensitive and high-throughput detection of the carcinoembryonic antigen (CEA) was constructed based on a CP/ZIF-8/PDA nanocomposite by self-supplied H_2O_2 generation (Scheme 1). ZIF-8 was used for the encapsulation of CP for the formation of ZIF-8/CP. After a layer of PDA membrane was formed onto the surface of ZIF-8/CP, the abundant amino of PDA was beneficial to the immobilization of biomolecules, such as antibodies or aptamers via the EDC-NHS reaction and glutaraldehyde cross-linking reaction. CP was the source of Cu^{2+} and H_2O_2 under the stimulation of the acidic condition. Thus, the CP/ZIF-8/PDA plays a key role as a "signal switch" and immunocarrier. The aminated CEA aptamer was modified on the surface of CP/ZIF-8/PDA to form an immunoprobe (CP/ZIF-8/PDA/Aptamer) via the glutaraldehyde cross-linking function between the aminated CEA aptamer and amino-rich PDA. In the presence of CEA, CP/ZIF-8/PDA/Aptamer and the modified CEA antibodies on the microtiter plate specifically recognized with CEA to form the sandwich immune complexes. Then, the structure of CP/ZIF-8/PDA was destroyed under acidic conditions, thus, releasing CP, which was rapidly converted to Cu^{2+} and H_2O_2 [16]. After that, Cu^{2+} underwent a Fenton-type reaction with H_2O_2 to produce ·OH and ·OH-oxidized o-phenylenediamine (OPD) to form 2,3-diaminophenothiazine (DPA) with fluorescence property. This method did not involve enzymes, metal ions and external H_2O_2 to induce Fenton-type reactions. Compared with an external H_2O_2-based sensing system, self-supplied H_2O_2 was more stable in the system solution and could effectively avoid the reduction in biomolecular activity [17]. The simultaneous production of H_2O_2 and Cu^{2+} could achieve enrichment in site to form higher concentrations of H_2O_2 and Cu^{2+}, which was conducive to producing a large amount of ·OH to oxidize OPD. The efficient high-throughput detection of CEA could be achieved by the FLISA sensing platform. Therefore, polydopamine-functionalized CP/ZIF-8 nanoparticle-based FLISA provided a new strategy for the sensitive and high-throughput detection of CEA.

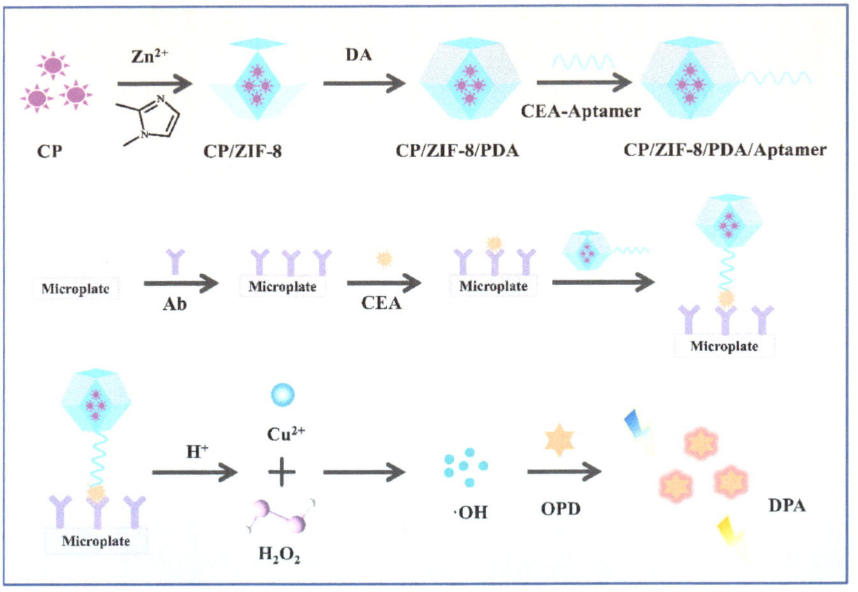

Scheme 1. Schematic illustration of CP/ZIF-8/PDA nanoparticle-based FLISA for the sensitive determination of the carcinoembryonic antigen by self-supplied H_2O_2 generation.

2. Experimental Section

2.1. Reagents and Materials

We purchased dopamine (DA), copper chloride dihydrate ($CuCl_2·2H_2O$), glacial acetic acid (HAc), sodium acetate (NaAc), sodium carbonate (Na_2CO_3), sodium bicarbonate ($NaHCO_3$), sodium dihydrogen phosphate (Na_2HPO_4) and disodium hydrogen phosphate (Na_2HPO_4) from Aladdin Reagent Co., Ltd. (Shanghai, China). Solarbio Science & Technology. Co., Ltd. (Beijing, China) provided bovine serum albumin (BSA, 97.0%) and Tween 20. From Jiuding Chemical Technology Co., Ltd. (Shanghai, China), we purchased methylimidazole and o-Diphenylamine (OPD). Sodium chloride (NaCl), potassium chloride (KCl) and sodium hydroxide (NaOH) were purchased from Sinopharm Chem. Re. Co., Ltd. (Shanghai, China). $Zn(NO_3)_2·6H_2O$ and anhydrous methanol were acquired from Xilong Scientific Co., Ltd. (Guangdong, China). All other chemicals were analytical grade. In these experiments, ultrapure water (Milli-Q, Millipore, ≥18.2 MΩ·cm) was used. Sangon Biotech (Shanghai, China) supplied the high-performance liquid chromatography (HPLC) purified CEA-binding aptamer. The following oligonucleotide sequence (5′ → 3′) was listed: H_2N-TTT TAT ACC AGC TTA TTC AAT T. Beijing Key-Bio Biotech Co., Ltd. (Beijing, China) provided the CEA monoclonal antibody (5 mg mL^{-1}), CEA (100 μg mL^{-1}), CA19-9 (50 KIU mL^{-1}), CA125 (50 KIU mL^{-1}), AFP (105 μg mL^{-1}) and PSA (100 μg mL^{-1}). The Guilin Hospital of Chinese Traditional and Western Medicine provided clinical human serum samples with different CEA concentrations based on the rules of the local ethical committee. The referenced values of CEA in clinical human serum samples were acquired from the commercialized Roche Cobas e601 automatic electrochemiluminescence immunoanalyzer (Basel, Switzerland)

2.2. Apparatus

Perkin-Elmer (USA) provided Fourier transform infrared (FTIR) spectrometer. Thermo Fisher (Thermo Fisher Scientific, Waltham, MA, USA) supplied field emission transmission electron microscopy (FE-TEM). The sample shaking was performed by Mini Shaker (Kylin-Bell Lab Instruments Co., Ltd., Haimen, China). Mapada (Shanghai, China) supplied the ultraviolet-vis absorption UV-6100S spectrophotometer. Malvern (Malvern, UK) supplied the Zetasizer Nano ZS90 Laser Particle Size/Potentiometer. Tecan (Männedorf, Switzerland) provided the TECAN microplate reader SPARK. The emission spectra at 565 nm wavelength were applied with the excitation spectra of DPA at 410 nm.

2.3. Synthesis of CP/ZIF-8/PDA

CP nanoparticles were prepared according to the literature synthesis method [16]. A total of 0.5 g of polyvinylpyrrolidone (PVP) and 0.01 g of $CuCl_2 \cdot 2H_2O$ were dissolved in 5 mL of aqueous solution. Then, 0.1 mL of H_2O_2 (30%) and 5 mL of NaOH (0.02 mol L^{-1}) were added to the above solution, and the color of the solution was changed from clear to brown by magnetic stirring for 30 min. The PVP-coated CP nanoparticles were washed by multiple centrifugations (12,000 rpm) with ultrapure water, and the centrifuged CP was dispersed with anhydrous methanol and stored in a refrigerator at 4 °C for use.

A total of 291 mg of $Zn(NO_3)_2 \cdot 6H_2O$ and 328 mg of 2-methylimidazole were sequentially dispersed in 25 mL of CP solution, stirred magnetically at room temperature for 24 h. CP/ZIF-8 was obtained by centrifugal washing (8000 rpm) and dried at 60 °C for 12 h.

A total of 100 mg of CP/ZIF-8 was weighed and dispersed in 25 mL of methanol. Subsequently, 10 mL of dopamine solution (DA, 1.0 mg mL^{-1}, dissolved in Tris-HCl, pH 8.5) was added under magnetic stirring and the reaction was continued for 2.5 h to form a polydopamine film (PDA) on the surface of CP/ZIF-8. CP/ZIF-8/PDA was obtained by centrifugal washing (8000 rpm) and dried at 60 °C for 12 h.

2.4. Synthesis of the CP/ZIF-8/PDA/Aptamer Immunoprobe

A total of 20 mg of CP/ZIF-8/PDA was dispersed into 20 mL of PBS buffer (10 mmol L^{-1}, pH 7.4), and 1 mL of 50% glutaraldehyde was added. The above resulting solution was incubated in a shaker at room temperature for 2 h, and the surface-activated CP/ZIF-8/PDA was washed by centrifugation at 10,000 rpm for 3 times and dispersed in 6.5 mL of PBS. Then, 0.5 OD of aminated CEA Aptamer was added and shaken at 37 °C for 16 h. After that, the immunoprobe (CP/ZIF-8/PDA/Aptamer) was obtained by adding BSA (final concentration of 1%) and shaking at room temperature for 1 h. The immunoprobe (CP/ZIF-8/PDA/Aptamer) was finally washed by centrifugation at 9500 rpm.

2.5. The 96-Well Plate Trimmed with CEA Antibody

First, the CEA antibody was diluted to 10 μg mL^{-1} with Na_2CO_3-$NaHCO_3$ coating buffer (100 mmol L^{-1}, pH 9.6). A total of 100 μL of CEA antibody was then added to each well of a 96-well plate and incubated for 12 h at 4 °C in the refrigerator. Then, the well plates were washed for three times with PBST wash buffer (10 mmol L^{-1}, pH 7.4, 27 mmol L^{-1} KCl, 135 mmol L^{-1} NaCl, 0.05% Tween 20), and then 200 μL of PBST buffer containing 1% BSA as blocking agent was added and incubated for 1 h at room temperature to block the remaining active sites. The above washing operation was repeated, and finally the 96-well plate was sealed with cling film and stored in a refrigerator at 4 °C.

2.6. FLISA Determination of CEA

A total of 200 μL of different concentrations of CEA were added to the microtiter wells and incubated at 37 °C for 1 h. Excess CEA was washed off with PBS (10 mmol L^{-1}, pH 7.4). Next, 100 μL of CP/ZIF-8/PDA/Aptamer immunoprobe was added and incubated at room temperature for 1 h. Then, excess immunoprobe was washed with PBS. Then, 100 μL of 4 mmol L^{-1} OPD (prepared with pH 5.5, 10 mmol L^{-1} HAc-NaAc buffer solution) was added for the reaction of 3 h. The fluorescence signal intensity at 565 nm was measured by a TECAN microplate reader SPARK.

3. Results and Discussion

3.1. Characterization of the CP and CP/ZIF-8/PDA/Aptamer

The morphology of the synthesized CP nanoparticles was characterized by TEM (Figure 1A). As seen in Figure 1A, the CP nanoparticles are small, spherical particles with a size of about 3 nm and uniform distribution. Figure 1B shows the UV-vis of CP nanoparticles in the aqueous solution; it can be seen that there is a shoulder peak at 345 nm, which is an absorption peak due to the charge transfer between O^{2-} and Cu(II) through a symmetrical bridging structure [18].

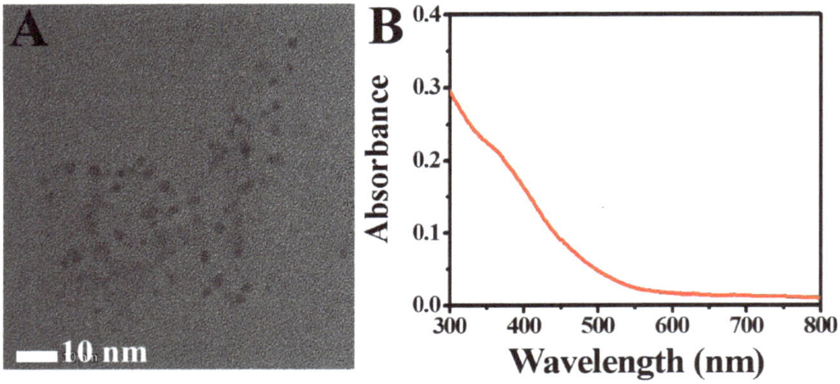

Figure 1. (**A**) TEM images of CP and (**B**) UV absorption peaks of CP.

Energy dispersive spectrometric spectra for the TEM images (EDS) were performed for ZIF-8 and CP/ZIF-8, respectively. Among them, Figure S1 shows the EDS mapping element analysis diagram of ZIF-8, from which ZIF-8 contained four elements, C, N, Zn and O. In contrast to Figure S2, the EDS mapping elemental analysis plot of CP/ZIF-8 showed that CP/ZIF-8 contained Cu elements in addition to the above four elements, indicating that CP nanoparticles were successfully encapsulated into ZIF-8. TEM of ZIF-8, CP/ZIF-8 and CP/ZIF-8/PDA were also performed (Figure 2) to examine the morphology of the materials after modification. In Figure 2A and B, after the CP nanoparticles were covered with ZIF-8, many CP nanoparticles were distributed in ZIF-8, and the shadow in the middle of the TEM image deepened. Additionally, the particle size of the nanoparticles became larger with about 150–200 nm, further indicating the successful synthesis of CP/ZIF-8 nanoparticles. With the coverage of PDA, the surface of CP/ZIF-8 nanoparticles gradually became rough (Figure 2C), and the shadow deepened again, indicating the successful modification of PDA on the surface of CP/ZIF-8.

The successful synthesis of CP/ZIF-8/PDA/Aptamer can be verified by FTIR. In Figure 2D, it could be seen that CP/ZIF-8/PDA had significant absorption bands at 1458 cm^{-1} and 1582 cm^{-1}, which were caused by the stretching vibrations of the aromatic ring skeleton in PDA [19], indicating the presence of PDA on the surface of CP/ZIF-8. Two absorption bands were observed between 3200–3500 cm^{-1}, which were attributed to the N-H stretching vibrations of $-NH_2$ on the PDA. When the CP/ZIF-8/PDA was modified with the aptamer by glutaraldehyde cross-linking method, the double peak between 3200–3500 cm^{-1} became a single peak, which was due to the N-H stretching vibration peak formed by glutaraldehyde crosslinking to produce amide bond, and $-NH_2$ became -NH. A clear absorption band at 1062 cm^{-1} appeared in the FTIR spectrum of CP/ZIF-8/PDA/Aptamer, which was mainly caused by the phosphodiester bond in the DNA strand [20], indicating the successful synthesis of CP/ZIF-8/PDA/Aptamer. Meanwhile, the zeta potentials of aptamer, CP/ZIF-8/PDA and CP/ZIF-8/PDA/Aptamer were also examined, as shown in Figure 2E. The average zeta potential carried by the CEA Aptamer was −9.36 mV and that of CP/ZIF-8/PDA was −11.02 mV. When CP/ZIF-8/PDA was modified with CEA Aptamer, the average zeta potential was −13.57 mV. These results further demonstrate the successful synthesis of CP/ZIF-8/PDA/Aptamer.

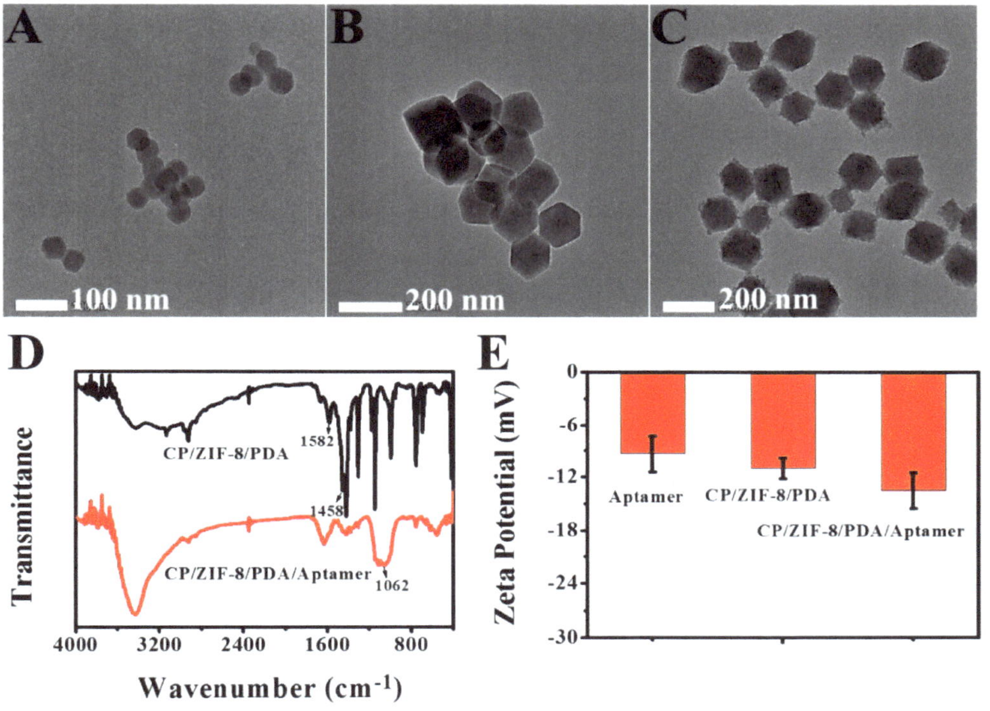

Figure 2. TEM image of (**A**) ZIF-8, (**B**) ZIF-8/CP, (C) CP/ZIF-8/PDA, and (**D**) FT-IR spectra of CP/ZIF-8/PDA, CP/ZIF-8/PDA/Aptamer; (**E**) Zeta potentials of CEA Aptamer, CP/ZIF-8/PDA and CP/ZIF-8/PDA/Aptamer.

3.2. Feasibility of the Proposed FLISA Platform

CP nanoparticles can be decomposed to form Cu^{2+} and H_2O_2 under acidic conditions, and Cu^{2+} and H_2O_2 will produce ·OH through the Fenton-type reaction. To verify the ability of CP/ZIF-8/PDA to catalyze the oxidation of 3,3′,5,5′-tetramethylbenzidine (TMB), H_2O, CP, CP/ZIF-8/and CP/ZIF-8/PDA were reacted with TMB under acidic conditions for 10 min. As can be seen in Figure 3A, under acidic conditions, H_2O cannot oxidize TMB with almost no UV absorption peak, while all CP, CP/ZIF-8 and CP/ZIF-8/PDA can oxidize TMB with different degrees of UV absorption peaks at 652 nm. The results indicate that CP still had the ability to catalyze the oxidation of TMB through the encapsulation and modification of ZIF-8 and PDA to produce ·OH. To further verify the generation of ·OH by CP/ZIF-8/PDA under acidic conditions, tert-butanol was used as a scavenger to react with CP/ZIF-8/PDA and TMB to identify ·OH. As shown in Figure 3B, the intensity of the UV-visible absorption peak decreases continuously with the increase in the tert-butanol concentration. The results indicate that CP/ZIF-8/PDA can generate ·OH under acidic conditions. Additionally, the fluorescence intensity of the FLISA platform after the introduction of CP/ZIF-8/PDA/Aptamer in the presence and absence of CEA was probed using OPD as the substrate (Figure S3). Comparing the absence of CEA, the fluorescence signal of the FLISA-sensing platform became significantly stronger in the presence of CEA. The above results show that the experiment is feasible.

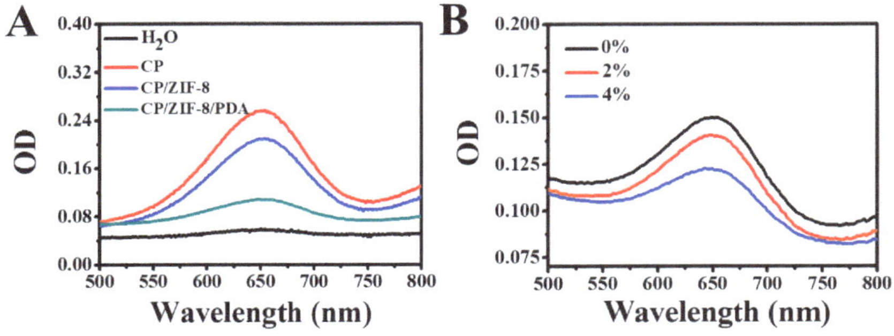

Figure 3. (**A**) Investigation of the catalytic oxidation of TMB by H_2O, CP, CP/ZIF-8, CP/ZIF-8/PDA, (**B**) investigation of ·OH by tert-butanol as a scavenger (0/2/4% is the volume ratio of tert-butanol in the solution).

3.3. Effect of Reaction Parameters

To obtain the optimal analytical performance of the FLISA platform, the influence of the pH value of OPD reaction, OPD concentration, incubation time of OPD and antibody concentration on the fluorescence signal were investigated, respectively. A concentration of 5 ng mL^{-1} CEA was used as the target concentration. The CP in CP/ZIF-8/PDA requires acidic release to produce ·OH to oxidize OPD, so the pH of the reaction is particularly important for the reaction system. HAc-NaAc buffer was used to dissolve OPD, and the pH of the buffer was optimized. The results in Figure 4A show that, in the pH range of 3.5 to 5.5, as the pH increased, the fluorescence signal was continuously enhanced, mainly because the oxidation of OPD was inhibited by over-acidic conditions [21]. In the pH 5.5–6.5 interval, the fluorescence signal continuously decreased with increasing pH, indicating that the dissolution of the ZIF-8 and PDA shell layers encapsulating CP slowed down after the pH reached 5.5, so pH 5.5 was chosen as the optimal pH. OPD concentration and reaction duration also have a large effect on the fluorescence signal. As shown in Figure 4B, as the OPD concentration increased from 1 mmol L^{-1} to 4 mmol L^{-1}, the DPA formed by ·OH oxidation of OPD increased and the fluorescence signal intensity of the reaction was enhanced. The fluorescence signal leveled off after the OPD concentration reached 4 mmol L^{-1}, indicating that the reaction reached equilibrium, thus, 4 mmol L^{-1} was chosen as the optimal OPD.

In Figure 4C, as the reaction time increased from 1 to 2.5 h, DPA increased, and the fluorescence intensity kept getting stronger. The fluorescence signal slowly enhanced with time growth after 2.5 h, thus, 2.5 h was chosen as the best reaction duration. Finally, the concentration of modified antibody (Ab) on the 96-well plate was optimized. As shown in Figure 4D, the concentration of Ab increased from 4 μg mL^{-1} to 10 μg mL^{-1}, the CEA captured by Ab kept increasing and the more CP/ZIF-8/PDA fixed onto the well plate, thus, the intensity of the fluorescence signal obtained was greater. When the concentration of Ab reached 10 μg mL^{-1}, the fluorescence signal reached a plateau, indicating that the amount of antibody binding to CEA reached saturation. Finally, 10 μg mL^{-1} of Ab was selected as the optimal concentration.

3.4. Analytical Performance

To examine the applicability of the FLISA analytical method for the sensitive detection of CEA, different concentrations of CEA were detected under optimal experimental conditions. In Figure 5A, the fluorescent signal was directly proportional to the logarithmic value of CEA concentrations in the range from 0.01 ng mL^{-1} to 20 ng mL^{-1}, corresponding with Figure 5B. The linear regression equation was $F = 2154.8 \times 1\,g\,C_{[CEA]}/ng\,mL^{-1} + 14439.75$ ($R^2 = 0.9967$). The detection limit was as low as 7.6 pg mL^{-1} ($3\sigma/K$, where σ stands for the standard deviation of 10 blank controls, K stands for the slope of regression equation). The analytical performances of other CEA assays [22–28] were listed in Table S1 and compared with our designed ELISA method. The results showed that the sensitivity of this FLISA method was comparable or even better than other analytical methods. This is because the sensing platform triggers a cascade reaction under acidic conditions to amplify the detection signal, and the simple steps eliminate many unnecessary errors. In addition, the self-supplied H_2O_2 is more stable and reacts more efficiently with Cu^{2+} than the applied H_2O_2. This FLISA platform has a low detection limit and can be used as an effective method for the clinical diagnosis of CEA.

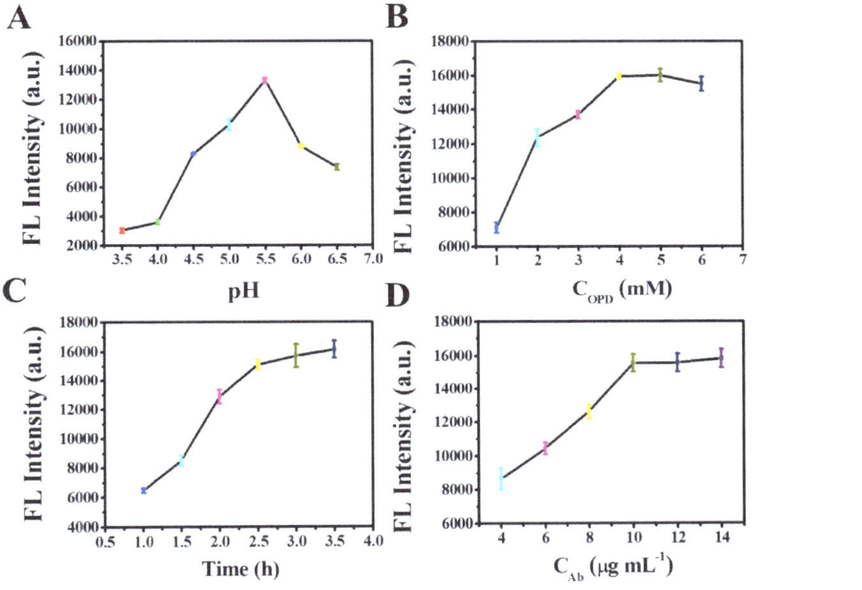

Figure 4. Effects of (**A**) pH of CP/ZIF-8/PDA reaction with OPD, (**B**) OPD concentration, (**C**) incubation time of OPD and (**D**) Ab concentration on the fluorescence intensity. A total of 2 mmol L^{-1} of OPD was used for pH optimization, 4 mmol L^{-1} of OPD was used for other optimization, and 5 ng mL^{-1} of CEA was used.

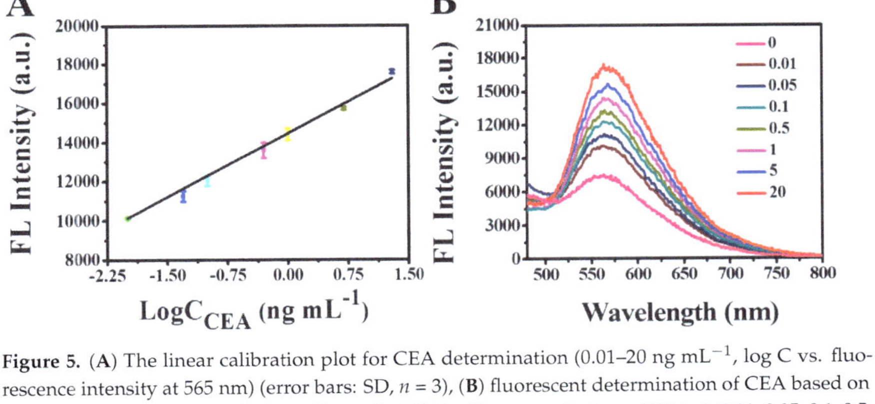

Figure 5. (**A**) The linear calibration plot for CEA determination (0.01–20 ng mL^{-1}, log C vs. fluorescence intensity at 565 nm) (error bars: SD, $n = 3$), (**B**) fluorescent determination of CEA based on CP/ZIF-8/PDA nanoparticles by self-supplied H_2O_2 (the concentration of CEA: 0, 0.01, 0.05, 0.1, 0.5, 1, 5, 10 and 20 ng mL^{-1}).

3.5. Selectivity and Stability Investigation

To examine the selectivity of this FLISA platform for CEA, coexisting substances that may interfere with CEA determination in actual samples such as prostate antigens (PSA, 25 ng mL^{-1}), alpha-fetoprotein (AFP, 25 ng mL^{-1}), glycoantigen 125 (CA125, 25 IU mL^{-1}) and glycoantigen 19-9 (CA19-9, 25 IU mL^{-1}) were investigated. The fluorescence signal intensity of 0.5 ng mL^{-1} CEA and interfering antigens and their mixtures were detected, respectively. The results are shown in Figure 6A. When comparing with other interferents, it was found that the fluorescence signal intensity of the experimental group with CEA was larger, and the fluorescence signal intensity of the interferent alone was consistent with that of the blank group. This was because the fluorescence immunosensor could not recognize other coexisting substances. When the CEA and high concentration of other tumor markers coexist in the sample, the fluorescence response results were in agreement with the results from the single presence of CEA. These results demonstrated that the fluorescence immunosensor could specifically distinguish CEA from its interfering species since the fluorescence signal was specifically triggered by the binding of the aptamer/antibody and CEA. So, the interferents did not interfere with the detection process. The results showed that the analytical method had good specificity for CEA. The reproducibility of the analytical method was investigated at three different concentrations of CEA. The coefficients of variation (CVs) of the three intra-batch measurements were 2.13%, 1.38% and 1.21% for the three CEA concentrations of 0.05, 0.5 and 5 ng mL^{-1}, respectively, and the CVs of the three inter-batch measurements were 5.09%, 6.10% and 6.17%, respectively. The results indicate that the FLISA platform has good reproducibility. The stability of the fluorescent immunosensor is a key factor which affects the practical application. As shown in Figure 6B, when the CP/ZIF-8/PDA/Aptamer was stored at 4 °C for 1, 6, 11 and 16 days, respectively, for the same concentration of CEA, the fluorescence signal change was not obvious. However, for the different concentrations of CEA, with the increasement in CEA concentration from 0.05 ng mL^{-1}, 0.5 ng mL^{-1} to 5 ng mL^{-1}, the fluorescence signal intensity gradually increased, which was consistent with the result of the linear calibration plot for CEA determination. The above results indicate that the method has good stability.

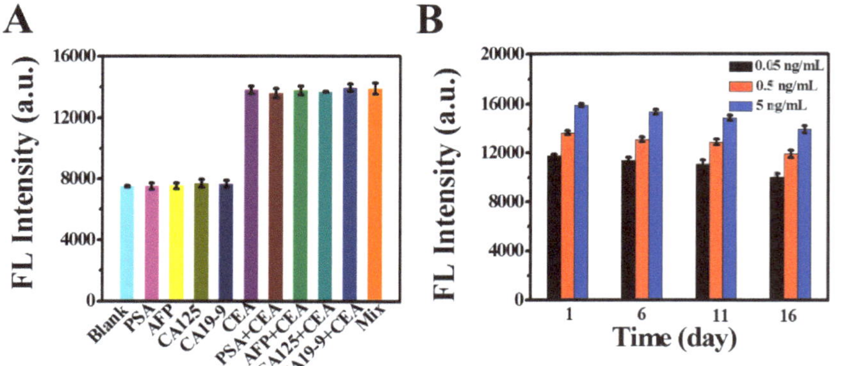

Figure 6. (**A**) Selectivity for the determination of CEA by monitoring the fluorescence intensity with different interfering species (error bars: SD, n = 3), (**B**) the stability of the FLISA platform for CEA (0.05, 0.5, 5 ng mL^{-1}) (error bars: SD, n = 3).

3.6. Analysis of Real Samples and Evaluation of Method Accuracy

The CEA referenced the concentrations of six clinical serum samples, which were 0.05, 0.5, 1.5, 2.1, 5.4 and 13.7 ng mL^{-1} for the follow-up FLISA determination, respectively. Among them, serum samples containing 0.05 ng mL^{-1} and 13.7 ng mL^{-1} CEA were obtained via dilution with PBS (10 mmol L^{-1}, pH 7.4). In Table 1, the FLISA results of the six different samples were consistent with the referenced values obtained from the commercialized Roche Cobas e601 automatic electrochemiluminescence immunoanalyzer. The maximum RSD of the FLISA method did not exceed 8.16%, indicating that the method has high accuracy and clinical application.

Table 1. Comparison of the results for the determination of CEA in clinical serum samples by using the self-supplied H_2O_2 fluorescent immunosensor and the referenced electrochemiluminescence method.

Sample Number	Found by the Electrochemiluminescence Method [ng mL^{-1}]	Found by Self-Supplied H_2O_2 Fluorescent Immunosensor [mean ± SD (RSD), ng mL^{-1}], n = 3
1	0.05	0.052 ± 0.001 (2.54%)
2	0.5	0.51 ± 0.03 (5.84%)
3	1.5	1.55 ± 0.13 (8.16%)
4	2.1	1.99 ± 0.14 (7.20%)
5	5.4	5.94 ± 0.17 (5.24%)
6	13.7	13.05 ± 0.68 (5.24%)

4. Conclusions

In summary, a self-supplied H_2O_2 of FLISA method was constructed based on CP/ZIF-8/PDA nanoparticles for the sensitive and high-throughput determination of CEA. The prepared CP/ZIF-8/PDA nanoparticles were modified by the aptamer to form an immunoprobe and were immobilized onto a 96-well plate by forming a sandwich complex with CEA and antibodies. Under acidic conditions, the cascade reaction was triggered to generate Cu^{2+} and H_2O_2 due to the CP dissociation. Both of them rapidly underwent a Fenton-type reaction to produce ·OH which oxidized OPD to form a strong fluorescent substance DPA for subsequent fluorescence detection. Because of the close distance between the simultaneously generated Cu^{2+} and H_2O_2, the concentration of them in the site was relatively high, and thus, a high reaction efficiency could be achieved without the addition of H_2O_2. Additionally, the signal output of this FLISA platform did not require enzymes. The sensitive determination of CEA could be achieved with a LOD value of 7.6 pg mL^{-1}, and the result of CEA determination in serum samples was satisfactory. Thus, the FLISA-sensing platform provided a new way of sensitive and large-scale screening biomarkers for the early diagnosis of clinical disease. Future work will focus on the point of care diagnosis for CEA via smartphone determination and explore app applications towards the biomarker-based FLISA method.

Supplementary Materials: The following supporting information can be downloaded at: https://www.mdpi.com/article/10.3390/bios12100830/s1, Figure S1: EDS mapping element analysis diagram of ZIF-8; Figure S2: EDS mapping element analysis diagram of CP/ZIF-8; Figure S3: Fluorescence intensity of FLISA platform with and without CEA; Table S1: Comparison of analytical performance for CEA determination by using different sensing methods.

Author Contributions: Conceptualization, J.H.; methodology, J.H., Y.Y. and Y.C.; investigation, J.H. and Y.Y.; writing—original draft preparation, J.H. and Y.Y.; writing—review and editing, Y.Y. and L.H.; supervision, L.H., T.L. and D.T.; project administration, L.H. and T.L.; funding acquisition, L.H. and T.L. All authors have read and agreed to the published version of the manuscript.

Funding: This research was funded by the Natural Science Foundation of Guangxi (2022GXNSFAA035475), the National Natural Science Foundation of China (21964004, 22164005).

Institutional Review Board Statement: Not applicable.

Informed Consent Statement: Not applicable.

Data Availability Statement: Not applicable.

Acknowledgments: The authors are grateful for financial support from the Natural Science Foundation of Guangxi and the National Natural Science Foundation of China.

Conflicts of Interest: The authors declare no conflict of interest.

References

1. Huang, L.; Chen, J.; Yu, Z.; Tang, D. Self-powered temperature sensor with seebeck effect transduction for photothermal-thermoelectric coupled immunoassay. *Anal. Chem.* **2020**, *92*, 2809–2814. [CrossRef] [PubMed]
2. Yu, Z.; Gong, H.; Xu, J.; Li, Y.; Zeng, Y.; Liu, X.; Tang, D. Exploiting photoelectric activities and piezoelectric properties of $NaNbO_3$ semiconductors for point-of-care immunoassay. *Anal. Chem.* **2022**, *94*, 3418–3426. [CrossRef]
3. Wei, Z.Y.; Xi, Z.; Vlasov, S.; Ayala, J.; Xia, X.H. Nanocrystals of platinum-group metals as peroxidase mimics for in vitro diagnostics. *Chem. Commun.* **2020**, *56*, 14962–14975. [CrossRef] [PubMed]
4. Peng, P.; Liu, C.; Li, Z.D.; Xue, Z.R.; Mao, P.; Hu, J.; Xu, F.; Yao, C.Y.; You, M.L. Emerging ELISA derived technologies for in vitro diagnostics. *TrAC Trends Anal. Chem.* **2022**, *152*, 116605. [CrossRef]
5. Ruan, Y.; Xu, H.H.; Yu, J.L.; Chen, Q.; Gu, L.H.; Guo, A.L. A fluorescence immunoassay based on CdTe: Zn/ZnS quantum dots for the rapid detection of bacteria, taking Delftia tsuruhatensis CM′13 as an example. *RSC Adv.* **2020**, *10*, 1042–1049. [CrossRef] [PubMed]
6. Wu, W.J.; Liu, X.Y.; Shen, M.F.; Shen, L.S.; Ke, X.; Cui, D.X.; Li, W.W. Multicolor quantum dot nanobeads based fluorescence-linked immunosorbent assay for highly sensitive multiplexed detection. *Sens. Actuators B* **2021**, *338*, 129827. [CrossRef]
7. Smitha, P.K.; Bathula, C.; Chandrashekara, K.N.; Das, M. Usage of graphene oxide in fluorescence quenching-linked immunosorbent assay for the detection of Cry2Ab protein present in transgenic plants. *J. Agric. Food Chem.* **2020**, *68*, 3656–3662. [CrossRef]
8. Lv, S.; Tang, Y.; Zhang, K.; Tang, D. Wet NH_3-triggered NH_2-MIL-125(Ti) structural switch for svisible fluorescence immunoassay impregnated on paper. *Anal. Chem.* **2018**, *90*, 14121–14125. [CrossRef]
9. Zhan, Y.J.; Yang, S.T.; Luo, F.; Guo, L.H.; Zeng, Y.B.; Qiu, B.; Lin, Z.Y. Emission wavelength switchable carbon dots combined with biomimetic inorganic nanozymes for a two-photon fluorescence immunoassay. *ACS Appl. Mater. Interfaces* **2020**, *12*, 30085–30094. [CrossRef]
10. Taron, W.; Phooplub, K.; Sanchimplee, S.; Piyanamvanich, K.; Jamnongkan, W.; Techasen, A.; Phetcharaburanin, J.; Klanrit, P.; Namwat, N.; Khuntikeo, N.; et al. Smartphone-based fluorescent ELISA with simple fluorescent enhancement strategy for Opisthorchis viverrini (Ov) antigen detection in urine samples. *Sens. Actuators B Chem.* **2021**, *348*, 130705. [CrossRef]
11. Lv, X.; Huang, Y.M.; Liu, D.F.; Liu, C.W.; Shan, S.; Li, G.Q.; Duan, M.L.; Lai, W.H. Multicolor and ultrasensitive enzyme-linked immunosorbent assay based on the fluorescence hybrid chain reaction for simultaneous detection of pathogens. *J. Agric. Food Chem.* **2019**, *67*, 9390–9398. [CrossRef] [PubMed]
12. Meng, J.S.; Liu, X.; Niu, C.J.; Pang, Q.; Li, J.T.; Liu, F.; Liu, Z.A.; Mai, L.Q. Advances in metal-organic framework coatings: Versatile synthesis and broad applications. *Chem. Soc. Rev.* **2020**, *49*, 3142–3186. [CrossRef]
13. Lv, S.; Zhang, K.; Zhu, L.; Tang, D. ZIF-8-assisted NaYF4:Yb, Tm@ZnO converter with exonuclease III-powered DNA walker for near-infrared light responsive biosensor. *Anal. Chem.* **2020**, *92*, 1470–1476. [CrossRef] [PubMed]
14. Guo, M.L.; Chi, J.T.; Zhang, C.; Wang, M.L.; Liang, H.; Hou, J.Y.; Ai, S.Y.; Li, X.Y. A simple and sensitive sensor for lactose based on cascade reactions in Au nanoclusters and enzymes co-encapsulated metal-organic frameworks. *Food Chem.* **2021**, *339*, 127863. [CrossRef] [PubMed]
15. Wang, H.Q.; Ma, Z.F. Copper peroxide/ZIF-8 self-producing H_2O_2 triggered cascade reaction for amperometric immunoassay of carbohydrate antigen 19-9. *Biosens. Bioelectron.* **2020**, *169*, 112644. [CrossRef]
16. Lin, L.S.; Huang, T.; Song, J.B.; Ou, X.Y.; Wang, Z.T.; Deng, H.Z.; Tian, R.; Liu, Y.J.; Wang, J.F.; Liu, Y.; et al. Synthesis of copper peroxide nanodots for H_2O_2 self-supplying chemodynamic therapy. *J. Am. Chem. Soc.* **2019**, *141*, 9937–9945. [CrossRef] [PubMed]
17. Li, X.J.; Du, Y.; Wang, H.; Ma, H.M.; Wu, D.; Ren, X.; Wei, Q.; Xu, J.J. Self-supply of H_2O_2 and O_2 by hydrolyzing CaO_2 to enhance the electrochemiluminescence of luminol based on a closed bipolar electrode. *Anal. Chem.* **2020**, *92*, 12693–12699. [CrossRef]
18. Ross, P.K.; Solomon, E.I. An electronic structural comparison of copper-peroxide complexes of relevance to hemocyanin and tyrosinase active sites. *J. Am. Chem. Soc.* **1991**, *113*, 3246–3259. [CrossRef]
19. Wan, W.; Han, Q.; Zhang, X.Q.; Xie, Y.M.; Sun, J.P.; Ding, M.Y. Selective enrichment of proteins for MALDI-TOF MS analysis based on molecular imprinting. *Chem. Commun.* **2015**, *51*, 3541–3544. [CrossRef]
20. Shuai, C.J.; Zan, J.; Deng, F.; Yang, Y.W.; Peng, S.P.; Zhao, Z.Y. Core-shell-structured ZIF-8@PDA-HA with controllable zinc ion release and superior bioactivity for improving a poly-l-lactic acid Scaffold. *ACS Sustain. Chem. Eng.* **2021**, *9*, 1814–1825. [CrossRef]
21. Fan, Y.C.; Xing, H.H.; Xue, Y.; Peng, C.; Li, J.; Wang, E.K. Universal platform for ratiometric sensing based on catalytically induced inner-filter effect by Cu^{2+}. *Anal. Chem.* **2020**, *92*, 16066–16071. [CrossRef] [PubMed]
22. Zheng, J.; Wang, J.P.; Song, D.D.; Xu, J.L.; Zhang, M. Electrochemical aptasensor of carcinoembryonic antigen based on concanavalin a-functionalized magnetic copper silicate carbon microtubes and gold-nanocluster-assisted signal amplification. *ACS Appl. Nano Mater.* **2020**, *3*, 3449–3458. [CrossRef]
23. Tang, Y.; Zhang, B.H.; Wang, Y.; Zhao, F.Q.; Zeng, B.Z. Electrochemiluminescence immunosensor for the detection of carcinoembryonic antigen based on oxygen vacancy-rich Co_3O_4 nanorods and luminol. *ACS Appl. Nano Mater.* **2021**, *4*, 7264–7271. [CrossRef]
24. Zhou, X.; Yang, C.T.; Xu, Q.S.; Lou, Z.C.; Xu, Z.F.; Thierry, B.; Gu, N. Gold nanoparticle probe-assisted antigen-counting chip using SEM. *ACS Appl. Mater. Interfaces* **2019**, *11*, 6769–6776. [CrossRef] [PubMed]
25. Jiang, Y.; Su, Z.Y.; Zhang, J.; Cai, M.J.; Wu, L.L. A novel electrochemical immunoassay for carcinoembryonic antigen based on glucose oxidase-encapsulated nanogold hollow spheres with a pH meter readout. *Analyst* **2018**, *143*, 5271–5277. [CrossRef]

26. Fan, Y.C.; Lv, M.M.; Xue, Y.; Li, J.; Wang, E.K. In situ fluorogenic reaction generated via ascorbic acid for the construction of universal sensing platform. *Anal. Chem.* **2021**, *93*, 6873–6880. [CrossRef]
27. Xu, S.H.; Feng, X.Y.; Gao, T.; Liu, G.F.; Mao, Y.N.; Lin, J.H.; Yu, X.J.; Luo, X.L. Aptamer induced multicoloured Au NCs-MoS_2 "switch on" fluorescence resonance energy transfer biosensor for dual color simultaneous detection of multiple tumor markers by single wavelength excitation. *Anal. Chim. Acta* **2017**, *983*, 173–180. [CrossRef]
28. Qiu, Z.; Shu, J.; Tang, D. Bioresponsive release system for visual fluorescence detection of carcinoembryonic antigen from mesoporous silica nanocontainers mediated optical color on quantum dot-enzyme-impregnated paper. *Anal. Chem.* **2017**, *89*, 5152–5160. [CrossRef]

biosensors

MDPI

Article

Viscosity-Sensitive Solvatochromic Fluorescent Probes for Lipid Droplets Staining

Mao-Hua Wang, Wei-Long Cui, Yun-Hao Yang and Jian-Yong Wang *

State Key Laboratory of Biobased Material and Green Papermaking, Key Laboratory of Paper Science and Technology of Ministry of Education, Faculty of Light Industry, Qi Lu University of Technology, Shandong Academy of Sciences, Jinan 250353, China

* Correspondence: wjy@qlu.edu.cn

Abstract: Lipid droplets (LDs) are simple intracellular storage sites for neutral lipids and exhibit important impact on many physiological processes. For example, the changes in the polar microenvironment inside LDs could affect physiological processes, such as lipid metabolism and storage, protein degradation, signal transduction, and enzyme catalysis. Herein, a new fluorescent chemosensor (Couoxo-LD) was formulated by our molecular design strategy. The probe could be applied to effectively label intracellular lipid droplets. Intriguingly, Couoxo-LD demonstrated positive sensitivity to both polarity and viscosity, which might be attributed to its D-π-A structure and the twisted rotational behavior of the carbon–carbon double bond (TICT). Additionally, Couoxo-LD was successfully implemented in cellular imaging due to its excellent selectivity, pH stability, and low biotoxicity. In HeLa cells, the co-localization curve between Couoxo-LD and commercial lipid droplet dyes overlapped at 0.93. The results indicated that the probe could selectively sense LDs in HeLa cells. Meanwhile, Couoxo-LD can be applied for in vivo imaging of zebrafish.

Keywords: lipid droplets; viscosity-sensitive; polarity-sensitive; solvatochromism

Citation: Wang, M.-H.; Cui, W.-L.; Yang, Y.-H.; Wang, J.-Y. Viscosity-Sensitive Solvatochromic Fluorescent Probes for Lipid Droplets Staining. *Biosensors* **2022**, *12*, 851. https://doi.org/10.3390/bios12100851

Received: 10 September 2022
Accepted: 7 October 2022
Published: 9 October 2022

Publisher's Note: MDPI stays neutral with regard to jurisdictional claims in published maps and institutional affiliations.

1. Introduction

Lipid droplets (LDs) serve as simple intracellular storage sites for neutral lipids and consist of a non-polar neutral lipid core [1]. Studies have shown that LDs are not only simple energy stores but also complex and dynamic multifunctional organelles [2]. For example, alterations in the polar microenvironment surrounding LDs affect physiological processes, such as lipid metabolism and storage, protein degradation, signal transduction, and enzyme catalysis [3]. Furthermore, previous studies reported that homeostasis of the LDs microenvironment was associated with diseases, such as obesity, cardiovascular disease, and diabetes mellitus [4–6]. Cancer cells exhibit a strong affinity for fatty acids and cholesterol, which are over-stored in lipid droplets. [7]. The high levels of LDs in tumors provide a potential means of monitoring and treating cancer [8]. Therefore, the tracking and monitoring of LDs are essential. In addition, LDs consist a unique structure (a single phospholipid membrane with a hydrophilic "head (group)" facing the cytoplasm and an internal storage of lipid cores, such as triglycerides and cholesterol esters [9,10]), i.e., a hydrophobic and viscous environment inside the lipid droplets. Theoretically, if the probe exhibits different fluorescence properties in these two microenvironments, they could be used as a tool for labeling LDs [11,12].

Fluorescence imaging has established itself as a beneficial tool for studying biological systems because of its high sensitivity, accessibility, non-invasiveness, and real-time and in situ detection of target molecules [13–19]. So far, several fluorescent probes for specifically imaging LDs have been reported [20–27]. For example, Yu et al., designed two heteroindole-based two-photon fluorescent probes and visualized the polarity of LDs at a cellular level and in zebrafish larvae [15]. However, there were only a few probes that were reported to respond positively to both the polarity and viscosity of lipid droplets. Therefore, there is a requirement to develop a strong color-changing LDs probe for sensing both polarity and viscosity [28,29].

Since Perkin et al. first synthesized artificial coumarins by chemical synthesis in 1868 [30], coumarin derivatives have been widely used in the design of small-molecule fluorescent sensors because of their good biocompatibility, controlled structure, and stable and strong fluorescence emission intensity [31–41]. Here, a solvent–chromogenic lipid droplets fluorescent probe, Couoxo-LD, was synthesized using benzoylglycine and coumarin fluorescent moiety. Couoxo-LD consisted an oxazolone and a coumarin derivative linked by a double bond. In optical characterization tests, the emission wavelength of the probe exhibited a significant red shift with increasing solvent polarity. The emission intensity of the probe increased with increasing solvent viscosity, showing regular polarity-viscosity-sensitive characteristics. The properties of good biocompatibility and pH stability were expressed in this probe. In addition, Couoxo-LD exhibited satisfactory lipid droplets targeting, possessed a high degree of overlap with commercial lipid droplets dye co-localization imaging, and had been successfully applied to cells and zebrafish imaging.

2. Synthesis of Probes

2.1. Reagents and Materials

The materials and reagents involved in the experiments were obtained commercially; furthermore, no secondary purification was carried out. The instruments used in the experiments were described in detail in the supporting materials.

2.2. Synthesis of Couoxo-LD

The synthesis of compound **1** was reported in detail in earlier work [42]. The synthetic design route of Couoxo-LD was shown in Scheme 1 and the specific synthesis was referred to in Ref. [43].

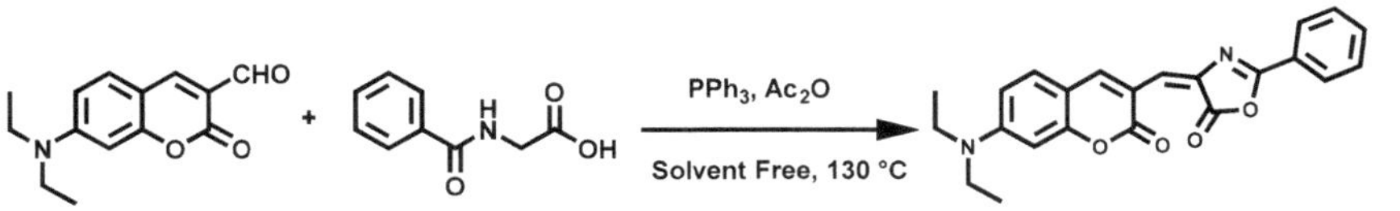

Scheme 1. Synthesis of Couoxo-LD.

Compound **1** (7-(diethylaMino)-2-oxo-2H-chroMene-3-carbaldehyde) (49.1 mg, 0.2 mmol), compound **2** (benzoylglycine) (39.42 mg, 0.22 mmol), and triphenylphosphine (5.25 mg, 0.022 mmol) were dissolved in anhydrous acetic anhydride (54.14 mg 50 μL, 0.53 mmol) and stirred for 4 h at 130 °C. After completion of the reaction (thin-layer chromatography monitoring), reaction mixture was cooled to room temperature. Ethyl alcohol (95%, 5.0 mL) was added to produce a large amount of precipitation. The product was obtained by filtration and recrystallization from ethanol (66.81 mg, 86 %). ^{1}H NMR (400 MHz, $CDCl_3$) δ 9.17 (s, 1H), 8.13–8.07 (m, 2H), 7.63–7.57 (m, 1H), 7.56–7.49 (m, 3H), 7.43 (d, J = 8.9 Hz, 1H), 6.64 (dd, J = 9.0, 2.4 Hz, 1H), 6.41 (d, J = 2.4 Hz, 1H), 3.45 (q, J = 7.1 Hz, 4H), 1.25 (t, J = 7.1 Hz, 6H). ^{13}C NMR (101 MHz, $CDCl_3$) δ 166.74, 162.76, 161.37, 157.09, 152.31, 146.86, 133.16, 131.94, 131.40, 128.94, 128.14, 125.67, 125.27, 113.79, 110.03, 109.86, 97.34, 45.30, 12.52.

3. Result and Discussion

3.1. Probe Design and Discussion

As mentioned above, LDs are distinguished from the surrounding viscous environment due to their unique internal structure, namely a phospholipid monolayer and a lipid core composed of fatty acids. The large amount of water and inorganic ions as components of the cytoplasmic solute provides a polar environment around LDs. Meanwhile, the characteristic environments such as viscosity and non-polarity exhibited inside LDs serve as a reliable theory to support our development of efficient lipid droplets fluorescent probes.

Based on the above principles, we designed and constructed a fluorescent probe Couoxo-LD with a structure sensitive to both viscosity and polarity. Couoxo-LD, consisting of coumarin derivatives and benzoylglycine, possessed strong electronic push–pull properties. In addition, the coumarin scaffold was selected as the monomeric component of Couoxo-LD, giving the probe a certain lipophilicity and therefore a higher sensitivity to lipid droplets. By introducing a diethylamine electron donor, Couoxo-LD was designed as a molecule with a typical D-π-A structure. In conclusion, the solvent-altering effect of Couoxo-LD was the main reason for illuminating intracellular lipid droplets. In an aqueous environment, such as a cytoplasmic solute, the twisted rotational behavior (TICT) around the carbon–carbon double bond hindered the emission behavior of Couoxo-LD [44]. In contrast to this, in non-polar media, such as LDs, it would release a strong signal through the locally excited (LE) state. That is, the vinyl structure in Couoxo-LD allowed the free rotation of the probe molecule that was restricted by the environment, which affected the emission behavior of the probe (Scheme 1). The probe Couoxo-LD was characterized by ^{1}H NMR and ^{13}C NMR (Supplementary Figures S1 and S2).

3.2. *Study of Photophysical Properties of Probes*

The molecular probe that possessed a donor (D)-π-acceptor (A) structure exhibited a pronounced solvatochromic effect and its photophysical properties varied with solvent polarity. Therefore, the absorption and emission behaviors of Couoxo-LD in different polar solvents were investigated (Figures 1 and S3), such as 1,4-dioxane, methanol (MeOH), dichloromethane (DCM), N,N-dimethylformamide (DMF), ethanol (EtOH), tetrahydrofuran (THF), toluene (Tol), dimethyl sulfoxide (DMSO), ethyl acetate (EtOAc), acetone, and acetonitrile (CH_3CN).

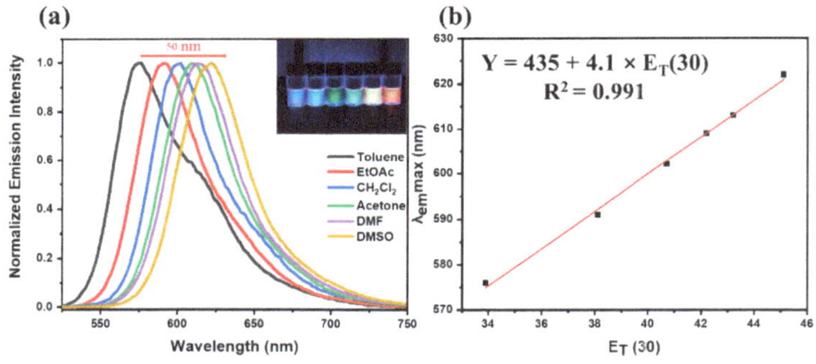

Figure 1. Normalized fluorescence spectra at 510 nm excitation. (**a**) Normalized fluorescence spectra in organic solvents with different polarities; a-Inner: Photographs of solvent discoloration with insertion of Couoxo-LD (10 μM). (**b**) For linearity between the maximum emission wavelength of the probe and the polarity of the solvent.

As shown in Figures 1 and S3, the spectral data of the probe in different polar solvents were firstly explored, and Couoxo-LD showed strong absorption and emission phenomena and exhibited red-shifted behavior. From the emission peak at 576 nm in toluene to 622 nm in dimethylsulfoxide, the emission peak of Couoxo-LD underwent a red shift of about 50 nm. Under 365 nm-UV irradiations, it was clearly seen that this probe with an electronic push–pull structure exhibited a clear solvent discoloration effect accompanied by a change in fluorescence color from blue (toluene) to orange-red (dimethylsulfoxide) (Figure 1a). The above presented results demonstrate the ICT effect of Couoxo-LD. As mentioned above, we evaluated the environment-sensitive and alike-solvent-discoloration effect of the Couoxo-LD by studying the emission behavior of Couoxo-LD under different solvent polarities. In particular, Couoxo-LD exhibited a strong positive solvent discoloration effect, which was consistent with the closely reported fluorophore [45]. Moreover, a gradual increase in the polarity parameter $E_T(30)$ from 33.9 kcal^{-1} to 55.4 kcal^{-1} molar concentration resulted in a continuous red shift of the maximum emission wavelength of the probe (Figure 1a). More significantly, we found that Couoxo-LD showed a satisfactory linear relationship between the emission peak in different polar solvents and the $E_T(30)$ of the solvent with their Pearson correlation coefficient of 0.991 (Figures 1b and S3) and a slope of 4.1 nm $E_T(30)$ units. This was caused by the function of the intra-molecular charge transfer effect (ICT) from the electron-giving N,N-diethyl unit to the electron-accepting oxazol-5(4H) one. These results indicated that the Couoxo-LD photophysical properties were closely related to solvent polarity.

To verify the above conjecture, we performed density general function theory (DFT) calculations for Couoxo-LD, and based on them we further optimized the electron cloud around Couoxo-LD using the Gaussian'09 program and DFT-derived Multiwfn and VMD program models. The distribution of Couoxo-LD in the more polar PBS-buffered solvent and the distribution of HOMO and LUMO in the less polar dioxane solvent were also calculated (Figure 2). The results show that the energy band gap ΔE of Couoxo-LD in aqueous solution is smaller than that in dioxane solvent. Obviously, the larger the energy band gap ΔE, the smaller the wavelength of the maximum absorption peak because the energy required for the electron leap is large. It was further verified that the wavelength of the absorption peak of Couoxo-LD in a low-polarity solvent environment (dioxane) was smaller than that of the absorption peak in a high-polarity solution (PBS buffer).

Since the Φ_F of Couoxo-LD depended strongly on the solvent's properties (Table 1), the Φ_F increased with increasing solvent polarity (Toluene to Acetone) and reached a maximum. The increase in Φ_F due to charge transfer (negative solvatokinetic effect) could be explained by several mechanisms, such as proximity effects and conformational changes. The decrease in Φ_F from DMSO to DMF (positive solvatokinetic effect) could be attributed to strong ICT interactions. As described in the literature [46], the fluorescence quantum yield of Couoxo-LD depended on the polarity of the solvent, as well as on specific solute–solvent interactions, such as hydrogen bonding and high photostability.

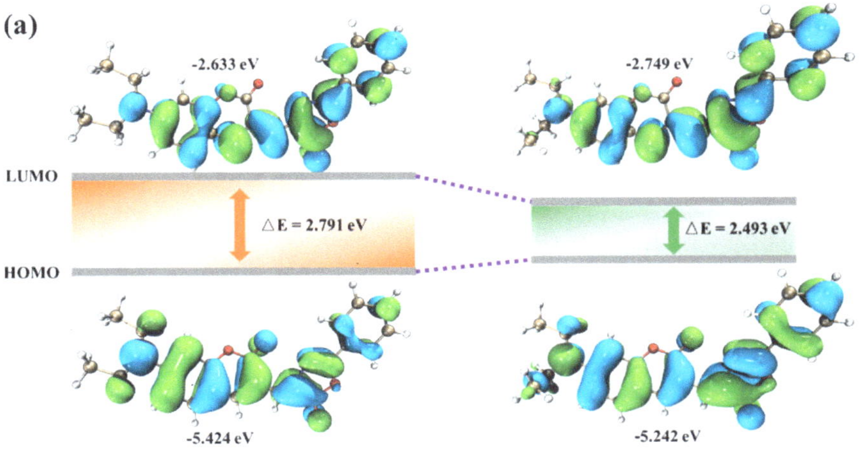

Figure 2. *Cont.*

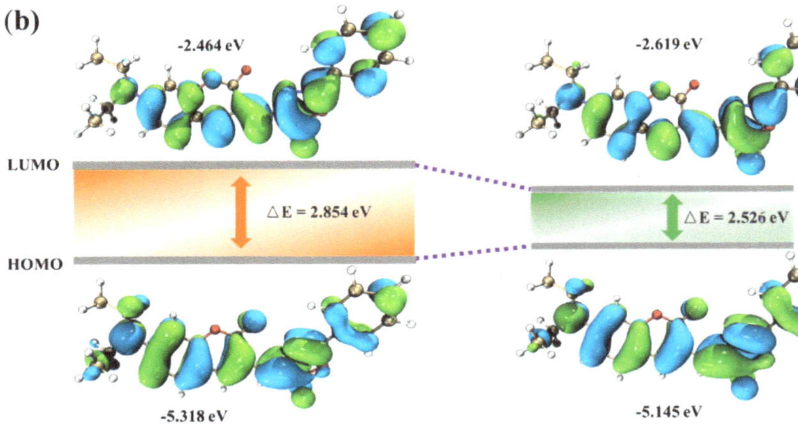

Figure 2. Molecular structure of Couoxo-LD and frontier molecular orbitals of the probe S_0 and S_1 states, calculated using Gaussian'09 at the TDB3L YP/6−31+G(d) level. (**a**) Used the PBS solvent model; (**b**) Used the Dioxane solvent model.

Table 1. Photophysical properties of Couoxo-LD in different solvents.

	Toluene	Dioxane	THF	EtOAc	DCM	Acetone	DMF	DMSO	MeCN
λ_{abs} (nm)	533	514	513	524	531	528	543	548	531
λ_{em} (nm)	576	581	592	591	602	609	613	622	607
$E_T(30)$ [1]	33.9	36	37.4	38.1	40.7	42.2	43.2	45.1	45.6
Stokes shift (nm/cm^{-1})	43/1401	67/2244	61/1940	67/2163	71/2221	81/2519	70/2103	74/2071	76/2358
Φ_F [2]	5.2%	6.1%	3.4%	3.4%	1.3%	19.0%	11.0%	11.3%	4.3%
(Log ε_{max}) [3]	4.74	4.33	4.78	4.76	4.74	4.24	4.75	4.78	4.76

[1] ET(30) is a solvent polarity parameter and means the molar transition energy in kcal·mol^{-1} [43]. [2] Fluorescence quantum yield was obtained by the reference method and applying equation $\Phi_x = \Phi_s (n_x/n_s)^2 (A_s/A_x) (F_x/F_s)$. [3] Molar extinction coefficients are calculated in the maximum of the highest peak.

3.3. *The Emission Behavior of the Probe in PBS and Dioxane*

Furthermore, we investigated the spectroscopic testing of the probe in different volume ratios of PBS buffer–dioxane mixtures (Figure 3). From the above experiments, we discovered that the emission behavior of the probe was related to the polarity of the medium; therefore, we further explored the lipid solubility test of the probe in different polar environments. In the binary solvent system of PBS buffer and dioxane, increasing the proportion of PBS buffer (f_W) from 40% to 100%, the fluorescence intensity of Couoxo-LD showed a trend of increasing and then rapidly decreasing. Its emission peak underwent a certain red shift, which we judged to be caused by the increased polarity of the solvent mixture. The emission of the probe decreased appreciably with the decrease of the lipid-soluble solvent. This might be due to the increasing content of the PBS buffer, in which the increased polarity of the solvent made Couoxo-LD more sensitive, masking the solubility of the probe to the lipid solvent. This results in a reduction in the emission intensity of Couoxo-LD and a red shift of the maximum emission peak. All these findings suggested that the optical properties of Couoxo-LD were closely related to the environmental polarity.

3.4. *Probe Emission Behavior in Viscous Environments*

We verified the viscosity-emitting behavior of Couoxo-LD in binary systems with different ratios of PBS buffer and glycerol. As shown in Figure 4a, the emission intensity gradually increased with the increase of glycerol content. We further explored the relationship between the probe in a medium with different volume ratios of PBS buffer–Glycerol and the maximum emission intensity. The emission intensity of Couoxo-LD increased exponentially with the viscosity of the system (Figure 4b), which was consistent with the D-π-A structure of the probe molecule we mentioned earlier. In a low-viscosity environment, the free intra-molecular rotation leads to a slackening of the excitation energy, which greatly attenuated the emission phenomenon. In high-viscosity environments, free rotation within the probe molecule was inhibited by the environment and the molecule released energy primarily by radiation, resulting in a significant enhancement of fluorescence emission from Couoxo-LD. Furthermore, the color change of Couoxo-LD in the mixed system was found to be affected by polarity in the study (Figure 4c,d). This could be due to the higher polarity of the solution in the pure-PBS-buffer environment. With the addition of glycerol, the polarity in the mixed system decreased. A shift from red to yellowish green occurred, while the spectrum underwent a blue shift. Finally, when the glycerol ration was increased to 100%, the polarity of the system was greater than the mixed polarity of both and a red shift in the spectrum occurred and the color changed back to red. These results showed that Couoxo-LD could successfully monitor changes in viscosity and respond to polarity.

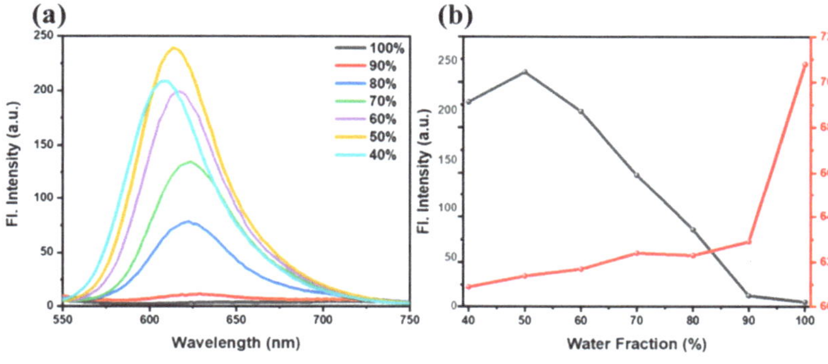

Figure 3. (**a**) Fluorescence emission spectra of Couoxo-LD (10 µM, λ_{ex} = 510 nm) in the binary system of PBS buffer and dioxane. (**b**) Folding graph of maximum absorption and emission wavelengths of Couoxo-LD.

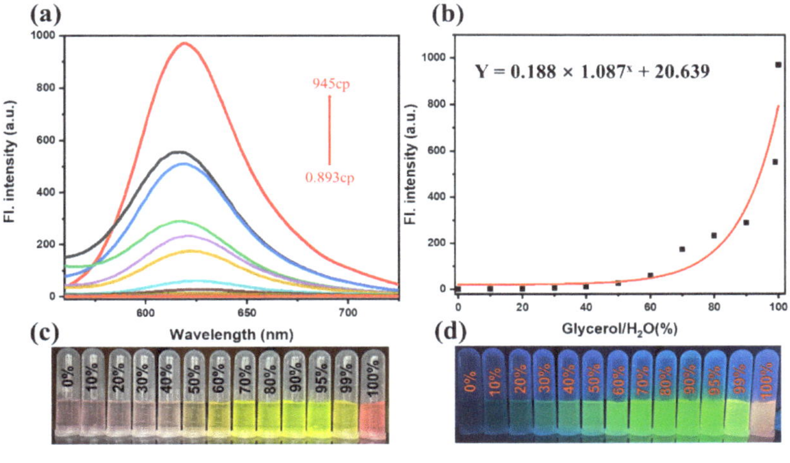

Figure 4. Couoxo-LD (10 µM, λ_{ex} = 510 nm) in solution viscosity (PBS buffer–Glycerol system) (2 mL) versus fluorescence spectrum (**a**) and maximum emission wavelength (**b**). Images of Couoxo-LD in different PBS buffer–Glycerol mixture systems under natural light (**c**). Images of Couoxo-LD in different PBS buffer–Glycerol mixes under 365 nm UV light (**d**).

3.5. Probe Stability Study

To verify whether the Couoxo-LD was adapted to the sophisticated environment inside living cells, we investigated the emission behavior of Couoxo-LD in phosphate buffer solutions over a wide range of pH values. The results showed that the fluorescence intensity of Couoxo-LD decreased slightly when decreasing the solution pH (Figure 5a). Overall, the fluorescence intensity of Couoxo-LD was observed to be almost identical in different pH environments, demonstrating that the probe was unaffected by pH. We then investigated the response of the probe Couoxo-LD under different interfering ion environments (Figure 5b). The ions and different small biomolecules were added to PBS buffer and dioxane solutions, respectively, and the results showed that the fluorescence intensity of the probe did not change significantly in the organic solvent dioxane and PBS buffer, indicating that the emission behavior of Couoxo-LD was almost unaffected by the other chemicals. Furthermore, the emission intensity of Couoxo-LD in PBS buffer remained almost unchanged after continuous irradiation with a 365 nm-UV lamp for 1000 s, while it remained at a high level in dioxane, although it decreased slightly (Figure 5c). The excellent photostability indicated that the probe Couoxo-LD possessed a strong resistance to photobleaching and photo bursts.

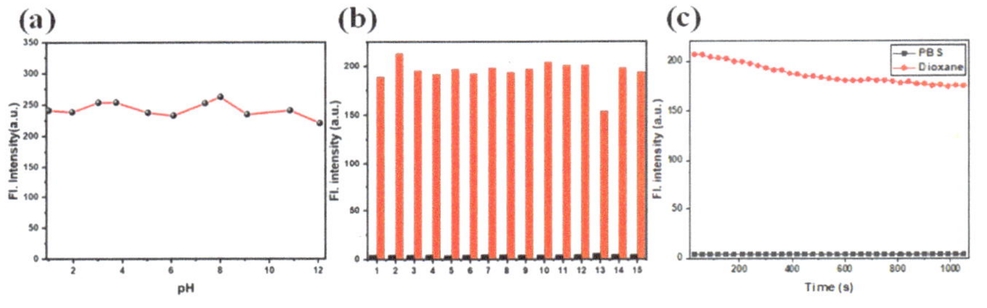

Figure 5. (**a**) The emission intensity of probe Couoxo-LD (10 μM, λ_{ex} = 510 nm) in PBS buffer at various pH (1–12). (**b**) In PBS buffer, the fluorescent intensity of Couoxo-LD in (10 μM, λ_{ex} = 510 nm, at 645 nm) and in Dioxane, Couoxo-LD (10 μM, λ_{ex} = 510 nm, at 580 nm) upon addition of various species (10 μM) including: 1, Blank; 2, HS^-; 3, NO_2^-; 4, CO_3^{2-}; 5, HCO_3^-; 6, TBHP; 7, SO_4^{2-}; 8, CI^-; 9, Ca^{2+}; 10, H_2O_2; 11, SO_3^{2-}; 12, HSO_3^-; 13, GSH; 14, Mg^{2+}; 15, Na^+. (**c**) Couoxo-LD (10 μM, λ_{ex} = 510 nm) emission pattern of continuous irradiation in PBS buffer and Dioxane solvent.

3.6. Probe Couoxo-LD for Bioimaging Applications

The cytotoxicity of the probe was tested by the standard MTT method (Supplementary Figure S5). Cells survival remained above 90% under incubation with a concentration of 20 μM Couoxo-LD, indicating that the cytotoxicity of Couoxo-LD was low and exhibited no significant impact on the cells testing. Encouraged by the excellent optical testing of the Couoxo-LD, we then estimated its cells imaging capabilities by laser scanning confocal (LSC) imaging techniques (Figure 6). BODIPY 493/503, a commercially available probe, was used for monitoring LDs as a control. We evaluated the absorption and emission spectra of Couoxo-LD and BODIPY (Supplementary Figure S6). Living HeLa cells were stimulated with oleic acid in order to generate more LDs. Figure 6d demonstrates the co-localization images of both dyes on intracellular lipid droplets. As anticipated, Couoxo-LD labeled intracellular lipid droplets well (Figure 6e,f). Furthermore, the Pearson's correlation coefficient between Couoxo-LD and the commercial dye BODIPY was as high as 0.93. These results suggested that Couoxo-LD possessed a good ability to sense LDs in living HeLa cells.

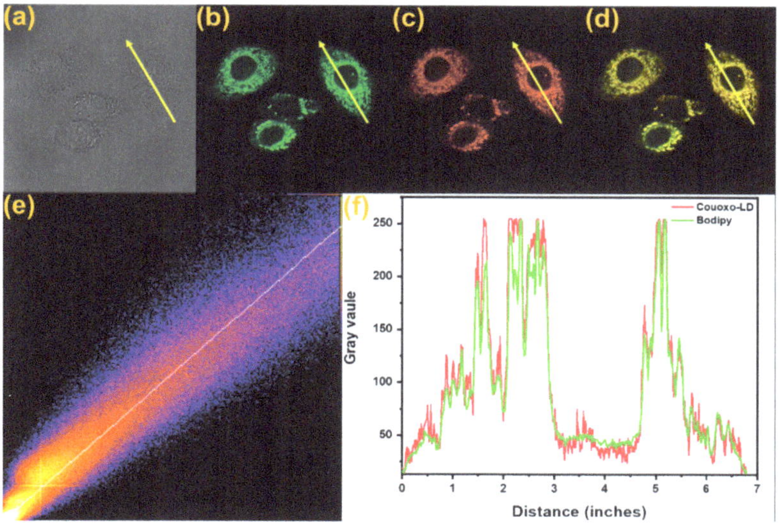

Figure 6. Fluorescence images of live HeLa cells incubated with Couoxo-LD (10 μM) and BODIPY (5 μM) for 30 min at 37 °C. (**a**) Bright-field image. (**b**) Emission of BODIPY (λ_{ex} = 405 nm, 480–510 nm). (**c**) Emission of Couoxo-LD (λ_{ex} = 405 nm, 600–650 nm). (**d**) Merged images. (**e**) The intensity scatter plot of (**b**) and (**c**). (**f**) Fluorescence intensity distribution within the rectangle.

Meanwhile, to investigate the response of the probe to intracellular polarity, we performed imaging tests on the probe. In the control group, HeLa cells were imaged after incubation with culture medium for 30 min. In the experimental group, the cells were further treated with a preconfigured H_2O_2 (500 μM, 20 μL) solution for 20 min on the basis of control cells. The H_2O_2 solution would kill the cells, leading to a decrease in the number of intracellular lipid droplets and resulting in a change in intracellular polarity. At this point, the fluorescence of the green channel diminished from bright green (Figure 7b,f), the fluorescence of the red channel was lighted up (Figure 7c,g), and the images of the combined channels also showed a change from green to orange fluorescence (Figure 7d,h).

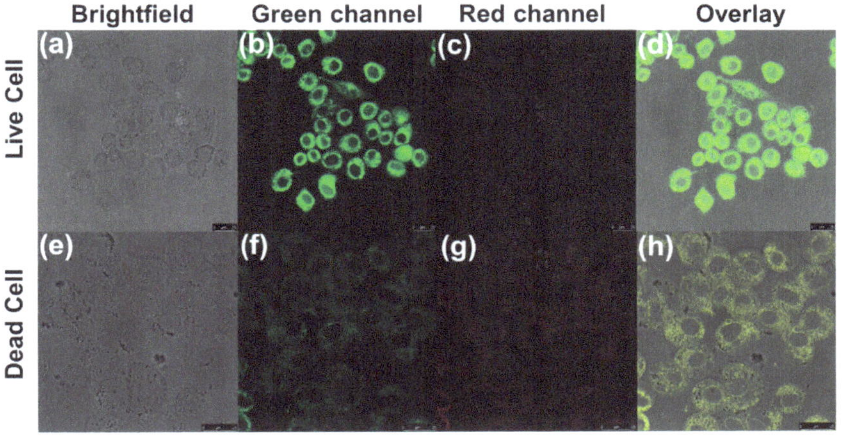

Figure 7. Fluorescent images of living (control) and dead (H_2O_2) HeLa cells treated with Couoxo-LD (10 μM). (**a**,**e**) Bright-field images; (**b**,**f**) Emission of BODIPY (λ_{ex} = 405 nm, λ_{em} 480–510 nm). (**c**,**g**) Couoxo-LD (λ_{ex} = 405 nm, λ_{em} 600–650 nm). (**d**,**h**) Merged images. Scale bar: 500 μm.

Zebrafish have similar digestive systems to humans, such as the liver and intestines. Additionally, their methods of digestion and nutrient absorption transport are highly similar to humans. Therefore, using live zebrafish to model intestinal lesions could help to further study human-related diseases. Approximately 70% of the zebrafish yolk sac fraction is neutral lipid, and by utilizing the lipid-specific fluorescent staining of the zebrafish yolk sac (Figures 8b and S8b), it was clearly observed that both dyes stained the zebrafish yolk sac. This indicates that Couoxo-LD can successfully label lipid droplets. Therefore, Couoxo-LD was informative for further studies on human physiology and pathology caused by abnormal expression of lipid droplets.

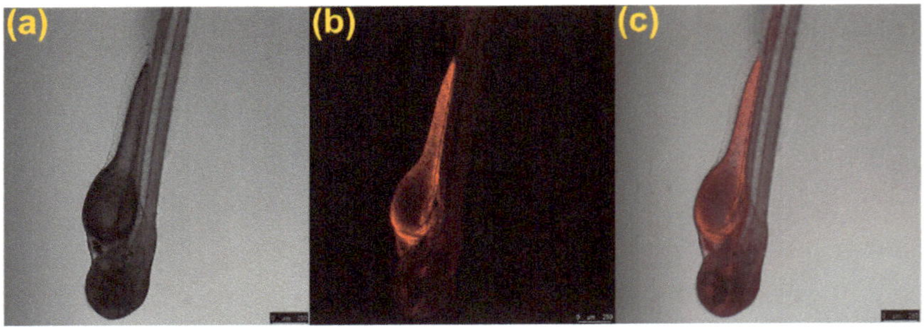

Figure 8. Fluorescent images of living zebrafish treated with Couoxo-LD (10 μM). (**a**) Bright field. (**b**) Red channel, λ_{ex} = 405 nm, λ_{em} = 600–650 nm. (**c**) Merge images. Scale bar: 500 μm.

4. Conclusions

In this work, we designed a polar viscosity-sensitive fluorescent probe for targeting LDs. The probe Couoxo-LD was extremely sensitive to the polarity and viscosity of different media and showed intense fluorescence in LDs. In addition, the probe possessed splendid selectivity, low biotoxicity, and photostability. By artificially altering the external environment, Couoxo-LD could be competently used for discriminating changes in LDs polarity between living and post-mortem HeLa cells. Couoxo-LD was successfully applied for zebrafish imaging. Furthermore, lipid droplets co-localization imaging illustrated the accurate targeting of intracellular lipid droplets by Couoxo-LD. We believe that Couoxo-LD could be a powerful tool to study the processes associated with LDs.

Supplementary Materials: The following supporting information can be downloaded at: https://www.mdpi.com/article/10.3390/bios12100851/s1, Figure S1: ^{1}H NMR ($CDCl_3$) spectrum of Couoxo-LD; Figure S2: ^{13}C NMR ($CDCl_3$) spectrum of Couoxo-LD; Figure S3: normalized (a) absorption and (b) fluorescence spectra of Couxox-LD (10 μM, λ_{ex} = 510 nm) in different solvents. (c,d) are photographs of Couoxo-LD (10 μM) in different solvents, under natural light, and with 365 nm handheld UV lamp irradiation, respectively; Figure S4: maximum emission wavelength of Couoxo-LD (10 μM, λ_{ex} = 510 nm) versus liquid viscosity (PBS buffers–Glycerol) system; Figure S5: cytotoxicity assays of probe Couoxo-LD at different concentrations (0 μM; 2 μM; 5 μM; 10 μM; 20 μM; 30 μM) for HeLa cells; Figure S6: absorption and emission spectra of Couoxo-LD (solid line) and BODIPY (dashed line) in dioxane; Figure S7: fluorescence images of live HeLa cells incubated with different concentrations of Couoxo-LD at 37 °C for 30 min; Figure S8: fluorescence images of live zebrafish treated with BODIPY (5 μM). (a) Bright-field view; (b) green channel, λ_{ex} = 405 nm, λ_{em} = 480–510 nm; (c) merged images. Scale bar. 500 μm.

Author Contributions: M.-H.W.: writing—original draft, methodology, investigation, fluorescence imaging, preparation, conceptualization and writing—review. W.-L.C.: fluorescence imaging, methodology, investigation, writing—original draft, preparation. Y.-H.Y.: fluorescence imaging, methodology, writing—original draft, preparation. J.-Y.W.: conceptualization, writing—review and editing, supervision. All authors have read and agreed to the published version of the manuscript.

Funding: This research was funded by the National Natural Science Foundation of China (21801145, 32000318).

Institutional Review Board Statement: Not applicable.

Informed Consent Statement: Not applicable.

Data Availability Statement: Not applicable.

Conflicts of Interest: The authors declare no conflict of interest.

References

1. Storey, S.M.; McIntosh, A.L.; Senthivinayagam, S.; Moon, K.C.; Atshaves, B.P. The phospholipid monolayer associated with perilipin-enriched lipid droplets is a highly organized rigid membrane structure. *J. Appl. Physiol. Endocrinol. Metab.* **2011**, *301*, E991–E1003. [CrossRef] [PubMed]
2. Krahmer, N.; Farese, R.V.; Walther, T.C. Balancing the fat: Lipid droplets and human disease. *EMBO Mol. Med.* **2013**, *5*, 973–983. [CrossRef] [PubMed]
3. Xiao, H.; Li, P.; Tang, B. Recent progresses in fluorescent probes for detection of polarity. *Coordin. Chem. Rev.* **2021**, *427*, 213582. [CrossRef]
4. Olzmann, J.A.; Richter, C.M.; Kopito, R.R. Spatial regulation of UBXD8 and p97/VCP controls ATGL-mediated lipid droplet turnover. *Proc. Natl. Acad. Sci. USA* **2013**, *110*, 1345–1350. [CrossRef]
5. Onal, G.; Kutlu, O.; Gozuacik, D.; Dokmeci Emre, S. Lipid Droplets in Health and Disease. *Lipids Health Dis.* **2017**, *16*, 128. [CrossRef]
6. Zheng, X.; Zhu, W.; Ni, F.; Ai, H.; Yang, C. A specific bioprobe for super-resolution fluorescence imaging of lipid droplets. *Sens. Actuat. B-Chem.* **2018**, *255*, 3148–3154. [CrossRef]
7. Beloribi-Djefaflia, S.; Vasseur, S.; Guillaumond, F. Lipid metabolic reprogramming in cancer cells. *Oncogenesis* **2016**, *5*, e189. [CrossRef]
8. Qiu, B.; Ackerman, D.; Sanchez, D.J.; Li, B.; Ochocki, J.D.; Grazioli, A.; Bobrovnikova-Marjon, E.; Diehl, J.A.; Keith, B.; Simon, M.C. HIF2α-Dependent Lipid Storage Promotes Endoplasmic Reticulum Homeostasis in Clear-Cell Renal Cell Carcinoma. *Cancer Discov.* **2015**, *5*, 652–667. [CrossRef]
9. Farese, R.V., Jr.; Walther, T.C. Lipid droplets finally get a little R-E-S-P-E-C-T. *Cell* **2009**, *139*, 855–860. [CrossRef]
10. Martin, S.; Parton, R.G. Lipid droplets: A unified view of a dynamic organelle. *Nat. Rev. Mol. Cell Biol.* **2006**, *7*, 373–378. [CrossRef]
11. Chowdhury, R.; Jana, B.; Saha, A.; Ghosh, S.; Bhattacharyya, K. Confocal microscopy of cytoplasmic lipid droplets in a live cancer cell: Number, polarity, diffusion and solvation dynamics. *Med. Chem. Comm.* **2014**, *5*, 536–539. [CrossRef]
12. Chen, S.; Hong, Y.; Zeng, Y.; Sun, Q.; Liu, Y.; Zhao, E.; Bai, G.; Qu, J.; Hao, J.; Tang, B.Z. Mapping live cell viscosity with an aggregation-induced emission fluorogen by means of two-photon fluorescence lifetime imaging. *Chem. Eur. J.* **2015**, *21*, 4315–4320. [CrossRef]
13. Biegas, K.J.; Swarts, B.M. Chemical probes for tagging mycobacterial lipids. *Curr. Opin. Chem. Biol.* **2021**, *65*, 57–65. [CrossRef]
14. Zhan, Z.; Zhuang, W.; Lei, Q.; Li, S.; Mao, W.; Chen, M.; Li, W. A smart probe for simultaneous imaging of the lipid/water microenvironment in atherosclerosis and fatty liver. *Chem. Commun.* **2022**, *58*, 4020–4023. [CrossRef]
15. Yu, C.; Guo, X.; Fang, X.; Chen, N.; Wu, Q.; Hao, E.; Jiao, L. Efficiently emissive, strongly solvatochromic and lipid droplet-specific, fluorescent probes for mapping polarity in vitro. *Dyes Pigments* **2022**, *197*, 109838. [CrossRef]
16. Tan, P.; Zhuang, W.; Li, S.; Zhang, J.; Xu, H.; Yang, L.; Liao, Y.; Chen, M.; Wei, Q. A lipid droplet targeted fluorescent probe for high-efficiency image-guided photodynamic therapy of renal cell carcinoma. *Chem. Commun.* **2021**, *57*, 1046–1049. [CrossRef]
17. Tian, H.; Sedgwick, A.C.; Han, H.H.; Sen, S.; Chen, G.R.; Zang, Y.; Sessler, J.L.; James, T.D.; Li, J.; He, X.P. Fluorescent probes for the imaging of lipid droplets in live cells. *Coord. Chem. Rev.* **2021**, *427*, 213577. [CrossRef]
18. Hu, R.; Chen, B.; Wang, Z.; Qin, A.; Zhao, Z.; Lou, X.; Tang, B.Z. Intriguing "chameleon" fluorescent bioprobes for the visualization of lipid droplet-lysosome interplay. *Biomaterials* **2019**, *203*, 43–51. [CrossRef]
19. Guo, L.; Tian, M.; Zhang, Z.; Lu, Q.; Liu, Z.; Niu, G.; Yu, X. Simultaneous Two-Color Visualization of Lipid Droplets and Endoplasmic Reticulum and Their Interplay by Single Fluorescent Probes in Lambda Mode. *J. Am. Chem. Soc.* **2021**, *143*, 3169–3179. [CrossRef]
20. Ye, M.; Hu, W.; He, M.; Li, C.; Zhai, S.; Liu, Z.; Wang, Y.; Zhang, H.; Li, C. Deep imaging for visualizing nitric oxide in lipid droplets: Discovering the relationship between nitric oxide and resistance to cancer chemotherapy drugs. *Chem. Commun.* **2020**, *56*, 6233–6236. [CrossRef]
21. Cho, M.K.; Seo, M.J.; Juvekar, V.; Jo, J.H.; Kim, W.; Choi, K.S.; Kim, H.M. Screening of Drug-Induced Steatosis and Phospholipidosis Using Lipid Droplet-Selective Two-Photon Probes. *Anal. Chem.* **2020**, *92*, 11223–11231. [CrossRef]
22. Jiang, G.; Jin, Y.; Li, M.; Wang, H.; Xiong, M.; Zeng, W.; Yuan, H.; Liu, C.; Ren, Z.; Liu, C. Faster and More Specific: Excited-State Intramolecular Proton Transfer-Based Dyes for High-Fidelity Dynamic Imaging of Lipid Droplets within Cells and Tissues. *Anal. Chem.* **2020**, *92*, 10342–10349. [CrossRef]
23. Zhang, X.; Yuan, L.; Jiang, J.; Hu, J.; du Rietz, A.; Cao, H.; Zhang, R.; Tian, X.; Zhang, F.; Ma, Y.; et al. Light-Up Lipid Droplets Dynamic Behaviors Using a Red-Emitting Fluorogenic Probe. *Anal. Chem.* **2020**, *92*, 3613–3619. [CrossRef]

24. Wang, K.; Ma, S.; Ma, Y.; Zhao, Y.; Xing, M.; Zhou, L.; Cao, D.; Lin, W. Aurone Derivative Revealing the Metabolism of Lipid Droplets and Monitoring Oxidative Stress in Living Cells. *Anal. Chem.* **2020**, *92*, 6631–6636. [CrossRef] [PubMed]
25. Wang, X.; Qi, F.; Jiang, Z.; Yan, M.; Xu, L. Highly fluorescent bisboron complexes in both solution and solid-state: Synthesis, photophysical properties and lipid droplet imaging in living cells. *Dyes Pigments* **2021**, *186*, 108999. [CrossRef]
26. Ren, W.; Wang, D.; Huang, W.; Li, J.; Tian, X.; Liu, Z.; Han, G.; Liu, B.; Han, M.Y.; Zhang, Z.; et al. Real-time tracking of lipid droplets interactions with other organelles by a high signal/noise probe. *Dyes Pigments* **2021**, *191*, 109366. [CrossRef]
27. Dong, B.; Song, W.; Lu, Y.; Sun, Y.; Lin, W. Revealing the Viscosity Changes in Lipid Droplets during Ferroptosis by the Real-Time and In Situ Near-Infrared Imaging. *ACS Sens.* **2021**, *6*, 22–26. [CrossRef] [PubMed]
28. Collot, M.; Fam, T.K.; Pichandi, A.; Faklaris, O.; Galli, T.; Danglot, L.; Klymchenko, A. Ultrabright and Fluorogenic Probes for Multicolor Imaging and Tracking of Lipid Droplets in Cells and Tissues. *J. Am. Chem. Soc.* **2018**, *140*, 5401–5411. [CrossRef]
29. Luo, J.; Xie, Z.; Lam, J.W.; Cheng, L.; Chen, H.; Qiu, C.; Kwok, H.S.; Zhan, X.; Liu, Y.; Zhu, D.; et al. Aggregation-induced emission of 1-methyl-1,2,3,4,5-pentaphenylsilole. *Chem. Commun.* **2001**, *18*, 1740–1741. [CrossRef]
30. Perkin, W.H., VI. On the artificial production of coumarin and formation of its homologues. *J. Chem. Soc.* **1868**, *21*, 53–63. [CrossRef]
31. Jiang, J.; Tian, X.; Xu, C.; Wang, S.; Feng, Y.; Chen, M.; Yu, H.; Zhu, M.; Meng, X. A two-photon fluorescent probe for real-time monitoring of autophagy by ultrasensitive detection of the change in lysosomal polarity. *Chem. Commun.* **2017**, *53*, 3645–3648. [CrossRef]
32. Bai, Y.; Shi, X.; Chen, Y.; Zhu, C.; Jiao, Y.; Han, Z.; He, W.; Guo, Z. Coumarin/BODIPY Hybridisation for Ratiometric Sensing of Intracellular Polarity Oscillation. *Chem. Eur. J.* **2018**, *24*, 7513–7524. [CrossRef]
33. Jimenez-Sanchez, A.; Lei, E.K.; Kelley, S.O. A Multifunctional Chemical Probe for the Measurement of Local Micropolarity and Microviscosity in Mitochondria. *Angew. Chem. Int. Ed.* **2018**, *57*, 8891–8895. [CrossRef]
34. Kahveci, B.; Menteşe, E. Microwave Assisted Synthesis of Coumarins: A Review From 2007 to 2018. *Curr. Micro. Chem.* **2019**, *5*, 162–178. [CrossRef]
35. Long, L.; Huang, M.; Wang, N.; Wu, Y.; Wang, K.; Gong, A.; Zhang, Z.; Sessler, J.L. A Mitochondria-Specific Fluorescent Probe for Visualizing Endogenous Hydrogen Cyanide Fluctuations in Neurons. *J. Am. Chem. Soc.* **2018**, *140*, 1870–1875. [CrossRef]
36. Chen, X.; Wang, H.; Ma, X.; Wang, M.; Zhang, Y.; Gao, G.; Liu, J.; Hou, S. Colorimetric and fluorescent probe for real-time detection of palladium (II) ion in aqueous medium and live cell imaging. *Dyes Pigments* **2018**, *148*, 286–291. [CrossRef]
37. Ding, Y.; Zhao, S.; Wang, Q.; Yu, X.; Zhang, W. Construction of a coumarin based fluorescent sensing platform for palladium and hydrazine detection. *Sens. Actuat. B-Chem.* **2018**, *256*, 1107–1113. [CrossRef]
38. Chen, W.; Pacheco, A.; Takano, Y.; Day, J.J.; Hanaoka, K.; Xian, M. A Single Fluorescent Probe to Visualize Hydrogen Sulfide and Hydrogen Polysulfides with Different Fluorescence Signals. *Angew. Chem. Int. Ed.* **2016**, *55*, 9993–9996. [CrossRef]
39. Yang, L.; Su, Y.; Geng, Y.; Zhang, Y.; Ren, X.; He, L.; Song, X. A Triple-Emission Fluorescent Probe for Discriminatory Detection of Cysteine/Homocysteine, Glutathione/Hydrogen Sulfide, and Thiophenol in Living Cells. *ACS Sens.* **2018**, *3*, 1863–1869. [CrossRef]
40. Kwon, N.; Hu, Y.; Yoon, J. Fluorescent Chemosensors for Various Analytes Including Reactive Oxygen Species, Biothiol, Metal Ions, and Toxic Gases. *ACS Omega* **2018**, *3*, 13731–13751. [CrossRef]
41. Wang, K.N.; Liu, L.Y.; Mao, D.; Xu, S.; Tan, C.P.; Cao, Q.; Mao, Z.W.; Liu, B. A Polarity-Sensitive Ratiometric Fluorescence Probe for Monitoring Changes in Lipid Droplets and Nucleus during Ferroptosis. *Angew. Chem. Int. Ed.* **2021**, *60*, 15095–15100. [CrossRef]
42. Cui, W.L.; Wang, M.H.; Chen, X.Q.; Zhang, Z.H.; Qu, J.; Wang, J.Y. A novel polarity-sensitive fluorescent probe for lighting up lipid droplets and its application in discriminating dead and living zebrafish. *Dyes Pigments* **2022**, *204*, 110433. [CrossRef]
43. Punna Rao, A.M.L.; Sridhar Rao, A.; Saratchandra Babu, M.; Krishnaji Rao, M. Triphenylphosphine (PPh3) Catalyzed Erlenmeyer Reaction for Azlactones under Solvent-free Conditions. *J. Heterocyclic. Chem.* **2017**, *54*, 429–435. [CrossRef]
44. Sasaki, S.; Drummen, G.P.C.; Konishi, G.i. Recent advances in twisted intramolecular charge transfer (TICT) fluorescence and related phenomena in materials chemistry. *J. Mater. Chem. C* **2016**, *4*, 2731–2743. [CrossRef]
45. Wu, C.J.; Li, X.Y.; Zhu, T.; Zhao, M.; Song, Z.; Li, S.; Shan, G.G.; Niu, G. Exploiting the Twisted Intramolecular Charge Transfer Effect to Construct a Wash-Free Solvatochromic Fluorescent Lipid Droplet Probe for Fatty Liver Disease Diagnosis. *Anal. Chem.* **2022**, *94*, 3881–3887. [CrossRef]
46. Marwani, H.M.; Asiri, A.M.; Khana, S.A. Spectral, stoichiometric ratio, physicochemical, polarity and photostability studies of newly synthesized chalcone dye in organized media. *J. Lumin.* **2013**, *136*, 296–302. [CrossRef]

biosensors

MDPI

Article

Voltammetric Determination of 3-Methylmorphine Using Glassy Carbon Electrode Modified with rGO and Bismuth Film

Ademar Wong [1], Anderson M. Santos [2], Camila A. Proença [2], Thaísa A. Baldo [2], Maria H. A. Feitosa [2], Fernando C. Moraes [2,*] and Maria D. P. T. Sotomayor [1]

[1] Institute of Chemistry, State University of São Paulo (UNESP), Araraquara 14801-970, Brazil
[2] Department of Chemistry, Federal University of São Carlos (UFSCar), São Carlos 13560-970, Brazil
* Correspondence: fcmoraes@ufscar.br

Abstract: This work reports the development and application of a simple, rapid and low-cost voltammetric method for the determination of 3-methylmorphine at nanomolar levels in clinical and environmental samples. The proposed method involves the combined application of a glassy carbon electrode modified with reduced graphene oxide, chitosan and bismuth film (Bi-rGO-CTS/GCE) via square-wave voltammetry using 0.04 mol L^{-1} Britton–Robinson buffer solution (pH 4.0). The application of the technique yielded low limit of detection of 24×10^{-9} mol L^{-1} and linear concentration range of 2.5×10^{-7} to 8.2×10^{-6} mol L^{-1}. The Bi-rGO-CTS/GCE sensor was successfully applied for the detection of 3-methylmorphine in the presence of other compounds, including paracetamol and caffeine. The results obtained also showed that the application of the sensor for 3-methylmorphine detection did not experience any significant interference in the presence of silicon dioxide, povidone, cellulose, magnesium stearate, urea, ascorbic acid, humic acid and croscarmellose. The applicability of the Bi-rGO-CTS/GCE sensor for the detection of 3-methylmorphine was evaluated using synthetic urine, serum, and river water samples through addition and recovery tests, and the results obtained were found to be similar to those obtained for the high-performance liquid chromatography method (HPLC)—used as a reference method. The findings of this study show that the proposed voltammetric method is a simple, fast and highly efficient alternative technique for the detection of 3-methylmorphine in both biological and environmental samples.

Keywords: reduced graphene oxide; bismuth; 3-methylmorphine; electrochemical sensor; electrodeposition

Citation: Wong, A.; Santos, A.M.; Proença, C.A.; Baldo, T.A.; Feitosa, M.H.A.; Moraes, F.C.; Sotomayor, M.D.P.T. Voltammetric Determination of 3-Methylmorphine Using Glassy Carbon Electrode Modified with rGO and Bismuth Film. *Biosensors* **2022**, *12*, 860. https://doi.org/10.3390/bios12100860

Received: 21 September 2022
Accepted: 9 October 2022
Published: 12 October 2022

1. Introduction

Over the past few decades, researchers from different fields have shown an increasingly growing interest in the determination of drugs present in clinical, environmental and pharmaceutical samples with a view to tackling the problems related to drug intoxication and wastewater contamination as well as ensuring the quality control of pharmaceutical products [1]. This has resulted in the development of several techniques with accurate tools targeted at the qualitative (simple yes/no analysis) and/or quantitative determination of drugs at trace levels in different samples (including urine, serum and water).

One of the drugs most commonly consumed for the treatment of pains is 3-methylmorphine; this opioid analgesic, which is also referred to as 3-methylmorphine (codeine), is obtained from the methylation of morphine derived from Papaver somniferum (poppy seeds) [2]. 3-methylmorphine has a sedative effect which helps to manage mild to moderate pain. Despite its proven efficiency in pain alleviation, 3-methylmorphine has some profoundly disturbing side effects; one of these side effects is that it can affect the central nervous system, causing mood swings. The analgesic effect of 3-methylmorphine can be potentiated when it is applied in combination with acetaminophen and caffeine [3].

Owing to its widespread application and rampant disposal in the environment, several studies reported in the literature have employed a wide range of analytical techniques for the determination/detection of 3-methylmorphine in different matrices; the techniques employed have included the following: colorimetry [4,5], spectrophotometry [6,7], chromatography [8,9], capillary electromigration [10,11], and electrochemical techniques [12,13]. Significant improvements have been increasingly made in the design of these techniques in order to ensure accurate quantification of drug species in complex matrices so as to avoid false positive or negative signals due to the high concentration levels of interferents in these matrices.

The use of electrochemical sensors for the detection of compounds of interest in complex matrices has attracted considerable interest among researchers because of their operational simplicity, sensibility and high degree of accuracy; in addition, the application of these sensors with rapid detection techniques such as square-wave voltammetry (SWV) allows one to identify the analytes of interest within an extremely short period of time (for instance, in seconds) [14]. The sensibility and selectivity of electrochemical sensors can be improved by modifying the devices with nanomaterials, such as reduced graphene oxide (rGO); this contributes toward enhancing the mechanical resistance of the sensors, as well as their thermal and electrical conductivity [15,16]. Furthermore, the use of suitable hybrid materials for the modification of working electrodes may help to improve the properties of the sensor. Bismuth film-based electrodes have been widely used in electroanalysis for the determination of analytes such as metals, pesticides and pharmaceutical products. Bismuth film can be used in combination with rGO in order to promote efficient synergistic effects which can effectively improve the sensor properties, leading to low residual current, good chemical stability, and high mechanical stability, apart from allowing the formation of "fused alloys" with heavy metals [17].

In the present work, we report the development and application of an electrochemical sensor, which was constructed using glassy carbon electrode (GCE) modified with rGO and bismuth film, for the determination of 3-methylmorphine. The clinical and environmental applicability of the sensor was also evaluated through the application of the device for the detection of 3-methylmorphine in synthetic urine, serum, and river water samples; the results obtained were then compared with those of HPLC used as a reference method.

2. Materials and Methods

2.1. Reagents and Apparatus

All the reagents (purity $\geq$ 98%) were used as purchased. 3-methylmorphine, ascorbic acid, caffeine, paracetamol, bismuth (Bi), chitosan (CTS), glutaraldehyde (Glu), KCl and bovine serum were purchased from Sigma-Aldrich (St. Louis, MO, USA). NaOH, $Na_2[B_4O_5(OH)_4$, (H_2SO_4), CH_3COOH and H_3BO_3 were acquired from Synth (São Paulo, Brazil). Graphene was acquired from the Graphene Supermarket (New York, NY, USA). All solutions were prepared using ultrapure water (with resistivity $\geq$ 18.0 MΩ cm) acquired from a Milli-Q system (Millipore®). Stock solution of 0.01 mol L^{-1} 3-methylmorphine was dissolved in ultrapure water.

The electrochemical experiments were performed using Autolab PGSTAT 302N potentiostat (Herisau, Switzerland) controlled by the NOVA 2.1 software and equipped with a conventional electrochemical cell of 10.0 mL volume which consisted of the following three electrodes: Ag/AgCl (3.0 mol L^{-1} KCl) used as reference electrode (RE); platinum coil employed as counter electrode (CE); and GCE used as working electrode (WE) (r = 0.15 cm).

Morphological characterization of the materials was performed using images acquired from Supra 35-VP microscope (Oberkochen, Germany) and the confocal optical microscope (Olympus, LEXT OLS 4000) controlled via the 2.2.6 software (CA, USA).

Fourier transform infrared spectroscopy analyses of the rGO and bismuth film were performed using FTIR–Vertex 70 spectrometer (Bruker, Karlsruhe, Germany) with spectral range of 4000 to 400 cm^{-1}.

The chromatographic analysis was conducted using Shimadzu model 10ATvp LC (Kyoto, Japan) system which consisted of two pumps (LC–10AT), column oven (CTO10A), and UV detector (SPD–10A). The chromatographic experiments were performed under the following conditions—mobile phase: 1% phosphoric acid (*v/v*) and acetonitrile in the ratio of 85:15 (*v/v*)); temperature: 30 °C; flow rate: 1.0 mL min^{-1}; detection wavelength: 220 nm [18].

2.2. Preparation of rGO and Modification of GCE with rGO and Bismuth Film

Initially, graphene oxide (GO) was synthesized, as described in the literature [19], using a mixture of H_2SO_4/HNO_3 concentrated in the proportion of 3:1 (*v/v*). Thus, an amount of 200 mg of graphene was mixed in the acidic solution, and the mixture was kept under stirring for 12 h at 25 °C. Subsequently, the GO suspension obtained was then filtered and washed with ultrapure water until a pH close to 7.0 was obtained. After that, the mixture was dried in an oven at 100 °C. A mass of 50 mg GO was diluted in ultrapure water (50 mL) to form a suspension of 1 mg/mL. Thereafter, 1.0 g $NaBH_4$ and 0.5 g $CaCl_2$ were added to a 50 mL GO suspension. The mixture was kept under stirring at 25 °C for 12 h to obtain the rGO. The solid material (rGO) obtained was filtered and washed with ultrapure water [20,21].

In the next step, the surface of the GCE was cleaned by polishing with alumina slurry (1.0 μm) on a polishing cloth. Subsequently, the GCE was subjected to ultrasonic bath with deionized water for 1 min. The Bi-rGO-CTS/GCE dispersion was prepared using 2.0 mg of rGO, 500 μL of CTS (0.1 % *v/v*), 500 μL of Glu (0.2 % *v/v*), and 1.0 mL of 0.01 mol L^{-1} sodium hydroxide. The material was subjected to ultrasonic agitation for 25 min. After that, 9.0 μL of the homogeneous dispersion was added on the electrode surface and was left to dry (25 °C) for 2 h. To prepare the Bi-rGO-CTS/GCE sensor, the bismuth film was electrochemically deposited on the electrode surface using an applied potential of −0.18 V vs. Ag/AgCl (3.0 mol L^{-1} KCl) for 200 s and a solution containing 0.02 mol L^{-1} $Bi(NO_3)_3$, 0.15 mol L^{-1} sodium citrate and 1.00 mol L^{-1} HCl [17].

2.3. Preparation of Synthetic Urine, Bovine Serum and River Water Samples

The synthetic urine sample was prepared based on the work conducted by Laube et al. [22] using 49, 10, 20, 15, 18, and 18 mmol L^{-1} of NaCl, $CaCl_2$, KCl, KH_2PO_4, NH_4Cl and urea, respectively. The serum samples were prepared using commercial bovine calf serum obtained from Sigma-Aldrich. River water samples were collected from the Jacaré-Guaçu river in the city of Araraquara—Sao Paulo, Brazil. The collection sites were recorded by GPS (21°51′23.3″ S 48°19′49.8″ W). The samples were subjected to conventional filtration to remove any solid material and were then stored in a 100.0 mL flask and kept in the refrigerator at 0 °C. The samples were enriched with 3-methylmorphine concentration. A dilution of 100 times was used in the electrochemical cell.

2.4. Analytical Procedure

First, the morphological and electrochemical characterizations of the electrodes were performed by scanning electron microscopy (SEM) and energy-dispersive X-ray (EDX).

The electrochemical oxidation of 3-methylmorphine was evaluated by cyclic voltammetry (CV) and the optimization of the experimental conditions was conducted using square-wave voltammetry (SWV). Under optimal conditions, analytical curves were constructed with the successive additions of 3-methylmorphine. The limit of detection (LOD) was calculated based on the following equation: $3 \times SD/m$, where "*SD*" stands for standard deviation for 10 blank solutions ($n = 10$) and "*m*" represents the slope. The precision and selectivity of the proposed method based on the application of the Bi-rGO-CTS/GCE sensor were evaluated through repeatability studies and analysis of potential interference. The proposed sensor (Bi-rGO-CTS/GCE) was applied for the detection of 3-methylmorphine in biological (bovine serum and urine synthetic) and environmental (river water) samples. Finally, the results obtained were compared with those of high-performance liquid chromatography (reference method).

3. Results and Discussion

3.1. Material Characterization

The morphology of the materials (rGO and Bi-rGO film) was characterized by scanning electron microscopy (SEM). The SEM image of the rGO material was characterized by agglomerated and thin "sheets" of different sizes (Figure 1A). Figure 1B,C show the SEM image of Bi on reduced rGO under different magnifications. As can be noted, bismuth characterized by loose leaf structure with well-defined morphology can be found uniformly distributed on both surfaces of the rGO sheets; this is clearly indicative of the successful formation of bismuth on the rGO material.

Figure 1. SEM images of reduced graphene oxide (**A**); bismuth on reduced graphene oxide (**B**,**C**); and EDX (**D**).

The EDX profile of the Bi-rGO material (Figure 1C) exhibited signals of carbon, oxygen and bismuth atoms; this pointed to the successful mixture of the rGO nanosheets with the Bismuth film.

The confocal optical microscope (LEXT OLS 4000) was used to obtain the images of the materials in 3D imaging format with great definition. Looking at Figure 2, one can clearly see a significant difference in the topographic images of the materials incorporated onto the surface of the proposed electrode. The thickness of the films was calculated using the Olympus software. The average values obtained for the thickness of the films were as follows: 3.1 ± 0.2 µm for rGO-CTS (Figure 2A) and 5.7 ± 0.2 µm for Bi-rGO-CTS (Figure 2B) (n = 5).

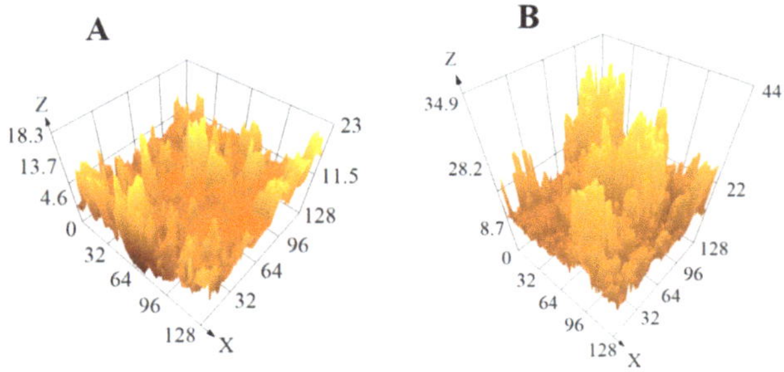

Figure 2. Topographic image of the (**A**) rGO-CTS and (**B**) Bi-rGO-CTS film surface obtained from confocal optical microscope (100 times of magnification). X and Z axes (scale, µm).

3.2. Analytical Response

Cyclic voltammetry plots were used to monitor the electrochemical profile of the modified electrodes. Figure 3 shows the electrochemical response (potential range: 0.5 V to 1.25 V, and ν = 50 mV s^{-1}) of the following electrodes: GCE, rGO-CTS/GCE and Bi-rGO-CTS/GCE in the presence of 3-methylmorphine and without the addition of 3-methylmorphine (inset).

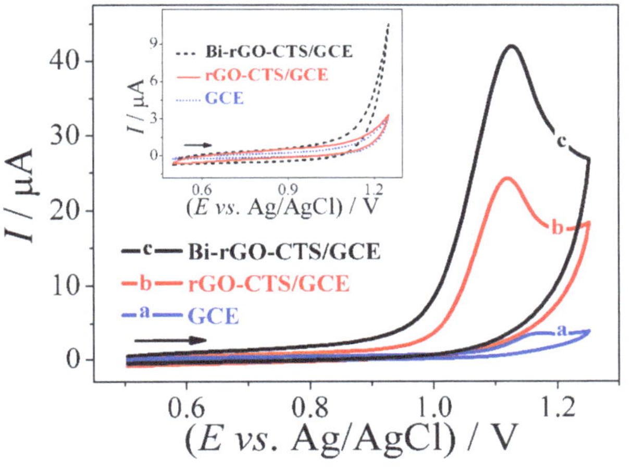

Figure 3. Electrochemical profiles of GCE, rGO-CTS/GCE and Bi-rGO-CTS/GCE. Analysis conditions-electrolyte solution: 0.04 mol L^{-1} Britton-Robinson buffer solution (pH 4); scan rate (ν): 50 mV s^{-1}; 3-methylmorphine concentration: 0.10 mmol L^{-1}. Inset: without the addition of 3-methylmorphine.

Figure 3a shows that the bare GCE did not exhibit any significant electrochemical response (voltammetric response) in the presence of 0.10 mmol L^{-1} standard solution of 3-methylmorphine. Figure 3b,c show the electrochemical profiles obtained for rGO-CTS/GCE and Bi-rGO-CTS/GCE, respectively, in the presence of a solution containing 3-methylmorphine; as can be observed, both electrodes exhibited high electrochemical response with well-defined irreversible oxidation peaks at +1.1 V, corresponding to the oxidation of 3-methylmorphine on the surface of the electrodes. The current values recorded for the bare GCE, rGO-CTS/GCE and Bi-rGO-CTS/GCE were 2.6 μA, 22 μA, and 39 μA, respectively. This result clearly points to the significant role played by Bi and rGO in the direct electrochemical detection of 3-methylmorphine. The analysis of the parameters mentioned above helped confirm that the modification of the electrode with Bi film led to an increase in both the surface area and the faradaic current.

Electrochemical studies were carried out in order to evaluate the electrochemical properties of the electrodes prepared in this study; with the aid of the electrochemical probe $[Fe(CN)_6]^{3-/4-}$ and through the application of cyclic voltammetry (CV), we were able to analyze the reversibility of the redox behavior, the change in conductivity, and the active surface area of the electrodes-see Figure 4. The scan rate of the bare GCE (Figure 4A), rGO-CTS/GCE (Figure 4B) and Bi-rGO-CTS/GCE (Figure 4C) was evaluated using 0.1 mol L^{-1} KCl as electrolyte solution with the redox probe $[Fe(CN)_6]^{3-/4-}$ (1.0 mmol L^{-1}) at scan rate of 10 to 500 mV s^{-1}.

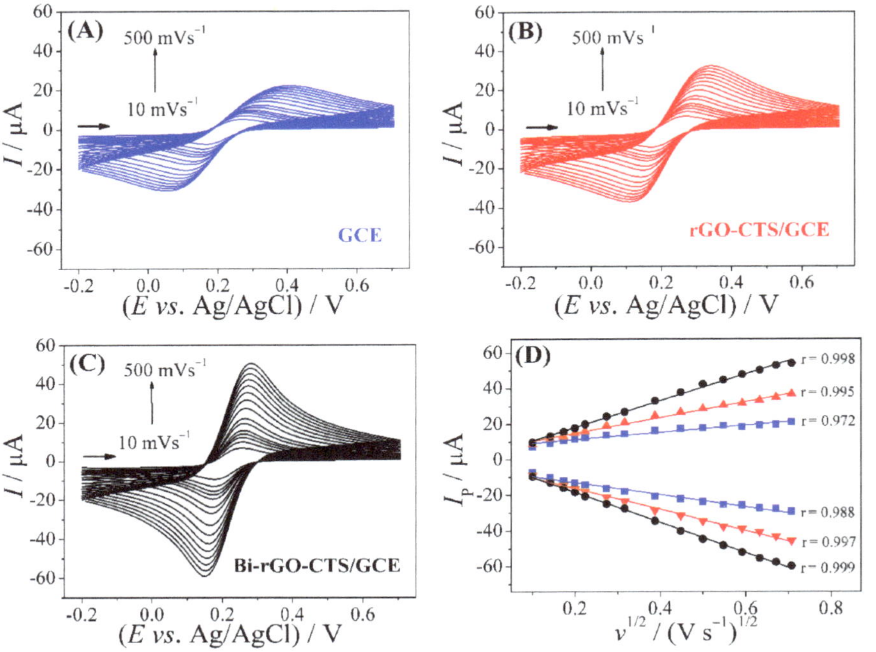

Figure 4. Analysis of the scan rate of the (**A**) GCE, (**B**) rGO-CCTS/GCE and (**C**) Bi-rGO-CTS/GCE electrodes using 0.1 mol L^{-1} KCl solution in the presence of redox probe $[Fe(CN)_6]^{3-/4-}$ (1.0 mmol L^{-1}) at different potential scan rates (10–500 mV s^{-1}). (**D**) I_p vs. $\nu^{1/2}$.

The Randles–Sevcik equation (Equation (1)) was used to calculate the active surface area of the sensors. The CVs obtained exhibited linear relationships between the anodic peak current (I_p) and the square root of the scan rate ($\nu^{1/2}$).

$$I_p = (2.69 \times 10^5)\, ACD^{1/2}n^{3/2}\nu^{1/2} \quad (1)$$

In Equation (1), I_p is the peak current (A), A is the electroactive area (cm^2), n is the number of electrons transferred, D is the diffusion coefficient of $[Fe(CN)_6]^{3-/4-}$ in the 0.10 mol L^{-1} KCl solution (7.6×10^{-6} cm^2 s^{-1}), ν is the potential scan rate (V s^{-1}), and C is the $[Fe(CN)_6]^{3-}$ concentration (mol cm^{-3}).

The surface area of the electrodes was calculated from the slope of I_p vs. $\nu^{1/2}$ (Figure 4D). The active surface area obtained for the bare GCE, rGO-CTS/GCE and Bi-rGO-CTS/GCE sensor was 0.041 cm^2, 0.077 cm^2 and 0.119 cm^2, respectively.

The heterogeneous electron transfer rate constant (k^0) was used to evaluate how the charge transfer kinetics is affected by the sensor modification. The rate constant values were found using the Nicholson Equation (2) [23].

$$\Psi = k^0\, [\pi D n \nu F/(RT)]^{-1/2} \quad (2)$$

In Equation (2), Ψ is the kinetic parameter, π = 3.1415, T = 298 K, and F and R are the Faraday (96,485 C mol^{-1}) and universal gas (8.314 J K^{-1} mol^{-1}) constants. The Ψ values were obtained as proposed by Lavagnini et al. [24], according to Equation (3), which correlates Ψ with ΔE_p.

$$\Psi = (-0.6288 + 0.0021\, n\Delta E_p)/(1 - 0.017\, n\Delta E_p) \quad (3)$$

With regard to the above equation, the slope of the equation of the line directly represents the value (obtained) of k^0, which in this case was equal to 1.7×10^{-3} cm s^{-1} for GCE, 3.8×10^{-3} cm s^{-1} for rGO-CTS/GCE, and 8.9×10^{-3} cm s^{-1} for Bi-rGO-CTS/GCE. Thus, the k^0 value obtained for Bi-rGO-CTS/GCE was about 5.2 times greater than that obtained for GCE. This result points to an improvement in the electronic transfer speed for Bi-rGO-CTS/GCE compared to the other electrodes.

A thorough analysis was also conducted in order to evaluate the influence of scan rate on the analytical response using 50.0 μmol L^{-1} 3-methylmorphine in phosphate-buffered solution (pH 6.0) and scan rates ranging from 10 to 400 mV s^{-1} (Figure 5A). The anodic peak currents (I_{pa}) of 3-methylmorphine in the phosphate-buffered solution (pH 6.0) shifted to more positive values as the scan rate increased; this is clearly indicative of an irreversible electrochemical reaction. The I_{pa} vs. $\nu^{1/2}$ plot was found to be linear (Figure 5B); this shows that the mass transport of the 3-methylmorphine on the electrode surface occurred by diffusion. Plots of log I_{pa} vs. log ν (Figure 5C) exhibited a slope of 0.60, which is typically characteristic of systems controlled by diffusion.

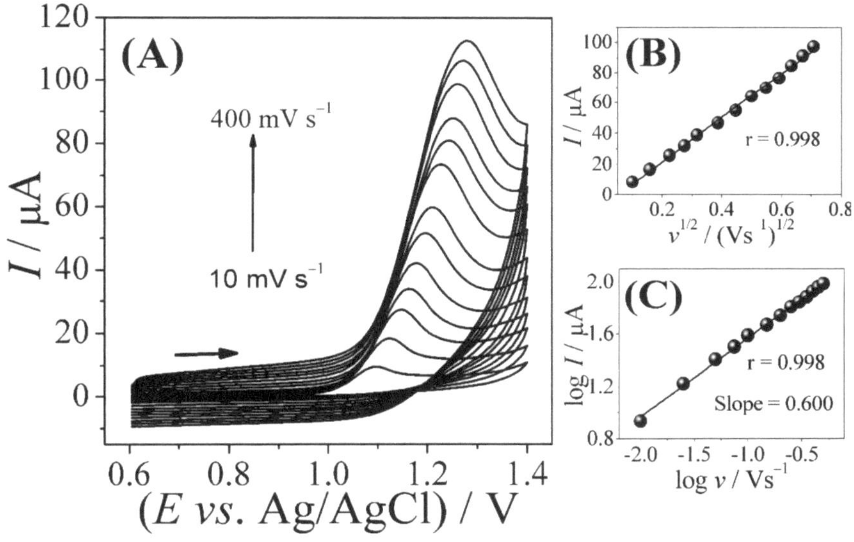

Figure 5. (**A**) Cyclic voltammograms at different scan rates (10–400 mV s^{-1}) based on the application of the Bi-rGO-CTS/GCE sensor in 0.20 mol L^{-1} phosphate-buffered solution (pH 6.0) in the presence of 50.0 μmol L^{-1} concentration of 3-methylmorphine. Plots: (**B**) I_{pa} vs. $\nu^{1/2}$ and (**C**) log I_{pa} vs. logν.

Using the cyclic voltammograms obtained at different potential scan rates, we were able to calculate the average value of $E_p - E_{1/2}$ (see Equation (4)). By inserting the value obtained (64 mV), we obtained the relation αn = 0.75. Considering α = 0.50 (value normally attributed to irreversible processes), we estimated the number of electrons (n) involved in the 3-methylmorphine irreversible oxidation on the electrode surface. The value of (n) was found to be approximately 1.5 (close to 2); this value is similar to the values obtained in other studies reported in the literature [25].

$$\left|E_p - E_{1/2}\right| = \frac{48}{\alpha n} \tag{4}$$

In Equation (4), E_p stands for peak potential, and $E_{1/2}$ is the half-wave potential obtained from the cyclic voltammetry graph (Figure 5).

The analysis of the influence of pH on the peak potential and the peak current related to 3-methylmorphine oxidation was conducted by square-wave voltammetry using 0.04 mol L^{-1} Britton–Robinson buffer solution (pH range of 2.0–7.0) as supporting electrolyte. As can be observed in Figure 6, a decrease in hydrogen ionic concentration of the electrolyte was found to cause a shift in peak potential for 3-methylmorphine oxidation toward less positive values; this outcome is attributed to deprotonation during the oxidation process, which is facilitated at higher pH values.

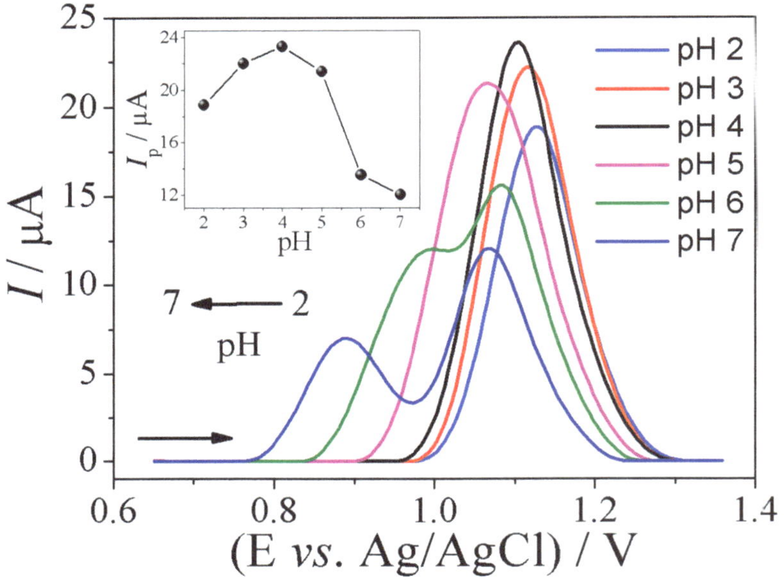

Figure 6. Results obtained from SWV analysis conducted based on the application of the Bi-rGO-CTS/GCE sensor in 0.04 mol L^{-1} Britton–Robinson buffer solution containing 5.0 µmol L^{-1} of 3-methylmorphine at pH ranging from 2.0 to 7.0. Parameters: f = 15 Hz; a = 50 mV; ΔE_s = 5 mV. Inset: I_p vs. pH plots.

Additionally, as can be verified, a single anodic peak was observed in the pH range of 2.0 to 5.0, and two anodic peaks were detected at pH = 6.0 and 7.0. Where peak one can be associated with the oxidation of the 6-hydroxy group with loss of 2 electrons and 1 proton, and peak two (varying of 0.95 V and 0.88 V) can be associated with the oxidation of the tertiary amine group involving the loss of 2 electrons and 2 protons [26]. For illustration purposes, Scheme 1 shows the electrochemical processes, which correspond to the oxidation processes of 3-methylmorphine [26].

The I_{pa} vs. pH plots (Figure 6 inset) for 3-methylmorphine shows an increase in the anodic peak current in the pH range of 2.0 to 4.0, with maximum peak current recorded at pH 4.0. The maximum value of the oxidation peak current for 3-methylmorphine was observed at pH 4.0; the value was found to decrease after that. In addition, an increase in pH from 2.0 to 7.0 led to a gradual shift in the oxidation peak potential to negative potential; this evidently showed that the electrochemical process was dependent on pH. Based on a comparison of the peak currents of 3-methylmorphine at different pH values, pH 4.0 was selected as the optimal pH level for the conduct of further experiments.

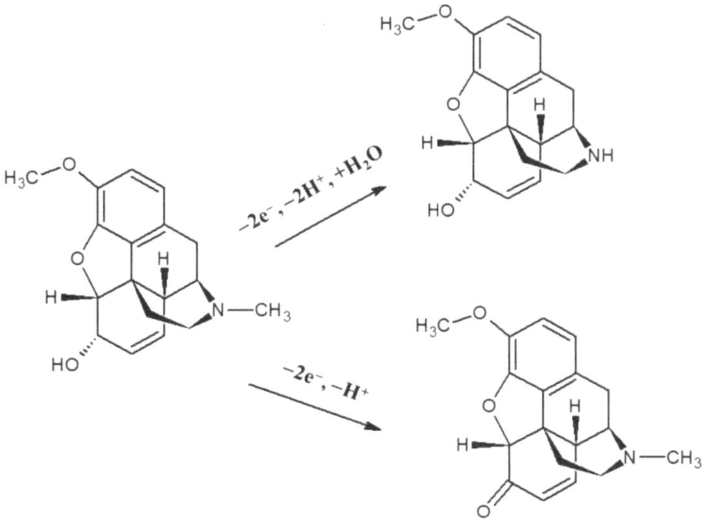

Scheme 1. Proposed mechanism for the electro-oxidation of 3-methylmorphine [26].

3.3. Calibration Curve

Under optimized experimental conditions, an increase in 3-methylmorphine concentration was found to lead to an increase in peak current (Figure 7). Figure 7 shows the SW voltammograms obtained from the application of Bi-rGO-CTS/GCE for the determination of 3-methylmorphine at concentrations ranging from 2.5×10^{-7} to 8.2×10^{-6} mol L^{-1} in 0.04 mol L^{-1} Britton–Robinson buffer solution (pH 4.0). The inset in Figure 7 shows the calibration plots that correspond to 3-methylmorphine concentrations. As can be observed, the calibration curve between peak current and 3-methylmorphine concentration exhibits a linear relationship (in the linear range) based on the following equation: I_p (μA) = 0.1 + 3.4 [3-methylmorphine], with correlation coefficient of 0.998. The LOD value obtained from the application of the proposed Bi-rGO-CTS/GCE was 24×10^{-9} mol L^{-1} (in triplicate experiments). Table 1 presents a comparative analysis of the LOD value obtained in this study and the values obtained in other studies reported in the literature related to the application of electrochemical sensors. As can be noted, the LOD value obtained in this study was satisfactorily comparable and, in some cases, better than the values reported in the literature; essentially, this points to the efficiency and viability of the method proposed in this study.

Table 1. Comparison of analytical parameters in the determination of 3-methylmorphine on different modified electrodes.

Electrode	Method	Linear Range (μmol L^{-1})	LOD (μmol L^{-1})	Sensitivity (μA mol^{-1} L)	Presence	Ref.
CD-HPC/CBPE [1]	SWV	0.5–38	0.0095	0.047	Dipyrone	[13]
AChE-TTF-SPCEs [2]	Chrono-amperometry	20–200	0.02	No	-	[27]
Zn_2SnO_4—GO/CPE [3]	DPV	0.02–15	0.009	2.938	Morphine	[28]
ND-DHP/GCE [4]	SWV	0.299–10.8	0.0545	0.0745	-	[29]
BDD [5]	DPV	0.1–60	0.08	0.155	-	[30]
HTP-MWCNT/CPE [6]	DPV	0.2–34.1 34.1–844.7	0.063	0.0288 0.009	Acetaminophen	[31]
CYP2D6-SPCEs [7]	Chrono-amperometry	5.0–35.0	4.9	-	-	[32]
Bi-rGO-CTS/GCE	SWV	0.25–8.2	0.02	3.5	Paracetamol and caffeine	**This work**

[1] Carbon black paste electrode modified with α-cyclodextrin and hierarchical porous carbon. [2] Tetrathiafulvalene/AChE- modified screen-printed carbon. [3] Carbon paste electrode (CPE) modified with Zn_2SnO_4-graphene nanocomposite. [4] Nanodiamond/dihexadecyl phosphate-modified glassy carbon. [5] Boron-doped diamond film electrode. [6] 4-hydroxy-2-(triphenylphosphonio)phenolate (HTP) and multiwall carbon nanotubes paste electrode. [7] CYP2D6-modified screen-printed carbon.

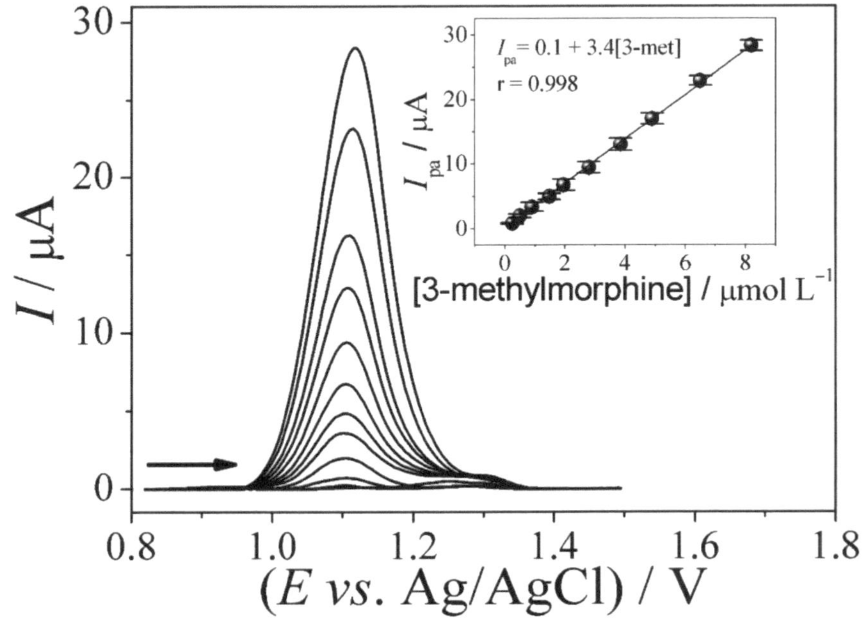

Figure 7. SW voltammograms obtained from the application of Bi-rGO-CTS/GCE for the determination of 3-methylmorphine (concentrations ranging from 2.5×10^{-7} to 8.2×10^{-6} mol L^{-1}) in 0.04 mol L^{-1} Britton–Robinson buffer solution (pH 4.0). Inset: Corresponding calibration plots related to 3-methylmorphine concentrations. Parameters: f = 15 Hz; a = 50 mV; ΔE_s = 5 mV.

To evaluate the selectivity of the Bi-rGO-CTS/GCE sensor in terms of 3-methylmorphine determination, a thorough analysis was performed in order to analyze the influence of some common interfering species. Under optimized conditions, the SWV experiments were performed in the potential range of 0.35 to +1.7 V vs. Ag/AgCl (3.0 mol L^{-1} KCl) using 0.1 mol L^{-1} phosphate buffer (pH 6.0), with sequential addition of 3-methylmorphine concentration in the range of 2.5×10^{-7} to 8.5×10^{-6} mol L^{-1}. The results obtained from the SWV experiments are shown in Figure 8. In all the SWV experiments, no overlap was observed between the 3-methylmorphine oxidation peak and the oxidation peaks of the interfering species. Based on the application of the proposed Bi-rGO-CTS/GCE sensor, the oxidation peak potentials obtained for paracetamol, 3-methylmorphine and caffeine were +0.50, +1.1, and +1.4 V, respectively.

Based on the comparative analysis, the linear concentration range and LOD obtained from the application of the Bi-rGO-CTS/GCE-based method was found to be comparable and sometimes better than those obtained in other studies reported in the literature.

A wide range of studies reported in the literature have employed similar oxidation reaction processes for the direct detection of 3-methylmorphine [27–32]. Carbon electrodes such as carbon paste, glassy carbon, and boron-doped diamond (BDD) have been mostly employed for the detection analysis (Table 1). Compared to metallic electrodes, the wider potential window exhibited by carbon electrodes allows one to effectively monitor redox species like 3-methylmorphine. This wider potential window is also desirable for the simultaneous determination of multiple analytes.

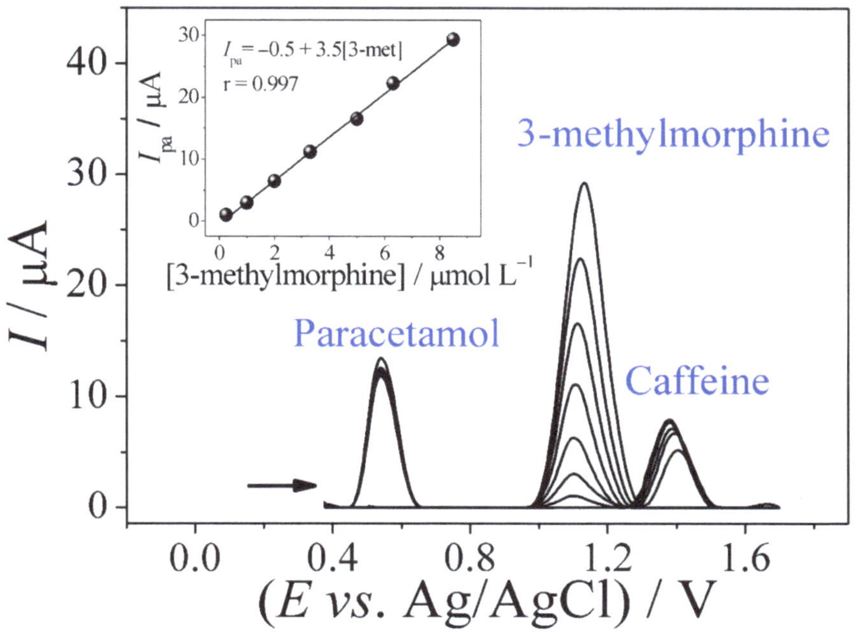

Figure 8. SW voltammograms obtained from the application of Bi-rGO-CTS/GCE for the determination of 3-methylmorphine in the presence of paracetamol and caffeine (at fixed concentration). Inset: Corresponding calibration plots related to 3-methylmorphine concentrations.

3.4. Study of Repeatability and the Influence of Possible Interferents

The analysis of repeatability of the Bi-rGO-CTS/GCE was conducted using 3.0×10^{-6} mol L^{-1} concentrations of 3-methylmorphine in 0.04 mol L^{-1} Britton–Robinson buffer solution (pH 4.0). As shown in Figure S1 (Supplementary Material), the RSD value obtained in 20 replicates was 3.1 % for 3-methylmorphine. The high repeatability of the proposed Bi-rGO-CTS/GCE can be attributed to the homogeneity of the electrode surface and the good conductivity derived from the application of Bi-rGO-CTS/GCE.

In addition, the influence of possible interferents, including silicon dioxide, povidone, cellulose, starch, croscarmellose, magnesium stearate, urea, ascorbic acid, and humic acid, in the ratio of 1:1 (analyte:possible interferent) was investigated. Based on the voltammograms obtained, the interferents were found to exert no influence on the determination of the analytes investigated.

3.5. Analysis of 3-Methylmorphine in Synthetic Urine, Bovine Serum and River Water Samples

The Bi-rGO-CTS/GCE sensor was applied for the quantification of 3-methylmorphine in samples of synthetic urine, bovine serum (commercial serum), and river water collected from the Jacaré-Guaçu River, Araraquara—São Paulo, Brazil; the samples were prepared as described in the experimental section. 3-methylmorphine determinations were performed in triplicate (n = 3) without any pre-treatment procedure. As can be observed in Table 2, the application of the proposed SWV-based method for 3-methylmorphine determination led to recovery percentages ranging from 95 % to 105 %.

Table 2. Results obtained from the analysis of river water, serum and artificial urine samples.

Samples	[3-Methylmorphine]/mol L^{-1}			Recovery ** (Sensor, %)
	Added	Found *	Comparative Method *	
Synthetic urine	1.0×10^{-7}	$(1.02 \pm 0.08) \times 10^{-7}$	$(1.05 \pm 0.05) \times 10^{-7}$	102
	1.0×10^{-6}	$(0.99 \pm 0.06) \times 10^{-6}$	$(0.95 \pm 0.04) \times 10^{-7}$	99.0
Serum	1.0×10^{-7}	$(1.03 \pm 0.09) \times 10^{-7}$	$(1.02 \pm 0.04) \times 10^{-6}$	103
	1.0×10^{-6}	$(1.05 \pm 0.07) \times 10^{-6}$	$(1.03 \pm 0.06) \times 10^{-6}$	105
River water	1.0×10^{-7}	$(1.00 \pm 0.09) \times 10^{-7}$	$(1.00 \pm 0.03) \times 10^{-7}$	100
	1.0×10^{-6}	$(0.95 \pm 0.06) \times 10^{-6}$	$(0.98 \pm 0.05) \times 10^{-6}$	95.0

* Average of 3 measured concentrations; ** Recovery percentage = [Found]/[Added] × 100.

Finally, the results obtained from the application of the two methods were compared using the paired Student's t test (95 % confidence level). For this study, the calculated experimental values (2.6) were found to be lower than the $t_{critical}$ value (0.2); that is, there was no significant difference between the results obtained from the two analytical methods. Thus, based on these results, it can be concluded that the proposed Bi-rGO-CTS/GCE-based method had no matrix effect; this clearly points to the great potential of the proposed technique when applied for the determination of 3-methylmorphine in urine, bovine serum and river waste samples.

4. Conclusions

The present work reported the successful development and application of Bi-rGO-CTS/GCE sensor for 3-methylmorphine detection. The modification of the working electrode (GCE) with rGO and bismuth film led to significant improvements in the electrochemical properties. The proposed sensor was successfully employed for 3-methylmorphine detection in synthetic urine and serum samples as well as in river water, and excellent recovery percentages were obtained (indicating the accuracy of the method). The results obtained in this study also showed that the Bi-rGO-CTS/GCE sensor has good repeatability, reproducibility and selectivity, and has great potential for application as an efficient tool for the analytical determination of 3-methylmorphine and other drugs in both biological and environmental samples.

Supplementary Materials: The following supporting information can be downloaded at: https://www.mdpi.com/article/10.3390/bios12100860/s1, Figure S1: Repeatability study of 3-methylmorphine with the Bi-Rgo-CTS/GCE sensor in 0.04 mol L^{-1} Britton-Robinson (pH 4.0). SWV parameters: f = 15 Hz, a = 50 mV, ΔEs = 5 mV.

Author Contributions: A.W.: Conceptualization, Methodology, Formal Analysis, Investigation, Validation, Writing—original draft. A.M.S.: Methodology, Investigation, Formal Analysis, Writing—Review and Editing. C.A.P.: Writing—Review and Editing. T.A.B.: Writing—Review and Editing. M.H.A.F.: Investigation, Writing—Review and Editing. F.C.M.: Resources, Writing—Review and Editing. M.D.P.T.S.: Project administration, Writing—Review and Editing, Funding acquisition. All authors have read and agreed to the published version of the manuscript.

Funding: This study was financed in part by the Coordenação de Aperfeiçoamento de Pessoal de Nível Superior—Brazil (CAPES)—Finance Code 001 (AUX/PE/PROEX Nº 0674/2018). The authors are also extremely indebted to the National Institute of Alternative Technologies for Detection, Toxicological Evaluation and Removal of Micropollutants and Radioactive Agents—INCTDATREN (FAPESP: 2014/50945-4; CNPq: 465571/2014-0) for the technical assistance provided in support of this work. The authors also gratefully acknowledge the financial support granted by FAPESP (grant numbers 2017/10118-0 and 2022/05454-9).

Institutional Review Board Statement: Not applicable.

Informed Consent Statement: Not applicable.

Data Availability Statement: Not applicable.

Conflicts of Interest: The authors have no conflict of interest to declare.

References

1. Verrinder, E.; Wester, N.; Leppänen, E.; Lilius, T.; Kalso, E.; Mikladal, B.r.; Varjos, I.; Koskinen, J.; Laurila, T. Electrochemical Detection of Morphine in Untreated Human Capillary Whole Blood. *ACS Omega* **2021**, *6*, 11563–11569. [CrossRef] [PubMed]
2. Pratiwi, R.; Noviana, E.; Fauziati, R.; Carrão, D.B.; Gandhi, F.A.; Majid, M.A.; Saputri, F.A. A Review of Analytical Methods for Codeine Determination. *Molecules* **2021**, *26*, 800. [CrossRef] [PubMed]
3. Khairy, M.; Mahmoud, B.G.; Banks, C.E. Simultaneous determination of codeine and its co-formulated drugs acetaminophen and caffeine by utilising cerium oxide nanoparticles modified screen-printed electrodes. *Sens. Actuators B Chem.* **2018**, *259*, 142–154. [CrossRef]
4. Mohseni, N.; Bahram, M. Mean centering of ratio spectra for colorimetric determination of morphine and codeine in pharmaceuticals and biological samples using melamine modified gold nanoparticles. *Anal. Methods* **2016**, *8*, 6739–6747. [CrossRef]
5. Lodha, A.; Pandya, A.; Sutariya, P.G.; Menon, S.K. A smart and rapid colorimetric method for the detection of codeine sulphate, using unmodified gold nanoprobe. *RSC Adv.* **2014**, *4*, 50443–50448. [CrossRef]
6. Saeed, A.; Ahmed, N.Q. Simultaneous spectrophotometric determination of paracetamol and codeine. *Eur. J. Pharm. Med. Res.* **2018**, *5*, 11–15.
7. Lotfi, A.; Karimi, S.; Hassanzadeh, J. Preconcentration of codeine in pharmaceutical and human urine samples by multi-walled carbon nanotubes and its spectrophotometric determination. *Can. J. Chem.* **2016**, *94*, 857–864. [CrossRef]
8. Grabenauer, M.; Moore, K.N.; Bynum, N.D.; White, R.M.; Mitchell, J.M.; Hayes, E.D.; Flegel, R. Development of a Quantitative LC–MS-MS Assay for Codeine, Morphine, 6-Acetylmorphine, Hydrocodone, Hydromorphone, Oxycodone and Oxymorphone in Neat Oral Fluid. *J. Anal. Toxicol.* **2018**, *42*, 392–399. [CrossRef]
9. Plotnikov, A.N.; Karpenko, Y.N.; Vikhareva, E.V.; Tyumina, E.A.; Richkova, M.I.; Selyaninov, A.A. Determination of Codeine Phosphate in the Culture Fluid of Rhodococcus by High-Performance Liquid Chromatography. *Mosc. Univ. Chem. Bull.* **2017**, *72*, 328–332. [CrossRef]
10. Abdul Keyon, A.S.; Miskam, M.; Ishak, N.S.; Mahat, N.A.; Mohamed Huri, M.A.; Abdul Wahab, R.; Chandren, S.; Abdul Razak, F.I.; Ng, N.-T.; Ali, T.G. Capillary electrophoresis for the analysis of antidepressant drugs: A review. *J. Sep. Sci.* **2019**, *42*, 906–924. [CrossRef]
11. Baciu, T.; Botello, I.; Borrull, F.; Calull, M.; Aguilar, C. Capillary electrophoresis and related techniques in the determination of drugs of abuse and their metabolites. *TrAC Trends Anal. Chem.* **2015**, *74*, 89–108. [CrossRef]
12. Santos, A.M.; Silva, T.A.; Vicentini, F.C.; Fatibello-Filho, O. Flow injection analysis system with electrochemical detection for the simultaneous determination of nanomolar levels of acetaminophen and codeine. *Arab. J. Chem.* **2020**, *13*, 335–345. [CrossRef]
13. Wong, A.; Riojas, A.C.; Baena-Moncada, A.M.; Sotomayor, M.D.P.T. A new electrochemical platform based on carbon black paste electrode modified with α-cyclodextrin and hierarchical porous carbon used for the simultaneous determination of dipyrone and codeine. *Microchem. J.* **2021**, *164*, 106032. [CrossRef]
14. Khairy, M. A synergetic effect of cerium oxide nanocubes and gold nanoparticles for developing a new photoelectrochemical sensor of codeine drug. *J. Electroanal. Chem.* **2021**, *895*, 115517. [CrossRef]
15. Wong, A.; Santos, A.M.; Fatibello-Filho, O. Determination of piroxicam and nimesulide using an electrochemical sensor based on reduced graphene oxide and PEDOT:PSS. *J. Electroanal. Chem.* **2017**, *799*, 547–555. [CrossRef]
16. Shao, Y.; Wang, J.; Wu, H.; Liu, J.; Aksay, I.A.; Lin, Y. Graphene Based Electrochemical Sensors and Biosensors: A Review. *Electroanalysis* **2010**, *22*, 1027–1036. [CrossRef]
17. De Figueiredo-Filho, L.C.S.; dos Santos, V.B.; Janegitz, B.C.; Guerreiro, T.B.; Fatibello-Filho, O.; Faria, R.C.; Marcolino-Junior, L.H. Differential Pulse Voltammetric Determination of Paraquat Using a Bismuth-Film Electrode. *Electroanalysis* **2010**, *22*, 1260–1266. [CrossRef]
18. Lau, I.-W.; Mok, C.-S. High-performance liquid chromatographic determination of active ingredients in cough-cold syrups with indirect conductometric detection. *J. Chromatogr. A* **1995**, *693*, 45–54. [CrossRef]
19. Wong, A.; Santos, A.M.; Silva, T.A.; Fatibello-Filho, O. Simultaneous determination of isoproterenol, acetaminophen, folic acid, propranolol and caffeine using a sensor platform based on carbon black, graphene oxide, copper nanoparticles and PEDOT:PSS. *Talanta* **2018**, *183*, 329–338. [CrossRef]
20. Yang, Z.-Z.; Zheng, Q.-B.; Qiu, H.-X.; Li, J.; Yang, J.-H. A simple method for the reduction of graphene oxide by sodium borohydride with CaCl2 as a catalyst. *New Carbon Mater.* **2015**, *30*, 41–47. [CrossRef]
21. Rowley-Neale, S.J.; Randviir, E.P.; Abo Dena, A.S.; Banks, C.E. An overview of recent applications of reduced graphene oxide as a basis of electroanalytical sensing platforms. *Appl. Mater. Today* **2018**, *10*, 218–226. [CrossRef]
22. Laube, N.; Mohr, B.; Hesse, A. Laser-probe-based investigation of the evolution of particle size distributions of calcium oxalate particles formed in artificial urines. *J. Cryst. Growth* **2001**, *233*, 367–374. [CrossRef]

23. Bao-Kang, J.; Jian-Rong, Z.; Zu-Xun, Z. Theory and application of cyclic voltammetry for measurement of fast electrode kinetics at microdisk electrode. *Chin. J. Chem.* **1996**, *14*, 338–347. [CrossRef]
24. Lavagnini, I.; Antiochia, R.; Magno, F. An Extended Method for the Practical Evaluation of the Standard Rate Constant from Cyclic Voltammetric Data. *Electroanalysis* **2004**, *16*, 505–506. [CrossRef]
25. Fatahi, A.; Malakooti, R.; Shahlaei, M. Electrocatalytic oxidation and determination of dexamethasone at an Fe3O4/PANI–CuII microsphere modified carbon ionic liquid electrode. *RSC Adv.* **2017**, *7*, 11322–11330. [CrossRef]
26. Garrido, J.M.P.J.; Delerue-Matos, C.; Borges, F.; Macedo, T.R.A.; Oliveira-Brett, A.M. Voltammetric Oxidation of Drugs of Abuse II. Codeine and Metabolites. *Electroanalysis* **2004**, *16*, 1427–1433. [CrossRef]
27. Asturias-Arribas, L.; Asunción Alonso-Lomillo, M.; Domínguez-Renedo, O.; Julia Arcos-Martínez, M. Screen-printed biosensor based on the inhibition of the acetylcholinesterase activity for the determination of codeine. *Talanta* **2013**, *111*, 8–12. [CrossRef]
28. Bagheri, H.; Khoshsafar, H.; Afkhami, A.; Amidi, S. Sensitive and simple simultaneous determination of morphine and codeine using a Zn2SnO4 nanoparticle/graphene composite modified electrochemical sensor. *New J. Chem.* **2016**, *40*, 7102–7112. [CrossRef]
29. Simioni, N.B.; Oliveira, G.G.; Vicentini, F.C.; Lanza, M.R.V.; Janegitz, B.C.; Fatibello-Filho, O. Nanodiamonds stabilized in dihexadecyl phosphate film for electrochemical study and quantification of codeine in biological and pharmaceutical samples. *Diam. Relat. Mater.* **2017**, *74*, 191–196. [CrossRef]
30. Švorc, L'.; Sochr, J.; Svítková, J.; Rievaj, M.; Bustin, D. Rapid and sensitive electrochemical determination of codeine in pharmaceutical formulations and human urine using a boron-doped diamond film electrode. *Electrochim. Acta* **2013**, *87*, 503–510. [CrossRef]
31. Garazhian, E.; Shishehbore, M.R. A New Sensitive Sensor for Simultaneous Differential Pulse Voltammetric Determination of Codeine and Acetaminophen Using a Hydroquinone Derivative and Multiwall Carbon Nanotubes Carbon Paste Electrode. *Int. J. Anal. Chem.* **2015**, *2015*, 783157. [CrossRef] [PubMed]
32. Asturias-Arribas, L.; Asunción Alonso-Lomillo, M.; Domínguez-Renedo, O.; Julia Arcos-Martínez, M. Cytochrome P450 2D6 based electrochemical sensor for the determination of codeine. *Talanta* **2014**, *129*, 315–319. [CrossRef] [PubMed]

biosensors

MDPI

Article

Electrochemical Sensor Based on Ni-Co Layered Double Hydroxide Hollow Nanostructures for Ultrasensitive Detection of Sumatriptan and Naproxen

Hadi Beitollahi [1],*, Zahra Dourandish [2], Somayeh Tajik [3], Fatemeh Sharifi [3] and Peyman Mohammadzadeh Jahani [4]

[1] Environment Department, Institute of Science and High Technology and Environmental Sciences, Graduate University of Advanced Technology, Kerman 7631885356, Iran
[2] Department of Chemistry, Faculty of Science, Shahid Bahonar University of Kerman, Kerman 76175-133, Iran
[3] Research Center of Tropical and Infectious Diseases, Kerman University of Medical Sciences, Kerman 7616913555, Iran
[4] School of Medicine, Bam University of Medical Sciences, Bam 7661771967, Iran
* Correspondence: h.beitollahi@yahoo.com

Abstract: In this work, Ni-Co layered double hydroxide (Ni–Co LDH) hollow nanostructures were synthesized and characterized by X-ray diffraction (XRD), field emission-scanning electron microscopy (FE-SEM), and Fourier-transform infrared spectroscopy (FT-IR) techniques. A screen-printed electrode (SPE) surface was modified with as-fabricated Ni–Co LDHs to achieve a new sensing platform for determination of sumatriptan. The electrochemical behavior of the Ni–Co LDH-modified SPE (Ni-CO LDH/SPE) for sumatriptan determination was investigated using voltammetric methods. Compared with bare SPE, the presence of Ni-Co LDH was effective in the enhancement of electron transport rate between the electrode and analyte, as well as in the significant reduction of the overpotential of sumatriptan oxidation. Differential pulse voltammetry (DPV) was applied to perform a quantitative analysis of sumatriptan. The linearity range was found to be between 0.01 and 435.0 μM. The limits of detection (LOD) and sensitivity were 0.002 ± 0.0001 μM and 0.1017 ± 0.0001 μA/μM, respectively. In addition, the performance of the Ni-CO LDH/SPE for the determination of sumatriptan in the presence of naproxen was studied. Simultaneous analysis of sumatriptan with naproxen showed well-separated peaks leading to a quick and selective analysis of sumatriptan. Furthermore, the practical applicability of the prepared Ni-CO LDH/SPE sensor was examined in pharmaceutical and biological samples with satisfactory recovery results.

Keywords: Ni-Co layered double hydroxide hollow nanostructures; screen-printed electrode; sumatriptan; naproxen

Citation: Beitollahi, H.; Dourandish, Z.; Tajik, S.; Sharifi, F.; Jahani, P.M. Electrochemical Sensor Based on Ni-Co Layered Double Hydroxide Hollow Nanostructures for Ultrasensitive Detection of Sumatriptan and Naproxen. *Biosensors* **2022**, *12*, 872. https://doi.org/10.3390/bios12100872

Received: 26 August 2022
Accepted: 6 October 2022
Published: 13 October 2022

Publisher's Note: MDPI stays neutral with regard to jurisdictional claims in published maps and institutional affiliations.

1. Introduction

Sumatripta (1-[3-(2-dimethylaminoethyl)-1H-indol-5-yl]-N-methylmethanesulfonamide) is one of the triptan drugs medically administrated to manage cluster headache and migraine [1,2]. Sumatriptan is a selective 5-hydroxytryptamine (5-$HT_{1B/1D}$) receptor agonist [3]. Its binding with serotonin type-1D receptors leads to extensively dilated cranial vessel vasoconstriction, thus reducing migraine pain [4]. Reportedly, extra dose of triptans can be followed by numerous complications, some of which are neck tension, seizures, sleepiness, paralysis, hypertension, leg or arm swelling, and feeling tremor [5,6]. The physiological significance of sumatriptan makes it necessary to quantitatively measure sumatriptan in different specimens, particularly biological samples in the disciplines of clinical diagnosis, pharmacology, and the life sciences. Naproxen (6-methoxy-a-methyl-2-naphthalene acetic acid) is widely used as an anti-inflammatory drug to manage numerous medical conditions such as degenerative joint disorder, rheumatoid arthritis, primary dysmenorrhea, ankylosing spondylitis, and acute gout [7]. Nevertheless, extra and long-term naproxen use can develop some complications such as gastrointestinal hemorrhage, stomach ulcers, nephrotoxicity, elevated heart disease risk, and kidney dysfunction [8,9]. The medical importance of naproxen in humans highlights the necessity of access to an appropriate rapid, facile, and sensitive analytical approach. In addition, the mode of complementary action of these two drugs suggests their combined use to obtain more effective clinical outcomes in the treatment of acute migraine than both drugs alone. Hence, simultaneous detection of these drugs in biological fluids and pharmaceutical formulations is very important [10].

Many strategies, including spectrophotometry [11,12], capillary electrophoresis [13,14], high-performance liquid chromatography [15,16], liquid chromatography–mass spectroscopy [17], spectrofluorimetry [18], chemiluminescence [19], and electrochemical techniques [20–22] have been applied for determination of these compounds. However, the widespread application of some of these methods has been limited by their complex operation and high cost. Electrochemical determinations have been shown to be more appropriate for analyte analysis [23–32], owing to commendable merits such as cost-effectiveness, narrow LOD, higher sensitivities, wide potential window, short analysis time, and ease to renew the surface.

Screen-printing technology has proven its effectiveness in making electroanalytical platforms with tailored purposes, some of which are point-of-care (POC) tools in biomedicine [33–35], and portable sensing systems in food industries [36,37] and environmental pollutant detection [38–40]. SPEs are potent materials for electroanalytical (bio)sensors [41–43] owing to their inexpensiveness and easy production process, especially for the fabrication of transducers required for on-site one-point measurements. The miniaturized SPEs are appropriate for on-site measurements during real-time analysis, and require small amounts of reagents and samples.

The application of nanomaterials in various fields is increasing rapidly [44–49] and offers promising prospects. In recent years, the advances in nanotechnology have been helpful to produce sensitive and selective (bio)sensors [50–56]. A variety of nanomaterials, such as metal and metal oxides nanoparticles, and carbon nanostructures, have been employed to fabricate electrochemical (bio)sensing platforms [57–65], with diverse performances such as biomolecule labeling or immobilization, the electrochemical process catalysis, electron transfer enhancement, and serving as reactant [66].

Layered double hydroxides (LDHs) have recently spurred extensive interest owing to multiple specific merits such as a layered nature, huge surface area, adjustable structure, cost-effectiveness, and environmental friendliness [67–69]. The LDHs containing transition metals are of great significance for catalyst, energy storage, and sensing [70–73]. One of the strategies to enhancing their electrochemical performance is the design of tunable porous nanostructures or architecture of LDHs with huge surface area [74–76]. Hierarchical hollow structures (HHSs) with well-defined micro- or nanostructures, mesoporous pore-size distribution, huge surface area, more active sites, and satisfactory charge transfer could potentially promote the electrochemical behavior of LDHs [77].

Among these materials, nickel–cobalt layered double hydroxides (Ni-Co LDHs) have attracted particular interest in electrochemical sensors because of their low cost, good redox activity, and eco-friendly properties. They have an inverse spinel crystal structure, where Ni^{2+} is distributed at the octahedral sites and Co^{3+} is distributed at both tetrahedral and octahedral sites. This composition offers higher conductivity than that of Ni-Co LDH, which in turn enhances the electron transfer and improves the performance of electrochemical sensors [78,79].

In this research, a simple strategy was used to design an electrochemical sensing platform based on SPE modification with Ni-Co LDH which was employed for the determination of sumatriptan in the presence of naproxen. The Ni-Co LDH-modified SPE demonstrated better sensor features with a low LOD, high sensitivity, and wide linear range. The sumatriptan and naproxen sensing platform was characterized by the successful measurement of these analytes in sumatriptan tablets, naproxen tablets, and urine samples.

2. Experimental

2.1. Equipment

A Metrohm Autolab PGSTAT 320N Potentiostat/Galvanostat Analyzer (Utrecht, The Netherlands) with GPES (General Purpose Electrochemical System-version 4.9) software was applied for all electrochemical determinations at ambient temperature. The electrochemical sensors were prepared using DRP-110 SPEs (DropSens, Oviedo, Spain) featuring a silver pseudo-reference electrode, graphite working electrode, and graphite auxiliary electrode. A Metrohm 713 pH meter with glass electrode (Herisau, Switzerland) was recruited to determine and adjust the solution pH. Direct-Q® 8 UV deionized water (Millipore, Darmstadt, Germany) was used to freshly prepare all solutions.

A Panalytical X'Pert Pro X-ray diffractometer (Almelo, The Netherlands) applying a Cu/Kα radiation (λ:1.54 Å) was used for XRD analysis, and a Bruker Tensor II spectrometer (Karlsruhe, Germany) was employed to capture the FT-IR spectra. An MIRA3 scanning electron microscope (Tescan, Brno, Czech Republic) was utilized for FE-SEM imaging.

2.2. Solvents and Chemicals

All solvents and chemicals applied in our protocol were of analytical grade and obtained from Merck and Sigma-Aldrich. Phosphate-buffered solution (PBS) was prepared using phosphoric acid and adjusted by NaOH to the desired pH value.

2.3. Synthesis of Ni-Co Layered Double Hydroxide Hollow Nanostructures

The Stöber method, with slight modification, was followed to prepare monodispersed silica (SiO_2) spheres [80]. To this end, tetraethyl orthosilicate (TEOS) (6 mL) was dissolved drop by drop in a solution containing ethanol (75 mL), deionized water (10 mL), and aqueous ammonia (3.15 mL), followed by stirring at an ambient temperature for 5 h. The centrifugation was performed to extract the SiO_2 spheres from the suspension, followed by rinsing by ethanol/deionized water. Finally, the obtained white precipitate was oven dried under vacuum condition at 65 °C for 12 h. Subsequently, the SiO_2@Ni-Co LDH core–shell structures were produced by following the protocol reported by Li and coworkers [77]. In brief, 200 mg of pre-synthesized silica spheres were dispersed in 100 mL ethanol under ultrasonication for 1 h. Then, 3 mmol $Ni(NO_3)_2.6H_2O$ (2.5 g) and 1.5 mmol $Co(NO_3)_2.6H_2O$ (5 g) were dissolved into the above suspension. After that, 23 mL of aqueous ammonia solution was dispersed drop by drop in the suspension containing SiO_2 spheres and metal salts while magnetically stirring for 1 h at room temperature. The co-precipitation process was carried out for deposition of hierarchical Ni-Co LDH nanosheets on SiO_2 sphere surface. The centrifugation was performed to extract the resulting precipitate, followed by thoroughly rinsing by ethanol/deionized water. The obtained precipitate was oven-dried at 80 °C for 12 h. Finally, Ni-Co LDH hollow structures were formed after removal of silica cores by etching SiO_2@Ni-Co LDH in 0.5 M KOH solution at for 1 h. The resulting product was centrifuged and rinsed thoroughly. The prepared Ni-Co LDH hollow structures were dried at 60 °C for 12 h.

2.4. Preparation of the Ni-Co LDH/SPE Sensor

A drop-casting technique was followed to fabricate the Ni-Co LDH/SPE. Thus, a certain amount of as-prepared Ni-Co LDH hollow nanostructures (1 mg) was subsequently dispersed in deionized water (1 mL) under 20 min ultrasonication. Then, the dispersed suspension (4 μL) was coated dropwise on the SPE surface and dried at the laboratory temperature.

2.5. Real Samples Preparation

Five sumatriptan tablets (labeled 50 mg of sumatriptan) purchased from a local pharmacy in Kerman (Iran) were pulverized together, and then 50 mg was dissolved in water (25 mL) under ultrasonication to prepare a sumatriptan solution. Then, variable volumes of diluted solution were poured into a 25 mL volumetric flask and brought to the final volume with PBS (pH = 7); the analyses were performed using a modified electrode.

Five naproxen tablets (labeled 500 mg of naproxen) purchased from a local pharmacy in Kerman (Iran) were pulverized together, and then 500 mg was dissolved in water (25 mL) under ultrasonication to prepare a naproxen solution. Then, variable volumes of diluted solution were poured into a 25 mL volumetric flask and brought to the final volume with PBS (pH = 7); the analyses were performed using modified electrode.

Moreover, 10 mL of refrigerated urine specimens were centrifuged at 1500 rpm for 20 min, followed by filtering the supernatant via 0.45 μm filter. Next, variable supernatant solution contents were placed in 25 mL volumetric flasks and diluted to the marks using PBS at the pH value of 7. Variable sumatriptan and naproxen contents were applied to spike the diluted urine specimens. At last, a standard addition method was followed to quantify the sumatriptan and naproxen.

3. Results and Discussion

3.1. Characterization of Ni-Co Layered Double Hydroxide Hollow Nanostructures

The surface morphologies of SiO_2 spheres, SiO_2@Ni-Co LDH core–shell structures, and Ni-Co LDH hollow structures were explored using FE-SEM. Figure 1a shows the FE-SEM images of SiO_2 spheres. The SiO_2 spherical particles showed good monodispersity, with a uniform size of approximately 170 nm. According to the FE-SEM images captured from SiO_2@Ni-Co LDH core–shell structures, it is clearly evident that, after the co-precipitation process, the hierarchical Ni-Co LDH nanosheets were well deposited on the surface of the silica spheres (Figure 1b). Subsequently, after KOH etching process to remove the SiO_2 cores, the Ni-Co LDH hollow structures were obtained and showed an obvious hollow structure (Figure 1c,d).

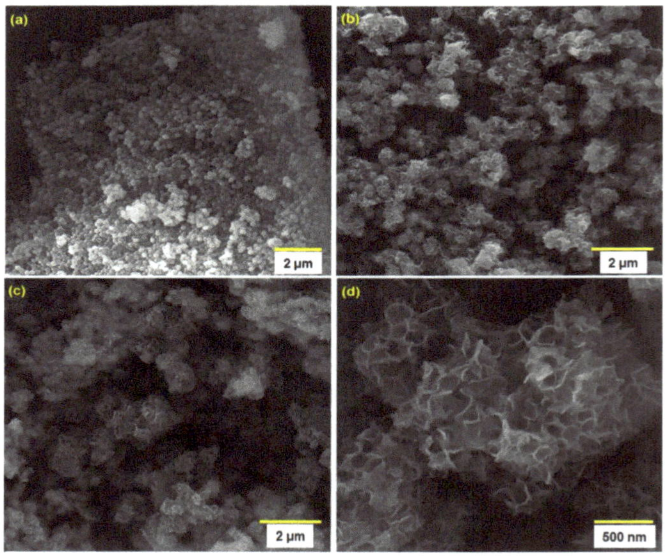

Figure 1. FE-SEM images of SiO_2 spheres (**a**), SiO_2@Ni-Co LDH core–shell structures (**b**), and Ni-Co LDH hollow structures (**c**,**d**).

The XRD pattern of Ni-Co LDH hollow structures is presented in Figure 2, showing the well-defined diffraction peaks observed at 2θ values of 11.4°, 23.0°, 34.2°, and 60.9° can be related to plane reflections of (003), (006), (012), and (110) for the hydrotalcite-like LDH phase. The XRD pattern of the synthesized sample is consistent with the XRD patterns reported in previous papers [81,82].

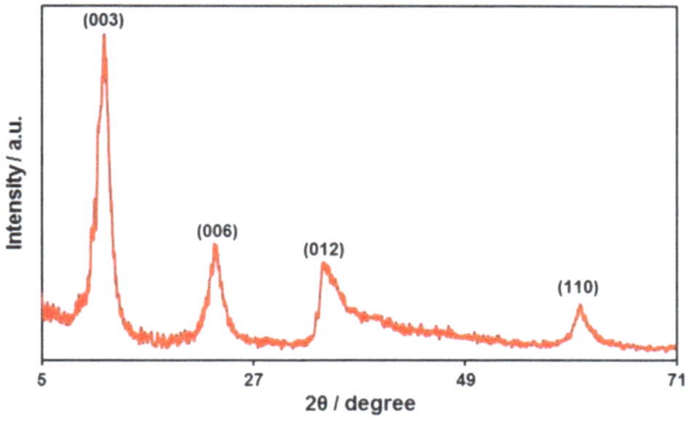

Figure 2. XRD pattern of Ni-Co LDH hollow structures.

FT-IR spectroscopy is a well-equipped tool to study the functional groups of the prepared samples. Figure 3 depicts the FT-IR spectra of SiO_2@Ni-Co LDH core–shell structures and Ni-Co LDH hollow structures. According to the FT-IR spectra captured from SiO_2@Ni-Co LDH, the distinctive adsorption peaks of SiO_2 were found at 467 cm^{-1}, 805 cm^{-1}, and 1101 cm^{-1}, corresponding to the bending vibration of Si–O–Si, stretching vibration of Si–O–Si, and asymmetric stretching vibration of Si–O–Si [83]. Below, the existence of characteristic absorption bands of Ni-Co LDH is mentioned. The broad vibration of hydroxyl groups (O–H stretching) of water molecules in the interlayer space of LDH was confirmed at 3459 cm^{-1}. The peak at 1637 cm^{-1} relates to the bending vibration of OH groups. The characteristic FT-IR band at 1383 cm^{-1} is generally assignable to the vibration of interlayer anions (CO_3^{2-} and NO_3^{-}) [82,84]. In addition, the peak at 642 cm^{-1} relates to the characteristic absorption band of M–O (metal-oxygen) vibrations. According to the FT-IR spectra captured for Ni-Co LDH hollow structures, following the etching process, disappearing of Si–O–Si characteristic peaks highlighted the silica template removal.

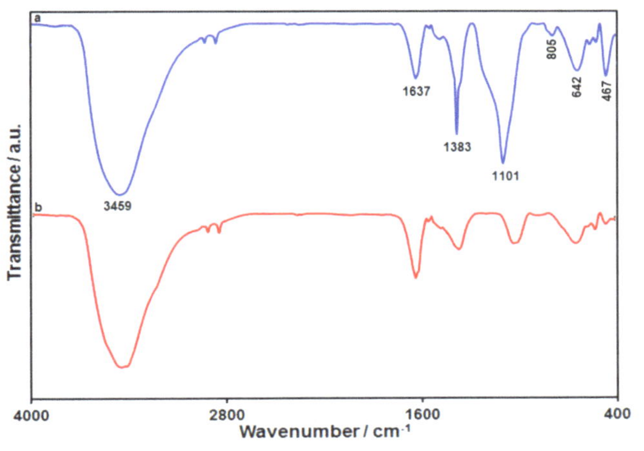

Figure 3. FT-IR spectra of (**a**) SiO_2@Ni-Co LDH, and (**b**) Ni-Co LDH.

3.2. Studying the Influence on the Structures on Voltammetric Detection of Sumatriptan Oxidation

The electrochemical response of sumatriptan oxidation in 0.1 M PBS adjusted to variable pH values (2.0 to 9.0) was explored to determine the influence of electrolyte solution pH. As shown in Figure 4, its electrochemical oxidation was dependent on the pH value of the solution, such that it reached a maximum with increasing pH up to 7.0

and then decreased with further pH values (Figure 5). Hence, the pH value of 7.0 was considered to be the optimum for subsequent electrochemical determinations.

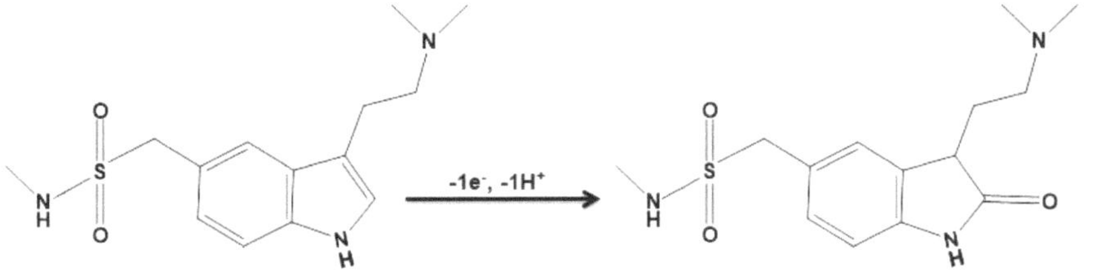

Figure 4. Proposed oxidation mechanism for sumatriptan.

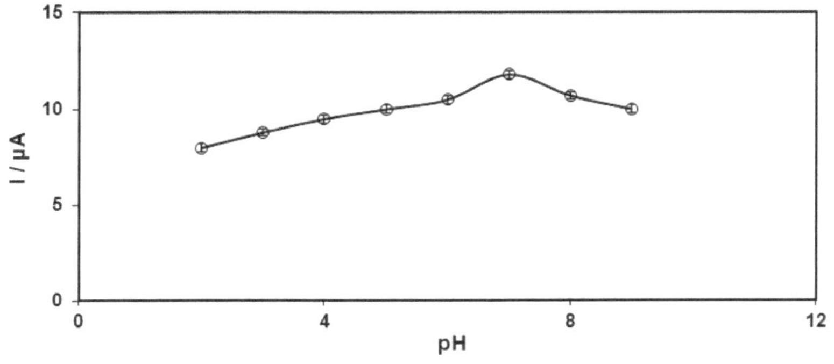

Figure 5. Plot of Ip vs. pH obtained from DPVs of Ni-Co LDH/SPE in a solution containing 100.0 µM of sumatriptan in 0.1 PBS with different pH (2.0, 3.0, 4.0, 5.0, 6.0, 7.0, 8.0, and 9.0).

Cyclic voltammetry (CV) was performed to clarify the electrochemical behavior of sumatriptan on unmodified (bare) and modified SPE surfaces. Figure 6 compares the bare SPE and Ni-Co LDH/SPE for 100.0 µM sumatriptan oxidation in 0.1 M PBS at the pH value of 7.0. The sumatriptan oxidation displayed a tiny and wide peak (2.9 µA) at the potential of 800 mV on the bare SPE surface. The Ni-Co LDH-modified SPE exhibited a shift in the peak current toward more negative potentials (610 mV) by raising the amount of current (11.8 µA). Such an improvement could have appeared because of the appreciable catalytic impact of Ni-Co LDH hollow nanostructures for sumatriptan oxidation.

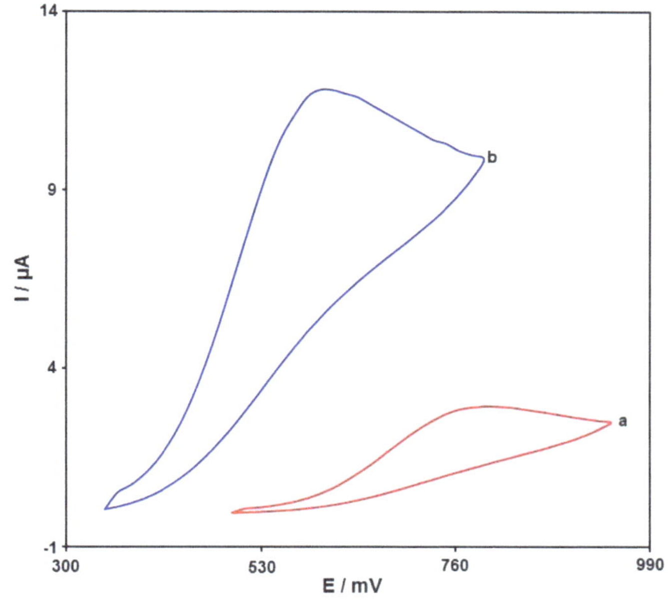

Figure 6. Cyclic voltammograms captured for oxidation of sumatriptan (100.0 μM) in PBS (0.1 M; pH = 7.0) on (a) unmodified SPE and (b) Ni-Co LDH/SPE with a scan rate of 50 mV/s.

3.3. *Effect of Scan Rate*

The linear sweep voltammograms (LSVs) were recorded for the oxidation of sumatriptan (100.0 μM) on the Ni-Co LDH/SPE under variable scan rates (Figure 7). There was an apparent gradual elevation in the oxidation peak by raising scan rate ranging from 10 to 400 mV/s. As seen in Figure 7 (Inset), the anodic peak current (Ipa) had a linear association with the scan rate square root ($\upsilon^{1/2}$). The regression equation was obtained as Ipa (μA) = 1.117 $\upsilon^{1/2}$ $(mV{\cdot}s^{-1})^{1/2}$ + 2.8278 (R^2 = 0.9986), indicating a controlled diffusion process of sumatriptan oxidation on the Ni-Co LDH/SPE.

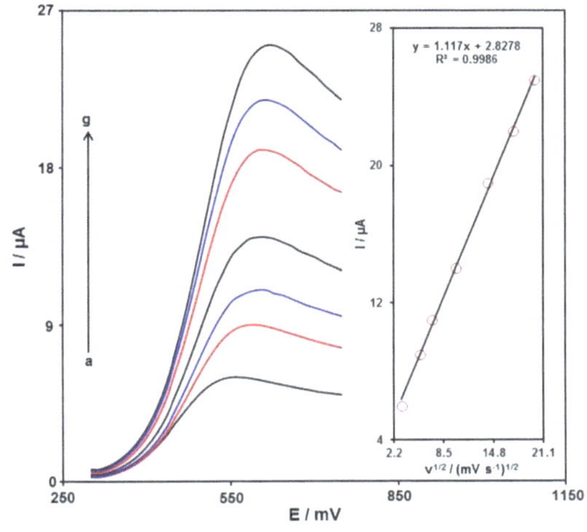

Figure 7. LSVs captured for the oxidation of sumatriptan (100.0 μM) on the Ni-Co LDH/SPE under variable scan rates (a–g: 10, 30, 50, 100, 200, 300, and 400 mV/s). Inset: correlation of Ipa with $\upsilon^{1/2}$.

A Tafel plot (Figure 8 (inset)) was achieved on the basis of data related to the rising domain of current–voltage curve at a low scan rate (10 mV/s) for sumatriptan (100.0 μM) to explore the rate-determining step. The linearity of the E vs. log I plot clarifies the involvement of electrode process kinetics. The slope from this plot could present the count of transferred electrons during the rate-determining step. On the basis of Figure 8 (inset), the Tafel slope was estimated to be 0.1393 V for the linear domain of the plot. The Tafel slope value revealed that the rate-limiting step was the one-electron transfer process considering a transfer coefficient (α) of 0.58.

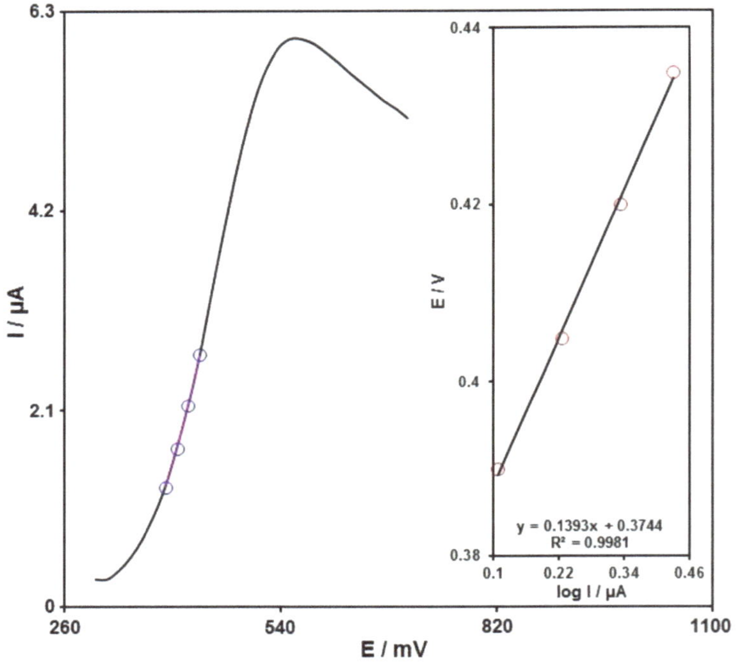

Figure 8. LSV for sumatriptan (100.0 μM) at the scan rate of 10 mV/s. Inset: The Tafel plot from the rising domain or the respective voltammogram.

3.4. Chronoamperometric Analysis

Chronoamperometry was recruited to explore the sumatriptan catalytic oxidation on the Ni-Co LDH/SPE surface. Chronoamperometric analysis was performed for variable sumatriptan contents on Ni-Co LDH/SPE at the working electrode potential of 660 mV. The chronoamperograms captured for variable sumatriptan contents on the Ni-Co LDH/SPE are shown in Figure 9. Cottrell's equation explains the current (I) for electrochemical reaction of an electroactive material with a D value (diffusion coefficient) under a mass transport limited condition. Figure 9A shows a linear relationship of the I value with t^{-12} for the oxidation of variable sumatriptan contents. The slopes from the obtained straight lines were plotted against variable sumatriptan contents (Figure 9B). The plotted slope and Cottrell equation estimated the D value to be 8.2×10^{-5} cm^2/s for sumatriptan.

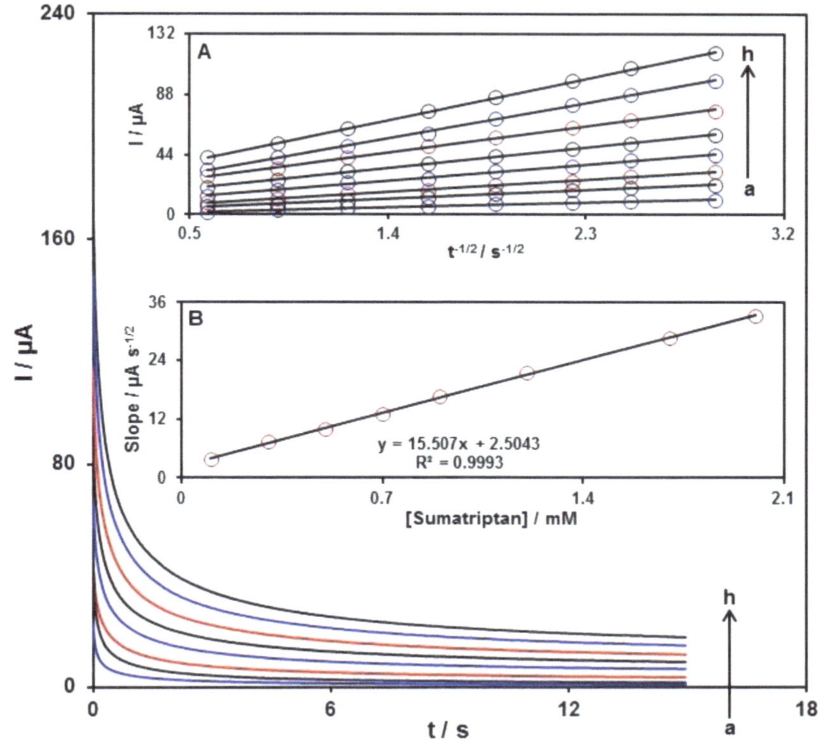

Figure 9. Chronoamperometric behavior of Ni-Co LDH/SPE in PBS (0.1 M; pH = 7.0) at potential of 660 mV for variable sumatriptan contents (a–h: 0.1, 0.3, 0.5, 0.7, 0.9, 1.2, 1.7, and 2.0 mM). Insets: (**A**) Plots of I vs. $t^{-1/2}$ and (**B**) plots of the slopes from the straight lines vs. sumatriptan level.

3.5. DPV Analysis of Sumatriptan

DPV analysis was performed for variable sumatriptan contents to explore the linear dynamic range, LOD, and sensitivity of the Ni-Co LDH/SPE under optimized experimental circumstances (Figure 10). As expected, the elevation in sumatriptan level enhanced the peak current. Figure 10 (Inset) shows a linear proportionality of the oxidation peak currents to variable sumatriptan contents (0.01 μM to 435.0 μM) with a linear regression equation of Ipa (μA) = 0.1017 ± 0.0001 $C_{sumatriptan}$ + 0.6849 (R^2 = 0.9995), and a sensitivity of 0.1017 μA/μM. In the equations of LOD = $3S_b/m$ and LOQ = $10S_b/m$, the S_b is the standard deviation of the response for blank solution, and m is the slope from the standard graph. The LOD and LOQ were estimated at 0.002 ± 0.0001 and 0.007 ± 0.0001 μM for sumatriptan determination on Ni-Co LDH/SPE.

Table 1 compares the efficiency of the sumatriptan sensor prepared by the Ni-Co LDH-modified SPE and other reported works.

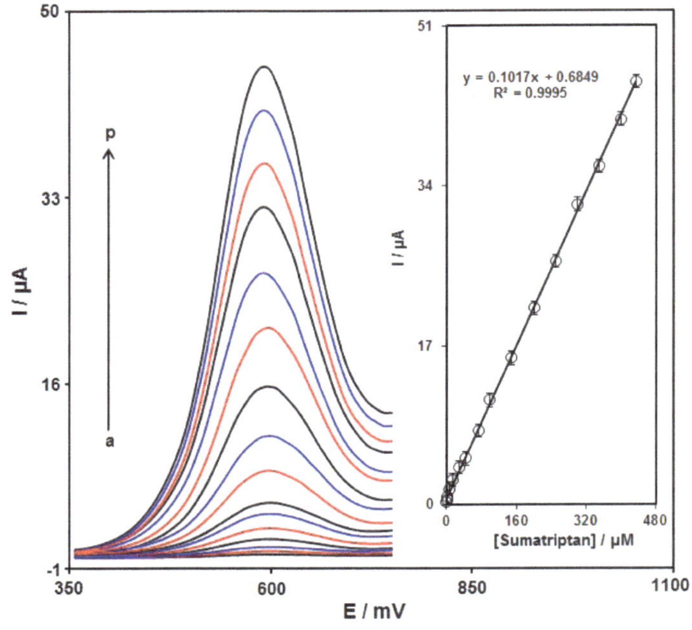

Figure 10. DPVs captured for the oxidation of variable sumatriptan contents on the Ni-Co LDH/SPE under variable contents (a–p: 0.01, 1.0, 2.5, 7.5, 15.0, 30.0, 45.0, 75.0, 100.0, 150.0, 200.0, 250.0, 300.0, 350.0, 400.0, and 435.0 μM). Inset: Calibration curve of voltammetric response (Ipa) against sumatriptan level.

Table 1. Comparison of the efficiency of the Ni-Co LDH/SPE sensor with other reported modified electrodes for sumatriptan determination.

Electrochemical Sensor	Electrochemical Method	Linear Range	LOD	Ref.
CuO/SPE	DPV	0.33–3.54 μM	0.066 μM	[4]
Cu nanoparticles (NPs)/poly-melamine/glassy carbon electrode	DPV	0.08–0.58 and 0.58–6.5 μM	0.025 μM	[85]
Multiwalled carbon nanotube (MWCNTs)decorated with silver NPs/pyrolytic graphite electrode	CV	0.08–100.0 μM	0.04 μM	[86]
MWCNTs and cobalt methyl-salophen complex/carbon paste electrode	DPV	1.0–1000.0 μM	0.3 μM	[21]
MWCNTs and polypyrrole doped with new coccine/glassy carbon electrode	LSV	0.02–10.0 μM	0.006 μM	[20]
Overoxidized poly(*p*-aminophenol) modified glassy carbon electrode	Square wave voltammetry	1.0–100.0 μM	0.294 μM	[87]
Ni-Co LDH/SPE	DPV	0.01–435.0 μM	0.002 μM	This work

3.6. DPV Analysis of Sumatriptan in the Presence of Naproxen

To confirm the ability of Ni-Co LDH/SPE for codetection of sumatriptan and naproxen, the electrochemical responses of these analytes were detected by simultaneously changing the concentration of both analytes in PBS at pH 7.0. As seen in Figure 11, with a concurrent change in their concentrations, two noninterference peaks were found on DPV curves. The peak currents of both sumatriptan and naproxen oxidation displayed a linear elevation with the respective concentrations (sumatriptan concentration range between 1.0 μM and 400.0 μM, and naproxen concentration range between 1.0 μM and 400.0 μM) (Figure 11A,B). The intensity of peak current showed good linearity with the target concentration change, highlighting the possibility of detecting sumatriptan and naproxen in the blended solution.

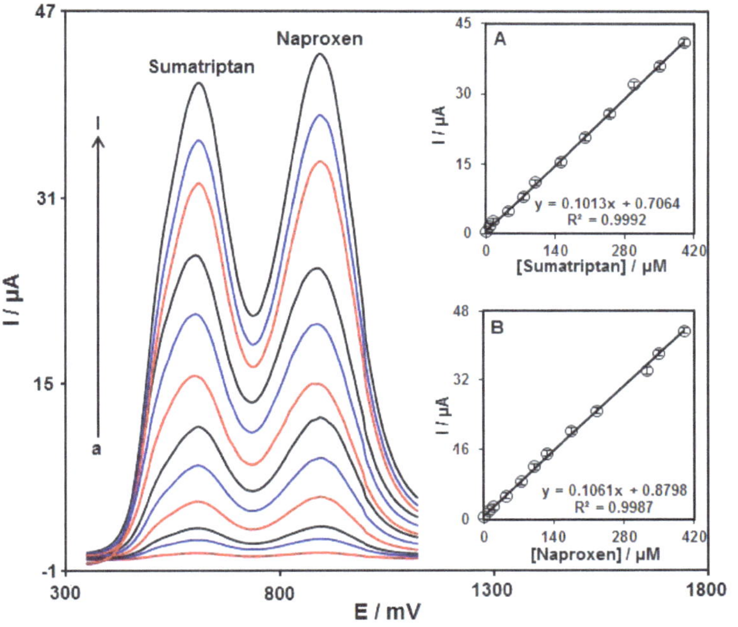

Figure 11. DPVs of Ni-Co LDH/SPE in 0.1 M PBS (pH 7.0) with various concentrations of sumatriptan (a–l: 1.0, 7.5, 15.0, 45.0, 75.0, 100.0, 150.0, 200.0, 250.0, 300.0, 350.0, and 400.0 μM) and naproxen (a–l: 1.0, 10.0, 20.0, 45.0, 75.0, 100.0, 125.0, 175.0, 225.0, 325.0, 350.0, and 400.0 μM). Insets: (**A**) The plot of peak current versus sumatriptan concentration, (**B**) the plot of peak current versus naproxen concentration.

3.7. Repeatability, Reproducibility, and Stability

The Ni-Co LDH/SPE was examined for repeatability through the measurement of the response of 40.0 μM sumatriptan on the surface of the same electrode 15 times. The relative standard deviation (RSD) of 3.9% for the current response of sumatriptan demonstrated the good repeatability of the proposed electrode.

To test the reproducibility, five Ni-Co LDH/SPE produced using the same procedures were applied to measure 40.0 μM sumatriptan under identical circumstances; the obtained RSD of 3.5% demonstrated commendable reproducibility.

To test the Ni-Co LDH/SPE stability, the current responses of sumatriptan were measured following 14 day storage of the sensor at ambient temperature. The decrease in peak current of sumatriptan to 4.2% of its original response demonstrated appreciable stability.

3.8. Selectivity Studies

The effects of some organic and inorganic species which commonly existed in pharmaceuticals and biological samples were examined on the analytical response of the proposed sensor (Ni-Co LDH/SPE). Therefore, a 50.0 μM solution of sumatriptan in the supporting electrolyte (PBS) was prepared. Various amounts of the interfering species were added to the sumatriptan solution. The voltammogram (DPV) of the sample was recorded in the presence of interfering species. The tolerance limit was defined as the maximum concentration of the interfering substance that caused an approximately ±5% relative error in the determination. The results revealed that 500-fold concentrations of Na^+, Mg^{2+}, Ca^{2+}, $NH_4{}^+$, and $SO_4{}^{2-}$, 300-fold concentrations of fructose, glucose, and lactose, 100-fold concentrations of histidine, phenyl alanine, methionine, and cysteine, and 20-fold concentrations of levodopa and uric acid did not show interference in determination (Table S1, Supplementary Materials). These results confirmed the suitable selectivity of the proposed sensor for determination.

3.9. Analysis of Real Specimens

The practical applicability of Ni-Co LDH/SPE was tested by sensing sumatriptan and naproxen in sumatriptan tablets, naproxen tablets, and urine specimens using the DPV procedure and a standard addition method, the results of which can be seen in Table 2. The recovery rate was between 96.4% and 102.5%, and all RSD values were ≤3.6%. According to the experimental results, the Ni-Co LDH/SPE sensor possesses a high potential for practical applicability.

Table 2. Voltammetric sensing of sumatriptan and naproxen in real specimens using Ni-Co LDH/SPE. All concentrations are in μA (n = 3).

Sample	Spiked (μM)		Found (μM)		Recovery (%)		RSD (%)	
	Sumatriptan	Naproxen	Sumatriptan	Naproxen	Sumatriptan	Naproxen	Sumatriptan	Naproxen
Sumatriptan Tablet	0	0	4.0	-	-	-	3.3	-
	1.0	4.0	4.9	4.1	98.0	102.5	1.9	2.3
	3.0	6.0	7.1	5.8	101.4	96.7	2.8	3.0
Naproxen Tablet	0	0	-	5.0	-	-	-	2.9
	5.0	1.0	5.1	5.9	102.0	98.3	3.0	2.2
	7.0	3.0	6.9	8.3	98.6	103.7	1.8	3.6
Urine	0	0	-	-	-	-	-	-
	4.5	5.5	4.6	5.3	102.2	96.4	2.5	2.8
	6.5	7.5	6.3	7.6	96.9	101.3	3.1	1.9

4. Conclusions

In this work, we reported the sensing application of Ni-Co LDH hollow nanostructures for electrochemical determination of sumatriptan. The sensing platform was fabricated via drop casting of a Ni-Co LDH hollow nanostructures dispersion on bare SPE. The electrochemical studies demonstrated efficient electrocatalytic activity of Ni-Co LDH hollow nanostructure-modified SPE for sensitive detection of sumatriptan. DPV findings showed an increase in the anodic peak currents with elevating sumatriptan contents (0.01–435.0 μM), with an LOD of 0.002 ± 0.0001 μM. Furthermore, for sensing sumatriptan in the presence of naproxen, the obtained voltammograms exhibited a desirable peak separation of about 300 mV potential differences. Moreover, the prepared sensor (Ni-Co LDH/SPE) was efficiently applied to detect sumatriptan and naproxen in in pharmaceutical and biological samples.

Supplementary Materials: The following supporting information can be downloaded at: https://www.mdpi.com/article/10.3390/bios12100872/s1, Table S1: Selectivity results for 50.0 μM sumatriptan in the presence of other interferences.

Author Contributions: Conceptualization, H.B. and S.T.; methodology, H.B.; validation, H.B. and Z.D.; formal analysis, Z.D., S.T., F.S. and P.M.J.; investigation, H.B.; writing—original draft preparation, Z.D., S.T., F.S. and P.M.J.; writing—review and editing, H.B., Z.D., S.T., F.S. and P.M.J. supervision, H.B.; project administration, H.B.; All authors have read and agreed to the published version of the manuscript.

Funding: The authors acknowledge the financial support provided for this paper by the Iran National Science Foundation (INSF).

Data Availability Statement: All the data are presented in the manuscript.

Conflicts of Interest: The authors declare no conflict of interest.

References

1. Coukell, A.J.; Lamb, H.M. Sumatriptan. *Pharmacoeconomics* **1997**, *11*, 473–490. [CrossRef] [PubMed]
2. Kerry, L.D.; Stephen, P.C. Sumatriptan-a review of its pharmacodynamic and pharmacokinetic properties and therapeutic efficacy in the acute treatment of migraine and cluster headache. *Drugs* **1992**, *43*, 776–791.
3. Wichitnithad, W.; Nantaphol, S.; Vicheantawatchai, P.; Kiatkumjorn, T.; Wangkangwan, W.; Rojsitthisak, P. Development and validation of liquid chromatography-tandem mass spectrometry method for simple analysis of sumatriptan and its application in bioequivalence study. *Pharmaceuticals* **2020**, *13*, 21. [CrossRef] [PubMed]
4. Aldawsari, A.M.; Khalifa, M.E.; Munshi, A.M.; Shah, R.; Keshk, A.A.; Saad, F.; El-Metwaly, N.M. Copper oxide based disposable sensors for sensitive voltammetric assay of sumatriptan. *Int. J. Electrochem. Sci.* **2021**, *16*, 210540. [CrossRef]
5. Goadsby, P.J.; Lipton, R.B.; Ferrari, M.D. Migraine-current understanding and treatment. *N. Engl. J. Med.* **2002**, *346*, 257–270. [CrossRef]
6. Dodick, D.W.; Lipton, R.B.; Goadsby, P.J.; Tfelt-Hansen, P.; Ferrari, M.D.; Diener, H.C.; Parsons, B. Predictors of migraine headache recurrence: A pooled analysis from the eletriptan database. *Headache* **2008**, *48*, 184–193. [CrossRef]
7. Todd, P.A.; Clissold, S.P. Naproxen. A reappraisal of its pharmacology, and therapeutic use in rheumatic diseases and pain states. *Drugs* **1990**, *40*, 91–137. [CrossRef]
8. Doomra, R.; Goyal, A. NSAIDs and self-medication: A serious concern. *J. Family Med. Prim. Care* **2020**, *9*, 2183. [CrossRef]
9. Sarhangzadeh, K. Application of multi wall carbon nanotube–graphene hybrid for voltammetric determination of naproxen. *J. Iran. Chem. Soc.* **2015**, *12*, 2133–2140. [CrossRef]
10. Smith, T.R.; Sunshine, A.; Stark, S.R.; Littlefield, D.E.; Spruill, S.E.; Alexander, W.J. Sumatriptan and naproxen sodium for the acute treatment of migraine. *Headache* **2005**, *45*, 983–991. [CrossRef]
11. Solanki, S.D.; Patel, P.U.; Suhagiya, B.N. Development and validation of spectrophotometric method for simultaneous estimation of sumatriptan succinate and naproxen sodium in pharmaceutical dosage form. *J. Pharm. Sci. Biosci. Res.* **2011**, *1*, 50–53.
12. Holzbecher, M.; Ellenberger, H.A.; Marsh, J.M.; Boudreau, S. An ultraviolet spectrophotometric procedure for the routine determination of naproxen. *Clin. Biochem.* **1979**, *12*, 66–67. [CrossRef]
13. Altria, K.D.; Filbey, S.D. Quantitative determination of sumatriptan by capillary electrophoresis. *Anal. Proc.* **1993**, *30*, 363–365.
14. Fillet, M.; Fotsing, L.; Bonnard, J.; Crommen, J. Stereoselective determination of S-naproxen in tablets by capillary electrophoresis. *J. Pharmaceut. Biomed. Anal.* **1998**, *18*, 799–805. [CrossRef]
15. Majithiya, R.J.; Majithiya, J.B.; Umrethia, M.L.; Ghosh, P.K.; Murthy, R.S.R. HPLC method for the determination of sumatriptan in plasma and brain tissue. *Ars. Pharm.* **2006**, *47*, 199–210.
16. Muneer, S.; Muhammad, I.N.; Abrar, M.A.; Munir, I.; Kaukab, I.; Sagheer, A.; Sultana, K. High Performance Liquid Chromatographic Determination of Naproxen in Prepared Pharmaceutical Dosage Form and Human Plasma and its Application to Pharmacokinetic Study. *J. Chromatogr. Sep. Tech.* **2017**, *8*, 1–5. [CrossRef]
17. Oxford, J.; Lant, M.S. Development and validation of a liquid chromatographic-mass spectrometric assay for the determination of sumatriptan in plasma. *J. Chromatogr. B Biomed. Sci. Appl.* **1989**, *496*, 137–146. [CrossRef]
18. Damiani, P.C.; Borraccetti, M.D.; Olivieri, A.C. Direct and simultaneous spectrofluorometric determination of naproxen and salicylate in human serum assisted by chemometric analysis. *Anal. Chim. Acta* **2002**, *471*, 87–96. [CrossRef]
19. Li, Y.; Lu, J. Flow injection chemiluminescence determination of naproxen based on $KMnO_4$–Na_2SO_3 reaction in neutral aqueous medium. *Anal. Chim. Acta* **2006**, *577*, 107–110. [CrossRef]
20. Shahrokhian, S.; Kamalzadeh, Z.; Saberi, R.S. Glassy carbon electrode modified with a bilayer of multi-walled carbon nanotube and polypyrrole doped with new coccine: Application to the sensitive electrochemical determination of Sumatriptan. *Electrochim. Acta* **2011**, *56*, 10032–10038. [CrossRef]
21. Amiri, M.; Pakdel, Z.; Bezaatpour, A.; Shahrokhian, S. Electrocatalytic determination of sumatriptan on the surface of carbonpaste electrode modified with a composite of cobalt/Schiff-base complex and carbon nanotube. *Bioelectrochemistry* **2011**, *81*, 81–85. [CrossRef] [PubMed]

22. Qian, L.; Thiruppathi, A.R.; Elmahdy, R.; van der Zalm, J.; Chen, A. Graphene-oxide-based electrochemical sensors for the sensitive detection of pharmaceutical drug naproxen. *Sensors* **2020**, *20*, 1252. [CrossRef] [PubMed]
23. Beitollahi, H.; Khalilzadeh, M.A.; Tajik, S.; Safaei, M.; Zhang, K.; Jang, H.W.; Shokouhimehr, M. Recent advances in applications of voltammetric sensors modified with ferrocene and its derivatives. *ACS Omega* **2020**, *5*, 2049–2059. [CrossRef] [PubMed]
24. Kamble, B.; Garadkar, K.M.; Sharma, K.K.; Kamble, P.; Tayade, S.; Ajalkar, B.D. Determination of 4-nitrophenol using MoO_3 loaded glassy carbon electrode via electrochemical sensing approach. *J. Electrochem. Sci. Eng.* **2021**, *11*, 143–159.
25. Mohanraj, J.; Durgalakshmi, D.; Rakkesh, R.A.; Balakumar, S.; Rajendran, S.; Karimi-Maleh, H. Facile synthesis of paper based graphene electrodes for point of care devices: A double stranded DNA (dsDNA) biosensor. *J. Colloid Interface Sci.* **2020**, *566*, 463–472. [CrossRef]
26. Montazarolmahdi, M.; Masrournia, M.; Nezhadali, A. A new electrochemical approach for the determination of phenylhydrazine in water and wastewater samples using amplified carbon paste electrode. *Chem. Methodol.* **2020**, *4*, 732–742.
27. Eren, T.; Atar, N.; Yola, M.L.; Karimi-Maleh, H. A sensitive molecularly imprinted polymer based quartz crystal microbalance nanosensor for selective determination of lovastatin in red yeast rice. *Food Chem.* **2015**, *185*, 430–436. [CrossRef]
28. Han, S.; Zhang, X.; Sun, H.; Wei, J.; Wang, H.; Wang, S.; Zhang, Z. Electrochemical behavior and voltammetric determination of chloramphenicol and doxycycline using a glassy carbon electrode modified with single-walled carbon nanohorns. *Electroanalysis* **2022**, *34*, 735–742. [CrossRef]
29. Shamsi, A.; Ahour, F. Electrochemical sensing of thioridazine in human serum samples using modified glassy carbon electrode. *Adv. J. Chem. A* **2020**, *4*, 22–31.
30. Elobeid, W.H.; Elbashir, A.A. Development of chemically modified pencil graphite electrode based on benzo-18-crown-6 and multi-walled CNTs for determination of lead in water samples. *Prog. Chem. Biochem. Res.* **2019**, *2*, 24–33.
31. Karimi-Maleh, H.; Khataee, A.; Karimi, F.; Baghayeri, M.; Fu, L.; Rouhi, J.; Boukherroub, R. A green and sensitive guanine-based DNA biosensor for idarubicin anticancer monitoring in biological samples: A simple and fast strategy for control of health quality in chemotherapy procedure confirmed by docking investigation. *Chemosphere* **2022**, *291*, 132928. [CrossRef] [PubMed]
32. Raoof, J.B.; Ojani, R.; Beitollahi, H. Electrocatalytic determination of ascorbic acid at chemically modified carbon paste electrode with 2,7-bis (ferrocenyl ethynyl) fluoren-9-one. *Int. J. Electrochem. Sci.* **2007**, *2*, 534–548.
33. Couto, R.A.S.; Lima, J.L.F.C.; Quinaz, M.B. Recent developments, characteristics and potential applications of screen-printed electrodes in pharmaceutical and biological analysis. *Talanta* **2016**, *146*, 801–814. [CrossRef] [PubMed]
34. Arduini, F.; Micheli, L.; Moscone, D.; Palleschi, G.; Piermarini, S.; Ricci, F.; Volpe, G. Electrochemical biosensors based on nanomodified screen-printed electrodes: Recent applications in clinical analysis. *Trends Anal. Chem.* **2016**, *79*, 114–126. [CrossRef]
35. Rama, E.C.; Costa-García, A. Screen-printed Electrochemical Immunosensors for the Detection of Cancer and Cardiovascular Biomarkers. *Electroanalysis* **2016**, *28*, 1700–1715. [CrossRef]
36. Smart, A.; Crew, A.; Pemberton, R.; Hughes, G.; Doran, O.; Hart, J.P. Screen-printed carbon based biosensors and their applications in agri-food safety. *TrAC Trends Anal. Chem.* **2020**, *127*, 115898. [CrossRef]
37. Vasilescu, A.; Nunes, G.; Hayat, A.; Latif, U.; Marty, J.L. Electrochemical affinity biosensors based on disposable screen-printed electrodes for detection of food allergens. *Sensors* **2016**, *16*, 1863. [CrossRef]
38. Barton, J.; García, M.B.G.; Santos, D.H.; Fanjul-Bolado, P.; Ribotti, A.; McCaul, M.; Magni, P. Screen-printed electrodes for environmental monitoring of heavy metal ions: A review. *Microchim. Acta* **2016**, *183*, 503–517.
39. Li, M.; Li, Y.T.; Li, D.W.; Long, Y.T. Recent developments and applications of screen-printed electrodes in environmental assays-A review. *Anal. Chim. Acta* **2012**, *734*, 31–44. [CrossRef]
40. Hayat, A.; Marty, J.L. Disposable screen printed electrochemical sensors: Tools for environmental monitoring. *Sensors* **2014**, *14*, 10432–10453. [CrossRef]
41. Tajik, S.; Dourandish, Z.; Garkani Nejad, F.; Aghaei Afshar, A.; Beitollahi, H. Voltammetric determination of isoniazid in the presence of acetaminophen utilizing MoS_2-nanosheet-modified screen-printed electrode. *Micromachines* **2022**, *13*, 369. [CrossRef] [PubMed]
42. Wang, X.; Zhang, Z.; Wu, G.; Xu, C.; Wu, J.; Zhang, X.; Liu, J. Applications of electrochemical biosensors based on functional antibody modified screen printed electrodes: A review. *Anal. Methods* **2021**, *14*, 7–16. [CrossRef] [PubMed]
43. Beitollahi, H.; Garkani Nejad, F.; Dourandish, Z.; Tajik, S. A novel voltammetric amaranth sensor based on screen printed electrode modified with polypyrrole nanotubes. *Environ. Res.* **2022**, *214*, 113725. [CrossRef] [PubMed]
44. Mirbaloochzehi, M.R.; Rezvani, A.; Samimi, A.; Shayesteh, M. Application of a novel surfactant-modified natural nano-zeolite for removal of heavy metals from drinking water. *Adv. J. Chem. A* **2020**, *3*, 612–620.
45. Fang, X.; Cao, J.; Shen, A. Advances in anti-breast cancer drugs and the application of nano-drug delivery systems in breast cancer therapy. *J. Drug Deliv. Sci. Technol.* **2020**, *57*, 101662. [CrossRef]
46. Karimi-Maleh, H.; Karaman, C.; Karaman, O.; Karimi, F.; Vasseghian, Y.; Fu, L.; Mirabi, A. Nanochemistry approach for the fabrication of Fe and N co-decorated biomass-derived activated carbon frameworks: A promising oxygen reduction reaction electrocatalyst in neutral media. *J. Nanostruct. Chem.* **2022**, *12*, 429–439. [CrossRef]
47. Mohammadi, S.S.; Ghasemi, N.; Ramezani, M. Bio-fabrication of silver nanoparticles using naturally available wild herbaceous plant and its antibacterial activity. *Eurasian Chem. Commun.* **2020**, *2*, 87–102.
48. Liu, Y.; Chen, Z.; Li, Z.; Zhao, N.; Xie, Y.; Du, Y.; Yang, Y. CoNi nanoalloy-Co-N_4 composite active sites embedded in hierarchical porous carbon as bi-functional catalysts for flexible Zn-air battery. *Nano Energy* **2022**, *99*, 107325. [CrossRef]

49. Khakyzadeh, V.; Rezaei-Vahidian, H.; Sediqi, S.; Azimi, S.; Karimi-Nami, R. Programming adsorptive removal of organic azo dye from aqueous media using magnetic carbon nano-composite. *Chem. Methodol.* **2021**, *5*, 324–330.
50. Mohammadzadeh Jahani, P.; Beitollahi, H.; Garkani Nejad, F.; Dourandish, Z.; Di Bartolomeo, A. Screen-printed graphite electrode modified with Co3O4 nanoparticles and 2D graphitic carbon nitride as an effective electrochemical sensor for 4-aminophenol detection. *Nanotechnology* **2022**, *33*, 395702. [CrossRef]
51. Miraki, M.; Karimi-Maleh, H.; Taher, M.A.; Cheraghi, S.; Karimi, F.; Agarwal, S.; Gupta, V.K. Voltammetric amplified platform based on ionic liquid/NiO nanocomposite for determination of benserazide and levodopa. *J. Mol. Liq.* **2019**, *278*, 672–676. [CrossRef]
52. Sengar, M.; Saxena, S.; Satsangee, S.; Jain, R. Silver nanoparticles decorated functionalized multiwalled carbon nanotubes modified screen printed sensor for the voltammetric determination of butorphanol. *J. Appl. Organomet. Chem.* **2021**, *1*, 95–108.
53. Pyman, H.; Roshanfekr, H.; Ansari, S. DNA-based electrochemical biosensor using chitosan–carbon nanotubes composite film for biodetection of Pirazon. *Eurasian Chem. Commun.* **2020**, *2*, 213–225.
54. Karimi-Maleh, H.; Sheikhshoaie, M.; Sheikhshoaie, I.; Ranjbar, M.; Alizadeh, J.; Maxakato, N.W.; Abbaspourrad, A. A novel electrochemical epinine sensor using amplified CuO nanoparticles and an-hexyl-3-methylimidazolium hexafluorophosphate electrode. *New J. Chem.* **2019**, *43*, 2362–2367. [CrossRef]
55. Tajik, S.; Beitollahi, H.; Torkzadeh-Mahani, M. Electrochemical immunosensor for the detection of anti-thyroid peroxidase antibody by gold nanoparticles and ionic liquid-modified carbon paste electrode. *J. Nanostruct. Chem.* **2022**, *12*, 581–588. [CrossRef]
56. Kouadio, K.E.; Kambiré, O.; Koffi, K.S.; Ouattara, L. Electrochemical oxidation of paracetamol on boron-doped diamond electrode: Analytical performance and paracetamol degradation. *J. Electrochem. Sci. Eng.* **2021**, *11*, 71–86.
57. Tajik, S.; Askari, M.B.; Ahmadi, S.A.; Garkani Nejad, F.; Dourandish, Z.; Razavi, R.; Beitollahi, H.; Di Bartolomeo, A. Electrochemical sensor based on $ZnFe_2O_4$/RGO nanocomposite for ultrasensitive detection of hydrazine in real samples. *Nanomaterials* **2022**, *12*, 491. [CrossRef]
58. Alavi-Tabari, S.A.; Khalilzadeh, M.A.; Karimi-Maleh, H. Simultaneous determination of doxorubicin and dasatinib as two breast anticancer drugs uses an amplified sensor with ionic liquid and ZnO nanoparticle. *J. Electroanal. Chem.* **2018**, *811*, 84–88. [CrossRef]
59. Sadeghi, H.; Shahidi, S.A.; Naghizadeh Raeisi, S.; Ghorbani-HasanSaraei, A.; Karimi, F. Electrochemical determination of folic acid in fruit juices samples using electroanalytical sensor amplified with CuO/SWCNTs and 1-butyl-2, 3-dimethylimidazolium hexafluorophosphate. *Chem. Methodol.* **2020**, *4*, 743–753.
60. Moshirian-Farahi, S.S.; Zamani, H.A.; Abedi, M. Nano-molar level determination of isoprenaline in pharmaceutical and clinical samples; A nanostructure electroanalytical strategy. *Eurasian Chem. Commun.* **2020**, *2*, 702–711. [CrossRef]
61. Karimi-Maleh, H.; Karimi, F.; Orooji, Y.; Mansouri, G.; Razmjou, A.; Aygun, A.; Sen, F. A new nickel-based co-crystal complex electrocatalyst amplified by NiO dope Pt nanostructure hybrid; a highly sensitive approach for determination of cysteamine in the presence of serotonin. *Sci. Rep.* **2020**, *10*, 11699. [CrossRef] [PubMed]
62. Karimi-Maleh, H.; Darabi, R.; Shabani-Nooshabadi, M.; Baghayeri, M.; Karimi, F.; Rouhi, J.; Karaman, C. Determination of D&C Red 33 and Patent Blue V Azo dyes using an impressive electrochemical sensor based on carbon paste electrode modified with ZIF-8/g-C_3N_4/Co and ionic liquid in mouthwash and toothpaste as real samples. *Food Chem. Toxicol.* **2022**, *162*, 112907. [PubMed]
63. Garkani Nejad, F.; Sheikhshoaie, I.; Beitollahi, H. Simultaneous detection of carmoisine and tartrazine in food samples using GO-Fe_3O_4-PAMAM and ionic liquid based electrochemical sensor. *Food Chem. Toxicol.* **2022**, *162*, 112864. [CrossRef] [PubMed]
64. Kanagavalli, P.; Andrew, C.; Veerapandian, M.; Jayakumar, M. In-situ redox-active hybrid graphene platform for label-free electrochemical biosensor: Insights from electrodeposition and electroless deposition. *TrAC Trends Anal. Chem.* **2021**, *143*, 116413. [CrossRef]
65. Kanagavalli, P.; Pandey, G.R.; Murugan, P.; Veerapandian, M. Electrochemical and DFT studies of andrographolide on electrochemically reduced graphene oxide for anti-viral herbaceutical sensor. *Anal. Chim. Acta* **2022**, *1209*, 339877. [CrossRef]
66. Luo, X.; Morrin, A.; Killard, A.J.; Smyth, M.R. Application of nanoparticles in electrochemical sensors and biosensors. *Electroanalysis* **2006**, *18*, 319–326. [CrossRef]
67. Chen, J.; Wang, X.; Wang, J.; Lee, P.S. Sulfidation of NiMn-layered double hydroxides/graphene oxide composites toward supercapacitor electrodes with enhanced performance. *Adv. Energy Mater.* **2016**, *6*, 1501745. [CrossRef]
68. Yang, J.; Yu, C.; Fan, X.; Qiu, J. 3D architecture materials made of NiCoAl-LDH Nanoplates coupled with NiCo-carbonate hydroxide nanowires grown on flexible graphite paper for asymmetric supercapacitors. *Adv. Energy Mater.* **2014**, *4*, 1400761. [CrossRef]
69. Wang, L.; Dong, Z.H.; Wang, Z.G.; Zhang, F.X.; Jin, J. Layered α-$Co(OH)_2$ nanocones as electrode materials for pseudocapacitors: Understanding the effect of interlayer space on electrochemical activity. *Adv. Funct. Mater.* **2013**, *23*, 2758–2764. [CrossRef]
70. Sarfraz, M.; Shakir, I. Recent advances in layered double hydroxides as electrode materials for high-performance electrochemical energy storage devices. *J. Energy Storage* **2017**, *13*, 103–122. [CrossRef]
71. Wang, Q.; O'Hare, D. Recent advances in the synthesis and application of layered double hydroxide (LDH) nanosheets. *Chem. Rev.* **2012**, *112*, 4124–4155. [CrossRef] [PubMed]

72. Sanati, S.; Abazari, R.; Morsali, A. Enhanced electrochemical oxygen and hydrogen evolution reactions using an NU-1000@NiMn-LDHS composite electrode in alkaline electrolyte. *Chem. Commun.* **2020**, *56*, 6652–6655. [CrossRef] [PubMed]
73. Baig, N.; Sajid, M. Applications of layered double hydroxides based electrochemical sensors for determination of environmental pollutants: A review. *Trends Environ. Anal. Chem.* **2017**, *16*, 1–15. [CrossRef]
74. Varadwaj, G.; Nyamori, V.O. Layered double hydroxide-and graphene-based hierarchical nanocomposites: Synthetic strategies and promising applications in energy conversion and conservation. *Nano Res.* **2016**, *9*, 3598–3621. [CrossRef]
75. Prevot, V.; Tokudome, Y. 3D hierarchical and porous layered double hydroxide structures: An overview of synthesis methods and applications. *J. Mater. Sci.* **2017**, *52*, 11229–11250. [CrossRef]
76. Zhang, X.; Lu, W.; Tian, Y.; Yang, S.; Zhang, Q.; Lei, D.; Zhao, Y. Nanosheet-assembled NiCo-LDH hollow spheres as high-performance electrodes for supercapacitors. *J. Colloid Interface Sci.* **2022**, *606*, 1120–1127. [CrossRef]
77. Li, M.; Yuan, P.; Guo, S.; Liu, F.; Cheng, J.P. Design and synthesis of Ni-Co and Ni-Mn layered double hydroxides hollow microspheres for supercapacitor. *Int. J. Hydrog. Energy* **2017**, *42*, 28797–28806. [CrossRef]
78. Amin, K.M.; Muench, F.; Kunz, U.; Ensinger, W. 3D NiCo-Layered double Hydroxide@ Ni nanotube networks as integrated free-standing electrodes for nonenzymatic glucose sensing. *J. Colloid Interface Sci.* **2021**, *591*, 384–395. [CrossRef]
79. Vilian, A.E.; Ranjith, K.S.; Lee, S.J.; Umapathi, R.; Hwang, S.K.; Oh, C.W.; Han, Y.K. Hierarchical dense Ni− Co layered double hydroxide supported carbon nanofibers for the electrochemical determination of metronidazole in biological samples. *Electrochim. Acta* **2020**, *354*, 136723. [CrossRef]
80. Fang, X.; Zang, J.; Wang, X.; Zheng, M.S.; Zheng, N. A multiple coating route to hollow carbon spheres with foam-like shells and their applications in supercapacitor and confined catalysis. *J. Mater. Chem. A* **2014**, *2*, 6191–6197. [CrossRef]
81. Soltani, R.; Marjani, A.; Shirazian, S. A hierarchical LDH/MOF nanocomposite: Single, simultaneous and consecutive adsorption of a reactive dye and Cr (vi). *Dalton Trans.* **2020**, *49*, 5323–5335. [CrossRef] [PubMed]
82. Cao, Y.; Khan, A.; Kurniawan, T.A.; Soltani, R.; Albadarin, A.B. Synthesis of hierarchical micro-mesoporous LDH/MOF nanocomposite with in situ growth of UiO-66-$(NH_2)_2$ MOF on the functionalized NiCo-LDH ultrathin sheets and its application for thallium (I) removal. *J. Mol. Liq.* **2021**, *336*, 116189. [CrossRef]
83. Bai, L.; Duan, S.; Jiang, W.; Liu, M.; Wang, S.; Sang, M.; Xuan, S. High performance polydopamine-functionalized mesoporous silica nanospheres for U (VI) removal. *Appl. Surf. Sci.* **2017**, *426*, 1121–1132. [CrossRef]
84. Palapa, N.R.; Mohadi, R.; Rachmat, A. Adsorption study of malachite green removal from aqueous solution using Cu/M^{3+} (M^{3+}= Al, Cr) layered double hydroxide. *Mediterr. J. Chem.* **2020**, *10*, 33–45. [CrossRef]
85. Karim-Nezhad, G.; Khanaliloo, S.; Khorablou, Z.; Dorraji, P.S. Signal amplification for sumatriptan sensing based on polymeric surface decorated with Cu nanoparticles. *J. Serb. Chem. Soc.* **2018**, *83*, 449–462. [CrossRef]
86. Ghalkhani, M.; Shahrokhian, S.; Ghorbani-Bidkorbeh, F. Voltammetric studies of sumatriptan on the surface of pyrolytic graphite electrode modified with multi-walled carbon nanotubes decorated with silver nanoparticles. *Talanta* **2009**, *80*, 31–38. [CrossRef]
87. Jalali-Sarvestani, M.R.; Madrakian, T.; Afkhami, A. Voltammetric Determination of Sumatriptan by an Overoxidized Poly (p-aminophenol) Modified Glassy Carbon Electrode. *Anal. Bioanal. Chem. Res.* **2021**, *8*, 245–259.

biosensors

MDPI

Article

Ultrasensitive Electrochemical Detection of Butylated Hydroxy Anisole via Metalloporphyrin Covalent Organic Frameworks Possessing Variable Catalytic Active Sites

Huacong Chu [1], Xin Sun [1], Xiaoqian Zha [1], Shifa Ullah Khan [2,*] and Yang Wang [1,*]

1 School of Chemistry and Chemical Engineering, Yangzhou University, Yangzhou 225002, China
2 The Institute of Chemistry, Faculty of Science, University of Okara, Renala Campus, Punjab 56300, Pakistan
* Correspondence: shifand786@gmail.com (S.U.K.); wangy@yzu.edu.cn (Y.W.)

Abstract: Three novel two-dimensional metalloporphyrin COFs (MPor−COF−366, M = Fe, Mn, Cu) were fabricated by changing the metal atoms in the center of the porphyrin framework. The physicochemical characteristics of MPor−COF−366 (M = Fe, Mn, Cu) composites were fully analyzed by diverse electron microscopy and spectroscopy. Under optimal conditions, experiments on determining butylated hydroxy anisole (BHA) at FePor−COF−366/GCE were conducted using differential pulse voltammetry (DPV). It is noted that the FePor−COF−366/GCE sensor showed excellent electrocatalytic performance in the electrochemical detection of BHA, compared with MnPor−COF−366/GCE and CuPor−COF−366/GCE. A linear relationship was obtained for 0.04–1000 μM concentration of BHA, with a low detection limit of 0.015 μM. Additionally, the designed sensor was successfully employed to detect BHA in practical samples, expanding the development of COF-based composites in electrochemical applications.

Keywords: electrochemical biosensor; covalent organic frameworks; metalloporphyrin; BHA

Citation: Chu, H.; Sun, X.; Zha, X.; Khan, S.U.; Wang, Y. Ultrasensitive Electrochemical Detection of Butylated Hydroxy Anisole via Metalloporphyrin Covalent Organic Frameworks Possessing Variable Catalytic Active Sites. *Biosensors* **2022**, *12*, 975. https://doi.org/10.3390/bios12110975

Received: 13 September 2022
Accepted: 4 November 2022
Published: 6 November 2022

1. Introduction

As a synthetic phenolic antioxidant, butylated hydroxy anisole (BHA) is frequently employed in food packaging, biofuels, and pharmaceutical formulations [1]. It has the effect of preventing or delaying the oxidative deterioration of food, thereby improving food stability and prolonging the shelf life of food [2,3]. If the content of BHA exceeds the specified level, it can cause serious health problems in humans, including nutrient loss and even toxic effects [4,5]. Hence, the negative impacts of BHA on human health restrict its application in food goods, and a quick, sensitive, and accurate method to quantify BHA is required.

Several techniques have been developed for determining BHA, including micellar electrokinetic capillary chromatography, gas chromatography, high-performance liquid chromatography, and spectrophotometry [6–10]. However, these methods often encounter difficulties associated with cumbersome or time-consuming procedures. Electrochemical technology has been utilized for the quantitative analysis of BHA owing to its advantages of simple operation, high sensitivity, and rapid response [11–14]. However, developing advanced materials to achieve high electrochemical performance remains a major research goal.

Covalent organic frameworks (COFs) are formed by powerful covalent links between light elements with a reticular network of molecular building blocks, and exhibit an exceptionally high surface area and high stability, displaying a wide range of applications [15–18]. Porphyrins are widely used as building blocks for the assembly of covalent organic frameworks because of their high stability, easy chelation with metal ions, and ease of modification [19]. Several porphyrin-based COFs (Por-COFs) have been constructed for use in electro chemiluminescence, heterogeneous catalysis, and photooxidization reactivity [20–22].

In addition, metalloporphyrin COFs (MPor-COFs) possess a unique N-C framework, which offers a novel platform for the preparation of self-supporting M-N-C sites with intrinsic metal coordination for use as porous electrocatalysts [23]. However, to our knowledge, the usage of MPor-COFs as electrode materials to improve electrochemical performance has rarely been investigated.

In this work, we synthesized a series of two-dimensional nanomaterials based on MPor−COF−366. Firstly, Por−COF−366 was produced by the condensation reaction of 5,10,5,20-tetra(4-aminophenyl)porphyrin (TAPP) and 1,3,5-benztriformaldehyde (BDA) under acidic conditions. Strong covalent bonds were constructed by using porphyrin macrocycle molecules as building blocks. Then, Por−COF−366 was impregnated with $FeSO_4{\cdot}6H_2O$, $MnSO_4{\cdot}4H_2O$ and $CuCl_2{\cdot}6H_2O$, respectively, to obtain the final composites MPor−COF−366 (M = Fe, Mn, Cu). Subsequently, to fabricate an electrochemical sensor, the obtained nanocomposites were fixed on the glass carbon electrode (GCE) surface. The sensor exhibited remarkable catalytic activity and selectivity towards BHA, as the synergistic effect of Por-COFs and metal atoms endowed the sensor with unique properties.

2. Materials and Methods

2.1. Materials

Chemicals were used exactly as bought and all were of analytical grade. The following substances were obtained from Shanghai Tansoole: 5,10,5,20-tetra(4-aminophenyl)porphyrin (TAPP), 1,3,5-benztriformaldehyde (BDA), 1,4-dioxane, ethanol, tetrahydrofuran, and acetone. Other chemicals were bought from Sinopharm Chemical Reagent Co., Ltd. (Shanghai, China), including $CuCl_2{\cdot}6H_2O$, $FeSO_4{\cdot}6H_2O$, $MnSO_4{\cdot}4H_2O$, CH_2Cl_2, methanol, acetic acid, and 1,3,5-trimethylbenzene. As a supporting electrolyte, phosphate buffer solution was prepared using 0.1 M NaH_2PO_4/Na_2HPO_4.

2.2. Synthesis of Por−COF−366, MPor−COF−366 (M = Fe, Mn, Cu)

Por−COF−366 was synthesized according to a previously reported method with some modifications [24]. TAPP (18 mg, 0.025 mM), BDA (10 mg, 0.075 mM), 1.0 mL ethanol, 1.0 mL mesitylene, and 0.25 mL acetic acid (6 M) were added into a 5 mL glass bottle. The mixed solution was ultrasonically dispersed for 5 min, followed by stirring at room temperature for 15 min, then transferred to a Teflon-sealed reactor and heated to 120 °C (72 h). Following the reaction, the product was filtered and collected, then washed with 1, 4-dioxane, THF, and acetone, respectively, and finally dried under vacuum at 40 °C for 12 h to obtain Por−COF−366.

Por−COF−366 (20.0 mg) and $FeSO_4{\cdot}7H_2O$ (25 mg, 0.1 mM) were dissolved in a mixture of 12 mL dichloromethane and methanol. The mixture was stirred continuously for 24 h. After centrifugation, the composites were washed with CH_2Cl_2 and were dried under vacuum at 60 °C for 12 h to obtain the product FePor−COF−366. After replacing the metal salts with $MnCl_2{\cdot}6H_2O$ and $CuCl_2{\cdot}6H_2O$, MnPor−COF−366 and CuPor−COF−366 were obtained by the same synthesis process.

2.3. Preparation of MPor−COF−366 (M = Fe, Mn, Cu) Modified Electrode

GCE was meticulously polished with α-Al_2O_3 before being sequentially rinsed in water and ethanol and then dried under N_2. The GCE surface was coated with 5.0 μL of FePor−COF−366 (1 mg/mL) suspension, and dried using infrared light. After that, 2.0 μL 5 % chitosan solution was selected as binder. MnPor−COF−366/GCE, CuPor−COF −366/GCE, and Por−COF−366/GCE were all modified using the same method.

2.4. Preparation of Samples

Rapeseed, corn, and peanut oils were bought at a nearby grocery. Then, 1.0 g of the material was dissolved in 10 mL of ethanol. The product was centrifuged after being oscillated for 10 min. The supernatant was collected, and filtered through a Millipore nylon filter with a 0.22 μm pore size. A 1.0 mL aliquot of the final product was put into 9.0 mL of pH 4.0 PBS for further analysis.

2.5. Characterizations

The morphology of the composites was examined using transmission electron microscopy (TEM, Tecnai 12, 120 KV, Amsterdam, Netherlands) and scanning electron microscopy (SEM, S-4800II, Tokyo, Japan). A Cary 610/670 infrared microspectrometer (Varian, Palo Alto, Santa Clara, CA, USA) was employed to obtain the FTIR spectra. Utilizing a D8 Advance X-ray diffractometer (Bruker Co., Karlsruhe, Germany), the X-ray diffraction (XRD) patterns were gathered from 1.5° to 80° at ambient temperature. XPS testing was carried out using an ESCALAB 250Xi XPS spectrometer from Thermo Scientific, Massachusetts, USA with an aluminum Kα X-ray source.

2.6. Electrochemical Analysis

A CHI852C electrochemical analyzer (Shanghai Chenhua Co., Shanghai, China) was applied for all electrochemical tests. A standard three-electrode system was employed to perform cyclic voltammetry (CV) and differential pulse voltammetry (DPV).

3. Results and Discussions

3.1. Characterizations of Synthesized Composites

The morphologies of Por−COF−366 and MPor−COF−366 composites were characterized using SEM and TEM measurements. In Figure 1A, it is shown that Por−COF−366 is a generally spherical nanomaterial with a size of 180 nm on average. After coordination with metal ions, the surfaces of MPor−COF−366 composites became rough, but the morphology did not change significantly.

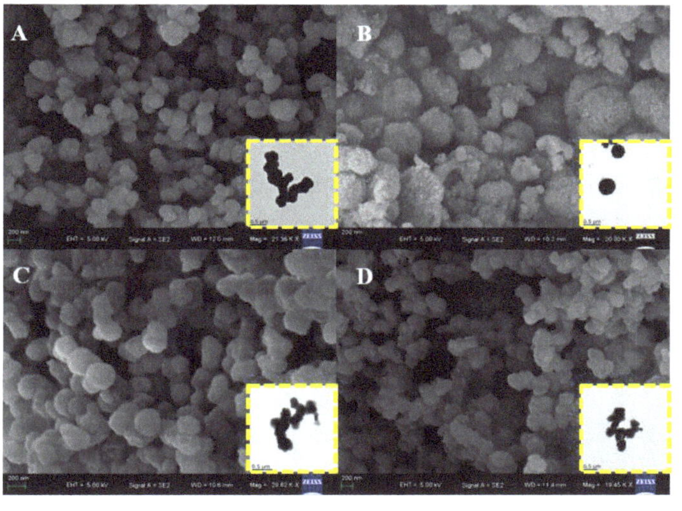

Figure 1. SEM and TEM (inset) images of (**A**) Por−COF−366, (**B**) FePor−COF−366, (**C**) MnPor−COF−366, and (**D**) CuPor−COF−366.

To further confirm the structure and crystallinity of the produced Por−COF−366 and MPor−COF−366, spectral PXRD analysis was performed. As displayed in Figure 2A, the obvious characteristic diffraction peak at 2θ = 3.5° was attributed to the (100) plane. Other weaker peaks appeared at 2θ = 6.1° and 7.9°, corresponding to the (110) and (200) planes, respectively, which were consistent with the reported data [25]. In addition, Por−COF−366 and MPor−COF−366 composites exhibited sharp pyrrole ring vibration at 797 cm^{-1}, and the characteristic stretching vibrations of C=N at 1623 cm^{-1} were seen in FTIR spectrum

(Figure 2B). There were also absorbance peaks at 1691 cm^{-1} and 3325 cm^{-1}, which were attributed, respectively, to the residual C=O and stretching N-H vibration of residual $-NH_2$ groups [26].

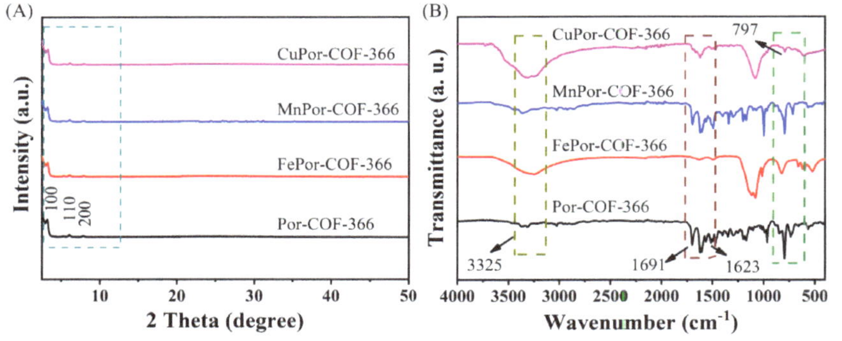

Figure 2. (**A**) XRD patterns and (**B**) FTIR spectra of Por−COF−366 and MPor−COF−366 (M = Fe, Mn, Cu).

Furthermore, Figure 3 shows the XPS measurements to evaluate the chemical state of the as-synthesized Por−COF−366 and MPor−COF−366 composites. The survey spectrum is shown in Figure 3A, revealing that the composites were composed of Fe, Mn, Cu, N, O, and C species. Figure 3B shows the spectrum of Fe 2p. The FePor−COF−366 was successfully synthesized, as evidenced by the four main peaks at 711.4, 715.9, 724.5, and 728.5 eV, which were consistent with the Fe^{2+} $2p_{3/2}$, Fe^{3+} $2p_{3/2}$, Fe^{2+} $2p_{1/2}$, and Fe^{3+} $2p_{1/2}$, respectively [27,28]. Additionally, Mn 2p in the XPS spectra shown in Figure 3C contained peaks at around 642.3 and 654.6 eV that were attributed to Mn $2p_{2/3}$ and Mn $2p_{1/2}$, respectively. Manganese ions were found to exist as Mn^{4+} with spin-orbital splitting of 12.3 eV [29]. Another peak at 646.7 eV was assigned to Mn^{3+}, indicating that the process may have created Mn in a lower oxidation state [30]. Additionally, as depicted in Figure 3D, the typical peaks at 954.2 and 934.4 eV corresponded to Cu $2p_{1/2}$ and Cu $2p_{3/2}$, respectively [31]. Typical peaks of the Cu(II) state were observed as two satellite peaks at about 963.12 and 943.46 eV [32]. These findings agreed with those of the XRD and FTIR analyses.

3.2. Electrochemical Properties

The changes between the electrode interface and the solution were investigated using electrochemical impedance spectroscopy (EIS) in 5.0 mM $[Fe(CN)_6]^{3-/4-}$ electrolyte [33]. The characteristic EIS plots of the different electrodes are shown, and an equivalent circuit (Randles model) is provided in the inset of Figure 4A. The Randles equivalent electrical circuit model consists of the electrode surface resistance (Rs), the element of interfacial electron transfer resistance (Rct), Warburg impedance (Zw), and the constant phase angle element (CPE). In the Nyquist impedance plot, a straight line at low frequencies was associated with a diffusion-limited process, and a small semicircle at high frequencies was associated with an electron-transfer-limited process. The semicircle is equal to the value of Rct, and the GCE revealed the largest Rct value (166.5 Ω). After immobilizing the Por−COF−366, the Rct value decreased to 77.2 Ω, indicating that the presence of Por−COF−366 lowered the charge transfer barrier, which may have been caused by the π-π stacking interaction between Por−COF−366. Because electrons were delocalized on all conjugated nanocomposites, the mobility of electrons was effectively enhanced. The Rct values of the MPor−COF−366 modified electrodes followed the orderFePor−COF−366 (88.6 Ω) < MnPor−COF−366 (96.7 Ω) < CuPor−COF−366 (104.1 Ω), indicating that MPor−COF−366 can effectively improve electron transfer and electrochemical activity.

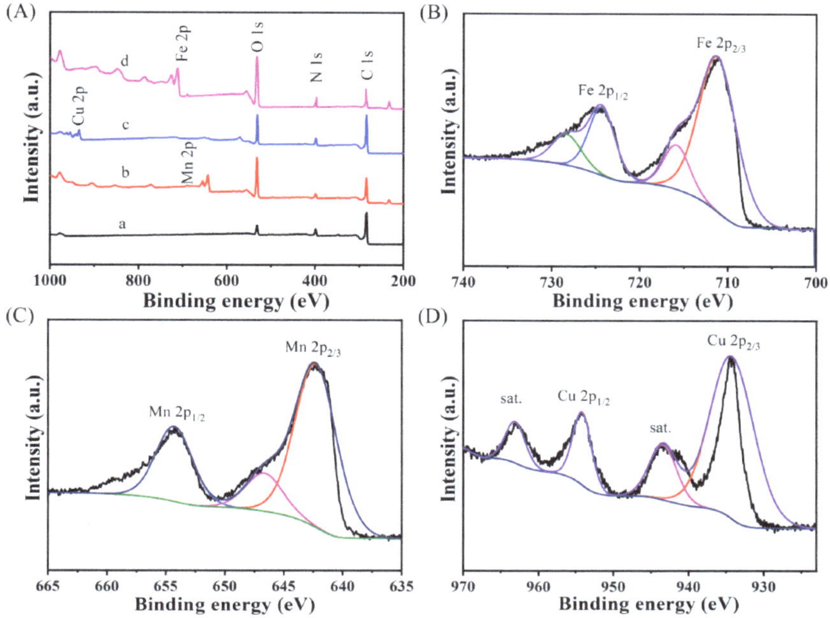

Figure 3. (**A**) Full spectrum diagram of (a) Por−COF−366, (b) MnPor−COF−366, (c) CuPor−COF−366, and (d) FePor−COF−366. XPS analysis of (**B**) Fe 2p, (**C**) Mn 2p, and (**D**) Cu 2p.

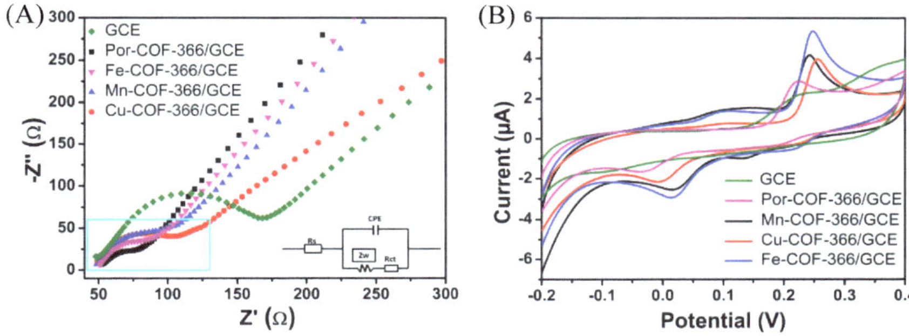

Figure 4. (**A**) EIS of bare GCE, Por−COF−366, and MPor−COF−366 (M = Fe, Mn, Cu). (Inset is the Randle circuit model). (**B**) CV values of bare GCE, Por−COF−366, and MPor−COF−366 (M = Fe, Mn, Cu) in 0.1 M PBS (pH 4.0) with the presence of 50 μM BHA at a scan rate of 100 mV s^{-1}.

The electrochemical behaviours of GCE, Por−COF−366/GCE, and MPor−COF−366/GCE were investigated through CV. As exhibited in Figure 4B, the redox peaks in the modified electrodes were well-defined and the electrochemical activity was significantly improved compared with the weak redox peaks on GCE. The anodic peaks located at 0.106 and 0.247 V belonged to butylated hydroquinone and BHA, respectively. The cathodic peak at 0.014 was assigned to *tert*-butylquinone (TBQ). Compared with CuPor−COF−366/GCE and MnPor−COF−366/GCE, a narrowed redox peak potential separation (ΔE_p) and enhanced peak current value were observed after modification with FePor−COF−366/GCE. The ΔE_p value in a cyclic voltammogram is one of the important factors for estimating the electron charge-transfer rate at the electrode–electrolyte interface. Generally, the ΔE_p value is inversely proportional to the charge-transfer rate constant. Here, the ΔE_p values decreased in the sequence CuPor−COF−366/GCE (263.5 mV) > MnPor−COF−366/GCE

(232.5 mV) > FePor−COF−366/GCE (229.5 mV). The results indicated that the engineered FePor−COF−366 largely minimized the electron-transfer barrier between the redox species and the electrode. The peak currents of the various modified electrodes followed the order FePor−COF−366 > MnPor−COF−366 > CuPor−COF−366 > Por−COF−366 > bare GCE, indicating that the introduction of Por−COF−366 and MPor−COF−366 can effectively improve the electrochemical activity of the electrode. Among all the investigated samples, the FePor−COF−366 exhibited the best electrochemical response. This phenomenon may be due to the fact that atomically dispersed Fe in the porphyrin macrocycle was coordinated as the catalytically active centre, which was fully exposed to the interaction with BHA, providing a greater specific surface area, increasing the quantity of open active sites, and improving the speed of electron transfer in the composite.

3.3. Optimization of the Experimental Parameters

Figure 5A depicts the results of investigation into the effect of pH values on the DPV response of BHA, for FePor−COF−366/GCE over a potential range of 0.1 to 0.8 V at various pH ranges from 2.0 to 6.0. With the increase of solution pH, the oxidation peak current of BHA showed first an increasing trend and then a decrease. The maximum current of BHA was reached at pH 4.0. At pH values between 4.0 and 6.0, a decrease was observed in the oxidation peak current value. Hence, PBS with pH value of 4.0 was selected for the analysis that followed. Furthermore, as the electrolyte pH increased, the anodic peak potential (E_{pa}) of BHA oxidation exhibited markedly negative movement, showing that protons were implicated in the electro-oxidation of BHA. The linear formula for the relationship between pH and the oxidation peak potential of BHA is: $E_{pa}(V) = -0.0572\ pH + 0.6832$ ($R^2 = 0.9949$). The peak potential and pH are related by:

$$\frac{dE_p}{dpH} = \frac{2.303mRT}{nF}$$

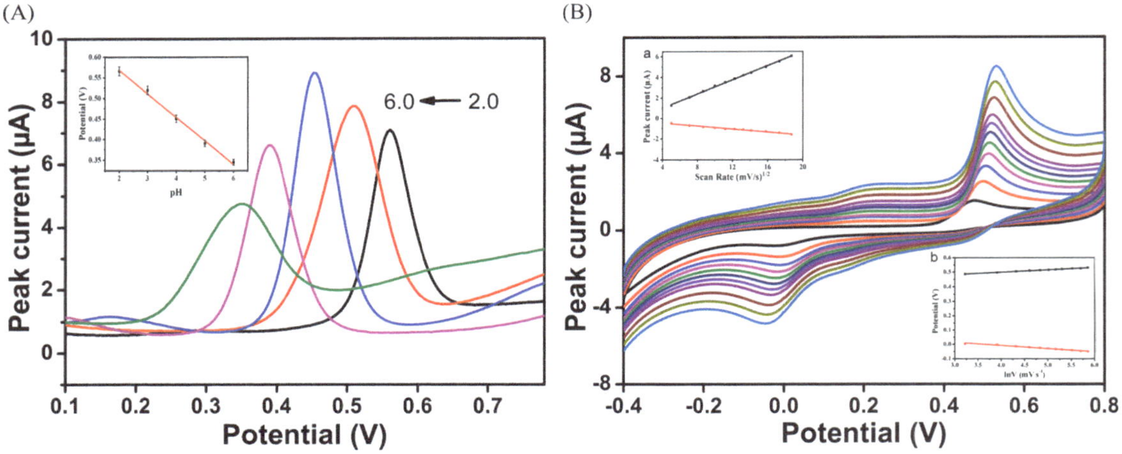

Figure 5. (**A**) Effect of pH on the peak currents of BHA, the inset is the linear relationship between pH values and electric potential; (**B**) CV values of the FePor−COF−366/GCE in 0.1 M PBS (pH 4.0) containing 50 μM BHA at different scan rates, the insets are (**a**) the linear relationship between currents and sweep-speed square root, and (**b**) the linear relationship between redox peak potentials (E_p) and lnv.

Substituting the equivalency of the slope values of the E_p versus pH plot resulted in an m/n ratio of 1, showing that the process involved an equal number of protons and electrons [34].

CV experiments at various scan speeds were carried out to investigate the electrochemical redox reaction kinetics of BHA on FePor−COF−366/GCE surfaces. Figure 5B indicates how the peak current grew progressively as the scan rate increased between 25 and 350 mV s^{-1}. The anodic and cathodic peak currents and the square root of the scan rates were linearly related, and followed the equations: I_{pa} (μA) = −0.3314 + 0.3415 $\nu^{1/2}$, (R^2 = 0.9986); I_{pc} (μA) = −0.1525 − 0.0737 $\nu^{1/2}$, (R^2 = 0.9834) (Inset a in Figure 5B). Diffusion was therefore primarily in charge of controlling the redox reaction process. The reduction peak potential (E_{pc}) shifted in a more negative direction as the scan rate increased, while the oxidation peak potential (E_{pa}) steadily moved in a more positive direction. The peak potentials and lnν can be represented by the linear equations: E_{pa} (V) = 0.4346 + 0.0163 lnν, (R^2 = 0.9967); E_{pc} (V) = 0.0812 − 0.0216 lnν, (R^2 = 0.9833) (Inset b in Figure 5B). The Laviron equation states:

$$E_{pa} = E^{\theta} + \left(\frac{RT}{\alpha nF}\right) \ln v \quad (1)$$

$$E_{pc} = E^{\theta} - \left[\frac{RT}{(1-\alpha)nF}\right] \ln v \quad (2)$$

The linear slopes of E_{pa} and E_{pc} relative to lnν are RT/αnF and RT/((1 − α)nF), respectively. n can be calculated as 2.1. Because the oxidation of BHA is a two-electron transfer process, two electrons and two protons are included in the redox reaction [35]. The reaction mechanism is shown in Figure 6.

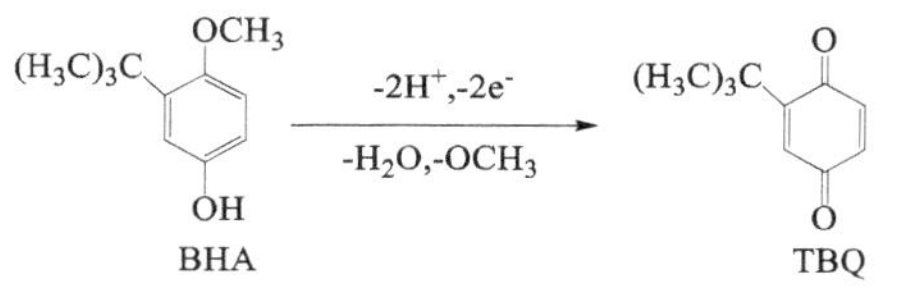

Figure 6. Electrochemical oxidation mechanism of BHA.

Furthermore, the accumulation step can be considered an effective method for improving the assay sensitivity. As shown in Figure 7A, the influence of accumulated potential on the peak current of BHA was tested from −0.3 to 0.2 V. The peak current reached its maximum value at −0.1 V, indicating that −0.1 V was favourable for the detection of BHA. As indicated in Figure 7B, considering the accumulation time, the peak current rose and reached its highest value at 60 s. As the amount of time increased, saturation of the BHA molecules occurred on the modified electrode. Therefore, 60 s was selected for subsequent studies.

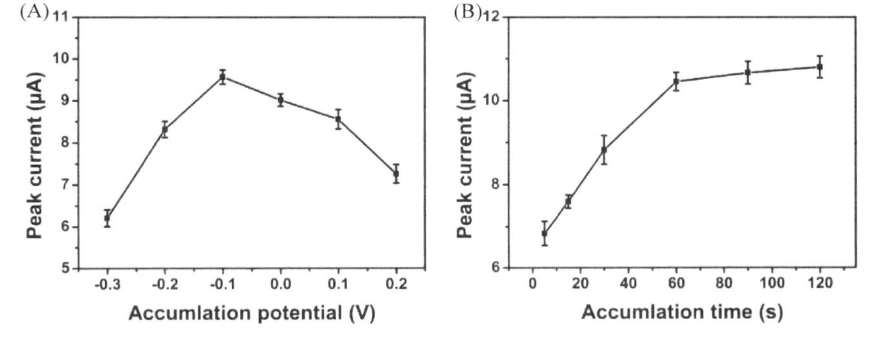

Figure 7. Influence of (**A**) accumulation potential and (**B**) time on the peak currents of BHA.

3.4. Analytical Performance

DPV was applied under optimal conditions to study the relationship between peak current and BHA concentration (Figure 8A). According to the linear regressions illustrated in Figure 8B, the peak currents of BHA were proportional to its concentrations between 0.04–100 and 100–1000 μM, with the linear regressions: I_{pa} (μA) = 0.3415 C − 0.3314 (C: 0.04–100 μM, R^2 = 0.998) and I_{pa} (μA) = 0.1525 C − 0.0737 (C: 100–1000 μM, R^2 = 0.997). The two linear regions in Figure 8B can be explained as follows: when the concentration of BHA was low, BHA was deposited on the electrode surface; at this time, the electrochemical oxidation process was mainly controlled by the adsorption process, and BHA molecules on the electrode surface were rapidly transformed. With the increase of BHA concentration, more BHA molecules were deposited on the surface of the modified electrode, and the mass transfer resistance increased, leading to the decrease of sensitivity. The detection limit was calculated as 0.015 μM, using a signal-to-noise ratio of 3 (S/N).

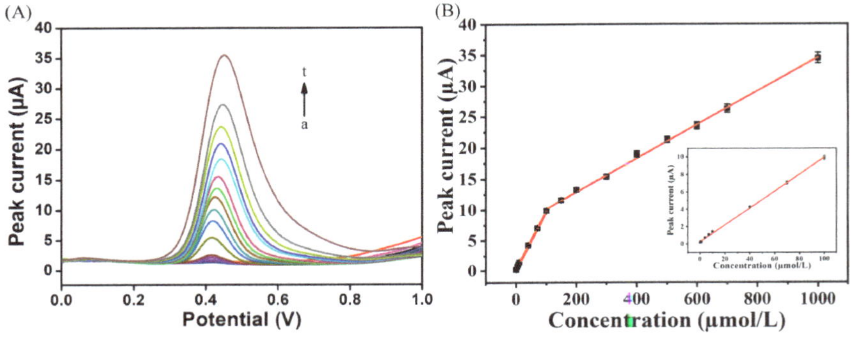

Figure 8. (**A**) DPV values for FePor−COF−366/GCE detection of BHA with different concentrations (from a to t: 0.04, 0.07, 0.1, 0.4, 0.7, 1.0, 4.0, 7.0, 10, 40, 70, 100, 150, 200, 300, 400, 500, 600, 700, 1000 μM) in 0.1 M pH 4.0 PBS. (**B**) Calibration plot of peak current versus BHA concentrations. Inset: trend of current intensity at low BHA concentrations.

According to the data in Table 1, comparing the performance of FePor−COF−366/GCE with other BHA sensors, FePor−COF−366/GCE had the lowest detection limit and the widest linear region. Several factors could account for its high analytical performance: (1) The metal porphyrins exist in a monomolecular state through the connection of metal ions, so that the metal ions are separated from each other and avoid aggregation, thereby increasing the catalytic active site and improving the catalytic activity; (2) the rigid and conjugated structure of Por-COFs at the molecular level endows them with inherent porous properties, enabling them to adsorb more BHA molecules; (3) Por-COFs have abundant π electron conjugated macrocycles, which can enhance affinity for BHA by π–π stacking and hydrogen bonding.

Table 1. Comparison of this work with other electrochemical materials for BHA detection.

Electrode Materials	Line Range (μM)	Detection Limit (μM)	Ref.
Poly carminic acid/MWCNT [1]	0.25–75	0.23	[36]
AuNPs [2]/GCE	0.55–8.3	0.22	[37]
SPE–MWCNT [3]	0.5–10	0.18	[38]
GCE/IrOxNPs [4]	1–280	0.60	[39]
ZnO TPHS [5]@GO [6] hybrid/GCE	0.3–60	0.04	[40]
Au-NP/Graphite	3.3–400	0.5	[41]
Surfactant/CPE [7]	1.1–10.2	0.07	[42]
Core-shell MIP [8]/GCE	0.6–300	1.62	[43]
FePor−COF−366/GCE	0.04–1000	0.015	This Work

[1] MWCNT: multiwalled carbon nanotube, [2] AuNPs: gold nanoparticles, [3] SPE-MWCNT: multi-walled carbon nanotube modified screen-printed electrodes, [4] IrOxNPs: iridium oxide nanoparticles, [5] TPHS: hierarchical triple-shelled porous hollow spheres, [6] GO: graphene oxide, [7] CPE: carbon paste electrode, [8] MIP: molecularly imprinted polymer.

3.5. Interference, Stability, and BHA Detection

A number of potential coexisting substances were evaluated by DPV in 50 μM BHA, to examine the selectivity of the sensor. These experiments showed that many inorganic ions, including Zn^{2+}, Mn^{2+}, Mg^{2+}, Cl^-, and SO_4^{2-}, did not obstruct the measurement of BHA. Furthermore, the BHA measurement was unaffected by ascorbic acid, glucose, and L-cysteine concentrations that were 20-fold greater. Butylated hydroxytoluene (BHT) is another phenolic antioxidant which shares the same hydroxyl group as BHA and can be oxidized close to the potential of BHA. It was found that BHT had no impact on the detection of BHA, even at a 20-fold concentration. Together, these findings demonstrate the great selectivity of the electrochemical sensor for BHA detection. Storage stability was assessed by storing a batch of freshly manufactured electrodes in a refrigerator at 4 °C and taking measurements every three days. After a lengthy storage period of 21 days, it was discovered that approximately 91.2% of the initial value was maintained. The outcomes amply demonstrated the sensor's ideal stability. Moreover, there was no evident decrease of current response observed after 35 cycles, and 95% of its original value was maintained. Three parallel measurements were taken for each sample, using the standard addition method, to assess the usefulness of the constructed electrochemical sensor. Table 2 showed that the recoveries ranged from 98.4 to 102.2%, and the RSD values ranged from 1.5% to 3.4%, indicating that the electrochemical sensor was capable of detecting the presence of BHA in actual samples.

Table 2. Determination of BHA in real samples by FePor−COF−366/GCE.

Samples	Added (μM)	Obtained (μM)	RSD (%)	Recovery (%)
Peanut oil	0	Not detected	-	-
	5.0	4.98	2.8	99.6
	50.0	50.6	3.4	101.2
Rapeseed oil	0	Not detected	-	-
	5.0	4.92	1.5	98.4
	50.0	49.4	2.3	100.4
Corn oil	0	Not detected	-	-
	5.0	5.11	3.0	102.2
	50.0	49.4	2.1	98.8

4. Conclusions

In conclusion, the pre-designed Por-COFs were successfully used as metallization modification matrixes, and the metal atoms were regulated in the form of coordinated metal-N4 as special functional modulators. The resulting FePor-COFs demonstrated distinctly higher electrocatalytic response for BHA in terms of wide linear range, low detection limit, high sensitivity, superior selectivity, and good reproducibility. This study not only describes an attractive and efficient catalyst electrode for BHA determination, but also opens a new avenue for the design and development of two-dimensional porphyrin COFs as sensors for electrochemical detection of small molecules.

Author Contributions: Conceptualization, methodology, data curation and writing—original draft, H.C.; methodology, data curation, X.S.; methodology, data curation, X.Z.; writing—review and editing, S.U.K.; funding acquisition, Y.W. All authors have read and agreed to the published version of the manuscript.

Funding: This research was funded by the National Natural Science Foundation of China (22174124), and a project funded by the Priority Academic Program Development of Jiangsu Higher Education Institutions (PAPD).

Institutional Review Board Statement: Not applicable.

Informed Consent Statement: Not applicable.

Data Availability Statement: Not available.

Conflicts of Interest: The authors declare no conflict of interest.

References

1. Wang, P.; Han, C.Y.; Zhou, F.Y.; Lu, J.S.; Han, X.G.; Wang, Z.W. Electrochemical determination of tert-butylhydroquinone and butylated hydroxyanisole at choline functionalized film supported graphene interface. *Sens. Actuators B Chem.* **2016**, *224*, 885–891. [CrossRef]
2. Wang, L.; Yang, R.; Wang, H.; Li, J.J.; Qu, L.B.; Harrington, P.B. High-selective and sensitive voltammetric sensor for butylated hydroxyanisole based on AuNPs-PVP-graphene nanocomposites. *Talanta* **2015**, *138*, 169–175. [CrossRef] [PubMed]
3. Yue, X.Y.; Song, W.S.; Zhu, W.X.; Wang, J.L.; Wang, Y.R. In situ surface electrochemical co-reduction route towards controllable construction of AuNPs/ERGO electrochemical sensing platform for simultaneous determination of BHA and TBHQ. *Electrochim. Acta* **2015**, *182*, 847–855. [CrossRef]
4. Freitas, K.H.; Fatibello-Filho, O. Simultaneous determination of butylated hydroxyanisole (BHA) and butylated hydroxytoluene (BHT) in food samples using a carbon composite electrode modified with $Cu_3(PO_4)_2$ immobilized in polyester resin. *Talanta* **2010**, *81*, 1102–1108. [CrossRef] [PubMed]
5. Manoranjitham, J.J.; Narayanan, S.S. Electrochemical sensor for determination of butylated hydroxyanisole (BHA) in food products using poly O-cresolphthalein complexone coated multiwalledcarbon nanotubes electrode. *Food Chem.* **2021**, *342*, 128246. [CrossRef] [PubMed]
6. Delgado-Zamarreño, M.M.; González-Maza, I.; Sánchez-Pérez, A.; Carabias Martínez, R. Analysis of synthetic phenolic antioxidants in edible oils by micellar electrokinetic capillary chromatography. *Food Chem.* **2007**, *100*, 1722–1727. [CrossRef]
7. Akkbik, M.; Assim, Z.B.; Ahmad, F.B. Optimization and Validation of RP-HPLC-UV/Vis Method for Determination Phenolic Compounds in Several Personal Care Products. *Int. J. Anal. Chem.* **2011**, *2011*, 858153. [CrossRef]
8. Chen, M.; Hu, X.J.; Tai, Z.G.; Qin, H.; Tang, H.N.; Liu, M.S.; Yang, Y.L. Determination of Four Synthetic Phenolic Antioxidants in Edible Oils by High-Performance Liquid Chromatography with Cloud Point Extraction Using Tergitol TMN-6. *Food Anal Methods* **2012**, *6*, 28–35. [CrossRef]
9. Davoli, E.; Bastone, A.; Bianchi, G.; Salmona, M.; Diomede, L. A simple headspace gas chromatography/mass spectrometry method for the quantitative determination of the release of the antioxidants butylated hydroxyanisole and butylated hydroxytoluene from chewing gum. *Rapid Commun. Mass Spectrom.* **2017**, *31*, 859–864. [CrossRef]
10. Capitan-Vallvey, L.F.; Valencia, M.C.; Nicolas, E.A. Monoparameter sensors for the determination of the antioxidants butylated hydroxyanisole and n-propyl gallate in foods and cosmetics by flow injection spectrophotometry. *Analyst* **2001**, *126*, 897–902. [CrossRef]
11. Shu, Y.; Li, B.; Xu, Q.; Gu, P.; Xiao, X.; Liu, F.P.; Yu, L.Y.; Pang, H.; Hu, X.Y. Cube-like $CoSn(OH)_6$ nanostructure for sensitive electrochemical detection of H_2O_2 in human serum sample. *Sens. Actuators B Chem.* **2017**, *241*, 528–533. [CrossRef]
12. Wang, J.; Li, N.; Xu, Y.X.; Pang, H. Two-Dimensional MOF and COF Nanosheets: Synthesis and Applications in Electrochemistry. *Chem. Eur. J.* **2020**, *26*, 6402–6422. [CrossRef] [PubMed]
13. Chen, Y.L.; Xie, Y.; Sun, X.; Wang, Y.; Wang, Y. Tunable construction of crystalline and shape-tailored Co_3O_4@TAPB-DMTP-COF composites for the enhancement of tert-butylhydroquinone electrocatalysis. *Sens. Actuators B Chem.* **2021**, *331*, 129438. [CrossRef]
14. Wang, J.; Xu, Q.; Xia, W.W.; Shu, Y.; Jin, D.Q.; Zang, Y.; Hu, X.Y. High sensitive visible light photoelectrochemical sensor based on in-situ prepared flexible Sn_3O_4 nanosheets and molecularly imprinted polymers. *Sens. Actuators B Chem.* **2018**, *271*, 215–224. [CrossRef]
15. Zhu, R.M.; Ding, J.W.; Jin, L.; Pang, H. Interpenetrated structures appeared in supramolecular cages, MOFs, COFs. *Coord. Chem. Rev.* **2019**, *389*, 119–140. [CrossRef]

16. Huang, Z.L.; Xu, Q.; Hu, X.Y. Covalent organic frameworks functionalized carbon fiber paper for the capture and detection of hydroxyl radical in the atmosphere. *Chin. Chem. Lett* **2020**, *31*, 2495–2498. [CrossRef]
17. Zhang, T.; Chen, Y.L.; Huang, W.; Wang, Y.; Hu, X.Y. A novel AuNPs-doped COFs composite as electrochemical probe for chlorogenic acid detection with enhanced sensitivity and stability. *Sens. Actuators B Chem.* **2018**, *276*, 362–369. [CrossRef]
18. Xie, Y.; Zhang, T.; Chen, Y.L.; Wang, Y.; Wang, L. Fabrication of core-shell magnetic covalent organic frameworks composites and their application for highly sensitive detection of luteolin. *Talanta* **2020**, *213*, 120843. [CrossRef]
19. Sun, Y.; Chen, C.Y.; Liu, J.B.; Liu, L.Z.; Tuo, W.; Zhu, H.T.Z.; Lu, S.; Li, X.P.; Stang, P.J. Self-Assembly of Porphyrin-Based Metallacages into Octahedra. *J. Am. Chem. Soc.* **2020**, *142*, 17903–17907. [CrossRef]
20. Cai, W.R.; Zeng, H.B.; Xue, H.G.; Marks, R.S.; Cosnier, S.; Zhang, X.J.; Shan, D. Enhanced Electrochemiluminescence of Porphyrin-Based Metal-Organic Frameworks Controlled via Coordination Modulation. *Anal. Chem.* **2020**, *92*, 1916–1924. [CrossRef]
21. Liu, M.J.; Cao, S.M.; Feng, B.Q.; Dong, B.X.; Ding, Y.X.; Zheng, Q.H.; Teng, Y.L.; Li, Z.W.; Liu, W.L.; Feng, L.G. Revealing the structure-activity relationship of two Cu-porphyrin-based metal-organic frameworks for the electrochemical CO_2-to-HCOOH transformation. *Dalton Trans.* **2020**, *49*, 14995–15001. [CrossRef] [PubMed]
22. Wu, L.T.; Han, C.; Wang, Z.J.; Wu, X.; Su, F.; Li, M.Y.; Zhang, Q.Y.; Jing, X.B. Porphyrin-Based Organoplatinum(II) Metallacycles With Enhanced Photooxidization Reactivity. *Front Chem.* **2020**, *8*, 262. [CrossRef] [PubMed]
23. Chen, R.F.; Wang, Y.; Ma, Y.; Mal, A.; Gao, X.Y.; Gao, L.; Qiao, L.J.; Li, X.B.; Wu, L.Z.; Wang, C. Rational design of isostructural 2D porphyrin-based covalent organic frameworks for tunable photocatalytic hydrogen evolution. *Nat. Commun.* **2021**, *12*, 1354. [CrossRef] [PubMed]
24. Wan, S.; Gándara, F.; Asano, A.; Furukawa, H.; Saeki, A.; Dey, S.K.; Liao, L.; Ambrogio, M.W.; Botros, Y.Y.; Duan, X.F.; et al. Covalent Organic Frameworks with High Charge Carrier Mobility. *Chem. Mater.* **2011**, *23*, 4094–4097. [CrossRef]
25. Wang, D.W.; Zhang, Z.; Lin, L.; Liu, F.; Wang, Y.B.; Guo, Z.P.; Li, Y.H.; Tian, H.Y.; Chen, X.S. Porphyrin-based covalent organic framework nanoparticles for photoacoustic imaging-guided photodynamic and photothermal combination cancer therapy. *Biomaterials* **2019**, *223*, 119459. [CrossRef]
26. Meng, F.L.; Qian, H.L.; Yan, X.P. Conjugation-regulating synthesis of high photosensitizing activity porphyrin-based covalent organic frameworks for photodynamic inactivation of bacteria. *Talanta* **2021**, *233*, 122536. [CrossRef]
27. Xu, Q.T.; Xue, H.G.; Guo, S.P. FeS2 walnut-like microspheres wrapped with rGO as anode material for high-capacity and long-cycle lithium-ion batteries. *Electrochim. Acta* **2018**, *292*, 1–9. [CrossRef]
28. Dai, L.X.; Li, W.L.; Zhou, K.H.; Tang, D.M.; Han, Y.; Wu, X.Y.; Wu, H.Y.; Diao, G.W.; Chen, M. Interfacial anchoring effect for enhanced lithium storage performance of sesame balls-like Fe_3O_4/C hollow nanospheres. *J. Electroanal. Chem.* **2019**, *855*, 113626. [CrossRef]
29. Shu, Y.; Xu, J.; Chen, J.Y.; Xu, Q.; Xiao, X.; Jin, D.Q.; Pang, H.; Hu, X.Y. Ultrasensitive electrochemical detection of H_2O_2 in living cells based on ultrathin MnO_2 nanosheets. *Sens. Actuators B Chem.* **2017**, *252*, 72–78. [CrossRef]
30. Wei, C.Z.; Cheng, C.; Ma, L.; Liu, M.N.; Kong, D.C.; Du, W.M.; Pang, H. Mesoporous hybrid NiO_x-MnO_x nanoprisms for flexible solid-state asymmetric supercapacitors. *Dalton Trans.* **2016**, *45*, 10789–10797. [CrossRef]
31. Gao, Y.J.; Yang, F.Y.; Yu, Q.H.; Fan, R.; Yang, M.; Rao, S.Q.; Lan, Q.C.; Yang, Z.J.; Yang, Z.Q. Three-dimensional porous Cu@Cu_2O aerogels for direct voltammetric sensing of glucose. *Mikrochim Acta* **2019**, *186*, 192. [CrossRef] [PubMed]
32. He, H.; Dong, J.; Li, K.; Zhou, M.; Xia, W.W.; Shen, X.S.; Han, J.R.; Zeng, X.H.; Cai, W.P. Quantum dot-assembled mesoporous CuO nanospheres based on laser ablation in water. *RSC Adv.* **2015**, *5*, 19479–19483. [CrossRef]
33. Jiang, J.J.; Ding, D.; Wang, J.; Lin, X.Y.; Diao, G.W. Three-dimensional nitrogen-doped graphene-based metal-free electrochemical sensors for simultaneous determination of ascorbic acid, dopamine, uric acid, and acetaminophen. *Analyst* **2021**, *146*, 964–970. [CrossRef] [PubMed]
34. Mao, A.R.; Li, H.B.; Yu, L.Y.; Hu, X.Y. Electrochemical sensor based on multi-walled carbon nanotubes and chitosan-nickel complex for sensitive determination of metronidazole. *J. Electroanal. Chem.* **2017**, *799*, 257–262. [CrossRef]
35. Yang, C.; Liu, M.M.; Bai, F.Q.; Guo, Z.Z.; Liu, H.; Zhong, G.X.; Peng, H.P.; Chen, W.; Lin, X.H.; Lei, Y.; et al. An electrochemical biosensor for sensitive detection of nicotine-induced dopamine secreted by PC12 cells. *J. Electroanal. Chem.* **2019**, *832*, 217–224. [CrossRef]
36. Ziyatdinova, G.; Guss, E.; Budnikov, H. Amperometric sensor based on MWNT and electropolymerized carminic acid for the simultaneous quantification of TBHQ and BHA. *J. Electroanal. Chem.* **2020**, *859*, 113885. [CrossRef]
37. Lin, X.Y.; Ni, Y.N.; Kokot, S. Glassy carbon electrodes modified with gold nanoparticles for the simultaneous determination of three food antioxidants. *Anal. Chim. Acta* **2013**, *765*, 54–62. [CrossRef] [PubMed]
38. Caramit, R.P.; de Freitas Andrade, A.G.; Gomes de Souza, J.B.; de Araujo, T.A.; Viana, L.H.; Trindade, M.A.G.; Ferreira, V.S. A new voltammetric method for the simultaneous determination of the antioxidants TBHQ and BHA in biodiesel using multi-walled carbon nanotube screen-printed electrodes. *Fuel* **2013**, *105*, 306–313. [CrossRef]
39. Roushani, M.; Sarabaegi, M. Electrochemical detection of butylated hydroxyanisole based on glassy carbon electrode modified by iridium oxide nanoparticles. *J. Electroanal. Chem.* **2014**, *717*, 147–152. [CrossRef]
40. Gan, T.; Zhao, A.X.; Wang, S.H.; Lv, Z.; Sun, J.Y. Hierarchical triple-shelled porous hollow zinc oxide spheres wrapped in graphene oxide as efficient sensor material for simultaneous electrochemical determination of synthetic antioxidants in vegetable oil. *Sens. Actuators B Chem.* **2016**, *235*, 707–716. [CrossRef]

41. Ng, K.L.; Tan, G.H.; Khor, S.M. Graphite nanocomposites sensor for multiplex detection of antioxidants in food. *Food Chem.* **2017**, *237*, 912–920. [CrossRef] [PubMed]
42. De Araujo, T.A.; Barbosa, A.M.; Viana, L.H.; Ferreira, V.S. Voltammetric determination of tert-butylhydroquinone in biodiesel using a carbon paste electrode in the presence of surfactant. *Colloids Surf. B* **2010**, *79*, 409–414. [CrossRef] [PubMed]
43. Zhao, P.N.; Hao, J.C. Tert-butylhydroquinone recognition of molecular imprinting electrochemical sensor based on core-shell nanoparticles. *Food Chem.* **2013**, *139*, 1001–1007. [CrossRef] [PubMed]

biosensors

MDPI

Article

Aptamer-Based Gold Nanoparticles–PDMS Composite Stamps as a Platform for Micro-Contact Printing

Amna Didar Abbasi [1], Zakir Hussain [1,*] and Kun-Lin Yang [2]

[1] Department of Materials Engineering, School of Chemical and Materials Engineering (SCME), National University of Sciences and Technology (NUST), Islamabad 44000, Pakistan
[2] Department of Chemical and Biomolecular Engineering, National University of Singapore, Engineering Drive 4, Singapore 117576, Singapore
* Correspondence: zakir.hussain@scme.nust.edu.pk

Abstract: In the present study, a functional template made up of in situ synthesised gold nanoparticles (AuNPs) is prepared on polydimethylsiloxane (PDMS) for patterning of target protein onto the desired solid substrates. Unlike previous studies in which bioreceptor probes are randomly attached to the PDMS stamp through electrostatic interactions, herein, we propose an AuNPs–PDMS stamp, which provides a surface for the attachment of thiol-modified biorecognition probes to link to the stamp surface through a dative bond with a single anchoring point based on thiol chemistry. By using this platform, we have developed the ability for microcontact printing (μCP) to selectively capture and transfer target protein onto solid surfaces for detection purposes. After μCP, we also investigated whether liquid crystals (LCs) could be used as a label-free approach for identifying transfer protein. Our reported approach provides promise for biosensing of various analytes.

Keywords: liquid crystals; DMOAP; AuNPs–PDMS composite; micro-contact printing

Citation: Abbasi, A.D.; Hussain, Z.; Yang, K.-L. Aptamer-Based Gold Nanoparticles–PDMS Composite Stamps as a Platform for Micro-Contact Printing. *Biosensors* **2022**, *12*, 1067. https://doi.org/10.3390/bios12121067

Received: 28 October 2022
Accepted: 16 November 2022
Published: 23 November 2022

1. Introduction

Patterning of biomolecules finds many applications in biomedical research, including biosensing, diagnostic immunoassays, DNA hybridisation and many others [1–4]. In biosensing, the most crucial step of sensor construction is the active placement of the biorecognition element on to the substrate. For active placement of biorecognition elements on surfaces, many procedures such as inkjet printing, micro arraying, electrospray and patterning through AFM tip demonstrate great promise. However, these procedures result in unclear patterns, protein aggregation and loss of biological activity [5]. In contrast, μCP is an efficient and simple procedure for direct patterning biorecognition elements (proteins, antibodies and oligos) on solid substrates without the need for photolithographic tools and a clean room facility [6,7].

Moreover, proteins cannot be synthesised on solid substrates but can be patterned. However, some proteins may not survive adsorption processes on surfaces. For these sensitive proteins, μCP is an ideal patterning option. In addition, μCP combines well with many biomedical procedures such as enzyme-linked immunosorbent assay (ELISA), microfluidic networks and fluorescence labelling [6].

The excellent properties exhibited by PDMS, including biocompatibility, bio- and oxidative stability, non-toxicity, optical transparency, flexibility, simple fabrication and good mechanical properties, make PDMS an ideal candidate for use as a stamp in μCP [8,9]. Despite these remarkable properties, the hydrophobicity of PDMS (water contact angle ~$108^\circ \pm 7^\circ$) often limits its application in sensor construction where the active placement of biomolecules is concerned. The undesired non-specific adsorption of proteins due to the hydrophobic nature of the PDMS surface results in compromised detection sensitivity [8–10].

Recently, in order to introduce biomolecules on the inert surface of PDMS, metal nanoparticle incorporation, especially AuNPs in a polymer matrix, has gained much interest in various fields, including in biosensing and optical devices [10–17]. As reported in earlier studies, thiol-containing molecules may directly interact with metal surfaces to form dative bonds [18]. Zhang et al. have proposed a simple in situ fabrication method to prepare an AuNPs–PDMS composite without the need for additional reducing or stabilising agents [12]. Since the inclusion of AuNPs on the PDMS surface makes it a suitable platform for attachment of biorecognition probes (aptamer, proteins or antibodies) with thiol modification [11,14,19], we propose this composite as a stamp material for attachment of a thiol-modified aptamer based on gold-thiol chemistry, which selectively binds with its target protein, and then transferring of target protein on the desired substrate through µCP. At the same time, aptamers are synthetic single-stranded oligonucleotides (RNA or DNA) that have been exploited as biorecognition probes in various bioassays due to their high specificity and selectivity [20–23]. In this study, we chose the B40t77 RNA aptamer as a biorecognition probe for HIV-1 surface glycoprotein gp-120. Initially, it was created for HIV-1 therapeutics [24]. We have recently used the B40t77 aptamer to develop bioassays for gp-120 [21,23]. The key steps involved in µCP are: (1) fabrication of PDMS stamps; (2) in situ synthesis of gold nanoparticles on the surface of an as-prepared PDMS stamp; (3) linking the stamp with a thiol-modified aptamer and incubating these stamps with the target protein before µCP; and (4) conformal contact of aptamer-conjugated AuNPs–PDMS stamps with the UV treated DMOAP-coated glass slide. In addition, liquid crystals (LCs) have recently drawn much attention due to their distinct optical characteristics and long range orientational order. Many researchers have exploited LCs as a signal transducer for label-free detection of biomolecules [25–30]. To analyse the fate of transferred protein, after µCP we utilized the LCs orientational response towards patterned protein on a DMOAP-coated slide, which yields optical readouts that are visible to the naked eye under crossed polarizers.

In the present study, we demonstrated a AuNPs–PDMS stamp as a stable platform for the attachment of a thiol-modified biorecognition probe based on thiol chemistry. We used a thiol-modified RNA aptamer (B40t77 aptamer) as a probe to check the interaction between probe oligos and stamp. Furthermore, the gp-120 protein was used as a target protein. Before conformal contact with the desired substrate, the aptamer bound AuNPs–PDMS stamp was incubated with a solution containing gp-120 protein. Then the PDMS stamp with the B40t77aptamer-gp-120 complex was conformally contacted with a UV exposed DMOAP-coated glass slide to transfer gp-120 to the DMOAP-coated glass substrate. This glass substrate with transferred protein was then used to fabricate LCs optical cells and analysed under the polarised optical microscope for results. Furthermore, by using 3′ 6 FAM-conjugated thiolated Apt 8 aptamer, we further confirmed that AuNPs–PDMS stamps are ideal stamp material for attachment of thiol-modified biorecognition probes.

2. Materials and Methods

2.1. Materials

PDMS monomer and curing agent (Sylgard 184) was purchased from Dow Corning (Midland, MI, USA). Glycoprotein-120 was obtained from Abcam (Cambridge, UK). FITC-hIgG was procured from Sigma Aldrich (Singapore). 2-fluoro pyrimidine substituted 77 nucleotides long B40t77 RNA aptamer was custom synthesised by Gene Link (Hawthorne, NY, USA). 2-fluoro pyrimidine substituted 23 nucleotides long Apt 8 RNA aptamer with 3′ 6-FAM and 5′ thiol modification was custom synthesised by IDT (Singapore). Tris EDTA (1× T.E., pH-8) and PBS buffer (10×, pH-7.4) were obtained from 1st BASE (Singapore). Decon-90 was purchased from VWR (Singapore). Microscopic glass slides were purchased from Merienfield (Berlin, Germany). *N*,*N*-dimethyl-*n*-octadecyl-3aminopropyltrimethoxysilyl chloride (DMOAP), RNase-free water and magnesium chloride were purchased from Sigma Aldrich (Singapore). Liquid crystals 4-cyano-4-pentylbiphenyl (5CB) were obtained from Merck (Tokyo, Japan). Mylar transparent film was procured from Infinite Graphics (Singapore). Gp-120 stock and working solutions were prepared in 0.01 M phosphate-buffered saline (PBS, pH-7.4). Aptamer stock solutions were prepared by dissolving B40t77 and Apt 8. The stock solutions were prepared in RNase-free water at room temperature and further dilutions were prepared in T.E. buffer with 100 mM of magnesium chloride. To ensure correct spatial folding, the aptamer solution was first heated to 95 °C for 3 min, followed by cooling at room temperature for 5 min, and lastly, the solution was cooled on ice. 0.2 μm filtered Milli-Q water (Milli-Q system, Billerica, MA, USA) was utilized to prepare all buffer solutions.

2.2. Glass Slides Cleaning and Surface Modification with DMOAP

Glass slides were twice cleaned with deionized water followed by overnight soaking in a solution containing 5 percent (*v*/*v*) Decon-90. Next, the glass slides were ultrasonically cleaned for 15 min with deionized water before being thoroughly rinsed with deionized water. For surface modification of glass slides with DMOAP, the cleaned glass slides were immersed for 5 min in a solution containing 0.1 percent (*v*/*v*) DMOAP. The glass slides were then rinsed with deionized water 5 times to eliminate remaining unreacted DMOAP from the surface and afterwards dried under a compressed stream of nitrogen gas. Lastly, to crosslink the immobilised DMOAP, these glass slides were heated to 100 °C in a vacuum oven for 15 min.

2.3. Preparation of PDMS Stamps

To prepare the PDMS stamp, the prepolymer mixture (silicon elastomer and curing agent) was first mixed thoroughly in a weight ratio of 10:1. After mixing, the air bubbles were removed by degassing the mixture in a vacuum desiccator for 1.5 h after pouring it on a cleaned Petri plate. After that, the prepolymer mixture was cured at 65 °C for 6–7 h. Finally, the cured PDMS was carefully peeled off the Petri plate and cut into the appropriate size. To clean the PDMS, it was soaked in absolute ethanol overnight, washed with deionised water and dried under a stream of nitrogen. Finally, the cleaned PDMS stamp was baked at 65 °C for 60 min to vaporise any ethanol trapped inside it.

2.4. U.V. Treatment of PDMS Stamp and DMOAP-Coated Slide

Before modification, the surface of the prepared PDMS stamp and DMOAP-coated glass slide was placed under a UV pen lamp (254 nm, model 11SC-1, Sigma-Aldrich, St. Louis, MO, USA) for durations of 5 and 1 min, respectively. The distance between the surface of the PDMS stamp and DMOAP-coated glass slide, and the UV pen lamp was kept constant at 1.5 cm throughout all experiments.

2.5. Preparation of AuNPs–PDMS Composite Stamp

PDMS–AuNPs composite film was prepared according to the procedure in the literature [12,19]. Briefly, prepared PDMS pieces were incubated with drops of 15 μL of 0.5% $HAuCl_4 \cdot 3H_2O$ solution in the form of the array at 37 °C in a self-made humidified chamber for 48 h. The PDMS pieces were then thoroughly rinsed with 0.22 μm filtered milli-Q water, followed by drying under a stream of nitrogen gas. They were then stored at 4 °C in the refrigerator for further use.

2.6. Immobilisation of B40t77 Aptamer on AuNPs–PDMS Stamp

Drops of 15 μL pretreated aptamer solution were dispensed on the cleaned and dried piece of PDMS–AuNPs composite stamp in the form of a small array with the help of a micropipette and then kept in a self-made humidified chamber at 4 °C for 12 h. It was reported previously that the kinetics of modification of AuNPs surfaces with thiols could take more than one hour to complete [18]. After incubation, they were thoroughly rinsed with RNAase-free water to remove any unbound or physically bound aptamer from the surface and then dried under a nitrogen gas stream. They were then stored at 4 °C until further use.

2.7. Characterisation of Prepared Solid Stamp Materials

The surface morphology of bare PDMS and AuNPs–PDMS stamps was analysed using a field emission scanning electron microscope (FE-SEM, J.S.M. 6700F) manufactured by JEOL. Samples for FE-SEM were kept in the dry box before sample preparation for analysis. These samples were attached to a sample holder using carbon tape and then coated with platinum for 90 s using a sputtering coater (JEOL LFC-1300) before the examination. To confirm the presence of AuNPs on the surface of the PDMS stamps, XRD analysis was performed using a Bruker D8 Advance Powder X-ray diffractometer. All solid samples of bare PDMS stamps (Section 2.3), AuNPs–PDMS stamps (Section 2.4) and AuNPs–PDMS stamps with aptamer immobilised on the surface (Section 2.6) were analysed by UV-Vis spectrometer (UV–Vis-NIR, Carry 5000, Varian New South Whales, Australia) in the spectral range of 200 nm–800 nm.

2.8. Patterning of Target Protein through Micro-Contact Printing

The areas of the PDMS stamp with immobilised aptamer were covered with the target gp-120 protein solution, followed by incubation at 37 °C for 1 h in a self-made humidified chamber. After incubation, the PDMS stamp was rinsed with RNase-free water and dried under a nitrogen gas stream. The aptamer and gp-120 protein complex immobilised PDMS stamp surface was brought into conformal contact with a UV treated DMOAP-coated slide. To ensure perfect contact weight, three glass slides were placed on top of this assembly during conformal contact. After 10 min of contact in a self-made humidified chamber, the stamp was peeled off from the DMOAP-coated glass slide surface. Then this glass slide surface was dried under a stream of nitrogen gas before LCs analysis.

2.9. Fabrication of LCs Optical Cell

An LCs optical cell was constructed by putting together a sample glass slide and a DMOAP-coated glass slide. At both ends of the two glass slides, mylar films of thickness 6 mm were employed to separate the two glass slides. Then these glass slides were secured with binder clips at both ends. Capillary force was used to fill the area between the two glass slides with around 10 µL of LCs. After 15 min, the constructed LCs optical cell was examined with a 1× objective lens using a transmission mode polarised optical microscope (Nikon ECLIPSE LV100POL, Tokyo, Japan).

2.10. Stability Check of the Prepared AuNPs–PDMS Stamp

To check the stability of the prepared PDMS–gold nanoparticles stamp, we used 3′ 6-FAM-conjugated thiolated aptamer (Apt 8). First, a drop of the aptamer solution was dispensed on the surface of the PDMS stamps with and without gold nanoparticles with the help of a micropipette and kept in a self-made humidified chamber at 4 °C for 12 h. After incubation, the stamps were thoroughly rinsed with RNAase-free water to remove any unbound or physically adsorbed aptamer. They were then dried under a nitrogen gas stream. Both PDMS stamps were brought into conformal contact with a DMOAP-coated glass slide separately. After 10 min of contact, the stamp was peeled off from the DMOAP-coated glass slide surface. Then this glass slide surface was dried under a stream of nitrogen gas in preparation for florescence analysis. The fluorescence signals were observed under a fluorescence microscope (Eclipse E200, Nikon, Tokyo, Japan) using a 6 s exposure time.

3. Results and Discussion

3.1. Proposed Scheme of µCP of Gp-120 Protein Based on AuNPs–PDMS Composite Stamp

In the proposed study, based on gold-thiol chemistry, an appropriate amount of thiol-modified B40t77 aptamer was allowed to immobilize onto the AuNPs–PDMS stamp. After immobilization, the B40t77 aptamer immobilized spot areas were incubated with gp-120 solution before performing µCP. Next, the AuNPs–PDMS stamp with complex was further used for gp-120 protein patterning on the glass slide surface. This patterning step was carried out in a Petri plate by creating handmade humid environment inside the Petri plate; we observed that the humid environment aids in μCP by allowing the protein to fall off from the stamp surface onto the glass slide surface. Finally, a glass slide patterned with target protein was used to fabricate LCs optical cells for imaging protein patterning results in the form of final optical responses of LCs due to their orientational transition. The LCs final optical images analysis showed that the presence of the target protein gp-120 transferred onto the surface of UV-treated DMOAP-coated glass slide leads to disruption of the LCs homeotropic alignment, which resulted in the bright image area. In contrast, the dark image areas were produced due to LCs homeotropic alignment induced by DMOAP (the orientational agent), which is coated on the glass slides surfaces (Figure 1).

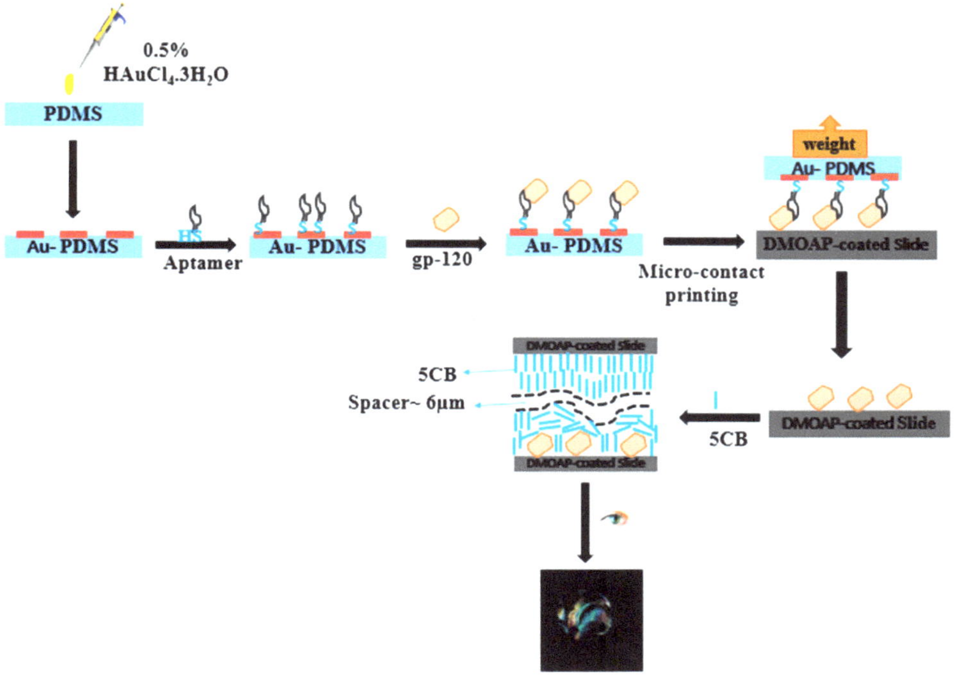

Figure 1. Schematic illustration of the fabrication of AuNPs–PDMS stamp and biorecognition process, including inking with probe aptamer, stamping and patterning target protein gp-120 on DMOAP-coated glass slide through μCP.

3.2. *Characterisation of AuNPs–PDMS Composite Stamp*

There was a visible difference in colour between the prepared bare PDMS stamps and the AuNPs–PDMS stamps. The bare PDMS stamps were transparent, but after surface modification, they turned red, which indicates the presence of AuNPs on the surface of the PDMS. Further, the AuNPs–PDMS stamps were characterized by UV-Vis spectroscopy, X-ray Diffraction (XRD) and Field emission Scanning Electron Microscope (FeSEM). Figure 2a shows UV-Vis spectra of bare PDMS with no absorbance peak, whereas AuNPs–PDMS exhibited an absorbance peak at 527 nm showing the characteristic peak of AuNPs, and the result was consistent with previous studies [12,19]. Figure 2c shows the scanning electron micrograph of bare PDMS and surface modified PDMS with AuNPs, which clearly indicates the presence of AuNPs on the surface of PDMS, as presented in Figure 2d. As can be observed in Figure 2b, the presence of AuNPs on the PDMS stamp surface was confirmed by XRD analysis. The obtained data was also corroborated with the UV-Vis spectroscopy analysis results. The XRD analysis of the AuNPs–PDMS stamp and obtained profiles were confirmed from the JCPDS (04-0784) with a face-centered cubic (F.C.C.) crystal system.

XRD peaks at 38.65, 44.86, 64.92 and 77.96, corresponding to 111, 200, 220 and 311 planes, respectively, confirmed the structure. The calculated interplanar spacing distances were 2.3331, 2.0199, 1.4339 and 1.225, respectively, which also corroborated previously reported results [15]. Furthermore, the prepared AuNPs–PDMS stamp also has long-term stability. After six months of storage in a refrigerator, the stamps still exhibited similar visible colour with little performance loss.

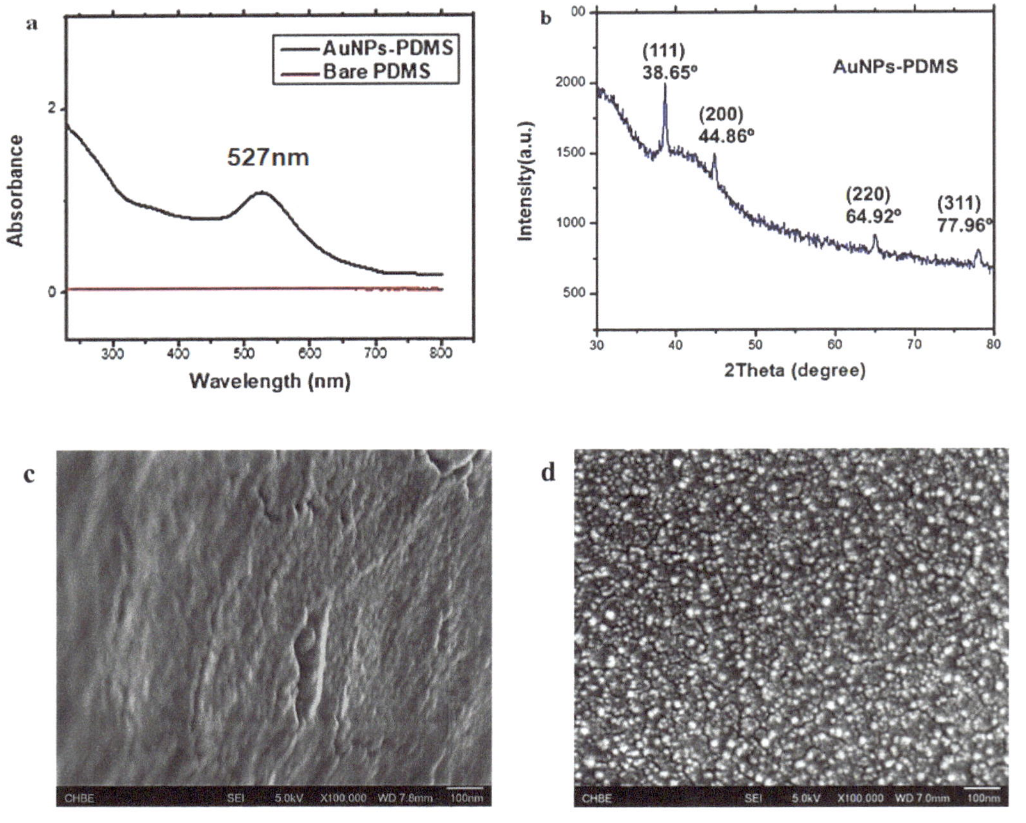

Figure 2. UV-vis absorption spectra. (**a**) UV-Vis spectra of bare PDMS and AuNPs–PDMS, (**b**) XRD of AuNPs–PDMS, (**c**) FeSEM image of bare PDMS and, (**d**) AuNPs–PDMS composite.

3.3. Characterisation of Aptamer Immobilised on PDMS–AuNPs Composite Stamps

Unlike the many previous studies in which bioreceptor probes are randomly attached to the PDMS stamps through electrostatic interactions, we proposed AuNPs–PDMS as a μCP stamp which provides the surface for attachment of thiol-modified biorecognition probes linked to the stamp surface through a dative bond with a single anchoring point based on thiol chemistry. To investigate whether the aptamers were successfully loaded on the surface of the AuNPs–PDMS stamps, UV-Vis absorption spectroscopy analysis was employed (Figure 3). After the attachment of aptamer on the surface of the AuNPs–PDMS stamp, the absorption value increased to 276 nm, indicating the presence of the aptamer, which has a characteristic peak at 260 nm. Hence, these results confirmed the successful attachment of aptamer at the stamp surface. The formation of aptamer bound on AuNPs–PDMS was further confirmed by the red shift in the absorption peak from 527 nm (AuNPs) to 537 nm, along with its increasing absorption intensity.

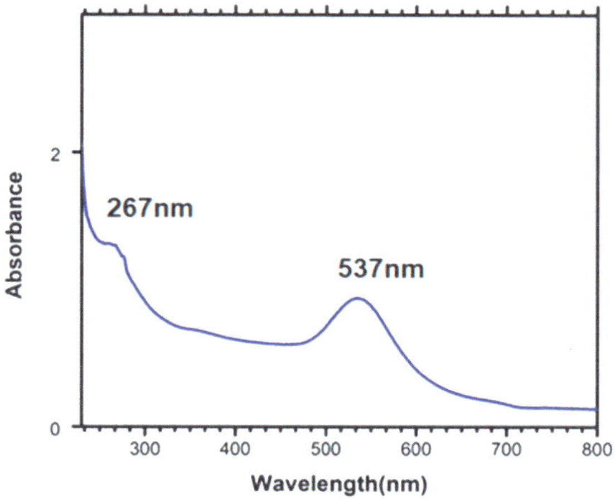

Figure 3. UV-Vis absorption spectra of AuNPs–PDMS with immobilized biorecognition aptamer probe.

Moreover, we also investigated the effect of AuNPs on the surface of the PDMS stamp on the immobilization of 3′ 6-FAM labelled Apt 8. In this case, we dispensed an Apt 8 solution drop on the AuNPs–PDMS stamp to attach a biorecognition probe on the stamp surface. Afterwards, this stamp was analyzed under a fluorescence microscope. As shown in Figure 4a, a green circular fluorescence signal could be observed, indicating successful attachment of the biorecognition probe (Apt 8) on the AuNPs–PDMS stamp. When μCP was carried out with this stamp, no fluorescence signal was observed, indicating no transfer of biorecognition probe onto the glass slide during μCP (Figure 4). Hence, the biorecognition probe is stable and not transferred from the stamp to the DMOAP-coated glass slide during μCP. These results indicated that AuNPs on the surface of the PDMS stamp play an important role in immobilizing the biorecognition probe on the stamp surface.

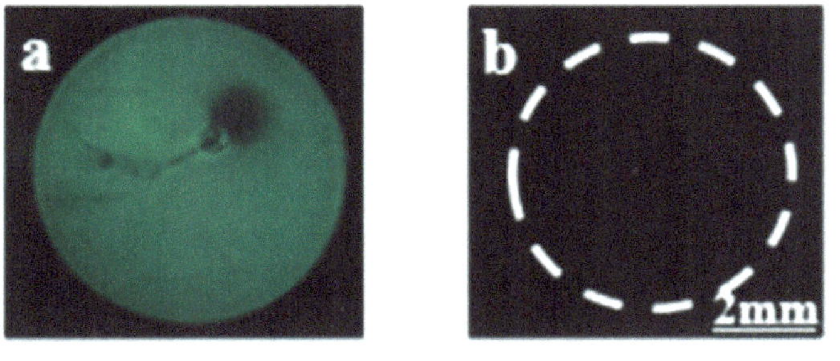

Figure 4. Fluorescence images showing (**a**) AuNPs–PDMS stamp surface with a circular region of immobilised RNA probe (3′ 6-FAM-conjugated Apt 8), and (**b**) DMOAP-coated glass slide surface after μCP. The RNA probe is stable and not transferred from the stamp to the DMOAP-coated glass slide during μCP.

3.4. Detection of Gp-120 Target Protein through μCP by Using LCs

After protein patterning by the AuNPs–PDMS stamp with the immobilized B40t77 aptamer, POM images of the LCs supported on DMOAP-coated slides were obtained and are shown in Figure 5. Before performing μCP, the stamps with attached aptamer were incubated with gp-120 solution. A droplet of a solution containing nonfluorescent gp-120 in different concentrations (8, 4, 2, 1, and 0.5 μg/mL) was placed on the already attached circular area on AuNPs–PDMS with attached B40t77 aptamer. The substrate was kept in a hand-made humid environment in a Petri plate for 1 h at room temperature. A humid environment provides enough time for the probe to interact with its target protein and avoid immediate drying of the solution. After rinsing and drying under a nitrogen stream, the AuNPs–PDMS stamp with the attached B40t77 aptamer and target protein gp-120 complex on its surface was used to print the target protein. Following the procedure, the DMOAP-coated glass slide was exposed to UV to study the transfer of target protein gp-120 during μCP. As previously reported, the UV treatment can strengthen the interactions between the target proteins and glass slides, and protein transfer efficiency strongly depends on interactions between the protein and the surface [7]. This DMOAP-coated glass slide with gp-120 protein printed on its surface was employed for the fabrication of the LCs optical cell for μCP. Bright LCs optical images in Figure 5 show the presence of transferred protein on the slide surface, which disrupt the LCs orientation, indicating successful transfer of the target protein gp-120 from the stamp surface to the DMOAP-coated glass slide. In contrast, the dark areas are due to the absence of the target protein or an insufficient amount of transferred protein which is unable to disrupt the LCs orientation, as in the case of 0 (just 1× PBS), 0.5 and 1 μg/mL concentrations of target protein gp-120 when dispensed on stamps containing the probe aptamer for biorecognition process followed by transferring to slide surface. Previously, John et al., reported an electrochemical aptasensor for HIV-1 gp-120 in which they proposed a dendrimer/streptavidin platform for B40 aptamer attachment and used this aptamer to electrochemically detect gp-120 with a limit of detection of 0.2 nM [31]. Despite of the fact that our proposed limit of detection (LoD) is higher than previously reported, the present biosensing technique is easy to operate, simple and allows detection of gp-120 from the protein mixture. Moreover, compared with our previously reported aptasensors for gp-120 detection [21,23], our present proposed sensing technique not only detects gp-120 from the protein mixture but also transfers the detected gp-120 protein onto a glass substrate for further qualitative analysis.

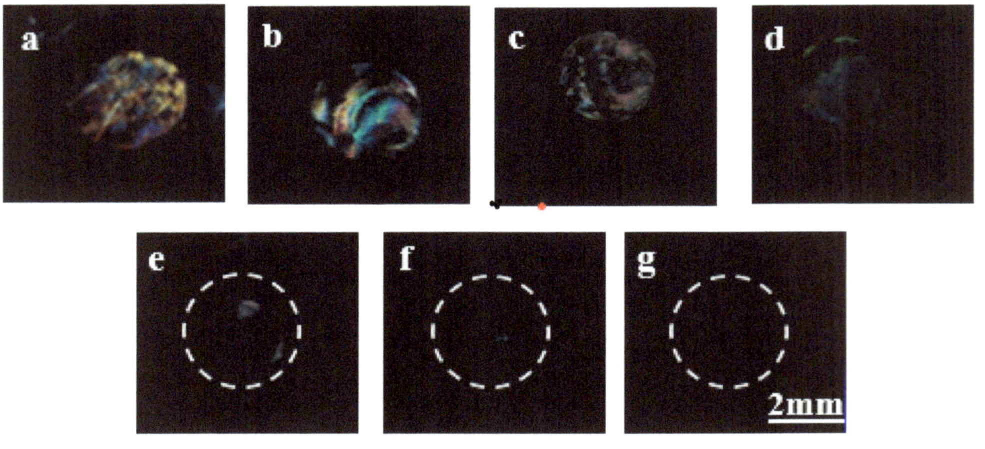

Figure 5. Polarized optical images of LCs supported on DMOAP-coated glass slide printed by AuNPs–PDMS stamp with immobilized probe (B40t77 aptamer) incubated with (**a**) 8 (μg/mL), (**b**) 6 (μg/mL), (**c**) 4 (μg/mL), (**d**) (2 μg/mL), (**e**) 1 (μg/mL), (**f**) 0.5 (μg/mL) and (**g**) (0 μg/mL) of target protein gp-120.

3.5. Capturing Desired Target Protein (Gp-120) in a Protein Mixture Solution

To test the selectivity of the proposed μCP setup, we further investigated the ability of this μCP stamp with aptamer (B40t77) fixed on its surface to capture the desired target protein gp-120 (nonfluorescent) from a solution containing a mixture of two proteins (target protein + one other protein labelled with a fluorophore). For this test, two solutions of gp-120 with a concentration of 8μg/mL were prepared. The solutions were mixed with

FITC labelled IgG with concentrations of 8 and 16 μg/mL. Then these two solutions were incubated with immobilized B40t77 aptamer on the surfaces of two different AuNPs–PDMS stamps. After incubation, the μCP results were analyzed under a polarized optical microscope and a fluorescence microscope. The LCs optical and fluorescence signal images captured are shown in Figure 6. Considering Figure 6a,b, the black fluorescence images with no signal indicates that FITC labelled IgG was not captured and transferred onto the glass slide by the AuNPs–PDMS stamp with B40t77 aptamer immobilized on its surface. For analyzing the transfer of gp-120, the LCs orientational response was exploited to predict the results (Figure 6c,d). These results show that the gp-120 target protein can be captured by the AuNPs–PDMS stamp with attached B40t77 aptamer, even if the gp-120 target protein solution is mixed with FITC labelled IgG solution.

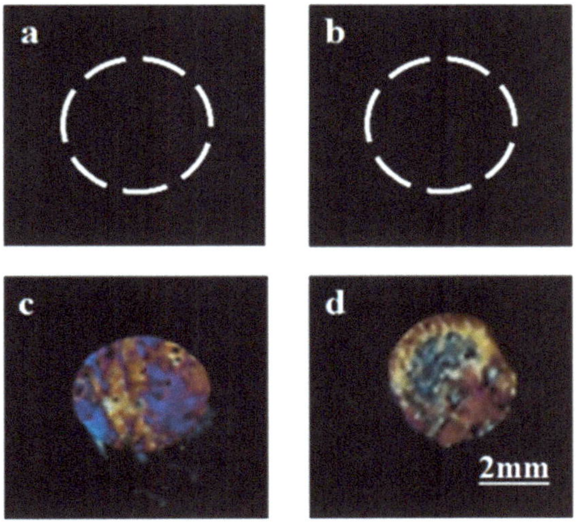

Figure 6. Specificity of the AuNPs–PDMS stamp with immobilized B40t77 aptamer probe after incubation with a solution of the mixture of two proteins (FITC-IgG + gp-120) is demonstrated by fluorescence images with no signals show no binding and transfer of FITC-IgG, (**a**) when 8 μg/mL quantity of gp-120 and FITC-IgG were mixed, and (**b**) when 8 μg/mL of gp-120 and 16 μg/mL of FITC-IgG were mixed. Bright optical images of LCs showing selective binding and transfer of gp-120 onto DMOAP-coated glass slide, (**c**) when 8 μg/mL quantity of gp-120 and FITC-IgG were mixed, and (**d**) when 8 μg/mL of gp-120 and 16 μg/mL of FITC-IgG were mixed.

4. Conclusions

A simple and cost-effective method of patterning desired protein for detection purposes is proposed in the present investigation. This patterning method captures a specific target protein, even from a mixture containing more than one protein, by a biorecognition probe attached to the stamp surface and then transferred to a desired solid substrate. AuNPs–PDMS stamps were exploited as a suitable stamp surface for attachment of the biorecognition probe. Furthermore, the AuNPs–PDMS stamps were fabricated by the already reported method for in situ modification of PDMS stamps with gold nanoparticles. Using this stamp material, it was found that protein patterning became much more effective by avoiding the possibility of protein degradation and aggregation, which usually occurs in the case of the direct adsorption of proteins on a solid substrate.

Furthermore, it could be seen that the dative bond (Au-SH) between AuNPs from the AuNPS–PDMS stamp surface and the thiol group of the biorecognition probe was strong enough to hold the probe with the stamp surface during μCP. The target protein can be separated from the probe during μCP and successfully transferred to a UV treated DMOAP-coated glass slide. The B40t77 aptamer was used as a biorecognition probe that has specificity for HIV-1 surface glycoprotein gp-120. Therefore, this patterning method allowed detection of gp-120 as the probe selectively binds to gp-120 from the solution and transfers it onto a glass slide. For analyzing the results of μCP, the LC's orientational response was exploited as a label-free approach for identifying transferred protein. The reported results demonstrate the significant potential of the developed procedure for the fabrication of bio-sensing systems for various analytes.

Author Contributions: Conceptualization, Z.H. and A.D.A.; methodology, A.D.A.; validation, A.D.A., Z.H. and K.-L.Y.; formal analysis, A.D.A.; investigation, A.D.A.; resources, Z.H. and K.-L.Y.; data curation, A.D.A.; writing—original draft preparation, A.D.A.; writing—review and editing, A.D.A., Z.H. and K.-L.Y.; visualization, Z.H.; supervision, Z.H. and K.-L.Y.; project administration, Z.H. and K.-L.Y.; funding acquisition, Z.H. and A.D.A. All authors have read and agreed to the published version of the manuscript.

Funding: This research was funded by the Higher Education Commission (HEC), Pakistan, grant number NRPU 20-3066.

Institutional Review Board Statement: Not applicable.

Informed Consent Statement: Not applicable.

Data Availability Statement: Not available from the authors.

Acknowledgments: A.D.A. acknowledges the financial assistance from the Higher Education Commission (H.E.C.) Pakistan under IRSIP fellowship (P.I.N.: IRSIP 40 Engg 01) and Indigenous 5000 PhD Scholarship (P.I.N.: 315-3443-2EG3-056). Moreover, Z.H. acknowledges the NRPU grant from H.E.C. under NRPU R&D grant No. 20-3066.

Conflicts of Interest: The authors declare no conflict of interest.

References

1. Sun, H.; Chen, G.Y.J.; Yao, S.Q. Review Recent Advances in Microarray Technologies for Proteomics. *Chem. Biol.* **2013**, *20*, 685–699. [CrossRef]
2. Hölz, K.; Schaudy, E.; Lietard, J.; Somoza, M.M. Multi-level patterning nucleic acid photolithography. *Nat. Commun.* **2019**, *10*, 1–8. [CrossRef]
3. Ogaki, R.; Alexander, M.; Kingshott, P. Chemical patterning in biointerface science Patterning of surfaces with different chemistries provides novel insights of new chemically patterned surfaces is highlighted. *Mater. Today* **2010**, *13*, 22–35. [CrossRef]
4. Delamarche, E.; Pereiro, I.; Kashyap, A.; Kaigala, G.V. Biopatterning: The Art of Patterning Biomolecules on Surfaces. *Langmuir* **2021**, *37*, 9637–9651. [CrossRef]
5. Juste-Dolz, A.; Avella-Oliver, M.; Puchades, R.; Maquieira, A. Indirect microcontact printing to create functional patterns of physisorbed antibodies. *Sensors* **2018**, *18*, 3163. [CrossRef]
6. Bernard, A.; Renault, J.P.; Michel, B.; Bosshard, H.R.; Delamarche, E. Microcontact printing of proteins. *Adv. Mater.* **2000**, *12*, 1067–1070. [CrossRef]
7. Chen, C.H.; Yang, K.L. Improving protein transfer efficiency and selectivity in affinity contact printing by using UV-modified surfaces. *Langmuir* **2011**, *27*, 5427–5432. [CrossRef]
8. Miranda, I.; Souza, A.; Sousa, P.; Ribeiro, J.; Castanheira, E.M.S.; Lima, R.; Minas, G. Properties and applications of PDMS for biomedical engineering: A review. *J. Funct. Biomater.* **2022**, *13*, 2. [CrossRef]
9. Gökaltun, A.; Kang, Y.B.; Yarmush, M.L.; Usta, O.B.; Asatekin, A. Simple Surface Modification of Poly(dimethylsiloxane) via Surface Segregating Smart Polymers for Biomicrofluidics. *Sci. Rep.* **2019**, *9*, 7377. [CrossRef]
10. Tu, Q.; Wang, J.C.; Zhang, Y.; Liu, R.; Liu, W.; Ren, L.; Shen, S.; Xu, J.; Zhao, L.; Wang, J. Surface modification of poly(dimethylsiloxane) and its applications in microfluidics-based biological analysis. *Rev. Anal. Chem.* **2012**, *31*, 177–192. [CrossRef]
11. Wang, W.; Wu, W.Y.; Zhong, X.; Wang, W.; Miao, Q.; Zhu, J.J. Aptamer-based PDMS-gold nanoparticle composite as a platform for visual detection of biomolecules with silver enhancement. *Biosens. Bioelectron.* **2011**, *26*, 3110–3114. [CrossRef] [PubMed]
12. Zhang, Q.; Xu, J.J.; Liu, Y.; Chen, H.Y. In-situ synthesis of poly(dimethylsiloxane)-gold nanoparticles composite films and its application in microfluidic systems. *Lab Chip* **2008**, *8*, 352–357. [CrossRef] [PubMed]
13. Wang, Q.; Zhou, J.; Wu, X.; Riaud, A. Optimization of synthesis conditions of gold nanoparticlespolydimethylsiloxane composite for ultrasound generation. *Nanotechnology* **2021**, *3*, 2850–2857.
14. Zhu, A.; Ali, S.; Xu, Y.; Ouyang, Q.; Chen, Q. A SERS aptasensor based on AuNPs functionalized PDMS film for selective and sensitive detection of Staphylococcus aureus. *Biosens. Bioelectron.* **2021**, *172*, 112806. [CrossRef]

15. Yao, J.Y.; Fostier, A.H.; Santos, E.B. In situ formation of gold and silver nanoparticles on uniform PDMS films and colorimetric analysis of their plasmonic color. *Colloids Surfaces A Physicochem. Eng. Asp.* **2020**, *607*, 125463. [CrossRef]
16. Pusty, M.; Shirage, P.M. Gold nanoparticle-cellulose/PDMS nanocomposite: A flexible dielectric material for harvesting mechanical energy. *RSC Adv.* **2020**, *10*, 10097–10112. [CrossRef]
17. SadAbadi, H.; Badilescu, S.; Packirisamy, M.; Wüthrich, R. Integration of gold nanoparticles in PDMS microfluidics for lab-on-a-chip plasmonic biosensing of growth hormones. *Biosens. Bioelectron.* **2013**, *44*, 77–84. [CrossRef]
18. Xue, Y.; Li, X.; Li, H.; Zhang, W. Quantifying thiol-gold interactions towards the efficient strength control. *Nat. Commun.* **2014**, *5*, 4348. [CrossRef]
19. Wu, W.Y.; Bian, Z.P.; Wang, W.; Wang, W.; Zhu, J.J. PDMS gold nanoparticle composite film-based silver enhanced colorimetric detection of cardiac troponin I. *Sens. Actuators B Chem.* **2010**, *147*, 298–303. [CrossRef]
20. Li, Y.; Liu, S.; Ling, L.; Xue, C.Y.; Chin, S.Y.; Khan, S.A.; Yang, K.L.; Tu, Q.; Wang, J.C.J.; Zhang, Y.; et al. Recent developments in aptasensors for diagnostic applications. *Langmuir* **2018**, *18*, 77–84. [CrossRef]
21. Abbasi, A.D.; Hussain, Z.; Yang, K.L. Aptamer laden liquid crystals biosensing platform for the detection of HIV-1 glycoprotein-120. *Molecules* **2021**, *26*, 2893. [CrossRef] [PubMed]
22. Wang, Q.; Yang, Q.; Wu, W. Graphene-Based Steganographic Aptasensor for Information Computing and Monitoring Toxins of Biofilm in Food. *Front. Microbiol.* **2020**, *10*, 3139. [CrossRef] [PubMed]
23. Abbasi, A.D.; Hussain, Z.; Liaqat, U.; Arif, D.; Yang, K.-L. Liquid Crystal Based Binding Assay for Detecting HIV-1 Surface Glycoprotein. *Front. Chem.* **2021**, *9*, 257. [CrossRef] [PubMed]
24. Dey, A.K.; Griffiths, C.; Lea, S.M.; James, W. Structural characterization of an anti-gp120 RNA aptamer that neutralizes R5 strains of HIV-1. *RNA* **2005**, *11*, 873–884. [CrossRef]
25. Chen, J.; Liu, Z.; Yang, R.; Liu, M.; Yao, J.; Zhang, M.; Li, N.; Yuan, Z.; Jin, M.; Shui, L. A label-free optical immunoassay based on birefringence of liquid crystal for insulin-like growth factor-I sensing. *Sens. Actuators B Chem.* **2022**, *352*, 131028. [CrossRef]
26. Lin, C.T.; Hsu, W.T.; Hwang, S.J. Real-time liquid crystal-based creatinine sensor using a micro-patterned flexible substrate. *Liq. Cryst.* **2021**, *48*, 1660–1670. [CrossRef]
27. Khan, M.; Liu, S.; Qi, L.; Ma, C.; Munir, S.; Yu, L.; Hu, Q. Liquid crystal-based sensors for the detection of biomarkers at the aqueous/LC interface. *TrAC Trends Anal. Chem.* **2021**, *144*, 116434. [CrossRef]
28. Hong, P.T.K.; Jang, C.H. Simple, sensitive technique for α-amylase detection facilitated by liquid crystal-based microcapillary sensors. *Microchem. J.* **2021**, *162*, 105864. [CrossRef]
29. Yin, F.; Cheng, S.; Liu, S.; Ma, C.; Wang, L.; Zhao, R.; Lin, J.M.; Hu, Q. A portable digital optical kanamycin sensor developed by surface-anchored liquid crystal droplets. *J. Hazard. Mater.* **2021**, *420*, 126601. [CrossRef]
30. SadAbadi, H.; Badilescu, S.; Packirisamy, M.; Wuthrich, R. PDMS-gold nanocomposite platforms with enhanced sensing properties. *J. Biomed. Nanotechnol.* **2012**, *8*, 539–549. [CrossRef]
31. John, S.V.; Khati, M.; Mamba, B.B.; Arotiba, O.; Rotherham, L.S. Towards HIV detection: Novel Poly (propylene) dendrimer-streptavidin platform for electrochemical DNA and gp-120 aptamer biosensors. *Int. J. Electrochem. Sci.* **2014**, *9*, 5425–5437.

biosensors

Article

Flexible Electrochemical Platform Coupled with In Situ Prepared Synthetic Receptors for Sensitive Detection of Bisphenol A

Chen-Yan Xu, Kang-Ping Ning, Zheng Wang, Yao Yao, Qin Xu * and Xiao-Ya Hu

School of Chemistry and Chemical Engineering, Yangzhou University, Yangzhou 225002, China
* Correspondence: xuqin@yzu.edu.cn

Abstract: A flexible electrochemical sensor based on the carbon felt (CF) functionalized with Bisphenol A (BPA) synthetic receptors was developed. The artificial Bisphenol A receptors were grafted on the CF by a simple thermal polymerization molecular imprinting process. Fourier-transform infrared spectroscopy (FTIR), scanning electron microscopy (SEM) and electrochemical characterizations were used to analyze the receptors. Characterization results demonstrated that the Bisphenol A synthetic receptors successfully formed on the CFs surface. Because the synthetic receptor and the porous CFs were successfully combined, the sensor displayed a better current response once Bisphenol A was identified. The sensor's linear range was determined to be from 0.5 to 8.0 nM and 10.0 to 300.0 nM, with a detection limit of 0.36 nM. Even after being bent and stretched repeatedly, the electrode's performance was unaffected, demonstrating the robustness, adaptability and viability of installing the sensor on flat or curved surfaces for on-site detection. The designed electrochemical sensor has been used successfully to identify Bisphenol A in milk samples with satisfactory results. This work provided a promising platform for the design of implantable, portable and miniaturized sensors.

Keywords: molecular imprinting; electrochemical; flexible; Bisphenol A

Citation: Xu, C.-Y.; Ning, K.-P.; Wang, Z.; Yao, Y.; Xu, Q.; Hu, X.-Y. Flexible Electrochemical Platform Coupled with In Situ Prepared Synthetic Receptors for Sensitive Detection of Bisphenol A. *Biosensors* **2022**, *12*, 1076. https://doi.org/10.3390/bios12121076

Received: 16 October 2022
Accepted: 21 November 2022
Published: 25 November 2022

1. Introduction

Endocrine disruptors (EDC) are exogenous chemicals that disrupt the human endocrine system and increase the risk of diabetes, obesity, cardiovascular disease and prostate cancer [1]. Bisphenol A (2,2-bis (4-hydroxyphenyl) propane, BPA) has been functionalized as a kind of EDC to disturb the human endocrine system due to its structural resemblance to estrogen. However, Bisphenol A is a precursor material for the synthesis of epoxy resins, polycarbonate and polysulfone resins [2]. It is frequently present in plastic products, the inner coating of canned food and food packaging bags. Long-term storage and heat treatment can lead to the penetration of Bisphenol A into food from packaging materials [3,4]. This has raised concerns about food safety [5]. Therefore, the analysis and detection of Bisphenol A are particularly important.

Several analytical methods have been reported for the detection of Bisphenol A, including high-performance liquid chromatography (HPLC) [6], liquid chromatography-mass spectrometry (LC-MS) [7], gas chromatography-mass spectrometry (GC-MS) [8], spectrophotometry [9] and enzyme-linked immunosorbent assay (ELISA) [10]. However, the chromatography and mass spectrometry-based methods are expensive, time-consuming and require complex sample pretreatment. ELISA method is readily available from many manufacturers, but the preparation of antibodies is labor-intensive and expensive. ELISA method also suffers from the instability of biomolecules and short signal stability [11]. Therefore, the development of new methods with simplicity, stability and sensitivity that can detect trace amounts of Bisphenol A residue has become a major research challenge.

Electrochemical analysis technology with low cost, fast response speed and high sensitivity is a promising option for the development of inexpensive, simple and convenient Bisphenol A sensors [12–14]. Bisphenol A has two phenolic hydroxyl groups, which can be oxidized easily on the electrode surface [15]. This character makes it possible to identify Bisphenol A using straightforward electrochemical techniques. It is crucial to choose a material to enhance the performance of the electrochemical sensor when an electrochemical approach is put into practice. Currently, several materials, such as carbon nanomaterials and transition metal oxides, have been utilized as electrode modifiers to improve the electrochemical sensitivity for Bisphenol A detection [16–18]. However, the oxidation products of Bisphenol A would produce non-conductive polymers on the electrode surface, deactivating the electrode surface and preventing further reactions [19]. When Bisphenol A was first recognized and accumulated on the electrode surface, the selectivity and sensitivity of the electrochemical sensor could be improved [20].

In recent years, synthetic receptors have been widely used in sensing platforms for recognition and accumulation. Molecularly imprinted polymers (MIPs) have become a research hotspot due to their superior selectivity over natural receptors, increased physico-chemical stability and better selectivity [21,22]. Hamed E M et al. have summarized the different production methods and the varied types of sensors that employed MIPs for Bisphenol A detection [23]. Nanomaterials with large surface areas are always modified on the rigid electrodes to construct the electrochemical sensors [24]. However, the growing demand for portable and wearable electronics demands is beyond the capacity of rigid electrodes. It is important and necessary to build flexible sensors with good sensitivity and selectivity. CFs have received a lot of interest as a flexible electrode material due to their rich porosity structure and effective adsorption. Compared with conventional rigid electrodes, CF electrodes can be simply regenerated by ultrasonic cleaning [25]. The CF electrode can further be modified to offer a larger specific surface area, more active sites and better recognition ability [26]. In this work, a molecularly imprinted polymer layer was deliberately modified on the surface of a CF electrode by a simple in situ thermal polymerization process to fabricate a novel and sensitive electrochemical sensor for the detection of Bisphenol A [27]. Figure 1 depicts the preparation and application of the MIP@CF sensor for the detection of Bisphenol A. The molecularly imprinted precursor solution was created using the ideal molar ratio of Bisphenol A to methacrylic acid (MAA) and then was thermally polymerized on the surface of the CF. After the release of the template molecule Bisphenol A, the imprinted cavity of Bisphenol A remained on the MIP@CF electrode as the layer with selectivity. The imprinted cavity reidentified the Bisphenol A and provided an electrochemical response signal during the detecting procedure. Cyclic voltammetry (CV) was utilized to evaluate the electro-oxidation behaviors of the identified Bisphenol A, and differential pulse voltammetry (DPV) was employed to quantify Bisphenol A in real samples.

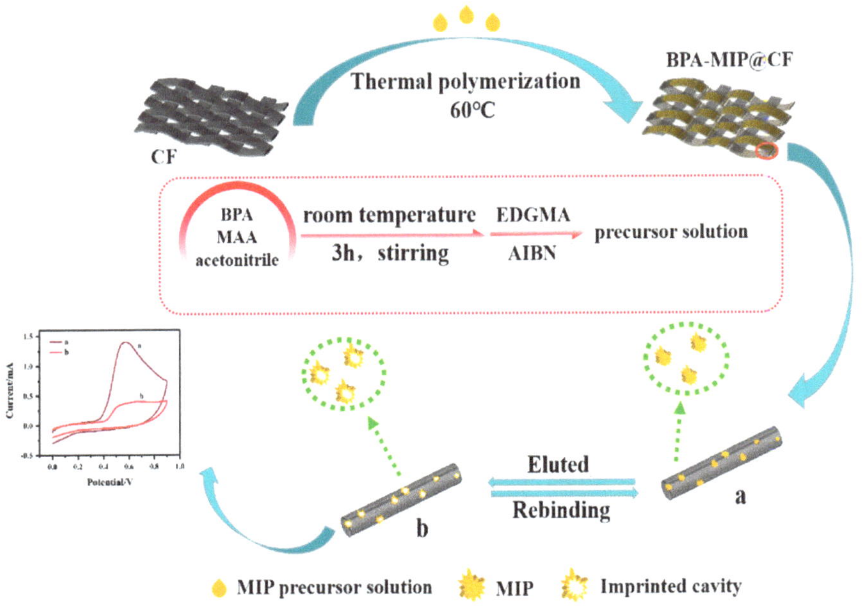

Figure 1. Schematic illustration of the preparation and application of MIP@CF sensor for the detection of BPA.

2. Experimental

2.1. Reagents and Apparatus

Graphite CFs, 2 mm thick, were purchased from Tianjin Carbon Factory (Tianjin, China). Acetonitrile (MW:41.05), ethanol (MW:46.07), methacrylic acid (MAA, MW: 86.09), Bisphenol A (MW:228.29), protocatechuic acid (PCA, MW:154.1), ascorbic acid (AA, MW:176.12), catechol (CC, MW:110.11) and hydroquinone (HQ, MW:110.11) with analytically purity were all purchased from Sinopharm Chemical Reagent Co., Ltd. (Shanghai, China). Ethylene glycol dimethacrylate (EGDMA, MW:198.22) was bought from Aladdin Industrial Corporation (Shanghai, China). Azobisisobutyronitrile (AIBN, MW:164.21) was obtained from Shanghai Shisi Hewei Chemical Co., Ltd. (Shanghai, China). All the reagents were of analytical grade and were used as received without further purification.

GeminiSEM 300 (German Carl Zeiss Co., Carl, Germany) was used to characterize the morphologies of the materials. The molecular structures of CF, Bisphenol A, Bisphenol A–MIP@CF and MIP@CF were checked and compared using a Cary 610/670 FTIR microscope (Varian, Palo Alto, CA, USA). The pH of solutions was adjusted using PHS-25 pH meter (Shanghai INESA Scientific Instrument Co., Ltd., Shanghai, China). A traditional three-electrode system was used for all the electrochemical experiments on a CHI-1040C workstation (Chenhua, Shanghai, China), using CF, MIP@CF and NIP@CF as working electrodes (working area = 1 cm^2), a platinum wire as the counter electrode and an Ag/AgCl electrode as the reference electrode, respectively.

2.2. Preparation of MIP or NIP Modified Electrodes

The CFs were cut into 1.0 cm × 1.5 cm before they were ultrasonically cleaned with ethanol and deionized water for 15 min, respectively, then they were dried in a vacuum oven at 60 °C for 2 h for further use.

The MIP precursor solution was prepared by dissolving template Bisphenol A (0.5 mmol) and functional monomer MAA (2.5 mmol) in 12.5 mL of acetonitrile solution, stirring at room temperature for 3 h to make a homogeneous mixture. Then, 10 mmol of crosslinking agent EGDMA and 40 mg of initiator AIBN were added in turn, and the mixture was further stirred for 10 min, followed by bubbling with nitrogen for 20 min. Finally, the mixture was heated in a water bath at 60 °C for 30 min to form a pre-polymerized solution [28,29].

To prepare MIP–modified electrodes, the CFs were saturated with a pre-polymerized solution (400 μL) before being put on an electric heating table at 60 °C for in situ thermal polymerizations for 4 h. The obtained electrodes were named Bisphenol A–MIP@CF before the elution operation. Lastly, the BPA–MIP@CF electrode was immersed in acetic acid/methanol (1:9, V:V) eluent until the template molecule was fully removed. The resulting electrode was named MIP@CF. For comparison, NIP@CF electrodes were fabricated by the same procedure, except that the template molecule Bisphenol A was not added to the precursor solution during the polymerization process.

2.3. Bisphenol A Electrochemical Measurement

The incubation solution was a mixture of acetonitrile and PBS with a volume ratio of 1:9 and Bisphenol A with various concentrations. The MIP@CF was incubated at room temperature for a period of time in the incubation solution. Then the obtained electrode was taken out, thoroughly washed with PBS three times, and then immersed in a blank PBS solution for electrochemical analysis. The difference of current in MIP@CF electrode before and after capturing Bisphenol A reflects the concentrations of Bisphenol A.

2.4. Real Sample Analysis

Pure milk samples were purchased from a local supermarket. 2 mL of milk was diluted with 10 mL of PBS and centrifuged at 12,000 rpm for 15 min. The supernatant was collected and centrifuged again. Then, a certain amount of supernatant was mixed with PBS for electrochemical analysis. A certain amount of Bisphenol A standard solutions were added to the milk samples to determine the recovery rates of the assay.

3. Results and Discussions

3.1. Characterization of CF, Bisphenol A–MIP@CF, MIP@CF

The surface morphologies of CF, Bisphenol A–MIP@CF and MIP@CF were analyzed and characterized by SEM. It was observed that the CF was composed of cross-linked networks (Figure 2A). The huge voids between carbon fibers were conductive to the transport of the incubation solution during the recognition process, while the cross-linked fibers were beneficial to the electron transfer to promote the electrocatalytic oxidation of Bisphenol A on the electrode [30–32]. In addition, there were abundant parallel cracks on the surface of carbon fiber (Figure 2B), and this roughness was beneficial for surface modification [33]. After the CF was modified by molecular imprinting technology, the diameter of the fiber was increased from 9.42 μm to 9.81 μm. The fiber surface became rough, and some bulk substances were loaded (Figure 2C,D), which may be the synthesized Bisphenol A molecularly imprinted polymers. For the MIP@CF electrode, as shown in Figure 2E,F, it could be clearly observed that the surface blocks were reduced, and the diameter of the fiber was slightly decreased to 9.76 μm. The elution process removed the template from the polymers. During this process, the loosely adhered polymer agglomerates were removed from the surface [34,35].

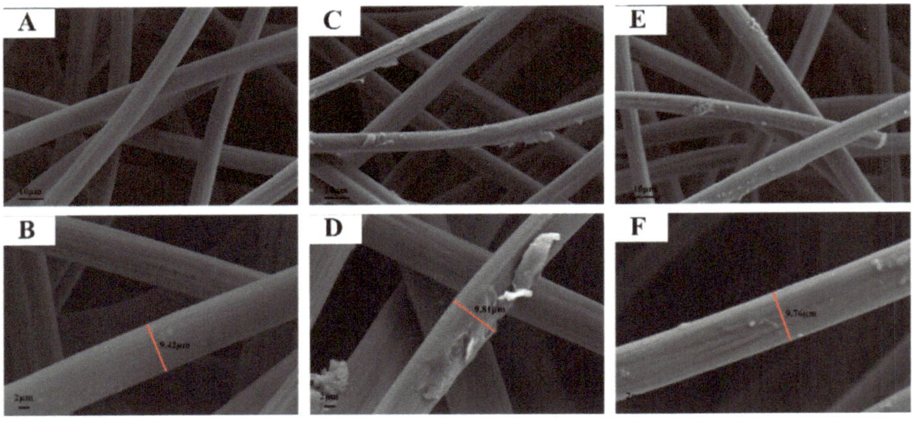

Figure 2. SEM images of bare CF (**A**,**B**), BPA–MIP@CF (**C**,**D**) and MIP@CF (**E**,**F**).

FTIR spectra of bare CF, Bisphenol A, BPA–MIP@CF and MIP@CF were studied and compared in Figure 3. The differences in these spectra confirmed the integration of new functional groups on the surface of the CF after imprinting. Compared with the bare CF (curve a) and Bisphenol A (curve b), the stretching vibration peak of the methylene group in the CF after Bisphenol A imprinting appeared at 2968 cm^{-1}, and the characteristic peak of the benzene ring appeared at 1500 cm^{-1}–1300 cm^{-1}, which proved that the Bisphenol A molecule was successfully imprinted on the surface of the CF (curve c). No peak in the range of 1500 cm^{-1}–1300 cm^{-1} was observed in this position of CF after elution of Bisphenol A, indicating that Bisphenol A has been successfully eluted from the electrode surface [36]. In addition, the characteristic absorption peaks at 1703 cm^{-1}, 1292 cm^{-1} and 1136 cm^{-1} were observed from curves c and d, which were attributed to the C=O vibration peaks of carboxyl group in MAA and the symmetric and asymmetric stretching vibration peaks of C-O of ester group in EGDMA. The methylene peak, which was attributed to the C-H stretching of the methylene group of EGDMA and AIBN, also appears at 2998 cm^{-1} [37–39]. These results indicated that Bisphenol A had been successfully imprinted on the surface of CF, and a molecular imprint cavity was formed after elution.

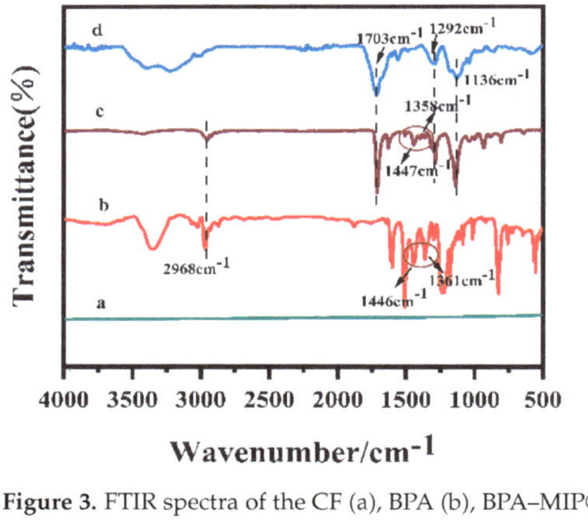

Figure 3. FTIR spectra of the CF (a), BPA (b), BPA–MIP@CF (c) and MIP@CF (d).

3.2. Flexibility Tests of MIP@CF

The MIP@CF electrodes express superior flexibility. CV images under different bending angles and twist times were studied. Figure 4A showed that the shapes of the CV curves were basically unchanged after the electrode bending angle was from 0 to 180°. In addition, CV curves remain almost unchanged after twisting 50 times (Figure 4B), indicating that bending and twisting did not affect the electrochemical performance of the electrode [40–42]. The insets of Figure 4A,B also confirmed the flexibility of the electrodes. Almost no change was observed in the oxidation and reduction peak current after bending and twisting (Figure 4C,D). The results showed that the electrode has good application potential in flexible electronics.

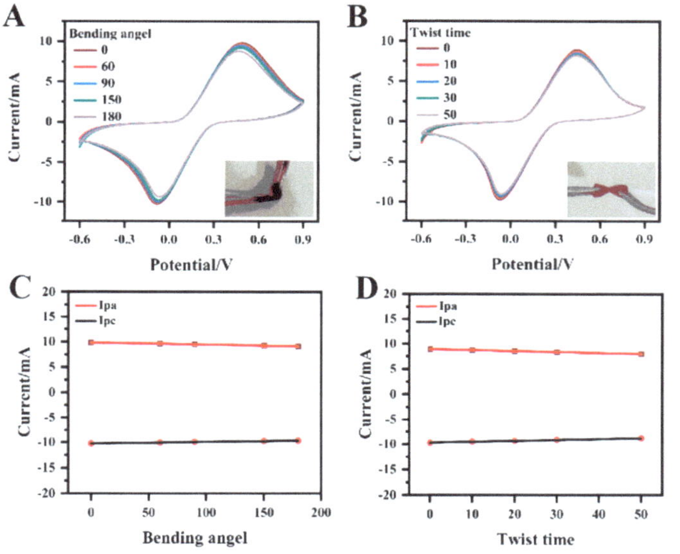

Figure 4. CVs of MIP@CF electrode with different bending angle (**A**) and twist time (**B**) in 5.0 mmol·L^{-1} $K_4[Fe(CN)_6]$ with 0.1 mol·L^{-1} KCl at the scan rate of 50 mV·s^{-1}; Relationship between the anodic (I_{pa}) and cathodic peak (I_{pc}) currents vs. bending angle (**C**) and twist time (**D**).

3.3. Electrochemical Behaviors of Different Modified Electrodes

The electrochemical behaviors of different electrodes were compared using the CV technique. Figure 5A showed the CV of the CF, BPA–MIP@CF, MIP@CF and NIP@CF in KCl solution containing 5 mM $[Fe(CN)_6]^{3-/4-}$. It can be seen from the figure that the bare CF electrode, without any modification, exhibits an obvious pair of well-separated redox peaks (curve a), indicating the good electron transfer ability of CF. Compared with the unmodified CF, the BPA–MIP@CF only showed an oxidation peak current at Epa of 605 mV. The redox peaks of $[Fe(CN)_6]^{3-/4-}$ disappeared (curve b). The polymer film on the CF inhibited the transfer of the $[Fe(CN)_6]^{3-/4-}$ to the electrode, so the redox peaks of $[Fe(CN)_6]^{3-/4-}$ were diminished. However, Bisphenol A imprinted in the polymer film could be oxidized at Epa of 605 mV [43]. The CV of NIP@CF supported this conclusion (curve d). No oxidation or reduction peak was observed on the NIP@CF. The formed non-conductive polymer layer on the CF hindered the electron transfer of $[Fe(CN)_6]^{3-/4-}$. The lack of the oxidation peak at around 605 mV also supported the fact that the peak was attributed to the oxidation of Bisphenol A [39]. The curve c in Figure 5A was the CV response of MIP@CF, which left the cavity on the polymer after the removal of Bisphenol A template. The redox peak of $[Fe(CN)_6]^{3-/4-}$ appeared again. After the elution of the template from the polymer, some cavities were left on the polymer. $[Fe(CN)_6]^{3-/4-}$ could diffuse through these cavities to the conductive CF surface, and electron transfer process was performed on the CF. The redox current on MIP@CF was slightly larger than that on the bared CF. This means that the extraction of template resulted in plenty of transport channels that were favorable to the redox reactions of $[Fe(CN)_6]^{3-/4-}$. The sensing performance of the MIP@CF for Bisphenol A detection was evaluated by comparing the oxidation current and oxidation potential of Bisphenol A on CF and MIP@CF. It can be seen from Figure 5B that Bisphenol A has obvious oxidation peak on both electrodes, and no corresponding reduction peak was found in the reverse scanning. This result indicates that the oxidation reaction of Bisphenol A on both electrodes was irreversible. Furthermore, the MIP@CF electrode (a) exhibited a significantly strong oxidation peak for Bisphenol A compared with bare CF (b). This was due to the specific recognition and accumulation of Bisphenol A by the imprinted sites [44]. The scan rate study is possible to understand whether the oxidation/reduction reaction mechanism of the analyte is controlled by adsorption or diffusion process. The CV responses of 1.0 μM of Bisphenol A on a MIP@CF were recorded at different scan rates (Figure S1). The oxidation peak current increased gradually with the increase in scan rate with a linear relationship in the range of 10–100 mVs^{-1}. The corresponding linear regression equation was: Ip (mA) = 0.00065 + 0.0227 υ (mVs^{-1}). This indicated that the oxidation of Bisphenol A on the MIP@CF was a typical adsorption-controlled process [45]. These results indicated that the MIP@CF can be used as an electrochemical tool for sensitive detection of Bisphenol A.

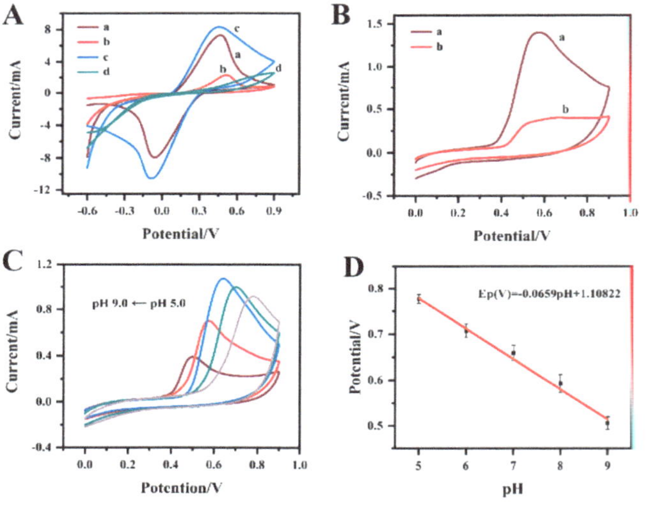

Figure 5. (**A**) CV responses of bare CF (a), BPA–MIP@CF (b), MIP@CF (c) and NIP@CF (d) in 5.0 $mmol \cdot L^{-1}$ $K_4[Fe(CN)_6]$ with 0.1 $mol \cdot L^{-1}$ KCl at the scan rate of 50 $mV \cdot s^{-1}$; (**B**) CV curves of bare CF (b), MIP@CF (a) after recognizing 1.0 μM BPA; (**C**) CV responses of MIP@CF electrode at different pH (5.0, 6.0, 7.0, 8.0 and 9.0) after recognizing 1.0 μM BPA at the scan rate of 50 $mV \cdot s^{-1}$; (**D**) The relationship between the peak potentials and pH values.

The pH value of the solution was an important parameter because it affected the oxidation potential of Bisphenol A. As shown in Figure 5C,D, the peak potential of Bisphenol A shifted negatively when the pH value increased from 5.0 to 9.0. The linear relationship between the oxidation peak potential (Ep) and pH was: Ep(V) = −0.0659·pH + 1.10822 (R^2 = 0.994). The slope of 66 mV/pH is close to the theoretical value of 59 mV/pH unit. The results showed that the oxidation reaction of Bisphenol A on the MIP@CF was accompanied by an equal amount of proton and electron transfer processes [46].

3.4. Parameters Optimization for Sensor Preparation

In order to obtain the optimal sensing performance of the MIP@CF sensor for Bisphenol A detection, various experimental parameters were studied and optimized. Since the molar ratio of the template molecule and the functional monomer played an important role in the structure and rebinding affinity of the polymer, the molar ratio of Bisphenol A and MAA was adjusted from 1:3 to 1:8, and the corresponding oxidation peak currents were measured respectively. As shown in Figure 6A, the oxidation current value was the largest when the Bisphenol A: MAA ratio was 1:5. On the one hand, when the molar ratio of MAA continued to increase, excessive MAA monomer hindered the recognition site of Bisphenol A and resulted in a decrease in oxidation current. On the other hand, when the proportion of template molecule Bisphenol A was too high, insufficient number of monomers led to insufficient polymerization, and it was difficult to form effective imprinting binding sites. The obtained MIP film would have lower imprinting efficiency [47]. Therefore, the optimal molar ratio of template molecule to functional monomer was set as 1:5 in the preparation of the MIP@CF sensor.

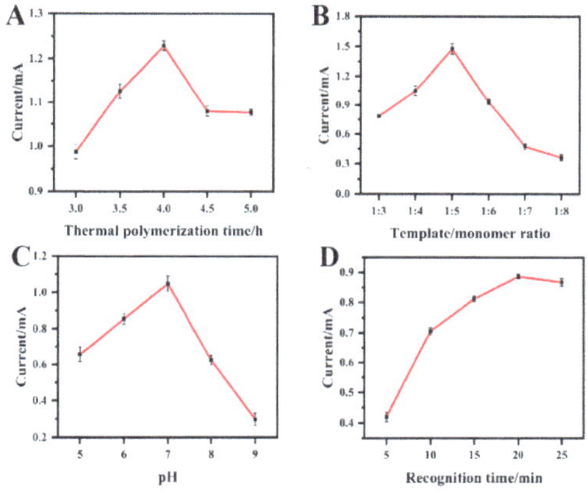

Figure 6. Plots of the oxidation currents of BPA versus (**A**) template/monomer ratio; (**B**) thermal polymerization time; (**C**) pH value of the detection solution and (**D**) elution time after the accumulation of 1.0 μM BPA.

The thermal polymerization time would influence the formation of the MIP layer on the CF surface and thus affect the performance of the MIP@CF sensor. It can be seen from Figure 6B that the oxidation current value gradually increases as the polymerization time increases and reaches the maximum current value at 4 h. However, the current value decreases as time continues to increase. This showed that the imprinted site had reached saturation after 4 h of thermal polymerization. Therefore, 4 h served as the optimal polymerization time for subsequent experiments.

Figure 6C shows that the current value of the oxidation peak is influenced by the pH of the detection solution. The peak current is the largest when pH 7.0 PBS is used. At higher pH, the decrease in peak current may be due to the electrostatic repulsion of the negatively charged anionic Bisphenol A to the sensor surface [48]. Therefore, a pH of 7.0 PBS was subsequently selected for the detection of Bisphenol A.

Figure 6D shows the effect of incubation time on the oxidative current after the MIP@CF sensor was incubated in the recognition solution, which contained Bisphenol A at different times. The oxidation current value gradually increased with the prolongation of incubation time and reached the maximum value after 20 min. So 20 min was selected as the optimum recognition time for Bisphenol A detection.

3.5. Sensing-Performance of MIP@CF in Bisphenol A Detection

The analytical performance of MIP@CF for Bisphenol A detection was evaluated by differential voltammetry pulse method (DPV). Figure 7A shows the response of different concentrations of Bisphenol A on the MIP@CF. It was observed that the peak currents increased linearly with the concentration of Bisphenol A. Figure 7B showed the linear relationship between DPV peak currents and Bisphenol A concentrations. Two calibration curves could be seen in the figure. In 0.5 nM to 8.0 nM range, the regression equation was: $I_P(\mu A) = 13.28 \cdot C\ (nM) + 14.18$. Bisphenol A was easily converted into its oxidation products on the electrode surface in this lower concentration range and had a high detection sensitivity. In 10.0 nM to 300.0 nM range, the regression equation was: $I_P(\mu A) = 0.843 \cdot C\ (\mu M) + 173.47$. The identification of the target and the diffusion of reaction products at high concentrations require more oxidation reaction time than when the concentration is low. Moreover, the surface of the MIP@CF electrode will be contaminated by high concentration of oxidation products. These facts result in relatively low detection sensitivity. The limit of detection (LOD) was as low as 0.36 nM based on $\frac{3s_{y/x}}{b}$, where $s_{y/x}$ was the standard deviation of y-intercepts of the regression line and b was the slope of the calibration curve in the low concentration range of Bisphenol A [49].

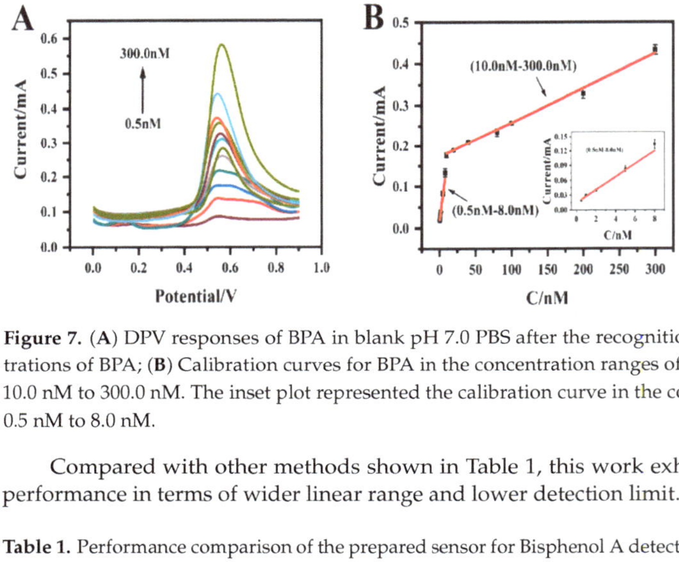

Figure 7. (**A**) DPV responses of BPA in blank pH 7.0 PBS after the recognition of different concentrations of BPA; (**B**) Calibration curves for BPA in the concentration ranges of 0.5 nM to 8.0 nM and 10.0 nM to 300.0 nM. The inset plot represented the calibration curve in the concentration ranges of 0.5 nM to 8.0 nM.

Compared with other methods shown in Table 1, this work exhibits better sensing performance in terms of wider linear range and lower detection limit.

Table 1. Performance comparison of the prepared sensor for Bisphenol A detection with other sensors.

Electrode	Method	Linear Range (μM)	LOD (nM)	Reference
MWCNTs-βCD/SPCE	LSV	0.125–2/2–30	13.7	[50]
CTAB/MIL-101(Cr)	DPV	0.02–0.35	10	[51]
$NiFe_2O_4$/SPCE	DPV	0.02–12.5	6	[52]
AgNPs-EG	SWV	5–100	230	[19]
AuNDs@CNFs/SPE	DPV	0.01–50	5	[53]
GO-MWCNT-βCD/SPE	LSV	0.05–5/5–30	6	[43]
BZPY/MCPE	DPV	2–18	29	[54]
MIP@CF	DPV	0.0005–0.008/0.01–0.3	0.36	This work

3.6. Selectivity

The selectivity of the method was evaluated by measuring the DPV responses of different interfering substances on the MIP@CF sensors. The electrodes were incubated in 0.5 μM of Bisphenol A or 5.0 μM of the interferences, including protocatechuic acid (PCA), ascorbic acid (AA), catechol (CC) and hydroquinone (HQ). The DPV responses of these compounds are shown in Figure S2. There was almost no oxidation peak of these interfering substances near the specific oxidation peak potential of Bisphenol A, and the high response signal to Bisphenol A verified the high selectivity of the sensor. Figure 8A shows the current responses of Bisphenol A and the interfering substances at about 0.6 V. It was shown that the presence of 10 times interferences does not have significant effects on the oxidation peak of Bisphenol A. The presence of the special binding site regions in MIP polymers ensures the specific recognition of Bisphenol A at the imprinted site through the hydrogen bonds and Van der Waals force and makes the sensor have good selectivity. The special oxidation peak potential also contributed to the good selectivity of the sensor for Bisphenol A detection.

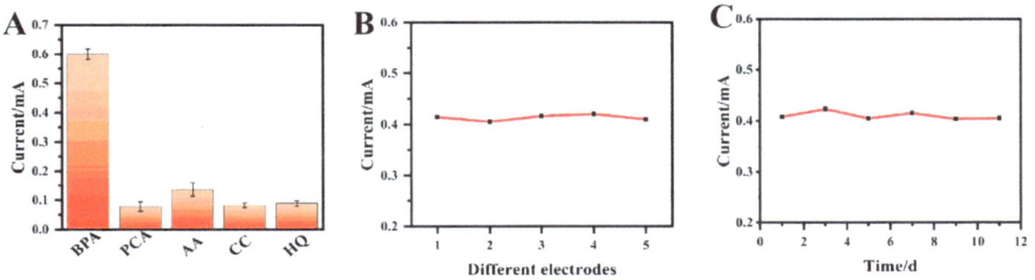

Figure 8. Investigating the selectivity of the MIP@CF sensor for 0.5 μM of Bisphenol A or 5.0 μM of the interferences (**A**), reproducibility (**B**) and stability (**C**) of MIP@CF electrodes for 5.0 μM of BPA.

3.7. Reproducibility and Stability of the Sensor

The reproducibility of the MIP@CF sensor for 0.5 μM of Bisphenol A was measured by measuring the current responses of five different modified electrodes to Bisphenol A under the optimal conditions. Figure 8B,C showed the result that the relative standard deviation (RSD) of the five electrodes was 3.7%, indicating the excellent reproducibility of the sensor. The prepared electrodes were stored in a refrigerator at 4 °C for 12 days, and the response of the electrodes to the same concentration of Bisphenol A was measured every 2 days to determine its stability. The results showed that the relative standard deviation (RSD) of the method was 4.1%, indicating the acceptable stability of the sensor.

3.8. Real Samples Analysis

To verify the analytical potential of the sensor in the detection of real samples, the sensor was applied to detect Bisphenol A in pure milk samples. Bisphenol A was not detected in pure milk samples. Therefore, the recovery experiments were performed. A certain amount of Bisphenol A was added to the pure milk samples, and the recoveries were tested (Figure S3). The analysis results are shown in Table 2. It could be seen that the recovery of Bisphenol A is between 91.26–112%. The results showed that MIP@CF can be used as a novel electrochemical sensor for the detection of Bisphenol A in real samples.

Table 2. Determination of BPA in real samples.

Sample	Original (nM)	Added (nM)	Found (nM)	Recovery (%)	RSD (%)
Milk	0.00	1.00	1.12	112	3.96
	0.00	50.00	51.28	102.5	4.16
	0.00	200.00	182.50	91.26	3.54

4. Conclusions

In this work, a Bisphenol A synthetic receptor was deliberately modified on a flexible CF by a simple thermal polymerization process to construct a Bisphenol A electrochemical sensor. The receptor and the flexible CF work together to give the sensor excellent sensitivity, stability, selectivity and flexibility. This sensor showed linear DPV current responses to

Bisphenol A from 0.5 to 8.0 nM and 10.0 to 300.0 nM. A good recovery result was obtained when this sensor was applied for the detection of Bisphenol A in milk samples. This work provides an effective way for the development of versatile electrochemical sensing platforms for the diagnosis of EDCs.

Supplementary Materials: The following supporting information can be downloaded at: https://www.mdpi.com/article/10.3390/bios12121076/s1. Figure S1. (A) CVs of 1.0 μmol·L^{-1} BPA at MIP@CF electrode at different scan rates; (B) The oxi-dation peak currents of 1.0 μmol·L^{-1} BPA against scan rates; Figure S2. DPV responses of the MIP@CF electrode in pH 7.0 PBS solution after the recognition of 0.5 μM of Bisphenol A or 5.0 μM of protocatechuic acid (PCA), ascorbic acid (AA), catechol (CC), and hydroquinone (HQ); Figure S3. Electrochemical response of MIP@CF for milk sample analysis.

Author Contributions: Conceptualization, C.-Y.X., Q.X. and X.-Y.H.; methodology, C.-Y.X. and K.-P.N.; software, Y.Y.; validation, C.-Y.X., K.-P.N. and Q.X.; formal analysis, C.-Y.X. and Z.W.; investigation, C.-Y.X. and Z.W.; resources, X.-Y.H.; data curation, C.-Y.X., K.-P.N. and Q.X; writing—original draft preparation, C.-Y.X., writing—review and editing, Q.X. and X.-Y.H.; visualization, C.-Y.X. and Z.W.; supervision, Q.X.; project administration, C.-Y.X., Y.Y. and Q.X.; funding acquisition, Q.X. All authors have read and agreed to the published version of the manuscript.

Funding: This research was funded by NSFC (22076161, 21675140 and 21705141), the Talent Support Program of Yangzhou University, the Open Project Program of Jiangsu Key Laboratory of Zoonosis (No. R2013), Yangzhou University Interdisciplinary Research Foundation for Chemistry Discipline of Targeted Support (yzuxk202009), the project funded by the PAPD and TAPP.

Institutional Review Board Statement: Not applicable.

Informed Consent Statement: Not applicable.

Data Availability Statement: Not applicable.

Acknowledgments: We thank the Test Center of Yangzhou University for the SEM and TEM characterizations.

Conflicts of Interest: The authors declare no conflict of interest.

Abbreviations

BPA: Bisphenol A; FTIR: Fourier-transform infrared spectroscopy; SEM: Scanning electron microscope; S/N: signal to noise; EDC: Endocrine disruptor; HPLC: high-performance liquid chromatography; LC-MS: liquid chromatography-mass spectrometry; GC-MS: gas chromatography-mass spectrometry; ELISA: enzyme-linked immunosorbent assay; MIP: Molecularly imprinted polymers; CF: carbon felt; MAA: methacrylic acid; DPV: differential voltammetry pulse method; EGDMA: ethylene glycol dimethacrylate; NIP: non-molecularly imprinted polymers; AIBN: Azobisisobutyronitrile; PBS: phosphate buffered saline; MW: molecular weight.

References

1. Xu, W.; Zhang, Y.; Yin, X.; Zhang, L.; Cao, Y.; Ni, X.; Huang, W. Highly sensitive electrochemical BPA sensor based on titanium nitride-reduced graphene oxide composite and core-shell molecular imprinting particles. *Anal. Bioanal. Chem.* **2021**, *413*, 1081–1090. [CrossRef] [PubMed]
2. Zhang, Y.; Mi, K.; Xue, W.; Wei, W.; Yang, H. Acute BPA exposure-induced oxidative stress, depressed immune genes expression and damage of hepatopancreas in red swamp crayfish Procambarus clarkii. *Fish Shellfish Immun.* **2020**, *103*, 95–102. [CrossRef] [PubMed]
3. Shim, K.; Kim, J.; Shahabuddin, M.; Yamauchi, Y.; Hossain, S.A.; Kim, J.H. Efficient wide range electrochemical bisphenol-A sensor by self-supported dendritic platinum nanoparticles on screen-printed carbon electrode. *Sens. Actuators B* **2018**, *255*, 2800–2808. [CrossRef]
4. Li, Y.; Gao, Y.; Cao, Y.; Li, H. Electrochemical sensor for bisphenol A determination based on MWCNT/melamine complex modified GCE. *Sens. Actuators B* **2012**, *171–172*, 726–733. [CrossRef]
5. Su, B.; Shao, H.; Li, N.; Chen, X.; Cai, Z.; Chen, X. A sensitive bisphenol A voltammetric sensor relying on AuPd nanoparticles/graphene composites modified glassy carbon electrode. *Talanta* **2017**, *616*, 126–132. [CrossRef] [PubMed]

6. Zhang, Y.; Yuan, Z.-L.; Deng, X.-Y.; Wei, H.-D.; Wang, W.-L.; Xu, Z.; Feng, Y.; Shi, X. Metal-organic framework mixed-matrix membrane-based extraction combined HPLC for determination of bisphenol A in milk and milk packaging. *Food Chem.* **2022**, *386*, 132753. [CrossRef]
7. Ma, W.; Wan, S.; Lin, C.; Lin, X.; Xie, Z. Towards online specific recognition and sensitive analysis of bisphenol A by using AuNPs@aptamer hybrid-silica affinity monolithic column with LC-MS. *Talanta* **2020**, *219*, 121275. [CrossRef]
8. Correia-Sá, L.; Norberto, S.; Delerue-Matos, C.; Calhau, C.; Domingues, V.F. Micro-QuEChERS extraction coupled to GC–MS for a fast determination of Bisphenol A in human urine. *J. Chromatogr. B* **2018**, *1072*, 9–16. [CrossRef]
9. Temel, N.K.; Gürkan, R. A micellar sensitized kinetic method for quantification of low levels of bisphenol A in foodstuffs by spectrophotometry. *Anal. Methods* **2017**, *9*, 1190. [CrossRef]
10. Lei, Y.; Fang, L.; Akash, M.S.H.; Liu, Z.; Shi, W.; Chen, S. Development and comparison of two competitive ELISAs for the detection of bisphenol A in human urine. *Anal. Methods* **2013**, *5*, 6106. [CrossRef]
11. Peng, Y.; Wang, J.; Wu, C. Determination of Endocrine Disruption Potential of Bisphenol A Alternatives in Food Contact Materials Using In Vitro Assays: State of the Art and Future Challenges. *J. Agric. Food Chem.* **2019**, *67*, 12613–12625. [CrossRef] [PubMed]
12. Gao, Y.; Wang, L.; Shen, C.; Wang, C.; Hu, X.; Wang, G. An electrochemical sensor on the hierarchically porous Cu-BTC MOF platform for glyphosate determination. *Sens. Actuators B* **2019**, *283*, 487–494.
13. Yu, L.; Zhang, Q.; Yang, B.; Xu, Q.; Xu, Q.; Hu, X. Electrochemical sensor construction based on Nafion/calcium lignosulphonate functionalized porous graphene nanocomposite and its application for simultaneous detection of trace Pb^{2+} and Cd^{2+}. *Sens. Actuators B* **2018**, *259*, 540–551. [CrossRef]
14. Mao, A.; Li, H.; Jin, D.; Yu, L.; Hu, X. Fabrication of electrochemical sensor for paracetamol based on multi-walled carbon nanotubes and chitosan–copper complex by self-assembly technique. *Talanta* **2015**, *144*, 252–257. [CrossRef] [PubMed]
15. Fu, S.; Zhu, Y.; Zhang, Y.; Zhang, M.; Zhang, Y.; Qiao, L.; Yin, N.; Song, K.; Liu, M.; Wang, D. Recent advances in carbon nanomaterials-based electrochemical sensors for phenolic compounds detection. *Microchem. J.* **2021**, *171*, 106776. [CrossRef]
16. Sanko, V.; Şenocak, A.; Tümay, S.O.; Orooji, Y.; Demirbas, E.; Khataee, A. An electrochemical sensor for detection of trace-level endocrine disruptor bisphenol A using $Mo_2Ti_2AlC_3$ MAX phase/MWCNT composite modified electrode. *Environ. Res.* **2022**, *212*, 113037. [CrossRef] [PubMed]
17. Teymoori, N.; Raoof, J.B.; Khalilzadeh, M.A.; Ojani, R. An electrochemical sensor based on CuO nanoparticle for simultaneous determination of hydrazine and bisphenol A. *J. Iran. Chem. Soc.* **2018**, *15*, 2271–2279. [CrossRef]
18. Akilarasan, M.; Kogularasu, S.; Chen, S.-M.; Chen, T.-W.; Lin, S.-H. One-step synthesis of reduced graphene oxide sheathed zinc oxide nanoclusters for the trace level detection of bisphenol A in tissue papers. *Ecotoxicol. Environ. Saf.* **2018**, *161*, 699–705. [CrossRef]
19. Tsekeli, T.R.; Sebokolodi, T.I.; Karimi-Maleh, H.; Arotiba, O.A. A Silver-Loaded Exfoliated Graphite Nanocomposite Anti-Fouling Electrochemical Sensor for Bisphenol A in Thermal Paper Samples. *ACS Omega* **2021**, *6*, 9401–9409. [CrossRef]
20. Ren, Y.; Ma, W.; Ma, J.; Wen, Q.; Wang, J.; Zhao, F. Synthesis and properties of bisphenol A molecular imprinted particle for selective recognition of BPA from water. *J. Colloid Interf. Sci.* **2012**, *367*, 355–361. [CrossRef]
21. Pei, D.-N.; Zhang, A.-Y.; Pan, X.-Q.; Si, Y.; Yu, H.-Q. Electrochemical Sensing of Bisphenol A on Facet-Tailored TiO2 Single Crystals Engineered by Inorganic-Framework Molecular Imprinting Sites. *Anal. Chem.* **2018**, *90*, 3165–3173. [CrossRef]
22. Jiang, H.; Jiang, D.; Shao, J.; Sun, X. Magnetic molecularly imprinted polymer nanoparticles based electrochemical sensor for the measurement of Gram-negative bacterial quorum signaling molecules (N-acyl-homoserine-lactones). *Biosens. Bioelectron.* **2016**, *75*, 411–419. [CrossRef] [PubMed]
23. Hamed, E.M.; Li, S.F.Y. Molecularly imprinted polymers-based sensors for bisphenol-A: Recent developments and applications in environmental, food and biomedical analysis. *Trends Environ. Anal. Chem.* **2022**, *35*, e00167. [CrossRef]
24. Karthika, P.; Shanmuganathan, S.; Viswanathan, S.; Delerue-Matos, C. Molecularly imprinted polymer-based electrochemical sensor for the determination of endocrine disruptor bisphenol-A in bovine milk. *Food Chem.* **2021**, *363*, 130287. [CrossRef] [PubMed]
25. Kim, M.; Song, Y.E.; Xiong, J.-Q.; Kim, K.-Y.; Jang, M.; Jeon, B.-H.; Kim, J.R. Electrochemical detection and simultaneous removal of endocrine disruptor, bisphenol A using a carbon felt electrode. *J. Electroanal. Chem.* **2021**, *880*, 114907. [CrossRef]
26. Pang, Y.-H.; Wang, Y.-Y.; Shen, X.-F.; Qiao, J.-Y. Covalent organic framework modified carbon cloth for ratiometric electrochemical sensing of bisphenol A and S. *Microchim. Acta* **2022**, *189*, 189. [CrossRef] [PubMed]
27. He, S.; Zhang, L.; Bai, S.; Yang, H.; Cui, Z.; Zhang, X.; Li, Y. Advances of molecularly imprinted polymers (MIP) and the application in drug delivery. *Eur. Polym. J.* **2021**, *143*, 110179. [CrossRef]
28. Kamel, A.H.; Jiang, X.; Li, P.; Liang, R. A paper-based potentiometric sensing platform based on molecularly imprinted nanobeads for determination of bisphenol A. *Anal. Methods* **2018**, *10*, 3890. [CrossRef]
29. Liu, K.; Song, Y.; Song, D.; Liang, R. Plasticizer-free polymer membrane potentiometric sensors based on molecularly imprinted polymers for determination of neutral phenols. *Anal. Chim. Acta* **2020**, *1121*, 50–56. [CrossRef]
30. Shi, J.; Chen, J.; Li, G.; An, T.; Yamashita, H. Fabrication of Au/TiO2 nanowires@carbon fiber paper ternary composite for visible-light photocatalytic degradation of gaseous styrene. *Catal. Today* **2017**, *281*, 621–629. [CrossRef]
31. Zuo, T.-T.; Wu, X.-W.; Yang, C.-P.; Yin, Y.-X.; Ye, H.; Li, N.-W.; Guo, Y.-G. Graphitized Carbon Fibers as Multifunctional 3D Current Collectors for High Areal Capacity Li Anodes. *Adv. Mater.* **2017**, *29*, 1700389. [CrossRef] [PubMed]

32. Wang, J.; Xu, Q.; Xia, W.W.; Shu, Y.; Jin, D.; Zang, Y.; Hu, X. High sensitive visible light photoelectrochemical sensor based on in-situ prepared flexible Sn_3O_4 nanosheets and molecularly imprinted polymers. *Anal. Bioanal. Chem.* **2018**, *271*, 215–224. [CrossRef]
33. Chen, S.; Chu, X.; Wu, L.; Foord, J.S.; Hu, J.; Hu, H.; Yang, J. Three-Dimensional PbO_2-Modified Carbon Felt Electrode for Efficient Electrocatalytic Oxidation of Phenol Characterized with In Situ ATR-FTIR. *J. Phys. Chem. C* **2022**, *126*, 912–921. [CrossRef]
34. Dong, L.; Liu, X.; Xu, H.; Wu, D.; Gao, S.; Zhong, L.; Kong, Y. Improved chiral electrochemical recognition of tryptophan enantiomers based on three- dimensional molecularly imprinted overoxidized polypyrrole/MnO_2/carbon felt composites. *Chirality* **2019**, *31*, 917–922. [CrossRef] [PubMed]
35. Yang, Z.; Qin, T.; Niu, Y.; Zhang, Y.; Zhang, C.; Li, P.; Zhu, M.; Jia, Y.; Li, Q. Flexible visible-light-driven photoelectrochemical biosensor based on molecularly imprinted nanoparticle intercalation-modulated graphene fiber for ultrasensitive urea detection. *Carbon* **2020**, *157*, 457–465. [CrossRef]
36. Xu, R.; Qian, X.; Zhang, Z.; Yuan, F.; Song, Y.; Liu, J.; Zhang, Q.; Wei, J. A molecularly Imprinted Electrochemical Sensor Based on NMWCNT/CPE for Highly Sensitive and Selective Detection of Bisphenol A. *Int. J. Electrochem. Sci.* **2022**, *17*, 220536. [CrossRef]
37. Metwally, M.G.; Shehab, O.R.; Ibrahim, H.; Nashar, R.M.E. Electrochemical detection of Bisphenol A in plastic bottled drinking waters and soft drinks based on molecularly imprinted polymer. *J. Environ. Chem. Eng.* **2022**, *10*, 107699. [CrossRef]
38. Pan, J.; Hu, W.; Dai, X.; Guan, W.; Zou, X.; Wang, X.; Huo, P.; Yan, Y. Molecularly imprinted polymers based on magnetic fly-ash-cenosphere composites for bisphenol A recognition. *J. Mater. Chem.* **2011**, *21*, 15741. [CrossRef]
39. Zhou, G.; Cao, Y.; Jin, Y.; Wang, C.; Wang, Y.; Hua, C.; Wu, S. Novel selective adsorption and photodegradation of BPA by molecularly imprinted sulfur doped nano-titanium dioxide. *J. Clean. Prod.* **2020**, *274*, 122929. [CrossRef]
40. Xu, M.; Song, Y.; Ye, Y.; Gong, C.; Shen, Y.; Wang, L.; Wang, L. A novel flexible electrochemical glucose sensor based on gold nanoparticles/polyaniline arrays/carbon cloth electrode. *Sens. Actuators B* **2017**, *252*, 1187–1193. [CrossRef]
41. Wu, D.; Zhong, W. A new strategy for anchoring a functionalized graphene hydrogel in a carbon cloth network to support a lignosulfonate/polyaniline hydrogel as an integrated electrode for flexible high arealcapacitance supercapacitors. *J. Mater. Chem. A* **2019**, *7*, 5819. [CrossRef]
42. Jiang, H.; Zhou, C.; Yan, X.; Miao, J.; You, M.; Zhu, Y.; Li, Y.; Zhou, W.; Cheng, X. Effects of various electrolytes on the electrochemistry performance of Mn_3O_4/carbon cloth to ultra-flexible all-solid-state asymmetric supercapacitor. *J. Energy Storage* **2020**, *32*, 101898. [CrossRef]
43. Alam, A.U.; Deen, M.J. Bisphenol A Electrochemical Sensor Using Graphene Oxide and β-Cyclodextrin-Functionalized Multi-Walled Carbon Nanotube. *Anal. Chem.* **2020**, *92*, 5532–5539. [CrossRef]
44. Zheng, W.; Xiong, Z.; Li, H.; Yu, S.; Li, G.; Niu, L.; Liu, W. Electrodeposited Pt@Molecularly imprinted polymer core-shell nanostructure: Enhanced sensing platform for sensitive and selective detection of bisphenol A. *Sens. Actuators B* **2018**, *272*, 655–661. [CrossRef]
45. Yücebaşa, B.B.; Yücebaşa, Y.T.; Bolat, G.; Özgür, E.; Uzun, L.; Abaci, S. Molecular imprinted polymer based electrochemical sensor for selective detection of paraben. *Sens. Actuators B* **2020**, *305*, 127368. [CrossRef]
46. Wang, Y.; Yin, C.; Zhuang, Q. An electrochemical sensor modified with nickel nanoparticle/nitrogen-doped carbon nanosheet nanocomposite for bisphenol A detection. *J. Alloys Compd.* **2020**, *827*, 154335. [CrossRef]
47. Rawool, C.R.; Srivastava, A.K. A dual template imprinted polymer modified electrochemical sensor based on Cu metal organic framework/mesoporous carbon for highly sensitive and selective recognition of rifampicin and isoniazid. *Sens. Actuators B* **2019**, *288*, 493–506. [CrossRef]
48. Wang, J.-Y.; Su, Y.-L.; Wu, B.-H.; Cheng, S.-H. Reusable electrochemical sensor for bisphenol A based on ionic liquid functionalized conducting polymer platform. *Talanta* **2016**, *147*, 103–110. [CrossRef]
49. Desimoni, E.; Brunetti, B. About estimating the limit of detection by the signal to noise approach. *Pharm. Anal.* **2015**, *6*, 355.
50. Ali, M.Y.; Alam, A.U.; Howlader, M.M.R. Fabrication of highly sensitive Bisphenol A electrochemical sensor amplified with chemically modified multiwall carbon nanotubes and β-cyclodextrin. *Sens. Actuators B* **2020**, *320*, 128319. [CrossRef]
51. Li, Z.; Hu, J.; Xiao, Y.; Zha, Q.; Zeng, L.; Zhu, M. Surfactant assisted Cr-metal organic framework for the detection of bisphenol A in dust from E-waste recycling area. *Anal. Chim. Acta* **2021**, *1146*, 174–183. [CrossRef] [PubMed]
52. Kesavan, G.; Nataraj, N.; Chen, S.-M.; Lin, L.-H. Hydrothermal synthesis of $NiFe_2O_4$ nanoparticles as an efficient electrocatalyst for the electrochemical detection of bisphenol A. *New J. Chem.* **2020**, *44*, 7698–7707. [CrossRef]
53. Lei, Y.; Zhang, Y.; Wang, B.; Zhang, Z.; Yuan, L.; Li, J. A lab-on-injector device with Au nanodots confined in carbon nanofibers for in situ electrochemical BPA sensing in beverages. *Food Control* **2022**, *134*, 108747. [CrossRef]
54. Naik, T.S.S.K.; Anil, A.G.; Swamy, B.E.K.; Singh, S.; Madhavi, V.; Raghavendra, S.M.; Ramamurthy, P.C. A novel electrochemical sensor based on 2,6-bis (2-benzimidazoyl) pyridine for the detection of Bisphenol A. *Mater. Chem. Phys.* **2022**, *275*, 125287. [CrossRef]

biosensors

Review

Current Status and Future Perspectives of Lactate Dehydrogenase Detection and Medical Implications: A Review

Yangzhe Zhou [1], Min Qi [1,*] and Minghui Yang [2,*]

1 Department of Plastic Surgery, Xiangya Hospital, Central South University, Changsha 410008, China
2 College of Chemistry and Chemical Engineering, Central South University, Changsha 410083, China
* Correspondence: qimin05@csu.edu.cn (M.Q.); yangminghui@csu.edu.cn (M.Y.); Tel.: +86-731-88879616 (M.Y.)

Abstract: The demand for glucose uptake and the accompanying enhanced glycolytic energy metabolism is one of the most important features of cancer cells. Unlike the aerobic metabolic pathway in normal cells, the large amount of pyruvate produced by the dramatic increase of glycolysis in cancer cells needs to be converted to lactate in the cytoplasm, which cannot be done without a large amount of lactate dehydrogenase (LDH). This explains why elevated serum LDH concentrations are usually seen in cancer patient populations. LDH not only correlates with clinical prognostic survival indicators, but also guides subsequent drug therapy. Besides their role in cancers, LDH is also a biomarker for malaria and other diseases. Therefore, it is urgent to develop methods for sensitive and convenient LDH detection. Here, this review systematically summarizes the clinical impact of lactate dehydrogenase detection and principles for LDH detection. The advantages as well as limitations of different detection methods and the future trends for LDH detection were also discussed.

Keywords: lactate dehydrogenase; quantum dots; fluorescence; nicotinamide adenine dinucleotide; tumor

Citation: Zhou, Y.; Qi, M.; Yang, M. Current Status and Future Perspectives of Lactate Dehydrogenase Detection and Medical Implications: A Review. *Biosensors* **2022**, *12*, 1145. https://doi.org/10.3390/bios12121145

Received: 11 November 2022
Accepted: 2 December 2022
Published: 7 December 2022

1. Introduction

Lactate dehydrogenase (LDH) is one of the most common enzymes in nature with an Enzyme Commission Number (EC)^{C} of 1. 1. 1. 27 [1,2]. Structurally, LDH is a tetrameric enzyme which mainly consists of two subunits, LDHA and LDHB. The two subunits are encoded by two independent genes, LDHA and LDHB, respectively [1,3]. LDHA and LDHB are expressed in skeletal muscle and cardiac muscle, respectively [4]. These two subunits can also be combined into five different forms to form homomers or heterotetramers in the human body, including: LDH−1, LDH−2, LDH−3, LDH−4 and LDH−5 [5]. The most critical reaction in which LDH is involved in the body is the conversion of lactic acid and oxidized nicotinamide adenine dinucleotide (NAD) to pyruvate and reduced nicotinamide adenine dinucleotide (NADH), respectively [6–8]. It is worth noting that LDHA has a higher affinity for pyruvate, which preferentially converts pyruvate to lactate while oxidizing NADH to NAD. Conversely, LDHB has a higher affinity for lactate, which preferentially convert lactate to pyruvate while reducing NAD to NADH [6,9–11]. This reversible reaction is sensitive to different conditions and was also applied in our previous LDH assay experiments [12].

It is known that the pyruvate produced in normal cells is mainly delivered to the inner mitochondrial membrane, where it is oxidized by pyruvate dehydrogenase complex to acetyl coenzyme A (CoA). Then CoA then enters the tricarboxylic acid cycle, where it is oxidatively phosphorylated (OXPHOS) for the efficient production of ATP for energy supply [11,13–15]. However, tumor cells are out of the ordinary. In the early 20th century, the German biochemist Otto Warburg explained the phenomenon that tumor cells produce large amounts of lactic acid even in the presence of sufficient oxygen. This is now known as the Warburg effect [11,13,16–20]. The lactate produced here is due to the dramatic increase of glycolysis rate in tumor cells, which occurs as the produced pyruvate is converted to lactate by the action of LDH without entering the mitochondria to be metabolized by oxidation [1,6,11,21]. This explains why LDH is usually elevated in cancer patients. Figure 1 illustrates the difference between the metabolism of normal cells and tumor cells in vivo. According to recent reports, dual inhibition of OXPHOS and glycolysis can disrupt the plasticity of tumor cells, curbing their energy supply and generating effective therapeutic options [22,23]. For example, Jennifer R. Molina et al. discovered IACS−010759, a substance that strongly inhibits tumor growth and induces apoptosis, which can be used to treat acute myeloid leukemia and brain cancer by inhibiting OXPHOS [24]. Other researchers also have found that by inhibiting OXPHOS in this pathway can be used to treat cancers such as pancreatic cancer and triple negative breast cancer [25,26]. In addition, Daan F. Boreel et al. found that targeting OXPHOS could be used to improve the efficacy of radiation and immunotherapy treatments, which may be a new approach for future tumor treatment [27].

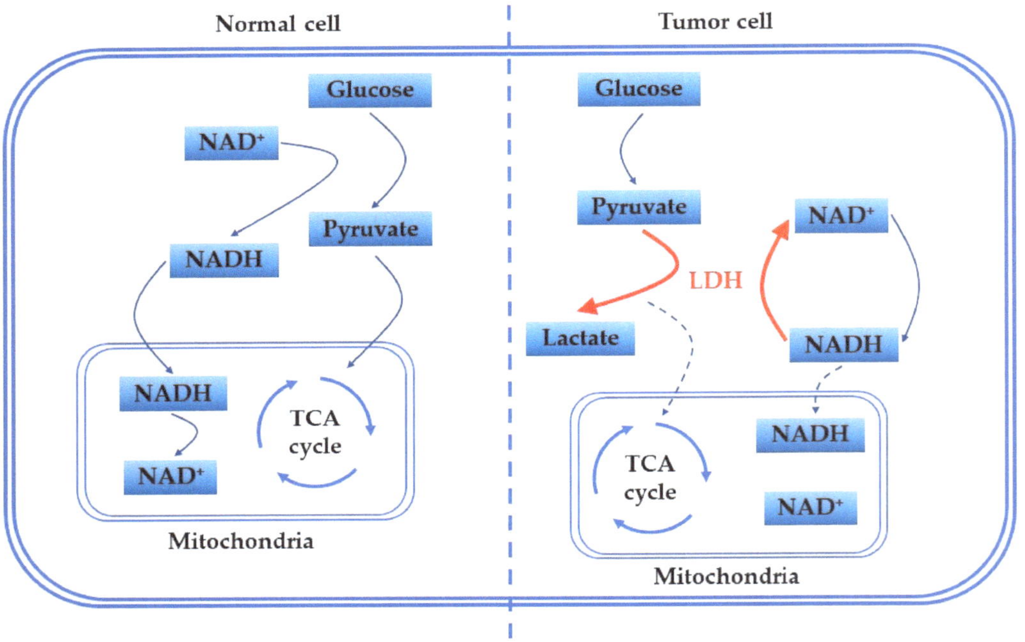

Figure 1. Difference of glucose metabolism between normal and tumor cells in vivo.

In healthy human serum, LDH concentration should not be higher than 200 U/L. This figure is the standard in routine clinical blood sampling for LDH testing. When elevated LDH levels are found, we are alerted to the possible presence of a tumor or residual tumor remaining after a tumor resection surgery. This is because LDH levels will drop significantly within one to two weeks after complete tumor removal [28]. At the same time, low levels of LDH are associated with good prognosis and complete tumor resection [29–31]. Currently, LDH has become a consensus as a biomarker for tumors, and plays a very important role as a prognostic survival indicator and guides drug treatment for cancer patients. For instance, numerous studies have reported LDH as a prognostic biomarker for colorectal cancer [32,33], lung cancer [34,35], melanoma [36,37], renal cell cancer [38,39] and hepatocellular carcinoma [40,41]. Many studies have also shown that LDH is also a biomarker for malaria [42–44]. Moreover, LDH activity is an important clinical guide to the choice of chemotherapy and helps to identify whether a pleural effusion is benign or malignant [45,46]. In addition, Sander Kelderman's team found that ipilimumab was less effective when serum LDH levels were higher than twice the normal value, and S. Diem et al. elucidated that LDH is predictive for melanoma patients receiving PD−1 therapy [47,48]. Since the outbreak of the novel coronavirus pneumonia in

2019 (COVID-19), it seriously endangers the survival and health of human beings. Some researchers have even discovered that elevated LDH is associated with an increased risk of severe COVID-19, which can be used as a survival indicator of poor prognosis [49–59].

In summary, LDH plays an important role in the medical field, especially as a biomarker for tumors. Figure 2 illustrates the applications of LDH in medicine. For these reasons, it is urgent to develop sensitive and convenient methods for LDH detection.

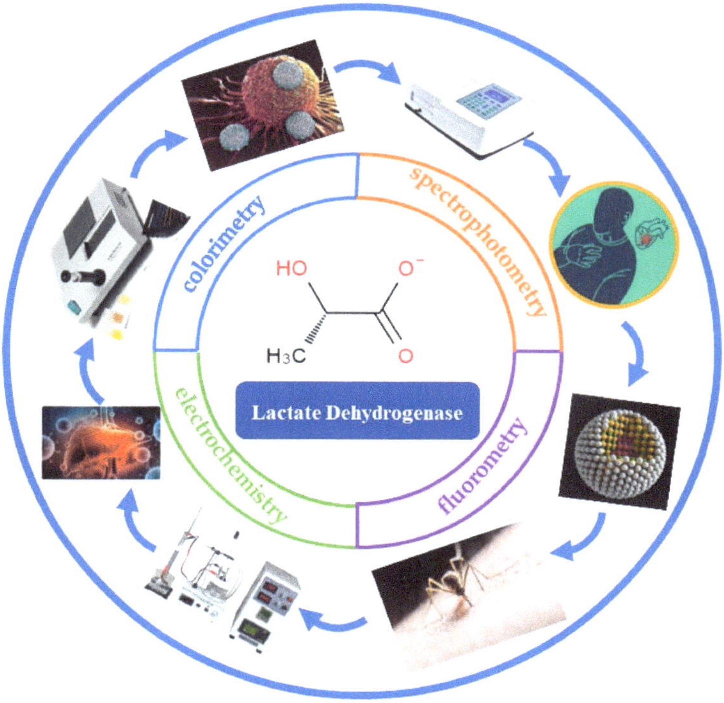

Figure 2. Lactate dehydrogenase applications in medical sectors.

Many methods have been reported for the quantitative detection of LDH, including colorimetry [60–69], spectrophotometry [70–74], electrochemistry [75–82] and fluorometry [12,61,74,83–87]. This article reviews the various methods for LDH detection, discusses their advantages as well as limitations and looks forward to future trends.

2. Various Substances Involved in LDH Testing

The most important reaction in which LDH involved is the conversion of lactate and NAD to pyruvate and NADH. The reaction equation is as follows:

$$\text{Lactate} + \text{NAD}^+ \xleftrightarrow{\text{LDH}} \text{Pyruvate} + \text{NADH} + \text{H}^+$$

Unlike other methods that directly detect the analyte, the detection of LDH is usually achieved through the indirect measurement of the substances involved in the redox reactions in which LDH participates. The two most critical substances in this reaction are NADH and its oxidized form NAD. Both play a crucial role in biological systems [88–90]. NADH is an essential oxidation reducer and is primarily responsible for bringing electrons from the tricarboxylic acid cycle into the electron transport chain to generate energy in the form of ATP [91]. Its oxidized form, NAD, is the best receptor for the reducing equivalents generated by the oxidation of various substrates in the cell. NAD receives reducing equivalents from the citric acid cycle and glycolytic processes, which also play a key role in cellular energy metabolism [92,93]. An imbalance in the ratio of NADH to NAD leads to a disruption of energy production by these pathways, which in turn leads to dysregulation of cellular metabolism [94,95].

3. LDH Detection Methods

3.1. Colorimetric Method

A colorimeter is a laboratory instrument that details the concentration represented by a color and is widely used for the routine measurement of substance concentrations because of advantages such as affordability, portability and direct visual observation of the results [64,96,97]. As shown in Figure 3, Kannan et al. developed a highly stable and mass-produced colorimetric biosensor for LDH detection, which can be operated in less than 5 min and has a detection limit as low as 13 U/L. The sensor was based on the principle of fixing Pullulan on paper. The serum containing LDH on the Pullulan-fixed paper is then added for colorimetric analysis. The sensor can be printed onto paper holes in a highly reproducible manner using an automated printing system, allowing the production of sensors in line with high-speed automated manufacturing. The color development after the addition of LDH can be seen with the naked eye and quantified with a digital camera and image processing software. The sensor uses an aggressively low volume of reagents, significantly reducing the costs associated with the assay. The paper wells can also be stored at room temperature for short periods of time (2–3 weeks) and in a refrigerator for at least five weeks. This eliminates the need for complex laboratory facilities, expensive reagent transport and storage and complicated sample handling. Therefore, the sensor can be used for rapid, inexpensive screening of large numbers of samples in resource-limited settings [98]. Moreover, Arias et al. reported a similar paper-based sensor which is based on a single-step magnetic immunoassay and can be performed in less than 20 min on an inexpensive and simple paper-based disposable device. The assay consists of a single incubation of the lysed whole blood sample with the reagent mixture for 5 min. The mixture is then pipetted directly into a single piece of paper-based equipment, manufactured using a low-cost process cuter. A detection limit of 0.39 U/L can be achieved by visual colorimetry [99]. In addition, Papaneophytou et al. indirectly detected LDH through the LDH-catalyzed production of NADH by reacting it with nitro blue tetrazolium (NBT) and phenazine methyl sulfate (PMS), resulting in a change in the color of the mixture followed by the formation of blue–purple beetles [69]. Furthermore, Halvorsen et al. reported an early prototype of a manually manufactured rapid paper-based point-of-care (POC) assay that required minute amounts of whole blood and used a smartphone camera to provide colorimetric LDH concentration measurements in less than 4 min. The POC analysis device involved lateral separation of whole blood into plasma on a set of filter papers, a colorimetric membrane using a dry chemical reaction on a filter membrane and analysis of the concentration using software on a smartphone. The ease of operation and smartphone-based readings make this POC platform particularly suitable for resource-limited environments. The smartphone provides real-time output of a colorimetric test that is completely self-contained, with a mobile application performing the data analysis. The smartphone can also be connected directly to an external computer. Mobile communication technology facilitates information management. Finally, POC systems are relatively inexpensive to manufacture and can be used as disposable device. These methods eliminate the need for complex laboratory facilities and complicated sample analyses, allowing for rapid, low-cost screening of large numbers of samples in resource-constrained medical settings [98,100]. However, the colorimetric method can only be applied to samples with relatively simple composition that are less susceptible to interference. Moreover, its relatively low sensitivity is obstacle to its wide application [60,96].

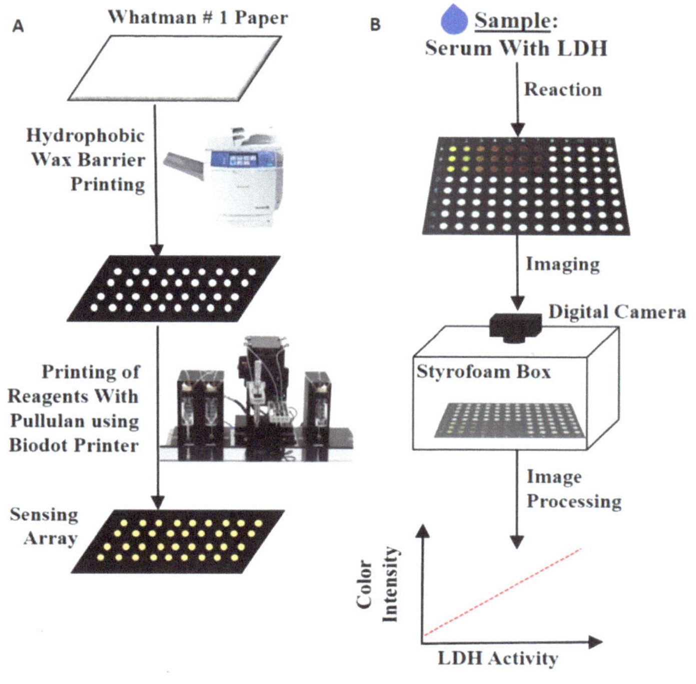

Figure 3. (**A**) Preparation of paper-based LDH sensors. (**B**) Assay for serum LDH [98]. Copyright 2015 American Chemical Society.

3.2. Spectrophotometric Method

The most commonly used spectrophotometry method is UV-spectrophotometry, which is easy to implement and often used for the detection of various substances [2,101]. NADH has one absorption peak at 260 nm and one at 340 nm, while NAD has only one absorption peak at 260 nm. This important property distinguishes the two, and is also the physical basis for measuring the metabolic rate in many metabolic tests [102]. Damaris et al. developed a facile spectroscopic assay for detection of LDH in saliva, where the assay is based on the interconversion of catalytic pyruvate and lactate in the presence of LDH, and the decrease in absorbance at 340 nm caused by NADH is proportional to the LDH activity in the sample [103]. In addition, Lee's team reported a microfluidic microplate-based immunoassay, which requires only a small amount of antibody for faster detection of LDH compared to traditional ELISA. The entire body of this microplate consists of a 96-well plate that has an inlet for pipette injection, an outlet open towards the absorbent pad and a microfluidic channel between them. The microfluidic microplate is a spiral microfluidic channel with 1.5 times larger surface area and 50 times larger surface area-to-volume ratio compared to conventional ELISA plates. The detection limit is as low as 6.25×10^{-3} U/L [43]. However, the UV spectrum also will be easily affected by the external environment, such as the color of the sample [70].

3.3. Electrochemical Measurement

Electrochemical detection is based on changes of the electrical signal due to the substance to be measured. This method is popular because of its low cost, fast response and ease of miniaturization. There are a wide variety of electrochemical biosensors based on different electrochemical techniques, such as cyclic voltammetry (CV), differential pulse voltammetry (DPV), stripping voltammetry, alternating current voltammetry (ACV), polarimetry, square wave voltammetry (SWV) and linear scanning voltammetry (LSV) [104,105].

As shown in Figure 4, Hong et al. fabricated an electrochemical sensor based on screen-printed electrodes for LDH detection. The reaction layer of the printed ink used for the working electrode consisted of graphite, electrodeposited 3,4−DHB, NAD and l−lactate, which were bound in a composite polymer binder. LDH, the target analyte, diffused from the supporting electrolyte into the reaction layer and reduced NAD to NADH. The generated NADH was then oxidized on the electrode surface, generating electrochemical current that is proportional to LDH activity. This method can detect LDH in the range of 50–500 U/L with a detection limit of 50 U/L, which is sufficient to meet the basic needs of the test [106]. In addition, Zhu et al. developed a novel detection platform combining microfluidics and electrochemical sensors arrays, which contains three parts: sample processing, detection and signal output. The processing analysis starts when the sample flows into the chip, uses lactic acid as the substrate. LDH catalyzes the reduction of NAD to NADH. Then, the concentration of LDH is assessed by electrochemical detection of NADH. It establishes a linear relationship in the range of 60–700 U/L with a detection limit of 25 U/L, which is much lower than the serum LDH concentration at the time of tumorigenesis [107]. Furthermore, researchers have tested a zwitterionic phenazine compound: 3−(1−methoxyphenazin−5−ium−5−yl)propane−1−sulfonate(mPPS), which acts as an electron mediator for the electrochemical oxidation of NADH. LDH was detected by the redox reaction of NAD to NADH with a detection limit as low as 0.5 U/L [108]. Gisela's team developed a point-of-care (POC) device that includes a single-use microfluidic paper, double-sided, screen-printed carbon electrode (MP-dsSPCE). The POC requires an optimized single-step immunoassay performed using magnetic beads and an immunomodified signal amplifier that is carried primarily in MP-dsSPCE. The system is capable of performing LDH assays in less than 20 min with minimal user intervention, and can provide a detection limit of 50 U/L [109]. In addition, Xu et al. developed an immunosensor for electrochemical detection of LDH. Firstly, multi-walled carbon nanotubes are assembled with gold nanoparticles onto electrode surface, which increases the surface area of the electrode and improves the conductivity of the electrode. LDH antibodies are then modified onto the electrode surface for capturing LDH. When LDH is immobilized on the electrode surface, it catalyzes the formation of pyruvate and NADH in the substrate solution, which enhances the current signal. Conversely, when the LDH concentration is low, the signal is weakened so that a linear relationship between LDH concentration and current intensity is established. The detection range of this method is 0.55–275 U/L, and the detection limit is 0.21 U/L [110–113].

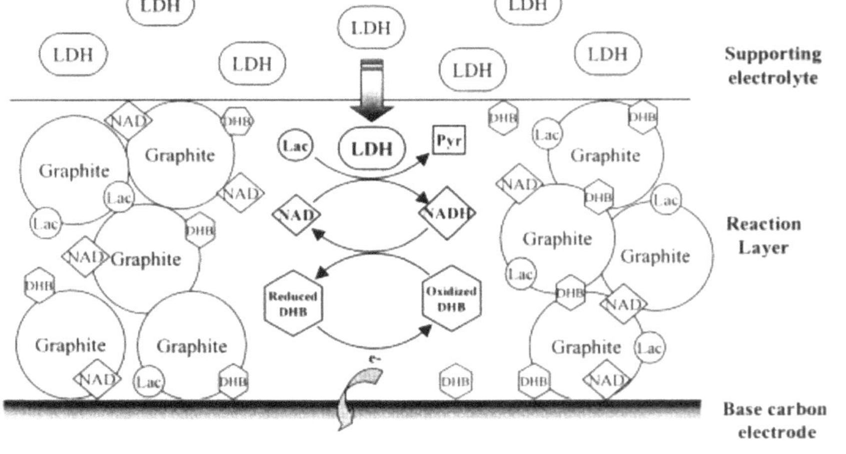

Figure 4. Response scheme for thick film LDH biosensors [106]. Copyright 2002 Elsevier.

Immunoassays have been widely used for the detection of LDH [114]. Figure 5 shows the work of Hemben et al., who developed an immunosensor using gold nanoparticles to enhance sensitivity. When humans are infected with malaria by mosquito bites, LDH rises rapidly in the blood. LDH is then detected by a gold nanoparticle-modified immunosensor during the active phase. In comparison with commercial kits, the sensor showed higher sensitivity and better reproducibility, allowing for immediate detection at low cost. The sensor is capable of detecting LDH in the range of 0–0.17 U/L with a detection limit as low as 0.45 U/L and can be used with a networked mobile device, which can greatly facilitate the detection process [115].

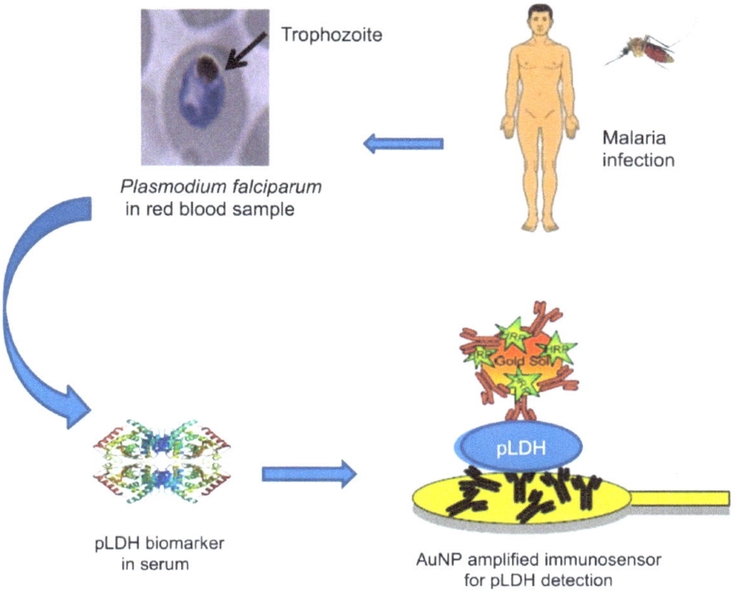

Figure 5. Schematic diagram of the process of LDH detection by immunosensor [115]. Copyright 2018 Elsevier.

3.4. Fluorometric Measurement

When a fluorescent substance is irradiated by incident light of a certain wavelength (usually UV light), the substances absorb light energy, enters the excited state, and immediately de-excites and emits light longer than the wavelength of the incident light (usually in the visible wavelength). Many fluorescent substances cease to emit light once the incident light stops, and the luminescence disappears immediately [116]. Compared with the previous detection methods, fluorescence analysis has advantages of high sensitivity, high throughput, wide linearity range, making it a good choice for LDH detection [117,118].

Back in 2010, Ren et al. reported the detection of LDH activity based on CdTe/CdS quantum dots. As shown in Figure 6, in this approach, the fluorescence of QDs was first quenched by NAD, caused by chemisorption or electrostatic diffusion of NAD onto the surface of QDs, which affecting the chemical bonds on QD surface. Then the fluorescence was gradually recovered with the addition of LDH, which was ascribed to the reduction of NAD to NADH catalyzed by LDH, thus consuming NAD and weakening the quenching effect of NAD on the fluorescence of QDs. The detection limit of this method was 75 U/L, and it has a good linearity in the range of 150–1500 U/L with a correlation coefficient of 0.996 [86]. One year later, they developed a CdTe QDs-based assay for LDH detection and successfully applied it to detect LDH in human serum samples, making up for the shortcomings of the previous work. The linear range of the assay is 250–6000 U/L, and the assay can be completed in 15 min [84]. Our team also developed silicon quantum dots (SiQDs) and sulfur quantum dots (SQDs) based assays for LDH detection, which have wide linear range for LDH detection. As shown in Figure 7, in the work on SiQDs, at an excitation wavelength of 350 nm, the synthesized SiQDs possess an emission peak at 450 nm. We found that no significant quenching effect was achieved when we mixed SiQDs with NAD. On the contrary, NADH has a significant quenching effect on SiQDs. The principle of fluorescence quenching is due to the diffusion of NADH onto the surface of SiQDs, leading to the electron transfer (ET) process on SiQDs. The detection limit is 970 U/L. To verify the feasibility of LDH detection by SiQDs, we conducted several control experiments. When NADH was added to the SiQDs solution, the fluorescence intensity of SiQDs was significantly reduced, indicating that NADH has a strong quenching effect on SiQDs. Then, the fluorescence intensity was restored after continuing to add LDH together with pyruvate to the mixed solution described above. However, when SiQDs were reacted with LDH alone, the fluorescence intensity did not change significantly. In addition, we also investigated the quenching of fluorescence intensity of SiQDs by NAD and the recovery of fluorescence with the addition of LDH. Although NAD can also quench the fluorescence of SiQDs, the quenching effect was not as significant as that of NADH. The fluorescence intensity was not recovered when LDH and pyruvate were added, indicating that LDH could not catalyze the conversion of NAD to NADH. Since LDH-catalyzed enzymatic reactions are reversible, we believe that the above phenomenon is attributed to the fact that it is much easier for LDH to catalyze pyruvate to lactate, while it is relatively difficult for LDH to catalyze lactate to pyruvate. Due to the presence of many interfering substances in human serum samples, the effect of various substances on the assay was also investigated to study the selectivity of the assay. The sensitivity of the method for LDH was much higher than that for other substances, indicating that the method has a high selectivity for the detection of LDH. As shown in Figure 8, in the experiments on SQDs, it was shown that NAD can increase the fluorescence intensity of SQDs, and its fluorescence intensity increased by about 30 times when 1 M of NAD was added. This is due to the fact that the emission wavelength of NAD is close to the excitation wavelength of SQDs, and the energy transfer between them enhances the fluorescence intensity of SQDs. A detection limit of 262.41 U/L can be achieved in the linear range of 0.5–40 $\times$ 10^3 U/L [119]. Compared to CdTe and CdTe/CdS QDs, SiQDs and SQDs have unique advantages such as good biocompatibility, low toxicity, good water solubility and photostability [12,119].

Wu's team used super-bright adenosine monophosphate (AMP)-covered gold nanoclusters (AuNCs@AMP) as a fluorescent probe to detect LDH. In contrast to the previous use of NAD or NADH as a quencher, they found that LDH could be used as a quencher for this probe. The mechanism of this quenching is due to the formation of Au-thiol complex by the free sulfhydryl groups in LDH in the microenvironment on the protein surface. When excited at 328 nm, AuNCs@AMP exhibited an intense fluorescence peak at 480 nm and their fluorescence emission was gradually quenched with time in the presence of 2.0 μM LDH. This resulted in an intensity loss of nearly 70% with a slight blue shift. The method was able to detect LDH linearly in the concentration range of 8–400 U/L with a detection limit of 0.8 U/L [120]. Building on their previous work, their team developed a quantitative assay for LDH based on Au−AgNCs@AMP. This time, LDH was instead used to enhance the fluorescence intensity of the probe, which could reach up to 5 times the initial intensity in the presence of 1.0 μM LDH. The mechanism suggests that the probe interacts with the structural domain of LDH near the active site and is driven by electrostatic interactions towards the free thiol group. The fluorescence enhancement was attributed to assembly-induced emission enhancement (AIEE) and hydrophobic transfer. Interestingly, they also introduced Al^{3+} in order to target specific LDH for detection. Al^{3+} can bind to CuNCs@GSH and AgNCs@GSH, forming aggregates through electrostatic interactions and producing strong emission enhancement effects. These results suggest that Al^{3+} are suitable promoters for improving the emission of nanoclusters and extending their application. Al^{3+} plays an important role in the specific detection of PvLDH by shielding the fluorescence response of Au−AgNCs@AMP to RLDH, PfLDH and

HLDH, but maintains the response to PvLDH. Therefore, specific LDH species can be detected more specifically [83].

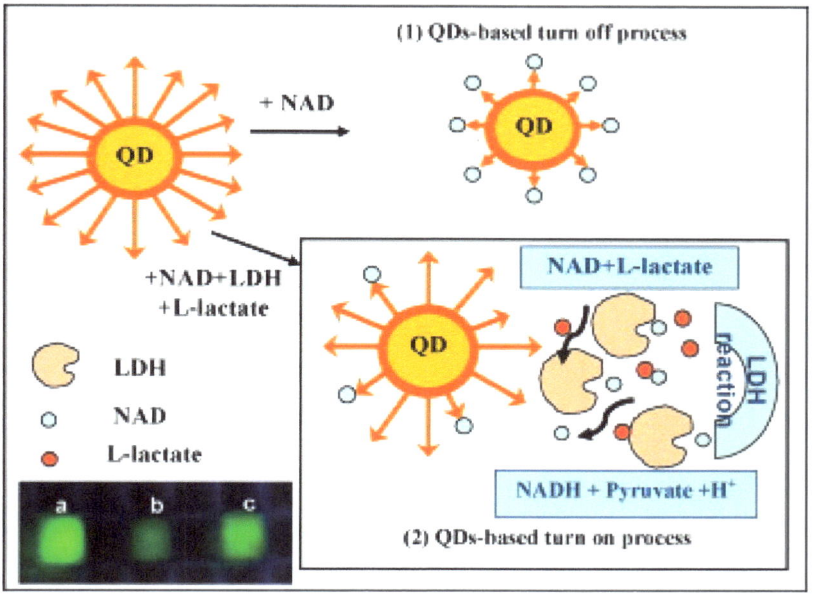

Figure 6. Schematic diagram of the principle of LDH determination, with the lower left panel showing the fluorescence image of the QDs in the absence (**a**) or presence (**b**) of NAD, and in the presence of LDH, NAD and LL (**c**) [86]. Copyright 2010 Elsevier.

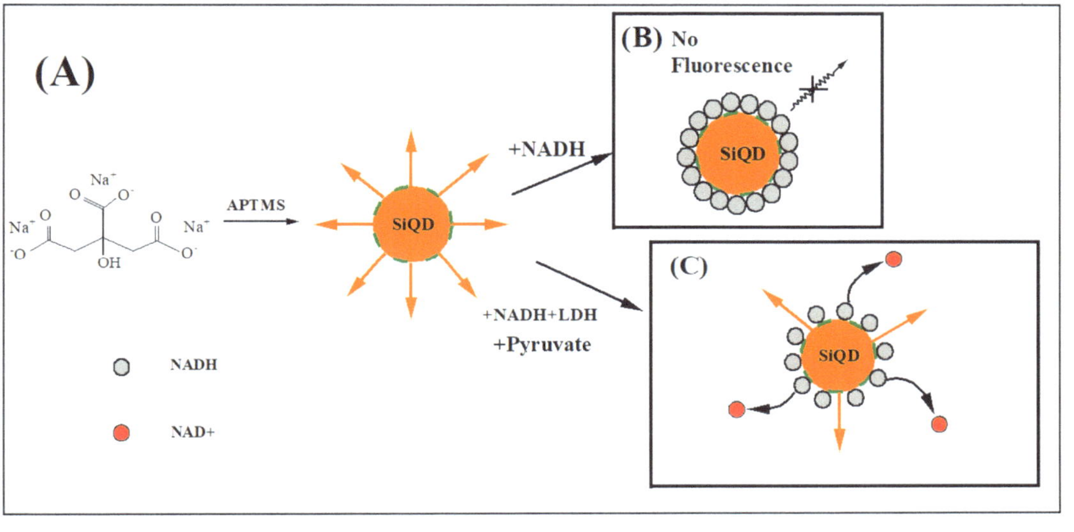

Figure 7. (**A**) The synthesis principle of SiQDs. (**B**) NADH quenches fluorescence of SiQDs. (**C**) Principle of LDH activity detection [12]. Copyright 2022 Elsevier.

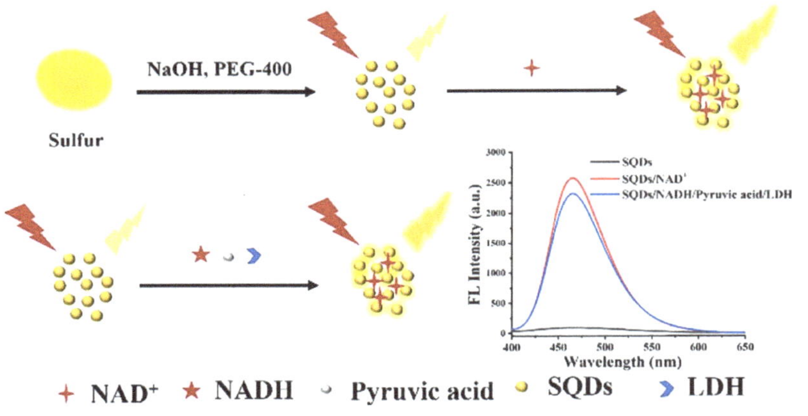

Figure 8. Schematic diagram of LDH detection by SQDs. Copyright 2022 Elsevier.

In addition, He et al. reported a fluorescence sensor based on CdSe quantum dots/polycaprolactone (PCL) composite electrospun fluorescent porous fiber membranes, where CdSe QDs were uniformly distributed within the PCL fiber as fluorescent probes, resulting in fluorescence quenching due to electron transfer (ET) between NAD and CdSe QDs. In order to accelerate the diffusion of analytes inside the fiber and to improve sensitivity, a porogenic agent was introduced to produce a secondary porous structure in the fiber. Compared with the fluorescence quenching sensor, this luciferase sensor can effectively reduce the background signal, avoiding the false signal interference caused by other quenching agents in the actual sample, which improves the sensitivity and selectivity of the sensor. Each assay takes only 10 min and is linear in the range of 200–2400 U/L [121]. Moreover, Kenry et al. reported a new application of monolayer MoS_2 nanosheets in the development of "catch-release" aptamer biomolecular sensors. Unlike chemisorption and electron transfer, the mechanism is as follows: the MoS_2 nanosheets first quench the LDH aptamer solution with fluorescent properties, and then the target LDH protein induces the release of the aptamer from the nanosheet surface to restore fluorescence. The linearity range is 0–578.13 U/L and detection limit is 5.09 U/L [122].

Unlike conventional quantum dots, Jenie developed a Resazurin based fluorescent probe to detect LDH. Resazurin, known as Alamar Blue, has weak fluorescent properties. With the transfer of electrons, the heterocyclic N−oxide group in Resazurin loses oxygen upon reduction and forms the strongly fluorescent product—resorufin. The porous silicon with microcavity structure enhanced the fluorescence signal. This is because the microcavities are multilayers consisting of intervening spacer layers and alternating porosity. The microcavities are capable of enhancing the emission of fluorophores confined in the porous structure according to the Purcell effect. The fluorescence intensity measured when LDH was present on porous silicon microcavities was ten and fives times higher than that of monolayers and detuned microcavities, respectively. The linear range of the assay system is $0.16–6.5 \times 10^3$ U/L and the detection limit is 80 U/L [123]. Dr. Minopoli presented a detection system based on plasma-enhanced fluorescence assay combining the plasma characteristics of AuNPs and a unique photochemical functionalization technique. The photochemical immobilization technique provides a fast and simple strategy to covalently tether antibodies to a gold surface, exposing fragment antigen binding sites to the surrounding environment. The assay is performed in a sandwich configuration where antibodies act as a capture bioreceptor and a fluorescently labeled aptamer binds to LDH from the top. AuNPs are used as fluorescent enhancers to improve the sensitivity of the assay [124]. Furthermore, Dr. Alpizar reported a paper-based fluorescent magnetic immunoassay that begins with a 5 min immunocapture in a test tube. The mixture is then transferred to the distal end of the paper device wash pad. When it is absorbed, 500 μL of PBS–T is added to the wash bath. This pushes the mixture into the magnetic particles' concentration fraction, where the magnet retains the magnetic particles while the unbound reagents flow toward the end absorbent pad. Finally, 50 μL of QuantaRed is added to the bottom of the wash pad and the device is incubated in the dark for 5 min. Fluorescence detection was achieved using a homemade portable fluorometer. The assay platform is capable of providing a detection limit of 0.225 U/L over a linear range of 0.2–3.13 U/L [99].

According to relevant studies, the normal range of LDH in human serum is 100–300 U/L. When the LDH level exceeds 1000 U/L, it suggests the possibility of related diseases or tumors. This indicates that the linear range of these current test are sufficient. Table 1 summaries the performance of recently reported assays for LDH.

Table 1. Performance of recently reported LDH detection methods.

Analytical Method	Materials	Detection Limit ($U\ L^{-1}$)	Linear Range ($U\ L^{-1}$)	Sensor Response Time (min)	Correlation Coefficient	Reference	Real Sample Test
Fluorescence	CdTe QDs	/	250–6000	15	0.996	[84]	Serum
Fluorescence	CdTe/CdS QDs	75	150–1500	20	0.996	[86]	/
Fluorescence	AuNCs@AMP	0.8	8.0–400	30	0.996	[120]	Serum
Fluorescence	Au−AgNCs@AMP	0.93	92.5–925	10	0.997	[83]	/
Fluorescence	SiQDs	970	$0.77–385 \times 10^3$	20	0.997	[12]	Serum
Fluorescence	SQDs	262.41	$0.5–40 \times 10^3$	60	0.991	[119]	Serum
Fluorescence	pSi	80	$0.16–6.5 \times 10^3$	10	0.984	[123]	/
Fluorescence	AuNPs	9.25×10^{-2}	$7.5 \times 10^{-2}–7.5 \times 10^4$	120	/	[124]	Blood
Fluorescence	CdSe QDs/PCL	/	200–2400	10	0.998	[121]	/
Fluorescence	Paper−based Sensor	0.225	0.2–3.13	<20	0.997	[99]	Blood
Fluorescence	Aptamer-Coated Magnetic Beads	1.51×10^{-2}	$9.25 \times 10^{-3}–9.25 \times 10^2$	60	0.990	[125]	/
Fluorescence	MoS_2 nanosheets	5.09	0–578.13	10	0.990	[122]	/
Fluorescence	Magnetic Beads	2.75×10^{-2}	0.025–6.25	15	/	[126]	Blood
Electrochemistry	Glassy Carbon Electrode	0.21	0.55–275	60	0.991	[110]	/
Electrochemistry	Screen-printed Electrode	50	50–500	10	0.998	[106]	/
Electrochemistry	$N-Mo_2C/SPE$	25	60–700	/	0.991	[107]	Plasma
Electrochemistry	MP−dsSPCE	50	3.13–25	<20	0.990	[109]	Blood
Electrochemistry	rGO−2DBioFET	7.22×10^{-5}	$7.22 \times 10^{-4}–9.25 \times 10^3$	/	0.990	[127]	Serum
Electrochemistry	mPPS	0.5	/	/	/	[108]	Serum
Colorimetry	Pullulan-Based Inks	13	0–225	5	/	[98]	Serum
Colorimetry	LFA	2.5	1.25–125	15	0.960	[128]	Serum
Colorimetry	mAb−functionalized Magnetic Beads	2.41×10^{-2}	$6.48 \times 10^{-2}–46.25$	15	/	[129]	Blood
Colorimetry	Magnetic Beads	0.24	/	30	/	[130]	/
Colorimetry	Paper-based Sensor	0.39	/	<20	/	[99]	Blood
Colorimetry	Aptamer-Coated Magnetic Beads	0.57	$9.25 \times 10^{-3}–9.25 \times 10^2$	60	0.990	[125]	/
Colorimetry	Magnetic Beads	0.03	0.1–6.25	20	/	[126]	Blood
Chemiluminescence	Magnetic Beads	5×10^{-3}	0.01–6.25	1	/	[126]	Blood
Immunoassay	TiO	12.8	1–100	30	0.996	[131]	/
Immunoassay	Microfluidic Microplate	6.25×10^{-3}	/	<90	/	[43]	Serum
Immunoassay	SPGE	0.45	0–0.17	120	0.991	[115]	Serum

4. Future Prospects

LDH plays an important role as a biomarker for different diseases such as cancers and malaria. In this review, we summarized various assays for LDH and discussed their advantages and limitations. Although different methods have been reported for LDH analysis, which generally meet most basic detection needs, further efforts are still needed to improve the performance for LDH detection for early screening of diseases. In the future, the main challenges we need to overcome are improvement of sensitivity, reduction of the cost and development of point-of-care testing (POCT). For example, by utilizing properties of nanomaterials, we can synthetize probes with dual fluorescence emissions to develop ratiometric assays for LDH. By using one emission peak as a built-in correction and the other emission as a signal, ratiometric probes can eliminate variability arising from environmental interference, leading to improved sensitivity and accuracy. Near-infrared fluorescence probes can also be developed for in vivo study and cellular imaging of LDH activity. Near-infrared fluorescence probes have been extensively used in biological detection and imaging due to their attractive properties, including large anti-Stokes shifts, good photostability, minimized autofluorescence and especially deep tissue penetration in biological samples. Based on electrochemistry, similar to a glucose testing strip, POCT for LDH can also be developed so that people can monitor LDH activity routine at home. Moreover, although many methods have been recently reported for LDH detection, the commercialization of these methods still needs further efforts, such as improving the reproducibility, selectivity as well as precision of the methods. More clinical samples need to be tested by the methods. More clinical data revealing the relationship of LDH with other diseases are required to find wider applications of LDH as clinical biomarkers for diseases diagnosis and therapeutic treatment.

Author Contributions: Y.Z.: Data curation, Investigation, Writing—original draft, Conceptualization, Methodology. M.Q.: Funding acquisition, Formal analysis, Software. M.Y.: Project administration, Supervision, Validation, Writing—review & editing. All authors have read and agreed to the published version of the manuscript.

Funding: This work was supported by the National Science Foundation of China (8207104259, 22174163), the Hunan Provincial Science and Technology Plan Project, China (No. 2019TP1001), and the Innovation Driven Project of Central South University (2020CX002).

Institutional Review Board Statement: Not applicable.

Informed Consent Statement: Not applicable.

Data Availability Statement: Not applicable.

Conflicts of Interest: The authors declare that they have no known competing financial inter-ests or personal relationships that could have appeared to influence the work reported in this paper.

References

1. Forkasiewicz, A.; Dorociak, M.; Stach, K.; Szelachowski, P.; Tabola, R.; Augoff, K. The usefulness of lactate dehydrogenase measurements in current oncological practice. *Cell. Mol. Biol. Lett.* **2020**, *25*, 35. [CrossRef] [PubMed]
2. Rattu, G.; Khansili, N.; Maurya, V.K.; Krishna, P.M. Lactate detection sensors for food, clinical and biological applications: A review. *Environ. Chem. Lett.* **2021**, *19*, 1135–1152. [CrossRef]
3. Rani, R.; Kumar, V. Recent Update on Human Lactate Dehydrogenase Enzyme 5 (hLDH5) Inhibitors: A Promising Approach for Cancer Chemotherapy. *J. Med. Chem.* **2016**, *59*, 487–496. [CrossRef] [PubMed]
4. Khan, A.A.; Allemailem, K.S.; Alhumaydhi, F.A.; Gowder, S.J.T.; Rahmani, A.H. The Biochemical and Clinical Perspectives of Lactate Dehydrogenase: An Enzyme of Active Metabolism. *Endocr. Metab. Immune Disord. Drug Targets* **2020**, *20*, 855–868. [CrossRef]
5. Mishra, D.; Banerjee, D. Lactate Dehydrogenases as Metabolic Links between Tumor and Stroma in the Tumor Microenvironment. *Cancers* **2019**, *11*, 750. [CrossRef]
6. Donnelly, R.P.; Finlay, D.K. Glucose, glycolysis and lymphocyte responses. *Mol. Immunol.* **2015**, *68*, 513–519. [CrossRef]
7. Alam, F.; RoyChoudhury, S.; Jalal, A.H.; Umasankar, Y.; Forouzanfar, S.; Akter, N.; Bhansali, S.; Pala, N. Lactate biosensing: The emerging point-of-care and personal health monitoring. *Biosens. Bioelectron.* **2018**, *117*, 818–829. [CrossRef]

8. Kucherenko, I.S.; Topolnikova, Y.V.; Soldatkin, O.O. Advances in the biosensors for lactate and pyruvate detection for medical applications: A review. *TrAC Trends Anal. Chem.* **2018**, *110*, 160–172. [CrossRef]
9. Urbańska, K.; Orzechowski, A. Unappreciated Role of LDHA and LDHB to Control Apoptosis and Autophagy in Tumor Cells. *Int. J. Mol. Sci.* **2019**, *20*, 2085. [CrossRef]
10. Kolappan, S.; Shen, D.L.; Mosi, R.; Sun, J.; McEachern, E.J.; Vocadlo, D.J.; Craig, L. Structures of lactate dehydrogenase A (LDHA) in apo, ternary and inhibitor-bound forms. *Acta Crystallogr. Sect. D Biol. Crystallogr.* **2015**, *71*, 185–195. [CrossRef]
11. Patel, C.H.; Powell, J.D. Warburg meets epigenetics. *Science* **2016**, *354*, 419–420. [CrossRef] [PubMed]
12. Zhou, Y.; Qi, M.; Yang, M. Fluorescence determination of lactate dehydrogenase activity based on silicon quantum dots. *Acta Part A Mol. Biomol. Spectrosc.* **2021**, *268*, 120697. [CrossRef] [PubMed]
13. Lu, J.; Tan, M.; Cai, Q. The Warburg effect in tumor progression: Mitochondrial oxidative metabolism as an anti-metastasis mechanism. *Cancer Lett.* **2015**, *356*, 156–164. [CrossRef] [PubMed]
14. Meyer, E.H.; Letts, J.A.; Maldonado, M. Structural insights into the assembly and the function of the plant oxidative phosphorylation system. *New Phytol.* **2022**, *235*, 1315–1329. [CrossRef] [PubMed]
15. Sukonina, V.; Ma, H.X.; Zhang, W.; Bartesaghi, S.; Subhash, S.; Heglind, M.; Foyn, H.; Betz, M.J.; Nilsson, D.; Lidell, M.E.; et al. FOXK1 and FOXK2 regulate aerobic glycolysis. *Nature* **2019**, *566*, 279–283. [CrossRef]
16. Martínez-Reyes, I.; Chandel, N.S. Cancer metabolism: Looking forward. *Nat. Rev. Cancer* **2021**, *21*, 669–680. [CrossRef]
17. Gentric, G.; Mieulet, V.; Mechta-Grigoriou, F. Heterogeneity in Cancer Metabolism: New Concepts in an Old Field. *Antioxid Redox Signal* **2017**, *26*, 462–485. [CrossRef]
18. Hamanaka, R.B.; Chandel, N.S. Warburg Effect and Redox Balance. *Science* **2011**, *334*, 1219–1220. [CrossRef]
19. Wortley, C. Warburg: Father of the metabolic approach to cancer. *Lancet Gastroenterol. Hepatol.* **2022**, *7*, 911. [CrossRef]
20. Sanderson, S.M.; Locasale, J.W. Revisiting the Warburg Effect: Some Tumors Hold Their Breath. *Cell Metab.* **2018**, *28*, 669–670. [CrossRef] [PubMed]
21. Hönigova, K.; Navratil, J.; Peltanova, B.; Polanska, H.H.; Raudenska, M.; Masarik, M. Metabolic tricks of cancer cells. *Biochim Biophys Acta Rev Cancer* **2022**, *1877*, 188705. [CrossRef] [PubMed]
22. Fang, X.; Yuan, M.; Dai, J.D.; Lin, Q.Y.; Lin, Y.H.; Wang, W.L.; Jiang, Y.F.; Wang, H.H.; Zhao, F.; Wu, J.Y.; et al. Dual inhibition of glycolysis and oxidative phosphorylation by aptamer-based artificial enzyme for synergistic cancer therapy. *Nano Res.* **2022**, *15*, 6278–6287. [CrossRef]
23. Gatenby, R.A.; Gillies, R.J. Why do cancers have high aerobic glycolysis? *Nat. Rev. Cancer* **2004**, *4*, 891–899. [CrossRef] [PubMed]
24. Molina, J.R.; Sun, Y.T.; Protopopova, M.; Gera, S.; Bandi, M.; Bristow, C.; McAfoos, T.; Morlacchi, P.; Ackroyd, J.; Agip, A.N.A.; et al. An inhibitor of oxidative phosphorylation exploits cancer vulnerability. *Nat. Med.* **2018**, *24*, 1036–1046. [CrossRef]
25. Xue, D.; Xu, Y.; Kyani, A.; Roy, J.; Dai, L.; Sun, D.; Neamati, N. Multiparameter Optimization of Oxidative Phosphorylation Inhibitors for the Treatment of Pancreatic Cancer. *J. Med. Chem.* **2022**, *65*, 3404–3419. [CrossRef] [PubMed]
26. Gao, W.; Zhang, J.; Wang, W.; Liu, Z.; Chen, M.; Hu, X.; Zeng, L.; Zheng, C.; Song, H.; Zhang, Q. Drug self-delivery nanorods enhance photodynamic therapy of triple-negative breast cancer by inhibiting oxidative phosphorylation. *Int. J. Pharm.* **2022**, *621*, 121775. [CrossRef] [PubMed]
27. Boreel, D.F.; Span, P.N.; Heskamp, S.; Adema, G.J.; Bussink, J. Targeting Oxidative Phosphorylation to Increase the Efficacy of Radio- and Immune-Combination Therapy. *Clin. Cancer. Res.* **2021**, *27*, 2970–2978. [CrossRef] [PubMed]
28. Comandatore, A.; Franczak, M.; Smolenski, R.T.; Morelli, L.; Peters, G.J.; Giovannetti, E. Lactate Dehydrogenase and its clinical significance in pancreatic and thoracic cancers. *Semin. Cancer Biol.* **2022**, *86*, 93–100. [CrossRef]
29. Dang, C.V. Links between metabolism and cancer. *Genes Dev.* **2012**, *26*, 877–890. [CrossRef]
30. Kopperschlager, G.; Kirchberger, J. Methods for the separation of lactate dehydrogenases and clinical significance of the enzyme. *J. Chromatogr. B* **1996**, *684*, 25–49. [CrossRef]
31. Schwartz, M.K. Lactic dehydrogenase. An old enzyme reborn as a cancer marker? *Am. J. Clin. Pathol.* **1991**, *96*, 441–443. [CrossRef] [PubMed]
32. Ma, Y.F.; Shi, L.L.; Lu, P.; Yao, S.; Xu, H.L.; Hu, J.J.; Liang, X.; Liang, X.J.; Wei, S.Z. Creation of a Novel Nomogram Based on the Direct Bilirubin-To-Indirect Bilirubin Ratio and Lactate Dehydrogenase Levels in Resectable Colorectal Cancer. *Front. Mol. Biosci.* **2021**, *8*, 751506. [CrossRef] [PubMed]
33. Ma, Y.F.; Lu, P.; Liang, X.J.; Wei, S.Z. Prognostic role of apolipoprotein and lactate dehydrogenase levels in resectable colorectal cancer. *J. Clin. Oncol.* **2021**, *39*, e15541. [CrossRef]
34. Zhang, Z.B.; Li, Y.; Yan, X.; Song, Q.; Wang, G.Q.; Hu, Y.; Jiao, S.C.; Wang, J.L. Pretreatment lactate dehydrogenase may predict outcome of advanced non small-cell lung cancer patients treated with immune checkpoint inhibitors: A meta-analysis. *Cancer Med.* **2019**, *8*, 1467–1473. [CrossRef] [PubMed]
35. Tjokrowidjaja, A.; Lord, S.J.; John, T.; Lewis, C.R.; Kok, P.S.; Marschner, I.C.; Lee, C.K. Pre- and on-treatment lactate dehydrogenase as a prognostic and predictive biomarker in advanced non-small cell lung cancer. *Cancer* **2022**, *128*, 1574–1583. [CrossRef] [PubMed]
36. Agarwala, S.S.; Keilholz, U.; Gilles, E.; Bedikian, A.Y.; Wu, J.; Kay, R.; Stein, C.A.; Itri, L.M.; Suciu, S.; Eggermont, A.M.M. LDH correlation with survival in advanced melanoma from two large, randomised trials (Oblimersen GM301 and EORTC 18951). *Eur. J. Cancer* **2009**, *45*, 1807–1814. [CrossRef]

37. Majithia, N.; Manana, A.V.; Yan, Y.; Kottschade, L.A.; Dronca, R.S.; Block, M.S.; Nevala, W.K.; Markovic, S. The prognostic role of preoperative serum lactate dehydrogenase (LDH) in patients with resected advanced melanoma. *J. Clin. Oncol.* **2017**, *35*, e21054. [CrossRef]
38. Girgis, H.; Masui, O.; White, N.M.A.; Scorilas, A.; Rotondo, F.; Seivwright, A.; Gabril, M.; Filter, E.R.; Girgis, A.H.A.; Bjarnason, G.A.; et al. Lactate Dehydrogenase A is a potential prognostic marker in clear cell renal cell carcinoma. *Mol. Cancer* **2014**, *13*, 101. [CrossRef]
39. Zhang, N.; Zhang, H.; Zhu, D.; JiRiGaLa; Yu, D.; Wang, C.; WuYunBiLiGe; Amin.; ZhiHong; Yu, H.; et al. Prognostic role of pretreatment lactate dehydrogenase in patients with metastatic renal cell carcinoma: A systematic review and meta-analysis. *Int. J. Surg.* **2020**, *79*, 66–73. [CrossRef]
40. Cui, Z.L.; Li, Y.; Gao, Y.N.; Kong, L.Y.; Lin, Y.F.; Chen, Y. Cancer-testis antigen lactate dehydrogenase C4 in hepatocellular carcinoma: A promising biomarker for early diagnosis, efficacy evaluation and prognosis prediction. *Aging* **2020**, *12*, 19455–19467. [CrossRef]
41. Wu, S.J.; Lin, Y.X.; Ye, H.; Xiong, X.Z.; Li, F.Y.; Cheng, N.S. Prognostic value of alkaline phosphatase, gamma-glutamyl transpeptidase and lactate dehydrogenase in hepatocellular carcinoma patients treated with liver resection. *Int. J. Surg.* **2016**, *36*, 143–151. [CrossRef] [PubMed]
42. de la Serna, E.; Arias-Alpizar, K.; Borgheti-Cardoso, L.N.; Sanchez-Cano, A.; Sulleiro, E.; Zarzuela, F.; Bosch-Nicolau, P.; Salvador, F.; Molina, I.; Ramirez, M.; et al. Detection of Plasmodium falciparum malaria in 1 h using a simplified enzyme-linked immunosorbent assay. *Anal. Chim. Acta* **2021**, *1152*, 338254. [CrossRef] [PubMed]
43. Lee, W.S.; Kang, T.; Kwak, K.J.; Park, K.; Yi, S.Y.; Lee, U.J.; Shin, Y.B.; Jeong, J. Simple, rapid, and accurate malaria diagnostic platform using microfluidic-based immunoassay of Plasmodium falciparum lactate dehydrogenase. *Nano Converg.* **2020**, *7*, 13. [CrossRef] [PubMed]
44. Lee, S.; Song, K.M.; Jeon, W.; Jo, H.; Shim, Y.B.; Ban, C. A highly sensitive aptasensor towards Plasmodium lactate dehydrogenase for the diagnosis of malaria. *Biosens. Bioelectron.* **2012**, *35*, 291–296. [CrossRef] [PubMed]
45. Zhang, F.; Hu, L.J.; Wang, J.J.; Chen, J.; Chen, J.; Wang, Y.M. Clinical value of jointly detection serum lactate dehydrogenase/pleural fluid adenosine deaminase and pleural fluid carcinoembryonic antigen in the identification of malignant pleural effusion. *J. Clin. Lab. Anal.* **2017**, *31*, e22106. [CrossRef] [PubMed]
46. Kim, J.H.; Oak, C.H.; Jung, M.H.; Jang, T.W.; Je, H.; Choi, J. High Level of ADA in Pleural Effusion, Diagnostic Value of CEA, NSE, Amylase, Glucose, and Lactic Dehydrogenase for the Differential Diagnosis of Pleural Effusion. *Chest* **2016**, *150*, 558A. [CrossRef]
47. Kelderman, S.; Heemskerk, B.; van Tinteren, H.; van den Brom, R.R.H.; Hospers, G.A.P.; van den Eertwegh, A.J.M.; Kapiteijn, E.W.; de Groot, J.W.B.; Soetekouw, P.; Jansen, R.L.; et al. Lactate dehydrogenase as a selection criterion for ipilimumab treatment in metastatic melanoma. *Cancer. Immunol. Immun.* **2014**, *63*, 449–458. [CrossRef]
48. Diem, S.; Kasenda, B.; Spain, L.; Martin-Liberal, J.; Marconcini, R.; Gore, M.; Larkin, J. Serum lactate dehydrogenase as an early marker for outcome in patients treated with anti-PD-1 therapy in metastatic melanoma. *Br. J. Cancer* **2016**, *114*, 256–261. [CrossRef]
49. Chen, X.Y.; Huang, M.Y.; Xiao, Z.W.; Yang, S.; Chen, X.Q. Lactate dehydrogenase elevations is associated with severity of COVID-19: A meta-analysis. *Crit. Care* **2020**, *24*, 459. [CrossRef]
50. Stoeckle, K.; Witting, B.; Kapadia, S.; An, A.; Marks, K. Elevated inflammatory markers are associated with poor outcomes in COVID-19 patients treated with remdesivir. *J. Med. Virol.* **2022**, *94*, 384–387. [CrossRef]
51. Zhang, S.; Guo, M.F.; Duan, L.M.; Wu, F.; Hu, G.R.; Wang, Z.H.; Huang, Q.; Liao, T.T.; Xu, J.J.; Ma, Y.L.; et al. Development and validation of a risk factor-based system to predict short-term survival in adult hospitalized patients with COVID-19: A multicenter, retrospective, cohort study. *Crit. Care* **2020**, *24*, 438. [CrossRef] [PubMed]
52. Gibellini, L.; De Biasi, S.; Meschiari, M.; Gozzi, L.; Paolini, A.; Borella, R.; Mattioli, M.; Lo Tartaro, D.; Fidanza, L.; Neroni, A.; et al. Plasma Cytokine Atlas Reveals the Importance of TH2 Polarization and Interferons in Predicting COVID-19 Severity and Survival. *Front. Immunol.* **2022**, *13*, 842150. [CrossRef] [PubMed]
53. Battaglini, D.; Lopes-Pacheco, M.; Castro-Faria-Neto, H.C.; Pelosi, P.; Rocco, P.R.M. Laboratory Biomarkers for Diagnosis and Prognosis in COVID-19. *Front. Immunol.* **2022**, *13*, 857573. [CrossRef]
54. Jin, Z.M.; Zheng, M.; Shi, J.C.; Ye, X.C.; Cheng, F.; Chen, Q.L.; Huang, J.P.; Jiang, X.G. Correlation Analysis Between Serum Uric Acid, Prealbumin Level, Lactate Dehydrogenase, and Severity of COVID-19. *Front. Mol. Biosci.* **2021**, *8*, 615837. [CrossRef]
55. Wu, M.Y.; Yao, L.; Wang, Y.; Zhu, X.Y.; Wang, X.F.; Tang, P.J.; Chen, C. Clinical evaluation of potential usefulness of serum lactate dehydrogenase (LDH) in 2019 novel coronavirus (COVID-19) pneumonia. *Respir. Res.* **2020**, *21*, 171. [CrossRef] [PubMed]
56. Han, Y.; Zhang, H.D.; Mu, S.C.; Wei, W.; Jin, C.Y.; Tong, C.Y.; Song, Z.J.; Zha, Y.F.; Xue, Y.; Gu, G.R. Lactate dehydrogenase, an independent risk factor of severe COVID-19 patients: A retrospective and observational study. *Aging* **2020**, *12*, 11245–11258. [CrossRef]
57. Poggiali, E.; Zaino, D.; Immovilli, P.; Rovero, L.; Losi, G.; Dacrema, A.; Nuccetelli, M.; Vadacca, G.B.; Guidetti, D.; Vercelli, A.; et al. Lactate dehydrogenase and C-reactive protein as predictors of respiratory failure in CoVID-19 patients. *Clin. Chim. Acta* **2020**, *509*, 135–138. [CrossRef]
58. Kermali, M.; Khalsa, R.K.; Pillai, K.; Ismail, Z.; Harky, A. The role of biomarkers in diagnosis of COVID-19—A systematic review. *Life. Sci.* **2020**, *254*, 117788. [CrossRef]

59. Neto, F.L.; Salzstein, G.A.; Cortez, A.L.; Bastos, T.L.; Baptista, F.V.D.; Moreira, J.A.; Lauterbach, G.P.; de Oliveira, J.C.; de Assis, F.C.; Aguiar, M.R.A.; et al. Comparative assessment of mortality risk factors between admission and follow-up models among patients hospitalized with COVID-19. *Int. J. Infect. Dis.* **2021**, *105*, 723–729. [CrossRef]
60. Chamchoy, K.; Pakotiprapha, D.; Pumirat, P.; Leartsakulpanich, U.; Boonyuen, U. Application of WST-8 based colorimetric NAD(P)H detection for quantitative dehydrogenase assays. *BMC Biochem.* **2019**, *20*, 4. [CrossRef]
61. Zhou, Y.; Xu, Z.; Yoon, J. Fluorescent and colorimetric chemosensors for detection of nucleotides, FAD and NADH: Highlighted research during 2004–2010. *Chem. Soc. Rev.* **2011**, *40*, 2222–2235. [CrossRef] [PubMed]
62. Hallaj, T.; Salari, R.; Amjadi, M. Morphology transition of Ag nanoprisms as a platform to design a dual sensor for NADH sensitive assay. *J. Photochem. Photobiol. A: Chem.* **2022**, *431*, 114043. [CrossRef]
63. Monosik, R.; dos Santos, V.B.; Angnes, L. A simple paper-strip colorimetric method utilizing dehydrogenase enzymes for analysis of food components. *Anal. Methods* **2015**, *7*, 8177–8184. [CrossRef]
64. Kjeld, M. An automated colorimetric method for the estimation of lactate dehydrogenase activity in serum. *Scand. J. Clin. Lab. Invest.* **1972**, *29*, 421–425. [CrossRef]
65. Babson, A.L.; Phillips, G.E. A rapid colorimetric assay for serum lactic dehydrogenase. *Clin. Chim. Acta.* **1965**, *12*, 210–215. [CrossRef]
66. Burd, J.F.; Usategui-Gomez, M. A colorimetric assay for serum lactate dehydrogenase. *Clin. Chim. Acta* **1973**, *46*, 223–227. [CrossRef]
67. Debnam, P.M.; Shearer, G. Colorimetric assay for substrates of NADP+−dependent dehydrogenases based on reduction of a tetrazolium dye to its soluble formazan. *Anal. Biochem.* **1997**, *250*, 253–255. [CrossRef]
68. Allain, C.C.; Henson, C.P.; Nadel, M.K.; Knoblesdorff, A.J. Rapid single-step kinetic colorimetric assay for lactate dehydrogenase in serum. *Clin. Chem.* **1973**, *19*, 223–227. [CrossRef]
69. Papaneophytou, C.; Zervou, M.E.; Theofanous, A. Optimization of a Colorimetric Assay to Determine Lactate Dehydrogenase B Activity Using Design of Experiments. *SLAS Discov. Adv. Sci. Drug Discov.* **2021**, *26*, 383–399. [CrossRef]
70. Veskoukis, A.S.; Margaritelis, N.V.; Kyparos, A.; Paschalis, V.; Nikolaidis, M.G. Spectrophotometric assays for measuring redox biomarkers in blood and tissues: The NADPH network. *Redox Rep.* **2018**, *23*, 47–56. [CrossRef]
71. Muthuchamy, M.; Muneeswaran, T.; Rajivgandhi, G.; Franck, Q.; Muthusamy, A.; Ji-Ming, S. Biologically synthesized copper and zinc oxide nanoparticles for important biomolecules detection and antimicrobial applications. *Mater. Toda. Commun.* **2020**, *22*, 100766. [CrossRef]
72. Stanko, A.P.; Soki, J.; Brkic, D.V.; Plecko, V. Lactate dehydrogenase activity in Bacteroides fragilis group strains with induced resistance to metronidazole. *J. Glob. Antimicrob. Resist.* **2016**, *5*, 11–14. [CrossRef] [PubMed]
73. Fernandes, E.N.; de Campos Moura, M.N.; Lima, J.; Reis, B.F. Automatic flow procedure for the determination of glycerol in wine using enzymatic reaction and spectrophotometry. *Microchem. J.* **2004**, *77*, 107–112. [CrossRef]
74. Moran, J.H.; Schnellmann, R.G. A rapid beta-NADH-linked fluorescence assay for lactate dehydrogenase in cellular death. *J. Pharmacol. Toxicol. Methods* **1996**, *36*, 41–44. [CrossRef] [PubMed]
75. Tang, L.; Zeng, G.M.; Shen, G.L.; Zhang, Y.; Li, Y.P.; Fan, C.Z.; Liu, C.; Niu, C.G. Highly sensitive sensor for detection of NADH based on catalytic growth of Au nanoparticles on glassy carbon electrode. *Anal. Bioanal. Chem.* **2009**, *393*, 1677–1684. [CrossRef]
76. Frattini, D.; Hyun, K.; Kwon, Y. Direct electrochemistry of lactate dehydrogenase in aqueous solution system containing L(+)−lactic acid, beta-nicotinamide adenine dinucleotide, and its reduced form. *J. Ind. Eng. Chem.* **2019**, *80*, 508–515. [CrossRef]
77. Di, J.W.; Cheng, J.J.; Xu, Q.A.; Zheng, H.I.; Zhuang, J.Y.; Sun, Y.B.; Wang, K.Y.; Mo, X.Y.; Bi, S.P. Direct electrochemistry of lactate dehydrogenase immobilized on silica sol-gel modified gold electrode and its application. *Biosens. Bioelectron.* **2007**, *23*, 682–687. [CrossRef]
78. O'Mahony, A.M.; Hardacre, C.; Compton, R.G. Selecting Room-Temperature Ionic Liquids to Optimize Voltammetric Responses: The Oxidation of NADH. *J. Electrochem. Soc.* **2010**, *157*, F49. [CrossRef]
79. Filip, J.; Sefcovicova, J.; Tomcik, P.; Gemeiner, P.; Tkac, J. A hyaluronic acid dispersed carbon nanotube electrode used for a mediatorless NADH sensing and biosensing. *Talanta* **2011**, *84*, 355–361. [CrossRef]
80. Arvinte, A.; Valentini, F.; Radoi, A.; Arduini, F.; Tamburri, E.; Rotariu, L.; Palleschi, G.; Bala, C. The NADH electrochemical detection performed at carbon nanofibers modified glassy carbon electrode. *Electroanalysis* **2007**, *19*, 1455–1459. [CrossRef]
81. Atta, N.F.; Galal, A.; Karagozler, A.E.; Russell, G.C.; Zimmer, H.; Mark, H.B., Jr. Electrochemistry and detection of some organic and biological molecules at conducting poly(3−methylthiophene) electrodes. *Biosens. Bioelectron.* **1991**, *6*, 333–341. [CrossRef] [PubMed]
82. Valentini, F.; Amine, A.; Orlanducci, S.; Terranova, M.L.; Palleschi, G. Carbon nanotube purification: Preparation and characterization of carbon nanotube paste electrodes. *Anal. Chem.* **2003**, *75*, 5413–5421. [CrossRef] [PubMed]
83. Zhang, C.X.; Tanner, J.A.; Li, H.W.; Wu, Y.Q. A novel fluorescence probe of Plasmodium vivax lactate dehydrogenase based on adenosine monophosphate protected bimetallic nanoclusters. *Talanta* **2020**, *213*, 120850. [CrossRef] [PubMed]
84. Yang, L.Q.; Ren, X.L.; Meng, X.W.; Li, H.B.; Tang, F.Q. Optical analysis of lactate dehydrogenase and glucose by CdTe quantum dots and their dual simultaneous detection. *Biosens. Bioelectron.* **2011**, *26*, 3488–3493. [CrossRef] [PubMed]
85. Jain, P.; Chakma, B.; Patra, S.; Goswami, P. Hairpin stabilized fluorescent silver nanoclusters for quantitative detection of NAD(+) and monitoring NAD(+)/NADH based enzymatic reactions. *Anal. Chim. Acta* **2017**, *956*, 48–56. [CrossRef] [PubMed]

86. Ren, X.L.; Yang, L.Q.; Tang, F.Q.; Yan, C.M.; Ren, J. Enzyme biosensor based on NAD-sensitive quantum dots. *Biosens. Bioelectron.* **2010**, *26*, 271–274. [CrossRef] [PubMed]
87. Koyappayil, A.; Kim, H.T.; Lee, M.H. An efficient and rapid synthesis route to highly fluorescent copper microspheres for the selective and sensitive excitation wavelength-dependent dual-mode sensing of NADH. *Sens. Actuators B Chem.* **2021**, *327*, 128887. [CrossRef]
88. Deore, L.A.; Freund, M.S. Reactivity of poly(anilineboronic acid) with NAD(+) and NADH. *Chem. Mater.* **2005**, *17*, 2918–2923. [CrossRef]
89. Gebicki, J.; Marcinek, A.; Zielonka, J. Transient species in the stepwise interconversion of NADH and NAD(+). *Acc. Chem. Res.* **2004**, *37*, 379–386. [CrossRef]
90. Titov, D.V.; Cracan, V.; Goodman, R.P.; Peng, J.; Grabarek, Z.; Mootha, V.K. Complementation of mitochondrial electron transport chain by manipulation of the NAD(+)/NADH ratio. *Science* **2016**, *352*, 231–235. [CrossRef]
91. Maynard, A.G.; Kanarek, N. NADH Ties One-Carbon Metabolism to Cellular Respiration. *Cell Metab.* **2020**, *31*, 660–662. [CrossRef] [PubMed]
92. Bilan, D.S.; Belousov, V.V. Genetically encoded probes for NAD(+)/NADH monitoring. *Free Radical Biol. Med.* **2016**, *100*, 32–42. [CrossRef] [PubMed]
93. Chu, X.G.; Raju, R.P. Regulation of NAD(+) metabolism in aging and disease. *Metabolism* **2022**, *126*, 154923. [CrossRef] [PubMed]
94. Liu, Y.; Landick, R.; Raman, S. A Regulatory NADH/NAD plus Redox Biosensor for Bacteria. *Acs. Synth. Biol.* **2019**, *8*, 264–273. [CrossRef] [PubMed]
95. Amjad, S.; Nisar, S.; Bhat, A.A.; Shah, A.R.; Frenneaux, M.P.; Fakhro, K.; Haris, M.; Reddy, R.; Patay, Z.; Baur, J.; et al. Role of NAD(+) in regulating cellular and metabolic signaling pathways. *Mol. Metab.* **2021**, *49*, 101195. [CrossRef]
96. Siu, V.S.; Lu, M.H.; Hsieh, K.Y.; Raines, K.; Asaad, Y.A.; Patel, K.; Afzali-Ardakani, A.; Wen, B.; Budd, R. Toward a Quantitative Colorimeter for Point-of-Care Nitrite Detection. *Acs. Omega* **2022**, *7*, 11126–11134. [CrossRef]
97. Lee, S.; Manjunatha, D.H.; Jeon, W.; Ban, C. Cationic Surfactant-Based Colorimetric Detection of Plasmodium Lactate Dehydrogenase, a Biomarker for Malaria, Using the Specific DNA Aptamer. *PLoS ONE* **2014**, *9*, e100847. [CrossRef]
98. Kannan, B.; Jahanshahi-Anbuhi, S.; Pelton, R.H.; Li, Y.F.; Filipe, C.D.M.; Brennan, J.D. Printed Paper Sensors for Serum Lactate Dehydrogenase using Pullulan-Based Inks to Immobilize Reagents. *Anal. Chem.* **2015**, *87*, 9288–9293. [CrossRef]
99. Arias-Alpízar, K.; Sánchez-Cano, A.; Prat-Trunas, J.; de la Serna Serna, E.; Alonso, O.; Sulleiro, E.; Sánchez-Montalvá, A.; Diéguez, A.; Baldrich, E. Malaria quantitative POC testing using magnetic particles, a paper microfluidic device and a hand-held fluorescence reader. *Biosens. Bioelectron.* **2022**, *215*, 114513. [CrossRef]
100. Halvorsen, C.P.; Olson, L.; Araujo, A.C.; Karlsson, M.; Nguyen, T.T.; Khu, D.T.K.; Le, H.T.T.; Nguyen, H.T.B.; Winbladh, B.; Russom, A. A rapid smartphone-based lactate dehydrogenase test for neonatal diagnostics at the point of care. *Sci. Rep.* **2019**, *9*, 9301. [CrossRef]
101. Passos, M.L.C.; Saraiva, M. Detection in UV-visible spectrophotometry: Detectors, detection systems, and detection strategies. *Measurement* **2019**, *135*, 896–904. [CrossRef]
102. Hikosaka, K.; Kim, J.; Kajita, M.; Kanayama, A.; Miyamoto, Y. Platinum nanoparticles have an activity similar to mitochondrial NADH: Ubiquinone oxidoreductase. *Colloids Surf. B Biointerfaces* **2008**, *66*, 195–200. [CrossRef] [PubMed]
103. dos Santos, D.R.; Souza, R.O.; Dias, L.B.; Ribas, T.B.; de Oliveira, L.C.F.; Sumida, D.H.; Dornelles, R.C.M.; Nakamune, A.; Chaves-Neto, A.H. The effects of storage time and temperature on the stability of salivary phosphatases, transaminases and dehydrogenase. *Arch. Oral Biol.* **2018**, *85*, 160–165. [CrossRef] [PubMed]
104. Kaya, H.O.; Cetin, A.E.; Azimzadeh, M.; Topkaya, S.N. Pathogen detection with electrochemical biosensors: Advantages, challenges and future perspectives. *J. Electroanal. Chem.* **2021**, *882*, 114989. [CrossRef] [PubMed]
105. Kucherenko, I.S.; Soldatkin, O.O.; Dzyadevych, S.V.; Soldatkin, A.P. Electrochemical biosensors based on multienzyme systems: Main groups, advantages and limitations—A review. *Anal. Chim. Acta* **2020**, *1111*, 114–131. [CrossRef]
106. Hong, M.Y.; Chang, J.Y.; Yoon, H.C.; Kim, H.S. Development of a screen-printed amperometric biosensor for the determination of L−lactate dehydrogenase level. *Biosens. Bioelectron.* **2002**, *17*, 13–18. [CrossRef]
107. Zhu, L.; Liu, X.X.; Yang, J.; He, Y.C.; Li, Y.C. Application of Multiplex Microfluidic Electrochemical Sensors in Monitoring Hematological Tumor Biomarkers. *Anal. Chem.* **2020**, *92*, 11981–11986. [CrossRef]
108. Lee, K.; Song, G.; Kwon, J.; Kim, J.; Yang, H. Electrochemical Detection of Glucose and Lactate Dehydrogenase Using a Zwitterionic Phenazine Compound as an Electron Mediator for NADH Oxidation. *Electroanalysis* **2022**, *34*, 1499–1506. [CrossRef]
109. Ruiz-Vega, G.; Arias-Alpízar, K.; de la Serna, E.; Borgheti-Cardoso, L.N.; Sulleiro, E.; Molina, I.; Fernàndez-Busquets, X.; Sánchez-Montalvá, A.; del Campo, F.J.; Baldrich, E. Electrochemical POC device for fast malaria quantitative diagnosis in whole blood by using magnetic beads, Poly-HRP and microfluidic paper electrodes. *Biosens. Bioelectron.* **2020**, *150*, 111925. [CrossRef]
110. Yi, Y.H.; Li, Y.; Li, W.Z.; Cheng, M.J.; Wu, M.S.; Miao, J.F.; Kang, W.; Xu, Y.Y. Electrochemical Immunosensor for Lactate Dehydrogenase Detection Through Analyte-driven Catalytic Reaction on Multi-walled Carbon Nanotubes and Gold Nanoparticle Modified Carbon Electrode. *Electroanal* **2022**, *34*, 1187–1192. [CrossRef]
111. Liu, J.H.; Zhou, W.H.; Zhao, R.Z.; Yang, Z.D.; Li, W.; Chao, D.L.; Qiao, S.Z.; Zhao, D.Y. Sulfur-Based Aqueous Batteries: Electrochemistry and Strategies. *J. Am. Chem. Soc.* **2021**, *143*, 15475–15489. [CrossRef] [PubMed]
112. Pumera, M. Electrochemistry of Graphene: New Horizons for Sensing and Energy Storage. *Chem. Rec.* **2009**, *9*, 211–223. [CrossRef]
113. Jaugstetter, M.; Blanc, N.; Kratz, M.; Tschulik, K. Electrochemistry under confinement. *Chem. Soc. Rev.* **2022**, *51*, 2491–2543. [CrossRef] [PubMed]

114. Byrnes, S.A.; Huynh, T.; Chang, T.C.; Anderson, C.E.; McDermott, J.J.; Oncina, C.I.; Weigl, B.H.; Nichols, K.P. Wash-Free, Digital Immunoassay in Polydisperse Droplets. *Anal. Chem.* **2020**, *92*, 3535–3543. [CrossRef] [PubMed]
115. Hemben, A.; Ashley, J.; Tothill, I.E. An immunosensor for parasite lactate dehydrogenase detection as a malaria biomarker—Comparison with commercial test kit. *Talanta* **2018**, *187*, 321–329. [CrossRef] [PubMed]
116. Li, C.Y.; Chen, G.C.; Zhang, Y.J.; Wu, F.; Wang, Q.B. Advanced Fluorescence Imaging Technology in the Near-Infrared-II Window for Biomedical Applications. *J. Am. Chem. Soc.* **2020**, *142*, 14789–14804. [CrossRef]
117. Lee, M.H.; Kim, J.S.; Sessler, J.L. Small molecule-based ratiometric fluorescence probes for cations, anions, and biomolecules. *Chem. Soc. Rev.* **2015**, *44*, 4185–4191. [CrossRef]
118. Gui, R.J.; Jin, H.; Bu, X.N.; Fu, Y.X.; Wang, Z.H.; Liu, Q.Y. Recent advances in dual-emission ratiometric fluorescence probes for chemo/biosensing and bioimaging of biomarkers. *Coord. Chem. Rev.* **2019**, *383*, 82–103. [CrossRef]
119. Fan, S.N.; Li, X.Q.; Ma, F.H.; Yang, M.H.; Su, J.; Chen, X. Sulfur quantum dot based fluorescence assay for lactate dehydrogenase activity detection. *J. Photoch. Photobio. A.* **2022**, *430*, 113989. [CrossRef]
120. Liu, J.; Li, H.W.; Wu, Y.Q. A highly selective and sensitive fluorescent probe for lactate dehydrogenase based on ultrabright adenosine monophosphate capped gold nanoclusters. *RSC Adv.* **2017**, *7*, 13438–13443. [CrossRef]
121. He, X.L.; Tan, L.F.; Wu, X.L.; Yan, C.M.; Chen, D.; Meng, X.W.; Tang, F.Q. Electrospun quantum dots/polymer composite porous fibers for turn-on fluorescent detection of lactate dehydrogenase. *J. Mater. Chem.* **2012**, *22*, 18471–18478. [CrossRef]
122. Kenry; Geldert, A.; Zhang, X.; Zhang, H.; Lim, C.T. Highly Sensitive and Selective Aptamer-Based Fluorescence Detection of a Malarial Biomarker Using Single-Layer MoS2 Nanosheets. *Acs. Sens.* **2016**, *1*, 1315–1321. [CrossRef]
123. Jenie, S.N.A.; Prieto-Simon, B.; Voelcker, N.H. Development of L-lactate dehydrogenase biosensor based on porous silicon resonant microcavities as fluorescence enhancers. *Biosens. Bioelectron.* **2015**, *74*, 637–643. [CrossRef] [PubMed]
124. Minopoli, A.; Della Ventura, B.; Campanile, R.; Tanner, J.A.; Offenhausser, A.; Mayer, D.; Velotta, R. Randomly positioned gold nanoparticles as fluorescence enhancers in apta-immunosensor for malaria test. *Microchim. Acta* **2021**, *188*, 88. [CrossRef]
125. Singh, N.K.; Jain, P.; Das, S.; Goswami, P. Dye Coupled Aptamer-Captured Enzyme Catalyzed Reaction for Detection of Pan Malaria and P. falciparum Species in Laboratory Settings and Instrument-Free Paper-Based Platform. *Anal. Chem.* **2019**, *91*, 4213–4221. [CrossRef]
126. Sanchez-Cano, A.; Ruiz-Vega, G.; Vicente-Gomez, S.; de la Serna, E.; Sulleiro, E.; Molina, I.; Sanchez-Montalva, A.; Baldrich, E. Development of a Fast Chemiluminescent Magneto-Immunoassay for Sensitive Plasmodium falciparum Detection in Whole Blood. *Anal. Chem.* **2021**, *93*, 12793–12800. [CrossRef]
127. Figueroa-Miranda, G.; Liang, Y.; Suranglikar, M.; Stadler, M.; Samane, N.; Tintelott, M.; Lo, Y.; Tanner, J.A.; Vu, X.T.; Knoch, J.; et al. Delineating charge and capacitance transduction in system-integrated graphene-based BioFETs used as aptasensors for malaria detection. *Biosens. Bioelectron.* **2022**, *208*, 114219. [CrossRef]
128. Cao, X.E.; Kim, J.; Mehta, S.; Erickson, D. Two-Color Duplex Platform for Point-of-Care Differential Detection of Malaria and Typhoid Fever. *Anal. Chem.* **2021**, *93*, 12175–12180. [CrossRef]
129. Markwalter, C.F.; Ricks, K.M.; Bitting, A.L.; Mudenda, L.; Wright, D.W. Simultaneous capture and sequential detection of two malarial biomarkers on magnetic microparticles. *Talanta* **2016**, *161*, 443–449. [CrossRef]
130. Markwalter, C.F.; Davis, K.M.; Wright, D.W. Immunomagnetic capture and colorimetric detection of malarial biomarker Plasmodium falciparum lactate dehydrogenase. *Anal. Biochem.* **2016**, *493*, 30–34. [CrossRef]
131. Sancho-Fornes, G.; Avella-Oliver, M.; Carrascosa, J.; Fernandez, E.; Brun, E.M.; Maquieira, Á. Disk-based one-dimensional photonic crystal slabs for label-free immunosensing. *Biosens. Bioelectron.* **2019**, *126*, 315–323. [CrossRef] [PubMed]

biosensors

MDPI

Article

Monitoring and Regulating Intracellular GPX4 mRNA Using Gold Nanoflare Probes and Enhancing Erastin-Induced Ferroptosis

Xiaoyan Liu [1,†], Qiangqiang Yang [1,†], Yanan Sui [1], Qiaoli Yue [1], Shuqing Yan [1], Chuan Li [1] and Min Hong [1,2,*]

[1] School of Chemistry and Chemical Engineering, Liaocheng University, Liaocheng 252059, China
[2] Shandong Harway Pharma Co., Ltd., Dongying 257000, China
* Correspondence: hongmin@lcu.edu.cn; Tel.: +86-635-823-9195; Fax: +86-635-823-9121
† These authors contributed equally to this work.

Abstract: Glutathione peroxidase 4 (GPX4) plays an important effect on ferroptosis. Down-regulating the expression of GPX4 mRNA can decrease the content of GPX4. In this work, a gold nanoflare (AuNF) probe loaded with anti-sense sequences targeting for GPX4 mRNA was designed to monitor and down-regulate intracellular GPX4 mRNA using fluorescence imaging in situ and using anti-sense technology. The results revealed that there was a marked difference for the expression of GPX4 mRNA in different cell lines, and the survival rate of cancer cells was not significantly effected when the relative mRNA and protein expression levels of GPX4 was down-regulated by AuNF probes. However, when co-treated with AuNF probes, the low expression of GPX4 strengthened erastin-induced ferroptosis, and this synergy showed a better effect on inhibiting the proliferation of cancer cells.

Keywords: ferroptosis; GPX4 mRNA; AuNF probe

Citation: Liu, X.; Yang, Q.; Sui, Y.; Yue, Q.; Yan, S.; Li, C.; Hong, M. Monitoring and Regulating Intracellular GPX4 mRNA Using Gold Nanoflare Probes and Enhancing Erastin-Induced Ferroptosis. *Biosensors* **2022**, *12*, 1178. https://doi.org/10.3390/bios12121178

Received: 3 November 2022
Accepted: 14 December 2022
Published: 17 December 2022

1. Introduction

In the foreseeable future, cancer may replace cardiovascular disease as the disease that is causing the most deaths worldwide. As a result, researchers have conducted a lot of research work for the anti-cancer field. The present progress of anti-cancer drugs, treatment strategies, surgical technology, and imaging technology has significantly improved the survival rate of cancer patients, and many treatment methods aside from chemotherapy, radiotherapy, and surgery have gradually emerged, such as immunotherapy and gene therapy [1,2]. However, there is still no effective treatment to cure most cancers. Thus, it is quite urgent to develop new and effective cancer treatment methods.

With the development of modern technology, people have a deeper understanding of cell death. In addition to apoptosis, researchers have also found some other abnormal forms of cell death, such as autophagy, necrosis, pyroptosis, ferroptosis, etc. Among them, ferroptosis, a regulatory and iron-dependent programmed cell death form, has received extensive attention in recent years. Studies revealed that ferroptosis has some unique characteristics, such as shrinking mitochondria, increasing membrane density of mitochondria, and aggregation of iron and a series of destructive reactive oxygen species (ROS, including hydroxyl free radicals, lipid free radicals, and lipid hydrogen peroxide free radicals) [3,4]. ROS can oxidize intracellular nucleic acids, proteins, lipids, and other macromolecules. Especially, they will cause the oxidation of cell membrane lipids, which will eventually promote ferroptosis [3,5]. An effective ferroptosis inducer such as erastin can inhibit the cysteine/glutamate transporter system Xc-, leading to the depletion of cysteine (Cys) and glutathione (GSH) [6–8], further disrupting cellular redox homeostasis, inducing ROS accumulation, and then leading to the occurrence of ferroptosis.

In addition to inducing ferroptosis in cancer cells by elevating intracellular ROS levels, numerous studies have demonstrated the effectiveness of cancer-killing by inducing ferroptosis through the inactivation of glutathione peroxidase 4 (GPX4). GPX4 is a membrane lipid repair enzyme that can reduce lipid peroxide in the process of transforming GSH into GSSG, so it is a key factor to protect cells from ferroptosis [9,10]. Therefore, the inhibition of GPX4 is a key step in the process of exploring the role that contributes to the ferroptosis of cancer cells. Many inhibitors of GPX4 have been widely reported, such as RSL3 [11], ML162 [12], and FINO2 [13]. However, these inhibitors have some obvious disadvantages, such as poor selectivity and poor pharmacokinetic properties. Thus, there is an urgent need to find a method with good biocompatibility, easy metabolism, and effective inhibition of GPX4 activity. Protein synthesis is all translated from mRNA. If gene regulation methods such as anti-sense oligonucleotide technology are used to silence and down-regulate the expression of intracellular GPX4 mRNA, the activity of GPX4 can be indirectly suppressed.

In light of the above description, in this study we designed an anti-sense oligonucleotide sequence that can silence GPX4 mRNA. The anti-sense sequence was loaded on gold nanoflare probes (AuNF probes), which was firstly developed by Mirkin's group [14]. After the endocytosis of AuNF probes, the fluorescent molecule Cy5-modified anti-sense sequence will directly hybridize with GPX4 mRNA and further down-regulate the expression of GPX4. Moreover, the inhibition of cancer cell proliferation was studied through down-regulation of GPX4 alone with AuNF probes or through the synergistic effect of erastin and AuNF probes. Based on this probe, we sought to achieve the purpose of visually down-regulating and in situ monitoring GPX4 mRNA. Meanwhile, in order to overcome the disadvantages of erastin's poor water solubility and avoiding its obvious toxic and side effects, we hope to explore a synergistic strategy that can more effectively induce ferroptosis in cancer cells under the effect of a low concentration of erastin.

2. Materials and Methods

2.1. Materials

Chloroauric acid ($HAuCl_4 \cdot 4H_2O$), potassium dihydrogen phosphate (KH_2PO_4), disodium hydrogen phosphate (Na_2HPO_4), sodium citrate ($C_6H_5Na_3O_7$), sodium chloride (NaCl), and potassium chloride (KCl) were purchased from Shanghai Chemical Reagent Company (Shanghai, China). Phosphate buffer (PBS) was freshly prepared (pH 7.4, 136.7 mM of NaCl, 2.7 mM of KCl, 8.72 mM of Na_2HPO_4, 1.41 mM of KH_2PO_4). Magnesium chloride solution (1 M), RNase inhibitor (DEPC), 3-(4,5-dimethylthiazole-2)-2,5- diphenyltetrazolium bromide (MTT), and Tween-20 were purchased from Sigma-Aldrich (St. Louis, MO, USA). Other reagents are commercially available. The DNA sequences used (Table 1) were synthesized and purchased from Sangon Biotechnology Co., Ltd. (Shanghai, China). A glutathione peroxide detection kit (NADPH method), BCA protein concentration assay kit, immunostaining reagents (including fixed solution, washing solution, and blocking solution), GPX4 primary antibody (GPX4 rabbit polyclonal antibody), and GPX4 secondary antibody [Cy3 labeled Goat anti rabbit IgG (H + L)], were all purchased from Shanghai Beyotime Biotechnology Co., Ltd. (Shanghai, China).

Table 1. DNA sequence list.

Names	Sequences
HS-DNA	5′-TCA GCG TAT CAA AAA AAA AA-SH-3′
HS-DNA′	5′-GAT CTA AGC TAA AAA AAA AA-SH-3′
Flare-DNA	5′-GAT ACG CTG AGT GTG GTT T-Cy5-3′
Flare-DNA′	5′-AGC TTA GAT CGC TAT GTA C-Cy5-3′
Flare-DNA-N	5′-GAT ACG CTG AGT GTG GTT T-3′
Target-DNA	5′-AAA CCA CAC TCA GCG TAT C-3′
Random-DNA	5′-GGC CGA TTG TGA ACA TGG A -3′
GPX4-F-primer	5′-AGA GAT CAA AGA GTT CGC CGC-3′
GPX4-R-primer	5′-TCT TCA TCC ACT TCC ACA GCG-3′
GAPDH-F-primer	5′-CTC AGA CAC CAT GGG GAA GGT GA-3′
GAPDH-R-primer	5′-ATG ATC TTG AGG CTG TTG TCA TA-3′

2.2. Instrumentation

Fluorescence spectra and intensities were obtained on an F-7000 spectrophotometer (Hitachi, Tokyo, Japan). Absorption spectral measurements were carried out on a UV-750 ultraviolet spectrophotometer (Perkin-Elmer, Waltham, MA, USA). Transmission electron microscope (TEM) images were taken using a JEM 2100 electron microscope (JEOL Ltd., Tokyo, Japan). The TEM was operated at an acceleration voltage of 200 kV, and a micro grid was used for sample suspension. Confocal laser scanning microscopic analysis (CLSM) was carried out on a ZEISS LSM 880 microscope. Reverse transcription fluorescence quantitative PCR (qTR-PCR) was performed on the Quantstudio 5 Applied Biosystems instrument (Waltham, MA, USA). An MTT assay was performed on a ELX808 microplate reader (BioTek, Winooski, VT, USA).

2.3. Experimental Sections

2.3.1. Preparation of Gold Nanoparticles and Gold Nanoflare Probes

Synthesis of gold nanoparticles (AuNPs) was conducted according to the literature [15]. We heated 50 mL of $HAuCl_4$ (1 mM) solution to 100 °C, and after full boiling, 10 mL of trisodium citrate (38.8 mM) solution was rapidly added. After that, the color of the solution will gradually turn dark red. After stirring for 15 min, the reaction was stopped, and the mixture was naturally dropped to room temperature and was stored at 4 °C for standby.

Gold nanoflare probes (AuNF probes) were synthesized according to the following steps. Add 100 μL of DEPC water to HS-DNA and Flare-DNA (or Flare-DNA-N) at 1 OD, respectively, and vortex to mix thoroughly. Then, add 1.5 μL (100 mM) of TCEP to the SH-DNA to activate the sulfhydryl groups, and let it stand for 1 h at room temperature in the dark. Mix HS-DNA and Flare-DNA (or Flare-DNA-N) at a ratio of 1:1.2 to form a nucleic acid hybridization solution. Then, the mixture was heated at 90 °C for 5 min and further cooled slowly. After the formation of the HS-DNA/Flare-DNA (or Flare-DNA-N) duplex, the solution was divided into 2 equal parts, and 1.2 mL of AuNPs solution was added to each part. After mixing at room temperature for 24 h in the dark, 10 μL of Tween diluent was firstly added, followed by adding 33 μL of PBS buffer solution containing 2 M of NaCl. Next, the passivation reaction was carried out by adding 33 μL of PBS buffer solution containing 2 M of NaCl three times every 2 h, and we continued to mix for 48 h. All operations were performed under dark conditions. Then, we centrifuged (13,500 rpm/30 min) the solution to remove the free DNA and continue to wash twice with PBS buffer solution. Finally, 1 mL of PBS buffer solution was used to dissolve the precipitate to form the AuNF probe solution, which was stored at 4 °C and should be used up in one month. The concentration of AuNF probe solution was calculated by the Lambert–Beer law ($\varepsilon = 2.7 \times 10^8$ L·mol^{-1}·cm^{-1}).

Gold nanoflare probe analogs (AuNF probe analogs) were also prepared using the same process mentioned above. However, the sequences of DNA duplexes (HS-DNA'/Flare-DNA') that was bound with AuNPs was different from those of HS-DNA/Flare-DNA but with the same amount of bases.

2.3.2. Incubation of AuNF Probes with Target-DNA or Cell Lysis Solution

AuNF probes (1.5 nM) and Target-DNA of different concentrations (0, 2.5, 5, 10, 50, 100, 200, and 300 nM) were mixed in PBS buffer solution to obtain 200 μL of solution, which was incubated at 37 °C for 4 h in the dark; we then measured the fluorescence intensity of each sample.

The cell lysis solution was prepared by using an RNA extraction kit to extract RNA from 2×10^6 HeLa cells and dissolve RNA in 100 μL of DEPC water. Then, take 20 μL cell lysis solution to mix with AuNF probes (1.5 nM) and produce up to 200 μL with DEPC water. Our mixture was incubated at 37 °C for 4 h in the dark and we then measured the fluorescence intensity.

2.3.3. Cell Culture, Confocal Imaging, and Flow Cytometric Analysis

The medium (DMEM, GIBCO), fetal bovine serum, and double antibody [penicillin (100 μg·mL^{-1}) and streptomycin (100 μg·mL^{-1})] were prepared according to the ratio of 9:1:0.1 to prepare the cell culture medium. The cells were cultured in a sterile environment at 37 °C humidified with 5% CO_2.

Confocal laser scanning microscopy (CLSM) was used to detect the content of intracellular GPX4 mRNA through in situ fluorescence imaging after incubating AuNF probes (1.5 nM) for 4 h in 6 different cancer cell lines, including the human cervical cancer cell line (HeLa), human lung cancer cell line (A549), human hepatoellular carcinomas (HepG2), human breast cancer cell lines (MCF-7 and MDA-MB-231), human osteosarcoma cell line (U2OS); two normal cell lines were also used, including human normal liver cell line (HL-7702) and human normal breast cell line (HBL-100). We thoroughly mixed 200 μL of cell suspension (1×10^6 mL^{-1}) and 2 mL of cell culture medium and then inoculated them in a confocal dish. After culturing for 24 h, the AuNF probes (1.5 nM) were added. After incubating the cell samples in the incubator for 4 h, CLSM fluorescence imaging was determined using a 633 nm excitation. Then, the medium was removed, and the cells were trypsinized, collected, and washed twice with pre-chilled PBS. Apoptotic cells were detected by a Guava easyCyte flow cytometer. For each experiment, 5000 cells were recorded per sample.

2.3.4. qRT-PCR Analysis

HeLa cells were seeded in 6-well plates, and three groups were set as the PBS group, control group (AuNF probe analogs), and AuNF probe group, which were prepared with Flare-DNA-N. The probes were added at different time points. After reaching the treatment time (12, 24, 36, 48 h), the culture medium was removed and thoroughly washed with PBS. Then, the cell samples were trypsinized and counted for extracting RNA according to the instruction of RNA extraction kit. After reverse transcription, they were stored in a −80 °C refrigerator for qRT-PCR detection. The contents of GPX4 mRNA were determined by qRT-PCR analysis using the SYBR Green Master Mix method. All primers used here are shown in Table 1.

2.3.5. Immunofluorescence Labeling

HeLa cells were seeded in laser confocal dishes for 24 h; AuNF probes or PBS were added for 48 h. Then, the control and the AuNF probe groups were individually fixed, blocked, and incubated with primary antibody and Cy3-labeled secondary antibody according to the instruction of the assay kit. Finally, two groups were washed and observed under a confocal laser scanning microscope. To avoid the disturbance of Cy5 labeled on the Flare-DNA, AuNF probes used here were prepared with Flare-DNA-N.

2.3.6. Determining the Content of Glutathione Peroxidases

HeLa cells were seeded in cell culture dishes for 24 h and then treated with AuNF probes at different concentrations (0, 0.5, 0.8, 1.2, and 1.5 nM) for 48 h. With the same method, AuNF probes (1.5 nM) were added into HeLa cell samples and incubated for 0, 12, 24, 36, or 48 h. After washing the cells with PBS, the cells were digested on a small dish with 0.02% EDTA, then lyzed with a glass homogenizer. The protein concentrations of the cell lysis were determined with a BCA protein concentration assay kit and then detected according to the instruction of the glutathione peroxidase assay kit.

2.3.7. Inhibiting the Proliferation of Cancer Cells through Erastin or the Synergistic Effect of Erastin and AuNF Probes

Five human cancer cell lines, including HeLa, A549, HepG2, MCF-7, and MDA-MB-231, were individually inoculated in 96-well plates (1×10^5 cells/well). After incubating for 24 h, different volumes of erastin dissolved in DMSO were added into cell samples with the final erastin concentrations of 0.1, 0.5, 1, 2, 5, 10, and 15 μM. The cell viability of each sample was determined using the MTT method after interacting with erastin for 48 h.

Similarly, five human cancer cell lines mentioned above were re-prepared in 96-well plates (1×10^5 cells/well). Each cell sample was prepared in triplicate. A different cell sample was treated with AuNF probes (1.5 nM), erastin dissolved in DMSO (2 μM), or AuNF probes (1.5 nM) + erastin (2 μM), respectively. The cell viabilities of the cell samples were determined using the MTT method after being treated with different systems for 12, 24, or 48 h.

2.3.8. Determining the Content Change of Intracellular ROS Caused by Erastin and AuNF Probes

Three HeLa cell samples cultured in confocal dishes were individually prepared by adding HeLa cell suspension (200 μL, 1×10^6 mL^{-1}) and 2 mL of cell culture medium. After culturing for 24 h, AuNF probes were added into two cell samples with the final concentration of 1.5 nM. After 6 h, the medium containing AuNF probes was removed, and 2 mL of new cell culture medium with or without erastin (2 μM) were added. These two cell samples were set as the "AuNF probes" or "AuNF probes + erastin" group, respectively. After culturing for 48 h, two cell samples were stained with 2′,7′-dichlorodihydrofluorescein diacetate (DCFH-DA, 50 mM) in the medium for 30 min and then washed for three times with PBS solution and observed under a CLSM. The DCF was excited with a 488 nm laser and the fluorescence intensity at 525 nm shows the amount of DCF, which was correlated with the content of ROS.

The CLSM imaging of intracellular ROS for the only erastin-treated cell group was also dealt with in a similar process as the "AuNF probes + erastin" group, but without the treatment of AuNF probes.

2.3.9. Statistical Analysis

All data were reported as the means ± standard deviation from at least three independent experiments. Statistical analyses were performed using the IBM SPSS Statistics 25 software package, and we assumed significance at $p < 0.05$ (*) and high significance at $p < 0.01$ (**).

3. Results and Discussion

3.1. Design of AuNF Probes and Mechanism of Enhancing Erastin-Induced Ferroptosis

Here, AuNF probes were prepared by the combination of HS-DNA/Flare-DNA duplexes and AuNPs through Au-S bonds. Flare-DNA was designed as the anti-sense sequences targeting for GPX4 mRNA and modified by the Cy5 fluorophore molecules at the 5′ end. Firstly, HS-DNA and Flare-DNA were co-incubated to form the HS-DNA/Flare-DNA duplexes, and the fluorescence signal of Cy5 was not effected. Once in the presence of AuNPs, AuNF probes were formed and the signal of Cy5 would be bleached due to the effect of the fluorescence resonance energy transfer (FRET). As shown in Figure 1, when the AuNF probes were endocytosed into the cell, intracellular GPX4 mRNA could competitively hybridize with Flare-DNA to form Flare-DNA/GPX4 mRNA hybridization systems, which will silence or degrade GPX4 mRNA and further down-regulate the expression of GPX4. At the same time, and the fluorescence signal of Cy5 is recovered due to the detachment of Flare-DNA from the surface of AuNPs. So, the above process of detecting and regulating GPX4 mRNA can be monitored by the fluorescence signal changes. In addition, the down-regulation of GPX4 was used to enhance erastin-induced ferroptosis for cancer cells.

3.2. Characterization of AuNPs and AuNF Probes

The size of AuNPs (~13 nm) was characterized by using TEM imaging (Figure 2A). The UV-vis spectra showed the characteristic absorption of AuNPs (~520 nm) and the DNA signal (~260 nm) for AuNF probes (Figure 2B). In addition, DLS was used to verify changes in the hydrodynamic size of AuNPs after binding with HS-DNA/Flare-DNA duplexes. As shown in Figure 2C,D, the hydrodynamic size of AuNF probes is 36 ± 0.3 nm, which is bigger than that of bare AuNPs (13.4 ± 0.4 nm). These results was basically consistent with relevant works [14,16].

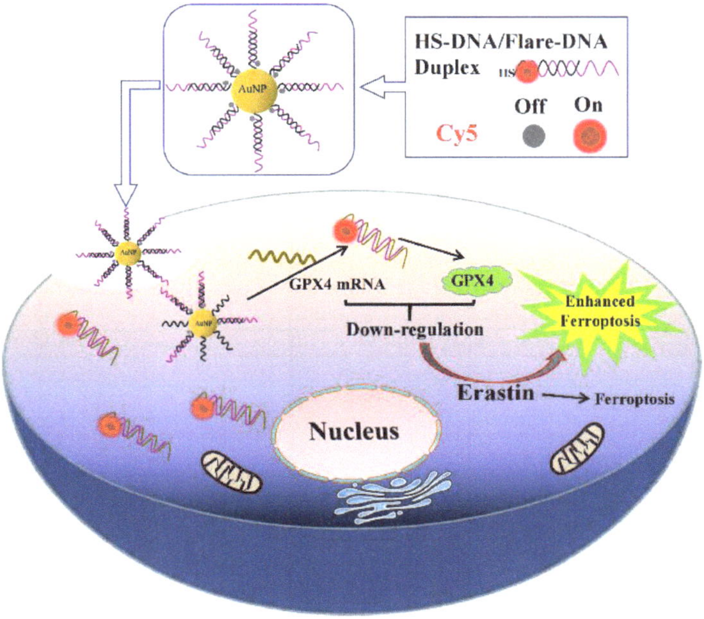

Figure 1. Schematic illustration of designation of AuNF probes, mechanism of down-regulating GPX4, and enhancement of erastin-induced ferroptosis for cancer cells.

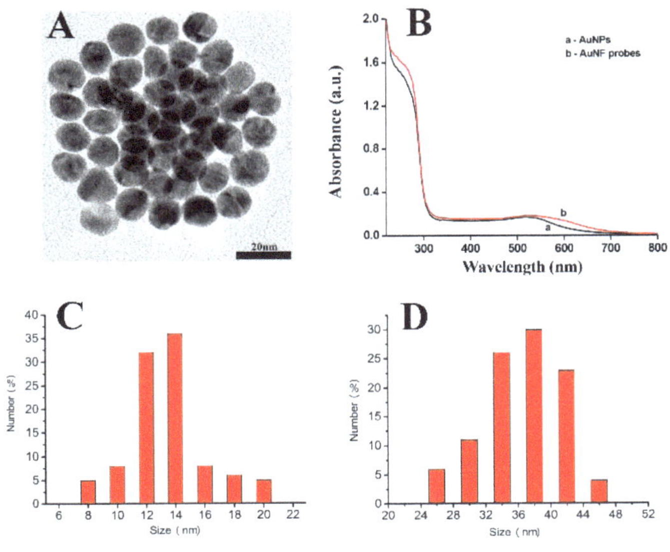

Figure 2. TEM images of AuNPs (**A**), UV-vis characterization of AuNPs and AuNF probes (**B**), hydrodynamic sizes of AuNPs (**C**) and AuNF probes (**D**) determined by DLS.

3.3. Fluorescence Response of AuNF Probes to Target-DNA and Cell Lysis Solution

According to the designation, the distance (-A_{10}-, ~3.4 nm) of Cy5 and AuNPs in the AuNF probes is in the range of FRET [17]. So, there is weak fluorescence signals of Cy5 in the solution of AuNF probes (Figure 3). When the AuNF probes were treated with different concentrations of Target-DNA that were designed to completely hybridize with the Flare-DNA, the fluorescence signals of Cy5 were recovered. With the increase in Target-DNA concentrations, the fluorescence intensity increased accordingly, and there was a plateau when the concentration of the Target-DNA was above 100 nM (Figure 3A). These results show that the competitive hybridization between the Target-DNA and the Flare-DNA would induce a Flare-DNA detach from AuNF probes and recover the Cy5 fluorescence. A similar fluorescence recovery also occurred to the experimental system of AuNF probes and HeLa cell lysis solution (Figure 3B). Furthermore, to determine the specificity of the AuNF probes to the Target-DNA, Random-DNAs with arbitrary sequences were used to test the fluorescence response of AuNF probes. Studies show that the fluorescence intensity is quite small when AuNF probes incubate with high concentrations of Random-DNA (Figure S2). In addition, AuNF probe analogs have also been prepared by binding arbitrary sequence HS-DNA'/Flare-DNA' hybridization duplexes on AuNPs. However, there was no obvious fluorescence response for the AuNF probe analogs to the Target-DNA (Figure S3A). All these results demonstrate the good specificity of the AuNF probes to GPX4 mRNA.

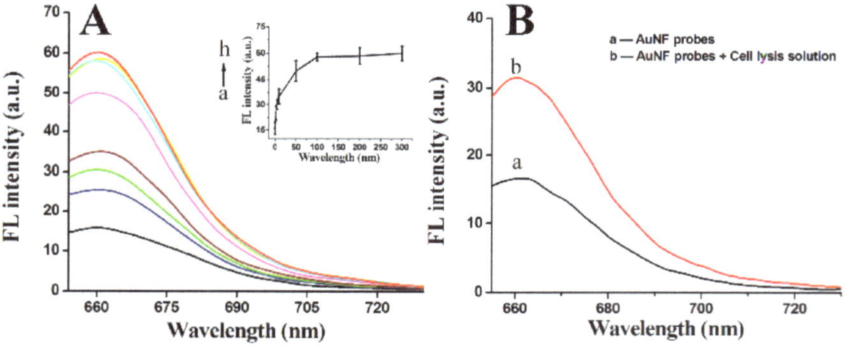

Figure 3. Fluorescence spectra of AuNF probes incubated with different concentrations of Target-DNA (0, 2.5, 5, 10, 50, 100, 200, and 300 nM) (**A**) or HeLa cell lysis solution (**B**).

3.4. Evaluation the Number of HS-DNA/Flare-DNA Duplexes Bound on AuNF Probes

According to the previous reports, the small molecule mercaptoethanol can compete with the nucleic acid hybridization duplexes and be bound to AuNPs by Au-S bonds. As AuNF probes (1.5 nM) were incubated with different concentrations of mercaptoethanol, and a sequentially enhanced fluorescence signal was observed with increasing mercaptoethanol concentration (Figure S1). This reveals that different amounts of HS-DNA/Flare-DNA duplexes were dissociated from the surface of the AuNPs. Based on the standard curve of HS-DNA/Flare-DNA duplexes, the amount of duplexes (~47) bound on each AuNF probe were evaluated by the fluorescence intensity of suspension solution of AuNF probes (1.5 nM) after being adequately treated with mercaptoethanol (50 μM) (Figure S1). Here, there are only 10 fully hybridized bases in the HS-DNA/Flare-DNA duplexes. Compared with previous similar gold nanoflare probes with 12 bases in the part of hybridization [16], the lower number of bound nucleic acid duplexes on the AuNF probe designed here should be due to the worse rigidity of the HS-DNA/Flare-DNA nucleic acid sequence caused by the lower number of hybridization bases.

3.5. Fluorescence Imaging of Intracellular GPX4 mRNA by CLSM

Next, the response of the AuNF probes to intracellular GPX4 mRNA was examined by confocal laser scanning microscopy (CLSM). Six types of cancer cell lines were tested, including A549, HeLa, HepG2, MCF-7, MDA-MB-231, and U2OS. As shown in Figure 4A, there were great differences in the expression of GPX4 mRNA in the different cancer cell lines. A549, HeLa, and HepG-2 presented high expression. Both of the two types of breast cancer cell lines showed low expression. Similarly, there were slight fluorescence signals for U2OS, showing the low expression of GPX4 mRNA. At the same time, the expression of GPX4 mRNA in two normal cell lines were also determined. Compared with the HepG2 cells, the human normal liver cell line (HL-7702) presented a low expression of GPX4 mRNA, but normal breast cells (HBL-100) had higher levels of expression than their breast cancer cell counterparts (MCF-7 and MDA-MB-231) (Figure 4B). In addition, to further verify this phenomena, parts of the cell samples were collected for flow cytometric analysis. The general fluorescence signals were in accordance with the results determined by CLSM (Figure 4C). The content of GPX4 mRNA in HeLa, A549, MCF-7, and HBL-100 have also been tested by qRT-PCR. The data show similar results to those determined by the CLSM and flow cytometric analyses (Figure S4).

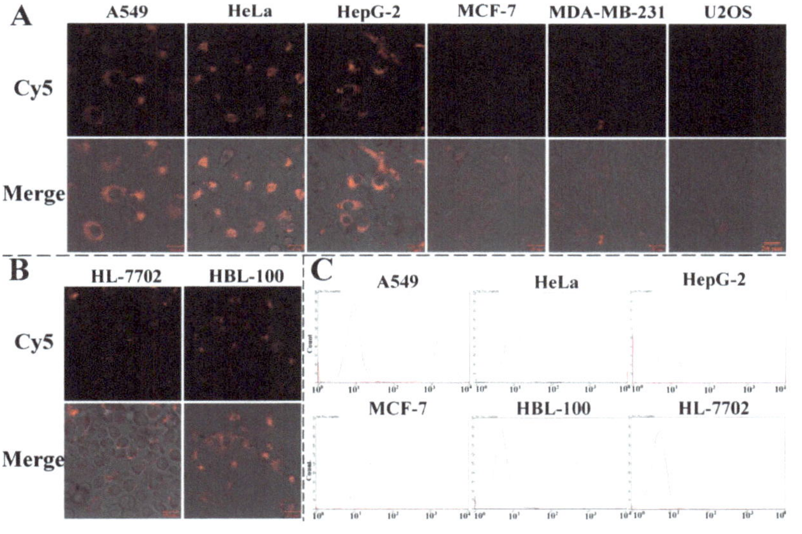

Figure 4. Fluorescence images of different cancer cell lines (**A**) and normal cell lines (**B**) that were treated with AuNF probes (1.5 nM) for 4 h. Flow cytometric analysis of different cell lines (**C**) collected after the CLSM determination.

Moreover, the intracellular non-specific response of AuNF probes can be excluded by CLSM imaging for HeLa cells that were treated with AuNF probe analogs (Figure S3B,C). Compared with the stronger fluorescence intensity of the AuNF probe treatment group, the AuNF probe analog group only showed slight background signals.

3.6. Expression Analysis of Intracellular GPX4 mRNA, GPX4 and Total GPX

Recently, Hangauer et al. reported that drug-resistant persistent cancer cells are sensitive to GPX4. Ferroptosis of resistant cancer cells can be induced by inhibiting GPX4 with a direct GPX4 inhibitor (RSL3) [18]. In addition, Ni et al. used siRNA to down-regulate GPX4 mRNA to synergistically enhance ferroptosis of cancer cells when reacted with iron oxide and platinum nanocomposite systems [19].

To further corroborate the critical role of GPX4, we down-regulated GPX4 mRNA expression in HeLa cells using anti-sense technology. The Flare-DNA sequences of AuNF probes were designed to completely hybridize with GPX4 mRNA. Once the AuNF probes enter the cells, Flare-DNA will hybridize with GPX4 mRNA to form a partial double-stranded structure. This hybridization will activate intracellular RNase H to hydrolyze GPX4 mRNA and reduce the content of intracellular GPX4 mRNA or to silence GPX4 mRNA to prevent its further translation into GPX4 protein [20]. To verify this effect, a qRT-PCR test was used to examine the content change of intracellular GPX4 mRNA when the AuNF probes reacted with HeLa cells for different times. With the increase in time, the GPX4 mRNA content showed a significant down-regulation trend, and it dropped to 61% at 48 h (Figure 5A). With a similar experimental operation, there was no obvious decrease for the expression of GPX4 mRNA in HeLa cells that were treated with the AuNF probe analogs. This result showed that AuNF probes could down-regulate the expression of intracellular GPX4 mRNA based on anti-sense technology.

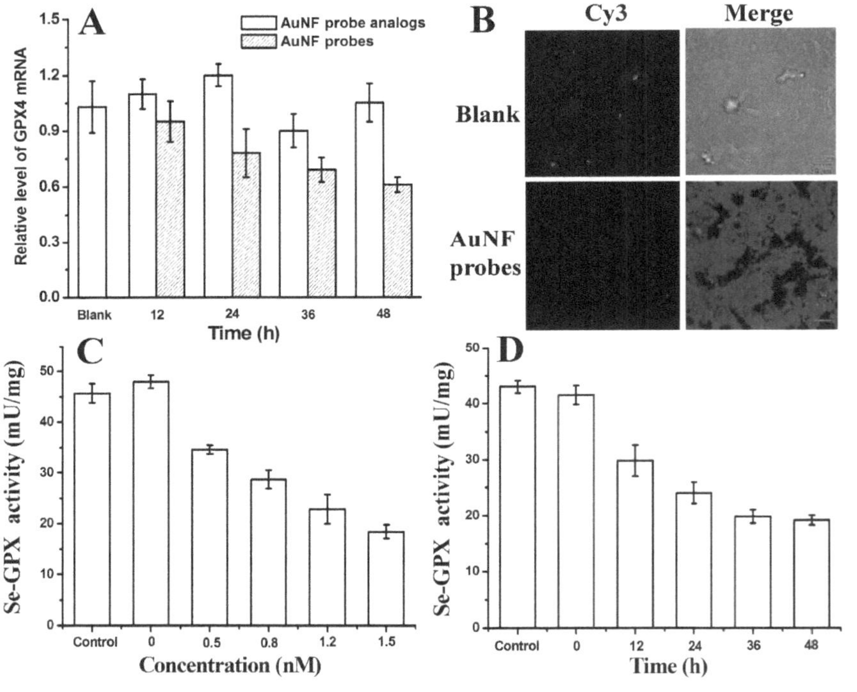

Figure 5. (**A**) qRT-PCR analysis of GPX4 mRNA in HeLa cells after being treated with AuNF probes (1.5 nM) or AuNF probe analogs (1.5 nM) for different times (12, 24, 36, and 48 h). (**B**) Immunofluorescence confocal imaging of GPX4 expression (purple) in HeLa cells after being treated with AuNF probes (1.5 nM) or PBS (Blank) for 48 h. (**C**) The activity of Se-GPX in HeLa cells after being treated with different concentrations of AuNF probes (0, 0.5, 0.8, 1.2, and 1.5 nM) for 48 h. (**D**) The activity of Se-GPX in HeLa cells after being treated with AuNF probes (1.5 nM) for different times (0, 12, 24, 36, and 48 h).

The protein level of GPX4 was observed using the immunofluorescence labeling method with CLSM. According to the Cy3 fluorescence signals labeled on secondary antibody (Figure 5B), it can be seen that the GPX4 content in the AuNF probe group is significantly lower than that of the PBS group. These results indicate that AuNF probes can effectively down-regulate the expression of GPX4 protein after regulating intracellular GPX4 mRNA.

Furthermore, the activity of total GPX containing selenium (Se-GPX) were determined using a glutathione peroxidase assay kit with the NADPH method. As shown in Figure 5C,D, with the increase in AuNF probe concentration and the prolongation of reaction time, the activity of Se-GPX showed a downward trend, which was consistent with the results of the down-regulation of the expression of GPX4.

3.7. Enhancing Erastin-Induced Ferroptosis Using the Synergistic Effect of AuNF Probes

Ferroptosis inducers can be divided into two types. The first type, such as erastin and sulfasalazine, acts by targeting cystinoglutamate anti-transporters (system Xc-). The second type, such as RSL3 and DP17, can directly inhibit GPX4 activity [21,22]. Both of these types of inducers can mediate intracellular ROS accumulation and inhibit the proliferation of cancer cells with the form of ferroptosis.

Here, we plan to study the inhibition of the proliferation of cancer cells while simultaneously using these two kinds of ferroptosis inducers. Erastin is a low molecular weight chemotherapy drug that can induce ferroptosis for different types of cancer cells [23–25]. Our experiments show that erastin exhibits obvious in vitro cytotoxicity against the five cancer cell lines, including HeLa, A549, HepG2, MCF-7, and MDA-MB-231 (Figure 6A). Especially for A549, the IC50 value is low, measuring at 2.32 μM. Overall, the inhibition rates of cell proliferation are dependent on the concentration of erastin. However, poor water solubility and renal toxicity have limited its clinical application. It was reported that the expression of GPX4 appeared to strongly correlate with the sensitivity to ferroptosis induction by erastin [12]. Compared with the above measured expression levels of GPX4 mRNA, we noticed that erastin showed significant cytotoxic activity at the lower concentration range (<5 μM) for HeLa, A549, and HepG2 cells, which presents a higher expression of GPX4 mRNA. The semi-inhibitory concentrations (IC50) are shown in Table S1. For MCF-7 and MDA-MB-231, both of their half-inhibitory concentrations were greater than 10 μM. All these results show the correlation of erastin-mediated ferroptosis with the expression of GPX4.

Considering the low water solubility of erastin, the authors aimed to explore some new strategies to conjugate with other ferroptosis promoters to maximize the anti-cancer effect of erastin at low concentrations. The inactivation of GPX4 is believed to lead to increased levels of uncontrolled lipid peroxidation culminating in cell death in vitro and in vivo [12,18]. To further test the possibility and the correlation between the expression of GPX4 and erastin, we treated five cancer cell lines with AuNF probes for different times. We found that the AuNF probes show a weaker inhibition for the proliferation of the five cancer cell lines by only down-regulating GPX4 (Figure 6B–F). Of course, the cytotoxicity of AuNPs itself can be excluded by determining the cell viability of HeLa cells that were treated with different concentrations of AuNF probe analogs for 48 h (Figure S3D).

Compared with the cell viability of different cancer cells treated with erastin (2 μM) alone for different times, the co-treatment with AuNF probes and erastin has better inhibition rates for the proliferation of all cancer cell lines except for A549 cells under these unoptimized conditions. According to the inhibition rate of different systems on the proliferation of different cancer cells at 48 h (shown in Table S2), we can see that there exists a drug synergism of "1 + 1 > 2". Thus, targeting system Xc- and directly inhibiting GPX4 activity can achieve a synergistic inhibition of cancer cell proliferation.

3.8. ROS Generation Caused by AuNF Probes and Erastin

To further demonstrate the relationship of cytotoxicity of AuNF probes or erastin and their ability to generate ROS, the content change of intracellular ROS was evaluated by the CLSM analysis using DCFH-DA as the fluorescence molecular probe.

A large number of studies have shown that ROS is the key point of ferroptosis in cells [3–5], and GPX4 shows the ability to clear intracellular ROS [9,10]. Therefore, the down-regulation of GPX4 should lead to intracellular ROS accumulation, which induces cell death. Figure 7 shows the confocal images of HeLa cells when the cells were treated by the AuNF probes. Unexpectedly, there is only a slight fluorescence signal of DCA, which demonstrates a lower content of intracellular ROS. Compared with the inhibition of GPX4 activity shown in Figure 5, we speculated that the above results should be due to the poor inhibition rate (40% at 48 h) of GPX4 activity by the AuNF probes, so that the remaining GPX4 could still clear most of the intracellular ROS. This may also explain why AuNF probes had less effect on the survival of the cancer cells tested (Figure 6B–F).

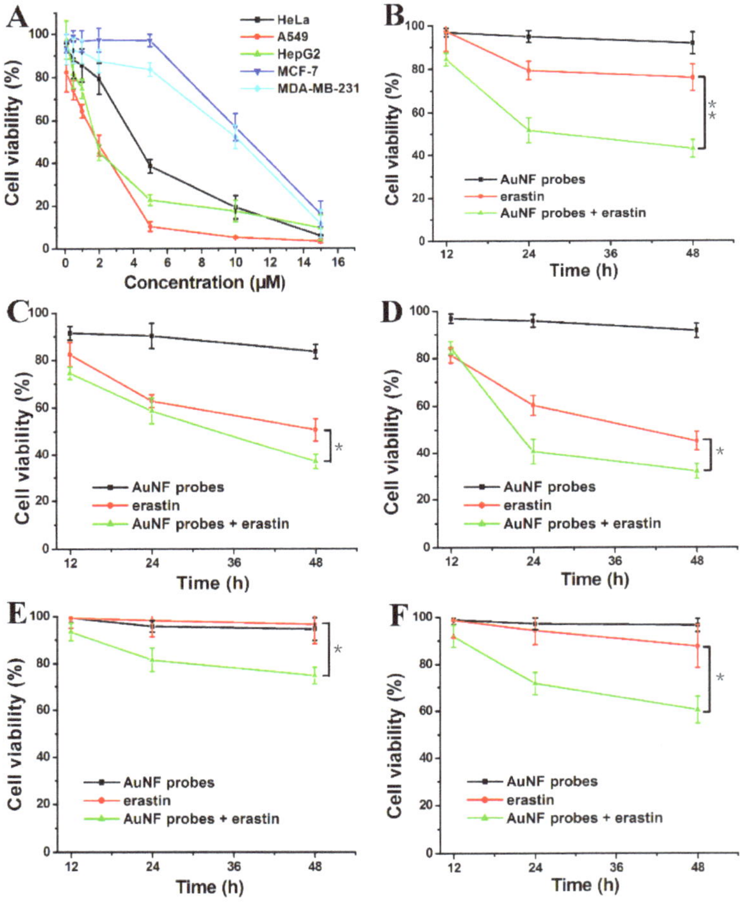

Figure 6. (**A**) Cell viabilities of different cancer cell lines when treated with different concentrations of erastin (0.1, 0.5, 1.0, 2.0, 5.0, 10.0, and 15.0 μM) for 48 h. Cell viability of HeLa (**B**), A549 (**C**), HepG2 (**D**), MCF-7 (**E**), and MDA-MB-231 cells (**F**) when treated with AuNF probes (1.5 nM), erastin (2.0 μM), and AuNF probes (1.5 nM) + erastin (2.0 μM) for different times (12, 24, and 48 h). $p < 0.05$ (*); $p < 0.01$ (**).

For the erastin-treated cell sample, the confocal images show obvious fluorescence signals of DCA, which verify the previous report that erastin can mediate intracellular ROS accumulation [23–25]. Apparently, the stronger fluorescence signals for AuNF probes and erastin co-treated cell sample demonstrated that the synergistic effect of the AuNF probes and erastin more significantly promoted the accumulation of large amounts of ROS in HeLa cells. These results are in accordance with the greater cell inhibitory activity shown in Figure 6B–F.

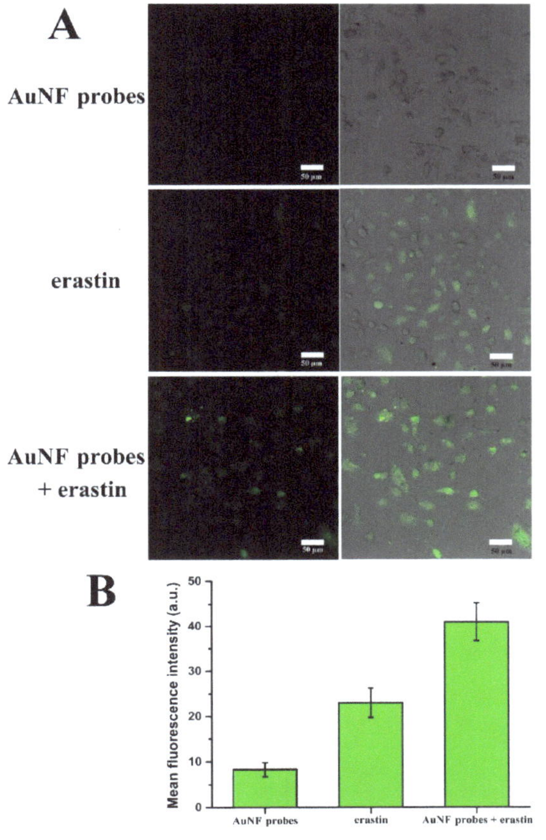

Figure 7. (**A**) Confocal images of HeLa cells showing the ROS response when cells were treated by different systems. (**B**) Mean fluorescence intensity of confocal images of HeLa cells shown in (**A**).

4. Conclusions

At present, there are more and more studies on ferroptosis, which are inseparable from the study of GPX4. In this study, we constructed a AuNF probe for monitoring and down-regulating intracellular GPX4 mRNA based on anti-sense oligonucleotide technology. Studies reveal that the proliferation of cancer cells was not obviously effected only by down-regulating the expression of GPX4. Nevertheless, ferroptosis induced by erastin can be enhanced by the synergistic effect of the AuNF probe. In conclusion, this study provides a new research idea for promoting the clinical application of erastin and finding better anti-cancer strategies.

Supplementary Materials: The following supporting information can be downloaded at: https://www.mdpi.com/article/10.3390/bios12121178/s1, Figure S1: (A): Fluorescence recovery of AuNF-probes incubated with different concentrations of mercaptoethanol (0, 0.2, 0.5, 1, 20, 50 μM). (B): Standard curves of the fluorescence intensity of HS-DNA/Flare-DNA duplexes (0, 5, 10, 20, 40, 60, 80, 100, 150 nM). Figure S2: Fluorescence spectra of AuNF-probes incubated with different concentrations of Random-DNA (0, 100, 300 nM). Figure S3 (A): Fluorescence spectra of AuNF-probe analogs incubated with different concentrations of Target-DNA (0, 100, 300 nM). (B): Confocal fluorescence images of HeLa cells that were treated with or without AuNF-probe analogs (1.5 nM) for 4 h. (C): Mean fluorescence intensity of HeLa cells that shown in Figure S2B and Figure 4A. (D): Viabilities of HeLa cells when treated with different concentrations of AuNF-probe analogs (0.5, 1.0, 1.5, 2.0 nM) for 48 h. Figure S4: qRT-PCR analysis of GPX4 mRNA content in HeLa, A549, MCF-7, and HBL-100 cells. Table S1: IC50 values of erastin against five different cancer cell lines. Table S2: Inhibition rate of different systems on the proliferation of different cancer cells at 48 h.

Author Contributions: Conceptualization, X.L. and Q.Y. (Qiangqiang Yang); methodology, X.L.; software, Q.Y. (Qiangqiang Yang); validation, Y.S., Q.Y. (Qiaoli Yue) and S.Y.; formal analysis, C.L.; investigation, M.H.; resources, C.L.; data curation, Q.Y. (Qiaoli Yue); writing—original draft preparation, X.L. and M.H.; writing—review and editing, M.H.; visualization, C.L.; supervision, M.H.; project administration, S.Y.; funding acquisition, M.H. All authors have read and agreed to the published version of the manuscript.

Funding: The work was supported by the Natural Science Foundation of Shandong Province (Grant No. ZR2015BM024), The Ministry of Education industry-school cooperative education project (Grant No. 220605042200513), the Tai-Shan Scholar Research Fund of Shandong Province, and Research Foundation of Liaocheng University (Grant No. 318012026).

Institutional Review Board Statement: Not applicable.

Informed Consent Statement: Not applicable.

Data Availability Statement: Not applicable.

Conflicts of Interest: The authors declare no conflict of interest.

References

1. Van Den Bulk, J.; Verdegaal, E.M.; De Miranda, N.F. Cancer immunotherapy: Broadening the scope of targetable tumours. *Open. Biol.* **2018**, *8*, 180037–180044. [CrossRef] [PubMed]
2. Goodall, G.J.; Wickramasinghe, V.O. RNA in cancer. *Nat. Rev. Cancer.* **2021**, *21*, 22–36. [CrossRef] [PubMed]
3. Scott, J.D.; Kathryn, M.L.; Michael, R.L.; Rachid, S.; Eleina, M.Z.; Caroline, E.G.; Darpan, N.P.; Andras, J.B.; Alexandra, M.C.; Wan, S.Y.; et al. Ferroptosis: An Iron-Dependent Form of Nonapoptotic Cell Death. *Cell* **2012**, *149*, 1060–1072
4. Fan, S.; Yang, Q.; Song, Q.; Hong, M.; Liu, X.; Chen, H.; Wang, J.; Li, C.; Cheng, S. Multi-pathway inducing ferroptosis by MnO_2-based nanodrugs for targeted cancer therapy. *Chem. Commun.* **2022**, *58*, 6486–6489. [CrossRef] [PubMed]
5. Wang, Y.; Qi, H.; Liu, Y.; Duan, C.; Liu, X.; Xia, T.; Chen, D.; Piao, H.; Liu, H. The double-edged roles of ROS in cancer prevention and therapy. *Theranostics* **2021**, *11*, 4839–4857. [CrossRef]
6. Li, Y.; Yan, H.; Xu, X.; Liu, H.; Wu, C.; Zhao, L. Erastin/sorafenib induces cis-platin-resistant non-small cell lung cancer cell ferroptosis through inhibition of the Nrf2/xCT pathway. *Oncol. Lett.* **2020**, *19*, 323–333.
7. Yang, W.S.; Kim, K.J.; Gaschler, M.M.; Patel, M.; Shchepinov, S.M.; Stockwell, B.R. Peroxidation of polyunsaturated fatty acids by lipoxygenases drives ferroptosis. *Proc. Natl. Acad. Sci. USA* **2016**, *113*, 4966–4975. [CrossRef]
8. Reed, J.C.; Pellecchia, M. Ironing out cell death mechanisms. *Cell* **2012**, *149*, 963–965. [CrossRef]
9. Seibt, T.M.; Proneth, B.; Conrad, M. Role of GPX4 in ferroptosis and its pharmacological implication. *Free. Radic. Biol. Med.* **2019**, *133*, 144–152. [CrossRef]
10. Krümmel, B.; Plötz, T.; Jörns, A.; Lenzen, S.; Mehmeti, I. The central role of glutathione peroxidase 4 in the regulation of ferroptosis and its implications for pro-inflammatory cytokine-mediated beta-cell death. *Biochim. Biophys. Acta-Mol. Basis Dis.* **2021**, *1867*, 166114–166136. [CrossRef]
11. Yang, W.S.; Stockwell, B.R. Synthetic lethal screening identifies compounds activating iron-dependent, nonapoptotic cell death in oncogenic-RAS-harboring cancer cells. *Chem. Biol.* **2008**, *15*, 234–245. [CrossRef]
12. Yang, W.S.; SriRamaratnam, R.; Welsch, M.E.; Shimada, K.; Skouta, R.; Viswanathan, V.S.; Cheah, J.H.; Clemons, P.A.; Shamj, A.F.; Clish, C.B.; et al. Regulation of ferroptotic cancer cell death by GPX4. *Cell* **2014**, *156*, 317–331. [CrossRef]
13. Gaschler, M.M.; Andia, A.A.; Liu, H.; Csuka, J.M.; Hurlocker, B.; Vaiana, C.A.; Heindel, D.W.; Zuckerman, D.S.; Bos, P.H.; Reznik, E.; et al. $FINO_2$ initiates ferroptosis through GPX4 inactivation and iron oxidation. *Nat. Chem. Biol.* **2018**, *14*, 507–515. [CrossRef]
14. Briley, W.E.; Bondy, M.H.; Randeria, P.S.; Dupper, J.; Mirkin, C.A. Quantification and real-time tracking of RNA in live cells using Sticky-flares. *Proc. Natl. Acad. Sci. USA* **2015**, *112*, 9591–9595. [CrossRef]
15. Grabar, K.C.; Freeman, R.G.; Hommer, M.B.; Natan, M.J. Preparation and Characterization of Au Colloid Monolayers. *Anal. Chem.* **1995**, *67*, 735–743. [CrossRef]
16. Hong, M.; Sun, H.; Yang, Q.; Cheng, S.; Yu, S.; Fan, S.; Li, C.; Cui, C.; Tan, W. A microRNA-21-responsive doxorubicin-releasing sticky-flare for synergistic anticancer with silencing of microRNA and chemotherapy. *Sci. China-Chem.* **2021**, *64*, 1009–1019. [CrossRef]
17. Stryer, L.; Haugland, R.P. Energy transfer: A spectroscopic ruler. *Proc. Natl. Acad. Sci. USA* **1967**, *58*, 719–726. [CrossRef]
18. Hangauer, M.J.; Viswanathan, V.S.; Ryan, M.J.; Bole, D.; Eato, J.K.; Matov, A.; Galeas, J.; Dhruv, H.D.; Berens, M.E.; Schreiber, S.L.; et al. Drug-tolerant persister cancer cells are vulnerable to GPX4 inhibition. *Nature* **2017**, *551*, 247–250. [CrossRef]

19. Zhang, Y.; Fu, X.; Jia, J.; Wikerholmen, T.; Xi, K.; Kong, Y.; Wang, J.; Chen, H.; Ma, Y.; Li, Z.; et al. Glioblastoma Therapy Using Codelivery of Cisplatin and Glutathione Peroxidase Targeting siRNA from Iron Oxide Nanoparticles. *ACS. Appl. Mater. Interfaces* **2020**, *12*, 43408–43421. [CrossRef]
20. Crooke, S.T.; Witztum, J.L.; Bennett, C.F.; Baker, B.F. RNA-targeted therapeutics. *Cell. Metab.* **2018**, *27*, 714–739. [CrossRef]
21. Zou, Y.; Palte, M.J.; Deik, A.A.; Li, H.; Eaton, J.K.; Wang, W.; Tseng, Y.; Deasy, R.; Kost-Alimova, M.; Dančík, V.; et al. A GPX4-dependent cancer cell state underlies the clear-cell morphology and confers sensitivity to ferroptosis. *Nat. Commun.* **2019**, *10*, 1617–1622. [CrossRef] [PubMed]
22. Liang, C.; Zhang, X.; Yang, M.; Dong, C. Recent progress in ferroptosis inducers for cancer therapy. *Adv. Mater.* **2019**, *31*, e1904197. [CrossRef] [PubMed]
23. Zhang, Y.; Tan, H.; Daniels, J.D.; Zandkarimi, F.; Liu, H.; Brown, L.M.; Uchida, K.; O'Connor, O.A.; Stockwell, B.R. Imidazole ketone erastin induces ferroptosis and slows tumor growth in a mouse lymphoma model. *Cell Chem. Biol.* **2019**, *26*, 623–633. [CrossRef] [PubMed]
24. Zhang, W.; Gai, C.; Ding, D.; Wang, F.; Li, W. Targeted p53 on small-molecules induced ferroptosis in cancers. *Front. Oncol.* **2018**, *8*, 507–511. [CrossRef]
25. Chen, L.; Qiao, L.; Bian, Y.; Sun, X. GDF15 knockdown promotes erastin-induced ferroptosis by decreasing SLC7A11 expression. *Biochem. Biophys. Res. Commun.* **2020**, *526*, 293–299. [CrossRef]

biosensors

MDPI

Article

Dual Classification Approach for the Rapid Discrimination of Metabolic Syndrome by FTIR

Kateryna Tkachenko [1], Isabel Esteban-Díez [1], José M. González-Sáiz [1], Patricia Pérez-Matute [2] and Consuelo Pizarro [1,*]

[1] Department of Chemistry, University of La Rioja, 26006 Logroño, Spain
[2] Infectious Diseases, Microbiota and Metabolism Unit, Infectious Diseases Department, Center for Biomedical Research of La Rioja (CIBIR), 26006 Logroño, Spain
* Correspondence: consuelo.pizarro@unirioja.es; Tel.: +34-941299626

Abstract: Metabolic syndrome is a complex of interrelated risk factors for cardiovascular disease and diabetes. Thus, new point-of-care diagnostic tools are essential for unambiguously distinguishing MetS patients, providing results in rapid time. Herein, we evaluated the potential of Fourier transform infrared spectroscopy combined with chemometric tools to detect spectra markers indicative of metabolic syndrome. Around 105 plasma samples were collected and divided into two groups according to the presence of at least three of the five clinical parameters used for MetS diagnosis. A dual classification approach was studied based on selecting the most important spectral variable and classification methods, linear discriminant analysis (LDA) and SIMCA class modelling, respectively. The same classification methods were applied to measured clinical parameters at our disposal. Thus, the classification's performance on reduced spectra fingerprints and measured clinical parameters were compared. Both approaches achieved excellent discrimination results among groups, providing almost 100% accuracy. Nevertheless, SIMCA class modelling showed higher classification performance between MetS and no MetS for IR-reduced variables compared to clinical variables. We finally discuss the potential of this method to be used as a supportive diagnostic or screening tool in clinical routines.

Keywords: metabolic syndrome; infrared spectroscopy; point of care; metabolic signatures; chemometrics; classification strategy; health and wellbeing monitoring

Citation: Tkachenko, K.; Esteban-Díez, I.; González-Sáiz, J.M.; Pérez-Matute, P.; Pizarro, C. Dual Classification Approach for the Rapid Discrimination of Metabolic Syndrome by FTIR. *Biosensors* **2023**, *13*, 15. https://doi.org/10.3390/bios13010015

Received: 4 November 2022
Revised: 12 December 2022
Accepted: 21 December 2022
Published: 23 December 2022

1. Introduction

The high prevalence of non-communicable diseases (NCD) in adults is reflected in increased costs for public health systems worldwide [1]. Among these NCD, metabolic syndrome (MetS) plays a significant role. MetS is often associated with an increased risk of diabetes and cardiovascular disease, resulting in increased incidence of morbidity and mortality and reduced quality of life [2–6]. Thus, the commensurate prevalence of metabolic syndrome burdens national health expenditure, representing a significant socio-economic problem, particularly in low- and middle-income countries [7–10]. However, MetS is a multifactorial disorder accompanied by conflicting opinions on its definition [11–13]. In particular, many different definitions have been proposed to describe MetS in adults. The main discrepancies were associated with inclusion and exclusion criteria adopted according to the World Health Organization (WHO), National Cholesterol Education Program (NCEP), Adult Treatment Panel III (ATPIII), and International Diabetes Federation (IDF). Finally, in 2009, the definition for metabolic syndrome was harmonised [14]: MetS is a disease formed by metabolic and vascular abnormalities, namely insulin resistance (IR), visceral adiposity, atherogenic dyslipidaemia, and oxidative and endothelial dysfunction. These risk factors easily predispose hyperglycaemia and hypertension, atherosclerotic vascular diseases and viral infection [15–18].

Given the complex and intertwined nature of MetS, it would be utopian to think that a single biomarker could define it unambiguously. Thus, parameters concerned around central obesity (waist circumference (WC)), hypertension (blood pressure), atherogenic dyslipidaemia (small low-density lipoprotein (LDL) and levels of high-density lipoprotein (HDL) cholesterol), and insulin resistance (fasting glucose levels) are usually measured to evaluate MetS diagnosis [19]. Due to the heterogeneity of these factors, people affected by metabolic syndrome are three times more likely to suffer acute myocardial infarction, cerebrovascular events, diabetes, or stroke. In addition, they have higher mortality rates [20]. Besides the economic impact, misdiagnosis or tardive diagnosis could lead not only to inefficient treatment outcomes but even to significant dysfunctions such as cancer [21,22]. Thus, early and proper diagnosis plays a crucial role in delaying the pathology's onset or progression as much as possible and improving a patient's condition.

Today, MetS diagnosis is based on several steps such as measuring metabolic markers of insulin resistance and other indices of metabolic syndrome (triglycerides, HDL cholesterol levels, and blood glucose) that are obtainable from routine clinical biochemistry laboratories, whereas blood pressure is measured in primary care [23]. The collection and analysis of samples also entails a waiting time for laboratory results and additional time for a new medical consultation. Although the proposed definition of MetS shares some common features, the clinical diagnosis lacks standardisation. On that basis, it was proposed that individuals showing a combination of any three out of these five simple clinical criteria were likely to be characterised by insulin resistance. Prospective analyses have also shown that any combination of these factors was predictive of an increased risk of both type 2 diabetes and cardiovascular disease. First, it is still challenging to identify a unified criteria for MetS applicable across all ethnicities. In addition, the contribution of each parameter seems to have different importance based on the evaluation adopted in each clinical environment (e.g., diagnosis focussed on glucose tolerance instead of obesity cut-offs). Moreover, there is variation in the cut-off values of diagnostic inclusion criteria ($\geq$140/90 mmHg according to WHO vs. $\geq$130/85 mmHg according to ATP III for blood pressure). The application of MetS diagnosis in clinical practice could also be compromised, since most patient registries have missing data, limiting a study's accuracy or leading to false-positive results. In addition, measurements such as WC, one of the predominant parameters for defining MetS, are not always feasible in patients because the diagnosis can often be limited by the patient's inability to perform a complete physical examination.

Given these perspectives, the need for standardised clinical diagnostic tools and protocols becomes imperative in the prevention and diagnosis of MetS. For this reason, analysing global metabolic profiles instead of disparate clinical measurements could be essential in shedding light on MetS disarrangements. A multifactorial and complex pathology such as MetS seems to require an approach from a holistic functional perspective, so an analysis of metabolic profiles reflecting the global clinical status of a patient could represent a suitable alternative.

By now, metabolomics plays a key role as a powerful analytical tool that has been widely applied to investigate plenty of disorders and disarrangements [24–26]. Metabolomics analysis has the potential to discover biomarkers and allow for the detection of a wide range of metabolites. In recent years, there has been a great interest in extracting biomarkers from biofluids and, considering that blood is a biofluid containing numerous valuable metabolic information, it seems that it in particular, it appropriately reflects metabolic changes and disarrangements during disease initiation or progression [27,28]. In this context, techniques based on vibrational spectroscopy are particularly suitable as sample preparation is simple, non-invasive, rapid, and low-cost [29]. Therefore, the Fourier transformed infrared spectroscopy (FTIR) technique has been established as a reliable analytical tool in metabolomic-based studies [30–34]. Moreover, another significant advantage resides in the fact that FTIR is ideally suitable for acquose matrices such as blood [35,36]; the instrument requires the collection of only one blood sample, with little or almost null pre-treatment. In this study, we proposed an FTIR-based method that investigates many components at a time, which are registered as spectral signatures. The development of a chemometric strategy capable of extrapolating the most significant infrared (IR) signatures plays a crucial role in this study, since each spectrum is unique for every patient and reflects their metabolic status. Non-targeted metabolomic studies, such as the one presented here, aim to extract the metabolic signatures instead of individual biomarkers with limited potential, and this permits the classification of patients according to their molecular patterns, reflecting clinical/pathological conditions such as MetS or no MetS.

This method could greatly support clinicians, capturing the complexity of the MetS metabolic profile when the clinical indicators are missing or lacking sufficient discriminative power, revealing the globality of physiological disturbances. We do not want to underestimate the importance of clinical diagnosis at any time. Still, our main aim is to propose an alternative analytical strategy that could be of great diagnostic relevance and support, limiting the time and cost of clinical measurements.

2. Materials and Methods

2.1. Study Population

A total of 105 plasma samples from anonymous donors were recruited from Infectious Disease Area, Center for Biomedical Research of La Rioja (Logroño, Spain). This study was approved by the Committee for Ethics in Drug Research in La Rioja (CEImLAR) (23 April 2013, reference number 121) and a written informed consent was achieved from all participants. The patients were evaluated by the NCEP-ATP-III scale and, if eligible, were assigned to a metabolic syndrome category. MetS was defined as the concomitant presence of at least three of the following risk factors: elevated TGL ($\geq$150 mg/dL), low concentrations of the fraction HDL cholesterol (<50 levels mg/dL in women or <40 mg/dL levels in men), increased WC ($\geq$88 cm in women or $\geq$102 cm in men), elevated blood pressure (>130/85 mmHg), and elevated fasting glucose (>110 mg/dL or diabetes) [37]. Thus, the patients were divided into two groups by the criteria of MetS: 19 patients tested as MetS positive and 86 as MetS negative. The patients enrolled in this study were also characterised by the presence of viral load through serological evidence of HIV or co-infection of HIV/HCV. A correct distribution between patients with and without infection in both categories has been ensured to not introduce bias in future models developed for diagnosing MetS.

2.2. Sample Collection

Once drawn, the venous blood samples were centrifuged at 2200$\times$ g for 15 min at 4 °C and the obtained plasma were transferred into a clean Eppendorf tube. Aliquots of 200 µL of each sample were stored at −80 °C until the day of the analysis. Before FTIR measurements, plasma samples were defrosted during the night according to the optimised ultrasound-based protocol for lipidomic analyses developed in our research group [38].

2.3. Method

FTIR spectroscopy measurements were performed by a Spectrum-One ABB Miracle Type MB3000 FT-IR Spectrophotometer using a PerkinElmer liquid cell (Omni Cell, Specac Ltd., Orpington, UK) with CaF_2 windows separated with a 50 µm Mylar spacer. The spectra from 25 µL of each plasma sample were recorded in the mid-IR region (4000–300 cm^{-1}) in triplicate. A mean spectrum was subsequently obtained from the replicates recorded for each plasma sample. The sample temperature was maintained at 23.0 $\pm$ 1.0 °C, and a constant N_2 purge was applied for atmospheric water vapour and CO_2 suppression. A resolution of 2 cm^{-1} was obtained using 32 scans. In order to monitor the stability and reproducibility of the analytical system, quality control (QC) samples were processed similarly to the actual samples and inserted regularly. In addition, the instrument performance was verified at the beginning of each day of data collection using PE-specific reference standards.

2.4. Data Analysis

After data acquisition, the processing and computational analysis of raw metabolic data was performed using Unscrambler (version X 11.0, Camo ASA, Oslo, Norway), V-Parvus (version PARVUS2011, Michele Forina, Genoa, Italy), and Matlab (MATLAB 9.4

R2018a). Two different regions of the mid-IR spectrum were analysed: the first region examined was the biochemical "fingerprint region" at 1500–1050 cm^{-1}, and the second was a higher region at 2950–2700 cm^{-1}. Remaining wavenumber ranges, as they were affected by signal saturation effects caused mainly by strong water absorptions or noise, were removed, and not considered for further analysis. Given the high dimensionality of biological spectral data, many disturbing factors influence the spectral data acquisition, such as random noise, baseline distortions, or light scattering. Thus, the pre-processing step is imperative in analysis to reduce these factors. To compensate for instrumental artefacts and sample to sample variations, different pre-processing methods were evaluated individually or in combination to minimise the adulterant-unrelated variability, namely derivatives (e.g., Savitzky–Golay (S–G) first and second derivatives), standard normal variate (SNV), and extended multiplicative scatter correction (EMSC). Thus, better resolution of overlapping peaks and decreased scatter effects were ensured after applying the combination (S–G) smoothing and SNV.

The entire data set was split into two independent subsets to develop and validate the classifications proposed: a training set with 95 samples (used to optimise and develop the classification rules and models) and a test set with ten samples (never used in the construction of the classification but to evaluate their actual predictive ability). The test set used was the same for all methods applied and classifications developed. As a result, the smoothed and normalised output tables were always centred before additional multivariate analysis and classification algorithms.

3. Results and Discussion

After careful pre-processing, FTIR measurements were submitted for further multivariate analysis. Thus, five measured clinical variables and a total of 838 spectra variables over the wavelength ranges of 1583–1050 cm^{-1} and 2973–2700 cm^{-1} collected from 105 patients were included. The two main categories of this study were patients with and without metabolic syndrome, i.e., MetS and no MetS, respectively.

3.1. Descriptive Statistics

Herein, an analysis was performed based on the distribution of five clinical parameters. It should be noted that one of the most critical clinical measurements, waist circumference, was not included in this study because most patients had missing data in the clinical register. Therefore, only parameters that were available for all patients have been used for the further comparative classification step. Thus, the descriptive statistics were calculated to analyse the distribution of clinical data in a box and whisker plot (Figure 1). The plot shows that TGL values seem to have more influence and variability between the two categories of patients; indeed, MetS patients have significantly higher values ranging from a minimum of 33 to 338 (mg/dL). The general distribution trend indicates that MetS patients also have slightly higher diastolic and systolic blood pressure values and glucose levels, whereas HDL values are lower, ranging from 25 to 95 (mg/dL). Table 1 shows the ranges of the collected values with the respective medians between the two categories.

Table 1. The distribution of the clinically measured parameters in MetS and no MetS patients expressed in mg/dL and in mmHg.

Category		MetS			No MetS	
Clinical Parameters	Max	Min	Mean	Max	Min	Mean
Systolic blood pressure	174	120	136	178	94	126
Diastolic blood pressure	109	75	87	115	61	79
Triglycerides	338	88	242	215	33	109
HDL	58	25	37	95	29	55
Glucose	164	82	114	123	63	91

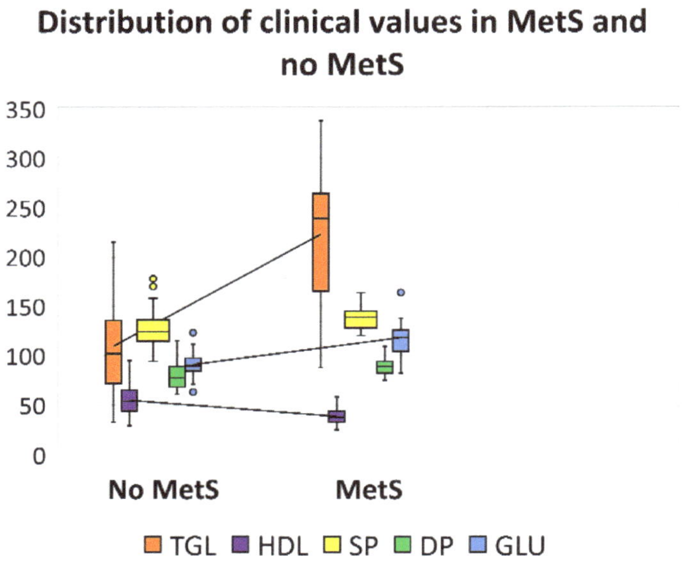

Figure 1. Box and whisker plot showing the distribution of clinical values levels in patients with MetS and no MetS. The line located in the middle of the box represents the median and is used to better visualise the differences between clinical parameters: triglycerides (TGL) levels are displayed in orange (■); high density lipoprotein (HDL) in violet (■); systolic pressure (SP) in yellow (■); diastolic pressure (DP) in green (■); and glucose (GLU) in blue (■).

3.2. Exploratory Analysis with PCA

An unsupervised pattern recognition method based on principal component analysis (PCA) was performed for the initial data overview and to investigate any possible clustering of samples based on five collected clinical parameters and 838 spectral variables, respectively.

The PCA score plot of clinical parameters, with 50.46% of explained variance by PC1, displays evident clustering according to known categories, delimitated by the parallel to the bisector of the second quadrant (Figure 2). Whereas PCA performed on pre-treated IR spectra accounted for 83.12% of explained variability on the PC1, evidenced by very subtle clustering between known categories (Figure 3).

In both cases, the first PCs explained most of the data's variability. The distribution of samples in principal component space suggests that it only seems possible to address subsequent, direct discrimination in the case of analysis of clinical parameters. Thus, parameters such as TGL and GLU majorly contributed to the segregation of no MetS from MetS and the values of HDL contributed to the separation of MetS from no MetS, as was shown in preliminary analysis by descriptive statistics. No evident clustering among the two main categories was observed performing PCA on spectral variables; only a few outliers were determined and excluded from further analysis. The high degree of overlapping features among the two classes was expected, as most blood components are common in all individuals. This also indicates the need to perform a selection of relevant spectral variables, closely related to clinicopathological parameters of prognostic importance in MetS. Therefore, other chemometric strategies were used to investigate and highlight metabolomic differences in metabolic syndrome using IR spectra.

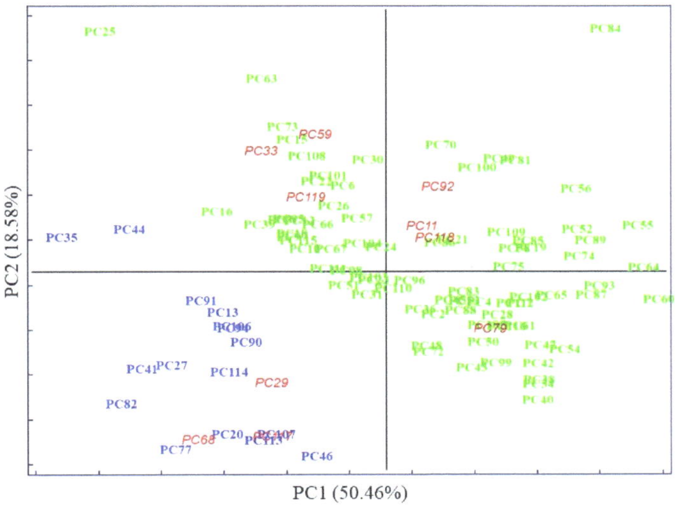

Figure 2. Scores for the plasma samples on the first two principal components explaining the variability in the dataset of five measured clinal parameters. The samples are labelled according to their specific pathology: no MetS (■), MetS (■), and external test samples (■).

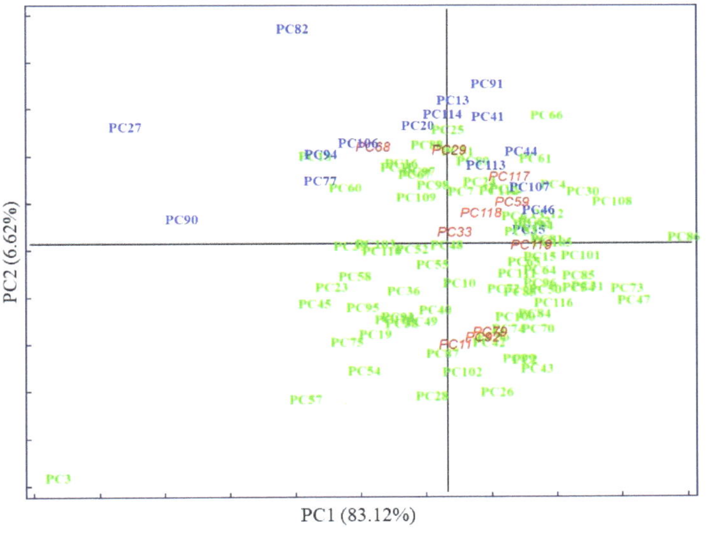

Figure 3. Scores for the plasma samples on the first two principal components explaining the variability in the IR spectral dataset. The samples are labelled according to their specific pathology: no MetS (■), MetS (■), and external test samples (■).

3.3. Supervised Techniques

The selection of variables in tandem with classification methods to extract reduced IR fingerprints that reflect the metabolic profiles of patients for a potential MetS diagnosis was studied. Therefore, a dual approach was applied based on a classification method on the one hand and a class modelling method on the other.

For its part, discriminant techniques focus on the differences between samples belonging to different categories, dividing the multidimensional space into as many subregions as the number of the considered classes. As a result of this work principle, every tested sample would always be assigned to one of the predefined categories, even in the case where an analysed sample truly belongs to a class not considered in the study. Regarding the above, it makes good sense to evaluate the application of a discriminant classification strategy in a two-class (binary) classification problem such as the one addressed in this paper. In particular, linear discriminant analysis (LDA), the most widely used classification algorithm, was used.

On the other hand, in contrast to class discrimination, class modelling approaches exploit similarities among inter-category samples to construct an individual model for every class independently from the others. Consequently, the developed class models may not entirely cover the original multivariate space. This fact opens the door to different assignment scenarios depending on whether a sample falls clearly into a single class region (so that it is assigned to that) or if it falls in overlapping regions (leading to a confusing classification in multiple classes), and, finally, when a sample falls outside every class model constructed (predicted as member of none of the considered categories). Therefore, due to their specific properties, modelling techniques, such as soft independent modelling by class analogy (SIMCA), are suitable for classification problems in which the emphasis is placed on a particular class of interest, as may be the case here with the MetS category.

3.3.1. SELECT

Considering that IR data presents high dimensionality, eliminating the futile features due to noise and identifying the relevant and important variables to be applied in the following classification steps was imperative. Thus, the stepwise orthogonalization of predictors (SELECT) algorithm [39,40] was prioritised among other variable selection techniques since it enabled us to optimise discrimination by simultaneously performing feature selection and classification. Moreover, thanks to its stepwise decorrelation procedure, SELECT also avoids the presence of redundant information in the subset of selected significant predictors. In addition, it has previously demonstrated its accurate prediction ability in selecting the most important variable for the discrimination of pathological status [41,42]. Thus, SELECT was applied to extract the most significant wavenumbers from the IR dataset, providing input features for a further dual-classification approach. Based on the commonly established rule, the number of training objects selected was always at least three times greater than the number of finally selected wavenumbers. An in-depth study of the literature is encouraged to understand the algorithm's rules [43].

3.3.2. LDA on Clinical Parameters

LDA is a well-known and extensively applied powerful supervised chemometric classification technique [44]. Based on LDA classification rules, the objects are always classified in one of the predefined classes.

LDA of five clinical parameters, built by leave one out (LOO) cross-validation, was performed to evaluate the feasibility of this classification methodology to differentiate between MetS and no MetS patients. Excellent discrimination among categories was achieved, providing a 100% level of correctly classified samples for no MetS subjects and patients with metabolic syndrome, respectively. Satisfactory external prediction performances ranging from 98.73% to 100% were achieved for both categories (within one no MetS subject classified as MetS), respectively (Table 2). Furthermore, a clear interclass separation achieved between these main categories can also be visually appreciated in the corresponding discriminative histogram (Figure 4). This classification performance was almost predictable since the PCA results already showed a clear clustering between the two groups.

The object belonging to the category MetS which was classified as no MetS was characterised by the following clinical parameters: 213 mg/mL of TGL, 76 mg/mL of HDL, 139 mmHg of SP, 83 mmHg of DP, and 102 mg/mL of GLU. As we can see, two out of five parameters have increased values, and the DP parameter is very close to the cut-off value, which is 85 mmHg based on the NCEP-ATP-III scale. Thus, this patient might instead be classified as MetS positive, presenting almost three out of five clinical parameters with augmented values. In addition, as we said above, the TGL parameter has a major contribution, among other parameters, to MetS classification. Thus, the plausible explanation could be that this subject, who has greater values of TGL, is more likely to be classified as MetS by LDA rather than no MetS. However, as we highlighted before, the eligibility criteria can be very insidious and create confusion and misassignment, worsening and delaying the patients' well-being.

Table 2. Results of LDA classification performance on clinical parameters.

Clinical Parameters	Classification (%)	External Prediction (%)	Total Rate (%)
MetS	100	100	100
No MetS	100	98.73 (1)[1]	99.36
Total rate	100	98.94	99.47

[1] The one corresponds to one misclassified subject in cross-validation.

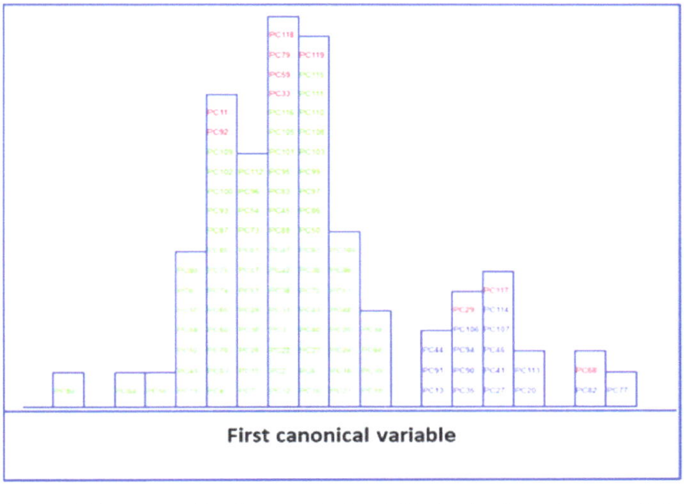

Figure 4. Histogram of the first canonical variable for the discrimination of MetS (■) and no MetS (■) patients within included (■) test set, after performing LDA in the stratification approach based on clinical parameters (*y*-axis indicates the maximum discrimination power between categories).

3.3.3. SELECT-LDA on IR Wavenumbers

Likewise, LDA on the IR dataset, containing 838 wavenumbers, was also performed. Before LDA analysis, as explained above, SELECT was applied to extract those predictor variables correlated with the discrimination between categories here considered. Therefore, based on the SELECT rules, 20 selected spectra variables were decorrelated from other signals and used for LDA. The 20 selected features showed an outstanding classification performance and the results were higher in performance than LDA results on clinical parameters, achieving 100% in classification and external prediction, respectively. The results of the SELECT LDA performance are displayed in Table 3. The suitability of the classification strategy applied to reduced IR plasma signatures can be visually appreciated in Figure 5. A discriminative histogram shows a clear group separation on the first canonical variable.

Table 3. Results of SELECT LDA classification performance on 20 IR selected spectral variables.

Clinical Parameters	Classification (%)	External Prediction (%)	Total Rate (%)
MetS	100	100	100
No MetS	100	100	100
Total rate	100	100	100

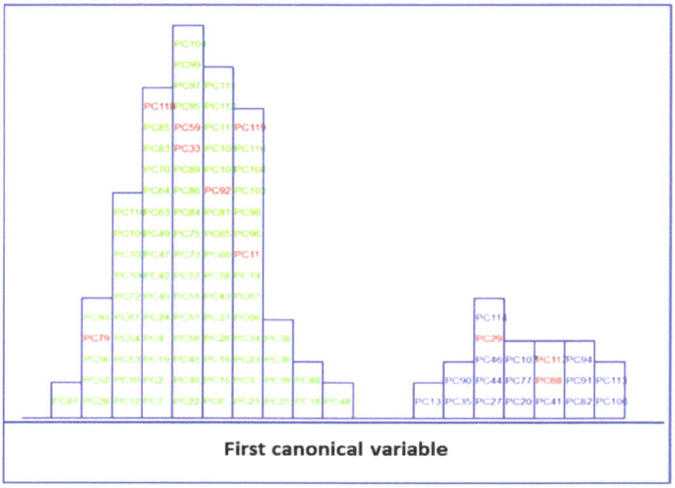

Figure 5. Histogram of the first canonical variable for the discrimination of MetS (■) and no MetS (■) patients within the included (■) test set, after performing SELECT-LDA in the stratification approach based on 20 IR variables (*y*-axis indicates the maximum discrimination power between categories).

3.3.4. SIMCA

In an attempt to go one step further in this classification strategy, it was decided to build optimised class models based on clinical parameters and the subset of reduced IR signatures selected by SELECT. SIMCA often outperforms other classification methods, where a new sample will always be classified in one of the predefined categories. Classification methods such as LDA are based on the development of classification rules and delimiters between classes, whereas in class models, significance limits are built for the specified classes. These limits define the membership parameters for each class; thus, an unknown sample can be classified as not belonging to any defined categories because it is not included in any of its class spaces. SIMCA class modelling uses the number of true/false positives and negatives and statistics, showing the ability of a classification model to recognise class members (*sensitivity* or true positive rate) and showing how good the model is for identifying strangers (*specificity* or true negative rate). Moreover, SIMCA class modelling is often used to describe the class structure of the data set, requiring little or no prior assumptions to build the model.

On applying SIMCA, independent PCA modelling is performed for each class; each sample is fitted in a PCA model to check the separation between classes [45]. This model uses the optimal number of principal components that best describes and groups an individual class. This model can then be used to classify new samples whose class is unknown. The principal components are obtained usually using the NIPALS (non-iterative partial least squares) algorithm after separate autoscaling of the data. Finally, the models built for the different classes are compared by studying their differences and analogies [46]. Each class is modelled independently; thus, it is sensitive to the quality of the data used to generate the principal component models for each class in the training set (at a 5% significance level).

SIMCA on Clinical Parameters

Herein, SIMCA modelling was performed on five clinical parameters (Table 4). A class modelling of five clinical parameters of MetS was built using 4PCs for the inner space of classes, achieving satisfactory results in both internal prediction (LOO) and external prediction 98.95%. SIMCA builds a mathematical model of the category with its principal components and a sample is accepted by the specific category if its distance to the model is not significantly different from the class residual standard deviation. The results of SIMCA modelling can be visually appreciated by a Cooman's Plot, representing the samples' distances against each of the two models. The Cooman's plots were built considering a 95% confidence level to define the class space and the unweighted augmented distance. This diagram is an effective visual representation that directly indicates the quality of the model constructed with the magnitude of the distance between categories. Thus, the distances to the principal component models and SIMCA approximation in a two-class problem for the class of MetS and no MetS are plotted in Figure 6. No clear outliers were observed, but several samples that fall into the joint space of both categories belong mainly to the MetS category. This relatively large number of samples plotted in the class-space common (overlapping) to the two models representing MetS and no MetS patients, as well as the considerable amount of no MetS samples located near their class boundary, suggest potential specificity problems associated with this classification approach based on clinical parameters. Therefore, the distribution of some samples from the MetS category in the area of relative indecision (small left quadrant) could be due to the unequivocal diagnostic parameters defining metabolic syndrome. In fact, these patients have three out of five altered parameters not necessarily similar. In addition, some parameters may be much less marked than others, confounding the decision about their location inside the model.

Table 4. The values of discriminant and modelling powers of clinical parameters after SIMCA class-modelling.

Clinical Parameters	Discriminant Power	Modelling Power	
		Category MetS	Category No MetS
Systolic blood pressure	1.99	0.70	0.73
Diastolic blood pressure	2.01	0.70	0.73
Triglycerides	2.18	0.94	0.96
HDL	2.34	0.79	0.94
Glucose	2.36	0.84	0.97

The data modelling power (MP) and discriminatory power (DP) of the SIMCA class modelling of clinical parameters are presented in Table 4. The MP describes how well a variable helps each principal component to model variation in the data, and discriminatory power (DP) describes how well a variable helps each principal component model to classify samples in a training set. The first detail that can be noticed is that, comparably, the MP in no MetS is consistently higher for all parameter pairs. This was expected as the distribution of the values of clinical parameters for each class of patients was significantly different. Nevertheless, the values of TGL have the highest modelling power in both MetS and no MetS categories, with values of 0.94 and 0.96, respectively. This ability of TGL to discriminate between the two groups is justified by previous studies, as metabolic syndrome patients should have significantly higher TGL values. This difference in modelling power is especially remarkable by the measured glucose (0.97 vs. 0.84) and HDL (0.94 vs. 0.79). In addition, clinical parameters such as glucose and HDL also showed significant discriminant power, with values of 2.63 and 2.58, respectively. These two parameters are also perfectly in line with the data collected from our patients. The MetS group is characterised by high glucose and low HDL values. These same parameters are often responsible for the presence or future development of comorbidities in patients such as diabetes, cardiac disease, and obesity. Other clinical parameters seem to contribute less to the principal component models; indeed, no significant difference was observed in the values distribution of SP or DP between the two categories.

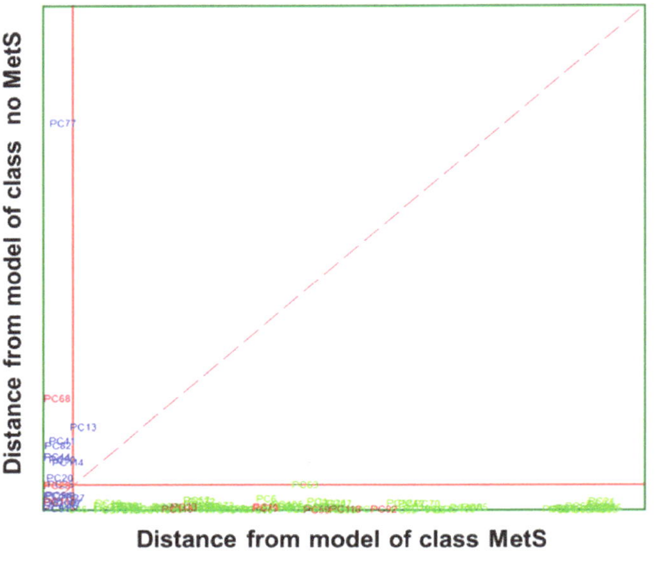

Figure 6. Cooman's plot displaying the results obtained by applying SIMCA class-modelling to clinical parameters: MetS (■) and no MetS (■) patients within the included (■) test set. The red solid line indicates a confidence level for class space at 95%. The red dashed line indicates equal class distance.

SELECT-SIMCA on IR Wavenumbers

The best recognition ability (percentage of the samples in training set correctly classified during the modelling step) afforded by SIMCA was achieved by only ten of 20 previously selected wavenumbers by SELECT, providing 98.94% in classification and 95.79% in external prediction, respectively. Interestingly, eight out of ten selected wavenumbers belong to the "fingerprint region", which reflects the production of characteristic perturbations in the metabolome and other such variations. The absorption pattern in this area is highly complex; that same inherent complexity makes it unique for each sample and reflects its pathophysiological status. Thus, eight of the selected IR spectral wavenumbers may reflect the current status of the organism and could be directly correlated with the presence or absence of the disease. The results of SIMCA performance applied to clinical variables and to reduced number of IR spectral variables are summarised in Table 5.

Table 5. The results of SIMCA class-modelling performance on clinical parameters and ten selected IR spectral variables.

Variables	Classification (%)	LOO (%)	CV Efficiency (%)	Efficiency Forced Model (%)	Total Rate (%)
5 clinical measurements	98.59	97.18	87.05	95.68	100
10 IR selected wavenumbers	97.18	94.37	87.92	97.86	100

A Cooman's plot is presented to show discrimination between the two MetS categories of IR variables (Figure 7), where the distance to the PC models for MetS and no MetS are displayed. Compared to the Cooman's plot of clinical parameters, it is observed that there is better separation and discrimination between categories. The Cooman's plot showed a high degree of interclass specificity and a patently clear separation between class models, with a significant improvement from the models constructed from available clinical parameters to those constructed from IR variables. The no MetS patients appear evidently segregated and concentrated forming a dense cluster at large distances from the model of MetS class. Likewise, the vast majority of MetS samples fall clearly and univocally into their class region, far from the class limit for the no MetS model. Furthermore, the single MetS sample located in the inconclusive classification region is virtually placed above the membership threshold.

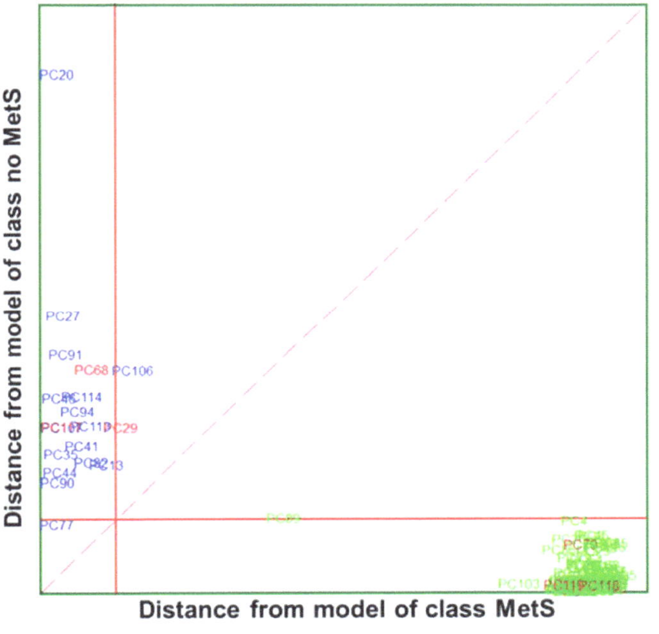

Figure 7. Cooman's plot displaying the results obtained by applying the SELECT-SIMCA class-modelling to ten selected IR signals: MetS (■) and no MetS (■) patients within included (■) test set. The red solid line indicates a confidence level for class space at 95%. The red dashed line indicates equal class distance.

From ten selected wavenumbers, the highest discriminant power (5.87) was obtained by the 1133.09 cm^{-1} spectra variable from the "fingerprint region" (Table 6), followed by 4.31 for 1557.40 cm^{-1} and 4.29 for 2948.94 cm^{-1} from the higher spectral region. The average discriminant power for IR variables is higher compared to DP values obtained with SIMCA modelling of clinical parameters, indicating the increased suitability of the method compared to those using values obtained from clinical measurements. Likewise, the contribution of IR variables to the model variation was of major strength compared to clinical parameters. Thus, all the selected variables contributed equally to marking the difference between MetS and no MetS with an MP equal to 1.00. Furthermore, the distance between classes was 5.19, significantly higher than in the case of SIMCA class modelling applied to clinical parameters (4.26). These results highlight that the proposed method outperformed in accuracy and specificity of the evaluation parameters used in clinical practice. Since the clinical diagnosis of metabolic syndrome lacks standardisation, the results of the obtained model capacity could greatly support clinical decisions, for example, in terms of exclusion and inclusion evaluation criteria for MetS discrimination.

Table 6. Discriminative and modelling powers of ten selected spectra variables after SELECT-SIMCA class modelling.

Wavenumber (cm^{-1})	Discriminant Power	Modelling Power	
		Category MetS	Category No MetS
2860.22	3.77	1.00	1.00
1423.36	4.23		
1562.22	3.66		
1578.61	3.75		
1108.98	3.70		
1316.32	3.64		
2948.94	4.29		
1557.40	4.31		
1133.09	5.86		
1247.85	3.58		

Our principal aim was to obtain optimal segregation between patients without additional clinical, physical, or ethnic data, and this goal was achieved.

3.3.5. Biochemical Reasoning of Ten Extracted Signals

Herein, we presented a simple, non-invasive, low-cost FTIR-based method for rapid discrimination between MetS and no MetS patients. The use of FTIR spectroscopy is gaining momentum for diagnosis of multiple disorders, from infectious diseases such as hepatitis C and B viruses or malaria to cancers [47–53]. Due to its ease of use and portability, the potential for using FTIR techniques in clinical environments is within reach. Our strategy extracted the metabolic signatures, instead of individual biomarkers with limited potential, that permit the classification of patients according to molecular patterns. Thus, the FTIR technique provided an overview of spectral changes associated with lipid, protein, or carbohydrate metabolisms.

Ten out of twenty previously selected wavenumbers showed higher discriminant power than clinical parameters. Thus, among these, influential bands at 1578.61, 1562.22, and 1557.40 cm^{-1} could be assigned to [δ (N-H) + ν (C-H)] of the amide II region of proteins. These discriminative signals may suggest some link with HDL lipoproteins, which showed significant influence among five clinical factors for the classification of MetS and no MetS subjects. Likewise, the higher absorbance in peaks at 2860.22 cm^{-1} and 2948.94 cm^{-1} could be attributed to CH3 and CH2 sym. stretching of lipids or carbohydrates, which is perfectly congruent with the formulated theories about MetS impairments and their possible implication in the disease. Moreover, as discussed above, TGL and GLU levels seemed to have more influence and variability between the two categories of patients; thus, these attempted assignments properly reflect the actual situation of the patient's metabolism. In addition, the variable at 1133.09 cm^{-1} could be associated with stretching C-O/C-O(H) of carbohydrates or proteins, since it was already shown that the parameters such as glucose or HDL have remarkable modelling and discriminant powers compared to other measured factors.

In this study, the selected spectral biomarkers perfectly reflect the clinical reality of the patient's metabolic profile. Thus, the explanation of the most significant spectral bands confirms the potential of FTIR spectroscopy to deal with such a complex disorder as MetS.

4. Conclusions

We firmly believe that this alternative analytical strategy could be of great diagnostic relevance and support for clinicians, limiting the time and cost of MetS diagnosis. Moreover, the evaluation of the metabolic profile captures the globality of physiological disturbances, whereas clinical indicators often lack sufficient discriminative power. The results indicate the possibility of rapid application of this strategy to screen for patients with metabolic syndrome. The LDA classifications and SIMCA developed models demonstrated that the spectral variables could provide the same discriminative results as measured clinical parameters. Therefore, why take five measurements when one measurement could provide the same classification ability, greatly stratifying categories of patients? The proposed FTIR method is quick, simple, and non-invasive, and it could be perfectly implemented for large scale-analysis in clinical routines. The principal limitation of this study resides in the relatively tiny sample size at our disposal. In addition, this is a cross-sectional study; therefore, no data on confounding factors (such as gender, age, or diet) were routinely included. The results of a more extensive data set would be required to strengthen the validity of the adopted classification strategy and lead to a firmer conclusion.

Author Contributions: Investigation and writing—original draft preparation, K.T.; methodology, J.M.G.-S.; data curation, I.E.-D.; resources, P.P.-M.; project administration, C.P. All authors have read and agreed to the published version of the manuscript.

Funding: This research was funded by the European Union's H2020 research grant (N· 801586) and Ministry of Science and Innovation (CTQ2011-26603).

Institutional Review Board Statement: The study was conducted in accordance with the Declaration of Helsinki, and approved by the Ethics Committee of San Pedro Hospital of La Rioja Province (CEImLAR, 23 April 2013, reference number 121).

Informed Consent Statement: Informed consent was obtained from all subjects involved in the study.

Data Availability Statement: Not applicable.

Conflicts of Interest: The authors declare no conflict of interest.

References

1. Saklayen, M.G. The Global Epidemic of the Metabolic Syndrome. *Curr. Hypertens. Rep.* **2018**, *20*, 12. [CrossRef] [PubMed]
2. Esposito, K.; Chiodini, P.; Capuano, A.; Bellastella, G.; Maiorino, M.I.; Giugliano, D. Metabolic Syndrome and Endometrial Cancer: A Meta-Analysis. *Endocrine* **2014**, *45*, 28–36. [CrossRef] [PubMed]
3. Mili, N.; Paschou, S.A.; Goulis, D.G.; Dimopoulos, M.-A.; Lambrinoudaki, I.; Psaltopoulou, T. Obesity, Metabolic Syndrome, and Cancer: Pathophysiological and Therapeutic Associations. *Endocrine* **2021**, *74*, 478–497. [CrossRef] [PubMed]
4. Esposito, K.; Chiodini, P.; Colao, A.; Lenzi, A.; Giugliano, D. Metabolic Syndrome and Risk of Cancer: A Systematic Review and Meta-Analysis. *Diabetes Care* **2012**, *35*, 2402–2411. [CrossRef] [PubMed]
5. Alexandra, K.; Konstantinos, I.; Konstantinos, S.; Alexandros, S.; Michalis, D.; Vasilios, A.; Katsimardou, A.; Imprialos, K.; Stavropoulos, K.; Sachinidis, A.; et al. Hypertension in Metabolic Syndrome: Novel Insights. *Curr. Hypertens. Rev.* **2019**, *16*, 12–18. [CrossRef]
6. Isomaa, B.; Almgren, P.; Tuomi, T.; Forsén, B.; Lahti, K.; Nissén, M.; Taskinen, M.R.; Groop, L. Cardiovascular Morbidity and Mortality Associated with the Metabolic Syndrome. *Diabetes Care* **2001**, *24*, 683–689. [CrossRef]
7. Federspil, G.; Nisoli, E.; Vettor, R. A Critical Reflection on the Definition of Metabolic Syndrome. *Pharmacol. Res.* **2006**, *53*, 449–456. [CrossRef]
8. Abebe, S.M.; Demisse, A.G.; Alemu, S.; Abebe, B.; Mesfin, N. Magnitude of Metabolic Syndrome in Gondar Town, Northwest Ethiopia: A Community-Based Cross-Sectional Study. *PLoS ONE* **2021**, *16*, e0257306. [CrossRef]
9. Motuma, A.; Gobena, T.; Roba, K.T.; Berhane, Y.; Worku, A. Metabolic Syndrome Among Working Adults in Eastern Ethiopia. *Diabetes Metab. Syndr. Obes. Targets Ther.* **2020**, *13*, 4941–4951. [CrossRef]
10. Misra, A.; Khurana, L. The Metabolic Syndrome in South Asians: Epidemiology, Determinants, and Prevention. *Metab. Syndr. Relat. Disord.* **2009**, *7*, 497–514. [CrossRef]
11. Huang, P.L. A Comprehensive Definition for Metabolic Syndrome. *Dis. Models Mech.* **2009**, *2*, 231–237. [CrossRef] [PubMed]
12. Punthakee, Z.; Goldenberg, R.; Katz, P. Definition, Classification and Diagnosis of Diabetes, Prediabetes and Metabolic Syndrome. *Can. J. Diabetes* **2018**, *42* (Suppl. 1), S10–S15. [CrossRef] [PubMed]
13. Alberti, K.G.M.M.; Zimmet, P.; Shaw, J. Metabolic Syndrome—A New World-Wide Definition. A Consensus Statement from the International Diabetes Federation. *Diabet. Med.* **2006**, *23*, 469–480. [CrossRef] [PubMed]
14. KG Alberti, R.E.S.G.P.Z.J.C.K.D.J.F.W.J.C.L.S.S. Harmonizing the Metabolic Syndrome: A Joint Interim Statement of the International Diabetes Federation Task Force on Epidemiology and Prevention. *Circulation* **2009**, *120*, 1640–1645. [CrossRef]
15. Reddy, P.; Leong, J.; Jialal, I. Amino Acid Levels in Nascent Metabolic Syndrome: A Contributor to the pro-Inflammatory Burden. *J. Diabetes Complicat.* **2018**, *32*, 465–469. [CrossRef] [PubMed]
16. Smith, M.; Honce, R.; Schultz-Cherry, S. Metabolic Syndrome and Viral Pathogenesis: Lessons from Influenza and Coronaviruses. *J. Virol.* **2020**, *94*, e00665-20. [CrossRef]

17. O'Neill, S.; O'Driscoll, L. Metabolic Syndrome: A Closer Look at the Growing Epidemic and Its Associated Pathologies. *Obes. Rev.* **2015**, *16*, 1–12. [CrossRef]
18. Lee, Y.H.; Pratley, R.E. The Evolving Role of Inflammation in Obesity and the Metabolic Syndrome. *Curr. Diabetes Rep.* **2005**, *5*, 70–75. [CrossRef]
19. Bovolini, A.; Garcia, J.; Andrade, M.A.; Duarte, J.A. Metabolic Syndrome Pathophysiology and Predisposing Factors. *Int. J. Sports Med.* **2021**, *42*, 199–214. [CrossRef]
20. Fanta, K.; Daba, F.B.; Asefa, E.T.; Chelkeba, L.; Melaku, T. Prevalence and Impact of Metabolic Syndrome on Short-Term Prognosis in Patients with Acute Coronary Syndrome: Prospective Cohort Study. *Diabetes Metab. Syndr. Obes.* **2021**, *14*, 3253–3262. [CrossRef]
21. Wiklund, P.K.; Pekkala, S.; Autio, R.; Munukka, E.; Xu, L.; Saltevo, J.; Cheng, S.; Kujala, U.M.; Alen, M.; Cheng, S. Serum Metabolic Profiles in Overweight and Obese Women with and without Metabolic Syndrome. *Diabetol. Metab. Syndr.* **2014**, *6*, 40. [CrossRef] [PubMed]
22. Esposito, K.; Chiodini, P.; Capuano, A.; Bellastella, G.; Maiorino, M.I.; Rafaniello, C.; Panagiotakos, D.B.; Giugliano, D. Colorectal Cancer Association with Metabolic Syndrome and Its Components: A Systematic Review with Meta-Analysis. *Endocrine* **2013**, *44*, 634–647. [CrossRef]
23. Lemieux, I.; Després, J.P. Metabolic Syndrome: Past, Present and Future. *Nutrients* **2020**, *12*, 3501. [CrossRef]
24. Shao, Y.; Le, W. Recent Advances and Perspectives of Metabolomics-Based Investigations in Parkinson's Disease. *Mol. Neurodegener.* **2019**, *14*, 3. [CrossRef] [PubMed]
25. González-Domínguez, R.; García-Barrera, T.; Gómez-Ariza, J.L. Combination of Metabolomic and Phospholipid-Profiling Approaches for the Study of Alzheimer's Disease. *J. Proteom.* **2014**, *104*, 37–47. [CrossRef] [PubMed]
26. Alonso, A.; Marsal, S.; Julià, A. Analytical Methods in Untargeted Metabolomics: State of the Art in 2015. *Front. Bioeng. Biotechnol.* **2015**, *3*, 23. [CrossRef]
27. Spalding, K.; Bonnier, F.; Bruno, C.; Blasco, H.; Board, R.; Benz-de Bretagne, I.; Byrne, H.J.; Butler, H.J.; Chourpa, I.; Radhakrishnan, P.; et al. Enabling Quantification of Protein Concentration in Human Serum Biopsies Using Attenuated Total Reflectance–Fourier Transform Infrared (ATR-FTIR) Spectroscopy. *Vib. Spectrosc.* **2018**, *99*, 50–58. [CrossRef]
28. Gika, H.G.; Wilson, I.D. Global Metabolic Profiling for the Study of Alcohol-Related Disorders. *Bioanalysis* **2014**, *6*, 59–77. [CrossRef]
29. Serkova, N.J.; Standiford, T.J.; Stringer, K.A. The Emerging Field of Quantitative Blood Metabolomics for Biomarker Discovery in Critical Illnesses. *Am. J. Respir. Crit. Care Med.* **2011**, *184*, 647–655. [CrossRef]
30. Finlayson, D.; Rinaldi, C.; Baker, M.J. Is Infrared Spectroscopy Ready for the Clinic? *Anal. Chem.* **2019**, *19*, 12117–12128. [CrossRef]
31. Lovergne, L.; Lovergne, J.; Bouzy, P.; Untereiner, V.; Offroy, M.; Garnotel, R.; Thiéfin, G.; Baker, M.J.; Sockalingum, G.D. Investigating Pre-Analytical Requirements for Serum and Plasma Based Infrared Spectro-Diagnostic. *J. Biophotonics* **2019**, *12*, e201900177. [CrossRef] [PubMed]
32. Maitra, I.; Morais, C.L.M.; Lima, K.M.G.; Ashton, K.M.; Date, R.S.; Martin, F.L. Attenuated Total Reflection Fourier-Transform Infrared Spectral Discrimination in Human Bodily Fluids of Oesophageal Transformation to Adenocarcinoma. *Analyst* **2019**, *144*, 7447–7456. [CrossRef] [PubMed]
33. Roy, S.; Perez-Guaita, D.; Bowden, S.; Heraud, P.; Wood, B.R. Spectroscopy Goes Viral: Diagnosis of Hepatitis B and C Virus Infection from Human Sera Using ATR-FTIR Spectroscopy. *Clin. Spectrosc.* **2019**, *1*, 100001. [CrossRef]
34. Kaznowska, E.; Depciuch, J.; Łach, K.; Kołodziej, M.; Koziorowska, A.; Vongsvivut, J.; Zawlik, I.; Cholewa, M.; Cebulski, J. The Classification of Lung Cancers and Their Degree of Malignancy by FTIR, PCA-LDA Analysis, and a Physics-Based Computational Model. *Talanta* **2018**, *186*, 337–345. [CrossRef]
35. Perez-Guaita, D.; Garrigues, S.; de la Miguel, G. Infrared-Based Quantification of Clinical Parameters. *TrAC Trends Anal. Chem.* **2014**, *62*, 93–105. [CrossRef]
36. Wang, X.; Wu, Q.; Li, C.; Zhou, Y.; Xu, F.; Zong, L.; Ge, S. A Study of Parkinson's Disease Patients' Serum Using FTIR Spectroscopy. *Infrared Phys. Technol.* **2020**, *106*, 103279. [CrossRef]
37. Baioumi, A.Y.A.A. Comparing Measures of Obesity: Waist Circumference, Waist-Hip, and Waist-Height Ratios. In *Nutrition in the Prevention and Treatment of Abdominal Obesity*; Elsevier: Amsterdam, The Netherlands, 2019; pp. 29–40.
38. Pizarro, C.; Arenzana-Rámila, I.; Pérez-del-Notario, N.; Pérez-Matute, P.; González-Sáiz, J.M. Thawing as a Critical Pre-Analytical Step in the Lipidomic Profiling of Plasma Samples: New Standardized Protocol. *Anal. Chim. Acta* **2016**, *912*, 1–9. [CrossRef]
39. Forina, M.; Lanteri, S.; Oliveros, M.C.C.; Millan, C.P. Selection of Useful Predictors in Multivariate Calibration. *Anal. Bioanal. Chem.* **2004**, *380*, 397–418. [CrossRef]
40. Pizarro, C.; Esteban-Díez, I.; Arenzana-Rámila, I.; González-Sáiz, J.M. Discrimination of Patients with Different Serological Evolution of HIV and Co-Infection with HCV Using Metabolic Fingerprinting Based on Fourier Transform Infrared. *J. Biophotonics* **2018**, *11*, e201700035. [CrossRef]
41. Pizarro, C.; Esteban-Díez, I.; Espinosa, M.; Rodríguez-Royo, F.; González-Sáiz, J.M. An NMR-Based Lipidomic Approach to Identify Parkinson's Disease-Stage Specific Lipoprotein-Lipid Signatures in Plasma. *Analyst* **2019**, *144*, 1334–1344. [CrossRef]
42. Tkachenko, K.; Espinosa, M.; Esteban-Díez, I.; González-Sáiz, J.M.; Pizarro, C. Extraction of Reduced Infrared Biomarker Signatures for the Stratification of Patients Affected by Parkinson's Disease: An Untargeted Metabolomic Approach. *Chemosensors* **2022**, *10*, 229. [CrossRef]

43. Cocchi, M.; Biancolillo, A.; Marini, F. Chemometric Methods for Classification and Feature Selection. In *Comprehensive Analytical Chemistry*; Elsevier B.V.: Amsterdam, The Netherlands, 2018; Volume 82, pp. 265–299, ISBN 9780444640444.
44. Forina, M.; Lanteri, S.; Armanino, C.; Oliveros, M.C.C.; Casolino, C. V-PARVUS. An Extendable Package of Programs for Explorative Data Analysis, Classification and Regression Analysis. Dip.Chimica e Tecnologie Farmaceutiche ed Alimentari, University of Genova, Genova (Italy) 2011. Available online: https://iris.unige.it/handle/11567/202703 (accessed on 3 November 2022).
45. Forina, M.; Oliveri, P.; Casale, M. Complete Validation for Classification and Class Modeling Procedures with Selection of Variables and/or with Additional Computed Variables. *Chemom. Intell. Lab. Syst.* **2010**, *102*, 110–122. [CrossRef]
46. Brown, S.; Tauler, R.; Walczak, B. *Comprehensive Chemometrics*; Elsevier: Amsterdam, The Netherlands, 2010; ISBN 9780444527011.
47. van der Greef, J.; Smilde, A.K. Symbiosis of Chemometrics and Metabolomics: Past, Present, and Future. *J. Chemom.* **2005**, *19*, 376–386. [CrossRef]
48. Martin, M.; Perez-Guaita, D.; Andrew, D.W.; Richards, J.S.; Wood, B.R.; Heraud, P. The Effect of Common Anticoagulants in Detection and Quantification of Malaria Parasitemia in Human Red Blood Cells by ATR-FTIR Spectroscopy. *Analyst* **2017**, *142*, 1192–1199. [CrossRef] [PubMed]
49. Tomasid, R.C.; Sayat, A.J.; Atienza, A.N.; Danganan, J.L.; Ramos, M.R.; Fellizar, A.; Notarteid, K.I.; Angeles, L.M.; Bangaoilid, R.; Santillan, A.; et al. Detection of Breast Cancer by ATR-FTIR Spectroscopy Using Artificial Neural Networks. *PLoS ONE* **2022**, *17*, e0262489. [CrossRef]
50. Sitnikova, V.E.; Kotkova, M.A.; Nosenko, T.N.; Kotkova, T.N.; Martynova, D.M.; Uspenskaya, M.v. Breast Cancer Detection by ATR-FTIR Spectroscopy of Blood Serum and Multivariate Data-Analysis. *Talanta* **2020**, *214*, 120857. [CrossRef]
51. Theophilou, G.; Lima, K.M.G.; Martin-Hirsch, P.L.; Stringfellow, H.F.; Martin, F.L. ATR-FTIR Spectroscopy Coupled with Chemometric Analysis Discriminates Normal, Borderline and Malignant Ovarian Tissue: Classifying Subtypes of Human Cancer. *Analyst* **2016**, *141*, 585–594. [CrossRef]
52. Banerjee, A.; Gokhale, A.; Bankar, R.; Palanivel, V.; Salkar, A.; Robinson, H.; Shastri, J.S.; Agrawal, S.; Hartel, G.; Hill, M.M.; et al. Rapid Classification of COVID-19 Severity by ATR-FTIR Spectroscopy of Plasma Samples. *Anal. Chem* **2021**, *93*, 10391–10396. [CrossRef]
53. el Khoury, Y.; Collongues, N.; de Sèze, J.; Gulsari, V.; Patte-Mensah, C.; Marcou, G.; Varnek, A.; Mensah-Nyagan, A.G.; Hellwig, P. Serum-Based Differentiation between Multiple Sclerosis and Amyotrophic Lateral Sclerosis by Random Forest Classification of FTIR Spectra. *Analyst* **2019**, *144*, 4647–4652. [CrossRef]

biosensors

Article

Selective Antibody-Free Sensing Membranes for Picogram Tetracycline Detection

Hamdi Ben Halima [1], Abdoullatif Baraket [1], Clara Vinas [2], Nadia Zine [1], Joan Bausells [3], Nicole Jaffrezic-Renault [1,*], Francesc Teixidor [2] and Abdelhamid Errachid [1]

1 Institut de Sciences Analytiques (ISA)-UMR 5280, Université Claude Bernard Lyon 1, 5 Rue de la Doua, 69100 Lyon, France
2 Inorganic Materials Laboratory, Institut de Ciencia de Materials de Barcelona (ICMAB-CSIC), Campus de la UAB, Bellaterra, 08193 Barcelona, Spain
3 Institute of Microelectronics of Barcelona (IMB-CNM, CSIC), Campus UAB, Bellaterra, 08193 Barcelona, Spain
* Correspondence: nicole.jaffrezic@univ-lyon1.fr; Tel.: +334-3742-3516

Abstract: As an antibody-free sensing membrane for the detection of the antibiotic tetracycline (TC), a liquid PVC membrane doped with the ion-pair tetracycline/θ-shaped anion $[3,3'\text{-Co}(1,2\text{-C}_2\text{B}_9\text{H}_{11})_2]^-$ ([*o*-COSAN]$^-$) was formulated and deposited on a SWCNT modified gold microelectrode. The chosen transduction technique was electrochemical impedance spectroscopy (EIS). The PVC membrane was composed of: the tetracycline/[*o*-COSAN]$^-$ ion-pair, a plasticizer. A detection limit of 0.3 pg/L was obtained with this membrane, using bis(2-ethylhexyl) sebacate as a plasticizer. The sensitivity of detection of tetracycline was five times higher than that of oxytetracycline and of terramycin, and 22 times higher than that of demeclocycline. A shelf-life of the prepared sensor was more than six months and was used for detection in spiked honey samples. These results open the way to having continuous monitoring sensors with a high detection capacity, are easy to clean, avoid the use of antibodies, and produce a direct measurement.

Keywords: tetracycline; electrochemical impedance spectroscopy; single-walled carbon nanotubes; tetracycline/[*o*-COSAN]$^-$ ion-pair complex; polyvinyl chloride; selectivity

Citation: Ben Halima, H.; Baraket, A.; Vinas, C.; Zine, N.; Bausells, J.; Jaffrezic-Renault, N.; Teixidor, F.; Errachid, A. Selective Antibody-Free Sensing Membranes for Picogram Tetracycline Detection. *Biosensors* **2023**, *13*, 71. https://doi.org/10.3390/bios13010071

Received: 23 November 2022
Revised: 18 December 2022
Accepted: 26 December 2022
Published: 31 December 2022

1. Introduction

Sensors are devices that detect and respond to some type of input from the physical environment. They are found everywhere in our daily lives and the aim is to make them practical and easy to handle, moving away from expensive complex instrumentation to small inexpensive systems that can be operated by anyone. In the chemical sensing field, potentiometry is the archetype of simplicity: simple sensing material commonly made of a membrane with standard and low-weight electronic equipment. The sensing material is responsible for interacting with the analyte, producing a change in the interfacial potential that is then transformed into a readable signal by the transducer [1–13]. For the popular ion-selective electrodes (ISE), the sensing membrane consists of a polymer matrix usually based on plasticized poly(vinyl chloride) and one electroactive additive, commonly a lipophilic salt [4,14,15]. The selectivity of these membranes has been related to the Hofmeister series, however when a such relationship is not sought, the membrane composition is complemented with a second electroactive additive or ionophore, usually a complexing selective ligand, e.g., valinomycin for K^+. Despite many ongoing theories about the mechanism of charge transfer at the interfaces, the reality is that the reasons behind the sensing and the selectivity are still not well inferred. Far less complex are the biosensors, for which the selectivity to the analyte is conceptually well interpreted because of the specific complementarity between enzyme or antibody, or aptamer and the analyte. It is generally accepted that an interaction between the analyte and the sensing material is a necessary condition for the feasibility of any chemical sensor. Non-covalent interactions may lead to a reversible or partially reversible response. Conversely, covalent bonding, which can provide high selectivity and sensitivity, often leads to an irreversible response. On the other hand, potentiometry, in practical terms, shows low detection limits near 10^{-6} or 10^{-7} M, whereas electrochemical antibodies or aptamer biosensors can go down to the pg [1,3,16]. As they are based on biological material, biosensors have associated problems such as the influence of the environment, temperature, and atmosphere, among others, which make them especially delicate and require rigorous storage conditions.

A few years ago we introduced, in potentiometry [1,3,5,17], the θ-shaped molecule $[3,3'\text{-Co}(1,2\text{-C}_2\text{B}_9\text{H}_{11})_2]^-$ ([*o*-COSAN]$^-$) into PVC membranes made of PVC: plasticizer in a ratio of 1:2. The principal motivation was that [NRR′R″R‴]$[3,3'\text{-Co}(1,2\text{-C}_2\text{B}_9\text{H}_{11})_2]$ salts are highly insoluble in water, whereas they are highly soluble in many organic solvents. At that time, we did not know much more about the physicochemical characteristics of the θ-shaped molecule $[3,3'\text{-Co}(1,2\text{-C}_2\text{B}_9\text{H}_{11})_2]^-$. The membrane did not contain any other components besides the cation of the analyte and the plasticizer. The potentiometry results on the macroscopic electrodes were excellent, with the membrane set on a solid support made of graphite powder mixed with Epoxy Resin, providing very good selectivity; whereas the low limit of detection (LOD) remained in most cases about 10^{-6} M. Electrodes for antibiotics [18], amino acids [19] and other biomolecules [20,21] were developed. The most surprising aspect was that such a simple membrane composition was able to discriminate a chiral amino acid from its enantiomer, with selectivity coefficients Kxy between 10^{-2} and 10^{-3} [22].

Since our first publications in the area of potentiometry with [NRR′R″R‴]$[3,3'\text{-Co}(1,2\text{-C}_2\text{B}_9\text{H}_{11})_2]$, we and others have contributed very much to the understanding of the physicochemical properties of $[\text{Co}(\text{C}_2\text{B}_9\text{H}_{11})_2]^-$. With regards to its sensing ability, we would construe that its tunable reversible redox potential [23–26], its self-assembling capacity in solid and aqueous solution [27,28], its capacity to produce hydrogen and dihydrogen bonds [29–31], its ability to dope conducting organic polymers [26,32], and its amphiphilic character [33], among others, are key points to explain the extraordinary performance of this anion. However, as stated above, it was not possible to overcome the Low LOD. We wondered if this problem could be solved with another electrochemical transducer, while keeping the same membrane composition so that this sort of electrochemical measurement could approach the biosensors in terms of low LOD. Hopefully, this would lead to the sensing materials being able to amalgamate the good points of potentiometry with the low LOD, and the selectivity of biosensors. We sought to use the electrochemical impedance spectroscopy (EIS) technique, and we decided to initiate our works with tetracycline. The aim was to discover if the impedance increased with increasing concentrations of analyte, and whether there was any selectivity with structurally closely related chemicals. For example, the set of tetracycline (TC) antibiotics, that would show the necessity for the analyte inside the membrane, whether there was a relationship between signal and concentration, and whether there was any real hope of getting a low value of the low detection limit.

As we shall see in this paper, the results are highly encouraging. We have made an ISE PVC membrane incorporating protonated tetracycline compensated by $[3,3'\text{-Co}(1,2\text{-C}_2\text{B}_9\text{H}_{11})_2]^-$ and, as an extra component, carbon nanotubes as an inner conductivity-enhancing layer. Tetracycline was chosen because there is a set of antibiotics all having the same tetracycline skeleton which, because of their very similar structures, will allow appreciation of the selectivity that can be achieved. Tetracycline is used to treat infections caused by bacteria in the respiratory tract, lymphatic, intestinal, genital and urinary systems, on the skin and eyes and certain other infections that are spread by infected animals [34–36]. The molecular structure of tetracycline and of selected similar molecules are shown in Figure 1a.

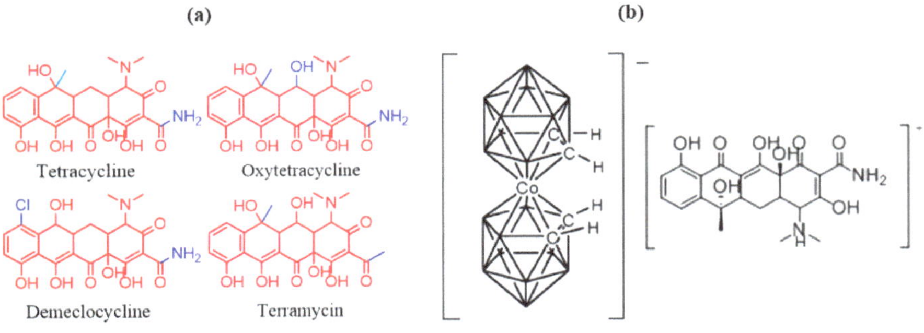

Figure 1. (**a**) Tetracycline and the chosen similar molecules and molecular structure of [*o*-COSAN]$^-$/tetracycline ion-pair complex (**b**).

Based on the literature, there are several analytical methods that have been described for tetracycline monitoring, including: high performance liquid chromatography coupled with mass spectrometry [37], capillary electrophoresis coupled with electrochemiluminescence [38], ELISA (enzyme-linked immune-sorbent assay) [39,40], liquid chromatography-mass spectroscopy (LC-MS) [41], and spectroscopy analysis [42,43]. However, these methods suffer from a lack of sensitivity compared to chromatographic techniques [44]. Commonly, the principal limitations of these steps lie in the lack of sensitivity, high cost, time-consuming implementation and the requirement for sophisticated technical skills [45]. New solutions need rapid, simple and accurate methods for the on-site screening of low TC residues without any supplementary steps such as extraction or clean-up. Owing to their advantages of high selectivity and rapid detection, several optical and electrochemical biosensors have been investigated [46]. For this reason, aptamer-based sensing techniques were widely used for the food safety determination. Above all, there is a growing rise in aptasensor fabrication for TC detection [47,48], with only some applications to honey samples [49,50]. Unfortunately, the mean disadvantage of these systems is their relatively low detection signals [51]. In addition, in some recent works, a molecularly imprinted sensor had been successfully applied to the analysis of antibiotic residues in honey samples [52,53]. However, these types of sensors suffer from reversibility and a short shelf-life when the imprinted membrane is fragile.

Tetracycline and tetracycline antibiotics (shown in Figure 1) all have a common linear fused tetracyclic nucleus that differ with the functional groups attached. All of them contain only one protonable amine group $N(Me)_2$. Therefore, the salt [tetracycline-H][3,3′-$Co(1,2\text{-}C_2B_9H_{11})_2$] was expected. It was thus anticipated that PVC/plasticizer/carbon nanotubes/[tetracycline-H][3,3′-$Co(1,2\text{-}C_2B_9H_{11})_2$] would be an excellent candidate to test the feasibility of the electrochemical sensor based on an EIS transducer. Indeed, this proved to be the case with a lower detection limit with excellent selectivity.

2. Materials and Methods

2.1. Materials and Chemicals

Tetracycline (TC), N-hydroxysuccinimide (NHS), 11-amino-1-undecanethiol, phosphate buffer solution (PBS) tablets, 1-ethyl-3-(3-dimethylaminopropyl) carbodiimide (EDC), Ethanol, sodium dodecyl sulfate (SDS), Polyvinyl chloride (PVC), o-nitro phenyl octyl ether (NPOE), di-octyl phthalate (DOP), dibutyl phthalate (DBP), dibutyl sebacate (DBS), hydrochloric acid (HCl), diethyl ether, tetrahydrofuran (THF), di-octyl phthalate, bis(2-ethyl hexyl) sebacate were purchased from Sigma-Aldrich (France). The standard solutions and buffers were prepared with Millipore Milli-Q nanopure water (resistivity > 18 MW cm) which is produced by a Millipore Reagent Water System (France). Epoxy resin EPO TEK H70E 2LC was from Epoxy Technology, France. Cs[*o*-COSAN], was synthesized from 1,2-*closo*-$C_2B_{10}H_{12}$ from Katchem Spol.sr.o (Kralupy nad Vltavou, Czech Republic), as reported in the literature [54]. The Na[*o*-COSAN] was obtained by means of cationic exchange resin from Cs[*o*-COSAN] following the previously described procedure [31].

2.2. Preparation of the Ion-Pair Complex [tetracycline-H][$Co(C_2B_9H_{11})_2$]

Tetracycline hydrochloride (40 mg, 0.083 mmol) was dissolved in diluted hydrochloric acid (~25 mL). After agitating and obtaining a clear solution, Na [3,3′-$Co(1,2-C_2B_9H_{11})_2$] (0.083 mmol) in 10 mL of diluted hydrochloric acid (1 or 3 M) were added. Almost instantly a precipitate appeared. The mixture was stirred for 5 min and left to rest for an additional 15 min. The orange solid was filtered through a Buchner funnel with filter paper. After rinsing first with 10 mL diluted hydrochloric acid (0.1 M) and then 2 × 10 mL of deionized water, the filter paper was carefully removed and placed in a round bottom flask with a ground glass joint for active 0.1–0.01 mm vacuum at room temperature. After 4–5 h, the solid was collected and was ready for the membrane preparation. The formula of the ion-pair complex is shown in Figure 2.

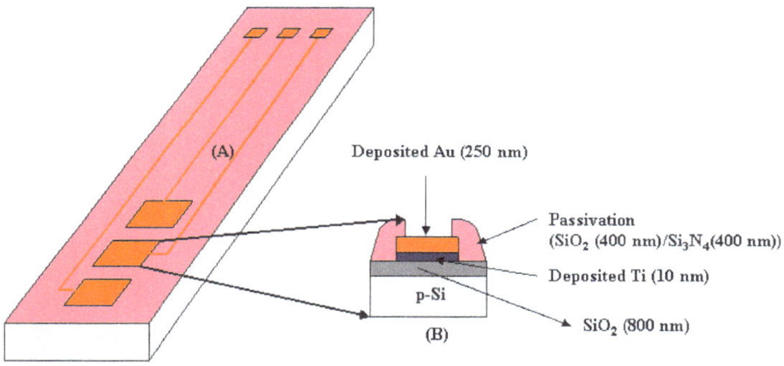

Figure 2. (**A**) Schematic view of microelectrodes based on silicon technology; (**B**) Cross section of one planar microelectrode.

2.3. Preparation of the Gold Microelectrode Modified by Single-Walled Carbon Nanotubes and Electrochemical Measurements

The microelectronics fabrication process for the microelectrodes was performed at Centro Nacional de Microelectronica (CNM). The process started with a thermal oxidation process to grow a thick oxide layer (8000 Å) on 100 mm diameter P-type <100> silicon wafers with a nominal thickness of 525 mm. The working microelectrode was made with a metal layer consisting of a thin titanium film (10 nm) promoting gold adhesion plus 250 nm of gold. After that, photoresist layer was spin-coated with a spinning speed of 3000 rpm and was exposed in UV light with a pattern mask. Etching away the exposed photoresist was performed with the developer OPD4262 from Fujifilm. The remaining photoresist corresponded exactly to the microelectrodes, and then the gold, unprotected by the photoresist, was etched away. The next step consisted of the deposition of two PECVD (Plasma-Enhanced Chemical Vapor Deposition) layers of SiO_2 (4000 Å) and Si_3N_4 (4000 Å), to act as a passivation layer. The second photolithographic process was performed to open the passivation on the active Au microelectrodes (300 μm × 300 μm; area: $9 \cdot 10^{-4}$ cm^2) and on the soldering pads. The structure of the microelectrodes is shown schematically in Figure 3. Wire bonding was performed using a Kulicke & Soffa 4523 A Digital instrument from Kulicke & Soffa, Singapore.

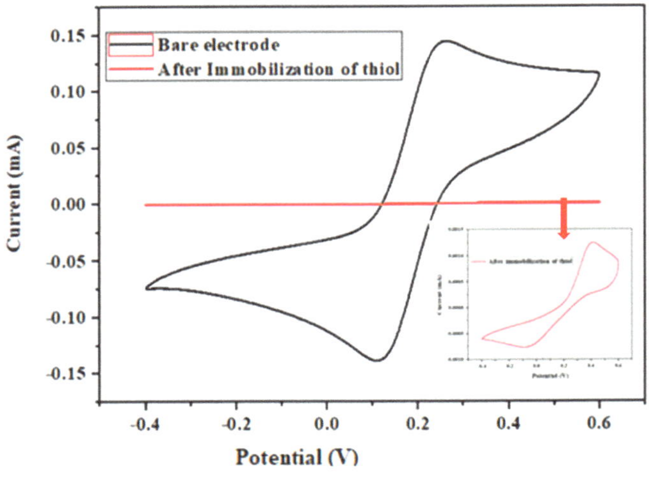

Figure 3. Characterization of gold microelectrode by cyclic voltammetry (CV), before and after immobilization of aminothiol in 5 mM $[Fe(CN)_6]^{3-/4-}$ and PBS, from −0.4 to +0.6 V at a scan rate of 80 mv/s. Insert: Enlarged cyclic voltammetry (CV) after immobilization of aminothiol.

Microelectrodes were exposed to UV using UV/ozone Procleaner^{TM} (BioForce, Germany) for 30 min for cleaning and activation by creating –OH groups. These were modified by adsorption of 11-amino-1-undecanthiol-HCl (aminothiol), by dipping the electrode in a 10 mM solution of the aminothiol in ethanol (EtOH) for one night at 4 °C. Then, the microelectrode was rinsed with ethanol to remove unbound thiols and dried with N_2. In a separate beaker, (0.17 g/L, 0.25 mL) of COOH-SWCNT (Carboxylic functionalized single-walled carbon nanotubes) were added to an aqueous solution of SDS (sodium dodecyl sulfate, 0.1 M), 0.1 M in NHS (N-hydroxysuccinimide) and 0.4 M in EDC (1-ethyl-3-(3-dimethylaminopropyl) carbodiimide-HCl), and the reagents were left in contact for one hour. Then, the microelectrode functionalized with aminothiol was placed with the solution containing the activated SWCNTs-COOH for 2 h at room temperature. Finally, the SWCNTs functionalized microelectrode was rinsed with ultrapure water to remove the unbounded SWCNTs and dried with N_2.

All electrochemical measurements were carried out in a Faraday cage at room temperature (r.t., 22 ± 2 °C). The electrochemical experiments were carried out using a VMP3 multichannel potentiostat (Biologic-EC-Lab, Seyssinet-Pariset, France). Data were acquired and analyzed using EC-Lab software V11.30. The EIS data were fitted by using the Randomize + Simplex method.

2.4. Liquid Membrane Preparation

In earlier potentiometric works related to $[3,3'\text{-}Co(1,2\text{-}C_2B_9H_{11})_2]^-$, it was found that the most suitable mix of electroactive material, plasticizer, and PVC powder for membranes was either 3% or 7% of the electroactive material, 63% of plasticizer and 34% or 30% of PVC powder, all in weight percentages that were dissolved in THF. The membrane solution was prepared as follows: 43 mg of PVC were dissolved by stirring in 1.5 mL of tetrahydrofuran (THF) until a viscous but clear solution was obtained. Then, 10 mg of [tetracycline-H]/$[3,3'\text{-}Co(1,2\text{-}C_2B_9H_{11})_2]$ and 90 mg of plasticizer were added. The resulting dispersing solution is deposited on the surface of the electrode body. In this paper, the results of three membranes (membranes named 1,2,3) are reported; the differences between the three membranes lie in the concentration of HCl in the preparation of the ion-pair complex (1 or 3 M) and in the plasticizer used: *Membrane 1*, HCl 1 M, 91.75 μL di-octylphthalate; *Membrane 2*, HCl 3 M, 91.75 μL di-octylphthalate, *Membrane 3*, HCl 3 M, 98.5 μL bis (2-ethyl hexyl) sebacate.

The membrane was drop-cast (2 μL drop) onto the gold microelectrode already functionalized with the SWCNTs layer, and the solvent was allowed to evaporate at an ambient temperature for 24 h [1,5]. Once the membrane had dried, the microelectrode was then immersed for 24 h at 4 °C in tetracycline at 10^{-3} M in order to achieve appropriate conditioning of the PVC tetracycline membrane. The microelectrodes were stored at room temperature for future use.

3. Results and Discussion

3.1. Characterization of the [o-COSAN]⁻/Tetracycline Ion-Pair Complex

In order to characterize our ion-pair complex, different techniques were used such as Proton Nuclear Magnetic Resonance (^{1}H NMR), Carbon Nuclear Magnetic Resonance (^{13}C{^{1}H} NMR), Fourier Transform Infrared Spectroscopy (FTIR), Matrix-Assisted Laser Desorption/Ionization-Time-of-Flight Mass Spectrometry (MALDI-TOF-MS) and Elemental Analysis.

The ratio [tetracycline-H]/[3,3′-Co(1,2-$C_2B_9H_{11}$)$_2$] (Figure 1b) is calculated from ^{1}H-NMR, in d_6-acetone, given the areas of the unquestionable key hydrogen atoms of the cation and the anion. The C-*H* resonances at the aromatic region of the spectrum (close to 7 ppm) correspond to the aromatic hydrogen atoms of the tetracycline cation, and the signal that appears close to 4 ppm corresponds to the four hydrogen atoms ($C_{cluster}$-*H*) of the anionic [3,3′-Co(1,2-$C_2B_9H_{11}$)$_2$] cluster. Additionally, the ^{1}H NMR spectrum displays the other signals corresponding to [*o*-COSAN]$^-$ and tetracycline that unambigously indicates the presence of the [*o*-COSAN]- anion, and the protonated tetracycline cation in the ion-pair complex. By integration, the ratio [tetracycline-H]/[*o*-COSAN]$^-$ in the ion-pair complex was calculated (Figure S1); there are three aromatic protons in each protonated tetracycline cation, and its integration area is around 3.5 per the 4 area corresponding to each [*o*-COSAN]$^-$ anion, which indicates that there is one molecule of [*o*-COSAN]$^-$ per one molecule of protonated tetracycline.

The ^{13}C{^{1}H} NMR spectrum (Figure S2) displays the signals of the different carbons of the protonated tetracycine molecules, which are represented with letters, as well as the signal of $C_{cluster}$-H vertices of the [*o*-COSAN]$^-$ that appear at 51.0 ppm. The signals without letters correspond to the solvent (deuterated acetone and other solvents used in the preparation).

On the other hand, to observe if in the sample there is [*o*-COSAN]$^-$, a signal should appear in the range 2520–2550 cm^{-1} in the IR spectrum (Figure S3).

FTIR spectrum of our ion-pair complex displays the ν(O-H) stretching vibration at 3611 cm^{-1}; ν(N-H; O-H) at 3370–3611 cm^{-1}; ν(C_{aryl}-H, $C_{cluster}$-H) at 3047 cm^{-1}; ν(B-H) at 2539 cm^{-1}; ν(C=O) at 1659 cm^{-1}; ν(C=O_{amide}; C=C; C=$C_{aromatic}$; N-H) at 1529–1612 cm^{-1}; ν(C-H) at 1322–1452 cm^{-1}; ν(C-O; C-C; C-N) at 1097–1322 cm^{-1}; ν(C-H; C=C) at 721–983 cm^{-1}.

In MALDI-TOF-MS we can see the mass spectrum of the negative and positive part of the sample, and know the molecular mass of both species. In Figure S4 we can see the positive part of the mass spectrum of our prepared ion-pair complexes; the negative part is the signal of [*o*-COSAN]$^-$ at 324 m/z. These spectra were created with a matrix. The Mass Spectra was recorded on a matrix, therefore not all signals correspond to the mass of the cation.

In addition, Table S1 shows the elemental analysis of our ion-pair complex, which contains the theoretical values of %C, %H, %N and %S and the real values. The real values are very similar to the theory, so we can confirm that our sample has the correct structure.

3.2. Electrochemical and Physical Characterization of the Gold Microelectrode

3.2.1. Characterization of the Gold Surface Before and After Immobilization of Amino Thiol Using Cyclic Voltammetry

The gold electrodes were characterized by cyclic voltammetry (CV) before and after the immobilization of the amino thiol. The CV scan was carried out from −0.4 to +0.6 V, with a scanning rate of 80 mV/s in 5 mL of PBS in the presence of the redox couple

$[Fe(CN)_6]^{3-/4-}$. Figure 3 shows that the oxido/reduction peaks of bare gold disappear after the immobilization of aminothiol. The disappearance of the oxido/reduction peaks was caused by the blocking of the aminothiol layer of the gold surface, thus creating a low electron transfer rate.

3.2.2. EIS Characterization of the Gold Microelectrode at the Different Stages of Modification

The electrode performance was studied by EIS by applying a potential of −0.4 V in a frequency range of 100 kHz to 6 Hz at each individual step of preparation: 1) to the bare Au electrode, 2) after functionalization with the aminothiol, and 3) after grafting of carbon nanotubes. Figure 4 shows the evolution of the conductance of the Nyquist diagrams at the different microelectrode functionalization stages. As expected, the most conductive is the bare electrode, whose conductivity decreases upon the addition of the aminothiol with its long carbon chain (C11). This conductivity is slightly increased by the grafting of carbon nanotubes.

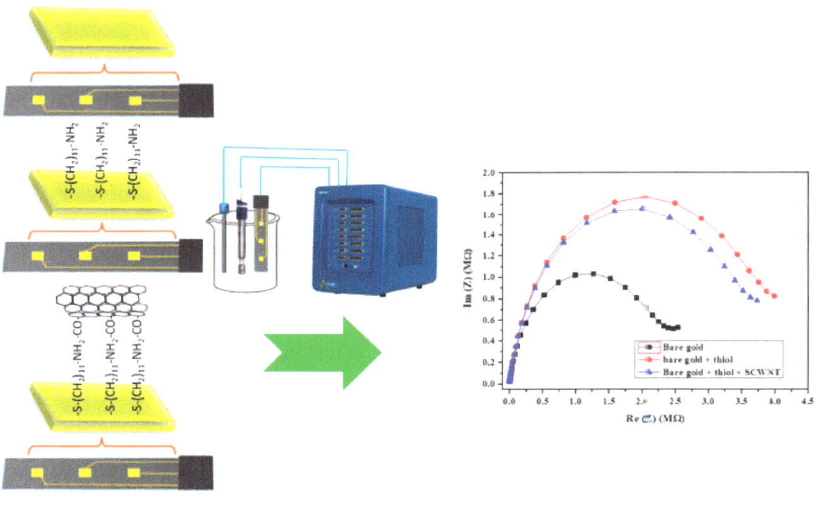

Figure 4. Nyquist diagram (-Im (Z) vs. Re (Z)) corresponding to the impedance measurements by applying a potential of −0.4 V and amplitude potential of 10 mV between a frequency range of 100 kHz to 6 Hz for the various grafted layers on the gold microelectrode. Bare gold (black), after the immobilization of the aminothiol (red) and after grafting SWCNTs (blue).

3.2.3. SEM Characterization

Scanning Electron Microscopy (SEM) was employed to investigate the surface morphology of the SWCNTs, fixed on to the gold microelectrode using FEI Quanta FEG 250. Figure 5 confirms the grafting of SWCNTs onto the gold microelectrode functionalized with amine thiol-SWCNTs. The immobilization of SWCNTs onto the microelectrode was confirmed by the small white rods, however, the black spots in the image indicate some spots not covered by SWCNTs.

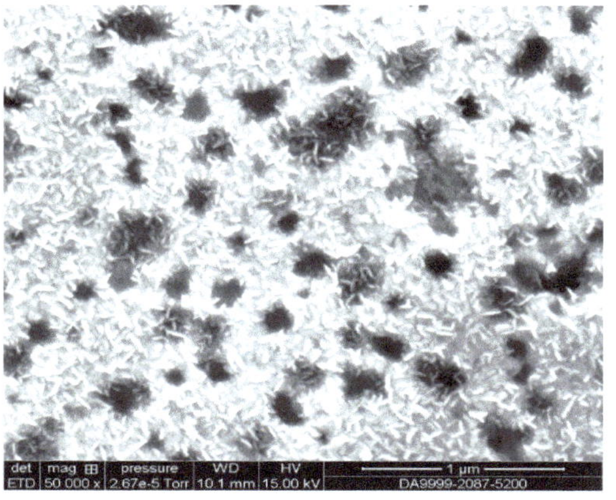

Figure 5. SEM image under functional microelectrode (amine thiol + SWCNTs).

3.2.4. Characterization by Fourier-Transform Infrared Spectroscopy (FTIR)

The presence of different compounds of our membrane, such as polymer matrix, plasticizer, and [*o*-COSAN]$^-$/tetracycline ion-pair complex, were confirmed after deposition onto the microelectrode. First, the FTIR spectrum of membrane 3 after deposition onto the microelectrode was carried out. In parallel, three different reference FTIR spectra of PVC, bis (2-ethylhexyl) sebacate and [*o*-COSAN]$^-$/tetracycline ion-pair complex were performed.

FTIR spectra of the different compounds of our membrane: bis (2-ethylhexyl) sebecate (plasticizer) (Figure S6), PVC polymer matrix (Figure S7), and ionophore of cobalt bis(dicarbollide) (Figure S8) were superimposed with the FTIR spectrum of membrane 3 after deposition onto the microelectrode (Figure 6).

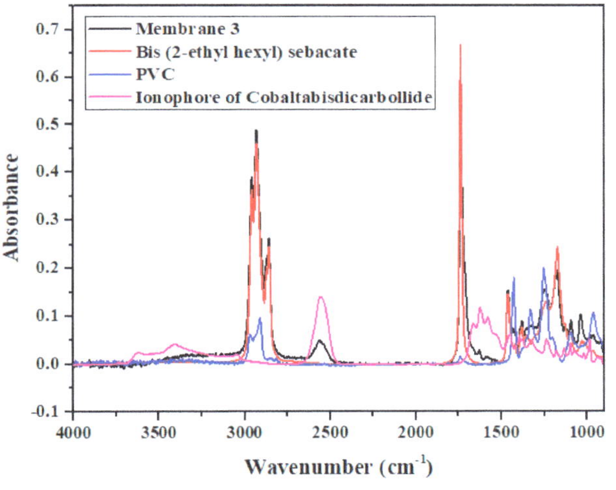

Figure 6. FTIR Spectrum of membrane 3 after deposition onto the microelectrode (black), bis (2-ehylhexyl) sebacate (red), PVC (blue), Ionophore of Cobalt bis(dicarbollide)–Tetracycline (pink).

Figure 6 confirms the presence of the different compounds of the membrane, since the absorption peaks are localized at the characteristic wavelengths of the membrane immobilized on the microelectrode. These results confirm the success of the immobilization of the membrane onto the microelectrode.

3.3. Detection of Tetracycline by Electrochemical Impedance Spectroscopy (EIS)

Impedance was measured using an electrochemical cell consisting of three electrodes (Figure 4). The reference electrode was a saturated calomel electrode, the auxiliary electrode was a platinum wire of 1 mm in diameter and the working electrode was the microelectrode described above. The measurements of tetracycline were carried out in a PBS buffer solution (pH = 7.4), varying the tetracycline concentration from 1 pg/L to 5 ng/L. All experiments were repeated three times to confirm the reproducibility of our biosensor. The experiments were performed in darkness to protect tetracycline, which is a fluorescent compound from degradation, and in a Faraday box to eliminate electrical interference. The Nyquist plots (-Im (Z)/Re (Z)) obtained with the three membranes were recorded while increasing the concentration of tetracycline. The reported results are the mean of the three sensing areas for one electrode (Figure 2).

From the Nyquist diagrams presented in Figure 7, it is noticeable that there is a marked response from the picogram quantities of tetracycline, as can be seen from the separation in the Nyquist plot between 0 pg and 1 pg. There is a steady increase of both Re(Z) and –Im(Z) in parallel with an increasing amount of analyte. In addition, we have noticed that the HCl concentration influences the preparation of [*o*-COSAN]$^{-}$/tetracycline ion-pair complex and of the used plasticizer.

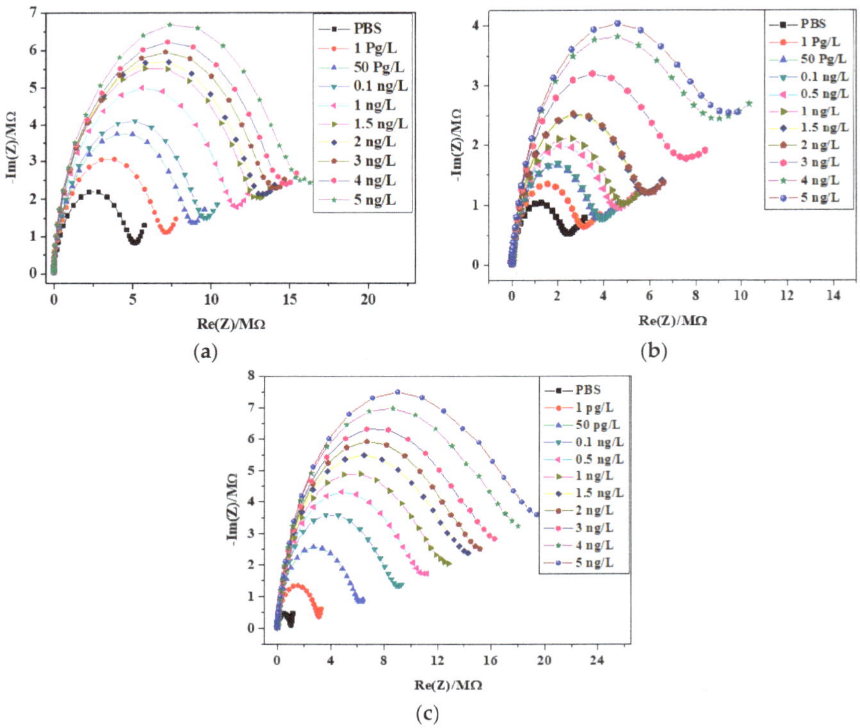

Figure 7. Nyquist diagrams of gold electrode/membrane interface (membrane 1, 2, 3) for different concentrations of tetracycline from 1 pg/L to 5 ng/L in a frequency range between 100 kHz and 6 Hz by applying a potential of −0.4 V and amplitude potential of 10 mV. (**a**) membrane 1; (**b**) membrane 2; (**c**) membrane 3.

Data fitting on EIS spectra was achieved by using the Randomize + Simplex method using Randles equivalent circuit model [R_s + Q_2/ (R_{ct} + Z_W)], in which R_s corresponds to the resistance of the electrolyte solution; Q_2 is the phase constant element that is in parallel with R_{ct}, which is the charge transfer resistance; and Z_W represents the Warburg impedance (Table S2). The standardization plots were obtained by drawing tetracycline concentrations vs. R_{ct} normalized data ($\Delta R/R = (R_{sample} - R_{Ref})/R_{Ref}$). The example of the calibration curve of the sensor with membrane 3 is presented in Figure S9. The reproducibility is 6%.

The sensitivity of the tetracycline sensor is the slope of the straight line. With membrane 3, the sensitivity is 1.83 ± 0.4 $ng^{-1}\cdot L$, as presented in Figure 8. The RSD of 22% is the repeatability obtained between 4 sensors. The sensitivity of the tetracycline sensor with membrane 1 is 0.5 ± 0.2 $ng^{-1}\cdot L$ and that of the sensor with membrane 2 is 0.5 ± 0.08 $ng^{-1}\cdot L$, showing the high influence of the used plasticizer and the negligible influence of the HCl concentration in the preparation of the [*o*-COSAN]$^-$/tetracycline ion-pair complex.

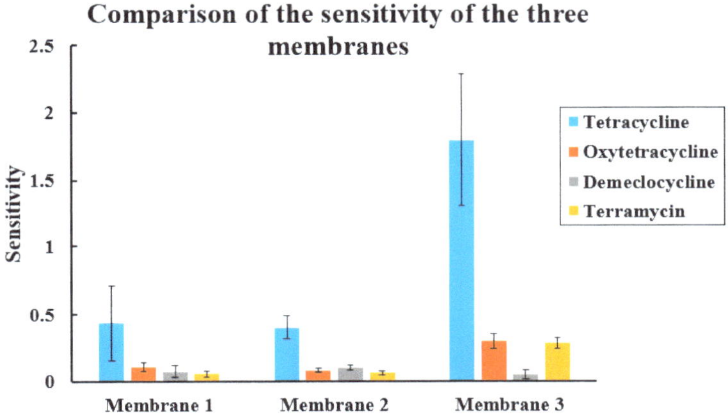

Figure 8. Sensitivities of membrane 1, membrane 2, membrane 3 sensors for tetracycline (blue) (concentrations ranging from 1 pg/L to 5 ng/L). Sensitivities for oxytetracycline (orange), demeclocycline (grey), and terramycin (yellow).

3.4. Selectivity of the Tetracycline Sensors

The selectivity of the tetracycline sensors was obtained by testing oxytetracycline, terramycin, and demeclocycline dissolved in a solution of PBS (pH = 7.4) in quantities ranging from 1 pg/L to 5 ng/L. The electrodes and the setup used were the same as those used for tetracycline, and the impedance of the system were measured. All experiments were done thrice to confirm the reproducibility of the sensor.

The slope of each curve of the three membranes for the detection of tetracycline, oxytetracycline, demeclocycline, and terramycin are drawn as bars in Figure 8. The sensitivity of detection of tetracycline was 5 times higher than that of oxytetracycline and of terramycin, and 22 times higher than that of demeclocycline.

Compared to the published tetracycline biosensors presented in Table 1, our tetracycline sensor presents a detection limit in the lower range. Most presented biosensors need aptamers/Ab TC and antibodies/Ag-TC, which are expensive compared to our developed approach.

In addition, the use of antibodies and antigen requires storage at 4 °C of their sensor; however, with our developed approach, our sensors could be stored at room temperature. Due to the reversibility of the interactions, there is a limit to the reusability of the tetracycline sensors. The shelf-life of the prepared sensor with membrane 3 was more than six months. For tetracycline biosensors based on antibody, aptamer or MIP, the resusability is quite limited. In Benvidi et al. [61], only five regenerations were possible without any loss of sensitivity.

Table 1. Comparison of the analytical performance of the prepared tetracycline sensor with membrane 3 and published electrochemical tetracycline biosensors.

Technique	Electrode	Immobilizing Biomolecules	Analyte	Linear Range (mol/L)	LOD (mol/L)	Refs
Photoelectrochemical aptasensor	cerium (Ce) doped CdS modified graphene (G)/BiYWO6	TC aptamer	Drug	4.5×10^{-10}–2.25×10^{-6}	2.25×10^{-11}	[47]
EIS	integrated bio micro-electromechanical system (Bio-MEMS) Au	Anti-TC Polyclonal antibody	Honey	2.25×10^{-13}–2.25×10^{-9}	2.7×10^{-12}	[16]
EIS	interdigitated array microelectrodes (IDAMs)	TC aptamer	Milk	1×10^{-10}–1×10^{-3}	3×10^{-9}	[55]
EIS	Au	TC aptamer	Milk	2.25×10^{-8}–6.75×10^{-6}	2.25×10^{-8}	[56]
EIS	Au	TC aptamer	Milk	1.12×10^{-8}–1.12×10^{-5}	2.25×10^{-9}	[57]
EIS	glassy carbon electrode (GCE) modified with graphene oxide nanosheets	TC aptamer	Tablet/serum	1×10^{-13}–1×10^{-5}	29×10^{-15}	[58]
EIS	nano-porous silicon (PS)	TC aptamer	-	2.1×10^{-9}–62.4×10^{-9}	2×10^{-9}	[59]
EIS	nanomaterial modified with pencil graphite electrode	TC aptamer	Milk	1×10^{16}–1×10^{-6}	3×10^{-17}	[60]
EIS	glassy carbon electrode	TC aptamer	Honey	1×10^{-16}–1×10^{-6}	3.7×10^{-17}	[61]
EIS	carbon paste electrode (CPE)	TC aptamer	Drug/Milk/Honey/blood serum	1×10^{-14}–1×10^{-6}	3.8×10^{-15}	[62]
EIS	interdigital array microelectrode (IDAM)	TC aptamer	Milk	1×10^{-9}– 1×10^{-3}	1×10^{-9}	[63]
differential pulse voltammetry (DPV)	Gold electrode	Anti-TC monoclonal antibody	Milk	1.8×10^{-10}–2.25×10^{-9}	7.22×10^{-11}	[64]
Amperometric immunosensor	Screen-printed dual carbon electrodes (SPdCEs)	Polyclonal sheep anti-TC antibody	Milk	1.12×10^{-12}–1.12×10^{-7}	1.93×10^{-9}	[65]
LSV	Gold electrode electropolymerization of PATP functionalized AuNPs	MIP	Honey	2.24×10^{-13}–2.24×10^{-8}	2.2×10^{-16}	[53]
EIS	PVC liquid membrane	$[o\text{-COSAN}]^{-}$/TC ion-pair complex	Honey	2.25×10^{-15}–1.12×10^{-11}	7.5×10^{-16}	This work

The proposed mechanism is based on the exchange process that occurs at the membrane interface with the analyte, and is a consequence of the geometry and chemical composition of the cobalt bis(dicarbollide) $[3,3'\text{-Co}(1,2\text{-C}_2\text{B}_9\text{H}_{11})_2]^-$, particularly the existence of B-H and $C_{cluster}$-H bonds that can generate hydrogen and dihydrogen bonds. We believe this is essential. In their absence, common ion exchangers will not generate strong interactions with either the ammonium cation or with the plasticizer solvent. Therefore, they have high mobility in the membrane. This is not the case with $[3,3'\text{-Co}(1,2\text{-C}_2\text{B}_9\text{H}_{11})_2]^-$ which does make these strong non-bonding interactions and generates more reticulate, and therefore more stability and a precise concentration of the analyte in the membrane. The cobalt bis(dicarbollide) has dimensions of 1.1 nm in length and 0.6 nm in width and is surrounded by hydrogens, the mentioned B-H bonds, that have considerable hydride character but not enough to be unstable in protic solvents. This sufficient hydride character of the B-H groups enables it to interact strongly with H-N units. The non-bonding interactions B-H···H-N are weak, but if there are many, as is the case, they become a strong interaction. This leads to a very stable concentration of the target analyte within the membrane, and allows for an adequate ion exchange thanks to the ion exchanger capacity of $[3,3'\text{-Co}(1,2\text{-C}_2\text{B}_9\text{H}_{11})_2]^-$, overall giving the appropriate stability and sensitivity.

The higher sensitivity for tetracycline compared to the similar molecules should be explained by a higher stability of the [*o*-COSAN]$^-$/tetracycline ion-pair complex, whereas the similar molecules present a primary amine group which is a prerequisite, apart from terramycin. Demeclocycline presents a chloride group and oxytetracycline, a supplementary OH group which can limit the ion-pair complex stability.

3.5. Determination of Tetracycline in Spiked Honey Samples with the Tetracycline Sensor (Membrane 3)

Honey solutions were prepared as follows: 1 g of honey was added to 1 mL of ethanol and maintained in an ultrasonic bath for 30 min. In order to reduce the matrix effect, the samples were then diluted 1/10 (v/v) with PBS and filtered through a 0.8 mm cellulose acetate filter. The samples were then spiked with standard TC solutions, and measurements were carried out through the standard addition method. Three replicates were made for each the two samples. Table 2 summarizes the tetracycline sensor results for the determination of TC in spiked honey samples. Recoveries of 100% to 107% were found, with an RSD of 6%. This indicates that the repeatability of the tetracycline sensor prepared with membrane 3 is acceptable in practice.

Table 2. Determination of TC in spiked honey samples through the standard addition method, using tetracycline sensor (membrane 3).

Samples	Added pg/L	Found pg/L	Recovery (%)
Honey 1	50	53.5	107 ± 6
	500	515	103 ± 2
Honey 2	50	51	102 ± 6
	500	500	100 ± 6

4. Conclusion

In this work, we have discussed the development process of a novel sensitive and highly selective tetracycline sensor, based on gold microelectrodes modified by single-walled carbon nanotubes (SWCNTs) and a plasticized PVC membrane doped with a [*o*-COSAN]$^-$/tetracycline ion-pair complex. The ion-pair complex was synthesized and characterized using different techniques such as Proton Nuclear Magnetic Resonance (^{1}H NMR), Carbon Nuclear Magnetic Resonance (^{13}NMR), Fourier-Transform Infrared Spectroscopy (FTIR), Matrix-Assisted Laser Desorption/Ionization- Time-of-Flight Mass Spectrometry (MALDI-TOF MS) and Elemental Analysis. The electrochemical detection was performed by electrochemical impedance spectroscopy (EIS) for three different membranes.

Membrane 3 with bis(2-ethyl hexyl) sebacate as the plasticizer, shows the highest sensitivity compared to the other membranes. The high selectivity of our prepared tetracycline sensors was demonstrated by analyzing solutions containing similar molecules, namely oxytetracycline, demeclocycline, and terramycin. The proposed approach presents a high sensitivity compared to published tetracycline biosensors; moreover, it is a cheap solution compared to those requiring aptamers or antibodies, and it is a durable solution because of the reversibility of the interactions with the [*o*-COSAN]$^-$/tetracycline ion-pair complex, and the shelf-life time of the prepared sensor was found to be more than six months. The work carried out in this paper aims at providing a solution to the food or drug control of improving portable instrumentations while incorporating new technologies.

Supplementary Materials: The following supporting information can be downloaded at: https://www.mdpi.com/article/10.3390/bios13010071/s1, Figure S1: ^{1}H NMR spectrum of [*o*-COSAN]$^-$/tetracycline (deuterated acetone); Figure S2: ^{13}C NMR spectrum of [*o*-COSAN]$^-$/tetracycline (deuterated acetone); Figure S3: FTIR spectrum of [*o*-COSAN]$^-$/tetracycline; Figure S4: MALDI-TOF spectrum of [*o*-COSAN]$^-$/tetracycline with matrix (positive part); Figure S5: FTIR spectrum of membrane 3; Figure S6: FTIR spectrum of bis (2-ethyl hexyl) sebacate; Figure S7: FTIR spectrum of PVC; Figure S8: FTIR spectrum of [*o*-COSAN]$^-$/tetracycline ion pair complex; Figure S9: Calibration curve of the tetracycline sensor with membrane 3; Table S1: Overview of elemental analysis in [*o*-COSAN]$^-$/tetracycline; Table S2: Fitting parameters obtained from Randles equivalent circuit model [R_s + Q_2/ (Rct + Z_W)] (membrane 3 detection of tetracycline).

Author Contributions: Writing—original draft preparation, formal analysis, methodology, validation, H.B.H.; formal analysis, A.B.; validation, C.V.; formal analysis, supervision N.Z.; Validation, J.B.; writing—review and editing, N.J.-R.; writing—original draft preparation, supervision, F.T.; supervision, A.E. All authors have read and agreed to the published version of the manuscript.

Funding: The authors acknowledge the financial support of the European Union's Horizon 2020 research and innovation program, Project NMBP–13-2017 KardiaTool (Grant agreement No. 768686), SEA-on-a-CHIP (FP7- OCEAN-2013) under the grant agreement No. 614168. The authors acknowledge the financial support from the POC4allergies project (Grant Agreement No. 768686), which received funding from ERA PerMed ERA-NET, and the financial support from the Bionanosens project (Grant Agreement No. 9[887]), which received funding from the European Union's Horizon 2020.

Institutional Review Board Statement: Not applicable.

Informed Consent Statement: Not applicable.

Data Availability Statement: The data that support the findings of this study are available from the corresponding author upon reasonable request.

Conflicts of Interest: The authors declare no conflict of interest.

References

1. Gallardo-Gonzalez, J.; Saini, A.; Baraket, A.; Boudjaoui, S.; Alcácer, A.; Streklas, A.; Teixidor, F.; Zine, N.; Bausells, J.; Errachid, A. A Highly Selective Potentiometric Amphetamine Microsensor Based on All-Solid-State Membrane Using a New Ion-Pair Complex, [3,3′-Co(1,2-Closo-C2B9H11)2]− [C9H13NH]+. *Sens. Actuators B Chem.* **2018**, *266*, 823–829. [CrossRef]
2. Ayroud, Z.; Gallardo-Gonzalez, J.; Baraket, A.; Hangouet, M.; Alcácer, A.; Streklas, A.; Bausells, J.; Errachid, A.; Zine, N. A Highly Sensitive Impedimetric Metamitron Microsensor Based on All-Solid-State Membrane Using a New Ion-Pair Complex, [3,3′-Co(1,2-Closo-C2B9H11)2]−[C10H11ON4]+. In Proceedings of the EUROSENSORS 2018, Graz, Austria, 9–12 September 2018; p. 1093.
3. Gallardo-Gonzalez, J.; Baraket, A.; Boudjaoui, S.; Clément, Y.; Alcácer, A.; Streklas, A.; Teixidor, F.; Zine, N.; Bausells, J.; Errachid, A. A Highly Sensitive Potentiometric Amphetamine Microsensor Based on All-Solid-State Membrane Using a New Ion-Par Complex, [3,3′-Co(1,2-Closo-C2B9H11)2]− C9H13NH+. In Proceedings of the Eurosensors 2017, Paris, France, 3–6 September 2017; p. 72017481.
4. Saini, A.; Gallardo-Gonzalez, J.; Baraket, A.; Fuentes, I.; Viñas, C.; Zine, N.; Bausells, J.; Teixidor, F.; Errachid, A. A Novel Potentiometric Microsensor for Real-Time Detection of Irgarol Using the Ion-Pair Complex [Irgarol-H]+[Co(C2B9H11)2]−. *Sens. Actuators B Chem.* **2018**, *268*, 164–169. [CrossRef]
5. Gallardo-González, J.; Baraket, A.; Bonhomme, A.; Zine, N.; Sigaud, M.; Bausells, J.; Errachid, A. Sensitive Potentiometric Determination of Amphetamine with an All-Solid-State Micro Ion-Selective Electrode. *Anal. Lett.* **2018**, *51*, 348–358. [CrossRef]

6. Sonu, A.; Baraket, A.; Boudjaoui, S.; Gallardo, J.; Zine, N.; Sigaud, M.; Hangouet, M.; Alcácer, A.; Streklas, A.; Bausells, J.; et al. Solid State Gas Sensor Based on Polyaniline Doped with [3,3′-Co(1,2-C2B9H11)2]−1 for Detection of Acetone: Diagnostic to Heart Failure Disease. In Proceedings of the EUROSENSORS 2018, Graz, Austria, 9–12 September 2018; p. 1086.
7. Ben Halima, H.; Bellagambi, F.G.; Alcacer, A.; Pfeiffer, N.; Heuberger, A.; Hangouët, M.; Zine, N.; Bausells, J.; Elaissari, A.; Errachid, A. A Silicon Nitride ISFET Based Immunosensor for Tumor Necrosis Factor-Alpha Detection in Saliva. A Promising Tool for Heart Failure Monitoring. *Anal. Chim. Acta* **2021**, *1161*, 338468. [CrossRef] [PubMed]
8. Ben Halima, H.; Bellagambi, F.G.; Brunon, F.; Alcacer, A.; Pfeiffer, N.; Heuberger, A.; Hangouët, M.; Zine, N.; Bausells, J.; Errachid, A. Immuno Field-Effect Transistor (ImmunoFET) for Detection of Salivary Cortisol Using Potentiometric and Impedance Spectroscopy for Monitoring Heart Failure. *Talanta* **2022**, 123802. [CrossRef]
9. Ben Halima, H.; Bellagambi, F.G.; Hangouët, M.; Alcacer, A.; Pfeiffer, N.; Heuberger, A.; Zine, N.; Bausells, J.; Elaissari, A.; Errachid, A. A Novel Electrochemical Strategy for NT-ProBNP Detection Using IMFET for Monitoring Heart Failure by Saliva Analysis. *Talanta* **2023**, *251*, 123759. [CrossRef] [PubMed]
10. Ben Halima, H.; Bellagambi, F.G.; Hangouët, M.; Alcacer, A.; Pfeiffer, N.; Heuberger, A.; Zine, N.; Bausells, J.; Errachid, A. A Novel IMFET Biosensor Strategy for Interleukin-10 Quantification for Early Screening Heart Failure Disease in Saliva. *Electroanalysis* **2022**, elan.202200141. [CrossRef]
11. Ben Halima, H.; Zine, N.; Bausells, J.; Jaffrezic-Renault, N.; Errachid, A. A Novel Cortisol Immunosensor Based on a Hafnium Oxide/Silicon Structure for Heart Failure Diagnosis. *Micromachines* **2022**, *13*, 2235. [CrossRef]
12. Halima, H.B.; Zine, N.; Gallardo-Gonzalez, J.; Aissari, A.E.; Sigaud, M.; Alcacer, A.; Bausells, J.; Errachid, A. A Novel Cortisol Biosensor Based on the Capacitive Structure of Hafnium Oxide: Application for Heart Failure Monitoring. In Proceedings of the 2019 20th International Conference on Solid-State Sensors, Actuators and Microsystems & Eurosensors XXXIII (TRANSDUCERS & EUROSENSORS XXXIII), Berlin, Germany, 23–27 June 2019; pp. 1067–1070.
13. Vozgirdaite, D.; Ben Halima, H.; Bellagambi, F.G.; Alcacer, A.; Palacio, F.; Jaffrezic-Renault, N.; Zine, N.; Bausells, J.; Elaissari, A.; Errachid, A. Development of an ImmunoFET for Analysis of Tumour Necrosis Factor-α in Artificial Saliva: Application for Heart Failure Monitoring. *Chemosensors* **2021**, *9*, 26. [CrossRef]
14. Hassani, N.E.A.E.; Baraket, A.; Neto, E.T.T.; Lee, M.; Salvador, J.P.; Colas, M.p.M.; Bausells, J.; Bari, N.E.; Bouchikhi, B.; Elaissari, A. Novel Strategy for Sulfapyridine Detection Using a Fully Integrated Bio-MEMS: Application to Honey Analysis. *Procedia Technol.* **2017**, *27*, 254–255. [CrossRef]
15. Hassani, N.E.A.E.; Baraket, A.; Neto, E.T.T.; Lee, M.; Salvador, J.-P.; Marco, M.; Bausells, J.; Bari, N.E.; Bouchikhi, B.; Elaissari, A.; et al. Novel Strategy for Sulfapyridine Detection Using a Fully Integrated Electrochemical Bio-MEMS: Application to Honey Analysis. *Biosens. Bioelectron.* **2017**, *93*, 282–288. [CrossRef]
16. El Alami El Hassani, N.; Baraket, A.; Boudjaoui, S.; Taveira Tenório Neto, E.; Bausells, J.; El Bari, N.; Bouchikhi, B.; Elaissari, A.; Errachid, A.; Zine, N. Development and Application of a Novel Electrochemical Immunosensor for Tetracycline Screening in Honey Using a Fully Integrated Electrochemical Bio-MEMS. *Biosens. Bioelectron.* **2019**, *130*, 330–337. [CrossRef] [PubMed]
17. Saini, A.; Fuentes, I.; Viñas, C.; Zine, N.; Bausells, J.; Errachid, A.; Teixidor, F. A Simple Membrane with the Electroactive [Sulfapyridine-H]+[Co(C2B9H11)2]- for the Easy Potentiometric Detection of Sulfonamides. *J. Organomet. Chem.* **2019**, *893*, 32–38. [CrossRef]
18. Stoica, A.-I.; Viñas, C.; Teixidor, F. Application of the Cobaltabisdicarbollide Anion to the Development of Ion Selective PVC Membrane Electrodes for Tuberculosis Drug Analysis. *Chem. Commun.* **2008**, 6492. [CrossRef] [PubMed]
19. Stoica, A.-I.; Viñas, C.; Teixidor, F. Cobaltabisdicarbollide Anion Receptor for Enantiomer-Selective Membrane Electrodes. *Chem. Commun.* **2009**, 4988. [CrossRef] [PubMed]
20. Stoica, A.-I.; Kleber, C.; Viñas, C.; Teixidor, F. Ion Selective Electrodes for Protonable Nitrogen Containing Analytes: Metallacarboranes as Active Membrane Components. *Electrochim. Acta* **2013**, *113*, 94–98. [CrossRef]
21. Bliem, C.; Fruhmann, P.; Stoica, A.-I.; Kleber, C. Development and Optimization of an Ion-Selective Electrode for Serotonin Detection. *Electroanalysis* **2017**, *29*, 1635–1642. [CrossRef]
22. Zdrachek, E.; Bakker, E. Describing Ion Exchange at Membrane Electrodes for Ions of Different Charge. *Electroanalysis* **2018**, *30*, 633–640. [CrossRef]
23. Hawthorne, M.F.; Andrews, T.D. Carborane Analogues of Cobalticinium Ion. *Chem. Commun. (Lond.)* **1965**, 443. [CrossRef]
24. Lever, A.B.P. Electrochemical Parametrization of Metal Complex Redox Potentials, Using the Ruthenium(III)/Ruthenium(II) Couple to Generate a Ligand Electrochemical Series. *Inorg. Chem.* **1990**, *29*, 1271–1285. [CrossRef]
25. Morris, J.H.; Gysling, H.J.; Reed, D. Electrochemistry of Boron Compounds. *Chem. Rev.* **1985**, *85*, 51–76. [CrossRef]
26. Núñez, R.; Romero, I.; Teixidor, F.; Viñas, C. Icosahedral Boron Clusters: A Perfect Tool for the Enhancement of Polymer Features. *Chem. Soc. Rev.* **2016**, *45*, 5147–5173. [CrossRef] [PubMed]
27. Bauduin, P.; Prevost, S.; Farràs, P.; Teixidor, F.; Diat, O.; Zemb, T. A Theta-Shaped Amphiphilic Cobaltabisdicarbollide Anion: Transition From Monolayer Vesicles to Micelles. *Angew. Chem.* **2011**, *123*, 5410–5412. [CrossRef]
28. Matějíček, P.; Cígler, P.; Procházka, K.; Král, V. Molecular Assembly of Metallacarboranes in Water: Light Scattering and Microscopy Study. *Langmuir* **2006**, *22*, 575–581. [CrossRef] [PubMed]
29. Hardie, M.J.; Raston, C.L. Solid State Supramolecular Assemblies of Charged Supermolecules (Na[2.2.2]Cryptate)+ and Anionic Carboranes with Host Cyclotriveratrylene. *Chem. Commun.* **2001**, 905–906. [CrossRef]

30. Fanfrlík, J.; Lepšík, M.; Horinek, D.; Havlas, Z.; Hobza, P. Interaction of Carboranes with Biomolecules: Formation of Dihydrogen Bonds. *ChemPhysChem* **2006**, *7*, 1100–1105. [CrossRef]
31. Zaulet, A.; Teixidor, F.; Bauduin, P.; Diat, O.; Hirva, P.; Ofori, A.; Viñas, C. Deciphering the Role of the Cation in Anionic Cobaltabisdicarbollide Clusters. *J. Organomet. Chem.* **2018**, *865*, 214–225. [CrossRef]
32. Masalles, C.; Borrós, S.; Viñas, C.; Teixidor, F. Are Low-Coordinating Anions of Interest as Doping Agents in Organic Conducting Polymers? *Adv. Mater.* **2000**, *12*, 1199–1202. [CrossRef]
33. Malaspina, D.C.; Viñas, C.; Teixidor, F.; Faraudo, J. Atomistic Simulations of COSAN: Amphiphiles without a Head-and-Tail Design Display "Head and Tail" Surfactant Behavior. *Angew. Chem. Int. Ed.* **2020**, *59*, 3088–3092. [CrossRef]
34. Nebot, C.; Guarddon, M.; Seco, F.; Iglesias, A.; Miranda, J.M.; Franco, C.M.; Cepeda, A. Monitoring the Presence of Residues of Tetracyclines in Baby Food Samples by HPLC-MS/MS. *Food Control* **2014**, *46*, 495–501. [CrossRef]
35. Jiang, X.; Shi, L. Distribution of Tetracycline and Trimethoprim/Sulfamethoxazole Resistance Genes in Aerobic Bacteria Isolated from Cooked Meat Products in Guangzhou, China. *Food Control* **2013**, *30*, 30–34. [CrossRef]
36. Antúnez, K.; Harriet, J.; Gende, L.; Maggi, M.; Eguaras, M.; Zunino, P. Efficacy of Natural Propolis Extract in the Control of American Foulbrood. *Vet. Microbiol.* **2008**, *131*, 324–331. [CrossRef] [PubMed]
37. Pokrant, E.; Maddaleno, A.; Araya, C.; San Martín, B.; Cornejo, J. In-House Validation of HPLC-MS/MS Methods for Detection and Quantification of Tetracyclines in Edible Tissues and Feathers of Broiler Chickens. *J. Braz. Chem. Soc.* **2017**. [CrossRef]
38. Deng, B.; Xu, Q.; Lu, H.; Ye, L.; Wang, Y. Pharmacokinetics and Residues of Tetracycline in Crucian Carp Muscle Using Capillary Electrophoresis On-Line Coupled with Electrochemiluminescence Detection. *Food Chem.* **2012**, *134*, 2350–2354. [CrossRef] [PubMed]
39. Gaurav, A.; Gill, J.P.S.; Aulakh, R.S.; Bedi, J.S. ELISA Based Monitoring and Analysis of Tetracycline Residues in Cattle Milk in Various Districts of Punjab. *Vet. World* **2014**, *7*, 26–29. [CrossRef]
40. Jeon, M.; Rhee Paeng, I. Quantitative Detection of Tetracycline Residues in Honey by a Simple Sensitive Immunoassay. *Anal. Chim. Acta* **2008**, *626*, 180–185. [CrossRef] [PubMed]
41. Desmarchelier, A.; Anizan, S.; Minh Tien, M.; Savoy, M.-C.; Bion, C. Determination of Five Tetracyclines and Their Epimers by LC-MS/MS Based on a Liquid-Liquid Extraction with Low Temperature Partitioning. *Food Addit. Contam. Part A* **2018**, *35*, 686–694. [CrossRef]
42. Qin, J.; Xie, L.; Ying, Y. Rapid Analysis of Tetracycline Hydrochloride Solution by Attenuated Total Reflection Terahertz Time-Domain Spectroscopy. *Food Chem.* **2017**, *224*, 262–269. [CrossRef]
43. Qin, J.; Xie, L.; Ying, Y. A High-Sensitivity Terahertz Spectroscopy Technology for Tetracycline Hydrochloride Detection Using Metamaterials. *Food Chem.* **2016**, *211*, 300–305. [CrossRef]
44. Mitra, S. *Sample Preparation Techniques in Analytical Chemistry*; John Wiley & Sons: New York, NY, USA, 2004; Volume 237, ISBN 0471457825.
45. Joshi, M.D.; Anderson, J.L. Recent Advances of Ionic Liquids in Separation Science and Mass Spectrometry. *RSC Adv.* **2012**, *2*, 5470–5484. [CrossRef]
46. Lan, L.; Yao, Y.; Ping, J.; Ying, Y. Recent Advances in Nanomaterial-Based Biosensors for Antibiotics Detection. *Biosens. Bioelectron.* **2017**, *91*, 504–514. [CrossRef] [PubMed]
47. Han, Q.; Wang, R.; Xing, B.; Chi, H.; Wu, D.; Wei, Q. Label-Free Photoelectrochemical Aptasensor for Tetracycline Detection Based on Cerium Doped CdS Sensitized BiYWO6. *Biosens. Bioelectron.* **2018**, *106*, 7–13. [CrossRef] [PubMed]
48. Tang, Y.; Zhang, J.; Liu, J.H.; Gapparov, I.; Wang, S.; Dong, Y.; Su, H.; Tan, T. The Development of a Graphene Oxide-Based Aptasensor Used for the Detection of Tetracycline in Honey. *Anal. Methods* **2017**, *9*, 1133–1140. [CrossRef]
49. Wang, S.; Yong, W.; Liu, J.; Zhang, L.; Chen, Q.; Dong, Y. Development of an Indirect Competitive Assay-Based Aptasensor for Highly Sensitive Detection of Tetracycline Residue in Honey. *Biosens. Bioelectron.* **2014**, *57*, 192–198. [CrossRef]
50. Wang, S.; Liu, J.; Yong, W.; Chen, Q.; Zhang, L.; Dong, Y.; Su, H.; Tan, T. A Direct Competitive Assay-Based Aptasensor for Sensitive Determination of Tetracycline Residue in Honey. *Talanta* **2015**, *131*, 562–569. [CrossRef] [PubMed]
51. Yi, X.; Li, L.; Peng, Y.; Guo, L. A Universal Electrochemical Sensing System for Small Biomolecules Using Target-Mediated Sticky Ends-Based Ligation-Rolling Circle Amplification. *Biosens. Bioelectron.* **2014**, *57*, 103–109. [CrossRef]
52. El Alami El Hassani, N.; Llobet, E.; Popescu, L.M.; Ghita, M.; Bouchikhi, B.; El Bari, N. Development of a Highly Sensitive and Selective Molecularly Imprinted Electrochemical Sensor for Sulfaguanidine Detection in Honey Samples. *J. Electroanal. Chem.* **2018**, *823*, 647–655. [CrossRef]
53. Bougrini, M.; Florea, A.; Cristea, C.; Sandulescu, R.; Vocanson, F.; Errachid, A.; Bouchikhi, B.; El Bari, N.; Jaffrezic-Renault, N. Development of a Novel Sensitive Molecularly Imprinted Polymer Sensor Based on Electropolymerization of a Microporous-Metal-Organic Framework for Tetracycline Detection in Honey. *Food Control* **2016**, *59*, 424–429. [CrossRef]
54. Hawthorne, M.F.; Young, D.C.; Andrews, T.D.; Howe, D.V.; Pilling, R.L.; Pitts, A.D.; Reintjes, M.; Warren, L.F.; Wegner, P.A. Pi.-Dicarbollyl Derivatives of the Transition Metals. Metallocene Analogs. *J. Am. Chem. Soc.* **1968**, *90*, 879–896. [CrossRef]
55. Xu, Q.-C.; Zhang, Q.-Q.; Sun, X.; Guo, Y.-M.; Wang, X.-Y. Aptasensors Modified by Antimony Tin Oxide Nanoparticle-Chitosan Based on Interdigitated Array Microelectrodes for Tetracycline Detection. *RSC Adv.* **2016**, *6*, 17328–17335. [CrossRef]
56. Le, T.H.; Pham, V.P.; La, T.H.; Phan, T.B.; Le, Q.H. Electrochemical Aptasensor for Detecting Tetracycline in Milk. *Adv. Nat. Sci. Nanosci. Nanotechnol.* **2016**, *7*, 015008. [CrossRef]

57. Chen, D.; Yao, D.; Xie, C.; Liu, D. Development of an Aptasensor for Electrochemical Detection of Tetracycline. *Food Control* **2014**, *42*, 109–115. [CrossRef]
58. Benvidi, A.; Tezerjani, M.D.; Moshtaghiun, S.M.; Mazloum-Ardakani, M. An Aptasensor for Tetracycline Using a Glassy Carbon Modified with Nanosheets of Graphene Oxide. *Microchim. Acta* **2016**, *183*, 1797–1804. [CrossRef]
59. Zhang, J.; Wu, Y.; Zhang, B.; Li, M.; Jia, S.; Jiang, S.; Zhou, H.; Zhang, Y.; Zhang, C.; Turner, A.P.F. Label-Free Electrochemical Detection of Tetracycline by an Aptamer Nano-Biosensor. *Anal. Lett.* **2012**, *45*, 986–992. [CrossRef]
60. Mohammad-Razdari, A.; Ghasemi-Varnamkhasti, M.; Rostami, S.; Izadi, Z.; Ensafi, A.A.; Siadat, M. Development of an Electrochemical Biosensor for Impedimetric Detection of Tetracycline in Milk. *J. Food Sci. Technol.* **2020**, *57*, 4697–4706. [CrossRef] [PubMed]
61. Benvidi, A.; Yazdanparast, S.; Rezaeinasab, M.; Tezerjani, M.D.; Abbasi, S. Designing and Fabrication of a Novel Sensitive Electrochemical Aptasensor Based on Poly (L-Glutamic Acid)/MWCNTs Modified Glassy Carbon Electrode for Determination of Tetracycline. *J. Electroanal. Chem.* **2018**, *808*, 311–320. [CrossRef]
62. Jahanbani, S.; Benvidi, A. Comparison of Two Fabricated Aptasensors Based on Modified Carbon Paste/Oleic Acid and Magnetic Bar Carbon Paste/Fe3O4@oleic Acid Nanoparticle Electrodes for Tetracycline Detection. *Biosens. Bioelectron.* **2016**, *85*, 553–562. [CrossRef]
63. Hou, W.; Shi, Z.; Guo, Y.; Sun, X.; Wang, X. An Interdigital Array Microelectrode Aptasensor Based on Multi-Walled Carbon Nanotubes for Detection of Tetracycline. *Bioprocess. Biosyst. Eng.* **2017**, *40*, 1419–1425. [CrossRef]
64. Liu, X.; Zheng, S.; Hu, Y.; Li, Z.; Luo, F.; He, Z. Electrochemical Immunosensor Based on the Chitosan-Magnetic Nanoparticles for Detection of Tetracycline. *Food Anal. Methods* **2016**, *9*, 2972–2978. [CrossRef]
65. Conzuelo, F.; Campuzano, S.; Gamella, M.; Pinacho, D.G.; Reviejo, A.J.; Marco, M.P.; Pingarrón, J.M. Integrated Disposable Electrochemical Immunosensors for the Simultaneous Determination of Sulfonamide and Tetracycline Antibiotics Residues in Milk. *Biosens. Bioelectron.* **2013**, *50*, 100–105. [CrossRef]

biosensors

Article

An Effective Electrochemical Platform for Chloramphenicol Detection Based on Carbon-Doped Boron Nitride Nanosheets

Jingli Yin [1,2], Huiying Ouyang [1,2], Weifeng Li [1] and Yumei Long [1,2,*]

1 College of Chemistry, Chemical Engineering and Materials Science, Soochow University, Suzhou 215123, China
2 The Key Lab of Health Chemistry and Molecular Diagnosis of Suzhou, Soochow University, Suzhou 215123, China
* Correspondence: yumeilong@suda.edu.cn

Abstract: Currently, accurate quantification of antibiotics is a prerequisite for health care and environmental governance. The present work demonstrated a novel and effective electrochemical strategy for chloramphenicol (CAP) detection using carbon-doped hexagonal boron nitride (C-BN) as the sensing medium. The C-BN nanosheets were synthesized by a molten-salt method and fully characterized using various techniques. The electrochemical performances of C-BN nanosheets were studied using cyclic voltammetry (CV) and electrochemical impedance spectroscopy (EIS). The results showed that the electrocatalytic activity of *h*-BN was significantly enhanced by carbon doping. Carbon doping can provide abundant active sites and improve electrical conductivity. Therefore, a C-BN-modified glassy carbon electrode (C-BN/GCE) was employed to determine CAP by differential pulse voltammetry (DPV). The sensor showed convincing analytical performance, such as a wide concentration range (0.1 μM–200 μM, 200 μM–700 μM) and low limit of detection (LOD, 0.035 μM). In addition, the proposed method had high selectivity and desired stability, and can be applied for CAP detection in actual samples. It is believed that defect-engineered *h*-BN nanomaterials possess a wide range of applications in electrochemical sensors.

Keywords: hexagonal boron nitride; carbon doping; chloramphenicol; differential pulse voltammetry; molten salt synthesis

Citation: Yin, J.; Ouyang, H.; Li, W.; Long, Y. An Effective Electrochemical Platform for Chloramphenicol Detection Based on Carbon-Doped Boron Nitride Nanosheets. *Biosensors* **2023**, *13*, 116. https://doi.org/10.3390/bios13010116

Received: 9 December 2022
Revised: 27 December 2022
Accepted: 4 January 2023
Published: 9 January 2023

1. Introduction

Chloramphenicol (CAP) is a nitrobenzene derivative and was first obtained from *Streptomyces venezuelae* in 1947 [1]. This compound is highly active against a variety of bacteria and thus has been widely employed as an antibiotic to treat various diseases [2,3]. However, the abuse of CAP has been a significant threat to environmental security and human health. For instance, CAP can easily accumulate to cause persistent environmental hazards due to its poor biodegradability [4,5]. Furthermore, these CAP residuals enter the human body through the food chain and result in serious diseases [6,7]. For this reason, the use of CAP in aquatic products and animal feeds has been prohibited in many countries [8]. Unfortunately, CAP residuals have still been detected in biological and environmental samples because of their low cost and high effectiveness [9–11]. Accordingly, it is imperative to develop an effective method for CAP detection.

Spectrophotometric and chromatographic methods have been developed for CAP detection [12–17]. Nevertheless, these methods are always time-consuming, expensive, and have cumbersome preprocessing [18]. Hence, most researchers have focused on electrochemical techniques due to their merits of simple preparation, fast response, good stability and in situ analysis [19,20]. Chloramphenicol is an electroactive substance and is able to produce an electrochemical signal when the redox reactions of CAP occur at the surface of an electrode. Generally, a bare GCE only gives rise to a poor electrochemical response toward CAP. To ensure high sensitivity and selectivity, chemical modification with appropriate electrocatalytic materials is necessary for GCEs. The surface modification is capable of accelerating charge migration and providing abundant active sites for further promoting electrochemical reactions. Consequently, tremendous efforts have been dedicated to preparing catalytic materials for the construction of electrochemical sensors.

Over the past decades, two-dimensional (2D) materials have been extensively exploited for the fabrication of electrochemical sensors based on their fascinating physiochemical performances [21–23]. Hexagonal boron nitride (*h*-BN) has a typical 2D layered structure and this compound has found widespread applications due to its high thermal conductivity, good lubricity, outstanding mechanical strength, and chemical stability [24–26]. Different from graphite of a semi-metallic nature, *h*-BN has a wide bandgap (~5.9 eV) because its π electrons are localized [27,28]. The applications of *h*-BN in functional devices have been limited due to its poor electron transport capability. Interestingly, the electronic properties of BN strongly rely on its microstructure, such as dimensionality, bond configuration and defects [29–31]. Among them, engineering defects is an effective way to ameliorate the photoelectrochemical properties of *h*-BN because they can form defect levels in the forbidden band to alter the energy band structure [32,33]. Recently, we demonstrated that point defects induced by heterovalent ion doping have great potential to improve sensing properties in *h*-BN systems [34,35]. Theoretical calculations also suggested that carbon doping can functionalize *h*-BN and makes it potentially applicable in the catalyst and sensor fields [36–38]. To the best of our knowledge, no studies on *h*-BN nanomaterials for the electrochemical detection of CAP have been reported.

In this work, a novel CAP electrochemical sensor was fabricated using C-BN nanosheets as the sensing medium. The C-BN nanosheets were prepared by a molten salt method and systematically characterized. Carbon doping enlarged the electrochemically active surface area (EASA) and facilitated electron transfer, which are favorable to the reaction kinetics. Therefore, enhanced electrochemical properties of *h*-BN were achieved upon doping with carbon. C-BN/GCE was employed for CAP detection, and its analytical performance was studied in detail. We also demonstrated the practical applications of the developed electrochemical sensor for CAP detection in real samples.

2. Materials and Methods

2.1. Materials

Cyanuramide ($C_3N_6H_6$), H_3BO_3, *h*-BN nanoparticles and chloramphenicol were obtained from Aladdin (Shanghai, China). Glucose, $ZnCl_2$, $CuCl_2{\cdot}2H_2O$, $NaNO_3$, $NaNO_2$, NaH_2PO_4, Na_2HPO_4, KCl, $K_3[Fe(CN)_6]$, and $K_4[Fe(CN)_6]$ were purchased from Sinopharm Chemical Reagent Co., Ltd. (Shanghai, China). Human serum, dopamine hydrochloride (DA), cysteine (Cys), chlortetracycline (CTC), and acetaminophen (AAP) were purchased from Shanghai Yuanye Bio-Technology Co., Ltd. Chloramphenicol eye drops (Runshu, Specification: 25 mg/10 mL) were produced by BAUSCH & LOMB Freda Pharmaceutical Co., Ltd. (Jinan, China). The Al_2O_3 polishing powder and glassy carbon electrodes (GCE, ϕ = 0.3 cm) were obtained from Shanghai Chenhua Co., Ltd. (Shanghai, China). All reagents were used as received. Deionized (DI) water was adopted for solution preparation throughout the work. Electrochemical tests were conducted in 0.2 M phosphate buffer saline (PBS), which were prepared by mixing an appropriate amount of NaH_2PO_4 (0.2 M) and Na_2HPO_4 (0.2 M).

2.2. Equipment

Powder X-ray diffraction (XRD) patterns were obtained on a D8 Bruker X-ray diffractometer with a Cu-$K_{\alpha 1}$ radiation source. Morphology, particle size and EDS analysis were investigated with an FEI Tecnai G20 transmission electron microscope (TEM). Fourier transform infrared (FTIR) spectra were acquired from a Nicolet 550 spectrometer. An F-4500 fluorescence spectrometer was used to test the photoluminescence (PL) properties of the samples. The UV–Vis diffuse reflectance spectroscopy (DRS) was recorded with a UV3600 spectrophotometer (Shimadzu Corporation, Japan).

A CHI 660E workstation (Shanghai China) was utilized for electrochemical tests, which were performed in conventional three-electrode installation at room temperature (25 ± 2 °C). The electrochemical installation consists of a saturated calomel electrode (SCE, reference electrode) and a platinum wire (counter electrode). Unmodified/modified GCEs were adopted as working electrodes. The CV and EIS measurements were carried out in a KCl solution (0.1 M) containing 1 mM $[Fe(CN)_6]^{3-/4-}$. The DPV plots were acquired in PBS solution. The scanning parameters were −0.8–−0.4 V for potential window, 50 mV for modulation amplitude, 4 mV for potential increment and 0.05 s for pulse width.

2.3. Synthesis of Carbon-Doped BN

Carbon-doped boron nitride (C-BN) was prepared modified from a previous report [35]. In general, cyanuramide aqueous solution is weakly basic, and H_3BO_3 is weakly acidic. They can interact with each other through acid-base interactions to form $C_3N_6H_6{\cdot}2H_3BO_3$ [39]. To successfully introduce carbon sources, excessive cyanuramide was added to the reaction. In a typical procedure for preparing carbon-doped BN nanosheets, 3.0 g H_3BO_3 was dissolved into 200 mL DI water and heated to 95 °C (water bath). After that, 5.0 g cyanuramide was added into the hot boric acid solution partially three times. The above mixture was vigorously stirred (95 °C) for 2 h and then subjected to the removal of water by vacuum rotary evaporation. The solid mixture was dried overnight at 80 °C and crushed to obtain a white precursor. Afterwards, 2 g precursor was well-mixed with 10 g NaCl/KCl (mass ratio of 45/55) eutectic matrix and transferred to a high-alumina crucible. The mixture was calcined at 900 °C for 3 h under a reducing atmosphere (carbon). To remove free carbon residues, the powders were washed with hot dilute nitric acid (0.2 M) and hot water several times. The final yellow powder was ground and dried and named C-BN.

2.4. Construction of the C-BN/GCE Sensor

The C-BN/GCE was obtained using a casting method. First, successive mirror polishing of the GCE was carried out on a suede using Al_2O_3 slurries (300 and 50 nm) and cleaned with ethanol and deionized water. Next, the cleaned GCE was treated in 1.0 M H_2SO_4 solution until a stable cyclic voltammetry curve was obtained. Meanwhile, C-BN aqueous suspensions (1 mg/mL) were obtained by dispersing C-BN nanosheets in deionized water under ultrasonication. Then, the suspension was spread on the surface of the pretreated GCE and dried naturally. Finally, the C-BN/GCE was carefully rinsed with deionized water to remove any unbound compounds. As a reference, an *h*-BN-modified GCE (*h*-BN/GCE) was constructed with the same method.

2.5. Sample Preparation

A CAP stock solution (10 mM) was freshly obtained by dissolving 0.323 g CAP in 100 mL deionized water and stored at 4 °C in the dark. The practicability of the proposed method was investigated by analyzing CAP in real samples (chloramphenicol eye drops and human serum), which were first diluted using 0.2 M PBS (pH 7.0). The final CAP concentrations were then examined by the C-BN/GCE sensor using a standard addition method.

3. Results and Discussion

3.1. Characterization

To identify the crystal structure of the products, powder XRD measurements were performed. Figure 1a displays the XRD patterns of C-BN and *h*-BN. As a reference, the standard diffraction data (JCPDS No. 34–0421) of hexagonal BN are also provided in Figure 1a. For both samples, all the diffraction peaks could be well indexed by standard diffraction data, and no additional diffraction peaks were found, indicating good phase purity. Different from *h*-BN, C-BN showed weak and broad diffraction peaks, suggesting that the hexagonal crystal structure was disturbed by carbon doping. A peak shift was also observed in the C-BN XRD pattern, which is attributed to the successful incorporation of carbon in the *h*-BN structure [39,40]. In the FTIR spectra (Figure 1b), both C-BN (red curve) and *h*-BN showed two strong absorption bands at 1389 and 805 cm^{-1}, which should be ascribed to the in-plane B-N stretching and out-of-plane B-N-B bending vibrations [41,42]. For the C-BN sample, some new IR absorption bands were visible. Several bands in the 1000–1210 cm^{-1} range can be assigned to C-N and C-B vibrations, and C = N vibrations contribute to the absorption band at 1662 cm^{-1} [40,42]. The results suggested that carbon atoms were successfully incorporated into the *h*-BN crystal. On the other hand, the existence of O-H and N-H was evidenced by absorption bands at 3428 cm^{-1} and 3182 cm^{-1}. These functional groups are beneficial for adsorbing target objects for further catalytic reactions.

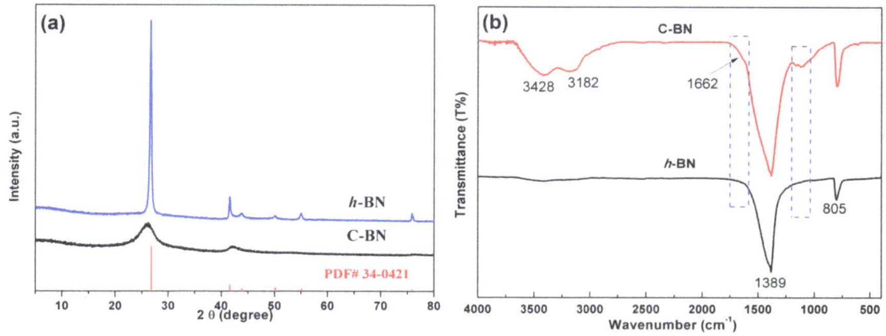

Figure 1. (**a**) XRD patterns and (**b**) FTIR spectra of *h*-BN and C-BN.

Figure 2 displays TEM images of the *h*-BN and C-BN samples. Both samples are composed of sheet-like materials. These nanosheets were highly transparent under the electron beam, implying their ultrathin features. Compared to that of *h*-BN (Figure 2a), C-BN (Figure 2b) exhibited a more uniform morphology and smaller size, which would provide a more active surface for adsorbing target compounds. Figure S1 shows the EDS profile of C-BN, confirming the presence of B, N and C.

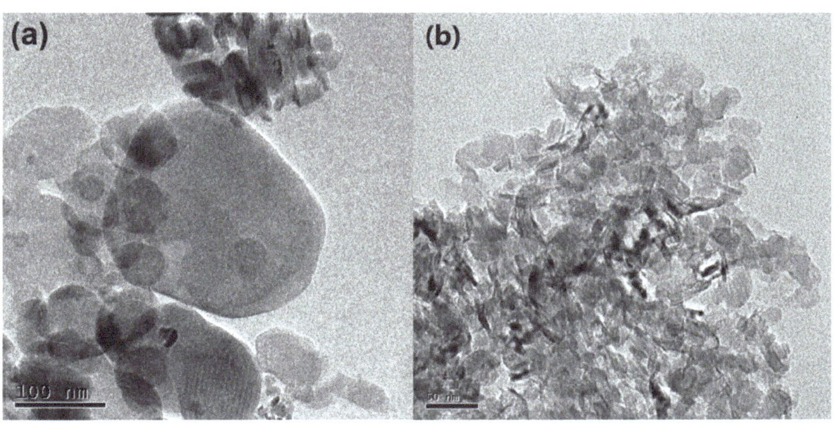

Figure 2. TEM images of (**a**) *h*-BN and (**b**) C-BN.

The band structure of semiconductors plays an important role in their electrocatalytic activity. The DRS was measured to investigate the influence of carbon doping on the band structure. As shown in Figure 3a, there was a drastic absorption in the ultraviolet window for both samples, which corresponds to the bandgap energy of a semiconductor. Compared to that of *h*-BN, the absorption edge of C-BN obviously shifts to the low energy direction. In addition, a wide absorption band appeared in the whole visible range for the C-BN sample. The results indicate that carbon doping has a significant impact on the band structure of *h*-BN and then modulates its photoelectronic properties. Figure 3b shows the PL spectra of the *h*-BN and C-BN samples under excitation at 230 nm. The C-BN shows multiple emission bands, which were not observed for *h*-BN. These emission peaks of C-BN were ascribed to carbon-linked defects, which modulated the behaviors of electrons [43].

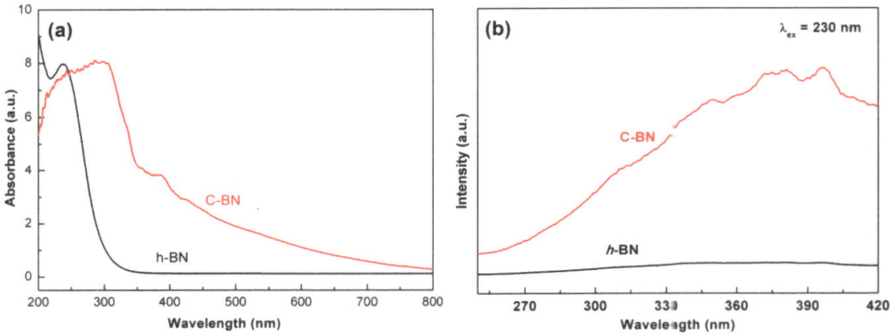

Figure 3. (**a**) UV–Vis DRS and (**b**) emission spectra (excited by 230 nm) of C-BN (red curve) and *h*-BN (black curve).

3.2. CV and EIS Studies

Figure 4a shows the CV response of different electrodes. Each electrode demonstrated a pair of well-defined redox peaks, which resulted from the redox of $[Fe(CN)_6]^{3-/4-}$. Among all the electrodes, C-BN/GCE exhibited the sharpest and highest redox peaks. This reveals that the electron transfer rate of C-BN/GCE is faster than that of the others. In addition, the EASA of the electrode was calculated based on the dependence of the CV on the scan rate (Figure S2a,c,e) using the Randles–Sevcik equation [44]

$$I_p = 2.69 \times 10^5 \cdot (n^{3/2})\, A \cdot D^{1/2} \cdot C \cdot \nu^{1/2} \quad (1)$$

where I_p is the peak current and ν is the scan rate (V/s). For 1 mM $[Fe(CN)_6]^{3-/4-}$ (C) in 0.1 M KCl solution, n equals 1 and the diffusion coefficient (D) is $6.5 \times 10^6 \cdot cm^2/s$. Accordingly, the EASA (A) of electrodes can be estimated based on the fitting results between I_p and $\nu^{1/2}$ (Figure S2b,d,f). The EASA values of GCE, *h*-BN/GCE and C-BN/GCE were 0.19 cm^2, 0.17 cm^2, and 0.23 cm^2, respectively. The good electron transfer ability and large EASA of C-BN/GCE should be ascribed to the defect structures induced by carbon-doping, as they not only modulate the electronic structure of *h*-BN but also provide more active sites for electrochemical reactions.

The EIS technique was utilized to study the interface nature of different electrodes, and the results are shown in Figure 4b. In the Nyquist plot, each curve consists of a semicircle part and a linear part. The semicircle diameter is proportional to the charge transfer resistance (R_{ct}) of an electrode, reflecting the electron transport ability at the electrode/electrolyte interface. As shown in Figure 4b, there are significant differences between the plots, revealing the distinct electrochemical behaviors of these electrodes. By fitting with the equivalent Randle circuit (inset of Figure 4b), the R_{ct} values are 843 Ω, 3383 Ω and 275 Ω for the GCE, *h*-BN/GCE and C-BN/GCE, respectively. The R_{ct} of C-BN/GCE was less than that of the other electrodes. This result is consistent with the above CV data, indicating the outstanding charge transferability of C-BN/GCE. Therefore, it is expected that carbon-doped BN has potential applications in electrochemical sensors.

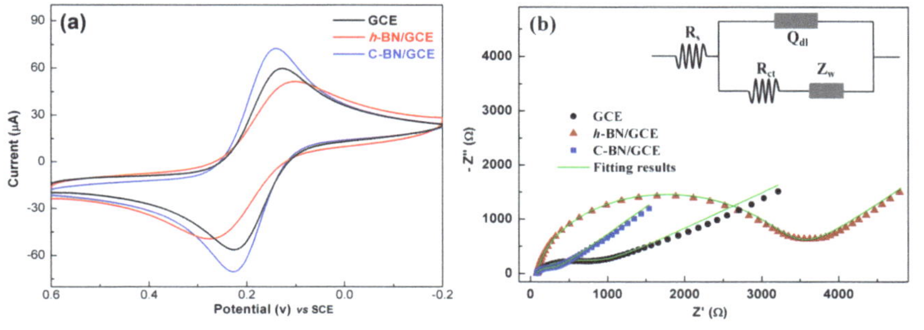

Figure 4. (**a**) CV and (**b**) EIS plots of 1 mM $[Fe(CN)_6]^{3-/4-}$ in 0.1 M KCl solution for GCE, *h*-BN/GCE and C-BN/GCE. Inset displaying the equivalent circuit for EIS tests.

3.3. Electrochemical Behaviors of CAP over C-BN/GCE

The electrochemical behaviors of CAP over different electrodes were first investigated by CV in 0.2 M PBS (pH = 7.0) with a scan rate of 100 mV/s. Figure 5a displays CV profiles obtained at the C-BN/GCE in the absence/presence of 100 μM CAP. No redox signals were discerned in the absence of CAP. In contrast, there was a pair of redox peaks in the presence of 100 μM CAP with an oxidation peak at −0.039 V and a reduction peak at −0.675 V (vs. SCE), respectively. The −0.675 V peak resulted from the nitro group reduction of CAP to hydroxylamine and the −0.039 V was ascribed to the oxidation of hydroxylamine to a nitroso group [45]. As depicted in Figure 5b, both the GCE and *h*-BN/GCE exhibited similar voltammetric signals toward CAP because they are electroactive. Interestingly, the redox signal produced by C-BN/GCE was remarkably larger than those obtained from bare GCE and *h*-BN/GCE. Its reduction peak current (88.9 μA) was approximately 2.6 times that of *h*-BN/GCE (34.1 μA) and 2.7 times that of GCE (32.9 μA). The favorable electrocatalytic effect of C-BN in the redox of CAP is attributed to its defect structure induced by carbon doping, which facilitates electron transfer and provides abundant active sites. Therefore, C-BN/GCE shows high sensitivity in the detection of CAP. As the reduction step of CAP at −0.68 V exhibited a better current response compared to the others, this peak was adopted for later experiments.

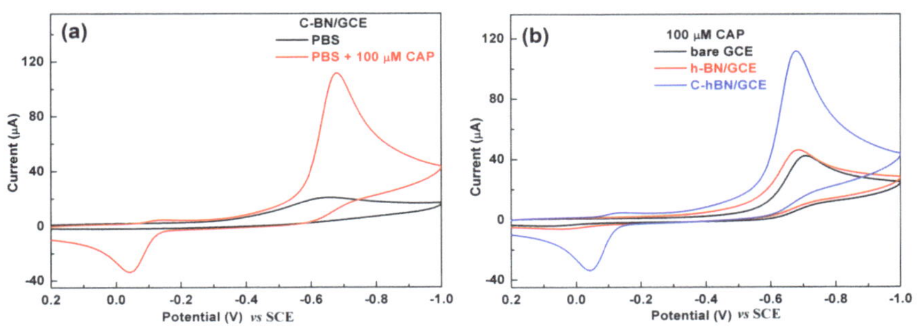

Figure 5. (**a**) CVs of C-BN/GCE in the absence/presence of 100 μM CAP; (**b**) CVs of GCE, *h*-BN/GCE and C-BN/GCE in the presence of 100 μM CAP. (Conditions: 0.2 M PBS with pH of 7.0 at a scan rate of 100 mV/s).

3.4. Effect of Scan Rate

Figure 6 illustrates the influence of the scan rate on the CV response of C-BN/GCE toward 100 μM CAP. The test conditions were 0.2 M PBS and pH 7.0. Figure 6a shows that the reduction peak current has a positive relation to the scan rate from 25 to 200 mV/s. A linear relationship between the reduction peak current and the square root of the scan rate ($\upsilon^{1/2}$) was observed (Figure 6b). The regression equation was defined as $I_{pc} = 9.477\ \upsilon^{1/2} - 6.788$.

The good linear relationship (R^2 = 0.998) revealed that electrochemical reduction of CAP over C-BN/GCE is a diffusion–controlled process [2].

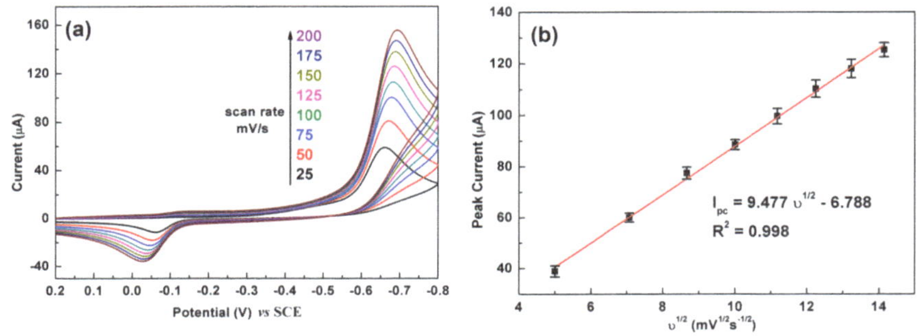

Figure 6. (**a**) The effect of scan rate on the CV response of the C-BN/GCE toward CAP (100 μM) and (**b**) corresponding linear plot of reduction currents against square root of scan rate. (Conditions: 0.2 M PBS with pH of 7.0).

3.5. Effect of pH

It is well known that the pH of the electrolytes plays an important role in an electrochemical reaction. Figure 7a presents the influence of pH on the CV response of C-BN/GCE toward CAP. The current response gradually increased with increasing pH from 5.0 to 7.0 and then significantly decreased with further increasing pH (Figure 7b, left). Meanwhile, the larger pH value makes the reduction peak potential shift to a more negative direction, implying that protons participated in the electrochemical reaction. As illustrated in Figure 7b (right), the peak potential showed a good linear relationship to pH with a regression equation of $I_{pc} = -0.0303$ pH–0.468 (R^2 = 0.994). The slope value of −0.0303 V/pH is very close to the theoretical value of −0.059/2 V/pH. Therefore, the electroredox of CAP over C-BN/GCE is an irreversible process involving one electron and two protons, which is consistent with previous reports [45,46]. According to literature [47], the pH value has an impact on the structure of CAP and further on its adsorption on the electrode surface. On the other hand, increasing the pH of the alkaline electrolyte could inhibit the reduction reaction of CAP. As a result, the best current response was achieved at pH 7.0, and was adopted for the later studies. The possible electrochemical reaction mechanism of CAP on C-BN/GCE can be described by Figure S3 [2].

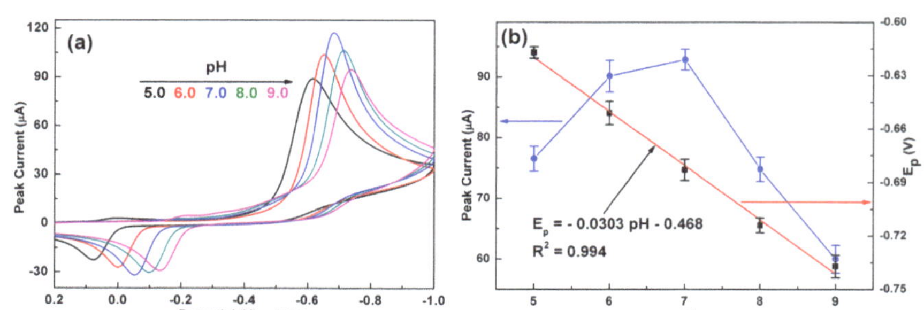

Figure 7. (**a**) CVs of C-BN/GCE at different pH (5.0, 6.0, 7.0, 8.0 and 9.0). (**b**) Dependence of the reduction peak current (left, blue curve) and E_{pc} (right, black dots) on pH, and corresponding fitting results (red plot). (Conditions: 0.2 M PBS containing 100 μM CAP at a scan rate of 100 mV/s).

3.6. Detection of CAP Using the DPV Method

The DPV technique is more effective and sensitive than CV for the quantification of organic substances and thus was employed for CAP detection. Figure 8a shows a series of

DPV curves of C-BN/GCE for different CAP concentrations. The current response monotonically increased with increasing CAP concentration from 0.1 μM to 700 μM. Figure 8b illustrates the linear calibration plots between the peak current and concentration. The regression equations can be described as I_{pc} (μA) = 0.422 C (μM) + 5.02 (R^2 = 0.994) in the low concentration region (0.1–200 μM) and I_{pc} (μA) = 0.160 C (μM) + 58.5 (R^2 = 0.989) in the high concentration region (200–700 μM). The limit of detection (LOD) was calculated using the equation LOD = 3σ/S according to the IUPAC definition [18], in which σ is the standard deviation of 5 blank measurements and S is the slope of the calibration plot. The proposed method shows a low LOD of 0.035 μM. The analytical performance of C-BN/GCE was compared with that of other electrodes reported previously. As summarized in Table 1, C-BN/GCE is a suitable platform for CAP detection. Moreover, the construction procedure of C-BN/GCE is facile, and all elements, including carbon, boron, and nitrogen, are earth-abundant.

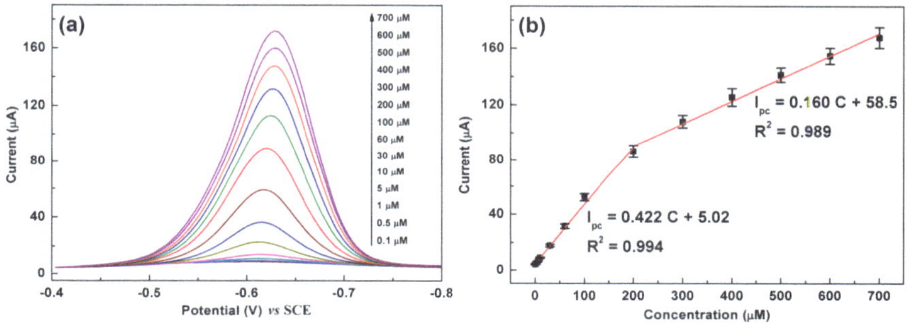

Figure 8. (**a**) DPV responses of C-BN/GCE toward different CAP concentrations; (**b**) corresponding linear plots of I_{pc} vs. CAP concentration (Circle dots are experimental data and red lines are fitting results). (Conditions: 0.2 M PBS with pH of 7.0, scanning window: −0.8–−0.4 V, modulation amplitude: 50 mV, potential increment: 4 mV, pulse width: 0.05 s).

Table 1. Comparison of the analytical performance of C-BN/GCE with previously reported electrochemical methods for CAP detection.

Electrode Material	Method	Linear Range (μM)	LOD (μM)	Refs
rGO@PDA@AuNPs	DPV	0.1–100	0.058	[2]
ZVO/SGN/LGE	DPV	0.005–325.5	0.0024	[44]
Co_3O_4@rGO	i-t	0.1–1500	0.1	[48]
Fe_3O_4/N-rGO	DPV	1–200	0.03	[49]
Sr-ZnO@rGO	LSV	0.19–2847.3	0.13	[50]
MoN@S-GCN	DPV	0.5–2450	0.0069	[51]
α-Fe_2O_3/SPCE	DPV	2.5–50	0.11	[52]
MoS_2-rGO	DPV	1–55	0.6	[53]
3D-printed CB/PLA	DPV	10–331	0.98	[54]
$NiCo_2O_4$@C	DPV	0.5–320	0.035	[55]
MoS_2-MWCNTs	DPV	1–35	0.4	[56]
MoS_2/f-MWCNTs	i-t	0.08–1392	0.015	[57]
Mn_2O_3 TNS/SPCE	DPV	0.015–1.28	0.00426	[58]
Eu_2O_3/GO	i-t	0.02–800.25	0.00132	[59]
C-BN	DPV	0.1–200, 200–700	0.035	Here

rGO: reduced graphene oxide; PDA: polydopamine; ZVO: ZnV_2O_8; SGN: sulfur doped carbon nitride; LGE: laser-induced graphene electrode; Sr-ZnO: Strontium doped zinc oxide; SPCE: screen-printed carbon electrode; S-GCN: sulfur-doped graphitic carbon nitride; CB/PLA: carbon-black integrated polylactic acid; MWCNT: multi-walled carbon nano-tube.

3.7. Selectivity and Stability

The selectivity of the C-BN/GCE for the electrocatalytic redox of CAP was investigated to ensure its practical applications. The DPV measurements were conducted using C-BN/GCE to detect 50 μM CAP in the absence/presence of various interfering compounds (IC), such as 10-fold concentrations of Zn^{2+}, Cu^{2+}, Fe^{3+}, NO_3^-, glucose, Cys, NO_2^-, DA, CTC, and AAP. Figure 9a displays the ratios of peak current before and after adding various interferents. All the investigated compounds had a negligible influence on the current response of CAP with a current deviation of less than 5.0%, indicating the good selectivity of the developed method. The satisfied selectivity of the proposed method should be ascribed to the π-π interaction and specific surface adsorption of CAP molecules by C-BN nanosheets.

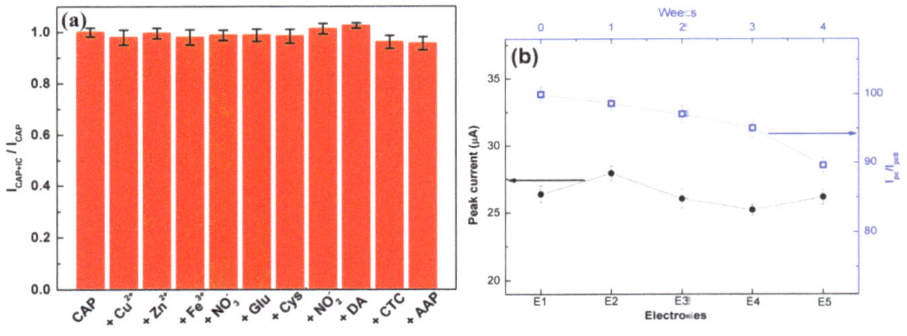

Figure 9. (**a**) Anti-interference ability of C-BN/GCE in the determination of CAP (50 μM CAP, 10-fold concentrations of Zn^{2+}, Cu^{2+}, Fe^{3+}, NO_3^-, glucose, Cys, NO_2^-, DA, CTC, and AAP; (**b**) Left (black curve): Reproducibility of the proposed sensor; Right (blue curve): Evaluation of the long-term stability of the proposed sensor for four weeks (Circle and square symbols are experimental data). All tests were performed in 0.2 M PBS at pH 7.0 using the DPV method.

Five different C-BN/GCEs were fabricated independently through the same process and used for CAP detection under the same conditions. All the electrodes displayed similar peak current values with a relative standard deviation (RSD) of 2.76% (Figure 9b, left), confirming the favorable reproducibility of the proposed method. Meanwhile, ten cycle tests were performed on the same C-BN/GCE in the presence of 50 μM CAP. The RSD of the current responses was 2.44%. Therefore, the modified electrode possesses prominent repeatability. For long-term stability, the current response of the same C-BN/GCE toward 50 μM CAP was measured every week. After 4 weeks of storage at room temperature, the retention rate of the peak current was 89.6% (Figure 9b, right), demonstrating that the sensor was stable.

3.8. Real Sample Analysis

To assess the practical feasibility of the proposed method, C-BN/GCE was applied to detect CAP in real samples using a standard addition method. Real samples (human serum and eye drops) were first diluted using 0.2 M PBS (pH 7.0) and spiked with defined CAP content. The final CAP concentrations were tested by the DPV technique, and three parallel tests were conducted for every concentration. As listed in Table 2, C-BN/GCE showed satisfactory recoveries and small RSDs (less than 5%), confirming that the proposed method is reliable and applicable.

Table 2. CAP quantification in real samples using C-BN/GCE sensor (n = 3).

Real Sample	Added (μM)	Founded (μM)	Recovery (%)	RSD (%)
Human serum	0	–	–	–
	20	19.75 ± 0.36	98.8 ± 1.8	3.05
	40	39.68 ± 1.07	99.2 ± 2.7	2.61
	60	61.02 ± 1.79	101.7 ± 2.9	2.93
Eye drops [1]	0	7.912 ± 0.31	102.3 ± 3.8	3.96
	20	27.65 ± 0.12	99.7 ± 0.4	0.43
	40	47.90 ± 0.82	100.3 ± 1.7	1.70
	60	65.79 ± 1.00	97.1 ± 1.4	1.52

[1] The stock solution of CAP eye drops is 7.737 μM.

4. Conclusions

In conclusion, a novel and efficient CAP detection method was developed using C-BN nanosheets as sensing materials. The C-BN nanosheets were successfully synthesized via a molten-salt process and well characterized. Modulation of the band structure in defective *h*-BN was observed by carbon doping, which improved the electrochemical performance of C-BN/GCE, such as conductivity and EASA. As a result, C-BN/GCE showed enhanced electrocatalytic activity in the redox of CAP compared to that of *h*-BN. Under the optimized conditions, the proposed method exhibited a low LOD of 0.035 μM and a wide linear range from 0.1 μM to 700 μM in the detection of CAP. Moreover, the good anti-interference ability, high stability, and desired reproducibility of C-BN/GCE were experimentally authenticated. The good recovery results obtained in real sample analyses suggested the practical feasibility of the as-fabricated sensor for CAP detection. Hence, defect-engineered *h*-BN nanomaterials are potential candidates for electrochemical sensing applications.

Supplementary Materials: The following supporting information can be downloaded at https://www.mdpi.com/article/10.3390/bios13010116/s1: Figure S1: EDS profile for C-BN; Figure S2: CVs of (a) bare GCE, (c) *h*-BN/GCE and (e) C-BN/GCE at different scan rate (25–200 mV/s). Plots of the redox peak current (I_p) versus the square root of the scan rate ($\upsilon^{1/2}$) at (b) bare GCE, (d) *h*-BN/GCE and (f) C-BN/GCE. All tests were conducted in 0.1 M KCl solution containing 1 mM $[Fe(CN)_6]^{3-/4-}$; Figure S3: The possible electrochemical reaction mechanism of CAP at C-BN/GCE.

Author Contributions: J.Y.: synthesis, collected data, data curation and writing—original draft; H.O.: synthesis, collected data, data curation; W.L.: data curation, validation and writing—review; Y.L.: resources, conceptualization, methodology, funding acquisition, formal analysis, writing—review and editing. All authors have read and agreed to the published version of the manuscript.

Funding: This research is supported by the Priority Academic Program Development of Jiangsu Higher Education Institutions and the Project of Scientific and Technologic Infrastructure of Suzhou (SZS201708).

Institutional Review Board Statement: Not applicable.

Informed Consent Statement: Not applicable.

Data Availability Statement: All the data are presented in the manuscript.

Conflicts of Interest: The authors declare no conflict of interest.

References

1. Nguyen, L.M.; Nguyen, N.T.T.; Nguyen, T.T.T.; Nguyen, T.T.; Nguyen, D.T.C.; Tran, T.V. Occurrence, toxicity and adsorptive removal of the chloramphenicol antibiotic in water: A review. *Environ. Chem. Lett.* **2022**, *20*, 1929–1963. [CrossRef] [PubMed]
2. Zhang, L.; Yin, M.; Wei, X.X.; Sun, Y.W.; Chen, Y.; Qi, S.Y.; Tian, X.X.; Qiu, J.X.; Xu, D.P. Synthesis of rGO@PDA@AuNPs for an effective electrochemical chloramphenicol sensor. *Diamond Relat. Mater.* **2022**, *128*, 109311. [CrossRef]
3. Hu, X.; Qin, J.Z.; Wang, Y.B.; Wang, J.J.; Yang, A.J.; Yang, A.J.; Tsang, Y.F.; Liu, B.J. Synergic degradation Chloramphenicol in photo-electrocatalytic microbial fuel cell over Ni/MXene photocathode. *J. Colloid Interface Sci.* **2022**, *628*, 327–337. [CrossRef] [PubMed]

4. Zhang, J.Y.; Zhao, R.X.; Cao, L.J.; Lei, Y.S.; Liu, J.; Feng, J.; Fu, W.J.; Li, X.Y.; Li, B. High-efficiency biodegradation of chloramphenicol by enriched bacterial consortia: Kinetics study and bacterial community characterization. *J. Hazard. Mater.* **2020**, *384*, 121344. [CrossRef] [PubMed]
5. Lin, J.; Zhang, K.T.; Jiang, L.K.; Hou, J.F.; Yu, X.; Feng, M.B.; Ye, C.S. Removal of chloramphenicol antibiotics in natural and engineered water systems: Review of reaction mechanisms and product toxicity. *Sci. Total Environ.* **2022**, *850*, 158059. [CrossRef]
6. Hu, X.; Deng, Y.; Zhou, J.T.; Liu, B.J.; Yang, A.J.; Jin, T.; Tsang, Y.F. N-and O self-doped biomass porous carbon cathode in an electro-Fenton system for Chloramphenicol degradation. *Sep. Purif. Technol.* **2020**, *251*, 1117376. [CrossRef]
7. Wei, L.Y.; Jiao, F.; Wang, Z.L.; Wu, L.; Dong, D.M.; Chen, Y.P. Enzyme-modulated photothermal immunoassay of chloramphenicol residues in milk and egg using a self-calibrated thermal imager. *Food Chem.* **2022**, *392*, 133232. [CrossRef]
8. Vilian, A.T.E.; Oh, S.Y.; Rethinasabapathy, M.; Umapathi, R.; Hwang, S.K.; Oh, C.W.; Park, B.; Huh, Y.S.; Han, Y.K. Improve conductivity of flower-like $MnWO_4$ on defect engineered graphitic carbon nitride as an efficient electrocatalyst for ultrasensitive sensing of chloramphenicol. *J. Hazard. Mater.* **2020**, *399*, 122868. [CrossRef]
9. Yang, J.W.; Ji, G.Z.; Gao, Y.; Fu, W.; Irfan, M.; Mu, L.; Zhang, Y.L.; Li, A.M. High-yield and high-performance porous biochar produced from pyrolysis of peanut shell with low-dose ammonium polyphosphate for chloramphenicol adsorption. *J. Clean. Prod.* **2020**, *264*, 121516. [CrossRef]
10. He, F.; Ma, W.C.; Zhong, D.; Yuan, Y.X. Degradation of chloramphenicol by α-FeOOH-activated two different double-oxidant systems with hydroxylamine assistance. *Chemosphere* **2020**, *250*, 126150. [CrossRef]
11. Sniegocki, T.; Sikorska, M.G.; Posyniak, A. Transfer of chloramphenicol from milk to commercial dairy products—Experimental proof. *Food Chem.* **2015**, *57*, 411–418. [CrossRef]
12. Elik, A.; Altunay, N. Chemometric approach for the spectrophotometric determination of chloramphenicol in various food matrices: Using natural deep eutectic solvents. *Spectrochim. Acta Part A* **2022**, *276*, 121198. [CrossRef]
13. Vuran, B.; Ulusoy, H.I.; Sarp, G.; Yilmaz, E.; Morgul, U.; Kabir, A.; Tartaglia, A.; Locatelli, M.; Soylak, M. Determination of chloramphenicol and tetracycline residues in milk samples by means of nanofiber coated magnetic particles prior to high-performance liquid chromatography-diode array detection. *Talanta* **2021**, *230*, 122307. [CrossRef]
14. Sharma, R.; Akshath, U.S.; Bhatt, P.; Raghavarao, K. Fluorescent aptaswitch for chloramphenicol detection—Quantification enabled by immobilization of aptamer. *Sens. Actuator B Chem.* **2019**, *290*, 110–117. [CrossRef]
15. Abnous, K.; Danesh, N.M.; Ramezani, M.; Emrani, A.S.; Taghdisi, S.M. A novel colorimetric sandwich aptasensor based on an indirect competitive enzyme-free method for ultrasensitive detection of chloramphenicol. *Biosen. Bioelectron.* **2016**, *78*, 80–86. [CrossRef]
16. Liu, T.S.; Xie, J.; Zhao, J.F.; Song, G.X.; Hu, Y.M. Magnetic chitosan nanocomposite used as cleanup material to detect chloramphenicol in milk by GC-MS. *Food Anal. Method.* **2014**, *7*, 814–819. [CrossRef]
17. Kim, N.; Park, I.S. Development of a chemiluminescent immunosensor for chloramphenicol. *Anal. Chim. Acta* **2006**, *578*, 19–24.
18. Demir, E.; Silah, H. Development of a new analytical method for determination of veterinary drug oxyclozanide by electrochemical sensor and its application to pharmaceutical formulation. *Chemosensors* **2020**, *8*, 25. [CrossRef]
19. Abhishek, K.J.; Sathish, R.; Shubha, A.; Lakshmi, B.; Deepak, K.; Naveen, C.S.; Harish, K.N.; Ramakrishna, S. A review on nanomaterial-based electrodes for the electrochemical detection of chloramphenicol and furazolidone antibiotics. *Anal. Methods* **2022**, *14*, 3228–3249.
20. Dong, X.Z.; Yan, X.H.; Li, M.; Liu, H.; Li, J.W.; Wang, L.; Wang, K.; Lu, X.; Wang, S.Y.; He, B.S. Ultrasensitive detection of chloramphenicol using electrochemical aptamer sensor: A mini review. *Electrochem. Commun.* **2020**, *120*, 106835. [CrossRef]
21. Tyagi, D.; Wang, H.D.; Huang, W.C.; Hu, L.P.; Tang, Y.F.; Guo, Z.N.; Ouyang, Z.B.; Zhang, H. Recent advances in two-dimensional-material-based sensing technology toward health and environmental monitoring applications. *Nanoscale* **2020**, *12*, 3535–3559. [CrossRef] [PubMed]
22. Li, T.; Shang, D.W.; Gao, S.W.; Wang, B.; Kong, H.; Yang, G.Z.; Shu, W.D.; Xu, P.L.; Wei, G. Two-dimensional Material-based electrochemical sensors/biosensors for food safety and biomolecular detection. *Biosensors* **2022**, *12*, 314. [CrossRef] [PubMed]
23. Li, M.N.; Huang, G.; Chen, X.J.; Yin, J.N.; Zhang, P.; Yao, Y.; Shen, J.; Wu, Y.W.; Huang, J. Perspectives on environmental applications of hexagonal boron nitride nanomaterials. *Nano Today* **2022**, *44*, 101486. [CrossRef]
24. Rohaizad, N.; Mayorga-Martinez, C.C.; Fojtu, M.; Latiff, N.M.; Pumera, M. Two-dimensional materials in biomedical, biosensing and sensing applications. *Chem. Soc. Rev.* **2021**, *50*, 619–657. [CrossRef] [PubMed]
25. Ihsanullah, I. Boron nitride-based materials for water purification: Progress and outlook. *Chemosphere* **2021**, *263*, 127970. [CrossRef]
26. Wang, J.M.; Zhang, L.Z.; Wang, L.F.; Lei, W.W.; Wu, Z.S. Two-dimensional boron nitride for electronics and energy applications. *Energy Environ. Mater.* **2021**, *5*, 10–44. [CrossRef]
27. Lopes, J.M.J. Synthesis of hexagonal boron nitride: From bulk crystals to atomically thin films. *Prog. Cryst. Growth Charact. Mater.* **2021**, *67*, 100522. [CrossRef]
28. Weng, Q.H.; Wang, X.B.; Wang, X.; Bando, Y.; Golberg, D. Functionalized hexagonal boron nitride nanomaterials: Emerging properties and application. *Chem. Soc. Rev.* **2016**, *45*, 3989–4012. [CrossRef]
29. Angizi, S.; Khalaj, M.; Alem, S.A.A.; Pakdel, A.; Willander, M.; Hatamie, A.; Simchi, A. Review—Towards the two-dimensional hexagonal boron nitride (2D h-BN) electrochemical sensing platforms. *J. Electrochem. Soc.* **2020**, *167*, 126513. [CrossRef]
30. Yin, J.; Li, J.D.; Hang, Y.; Yu, J.; Tai, G.A.; Li, X.M.; Zhang, Z.H.; Guo, W.L. Boron nitride nanostructures: Fabrication, functionalization and applications. *Small* **2016**, *12*, 2942–2968. [CrossRef]

31. Gottscholl, A.; Diez, M.; Soltamov, V.; Kasper, C.; Sperlich, A.; Kianinia, M.; Bradac, C.; Aharnonvich, I.; Dyakonov, V. Room temperature coherent control spin defects in hexagonal boron nitride. *Sci. Adv.* **2021**, *7*, eabf3630. [CrossRef]
32. Legesse, M.; Rashkeev, S.N.; Saidaoui, H.; Mellouhi, F.E.; Ahzi, S.; Alharbi, F.H. Band gap tuning in aluminum doped two-dimensional hexagonal boron nitride. *Mater. Chem. Phys.* **2020**, *250*, 123176. [CrossRef]
33. Zhang, J.J.; Sun, R.; Ruan, D.L.; Zhang, M.; Li, Y.X.; Zhang, K.; Cheng, F.L.; Wang, Z.C.; Wang, Z.M. Point defects in two-dimensional hexagonal boron nitride: Perspective. *J. Appl. Phys.* **2020**, *128*, 100902. [CrossRef]
34. Shen, Y.L.; Ouyang, H.Y.; Li, W.F.; Long, Y.M. Defect-enhanced electrochemical property of h-BN for Pb^{2+} detection. *Microchim. Acta* **2021**, *188*, 40. [CrossRef]
35. Ouyang, H.Y.; Li, W.F.; Long, Y.M. Carbon-doped h-BN for the enhanced electrochemical determination of dopamine. *Electrochim. Acta* **2021**, *369*, 137682. [CrossRef]
36. Esrafili, M.D.; Rad, F.A. Carbon-doped boron nitride nanosheets as highly sensitive materials for detection of toxic NO and NO_2 gases: A DFT study. *Vacuum* **2019**, *166*, 127–134. [CrossRef]
37. Mudchimo, T.; Namuangruk, S.; Kungwan, N.; Jungsuttiwong, S. Carbon-doped boron nitride nanosheet as a promising metal-free catalyst for NO reduction: DFT mechanistic study. *Appl. Catal. A Gen.* **2018**, *557*, 79–88. [CrossRef]
38. Gao, M.; Adachi, M.; Lyalin, A.; Taketsugu, T. Long range functionalization of h-BN monolayer by carbon doping. *J. Phys. Chem. C* **2016**, *120*, 15993–16001. [CrossRef]
39. Tippo, P.; Singjai, P.; Choopun, S.; Sakulsermsuk, S. Preparation and electrical properties of nanocrystalline BCNO. *Mater. Lett.* **2018**, *211*, 51–54. [CrossRef]
40. Lei, W.W.; Portehault, D.; Dimova, R.; Antonietti, M. Boron carbon nitride nanostructures from salt melts: Tunable water-soluble phosphors. *J. Am. Chem. Soc.* **2011**, *133*, 7121–7127. [CrossRef]
41. Huang, C.J.; Chen, C.; Zhang, M.W.; Lin, L.H.; Ye, X.X.; Lin, S.; Antonietti, M.; Wang, X.C. Carbon-doped BN nanosheets for metal-free photoredox catalysis. *Nat. Commun.* **2015**, *6*, 7698. [CrossRef] [PubMed]
42. Chen, S.R.; Li, P.; Xu, S.T.; Pan, X.L.; Fu, Q.; Bao, X.H. Carbon doping of hexagonal boron nitride porous materials toward CO_2 capture. *J. Mater. Chem. A* **2018**, *6*, 1832–1839. [CrossRef]
43. Vokhmintsev, A.; Weinstein, I.; Zamyatin, D. Electron-phonon interactions in subband excited photoluminescence of hexagonal boron nitride. *J. Lumin.* **2019**, *208*, 363–370. [CrossRef]
44. Rajaji, U.; Govindasamy, M.; Sha, R.; Alsjgari, R.A.; Juang, R.S.; Liu, T.Y. Surface engineering of 3D spinel $Zn_3V_2O_8$ wrapped on sulfur doped graphitic nitride composites: Investigation on the dual role of electrocatalyst for simultaneous detection of antibiotic drugs in biological fluids. *Compos. Part B* **2022**, *242*, 110017. [CrossRef]
45. Kokulnathan, T.; Sharma, T.S.K.; Chen, S.M.; Chen, T.W.; Dinesh, B. Ex-situ decoration of graphene oxide with palladium nanoparticles for the highly sensitive and selective electrochemical determination of chloramphenicol in food and biological samples. *J. Taiwan Inst. Chem. Eng.* **2018**, *89*, 26–38. [CrossRef]
46. Kokulnathan, T.; Sharma, T.S.K.; Chen, S.M.; Yu, Y.H. Synthesis and characterization of Zirconium dioxide anchored carbon nanofiber composite for enhanced electrochemical determination of chloramphenicol in food samples. *J. Electrochem. Soc.* **2018**, *165*, B281–B288. [CrossRef]
47. Zhao, H.M.; Chen, Y.Q.; Tian, J.P.; Yu, H.T.; Quan, X. Selectively electrochemical determination of chloramphenicol in aqueous solution using molecularly imprinted polymer-carbon nanotubes-gold nanoparticles modified electrode. *J. Electrochem. Soc.* **2012**, *159*, J231–J236. [CrossRef]
48. Adav, M.; Ganesan, V.; Gupta, R.; Yadav, D.K.; Sonkar, P.K. Cobalt oxide nanocrystals anchored on graphene sheets for electrochemical determination of chloramphenicol. *Microchem. J.* **2019**, *146*, 881–887. [CrossRef]
49. Qi, X.; Teng, Z.X.; Yu, J.H.; Jia, D.L.; Zhang, Y.F.; Pan, H.Z. A simple one-step synthesis of Fe_3O_4/N-rGO nanocomposite for sensitive electrochemical detection of chloramphenicol. *Mater. Lett.* **2023**, *330*, 133350. [CrossRef]
50. Selvi, S.V.; Nataraj, N.; Chen, S.M. The electro-catalytic activity of nanosphere strontium doped zinc oxide with rGO layers screen-printed carbon electrode for the sensing of chloramphenicol. *Microchem. J.* **2020**, *159*, 105580. [CrossRef]
51. Jaysiva, G.; Manavalan, S.; Chen, S.M.; Veerakumar, P.; Keerthi, M.; Tu, H.S. MoN nanorod/sulfur-doped graphitic carbon nitride for electrochemical determination of chloramphenicol. *ACS Sustain. Chem. Eng.* **2020**, *8*, 11088–11098. [CrossRef]
52. Huyen, N.N.; Dinh, N.X.; Doan, M.Q.; Vu, N.P.; Das, R.; Le, M.T.; Thang, D.; Le, A.T. Unraveling the roles of morphology and steric hindrance on electrochemical analytical performance of α-Fe_2O_3 nanostructures-based nanosensors towards chloramphenicol antibiotic in shrimp samples. *J. Electrochem. Soc.* **2022**, *169*, 026507. [CrossRef]
53. Gao, S.; Yang, Z.M.; Zhang, Y.Q.; Zhao, L.; Xing, Y.P.; Fei, T.; Liu, S.; Zhang, T. The synergistic effects of MoS_2 and reduced graphene oxide on sensing performance for electrochemical chloramphenicol sensor. *FlatChem* **2022**, *33*, 100364. [CrossRef]
54. Oliveira, M.D.; Rocha, R.G.; Faria, L.V.D.; Richter, E.M.; Munoz, R.A.A. Carbon-black integrated polylactic acid electrochemical sensor for chloramphenicol determination in milk and water samples. *J. Electrochem. Soc.* **2022**, *169*, 047517. [CrossRef]
55. Niu, X.; Bo, X.J.; Guo, L.P. MOF-derived hollow $NiCo_2O_4$/C composite for simultaneous electrochemical determination of furazolidone and chloramphenicol in milk and honey. *Food Chem.* **2021**, *364*, 130368. [CrossRef]
56. Gao, S.; Zhang, Y.P.; Yang, Z.M.; Fei, T.; Liu, S.; Zhang, T. Electrochemical chloramphenicol sensors-based on trace MoS_2 modified carbon nanomaterials: Insight into carbon supports. *J. Alloys Compd.* **2021**, *872*, 159687. [CrossRef]

57. Govindasamy, M.; Chen, S.M.; Mani, V.; Devasenathipathy, R.; Umamaheswari, R.; Santhanaraj, K.J.; Sathiyan, A. Molybdenum disulfide nanosheets coated multiwalled carbon nanotubes composite for highly sensitive determination of chloramphenicol in food samples milk, honey and powdered milk. *J. Colloid Interface Sci.* **2017**, *485*, 129–136. [CrossRef]
58. Rajaji, U.; Muthumariappan, A.; Chen, S.M.; Chen, T.W.; Tseng, T.W.; Wang, K.; Qi, D.D.; Jiang, H.Z. Facile sonochemical synthesis of porous and hierarchical manganese(III) oxide tiny nanostructures for super sensitive electrocatalytic detection of antibiotic (chloramphenicol) in fresh milk. *Ultrason. Sonochem.* **2019**, *58*, 104648. [CrossRef]
59. Rajaji, U.; Manavalan, S.; Chen, S.M.; Govindasamy, M.; Maiyalagan, T. Microwave-assisted synthesis of europium(III) oxide decorated reduced graphene oxide nanocomposite for detection of chloramphenicol in food samples. *Compos. Part B* **2019**, *161*, 29–36. [CrossRef]

biosensors

MDPI

Article

A Portable Wireless Intelligent Nanosensor for 6,7-Dihydroxycoumarin Analysis with A Black Phosphorene and Nano-Diamond Nanocomposite-Modified Electrode

Xiaoqing Li [1,2], Lisi Wang [1], Lijun Yan [1], Xiao Han [1], Zejun Zhang [1], Xiaoping Zhang [1] and Wei Sun [1,*]

[1] Key Laboratory of Functional Materials and Photoelectrochemistry of Haikou, Key Laboratory of Laser Technology and Optoelectronic Functional Materials of Hainan Province, College of Chemistry and Chemical Engineering, Hainan Normal University, Haikou 571158, China
[2] College of Health Sciences, Shandong University of Traditional Chinese Medicine, Jinan 250355, China
* Correspondence: sunwei@hainnu.edu.cn

Abstract: In this work, a novel portable and wireless intelligent electrochemical nanosensor was developed for the detection of 6,7-dihydroxycoumarin (6,7-DHC) using a modified screen-printed electrode (SPE). Black phosphorene (BP) nanosheets were prepared via exfoliation of black phosphorus nanoplates. The BP nanosheets were then mixed with nano-diamond (ND) to prepare ND@BP nanocomposites using the self-assembly method, achieving high environmental stability. The nanocomposite was characterized by SEM, TEM, Raman, XPS and XRD. The nanocomposite was used for the modification of SPE to improve its electrochemical performances. The nanosensor displayed a wide linear range of 0.01–450.0 μmol/L with a low detection limit of 0.003 μmol/L for 6,7-DHC analysis. The portable and wireless intelligent electrochemical nanosensor was applied to detect 6,7-DHC in real drug samples by the standard addition method with satisfactory recoveries, which extends the application of BP-based nanocomposite for electroanalysis.

Keywords: portable wireless intelligent electrochemical nanosensor; black phosphorene nanosheets; nano-diamond; screen-printed electrode; electrochemistry; 6,7-Dihydroxycoumarin

Citation: Li, X.; Wang, L.; Yan, L.; Han, X.; Zhang, Z.; Zhang, X.; Sun, W. A Portable Wireless Intelligent Nanosensor for 6,7-Dihydroxycoumarin Analysis with A Black Phosphorene and Nano-Diamond Nanocomposite-Modified Electrode. *Biosensors* **2023**, *13*, 153. https://doi.org/10.3390/bios13020153

Received: 10 December 2022
Revised: 14 January 2023
Accepted: 16 January 2023
Published: 18 January 2023

1. Introduction

Coumarins are the general term for o-hydroxycinnamic acid lactones with the basic skeleton of a benzo α-pyrone nucleus. Most coumarin compounds feature oxygen functional groups at the C-7 positions. 6,7-dihydroxycoumarin (6,7-DHC), known as esculetin, is a coumarin derivative (its structure is shown in Figure S1), which is widely present in various plants such as artemisia annua, lemon, etc. It has attracted much attention from researchers, particularly in the medical field, due to its strong antioxidant property [1,2]. Different techniques have been used to detect 6,7-DHC in biological and pharmaceutical samples [3–6]. Among these methods, electrochemical techniques have also been proposed for the detection of 6,7-DHC with high sensitivity and reliability. For example, Sheng et al., developed a voltammetric sensor based on the electropolymerization L-lanthionine on glassy carbon electrode (GCE) for esculetin measurement [7]. Li et al., prepared a sensitive electrochemical sensor based on TiO_2NPs-coated poly (diallyldimethylammonium chloride)-functionalized graphene nanocomposites and used them to detect 6,7-DHC in real samples, yielding satisfactory results [8]. Therefore, developing sensitive electrochemical methods for the detection of 6,7-DHC is a meaningful and important task.

The working electrode is the core of electrochemical sensors. In recent years, two-dimensional (2D) nanomaterials with excellent properties have been used in the design of chemically modified electrodes [9]. Due to their large surface area, excellent conductivity, abundant catalytic activity sites, and easy functionalization, 2D nanomaterial modified electrodes can be used in different types of electrochemical biosensors [10]. Among them, black phosphorus nanoplates (BPNPs) are a new type of 2D nanomaterial with unique physical and structural properties [11], such as a large specific surface area, low cytotoxicity, and high thermal stability and biocompatibility [12–14]. They have already been used in biosensing [15,16], photothermal therapy [17], drug delivery [18], and other fields. However, BPNPs are unstable and easily biodegrade into PxOy in an environment containing oxygen and water [19]. Therefore, the modification of the surface of BPNPs is one of the most effective ways to avoid or reduce their degradation and to improve their performance [20,21]. Li et al., constructed a novel label-free electrochemical aptasensor based on a carbon dots–black phosphorus nanohybrid to carry out ultrasensitive detection of ochratoxins A [22]. Liu et al., prepared a sandwich-like BP@AuNPs@aptamer biosensor to carry out the highly efficient detection of circulating tumor cells in whole blood samples [23]. Li et al., developed a new electrochemical horseradish peroxidase (HRP) biosensor based on nanocomposites of black phosphorene (BP) nanosheets and single-walled carbon nanotubes for the detection of trichloroacetic acid, sodium nitrite, and hydrogen peroxide [24].

Nano-diamond (ND) is a unique type of carbon nanoparticle with many excellent advantages, including good electronic properties, excellent biocompatibility, low dielectric constant, a wide electrochemical potential window, and excellent chemical stability, earning it an increasing amount of attention [25]. Furthermore, diamond-based materials have been used in supercapacitors [26], pharmaceutical analysis [27], food detection [28], and biosensors [29]. Chen et al., prepared an ND-modified electrode for the oxidation of nitrite via the oxidation mechanism previously discussed [30]. Shahrokian et al., used an ND–graphite/chitosan membrane-modified GCE and applied it in the electrochemical detection of azathioprine [31]. Simioni et al., proposed an ND and dihexadecyl phosphate-modified GCE for the determination of codeine in different sample matrices via cyclic voltammetry [32]. Xie et al., designed an electrochemical biosensor with excellent electrocatalytic performance using myoglobin and an ND-modified electrode to detect trichloroacetic acid, sodium nitrite, and hydrogen peroxide [33].

Electrochemical sensors have many merits, such as their remarkable sensitivity, good selectivity, simple operation, low cost, easy miniaturization, etc., which can achieve both qualitative and quantitative analysis of targets. It has wide practical applications in environmental monitoring, food safety, medical analysis, and so on [34–36]. Govindasamy et al., applied La-based, perovskite-type lanthanum aluminate nanorod-incorporated graphene-oxide nanosheets for the sensitive determination of nitrite in samples of meat and drinking water [37]. With the rapid development of microelectronic technology, portable electrochemical sensors have been designed, which feature characteristics such as miniaturization, intelligence, and wireless transmission [38–41]. These kinds of sensors can be small, light, mobile, low-cost, and energy-efficient, and are designed to meet practical needs using disposable electrodes. Ge et al., developed a portable and wireless intelligent electrochemical sensor for the on-site detection of terbutaline residue in meat products [42]. Gao et al., fabricated a highly sensitive portable electrochemical immunosensor based on the dual-functionalized AuNPs for rapid point-of-care testing of genetically modified crops [43]. Chinnapaiyan et al., fabricated a thulium (III) metal–organic framework based on a smartphone sensor for the purpose of detecting roxarsone in real food samples [44]. Portable sensing devices can overcome the disadvantages of conventional laboratory-based electrochemical analysis, such as the high cost, bulkiness, and complexity. In contrast, portable sensing devices are portable, easy to operate, multifunctional, consume less power, and are practical in situ applications [45–47].

In this paper, ND was used as a protective agent and functional modifier of BP. ND@BP nanocomposite was synthesized via the self-assembly method and further used for construction of electrochemical sensor. The electrochemical behavior of 6,7-DHC was studied by using a portable wireless intelligent electrochemical workstation. The schematic diagram of the preparation process of this electrochemical sensor is shown in Scheme 1.

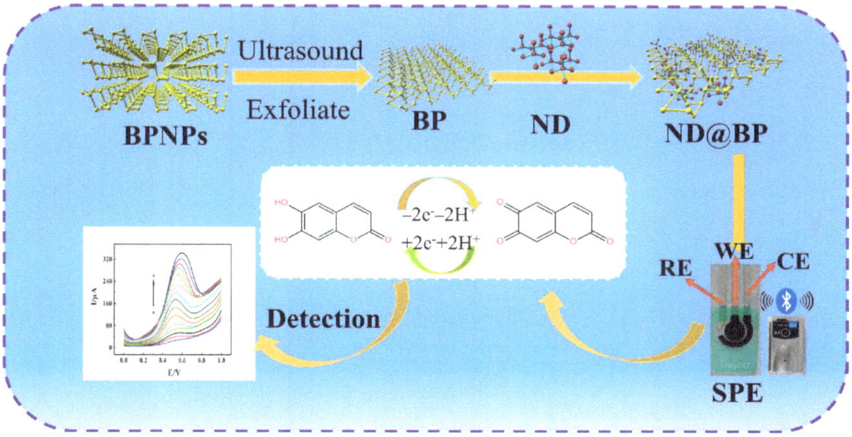

Scheme 1. Preparation process of the electrochemical sensor (BPNPs: black phosphorus nanoplates; BP: black phosphorene; ND: nano-diamond; SPE: screen-printed electrode; WE: working electrode; RE: reference electrode; CE: counter electrode).

2. Experimental

2.1. Reagents

The BPNPs (Nanjing XFNANO Materials Tech. Co., Ltd., Nanjing, China), isopropanol (IPA, Xilong Scientific Co., Ltd., Guangdong, China), ND powder (diameter: 5–10 nm, Nanjing XFNANO Materials Tech. Co., Ltd., Nanjing, China), and 6,7-DHC (Shanghai Aladdin Co., Shanghai, China) were used without further purification.

Phosphate buffer solution (PBS, 0.1 mol/L) was prepared as the supporting electrolyte with various pH from 2.0 to 6.0. Ultrapure water (IQ-7000, Milli-Q, Boston, MA, USA) was used throughout, and all other reagents were of analytical grade.

2.2. Synthesis of ND@BP

ND power (1.0 mg) was dispersed in 1.0 mL N_2-satured ultrapure water and then sonicated for 4 h at room temperature to obtain a 1.0 mg/mL ND dispersion.

Based on the reported procedure with little modification [48], the BP was exfoliated with the following procedure. Firstly, the BPNPs solution was dispersed in water to obtain a 0.2 mg/mL solution. The BPNPs solution (20.0 mL) was mixed with 5.0 mL of IPA, which was sealed in a brown centrifuge tube and sonicated in an ice bath for 12 h. After that, the solution was centrifuged twice at 10,000 rpm for 20 min. Finally, BP nanosheets were obtained and stored in a 4 °C refrigerator.

ND@BP nanocomposite was prepared by undergoing an electrostatic assembly process [49]. In brief, 1.0 mg/mL ND dispersion and 0.16 mg/mL BP solution were mixed in a volume ratio of 1:1 and stirred vigorously for 10 min. After sonication in an ice bath for 4 h, the resultant mixture solution was incubated in the 4 °C refrigerator and kept for 24 h under the protection of nitrogen to obtain ND@BP nanocomposite suspension.

2.3. Materials Characterization

X-ray diffraction (XRD) experiments were conducted using a D/Max-2500V X-ray diffractometer (Rigaku, Tokyo, Japan) with Cu-Kα (λ = 0.15418 nm) radiation at a scan rate of 2°/min in the 2 θ from 5 to 80°. X-ray photoelectron spectroscopy (XPS) was performed using an AXIS HIS 165 spectrophotometer (Kratos Analytical, Manchester, UK) equipped with a microfocused Al Kα X-ray beam (500 μm) and a photoelectron take off angle of 90°.

The adventitious carbon C1s peak at 284.8 eV was used for peak calibration. XPS survey spectra were analyzed by using XPSPEAK software. Scanning electron microscopy (SEM) was recorded on a JSM-7100F scanning electron microscope (JEOL Electron Co.,

Tokyo, Japan) with the operation voltage of 5.0 kV and focal length of 10 mm. Transmission electron microscopy (TEM) was performed on a JEM-2010F transmission electron microscope (JEOL Electron Co., Tokyo, Japan) at an operation voltage of 200 kV.

2.4. Construction of the Modified Electrodes

The modified working electrode was prepared via the drop-coating method with a 15.0 μL ND@BP solution coated on the surface of SPE and dried at room temperature in a glove box filled with nitrogen. Similarly, other modified electrodes were prepared in the same procedure for comparison.

2.5. Electrochemical Investigations

Voltammetric experiments were carried out on a portable electrochemical workstation (EmStat3 + Blue, Palmsens BV, Houten, The Netherlands) with electrochemical impedance spectroscopy (EIS) on a CHI 660E electrochemical workstation (Shanghai CH Instrument, Shanghai, China). A three-electrode system of SPE (φ = 5 mm, Qingdao Poten Technology Co., Ltd., Qingdao, China) was utilized with ND@BP as the modifier on the working electrode.

Cyclic voltammetry (CV) was performed to test the electrochemical performances of different working electrodes using a 1.0 mmol/L $K_3[Fe(CN)_6]$ and 0.5 mol/L KCl mixture solution in the potential window from −0.3 to 0.7 V. EIS was recorded using a 10.0 mmol/L $K_3[Fe(CN)_6]/K_4[Fe(CN)_6]$ and 0.1 mol/L KCl mixture solution in the frequency range of 10^5–10^{-2} Hz and an amplitude of 5 mV.

Electrochemical behaviors of 0.1 mmol/L 6,7-DHC were investigated by CV in the potential window from 0.0 to 1.0 V in 0.1 mol/L pH 3.0 PBS. Differential pulse voltammetry (DPV) was used for the electrochemical detection of different concentrations of 6,7-DHC in 0.1 mol/L pH 3.0 PBS with the following parameters: potential window from 0.0 to 1.0 V, pulse amplitude of 0.2 V, pulse width of 0.02 s, pulse period of 0.1 s, and quiet time of 2.0 s.

3. Results and Discussions

3.1. Characterization of ND@BP Nanocomposite

SEM and TEM were carried out to characterize the morphology of the BPNPs, BP, ND, and ND@BP nanocomposite, with the results presented in Figure 1. In Figure 1A, the inhomogenous and discontinuous irregular multilayer flake morphology of the BPNPs can be observed. Figure 1B illustrates the small grains and uniform distribution of the ND, which tends to form little agglomerations. As illustrated in Figure 1C, the morphology of the ND@BP nanocomposites can be observed as having granular ND particles adhered to and uniformly dispersed on the surface of the BP.

TEM images show that the BP is only a few layers thick (Figure 1D). As shown in Figure 1E, the ND is present as spherical tiny particles with uniform dispersion. The electron diffraction pattern of ND demonstrated the crystallinity of the structure. In the image of the ND@BP nanocomposites, the ND particles were uniformly dispersed on the BP nanosheets with particle sizes less than 10 nm (Figure 1F).

The chemical states and surface compositions of the elements in ND@BP were analyzed via XPS. The survey spectrum of BP and ND@BP showed the presence of C, O, and P elements clearly (Figure 2A). Figure 2B,D analyzed the high-resolution XPS C 1s spectrum of the BP and ND@BP. The binding energies at 284.7, 285.7, and 287.0 eV for BP were assigned to the C-C, C-C-O and O-C-C=O bonds, respectively [50]. For ND@BP, the binding energies were slightly shifted to the left, and the P-C bond appeared at 282.3 eV, which was attributed to the electrostatic interaction between ND and BP [51]. For Figure 2C, three peaks can be seen at 129.7, 130.5, and 133.8 eV, which belonged to P 2p3/2, P 2p1/2, and P-Ox bonds, respectively [52,53]. In the ND@BP nanocomposite, there was a new C-P bond at 132.1 eV, which indicated that ND was successfully combined with BP (Figure 2E).

The crystallographic properties of ND, BP, and ND@BP were analyzed via XRD, and the results are shown in Figure 2F. The XRD patterns of BP showed three diffraction peaks at 16.9°, 34.2° and 52.3°, respectively, which correspond to the (020), (040), and (060) planes, respectively, indicating the perfect crystal structure of black phosphorus [54]. However, the disappearance of the (020) and (040) characteristic diffraction peaks of the layer spacing of ND@BP suggested that the BPNPs had been stripped [55]. The diffraction peak at 43.9° in ND and ND@BP corresponded to the (111) diffraction peak of ND [56].

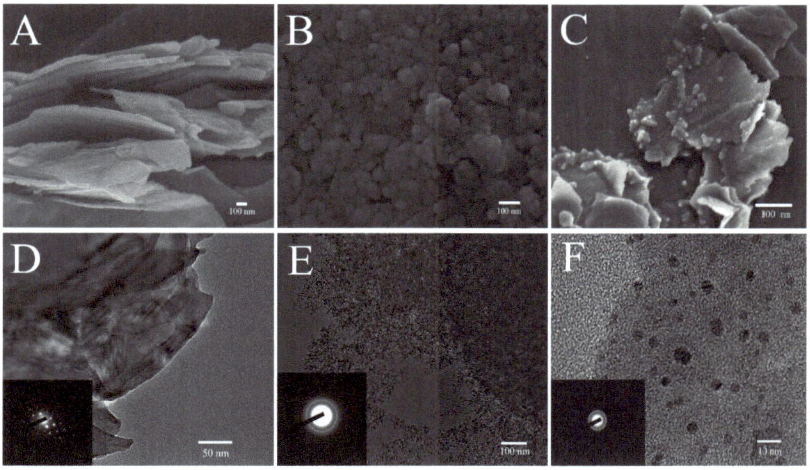

Figure 1. SEM images of BPNPs (**A**), ND (**B**), ND@BP (**C**). TEM images of BP (**D**), ND (**E**), and ND@BP (**F**) with electron diffraction patterns (inset).

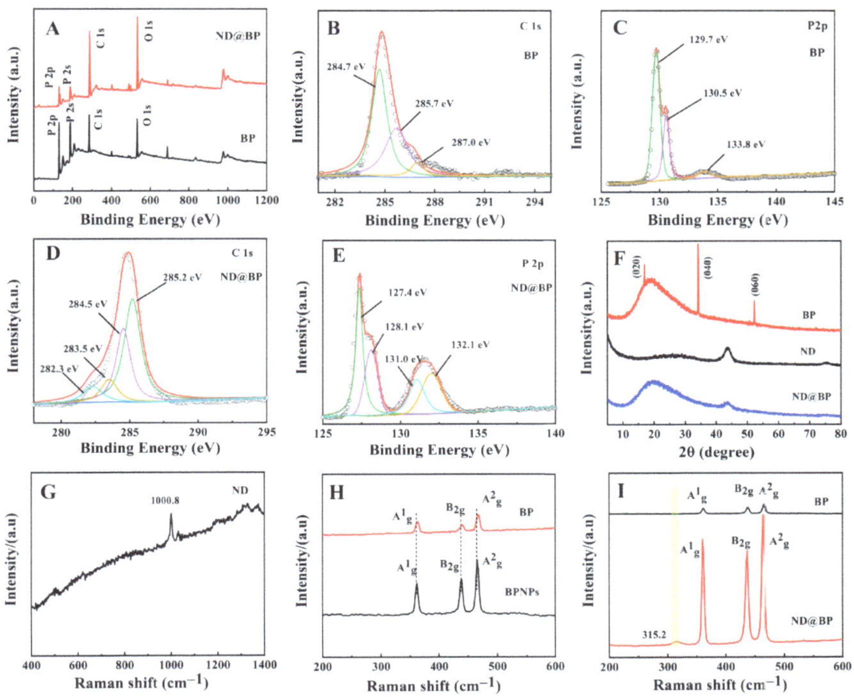

Figure 2. XPS survey spectrum of BP and ND@BP (**A**); high-resolution XPS spectra of the C 1s (**B**) and P2p (**C**) signal for BP; high-resolution XPS spectra of the C 1s (**D**) and P 2p (**E**) signal for ND@BP; XRD patterns of ND, BP, and ND@BP (**F**); Raman spectra of (**G**) ND, (**H**) BPNPs and BP, and (**I**) BP and ND@BP.

The recorded Raman spectra are shown in Figure 2 (G–I). A typical peak of ND appeared at 1000.8 cm^{-1}, attributed to the aromatic C-C stretching (Figure 2G). BPNPs showed three typical peaks at 361.2, 438.0, and 465.7 cm^{-1}, which were attributed to the $A^1{}_g$ (out of plane), B_{2g} (in plane), and $A^2{}_g$ (in plane) modes of BPNPs, respectively, [57] and are consistent with other research results [58]. The Raman peaks of ND@BP were blue-shifted compared with those of the BPNPs, and the $A^1{}_g$, B_{2g}, and $A^2{}_g$ peaks shifted to 363.2, 439.8, and 468.1 cm^{-1}, which was attributed to a reduced number of BPNPs layers (Figure 2H) [53]. As shown in Figure 2I, a new peak appeared at 315.2 cm^{-1} for ND@BP, indicating the interaction between BP and ND with the formation of the P-C bond. Furthermore, the vibration of the P atoms was facilitated, which could increase the Raman scattering energy and slowed down the oxidation of BP.

3.2. Electrochemical Characterizations

EIS is a commonly used method to evaluate the interfacial information of modified electrodes. The results of EIS reflect the electron transfer ability of the electrode's surface. The value of the electron transfer resistance (Ret) can be used to describe the interface properties of the modified surface, which can be expressed by the semicircle's diameter [59]. Figure 3 shows the EIS of different modified electrodes in a solution of 0.1 mol/L KCl containing 10.0 mmol/L $[Fe(CN)_6]^{3-/4-}$ with the frequency ranging from 10^5 to 10^{-2} Hz. The Ret of ND/SPE (curve b) was smaller than SPE only (curve a), indicating the excellent conductivity of ND. The Nyquist plots of BP/SPE (curve c) and ND@BP/SPE (curve d) displayed almost straight lines at all frequencies. Furthermore, the slope of ND@BP/SPE was larger than that of BP/SPE, demonstrating the excellent conductivity of ND@BP as a result of the synergistic enhancement effect of the two components.

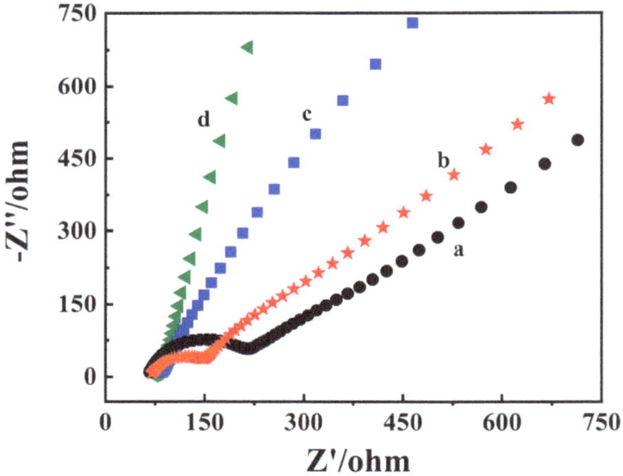

Figure 3. EIS of (**a**) SPE, (**b**) ND/SPE, (**c**) BP/SPE, and (**d**) ND@BP/SPE in a 10.0 mmol/L $[Fe(CN)_6]^{3-/4-}$ and 0.1 mol/L KCl mixture, with frequencies ranging from 10^5 to 10^{-2} Hz and an amplitude of 5 mV.

Figure 4 shows the CV of different modified electrodes at various scan rates and the corresponding relationships between the current and the scan rate. Both oxidation and reduction peak currents increased with the increase in scan rate, and the currents against the square root of the scan rate ($\upsilon^{1/2}$) exhibited a good linear relationship. According to the Randles–Sevcik equation [60], $Ip = 2.69 \times 10^5 AD^{1/2}n^{3/2}\upsilon^{1/2}C$, the electrochemical effective areas (A) of different modified electrodes were calculated by the $I–\upsilon^{1/2}$ curve with values of 0.131 cm^2 for SPE (Figure 4A), 0.146 cm^2 for ND/SPE (Figure 4B), 0.153 cm^2 for BP/SPE (Figure 4C), and 0.181 cm^2 for ND@BP/SPE (Figure 4D). Therefore, the presence of ND@BP on the electrode's surface provides a large effective area, which is a very important factor in improving the responsiveness of an electrochemical sensor.

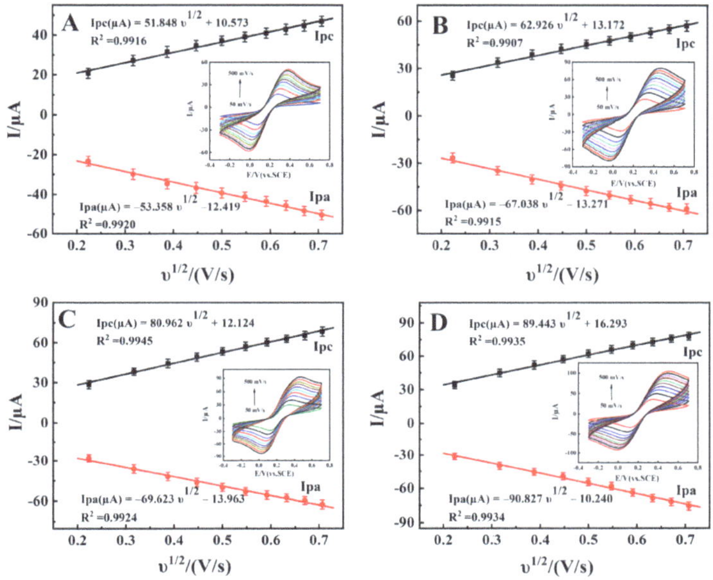

Figure 4. Relationship of Ip and $\upsilon^{1/2}$ of (**A**) SPE, (**B**) ND/SPE, (**C**) BP/SPE, and (**D**) ND@BP/SPE. Inset is the CV of different electrodes in a 1.0 mmol/L $K_3[Fe(CN)_6]$ and 0.5 mol/L KCl mixture with different scan rates from 50 to 500 mV/s.

The electrochemical behaviors of 0.1 mmol/L 6,7-DHC on different modified electrodes were investigated in 0.1 mol/L pH 3.0 PBS, with the curves shown in Figure 5. On SPE, a pair of obvious redox peaks of 6,7-DHC were recorded (curve a), with the anodic peak current (Ipa) and cathodic peak current (Ipc) being 9.99 and 5.98 μA, respectively. On ND/SPE (curve b), the redox peak current was increased, which could be due to the presence of ND with good conductivity. On BP/SPE (curve c), electrochemical response was further improved, while on ND@BP/SPE (curve d), the peak currents increased to 17.32 μA (Ipa) and 8.08 μA (Ipc), which were 1.73 times and 1.35 times larger than that on SPE. The drastic increase in the redox responses could be attributed to the synergistic effect between ND and BP. ND could increase the electric conductivity, and BP supplied a large surface area and active catalytic ability.

The effect of different scan rates on the redox process of 0.1 mmol/L 6,7-DHC was investigated in the range from 50 to 1000 mV/s on ND@BP/SPE; the results are shown in Figure S2A. As the scan rate increased, the redox peak currents and potential gradually increased, indicating that 6,7-DHC underwent a quasi-reversible electrochemical reaction. A linear relationship between Ip and υ (Figure S2B) demonstrated that the electrode reaction of 6,7-DHC on the surface of ND@BP/SPE was an adsorption control process. Furthermore, Figure S2C displayed a good linear relationship between Ep and lnυ. According to Laviron's formula [61,62], the value of the electron transfer coefficient (α) and the electron transfer number (n) of 6,7-DHC were generally assumed to be 0.50 and 2.04, respectively. Meanwhile, the electron transfer rate constant (k_s) was calculated as 5.61 s^{-1}.

The influence of the solution pH on the electrochemical response of 6,7-DHC was further investigated. As shown in Figure S3, the formal potential ($E^{0'}$) was negatively shifted with the increase of the pH in the range from 2.0 to 6.0 (Figure S3A), indicating that protons were involved in electrode reaction. There was a good linear relationship between $E^{0'}$ and pH, with the following linear regression equation: $E^{0'}$ (V) = −0.0535 pH + 0.7408 (R^2 = 0.9977)

(Figure S3B, curve a). The slope was close to the theoretical value of -59 mV/pH at 25 °C [63], which demonstrated that the same amounts of protons and electrons were involved in the electrode reaction of 6,7-DHC. The peak current changed with the increase in pH, and the maximum value was obtained at pH 3.0 (Figure S3B, curve b). Therefore, in the experiment, pH 3.0 PBS was selected as the supporting electrolyte.

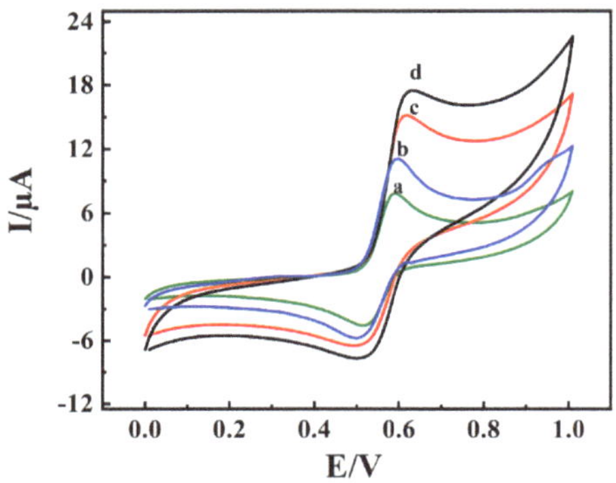

Figure 5. Electrochemical behaviors of 0.1 mmol/L 6,7-DHC on (a) SPE, (b) ND/SPE, (c) BP/SPE and (d) ND@BP/SPE in pH 3.0 PBS at the scan rate of 100 mV/s.

3.3. Calibration Curve

DPV was used to establish the calibration curve, and the results are shown in Figure 6. It can be clearly observed from Figure 6A that the Ipa value increased gradually with the addition of 6,7-DHC. The linear relationships of the three sections are 0.01–10.0, 10.0–100.0, and 100.0–450.0 μmol/L, and their linear regression equations are Ipa (μA) = 4.858C (μmol/L) + 4.313 (R^2 = 0.9912), Ipa (μA) = 1.049C (μmol/L) + 46.41 (R^2 = 0.9920), and Ipa (μA) = 0.1803C (μmol/L) + 128.2 (R^2 = 0.9915), respectively (Figure 6B). The detection limit was 0.003 μmol/L ($3S_0/S$). For the adsorption-controlled process of 6,7-DHC, different calibration curves were ascribed to the differences in the activity of the modified electrode surface when exposed to high and low 6,7-DHC concentrations. The presence of the ND@BP nanocomposite, with its high conductivity and large effective surface, resulted in more active sites. At low concentrations of 6,7-DHC, the surface of ND@BP/SPE had a larger number of active sites, in which the adsorption-controlled kinetics were dramatically bigger than at higher concentrations.

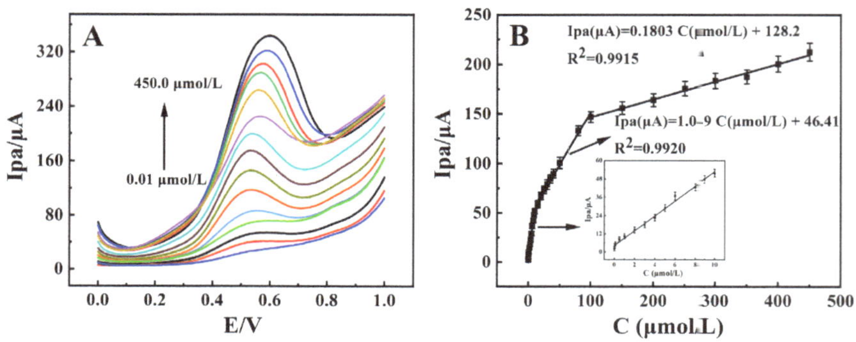

Figure 6. (**A**) DPV curves of different concentrations of 6,7-DHC from 0.01 to 450.0 μmol/L in pH 3.0 PBS; (**B**) Linear relationship between Ipa and 6,7-DHC concentration (n = 3).

3.4. Samples Analysis

In order to evaluate the analytical applicability, this analytical method was applied to detect 6,7-DHC analysis in Qinpi Jiegu Capsules (Sample 1) and Bawei Qinpi Wan (Sample 2). The preparation method was as follows: three tablets were added in a mortar and then ground for 20 min. Then, the sample was transferred to a centrifuge tube and dispersed in 50 mL methanol. After being sealed carefully, the centrifuge tube was sonicated in an ice-bath for 4 h. The dispersion was centrifuged at 8000 and 10,000 rpm for 5 min each, and the supernatant was then removed and diluted to 50 mL with methanol; this was the final sample and was used for DPV measurement. The results are summarized in Table 1. It was found that the recoveries of 95.0–101.5% and 94.7–104.4% were obtained from these spiked samples with satisfactory accuracy. Furthermore, the relative standard deviations (RSD) were 1.41–3.28% with ND@BP/SPE. All the RSD values were less than 5.0%, demonstrating that ND@BP/SPE has a potential application for real sample analysis and could be a novel method for rapid on-site detection of 6,7-DHC in medicinal samples.

Table 1. Detection results of 6,7-DHC in real samples (n = 3).

Samples	Detected (μmol/L)	Added (μmol/L)	Total (μmol/L)	RSD (%)	Recovery (%)
Sample 1 (Z20026310)	1.96	5.00	6.71	3.28	95.0
		10.00	12.11	3.11	101.5
		15.00	16.24	2.07	95.2
Sample 2 (Z54020022)	2.60	5.00	7.82	3.10	104.4
		10.00	12.58	1.41	99.8
		15.00	16.81	2.55	94.7

3.5. Interference

The selectivity of ND@BP/SPE was investigated by DPV in the presence of 100-fold concentrations of inorganic ions (K^+, Na^+, Ca^{2+}, Cu^{2+}) and 10-fold concentrations of amino acids (L-cysteine, aspartic acid), glucose, urea, and uric acid as interference agents with 10.0 μmol/L 6,7-DHC. As shown in Figure 7, the coexisting interferents did not have a significant impact on the detection of 6,7-DHC, with an RSD below 5.0%. Therefore, ND@BP/SPE has good anti-interference performance.

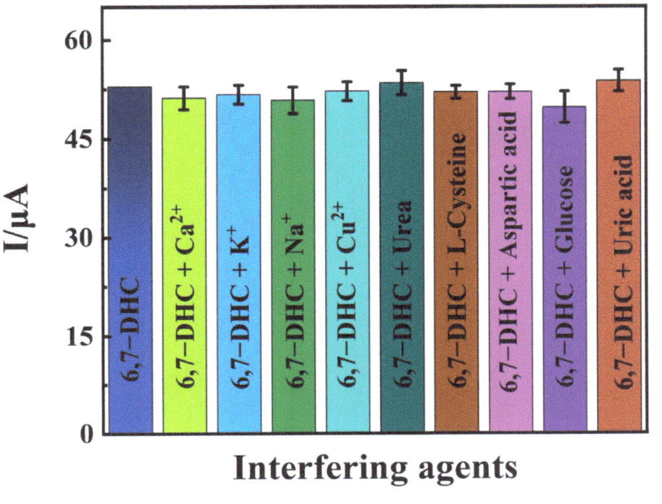

Figure 7. Influence of co-existing substances for 10.0 μmol/L 6,7-DHC analysis (n = 3) for 100-fold concentrations of K^+, Na^+, Ca^{2+}, and Cu^{2+} and 10-fold concentrations of L-cysteine, aspartic acid, glucose, urea, and uric acid.

3.6. Stability, Repeatability, and Reproducibility

The stability of the different modified electrodes was investigated carefully by a continuous cyclic voltammetry test with a scan rate of 100 mV/s. As shown in Figure 8, the RSD of the anodic and cathodic peak currents for the three modified electrodes were 1.22% and 1.09% (Figure 8A), 3.69% and 4.89% (Figure 8B), and 1.68% and 2.05% (Figure 8C), respectively. Therefore, ND not only had a certain protective effect on BP, but it also interacted with BP, which improved the ND@BP/SPE's electrochemical cycle stability. In addition, the reduction peak current was still higher than 83.4% of the initial signal value after storing the sample in a refrigerator at 4 °C for 10 days, which indicated that ND@BP/SPE has good storage stability.

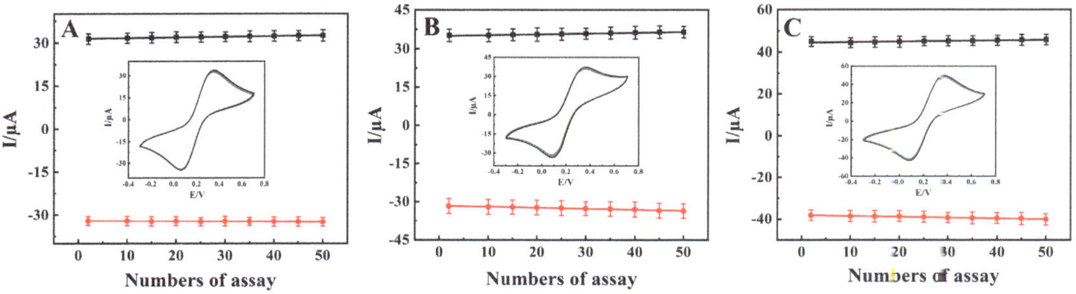

Figure 8. Stability of different electrodes with successive CV scans of (**A**) ND/SPE, (**B**) BP/SPE, and (**C**) ND@BP/SPE (n = 3). (Inset: Multi−scan CV curves of (**A**) ND/SPE, (**B**) BP/SPE, and (**C**) ND@BP/SPE in 1.0 mmol/L $K_3[Fe(CN)_6]$ and 0.5 mol/L KCl mixture at a scan rate of 100 mV/s.).

Five parallel electrodes (ND@BP/SPE) were fabricated and used to detect 0.1 mmol/L 6,7-DHC by CV. The RSD peak currents were 3.86% and 4.34%, respectively (Figure 9A), which demonstrates an excellent repeatability of ND@BP/SPE. Furthermore, nine parallel detections of the same ND@BP/SPE in 0.1 mmol/L 6,7-DHC were checked with the RSD of redox peak currents as 1.69% and 1.41% (Figure 9B), which reflected a good reproducibility.

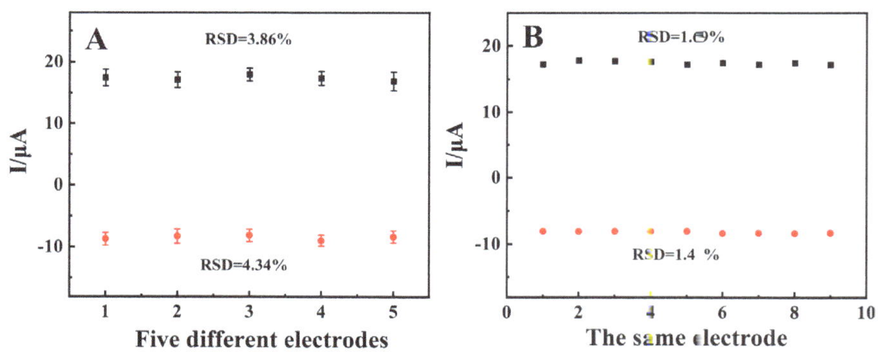

Figure 9. (**A**) Repeatability and (**B**) reproducibility of ND@BP/SPE for the detection of 0.1 mmol/L 6,7-DHC in pH 3.0 PBS at a scan rate of 100 mV/s by CV.

4. Conclusions

In this paper, a novel portable and wireless intelligent electrochemical nanosensor was prepared based on ND@BP nanocomposite-modified SPE. The prepared nanosensor exhibited excellent electrochemical performance in detecting 6,7-DHC as a result of the synergistic effects of ND and BP. Our portable and wireless intelligent electrochemical nanosensor can overcome the disadvantages of conventional laboratory-based electrochemical analysis, such as the high cost, bulkiness, and complexity. In contrast, our sensing device is portable, easy to operate, multifunctional, consumers power, and is practical in in situ practical applications. However, the modification of SPE is the key step in the electrode's preparation, an such modification should be carried out carefully. In addition, the disposable usage of the modified SPE results in an increase in the cost of each measurement. Nevertheless, the presence of ND on the BP nanosheet results in good stability of the nanocomposite, which was further used for the modification of SPE, achieving excellent electrochemical performance. The fabricated modified electrode was connected to a portable electrochemical workstation to carry out the sensitive detection of 6,7-DHC in real drug samples. Our results extend the application of BP-based nanocomposite to intelligent sensing platforms.

Supplementary Materials: The following supporting information can be downloaded at: https://www.mdpi.com/article/10.3390/bios13020153/s1, Figure S1. The structural diagram of 6,7-DHC; Figure S2. (A) CV curves of 0.1 mmol/L 6,7-DHC in 0.1 mol/L pH 3.0 PBS on ND@BP/SPE at different scan rates (from a to k: 0.05, 0.1, 0.2, 0.3, 0.4, 0.5, 0.6, 0.7, 0.8, 0.9, 1.0 V/s). (B) Linear relationship of Ip versus υ (n = 3). (C) Linear relationship of Ep versus lnυ (n = 3); Figure S3. (A) CV curves of 0.1 mmol/L 6,7-DHC in 0.1 mol/L pH 3.0 PBS on ND@BP/SPE at different pH values (from a to e: 2.0, 3.0, 4.0, 5.0, and 6.0) at the scan rate of 100 mV/s; (B) The relationship between $E^{0'}$ versus pH (a) and the plot of Ipa versus pH (b) (n = 3).

Author Contributions: X.L.: writing—original draft, methodology, investigation, data curation, and conceptualization. L.W.: investigation, validation, and software. L.Y.: methodology and validation. X.H.: software and formal analysis. Z.Z.: formal analysis and supervision. X.Z.: conceptualization and investigation. W.S.: project administration and funding acquisition. All authors have read and agreed to the published version of the manuscript.

Funding: This project was financially supported by the National Natural Science Foundation of China (21964007), the Specific Research Fund of the Innovation Platform for Academicians of Hainan Province (YSPTZX202126), the Innovation Platform for Academicians of Hainan Province, the Key Research and Development Program of Hainan Province–Social Development Direction (ZDYF2020204), the Open Foundation of Key Laboratory of Laser Technology and Optoelectronic Functional Materials of Hainan Province (2022LTOM01), and the Graduate Student Innovation Project of the College of Chemistry and Chemical Engineering of Hainan Normal University (Hgb 202204).

Institutional Review Board Statement: Not applicable.

Informed Consent Statement: Not applicable.

Data Availability Statement: Not applicable.

Conflicts of Interest: The authors declare no conflict of interest.

References

1. Madureira, J.; Botelho, M.L.; Cooper, W.J.; Leal, J.P.; Melo, R. Aqueous degradation of esculetin (6,7-dihydroxycoumarin) using gamma radiation. *Desalin. Water Treat.* **2020**, *181*, 385–390. [CrossRef]
2. Prajapati, S.; Tomar, B.; Srivastava, A.; Narkhede, Y.B.; Gaikwad, A.N.; Lahiri, A.; Mulay, S.R. 6,7-Dihydroxycoumarin ameliorates crystal-induced necroptosis during crystal nephropathies by inhibiting MLKL phosphorylation. *Life Sci.* **2021**, *271*, 119193. [CrossRef] [PubMed]
3. Zhao, X.J.; Zhang, F.Y.; Lu, D.B.; Du, Y.L.; Ye, W.C.; Wang, C.M. Fabrication of a CuS/graphene nanocomposite modified electrode and its application for electrochemical determination of esculetin. *Anal. Methods.* **2013**, *5*, 3992–3998. [CrossRef]
4. Sproll, C.; Ruge, W.; Andlauer, C.; Godelmann, R.; Lachenmeier, D.W. HPLC analysis and safety assessment of coumarin in foods. *Food Chem.* **2008**, *109*, 462–469. [CrossRef] [PubMed]
5. Li, Y.Y.; Song, Y.Y.; Liu, C.H.; Huang, X.T.; Zheng, X.; Li, N.; Xu, M.L.; Mi, S.Q.; Wang, N.S. Simultaneous determination of esculin and its metabolite esculetin in rat plasma by LC–ESI-MS/MS and its application in pharmacokinetic study. *J. Chromatogr. B* **2012**, *907*, 27–33. [CrossRef] [PubMed]
6. Li, J.F.; Wan, C.J.; Wang, C.; Zhang, H.; Chen, X.D. 2D material chemistry: Graphdiyne-based biochemical sensing, *Chem. Res. Chin. Univ.* **2020**, *36*, 1–4. [CrossRef]
7. Sheng, K.; Li, T.T.; Zhang, Q.; Wang, Y.L. A highly sensitive sensor based on electropolymerization for electrochemical detection of esculetin. *Microchem. J.* **2020**, *159*, 105368. [CrossRef]

8. Li, H.C.; Wang, L.; Sheng, K.; Zoua, L.; Yea, B.X. Highly sensitive determination of esculetin on TiO_2NFs-coated poly (diallyldimethylammonium chloride)-functionalized graphene modified electrode. *Talanta* **2016**, *161*, 838–846. [CrossRef]
9. Huang, Y.H.; Yan, L.J.; Wang, B.; Zhu, L.; Shao, B.; Niu, Y.Y.; Zhang, X.P.; Yin, P.; Ge, Y.Q.; Sun, W.; et al. Recent applications of black phosphorus and its related composites in electrochemistry and bioelectrochemistry: A mini review. *Electrochem. Commun.* **2021**, *129*, 107095. [CrossRef]
10. Khushaim, W.; Mani, V.; Peramaiya, K.; Huang, K.W.; Salama, K.N. Ruthenium and nickel molybdate-decorated 2D porous graphitic carbon nitrides for highly sensitive cardiac troponin biosensor. *Biosensors* **2022**, *12*, 783. [CrossRef]
11. Zhu, X.Y.; Lin, L.; Wu, R.M.; Zhu, Y.F.; Sheng, Y.Y.; Nie, P.C.; Liu, P.; Xu, L.L.; Wen, Y.P. Portable wireless intelligent sensing of ultra-trace phytoregulator α-naphthalene acetic acid using self-assembled phosphorene/Ti_3C_2-MXene nanohybrid with high ambient stability on laser induced porous graphene as nanozyme flexible electrode. *Biosens. Bioelectron.* **2021**, *179*, 113062. [CrossRef]
12. Tao, W.; Zhu, X.B.; Yu, X.H.; Zeng, X.W.; Xiao, Q.L.; Zhang, X.D.; Ji, X.Y.; Wang, X.S.; Shi, J.J.; Zhang, H.; et al. Black phosphorus nanosheets as a robust delivery platform for cancer theranostics. *Adv. Mater.* **2017**, *29*, 1603276. [CrossRef] [PubMed]
13. Sun, Z.B.; Xie, H.H.; Tang, S.Y.; Yu, X.F.; Guo, Z.N.; Shao, J.D.; Zhang, H.; Huang, H.; Wang, H.Y.; Chu, P.K. Ultrasmall black phosphorus quantum dots: Synthesis and use as photothermal agents. *Angew. Chem. Int. Edit.* **2015**, *54*, 11526–11530. [CrossRef] [PubMed]
14. Lee, H.U.; Park, S.Y.; Lee, S.C.; Choi, S.; Seo, S.; Kim, H.; Won, J.; Choi, K.; Kang, K.S.; Park, H.G. Black Phosphorus (BP) nanodots for potential biomedical applications. *Small* **2016**, *12*, 214–219. [CrossRef] [PubMed]
15. Kumar, V.; Brent, J.R.; Shorie, M.; Kaur, H.; Chadha, G.; Thomas, A.G.; Lewis, E.A.; Rooney, A.P.; Nguyen, L.; Zhong, X.L.; et al. Nanostructured aptamer-functionalized black phosphorus sensing platform for label-free detection of myoglobin, a cardiovascular disease biomarker. *ACS Appl. Mater. Inter.* **2016**, *8*, 22860–22868. [CrossRef] [PubMed]
16. Yew, Y.T.; Sofer, Z.; Mayorga-Martinez, C.C.; Pumera, M. Black phosphorus nanoparticles as a novel fluorescent sensing platform for nucleic acid detection. *Mater. Chem. Front.* **2017**, *1*, 1130–1136. [CrossRef]
17. Yang, G.C.; Liu, Z.M.; Li, Y.; Hou, Y.Q.; Fei, X.X.; Su, C.K.; Wang, C.M.; Zhuang, Z.F.; Guo, Z.Y. Facile synthesis of black phosphorus–Au nanocomposites for enhanced photothermal cancer therapy and surface-enhanced Raman scattering analysis. *Biomater. Sci.* **2017**, *5*, 2048–2055. [CrossRef] [PubMed]
18. Qiu, M.; Wang, D.; Liang, W.Y.; Liu, L.P.; Zhang, Y.; Chen, X.; Sang, D.K.; Xing, C.Y.; Li, Z.J.; Dong, B.Q.; et al. Novel concept of the smart NIR-light–controlled drug release of black phosphorus nanostructure for cancer therapy. *Proc. Natl. Acad. Sci. USA* **2018**, *115*, 501–506. [CrossRef]
19. Huang, Y.; Qiao, J.; He, K.; Bliznakov, S.; Sutter, E.; Chen, X.J.; Luo, D.; Meng, F.K.; Su, D.; Decker, J.; et al. Interaction of black phosphorus with oxygen and water. *Chem. Mater.* **2016**, *28*, 8330–8339. [CrossRef]
20. Gusmao, R.; Sofer, Z.; Pumera, M. Black phosphorus rediscovered: From bulk to monolayer. *Angew. Chem.* **2017**, *56*, 8052–8072. [CrossRef]
21. Ryder, C.R.; Wood, J.D.; Wells, S.A.; Yang, Y.; Jariwala, D.; Marks, T.J.; Schatz, G.C.; Hersam, M.C. Covalent functionalization and passivation of exfoliated black phosphorus via aryl diazonium chemistry. *Nat. Chem.* **2016**, *8*, 597–602. [CrossRef] [PubMed]
22. Li, K.X.; Qia, X.J.; Zhao, H.Y.; He, Y.P.; Sheng, Q.L.; Yue, T.L. Ultrasensitive and label-free electrochemical aptasensor based on carbon dots-black phosphorus nanohybrid for the detection of ochratoxins A. *Microchem. J.* **2021**, *168*, 106378. [CrossRef]
23. Liu, S.P.; Luo, J.J.; Jiang, X.X.; Li, X.Q.; Yang, M.H. Gold nanoparticle–modified black phosphorus nanosheets with improved stability for detection of circulating tumor cells. *Microchim. Acta* **2020**, *187*, 397. [CrossRef] [PubMed]
24. Li, X.Q.; Wang, L.S.; Wang, B.L.; Zhang, S.Y.; Jiang, M.; Fu, W.T.; Sun, W. Preparation and application of electrochemical horseradish peroxidase sensor based on a black phosphorene and single-walled carbon nanotubes nanocomposite. *Molecules* **2022**, *27*, 8064. [CrossRef]
25. Zhang, Y.H.; Rhee, K.Y.; Hui, D.; Park, S.J. A critical review of nanodiamond based nanocomposites: Synthesis, properties and applications. *Compos. Part B* **2018**, *143*, 19–27. [CrossRef]
26. Bogdanowicz, R.; Dettlaff, A.; Skiba, F.; Trzcinski, K.; Szkoda, M.; Sobaszek, M.; Ficek, M.; Dec, B.; Macewicz, L.; Wyrębski, K.; et al. Enhanced charge storage mechanism and long-term cycling stability in diamondized titania nanocomposite supercapacitors operating in aqueous electrolytes. *J. Phys. Chem. C* **2020**, *124*, 15698–15712. [CrossRef]
27. Yence, M.; Cetinkaya, A.; Ozcelikay, G.; Kaya, S.I.; Ozkan, S.A. Boron-doped diamond electrodes: Recent developments and advances in view of electrochemical drug sensors. *Crit. Rev. Anal. Chem.* **2022**, *52*, 1122–1138. [CrossRef] [PubMed]
28. Sarakhman, O.; Švorc, L. A review on recent advances in the applications of boron-doped diamond electrochemical sensors in food analysis. *Crit. Rev. Anal. Chem.* **2020**, *52*, 791–813. [CrossRef]
29. Krecmarová, M.; Gulka, M.; Vandenryt, T.; Hrubý, J.; Fekete, L.; Hubík, P.; Taylor, A.; Mortet, V.; Thoelen, R.; Bourgeois, E.; et al. A label-free diamond microfluidic DNA sensor based on active nitrogen-vacancy center charge state control. *ACS Appl. Mater. Interfaces* **2021**, *13*, 18500–18510. [CrossRef] [PubMed]
30. Chen, L.H.; Zang, J.B.; Wang, Y.H.; Bian, L.Y. Electrochemical oxidation of nitrite on nanodiamond powder electrode. *Electrochim. Acta* **2008**, *53*, 3442–3445. [CrossRef]
31. Shahrokhian, S.; Ghalkhani, M. Glassy carbon electrodes modified with a film of nanodiamond–graphite/chitosan: Application to the highly sensitive electrochemical determination of Azathioprine. *Electrochim. Acta* **2010**, *55*, 3621–3627. [CrossRef]

32. Simioni, N.B.; Oliveira, G.G.; Vicentini, F.C.; Lanza, M.; Janegitz, B.C.; FatibelloFilho, O. Nanodiamonds stabilized in dihexadecyl phosphate film for electrochemical study and quantification of codeine in biological and pharmaceutical samples. *Diam. Relat. Mater.* **2017**, *74*, 191–196. [CrossRef]
33. Xie, H.; Li, X.Y.; Luo, G.L.; Niu, Y.Y.; Zou, R.Y.; Yin, C.X.; Huang, S.M.; Sun, W.; Li, G.J. Nano-diamond modified electrode for the investigation on direct electrochemistry and electrocatalytic behavior of myoglobin. *Diam. Relat. Mater.* **2019**, *97*, 107453. [CrossRef]
34. Dhamu, V.N.; Poudyal, D.C.; Muthukumar, S.; Prasad, S. A highly sensitive electrochemical sensor system to detect and distinguish between glyphosate and glufosinate. *J. Electrochem. Soc.* **2021**, *168*, 057531. [CrossRef]
35. Yan, L.J.; Hu, T.X.; Li, X.Q.; Ding, F.Z.; Wang, B.; Wang, B.L.; Zhang, B.X.; Shi, F.; Sun, W. Graphdiyne and ionic liquid composite modified gold electrode for sensitive voltammetric analysis of rutin. *Electroanalysis* **2022**, *34*, 286–293. [CrossRef]
36. Saylan, Y.; Erdem, O.; Ünal, S.; Denizli, A. An alternative medical diagnosis method: Biosensors for virus detection. *Biosensors* **2019**, *9*, 65. [CrossRef]
37. Govindasamy, M.; Wang, S.F.; Huang, C.H.; Alshgari, R.A.; Ouladsmane, M. Colloidal synthesis of perovskite-type lanthanum aluminate incorporated graphene oxide composites: Electrochemical detection of nitrite in meat extract and drinking water. *Microchim. Acta* **2022**, *189*, 210. [CrossRef]
38. Shi, F.; Wang, B.L.; Yan, L.J.; Wang, B.; Niu, Y.Y.; Wang, L.S.; Sun, W. In-site growth of nitrogen-doped carbonized polymer dots on black phosphorus for electrochemical DNA biosensor of *Escherichia coli* O157: H7. *Bioelectrochemistry* **2022**, *148*, 108226. [CrossRef]
39. Ozer, T.; Henry, C.S. Review-recent advances in sensor arrays for the simultaneous electrochemical detection of multiple analytes. *J. Electrochem. Soc.* **2021**, *168*, 057507. [CrossRef]
40. Chen, Y.X.; Sun, Y.X.; Niu, Y.Y.; Wang, B.L.; Zhang, Z.J.; Zeng, L.N.; Li, L.; Sun, W. Portable electrochemical sensing of indole-3-acetic acid based on self-assembled Mxene and carbon nanotube composite modified screen-printed electrode. *Electroanalysis* **2022**, *34*, 1–10. [CrossRef]
41. Hoilett, O.S.; Walker, J.F.; Balash, B.M.; Jaras, N.J.; Boppana, S.; Linnes, J.C. KickStat: A coin-sized potentiostat for high-resolution electrochemical analysis. *Sensors* **2020**, *20*, 2407. [CrossRef] [PubMed]
42. Ge, Y.; Liu, P.; Xu, L.J.; Qu, M.G.; Hao, W.X.; Liang, H.; Sheng, Y.Y.; Zhu, Y.F.; Wen, Y.P. A portable wireless intelligent electrochemical sensor based on layer-by-layer sandwiched nanohybrid for terbutaline in meat products. *Food Chem.* **2022**, *371*, 131140. [CrossRef] [PubMed]
43. Gao, H.F.; Wen, L.K.; Tian, J.; Wu, Y.H.; Liu, F.; Lin, Y.J.; Hua, W.; Wu, G. A portable electrochemical immunosensor for highly sensitive point-of-care testing of genetically modified crops. *Biosens. Bioelectron.* **2019**, *142*, 111504. [CrossRef]
44. Chinnapaiyan, S.; Rajaji, U.; Chen, S.M.; Liu, T.Y.; Filho, J.I.O.; Chang, Y.S. Fabrication of thulium metal-organic frameworks based smartphone sensor towards arsenical feed additive drug detection: Applicable in food safety analysis. *Electrochim. Acta* **2022**, *401*, 139487. [CrossRef]
45. Li, J.J.; Bo, X.J. Laser-enabled flexible electrochemical sensor on finger for fast food security detection. *J. Hazard. Mater.* **2022**, *423*, 127014. [CrossRef]
46. Shao, B.; Ai, Y.J.; Yan, L.J.; Wang, B.; Huang, Y.H.; Zou, Q.W.; Fu, H.Y.; Niu, X.L.; Sun, W. Wireless electrochemical sensor for the detection of phytoregulator indole-3-acetic acid using gold-nanoparticles and three-dimensional reduced graphene oxide modified screen printed carbon electrode. *Talanta* **2023**, *253*, 124030. [CrossRef]
47. Shi, F.; Ai, Y.J.; Wang, B.L.; Yao, Y.C.; Zhang, Z.J.; Zhou, J.; Wang, X.H.; Sun, W. Portable wireless intelligent electrochemical sensor for the ultrasensitive detection of rutin using functionalized black phosphorene nanocomposite. *Molecules* **2022**, *27*, 6603. [CrossRef]
48. Lin, S.H.; Liu, S.H.; Yang, Z.B.; Li, Y.Y.; Ng, W.T.; Xu, Z.Q.; Bao, Q.L.; Hao, J.H.; Lee, S.H.; Surya, C.; et al. Solution-processable ultrathin black phosphorus as an effective electron transport layer in organic photovoltaics. *Adv. Funct. Mater.* **2016**, *26*, 864–871. [CrossRef]
49. Yang, J.; Pan, Z.H.; Zhong, J.; Li, S.; Wang, J.; Chen, P.Y. Electrostatic self-assembly of heterostructured black phosphorus-Mxene nanocomposites for flexible microsupercapacitors with high rate performance. *Energy Storage Mater.* **2021**, *36*, 257–264. [CrossRef]
50. Machabaphala, M.K.; Hlekelele, L.; Dlamini, L.N. A heterostructure of black phosphorus and zirconium-based MOF as a photocatalyst for photocatalytic applications. *Mater. Lett.* **2020**, *281*, 128660. [CrossRef]
51. Tan, W.C.; Cai, Y.Q.; Ng, R.J.; Huang, L.; Fang, X.W.; Zhang, G.; Zhang, Y.W.; Nijhuia, C.A.; Liu, X.K.; Ang, K.W. Few-layer black phosphorus carbide field-effect transistor via carbon doping. *Adv. Mater.* **2017**, *29*, 1700503–1700509. [CrossRef] [PubMed]
52. Reddy, D.A.; Kim, E.H.; Gopannagari, M.; Kim, Y.; Kumar, D.P.; Kim, T.K. Few layered black phosphorus/MoS_2 nanohybrid: A promising co-catalyst for solar driven hydrogen evolution. *Appl. Catal. B Environ.* **2019**, *241*, 491–498. [CrossRef]
53. Favron, A.; Gaufrès, E.; Fossard, F.; Phaneuf-L'Heureux, A.L.; Tang, N.Y.W.; Lévesque, P.L.; Loiseau, A.; Leonelli, R.; Francoeur, S.; Martel, R. Photooxidation and quantum confinement effects in exfoliated black phosphorus. *Nat. Mater.* **2015**, *14*, 826–832. [CrossRef] [PubMed]
54. He, L.D.; Lu, Q.J.; Yang, Y.; Liu, Y.Q.; Zhu, Y.Z.; Mei, Y. Facile synthesis of holey phosphorene via low temperature electrochemical exfoliation for electrocatalytic nitrogen reduction. *ChemistrySelect* **2021**, *6*, 5021–5026. [CrossRef]
55. Li, C.; Cui, X.M.; Gao, X.; Liu, S.Z.; Sun, Q.; Lian, H.Q.; Zu, L.; Liu, Y.; Wang, X.D.; Cui, X.G. Electrochemically prepared black phosphorene micro-powder as flame retardant for epoxy resin. *Compos. Interface* **2020**, *28*, 693–705. [CrossRef]

56. dos Santos, A.J.; Fortunato, G.V.; Kronka, M.S.; Vernasqui, L.G.; Ferreira, N.G.; Lanza, M.R.V. Electrochemical oxidation of ciprofloxacin in different aqueous matrices using synthesized boron-doped micro and nano-diamond anodes. *Environ. Res.* **2022**, *204*, 112027. [CrossRef]
57. Wang, L.J.; Hu, Y.Y.; Qi, F.; Ding, L.; Wang, J.M.; Zhang, X.Y.; Liu, Q.W.; Liu, L.Z.; Sun, H.Z.; Qu, P. Anchoring black phosphorus nanoparticles onto ZnS porous nanosheets: Efficient photocatalyst design and charge carrier dynamics. *ACS Appl. Mater. Interfaces* **2020**, *12*, 8157–8167. [CrossRef]
58. Barman, S.C.; Sharifuzzaman, M.; Zahed, M.A.; Park, C.; Yoon, S.H.; Zhang, S.P.; Kim, H.; Yoon, H.; Park, J.Y. A highly selective and stable cationic polyelectrolyte encapsulated black phosphorene based impedimetric immunosensor for interleukin-6 biomarker detection. *Biosens. Bioelectron.* **2021**, *186*, 113287. [CrossRef]
59. Ribeiro, D.; Abrantes, J. Application of electrochemical impedance spectroscopy (EIS) to monitor the corrosion of reinforced concrete: A new approach. *Constr. Build. Mater.* **2016**, *111*, 98–104. [CrossRef]
60. Niu, X.L.; Wen, Z.R.; Li, X.B.; Zhao, W.S.; Li, X.Y.; Huang, Y.Q.; Li, Q.T.; Li, G.J.; Sun, W. Fabrication of graphene and gold nanoparticle modified acupuncture needle electrode and its application in rutin analysis. *Sens. Actuators B Chem.* **2018**, *255*, 471–477. [CrossRef]
61. Laviron, E. General expression of the linear potential sweep voltammogram in the case of diffusionless electrochemical systems. *J. Electroanal. Chem.* **1979**, *101*, 19–28. [CrossRef]
62. Laviron, E. Adsorption, autoinhibition and autocatalysis in polarography in linear potential sweep voltammetry. *J. Electroanal. Chem.* **1974**, *52*, 355–393. [CrossRef]
63. Nicholson, R.S.; Shain, I. Theory of stationary electrode polarography for a chemical reaction coupled between two charge transfers. *Anal. Chem.* **1965**, *37*, 178–190. [CrossRef]

biosensors

MDPI

Article

Porphyrin-Based Covalent Organic Frameworks with Donor-Acceptor Structure for Enhanced Peroxidase-like Activity as a Colorimetric Biosensing Platform

Qian Wang [1,†], Liang Lv [2,†], Wenhao Chi [1], Yujiao Bai [1], Wenqing Gao [1], Peihua Zhu [1,*] and Jinghua Yu [1]

[1] School of Chemistry and Chemical Engineering, University of Jinan, Jinan 250022, China
[2] Jinan Agricultural Product Quality and Safety Center, Jinan 250316, China
* Correspondence: chm_zhuph@ujn.edu.cn
† These authors equally contributed to this work.

Abstract: Hydrogen peroxide (H_2O_2) and glucose play a key role in many cellular signaling pathways. The efficient and accurate in situ detection of H_2O_2 released from living cells has attracted extensive research interests. Herein, a new porphyrin-based porous covalent organic framework (TAP-COF) was fabricated via one-step condensation of 1,6,7,12-tetrachloroperylene tetracarboxylic acid dianhydride and 5,10,15,20-tetrakis (4-aminophenyl)porphyrin iron(III). The obtained TAP-COF has high surface areas, abundant surface catalytic active sites, and highly effective electron transport due to its precisely controllable donor–acceptor arrangement and 3D porous structure. Then, the new TAP-COF exhibited excellent peroxidase-like catalytic activity, which could effectively catalyze oxidation of the substrate 3,3′,5,5′-tetramethylbenzidine by H_2O_2 to produce a typical blue-colored reaction. On this basis, simple, rapid and selective colorimetric methods for in situ H_2O_2 detection were developed with the detection limit of 2.6 nM in the wide range of 0.01 to 200 μM. The colorimetric approach also could be used for in situ detection of H_2O_2 released from living MCF-7 cells. This portable sensor based on a COF nanozyme not only opens a new path for point-of-care testing, but also has potential applications in the field of cell biology and clinical diagnosis.

Keywords: porphyrin; covalent organic framework; peroxidase-like activity; cells; H_2O_2

Citation: Wang, Q.; Lv, L.; Chi, W.; Bai, Y.; Gao, W.; Zhu, P.; Yu, J. Porphyrin-Based Covalent Organic Frameworks with Donor-Acceptor Structure for Enhanced Peroxidase-like Activity as a Colorimetric Biosensing Platform. *Biosensors* **2023**, *13*, 188. https://doi.org/10.3390/bios13020188

Received: 4 January 2023
Revised: 23 January 2023
Accepted: 23 January 2023
Published: 26 January 2023

1. Introduction

Small molecule biomarkers are a significant part of sustaining life processes and biological metabolism [1]. Among them, hydrogen peroxide (H_2O_2), one of the frequent reactive oxygen species in biological systems, is essential for promoting cellular proliferation, differentiation, and emigration [2,3]. Various illnesses including cancer [4], diabetes mellitus [5], neurological disorders [6], and angiocardiopathy [7] are tightly accompanied by abnormal H_2O_2 production. For example, cancer cells have considerably greater amounts of H_2O_2 (6.0 μM–1.2 mM) than normal cells (less than 0.72 μM) [8]. Therefore, the need for a reliable and accurate approach for the extremely hypersensitive identification of H_2O_2 in living creatures is critical. The examination of H_2O_2 has been reported using a variety of strategies, including electrochemical [9], fluorescence [10,11], electrochemiluminescence [12], spectrophotometry [13], etc., among which colorimetric tests [14] are considered to be an effective and straightforward method due to its affordability and usefulness.

Peroxidases, like horseradish peroxidase (HRP) and thyroid peroxidase, are found in plankton and flora and fauna, which are commonly utilized because of their high particularity and effectiveness under mild settings [15]. However, due to their inherent nature as proteins, the majority of natural enzymes have many inevitable shortcomings, such as intrinsic instability, costly preparation, and easy denaturation under adverse conditions, which severely restricts their real applications [16]. Then, a lot of work has been put into researching nanozymes, such as porphyrins and metal complexes, with the aim of simulating the structures and activities of naturally produced enzymes. Due to their affordability, a large-scale production capacity, superior stability and robustness compared to natural enzymes, the interests in nanozymes with unexpected enzyme-like activity have grown recently.

Porphyrin-based nanozymes are currently receiving more focus for their biological similarity to inartificial enzymes like cytochrome P450 and hemoglobin [17]. Various iron porphyrins, including hemin [18] and 5,10,15,20-tetrakis-(4′-bromophenyl)porphyrinato iron(III) [19] are used to simulate HRP activity. Free iron porphyrins cannot, however, be used in standard immunoassay platforms because of their poorer imitation of peroxidase activity and facile aggregation property [20]. Recently, a lot of interest has been focused on porphyrin-based covalent organic frameworks (COFs), which are produced from porphyrin or metalloporphyrin through strong covalent bonding, as biomimetic catalysts, electrocatalysts, and photocatalysts [21–23] due to their superior catalytic operations and close biological affinity to inartificial enzymes. These possess COFs′ intrinsic benefits, including their high surface area, tunable pore topologies, functionalization, and excellent thermal and chemical durability. By incorporating metalloporphyrin into COFs, high density active sites would be added to the skeleton of the COFs, and the catalytic deactivation brought on by porphyrin aggregation might also be prevented. Typically, Song's group proposed a series of ultrastable iron porphyrin-based COFs with 3D channels as biomimetic catalysts [24]. Lin's group created a 3D porphyrin-based COFs through the co-condensation of tetrahedrons with parallelogram construction blocks to enhance catalytic performance [25]. Additionally, to produce small band gaps for the dissociation and transportation of electron-hole pairs, the appropriate combination and arrangement of complementary donor–acceptor pairs inside the COFs skeleton have been shown to be an effective strategy to generate higher catalytic activity. Meanwhile, the unitarity of the COFs would be advantageous to boost the efficiency of enzymatic processes by improving the stable enzymes and decreasing the diffusion barrier.

Inspired by the above discussion, a donor–acceptor COFs (defined as TAP-COF) with 5,10,15,20-tetrakis (4-aminophenyl)porphyrine iron(III) chloride (FeTAPP) as the donor and 1,6,7,12-tetrachloroperylene tetracarboxylic acid dianhydride (TAD) as the acceptor was synthesized and employed as a peroxidase-mimicking nanozyme. The produced D-A COFs had a high degree of crystallinity, a sizable amount of porosity, excellent electron transport, and a significant number of catalytic active sites. Moreover, TAP-COF was a metalloporphyrin-based COFs with a built-in catalyst that could successfully prevent deactivation, which was frequently linked to porphyrin aggregation. TAP-COF was utilized to build a colorimetric assay in viable cells for in situ analysis of H_2O_2 using of 3,3′,5,5′-tetramethylbenzidine (TMB), which offered a novel method of cell monitoring and has promising uses in the field of fast clinical detection.

2. Materials and Methods

2.1. *Peroxidase-like Performance of TAP-COF*

By catalytically oxidizing the peroxidase zymolyte TMB in the absence of H_2O_2, the peroxidase-like activity of TAP-COF was examined. The measurements were made by monitoring the absorbance information in time-scan patterns at 652 nm. In the typical experiment, 200 μL 0.32 mg·mL^{-1} TAP-COF dispersed in water, 200 μL H_2O_2 (30 mM), and 200 μL TMB (0.80 mM) as the substrate were added to a HAc–NaAc buffer solution (pH 3.8).

In a typical colorimetric experiment, 200 μL TAP-COF (0.32 mg·mL^{-1}) dispersed in water, 200 μL H_2O_2 (30 mM), and 200 μL *tert*-butyl alcohol (1 mM) were added into 1200 μL of a HAc–NaAc buffer solution (pH 3.8). Next, the above solution was incubated at 25 °C for 0.5 h and 200 μL TMB (0.80 mM) was added before incubating for another 10 min. Subsequently, UV–vis absorption spectroscopies were used to gauge the final reaction solution. In comparison, the same experimental conditions were used to investigate the expression system of TAP-COF + H_2O_2 + TMB + buffer by omitting the *tert*-butyl alcohol.

2.2. Condition Optimization

The influences of the pH values (1.7–8.0), temperatures (25–70 °C), and various concentrations of H_2O_2 (1–400 mM) and TMB (0.20–2.5 mM) were investigated to evaluate the peroxidase-like activity of the TAP-COF composites. The effect of the catalyst amount was also studied.

2.3. Detection of H_2O_2 in Cells

A total of 200 μL phorbol 12-myristate-13-acetate (PMA) (50 g·mL^{-1}) was injected to a counting plate containing 0, 100, 200, 500, 1000, 3000, or 5000 cells·mL^{-1} MCF-7 cells in order to determine the amount of H_2O_2 released by the cells. To facilitate color development, 200 μL TMB (0.8 mM) and 200 μL TAP-COF (0.32 mg·mL^{-1}) were injected to each counting plate after the specimens had been incubated for 10 min in 0.2 M HAc–NaAc (pH 4.0). A microplate reader was used to measure each well's absorbance at 652 nm, which was believed to be related to the emission of H_2O_2 in cells.

3. Results and Discussions

3.1. Structure and Morphology Characterization

TAP-COF was synthesized by condensation between FeTAPP with TAD in the presence of 4 Å molecular sieves under solvothermal conditions (Scheme 1). To confirm the formation of the diimide linkages between FeTAPP and TAD units in the TAP-COF, Fourier transform infrared (FT-IR) spectroscopy was performed (Figure 1A). FT-IR spectra indicated that the peak of the N-H stretching of -NH_2 (at 3364 cm^{-1}) from FeTAPP disappeared after polymerization. Meanwhile, the C = O amide stretching vibrations correspond to the absorptions at 1666 cm^{-1}. The peak at 1323 cm^{-1} was attributed to the C-H in phenyl rings' in-plane bending vibration. In addition, with the presence of C-N-C, a clear band was displayed at 1594 cm^{-1} [26]. However, the peaks belonging to anhydride were not accurately identified due to the synthesis of imide between amino groups and anhydride. The X-ray diffraction (XRD) measurement indicated that TAP-COF was a well crystalline material as illustrated in Figure 1B. The XRD diagram of TAP-COF demonstrated two peaks at 1.26 nm and 0.72 nm, which were ascribed to diffraction from the (200) and (002) planes, respectively. In addition, the XRD pattern also displayed three higher order refractions at 0.88, 0.42, and 0.38 nm, which were ascribed to diffraction from the (300), (600), and (800) planes, respectively, revealing the high molecular ordering nature of this COF along the a-axis of the unit cell [27]. The (001) plane gave its higher order diffractions at 0.34 nm in the wide-angle range of the XRD pattern. These diffraction results could be assigned to refraction from a rectangular lattice with the cell parameter of a = 2.52 nm, b = 2.52 nm, and c = 1.44 nm (Figure S1 in Supplementary Materials). In addition, the XRD pattern of TAP-COF exhibited additional refraction at 0.34 nm, which was attributed to the π-π stacking distance between (100) planes [28–30].

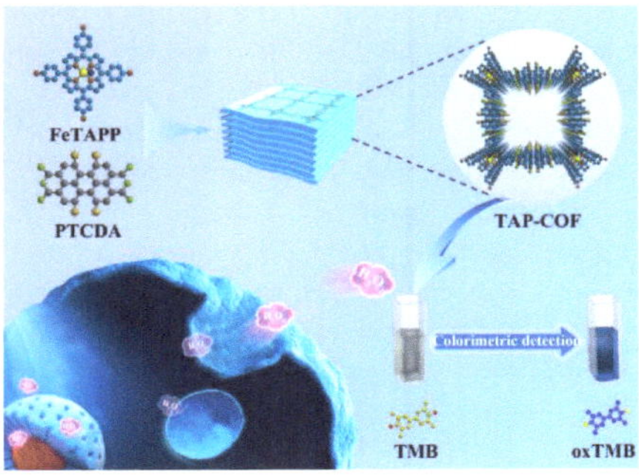

Scheme 1. Diagrammatic drawing for the synthesis of TAP-COF and in situ detection of H_2O_2 released from MCF-7 cells.

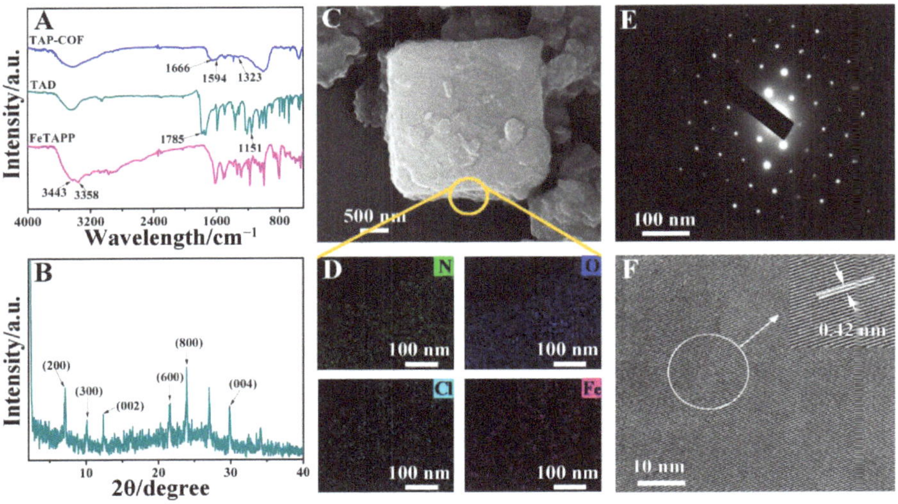

Figure 1. (**A**) FT-IR spectra of FeTAPP, TAD, and TAP-COF; (**B**) XRD spectra of the TAP-COF; (**C**) SEM image; (**D**) N, O, Cl, Fe element mapping; (**E**) SAED pattern from the crystalline domain; (**F**) HRTEM image of TAP-COF.

The morphologies of TAP-COF were analyzed using scanning electron microscopy (SEM). The SEM image (Figure 1C) revealed that TAP-COF had a distinct lamellar structure. In addition, the energy-dispersive spectrometry mapping was exhibited in Figure 1D. Atoms of the expected composition (N, O, Cl, Fe) of the compound could be detected, suggesting that TAP-COF was synthesized successfully. In order to further investigate the morphology of TAP-COF, high-resolution transmission electron microscopy (HRTEM) was also performed. The sample's crystalline nature was confirmed by the regular spot pattern that was revealed by the selected area electron diffraction (SAED) in the chosen location, as exhibited in Figure 1E. The HRTEM image indicated that the lattice space of the TAP-COF was calculated to be 0.42 nm (Figure 1F), which corresponded to (600) plane of TAP-COF as seen in the XRD profile.

To further confirm the thermal stability of TAP-COF, thermal gravimetric analysis (TGA) under air conditions of the sample was conducted. As illustrated in Figure S2, the TGA presented a rapid weight loss of ~12.2% in the temperature range of 25–150 °C, which seemed to be ascribed to the loss of the solvent absorbed in the porous sample [31]. Then, the TGA profile of the TAP-COF did not exhibit any weight loss in the temperature range of 150–380 °C. In addition, TAP-COF showed an observable weight loss over 380 °C, which was attributed to the decomposition of the frameworks. Based on the above analysis, the TGA curves revealed that the TAP-COF was thermally stable up to 380 °C.

The permanent porosity of TAP-COF was checked by N_2 adsorption isotherms collected at 77 K (Figure S3). The sorption isotherm fit well with the type IV characters, suggesting the permanent mesoporous nature of TAP-COF. According to the Figure S3, the COF had a total pore volume of 0.124 $cm^3 \cdot g^{-1}$ (P/P_0 = 0.99). In addition, the BET and Langmuir surface areas of TAP-COF were calculated to be 66.38 and 140.12 $m^2 \cdot g^{-1}$, respectively, which indicated the small surface area of TAP-COF compared to other COFs. It might be related to the disorganized stacking of the TAP-COF layers caused by the steric hindrance of TAD and the twisting configuration of the phenyl units in FeTAPP.

3.2. Peroxidase-like Activity of TAP-COF

The peroxide simulation properties of TAP-COF were examined by oxidizing the colorimetric primer TMB to oxTMB, which could be easily monitored by the UV–vis absorbance spectroscopy or observed by the naked eyes. Figure 2A shows the UV–vis absorption spectra for the different test solutions. The absence experiments were performed to determine the components needed to trigger the reaction. When both TAP-COF and H_2O_2 were present in the reaction system, there was a higher absorption peak at 652 nm, indicating that TAP-COF and H_2O_2 were required for the catalytic process. Additionally, Figure 2B shows the variations of absorbance over time for TAP-COF + H_2O_2 + TMB and H_2O_2 + TMB + FeTAPP. The absorbance at 652 nm rose for each system as the reaction proceeded, but the TAP-COF + H_2O_2 + TMB system responded much more quickly than the other systems. After only five minutes of reaction, the absorbance of TAP-COF + H_2O_2 + TMB exceeded that of FeTAPP + H_2O_2 + TMB, indicating that TAP-COF had a greater catalytic activity than FeTAPP. These findings unequivocally indicated that TAP-COF had a better sensitivity for accelerating the oxidation of TMB. The inherent skeleton structure, high N content, and D-A structure of TAP-COF might be the origin of its porosity characteristics, electron cloud density, and greater surface catalytic activity sites. These would hasten the electron exchange between reactive oxygen species generated by catalyzing hydrogen peroxide decomposition and TMB, thus increasing the catalytic activity. They also had electrostatic interaction that bind the catalytically active sites of TAP-COF to the positively charged TMB.

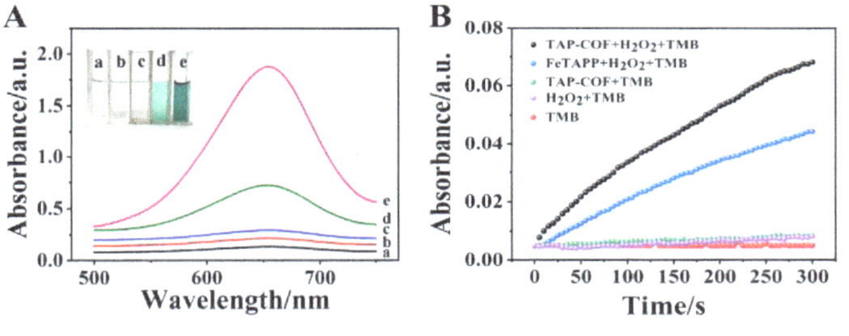

Figure 2. The absorbance changes of TMB, H_2O_2 + TMB, TAP-COF + TMB, FeTAPP + H_2O_2 + TMB, and TAP-COF + H_2O_2 + TMB are illustrated in (**A**) UV–vis spectra and (**B**) the impact of time on these changes. TMB reaction mixture's color variations are seen in the inset.

3.3. Optimization of Experimental Conditions

Similar to HRP and other nanomaterial-based peroxidase mimics, the pH, temperature, and substrate concentration during the color development process also influenced the catalytic performance of TAP-COF. Then, the peroxidase-like activity of TAP-COF was investigated by varying the pH (1.7–8), temperature (25–70 °C), H_2O_2 concentration (0–200 M), and TMB concentration (0.2–2.5 mM) as shown in Figure S4. These data suggested that the ideal experimental settings were pH 3.8, 25 °C, the relative lowest necessary concentration of H_2O_2 (200 μM) and 0.8 nM of TMB. Moreover, the catalytic activity of TAP-COF was also related to its amount. The catalytic activity increased with a rapid increase in catalyst concentration from 0 to 1.0 mg·mL^{-1} and then there was a downtrend from 1.0 to 1.5 mg·mL^{-1}. Thus, the optimal concentration of TAP-COF was chosen as 1.0 mg·mL^{-1}.

3.4. Steady-State Kinetic Analysis

Typical Michaelis–Menten curves were produced under ideal conditions over a range of TMB or H_2O_2 concentrations to quantitatively explore the catalytic activities of the TAP-COF (Figure S5). Under the guidance of the Lineweaver–Burk equation, the Michaelis constant (K_m) and the maximum response velocity (V_{max}) were calculated (Table 1). In general, K_m describes how well an enzyme binds to its substrates, with a lower K_m value indicating a stronger binding. The K_m of TAP-COF (0.78 mM) for H_2O_2 as primer was significantly lower than that of HRP (3.7 mM), demonstrating that the affinity of TAP-COF for H_2O_2 was stronger than that of HRP. This might be because of the increased electrostatic contact between TMB and TAP-COF produced by its high N content. Furthermore, the TAP-COF had a K_m of 0.18 mM with TMB as the substrate, which was lower than the HRP's K_m of 0.43 mM, indicating that TAP-COF also had a high affinity for the TMB mediator. These results implied that TAP-COF had promising peroxidase-like activity, which was in line with other discoveries using various peroxidase mimics based on nanomaterials (Table 1).

Table 1. Comparison of maximal reaction rate (V_{max}) and apparent Michaelis-Menten constant (K_m).

Catalyst	K_m/mM		$V_{max}/10^{-8}$ M s^{-1}		Reference
	H_2O_2	TMB	H_2O_2	TMB	
FePor-TFPA-COP	2.77	0.027	83.3	1.76	[32]
H_2TCPP-Co_3O_4	3.70	0.43	8.71	10.0	[33]
Fe-MIL-88NH_2	6.10	0.028	0.710	0.670	[34]
C-Dots	39.1	0.011	1.40	4.80	[35]
TAP-COF	0.780	0.18	2.94	4.78	This work

3.5. Colorimetric Sensing

The catalysis-based colorimetric test revealed extremely sensitive changes in absorption with varying H_2O_2 concentrations. Figure 3A demonstrates that the immunoassay solutions gradually darkened from colorless to blue as the H_2O_2 content increased. Additionally, Figure 3B and C illustrates a superb linear correlation between the logarithm of H_2O_2 concentration in the range of 0.01 μM to 200 μM and the absorbance of oxidized TMB at 652 nm. The TAP-COF-based sensor had a lower limit of detection (LOD) and larger linear range compared to other previously published outstanding peroxidase-based sensors (see Table S1). Additionally, our proposed sensor's LOD was lower than that of some of the other electrochemical and fluorescent sensors, which might be explained by the fact that the interlinking of electron acceptor and donor of TAP-COF could improve the electron transport rate, as well as allow exposure to high concentrations of catalytic active sites, thus contributing to its increased catalytic activity.

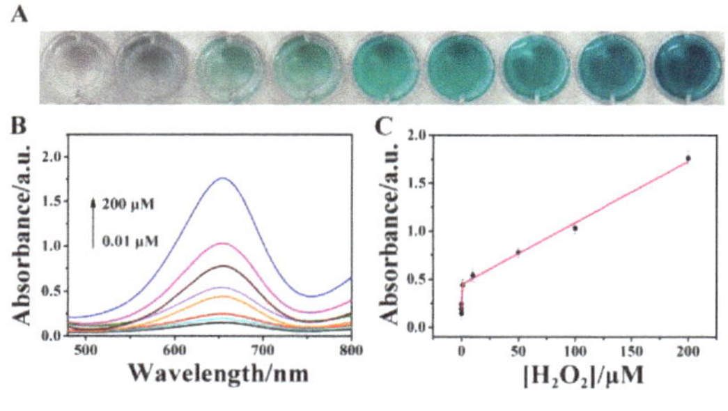

Figure 3. Photographs (**A**) of the TMB + H_2O_2 colorimetric reaction solutions facilitated by various H_2O_2 concentrations in the range of 0.01–200 μM and (**B**) UV–vis absorption spectra. (**C**) Relationship between absorbance at 650 nm in reaction solutions' UV–vis absorption spectra and H_2O_2 concentration. Standard deviations are shown by error bars (n = 5).

Moreover, H_2O_2 was also a major byproduct of other metabolic processes, including the oxidation of uric acid by the enzyme uricase as well as the oxidation of glucose by the enzyme glucose oxidase (GOD). Consequently, the measurement of H_2O_2 might be used to indirectly infer the quantities of living components. Figure 4A displays the TMB + TAP-COF + GOD photographs at various glucose concentrations. As the glucose concentration increased linearly, the reaction became progressively darker blue in the presence of TAP-COF and GOD. The standard curves for these reactions are shown in Figure 4C, where the linear range of glucose was 0.3–800 μM, and the LOD was 0.14 μM, which was comparable to that of other reported fluorescent probes (Table S2).

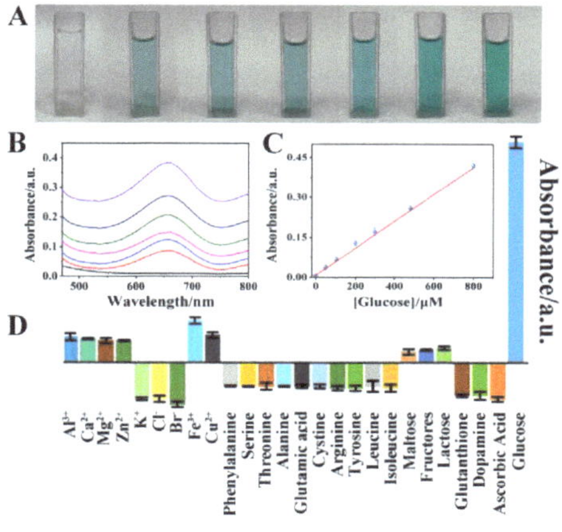

Figure 4. (**A**) A picture of the reaction mixture with various concentrations of glucose present. (**B**) The UV–vis absorption spectra. (**C**) Linear calibration of glucose detection. (**D**) The absorption of TMB-TAP-COF for different substances. (The concentration of potentially interfering substances was 50 μM, and the concentration of glucose was 0.5 μM, TMB was 0.8 mM and TAP-COF was 1.0 mg·mL^{-1}).

3.6. Selectivity for the Detection of Glucose

Several interference experimental studies were conducted in the presence of interfering chemicals to determine the selectivity of TAP-COF-based colorimetric sensors for H_2O_2 detection (Figure 4D). The response of glucose (2×10^{-4} mmol·L^{-1}) was up to 100 times greater than that of other substances (2×10^{-2} mol·L^{-1}), none of which interfered with the measurement of glucose. These outcomes demonstrated the TAP-COF-based sensor exhibited superior glucose detection selectivity.

3.7. Colorimetric Detection Mechanism for H_2O_2

The cascade reaction-based colorimetric glucose assay is depicted in Figure 5A. The peroxidase-like activity of TAP-COF was derived from its catalytic H_2O_2 decomposition to produce ROS, mainly including singlet oxygen ($^1O_2$), hydroxyl radical (·OH), and superoxide anion ($O_2^{\bullet -}$), where ·OH was a possible source of the catalytic process of TMB [36,37]. Next, capture experiments were used to test the ROS, and the catalytic mechanism and specificity of TAP-COF was further studied. As demonstrated in Figure 5B, the TMB + H_2O_2 + TAP-COF system showed different absorbance intensity with or without the addition of D-histidine ($^1O_2$ scavenger), *tert*-butyl alcohol (·OH scavenger), and benzoquinone ($O_2^{\bullet -}$ scavenger). The absorbance significantly decreased after the addition of *tert*-butyl, suggesting that the generated ·OH was scavenged by *tert*-butyl alcohol. Meanwhile, the system's absorbance value exhibited no evident change after the addition of D-histidine and benzoquinone, indicating the specificity of TAP-COF peroxide-mimicking enzymes [38]. In summary, the peroxide-mimicking enzyme activity of TAP-COF mostly resulted from the ·OH, which was at odds with earlier studies on certain inorganic nanomaterials that were thought to originate via electron transport [39]. The electrocatalytic behavior of the TAP-COF and the FeTAPP modified GCE towards the electrochemical reduction of H_2O_2 was examined in further detail utilizing amperometric responses. Compared with the FeTAPP-modified electrode, the TAP-COF-modified electrode exhibited an enhanced reduction peak current, suggesting that TAP-COF exhibited higher electrocatalytic activity in the reduction of H_2O_2, which indicated that the TAP-COF had the better ability to accelerate the electron transfer (Figure 5C). Based on the above discussion, the possible explanation of TMB substrate oxidation by the TAP-COF catalyst in the presence of H_2O_2 might be proposed as follows: first, H_2O_2 entered the pores of TAP-COF or attached to the surface, and then the O–O bond of H_2O_2 broke to form ·OH. Second, ·OH oxidized TMB to produce a blue-colored byproduct. These phenomena indicated that the blue signal came from the charge transfer complex (chromogen). Electrochemical studies have indicated that there was a high degree of communication between the donor and acceptor of TAP-COF and the entire molecule's electrical activity was governed by this connection. Moreover, the catalytic oxidation of TMB by H_2O_2 was aided by the D-A ordered structure of TAP-COF's capacity to speed electron transport. Therefore, the design of TAP-COF with D-A properties was of great significance for the construction of novel colorimetric sensors.

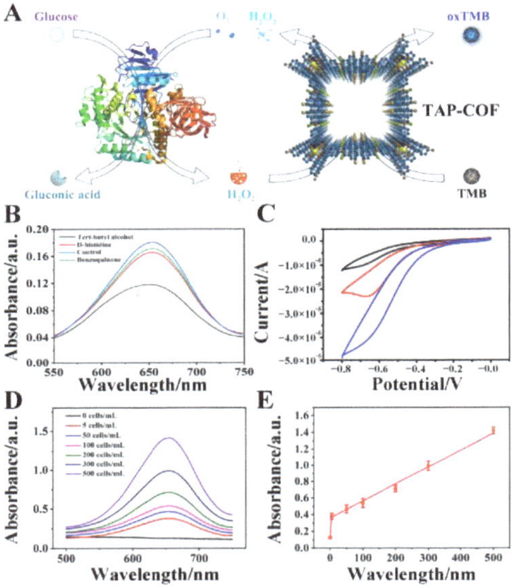

Figure 5. (**A**) Diagrammatic representation of the blue-colored oxTMB produced in cells by the INAzyme-catalyzed cascade oxidation of glucose and TMB. (**B**) ROS trapping experiment. The TMB oxidation process in the presence of D-histidine ($^1O_2$ scavenger), tert-butyl alcohol (·OH scavenger), and benzoquinone ($O_2^{\bullet -}$ scavenger). (**C**) CV curves of different electrodes, GCE electrode (black line), FeTAPP/GCE (red line), TAP-COF/GCE (blue line). TAP-COF-based colorimetric examination of MCF-7 cells in binding buffer for quantitative analysis of (**D**) UV-vis absorption spectroscopy and (**E**) linear calibration of MCF-7 cells detection.

3.8. Colorimetric Detection of Cancer Cells

H_2O_2 was an important component of several metabolic processes, including the metabolism of proteins and carbohydrates, and it may be used as a marker for a number of serious diseases, including cancer [40]. Then, samples with various MCF-7 cells ranging from 0–3000 cells·mL^{-1} were identified in 100 μL serial dilutions of binding buffer using TAP-COF as described above. Without including MCF-7 cells, the absorbance of the background was estimated using a similar approach (Figure 5D), and the absorbance at 650 nm was gathered for profiling. As demonstrated in Figure 5E, with the minimum cell concentration of 25 cells·mL^{-1} observed in our actual studies, a strong linear correlation was displayed between the colorimetric signal and the cell concentration ($R^2 > 0.98$). This LOD was significantly lower than that attained using the currently available colorimetric cancer cell detection techniques, and even on par with that attained using electrochemical or fluorescent assays that are more sensitive [41–43]. The above results indicated that the TAP-COF-based colorimetric biosensor displayed a powerful ability to detect H_2O_2

released from cancer cells, and thus our proposed simple and convenient colorimetric assay could be used for the early diagnosis of cancer cells.

4. Conclusions

In conclusion, a new functional TAP-COF nanozyme with a donor–acceptor arrangement has been constructed through one-step condensation. The TAP-COF demonstrated increased enzyme mimicking activities for the direct catalysis of TMB to create blue oxTMB due to very efficient charge transfer, porous strutted structure, and the abundance of iron centers. Correspondingly, an H_2O_2-sensitive colorimetric detection method based on the TAP-COF sensing system was established by facilitating the electron change between H_2O_2 and TMB. The established colorimetric biosensor has high selectivity, excellent compatibility, and outstanding sensitivity, which was successfully used for the visual detection of H_2O_2 released from living cells. Our work provides a portable COF nanozyme-based sensing platform for the accurate detection of small molecule biomarkers, and a new perspective for point-of-care testing and medical diagnosis.

Supplementary Materials: The following supporting information can be downloaded at: https://www.mdpi.com/article/10.3390/bios13020188/s1. Figure S1: Schematic representation of the unit cell in the TAP-COF (A) top view and (B) side view; Figure S2: TGA data of TAP-COF; Figure S3: N2 adsorption-desorption isotherm curve of TAP-COF at 77 K; Figure S4: Exploring the impacts of this detection system, including (A) temperature, (B) pH, (C) TMB concentration, (D) COF concentration, and (E) H2O2 concentration; Figure S5: Steady-state kinetic analysis using the Michaelis-Menten model (A, B) and Lineweaver-Burk model (C, D) for TAP-COF. The concentration of H2O2 was 60 mM and TMB concentration was varied (A, C). The concentration of TMB was 0.80 mM and H2O2 concentration was varied (B, D); Table S1: Detection limit comparison of TAP-COF for H2O2 with some recently reported peroxidase mimics; Table S2: Comparison of analytical parameters for the determination of glucose. References [44–59] are cited in Supplementary Materials.

Author Contributions: Experiment design, data analysis, and manuscript preparation, Q.W. and L.L.; experiments, W.C.; investigation and data analysis, Y.B. and W.G.; supervision, funding acquisition, and formal analysis, P.Z. and J.Y. All authors have read and agreed to the published version of the manuscript.

Funding: This research was funded by the National Natural Science Foundation of China (22276069) and the Shandong Provincial Natural Science Foundation (ZR2021MB047).

Institutional Review Board Statement: Not applicable.

Informed Consent Statement: Not applicable.

Data Availability Statement: Not applicable.

Conflicts of Interest: The authors declare no conflict of interest.

References

1. Liu, Y.; Zhang, H.; Li, B.; Liu, J.; Jiang, D.; Liu, B.; Sojic, N. Single Biomolecule Imaging by Electrochemiluminescence. *J. Am. Chem. Soc.* **2021**, *143*, 17910–17914. [CrossRef]
2. Jiao, L.; Xu, W.; Yan, H.; Wu, Y.; Liu, C.; Du, D.; Lin, Y.; Zhu, C. Fe-N-C Single-Atom Nanozymes for the Intracellular Hydrogen Peroxide Detection. *Anal. Chem.* **2019**, *91*, 11994–11999. [CrossRef]
3. Teodoro, K.B.R.; Migliorini, F.L.; Christinelli, W.A.; Correa, D.S. Detection of Hydrogen Peroxide (H_2O_2) Using a Colorimetric Sensor Based on Cellulose Nanowhiskers and Silver Nanoparticles. *Carbohydr. Polym* **2019**, *212*, 235–241. [CrossRef]
4. Chen, Y.; Chen, Z.; Fang, L.; Weng, A.; Luo, F.; Guo, L.; Qiu, B.; Lin, Z. Electrochemiluminescence Sensor for Cancer Cell Detection Based on H_2O_2-Triggered Stimulus Response System. *J. Anal. Test.* **2020**, *4*, 128–135. [CrossRef]
5. Chen, Y.; Shi, X.; Lu, Z.; Wang, X.; Wang, Z. A Fluorescent Probe for Hydrogen Peroxide in Vivo Based on the Modulation of Intramolecular Charge Transfer. *Anal. Chem.* **2017**, *89*, 5278–5284. [CrossRef] [PubMed]
6. Jain, V.; Bhagat, S.; Singh, S. Bovine Serum Albumin Decorated Gold Nanoclusters: A Fluorescence-Based Nanoprobe for Detection of Intracellular Hydrogen Peroxide. *Sens. Actuators B* **2021**, *327*, 128886. [CrossRef]
7. Ye, S.; Hu, J.J.; Yang, D. Tandem Payne/Dakin Reaction: A New Strategy for Hydrogen Peroxide Detection and Molecular Imaging. *Angew. Chem. Int. Ed.* **2018**, *57*, 10173–10177. [CrossRef]

8. Kumar, R.; Han, J.; Lim, H.J.; Ren, W.X.; Lim, J.Y.; Kim, J.H.; Kim, J.S. Mitochondrial Induced and Self-Monitored Intrinsic Apoptosis by Antitumor Theranostic Prodrug: In Vivo Imaging and Precise Cancer Treatment. *J. Am. Chem. Soc.* **2014**, *136*, 17836–17843. [CrossRef] [PubMed]
9. Bhunia, S.K.; Dolai, S.; Sun, H.; Jelinek, R. "On/Off/On" Hydrogen-Peroxide Sensor with Hemoglobin-Functionalized Carbon Dots. *Sens. Actuators B* **2018**, *270*, 223–230. [CrossRef]
10. Shi, Y.; Wu, Q.; Li, W.; Lin, L.; Qu, F.; Shen, C.; Wei, Y.; Nie, P.; He, Y.; Feng, X. Ultra-Sensitive Detection of Hydrogen Peroxide and Levofloxacin Using a Dual-Functional Fluorescent Probe. *J. Hazard. Mater.* **2022**, *432*, 128605. [CrossRef]
11. Rajendran, S.; Zichri, S.B.; Usha Vipinachandran, V.; Jelinek, R.; Bhunia, S.K. Triphenylphosphonium-Derived Bright Green Fluorescent Carbon Dots for Mitochondrial Targeting and Rapid Selective Detection of Tetracycline. *ChemNanoMat* **2021**, *7*, 545–552. [CrossRef]
12. Kitte, S.A.; Gao, W.; Zholudov, Y.T.; Qi, L.; Nsabimana, A.; Liu, Z.; Xu, G. Stainless Steel Electrode for Sensitive Luminol Electrochemiluminescent Detection of H_2O_2, Glucose, and Glucose Oxidase Activity. *Anal. Chem.* **2017**, *89*, 9864–9869. [CrossRef]
13. Zhao, J.; Dong, W.; Zhang, X.; Chai, H.; Huang, Y. FeNPs@Co_3O_4 Hollow Nanocages Hybrids as Effective Peroxidase Mimics for Glucose Biosensing. *Sens. Actuators B* **2018**, *263*, 575–584. [CrossRef]
14. UshaVipinachandran, V.; Rajendran, S.; Ali, H.; Ashokan, I.; Bhunia, S.K. Citrate Capped Silver Nanoparticles as an Instantaneous Colorimetric Selective Sensor for Neomycin and Thiamine in Wastewater. *New J. Chem.* **2022**, *46*, 14081–14090. [CrossRef]
15. Liu, H.; Ding, Y.; Yang, B.; Liu, Z.; Liu, Q.; Zhang, X. Colorimetric and Ultrasensitive Detection of H_2O_2 Based on Au/Co_3O_4-CeOx Nanocomposites with Enhanced Peroxidase-Like Performance. *Sens. Actuators B* **2018**, *271*, 336–345. [CrossRef]
16. Lin, A.; Liu, Q.; Zhang, Y.; Wang, Q.; Li, S.; Zhu, B.; Miao, L.; Du, Y.; Zhao, S.; Wei, H. A Dopamine-Enabled Universal Assay for Catalase and Catalase-Like Nanozymes. *Anal. Chem.* **2022**, *94*, 10636–10642. [CrossRef]
17. Fan, X.; Tian, R.; Wang, T.; Liu, S.; Wang, L.; Xu, J.; Liu, J.; Ma, M.; Wu, Z. An Ultrathin Iron-Porphyrin Based Nanocapsule with High Peroxidase-Like Activity for Highly Sensitive Glucose Detection. *Nanoscale* **2018**, *10*, 22155–22160. [CrossRef]
18. Liu, J.; He, X.; Wang, K.; He, D.; Wang, Y.; Mao, Y.; Shi, H.; Wen, L. A Highly Sensitive Electrochemiluminescence Assay for Protein Kinase Based on Double-Quenching of Graphene Quantum Dots by G-quadruplex-Hemin and Gold Nanoparticles. *Biosens. Bioelectron.* **2015**, *70*, 54–60. [CrossRef]
19. Li, Y.; Fang, Y.; Gao, W.; Guo, X.; Zhang, X. Porphyrin-Based Porous Organic Polymer as Peroxidase Mimics for Sulfide-Ion Colorimetric Sensing. *ACS Sustain. Chem. Eng.* **2020**, *8*, 10870–10880. [CrossRef]
20. Liu, Q.; Chen, P.; Xu, Z.; Chen, M.; Ding, Y.; Yue, K.; Xu, J. A Facile Strategy to Prepare Porphyrin Functionalized ZnS Nanoparticles and Their Peroxidase-Like Catalytic Activity for Colorimetric Sensor of Hydrogen Peroxide and Glucose. *Sens. Actuators B* **2017**, *251*, 339–348. [CrossRef]
21. Li, D.; Fang, Y.; Zhang, X. Bacterial Detection and Elimination Using a Dual-Functional Porphyrin-Based Porous Organic Polymer with Peroxidase-Like and High Near-Infrared-Light-Enhanced Antibacterial Activity. *ACS Appl. Mater. Interfaces* **2020**, *12*, 8989–8999. [CrossRef]
22. Xiang, Z.; Xue, Y.; Cao, D.; Huang, L.; Chen, J.F.; Dai, L. Highly Efficient Electrocatalysts for Oxygen Reduction Based on 2D Covalent Organic Polymers Complexed with Non-Precious Metals. *Angew. Chem. Int. Ed.* **2014**, *53*, 2433–2437. [CrossRef]
23. Zhu, P.; Li, S.; Zhou, S.; Ren, N.; Ge, S.; Zhang, Y.; Wang, Y.; Yu, J. In Situ Grown COFs on 3D Strutted Graphene Aerogel for Electrochemical Detection of NO Released from Living Cells. *Chem. Eng. J.* **2021**, *420*, 127559. [CrossRef]
24. Xie, Y.; Xu, M.; Wang, L.; Liang, H.; Wang, L.; Song, Y. Iron-Porphyrin-Based Covalent-Organic Frameworks for Electrochemical Sensing H_2O_2 and pH. *Mater. Sci. Eng. C* **2020**, *112*, 110864. [CrossRef]
25. Li, G.; Tian, W.; Zhong, C.; Yang, Y.; Lin, Z. Construction of Donor-Acceptor Heteroporous Covalent Organic Frameworks as Photoregulated Oxidase-Like Nanozymes for Sensing Signal Amplification. *ACS Appl. Mater. Interfaces* **2022**, *14*, 21750–21757. [CrossRef]
26. Liu, K.; Qi, H.; Dong, R.; Shivhare, R.; Addicoat, M.; Zhang, T.; Sahabudeen, H.; Heine, T.; Mannsfeld, S.; Kaiser, U.; et al. On-Water Surface Synthesis of Crystalline, Few-Layer Two-Dimensional Polymers Assisted by Surfactant Monolayers. *Nat. Chem.* **2019**, *11*, 994–1000. [CrossRef]
27. Han, B.; Ding, X.; Yu, B.; Wu, H.; Zhou, W.; Liu, W.; Wei, C.; Chen, B.; Qi, D.; Wang, H.; et al. Two-Dimensional Covalent Organic Frameworks with Cobalt(II)-Phthalocyanine Sites for Efficient Electrocatalytic Carbon Dioxide Reduction. *J. Am. Chem. Soc.* **2021**, *143*, 7104–7113. [CrossRef]
28. Ma, P.; Kan, J.; Zhang, Y.; Hang, C.; Bian, Y.; Chen, Y.; Kobayshi, N.; Jiang, J. The First Solution-Processable n-Type Phthalocyaninato Copper Semiconductor: Tuning the Semiconducting Nature Via Peripheral Electron-Withdrawing Octyloxycarbonyl Substituents. *J. Mater. Chem.* **2011**, *21*, 18552–18559. [CrossRef]
29. Chandra, S.; Kundu, T.; Dey, K.; Addicoat, M.; Heine, T.; Banerjee, R. Interplaying Intrinsic and Extrinsic Proton Conductivities in Covalent Organic Frameworks. *Chem. Mater.* **2016**, *28*, 1489–1494. [CrossRef]
30. Zhao, C.; Zhang, L.; Wang, Q.; Zhang, L.; Zhu, P.; Yu, J.; Zhang, Y. Porphyrin-Based Covalent Organic Framework Thin Films as Cathodic Materials for "On–Off–On" Photoelectrochemical Sensing of Lead Ions. *ACS Appl. Mater. Interfaces* **2021**, *13*, 20397–20404. [CrossRef]
31. Lu, D.; Li, J.; Wu, Z.; Yuan, L.; Fang, W.; Zou, P.; Ma, L.; Wang, X. High-Activity Daisy-Like Zeolitic Imidazolate Framework-67/Reduced Grapheme Oxide-Based Colorimetric Biosensor for Sensitive Detection of Hydrogen Peroxide. *J. Colloid Interface Sci.* **2022**, *608*, 3069–3078. [CrossRef]

32. Deng, X.; Fang, Y.; Lin, S.; Cheng, Q.; Liu, Q.; Zhang, X. Porphyrin-Based Porous Organic Frameworks as a Biomimetic Catalyst for Highly Efficient Colorimetric Immunoassay. *ACS Appl. Mater. Interfaces* **2017**, *9*, 3514–3523. [CrossRef]
33. Xie, J.; Cao, H.; Jiang, H.; Chen, Y.; Shi, W.; Zheng, H.; Huang, Y. Co_3O_4-Reduced Graphene Oxide Nanocomposite as an Effective Peroxidase Mimetic and Its Application in Visual Biosensing of Glucose. *Anal. Chim. Acta* **2013**, *796*, 92–100. [CrossRef]
34. Liu, Y.L.; Zhao, X.J.; Yang, X.X.; Li, Y.F. A Nanosized Metal-Organic Framework of Fe-MIL-88NH_2 as a Novel Peroxidase Mimic Used for Colorimetric Detection of Glucose. *Analyst* **2013**, *138*, 4526–4531. [CrossRef] [PubMed]
35. Shi, W.; Wang, Q.; Long, Y.; Cheng, Z.; Chen, S.; Zheng, H.; Huang, Y. Carbon Nanodots as Peroxidase Mimetics and Their Applications to Glucose Detection. *Chem. Commun.* **2011**, *47*, 6695–6697. [CrossRef] [PubMed]
36. Zheng, D.J.; Yang, Y.S.; Zhu, H.L. Recent Progress In the Development of Small-Molecule Fluorescent Probes for the Detection of Hydrogen Peroxide. *TrAC Trends Anal. Chem.* **2019**, *118*, 625–651. [CrossRef]
37. Zhang, Y.; Xu, J.; Zhou, S.; Zhu, L.; Lv, X.; Zhang, J.; Zhang, L.; Zhu, P.; Yu, J. DNAzyme-Triggered Visual and Ratiometric Electrochemiluminescence Dual-Readout Assay for Pb(II) Based on an Assembled Paper Device. *Anal. Chem.* **2020**, *92*, 3874–3881. [CrossRef]
38. Hu, F.X.; Miao, J.; Guo, C.; Yang, H.B.; Liu, B. Real-Time Photoelectrochemical Quantification of Hydrogen Peroxide Produced by Living Cells. *Chem. Eng. J.* **2021**, *407*, 127203. [CrossRef]
39. Dong, Y.L.; Zhang, H.G.; Rahman, Z.U.; Su, L.; Chen, X.J.; Hu, J.; Chen, X.G. Graphene Oxide-Fe_3O_4 Magnetic Nanocomposites with Peroxidase-Like Activity for Colorimetric Detection of Glucose. *Nanoscale* **2012**, *4*, 3969–3976. [CrossRef]
40. Zhu, D.; Kong, H.; Yang, G.; He, P.; Luan, X.; Guo, L.; Wei, G. Peptide Nanosheet-Inspired Biomimetic Synthesis of CuS Nanoparticles on Ti_3C_2 Nanosheets for Electrochemical Biosensing of Hydrogen Peroxide. *Biosensors* **2022**, *13*, 14. [CrossRef]
41. Lu, W.; Arumugam, S.; Senapati, D.; Khan, S.A.; Yu, H.; Ray, P. Multifunction Oval-Shaped Gold-Nanoparticle-Based Selective Detection of Breast Cancer Cells Using Simple Colorimetric and Highly Sensitive Two-Photon Scattering Assay. *ACS Nano* **2010**, *4*, 1739–1749. [CrossRef]
42. Sun, H.; Gao, Y.; Hu, N.; Zhang, Y.; Guo, C.; Gao, G.; Ma, Z.; Ivan Ivanovich, K.; Qiu, Y. Electronic Coupling Between Molybdenum Disulfide and Gold Nanoparticles to Enhance the Peroxidase Activity for the Colorimetric Immunoassays of Hydrogen Peroxide and Cancer Cells. *J. Colloid Interface Sci.* **2020**, *578*, 366–378. [CrossRef]
43. Li, M.; Lao, Y.H.; Mintz, R.L.; Chen, Z.; Shao, D.; Hu, H.; Wang, H.X.; Tao, Y.; Leong, K.W. A Multifunctional Mesoporous Silica-Gold Nanocluster Hybrid Platform for Selective Breast Cancer Cell Detection Using a Catalytic Amplification-Based Colorimetric Assay. *Nanoscale* **2019**, *11*, 2631–2636. [CrossRef]
44. Abdinejad, M.; Dao, C.; Zhang, X.A.; Kraatz, H.B. Enhanced Electrocatalytic Activity of Iron Amino Porphyrins Using a Flow Cell for Reduction of CO_2 to CO. *J. Energy Chem.* **2021**, *58*, 162–169. [CrossRef]
45. Pegis, M.L.; Martin, D.J.; Wise, C.F.; Brezny, A.C.; Johnson, S.I.; Johnson, L.E.; Kumar, N.; Raugei, S.; Mayer, J.M. Mechanism of Catalytic O_2 Reduction by Iron Tetraphenylporphyrin. *J. Am. Chem. Soc.* **2019**, *141*, 8315–8326. [CrossRef] [PubMed]
46. Hartnett, P.E.; Mauck, C.M.; Harris, M.A.; Young, R.M.; Wu, Y.L.; Marks, T.J.; Wasielewski, M.R. Influence of Anion Delocalization on Electron Transfer in a Covalent Porphyrin Donor-Perylenediimide Dimer Acceptor System. *J. Am. Chem. Soc.* **2017**, *139*, 749–756. [CrossRef]
47. Zhang, C.; Zhang, S.; Yan, Y.; Xia, F.; Huang, A.; Xian, Y. Highly Fluorescent Polyimide Covalent Organic Nanosheets as Sensing Probes for the Detection of 2,4,6-Trinitrophenol. *ACS Appl. Mater. Interfaces* **2017**, *9*, 13415–13421. [CrossRef]
48. Cui, C.; Wang, Q.; Liu, Q.; Deng, X.; Liu, T.; Li, D.; Zhang, X. Porphyrin-Based Porous Organic Framework: An Efficient and Stable Peroxidase-Mimicking Nanozyme for Detection of H_2O_2 and Evaluation of Antioxidant. *Sens. Actuators, B* **2018**, *277*, 86–94. [CrossRef]
49. Wang, N.; Sun, J.; Chen, L.; Fan, H.; Ai, S. A $Cu_2(OH)_3Cl$-CeO_2 Nanocomposite with Peroxidase-Like Activity, and Its Application to the Determination of Hydrogen Peroxide, Glucose and Cholesterol. *Microchim. Acta* **2015**, *182*, 1733–1738. [CrossRef]
50. Socaci, C.; Pogacean, F.; Biris, A.R.; Coros, M.; Rosu, M.C.; Magerusan, L.; Katona, G.; Pruneanu, S. Graphene Oxide vs. Reduced Graphene Oxide as Carbon Support in Porphyrin Peroxidase Biomimetic Nanomaterials. *Talanta* **2016**, *148*, 511–517. [CrossRef]
51. Yang, H.; Zha, J.; Zhang, P.; Xiong, Y.; Su, L.; Ye, F. Sphere-Like CoS with Nanostructures as Peroxidase Mimics for Colorimetric Determination of H_2O_2 and Mercury Ions. *RSC Adv.* **2016**, *6*, 66963–66970. [CrossRef]
52. Liu, Q.; Yang, Y.; Lv, X.; Ding, Y.; Zhang, Y.; Jing, J.; Xu, C. One-Step Synthesis of Uniform Nanoparticles of Porphyrin Functionalized Ceria with Promising Peroxidase Mimetics for H_2O_2 and Glucose Colorimetric Detection. *Sens. Actuators, B* **2017**, *240*, 726–734. [CrossRef]
53. Chen, H.; Shi, Q.; Deng, G.; Chen, X.; Yang, Y.; Lan, W.; Hu, Y.; Zhang, L.; Xu, L.; Li, C.; et al. Rapid and Highly Sensitive Colorimetric Biosensor For the Detection of Glucose and Hydrogen Peroxide Based on Nanoporphyrin Combined with Bromine as a Peroxidase-Like Catalyst. *Sens. Actuators, B* **2021**, *343*, 130104. [CrossRef]
54. Jin, L.; Shang, L.; Guo, S.; Fang, Y.; Wen, D.; Wang, L.; Yin, J.; Dong, S. Biomolecule-Stabilized Au Nanoclusters as a Fluorescence Probe for Sensitive Detection of Glucose. *Biosens. Bioelectron.* **2011**, *26*, 1965–1969. [CrossRef] [PubMed]
55. Liu, L.M.; Wen, J.; Liu, L.; He, D.; Kuang, R.Y.; Shi, T. A Mediator-Free Glucose Biosensor Based on Glucose Oxidase/Chitosan/ Alpha-Zirconium Phosphate Ternary Biocomposite. *Anal. Biochem.* **2014**, *445*, 24–29. [CrossRef]
56. Wang, L.; Zhu, W.; Lu, W.; Shi, L.; Wang, R.; Pang, R.; Cao, Y.; Wang, F.; Xu, X. One-Step Electrodeposition of AuNi Nanodendrite Arrays as Photoelectrochemical Biosensors for Glucose and Hydrogen Peroxide Detection. *Biosens. Bioelectron.* **2019**, *142*, 111577. [CrossRef]

57. Xia, Y.; Huang, W.; Zheng, J.; Niu, Z.; Li, Z. Nonenzymatic Amperometric Response of Glucose on A Nanoporous Gold Film Electrode Fabricated by A Rapid and Simple Electrochemical Method. *Biosens. Bioelectron.* **2011**, *26*, 3555–3561. [CrossRef]
58. Aksorn, J.; Teepoo, S. Development of The Simultaneous Colorimetric Enzymatic Detection of Sucrose, Fructose and Glucose Using a Microfluidic Paper-based Analytical Device. *Talanta* **2020**, *207*, 120302. [CrossRef]
59. Wang, K.; Li, N.; Zhang, J.; Zhang, Z.; Dang, F. Size-Selective QD@MOF Core-Shell Nanocomposites for the Highly Sensitive Monitoring of Oxidase Activities. *Biosens. Bioelectron.* **2017**, *87*, 339–344. [CrossRef]

biosensors

MDPI

Article

Highly Stable InSe-FET Biosensor for Ultra-Sensitive Detection of Breast Cancer Biomarker CA125

Hao Ji [1], Zhenhua Wang [1], Shun Wang [1], Chao Wang [1], Kai Zhang [1], Yu Zhang [1,*] and Lin Han [1,2,3,*]

[1] Institute of Marine Science and Technology, Shandong University, Qingdao 266237, China
[2] Shenzhen Research Institute of Shandong University, Shenzhen 518057, China
[3] Shandong Engineering Research Center of Biomarker and Artificial Intelligence Application, Ji'nan 250100, China
* Correspondence: yuzhang@sdu.edu.cn (Y.Z.); hanlin@sdu.cn (L.H.)

Abstract: Two-dimensional materials-based field-effect transistors (FETs) are promising biosensors because of their outstanding electrical properties, tunable band gap, high specific surface area, label-free detection, and potential miniaturization for portable diagnostic products. However, it is crucial for FET biosensors to have a high electrical performance and stability degradation in liquid environments for their practical application. Here, a high-performance InSe-FET biosensor is developed and demonstrated for the detection of the CA125 biomarker in clinical samples. The InSe-FET is integrated with a homemade microfluidic channel, exhibiting good electrical stability during the liquid channel process because of the passivation effect on the InSe channel. The InSe-FET biosensor is capable of the quantitative detection of the CA125 biomarker in breast cancer in the range of 0.01–1000 U/mL, with a detection time of 20 min. This work provides a universal detection tool for protein biomarker sensing. The detection results of the clinical samples demonstrate its promising application in early screenings of major diseases.

Keywords: field-effect transistor; InSe; biosensor; CA125; biomarker detection; liquid gate

Citation: Ji, H.; Wang, Z.; Wang, S.; Wang, C.; Zhang, K.; Zhang, Y.; Han, L. Highly Stable InSe-FET Biosensor for Ultra-Sensitive Detection of Breast Cancer Biomarker CA125. *Biosensors* **2023**, *13*, 193. https://doi.org/10.3390/bios13020193

Received: 29 November 2022
Revised: 15 January 2023
Accepted: 19 January 2023
Published: 28 January 2023

1. Introduction

Biomarkers play a key role in disease diagnosis and treatment [1–3]. The early detection of major diseases such as cancer can be achieved by detecting biomarkers in the blood. Various immunoassays have been proposed and investigated to meet the growing demand for assays while ensuring sensitivity and specificity [4,5]. As an example, fluorescent immunoassays can provide highly sensitive and reliable detection in aqueous or cellular environments, but their cumbersome steps cause a reduction in efficiency [6]. In recent years, field-effect transistor (FET) biosensors, as a promising label-free and fast biomolecular detection method, have attracted much attention due to their low power consumption, scalability to on-chip integration, and low processing cost [7–10]. Biosensors constructed with various nanomaterials and nanostructures, including silicon nanowires and carbon nanotubes, have shown potential in improving the detection sensitivity of biomolecules [11–15]. In addition, two-dimensional (2D) nanomaterials, such as graphene and transition metal sulfide (TMD) compounds, show ultrasensitive properties for existing detection methods. The active layer of the 2D material allows for a highly specific surface area, resulting in superior charge sensitivity [16–18]. In particular, unlike zero-bandgap graphene, the presence of a bandgap in TMDs is critical for FET-based platforms because the modulation of two-dimensional channel carrier transport is triggered by binding processes between the surface and biomolecules [19]. Recently, MoS_2-based FETs have been used for the immediate diagnosis of nucleic acid molecules and biomolecules, such as proteins, demonstrating the potential of two-dimensional semiconductor materials for applications [20,21].

For FET-based biosensors, a liquid gate (Ag/AgCl reference electrode) is commonly used. Liquid gate modulated FETs with low power consumption and easy detection are widely used in graphene- and MoS_2-based FET biosensors. However, in ionic liquids, the liquid gate modulation process is similar to that of an electrochemical reaction, and the operating voltage is prone to the electrolysis of the aqueous solution, causing the decomposition of the channel material. Specifically, some potential 2D materials for biomolecular sensing require further evaluation of the impact of liquid gate modulation. High-mobility 2D materials have natural advantages in sensor construction and can achieve highly sensitive detection [22]. As a III–VI two-dimensional semiconductor compound, InSe has a direct and moderate band gap of 1.26 eV with an ultra-high Hall mobility of over 1000 $cm^2 V^{-1} s^{-1}$ at room temperature due to the light electron effective mass ($\sim$0.143 m_0), making it an ideal material for biomolecular sensing [23].

In this study, we propose an InSe-FET biosensor and investigate its electrical stability in the liquid channel. In addition to the commonly used liquid gate electrode, the back gate working mode is applied to conduct the biomarker detection. The liquid channel works as a passivation layer and effectively improves the stability of InSe-FETs. even though the field-effect mobility degrades slightly due to the scattering of ions in the liquid channel. The proposed InSe-FET biosensor achieves an extra-large quantitative linear and selective detection range during CA125 biomarker sensing. The detection capability of the CA125 biomarker in clinical serum samples demonstrates its potential applications in the screening of major diseases.

2. Materials and Methods

2.1. Materials and Regents

Bulk InSe was bought from XFNANO Materials Tech Co., Ltd., Shenzhen, China. Silicon wafers were purchased from Saibang Electronic Technology Co. Ltd., Kunshan, China. Phosphate-buffered saline (PBS) was purchased from Corning. 3-aminopropyltrimethoxisylane (APTES, >99%) was purchased from Sigma-Aldrich. Capture antibodies, recombinant proteins, and detection antibodies, CA125 and CA199, were purchased from Fitzgerald (America). The types and purity of all antigens and antibodies used in the experiments are summarized in Table 1. Clinical serum specimens were collected at Qilu Hospital of Shandong University. The ultrapure water (18.25 MU/cm^3) used throughout all experiments was made by a Millipore system. All chemicals used in this work were of analytical grade or highest purity available and used directly without further purification. Acetone (>99.9%) and isopropyl alcohol (>99.7%) were purchased from Sinopharm.

Table 1. Material information of detection regents.

Name	Function	Host/Source	Type	Subclass	Purity
CA125	Antibody	Mouse	Monoclonal	IgG1	>90%
	Protein	Mouse	Monoclonal	IgG1	>95%
CA199	Protein	Mouse	Monoclonal	IgG1	>90%

2.2. Clinical Samples Preparation

First, 1–2 mL of venous blood from healthy person and patient is collected in a non-anticoagulant tube and is kept standing for 2 h. Then, the supernatant is transferred into a new Eppendorf tube and centrifuged at 3000 rpm for 10 min. Finally, 100 μL of serum supernatant is collected and stored at −80 or −20 °C for testing.

2.3. InSe Material Characterization

InSe is examined using high-resolution transmission electron microscope (TEM) (FEI-G20, Thermo Fisher Scientific Inc., Waltham, MA, USA). The thickness of the InSe is determined using atomic force microscopy (AFM) on a Smart SPM AFM system. Raman spectra and photoluminescence of InSe are determined by a Renishaw inVia Raman microscope at room temperature with a 532 nm laser as an excitation source.

2.4. Fabrication of InSe-FET-Based Biosensor

A schematic diagram of the InSe-FET biosensor fabrication process is shown in Figure S1. The p-doped Si substrate with 100 nm SiO_2 is cleaned using acetone, methanol, and deionized water. Then a high-quality layer of HfO_2 is grown on the Si/SiO_2 substrate by thermal ALD (atomic layer deposition) as a way to reduce the scattering centers generated by fixed groups and defects on the SiO_2 surface [24,25]. The PE-ALD Beneq TFS200 Instrument is used for the deposition of HfO_2 films from tetrakis-dimethylamino hafnium (TDMAHf) precursor. The TDMAHf is heated at 75 °C and N_2 carrier gas is flown at 40 sccm to improve the hafnium precursor transport from the bubbler to the reactor chamber. The oxidation step is performed using water vapor in case of the thermal ALD at 250 °C, while the number of cycles ranges from 125 (growth rate is about 0.08 nm/cycle) for the T-ALD process. The InSe with few layers is transferred from the bulk material to the substrate by mechanical exfoliation. Ti/Au (10 nm/20 nm) source/drain electrode is formed by high-vacuum (4×10^{-5} Pa) electron beam evaporation using a shadow mask to avoid the introduction of contamination by the photoresist, creating a good ohmic contact. The linearity of the output curve in Figure S2 illustrates the good ohmic contact of the device at room temperature. A microfluidic chip with a channel of approximately $5000 \times 40 \times 20\ \mu m^3$ (the width of the flow channel is 20 μm, the height is 50 μm, and the length is 5000 μm) is designed and aligned with the InSe sensing channel to form a biosensor, allowing solutions to flow into the channel and interact with the InSe channel. The dimensions of the PDMS device are $10 \times 5 \times 2\ mm^3$ (see in Figure S3). The schematic cross-sectional structure of the InSe biosensor is shown in Figure 1a. The channel current can be regulated by the voltage applied to the back gate or liquid gate (Ag/AgCl reference electrode) mode. The electrical performance and sensing tests are performed using a Keithley 2636B system for I–V testing. The fabrication process' time consumption schematic graphs and photograph of fabricated InSe-FET biosensor are shown in Figure S3.

2.5. Capture Antibody-CA125 Immobilization and CA125 Antigen Sensing

APTES solution is made of 0.4 mL APTES with ethanol and DI water (19 mL/1 mL), and then injected into the flow channel and incubated with InSe for 1 h. After incubation, InSe channel is cleaned by inflowing ethanol and DI water. A concentration of 100 ug/L antibody-CA125 is loaded into the microfluidic channel and incubated for 2 h to achieve adequate immobilization on InSe channel. After cleaning, CA125 antigen sample is loaded in the detection channel and the reaction time is 20 min. Finally, the output current is detected after cleaning of reaction residues using 1 × PBS.

2.6. Electrical Characterization of InSe-FET

The $I_{DS}-V_{DS}$ output characteristics and $I_{DS}-V_{GS}$ transfer characteristics of the InSe-FET are measured by a Keithley 2636 under ambient and liquid conditions. The working mode includes back gate and liquid gate modes. For the back gate mode, V_{BG} is swept from −10 V to 10 V at V_{DS} = 0.1 V to obtain transfer characteristic. For liquid gate, V_{LG} is swept from −1 V to 1 V at V_{DS} = 0.1 V. The details of circuit connections and the experimental setup are shown in Figure 2a. For the InSe-FET stability test, the same parameters as for the electrical test are selected, and the transfer curves are tested at 15 min intervals under atmospheric and liquid environments. The transfer characteristics are measured for the fabricated InSe-FET biosensor after antibody immobilization and after CA125 capture antibody immobilization, respectively.

3. Results

3.1. Characterization of the InSe

The schematic structure of the InSe-FET biosensor is shown in Figure 1a. The back gate InSe-FET is integrated with the homemade microfluidic channel. High-quality ALD-grown HfO_2 is applied to effectively improve field-effect mobility and reduce hysteresis by reducing the parasitic capacitance and shielding the interfacial Coulomb scattering [26]. As shown in Figure S4, the hysteresis of the transistor with the HfO_2/SiO_2 substrate is about 0.5 V, which is significantly less than its value, 3.8 V, with the SiO_2 substrate only. APTES as a coupling agent is functionalized directly on the InSe surface by the chemical reaction of C-Si bonds and Se vacancy defects [27]. Although the InSe lacks suspension bonds, the highly reactive sites in the Se vacancies can react with APTES to form In-O bonds [28]. The atomic force microscopy characterization in Figure S5 demonstrates that the surface roughness Ra increases from 0.1264 nm to 0.5062 nm after the APTES functionalization on InSe. In order to confirm the successful functionalization of APTES and the immobilization of anti-CA125, Raman spectroscopy was conducted after anti-CA125 immobilization on APTES-modified InSe in comparison to that on bare InSe. A random spot in the channel and the channel area's scanned Raman spectra are shown in Figure S6. They show representative peaks of anti-CA125 at 1096 cm^{-1} and 1407 cm^{-1} on the APTES-modified InSe channel, and the scanned Raman spectra present a uniform distribution of anti-CA125, indicating the successful functionalization of APTES and the immobilization of anti-CA125. APTES is functionalized directly on the InSe surface as a coupling agent without an additional top passivation structure, allowing the surface charge of the detection antigen to act directly on the channel surface to introduce changes in output current, effectively improving the detection sensitivity [29]. The electrical properties of the FET devices are determined by the quality of the InSe films, which strongly influences the sensing performance of the biosensors. In order to characterize the quality of the InSe material, the atomic structure of the InSe is determined by high-resolution electron scanning transmission microscopy as shown in Figure 1b. The multilayered InSe shows a complete honeycomb structure with alternating rows of In and Se atoms with different brightnesses corresponding to different atomic numbers. The atomic structure indicates that the InSe used in the experiment is in the γ-conformation with the lattice constant a = b = 0.34 nm, which is consistent with previous reports [30]. It has been shown that a higher device performance and smaller thickness of the material are beneficial in improving the sensitivity of the sensor [31]. FETs constructed with 20–30 nm thick InSe have the best performance [22,26] and are conducive to improving the device response to the biosensor. Thus, an InSe film with a thickness of about 20 nm is selected to construct the InSe-FET, and the thickness conducted by AFM is shown in Figure 1c. As shown in Figure 1d, three peaks can be observed at 114 cm^{-1}, 176 cm^{-1}, and 226 cm^{-1} for the InSe layers, which are consistent with the Raman modes of γ-InSe [32]. The band gap of the transferred InSe can be inferred from the photoluminescence (PL) spectrum in Figure 1e, and the peak at 990 nm corresponding to the band gap of the multilayer InSe is calculated to be about 1.25 eV. The typical multilayer InSe material properties are presented by the Raman and PL spectra of the InSe, indicating the successful transfer of InSe flakes to the substrate.

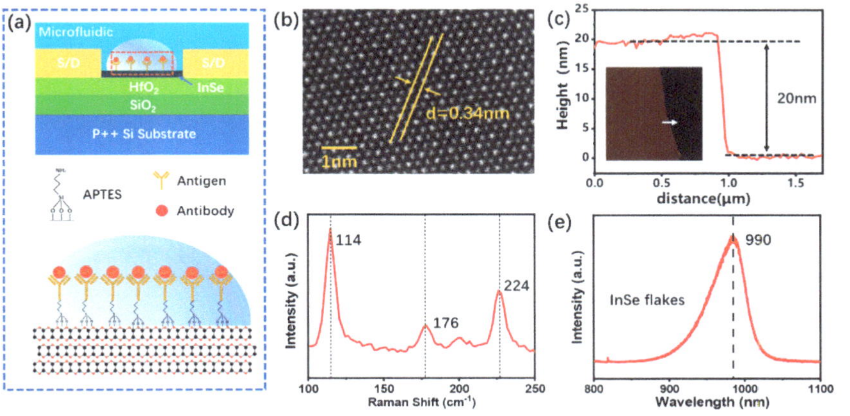

Figure 1. (**a**) Schematic of the InSe-FET biosensor for biomarker CA125 detection. (**b**) High-resolution transmission electron microscopy of InSe. (**c**) Thickness of typical InSe films used in FET determined by AFM. (**d**,**e**) Raman spectra and photoluminescence spectra of transferred InSe flakes.

3.2. Electrical Characterization of InSe-FET

The working mode of InSe-FET plays an essential role in biosensing performance. Here, the performances of the InSe-FETs working in the back gate liquid channel and top liquid gate are compared with those of the fabricated InSe-FET. As shown in Figure 2a, the fabricated device tested in the air (dry channel) is used as a benchmark to evaluate the difference in the electrical performance between the two working modes in a liquid environment. Figure 2b and c shows the transfer characteristics of the InSe-FET working in the back gate liquid channel on linear and logarithmic scales in comparison with the fabricated InSe-FET, and the slight decrease in the slope implies a slight decrease in the field-effect mobility from 453.42 to 418.42 $cm^2/V^{-1}s^{-1}$, which may be caused by the adsorption of the ion group in the PBS solution on the InSe surface. The field-effect mobility of multilayer InSe-FETs can be extracted from the transfer curve using the following equation: $\mu = [L/WC_iV_{DS}] \times [dI_{DS}/dV_G]$, where L is channel length of 30 μm, W is the channel width of 20 μm, and Ci is the series capacitance of 100 nm SiO_2 and 10 nm HfO_2, where the dielectric constants of both are 3.9 and 15, respectively. The negative shift of −0.9 V in the threshold voltage in the logarithmic coordinate system in Figure 2c indicates that the InSe-FET appears to be n-doping in the PBS solution [19], which is possibly due to the charge transfer of the OH-groups on the InSe surface [6]. The InSe-FET in the liquid gate working mode switches between the on and off states in the 2 V range with a higher regulation efficiency compared with that of the back gate mode, as shown in Figure 2d. However, a calculated mobility of 211.22 $cm^2/V^{-1}s^{-1}$ is obtained, which decreases by 50% compared to that of the back gate working mode. The larger mobility drop is probably due to the smaller electron effective mass of InSe, which is exceptionally sensitive to interfacial Coulomb scattering [26]. In the liquid gate working mode, it forms a 2~3 nm charge distribution layer at the InSe–liquid interface, as shown in Figure S7, in which the thickness range of the induced charges is only due to the direct formation of a double electric layer on the surface of InSe by liquid gate regulation [33]. The formation of a bilayer on the InSe surface by the liquid gate modulates the channel current while increasing the scattering chance of the induced charge near the surface of the InSe. Figure 2e shows that the device has clearly ambipolar behavior in the liquid gate working mode. A similar phenomenon was previously observed in liquid gate MoS_2 layers of more than 10 nm [30], but this is the first time that it has been found in the InSe materials. This indicates that the occurrence of ambipolar phenomena is most likely related to the efficient modulation of the charge transport properties of the InSe surface layer by the liquid gate and not the intrinsic properties of the InSe material. The gate leakage current increases when the liquid gate bias voltage is negative, as shown in Figure 2e. The basic FET performance indicates that the back gate working mode is appropriate for biomarker sensing.

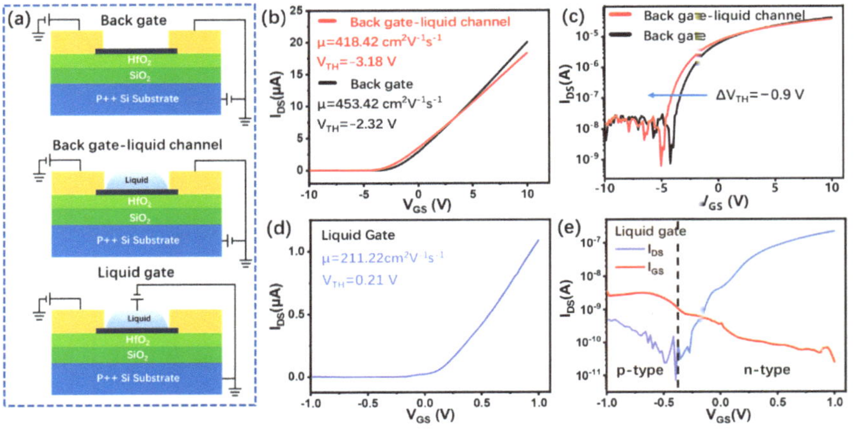

Figure 2. (**a**) Schematic diagram of the three operating modes of InSe-FET. Comparison of InSe-FET performance before and after filling the microfluidic channel with liquid in (**b**) linear and (**c**) logarithmic coordinates. Transfer characteristic curves of InSe-FET regulated by liquid gate in (**d**) linear and (**e**) logarithmic coordinate systems.

3.3. Electrical Stability of InSe-FET

During biomarker detection, the electrical stability during the test cycles determines the reliability and sensitivity of the InSe-FET biosensor. Here, in order to test the electrical stability, the transfer characteristics are scanned eight times at an interval of 15 min for the InSe-FETs in the different working modes. However, the InSe-FETs usually fail on the seventh test. Figure 3a and b shows the transfer curves of the fabricated InSe-FET in the air corresponding to the linear and logarithmic coordinates. As the number of tests increases, the output current slightly drops and threshold voltage drifts are observed, which are probably due to the desorption of water and oxygen on the surface of the InSe in the atmosphere. This non-stationarity is more pronounced in the first few tests of the freshly processed device, indicating the rapid desorption of water and oxygen on the InSe surface in the atmosphere. When the devices are tested in a liquid environment through back gate modulation, an extremely high stability was observed, as shown in Figure 3c,d, and there were no observable changes in the transfer curves over the eight repeating tests. The liquid in the flow channel actually plays a similar role to passivation, avoiding the creation of unstable states caused by adsorbed substances on the InSe surface. The excellent electrical stability is especially important for biosensors for ultra-low-concentration biomolecule detection [31]. When the InSe-FET is working with the liquid gate, the transfer curve shows a significant drop in current in several of the repeating tests and is irrecoverable. This irreversible property degradation is probably due to the change in material properties induced by the liquid electrode. Due to the reactivity of InSe [28], an electrochemical etching-like reaction probably occurs on the InSe surface, causing irreversible electrical property degradation of the device, even if the voltage applied to the liquid gate is small. Figure 3g shows the negligible change in output current for the InSe-FET in the back gate and liquid channel working modes. The output current increased 1.11 times after eight replications for InSe-FET in the back gate and dry channel working modes, while the output current decreased to 0.79 after six replications for the top liquid gate working mode. Figure 3h shows slight threshold voltage shifts of 0.032 V after eight scans for InSe-FET in the back gate and liquid channel working modes, positive threshold voltage shifts up to 1.43 V after eight scans for InSe-FET in the back gate and dry channel working modes, and positive threshold voltage shifts up to 0.24 V after six scans for InSe-FET in the top liquid gate working mode. Even though the threshold voltage shifts in the liquid gate working mode, the devices fail quickly after six repeated scans. Cyclic tests indicate that the InSe-FET working in the back gate mode with a liquid channel has the best stability under gate electrical stress and is appropriate for biosensing.

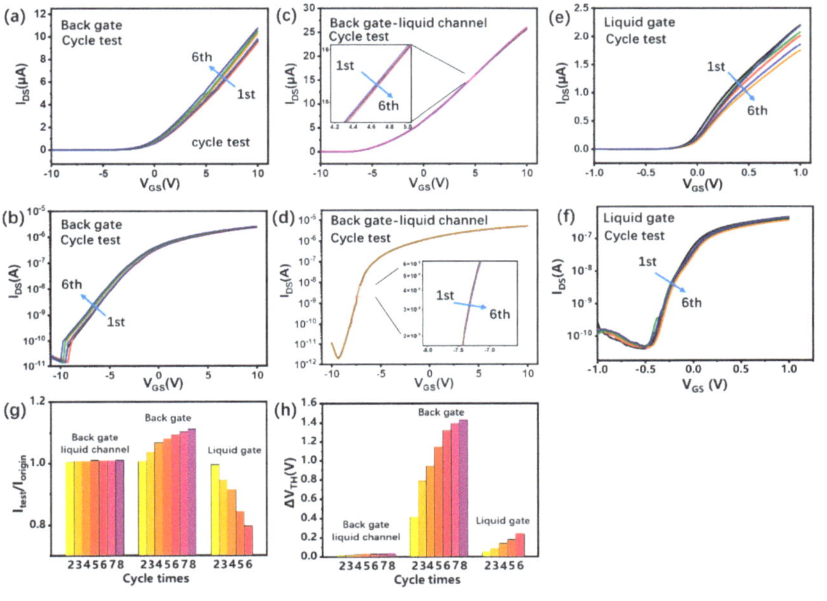

Figure 3. Results of cyclic tests in three modes. The transfer characteristics cycle tests of InSe-FET in (**a**,**b**) back gate dry channel, (**c**,**d**) back gate liquid channel, (**e**,**f**) and liquid gate in linear and logarithmic coordinates. The current (**g**) and threshold voltage (**h**) evolution of three modes with cycle times.Note: In back gate working mode, the point at V_{DS} = 0.1 V, V_{GS} = 10 V is selected to compare output current change and threshold shift; in liquid gate working mode, the point at V_{DS} = 0.1 V, V_{GS} = 1 V is selected to compare output current change and threshold shift.

In order to further characterize the stability of the InSe-FET biosensor, the storage stability is tested. The InSe-FETs are stored in atmosphere and liquid, respectively. Figure 4a,b shows the changes in PL spectra after five days. The peak intensity of the InSe immersed in the 0.1×PBS solution declined by 7.43% after five days, while that of the InSe exposed to atmosphere declined by 35.68%. The decrease in the PL peak intensity is probably due to the formation of InSe oxides [34]. The InSe in the atmosphere is exposed to an environment containing both oxygen and water, subjected to oxidation by oxygen. In the liquid channel, the oxidation process is significantly slowed due to the extremely low dissolved oxygen in the liquid. Figure 4c,d shows that the transfer curves shift less in the InSe-FET with the liquid channel than they do with dry channel during the back gate working mode after being stored for five days. When the FET channel is filled with liquid, the threshold voltage shift increases to 1.72 V, and when the channel is exposed to the atmosphere, the threshold voltage shift decreases to 2.59 V. The output current also shows faster degradation for the InSe-FET exposed to the atmosphere than that filled with liquid, as shown in Figure 4f. The electrical properties' degradation during storage when exposed to air is probably because of the decrease in the intrinsic carrier concentration due to the occurrence of oxidation reactions, leading to the conversion of the material from InSe to $InSe_xO_y$ [24]. The liquid channel protects the damage of InSe from oxidation, and a negative shift in V_{TH} suggests that the InSe is n-doped by ionic adsorbates in liquid.

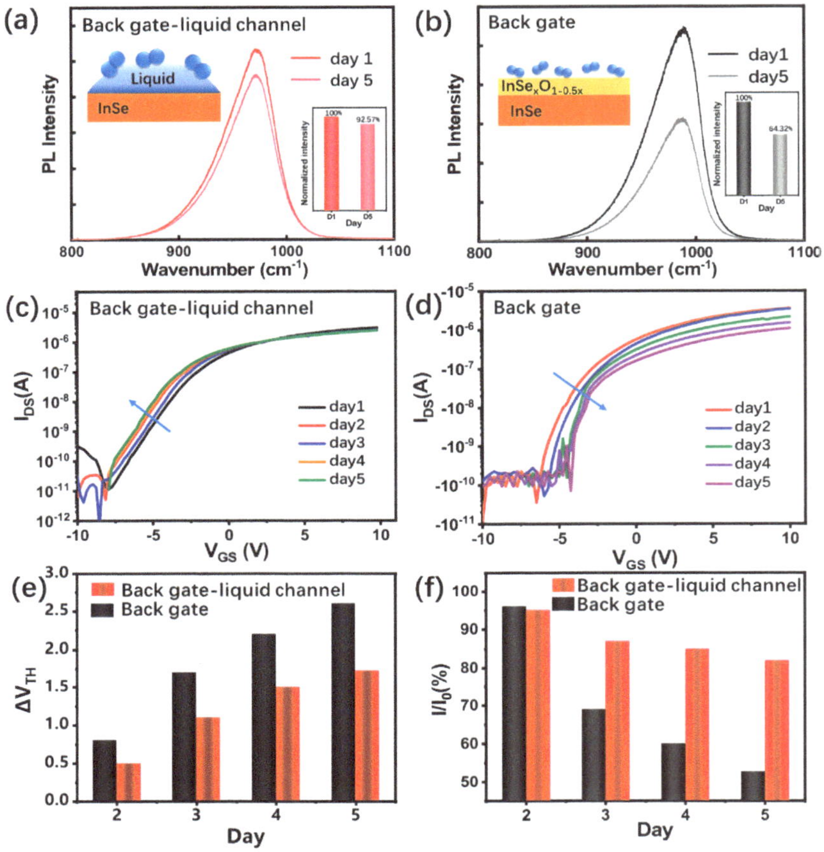

Figure 4. PL spectra of InSe before and after storage for five days in (**a**) liquid and (**b**) atmosphere. Transfer curves before and after storage for five days in (**c**) air and (**d**) liquid. (**e**) The threshold voltage shift and (**f**) the output current change at V_{GS} = 10 V for the InSe-FETs over storage time.

3.4. Subsection

The specific detection of the cancer marker CA125 is based on the specific binding of antigen–antibody. As shown in Figure 5a, before the detection process, the CA125 capture antibody is immobilized on the InSe channel, and a large number of amino and carboxyl residues on the antibody molecules present charges in the liquid and induce a negative shift in the threshold voltage, as shown in Figure 5b. It is important to note that the scattering effect on the channel carriers induced by the target antigen is limited because of its larger molecule size. The field-effect mobility response for target CA125 is less than 5% and does not show a clear linear relationship with concentration, as shown in Figure S8. Since there is not a dielectric layer between the antibody and the channel material surface, the InSe channel electrons can be directly induced by the antibody, which can be explained from $\Delta V_{TH} = -Q_F$, where Q_F represents the effective charges that can induce the change in the conductivity of the InSe channel [35]. In the meantime, the field-effect mobility drops from 425 $cm^2/V^{-1}s^{-1}$ to 102 $cm^2/V^{-1}s^{-1}$ after the antibody immobilization of InSe, because the charge in direct contact with the surface introduces the Coulomb scattering centers, and the channel electrons are affected by the scattering, which results in a decrease in the field-effect mobility. Once the target sample is loaded into the microfluidic channel, the antigen CA125 is bonded with anti-CA125 and these charges decrease, inducing a positive shift in threshold voltage. The antigen CA125 neutralizes a portion of the antibody-induced gate voltage effect, which is equivalent to applying a small voltage to the InSe surface, visualized as a positive shift in the threshold voltage, and the positive drift of threshold voltage increases with the increasing antigen concentration. Figure 5c shows the linear relationship between the threshold voltage shift and antigen CA125 concentration in the semi-log scale with a detectable range of 0.01–1000 U/mL. Each concentration is tested three times using the same batch devices, and these devices present a standard deviation of 9.05%, as shown in Figure S9. In addition, CA125 can be detected at concentrations as low as 0.01 U/mL, which are much lower those previously reported (0.1 U/mL) [21]. Meanwhile, a comparison of information, such as the detection limits and detection times, between this study and other methods is shown in Table S1. The major contributing factor for the ultralow detection concentration is the ultrasensitive InSe scattering of carriers with small mass and the simple channel treatment. Unlike previously reported protein assays, APTES is used to immobilize antibodies on the surface of InSe without any additional dielectric layer, which greatly improves the ability of the target molecules to modulate the channel current. In addition, the excellent stability of InSe-FETs in liquid is also favorable for the low detection limit.

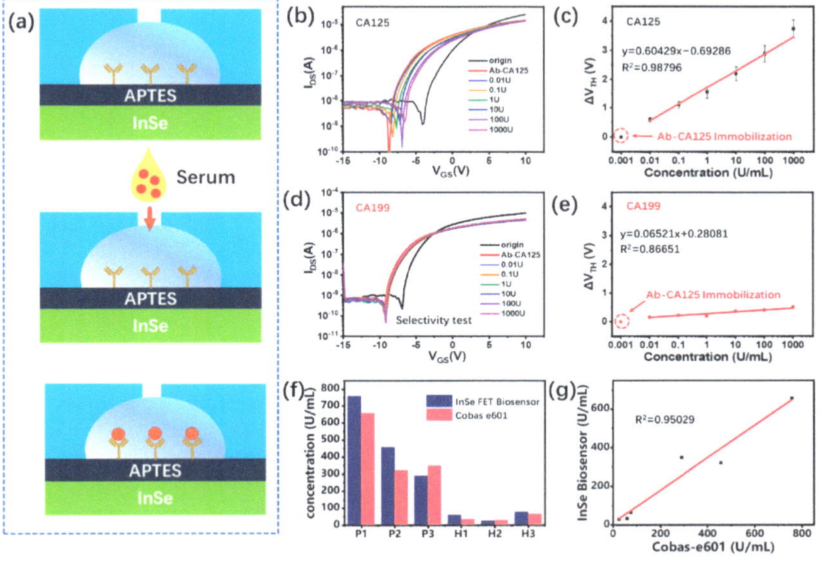

Figure 5. (**a**) Schematic diagram of InSe-FET biosensor for biomarker detection. (**b**) Transfer characteristic responses to target CA125 antigen in different concentrations. (**c**) Quantitative relationship between threshold voltage shift and antigen CA125 concentration. (**d**) Transfer characteristics responses to the non-target antigen CA199, and (**e**) threshold voltage shift corresponding to different concentrations of antigen CA199. (**f**) Comparison of the results of antigen CA125 detection by InSe-FET biosensor and Cobas Roche electroluminescence assay on serum samples from breast cancer patients and healthy individuals. P1~3 are serum samples from three breast cancer patients, and H1~3 are serum samples from three healthy people. (**g**) Correlation of the results of the proposed InSe-FET biosensor with those of the Cobas e601 electroluminescence device.

The specificity of CA125 detection is performed by loading different concentrations of CA199 to the CA125 antibody immobilized channel. Figure 5d shows the negligible response of the InSe-FET to loaded CA199, which demonstrates a nonspecific reaction of

CA199 to the antibody CA125. To confirm the ability of the InSe-FET biosensor to detect clinical samples, we collected and tested serum samples from three breast cancer patients and three healthy volunteers. The detected transfer curves and the derived ΔV_{TH} histogram are shown in Figure S10, which presents a much larger transfer curve shift and threshold voltage change in the patients' serum versus the healthy people's serum. According to the linear relationship in Figure 5c, the detected concentration of CA125 is derived and shown in Figure 5f in comparison to the results detected by Cobas e 601 from Roche. The proposed InSe-FET biosensor presents a high correlation of $R^2 > 0.95$ with commercial Cobas e 601. In addition, the proposed InSe-FET biosensor presents a lower detection limit and rapid detection speed than other representative biosensors, as listed in Table S1. The detection performance indicates the promising application of the proposed InSe-FET biosensor.

4. Conclusions

In summary, we tested an InSe-FET biosensor to perform the ultrasensitive, specific, fast, and label-free detection of the breast cancer biomarker CA125. Through systematic experiments and an analysis of InSe-FETs' electrical characteristics and stability in different working modes, we reveal that the back gate working mode is favorable for conducting biomolecules with filled liquid in the InSe channel. It also indicates that the liquid gate electrode is not suitable for a reactive InSe-based FET biosensor. The proposed biosensor is capable of detecting an ultra-large range of the antigen CA125 from 0.01 to 1000 U/mL with a standard error under 8.78%. The detection of clinical samples has shown that InSe-FET biosensors hold great promise for practical applications, such as the early diagnosis and prognosis of cancer, the study of the pathogenesis of major diseases, and the real-time monitoring of health.

Supplementary Materials: The references in Supplementary Materials were cited in Refs. [36–38]. The following supporting information can be downloaded at: https://www.mdpi.com/article/10.3390/bios13020193/s1, Figure S1: Fabrication process of InSe-field transistor biosensor; Figure S2: Typical I_{DS} $-V_{DS}$ curves at 300K for the InSe FET. The device exhibits linear output characteristics at all the temperatures measured.; Figure S3: (a) Real InSe-FET biosensor and detection schematic. (b) Top view of the optical microscope of the biosensor. Figure S4: Comparison of InSe-FET hysteresis using SiO_2 (a) and HfO_2/SiO_2¬ (b) substrates. Figure S5: Surface morphology of (d) bare InSe and (e) APTES modified InSe characterized by AFM. Figure S6: Raman spectra of anti-CA125 on (a) bare InSe and (b) APTES modified InSe, (c) Raman spectra mapping. Figure S7: (a,b) show the distribution of channel electrons after applying gate pressure to the liquid-gate and back-gate FETs, respectively. (c) Schematic diagram of the liquid gate regulated trench inversion layer in the device on state. Figure S8: (a) Evolution of linear coordinate system transfer curve with CA125 concentration. Evolution of mobility with CA125 concentration. Figure S9: Detection curves of the three independent sensors (a), (b), and (c) for the same concentration of CA125, and (d) the threshold voltage shift of the three independent devices. Figure S10: (a) Serum samples from patients and healthy individuals tested using three devices. The threshold voltage offsets extracted for the three patient and healthy human serum samples are shown in (b). Table S1: The detection performance of InSe-FET biosensor and other representative biosensors.

Author Contributions: H.J.: Writing: original draft, carried out the experiments, and wrote this manuscript draft. Y.Z.: Writing: review and editing, designed the study. Z.W.: provided some of the samples needed in the experiment. S.W.: helped perform the experiments. K.Z.: helped to perform the Raman tests. C.W.: helped to perform the experiments. L.H.: Writing: review and editing, brought up the idea. All authors have read and agreed to the published version of the manuscript.

Funding: This research was funded by Key Fundamental Research Program of Shenzhen (JCYJ20210324115601004), the Major Scientific and Technological Innovation Project of Shandong Province (2021CXGC010603), Shandong University Integrated Research and Cultivation Project (Grant No. 2022JC001), the National Science Foundation of China (Grant Nos. 62174101 and 32001018), the Fundamental Research Funds of Shandong University (2020QNQT001), and the Collaborative Innovation Center of Technology and Equipment for Bio-logical Diagnosis and Therapy at the Universities of Shandong Province.

Institutional Review Board Statement: The study was conducted in accordance with the Declaration of Helsinki, and approved by the Scientific Research Ethics Committee of Qilu Hospital of Shandong University (protocol code KYLL-2019-122 and date of approval).

Informed Consent Statement: Informed consent was obtained from all subjects involved in the study. Written informed consent has been obtained from the patients to the publication of this paper.

Data Availability Statement: Not applicable.

Acknowledgments: The authors would like to thank Xiaoju Li from the State Key Laboratory of Microbial Technology, Shandong University for their assistance with SEM.

Conflicts of Interest: The authors declare no conflict of interest.

References

1. Rao, C.N.; Gopalakrishnan, K.; Maitra, U. Comparative Study of Potential Applications of Graphene, MoS2, and Other Two-Dimensional Materials in Energy Devices, Sensors, and Related Areas. *ACS Appl. Mater. Interfaces* **2015**, *7*, 7809–7832. [CrossRef] [PubMed]
2. Wu, X.; Mu, F.; Wang, Y.; Zhao, H. Graphene and Graphene-Based Nanomaterials for DNA Detection: A Review. *Molecules* **2018**, *23*, 2050. [CrossRef] [PubMed]
3. Wang, C.; Zhang, Y.; Tang, W.; Wang, C.; Han, Y.; Qiang, L.; Gao, J.; Liu, H.; Han, L. Ultrasensitive, high-throughput and multiple cancer biomarkers simultaneous detection in serum based on graphene oxide quantum dots integrated microfluidic biosensing platform. *Anal. Chim. Acta* **2021**, *1178*, 338791. [CrossRef] [PubMed]
4. Wang, C.; Wang, C.; Qiu, J.; Gao, J.; Liu, H.; Zhang, Y.; Han, L. Ultrasensitive, high-throughput, and rapid simultaneous detection of SARS-CoV-2 antigens and IgG/IgM antibodies within 10 min through an immunoassay biochip. *Mikrochim Acta* **2021**, *188*, 262. [CrossRef] [PubMed]
5. Makowski, M.S.; Ivanisevic, A. Molecular analysis of blood with micro-/nanoscale field-effect-transistor biosensors. *Small* **2011**, *7*, 1863–1875. [CrossRef] [PubMed]
6. Gao, J.; Gao, Y.; Han, Y.; Pang, J.; Wang, C.; Wang, Y.; Liu, H.; Zhang, Y.; Han, L. Ultrasensitive Label-free MiRNA Sensing Based on a Flexible Graphene Field-Effect Transistor without Functionalization. *ACS Appl. Electron. Mater.* **2020**, *2*, 1090–1098. [CrossRef]
7. Zheng, C.; Huang, L.; Zhang, H.; Sun, Z.; Zhang, Z.; Zhang, G.J. Fabrication of Ultrasensitive Field-Effect Transistor DNA Biosensors by a Directional Transfer Technique Based on CVD-Grown Graphene. *ACS Appl Mater Interfaces* **2015**, *7*, 16953–16959. [CrossRef]
8. Hwang, M.T.; Wang, Z.; Ping, J.; Ban, D.K.; Shiah, Z.C.; Antonschmidt, L.; Lee, J.; Liu, Y.; Karkisaval, A.G.; Johnson, A.T.C.; et al. DNA Nanotweezers and Graphene Transistor Enable Label-Free Genotyping. *Adv. Mater.* **2018**, *30*, e1802440. [CrossRef]
9. Xu, S.; Jiang, S.; Zhang, C.; Yue, W.; Zou, Y.; Wang, G.; Liu, H.; Zhang, X.; Li, M.; Zhu, Z.; et al. Ultrasensitive label-free detection of DNA hybridization by sapphire-based graphene field-effect transistor biosensor. *Appl. Surf. Sci.* **2018**, *427*, 1114–1119. [CrossRef]
10. Gao, A.; Lu, N.; Wang, Y.; Dai, P.; Li, T.; Gao, X.; Wang, Y.; Fan, C. Enhanced sensing of nucleic acids with silicon nanowire field effect transistor biosensors. *Nano Lett.* **2012**, *12*, 5262–5268. [CrossRef]
11. Hong, E.-K.; Cho, W.-J. High sensitivity In-Ga-Zn-O nanofiber-based double gate field effect transistors for chemical sensing. *Sens. Actuators B Chem.* **2021**, *326*, 128827. [CrossRef]
12. Liang, Y.; Xiao, M.; Wu, D.; Lin, Y.; Liu, L.; He, J.; Zhang, G.; Peng, L.M.; Zhang, Z. Wafer-Scale Uniform Carbon Nanotube Transistors for Ultrasensitive and Label-Free Detection of Disease Biomarkers. *ACS Nano* **2020**, *14*, 8866–8874. [CrossRef]
13. Zhao, C.; Liu, Q.; Cheung, K.M.; Liu, W.; Yang, Q.; Xu, X.; Man, T.; Weiss, P.S.; Zhou, C.; Andrews, A.M. Narrower Nanoribbon Biosensors Fabricated by Chemical Lift-off Lithography Show Higher Sensitivity. *ACS Nano* **2021**, *15*, 904–915. [CrossRef] [PubMed]
14. Li, F.; Peng, J.; Wang, J.; Tang, H.; Tan, L.; Xie, Q.; Yao, S. Carbon nanotube-based label-free electrochemical biosensor for sensitive detection of miRNA-24. *Biosens. Bioelectron.* **2014**, *54*, 158–164. [CrossRef] [PubMed]
15. Park, H.; Han, G.; Lee, S.W.; Lee, H.; Jeong, S.H.; Naqi, M.; AlMutairi, A.; Kim, Y.J.; Lee, J.; Kim, W.-j.; et al. Label-Free and Recalibrated Multilayer MoS2 Biosensor for Point-of-Care Diagnostics. *ACS Appl. Mater. Interfaces* **2017**, *9*, 43490–43497. [CrossRef] [PubMed]
16. Forster, A.; Gemming, S.; Seifert, G.; Tomanek, D. Chemical and Electronic Repair Mechanism of Defects in MoS_2 Monolayers. *ACS Nano* **2017**, *11*, 9989–9996. [CrossRef] [PubMed]
17. Moudgil, A.; Singh, S.; Mishra, N.; Mishra, P.; Das, S. MoS 2/TiO2 Hybrid Nanostructure-Based Field-Effect Transistor for Highly Sensitive, Selective, and Rapid Detection of Gram-Positive Bacteria. *Adv. Mater. Technol.* **2019**, *5*, 1900615. [CrossRef]
18. Majd, S.M.; Salimi, A.; Ghasemi, F. An ultrasensitive detection of miRNA-155 in breast cancer via direct hybridization assay using two-dimensional molybdenum disulfide field-effect transistor biosensor. *Biosens. Bioelectron.* **2018**, *105*, 6–13. [CrossRef]
19. Liu, J.; Chen, X.; Wang, Q.; Xiao, M.; Zhong, D.; Sun, W.; Zhang, G.; Zhang, Z. Ultrasensitive Monolayer MoS2 Field-Effect Transistor Based DNA Sensors for Screening of Down Syndrome. *Nano Lett.* **2019**, *19*, 1437–1444. [CrossRef]

20. Jiao, S.; Liu, L.; Wang, J.; Ma, K.; Lv, J. A Novel Biosensor Based on Molybdenum Disulfide (MoS_2) Modified Porous Anodic Aluminum Oxide Nanochannels for Ultrasensitive microRNA-155 Detection. *Small* **2020**, *16*, e2001223. [CrossRef]
21. Mudd, G.W.; Svatek, S.A.; Ren, T.; Patane, A.; Makarovsky, O.; Eaves, L.; Beton, P.H.; Kovalyuk, Z.D.; Lashkarev, G.V.; Kudrynskyi, Z.R.; et al. Tuning the bandgap of exfoliated InSe nanosheets by quantum confinement. *Adv. Mater.* **2013**, *25*, 5714–5718. [CrossRef] [PubMed]
22. Feng, W.; Zheng, W.; Cao, W.; Hu, P. Back gated multilayer InSe transistors with enhanced carrier mobilities via the suppression of carrier scattering from a dielectric interface. *Adv. Mater.* **2014**, *26*, 6587–6593. [CrossRef] [PubMed]
23. Tsai, T.H.; Yang, F.S.; Ho, P.H.; Liang, Z.Y.; Lien, C.H.; Ho, C.H.; Lin, Y.F.; Chiu, P.W. High-Mobility InSe Transistors: The Nature of Charge Transport. *ACS Appl. Mater. Interfaces* **2019**, *11*, 35969–35976. [CrossRef] [PubMed]
24. Wu, L.; Shi, J.; Zhou, Z.; Yan, J.; Wang, A.; Bian, C.; Ma, J.; Ma, R.; Liu, H.; Chen, J.; et al. InSe/hBN/graphite heterostructure for high-performance 2D electronics and flexible electronics. *Nano Res.* **2020**, *13*, 1127–1132. [CrossRef]
25. Jiang, J.; Li, J.; Li, Y.; Duan, J.; Li, L.; Tian, Y.; Zong, Z.; Zheng, H.; Feng, X.; Li, Q.; et al. Stable InSe transistors with high-field effect mobility for reliable nerve signal sensing. *Npj 2d Mater. Appl.* **2019**, *3*, 29. [CrossRef]
26. Lei, S.; Ge, L.; Najmaei, S.; George, A.; Kappera, R.; Lou, J.; Chhowalla, M.; Yamaguchi, H.; Gupta, G.; Vajtai, R.; et al. Evolution of the Electronic Band Structure and Efficient Photo-Detection in Atomic Layers of InSe. *ACS Nano* **2014**, *8*, 1263–1272. [CrossRef]
27. Shi, L.; Zhou, Q.; Zhao, Y.; Ouyang, Y.; Ling, C.; Li, Q.; Wang, J. Oxidation Mechanism and Protection Strategy of Ultrathin Indium Selenide: Insight from Theory. *J. Phys. Chem. Lett.* **2017**, *8*, 4368–4373. [CrossRef]
28. Zhang, Y.J.; Ye, J.T.; Yomogida, Y.; Takenobu, T.; Iwasa, Y. Formation of a stable p-n junction in a liquid-gated MoS_2 ambipolar transistor. *Nano Lett.* **2013**, *13*, 3023–3028. [CrossRef]
29. Arora, H.; Erbe, A. Recent progress in contact, mobility, and encapsulation engineering of InSe and GaSe. *InfoMat* **2020**, *3*, 662–693. [CrossRef]
30. Masurkar, N.; Thangavel, N.K.; Yurgelevic, S.; Varma, S.; Auner, G.W.; Reddy Arava, L.M. Reliable and highly sensitive biosensor from suspended MoS_2 atomic layer on nano-gap electrodes. *Biosens. Bioelectron.* **2021**, *172*, 112724. [CrossRef]
31. Hu, T.; Zhang, L.; Wen, W.; Zhang, X.; Wang, S. Enzyme catalytic amplification of miRNA-155 detection with graphene quantum dot-based electrochemical biosensor. *Biosens. Bioelectron.* **2016**, *77*, 451–456. [CrossRef] [PubMed]
32. Ho, P.H.; Chang, Y.R.; Chu, Y.C.; Li, M.K.; Tsai, C.A.; Wang, W.H.; Ho, C.H.; Chen, C.W.; Chiu, P.W. High-Mobility InSe Transistors: The Role of Surface Oxides. *ACS Nano* **2017**, *11*, 7362–7370. [CrossRef] [PubMed]
33. Arora, H.; Jung, Y.; Venanzi, T.; Watanabe, K.; Taniguchi, T.; Hubner, R.; Schneider, H.; Helm, M.; Hone, J.C.; Erbe, A. Effective Hexagonal Boron Nitride Passivation of Few-Layered InSe and GaSe to Enhance Their Electronic and Optical Properties. *ACS Appl. Mater. Interfaces* **2019**, *11*, 43480–43487. [CrossRef]
34. Wells, S.A.; Henning, A.; Gish, J.T.; Sangwan, V.K.; Lauhon, L.J.; Hersam, M.C. Suppressing Ambient Degradation of Exfoliated InSe Nanosheet Devices via Seeded Atomic Layer Deposition Encapsulation. *Nano Lett.* **2018**, *18*, 7876–7882. [CrossRef]
35. Lin, P.; Luo, X.; Hsing, I.M.; Yan, F. Organic electrochemical transistors integrated in flexible microfluidic systems and used for label-free DNA sensing. *Adv. Mater.* **2011**, *23*, 4035–4040. [CrossRef] [PubMed]
36. Xie, Y.; Zhi, X.; Su, H.; Wang, K.; Yan, Z.; He, N.; Zhang, J.; Chen, D.; Cui, D. A Novel Electrochemical Microfluidic Chip Combined with Multiple Biomarkers for Early Diagnosis of Gastric Cancer. *Nanoscale Res. Lett.* **2015**, *10*, 477. [CrossRef]
37. Li, X.-H.; Sun, W.-M.; Wu, J.; Gao, Y.; Chen, J.-H.; Chen, M.; Ou, Q.-S. An ultrasensitive fluorescence aptasensor for carcino-embryonic antigen detection based on fluorescence resonance energy transfer from upconversion phosphors to Au nanoparticles. *Anal. Methods* **2018**, *10*, 1552–1559. [CrossRef]
38. Terenghi, M.; Elviri, L.; Careri, M.; Mangia, A.; Lobinski, R. Multiplexed Determination of Protein Biomarkers Using Metal-Tagged Antibodies and Size Exclusion Chromatography−Inductively Coupled Plasma Mass Spectrometry. *Anal. Chem.* **2009**, *81*, 9440–9448. [CrossRef]

biosensors

MDPI

Communication

Complementary DNA Significantly Enhancing Signal Response and Sensitivity of a Molecular Beacon Probe to Aflatoxin B1

Chao Wang [1], Kexiao Zhu [2], Jie Yu [2] and Pengfei Shi [1,*]

[1] College of Medicine, Linyi University, Linyi 276005, China
[2] College of Chemistry and Chemical Engineering, Linyi University, Linyi 276005, China
* Correspondence: shipengfei@lyu.edu.cn

Abstract: This paper reported an improved molecular beacon method for the rapid detection of aflatoxin B1 (AFB1), a natural mycotoxin with severe carcinogenicity. With the assistance of a complementary DNA (cDNA) chain, the molecular beacon which consists of a DNA aptamer flanked by FAM and BHQ1 displayed a larger fluorescent response to AFB1, contributing to the sensitive detection of AFB1. Upon optimization of some key experimental factors, rapid detection of AFB1 ranging from 1 nM to 3 μM, within 20 min, was realized by using this method. A limit of detection (LoD) of 1 nM was obtained, which was lower than the LoD (8 nM) obtained without cDNA assistance. This aptamer-based molecular beacon detection method showed advantages in easy operation, rapid analysis and larger signal response. Good specificity and anti-interference ability were demonstrated. This method showed potential in real-sample analysis.

Keywords: aflatoxin B1; aptamer; molecular beacon; complementary DNA; enhanced signal

Citation: Wang, C.; Zhu, K.; Yu, J.; Shi, P. Complementary DNA Significantly Enhancing Signal Response and Sensitivity of a Molecular Beacon Probe to Aflatoxin B1. *Biosensors* **2023**, *13*, 195. https://doi.org/10.3390/bios13020195

Received: 22 December 2022
Revised: 20 January 2023
Accepted: 23 January 2023
Published: 28 January 2023

1. Introduction

A simple, rapid and on-site detection method protocol is more desired for the sensing of toxins and hazardous chemicals. Aflatoxin B1 (AFB1) is a natural mycotoxin commonly found in fungus colonized cereals (e.g., corn, barley, wheat, peanut, bean, etc.) and their products [1]. The International Agency for Research on Cancer (IARC) has classified AFB1 as a Group 1 carcinogen due to its definite carcinogenesis [2,3]. AFB1 monitoring is an effective way to prevent contaminated samples from entering the human food chain and to reduce AFB1 exposure [4,5]. To date, however, routine detection methods for AFB1 including HPLC and LC-MS have been limited to complex sample pretreatment, sophisticated instruments and professional staff [6,7], resulting in these methods being unserviceable in some practical on-site and rapid analysis cases. The antibody-based immunoassay technique provides an alternative, which allows for the simple, rapid and in situ detection of AFB1 [8,9]. However, the preparation of antibodies is time-consuming and costly, and antibodies are vulnerable to denaturing. It is more difficult to prepare antibodies specific to toxic matter than non-toxic substances. These shortcomings restrict the wide application of immunoassays in toxins and hazardous chemical sensing. Therefore, it is desirable to develop new AFB1 detection methods with characteristics of easy operation and rapid and low-cost analysis.

Aptamer is an oligonucleotide that holds high specificity and binding affinity toward a wide range of targets, which is isolated from a combinatorial DNA library via a selection technology named systemic evolution of ligands by exponential enrichment (SELEX) [10–12]. Aptamer provides an ideal alternative of antibody and shows many advantages including chemical synthesis, thermal stability, low cost and easy modification over other antibodies [13,14]. The development of novel aptamer-based detection methods is a future direction for toxins and hazardous chemical sensing [15–19]. In the last decades, many aptamer-based assays/biosensors for AFB1 have been developed [20–22].

However, special reagents, repeated incubation and washing are usually needed in these assay/biosensor methods, which might retard their practical use ability in simple and rapid analysis.

The molecular beacon (MB) has been attractive for bioassays for nucleic acids, showing advantages in simplicity, rapidity and sensitivity [23–25]. The use of aptamer has allowed one to develop aptamer-based fluorescent molecular beacon assays for the detection of proteins and small molecules [26,27], e.g., Tat protein of HIV-1 [28], thrombin [29,30], cocaine [31], adenosine triphosphate (ATP) [32,33], etc., based on binding-induced structure change and the subsequent fluorescence decrease or fluorescence increase. Previously, we have reported a "signal-off" molecular beacon strategy for the fluorescence detection of AFB1 [34], which has shown advantages in simple operation and rapid analysis. However, we found that there were two facts which would reduce detection performance of this detection strategy: for one, the initial hairpin structure of the molecular beacon (MB) molecule without AFB1 would lead to a low starting fluorescence intensity of MB before adding AFB1; for the other, the intermolecular base-pairing reaction of MB molecules would cause fluorescence quenching before adding AFB1 and invalidate the "signal-off" response of MB to AFB1. Therefore, it is necessary to make some advancements to this detection proposal.

Herein, we have presented a complementary DNA (cDNA)-chain-assisted molecular beacon method for the rapid detection of AFB1. As illustrated in Scheme 1, in the absence of AFB1, aptamer hybridized with its complementary DNA (cDNA) strand to form a duplex structure, resulting in fluorescein (FAM) and fluorescent quencher (BHQI), is completely separated, and no fluorescence quenching occurs. Upon the addition of AFB1, aptamer prefers to bind with AFB1 rather than the cDNA strand, forming a stem-loop structure in which FAM and BHQI are in close proximity, and fluorescence quenching occurs. The cDNA strand could reduce fluorescence quenching before the addition of AFB1 and enhance the initial fluorescence strength of the MB probe. With the assistance of a cDNA strand, MBs signal change caused by AFB1 was enhanced and detection performance was improved. Under optimized conditions, we quantitatively achieved the detection of AFB1 in a range of 1 nM to 3 μM, with a detection limit of 1 nM which was lower than that (8 nM) obtained without cDNA's assistance. The specificity of this method for AFB1 was demonstrated. We successfully detected that AFB1 spiked in 50-fold diluted beer, 50-fold diluted serum and 10-fold diluted tap water by using this method, respectively. This shows its applicable potential in real-sample analysis.

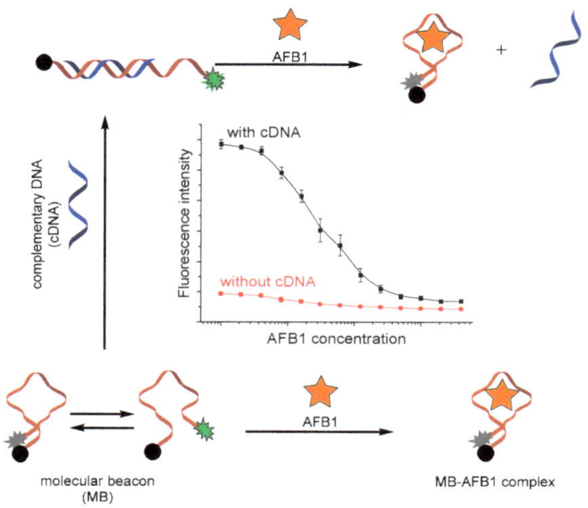

Scheme 1. Picture illustrating the principle of this proposal. With the assistance of a complementary DNA (cDNA) strand, the molecular beacon (MB) hybridizes with the cDNA to form a duplex structure,

in which FAM labeled at its 3′ end is apart from BHQI labeled at its 5′ end, and there is no fluorescence quenching. Upon the addition of AFB1, the MB prefers to bind with AFB1 rather than cDNA, and adapts into a hairpin structure in which FAM and BHQI are in close proximity, and fluorescence quenching occurs. Detection of AFB1 can be rapidly achieved by measurement of fluorescence intensity decline. Without cDNA, signal response and sensitivity of the MB probe are weak.

2. Materials and Methods

2.1. Materials and Reagents

DNA chains listed in Table 1 were synthesized and purified via HPLC by Sangon Bioteh Co., Ltd. (Shanghai, China, https://www.sangon.com/ (accessed on 28 January 2022)). Target mycotoxin aflatoxin B1 (AFB1) and non-target mycotoxins including ochratoxin A (OTA), zearalenone (ZAE) and deoxynivalenol (DON) were obtained from Sangon Bioteh Co., Ltd. (Shanghai, China, https://www.sangon.com/ (accessed on 28 January 2022)). Normal human serum was purchased from Beijing Solarbio Life Science Co., Ltd. (Beijing, China, https://www.solarbio.com/ (accessed on 28 January 2022)). Beer was bought from a local supermarket. Assay buffer (pH 7.5) contained 20 mM Tris-HCl, 200 mM NaCl and 20 mM $MgCl_2$, and 0.01% (*v*/*v*) tween20 was used for analysis. Aqueous solutions were prepared with ultrapure water where resistance was more than 18.2 MΩ·cm.

Table 1. Information of different DNA chains.

Name	Base Sequence (5′ to 3′)	Labels
MB	CGTGTTGTCTCTCTGTGT CTCG	5′-BHQ1, 3′-FAM
FDNA	CGTGTTGTCTCTCTGTGT CTCG	3′-FAM
BDNA	CGTGTTGTCTCTCTGTGT CTCG	5′-BHQ1
cDNA	GAGACAACACG	no

BHQ1 is the abbreviation for black hole quencher I; FAM is the abbreviation for fluorescein.

2.2. Analysis Procedures for AFB1 Detection

Firstly, the AFB1 sample, MB probe and cDNA chain were mixed in assay buffer solution. Unless otherwise specified, final concentrations of MB probe and cDNA were 20 nM. After incubation for 20 min at 4 °C, 200 μL of the reaction mixture solution was transferred into a microvolume quartz cuvette with 10 mm pathway length, and fluorescence intensity was measured immediately using a fluorescence spectrophotometer (Edinburgh FLS980-STM, U.K.). For fluorescence intensity measurement, the excitation wavelength was 485 nm and the emission wavelength was 518 nm. Three repeated measurements were carried out and the average data were used for quantitative analysis of AFB1.

2.3. Specificity Tests

To assess specificity of this proposal method, non-target mycotoxins including OTA, ZAE, DON and their mixtures were tested as interferences. In detail, these non-target mycotoxins were mixed with MB and cDNA. After 20 min incubation at 4 °C, the fluorescence intensity of MB was measured. The analysis procedures and conditions were the same as described above for the AFB1 detection.

2.4. Detection of AFB1 Spiked in Complex Matrix

We detected different concentrations of AFB1 spiked in 50-fold diluted beer, 50-fold diluted human serum and 10-fold diluted tap water, respectively. Prior to analysis, the beer, serum and tap water were ultrasonicated to degas, and then filtered through a syringe filter (0.22 μm) before dilution with the assay buffer. After dilution via the assay buffer and adulteration with standard AFB1 solution, the filtrate was used as a complex matrix sample for analysis.

3. Results and Discussions

3.1. Inspiration and Feasibility of This Proposal

According to the previous reports, the anti-AFB1 aptamer selected by Le et al. has a hairpin structure that is critical to its binding function; we have fabricated a molecular beacon (MB) capable of rapid analysis of AFB1, using a deliberately designed aptamer with a hairpin structure. This MB could bind with the AFB1 molecule through chemical forces including hydrogen bond, π-π stacking, electrostatic, van der Waals and hydrophobic interactions, and an MB conformational change occurred. As an assumption (bottom section of Scheme 1), the presence of AFB1 would cause the MB molecule to adjust into a stable stem-loop structure, in which FAM would be close to BHQ1 and fluorescence quenching would occur so that AFB1 could be detected by measuring the reduction in fluorescence intensity. However, we found that a large proportion (~90%) of FAM fluorescence had been quenched before the addition of AFB1, causing poor efficiency. This may have resulted from two possibilities: (a) most of the MB molecules stayed in a tertiary structure which brought FAM and BHQ1 to be adjacent in space, rather than in a formation that separated FAM and BHQ1, in the absence of AFB1; (b) intermolecular hybridization occurred among MBs, which caused FAM from one MB molecule and BHQ1 from another MB molecule to come into a close, causing fluorescence quenching. To verify or falsify these possibilities, we measured the fluorescence intensities of samples containing different reagents (Figure 1). The fluorescence intensity of the sample containing only MB (Figure 1c) is much lower than that of the sample containing only FDNA (Figure 1a), which is similar to that of the sample containing FDNA and BDNA (Figure 1b). These results imply that MB molecules had already stayed in a hairpin structure before the addition of AFB1, which is the major reason that caused low initial fluorescence intensity. This major reason further resulted in less of a reduction in fluorescence intensity upon the addition of AFB1. To resolve the problem and enhance the fluorescence response, we employed a cDNA chain to hybridize with MB and hoped to increase the initial fluorescence intensity before the addition of AFB1. As a result, fluorescence intensity of the sample containing only MB and cDNA (Figure 1e) was obviously higher than that of the sample containing only MB (Figure 1c), meaning this resolution is viable. Then, AFB1 was added. The fluorescence intensity of the sample containing MB, cDNA and AFB1 (Figure 1f) decreased in comparison to that of the sample containing MB and cDNA (Figure 1e). This means that AFB1 detection could be achieved via this proposal.

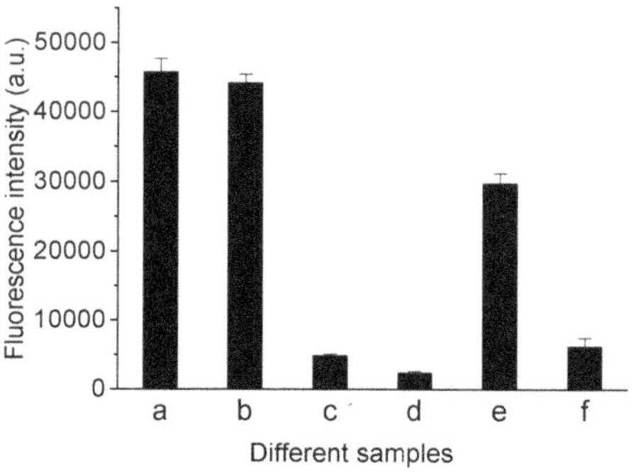

Figure 1. Fluorescence intensities of samples containing different reagents: (**a**) FDNA; (**b**) FDNA and BDNA; (**c**) MB; (**d**) MB and AFB1; (**e**) MB and cDNA and (**f**) MB, cDNA and AFB1. Concentrations of FDNA, BDNA and MB were 20 nM. Concentration of AFB1 was 500 nM. Excitation and emission wavelengths set for fluorescence measurements were 485 nm and 518 nm.

3.2. Optimization of Experimental Conditions

We optimized some key factors of this experiment. Figure 2a depicts the influences of the concentration ratio between cDNA and MB (C_{cDNA}:C_{MB}) on the detection performance of this proposal. The fluorescence intensity of the sample without AFB1 (F_{blank}) and the fluorescence intensity of the sample containing AFB1 (F_{AFB1}) both increased by increasing the ratio of C_{cDNA}:C_{MB}, demonstrating that much more MB hybridized with cDNA. However, too much cDNA might increase the hybridization reaction possibility of MB and subsequently decrease its binding with AFB1. To obtain a better competition between cDNA and AFB1 for MB, the optimal C_{cDNA}:C_{MB} was determined to be 20 nM:20 nM, as the signal descent degree $(1 - (F_{AFB1}/F_{blank})) \times 100\%$ caused by AFB1 approached the maximum at this ratio (Figure S1 in the electronic Supplementary Material (ESM)).

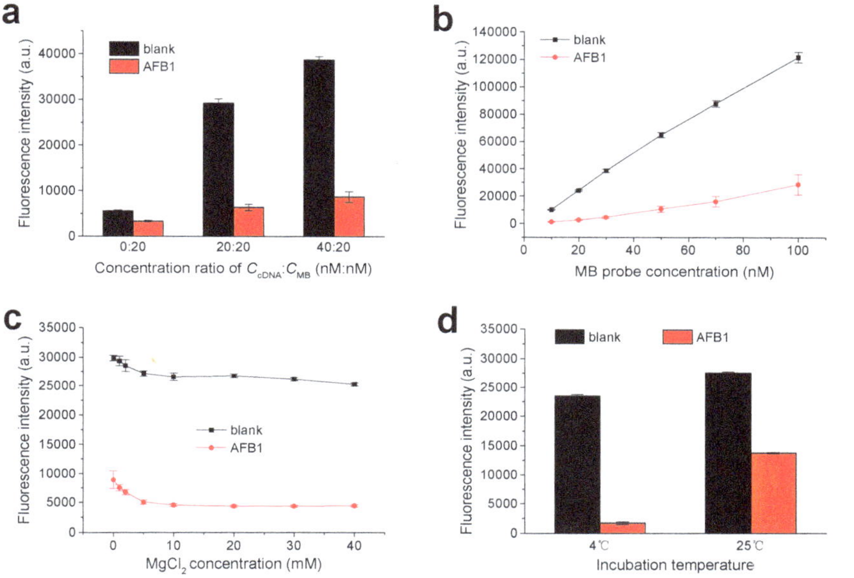

Figure 2. Optimizations of experimental conditions. (**a**) Fluorescence intensity of samples with or without AFB1 obtained with different concentration ratios between cDNA and MB; (**b**) fluorescence intensity of samples with or without AFB1 obtained with different concentrations of MB, when ratio value of C_{cDNA}:C_{MB} was fixed at 1:1; (**c**) fluorescence intensity of samples with or without AFB1 using assay buffer containing different concentrations of $MgCl_2$ and (**d**) fluorescence intensity of samples with or without AFB1 under different incubation temperatures.

The influences of molecular beacon (MB) probe concentration on detection performance were researched under the C_{cDNA}:C_{MB} ratio value fixed at 1:1. The results are shown in Figure 2b. F_{blank}, F_{AFB1} and fluorescence reduction (F_{blank}-F_{AFB1}) caused by AFB1 all increased with an increase in MB concentration. Finally, 20 nM MB was chosen to be used, since signal descent degree $(1 - (F_{AFB1}/F_{blank})) \times 100\%$ was largest at this concentration (Figure S2 in ESM).

Effects of cations (Mg^{2+} and Na^{+}) in assay buffer were also tested. Figure 2c shows the effects of Mg^{2+} cations in assay buffer on detection performance. In a lower $MgCl_2$ concentration range, F_{blank} and F_{AFB1} all decreased with an increasing concentration of $MgCl_2$. This might be due to the fact that Mg^{2+} can induce MB to form a stable hairpin structure, which causes fluorescence quenching. When 20 mM $MgCl_2$ was used, the value of $(1 - (F_{AFB1}/F_{blank})) \times 100\%$ was the largest (Figure S3 in ESM), meaning better sensitivity. Therefore, assay buffer containing 20 mM $MgCl_2$ was used in this study. Figure S4 in ESM shows the effects of NaCl in assay buffer containing 20 mM $MgCl_2$. When AFB1 was absent, a proper amount of NaCl could reduce the non-specific adsorption. Finally, NaCl at 200 mM in the assay buffer was chosen and applied to the further experiments.

The experimental temperature, which could transform DNA conformation and subsequently alter the binding affinity of the aptamer target, was also optimized. A certain concentration of AFB1 was detected under different incubation temperatures of 4 °C and 25 °C, respectively (Figure 2d). A larger signal change (F_{blank}-F_{AFB1}) was obtained at 4 °C, meaning better sensitivity.

3.3. Sensitivity Analysis of Novel Aptameric Sensor against AFB1

Under optimal conditions obtained above, we detected different concentrations of AFB1 using this MB with or without cDNA, respectively. As Figure 3 shows, with the assistance of the cDNA chain, the MB generated a much higher initial fluorescence intensity before the addition of AFB1, in comparison to that without cDNA assistance. Additionally, a more-than-ten-times-larger decrease in fluorescence intensity caused by the addition of AFB1 was observed when cDNA was used, which means an enhanced response and better sensitivity. In the concentration range of 1 nM to 125 nM, a linear relationship between fluorescence intensity and AFB1 concentration (Y = 12841X + 35922, R^2 = 0.99685, where Y is the fluorescence intensity and X is the logarithm of AFB1 concentration) was obtained with cDNA assistance. The limit of detection (LOD) determined by signal/noise being more than three (S/N > 3) was 1 nM, lower than the LOD (8 nM) obtained without cDNA assistance. This detection performance is better than or comparable to some of the previous literature listed in Table S1. Additionally, this aptasensor shows advantages in easy operation, rapid analysis and large signal change.

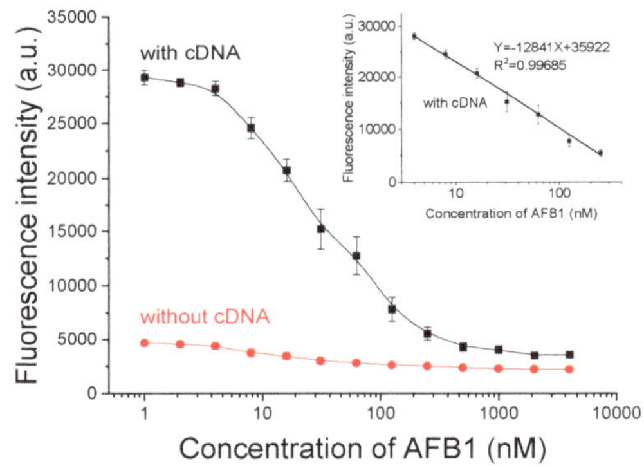

Figure 3. Fluorescence intensities corresponding to different concentrations of AFB1, with or without cDNA assistance. Assay buffer (pH 7.5) containing 20 mM Tris-HCl, 20 mM $MgCl_2$, 200 mM NaCl and 0.1% (*v*/*v*) tween20 was used. MB probe concentration was 20 nM. cDNA concentration was 20 nM. Excitation/emission was 485 nm/518 nm. Incubation was carried out at 4 °C for 20 min.

3.4. Specificity of This Detection Method

To demonstrate the specificity of this method for AFB1 detection, some non-target mycotoxins were also detected using this method. Results are shown in Figure 4. In comparison with the blank sample, AFB1 addition caused an obvious decrease in the fluorescence intensity. In contrast, the addition of these non-target mycotoxins OTA, ZAE and DON caused a neglectful change in fluorescence intensity, compared with the blank sample. Co-exists of these non-target mycotoxins had no interference on AFB1 detection. These results imply a good specificity of this method toward AFB1

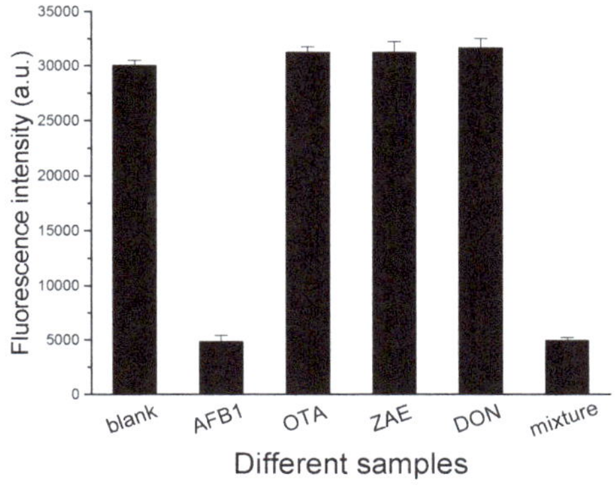

Figure 4. Fluorescence intensities corresponding to samples containing different mycotoxins. Assay buffer (pH 7.5) containing 20 mM Tris-HCl, 20 mM $MgCl_2$, 200 mM NaCl and 0.1% (v/v) tween20 was used. Concentrations of MB probe and cDNA were 20 nM. AFB1 concentration was 500 nM. Concentrations of OTA, ZAE and DON were 1 μM. Excitation/emission was 485 nm/518 nm. Incubation was carried out at 4 °C for 20 min.

3.5. Complex Matrix Interference Tests

To assess the application ability of this detection protocol in a complex matrix, different concentrations of AFB1 spiked in 50-fold diluted beer, 50-fold diluted serum and 10-fold diluted tap water were detected by using this method, respectively (Figure 5). Corresponding to any of these matrixes, fluorescence intensity declined with the increasing spiked amount of AFB1. A dynamic range of 1 nM to 3 μM and LOD of 1 nM were also achieved. These detection performances were comparable to those in a pure assay buffer system. These results imply the good anti-interference ability of this method, and its application potential in real-sample analysis.

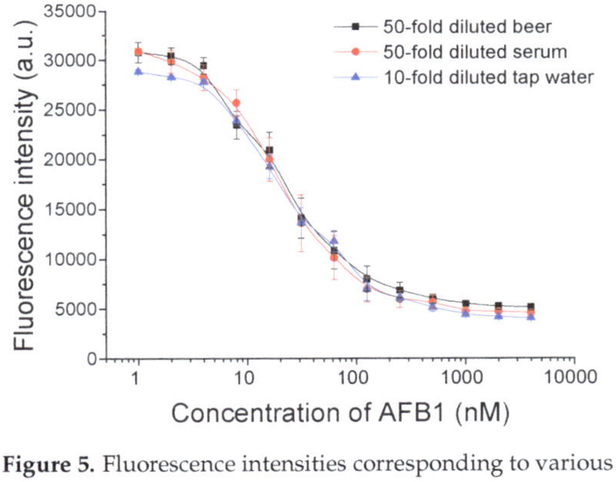

Figure 5. Fluorescence intensities corresponding to various concentrations of AFB1 spiked in 50-fold diluted beer, 50-fold diluted serum and 10-fold diluted tap water, respectively.

4. Conclusions

We have developed a simple aptamer-based molecular beacon method for the rapid detection of AFB1, in which a cDNA chain was employed to increase the initial fluorescence intensity of a molecular beacon (MB) probe before the addition of AFB1. Compared with the use of the MB probe alone, a larger fluorescence signal change caused by AFB1 and a lower detection limit were obtained, with the assistance of a cDNA chain. The detection range of this proposed method was 1 nM to 3 μM AFB1. This method showed good specificity toward AFB1, and resistance ability to the complex matrix interference.

Supplementary Materials: The following supporting information can be downloaded at https://www.mdpi.com/article/10.3390/bios13020195/s1: Figure S1: selection of concentration ratio between molecular beacon and complementary DNA strand; Figure S2: selection of MB concentration; Figure S3: optimization of MgCl2 concentration in assay buffer; Figure S4: effects of NaCl concentration in assay buffer on detection performance of this proposal [35–41].

Author Contributions: Conceptualization and methodology, C.W. and P.S.; data collection and analysis, writing—original draft preparation, C.W., K.Z. and J.Y.; writing—review and editing, C.W. and P.S.; supervision, interpretation of the draft, P.S.; funding acquisition, C.W. All authors have read and agreed to the published version of the manuscript.

Funding: This research was supported by Shandong Provincial Natural Science Foundation of China, grant number ZR2021QB206.

Conflicts of Interest: The authors declare no conflict of interest.

References

1. Rushing, B.R.; Selim, M.I. Aflatoxin B1: A review on metabolism, toxicity, occurrence in food, occupational exposure, and detoxification methods. *Food Chem. Toxicol.* **2019**, *124*, 81–100. [CrossRef] [PubMed]
2. Ostry, V.; Malir, F.; Toman, J.; Grosse, Y. Mycotoxins as human carcinogens-the IARC monographs classification. *Mycotoxin Res.* **2017**, *33*, 65–73. [CrossRef] [PubMed]
3. Marchese, S.; Polo, A.; Ariano, A.; Velotto, S.; Costantini, S.; Severino, L. Aflatoxin B1 and M1: Biological properties and their involvement in cancer development. *Toxins* **2018**, *10*, 214. [CrossRef] [PubMed]
4. Ali, N. Aflatoxins in rice: Worldwide occurrence and public health perspectives. *Toxicol. Rep.* **2019**, *6*, 1188–1197. [CrossRef] [PubMed]
5. Mahato, D.K.; Lee, K.E.; Kamle, M.; Devi, S.; Dewangan, K.N.; Kumar, P.; Kang, S.G. Aflatoxins in food and feed: An overview on prevalence, detection and control strategies. *Front. Microbiol.* **2019**, *10*, 2266. [CrossRef]
6. Yao, H.; Hruska, Z.; Diana Di Mavungu, J. Developments in detection and determination of aflatoxins. *World Mycotoxin J.* **2015**, *8*, 181–191. [CrossRef]
7. Abbas, H.K.; Shier, W.T.; Horn, B.W.; Weaver, M.A. Cultural methods for aflatoxin detection. *J. Toxicol. -Toxin. Rev.* **2004**, *23*, 295–315. [CrossRef]
8. Li, P.; Zhang, Q.; Zhang, W. Immunoassays for aflatoxins. *TrAC Trends Anal. Chem.* **2009**, *28*, 1115–1126. [CrossRef]
9. Qileng, A.; Huang, S.; He, L.; Qin, W.; Liu, W.; Xu, Z.; Liu, Y. Composite films of CdS nanoparticles, MoS2 nanoflakes, reduced graphene oxide, and carbon nanotubes for ratiometric and modular immunosensing-based detection of toxins in cereals. *ACS Appl. Nano Mater.* **2020**, *3*, 2822–2829. [CrossRef]
10. Ellington, A.D.; Szostak, J.W. In vitro selection of RNA molecules that bind specific ligands. *Nature* **1990**, *346*, 818–822. [CrossRef]
11. Tuerk, C.; Gold, L. Systematic evolution of ligands by exponential enrichment: RNA ligands to bacteriophage T4 DNA polymerase. *Science* **1990**, *249*, 505–510. [CrossRef] [PubMed]
12. Stoltenburg, R.; Reinemann, C.; Strehlitz, B. SELEX—A (r)evolutionary method to generate high-affinity nucleic acid ligands. *Biomol. Eng.* **2007**, *24*, 381–403. [CrossRef] [PubMed]
13. Bognar, Z.; Gyurcsanyi, R.E. Aptamers against immunoglobulins: Design, selection and bioanalytical applications. *Int. J. Mol. Sci.* **2020**, *21*, 5748. [CrossRef] [PubMed]
14. Toh, S.Y.; Citartan, M.; Gopinath, S.C.; Tang, T.H. Aptamers as a replacement for antibodies in enzyme-linked immunosorbent assay. *Biosens. Bioelectron.* **2015**, *64*, 392–403. [CrossRef] [PubMed]
15. Xie, M.; Zhao, F.; Zhang, Y.; Xiong, Y.; Han, S. Recent advances in aptamer-based optical and electrochemical biosensors for detection of pesticides and veterinary drugs. *Food Control* **2022**, *131*, 108399. [CrossRef]
16. Liu, L.S.; Wang, F.; Ge, Y.; Lo, P.K. Recent Developments in Aptasensors for Diagnostic Applications. *ACS Appl. Mater. Interfaces* **2021**, *13*, 9329–9358. [CrossRef]
17. Lv, X.; Zhu, L.; Rong, Y.; Shi, H.; Zhang, L.; Yu, J.; Zhang, Y. Ratiometric electrochemiluminescence lab-on-paper device for DNA methylation determination based on highly conductive copper paper electrode. *Biosens. Bioelectron.* **2022**, *214*, 114522. [CrossRef]
18. Zhou, Y.; Lv, S.; Wang, X.Y.; Kong, L.; Bi, S. Biometric photoelectrochemical-visual multimodal biosensor based on 3D hollow HCdS@Au nanospheres coupled with target induced ion exchange reaction for antigen detection. *Anal. Chem.* **2022**, *94*, 14492–14501. [CrossRef]
19. Shen, Y.; Guan, J.; Ma, C.; Shu, Y.; Xu, Q.; Hu, X. Competitive displacement triggering DBP photoelectrochemical aptasensor via cetyltrimethylammonium bromide bridging aptamer and perovskite. *Anal. Chem.* **2022**, *94*, 1742–1751. [CrossRef]
20. Hou, Y.; Jia, B.; Sheng, P.; Liao, X.; Shi, L.; Fang, L.; Zhou, L.; Kong, W. Aptasensors for mycotoxins in foods: Recent advances and future trends. *Compr. Rev. Food Sci. Food Saf.* **2022**, *21*, 2032–2073. [CrossRef]
21. Xue, Z.; Zhang, Y.; Yu, W.; Zhang, J.; Wang, J.; Wan, F.; Kim, Y.; Liu, Y.; Kou, X. Recent advances in aflatoxin B1 detection based on nanotechnology and nanomaterials–A review. *Anal. Chim. Acta* **2019**, *1069*, 1–27. [CrossRef] [PubMed]

22. Beitollahi, H.; Tajik, S.; Dourandish, Z.; Zhang, K.; Le, Q.V.; Jang, H.W.; Kim, S.Y.; Shokouhimehr, M. Recent advances in the aptamer-based electrochemical biosensors for detecting aflatoxin B1 and its pertinent metabolite aflatoxin M1. *Sensors* **2020**, *20*, 3256. [CrossRef] [PubMed]
23. Tyagi, S.; Kramer, F.R. Molecular beacons: Probes that fluoresce upon hybridization. *Nat. Biotechnol.* **1996**, *14*, 303–308. [CrossRef]
24. Bidar, N.; Amini, M.; Oroojalian, F.; Baradaran, B.; Hosseini, S.S.; Shahbazi, M.-A.; Hashemzaei, M.; Mokhtarzadeh, A.; Hamblin, M.R.; de la Guardia, M. Molecular beacon strategies for sensing purpose. *TrAC Trends Anal. Chem.* **2021**, *134*, 116143. [CrossRef]
25. Moutsiopoulou, A.; Broyles, D.; Dikici, E.; Daunert, S.; Deo, S.K. Molecular aptamer beacons and their applications in sensing, imaging, and diagnostics. *Small* **2019**, *15*, 1902248. [CrossRef] [PubMed]
26. Zheng, J.; Yang, R.; Shi, M.; Wu, C.; Fang, X.; Li, Y.; Li, J.; Tan, W. Rationally designed molecular beacons for bioanalytical and biomedical applications. *Chem. Soc. Rev.* **2015**, *44*, 3036–3055. [CrossRef] [PubMed]
27. Goulko, A.A.; Li, F.; Chris Le, X. Bioanalytical applications of aptamer and molecular-beacon probes in fluorescence-affinity assays. *TrAC Trends Anal. Chem.* **2009**, *28*, 878–892. [CrossRef]
28. Yamamoto, R.; Baba, T.; Kumar, P.K. Molecular beacon aptamer fluoresces in the presence of Tat protein of HIV-1. *Genes Cells* **2000**, *5*, 389–396. [CrossRef]
29. Hamaguchi, N.; Ellington, A.; Stanton, M. Aptamer beacons for the direct detection of proteins. *Anal. Biochem.* **2001**, *294*, 126–131. [CrossRef]
30. Li, J.J.; Fang, X.; Tan, W. Molecular aptamer beacons for real-time protein recognition. *Biochem. Biophys. Res. Commun.* **2002**, *292*, 31–40. [CrossRef]
31. Stojanovic, M.N.; de Prada, P.; Landry, D.W. Aptamer-based folding fluorescent sensor for cocaine. *J. Am. Chem. Soc.* **2001**, *123*, 4928–4931. [CrossRef] [PubMed]
32. Tang, Z.; Mallikaratchy, P.; Yang, R.; Kim, Y.; Zhu, Z.; Wang, H.; Tan, W. Aptamer switch probe based on intramolecular displacement. *J. Am. Chem. Soc.* **2008**, *130*, 11268–11269. [CrossRef] [PubMed]
33. Li, N.; Ho, C.M. Aptamer-based optical probes with separated molecular recognition and signal transduction modules. *J. Am. Chem. Soc.* **2008**, *130*, 2380–2381. [CrossRef]
34. Wang, C.; Sun, L.; Zhao, Q. A simple aptamer molecular beacon assay for rapid detection of aflatoxin B1. *Chinese Chem. Lett.* **2019**, *30*, 1017–1020. [CrossRef]
35. Chen, L.; Wen, F.; Li, M.; Guo, X.; Li, S.; Zheng, N.; Wang, J. A simple aptamer-based fluorescent assay for the detection of aflatoxin B1 in infant rice cereal. *Food Chemistry* **2017**, *215*, 377–382. [CrossRef] [PubMed]
36. Jia, Y.; Wu, F.; Liu, P.; Zhou, G.; Yu, B.; Lou, X.; Xia, F. A label-free fluorescent aptasensor for the detection of aflatoxin B1 in food samples using AIEgens and graphene oxide. *Talanta* **2019**, *198*, 71–77. [CrossRef]
37. Sabet, F.S.; Hosseini, M.; Khabbaz, H.; Dadmehr, M.; Ganjali, M.R. FRET-based aptamer biosensor for selective and sensitive detection of aflatoxin B1 in peanut and rice. *Food Chemistry* **2017**, *220*, 527–532. [CrossRef]
38. Xia, X.; Wang, Y.; Yang, H.; Dong, Y.; Zhang, K.; Lu, Y.; Deng, R.; He, Q. Enzyme-free amplified and ultrafast detection of aflatoxin B1 using dual-terminal proximity aptamer probes. *Food Chemistry* **2019**, *283*, 32–38. [CrossRef]
39. Lu, Z.; Chen, X.; Wang, Y.; Zheng, X.; Li, C.M. Aptamer based fluorescence recovery assay for aflatoxin B1 using a quencher system composed of quantum dots and graphene oxide. *Microchimica Acta* **2014**, *182*, 571–578. [CrossRef]
40. Zhao, Z.; Yang, H.; Deng, S.; Dong, Y.; Yan, B.; Zhang, K.; Deng, R.; He, Q. Intrinsic conformation response-leveraged aptamer probe based on aggregation-induced emission dyes for aflatoxin B1 detection. *Dye. Pigment.* **2019**, *171*, 107767. [CrossRef]
41. Mousivand, M.; Javan-Nikkhah, M.; Anfossi, L.; Di Nardo, F.; Salina, M.; Bagherzadeh, K. High performance aptasensing platform development through in silico aptamer engineering for aflatoxin B1 monitoring. *Food Control* **2023**, *145*, 109418. [CrossRef]

biosensors

MDPI

Communication

A Single Aptamer-Dependent Sandwich-Type Biosensor for the Colorimetric Detection of Cancer Cells via Direct Coordinately Binding of Bare Bimetallic Metal–Organic Framework-Based Nanozymes

Yuhui Zhu [1,†], Xueting Fang [1,†], Xiaofei Lv [1], Meijun Lu [1], Hui Xu [2], Shengqiang Hu [1], Shulin Zhao [1] and Fanggui Ye [1,*]

[1] State Key Laboratory for the Chemistry and Molecular Engineering of Medicinal Resources, School of Chemistry and Pharmaceutical Sciences, Guangxi Normal University, Guilin 541004, China
[2] Nanxian Inspection and Testing Center of Yiyang City in Hunan Province, Yiyang 413299, China
* Correspondence: fangguiye@163.com
† These authors contributed equally to this work.

Abstract: A typical colorimetric sandwich-type sensor relies on dual antibodies/aptamers to specifically visualize the targets. The requirement of dual antibodies/aptamers and low signal intensity inevitably increases the design difficulty and compromises the sensing sensitivity. In this work, a novel sandwich-type aptasensor was developed using single aptamer-functionalized magnetic nanoparticles as a specific recognition unit to target cancer cells and a bimetallic metal–organic frameworks (MOFs)-based nanozymes as a colorimetric signal amplification unit. The well-defined crystalline structure of UIO-66 MOFs enabled the introduction of Fe/Zr bimetal nodes, which possessed integrated properties of the peroxidase-like nanozyme activity and direct coordinately binding to the cell surface. Such a novel construction strategy of sandwich-type aptasensors achieved simple, sensitive, and specific detection of the target cancer cells, which will inspire the development of biosensors.

Keywords: metal–organic frameworks; sandwich-type; single aptamer; coordination bond; colorimetric detection

Citation: Zhu, Y.; Fang, X.; Lv, X.; Lu, M.; Xu, H.; Hu, S.; Zhao, S.; Ye, F. A Single Aptamer-Dependent Sandwich-Type Biosensor for the Colorimetric Detection of Cancer Cells via Direct Coordinately Binding of Bare Bimetallic Metal–Organic Framework-Based Nanozymes. *Biosensors* **2023**, *13*, 225. https://doi.org/10.3390/bios13020225

Received: 16 January 2023
Revised: 30 January 2023
Accepted: 30 January 2023
Published: 3 February 2023

1. Introduction

Cancer remains a major threaten to human life in the world [1]. Current clinical practice suggests that early detection of cancer cells contributes to monitoring the progressions of cancer and conducting precise therapy [2]. Consequently, it is desirable to develop simple, sensitive, and specific technologies to quantify cancer cells. Among all the developed detection strategies, colorimetric sandwich-type methods have attracted considerable attention due to the advantages of high specificity, easy operation, as well as bare-eye-based readout [3]. In general, the colorimetric sandwich-type assay is carried out by using two specific antibodies/aptamers to recognize the targets and using appropriate labels to provide detectable signals [4,5]. However, the requirement of two antibodies/aptamers inevitably increases the design difficulty. In fact, the use of single high-affinity antibody/aptamer fully meets the need of specific recognition of target cells [6–8]. In addition, low colorimetric signals greatly compromise the response sensitivity of visual analysis.

Metal–organic frameworks (MOFs) are a type of highly ordered porous crystalline material constructed by reasonable self-assembly of metal ions and organic ligands [9,10]. The well-defined crystalline structure of MOFs offers an excellent alternative to design metal nodes to acquire metal-related physicochemical properties, which holds great potential in colorimetric sandwich-type analysis. For instance, the introduction of Fe atoms might endow the resulting Fe-containing MOFs with intrinsic peroxidase-like catalytic activity [11], which could significantly increase the colorimetric analysis sensitivity via accelerating H_2O_2-mediated oxidation of chromogenic substrates. The coordination affinity between Zr^{4+} and PO_4^{3-} contributes to identifying the cellular phospholipid bilayer using Zr-containing MOFs [12]. Correspondingly, developing bimetallic MOFs with a given ionic ratio offers an interesting candidate to obtain integrated properties from different metal nodes.

Herein, a novel sandwich-type aptasensor was prepared for colorimetric analysis of cancer cells using single aptamer-modified magnetic nanoparticles to selectively capture and separate target cells, followed by direct binding of bare bimetallic metal–organic-framework-based nanozymes. By adjusting the added amount of metal precursors among the synthesis of a given UIO-66 MOF, UIO-66(Fe/Zr) bimetallic MOFs were prepared with integrated properties, including the peroxidase-like nanozyme activity from Fe nodes and direct binding to cell surface based on the Zr-involved coordination bond. Such a single aptamer-dependent colorimetric sandwich-type aptasensor facilitated simple, sensitive, and specific detection of target cancer cells even in complex biological systems.

2. Experimental Section

2.1. Materials

Ferric chloride hexahydrate ($FeCl_3{\cdot}6H_2O$) was purchased from Shanghai Yien Chemical Technology Co., Ltd. (Shanghai, China). Zirconium tetrachloride ($ZrCl_4$), ferric oxide nanoparticles (Fe_3O_4 NPs), and (1-(3-dimethylamino-propyl)-3-ethylcarbon diamide hydrochloride (EDC) were obtained from Aladdin Reagent Co., Ltd. (Shanghai, China). Ethyl orthosilicate (TEOS) was purchased from Sinopharm Chemical Reagent Co., Ltd (Shanghai, China); 3-Aminopropyltriethoxysilane (APTES) and N, N-dimethylformamide (DMF) were attained from Titan Technology Co., Ltd. (Shanghai, China). Hydrogen peroxide (H_2O_2, 30 wt%), acetic acid, ammonia ($NH_3{\cdot}H_2O$), and 2-propanol were provided by Xilong Chemical Co., Ltd. (Guangzhou, China); 3,3′,5,5′-Tetramethylbenzidine (TMB) was purchased from Tokyo Chemical Industry Co., Ltd. (Shanghai, China). Dulbecco's modified Eagle's medium (DMEM) and 1640 cell culture media were acquired from Thermo Fisher Scientific Inc. (Cleveland, OH, USA); 2-(N-morpholino)ethanesulfonic acid (MES) buffer (100 mM, pH 4.8), Cy3 dyes, and carboxylated AS14111 aptamer (from 5′ to 3′: HOOC-TTT TTT TTT GGT GGT GGT TGT GGT GGT GGT GG) were provided by Sangon Biotech Co., Ltd. (Shanghai, China). The human serum sample was kindly provided from a healthy volunteer, and all related experiments were approved by the Institutional Ethical Committee (IEC) of Guangxi Normal University.

2.2. Instrumentation

Transmission electron microscopy (TEM) images were acquired from a Talos 200S transmission electron microscope (Thermo Fisher Scientific Inc., Cleveland, OH, USA). Scanning electron microscope (SEM) and energy dispersive spectroscopy (EDS) were performed with a FEI Quanta 200 scanning electron microscope (Hitachi, Japan). Inductively coupled plasma mass spectrometry (ICP-MS) was conducted on Agilent 7700ce (Agilent, Santa Clara, CA, USA). Fourier transform infrared (FT-IR) spectrum was obtained using a Spectrum Two UATR FT-IR spectrometer (Perkin-Elmer, Waltham, MA, USA). Zeta potential was measured using a Nano ZS-90 Malvern Particle size potentiometer (Malvern Instruments, Worcestershire, UK). X-ray diffraction (XRD) patterns were collected on Rigaku D/max 2550 VB/PC X-ray diffractometer (Rigaku Company, Tokyo, Japan). Ultraviolet−visible (UV−vis) absorption spectra were collected on a Cary 60 spectrometer (Agilent, Santa Clara, CA, USA). Cell imaging was performed using a LSM 710 confocal laser scanning microscope (Zeiss, Jena, Germany).

2.3. Preparation of UIO-66(Fe/Zr)

The UIO-66(Fe/Zr) NPs were synthesized following a previously reported method [13]. In brief, 0.8 mM $ZrCl_4$, 0.8 mM $FeCl_3{\cdot}6H_2O$, and 0.8 mM H_2BDC were mixed into 25 mL of DMF, followed by the addition of 5 mL of acetic acid. After ultrasonic treatment at ambient temperature, the mixture was transferred to a 50-mL Teflon-lined hydrothermal reactor and reacted at 120 °C for 12 h. Subsequently, the obtained products were washed with ultrapure water and ethanol three times, respectively. After drying the products at 60 °C overnight, UIO-66(Fe/Zr) powder was collected for further use. For the preparation of Cy3-dye-incorporated UIO-66(Fe/Zr), 9.7 mg of UIO-66(Fe/Zr) NPs were mixed with 0.03 mM Cy3 dyes, followed by stirring for 24 h.

2.4. Preparation of AS1411-Functionalized Fe_3O_4@SiO_2

First, Fe_3O_4@SiO_2 core-shell NPs were synthesized using a reported sol-gel strategy [14]. Typically, 0.15 g of Fe_3O_4 NPs were ultrasonically dispersed into a mixture solution containing 70 mL of ultrapure water, 280 mL of ethanol, and 5.0 mL of $NH_3{\cdot}H_2O$ (28 wt%), followed by a slow addition of 4.0 mL of TEOS. After stirring for 10 h, the resulting Fe_3O_4@SiO_2 NPs were collected after magnetic separation and washing with ultrapure water and ethanol.

Then, 0.3 g of Fe_3O_4@SiO_2 NPs were mixed with 303 mL of 2-propanol containing 3 mL of APTES. After stirring at 70 °C for 6 h, Fe_3O_4@SiO_2-NH_2 NPs were obtained by magnetic separation. Subsequently, the carboxylated aptamer AS1411 and Fe_3O_4@SiO_2-NH_2 (250 μL, 5 mg/mL) were added to 500 μL of MES (100 mM, pH 4.8) solution containing 9.6 mg of EDC. After incubation at 37 °C overnight, the obtained AS1411-functionalized Fe_3O_4@SiO_2 (Fe_3O_4@SiO_2-Apt) NPs were magnetically isolated and resuspended in Tris-HCl buffer (10 mM, pH 7.4) for future use.

2.5. Assessment of Peroxidase-Like Nanozyme Activity

To investigate the peroxidase-like nanozyme activity, UIO-66(Fe/Zr) or Fe_3O_4@SiO_2-Apt were added into 200 μL of NaAc-HAc buffer (10 mM, pH 3.5) containing TMB (50 μL, 10 mM) and H_2O_2 (50 μL, 20 mM). After reacting at 37 °C for 30 min, the catalytic reaction was terminated using 1 M H_2SO_4. The resulting solution was photographed and measured using an UV-vis spectrophotometer.

2.6. Cell Culture

Human cervical cancer HeLa cells were cultured in 1640 medium containing 10% FBS and 1% penicillin-streptomycin. For the cases of human hepatoma HepG2 cells and mouse fibroblast L929 cells, DMEM medium was used. All the cell lines were incubated in a humidified atmosphere of 5% CO_2 at 37 °C.

2.7. Colorimetric Detection of Cancer Cells

First, HeLa cells with different concentrations (5×10^2, 1×10^3, 2×10^3, 3×10^3, 5×10^3, 8×10^3, 9×10^3, 1×10^4, 2×10^4, 3×10^4 cells/mL) were cultured in 1640 medium overnight. Then, fresh cell medium containing Fe_3O_4@SiO_2-Apt was added and incubated for 2 h. After washing with PBS (10 mM, pH 7.4), the cells were digested and magnetically collected. Subsequently, UIO-66(Fe/Zr) were added into the resulting cells and incubated at 37 °C for 30 min. After removing free UIO-66(Fe/Zr), 200 μL of NaAc-HAc buffer (10 mM, pH 3.5) containing TMB (50 μL, 10 mM) and H_2O_2 (50 μL, 20 mM) was added and reacted at 37 °C for 30 min. After terminating the reaction using 1 M H_2SO_4, the absorbance spectra were collected by an UV-vis spectrophotometer. To investigate the potential in detecting real samples, HeLa cells with different concentrations (5×10^3, 8×10^3 1×10^4 cells/mL) were spiked into 1000-fold diluted human blood, respectively, followed by colorimetric analysis.

3. Results and Discussion

3.1. Principle of Colorimetric Sandwich-Type Detection of Cancer Cells

As shown in Figure 1A, UIO-66(Fe/Zr) NPs were synthesized by a one-step hydrothermal method using $ZrCl_4$ and $FeCl_3{\cdot}6H_2O$ as metal precursors and H_2BDC as organic ligands. In order to guarantee selective detection of target cancer cells, a novel sandwich-type aptasensor was proposed using aptamer-functionalized $Fe_3O_4@SiO_2$ core-shell NPs to selectively capture and magnetically separate cancer cells from complex biological systems (Figure 1B). Considering bare Fe_3O_4 NPs also own peroxide-like nanozyme activity, $Fe_3O_4@SiO_2$ core-shell NPs originating from the silylanization of Fe_3O_4 NPs were adopted here to improve the signal-to-noise ratio toward cancer cell analysis. Taking advantage of the Zr-O-P coordination bond between phosphate units in the phospholipid bilayer of the cell membrane and zirconium nodes on MOFs, UIO-66(Fe/Zr) could directly bind to the cell surface. In addition, UIO-66(Fe/Zr) possesses a strong peroxidase-like nanozyme activity, which can catalyze H_2O_2-mediated oxidation of colorless TMB to form yellow-colored oxidized ones (oxTMB) with the aid of 1 M H_2SO_4, providing an excellent alternative for colorimetric analysis of cancer cells.

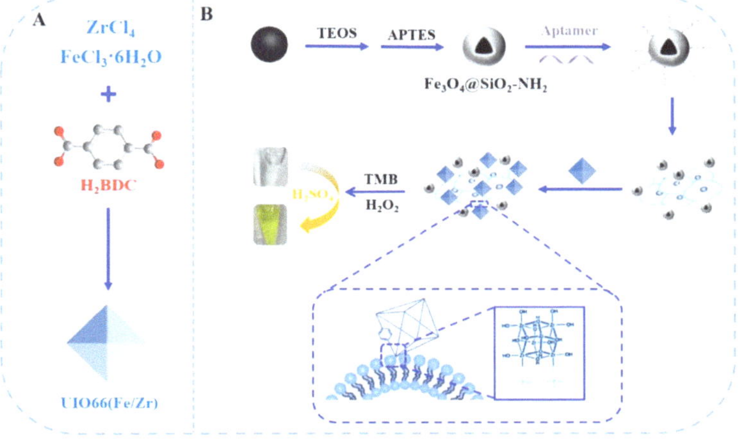

Figure 1. Schematic illustration of (**A**) UIO-66(Fe/Zr) synthesis and (**B**) their application in constructing a single aptamer-dependent sandwich-type biosensor for colorimetric detection of cancer cells.

3.2. Characterization of UIO-66(Fe/Zr) and $Fe_3O_4@SiO_2$-Apt

After a simple hydrothermal process, UIO-66(Fe/Zr) NPs were successfully prepared with a size ranging from 70 nm to 96 nm (Figure 2A). Observed from the EDS (Figure 2B) and elemental imaging (Figure 2C), the elements of C, O, Zr, and Fe were uniformly distributed among the bimetallic MOFs. Furthermore, the molar ratio of Fe and Zr was calculated to be 1:32 according to the ICP-MS analysis. Observed from the FT-IR (Figure 2D), the stretching vibration of the C=O bond in the H_2BDC organic ligand displayed a characteristic absorption peak at 1660 cm^{-1} [15], while the vibration of Zr-O-Zr yielded triple peaks at 766, 662, 482 cm^{-1} [16]. Powder XRD results demonstrated the prepared UIO-66(Fe/Zr) exhibited a similar crystal structure to that of simulated UIO-66, as evidenced by the characteristic diffraction peaks at 7.36°, 8.48°, and 25.72° assigned to the (111), (002), and (224) crystal planes (CCDC-1507786), respectively (Figure 2E) [17]. These results indicated that the introduction of additional Fe nodes was unable to disturb the intact metal–organic framework of UIO-66(Zr), consistent with that reported previously [13]. Taken all the above results into consideration, it can be concluded that a part of the Zr nodes in the UIO-66 framework were carefully displaced by Fe nodes, and the similar Fe-O-Fe bond to Zr-O-Zr bond ensured a comparable spatial configuration of the UIO-66(Fe/Zr) to that of UIO-66(Zr).

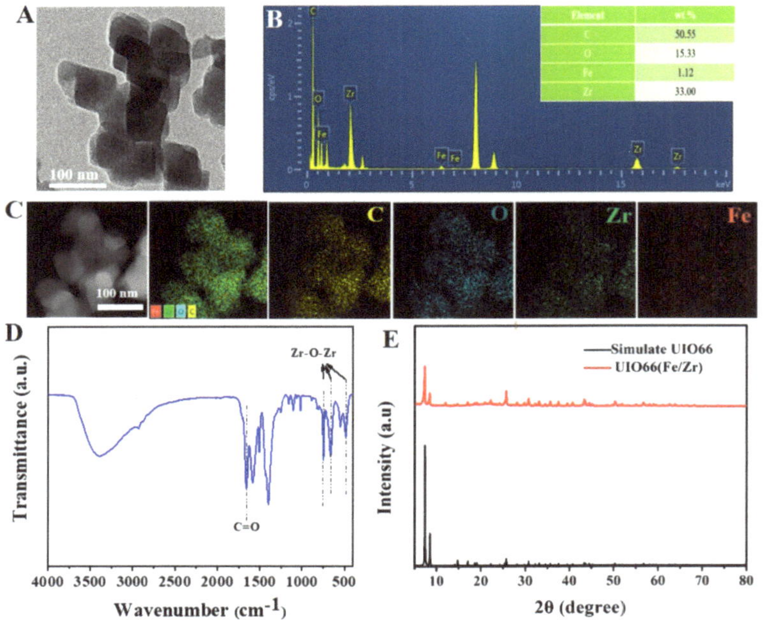

Figure 2. (**A**) Transmission electron microscopy (TEM) image, (**B**) elemental analysis, (**C**) EDS images, and (**D**) FT-IR spectrum of the prepared UIO-66(Fe/Zr). The inset shows the corresponding percentage of elements C, O, Fe, and Zr. (**E**) XRD patterns of the prepared UIO-66(Fe/Zr) and simulate UIO-66.

Next, Fe_3O_4@SiO_2 core-shell NPs were prepared via sonication-assisted coating of a silica shell on Fe_3O_4 NPs core. As shown in Figure S1, spherical Fe_3O_4 NPs had a mean diameter of approximately 320 nm. After being coated with a thin silica shell (approximately 32 nm), an obvious core-shell structure was observed for the obtained Fe_3O_4@SiO_2 NPs. Strong asymmetric stretching peaks of the Si-O-Si bond at 1207 and 1079 cm^{-1} appeared [18], accompanied by the disappearance of the characteristic peak of the Fe-O-Fe bond at 593 cm^{-1} [19], which indicated the successful coating of SiO_2 on the surface of Fe_3O_4 NPs (Figure S2A). For the formed Fe_3O_4@SiO_2 core-shell NPs, aminated treatment gave rise to a positive zeta potential (Figure S2B). However, further functional modification with aptamer AS1411 molecules induced the resultant Fe_3O_4@SiO_2-Apt NPs to be negative-charged.

3.3. Assessment of Peroxidase-Like Nanozyme Activity

The peroxidase-like nanozyme activity was visually evaluated using the typical TMB-H_2O_2 colorimetric system [7]. As shown in Figure 3A, under the condition of the acetate buffer, the mixture of TMB and H_2O_2 almost remained colorless with a low absorption peak intensity at 450 nm, indicating a relatively slow reaction efficiency (inset b). After introduction of UIO-66(Fe/Zr) (inset d), the solution color of the TMB+H_2O_2 system turned to be yellow. Since UIO-66(Fe/Zr) NPs alone exhibited negligible absorption at 450 nm (inset c), the above-mentioned yellow solution was, thus, attributed to the catalytic decomposition of H_2O_2 by UIO-66(Fe/Zr) NPs to promote the oxidation of TMB (Figure 3B). Furthermore, the absorption peak intensity was positively related to the added amount of UIO-66(Fe/Zr) NPs (Figure 3C), revealing the feasibility of visual analysis using a UIO-66(Fe/Zr)-mediated colorimetric system.

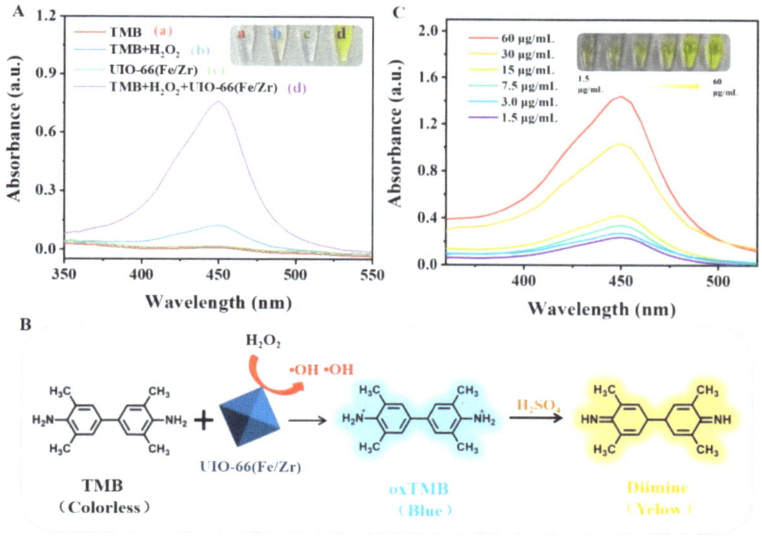

Figure 3. (**A**) UV-vis absorbance spectra of different solution. The insets showing the images of their corresponding solution. (**B**) Schematic illustration of UIO-66(Fe/Zr)-mediated catalytic oxidation of TMB-H_2O_2 system. (**C**) UIO-66(Fe/Zr) concentration-dependent UV-vis absorbance spectra of TMB-H_2O_2 system. The insets show the images of their corresponding solution.

In order to obtain a low background signal, Fe_3O_4@SiO_2-Apt were expected without peroxidase-like activity. As anticipated, bare Fe_3O_4 NPs did catalyze the decomposition of hydrogen peroxide to accelerate TMB oxidization (inset b, Figure S3). For the case of Fe_3O_4@SiO_2-Apt, effective coverage of Fe_3O_4 with a thin shell of SiO_2 resulted in a significant decrease in the number of catalytic sites, and the peroxidase-like activity was almost lost (insets c and d, Figure S3).

3.4. Construction of a Single-Aptamer-Based Sandwich-Type Biosensor

The direct binding of bare UIO-66(Fe/Zr) to target cancer cells was of vital importance for constructing a single-aptamer-based sandwich-type biosensor. To this end, Cy3-dyes-incorporated UIO-66(Fe/Zr) NPs were prepared for fluorescence-assisted positioning of their binding sites (Figure S4). As a proof-of-concept assay, we attempted to incubate human cervical cancer HeLa cells with Cy3-dyes-incorporated UIO-66(Fe/Zr) NPs. As imaged by confocal laser scanning microscopy (CLSM), the outline of the HeLa cell was clearly lighted up, taking advantage of the luminance of Cy3 (Figure 4A), which proved the successful binding of bare UIO-66(Fe/Zr) NPs to the surface of HeLa cells. The resultant HeLa cell maintained the original morphology, which demonstrates the excellent biocompatibility of UIO-66(Fe/Zr) NPs.

Inspired by the direct binding of bare UIO-66(Fe/Zr) NPs to HeLa cells, a colorimetric sandwich-type aptasensor was then designed using Fe_3O_4@SiO_2-Apt to selectively capture and magnetically separate HeLa cells through specific recognition of aptamer AS1411 to the overexpressed nucleolin [20,21]. To achieve the colorimetric detection of HeLa cells, Fe_3O_4@SiO_2-Apt were first incubated with HeLa cells at 37 °C for 2 h, followed by the addition of UIO-66(Fe/Zr) NPs and incubation at 37 °C for another 30 min. After magnetic separation of the free nanozymes, the formed sandwich-type structure was added into the TMB+H_2O_2 system. Without HeLa cells or UIO-66(Fe/Zr), the solution color of TMB+H_2O_2 system in the presence of Fe_3O_4@SiO_2-Apt was nearly colorless, and weak absorption at 450 nm was generated (curves a and b, Figure 4B). For the colorimetric system treated with a stable sandwich-type structure (curve c, Figure 4B), the yellow solution appeared with an over 7-fold absorbance intensity compared to that of the blank sample (inset c, Figure 4B), which demonstrates the feasibility of colorimetric detection of HeLa cells.

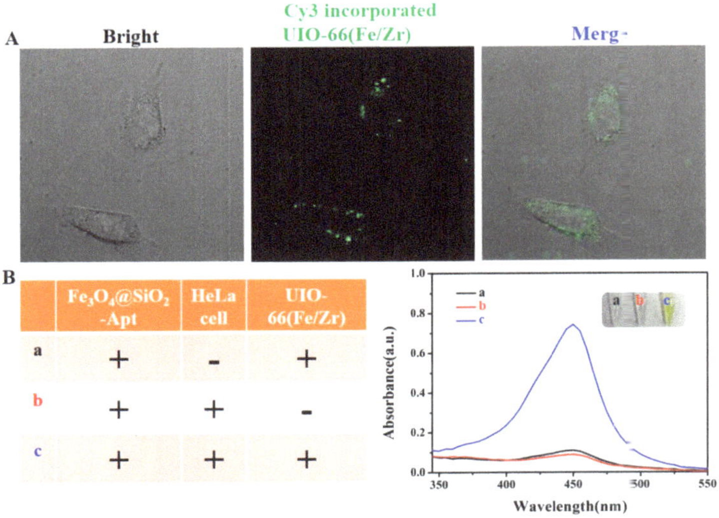

Figure 4. (**A**) Bright-field image, fluorescence image, and overlay of the fluorescence and bright-field images of HeLa cells after incubation with bare Cy3 dyes-incorporated UIO-66(Fe/Zr) NPs. (**B**) UV-vis absorbance spectra of TMB-H_2O_2 system after different treatments. The insets show the images of their corresponding solution.

3.5. Colorimetric Detection of Cancer Cells

As the concentration of HeLa cells increased from 0 to 2 × 10^4 cells/mL, the colorimetric system gradually darkened in solution color (the inset, Figure 5A). By measuring the absorption intensity at 450 nm, quantitative analysis further confirmed a positive correlation between HeLa cell concentration and the oxidation efficiency of TMB (Figure 5A). These results were originated from the fact that the presence of more HeLa cells meant more cell membrane-bound UIO-66(Fe/Zr) nanozymes, resulting in a higher catalytic reaction dynamic of TMB oxidation by H_2O_2. As illustrated in Figure 5B, a good linear relationship was attained between the absorption intensity at 450 nm and the cell concentration ranging from 10^3 to 10^4 cells/mL. The detection limit (LOD) was calculated to be 481 cells/mL based on $3\sigma/k$ (σ and k representing the standard deviation of blank signal and the curve slope, respectively).

To further verify the sensing selectivity toward HeLa cells, a same concentration of human hepatoma HepG2 cells and mouse fibroblast L929 cells were tested with the proposed colorimetric sandwich-type aptasensor, respectively. When compared with the blank sample, the group of L929 cells caused no apparent color change, while a significant yellow solution was produced for the samples of HepG2 and HeLa cells (Figure 5C). Higher response sensitivity toward cancer cells might be ascribed to a higher expression level of nucleolin on cancer cells than that on normal cells [8], which verifies the sensing specificity of the colorimetric sandwich-type aptasensor. Furthermore, this aptasensor could accurately distinguish HeLa cells from an analogue biological environment via mixing HeLa and L929 cells (light-blue column, Figure 5C).

Encouraged by the excellent selectivity, the developed aptasensor was employed for the analysis of real samples by spiking different concentrations of HeLa cells into 1000-fold diluted healthy human blood. A satisfactory recovery rate between 96.8% and 106.0% with an acceptable relative standard deviation (RSD) value was obtained (Figure 5D), which fully demonstrated the accuracy and reliability of the developed aptasensor in analyzing target cancer cells in the complex biological system.

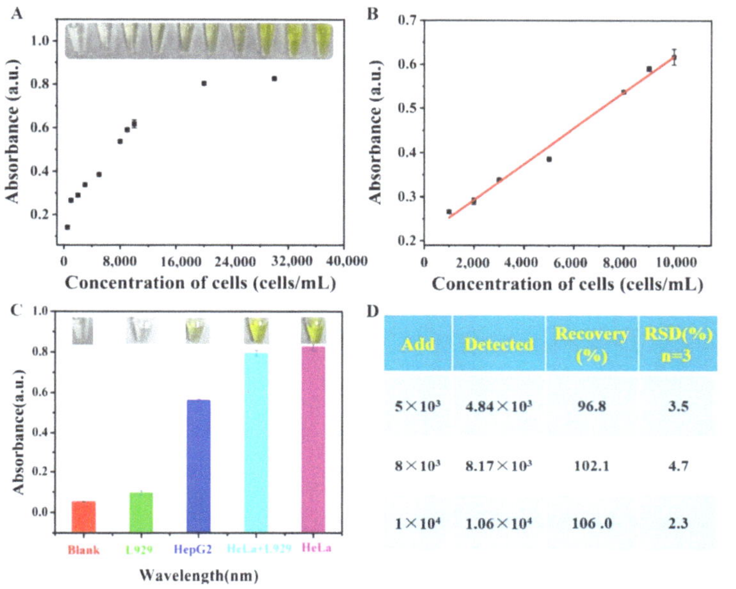

Figure 5. (**A**) The absorbance intensity response of TMB-H_2O_2 system at 450 nm toward various concentrations of HeLa cells (5×10^2, 1×10^3, 2×10^3, 3×10^3, 5×10^3, 8×10^3, 9×10^3, 1×10^4, 2×10^4, 3×10^4 cells/mL). The insets show the images of their corresponding solution. (**B**) Linear relationship of the absorbance intensity versus HeLa cell concentration from 10^3 to 10^4 cells/mL. (**C**) Specificity of the colorimetric sandwich-type aptasensor toward different interference cells. The insets show the images of their corresponding solution. (**D**) The recovery assays.

4. Conclusions

In summary, a novel single aptamer-dependent sandwich-type biosensor was proposed for simple, sensitive, and specific detection of HeLa cells using Fe_3O_4@SiO_2-Apt to selectively capture and magnetically separate target cancer cells, and using UIO-66(Fe/Zr) NPs to output amplified colorimetric signals. Through designing appropriate metal nodes in the given MOFs, the prepared UIO-66(Fe/Zr) NPs show desirable integrated properties, that is, the intrinsic peroxidase-like activity originated from Fe nodes, and direct binding to the cell surface using the Zr-O-P coordination bond between phosphate units in the phospholipid bilayer of the cell membrane and zirconium nodes. These unique properties ensured high sensing sensitivity and broke through the limitation of the requirement of labeling two antibodies or aptamers. The constructed colorimetric aptasensor could achieve visual detection of HeLa cells in the range of 10^3-10^4 cells/mL with a detection limit of 481 cells/mL. Such a novel single aptamer-dependent colorimetric sandwich-type biosensor has great potential in the diagnosis and treatment evaluation of cancer.

Supplementary Materials: The following supporting information can be downloaded at: https://www.mdpi.com/article/10.3390/bios13020225/s1, Figure S1: TEM images of Fe_3O_4 and Fe_3O_4@SiO_2 NPs; Figure S2. (A) FT-IR spectra of Fe_3O_4, Fe_3O_4@SiO_2, Fe_3O_4@SiO_2-NH_2. (B) Zeta potential of Fe_3O_4@SiO_2, Fe_3O_4@SiO_2-NH_2, Fe_3O_4@SiO_2-Apt; Figure S3. UV-vis absorbance spectra of the TMB-H_2O_2 system after different treatments; Figure S4. Fluorescence emission spectra of UIO-66(Fe/Zr) before and after incubation with Cy3 dyes.

Author Contributions: Y.Z. and X.F.: methodology, investigation, writing—original draft. X.L. and M.L.: investigation and validation. H.X. and S.Z.: methodology and writing—review and editing.

S.H. and F.Y.: supervision, writing—review and editing, and funding acquisition. All authors have read and agreed to the published version of the manuscript.

Funding: Financial support from the National Natural Science Foundation of China (No. 22064004), Natural Science Foundations of Guangxi Province (No. 2019GXNSFBA245035), and Guangxi Provincial Science and Technology Bases and Special Fund for Talented Persons (No. GUIKEAD20159072) is gratefully acknowledged.

Institutional Review Board Statement: The study was conducted in accordance with the Declaration of Helsinki, and approved by the Institutional Ethical Committee (IEC) of Guangxi Normal University (protocol code 202207-003).

Informed Consent Statement: Informed consent was obtained from all subjects involved in the study.

Data Availability Statement: The data presented in this study are available within the article and its Supplementary Materials. Other data that support the findings of this study are available upon request from the corresponding author.

Conflicts of Interest: The authors declare no conflict of interest.

References

1. Mao, J.J.; Pillai, G.G.; Andrade, C.J.; Ligibel, J.A.; Basu, P.; Cohen, L.; Khan, I.A.; Mustian, K.M.; Puthiyedath, R.; Dhiman, K.S.; et al. Integrative oncology: Addressing the global challenges of cancer prevention and treatment. *CA Cancer J. Clin.* **2022**, *72*, 144–164. [CrossRef] [PubMed]
2. Crosby, D.; Bhatia, S.; Brindle, K.M.; Coussens, L.M.; Dive, C.; Emberton, M.; Esener, S.; Fitzgerald, R.C.; Gambhir, S.S.; Kuhn, P.; et al. Early detection of cancer. *Science* **2022**, *375*, eaay9040. [CrossRef]
3. Kaur, B.; Kumar, S.; Kaushik, B.K. Recent advancements in optical biosensors for cancer detection. *Biosens. Bioelectron.* **2022**, *197*, 113805. [CrossRef]
4. Yang, L.; Cui, M.; Zhang, Y.; Jiang, L.; Liu, H.; Liu, Z. A colorimetric aptasensing assay with adjustable color mutation points for threshold-readout detection of carcinoembryonic antigen. *Sens. Actuators B* **2022**, *350*, 130857. [CrossRef]
5. Li, G.; Chen, M.; Wang, B.; Wang, C.; Wu, G.; Liang, J.; Zhou, Z. Dual-signal sandwich-type aptasensor based on H-rGO-Mn_3O_4 nanozymes for ultrasensitive Golgi protein 73 determination. *Anal. Chim. Acta* **2022**, *1221*, 340102. [CrossRef]
6. Hu, S.; Tong, L.; Wang, J.; Yi, X.; Liu, J. NIR Light-Responsive Hollow Porous Gold Nanospheres for Controllable Pressure-Based Sensing and Photothermal Therapy of Cancer Cells. *Anal. Chem.* **2019**, *91*, 15418–15424. [CrossRef] [PubMed]
7. Wei, Z.; Yu, Y.; Hu, S.; Yi, X.; Wang, J. Bifunctional Diblock DNA-Mediated Synthesis of Nanoflower-Shaped Photothermal Nanozymes for a Highly Sensitive Colorimetric Assay of Cancer Cells. *ACS Appl. Mater. Interfaces* **2021**, *13*, 16801–16811. [CrossRef] [PubMed]
8. Fang, X.; Zhang, X.; Zhang, Z.; Hu, S.; Ye, F.; Zhao, S. Complementary atomic flame/molecular colorimetry dual-mode assay for sensitive and wide-range detection of cancer cells. *Chem. Commun.* **2021**, *57*, 3327–3330. [CrossRef] [PubMed]
9. Hou, L.; Qin, Y.; Lin, T.; Sun, Y.; Ye, F.; Zhao, S. Michael reaction-assisted fluorescent sensor for selective and one step determination of catechol via bifunctional Fe-MIL-88NH_2 nanozyme. *Sens. Actuators B* **2020**, *321*, 128547. [CrossRef]
10. Cheng, Y.; Liang, L.; Ye, F.; Zhao, S. Ce-MOF with Intrinsic Haloperoxidase-Like Activity for Ratiometric Colorimetric Detection of Hydrogen Peroxide. *Biosensors* **2021**, *11*, 204. [CrossRef]
11. Sun, Y.; Lin, T.; Zeng, C.; Jiang, G.; Zhang, X.; Ye, F.; Zhao, S. A self-correcting fluorescent assay of tyrosinase based on Fe-MIL-88B-NH_2 nanozyme. *Microchim. Acta* **2021**, *188*, 158. [CrossRef]
12. Ji, Z.; Zhang, H.; Liu, H.; Yaghi, O.M.; Yang, P. Cytoprotective metal-organic frameworks for anaerobic bacteria. *Proc. Natl. Acad. Sci. USA* **2018**, *115*, 10582–10587. [CrossRef] [PubMed]
13. Li, X.; Liu, P.; Niu, X.; Ye, K.; Ni, L.; Du, D.; Pan, J.; Lin, Y. Tri-functional Fe–Zr bi-metal–organic frameworks enable high-performance phosphate ion ratiometric fluorescent detection. *Nanoscale* **2020**, *12*, 19383–19389. [CrossRef]
14. Miao, J.; Zhao, X.; Zhang, Y.-X.; Liu, Z.-H. Feasible synthesis of hierarchical porous MgAl-borate LDHs functionalized $Fe_3O_4@SiO_2$ magnetic microspheres with excellent adsorption performance toward congo red and Cr(VI) pollutants. *J. Alloys Compd.* **2021**, *861*, 157974. [CrossRef]
15. Cavka, J.H.; Jakobsen, S.; Olsbye, U.; Guillou, N.; Lamberti, C.; Bordiga, S.; Lillerud, K.P. A New Zirconium Inorganic Building Brick Forming Metal Organic Frameworks with Exceptional Stability. *J. Am. Chem. Soc.* **2008**, *130*, 13850–13851. [CrossRef] [PubMed]
16. He, X.; Deng, F.; Shen, T.; Yang, L.; Chen, D.; Luo, J.; Luo, X.; Min, X.; Wang, F. Exceptional adsorption of arsenic by zirconium metal-organic frameworks: Engineering exploration and mechanism insight. *J. Colloid Interface Sci.* **2019**, *539*, 223–234. [CrossRef]
17. Peterson, G.W.; Mahle, J.J.; DeCoste, J.B.; Gordon, W.O.; Rossin, J.A. Extraordinary NO_2 Removal by the Metal–Organic Framework UiO-66-NH_2. *Angew. Chem.* **2016**, *128*, 6343–6346. [CrossRef]

18. Morel, A.-L.; Nikitenko, S.I.; Gionnet, K.; Wattiaux, A.; Lai-Kee-Him, J.; Labrugere, C.; Chevalier, B.; Deleris, G.; Petibois, C.; Brisson, A.; et al. Sonochemical Approach to the Synthesis of $Fe_3O_4@SiO_2$ Core−Shell Nanoparticles with Tunable Properties. *ACS Nano* **2008**, *2*, 847–856. [CrossRef]
19. Jermoumi, T.; Hafid, M.; Toreis, N. Density, thermal and FTIR analysis of (50-x)BaO.xFe_2O_2.50P_2O_5 glasses. *Phys. Chem. Glas.* **2002**, *43*, 129–132.
20. Li, Q.; Zhao, D.; Shao, X.; Lin, S.; Xie, X.; Liu, M.; Ma, W.; Shi, S.; Lin, Y. Aptamer-Modified Tetrahedral DNA Nanostructure for Tumor-Targeted Drug Delivery. *ACS Appl. Mater. Interfaces* **2017**, *9*, 36695–36701. [CrossRef] [PubMed]
21. Jing, Y.; Cai, M.; Zhou, L.; Jiang, J.; Gao, J.; Wang, H. Aptamer AS1411 utilized for super-resolution imaging of nucleolin. *Talanta* **2020**, *217*, 121037. [CrossRef] [PubMed]

biosensors

MDPI

Article

Visual/Photoelectrochemical Off-On Sensor Based on Cu/Mn Double-Doped CeO_2 and Branched Sheet Embedded Cu_2O/CuO Nanocubes

Huihui Shi [1], Yanfei Che [1], Yumeng Rong [1], Jiajun Wang [1], Yanhu Wang [2], Jinghua Yu [1] and Yan Zhang [1,3,*]

1 School of Chemistry and Chemical Engineering, University of Jinan, Jinan 250022, China
2 Shandong Analysis and Test Center, Qilu University of Technology (Shandong Academy of Sciences), Jinan 250014, China
3 Key Laboratory of Optic-Electric Sensing and Analytical Chemistry for Life Science, MOE, Qingdao University of Science and Technology, Qingdao 266042, China
* Correspondence: chm_zhangyan@hotmail.com

Abstract: An integrated dual-signal bioassay was devised to fulfil thrombin (TB) ultrasensitive detection by integrating visualization with the photoelectrochemical technique based on G-quadruplex/hemin. During the process, branched sheet embedded copper-based oxides prepared with illumination and alkaline condition play a vital role in obtaining the desirable photocurrent. The switchover of photoelectrochemical signal was realized by the adjustable distance between electron acceptor G-quadruplex/hemin and interface materials due to dissociation of the Cu/Mn double-doped cerium dioxide ($CuMn@CeO_2$)/DNA caused by the addition of TB. Then, $CuMn@CeO_2$ transferred onto visual zones triggered catalytic reactions under the existence of 3,3′,5,5′-tetramethylbenzidine and hydrogen peroxide, making a variation in color recognized by the naked eye and providing visual prediction. Under optimized conditions, this bioassay protocol demonstrated wide linear ranges (0.0001–50 nM), high selectivity, stability, and reproducibility. More importantly, the proposed visual/photoelectrochemical transduction mechanism platform exhibits a lower background signal and more reliable detection results, which also offers an effective way for detecting other proteins and nucleic acids.

Keywords: photoelectrochemical; colorimetric; thrombin; paper-based device; Cu_2O/CuO; $CuMn@CeO_2$

Citation: Shi, H.; Che, Y.; Rong, Y.; Wang, J.; Wang, Y.; Yu, J.; Zhang, Y. Visual/Photoelectrochemical Off-On Sensor Based on Cu/Mn Double-Doped CeO_2 and Branched Sheet Embedded Cu_2O/CuO Nanocubes. *Biosensors* **2023**, *13*, 227. https://doi.org/10.3390/bios13020227

Received: 23 November 2022
Revised: 27 January 2023
Accepted: 2 February 2023
Published: 4 February 2023

1. Introduction

Thrombin (TB), as a member of the serine protease group, has been identified to be expressed in various cardiovascular diseases including cerebral ischemia and infarction. Considering the importance of TB assays, an efficient sensing strategy for fast and sensitive TB detection is highly desired [1,2]. Up to now, numerous analytical methods were reported, including chemiluminescence [3], electrochemiluminescence [4,5], fluorescence [6], electrochemistry [7,8], and photoelectrochemistry (PEC) [9,10]. Among these protocols, PEC, as an emerged analytical method with high sensitivity and a low background signal, has received substantial attention [11,12]. However, most of them relied on a signal output model, which still need to be improved to decrease the influence of systematic error. To achieve a more accurate and reliable analysis, an alternative method is to combine a naked-eye colorimetric strategy with the PEC transduction mechanism.

As an essential portion in colorimetric reaction and the PEC process, nanomaterials with catalytic and photoactive activity have a great impact on the obtained signal strength [13–16]. Cerium oxide (CeO_2), simultaneously existing Ce^{3+} and Ce^{4+} oxidation states on the lattice surface, could catalyze 3,3′,5,5′-tetramethylbenzidine (TMB) to convert colorless to blue with assistance of H_2O_2 due to oxygen vacancy [17,18]. However, the capacity is relatively low to be used for low-concentration analytes detection. It is a good method to dope CeO_2 with metal to expand their application as colorimetric reaction probes [19,20]. As a consequence, double-doped CeO_2 (CuMn@CeO_2) was prepared with superior catalytic performance for more accurate detection. Furthermore, to obtain a desirable PEC signal, a cuprous oxide–cupric oxide (Cu_2O/CuO) composite was introduced as a photoactive material owing to its abundance, affordable price, environmental acceptability, and low band-gap energy [21,22]. The fabrication process was carried out under normal temperature and pressure, with low-cost and high-quality products. The introduction of an external stimulus can promote the reduction process significantly, and the obtained Cu_2O in such a strong redox photochemical surrounding possessed the intrinsic superiority to endure photocorrosion in a kinetically more favorable way [23]. Furthermore, the well-designed overall architecture of the branched sheet embedded nanocubic Cu_2O/CuO complex not only provides more active sites and a larger surface area, but also helps Cu_2O to overcome its high carrier recombination caused by short electron diffusion length (20–100 nm) [24,25].

Apart from desirable initial PEC performance, an efficient electron transfer-regulated strategy is also of vital importance in broadening the application range of the PEC system [9,26]. G-quadruplex/hemin, which forms between single-stranded guanine-rich aptamer and hemin, has been utilized to act as an electron acceptor and mediate the catalytic reduction of dissolved oxygen [9,27]. Thus, this special structure could be employed as an effective way to realize off-on transformation in the PEC detecting process. In addition, to operate the whole process, an easy-to-operate platform was necessary. Cellulose paper, owning virtues of unique structure, low cost, large specific surface area, and portability, has aroused widespread interest [28–30]. Concretely, it is easy to be folded to satisfy diverse demand when applied as micro-reactors. Combined with the abovementioned factors, an integrated paper-based platform was fabricated to functionalize the working electrode and implement a multi-module microfluidic device.

Hence, a dual-signal output paper-based sensing strategy integrating the PEC technique with visualization in two spatially separated working areas, gold nanoparticles (AuNPs)-modified PEC working electrodes (PWE) and visual areas, was proposed for TB detection (Scheme 1). Concretely, TMB and H_2O_2 were applied on the visual areas for chromogenic reactions, while branched sheet-like nanocubic Cu_2O/CuO and hemin-DNA_1 were applied on the PWE surface with excellent PEC performance due to G-quadruplex/hemin. With addition of Cu/Mn-doped cerium dioxide/DNA_2 (CuMn@CeO_2-DNA_2), the specific double-stranded structure based on hybridization of DNA_1 and DNA_2 and steric hindrance of CuMn@CeO_2 could dramatically decrease the photocurrent, thus the "signal-off" trend for PEC signal can be triggered. After that, with the application of TB, DNA_2 could hybridize with TB, causing free CuMn@CeO_2-DNA_2. Then, the released CuMn@CeO_2 could catalyze H_2O_2 on visual areas, forming the hydroxyl radical that can make TMB effectively develop colors, which offers visual prediction for thrombin concentration. Meanwhile, due to a DNA_2-TB binding event, dissociation of CuMn@CeO_2 facilitated the electrons transfer and G-quadruplex/hemin could accept electrons from the illuminated Cu_2O/CuO, resulting in the increase of PEC intensity and switchover from off to on mode of the PEC signal. Such a distance-based PEC sensing system between electron acceptor and interface materials endows detection methods with lower background signal. Accordingly, the paper-based analytical device with a dual-signal output sensing mechanism was successfully fabricated, which could be introduced as quantitative platform for screening of other proteins and nucleic acids.

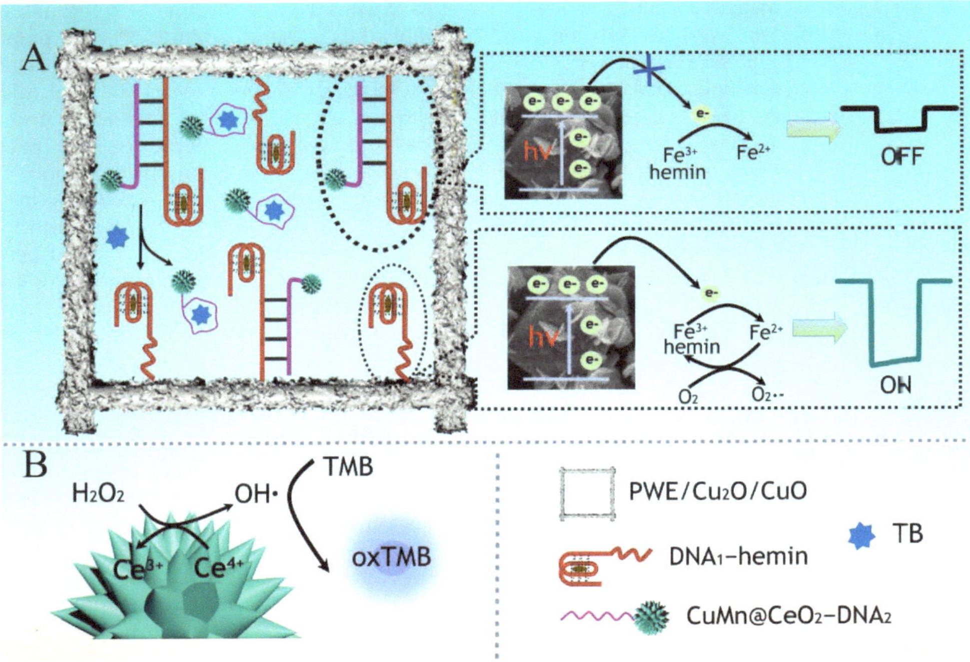

Scheme 1. Schematic illustration of (**A**) off−on switchover of PEC areas and (**B**) visual detection of TB.

2. Experimental Section

2.1. Design of Paper-Based Device

The prepared whole device was made with four units: detection tab, photoreaction auxiliary tab, fluid transfer tab, and electrode tab, as depicted in Figure 1A,B. Two working zones (8 mm in diameter) for the establishment of the sensing platform were patterned on the detection tab (I for PEC system, II for chromogenic reaction). In order to form a densely distributed nanosized oxide layer, more growth solution that exceeded the capacity of a single working area was required. Thus, the photoreaction auxiliary tab, where the white area was designed to be hollow while the green area is hydrophobic, was added to the device to hold more solutions in the light reaction process after folding (Figure 1C). When the device was folded as shown in Figure 1D, the branch channel (3.0 mm × 5.0 mm) on the fluid transfer tab is favorable to connect two working zones which are ready for fluid transfer. The carbon working electrode (WE) was printed on the back of the detection zone, while the carbon counter electrode (CE) and Ag/AgCl reference electrode (RE) are on the hydrophilic zone (8 mm in diameter) of the electrode tab. Finally, a three-dimensional device for visual and PEC detection was successfully constructed after the electrode tab folded under the detection tab, as displayed in Figure 1E.

2.2. Preparation of CuMn@CeO$_2$-DNA$_2$

Firstly, 0.1 g CuMn@CeO$_2$ (the synthesis process can be found in the Supporting Information) was redispersed in 10 mL ethanol and stirred for 30 min. Then, 0.1 mL 3-aminopropyltriethoxysilane was injected into the above solution and refluxed at 70 °C for 90 min. After the obtained solution was centrifugated and washed with ethanol and deionized water, 100 μL glutaraldehyde (1%, *w*/*w*) was added to activate the amino groups on the CuMn@CeO$_2$ surface. To obtain CuMn@CeO$_2$-modified DNA$_2$, 15 μL of DNA$_2$ was first added in the dispersed 2 mL CuMn@ CeO$_2$ solution with stirring for 12 h at 4 °C.

Next, the above solution was diluted with 25 mM Tris-HCl and 0.3 M NaCl, followed by centrifugation at 12,000 rpm for 10 min to remove unconjugated DNA. The resultant CuMn@CeO_2-DNA_2 was stored at 4 °C when not in use.

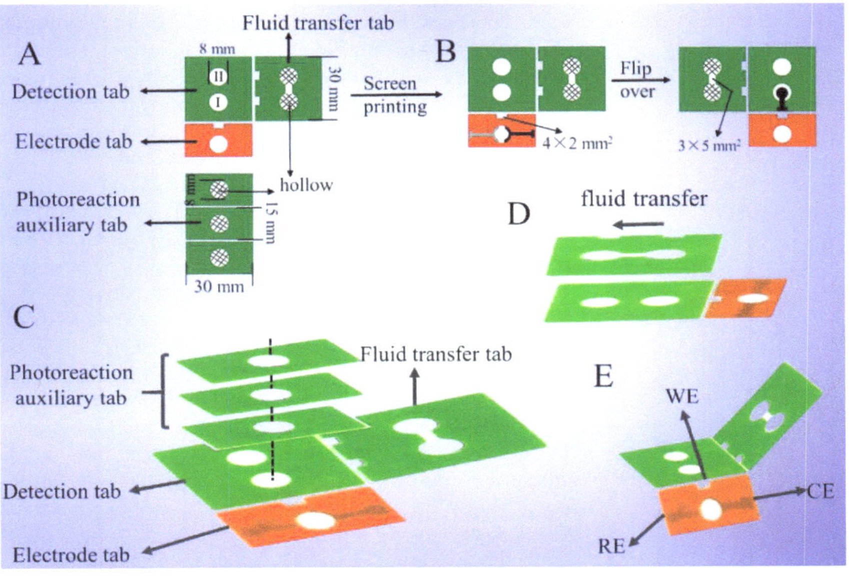

Figure 1. Schematic layout of (**A**) prepared device and (**B**) the fabricated device with screen−printed electrodes. Assembly illustration of prepared device during (**C**) photoreaction, (**D**) fluid transfer, and (**E**) sensing platform.

2.3. Fabrication of Dual-Signal Sensing Platform

PWE were prepared as shown in the Supporting Information. In order to achieve a desirable PEC signal, nanocubic Cu_2O was first coated on PWE by the light-induced photochemical synthesis approach [23]. Briefly, 1 mL as-prepared copper tungstate ($CuWO_4$) solution (0.15 M, the preparation process can be found in the Supporting Information) was coated onto PWE by a spin coater, then 50 μL solution containing 0.3 M NaOH and 0.1 M glucose was dropped onto the surface of PWE with a photoreaction auxiliary unit. Next, the assembled photochemical reaction device was irradiated by a 300W Xe lamp for 60 min and then rinsed with distilled water. After that, the electrode dealt with 50 μL 0.2 M NaOH for 30 min. Subsequently, 50 μL chitosan aqueous solution (0.08 wt%) in 1% acetic acid was applied onto the working electrode and reacted for 60 min. After drying, 5 wt% glutaraldehyde was added to activate the electrode, and the final electrode was denoted as PWE/Cu_2O/CuO. Finally, 20 μL acetic acid (pH 4.5), 20 mM TMB, and 20 μL of 0.5 M H_2O_2 were dropped onto the II visual area.

2.4. Analysis Process of TB

The dual-signal mechanism detection was performed by the following procedure. Firstly, 50 μL of 1 μM DNA_1 and 1 μM hemin was dropped onto the PWE and kept for 80 min, followed by adding 50 μL of 1 μM CuMn@CeO_2-DNA_2. After the PEC electrode was incubated with different concentrations of TB for 50 min, the device was folded as depicted in Figure 1D and the above mixed solution's fluid flow from PWE to the II visual zone could be generated along the hydrophilic channel on the fluid transfer tab. The electrode was carefully cleaned with 0.01 M phosphate buffer saline (PBS, pH 7.4) after each step. Finally, color variance could be signaled and the PEC signal was recorded by a three-electrode system (−0.1 V).

3. Results and Discussion

3.1. Structure Characterization

The morphology evolution of nanomaterials was investigated by scanning electron microscope (SEM) as illustrated in Figure 2. Clearly, it could be found that interconnected fibers of the bare sample zone displayed porous architecture with high surface area (Figure 2A), which provided a special micro-environment for AuNPs' growth. Pyknotic AuNPs were modified onto the cellulose by an in situ growth technique and they were distributed onto the fibers' surface uniformly (Figure 2B). Moreover, the conductivity of the paper-based electrode was measured by a four-probe tester (Agilent U1251B multimeter), with the results shown in Figure S1. Significantly, PWE displayed lower resistance (0.385 Ω) than that of new clean FTO (2.017 Ω) and ITO (1.337 Ω). Then, deposition of Cu_2O/CuO on the PWE was conducted by an in situ synthesis approach with two steps. To better survey the preparation process of the Cu_2O, samples were gathered at diverse stages of photochemical reaction time. At first, layered substances (Figure S2A) with a particle structure were clearly observed adsorbing on the surface of the PWE. After 15 min irradiation, a small amount of crystals similar to microcubes were sparsely distributed on the fibers (Figure S2B). To additionally probe the existing elements, energy-dispersive X-ray spectroscopy (EDS) and element mapping analysis were performed. Clearly, it could be seen that peaks of tungsten are observed in the EDS spectra (Figure S3) for working electrodes and Figure S2B_1–B_4 reveals that Cu, W, O, and C elements were uniformly distributed on the fibers. Hence, layered substances were presumed to be $CuWO_4$. In addition, with the increase of illumination time, the distribution of cubes became more and more intensive until it reached optimum at the photoreaction time of 60 min (Figures S2C,D and 2D). Only Cu, O, and C peaks were collected for PWE/Cu_2O (Figure 2C) and they were evenly distributed on the fibers (Figure 2E). The fact that no spectrum pertaining to the tungsten element was found in EDS patterns of electrodes with 60 min photoreaction time demonstrated that $CuWO_4$ was gradually converted to oxides of microcubes with the increasement of photoreaction time. The more microscopic morphology of cubes was observed by zoom-in SEM with ~400 nm diameter (inset in Figure 2D). Moreover, there was a sparse distribution of cubes on the PWE (Figure S2E) without the photoreaction auxiliary unit for photochemical processes, verifying that the existence of the photoreaction auxiliary unit was necessary. The separate working electrode only has a growth fluid capacity of 50 μL, whereas the electrode with the photoreaction auxiliary unit could hold up to 1.0 mL. Moreover, the SEM image of the working electrode without illumination showed unsatisfactory performance (Figure S2F), illustrating the necessity of illumination. In the subsequent reaction process under alkaline condition, Cu_2O in cubes was gradually partly transformed into intersecting sheet CuO (Figure 2F).

The phase structure and the crystallization details of the sample were further researched with X-ray diffraction (XRD) and Raman spectroscopy. A peak at 22.89° gathered from the original working zone of the paper-based device was assigned to (002) planes of cellulose [31], as depicted in Figure 3A. For PWE, four additional peaks at 38.18, 44.39, 64.58, and 77.55° were well-matched with characteristic peaks of Au NPs (JCPDS 04-0784). Except for naked papers and Au NPs peaks, other distinguishable peaks in curve c could be well-indexed to the pristine cubic phase of Cu_2O shown in Figure S4A. The PWE/Cu_2O/CuO electrode exhibited apparent diffractions of cubic Cu_2O and additional diffraction peaks at 35.49° of CuO, illustrating successful preparation of Cu_2O/CuO. Furthermore, Raman spectra of above electrodes were shown in Figure 3B. Similar to that of cellulose paper, no obvious characteristic peaks were recorded in the Raman pattern of PWE. After the first photoreaction, three peaks were collected for the Cu_2O-functionalized PWE. Peaks at 217 and 636 cm^{-1} resulted from the second-order Raman-allowed mode and the red-allowed mode of Cu_2O, meanwhile another presented at 413 cm^{-1} corresponding to the four-phonon mode. Both Cu_2O and CuO peaks were recorded simultaneously for PWE/Cu_2O/CuO. Apart from the peak of Cu_2O at 217 cm^{-1}, the other three peaks at 280, 326, and 617 cm^{-1}

were assigned to Ag, Bg, and Bg mode of CuO. All these peaks exhibited sharpness, indicating preferable crystallization of the synthesized materials.

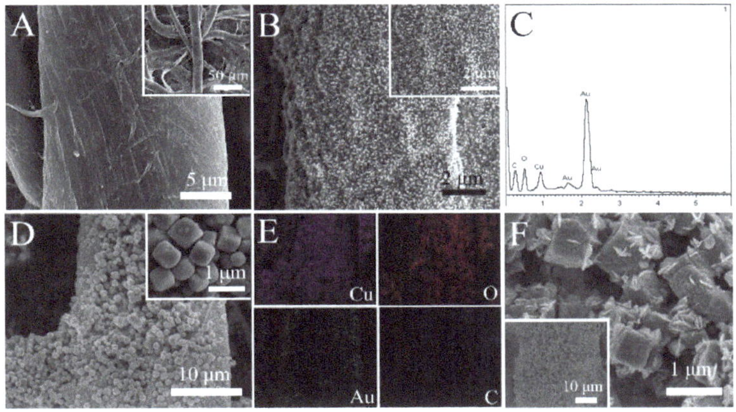

Figure 2. SEM images of (**A**) sample zone, (**B**) PWE, (**D**) PWE/Cu_2O, and (**F**) PWE/Cu_2O/CuO at different magnifications. (**C**) EDS and (**E**) elemental mappings of PWE/Cu_2O.

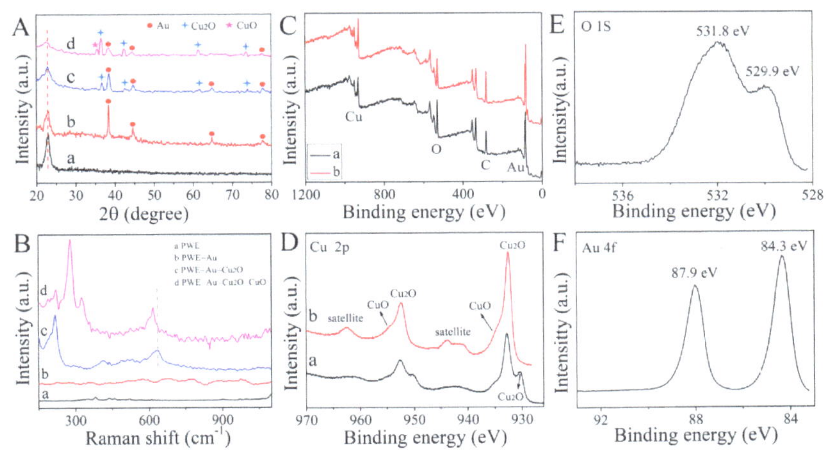

Figure 3. (**A**) XRD patterns and (**B**) Raman spectra (a) cellulose paper, (b) PWE, (c) PWE/Cu_2O, and (d) PWE/Cu_2O/CuO. (**C**) XPS survey scan of (a) PWE/Cu_2O and (b) PWE/Cu_2O/CuO. (**D**) High−resolution XPS spectra of Cu 2p of (a) PWE/Cu_2O and (b) PWE/Cu_2O/CuO. High−resolution XPS spectra of (**E**) O 1S and (**F**) Au 4f of PWE/Cu_2O/CuO.

X-ray photoelectron spectroscopy experiments were further operated to gain chemical composition information of the PWE/Cu_2O and PWE/Cu_2O/CuO. As shown in Figure 3C, the presence of Cu, O, and Au elements could be attributed to the product of the photochemical reaction and conductive Au NPs. To explore the chemical state of the Cu element, the Cu 2p core-level spectra of PWE/Cu_2O and PWE/Cu_2O/CuO were compared as displayed in Figure 3D. For Cu_2O-modified PWE, the characteristic peaks at the binding energies of 932.8 and 952.6 eV were associated with the Cu 2p3/2 and Cu 2p1/2 of Cu_2O, respectively [32]. As a contrast, PWE/Cu_2O/CuO showed two additional obvious peaks at 934.8 and 954.6 eV, which corresponded to Cu 2p3/2 and Cu 2p1/2 of CuO. Additionally, three weak satellite peaks at 940.8, 943.8, and 962.3 eV originated from multiple excitations in copper oxides and were attributed to the open $3d^9$ shell of Cu^{2+} in CuO [33]. As shown in

Figure 3E, the peak at 529.9 eV was related to the lattice oxygen from both Cu_2O and CuO phases and the other peak at 531.8 eV marked the surface hydroxyl group, which was consistent with the reported literature [34]. Moreover, Au 4f 7/2 and Au 4f 5/2 peaks presented at 84.3 and 87.9 eV were observed (Figure 3F), which were similar to the counterpart for the standard Au (0) [35]. Moreover, the solution's color change in the whole process was also favorable evidence for the successful preparation of Cu_2O/CuO (Figure S4B). All of these results matched each other, demonstrating the successful fabrication of PWE/Cu_2O/CuO.

3.2. Ce-Based Materials Characterization

The SEM technique was characterized to explore the morphology of the beacon in colorimetric reactions. Obviously, uniform-size nanospheres with a rough surface were found and the average size of CuMn@CeO_2 was around 60 nm (Figure 4A). To probe the elements, EDS and mapping were conducted, with the results displayed in Figure 4B,C, respectively. It could be observed that Ce, Cu, Mn, and O elements were collected and they were evenly distributed in the nanospheres. Furthermore, the main peak of CeO_2 in Raman spectra at about 463 cm^{-1} was attributed to the F_{2g} Raman mode (Figure 4D). Compared with that of pure CeO_2, the peak shifts to lower frequency at around 459 cm^{-1}, indicating occurrence of defects in the CeO_2 lattice structure after doping with Cu and Mn metals [36]. Meanwhile, additional peaks at 368 and 544 cm^{-1} were collected, which correspond to oxygen vacancies for CuMn@CeO_2, confirming that extra oxygen vacancies existed in CuMn@CeO_2 with excellent catalytic performance.

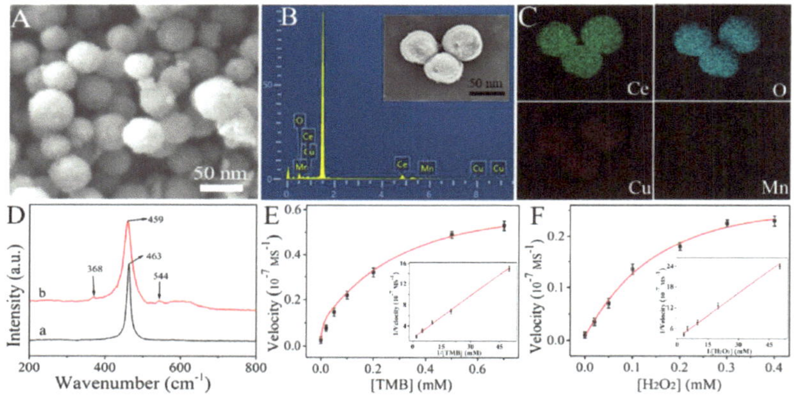

Figure 4. (**A**) SEM image, (**B**) EDS, and (**C**) mapping of CuMn@CeO_2. (**D**) Raman spectra of (a) CeO_2 and (b) CuMn@CeO_2. Michaelis–Menten curves at (**E**) fixed concentration of H_2O_2 (2 mM) and various concentrations of TMB (0–0.7 mM) and (**F**) fixed concentration of TMB (0.6 mM) and various concentrations of H_2O_2 (0−0.4 mM) (inset: corresponding Lineweaver−Burk plots).

Additionally, steady-state kinetic constants measurement experiments were carried out to verify catalytic performances of CuMn@CeO_2 (Figure 4E,F). During the process, TMB and H_2O_2 concentration were variables. Obtained data were fitted by the Michaelis–Menten kinetic equation, $v = (v_{max} \times [S])/(Km + [S])$, where a smaller Km value means a higher affinity of enzyme to substrate. Compared with that of horseradish peroxidase (Km(H_2O_2) = 3.7 mM, Km(TMB) = 0.434 mM), the values for CuMn@CeO_2 towards H_2O_2 and TMB were 0.15 mM and 0.12 mM, indicating that CuMn@CeO_2 had a stronger affinity for H_2O_2 and TMB than enzyme and provided accessible active sites for analytes.

3.3. EIS and PEC Behavior

Electrochemical impedance spectroscopy (EIS) was an effective technique to verify the manufacturing procedure of the sensing platform [37]. As magnified in Figure 5A,

PWE showed a small electron-transfer resistance (R_{et}) due to great conductivity of AuNPs (curve a). With the nanomaterial modification progress, the *Ret* increased dramatically for PWE/Cu_2O and PWE/Cu_2O/CuO (curves b and c) due to low conductivity of Cu_2O and CuO. As expected, the R_{et} showed an upward trend when hemin-DNA_1 and CuMn@CeO_2-DNA_2 were attached onto the PWE surface in turn (curves d and e). It could be explained as the steric hindrance of hemin-DNA_1 and CuMn@CeO_2-DNA_2 that prevented the charge-transfer rate of redox. Furthermore, the transient photocurrent responses of modified PWE were monitored. As revealed in Figure 5B, PWE/Cu_2O/CuO (curve b) displayed an obvious enhancement of PEC recovery compared with PWE/Cu_2O (curve a) due to the heterojunction formation of Cu_2O/CuO. Significantly, PEC performance of a paper-based electrode with greater immobilization capacity for photoelectric materials caused by special three-dimensional interlaced fibers was superior to FTO/ITO-based electrodes (Figure S5). As the electrode incubated with hemin-DNA_1, the value of photocurrent increased dramatically due to the presence of G-quadruplex/hemin (curve c). With the addition of CuMn@CeO_2-DNA_2, a reduced photocurrent was acquired (curve d) caused by enhanced steric hindrance and reduced electron transfer efficiency, indicating that the biosensor was successfully prepared as expected.

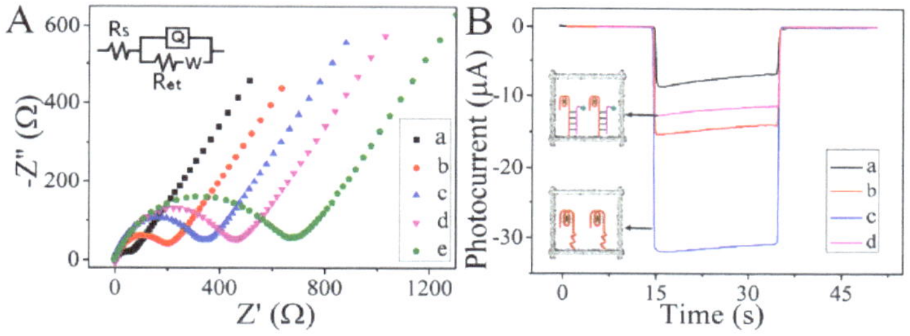

Figure 5. (**A**) EIS spectra of (a) PWE, (b) PWE/Cu_2O, (c) PWE/Cu_2O/CuO, (d) PWE/Cu_2O/CuO/hemin−DNA_1, and (e) PWE/Cu_2O/CuO/hemin−DNA_1/CuMn@CeO_2−DNA_2 in 5 mM $[Fe(CN)_6]^{3-/4-}$ solution containing 0.1 M KCl. (**B**) Photocurrent responses of (a) PWE/Cu_2O, (b) PWE/Cu_2O/CuO, (c) PWE/Cu_2O/CuO/hemin−DNA_1, and (d) PWE/Cu_2O/CuO/hemin−DNA_1/CuMn@CeO_2−DNA_2 in 0.01 M PBS solution.

3.4. Analytical Performance

The well-designed signal output paper-based device was employed to realize the quantitative detection by analyzing the PEC signals and color intensity with the variety of TB concentrations. Excellent catalytic performances toward H_2O_2 of CuMn@CeO_2 can offer visual prediction for analytes. Under optimal conditions (Figure S6), color intensity and photocurrent witnessed an upward trend along with the elevated concentration of TB, and a dynamic range was obtained from 0.0001 to 50 nM (Figure 6A,B). Meanwhile, the excellent linear relationship between logarithmic value of TB concentrations (Figure 6C) and photocurrent response could be fitted and the linear regression was $-\Delta I_{PEC}$ (μA) = 2.30 lgc + 10.99 (R = 0.994), and the detection limit was calculated as 0.035 pM (S/N = 3). Compared with single-signal readout strategies (Table 1), the paper-based visual/PEC biosensor exhibited apparent merits in wide response range and dual-signal sensing mode, providing a more accurate and effective method for TB detection.

3.5. Specificity, Stability, and Reproducibility

To further assess the feasibility of the present protocol, three significant criterions for biosensors, the selectivity, stability, and reproducibility, were explored. The specificity of the dual-signal bioassay was investigated by incubation with 0.1 nM TB, human immunoglobulin G (HIgG), carcinoembryonic antigen (CEA), bovine serum albumin (BSA), and their mixture. The highest photocurrent response was obtained for samples containing

TB (Figure 7A), demonstrating gratifying specificity. The stability of the as-prepared sensing device was measured by applying 0.1 nM TB as samples and monitoring the photocurrent responses intermittently (every 2 days). As expected, experimental results revealed acceptable stability (Figure 7B). Furthermore, the reproducibility was investigated by testing five independent electrodes, and the relative standard deviation of photocurrent response was 4.3%, demonstrating acceptable precision and repeatability.

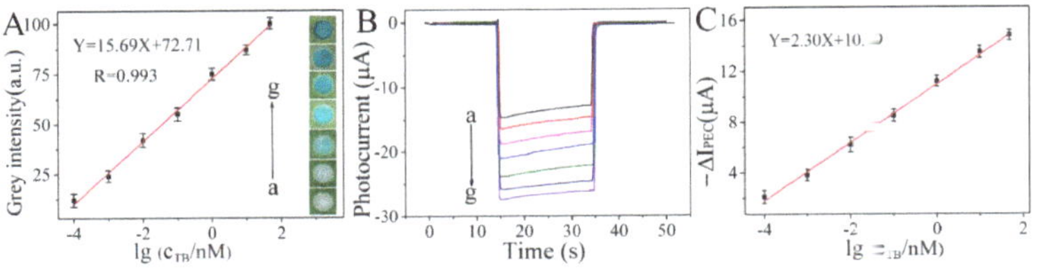

Figure 6. (**A**) Calibration curve between color intensity and logarithm of TB concentration (inset: color intensity). (**B**) Photocurrent of sensing device at different TB concentrations (0.0001, 0.001, 0.01, 0.1, 1, 10, 50 nM) and (**C**) corresponding calibration curve between $-\Delta I_{PEC}$ and logarithm of TB concentration.

Table 1. Comparison of the biosensor with other analytical strategies.

Method	Materials	Liner Range (nM)	Detection Limit (pM)	References
Electrochemiluminescence	Graphene oxide and carbon nanotubes	0.001–5	0.23	[38]
Electrochemical	Au NPs	0.005–50	1.1	[39]
Electrochemical	Au electrode and methylene blue	0.005–1	1.7	[40]
Fluorescence	Zinc selenide quantum dots	0.1–20	25	[41]
PEC	Au-ZnO Nanoflowers	0.001–30	0.37	[7]
Visual/PEC	Cu_2O/CuO and $CuMn@CeO_2$	0.0001–50	0.035	This work

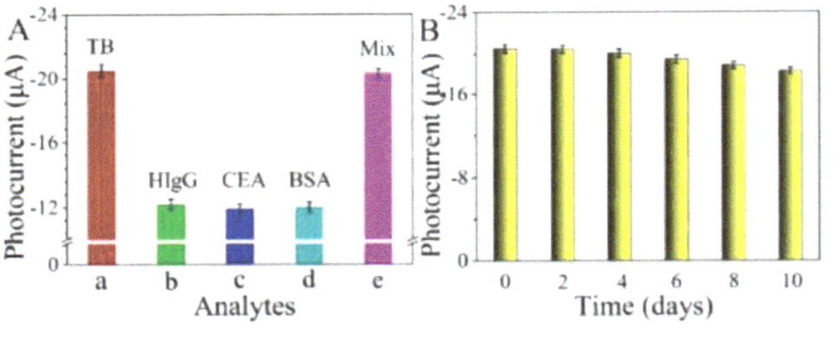

Figure 7. (**A**) Photocurrent of the bioassay in the presence of 0.1 nM sample. (**B**) Photocurrent responses of constructed PEC sensing platform with 0.1 nM TB.

4. Conclusions

Herein, Cu/Mn-doped CeO_2 and branched sheet embedded Cu-based nanocubes were prepared for fabrication of an effective visual/PEC paper-based sensing platform to obtain sensitive analysis of TB. The rapid chromogenic reactions were achieved by the release of $CuMn@CeO_2$ with the addition of TB, which provides a simple visual prediction. Meanwhile, the presence of TB led to dissociation of $CuMn@CeO_2$ and the electrons-acceptable distance of G-quadruplex/hemin, which then contributed to the switchover of the original PEC signal caused by branched sheet embedded nanocubic Cu_2O/CuO. Undoubtedly, such dual-signal output strategy with good performance was capable of making detection results more sensitive and accurate. Additionally, it was anticipated that the approach shows potentiality in designing numerous biosensors with excellent performance for analytes detection.

Supplementary Materials: The following supporting information can be downloaded at: https://www.mdpi.com/article/10.3390/bios13020227/s1, Figure S1: Physical photos of (A) FTO, (B) ITO, and (C) PWE sheet resistance; Figure S2: SEM images of PWE with photoreaction progresses at (A) 0 min, (C) 30 min, (D) 45 min. (B–B4) Elemental mappings of PWE with photoreaction progresses at 15 min. SEM images of (E) PWE/Cu_2O without photoreaction auxiliary unit and (F) the electrode without illumination; Figure S3: EDS spectra of PWE with photoreaction progresses at 15 min; Figure S4: (A) XRD pattern of Cu_2O and CuO. (B) Photographs of photoreaction progresses at (a) 0 min, (b) 15 min, (c) 30 min, (d) 45 min and (e) 60 min; and (f) the obtained Cu_2O/CuO photograph; Figure S5: Photocurrent response of (a) FTO/Cu_2O/CuO, (b) ITO/Cu_2O/CuO, and (c) PWE/Cu_2O/CuO; Figure S6: Effect of TB incubation time. References [42,43] are cited in the Supplementary Materials.

Author Contributions: H.S.: Conceptualization, methodology, investigation, and writing—original draft. Y.C.: Investigation and data curation. Y.R.: Methodology and investigation. J.W.: Data curation. Y.W.: Investigation, formal analysis. Y.Z.: Investigation, writing—reviewing and editing, resources. J.Y.: Funding acquisition, project administration. All authors have read and agreed to the published version of the manuscript.

Funding: This work was financially supported by the National Natural Science Foundation of China (21874055, 21904047), Taishan Scholars Program (tsqn202103082), the Excellent Youth Innovation Team in Universities of Shandong (2021KJ021), Science and Technology Program of University of Jinan (XKY2203), and the Open Fund of Key Laboratory of Optic-electric Sensing and Analytical Chemistry for Life Science, MOE, Qingdao University of Science and Technology (M2023-5).

Institutional Review Board Statement: Not applicable.

Informed Consent Statement: Not applicable.

Data Availability Statement: Not applicable.

Conflicts of Interest: The authors declare no conflict of interest.

References

1. Li, M.; Zheng, Y.; Liang, W.; Yuan, Y.; Chai, Y.; Yuan, R. An ultrasensitive "on–off–on" photoelectrochemical aptasensor based on signal amplification of a fullerene/CdTe quantum dots sensitized structure and efficient quenching by manganese porphyrin. *Chem. Commun.* **2016**, *52*, 8138–8141. [CrossRef] [PubMed]
2. Shao, C.; Chi, J.; Chen, Z.; Cai, L.; Zhao, Y. Superwettable colloidal crystal micropatterns on butterfly wing surface for ultrasensitive detection. *J. Colloid Interf. Sci.* **2019**, *546*, 122–129. [CrossRef]
3. Sun, Y.; Zhu, X.; Liu, H.; Dai, Y.; Han, R.; Gao, D.; Luo, C.; Wang, X.; Wei, Q. Novel chemiluminescence sensor for thrombin detection based on dual-aptamer biorecognition and mesoporous silica encapsulated with iron porphyrin. *ACS Appl. Mater. Interfaces* **2020**, *12*, 5569–5577. [CrossRef]
4. Fang, Y.; Wang, H.M.; Gu, Y.X.; Yu, L.; Wang, A.J.; Yuan, P.X.; Feng, J.J. Highly enhanced electrochemiluminescence luminophore generated by zeolitic imidazole framework-8-linked porphyrin and its application for thrombin detection. *Anal. Chem.* **2020**, *92*, 3206–3212. [CrossRef] [PubMed]
5. Shao, K.; Wang, B.; Ye, S.; Zuo, Y.; Wu, L.; Li, Q.; Lu, Z.; Tan, X.; Han, H. Signal-amplified near-infrared ratiometric electrochemiluminescence aptasensor based on multiple quenching and enhancement effect of graphene/gold nanorods/G-quadruplex. *Anal. Chem.* **2016**, *88*, 8179–8187. [CrossRef] [PubMed]
6. Zhao, X.; Li, S.; Xu, L.; Ma, W.; Wu, X.; Kuang, H.; Wang, L.; Xu, C. Up-conversion fluorescence "off-on" switch based on heterogeneous core-satellite assembly for thrombin detection. *Biosens. Bioelectron.* **2015**, *70*, 372–375. [CrossRef]
7. Yang, Z.H.; Zhuo, Y.; Yuan, R.; Chai, Y.Q. Amplified thrombin aptasensor based on alkaline phosphatase and hemin/G-quadruplex-catalyzed oxidation of 1-naphthol. *ACS Appl. Mater. Interfaces* **2015**, *7*, 10308–10315. [CrossRef] [PubMed]
8. Heydari-Bafrooei, E.; Amini, M.; Ardakani, M.H. An electrochemical aptasensor based on TiO_2/MWCNT and a novel synthesized Schiff base nanocomposite for the ultrasensitive detection of thrombin. *Biosens. Bioelectron.* **2016**, *85*, 828–836. [CrossRef]
9. Wang, G.L.; Shu, J.X.; Dong, Y.M.; Wu, X.M.; Zhao, W.W.; Xu, J.J.; Chen, H.Y. Using G-quadruplex/hemin to "switch-on" the cathodic photocurrent of p-type PbS quantum dots: Toward a versatile platform for photoelectrochemical aptasensing. *Anal. Chem.* **2015**, *87*, 2892–2900. [CrossRef] [PubMed]
10. Xu, F.; Zhu, Y.C.; Ma, Z.Y.; Zhao, W.W.; Xu, J.J.; Chen, H.Y. An ultrasensitive energy-transfer based photoelectrochemical protein biosensor. *Chem. Commun.* **2016**, *52*, 3034–3037. [CrossRef]
11. Wang, Z.J.; Li, Q.; Tan, L.L.; Liu, C.G.; Shang, L. Metal-organic frameworks-mediated assembly of gold nanoclusters for sensing applications. *J. Anal. Test.* **2022**, *6*, 163–177. [CrossRef] [PubMed]

12. Wang, F.; Fan, Q.; Wang, Y.; Wang, Y.; Ge, S.; Yan, M.; Yu, J. A paper-supported photoelectrochemical sensing platform based on surface plasmon resonance enhancement for real-time H_2S determination. *J. Anal. Test.* **2019**, *3*, 89–93. [CrossRef]
13. Shi, X.M.; Fan, G.C.; Shen, Q.; Zhu, J.J. Photoelectrochemical DNA biosensor based on dual-signal amplification strategy integrating inorganic-organic nanocomposites sensitization with λ-exonuclease-assisted target recycling. *ACS Appl. Mater. Interfaces* **2016**, *8*, 35091–35098. [CrossRef] [PubMed]
14. Pravin, S.; Xiaoniu, P.; Jue, W.; Yanxiao, M.; Louis, M.; Nathan, H.; Arunava, G.; Shanlin, P. Rapid screening of photoanode materials using scanning photoelectrochemical microscopy technique and formation of Z-scheme solar water splitting system by coupling p- and n-type heterojunction photoelectrodes. *ACS Appl. Energy Mater.* **2018**, *1*, 2283–2294.
15. Wang, B.; Mei, L.P.; Ma, Y.; Xu, Y.T.; Ren, S.W.; Cao, J.T.; Liu, Y.M.; Zhao, W.W. Photoelectrochemical-chemical-chemical redox cycling for advanced signal amplification: Proof-of-concept toward ultrasensitive photoelectrochemical bioanalysis. *Anal. Chem.* **2018**, *90*, 12347–12351. [CrossRef] [PubMed]
16. Jessica, C.B.; Charles, R.M. Scalable methods for device patterning as an outstanding challenge in translating paper-based microfluidics from the academic benchtop to the point-of-care. *J. Anal. Test.* **2019**, *3*, 50–60.
17. Tang, L.; Li, J. Plasmon-based colorimetric nanosensors for ultrasensitive molecular diagnostics. *ACS Sens.* **2017**, *2*, 857–875. [CrossRef]
18. Kong, Q.; Cui, K.; Zhang, L.; Wang, Y.; Sun, J.; Ge, S.; Zhang, Y.; Yu, J. "On−Off−On" Photoelectrochemical/Visual Lab-on-Paper Sensing via Signal Amplification of CdS Quantum Dots@Leaf-Shape ZnO and Quenching of Au-Modified Prism-Anchored Octahedral CeO_2 Nanoparticles. *Anal. Chem.* **2018**, *90*, 11297–11304. [CrossRef]
19. Cheng, X.; Huang, L.; Yang, X.; Elzatahry, A.; Alghamdi, A.; Deng, Y. Rational design of a stable peroxidase mimic for colorimetric detection of H_2O_2 and glucose: A synergistic CeO_2/Zeolite Y nanocomposite. *J. Colloid Interf. Sci.* **2019**, *535*, 425–435. [CrossRef]
20. Liu, W.; Liu, X.; Feng, L.; Guo, J.; Xie, A.; Wang, S.; Zhang, J.; Yang, Y. The synthesis of CeO_2 nanospheres with different hollowness and size induced by copper doping. *Nanoscale* **2014**, *6*, 10693–10700. [CrossRef]
21. Yang, Z.; Kang, T.; Ji, Y.; Li, J.; Zhu, Y.; Liu, H.; Jiang, X.; Zhong, Z.; Su, F. Architectural Cu_2O@ CuO mesocrystals as superior catalyst for trichlorosilane synthesis. *J. Colloid Interf. Sci.* **2021**, *589*, 198–207. [CrossRef] [PubMed]
22. Wu, S.; Fu, G.; Lv, W.; Wei, J.; Chen, W.; Yi, H.; Gu, M.; Bai, X.; Zhu, L.; Tan, C.; et al. A single-step hydrothermal route to 3D hierarchical Cu_2O/CuO/rGO nanosheets as high-performance anode of lithium-ion batteries. *Small* **2018**, *14*, 1702667. [CrossRef] [PubMed]
23. Zhang, L.; Jing, D.; Guo, L.; Yao, X. In situ photochemical synthesis of Zn-doped Cu_2O hollow microcubes for high efficient photocatalytic H_2 production. *ACS Sustain. Chem. Eng.* **2014**, *2*, 1446–1452. [CrossRef]
24. Jiang, D.; Xue, J.; Wu, L.; Zhou, W.; Zhang, Y.; Li, X. Photocatalytic performance enhancement of CuO/Cu_2O heterostructures for photodegradation of organic dyes: Effects of CuO morphology. *Appl. Catal. B Environ.* **2017**, *211*, 199–204. [CrossRef]
25. Huang, Q.; Kang, F.; Liu, H.; Li, Q.; Xiao, X. Highly aligned Cu_2O/CuO/TiO_2 core/shell nanowire arrays as photocathodes for water photoelectrolysis. *J. Mater. Chem. A* **2013**, *1*, 2418–2425. [CrossRef]
26. Zhao, W.W.; Xu, J.J.; Chen, H.Y. Photoelectrochemical bioanalysis: The state of the art. *Chem. Soc. Rev.* **2015**, *44*, 729–741. [CrossRef]
27. Shu, J.; Tang, D.P. Recent advances in photoelectrochemical sensing: From engineered photoactive materials to sensing devices and detection modes. *Anal. Chem.* **2020**, *92*, 363–377. [CrossRef]
28. Zhang, Y.; Zhang, L.; Cui, K.; Ge, S.; Cheng, X.; Yan, M.; Yu, J.; Liu, H. Flexible electronics based on micro/nanostructured paper. *Adv. Mater.* **2018**, *30*, 1801588. [CrossRef] [PubMed]
29. Hou, Y.; Lv, C.C.; Guo, Y.L.; Ma, X.H.; Liu, W.; Jin, Y.; Yao, S.Y. Recent advances and applications in paper-based devices for point-of-care testing. *J. Anal. Test.* **2022**, *6*, 247–273. [CrossRef]
30. Zhu, L.; Lv, X.; Yu, H.; Tan, X.; Rong, Y.; Feng, W.; Zhang, L.; Yu, J.; Zhang, Y. Paper-Based Bipolar Electrode Electrochemiluminescence Platform Combined with Pencil-Drawing Trace for the Detection of M.SssI Methyltransferase. *Anal. Chem.* **2022**, *94*, 8327–8334. [CrossRef]
31. Zhang, Y.; Li, Y.L.; Cui, S.H.; Wen, C.Y.; Li, P.; Yu, J.F.; Zeng, J.B. Distance-based detection of Ag^+ with gold nanoparticles coated microfluidic paper. *J. Anal. Test.* **2021**, *5*, 11–18. [CrossRef]
32. Shyamal, S.; Maity, A.; Satpati, A.K.; Bhattacharya, C. Amplification of PEC hydrogen production through synergistic modification of Cu_2O using cadmium as buffer layer and dopant. *Appl. Catal. B Environ.* **2019**, *246*, 111–119. [CrossRef]
33. Cheng, S.; DelaCruz, S.; Chen, C.; Tang, Z.; Shi, T.; Carraro, C.; Maboudian, R. Hierarchical Co_3O_4/CuO nanorod array supported on carbon cloth for highly sensitive non-enzymatic glucose biosensing. *Sens. Actuators B* **2019**, *298*, 126860. [CrossRef]
34. Li, H.; Su, Z.; Hu, S.; Yan, Y. Free-standing and flexible Cu/Cu_2O/CuO heterojunction net: A novel material as cost-effective and easily recycled visible-light photocatalyst. *Appl. Catal. B: Environ.* **2017**, *207*, 134–142. [CrossRef]
35. Yao, W.; Li, F.L.; Li, H.X.; Lang, J.P. Fabrication of hollow Cu_2O@CuO-supported Au-Pd alloy nanoparticles with high catalytic activity through the galvanic replacement reaction. *J. Mater. Chem. A* **2015**, *3*, 4578–4585. [CrossRef]
36. Xue, S.; Li, Q.; Wang, L.; You, W.; Zhang, J.; Che, R. Copper- and cobalt-codoped CeO_2 nanospheres with abundant oxygen vacancies as highly efficient electrocatalysts for dual-mode electrochemical sensing of MicroRNA. *Anal. Chem.* **2019**, *91*, 2659–2666. [CrossRef]

37. Zhang, Y.; Xu, J.; Zhou, S.; Zhu, L.; Lv, X.; Zhang, J.; Zhang, L.; Zhu, P.; Yu, J. DNAzyme-triggered visual and ratiometric electrochemiluminescence dual-readout assay for Pb(II) based on an assembled paper device. *Anal. Chem.* **2020**, *92*, 3874–3881. [CrossRef] [PubMed]
38. Wu, D.; Xin, X.; Pang, X.; Pietraszkiewicz, M.; Hozyst, R.; Sun, X.; Wei, Q. Application of europium multiwalled carbon nanotubes as novel luminophores in an electrochemiluminescent aptasensor for thrombin using multiple amplification strategies. *ACS Appl. Mater. Interfaces* **2015**, *7*, 12663–12670. [CrossRef]
39. Qiu, F.; Gan, X.; Jiang, B.; Yuan, R.; Xiang, Y. Electrode immobilization-free and sensitive electrochemical sensing of thrombin via magnetic nanoparticle-decorated DNA polymers. *Sens. Actuators B Chem.* **2021**, *331*, 129395. [CrossRef]
40. Yang, J.; Dou, B.; Yuan, R.; Xiang, Y. Aptamer/protein proximity binding-triggered molecular machine for amplified electrochemical sensing of thrombin. *Anal. Chem.* **2017**, *89*, 5138–5143. [CrossRef]
41. Tian, Y.; Xin, C.; Liu, S.; Liu, Y.; Liu, S. Affinity binding-induced Hg^{2+} release and quantum dot doping for general, label-free, and homogenous fluorescence protein assay. *ACS Sens.* **2018**, *3*, 1401–1408. [CrossRef] [PubMed]
42. Wang, H.; Jian, Y.; Kong, Q.; Liu, H.; Lan, F.; Liang, L.; Ge, S.; Yu, J. Ultrasensitive electrochemical paper-based biosensor for microRNA via strand displacement reaction and metal-organic frameworks. *Sens. Actuators B Chem.* **2018**, *257*, 561–569. [CrossRef]
43. Li, X.; Zhang, H.; Tang, Y.; Wu, P.; Xu, S.; Zhangg, X. A both-end blocked peroxidase-mimicking DNAzyme for low-background chemiluminescent sensing of miRNA. *ACS Sens.* **2017**, *2*, 810–816. [CrossRef]

biosensors

MDPI

Review

Application of Nanozymes in Environmental Monitoring, Management, and Protection

Miaomiao Wang [1,2], Ping Zhu [1,2], Shuge Liu [1,2], Yating Chen [1,2], Dongxin Liang [1], Yage Liu [1,2], Wei Chen [1,2,*], Liping Du [1,2,*] and Chunsheng Wu [1,2,*]

[1] Institute of Medical Engineering, Department of Biophysics, School of Basic Medical Sciences, Health Science Center, Xi'an Jiaotong University, Xi'an 710061, China

[2] Key Laboratory of Environment and Genes Related to Diseases, Xi'an Jiaotong University, Ministry of Education of China, Xi'an 710061, China

[*] Correspondence: weiwcchen@xjtu.edu.cn (W.C.); duliping@xjtu.edu.cn (L.D.); wuchunsheng@xjtu.edu.cn (C.W.)

Abstract: Nanozymes are nanomaterials with enzyme-like activity, possessing the unique properties of nanomaterials and natural enzyme-like catalytic functions. Nanozymes are catalytically active, stable, tunable, recyclable, and versatile. Therefore, increasing attention has been paid in the fields of environmental science and life sciences. In this review, we focused on the most recent applications of nanozymes for environmental monitoring, environmental management, and environmental protection. We firstly introduce the tuning catalytic activity of nanozymes according to some crucial factors such as size and shape, composition and doping, and surface coating. Then, the application of nanozymes in environmental fields are introduced in detail. Nanozymes can not only be used to detect inorganic ions, molecules, organics, and foodborne pathogenic bacteria but are also involved in the degradation of phenolic compounds, dyes, and antibiotics. The capability of nanozymes was also reported for assisting air purification, constructing biofuel cells, and application in marine antibacterial fouling removal. Finally, the current challenges and future trends of nanozymes toward environmental fields are proposed and discussed.

Keywords: nanozyme; sensing; monitoring; environmental pollutant; catalytic activity

Citation: Wang, M.; Zhu, P.; Liu, S.; Chen, Y.; Liang, D.; Liu, Y.; Chen, W.; Du, L.; Wu, C. Application of Nanozymes in Environmental Monitoring, Management, and Protection. *Biosensors* **2023**, *13*, 314. https://doi.org/10.3390/bios13030314

Received: 4 January 2023
Revised: 10 February 2023
Accepted: 16 February 2023
Published: 24 February 2023

1. Introduction

Most life activities in nature involve enzymes. Natural enzymes are macromolecular biocatalysts composed of most proteins and a few nucleic acids that run through the metabolism of life [1]. They have high catalytic efficiency, good substrate specificity, and biocompatibility [2]. Therefore, they are widely used in various fields, including disease diagnosis, clinical treatment, agricultural engineering, paper and leather, textile industry, and food processing. However, most natural enzymes are easy to inactivate or their activities are inhibited under nonphysiological conditions, which severely limits the wide application of enzymes. In addition, enzymes also have defects such as storage stability and recovery difficulties, complex production as well as purification processes, and high costs [3,4]. With the rapid development of nanoscience and life science, simulating the structure and catalytic activity of natural enzymes to construct substitute products has gradually become a new direction in which to expand the application of natural enzymes.

The term "artificial enzymes" was coined by Ronald Breslow for enzyme mimics. An "artificial enzyme" combines a metal catalytic group and a hydrophobic binding cavity [5]. In 2004, Scrimin and his colleagues created the term "nanozyme" and used gold nanoparticles functionalized by triazetidine as the catalyst for the transphosphorylation reaction [6]. In 2007, Yan's team found that magnetic nanoparticles (Fe_3O_4 MNPs) have a catalytic activity similar to horseradish peroxidase (HRP), indicating that some inorganic nanoparticles can also have peroxidase-like properties [7]. In 2013, Wei and Wang used the term "nanozyme" to describe some nanoscale materials with enzyme-like characteristics, namely, nanozyme is a kind of nanomaterial with similar natural enzyme catalytic activity and enzymatic reaction kinetics [8]. In the ten years since then, nanozyme experienced a period of rapid development and application in various fields. Based on advanced nanotechnology, a variety of nanozymes with catalytic activity comparable to that of natural enzymes have been explored. Compared with natural enzyme, nanozyme has the following obvious advantages: (I) High stability: inorganic nanomaterials are less fragile than natural enzymes, which enables the use of nanozymes under a wide range of pH (3–12) and temperature (4–90 °C) conditions. In contrast, natural enzymes are usually inactivated under extreme pH and temperature conditions. (II) Low cost: the production process of enzymes is usually complex and expensive, while inorganic nanomaterials are easy to produce, with high efficiency and low cost. (III) Recycling: Nanozymes are recyclable, and there is no substantial loss of catalytic activity in subsequent cycles. (IV) Easily multifunctional: Nanozymes have sufficient surface area to allow them to be coupled with multiple ligands to achieve multifunctionability [9]. (V) High catalytic activity: The level of activity is comparable to that of biological enzymes with the help of advanced nanotechnology. At the same time, a variety of factors influence the level of activity such as size, shape, composition, crystal surface, charge, and hydrophilicity. Although nanozymes have the above advantages, some nanozymes have the disadvantages of toxicity, low specificity, and poor dispersion. During the degradation and protection of environmental pollutants by nanozymes, nanozymes inevitably contact with water, animals and plants, soil, and air. Therefore, the safety of nanozymes is crucial. For example, some heavy metal nanozymes (Au, Cu, Ce, Fe, etc.) will be absorbed into soil and water, causing ecological pollution [10]. The continuous enrichment of heavy metals will eventually endanger human health through the food chain. In addition, graphene, quantum dots, copper–carbon dots and other nanomaterials have their own toxicity, and their dispersion will be strengthened during trial, which will make them more easily absorbed by aquatic plants and water bodies [11]. Toxicity can be reduced by reducing the size of nanozyme [12]. In addition, we can control the surface charge of nanozyme to regulate its permeability to cells in the human body [13]. The metal core is the source of toxicity, so it can be prevented from leakage by mixing other metal ions and chemical sealing [14,15].

The International Enzyme Commission (I.E.C.) specifies that natural enzymes can be classified into six categories by the enzymatic reaction as oxidoreductases, hydrolases, isomerases, lyases, ligases, transferases. Since nanozymes are a class of nanomaterials that mimic the catalytic activity and enzyme kinetic characteristics of natural enzymes, the categories also similar to that of natural enzymes could divide nanozymes into the following six categories: redox nanozyme, hydrolyzing nanozyme, lytic nanozyme, transfer nanozyme, isomeric nanozyme, and linked nanozyme. Currently, the reported nanozymes are mainly in the redox nanozyme family, whose members are oxidase (OXD) [16], peroxidase (POD) [17], catalase (CAT) [18], and superoxide dismutase (SOD) [19] (Figure 1). Oxidase catalyzes the oxidation of substrates using oxygen as the electron acceptor. Subsequently, O_2 is reduced to water or hydrogen peroxide. Peroxidase nanozyme can catalyze peroxides whose substrate is usually used as an electron donor. In biomedicine, it defends against pathogens and removes the toxicity of reactive oxygen [20]. Catalase often catalyzes H_2O_2 to produce oxygen and water. It is found that many metal materials and even metal oxides have catalase-like activity. Superoxide dismutase disproportionates superoxide radicals into oxygen and hydrogen peroxide and alleviates oxidative stress generated from cell metabolism. There is also a family of hydrolyzing nanozymes including phosphatases [21], nucleases, and proteases. They catalyze the separation of phosphate groups and the hydrolysis of phosphate diester bonds and peptide bonds. The family of lytic nanozyme comprises the carbonic anhydrase [19]. Nanozymes mostly catalyze the optical signal transmission of chromogenic substrates, so that other three types of nanozymes are rarely reported. It is hoped that the design of nanozymes can break through more types and functions limitations.

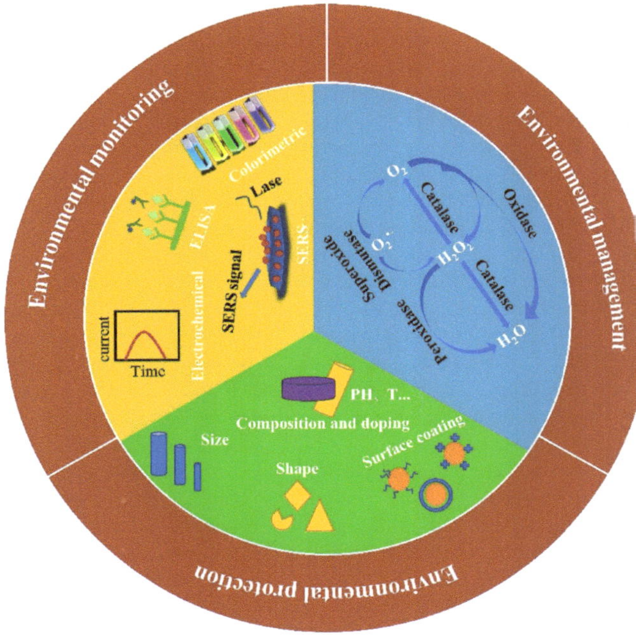

Figure 1. Nanozyme as a promising tool from environmental monitoring and environmental management to environmental protection.

The reported nanozymes can be classified into four categories according to the materials, such as metal-based, metal oxide-based, carbon-based, MOF-based, and other material-based. Currently reported carbon-based nanozymes include fullerenes, carbon nanotubes (CNTs), graphene, graphene oxide (GO), carbon dots (CDS), graphitic quantum dots (GQDS), and carbon nitrides [22]. Because of their special electronic and geometric properties, they can mimic the catalytic center of natural enzymes and possess catalytic activities such as oxidase, peroxygenase, superoxide dismutase, and catalase. On account of the intrinsic enzymatic activity of carbon nanomaterials, they can be combined with other materials or modified and functionalized to enhance enzymatic activity. Ye et al. [23] reported a highly specific N-doped nanozyme with HBF as a porous carbonaceous and nitrogen-containing precursor to prepare N-doped carbon nanozymes named HBF-1-c800 by high temperature pyrolysis with N-doping efficiency up to 5.48%, which is higher than the value of most reported N-doped carbon nanozymes. As a result, they found that the apparent POD activity of HBF-1-c800 shows a three to seven-fold enhancement over traditional carbon nanozymes and a five-fold enhancement over the reported N-doped graphene. Consequently, it has been widely used in environmental monitoring and environmental remediation. Although slightly inferior to metal-based nanozymes in terms of catalytic activity, its catalytic activity has been improved, and some excellent designed carbon-based nanozymes show comparable or even better results than that of natural enzymes.

Metal-based nanozymes are one of the most widely used nanozymes. They have unique optical and electrical properties at the nanoscale as well as excellent catalytic properties [24] and exhibit good activity tunability and high stability. They have been found to exhibit a variety of enzyme-like properties, including oxidase-, peroxidase-, catalase-, and/or superoxide dismutase-like activities [25]. Metallic nanomaterials commonly include Au, Ag, Pt, PD, Rh, Ru, and Ir. For metal-based nanozymes, the catalytic mechanism arises from the adsorption, activation, and electron transfer of the substrate onto the metal surface, in contrast to the mechanisms occurring by changes in the metal valence of the nanomaterial, as in the case of other metal compound-based nanozymes [22]. Studies have found that the surface state of metal-based nanomaterials is one of the important factors affecting their catalytic activity. Therefore, much attention has been focused on how to optimally control the surface of metal-based nanozymes for high electrical conductivity. It can be classified into monometallic nanozymes, bimetallic nanozymes, and multimetallic nanozymes, which are distinguished, as the name implies, by having several metal cores. Bimetallic nanomaterials (BNMS) usually exhibit stronger catalytic performance than monometallic nanomaterials due to the synergistic effect [26], among which platinum-based BNMS have been extensively studied in the field of catalysis for many years. Multimetallic NPs composed of at least three different metals have more possibilities in modulating the activity, selectivity, and stability of surface catalyzed reactions. Hence, the rational design and controllable synthesis of multimetal nanozymes are of great significance.

Transition metals other than noble metals (Ti, V, Cr, Mn, Fe, Co, Ni, Zn, Al, Mo, and W) usually exist stably in the form of their complex ions, among them the oxide form, thus constituting metal oxide-based nanozymes. Metal oxide-based nanozymes including CeO_2, Fe_2O_3, Fe_3O_4, Co_3O_4, Mn_2O_3, and Mn_3O_4 all exhibit multienzyme activities, such as peroxidases, oxidases, hydrolases, and catalases. In addition, they show other physicochemical properties such as fluorescence quenching, dielectric properties, and magnetism [27]. It is worth noting that most metal oxide nanomaterials exhibit lower Km than HRP, which provides the possibility for a wide range of applications. Metal-organic frameworks (MOFs) are porous coordination crystalline materials formed by the self-assembly of metal ions (or metal clusters) and organic ligands through the principles of coordination chemistry [28]. In recent years, some MOFs exhibit their own good enzyme mimicking properties, mimicking the functions of a variety of enzymes, including oxidases, peroxidases, catalases, superoxide dismutases, and hydrolases. The high specific surface area, homogeneously dispersed active sites, structural diversity, and pore size tunability of MOFs can facilitate the efficient contact of reaction substrates to the catalytic sites and, in turn, enhance the catalytic efficiency of subsequent processes. Therefore, its pore size, size, modification, and composition are important factors for regulating enzyme activity. In addition, it can also participate in regulation by external conditions pH, temperature, H_2O_2, and so on [29]. The development of other nanomaterials with different structures and properties has provided new sources for artificial enzyme research, such as perovskites, metal sulfides, metal, dichalcogenides, methyl hydroxides, metal phosphates, and polymeric nanostructures [22]. The development of other types of nanomaterials has provided new sources for artificial enzyme research, such as perovskites, metal sulfides, metal, dichalcogenides, methyl hydroxides, metal phosphates, polymeric nanostructures, and others. They can also mimic the enzymatic activity properties of the oxidoreductase family and have received much attention for their unique structures or properties that are different from those of carbon-based nanomaterials, metal-based nanomaterials, and metal oxide nanomaterials and applications in environmental monitoring and remediation. As a carbon nitride, MXENEs have large surface areas, metallic conductivity, antimicrobial activity, and biocompatibility [30] and have been found to possess intrinsic peroxidase-like and oxidase-like activities and can be enhanced by single stranded DNA (ssDNA) adsorbed onto nanosheets. Li et al. constructed a simple label-free colorimetric sensing platform for TB-selective detection based on Ti_3C_2@ssDNA. The sensor exhibited good selectivity and sensitivity with a wide linear range of 1.0×10^{-11} to 1.0×10^{-8} m and a low detection limit of 1.0×10^{-11} M [31].

In this review, we summarize the most recent applications of nanozymes for environmental monitoring, environmental management, and environmental protection (Figure 1). We firstly introduce the tuning catalytic activity of nanozymes according to some crucial factors such as size and shape, composition and doping, and surface coating. Then, the application of nanozymes in environmental fields is introduced in detail. Nanozymes can not only be used to detect inorganic ions, molecules, organics and foodborne pathogenic bacteria but are also involved in the degradation of phenolic compounds, dyes, and antibiotics. The capability of nanozymes was also reported for assisting air purification, constructing biofuel cells, and application in marine antibacterial fouling removal. Finally, the current challenges and future trends of nanozymes toward environmental fields is proposed and discussed.

2. Tuning Catalytic Activity

Nanozymes are alternatives to natural enzymes but remain slightly inferior in catalytic activity. Thus, we need to focus on several important factors that affect the enzymatic activity of nanozymes as well as current strategies to enhance activity, thereby laying a theoretical foundation for the design of nanozymes.

One of the distinct features of enzymes are their ultrahigh reaction rate. Correspondingly, nanozymes with comparable or even superior activity are long-standing pursuits. Two strategies are discussed here to improve the activity of nanozymes: (I) increasing the inherent activity by delicate design and (II) boosting the activity by confinement effect or external stimulators [32]. The main factors affecting the intrinsic activity of nanozyme are size, composition, doping, shape, and surface modification. External stimulus factors, such as pH, substrate concentration, temperature, and light, affect the catalytic activity. Factors determining the activity of nanozymes need to be optimized for specific conditions in order to achieve maximum efficiency in applications involving detection of target analytes.

2.1. Size and Shape

Size, shape, and atomic arrangement can lead to changes in the catalytic performance of materials. It was found that the catalytic activity and stability of nanozyme increased with the increase in surface volume ratio. For example, Valden et al. [33] prepared gold clusters with a diameter of 1 to 6 nm on the single crystal surface of titanium dioxide under ultrahigh vacuum to investigate the size dependence of their low-temperature catalytic oxidation of carbon monoxide. It was found that the gold cluster with the largest carbon monoxide oxidation activity was 3 nm. In another case, Zhou et al. [34] used Au nanoparticles with various sizes (2–15 nm) to catalyze the reduction of resazurin, showing that Au nanoparticles of 6 nm exhibited the highest activity. However, small gold nanoparticles tend to aggregate and lose their activity. Scientists often anchor gold nanoparticles to carbon, silica, graphene, and other supporting materials to improve the dispersion of bare Au. Kalantari et al. [35] adjusted the delayed addition time of the thiolated organosilica precursor to control the nanostructure and the thiol density. Moreover, for the first time, they demonstrated that the peroxidase-like activity of T-Dendritic Mesoporous Silica Microspheres (DMSNs)-Au depended on nano-Au size. In addition, the highest activity was achieved at the Au particle size of 1.9 nm (Figure 2).

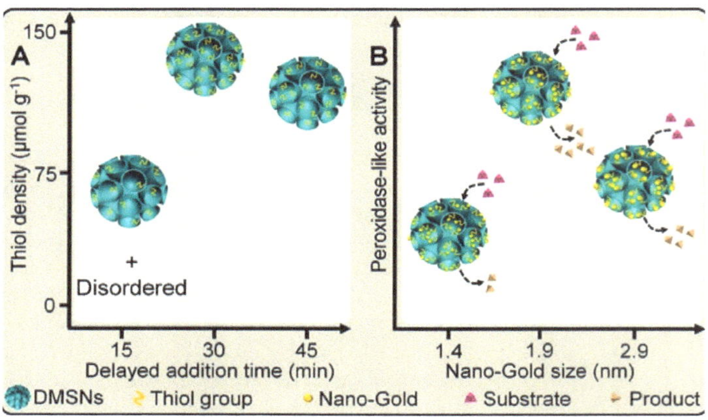

Figure 2. Schematic illustration of the effects of (**A**) delayed addition time on the structures of the final product and (**B**) Nano-Au size on the peroxidase-like activity [35]. Reproduced with permission from Ref. [35], Copyright 2019 American Chemical Society.

The catalytic performance of nanozymes can also be modulated by adjusting the shape of the nanostructures. Biswas et al. [36] compared the catalytic efficiency of gold nanorods (GNRs), gold nanoparticles (GNPs), and horseradish peroxidase (HRP). It was proved that the peroxidase activity of gold nanorods with a length diameter ratio of 2.8 was 2.5 times higher than that of HRP and gold nanoparticles, which showed stability in a wide range of pH and temperature. Based on this, a colorimetric sensor for malathion was developed, whose sensitivity of the assay was 1.78 μg/mL. A comparative study of VO_2 nanoparticles with different morphologies (nanofibers, nanosheets, and nanorods) was conducted and applied to the sensitive colorimetric detection of H_2O_2 and glucose by Tian et al. [37]. According to the typical Michaelis–Menten curve obtained for VO_2 nanozymes, the apparent K_M values of VO_2 nanofibers with H_2O_2 as the substrate were lower than that of VO_2 nanorods and VO_2 nanosheets. It shows that the VO_2 nanofibers have a higher affinity for H_2O_2 compared with VO_2 nanosheets and VO_2 nanorods. Moreover, compared with VO_2 nanorods and VO_2 nanosheets, the VO_2 nanofibers demonstrated the most sensitive response during the H_2O_2 and glucose sensing.

2.2. Composition and Doping

Some researchers have shown, based on the synergistic effect, that combining a variety of nanomaterials or conjugating several nanomaterials to form a hybrid can provide a catalytic center [38], improve the electron transfer between the nanozyme and the substrate, and generate additional active sites, which can adjust the catalytic activity of the catalyst.

Zhu et al. [39] combined TiO_2, $CuInS_2$, and CuS into a ternary metal sulfide-based hybrid. Owing to the synergistic effect among TiO_2, $CuInS_2$, and CuS components, compared with the control sample of Fe_3O_4/rGO, TiO_2/rGO, Fe_3O_4, TiO_2, and rGO, the prepared $TiO_2/CuInS_2/CuS$ nanofibers showed excellent peroxidase (POD)-like activity. They subsequently developed a sensor for the detection of dopamine with a detection limit of 1.2 μM. Wang et al. [40] incorporated iron oxide nanoparticles (Fe_3O_4NPs) into the heterodimer composed of gold and platinum to form a hybrid nanomaterial with good peroxidase-like activity. The formation of an alloy between platinum and gold can significantly improve the activity and selectivity of platinum-based catalysts. The nature of the peroxidase-like activity of the Fe_3O_4@Au-Pt hybrid nanomaterial originates from their ability to transfer electrons between the reducing substances and H_2O_2. The colorimetric sensor with a lower detection limit of 0.0018 μM was developed for glucose. Another form of composition is loading. Zhao et al. [41] covalently fixed the carbon point (C-dots) on the inner surface of the amino terminated dendritic silica sphere (dSs) while coupling the gold nanoclusters (Au NCs) on the outer surface. It not only maintains the superoxide dismutase-like enzyme activity of the carbon point but also improves the peroxidase-like-activity of the gold nanoparticles. Furthermore, adjusting the loading ratio of the two kinds of nanozymes can meet different functional requirements.

2.3. Surface Coating

The surface modification of nanozyme not only plays a connecting role in the combination of nanomaterials but also is of great importance to the regulation of catalytic activity. The surface catalytic reaction process can be described by several basic reaction steps, including substrate adsorption, substrate diffusion on the surface, chemical reaction, and then product desorption to regenerate the active site [42]. Each step will be affected by surface modification. Thus, some general strategies can be adopted for surface modification, such as changing the electronic structure of the surface, regulating the surface acidity, blocking surface contact, promoting product desorption, mediating the exposure of active sites to regulate substrate binding, and applying effective methods for surface electronic structure.

Surface modifiers can be divided into three categories: ions, small molecules, and macromolecules. Lee et al. [43] introduced Mn(acetate)$_2$ during the synthetic step of N-doped carbon dots to improve the enzymatic properties of metal-induced N-doped carbon dots (N-CDs). Its influence on the enzymatic properties of Mn-induced N-CDs (Mn:N-CDs) was investigated. Finally, the addition of Mn(acetate)$_2$ to the reaction solution seemed to generate more functional groups at the edge of carbogenic domains in Mn:N-CDs than in N-CDs, resulting in improved peroxidase-like properties (Figure 3a). Mn:N-CDs with strong enzymatic effects can be applied as a colorimetric sensor probe for the detection of gamma-aminobutyric acid (GABA). Surface modification can also change the intrinsic enzyme activity of nanomaterials. Zeolitic imidazolate framework-8 (ZIF-8) is a monatomic nanozyme with peroxidase activity. Sun et al. [44] introduced amino acid (AA) to regulate the growth of ZIF-8 crystal, thus simulating the structure and function of natural carbonic anhydrase (CA). Amino acid as a capping agent regulates the shape and size of ZIF-8 and forms a hydrophobic region on the surface of ZIF-8 to simulate the hydrophobic pocket of natural carbonic anhydrase (Figure 3b). Compared with natural carbonic anhydrase, Val-ZIF-8 not only has excellent esterase activity but also has better hydrothermal stability. Surface coating may also weaken or even lead to loss of enzyme activity. Jain et al. [45] reported the replacement of cetyl trimithyl ammonium bromide (CTAB) by 11-MUA from the surface of Au-core CeO$_2$-shell NP-based nanozyme studied for exhibiting multiple enzyme-like activities such as peroxidase, catalase, and superoxide dismutase. They found that 11-MUA coating AuNPs lost the SOD and catalase-like activity, which compromise the multifunctional property of chitosan nanoparticles (CSNPs).

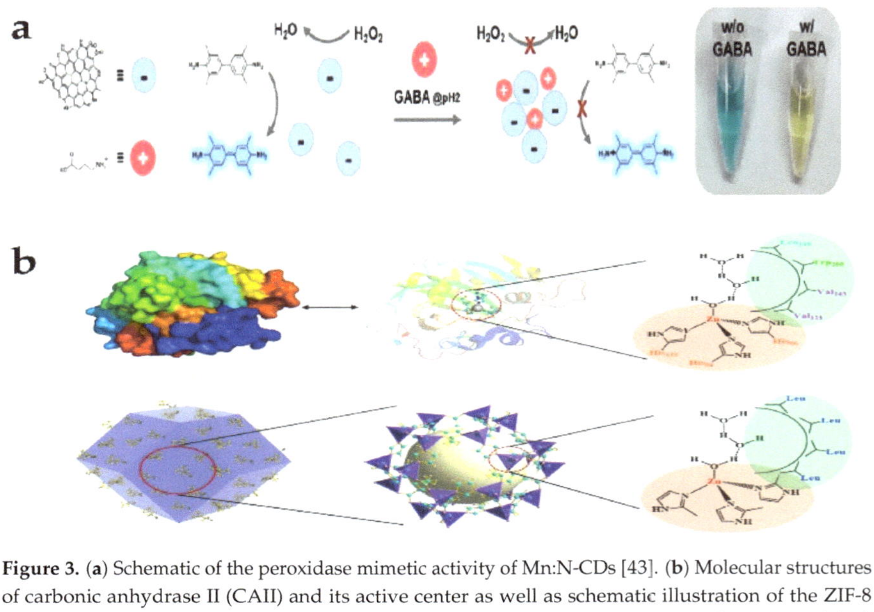

Figure 3. (**a**) Schematic of the peroxidase mimetic activity of Mn:N-CDs [43]. (**b**) Molecular structures of carbonic anhydrase II (CAII) and its active center as well as schematic illustration of the ZIF-8 structure and its CA-mimetic active center [44]. Reproduced with permission from (**a**) Ref. [43], Copyright 2021 Multidisciplinary Digital Publishing Institute; (**b**) Ref. [44], Copyright 2022 American Chemical Society.

2.4. Other Factors

Except regulating the intrinsic enzyme activity of nanozyme to control catalytic activity, external factors can also affect the final enzyme activity. The pH and temperature are the main external influencing factors. A lot of studies have confirmed that acidic conditions are suitable for peroxidase-like activity, while neutral and alkaline conditions are favorable for superoxide dismutase and catalase. For example, the esterase activity of Val-ZIF-8 synthesized by Sun et al. [44] would greatly increase with the increase in temperature. The enzyme activity at 80 °C was about 25 times higher than that at 25 °C. Gao et al. [46] reported a new strategy for controlling plaque biofilm with a peroxidase-like nanozyme (CAT-NP).

CAT-NP showed a strong dependence on acidic conditions. It killed 99% of bacteria in the acidic microenvironment simultaneously in a short time for biofilm control and prevention of dental caries. However, several studies have broken through the limitation of optimal pH for different nanozymes. Li et al. [47] developed the copper-based nanozyme ($CuCo_2S_4$), which showed enhanced peroxidase-like activity and antibacterial ability under neutral conditions (Figure 4). This would be used for infected wounds with pH close to neutral.

3. Improvement of the Specificity

At present, the research of nanozyme is not only to improve catalytic activity but also to improve specific recognition ability to realize the replacement of natural enzyme. The activity change of nanozyme is related to the action mechanism of nanozyme, while the catalytic specificity of nanozyme affects the accuracy of target capture. Most applications of nanozymes are based on the discovery and simulation of nanozyme activity, but there is still a lack of catalytic specificity. The simulation of catalytic activity mostly comes from the functional replica rather than the remodeling of the active center structure of natural enzyme, so the catalytic specificity is greatly reduced. In the years of rapid development of nanozyme, some strategies have been explored to solve these problems: (I) to simulate the active center and binding site of natural enzyme more precisely from the chemical structure at the design and construction of nanozyme and (II) to combine some specific molecular-assisted recognition.

The premise for natural enzyme to work is to combine it with the substrate, that is, to capture the substrate. It mostly depends on the primary structure and spatial configuration of protein or RNA to realize the complementarity with specific substrate. Therefore, the research on the specificity of nanozyme can start from this point. Currently, there have been many reports on biomimetic research of nanozyme [48–50]. For example, Zhou et al. [48] proposed a chiral COF nanozyme with highly ordered active centers and substrate binding sites that is mainly used to simulate horseradish peroxidase (HRP). The active site of HRP contains porphyrin heme as the catalytic active center and the distal L-histidine (L-His) residue as the binding site. Biomimetic COF enzyme is mainly constructed by mixing iron 5, 10, 15, 20-tetra (4′-tetraphenylamino) porphyrin unit (Fe-ATPP) into the COF skeleton as the active center and modifying L-His as the substrate binding site for chiral recognition. COF can be used as the skeleton of nanozyme. The well-dispersed Fe-ATPP unit in the skeleton endows COF nanozyme with high enzyme-like activity, which is 21.7 times higher than HRP. At the same time, the incorporation of L-/D-Hiss imparts the COF nanozyme with enantioselectivity in the oxidation of L-/D-dopa enantiomers and displays a preference for dopa. Changing the content of L-/D-Hiss can also optimize the selectivity of COF nanozyme. This work can easily adjust the activity and stereospecificity of COF chiral nanozyme by changing the doped amino acid and its content.

Although the biomimetic simulation of active centers and binding sites can be carried out according to the analysis of the three-dimensional structure of natural enzymes, the biological affinity and structural simulation of nanomaterials are still limited, which still need to be assisted by biological molecules with specific recognition ability. In this method, the biological recognition element and nanozyme are coupled to achieve the dual improvement of catalytic activity and specificity. Biorecognition elements mainly include antibody, DNA, aptamer, molecularly imprinted polymer (MIP), and biological enzyme. Molecularly imprinted polymer (MIP) is a polymer processed by molecular imprinting technology, which leaves a cavity in the polymer matrix and has affinity for selected "template" molecules [51]. Molecular recognition sites of specific target molecules are created in MIP to obtain solid materials with high selectivity for specific target molecules. Zhang et al. [52] initiated polymerization on the surface of the nanozyme substrate conjugate by adding a variety of polymerization monomers, thus forming a molecular imprinted hydrogel layer. Remove the imprinted substrate molecule to obtain the substrate specific recognition site constructed on the periphery of the nanozyme (Figure 4a). In addition, the incorporation of functional monomers and charges further improved the activity and specificity of nanozyme. Under the optimum conditions, the specificity can reach 100 times. A variety of nanozyme materials, including ferric oxide, gold, cerium dioxide, and other nanomaterials with peroxidase or oxidase activity, can be significantly improved by molecular imprinting.

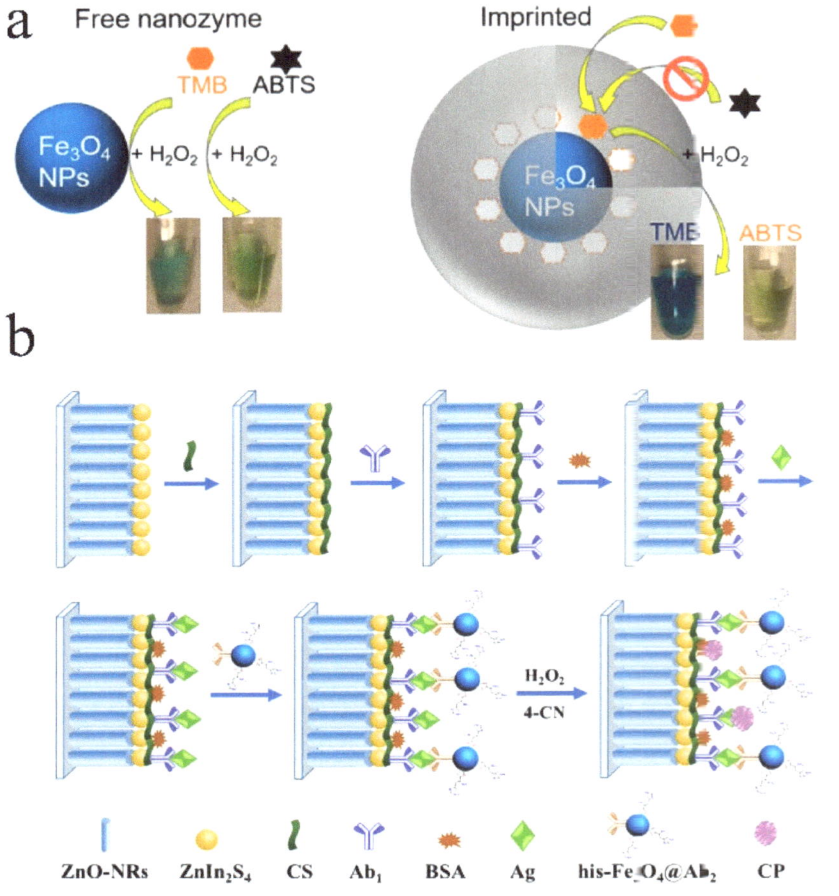

Figure 4. (**a**) Schematic diagram of molecularly imprinted nanozyme preparation [52]. (**b**) Schematic diagram of the PEC immunoassay using high-activity Fe_3O_4 nanozyme as signal amplifier. Reproduced with permission from (**a**) Ref. [52], Copyright 2017 American Chemical Society; (**b**) Ref. [53], Copyright 2019 Elsevier.

The biological enzyme–nanozyme cascade sensing system combined with biological enzyme is mostly used in the application of peroxide–nanozyme. Biological enzymes combine to oxidize specific substrates and produce H_2O_2, which in turn triggers peroxide nanozyme color or fluorescence reaction to realize signal sensing. Although biological enzymes can selectively capture target molecules, just like the reason for the birth of nanozymes, biological enzymes are limited by pH, temperature, reaction system, and other factors. At present, two kinds of enzymes are often linked on the same carrier for a cascade reaction. The ratio of enzyme content, immobilization method, and intermediate loss all affect the catalytic efficiency and sensing accuracy. The highly variable region of the antibody endows the antibody with the ability to recognize antigen specifically. Researchers often use nanozyme to replace the biological enzyme in ELISA to provide color signals. In addition to recording optical (colorimetric, fluorescent, chemiluminescence) signals for analysis, nanozyme catalysis can also trigger changes in temperature, volume, mass and pressure to form a new sensing mode. The detection methods are not confined to colorimetric detection but can also use electrochemical detection and Raman analysis. Li et al. [53] used high-activity Fe_3O_4 nanozyme as a signal amplifier to develop an ultrasensitive photoelectronic (PEC) immunoassay. In short, ZnO nanorods (ZnO-NRs) growing vertically on a bar indium–tin oxide (ITO) electrode were dispersed with $ZnIn_2S_4$ nanocrystals, producing a $ZnIn_2S_4$/ZnO-NRs/ITO photoelectronic as the PEC material mix to modify and capture PSA antibodies (Ab_1). Histidine-modified Fe_3O_4 (His-Fe_3O_4) nanozyme acts as a signal amplifier and connects with the signal PSA antibody (Ab_2) to form His-Fe_3O_4@Ab_2 conjugate, which is anchored by a specific sandwich immune reaction (Figure 4b). Labeled His-Fe_3O_4 nanozyme as a peroxidase induced the production of insoluble and insulating precipitation, resulting in a significant reduction of photocurrent signal. Finally, the ultralow detection limit of prostate specific antigen (PSA) 18 fg/mL was achieved. Recently, the methods that can be used for the biological coupling of nanozyme and an antibody are still limited. The technology that can effectively biocouple nanozyme with an antibody or antigen is not mature [19], and the reproducibility of nanozyme-labeled immunosensor is not good, and the large-scale commercial application technology still needs to be developed.

Deoxyribonucleic acid (DNA) and aptamers are other biological recognition elements that assist nanozymes to achieve high specificity. DNA, which follows the principle of complementary base pairing and has a specific sequence, can accurately identify the target, which is widely used in biological and medical fields. However, there are some problems to be solved in the nanozyme labeled with single-stranded DNA (ssDNA). ssDNA endows the surface of nanozyme with more negative charges, which will further affect the adsorption kinetics of substrate and the catalytic activity of nanozyme. It has been found that the modification of DNA changes the surface charge state of Fe_3O_4NPs, thus promoting the combination of nanozyme and substrate [54]. In addition, chemical modification or physical adsorption of DNA may also block the active site of nanozyme, resulting in the reduction of catalytic efficiency. Therefore, the length, concentration, and two-dimensional structure of DNA and the surface charge distribution of nanozyme affect the catalytic activity of nanozyme [55]. Although there have been many research outputs of DNA-modified nanozymes, we still do not know enough about the mechanism of the interaction between the DNA chain on the surface of the nanozyme and the nanozyme. Most of the results are due to the trial and error of scientific researchers. Understanding the mechanism may better control the application of DNA in the field of nanozymes. The aptamer is a short DNA sequence screened in vitro, which is also highly specific. There are many aptamer sensors based on nanozyme. A simple and low-cost colorimetric analysis was established for a highly sensitive determination of Kanamycin (KAN) through integrating boron nitrate quantum dots-anchored porous CeO_2 nanorods (BNQDs/CeO_2) and aptamer by Zhu et al. [56] due to the large specific surface area and synergistic interaction between BNQDs and CeO_3, which can effectively catalyze the oxidation of 3,5′,5,2′-tetramethylbenzidine (TMB). In addition, the catalytic activity of BNQDs/CeO_2 nanozyme was significantly enhanced because of the dispersion of BNQDs/CeO_2 nanozyme and the increase in substrate affinity after the substrate was combined with KAN-specific aptamer. KAN can combine with the aptamer to reduce the catalytic efficiency. The proposed colorimetric method realized the low detection limit of 4.6 pM. DNA and aptamer are superior to other biological recognition molecules in cost and stability, and aptamer can also further improve the detection specificity. However, there is also a complex interaction between the aptamer and the catalytic performance of nanozyme. Mechanism analysis and rule summary need to be obtained.

4. Environmental Monitoring

Over recent years, with the development of industry, environmental pollution has become increasingly serious, especially water pollution, which directly threatens human health. Common pollutants include heavy metal ions, dyes, phenols, pesticides, antibiotics, drugs, plasticizers, and other organic substances due to biological pollutants such as biological pathogenic bacteria or toxin pollution (Table 1). Hence, monitoring environmental pollutants has become a global concern.

4.1. Toxic Ions

With the rapid development of manufacturing industry around the world, overexploitation of minerals and groundwater, and industrial wastewater discharge, toxic ion pollution has become an unavoidable environmental problem. Toxic ions mainly refer to heavy metal ions, including mercury, cadmium, lead, chromium, arsenic, and other elements with potential biological toxicity. Because it cannot be degraded, it can only be converted into different chemicals through abiotic or biological mechanisms and can be amplified through the food chain, posing a serious threat to the ecosystem and human health [57]. At present, many methods based on nanozyme detection of toxic ions have been explored. Most of the sensors are miniaturized and portable to be used in point-of-care testing (POCT).

Wang et al. [58] loaded Au NPs onto HS-rGO to modify a glass carbon electrode (GCE) as a sensing platform. Au Pd-modified zirconium metal organic skeleton (AuPd@UiO-67) labeled with signal chain (Apt2) is used as signal enhancer to capture Hg^{2+} based on T-Hg (II)-T structure (Figure 5a). With the increase in modified Hg^{2+} concentration, the amount of Apt2-AuPd@UiO-67 is increased, thus realizing the detection of Hg^{2+}. The electrochemical sensor has a wide linear range (1.0 nmol/L–1.0 mmol/L) and a low detection limit (0.16 nmol/L). Except for electrochemical detection methods, colorimetry is the most commonly used method to detect Hg^{2+}. A graphene oxide nanosheet (CGO) based on L-cysteine functionalization was found to have a strong peroxidase-like activity compared with graphene oxide [59]. The introduction of more S and N species can effectively produce more surface defects and active sites, thus endowing carbon high peroxidase-like properties. The nanozyme can be used to realize the microdetection of Hg^{2+}, and its sensing principle is mainly based on the competitive adsorption between Hg^{2+} and photothermal properties of 3,3′,5,5′-tetramethylbenzidine (TMB). Because Hg^{2+} hinders the combination of TMB and CGO, TMB is catalyzed by H_2O_2 to produce more colored oxidation products, resulting in a more significant colorimetric response (Figure 5b), meaning, therefore, good detection of Hg^{2+}. Zhong et al. [60] used the peroxidase-like activity of iron hydroxide (FeOOH) nanorods to detect As(V) by colorimetry. Unlike TMB, the catalytic substrate is ABST, and its oxidation product is green and reaches the maximum absorption peak at 418 nm. As(V) can be adsorbed onto FeOOH nanorods through electrostatic interaction and an As-O bond, so the oxidation is gradually embedded. Finally, the colorimetric determination with response of 0–8 ppb and 8–200 ppb and detection limit of 0.1 ppb is realized. Ag^+ is also a heavy metal ion. Zhang et al. [61] utilized the excellent colorimetric and TMB to construct photothermal and colorimetric double-readout sensors for Ag+ analysis. MnO_2 nanosheets (NSs) were used to catalyze the oxidation of TMB to oxTMB. However, the reduction of MnO_2 NSs by glutathione (GSH) can reduce the catalytic capacity of MnO_2 NSs (Figure 5c). In this method, a specific combination of Ag^+ and GSH is utilized to inhibit this reduction process. According to this principle, the Ag^+ concentration can be converted into temperature and color signals. Consequently, the Ag^+ content can be determined both with the naked eye and with a portable thermometer. It is very suitable for POCT in the process of environmental detection.

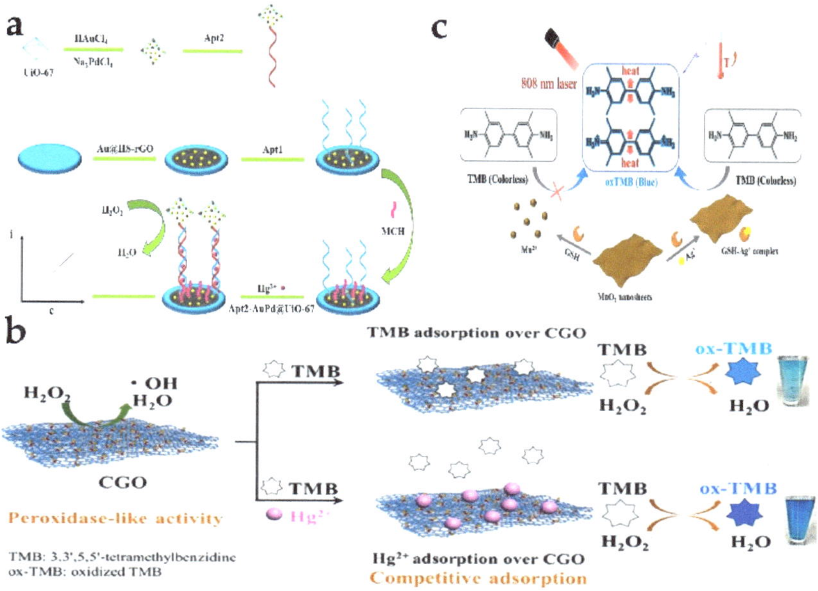

Figure 5. (**a**) Schematic diagram for fabricating the designed electrochemical aptasensor to detect Hg^+ [58]. (**b**) Schematic of CGO preparation and the principle of color rendering [59]. (**c**) Schematic of the proposed MnO_2 NSs-GSH-TMB colorimetric and photothermal platform for the Ag^+ analysis [61]. Reproduced with permission from (**a**) Ref. [58], Copyright 2022 Elsevier; (**b**) Ref. [59], Copyright 2022 Elsevier; (**c**) Ref. [61], Copyright 2019 Springer.

4.2. Organic Pollutants

Herbicides and pesticides used in agriculture to increase food production will cause serious pollution of soil and water quality. At present, most pesticides on the market are not degradable organophosphorus pesticides. If the metabolites or degradation products of pesticides exceed the maximum residue, they will cause pollution. In addition, antibiotic residues are also considered as typical organic pollution. Because of their longer half-life, they are more threatening to humans and other organisms. Another common organic pollutant, phenolic compounds, are common in dye, pharmaceutical, photo development and other industrial fields and is difficult to degrade in the aquatic ecological environment.

Parathion is an organophosphate (OP) insecticide, and it is also an irreversible inhibitor of nervous system function. Chen et al. [62] designed a bimetallic nanozyme (Au@Pt: gold@platinum) catalytic competitive sensitive biological bar code immunoassay. This novel biobarcode immunoassay contained three types of probes: (I) mAbs and ssDNA-labeled AuNP probes, (II) parathion OVA-hapten-modified immunomagnetic nanoparticle (MNP) probes, and (III) C-ssDNA-labeled Au@Pt probes (Figure 6a). The Au@Pt probe reacts with the AuNP probe through complementary base pairing. Parathion then competes with the MNP probe to bind mAb onto the AuNP probe. Finally, Au@Pt nanozyme is released from the complex to catalyze the color development of TMB. The most common and efficient platform to detect OPs is the enzyme biosensor, which is based on OP_S to inhibit acetylcholinesterase (AChE) activity. In another work, lactate dehydrogenase (LDH)-based ZIF-8 nanocomposite was prepared by Bagheria et al. [63], utilizing the formation of a simple complex between Zn^{2+} and 2-methylimidazole. This process leads to the formation of highly dispersed ZIF-8 nanostructures on the surface of ZnFe-LDH nanosheets (LDH@ZIF-8). Moreover, the peroxidase-mimicking behavior of the prepared nanocomposites is improved compared to pristine LDHs and MOFs. LDH@ZIF-8 significantly contributes to the CL emission intensity of the H_2O_2 rhodamine B (RhoB) system.

When organophosphorus pesticides (OPs) were present, AChE activity was inhibited, and reduced production of H_2O_2 eventually led to the attenuated chemiluminescence of RhoB (Figure 6c). This detection method achieved highly sensitive sensing of OPs indirectly by detecting the change of the hydrogen peroxide content. Apart from the chemiluminescence strategy, photoelectrochemical biosensors are another emerging analytical method. Song et al. [64] used the double amplification strategy to construct the photoelectrochemical aptamer sensor. The sensor realizes signal sensing mainly based on Co_9S_8@In-CdS NTs and PtNi nanozymes acting as signal amplifiers. Because sulfamethazine (SMZ) has a stronger affinity with the aptamer, the aptamer tends to combine with SMZ and escape from the electrode through the dissociation of the double chain structure, generating a photocurrent response. SA-modified PtNi nanostructures further increase the spatial steric resistance of the electrode surface. The insoluble 4-CD precipitate formed by incubation with 4-CN in the presence of H_2O_2 seriously hinders electron transfer and again changes the photocurrent response to achieve sensing of SMZ (Figure 6b).

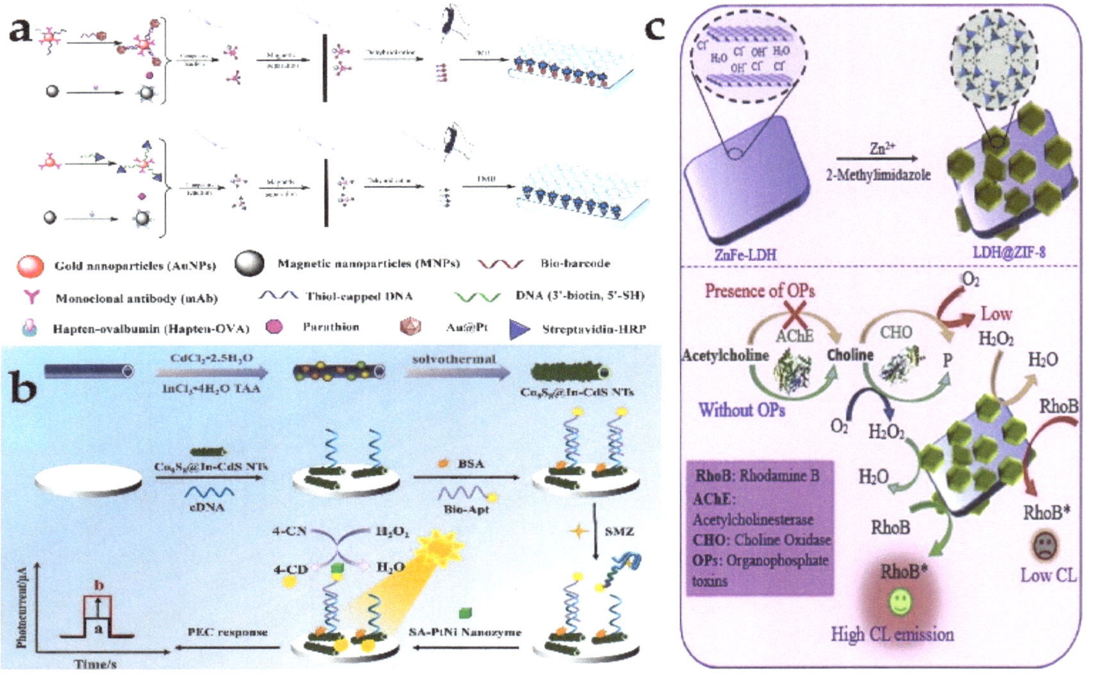

Figure 6. (**a**) Schematic presentation of the colorimetric biobarcode immunoassays for parathion based on amplification using Au@Pt and HRP [62]. (**b**) Schematic illustration of the preparation process of Co9S8@In-CdS NTs and the proposed PEC aptasensor for SMZ detection [64]. (**c**) Detection of OPs by LDH@ZIF-8-assisted RhoB-H_2O_2 CL-based system [63]. Reproduced with permission from (**a**) Ref. [62], Copyright 2020 American Chemical Society; (**b**) Ref. [64] Copyright 2022 Elsevier; (**c**) Ref. [63], Copyright 2019 Elsevier.

In order to make detection and environmental assessment portable and fast, some researchers bind sensors and smartphones to achieve mobile data transmission. This kind of combination can be used for POCT. Sun et al. [65] synthesized a three-layer FeOx@ZnMnFeOy @Fe Mn bimetallic organism with excellent multienzyme activity (peroxidase, oxidase, and catalase). Therefore, this sensing platform can complete the four functions of detecting H_2O_2, citric acid (CA), norfloxacin (NOR), and gallic acid. Moreover, smartphones are also used for automatic quantitative detection of CA and NOR, and the detection limit of norfloxacin is as low as 52 nM. Nanozyme with multienzyme activity was also used in the detection of phenolic compounds. A highly efficient mimic catalyst of $Co_{1.5}Mn_{1.5}O_4$ with four enzyme activities (peroxidase, oxidase, catalase, laccase) was used to detect dihydroxybenzene isomers. Finally, a dual function colorimetric sensor was constructed using TMB [66]. Ye et al. [67] constructed $NiCo_2O_4@MnO_2$ with p–n junctions, which not only has the photoelectric effect brought by the p–n junctions but also has inherent oxidase and peroxidase-like activities. The result is an excellent minimum detection limit of 0.0042 μM for hydroquinone.

4.3. Foodborne Pathogens

Biological pollution mainly refers to environmental pollution caused by various organisms that pose a threat to human health. The biological pollution in the water and soil environment mainly comes from untreated domestic sewage, industrial wastewater, garbage, and feces, which eventually leads to the excessive content of foodborne pathogens. In history, Vibrio cholerae once polluted the water environment as a foodborne pathogen, which eventually led to the global epidemic of cholera in the 1930s. Intestinal bacteria such as Escherichia coli, Streptococcus faecalis and Clostridium are the main bacteria that pollute water. Hepatitis also erupts through fecal sewage.

The first microorganism to be detected in drinking water is the content of Escherichia coli (*E. coli*). Thus, the detection of foodborne pathogens is increasingly urgent. In the detection of *E. coli*, β-Galactosidase (β-Gal) is applied. A multicolor colorimetric platform triggered by a designed enzyme nanozyme cascade reaction was designed and prepared [68]. MnO_2 nanoparticles with oxidase-like activity can catalyze the oxidation of TMB. Then TMB^{2+} quickly etched the gold nanorods (Au NRs), and the longitudinal local surface plasmon resonance peak appeared as an obvious blue shift and the polychromatic change of Au NRs. The presence of *E. coli* will hydrolyze p-aminophenyl β-d-galactopyranoside (PAPG) to produce p-aminophenol (PAP) through β-galactose, thereby mediating the reduction of MnO_2 nanosheets, destroying their oxidase-mimicking activity, and affecting the production of TMB^{2+}. Consequently, sensing systems that exhibit different colors can be easily observed for different concentrations of *E. coli*. A colorimetric sensor based on Ps-Pt nanozymes for the detection of Salmonella typhimurium was reported by Jiao et al. [69]. The sensor combines immunosensing and magnetic separation techniques. They covalently bound streptavidin first to Ps-Pt. Then the bacteria were recognized by coupling a biotinylated antibody of *S. typhi* onto PS Pt through the high affinity between streptavidin and biotin. Moreover, the magnetic beads conjugated antibodies were also prepared to facilitate the subsequent bacteria separation test, thus making the detection simple and fast. Similarly, Ps-Pt nanozyme with peroxidase activity has also been used to detect *Staphylococcus aureus* (*S. aureus*) on paper-based analytical equipment with a detection limit of 9.56 ng/mL [70]. Targeting *S. aureus*, Luo et al. [71] constructed a PEC sensor for *S. aureus* with a wide linear range between 10 and 10^8 CFU/mL and a limit of detection (LOD) as low as 3.40 CFU/mL based on "signal off" using the $Cu\text{-}C_3N_4\text{-}TIO_2$ heterostructure as the photoactive material and $Cu\text{-}C_3N_4$ peroxidase-like nanozymes as signal amplifiers. During the detection, $Cu\text{-}C_3N_4$ ($Cu\text{-}C_3N_4@Apt$) and benzo-4-chlorohexanedione (4-CD) produced by the oxidation of 4-chloro-1-naphthol (4-CN) in the presence of hydrogen peroxide participated in decreasing the photocurrent signal (Figure 7b). Polyoxometalates (POMS) of different structures have also been noted to possess peroxidase activity, among which $P_2Fe_4W_{18}$ enzyme activity was reported by Zhang et al. [72] Polydopamine (PDA) as an emerging biomimetic adhesive polymer combines with $P_2Fe_4W_{18}$ to enhance enzyme activity (Figure 7a). Ultimately, $Fe_4P_2W_{18}$/PDA achieved the detection of *E. coli* O157:H7 with a detection limit of 4.2×10^2 CFU/mL.

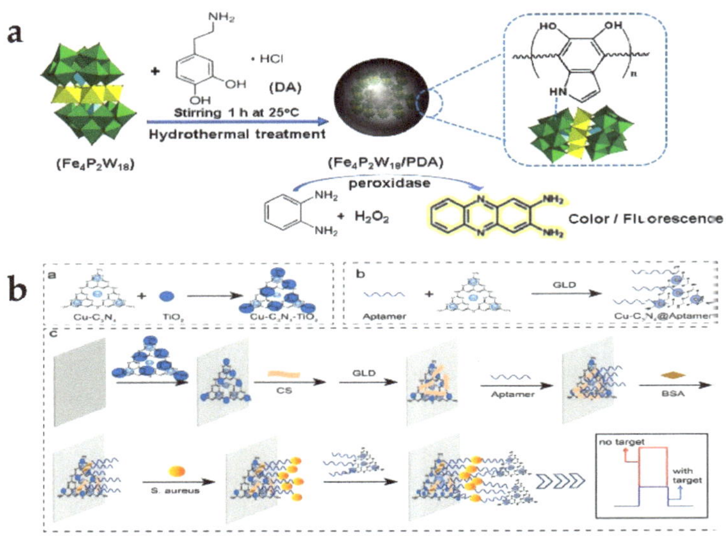

Figure 7. (**a**) Construction of the PEC sensor and detection of S. aureus [72]. (**b**) Schematic illustration of the synthesis procedure and the peroxidase activity of the P2W18Fe4/PDA nanozyme [71]. Reproduced with permission from (**a**) Ref. [72], Copyright 2022 Elsevier; (**b**) Ref. [71], Copyright 2021 Elsevier.

Table 1. Summary of the application of nanozymes in environmental monitoring.

Category	Analyte	Nanozyme	Activity	Detection Mode	Detection Range	LOD	Ref.
Toxic ions	Fe^{2+}/Pb^{2+}	MnO_2	CAT	Colorimetric	0.001~0.02 mmol/L 0.05~0.4 mmol/L	0.5 µmol/L 2 µmol/L	[73]
	F^-	AgPt-Fe_3O_4	POD	Colorimetric	50~2000 µM	13.73 µM	[74]
	Nitrite	AuNP-CeO_2 NP@GO	OXD	Colorimetric	100~5000 µM	4.6 M	[75]
	Cl^-, Br^-, I^-	Ag_3Cit	OXD	Colorimetric	/	26, 12, 7 nM	[76]
	Cu^{2+}	E-ChlCu/ZnO	POD	Colorimetric	0–1/1–15 µM	0.024 µM	[77]
	As^{3+}	Pd-DTT	OXD	Colorimetric	33~3.333 × 10^5 ng/L	35 ng/L	[78]
	Fe^{2+}	C-dots/Mn_3O_4 NCs	OXD	Colorimetric	0.03~0.83 µM	0.03 µM	[79]
	Nitrite	His@AuNCs/RGO	POD	Electrochemical	2.5~5700 µM	0.5 µM	[80]
	Hg^{2+}	MXene/DNA/Pt NCs	POD	Colorimetric	50~250 nM	9.0 nM	[81]
	Fe^{3+}	NCD/UiO-66 NCs	SOD POD	Colorimetric	0~0.1 mM	/	[82]
	Cr^{6+}	PEI-AgNCs	OXD	Colorimetric	/	1.1 µM	[83]
	Fe^{2+}	AuRu aerogels	OXD POD	Colorimetric	5~250 µmol/L	0.7 µmol/L	[84]
	Hg^{2+}	CS-$MoSe_2$NS	POD OXD	Colorimetric	0.1~4.0 µM	3.5 nM	[85]
	Fe^{3+}	$MoSe_2$@Fe	POD	Colorimetric	25~300 µM	1.97 µM	[86]
	F^-	R-$MnCo_2O_4$/Au NTs	POD	SERS	0.1~10 nM	0.1 nM	[87]
	Sn^{2+}	nano-UO_2	POD	Colorimetric	0.5–100 µM	0.36 µM	[88]
	PO_4^{3-}	MB@ZrHCF	POD	Colorimetric	10~200 µM	2.25 µM	[89]
	Cr^{3+}	GdOOH	Phospholipase	Colorimetric	5.0~200 µM	0.34 µM	[90]

Table 1. *Cont.*

Category	Analyte	Nanozyme	Activity	Detection Mode	Detection Range	LOD	Ref.
Toxic ions	Hg^{2+}	AuPd@UiO-67	POD	Electrochemical	1~10^6 mM	0.16 nmol/l	[58]
	Al^{3+}	Single atom Ce-N-C	Laccase	Colorimetric	5–25 μg/mL	22.89 ng/mL	[91]
	Cr^{6+}	CD/g-C_3N_4	POD	Colorimetric	0.3–1.5 μM	0.31 μM	[92]
	Hg^{2+}	CuS HNS	POD	Colorimetric	50~4 × 10^5 ng/mL	50 ng/L	[93]
	As^{3+}	CoOOH	POD	Electrochemical	0.1~200 μg/L	56.1 ng/L	[94]
	Cr^{6+}	Cu-PyC MOF	POD	Colorimetric	0.5–50 μM	0.051 μM	[95]
	Cr^{6+}	Ni/Al LDH (Ni/Al–Fe$(CN)_6$ LDH)	POD	Colorimetric	0.067~10 mM	0.039 mM	[96]
	Pb^{2+}	Tannic Acid@Au NPs	POD	Colorimetric	25~500 ng/mL	11.3 ng/mL	[97]
	S^{2-}	MoS_2/g-C_3N_4HNs	POD	Colorimetric	0.1~10 μM	37 nM	[98]
	S^{2-}	PDA@Co_3O_4NPs	CAT	Colorimetric	4.3~200 μM	4.3 μM	[99]
	As^{3+}	AuNPs	POD	Colorimetric	0.01~11.67 mg/L	0.008 mg/L	[100]
	S^{2-}	GMP-Cu	Laccase	Colorimetric	0~220 μmol/L	0.67 μmol/L	[101]
	Hg^{2+}	Ag_2S@GO	OXD	Colorimetric	5.0~120.0 × 10^{-8} M	9.8 × 10^{-9} mol/L	[102]
	Cu^{2+}	MMoO	POD	Colorimetric	0.1~24 μM	0.024 μM	[103]
	Cr^{6+}	MOF	OXD	Colorimetric	0.1~30 μM	20 nM	[104]
	Cr^{6+}	CuS-frGO	POD	Colorimetric	0–200 nM	26.60 nM	[105]
	Cr^{6+}	SA-Fe/NG	POD	Colorimetric	30~3 μM	3 nM	[106]
	Cr^{3+}	$CuFe_2O_4$/rGO	POD	Colorimetric	0.1~25 μM	35 nM	[107]
	Hg^{2+}	L-cysteine@GO	POD	Colorimetric	0~200 μg/L	5 μg/L	[59]
	Hg^{2+}	PtNPs	POD	Colorimetric	20~3000 nM	10.5 nM	[108]
	Hg^{2+}	Au-HBNz	POD	Colorimetric	0.008~20 μg/mL	1.10 ng/mL	[109]
	Hg^{2+}	AuPt@DSN	POD	Colorimetric	0.1~10^3 nM	8.58 pM	[110]
	Hg^{2+}	MVC-MOF	OXD	Colorimetric	0.05~6 μM	10.5 nM	[111]
	Hg^{2+}	Citrate-capped Cu NPs	POD	Colorimetric	0.100~6.000 μM	0.052 μM	[112]
	Hg^{2+}	Fe-MoS_2@AuNPs	POD	Electrochemical	0.5~200 nM	0.2 nM	[113]
	Hg^{2+}	Ag NWs	OXD	Colorimetric	25~5000 μg/L	19.9 ng/L	[114]
	Hg^{2+}	Cys-Fe_3O_4	POD	Colorimetric	0.02–90 nM	5.9 pM	[115]
	Hg^{2+}	His-AuNCs	OXD	Colorimetric	0.05–0.8 μM	8 nM	[116]
	Ag^{+}	MnO_2 NSs	OXD	Colorimetric	0.02~1.0 μM	6.7 nM	[61]
	As^{5+}	FeOOH	POD	Electrochemical	0.04~200 μg/L	12 ng/L	[60]
	Al^{3+}	Nanoceria	Phosphatase	Electrochemical	30~3.5 × 10^3 nM	10 nM	[117]
	H_2O_2	MA-Hem/Au-Ag	POD	Colorimetric	0.010–2.50 mM	2.5 μM	[118]
	H_2O_2	Pt/CeO_2/NCNFs	CAT	Electrochemical	0.0005–15 mM	0.049 μM	[119]
Phenolic	Phenol Compounds	1-Methylimidazole/Cu Nanozyme	Laccase	Colorimetric	0.5~4 μg/mL	0.57 μg/ml	[120]
	2,4-dinitrophenol	polymer-Fe-doped ceria/Au NC	POD	Colorimetric	1~100 μg/mL	2.4 μM	[121]
	Hydroquinone	$NiCo_2O_4$@MnO_2	POD OXD	Colorimetric	0~24 μM	0.042 μM	[67]
	Hydroquinone	$Co_{1.5}Mn_{1.5}O_4$	OXD	Colorimetric	0.05~100μM	0.04μM	[66]
	2,4,6-TNT	2H–MoS_2/Co_3O_4	OXD	Electrochemical	/	1 pM	[122]
	Hydroqui-none	Fe_3O_4@COF	POD	Colorimetric	0.5~300 μmol L	0.12 μmol L	[123]
	2,4-DP	AMP-Cu	Laccase	Colorimetric	0.1~100 μmol/L	0.033 μmol/L	[124]
	2,4-DP	MnCo@C NCs	Laccase	Electrochemical	3.1~122.7 μM	0.76 μM	[125]
	2,4-DP	$NiFe_2O_4$	POD	Colorimetric	0.218~3.282 μg/mL	0.311 μg/mL	[126]

Table 1. *Cont.*

Category	Analyte	Nanozyme	Activity	Detection Mode	Detection Range	LOD	Ref.
OPs	Carbendazim	MoS_2/MWCNTs	OXD	Electrochemical	0.04~100 μM	7.4 nM	[127]
	parathion ethyl	C-Au NPs	POD	Colorimetric	11.6~92.8 ng/mL	5.8 ng/mL	[128]
	Dichlorvos	γ-MnOOH NWs	OXD	Colorimetric	0~15 ng/mL	3 ng/ml	[129]
	Diazinon	LDH@ZIF-8	POD	Colorimetric	0.5~300 nM	0.22 nM	[63]
	Paraoxon	2D MnO_2	OXD POD	Electrochemical	0.1~20 ng/mL	0.025 ng/mL	[130]
	Benomyl	AgNPs/MWCNTs/GO	OXD	Electrochemical	0.2~122.2 μM	/	[131]
	Dimethoate	Pt NPs	POD	Colorimetric	0.5~9 μg/mL	0.15 μg/mL	[132]
	Naphthalene acetic acid	Ti_3C_2-MXene/BP	OXD	Electrochemical	0.02~40 μM	1.6 nM	[133]
	Parathion	NiO-SPE	OXD	Electrochemical	0.1~30 μM	0.024 μM	[134]
	MeHg	NA-CDs/AuNPs	POD	Colorimetric	0.375~75 μg L	0.06 μg L	[135]
	Chlorpyrifos	Ag-Nanozyme	POD	Colorimetric	35~210 ppm	11.3 ppm	[136]
	Omethoate	SACe-N-C	POD	Colorimetric	100~700 μg/mL	55.83 ng/mL	[137]
	Methyl-paraoxon	Nanoceria	Laccase	Colorimetric	0.42~126 μM	0.42 μmol/L	[138]
	Methyl-paraoxon	CeO_2	POD OXD	Electrochemical	0.1~100 μmol/L	0.06 μmol/L	[139]
	Methyl-parathion	Fe_3O_4/C-dots@Ag-MOFs	/	Electrochemical	5×10^{-11}~2×10^{-9} mol/L	1.16×10^{-11} mol/L	[140]
	Atrazine	Fe_3O_4-TiO_2/rGO	POD	Colorimetric	2~20 mμ g/L	2.98 μg/L	[141]
	Glyphosate	Au@PN	POD	Colorimetric	0.5~20 nM	0.24 nM	[142]
	Glyphosate	Porous Co_3O_4	POD	Colorimetric	8~80 μg/L	2.37 μg/L	[143]
	Glyphosate	Fe_3O_4@C_7/PB	POD	Colorimetric	0.125~15 μg/mL	0.1 μg/mL	[144]
	Carbaryl	NH_2-MIL-101(Fe)	POD	Colorimetric	2~100 ng/mL	1.45 ng/mL	[145]
	Chlorophenols	Fe_3O_4@MnOx	OXD	Colorimetric	10~1600 μM	0.85 μM	[146]
	Fipronil	ZIF-8	POD	Colorimetric	0.2~4μM	0.036 μM	[147]
	Malathion	Fe-N/C SAzyme	OXD	Colorimetric	0.5~10 nM	0.42 nM	[148]
Antibiotic residues	Sulfamethazine	PtNi NCs	POD	Photoelectrochemical	0.05~10^3 pg/mL	37.2 fg/mL	[64]
	Sulfonamides	2D Cu-TCPP (Fe)	POD	Electrochemical	1.186~28.051 ng/mL	0.395 ng/mL	[149]
	Streptomycin	Au@Pt NPs	POD	Lateral Flow Immunoassays	0.062~0.271 ng/mL	1 ng/mL	[150]
	Tetracycline	Cu-doped-g-C_3N_4	POD	Colorimetric	0.1~50 μM	31.51 nM	[151]
	Tetracycline	Fe_3O_4@MIP	POD	Colorimetric	2~225 μM	0.4 μM	[152]
	Tetracycline	MIL-101(Fe/Co)	POD	Colorimetric	1–8 μM	0.24 μM	[153]
	Norfloxacin	FO@ZMFO@FM-MOG	CAT OXD POD	Colorimetric	0.415–6.21 μM	52 nM	[65]
	Kanamycin	$CoFe_2O_4$NPs	POD	Electrochemical	1~10^{-6} μM	0.5 pM	[62]
	Chloramphenicol	Co_3O_4	POD	Electrochemiluminescence	5×10^{-13}~4×10^{-10} mol/L	1.18×10^{-13} mol/L	[154]
	Kanamycin	WS_2 Nanosheets	POD	Colorimetric	0.1–0.5 μM	0.06 μM	[155]
	Metronidazole	MIL-53 (Fe)@ molecularly imprinted polymer (MIP)	POD	Colorimetric	1~200 μM	53.4 nM	[156]
Foodborne pathogens	*Staphylococcus aureus*	Cu-C_3N_4-TiO_2	POD	Photoelectrochemical	10~10^8 CFU/mL	3.40 CFU/mL	[71]
	Staphylococcus aureus	Pd@Pt NPs	POD	Lateral Flow Immunoassays	10–300 ng/mL	9.56 ng/mL	[70]
	Salmonella typhimurium	IPs-Pt	POD	Colorimetric	10^4~10^6 CFU/mL	10^3 CFU/mL	[69]
	Escherichia coli	Au NRs	OXD	Colorimetric	1.0×10^2~1.0×10^5 CFU/mL	22 CFU/mL	[68]
	E. coli O157:H7	Au-Pt dumbbell NPs	POD	Colorimetric	10~10^7 CFU/mL	2 CFU/mL	[157]
	E. coli O157:H7	man-Pediatric lead (PB)	POD	Lateral Flow Immunoassays	10^2~10^8 CFU/mL	10^2 CFU/mL	[158]
	E. coli O157:H7	$P_2W_{18}Fe_4$/PDA	POD	Colorimetric	10^3~10^6 CFU/mL	4.2×10^2 CFU/mL	[72]

5. Environmental Management

Over these years, industrial development and natural resource exploitation have brought economic prosperity, but at the same time, all over the world, it has also faced serious environmental governance challenges. Most pollutant residues are found in water sources and soils on which mankind depends for survival. Furthermore, the common degradation methods of pollutants are of three types: (I) biodegradation, (II) physical adsorption, and (III) chemical oxidation. Nanozymes as an emerging research outcome in the 21st century exhibit excellent qualities in environmental governance. They (I) can handle compounds that are often difficult to biodegrade, (II) can operate independently of pollutant concentration, (III) can operate over a wide range of pH, temperature, and salinity, (IV) are not inhibited by biofouling, (V) are relatively simple and easy to control, and (VI) are highly stable and recyclable [57]. Environmental monitoring Table 2 demonstrates the application of nanomaterials in the degradation of various pollutants.

Table 2. Summary of the application of nanozymes in environmental management.

Category	Pollutant	Activity	Nanozyme	Removal Efficiency	Ref.
Dyes	RhB	POD	Sulfur-doped graphdiyne	>98%	[159]
	Methyl orange	POD	CNZ	93%	[160]
	Rhodamine B	OXD	FeBi-NC SAzyme	99%	[161]
	Methylene Blue	POD	ZnNi-MOF/GO NCs	95%	[162]
	Methylene Blue	POD	Cu^{2+}-HCNSs-COOH	80.7%	[163]
	Methylene Blue	POD OXD	PdNPs/PCNF	99.64%	[151]
	Amido Black	Laccase	Cu/H_3BTC MOF	60%	[164]
	Malachite green	Laccase	Fe_3O_4@C-Cu^{2+}	99%	[165]
	Organic dyes	POD	Fe_3O_4@Gel	99%	[166]
Antibiotics	Tetracycline	POD	Sulfur-doped graphdiyne	>69%	[159]
Toxic ions	Cr^{6+}/As^{3+}	CAT	NanoMn_3O_4	>98%	[167]
	Hg^{2+}/Cl^-	POD	AgRu@β-CD-co-GO	94.9% 93.8%	[168]
	H_2O_2	CAT POD	DMNS@AuPtCo	>95%	[169]
Phenolic	Hydroquinone	Laccase	Aminopropyl-functionalized copper containing phyllosilicate (ACP)	100%	[170]
	Phenol	POD	MNP@CTS	>95%	[171]
	Phenol	CAT POD	DMNS@AuPtCo	90%	[169]
	2,4-DP	Laccase	Fe1@CN-20	65%	[172]
	2,4-DP	Laccase	AMP-Cu	65%	[124]
	2,4-DP	Laccase	CH-Cu	82%	[173]
	2,4-DP	Laccase	Cu-Cys@COF-OMe	>75%	[174]
	2,4-DP	Laccase	CA-Cu NPs	90%	[175]
	DEHP phthalic acid esters	Hydrolase	Zn-heptapeptide bionanozymes	86.80%	[176]
	Microplastics	POD	Fe_3O_4NPs	100%	[177]
Pathogens	*Escherichia coli*	Phospholipase	PAA-Cnp	>80%	[178]
	Escherichia coli	POD	Au-Pt dumbbell NPs	95%	[157]
	Escherichia coli	OXD	w-SiO_2/CuO	90%	[179]
	Gram-negative bacteria	POD	SA-Pt/g-C_3N_4-K	>99.99%	[180]
OPs	Simazine	POD	Fe_3O_4/DG	99%	[181]
	Atrazine	POD	Fe_3O_4-TiO_2/rGO	98%	[141]
	Cinosulfuron	POD	CP@CA	96.25%	[182]

Currently, most of the substances that nanozymes are able to degrade are organic such as phenolic compounds, herbicides, insecticides, dyes, and antibiotics. A few nanozymes are used to degrade inorganic heavy metal ions. Citric acid-modified copper peroxide nanodots (CP@CA) synthesized by Yang et al. [182] are an autocatalytic nanozyme. Under acidic conditions, they can decompose into H_2O_2 and Cu^{2+} in water or soil, while H_2O_2 would further decompose into •OH, capable of degrading nicosulfuron, based on a Fenton-like reaction. Its degradation rate can reach 97.58% within 1 h. Furthermore, after CP@CA was involved in pollutant degradation, the ecotoxicity of most degradation intermediates was reduced to a lower level compared with nicosulfuron. Moreover, CP@CA had little effect on the active components of the soil bacterial community. Photocatalytic degradation is another pathway for the degradation of pollutants. Baruah et al. synthesized magnetic Fe_3O_4 NPs on the surface of polydopamine functionalized RGO sheets (FDGs) for photocatalytic degradation of the hazardous pesticide simazine. Due to its excellent photocatalytic activity and magnetic separability, this makes the degradation rate of this nanozyme 99% and highly sustainable. The specific mechanism of its degradation relies on the formation of •OH under photocatalysis. The graphene sheets with good optical properties enhanced Fe_3O_4 with very high electron hole pair recombination characteristics. Dopamine-functionalized graphene sheets (DG) have high electron carrier capacity through their π-bond network, resulting in FDG nanozymes with high photocatalytic activity. Upon light irradiation, nanozymes absorb photons and undergo redox reactions by elevating electrons from the valence band (VB) to the conduction band (CB) (Eq^n 8). The electrons in the CB can be easily transferred to the DG surface, forming a hole in the VB (Eq^n 8). The electrons on the surface of DG simultaneously trap dissolved molecular oxygen and lead to the formation of superoxide radical anions ($O•^{2-}$) (Eq^n 9). Superoxide radical anions directly interact with water molecules to form •OH (Eq^n 10–12). Similarly, the holes (H^+) can be in contact with water molecules producing •OH (Eq^n 13). •OH decomposes simazine pesticides into nontoxic inorganic molecules and ions [181].

There are few recent studies on the removal of heavy metal ions by nanozyme, but there are still several methods with high removal efficiency. AgRu bimetal mesoporous nanozyme costabilized by β-CD and GO (AgRu@ β-CD co GO) was first constructed [168]. The nanozyme has a porous microstructure and a large number of hydroxyl and GO aromatic rings, which can enrich and adsorb a large amount of Hg^+ and Cl^- in water. The authors, through a 0.22 μm commercial millipore filtration membrane, repeatedly filtered the mixed solution containing Hg^+, Cl^-, and nanozyme three times, and the $Hg2^+$ and Cl^- removal efficiency reached more than 95.4% and 93.8%, respectively. In another work, Su et al. [167] investigated the microbial sensitivity regulation mechanism (MSRM) on typical paddy field heavy metal pollution (As^{3+} and Cr^{6+}) using nanozyme nanoMn_3O_4-coated microbial populations (NMCMP) and proved that Flavoisolibacter and Arthrobacter were two main bacteria related to heavy metal (As^{3+} and Cr^{6+}) pollution remediation. In addition, NMCMP can enhance the reduction of Cr^{6+} level and inhibit the release and rapid oxidation of As^{3+} during the repair process of $As_2H_2S_3$ (Figure 8b). Methyl orange is a typical dye in industrial wastewater selected as a typical dye pollutant because it is not easily degraded. CNZ was applied to the degradation of methyl orange pollutants [160]. At a high temperature of 60 °C and pH value of 3.93%, the degradation rate can be obtained in less than 10 minutes. In addition, the nanozyme showed excellent reusability and storage stability. However, Pd@ZnNi-MOF/GO nanocomposites with high peroxidase-like activity took only 8 min to completely degrade the methyl blue dye, and the catalytic degradation efficiency was as high as 95% [162].

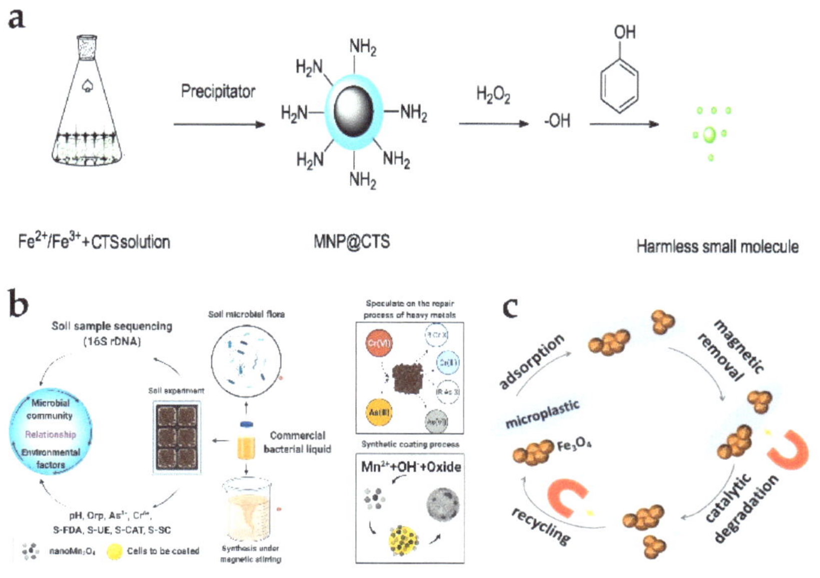

Figure 8. (**a**) Schematic diagram of MNP@CTS nanozyme-catalyzed degradation of phenol [171]. (**b**) Experimental design and process on NMCMP (NMCMP: nanozyme nanoMn_3O_4-coated microbial populations) [167]. (**c**) Schematic diagram of Fe_3O_4 nanozyme-catalyzed degradation of microplastics [177]. Reproduced with permission from (**a**) Ref. [171], Copyright 2018 Elsevier; (**b**) Ref. [167], Copyright 2022 Elsevier; (**c**) Ref. [177], Copyright 2022 Wiley Online Library.

Decomposition of phenol and phenolic compound purification of the environment is the focus of social attention. The degradation of phenol and phenolic compounds using ferromagnetic nanoparticles (MNPs) has many advantages. Ferromagnetic chitosan nanozymes (MNP@CTS) have the ability to catalyze the production of reactive oxygen species from hydrogen peroxide. Under the action of reactive oxygen species, the substrate phenol can be rapidly oxidized into various small molecules. Meanwhile, CTS can improve the catalytic efficiency and increase the degradation rate and degradation effect (Figure 8a). The removal efficiency is higher than 95% within 5 h [171]. Microplastics have a high surface-to-volume ratio, and they can act as a carrier for invading microorganisms, heavy metals, and other contaminants. Some of the long-term deleterious effects of microplastics include infertility, degradation of microplastics, and cancer. Therefore, it is crucial to remove and degrade microplastics in water resources. Hydrophilic bare Fe_3O_4 nanoaggregates allowed efficient removal of the most common microplastics including high-density polyethylene, polypropylene, polyvinyl chloride, polystyrene, and polyethylene terephthalate [177]. The bare Fe_3O_4 nanoaggregates with peroxidase-like activity further catalyzed the degradation of microplastics with nearly 100% efficiency by adsorbing to microplastics via hydrogen bonding (Figure 8c).

6. Other Environmental Protection Applications

6.1. Air Purification

Air pollution does great harm to people who breathe with their lungs. Formaldehyde is particularly harmful. Newly decorated rooms often face the problem of too much formaldehyde and cannot be occupied. Ecological nanozymes can catalyze the decomposition of formaldehyde. The average purification rate of formaldehyde in two hours was 91.9% [57]. The composite material is made of activated carbon fiber (ACF) and porous polymer composite and is also equipped with an antibacterial agent. When formaldehyde molecules are sucked into the nanospace, the nanozyme will react with the oxygen in the air to produce highly active superoxide ions with active oxygen structure. Due to the large contact area between the ecological enzyme catalyst and the adsorbed formaldehyde molecules in the nanopores, the active catalyst molecules quickly combine with formaldehyde molecules. After a series of oxidation-reduction enzyme catalyzed reactions, different peroxyintermediate oxidation molecules are formed. Finally, the formaldehyde molecules are oxidized into water and carbon dioxide molecules [57]. However, the ecological enzyme quickly returned to its original state and combined with oxygen molecules in the air again. The process of "combination with oxygen molecules– formation of active oxygen molecules–combination with formaldehyde molecules–enzymatic decomposition" keeps repeating so as to remove formaldehyde, bacteria, and other organic molecules in the air. This can keep the composite material clean for a long time.

6.2. Antibacterial and Antifouling Agent

Marine biofouling refers to the process in which marine microorganisms, animals, and plants continuously enrich and grow on artificial surfaces to form biofouling, which is a worldwide problem affecting maritime transport and communication facilities and coastal power plants [183]. The microfouling organisms on the substrate surface form a heterogeneous biofilm, which is composed of a variety of heterotrophic bacteria, cyanobacteria, diatoms, protozoa, and fungi [184]. Biofouling can increase hull roughness and weight, increase navigation resistance, greatly increase fuel consumption, cause economic losses of billions of dollars every year, increase carbon dioxide emissions, and intensify the room temperature effect [185]. In addition, organisms attached to distant ships will enter different sea areas, causing potential "species invasion" and affecting the marine ecological balance. When they block the mesh of mariculture cages, they can cause large-scale death of fish and shrimp. For a long time, the method to solve the pollution of marine organisms mainly depended on the toxic effect of heavy metal ions, which also causes serious pollution of the marine environment. In recent years, the research and development of nanozyme provides a new solution to prevent and remove marine biological fouling.

A semiconducting nanozyme consisting of chromium single atoms coordinated on carbon nitride (Cr-SA-CN) that performs bifunctional roles of nonsacrificial H_2O_2 photosynthesis and haloperoxidase-mimicking activity for antibiofouling was constructed [186]. The bifunctional Cr-SA-CN nanoplatform promotes the sustainable formation of HOBr under visible light radiation, so it has excellent antibacterial ability. Moreover, the nanozyme can continuously produce H_2O_2 from underwater and oxygen under visible light irradiation for enzymatic reaction. Field tests in seawater show that Cr-SA-CN, as an antibacterial additive for environmental protection coatings, can prevent the colonization of marine microorganisms on inert surfaces. In addition, the disinfection efficiency of Cr-SA-CN + Br^- against Escherichia coli, Staphylococcus aureus, Pseudomonas aeruginosa, and Vibrio vulnificus was, respectively, 97%, 96%, 92%, and 95%. This study not only proved the ability of monatomic nanozyme to resist biological pollution and sterilization but also provided a strategy for designing more innovative nanozymes with multifunctionality. Nanozyme antifouling agents with the same principle also include photothermal nanozyme with Mo single atom as the active site (Mo SA-N/C), which also has halogen oxide enzyme-like activity [187]. It catalyzes the oxidation of Br^- and H_2O_2 to produce cytotoxic HOBr. At the same time, the photothermal effect induced by visible light greatly accelerates the reaction process. Attapulgite (ATP) is a kind of natural and available nanomineral that has a special layered chain structure, large specific surface area, strong adsorption capacity, and surface dynamic properties. It provides rapid mass transfer and abundant accessible sites for efficient catalytic reactions [188]. Feng et al. [189] synthesized iron and copper-doped ATP (Fe Cu/ATP) with POD-like activity. The addition of Fe and Cu improves the conversion efficiency of H_2O_2, thus showing enhanced POD-like activity. The bactericidal mechanism is to produce reactive oxygen species to attack bacterial populations (Figure 9a). The antibacterial rate of Fe Cu/ATP against Escherichia coli and Staphylococcus aureus is 100% and has a long-term effect. Wei et al. [190] synthesized Fe_3O_4@MoS_2-Ag made great efforts in bacterial adsorption and toxic attack. Topological structure of MoS_2 nanosheets and S-vacancy enabled $Fe_3O_4@MoS_2$-Ag to have strong adhesion with bacteria by forming chemical bonds, which shortens the diffusion distance of free radicals and enhances the antibacterial effect (Figure 9b). The nanocomposite has the characteristics of peroxidase simulation and can catalyze H_2O_2 to produce living oxygen to attack bacteria. In addition, the released Ag^+ plays an auxiliary role while attacking the bacterial membrane. Under near-infrared radiation, local hyperthermia and peroxidase simulation can further enhance the sterilization effect. Magnetism also makes it reusable. The method has broad-spectrum antibacterial performance against Gram-negative bacteria, Gram-positive bacteria, drug-resistant bacteria, and fungal bacteria.

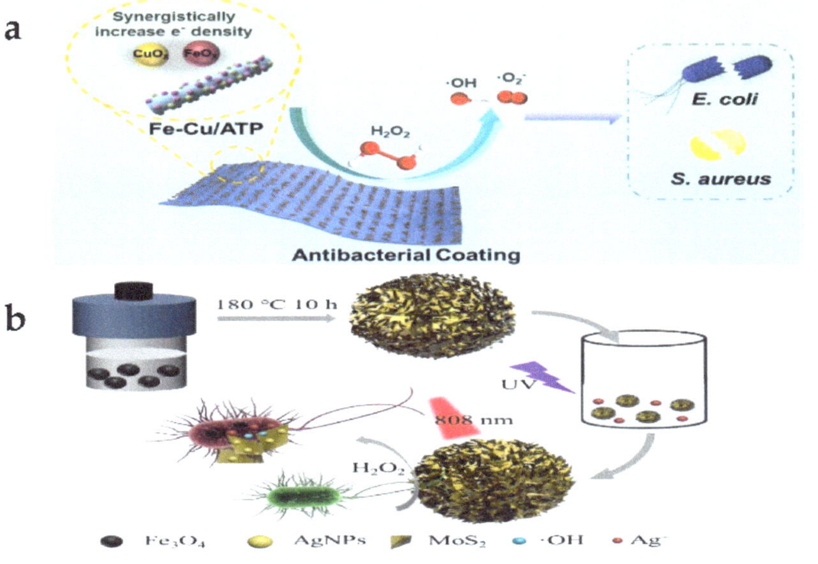

Figure 9. (**a**) Attapulgite doped with Fe and Cu nanooxides as peroxidase nanozymes for antibacterial coatings [189]. (**b**) The schematic preparation of $Fe_3O_4@MoS_2$-Ag with antibacterial function [190]. Reproduced with permission from (**a**) Ref. [189], Copyright 2022 American Chemical Society; (**b**) Ref. [190], Copyright 2021 Elsevier.

6.3. Enzyme-like Nanomaterial (Nanozyme)-Based Biofuel Cells

Human beings are facing environmental problems caused by the excessive exploitation of fossil energy. In recent years, countries all over the world have focused on sustainable and environment-friendly new energy. Biofuel cells have become an alternative energy conversion device [191]. Biofuel cells can be divided into three categories: microbial biofuel cells (MBFC), enzyme biofuel cells (EBFC), and enzyme-like nanomaterial (nanozymes)-based biofuel cells (NBFC). MBFC have many advantages in waste treatment and environmental protection [192], but the key disadvantage limiting their wide application and commercialization is that their power output is significantly low, and they are extremely difficult to control the internal electron transfer of microorganisms. Unlike MBFC, EBFC catalyze the oxidation of biofuels to generate electricity with the help of natural enzymes. The biofuels of EBFC are usually sugar relatives, such as glucose, sucrose, fructose, alcohols (including ethanol and methanol), organic acids, and organic salts (such as sulfite). However, glucose-based EBFC have many limitations derived from natural enzymes, such as variability and instability, high production cost, and difficult electron transfer [193,194]. In this case, compared with natural enzymes, nanozymes have become potential catalytic materials for developing glucose biofuel cells due to their inherent characteristics (such as long-term stability, easy synthesis, low cost, and adjustable enzyme mimic activity). Gu et al. [195] found that black phosphorus (BP) showed glucose dehydrogenase (GDH)-like activity and could catalyze the oxidation of glucose without any by-products. BP is a strong alternative candidate for sustainable biofuel cells. Finally, nanozyme-based EBFC consisting of BP anodes for glucose oxidation and Cu^{2+}/carbon nanotube (Cu^{2+}/CNT) cathodes for reducing O_2 under natural conditions were successfully constructed (Figure 10a). The power output of BP-based EBFC is higher than that of GDH-based EBFC. BP nanosheets maintained structural integrity before 360 °C, while the protein structure of GDH was destroyed after 250 °C. More important, EBFC based on nanozyme still showed high stability after 30 days of operation. This work provides more possibilities for the application of BP in the field of nanozyme. Compared with metal-based nanozymes, metal oxide nanozymes can be easily synthesized at low cost. Ho et al. [196] proposed interesting NBFC based on metal oxides, in which $CoMn_2O_4$/carbon was used as a GOx-like anode catalyst in biofuel cell systems. NBFC show a power output of 2.372 mW/cm^2, which is comparable to the commercial platinum/carbon-based biofuel cell system.

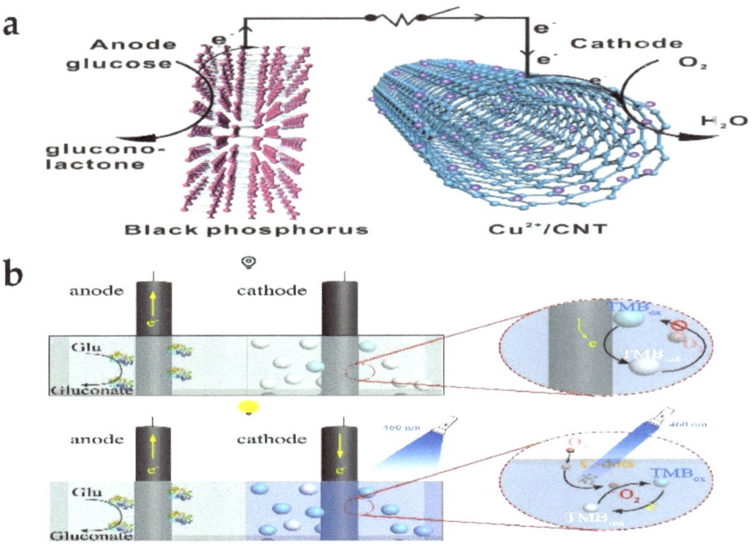

Figure 10. (**a**) Schematic illustration of the as-constructed EBFCs based on BP anode for glucose oxidation and Cu^{2+}/CNT cathode for O_2 reduction [195]. (**b**) Diagram of the photo-switch working mechanism based on C-dots and TMB [197]. Reproduced with permission from (**a**) Ref. [195], Copyright 2020 American Chemical Society; (**b**) Ref. [197], Copyright 2022 Elsevier.

In the traditional design of EBFC, once the fuel is added to the anode, the generation of electric energy will start and continue until the fuel is exhausted or the circuit is cut off. This uncontrollable way will lead to energy waste when EBFC are not used. Li et al. [197] proposed a novel optical switch for EBFC by controlling the electron acceptor in the cathode. When the stable TMBred was oxidized by singlet oxygen activated by the C point under light conditions, the medium can accept the electrons generated by the enzyme anode, thus leading to the formation of a path for the external circuit. Without radiation, TMBred cannot be converted into TMB198ox. A limited amount of electron acceptors is rapidly depleted, resulting in almost zero current and power density. Here, the C-point nanozyme was used as a photosensitizer of oxygen, and TMB is added to the cathode chamber as an electron acceptor. Theoretically, without lighting, there should be no current in the external circuit, showing a completely "off" state. The EBFC can be precisely and easily adjusted by the optical switch with light as the input signal (Figure 10b). Therefore, it avoids the damage to the anode enzyme caused by the traditional pH switch. This optical switch can adjust the power output of EBFC according to specific optical signals and promote the "intelligent" application of EBFC.

Enzymes such as nanomaterial (nanozyme)-based biofuel cells mainly use noble metal-based nanozymes, metal oxide-based nanozymes, and electron-receiving laccase that can mimic natural GOx and catalase. NBFC are stable and have a long service life. The high catalytic activity of glucose oxidation can output enough power, which can be synthesized on a large scale and has low production cost. The use of nanozymes in glucose biofuel cell systems has significantly improved the power generation performance. The utilization of NBFC will continue to improve.

7. Conclusions and Prospects

To summarize, we reviewed the application of nanozyme in the environmental field from three aspects: environmental monitoring, environmental management, and other environmental applications. The influencing factors of nanozyme catalytic activity were briefly summarized as well. The size, structure, composition combination, doping, and surface modification of nanozyme can adjust the catalytic activity of nanozyme. The specificity of nanozyme is mainly improved by biological recognition molecules (biological enzyme, MIP, DNA, antibody, aptamer) and the simulation of the active center and binding site of the biological enzyme. In addition to its own intrinsic enzyme activity, it can also affect the catalytic efficiency through pH, temperature, light stimulation, and so on. In terms of environmental monitoring and treatment, the detection and degradation of heavy metal ions, phenolic compounds, dyes, plasticizers, pesticides, and antibiotics all involve nanozymes. Moreover, nanozymes can be used as antibacterial and antifouling agents and biofuel cells to indirectly protect the environment. Great development and applications of nanozyme have been promoted in the field of environmental science.

Nanozyme has been developed for more than 20 years. Nanomaterials with new enzyme activities have been continuously explored. The design strategy of nanozymes has been constantly improved. However, nanozyme still has some limitations regarding the direction of development in the future.

- At present, the types of nanozymes are still too few, and they are mainly concentrated in the oxidoreductase family and hydrolase family. Compared with the six categories of natural enzymes, there is still an urgent need to unlock more simulated enzymes with different catalytic activities to expand the scope of application.
- Nanozymes are a succedaneum for natural enzymes, but the catalytic activity of most nanozymes is far inferior to natural enzymes. Hence, strategies need to be continuously explored to improve their catalytic activity.
- Nanozymes can show good performance in the laboratory. Nevertheless, they are still disadvantaged because they cannot be used on a large scale for the actual pollutant treatment industry, such as catalytic devices requiring high-precision technology, short service life, and higher cost than traditional environmental treatment methods.
- Although some nanozymes that break through the restriction of pH have appeared, most nanozymes are still limited by pH with narrow range. Technological breakthroughs are still needed in this regard so that the catalytic activity of most nanozymes is no longer limited by pH.
- Nanozymes are intrinsically toxic. It is vital to design low-toxicity nanozymes by adjusting their physical and chemical properties such as size, shape, surface properties, surface charge, and chemical composition to avoid secondary contamination.
- In recent years, nanozymes with multienzyme activity have been continuously developed, which can be used for multifunctional applications. However, at the same time, facing the challenge that the selectivity of nanozymes with multienzyme activity is lower than that of single-enzyme live nanozymes, will

challenge researchers to balance the relationship between "multifunction" and "high selectivity" as well as achieve a win–win situation.

- Recently, the specificity of nanozyme is much lower than that of natural enzyme. The design of nanozyme should be committed to better a bionic biological enzyme's active center and binding site, and the recognition element should be stably and effectively connected to the nanozyme. In addition, it is critical to explore the mechanism and law of interactions between nanozyme and a recognition element. Meanwhile, it is still necessary to improve the performance of nanozyme sensors by combining the research results of specific recognition in other fields and sensing technologies.
- Nanozyme detection mostly relies on colorimetric sensing, but colorimetric sensing has the problems of large interference and low sensitivity. In addition to electrochemistry, photoelectrochemistry, and surface-enhanced Raman scattering (SERS), adding more detection modes can have unexpected effects on environmental monitoring.

Author Contributions: M.W.: Conceptualization, Writing—Original Draft Preparation; P.Z.: Writing—Review and Editing, Visualization; S.L.: Writing—Review and Editing; Y.C.: Writ-ing—Review and Editing; D.L.: Conceptualization, Writing—Review and Editing; Y.L.: Writing—Review and Editing; W.C.: Concep-tualization, Writing—Review and Editing; L.D.: Conceptualization, Writing—Review and Edit-ing; Supervision, Funding Acquisition; C.W.: Writing—Review and Editing; Supervision, Funding Acquisition; Project Administration. All authors have read and agreed to the published version of the manuscript.

Funding: This research was funded by the Key Research and Development Program of Shaanxi Province-International Science and Technology Cooperation General Project (Grant No. 2022KW-23), the National Natural Science Foundation of China (Grant No. 32071370, 51861145307, and 32271427), and the Chunhui Research Grant of the Ministry of Education of China.

Institutional Review Board Statement: No human participants and/or animals were used for the research.

Informed Consent Statement: No human participants and/or animals were used for the research.

Conflicts of Interest: The authors declare that they have no known competing financial interest or personal relationships that could have appeared to influence the work reported in this paper.

References

1. Van Beilen, J.B.; Li, Z. Enzyme technology: An overview. *Curr. Opin. Biotechnol.* **2002**, *13*, 338–344. [CrossRef]
2. Jaeger, K.E.; Eggert, T. Enantioselective biocatalysis optimized by directed evolution. *Curr. Opin. Biotechnol* **2004**, *15*, 305–313. [CrossRef]
3. Huang, W.; Huang, S.M.; Chen, G.S.; Ouyang, G.F. Biocatalytic Metal-Organic Frameworks: Promising Materials for Biosensing. *Chembiochem* **2022**, *23*, e202100567. [CrossRef] [PubMed]
4. Garcia-Galan, C.; Berenguer-Murcia, A.; Fernandez-Lafuente, R.; Rodrigues, R.C. Potential of Different Enzyme Immobilization Strategies to Improve Enzyme Performance. *Adv. Synth. Catal.* **2011**, *353*, 2885–2904. [CrossRef]
5. Breslow, R.; Baldwin, S.W. Conversion of cholestanol to 12-ketocholestanol and to delta14-cholestenol and delta8(14)-cholestenol by remote oxidation. *J. Am. Chem. Soc.* **1970**, *92*, 732–734. [CrossRef]
6. Manea, F.; Houillon, F.B.; Pasquato, L.; Scrimin, P. Nanozymes: Gold-nanoparticle-based transphosphorylation catalysts. *Angew. Chem. Int. Ed.* **2004**, *43*, 6165–6169. [CrossRef] [PubMed]
7. Gao, L.Z.; Zhuang, J.; Nie, L.; Zhang, J.B.; Zhang, Y.; Gu, N.; Wang, T.H.; Feng, J.; Yang, D.L.; Perrett, S.; et al. Intrinsic peroxidase-like activity of ferromagnetic nanoparticles. *Nat. Nanotechnol.* **2007**, *2*, 577–583. [CrossRef]
8. Wei, H.; Wang, E.K. Nanomaterials with enzyme-like characteristics (nanozymes): Next-generation artificial enzymes. *Chem. Soc. Rev.* **2013**, *42*, 6060–6093. [CrossRef]
9. Jiang, B.; Duan, D.M.; Gao, L.Z.; Zhou, M.J.; Fan, K.L.; Tang, Y.; Xi, J.Q.; Bi, Y.H.; Tong, Z.; Gao, G.F.; et al. Standardized assays for determining the catalytic activity and kinetics of peroxidase-like nanozymes. *Nat. Protoc.* **2018**, *13*, 1506–1520. [CrossRef]
10. Iswarya, V.; Johnson, J.B.; Parashar, A.; Pulimi, M.; Chandrasekaran, N.; Mukherjee, A. Modulatory effects of Zn^{2+} ions on the toxicity of citrate-and PVP-capped gold nanoparticles towards freshwater algae, Scenedesmus obliquus. *Environ. Sci. Pollut. Res.* **2017**, *24*, 3790–3801. [CrossRef]

11. Lv, X.; Yang, Y.; Tao, Y.; Jiang, Y.; Chen, B.; Zhu, X.; Cai, Z.; Li, B. A mechanism study on toxicity of graphene oxide to Daphnia magna: Direct link between bioaccumulation and oxidative stress. *Environ. Pollut.* **2018**, *234*, 953–959. [CrossRef] [PubMed]
12. Albanese, A.; Tang, P.S.; Chan, W.C.W. The Effect of Nanoparticle Size, Shape, and Surface Chemistry on Biological Systems. *Annu. Rev. Biomed. Eng.* **2012**, *14*, 1–16. [CrossRef] [PubMed]
13. Liu, Y.; Li, W.; Lao, F.; Liu, Y.; Wang, L.; Bai, R.; Zhao, Y.; Chen, C. Intracellular dynamics of cationic and anionic polystyrene nanoparticles without direct interaction with mitotic spindle and chromosomes. *Biomaterials* **2011**, *32*, 8291–8303. [CrossRef] [PubMed]
14. Rasmussen, M.K.; Pedersen, J.N.; Marie, R. Size and surface charge characterization of nanoparticles with a salt gradient. *Nat. Commun.* **2020**, *11*, 2337. [CrossRef] [PubMed]
15. Gahlawat, G.; Choudhury, A.R. A review on the biosynthesis of metal and metal salt nanoparticles by microbes. *RSC Adv.* **2019**, *9*, 12944–12967. [CrossRef]
16. Zhang, M.; Qu, Y.; Li, D.X.; Liu, X.Y.; Niu, Y.S.; Xu, Y.H. To Love and to Kill: Accurate and Selective Colorimetry for Both Chloride and Mercury Ions Regulated by Electro-Synthesized Oxidase-like SnTe Nanobelts. *Anal. Chem.* **2021**, *93*, 10132–10140. [CrossRef]
17. Liang, X.; Han, L. White Peroxidase-Mimicking Nanozymes: Colorimetric Pesticide Assay without Interferences of O-2 and Color. *Adv. Funct. Mater.* **2020**, *30*, 2001933. [CrossRef]
18. Xu, D.; Wu, L.; Yao, H.; Zhao, L. Catalase-Like Nanozymes: Classification, Catalytic Mechanisms, and Their Applications. *Small* **2022**, *18*, 2203400. [CrossRef] [PubMed]
19. Li, X.; Zhu, H.; Liu, P.; Wang, M.; Pan, J.; Qiu, F.; Ni, L.; Niu, X. Realizing selective detection with nanozymes: Strategies and trends. *TrAC Trends Anal. Chem.* **2021**, *143*, 116379. [CrossRef]
20. Golchin, J.; Golchin, K.; Alidadian, N.; Ghaderi, S.; Eslamkhah, S.; Eslamkhah, M.; Akbarzadeh, A. Nanozyme applications in biology and medicine: An overview. *Artif. Cells Nanomed. Biotechnol.* **2017**, *45*, 1069–1076. [CrossRef]
21. Jiang, L.; Sun, Y.; Chen, Y.; Nan, P. From DNA to Nerve Agents—The Biomimetic Catalysts for the Hydrolysis of Phosphate Esters. *Chemistryselect* **2020**, *5*, 9492–9516. [CrossRef]
22. Ballesteros, C.A.S.; Mercante, L.A.; Alvarenga, A.D.; Facure, M.H.M.; Schneider, R.; Correa, D.S. Recent trends in nanozymes design: From materials and structures to environmental applications. *Mater. Chem. Front.* **2021**, *5*, 7419–7451. [CrossRef]
23. Ye, N.R.; Huang, S.M.; Yang, H.S.; Wu, T.; Tong, L.J.; Zhu, F.; Chen, G.S.; Ouyang, G.F. Hydrogen-Bonded Biohybrid Framework-Derived Highly Specific Nanozymes for Biomarker Sensing. *Anal. Chem.* **2021**, *93*, 13981–13989. [CrossRef]
24. Cai, R.; Liu, J.B.; Wu, X.C. Research Progress of Noble Metal-based Nanozymes. *Chem. J. Chin. Univ. Chin.* **2021**, *42*, 1188–1201. [CrossRef]
25. Cai, S.; Yang, R. Noble Metal-Based Nanozymes. In *Nanozymology. Nanostructure Science and Technology*; Springer: Singapore, 2020; pp. 331–365.
26. Cai, Z.; Kuang, Y.; Qi, X.; Wang, P.; Zhang, Y.; Zhang, Z.; Sun, X. Ultrathin branched PtFe and PtRuFe nanodendrites with enhanced electrocatalytic activity. *J. Mater. Chem. A* **2015**, *3*, 1182–1187. [CrossRef]
27. Liu, Q.; Zhang, A.; Wang, R.; Zhang, Q.; Cui, D. A Review on Metal- and Metal Oxide-Based Nanozymes: Properties, Mechanisms, and Applications. *Nano-Micro Lett.* **2021**, *13*, 154. [CrossRef]
28. Liang, B.; Li, B.; Li, Z.Q.; Chen, B.L. Progress in Multifunctional Metal-Organic Frameworks/Polymer Hybrid Membranes. *Chem. A Eur. J.* **2021**, *27*, 12940–12952. [CrossRef] [PubMed]
29. Huang, X.; Zhang, S.T.; Tang, Y.J.; Zhang, X.Y.; Bai, Y.; Pang, H. Advances in metal-organic framework-based nanozymes and their applications. *Coord. Chem. Rev.* **2021**, *449*, 214216. [CrossRef]
30. Meng, W.; Liu, X.; Song, H.; Xie, Y.; Shi, X.; Dargusch, M.; Chen, Z.-G.; Tang, Z.; Lu, S. Advances and challenges in 2D MXenes: From structures to energy storage and conversions. *Nano Today* **2021**, *40*, 101273. [CrossRef]
31. Li, M.M.; Peng, X.Y.; Han, Y.J.; Fan, L.F.; Liu, Z.G.; Guo, Y.J. Ti_3C_2 MXenes with intrinsic peroxidase-like activity for label-free and colorimetric sensing of proteins. *Microchem. J.* **2021**, *166*, 106238. [CrossRef]
32. Li, S.R.; Zhang, Y.H.; Wang, Q.; Lin, A.Q.; Wei, H. Nanozyme-Enabled Analytical Chemistry. *Anal. Chem.* **2022**, *94*, 312–323. [CrossRef]
33. Valden, M.; Lai, X.; Goodman, D.W. Onset of catalytic activity of gold clusters on titania with the appearance of nonmetallic properties. *Science* **1998**, *281*, 1647–1650. [CrossRef] [PubMed]
34. Zhou, X.C.; Xu, W.L.; Liu, G.K.; Panda, D.; Chen, P. Size-Dependent Catalytic Activity and Dynamics of Gold Nanoparticles at the Single-Molecule Level. *J. Am. Chem. Soc.* **2010**, *132*, 138–146. [CrossRef]
35. Kalantari, M.; Ghosh, T.; Liu, Y.; Zhang, J.; Zou, J.; Lei, C.; Yu, C.Z. Highly Thiolated Dendritic Mesoporous Silica Nanoparticles with High-Content Gold as Nanozymes: The Nano-Gold Size Matters. *Acs Appl. Mater. Interfaces* **2019**, *11*, 13264–13272. [CrossRef]
36. Biswas, S.; Tripathi, P.; Kumar, N.; Nara, S. Gold nanorods as peroxidase mimetics and its application for colorimetric biosensing of malathion. *Sens. Actuators B-Chem.* **2016**, *231*, 584–592. [CrossRef]
37. Tian, R.; Sun, J.H.; Qi, Y.F.; Zhang, B.Y.; Guo, S.L.; Zhao, M.M. Influence of VO_2 Nanoparticle Morphology on the Colorimetric Assay of H_2O_2 and Glucose. *Nanomaterials* **2017**, *7*, 347. [CrossRef] [PubMed]
38. Wu, J.J.X.; Wang, X.Y.; Wang, Q.; Lou, Z.P.; Li, S.R.; Zhu, Y.Y.; Qin, L.; Wei, H. Nanomaterials with enzyme-like characteristics (nanozymes): Next-generation artificial enzymes (II). *Chem. Soc. Rev.* **2019**, *48*, 1004–1076. [CrossRef]

39. Zhu, W.; Cheng, Y.; Yan, S.; Chen, X.; Wang, C.; Lu, X. A general cation-exchange strategy for constructing hierarchical TiO_2/$CuInS_2$/CuS hybrid nanofibers to boost their peroxidase-like activity toward sensitive detection of dopamine. *Microchem. J.* **2022**, *183*, 108090. [CrossRef]
40. Feng, X.Y.; Fu, H.; Bai, Z.Y.; Li, P.; Song, X.L.; Hu, X.P. Colorimetric detection of glucose by a hybrid nanomaterial based on amplified peroxidase-like activity of ferrosoferric oxide modified with gold-platinum heterodimer. *New J. Chem.* **2021**, *46*, 239–249. [CrossRef]
41. Zhao, L.; Ren, X.; Zhang, J.; Zhang, W.; Chen, X.; Meng, X. Dendritic silica with carbon dots and gold nanoclusters for dual nanozymes. *New J. Chem.* **2020**, *44*, 1988–1992. [CrossRef]
42. Liu, B.W.; Liu, J.W. Surface modification of nanozymes. *Nano Res.* **2017**, *10*, 1125–1148. [CrossRef]
43. Lee, A.H.Y.; Kang, W.S.; Choi, J.S. Highly Enhanced Enzymatic Activity of Mn-Induced Carbon Dots and Their Application as Colorimetric Sensor Probes. *Nanomaterials* **2021**, *11*, 3046. [CrossRef] [PubMed]
44. Sun, S.X.; Zhang, Z.J.; Xiang, Y.; Cao, M.W.; Yu, D.Y. Amino Acid-Mediated Synthesis of the ZIF-8 Nanozyme That Reproduces Both the Zinc-Coordinated Active Center and Hydrophobic Pocket of Natural Carbonic Anhydrase. *Langmuir* **2022**, *38*, 1621–1630. [CrossRef] [PubMed]
45. Jain, V.; Bhagat, S.; Singh, M.; Bansal, V.; Singh, S. Unveiling the effect of 11-MUA coating on biocompatibility and catalytic activity of a gold-core cerium oxide-shell-based nanozyme. *RSC Adv.* **2019**, *9*, 33195–33206. [CrossRef] [PubMed]
46. Gao, L.Z.; Liu, Y.; Kim, D.; Li, Y.; Hwang, G.; Naha, P.C.; Cormode, D.P.; Koo, H. Nanocatalysts promote Streptococcus mutans biofilm matrix degradation and enhance bacterial killing to suppress dental caries in vivo. *Biomaterials* **2016**, *101*, 272–284. [CrossRef]
47. Li, D.D.; Guo, Q.Q.; Ding, L.M.; Zhang, W.; Cheng, L.; Wang, Y.Q.; Xu, Z.B.; Wang, H.H.; Gao, L.Z. Bimetallic $CuCo_2S_4$ Nanozymes with Enhanced Peroxidase Activity at Neutral pH for Combating Burn Infections. *Chembiochem* **2020**, *21*, 2620–2627. [CrossRef]
48. Chen, J.X.; Huang, L.; Wang, Q.Q.; Wu, W.W.; Zhang, H.; Fang, Y.X.; Dong, S.J. Bio-inspired nanozyme: A hydratase mimic in a zeolitic imidazolate framework. *Nanoscale* **2019**, *11*, 5960–5966. [CrossRef]
49. Li, M.H.; Chen, J.X.; Wu, W.W.; Fang, Y.X.; Dong, S.J. Oxidase-like MOF-818 Nanozyme with High Specificity for Catalysis of Catechol Oxidation. *J. Am. Chem. Soc.* **2020**, *142*, 15569–15574. [CrossRef]
50. Xu, B.L.; Wang, H.; Wang, W.W.; Gao, L.Z.; Li, S.S.; Pan, X.T.; Wang, H.Y.; Yang, H.L.; Meng, X.Q.; Wu, Q.W.; et al. A Single-Atom Nanozyme for Wound Disinfection Applications. *Angew. Chem.-Int. Ed.* **2019**, *58*, 4911–4916. [CrossRef]
51. Chen, L.X.; Xu, S.F.; Li, J.H. Recent advances in molecular imprinting technology: Current status, challenges and highlighted applications. *Chem. Soc. Rev.* **2011**, *40*, 2922–2942. [CrossRef]
52. Zhang, Z.J.; Zhang, X.H.; Liu, B.W.; Liu, J.W. Molecular Imprinting on Inorganic Nanozymes for Hundred-fold Enzyme Specificity. *J. Am. Chem. Soc.* **2017**, *139*, 5412–5419. [CrossRef] [PubMed]
53. Li, W.S.; Fan, G.C.; Gao, F.X.; Cui, Y.G.; Wang, W.; Luo, X.L. High-activity Fe_3O_4 nanozyme as signal amplifier: A simple, low-cost but efficient strategy for ultrasensitive photoelectrochemical immunoassay. *Biosens. Bioelectron.* **2019**, *127*, 64–71. [CrossRef] [PubMed]
54. Chau, L.Y.; He, Q.; Qin, A.; Yip, S.P.; Lee, T.M.H. Platinum nanoparticles on reduced graphene oxide as peroxidase mimetics for the colorimetric detection of specific DNA sequence. *J. Mater. Chem. B* **2016**, *4*, 4076–4083. [CrossRef] [PubMed]
55. Liu, B.; Liu, J. Accelerating peroxidase mimicking nanozymes using DNA. *Nanoscale* **2015**, *7*, 13831–13835. [CrossRef] [PubMed]
56. Zhu, X.; Tang, L.; Wang, J.J.; Peng, B.; Ouyang, X.L.; Tan, J.S.; Yu, J.F.; Feng, H.P.; Tang, J.L. Enhanced peroxidase-like activity of boron nitride quantum dots anchored porous CeO_2 nanorods by aptamer for highly sensitive colorimetric detection of kanamycin. *Sens. Actuators B-Chem.* **2021**, *330*, 129318. [CrossRef]
57. Meng, Y.T.; Li, W.F.; Pan, X.L.; Gadd, G.M.G. Applications of nanozymes in the environment. *Environ. Sci. Nano* **2020**, *7*, 1305–1318. [CrossRef]
58. Wang, Y.; Wang, Y.; Wang, F.; Chi, H.; Zhao, G.; Zhang, Y.; Li, T.; Wei, Q. Electrochemical aptasensor based on gold modified thiol graphene as sensing platform and gold-palladium modified zirconium metal-organic frameworks nanozyme as signal enhancer for ultrasensitive detection of mercury ions. *J. Colloid Interface Sci.* **2022**, *606*, 510–517. [CrossRef]
59. Tian, H.; Liu, J.; Guo, J.; Cao, L.; He, J. L-Cysteine functionalized graphene oxide nanoarchitectonics: A metal-free Hg^{2+} nanosensor with peroxidase-like activity boosted by competitive adsorption. *Talanta* **2022**, *242*, 123320 [CrossRef]
60. Zhong, X.-L.; Wen, S.-H.; Wang, Y.; Luo, Y.-X.; Li, Z.-M.; Liang, R.-P.; Zhang, L.; Qiu, J.-D. Colorimetric and electrochemical arsenate assays by exploiting the peroxidase-like activity of FeOOH nanorods. *Microchim. Acta* **2019**, *186*, 732. [CrossRef]
61. Zhang, Y.; Zhao, T.; Zhang, X.; Akhtar, M.H.; Zhang, Q.; Li, M.; Yu, C. Photothermal and colorimetric dual-readout silver ions determination utilizing the oxidase-mimicking activity of MnO_2 nanosheets. *Sens. Actuators B Chem.* **2021**, *346*, 130494. [CrossRef]
62. Chen, G.; Jin, M.; Ma, J.; Yan, M.; Cui, X.; Wang, Y.; Zhang, X.; Li, H.; Zheng, W.; Zhang, Y.; et al. Competitive Bio-Barcode Immunoassay for Highly Sensitive Detection of Parathion Based on Bimetallic Nanozyme Catalysis. *J. Agric. Food Chem.* **2020**, *68*, 660–668. [CrossRef] [PubMed]
63. Bagheri, N.; Khataee, A.; Hassanzadeh, J.; Samaei, L. Highly sensitive chemiluminescence sensing system for organophosphates using mimic LDH supported ZIF-8 nanocomposite. *Sens. Actuators B Chem.* **2019**, *284*, 220–227. [CrossRef]
64. Song, P.; Wang, M.; Xue, Y.; Wang, A.-J.; Mei, L.P.; Feng, J.-J. Bimetallic PtNi nanozyme-driven dual-amplified photoelectrochemical aptasensor for ultrasensitive detection of sulfamethazine based on Z-scheme heterostructured Co9S8@In-CdS nanotubes. *Sens. Actuators B Chem.* **2022**, *371*, 132519. [CrossRef]

65. Sun, M.M.; He, M.X.; Jiang, S.J.; Wang, Y.Y.; Wang, X.X.; Liu, T.; Song, C.; Wang, S.N.; Rao, H.B.; Lu, Z.W. Multi-enzyme activity of three layers FeOx@ZnMnFeOy@Fe-Mn organogel for colorimetric detection of antioxidants and norfloxacin with smartphone. *Chem. Eng. J.* **2021**, *425*, 131823. [CrossRef]
66. Liu, X.; Yang, J.; Cheng, J.; Xu, Y.; Chen, W.; Li, Y. Facile preparation of four-in-one nanozyme catalytic platform and the application in selective detection of catechol and hydroquinone. *Sens. Actuators B-Chem.* **2021**, *337*, 129763. [CrossRef]
67. Ma, Y.; Zhu, M.; He, Q.; Zhao, M.; Cui, H. Photoenhanced Oxidase-Peroxidase-like $NiCo_2O_4$@MnO_2 Nanozymes for Colorimetric Detection of Hydroquinone. *ACS Sustain. Chem. Eng.* **2022**, *10*, 5651–5658. [CrossRef]
68. Zhou, J.; Tian, F.; Fu, R.; Yang, Y.; Jiao, B.; He, Y. Enzyme-Nanozyme Cascade Reaction-Mediated Etching of Gold Nanorods for the Detection of Escherichia coli. *ACS Appl. Nano Mater.* **2020**, *3*, 9016–9025. [CrossRef]
69. Hu, J.; Tang, F.; Wang, L.; Tang, M.; Jiang, Y.-Z.; Liu, C. Nanozyme sensor based-on platinum-decorated polymer nanosphere for rapid and sensitive detection of Salmonella typhimurium with the naked eye. *Sens. Actuators B Chem.* **2021**, *346*, 130560. [CrossRef]
70. Wu, S.-C.; Tsai, T.-T.; Li, T.-H.; Tung, C.-Y.; Chiu, P.-Y.; Lin, J.-H.; Chen, C.-F. Palladium-platinum bimetallic nanomaterials and their application in Staphylococcus aureus detection on paper-based devices. *Biosens. Bioelectron.* **2022**, *216*, 114669. [CrossRef]
71. Luo, S.; Liu, F.; Gu, S.; Chen, K.; Yang, G.; Gu, Y.; Cao, J.; Qu, L.-L. Nanozyme-mediated signal amplification for ultrasensitive photoelectrochemical sensing of Staphylococcus aureus based on Cu-C_3N_4-TiO(2)heterostructure. *Biosens. Bioelectron.* **2022**, *216*, 114593. [CrossRef]
72. Zhang, B.; Zou, H.; Qi, Y.; Zhang, X.; Sheng, R.; Zhang, Y.; Sun, R.; Chen, L.; Lv, R. Assembly of polyoxometalates/polydopamine nanozymes as a multifunctional platform for glutathione and *Escherichia coli* O157:H7 detection. *Microchem. J.* **2021**, *164*, 106013. [CrossRef]
73. Xie, D.; Liu, Q.; Wang, J.; Zhang, X.; Gu, K.; Sun, K. Chromogenic reaction of ABTS catalyzed by MnO_2 nanozyae and its application in the visual detection of Fe^{2+} and Pb^{2+}. *Huanjing Huaxue-Environ. Chem.* **2019**, *38*, 2843–2850. [CrossRef]
74. Qiu, Z.W.; Duan, W.; Cao, S.F.; Zeng, T.; Zhao, T.Y.; Huang, J.K.; Lu, X.Q.; Zeng, J.B. Highly Specific Colorimetric Probe for Fluoride by Triggering the Intrinsic Catalytic Activity of a AgPt-Fe_3O_4 Hybrid Nanozyme Encapsulated in SiO_2 Shells. *Environ. Sci. Technol.* **2022**, *56*, 1713–1723. [CrossRef] [PubMed]
75. Adegoke, O.; Zolotovskaya, S.; Abdolvand, A.; Daeid, N.N. Rapid and highly selective colorimetric detection of nitrite based on the catalytic-enhanced reaction of mimetic Au nanoparticle-CeO_2 nanoparticle-graphene oxide hybrid nanozyme. *Talanta* **2021**, *224*, 121875. [CrossRef] [PubMed]
76. Wang, L.J.; Xu, X.C.; Liu, P.; Wang, M.Z.; Niu, X.H.; Pan, J.M. A single-nanozyme colorimetric array based on target-induced differential surface passivation for quantification and discrimination of Cl^-, Br^- and I^- ions. *Anal. Chim. Acta* **2021**, *1160*, 338451. [CrossRef]
77. Hu, S.Y.; Yan, G.H.; Zhang, L.Y.; Yi, S.S.; Zhang, Z.T.; Wang, Y.; Chen, D.L. Highly Selective Colorimetric Detection of Cu^{2+} Using EDTA- Complexed Chlorophyll-Copper/ZnO Nanorods with Cavities Specific to Cu^{2+} as a Light-Activated Nanozyme. *Acs Appl. Mater. Interfaces* **2022**, *14*, 37716–37726. [CrossRef]
78. Xu, X.C.; Wang, L.J.; Zou, X.B.; Wu, S.W.; Pan, J.M.; Li, X.; Niu, X.H. Highly sensitive colorimetric detection of arsenite based on reassembly-induced oxidase-mimicking activity inhibition of dithiothreitol-capped Pd nanozyme. *Sens. Actuators B Chem.* **2019**, *298*, 126876. [CrossRef]
79. Honarasa, F.; Peyravi, F.; Amirian, H. C-dots/Mn_3O_4 nanocomposite as an oxidase nanozyme for colorimetric determination of ferrous ion. *J. Iran. Chem. Soc.* **2020**, *17*, 507–512. [CrossRef]
80. Liu, L.; Du, J.; Liu, W.E.; Guo, Y.L.; Wu, G.F.; Qi, W.N.; Lu, X.Q. Enhanced His@AuNCs oxidase-like activity by reduced graphene oxide and its application for colorimetric and electrochemical detection of nitrite. *Anal. Bioanal. Chem.* **2019**, *411*, 2189–2200. [CrossRef]
81. Shi, Y.; Liu, Z.; Liu, R.; Wu, R.; Zhang, J.J. DNA-encoded MXene-Pt nanozyme for enhanced colorimetric sensing of mercury ions. *Chem. Eng. J.* **2022**, *442*, 136072. [CrossRef]
82. Zhao, J.; Gong, J.W.; Wei, J.N.; Yang, Q.; Li, G.J.; Tong, Y.P.; He, W.W. Metal organic framework loaded fluorescent nitrogen-doped carbon nanozyme with light regulating redox ability for detection of ferric ion and glutathione. *J. Colloid Interface Sci.* **2022**, *618*, 11–21. [CrossRef] [PubMed]
83. Xue, Q.S.; Li, X.; Peng, Y.X.; Liu, P.; Peng, H.B.; Niu, X.H. Polyethylenimine-stabilized silver nanoclusters act as an oxidoreductase mimic for colorimetric determination of chromium(VI). *Microchim. Acta* **2020**, *187*, 263. [CrossRef] [PubMed]
84. Xu, R.X.; Wang, Z.Y.; Liu, S.Q.; Li, H. Bimetallic AuRu aerogel with enzyme-like activity for colorimetric detection of Fe2+ and glucose. *Chin. Chem. Lett.* **2022**, *33*, 4683–4686. [CrossRef]
85. Huang, L.J.; Zhu, Q.R.; Zhu, J.; Luo, L.P.; Pu, S.H.; Zhang, W.T.; Zhu, W.X.; Sun, J.; Wang, J.L. Portable Colorimetric Detection of Mercury(II) Based on a Non-Noble Metal Nanozyme with Tunable Activity. *Inorg. Chem.* **2019**, *58*, 1638–1646. [CrossRef]
86. Lin, L.; Chen, D.X.; Lu, C.F.; Wang, X.X. Fluorescence and colorimetric dual-signal determination of Fe^{3+} and glutathione with $MoSe_2$@Fe nanozyme. *Microchem. J.* **2022**, *177*, 107283. [CrossRef]
87. Wen, S.S.; Zhang, Z.W.; Zhang, Y.P.; Liu, H.; Ma, X.W.; Li, L.J.; Song, W.; Zhao, B. Ultrasensitive Stimulation Effect of Fluoride Ions on a Novel Nanozyme-SERS System. *ACS Sustain. Chem. Eng.* **2020**, *8*, 11906–11913. [CrossRef]

88. Yang, Z.P.; Liu, Y.Q.; Liu, Y.; Wang, Y.Y.; Rao, H.B.; Liu, Y.; Yin, J.J.; Yue, G.Z.; Wu, C.M.; Li, H.; et al. Preparation of porous uranium oxide hollow nanospheres with peroxidase mimicking activity: Application to the colorimetric determination of tin(II). *Microchim. Acta* **2019**, *186*, 501. [CrossRef]
89. Zhang, G.Y.; Yu, K.; Zhou, B.J.; Wang, J.Y.; Zheng, C.; Qu, L.J.; Chai, H.N.; Zhang, X.J. Magnetic zirconium-based Prussian blue analog nanozyme: Enhanced peroxidase-mimicking activity and colorimetric sensing of phosphate ion. *Microchim. Acta* **2022**, *189*, 220. [CrossRef]
90. Xiong, Y.H.; Su, L.J.; Ye, F.G.; Zhao, S.L. Porous Oxyhydroxide Derived from Metal-Organic Frameworks as Efficient Triphosphatase-like Nanozyme for Chromium(III) Ion Colorimetric Sensing. *Acs Appl. Bio Mater.* **2021**, *4*, 6962–6973. [CrossRef]
91. Song, G.C.; Li, J.C.; Majid, Z.; Xu, W.T.; He, X.Y.; Yao, Z.Y.; Luo, Y.B.; Huang, K.L.; Cheng, N. Phosphatase-like activity of single-atom Ce-N-C nanozyme for rapid detection of Al^{3+}. *Food Chem.* **2022**, *390*, 133127. [CrossRef]
92. Goswami, J.; Saikia, L.; Hazarika, P. Carbon Dots-Decorated g-C_3N_4 as Peroxidase Nanozyme for Colorimetric Detection of Cr(VI) in Aqueous Medium. *Chemistryselect* **2022**, *7*, e202201963. [CrossRef]
93. Fang, Y.M.; Zhang, Y.; Cao, L.G.; Yang, J.Z.; Hu, M.H.; Pang, Z.L.; He, J.H. Portable Hg^{2+} Nanosensor with ppt Level Sensitivity Using Nanozyme as the Recognition Unit, Enrichment Carrier, and Signal Amplifier. *Acs Appl. Mater. Interfaces* **2020**, *12*, 11761–11768. [CrossRef]
94. Wen, S.H.; Zhong, X.L.; Wu, Y.D.; Liang, R.P.; Zhang, L.; Qiu, J.D. Colorimetric Assay Conversion to Highly Sensitive Electrochemical Assay for Bimodal Detection of Arsenate Based on Cobalt Oxyhydroxide Nanozyme via Arsenate Absorption. *Anal. Chem.* **2019**, *91*, 6487–6497. [CrossRef]
95. Kulandaivel, S.; Lo, W.C.; Lin, C.H.; Yeh, Y.C. Cu-PyC MOF with oxidoreductase-like catalytic activity boosting colorimetric detection of Cr(VI) on paper. *Anal. Chim. Acta* **2022**, *1227*, 340335. [CrossRef]
96. Amini, R.; Rahimpour, E.; Jouyban, A. An optical sensing platform based on hexacyanoferrate intercalated layered double hydroxide nanozyme for determination of chromium in water. *Anal. Chim. Acta* **2020**, *1117*, 9–17. [CrossRef]
97. Serebrennikova, K.V.; Komova, N.S.; Berlina, A.N.; Zherdev, A.V.; Dzantiev, B.B. Tannic Acid-Capped Gold Nanoparticles as a Novel Nanozyme for Colorimetric Determination of Pb2+ Ions. *Chemosensors* **2021**, *9*, 332. [CrossRef]
98. Liu, X.N.; Huang, L.J.; Wang, Y.P.; Sun, J.; Yue, T.L.; Zhang, W.T.; Wang, J.L. One-pot bottom-up fabrication of a 2D/2D heterojuncted nanozyme towards optimized peroxidase-like activity for sulfide ions sensing. *Sens. Actuators B-Chem.* **2020**, *306*, 127565. [CrossRef]
99. Vu, T.H.; Nguyen, P.T.; Kim, M.I. Polydopamine-Coated Co_3O_4 Nanoparticles as an Efficient Catalase Mimic for Fluorescent Detection of Sulfide Ion. *Biosensors* **2022**, *12*, 1047. [CrossRef]
100. Xue, Q.S.; Niu, X.H.; Liu, P.; Wang, M.Z.; Peng, Y.X.; Peng, H.B.; Li, X. Analyte-triggered citrate-stabilized Au nanoparticle aggregation with accelerated peroxidase-mimicking activity for catalysis-based colorimetric sensing of arsenite. *Sensors and Actuators B-Chem.* **2021**, *334*, 129650. [CrossRef]
101. Huang, H.; Li, M.N.; Hao, M.W.; Yu, L.L.; Li, Y.X. A novel selective detection method for sulfide in food systems based on the GMP-Cu nanozyme with laccase activity. *Talanta* **2021**, *235*, 122775. [CrossRef] [PubMed]
102. Zhao, Q.; Gou, W.X.; Zhang, X.T.; Zhang, M.Y.; Bu, Y.R.; Wang, L.J.; Hu, L.; Yao, W.L.; Yan, Z.Q. Hg^{2+}-activated oxidase-like activity of Ag_2S@graphene oxide nanozyme and its naked-eye monitoring Hg^{2+} application with obvious hyperchromic effect. *Appl. Surf. Sci.* **2021**, *545*, 148973. [CrossRef]
103. Luo, L.P.; Su, Z.H.; Zhuo, J.C.; Huang, L.J.; Nian, Y.; Su, L.H.; Zhang, W.T.; Wang, J.L. Copper-Sensitized "Turn On" Peroxidase-Like Activity of $MMoO_4$ (M = Co, Ni) Flowers for Selective Detection of Aquatic Copper Ions. *Acs Sustain. Chem. Eng.* **2020**, *8*, 12568–12576. [CrossRef]
104. Shi, W.; He, M.Q.; Li, W.T.; Wei, X.; Bui, B.; Chen, M.L.; Chen, W. Cu-Based Metal-Organic Framework Nanoparticles for Sensing Cr(VI) Ions. *Acs Appl. Nano Mater.* **2021**, *4*, 802–810. [CrossRef]
105. Borthakur, P.; Das, M.R.; Szunerits, S.; Boukherroub, R. CuS Decorated Functionalized Reduced Graphene Oxide: A Dual Responsive Nanozyme for Selective Detection and Photoreduction of Cr(VI) in an Aqueous Medium. *Acs Sustain. Chem. Eng.* **2019**, *7*, 16131–16143. [CrossRef]
106. Mao, Y.; Gao, S.J.; Yao, L.L.; Wang, L.; Qu, H.; Wu, Y.; Chen, Y.; Zheng, L. Single-atom nanozyme enabled fast and highly sensitive colorimetric detection of Cr(VI). *J. Hazard. Mater.* **2021**, *408*, 124898. [CrossRef] [PubMed]
107. Yi, W.W.; Zhang, P.; Wang, Y.P.; Li, Z.P.; Guo, Y.J.; Liu, M.; Dong, C.; Li, C.F. Copper ferrite nanoparticles loaded on reduced graphene oxide nanozymes for the ultrasensitive colorimetric assay of chromium ions. *Anal. Methods* **2022**, *14*, 3434–3443. [CrossRef]
108. Li, W.J.; Liu, C.; Liu, D.; Liu, S.D.; You, T.Y. Ratiometric fluorescent sensing of mercury (II) ion based on the Pt nanozyme-triggered fluorescence resonance energy transfer between Si quantum dots and 2,3-diaminophenazine. *Sens. Actuators A-Phys.* **2021**, *331*, 112976. [CrossRef]
109. Ko, E.; Hur, W.; Son, S.E.; Seong, G.H.; Han, D.K. Au nanoparticle-hydrogel nanozyme-based colorimetric detection for on-site monitoring of mercury in river water. *Microchim. Acta* **2021**, *188*, 382. [CrossRef] [PubMed]
110. Zhou, J.J.; Zhu, W.J.; Lv, X.Q.; Du, X.P.; He, J.L.; Cai, J. Dendritic Silica Nanospheres with Au-Pt Nanoparticles as Nanozymes for Label-Free Colorimetric Hg^{2+} Detection. *Acs Appl. Nano Mater.* **2022**, *5*, 18885–18893. [CrossRef]
111. Wang, C.H.; Tang, G.G.; Tan, H.L. Colorimetric determination of mercury(II) via the inhibition by ssDNA of the oxidase-like activity of a mixed valence state cerium-based metal-organic framework. *Microchim. Acta* **2018**, *185*, 475. [CrossRef] [PubMed]

112. Li, Q.; Wu, F.; Mao, M.; Ji, X.; Wei, L.Y.; Li, J.Y.; Ma, L. A dual-mode colorimetric sensor based on copper nanoparticles for the detection of mercury-(ii) ions. *Anal. Methods* **2019**, *11*, 4014–4021. [CrossRef]
113. Dai, J.W.; Wang, L.W.; Xu, F.A.; Ma, G.R. Dual-functional Fe-MoS_2@AuNPs for simple and sensitive colorimetric-electrochemical coupled detection of Hg^{2+}. *Ionics* **2022**, *28*, 5251–5255. [CrossRef]
114. Cao, L.G.; Fang, Y.M.; Zhang, Y.; Yang, J.Z.; He, J.H. Colorimetric Detection of Hg^{2+} Based on the Promotion of Oxidase-Like Catalytic Activity of Ag Nanowires. *Int. J. Nanosci.* **2020**, *19*, 2050004. [CrossRef]
115. Niu, X.H.; He, Y.F.; Li, X.; Zhao, H.L.; Pan, J.M.; Qiu, F.X.; Lan, M.B. A peroxidase-mimicking nanosensor with Hg^{2+}-triggered enzymatic activity of cysteine-decorated ferromagnetic particles for ultrasensitive Hg^{2+} detection in environmental and biological fluids. *Sens. Actuators B-Chem.* **2019**, *281*, 445–452. [CrossRef]
116. Fu, M.L.; Li, L.; Yang, D.Y.; Tu, Y.F.; Yan, J.L. Colorimetric detections of iodide and mercuric ions based on a regulation of an Enzyme-Like activity from gold nanoclusters. *Spectrochim. Acta Part A Mol. Biomol. Spectrosc.* **2022**, *279*, 121450. [CrossRef]
117. Tian, X.; Liao, H.; Wang, M.; Feng, L.Y.; Fu, W.S.; Hu, L.Z. Highly sensitive chemiluminescent sensing of intracellular Al^{3+} based on the phosphatase mimetic activity of cerium oxide nanoparticles. *Biosens. Bioelectron.* **2020**, *152*, 112027. [CrossRef]
118. Liu, H.; Hua, Y.; Cai, Y.Y.; Feng, L.P.; Li, S.; Wang, H. Mineralizing gold-silver bimetals into hemin-melamine matrix: A nanocomposite nanozyme for visual colorimetric analysis of H_2O_2 and glucose. *Anal. Chim. Acta* **2019**, *1092*, 57–65. [CrossRef]
119. Guan, H.J.; Zhao, Y.F.; Cheng, J.H.; Zhang, Y.D.; Yang, Q.; Zhang, B. Fabrication of Pt/CeO_2/NCNFs with embedded structure as high-efficiency nanozyme for electrochemical sensing of hydrogen peroxide. *Synth. Met.* **2020**, *270*, 116604. [CrossRef]
120. Lei, Y.; He, B.; Huang, S.J.; Chen, X.Y.; Sun, J. Facile Fabrication of 1-Methylimidazole/Cu Nanozyme with Enhanced Laccase Activity for Fast Degradation and Sensitive Detection of Phenol Compounds. *Molecules* **2022**, *27*, 4712. [CrossRef]
121. Adegoke, O.; Daeid, N.N. Polymeric-coated Fe-doped ceria/gold hybrid nanocomposite as an aptasensor for the catalytic enhanced colorimetric detection of 2,4-dinitrophenol. *Colloids Surf. A-Physicochem. Eng. Asp.* **2021**, *627*, 127194. [CrossRef]
122. Ma, Y.J.; Deng, M.S.; Wang, X.F.; Gao, X.H.; Song, H.X.; Zhu, Y.H.; Feng, L.Y.; Zhang, Y. 2H-MoS_2/Co_3O_4 nanohybrid with type I nitroreductase-mimicking activity for the electrochemical assays of nitroaromatic compounds. *Anal. Chim. Acta* **2022**, *1221*, 340078. [CrossRef]
123. Sun, X.; Xie, Y.; Chu, H.C.; Long, M.; Zhang, M.Y.; Wang, Y.; Hu, X.Y. A highly sensitive electrochemical biosensor for the detection of hydroquinone based on a magnetic covalent organic framework and enzyme for signal amplification. *New J. Chem.* **2022**, *46*, 11902–11909. [CrossRef]
124. Huang, H.; Lei, L.L.; Bai, J.; Zhang, L.; Song, D.H.; Zhao, J.Q.; Li, J.L.; Li, Y.X. Efficient elimination and detection of phenolic compounds in juice using laccase mimicking nanozymes. *Chin. J. Chem. Eng.* **2021**, *29*, 167–175. [CrossRef]
125. Zhu, X.; Tang, J.; Ouyang, X.L.; Liao, Y.B.; Feng, H.P.; Yu, J.F.; Chen, L.; Lu, Y.T.; Yi, Y.Y.; Tang, L. Multifunctional MnCo@C yolk-shell nanozymes with smartphone platform for rapid colorimetric analysis of total antioxidant capacity and phenolic compounds. *Biosens. Bioelectron.* **2022**, *216*, 114652. [CrossRef]
126. Omar, N.A.; Jabbar, H.S. $NiFe_2O_4$ nanoparticles as nanozymes, a new colorimetric probe for 2,4-dichlorophenoxyacetic acid herbicide detection. *Inorg. Chem. Commun.* **2022**, *146*, 110104. [CrossRef]
127. Zhu, X.Y.; Liu, P.; Ge, Y.; Wu, R.M.; Xue, T.; Sheng, Y.Y.; Ai, S.R.; Tang, K.J.; Wen, Y.P. MoS_2/MWCNTs porous nanohybrid network with oxidase-like characteristic as electrochemical nanozyme sensor coupled with machine learning for intelligent analysis of carbendazim. *J. Electroanal. Chem.* **2020**, *862*, 113940. [CrossRef]
128. Shah, M.M.; Ren, W.; Irudayaraj, J.; Sajini, A.A.; Ali, M.I.; Ahmad, B. Colorimetric Detection of Organophosphate Pesticides Based on Acetylcholinesterase and Cysteamine Capped Gold Nanoparticles as Nanozyme. *Sensors* **2021**, *21*, 8050. [CrossRef]
129. Huang, L.J.; Sun, D.W.; Pu, H.B.; Wei, Q.Y.; Luo, L.P.; Wang, J.L. A colorimetric paper sensor based on the domino reaction of acetylcholinesterase and degradable gamma-MnOOH nanozyme for sensitive detection of organophosphorus pesticides. *Sens. Actuators B-Chem.* **2019**, *290*, 573–580. [CrossRef]
130. Wu, J.H.; Yang, Q.T.; Li, Q.; Li, H.Y.; Li, F. Two-Dimensional MnO_2 Nanozyme-Mediated Homogeneous Electrochemical Detection of Organophosphate Pesticides without the Interference of H_2O_2 and Color. *Anal. Chem.* **2021**, *93*, 4084–4091. [CrossRef]
131. Xu, L.L.; Xiong, Y.; Wu, R.M.; Geng, X.; Li, M.H.; Yao, H.; Wang, X.; Wen, Y.P.; Ai, S.R. An Emerging Machine Learning Strategy for the Fabrication of Nanozyme Sensor and Voltametric Determination of Benomyl In Agro-Products. *J. Electrochem. Soc.* **2022**, *169*, 047506. [CrossRef]
132. Li, F.; Jiang, J.M.; Peng, H.; Li, C.X.; Li, B.; He, J.B. Platinum nanozyme catalyzed multichannel colorimetric sensor array for identification and detection of pesticides. *Sens. Actuators B-Chem.* **2022**, *369*, 132334. [CrossRef]
133. Zhu, X.Y.; Lin, L.; Wu, R.M.; Zhu, Y.F.; Sheng, Y.Y.; Nie, P.C.; Liu, P.; Xu, L.L.; Wen, Y.P. Portable wireless intelligent sensing of ultra-trace phytoregulator alpha-naphthalene acetic acid using self-assembled phosphorene/Ti_3C_2-MXene nanohybrid with high ambient stability on laser induced porous graphene as nanozyme flexible electrode. *Biosens. Bioelectron.* **2021**, *179*, 113062. [CrossRef] [PubMed]
134. Khairy, M.; Ayoub, H.A.; Banks, C.E. Non-enzymatic electrochemical platform for parathion pesticide sensing based on nanometer-sized nickel oxide modified screen-printed electrodes. *Food Chem.* **2018**, *255*, 104–111. [CrossRef]
135. Li, Q.L.; Li, H.; Li, K.X.; Gu, Y.; Wang, Y.J.; Yang, D.Z.; Yang, Y.L.; Gao, L. Specific colorimetric detection of methylmercury based on peroxidase-like activity regulation of carbon dots/Au NPs nanozyme. *J. Hazard. Mater.* **2023**, *441*, 129919. [CrossRef]

136. Weerathunge, P.; Behera, B.K.; Zihara, S.; Singh, M.; Prasad, S.N.; Hashmi, S.; Mariathomas, P.R.D.; Bansal, V.; Ramanathan, R. Dynamic interactions between peroxidase-mimic silver NanoZymes and chlorpyrifos-specific aptamers enable highly-specific pesticide sensing in river water. *Anal. Chim. Acta* **2019**, *1083*, 157–165. [CrossRef]
137. Song, G.C.; Zhang, J.J.; Huang, H.X.; Wang, X.; He, X.Y.; Luo, Y.B.; Li, J.C.; Huang, K.L.; Cheng, N. Single-atom Ce-N-C nanozyme bioactive paper with a 3D-printed platform for rapid detection of organophosphorus and carbamate pesticide residues. *Food Chem.* **2022**, *387*, 132896. [CrossRef] [PubMed]
138. Wei, J.C.; Yang, L.L.; Luo, M.; Wang, Y.T.; Li, P. Nanozyme-assisted technique for dual mode detection of organophosphorus pesticide. *Ecotoxicol. Environ. Saf.* **2019**, *179*, 17–23. [CrossRef] [PubMed]
139. Sun, Y.Z.; Wei, J.C.; Zou, J.; Cheng, Z.H.; Huang, Z.M.; Gu, L.Q.; Zhong, Z.F.; Li, S.L.; Wang, Y.T.; Li, P. Electrochemical detection of methyl-paraoxon based on bifunctional cerium oxide nanozyme with catalytic activity and signal amplification effect. *J. Pharm. Anal.* **2021**, *11*, 653–660. [CrossRef] [PubMed]
140. Li, S.H.; Pang, C.H.; Ma, X.H.; Zhang, Y.L.; Xu, Z.; Li, J.P.; Zhang, M.; Wang, M.Y. Microfluidic paper-based chip for parathion-methyl detection based on a double catalytic amplification strategy. *Microchim. Acta* **2021**, *188*, 438. [CrossRef] [PubMed]
141. Boruah, P.K.; Das, M.R. Dual responsive magnetic Fe_3O_4-TiO_2/graphene nanocomposite as an artificial nanozyme for the colorimetric detection and photodegradation of pesticide in an aqueous medium. *J. Hazard. Mater.* **2020**, *385*, 121516. [CrossRef]
142. Wang, Z.H.; Shu, Y.Y.; Li, J.J.; Liang, A.H.; Jiang, Z.L. Silver nanosol RRS aptamer assay of trace glyphosate based on gold-doped. *Microchem. J.* **2022**, *176*, 107252. [CrossRef]
143. Wu, Z.; Hu, Y.; Pan, X.L.; Tang, Y.; Dai, Y.F.; Wu, Y.G. A liquid colorimetric chemosensor for ultrasensitive detection of glyphosate residues in vegetables using a metal oxide with intrinsic peroxidase catalytic activity. *Food Addit. Contam. Part A Chem. Anal. Control. Expo. Risk Assess* **2022**, *39*, 710–723. [CrossRef] [PubMed]
144. Chen, D.; Wang, C.Q.; Yang, D.Z.; Deng, H.M.; Li, Q.L.; Chen, L.; Zhao, G.K.; Shi, J.L.; Zhang, K.; Yang, Y.L. A portable smartphone-based detection of glyphosate based on inhibiting peroxidase-like activity of heptanoic acid/Prussian blue decorated Fe_3O_4 nanoparticles. *RSC Adv.* **2022**, *12*, 25060–25067. [CrossRef] [PubMed]
145. Liu, P.; Li, X.; Xu, X.C.; Ye, K.; Wang, L.J.; Zhu, H.J.; Wang, M.Z.; Niu, X.H. Integrating peroxidase-mimicking activity with photoluminescence into one framework structure for high-performance ratiometric fluorescent pesticide sensing. *Sens. Actuators B-Chem.* **2021**, *328*, 129024. [CrossRef]
146. Xu, X.C.; Wu, S.W.; Guo, D.Z.; Niu, X.H. Construction of a recyclable oxidase-mimicking Fe_3O_4@MnOx-based colorimetric sensor array for quantifying and identifying chlorophenols. *Anal. Chim. Acta* **2020**, *1107*, 203–212. [CrossRef]
147. Zhang, X.; Zhou, Y.; Huang, X.; Hu, X.; Huang, X.; Yin, L.; Huang, Q.; Wen, Y.; Li, B.; Shi, J.; et al. Switchable aptamer-fueled colorimetric sensing toward agricultural fipronil exposure sensitized with affiliative metal-organic framework. *Food Chem.* **2022**, *407*, 135115. [CrossRef]
148. Ge, J.; Yang, L.K.; Li, Z.H.; Wan, Y.; Mao, D.S.; Deng, R.J.; Zhou, Q.; Yang, Y.; Tan, W.H. A colorimetric smartphone-based platform for pesticides detection using Fe-N/C single-atom nanozyme as oxidase mimetics. *J. Hazard. Mater.* **2022**, *436*, 129199. [CrossRef]
149. Xiao, J.X.; Hu, X.L.; Wang, K.; Zou, Y.M.; Gyimah, E.; Yakubu, S.; Zhang, Z. A novel signal amplification strategy based on the competitive reaction between 2D Cu-TCPP(Fe) and polyethyleneimine (PEI) in the application of an enzyme-free and ultrasensitive electrochemical immunosensor for sulfonamide detection. *Biosens. Bioelectron.* **2020**, *150*, 111883. [CrossRef]
150. Wei, D.; Zhang, X.; Chen, B.; Zeng, K. Using bimetallic Au@Pt nanozymes as a visual tag and as an enzyme mimic in enhanced sensitive lateral-flow immunoassays: Application for the detection of streptomycin. *Anal. Chim. Acta* **2020**, *1126*, 106–113. [CrossRef]
151. Shen, Y.Z.; Wei, Y.L.; Liu, Z.M.; Nie, C.; Ye, Y.W. Engineering of 2D artificial nanozyme-based blocking effect-triggered colorimetric sensor for onsite visual assay of residual tetracycline in milk. *Microchim. Acta* **2022**, *189*, 233. [CrossRef]
152. Liu, B.X.; Zhu, H.J.; Feng, R.L.; Wang, M.Z.; Hu, P.W.; Pan, J.M.; Niu, X.H. Facile molecular imprinting on magnetic nanozyme surface for highly selective colorimetric detection of tetracycline. *Sens. Actuators B Chem.* **2022**, *370*, 132451. [CrossRef]
153. Zhu, B.C.; Dong, S.M.; Liu, Z.C.; Gao, Y.; Zhu, X.X.; Xie, M.; Liu, Q.Y. Enhanced peroxidase-like activity of bimetal (Fe/Co) MIL-101 for determination of tetracycline and hydrogen peroxide. *New J. Chem.* **2022**, *46*, 21834–21844. [CrossRef]
154. Li, S.H.; Ma, X.H.; Pang, C.H.; Wang, M.Y.; Yin, G.H.; Xu, Z.; Li, J.P.; Luo, J.H. Novel chloramphenicol sensor based on aggregation-induced electrochemiluminescence and nanozyme amplification. *Biosens. Bioelectron.* **2021**, *176*, 112944. [CrossRef] [PubMed]
155. Tang, Y.; Hu, Y.; Zhou, P.; Wang, C.X.; Tao, H.; Wu, Y.G. Colorimetric Detection of Kanamycin Residue in Foods Based on the Aptamer-Enhanced Peroxidase-Mimicking Activity of Layered WS2 Nanosheets. *J. Agric. Food Chem.* **2021**, *69*, 2884–2893. [CrossRef]
156. Zhang, Z.P.; Liu, Y.; Huang, P.C.; Wu, F.Y.; Ma, L.H. Polydopamine molecularly imprinted polymer coated on a biomimetic iron-based metal-organic framework for highly selective fluorescence detection of metronidazole. *Talanta* **2021**, *232*, 122411. [CrossRef]
157. Lu, C.; Tang, L.H.; Gao, F.; Li, Y.Z.; Liu, J.W.; Zheng, J.K. DNA-encoded bimetallic Au-Pt dumbbell nanozyme for high-performance detection and eradication of *Escherichia coli* O157:H7. *Biosens. Bioelectron.* **2021**, *187*, 113327. [CrossRef]
158. Liu, J.C.; Xing, Y.P.; Xue, B.Y.; Zhou, X.H. Nanozyme enhanced paper-based biochip with a smartphone readout system for rapid detection of cyanotoxins in water. *Biosens. Bioelectron.* **2022**, *205*, 114099. [CrossRef]

159. Zhang, J.C.; Bai, Q.; Bi, X.L.; Zhang, C.H.; Shi, M.Y.; Yu, W.W.; Du, F.L.; Wang, L.A.; Wang, Z.B.; Zhu, Z.L.; et al. Piezoelectric enhanced peroxidase-like activity of metal-free sulfur doped graphdiyne nanosheets for efficient water pollutant degradation and bacterial disinfection. *Nano Today* **2022**, *43*, 101429. [CrossRef]
160. Geng, X.; Xie, X.N.; Liang, Y.C.; Li, Z.Q.; Yang, K.; Tao, J.; Zhang, H.; Wang, Z. Facile Fabrication of a Novel Copper Nanozyme for Efficient Dye Degradation. *Acs Omega* **2021**, *6*, 6284–6291. [CrossRef]
161. Chen, Q.M.; Liu, Y.; Lu, Y.W.; Hou, Y.J.; Zhang, X.D.; Shi, W.B.; Huang, Y.M. Atomically dispersed Fe/Bi dual active sites single-atom nanozymes for cascade catalysis and peroxymonosulfate activation to degrade dyes. *J. Hazard. Mater.* **2022**, *422*, 126929. [CrossRef]
162. Su, C.; Wang, B.X.; Li, S.L.; Wie, Y.B.; Wang, Q.L.; Li, D.J. Fabrication of Pd@ZnNi-MOF/GO Nanocomposite and Its Application for H_2O_2 Detection and Catalytic Degradation of Methylene Blue Dyes. *Chemistryselect* **2021**, *6*, 8480–8489. [CrossRef]
163. Zhu, J.L.; Luo, G.; Xi, X.X.; Wang, Y.J.; Selvaraj, J.N.; Wen, W.; Zhang, X.H.; Wang, S.F. Cu^{2+}-modified hollow carbon nanospheres: An unusual nanozyme with enhanced peroxidase-like activity. *Microchim. Acta* **2021**, *188*, 8. [CrossRef]
164. Shams, S.; Ahmad, W.; Memon, H.; Wei, Y.; Yuan, Q.P.; Liang, H. Facile synthesis of laccase mimic Cu/H3BTC MOF for efficient dye degradation and detection of phenolic pollutants. *RSC Adv.* **2019**, *9*, 40845–40854. [CrossRef] [PubMed]
165. Li, Z.G.; Chen, Z.M.; Zhu, Q.P.; Song, J.J.; Li, S.; Liu, X.H. Improved performance of immobilized laccase on Fe_3O_4@C-Cu^{2+} nanoparticles and its application for biodegradation of dyes. *J. Hazard. Mater.* **2020**, *399*, 123088. [CrossRef]
166. Zha, J.Q.; Wu, W.G.; Xie, P.; Han, H.H.; Fang, Z.; Chen, Y.T.; Jia, Z.F. Polymeric Nanocapsule Enhances the Peroxidase-like Activity of Fe_3O_4 Nanozyme for Removing Organic Dyes. *Catalysts* **2022**, *12*, 614. [CrossRef]
167. Su, H.F.; Zhang, Y.Z.; Lu, Z.C.; Wang, Q.Y. A mechanism of microbial sensitivity regulation on interventional remediation by nanozyme manganese oxide in soil heavy metal pollution. *J. Clean. Prod.* **2022**, *373*, 133470. [CrossRef]
168. Yan, Z.Q.; Xing, L.; Zhao, L.; Zhang, X.Y.; Zhang, Y.F.; Tang, Y.L.; Zhou, X.M.; Hu, L.; Zhu, N.L. beta-Cyclodextrin and graphene oxide co-strengthened AgRu bimetal mesoporous nanozyme: An efficient strategy for visual detection and removal of toxic Hg^{2+} and Cl. *J. Environ. Chem. Eng.* **2022**, *10*, 108242. [CrossRef]
169. Wang, F.M.; Zhang, Y.; Liu, Z.W.; Ren, J.S.; Qu, X.G. A mesoporous encapsulated nanozyme for decontaminating two kinds of wastewater and avoiding secondary pollution. *Nanoscale* **2020**, *12*, 14465–14471. [CrossRef]
170. Lv, R.; Sun, S.Y.; Wang, K.; Golubev, Y.A.; Dong, F.Q.; Kotova, O.B.; Liu, J.; Liu, M.X.; Tan, D.Y. Design and construction of copper-containing organophyllosilicates as laccase-mimicking nanozyme for efficient removal of phenolic pollutants. *J. Mater. Sci.* **2022**, *57*, 10084–10099. [CrossRef]
171. Jiang, J.F.; He, C.Y.; Wang, S.; Jiang, H.; Li, J.D.; Li, L.S. Recyclable ferromagnetic chitosan nanozyme for decomposing phenol. *Carbohydr. Polym.* **2018**, *198*, 348–353. [CrossRef] [PubMed]
172. Lin, Y.M.; Wang, F.; Yu, J.; Zhang, X.; Lu, G.P. Iron single-atom anchored N-doped carbon as a 'laccase-like' nanozyme for the degradation and detection of phenolic pollutants and adrenaline. *J. Hazard. Mater.* **2022**, *425*, 127763. [CrossRef] [PubMed]
173. Wang, J.H.; Huang, R.L.; Qi, W.; Su, R.X.; Binks, B.P.; He, Z.M. Construction of a bioinspired laccase-mimicking nanozyme for the degradation and detection of phenolic pollutants. *Appl. Catal. B Environ.* **2019**, *254*, 452–462. [CrossRef]
174. Tang, Y.; Jiang, S.L.; Li, W.Y.; Shah, S.J.; Zhao, Z.X.; Pan, L.; Zhao, Z.X. Confined construction of COF@Cu-nanozyme with high activity and stability as laccase biomimetic catalyst for the efficient degradation of phenolic pollutants. *Chem. Eng. J.* **2022**, *448*, 137701. [CrossRef]
175. Xu, X.J.; Wang, J.H.; Huang, R.L.; Qi, W.; Su, R.X.; He, Z.M. Preparation of laccase mimicking nanozymes and their catalytic oxidation of phenolic pollutants. *Catal. Sci. Technol.* **2021**, *11*, 3402–3410. [CrossRef]
176. Liang, S.; Wu, X.L.; Zong, M.H.; Lou, W.Y. Construction of Zn-heptapeptide bionanozymes with intrinsic hydrolase-like activity for degradation of di(2-ethylhexyl) phthalate. *J. Colloid Interface Sci.* **2022**, *622*, 860–870. [CrossRef] [PubMed]
177. Zandieh, M.; Liu, J.W. Removal and Degradation of Microplastics Using the Magnetic and Nanozyme Activities of Bare Iron Oxide Nanoaggregates. *Angew. Chem. Int. Ed.* **2022**, *61*, e202212013. [CrossRef]
178. Khulbe, K.; Karmakar, K.; Ghosh, S.; Chandra, K.; Chakravortty, D.; Mugesh, G. Nanoceria-Based Phospholipase-Mimetic Cell Membrane Disruptive Antibiofilm Agents. *ACS Appl. Bio Mater.* **2020**, *3*, 4316–4328. [CrossRef]
179. Fuentes, K.M.; Onna, D.; Rioual, T.; Huvelle, M.A.L.; Britto, F.; Simian, M.; Sanchez-Dominguez, M.; Soler-Illia, G.; Bilmes, S.A. Copper upcycling by hierarchical porous silica spheres functionalized with branched polyethylenimine: Antimicrobial and catalytic applications. *Microporous Mesoporous Mater.* **2021**, *327*, 111391. [CrossRef]
180. Fan, Y.F.; Gan, X.R.; Zhao, H.M.; Zeng, Z.X.; You, W.J.; Quan, X. Multiple application of SAzyme based on carbon nitride nanorod-supported Pt single-atom for H_2O_2 detection, antibiotic detection and antibacterial therapy. *Chem. Eng. J.* **2022**, *427*, 131572. [CrossRef]
181. Boruah, P.K.; Darabdhara, G.; Das, M.R. Polydopamine functionalized graphene sheets decorated with magnetic metal oxide nanoparticles as efficient nanozyme for the detection and degradation of harmful triazine pesticides. *Chemosphere* **2021**, *268*, 129328. [CrossRef]
182. Yang, D.C.; Huo, J.Q.; Zhang, Z.; An, Z.X.; Dong, H.J.; Wang, Y.N.; Duan, W.D.; Chen, L.; He, M.X.; Gao, S.T.; et al. Citric acid modified ultrasmall copper peroxide nanozyme for in situ remediation of environmental sulfonylurea herbicide contamination. *J. Hazard. Mater.* **2023**, *443*, 130265. [CrossRef] [PubMed]
183. Pan, J.; Ai, X.; Ma, C.; Zhang, G. Degradable Vinyl Polymers for Combating Marine Biofouling. *Acc. Chem. Res.* **2022**, *55*, 1586–1598. [CrossRef]

184. Dobretsov, S.; Abed, R.M.M.; Teplitski, M. Mini-review: Inhibition of biofouling by marine microorganisms. *Biofouling* **2013**, *29*, 423–441. [CrossRef] [PubMed]
185. Cao, S.; Wang, J.; Chen, H.; Chen, D. Progress of marine biofouling and antifouling technologies. *Chin. Sci. Bull.* **2011**, *56*, 598–612. [CrossRef]
186. Luo, Q.; Li, Y.; Huo, X.; Li, L.; Song, Y.; Chen, S.; Lin, H.; Wang, N. Atomic Chromium Coordinated Graphitic Carbon Nitride for Bioinspired Antibiofouling in Seawater. *Adv. Sci.* **2022**, *9*, 2105346. [CrossRef] [PubMed]
187. Wang, W.; Luo, Q.; Li, J.Y.; Li, L.Q.; Li, Y.H.; Huo, X.B.; Du, X.W.; Li, Z.S.; Wang, N. Photothermal-Amplified Single Atom Nanozyme for Biofouling Control in Seawater. *Adv. Funct. Mater.* **2022**, *32*, 2205461. [CrossRef]
188. You, J.; Chen, F.; Zhao, X.; Chen, Z. Preparation, characterization and catalytic oxidation property of CeO_2/Cu^{2+}-attapulgite (ATP) nanocomposites. *J. Rare Earths* **2010**, *28*, 347–352. [CrossRef]
189. Feng, F.; Zhang, X.; Mu, B.; Wang, P.X.; Chen, Z.S.; Zhang, J.H.; Zhang, H.F.; Zhuang, J.L.; Zhao, L.; An, Q.; et al. Attapulgite Doped with Fe and Cu Nanooxides as Peroxidase Nanozymes for Antibacterial Coatings. *ACS Appl. Nano Mater.* **2022**, *5*, 16720–16730. [CrossRef]
190. Wei, F.; Cui, X.; Wang, Z.; Dong, C.; Li, J.; Han, X. Recoverable peroxidase-like Fe_3O_4@MoS_2-Ag nanozyme with enhanced antibacterial ability. *Chem. Eng. J.* **2021**, *408*, 127240. [CrossRef]
191. Hao, S.; Sun, X.; Zhang, H.; Zhai, J.; Dong, S. Recent development of biofuel cell based self-powered biosensors. *J. Mater. Chem. B* **2020**, *8*, 3393–3407. [CrossRef]
192. Lutterbeck, C.A.; Machado, E.L.; Sanchez-Barrios, A.; Silveira, E.O.; Layton, D.; Rieger, A.; Lobo, E.A. Toxicity evaluation of hospital laundry wastewaters treated by microbial fuel cells and constructed wetlands. *Sci. Total Environ.* **2020**, *729*, 138816. [CrossRef]
193. Torres Castillo, N.E.; Melchor-Martinez, E.M.; Ochoa Sierra, J.S.; Mariana Ramirez-Torres, N.; Eduardo Sosa-Hernandez, J.; Iqbal, H.M.N.; Parra-Saldivar, R. Enzyme mimics in-focus: Redefining the catalytic attributes of artificial enzymes for renewable energy production. *Int. J. Biol. Macromol.* **2021**, *179*, 80–89. [CrossRef]
194. Le, P.G.; Kim, M.I. Research Progress and Prospects of Nanozyme-Based Glucose Biofuel Cells. *Nanomaterials* **2021**, *11*, 2116. [CrossRef]
195. Gu, C.; Kong, X.; Yan, S.; Gai, P.; Li, F. Glucose Dehydrogenase-like Nanozyme Based on Black Phosphorus Nanosheets for High-Performance Biofuel Cells. *ACS Sustain. Chem. Eng.* **2020**, *8*, 16549–16554. [CrossRef]
196. Ho, J.; Li, Y.; Dai, Y.; Kim, T.; Wang, J.; Ren, J.; Yun, H.; Liu, X. Ionothermal synthesis of N-doped carbon supported $CoMn_2O_4$ nanoparticles as ORR catalyst in direct glucose alkaline fuel cell. *Int. J. Hydrogen Energy* **2021**, *46*, 20503–20515. [CrossRef]
197. Li, Z.; Kang, Z.; Zhu, Z. A photo-switch for enzymatic biofuel cells based on the photo-oxidization of electron acceptor in cathode by C-dots nanozyme. *Chem. Eng. J.* **2022**, *428*, 13125. [CrossRef]

biosensors

MDPI

Article

Dual-Mode Sensing Platform for Cancer Antigen 15-3 Determination Based on a Silica Nanochannel Array Using Electrochemiluminescence and Electrochemistry

Jie Huang [1,†], Tongtong Zhang [2,†], Yanyan Zheng [1] and Jiyang Liu [1,*]

1 Key Laboratory of Surface & Interface Science of Polymer Materials of Zhejiang Province, Department of Chemistry, Zhejiang Sci-Tech University, Hangzhou 310018, China
2 Key Laboratory of Integrated Oncology and Intelligent Medicine of Zhejiang Province, Department of Hepatobiliary and Pancreatic Surgery, Affiliated Hangzhou First People's Hospital, Zhejiang University School of Medicine, Hangzhou 310006, China
* Correspondence: liujy@zstu.edu.cn
† These authors contributed equally to this work.

Abstract: An electrochemiluminescence-electrochemistry (ECL-EC) dual-mode sensing platform based on a vertically-ordered mesoporous silica films (VMSF) modified electrode was designed here for the sensitive and selective determination of cancer antigen 15-3 (CA 15-3), a specific biomarker of breast cancer. VMSF was assembled through a rapid electrochemically assisted self-assembly (EASA) method and plays a crucial role in signal amplification via a strong electrostatic interaction with the positively charged bifunctional probe $Ru(bpy)_3^{2+}$. To construct the biorecognition interface, epoxy functional silane was linked to the surface of VMSF for further covalent immobilization of the antibody. As a benefit from the specific combination of antigen and antibody, a non-conductive immunocomplex layer was formed in the presence of CA 15-3, leading to the hinderance of the mass and electron transfer of the probes. Based on this strategy, the dual-mode determination of CA 15-3 ranging from 0.1 mU/mL to 100 mU/mL with a LOD of 9 μU/mL for ECL mode, and 10 mU/mL to 200 U/mL with a LOD of 5.4 mU/mL for EC mode, was achieved. The proposed immunosensor was successfully employed for the detection of CA 15-3 in human serum without tedious pretreatment.

Keywords: dual-mode detection; electrochemiluminescence; electrochemistry; cancer antigen 15-3; vertically ordered mesoporous silica film; immunosensor

Citation: Huang, J.; Zhang, T.; Zheng, Y.; Liu, J. Dual-Mode Sensing Platform for Cancer Antigen 15-3 Determination Based on a Silica Nanochannel Array Using Electrochemiluminescence and Electrochemistry. *Biosensors* **2023**, *13*, 317. https://doi.org/10.3390/bios13030317

Received: 17 January 2023
Revised: 15 February 2023
Accepted: 21 February 2023
Published: 24 February 2023

1. Introduction

Cancer has become the second major cause of death worldwide [1]. According to the latest statistics from the International Agency for Research on Cancer (IARC), breast cancer occurs with the highest incidence and is also the major cause of cancer-related mortality in females [2], indicating that it is the most common and high-risk cancer globally. The diagnosis of breast cancer mainly relies on clinical screening or imaging analysis that often requires invasive means to obtain tissue samples and may bring sufferings to the patients [3]. The establishment of a noninvasive/minimally invasive, non-radioactive, simple, and rapid technique for early diagnosis remains a serious challenge and an unmet demand. Tumor markers, a kind of biological macromolecules existing in accessible body fluids (e.g., blood, serum, and urine) [4], provide a new standard for cancer diagnosis, health monitoring and personalized treatment. Cancer antigen 15-3 (CA 15-3) is an acidic glycoprotein with a molecular weight of 300–450 kDa that is often found in those who suffer from breast cancer [5,6]. The serum content of CA 15-3 in healthy people is usually less than 30 U/mL [7]. Abnormally elevated serum CA 15-3 levels indicate a higher risk of breast cancer. Simple but sensitive identification and quantitation of serum CA 15-3 level is critical for cancer screening, determination of disease progression, and treatment effect assessment.

Enzyme-linked immunosorbent assay (ELISA) [8], fluorescent immunoassay (FIA) [9], and radioimmunoassay (RIA) [10] have been reported for the quantitation of CA 15-3, but cumbersome instruments, trained operators and time-consuming preprocesses are always required. The electrochemical (EC) sensing technique has proven itself to be a powerful tool because of the fast response, simple operation, and ease of integration [11,12] compared with the traditional immunoassay, despite the hidden risk of inaccurate test results. To guarantee the accuracy and reliability of the diagnosis, dual-mode bioassays with different mechanisms and independent signal conversion have been developed. To date, a series of dual-mode biosensors which incorporate colorimetric-electrochemiluminescence (ECL) [13], ECL-electrochemical impedance spectroscopy (EIS) [14], fluorescence-colorimetry [15], and ECL-EC [16,17] have been reported. Among them, the ECL-EC dual-mode that combines the wider linear range and zero background signal of the former and the fast and easy operation characteristics of the latter with highly sensitivity, has attracted much attention. Generally, two different probes for ECL and EC are involved in the reported dual-mode detection strategy, which requires more laborious procedures. Tris(2,2′-bipyridyl) ruthenium ($Ru(bpy)_3^{2+}$), as an excellent bifunctional probe with both outstanding EC and ECL response, is a good candidate for constructing an ECL-EC dual-mode sensor, not only simplifying the manual work, but also reducing the analysis error effectively [17,18].

In the recent three decades, functional mesoporous materials possessing abundant unique mesostructures have exhibited tremendous potential in the fields of adsorption, separation, catalysis, and sensing [19–21]. Among them, the vertically-ordered mesoporous silica films (VMSF) made of numerous perpendicularly aligned nanochannels with a ultrasmall and uniform diameter of 2~3 nm has attracted much attention [22,23]. The deprotonation of plentiful silanols with pK_a of ~2 on the VMSF surface endows its negatively charged property [24,25]. All these characteristics endow VMSF with a good capability of prominent charge-based permselectivity, high molecular permeability, and outstanding antifouling ability [26–28]. Therefore, VMSF has been widely used in the construction of sensing platforms to serve certain functions (e.g., preconcentration through electrostatic interaction [29], hydrogen bonding [30] or hydrophobic extraction [31], specific recognition by tailoring recognitive molecules [32], and antifouling ability through size exclusion effect of nanopores [33]). As a monolayer of modification material with a nanoscale thickness, diverse VMSF-based sensors have been developed for the determination of biomolecules (DNA, antibodies, and antigens) [17,34], pharmaceutical molecules [35,36], small organic pollutants [30,37], and metal ions [38,39] in biofluids (blood, serum, urine, and sweat) and environmental samples. Su's group observed a two-orders-of-magnitude ECL intensity enhancement when equipping indium tin oxide (ITO) electrode with a layer of VMSF due to the strong electrostatic attraction between the negatively charged nanochannels and the cationic ECL luminophore $Ru(bpy)_3^{2+}$ and obtained a high ECL intensity even with a very low concentration [40]. On the other hand, the existence of silanols with excellent covalent reactivity towards silane coupling reagents provides the precondition to construct a recognition interface by introducing certain specific molecules to design a gated sensing scheme [41,42]. VMSF has become a fascinating material in the construction of sensing interfaces, demonstrating a great potential for direct and sensitive detection of targets in complex samples without cumbersome pretreatment processes including filtration, separation, and pre-enrichment.

In this work, an ECL-EC dual-mode immunosensor constructed on an ITO substrate composed of a layer of mesoporous silica membrane as the signal amplifier, and a monomolecular layer of anti-CA 15-3 antibody as the recognition element was designed for the sensitive and selective detection of the breast cancer biomarker CA 15-3. Compared with medical imaging and pathological biopsy applied to the clinical diagnosis of breast cancer, this method offers more hope for the early detection of breast cancer rapidly and conveniently. As shown in Scheme 1, VMSF was first assembled on ITO through the electrochemical assisted self-assembly (EASA) method. The construction of the immune recognition interface for further identifying and capturing targets specifically was achieved by immobilizing the antibodies with the assistance of (3-glycidyloxypropyl)trimethoxysilane (GPTMS) linked to the outer surface of VMSF beforehand. A layer of immunocomplex with poor conductivity and large size would form in the presence of the CA 15-3 antigen and hamper the mass transfer and electron transfer of the bifunctional probe $Ru(bpy)_3^{2+}$, leading to decreased ECL-EC signal and the final signal-off detection of CA 15-3. The benefits of the immunosensor lie in the easy fabrication of the modification material which serves as both a signal amplifier to enrich $Ru(bpy)_3^{2+}$ and an anti-fouling filter, as well as in the generalizability of the detection strategy which is suitable for the quantitation of various targets by simply changing the immobilized bio-receptors.

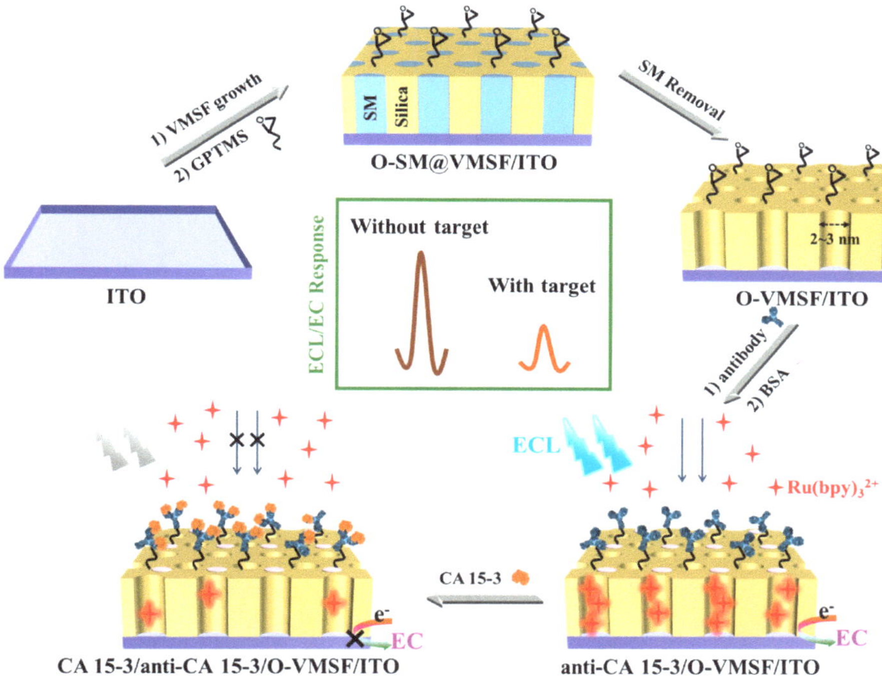

Scheme 1. Illustration of the preparation of the immunosensor and detection of CA 15-3.

2. Materials and Methods

2.1. Reagents and Solution

CA 15-3, anti-CA 15-3 antibody, carcinoembryonic antigen (CEA), and prostate specific antigen (PSA) were from Beijing KEY-BIO Biotech. Tetraethyl orthosilicate (TEOS), hexadecyl trimethyl ammonium bromide (CTAB), sodium nitrate ($NaNO_3$), sodium hydroxide (NaOH), potassium ferrocyanide ($K_4[Fe(CN)_6]$), potassium ferricyanide ($K_3[Fe(CN)_6]$), bovine serum albumin (BSA), and (3-glycidoxypropyl)methyldiethoxysilane (GPTMS) were from Aladdin Chemistry Co. Ltd. (China). Tris(2,2-bipyridyl) dichlororuthenium (II) hexahydrate ($Ru(bpy)_3Cl_2{\cdot}6H_2O$) and hexaammineruthenium (III) chloride ($Ru(NH_3)_6Cl_3$) were purchased from Sigma-Aldrich. Ethanol (99.8%), hydrochloric acid (HCl, 38%), and acetone were from Hangzhou Shuanglin Chemical Reagent Co. Ltd. Phosphate buffered saline (PBS, 0.01 M, pH = 7.4) was obtained by mixing NaH_2PO_4 and Na_2HPO_4 in a certain ratio. All chemicals and reagents were used as received without any further purification. All the aqueous solutions used here were prepared using ultrapure water (18.2 MΩ cm) from Milli-Q Systems (Millipore Inc., Massachusetts, America). Serum of healthy men was obtained from the Center of Disease Control and Prevention (Hangzhou, China) for real sample analysis.

2.2. Apparatus and Electrodes

An MPI-E II ECL analytical system (Xi'an Remex Analytical Instrument Ltd., Xi'an, China) was used to perform ECL measurements. EC measurements, including cyclic voltammetry (CV), differential pulse voltammetry (DPV), and electrochemical impedance spectroscopy (EIS) were performed on an Autolab electrochemical workstation (PGSTAT302N, Metrohm, Switzerland). A traditional three-electrode system was adopted. Bare or modified ITO, platinum wire or platinum plate, and Ag/AgCl electrode (saturated with KCl solution) were used as the working electrode, counter electrode and reference electrode, respectively. Zhuhai Kaivo Optoelectronic Technology provided ITO glasses (<17 Ω/square, thickness: 100 ± 20 nm), which were cleaned ultrasonically with 1 M NaOH, acetone, ethanol and deionized water, respectively. The geometric surface area of the electrode was fixed to 1 cm × 0.5 cm.

The morphology of VMSF/ITO was investigated on a JEM-2100 (JEOL Ltd., Tokyo, Japan) transmission electron microscopy (TEM). The thickness of VMSF/ITO was characterized with a SU8010 (Hitachi, Tokyo, Japan) scanning electron microscopy (SEM).

2.3. Procedures

2.3.1. Preparation of VMSF/ITO Electrode

According to the reported electrochemically assisted self-assembly (EASA) method, an easy and rapid modification of VMSF on the ITO electrode was achieved [43]. A sol for film electrodeposition was prepared as follows: first mix 20 mL of ethanol and 20 mL of $NaNO_3$ (0.1 M, adjusted to pH = 2.6 by HCl) together, then add 1.585 g of CTAB and 3050 μL of TEOS. After their complete dissolution, the solution was stirred for 2.5 h at room temperature for hydrolyzation. The assembly of VMSF was conducted under a galvanostatic condition ($j = -0.70$ mA/cm^2) for 10 s after immersing the ITO in the hydrolyzed sol, using an Ag/AgCl (saturated KCl) as reference electrode and a platinum plate (2 cm × 4 cm) as counter electrode. The as-prepared electrode was rapidly removed from the sol and washed with copious deionized water, dried with N_2 stream, and further treated overnight at 120 °C. The obtained electrodes were denoted SM@VMSF/ITO, which retained surfactant micelles (SM) templates in the nanochannels. The templates can be conveniently removed using 0.1 M HCl-ethanol by stirring for 5 min and the obtained electrode modified with open nanochannels was termed VMSF/ITO.

2.3.2. Characterization of VMSF/ITO Electrode

The morphology of VMSF was characterized with a JEM-2100 TEM with an acceleration voltage of 100 kV. The TEM specimen was prepared by dispersing the VMSF peeled off from the VMSF/ITO in ethanol through ultrasonication, and finally dropped onto the copper grids for investigation. The thickness of VMSF was characterized using a SU8010 SEM with an acceleration voltage of 5 kV. Before observation, a fresh cross section was obtained by breaking up the VMSF/ITO and spraying it with gold.

The integrity and permeability of VMSF were investigated with the CV technique using $Fe(CN)_6^{3-}$ and $Ru(NH_3)_6^{3+}$, two types of standard redox probes with different charges.

2.3.3. Fabrication of Immunosensor towards CA 15-3

For the immunosensor fabrication, the construction of a biorecognition interface is of great significance. To immobilize the antibody on the outer surface of the VMSF/ITO electrode, the bifunctional reagent GPTMS with both epoxy and silyl groups was selected as the crosslinker [34]. Typically, the SM@VMSF/ITO was immersed in GPTMS (2.26 mM in ethanol) for 60 min, then the GPTMS was covalently linked to the outer surface instead of the inner channels of the VMSF through silane reaction. The electrode was then treated with 0.1 M HCl-ethanol to remove the SM templates. The electrode modified with epoxy-functionalized nanochannels was named O-VMSF/ITO. Anti-CA 15-3 antibodies (anti-CA 15-3) were further immobilized by covering the O-VMSF/ITO with 50 μL of anti-CA 15-3 solution (10 μg/mL in PBS) for 90 min at 37 °C via the ring opening reaction between active epoxy groups on the VMSF and amino residues of the anti-CA 15-3. After rinsing with PBS, it was incubated with s BSA solution (1%, *w*/*w*) for 90 min to block the nonspecific binding sites. The immunosensor termed as anti-CA 15-3/O-VMSF/ITO was obtained and was stored at 4 °C for further tests. The procedure of the immunosensor fabrication is illustrated in Scheme 1.

2.3.4. Dual-Mode Tests

Before the ECL and EC tests, the immunosensor was incubated with different concentrations of CA 15-3 (50 μL in PBS) for 45 min at 37 °C and was noted as CA 15-3/anti-CA 15-3/O-VMSF/ITO, followed by thoroughly washing with PBS. For ECL measurement, the CA 15-3/anti-CA 15-3/O-VMSF/ITO was placed in PBS containing 10 μM $Ru(bpy)_3{}^{2+}$ and 3 mM TPA. The ECL process was triggered by a continuous CV procedure with the potential range of 0 to 1.25 V at 0.1 V/s, which was recorded simultaneously along with the ECL signals. The voltage of the photomultiplier tube (PMT) was set at 400 V. For the EC measurement, the CA 15-3/anti-CA 15-3/O-VMSF/ITO was immersed in PBS including 10 μM $Ru(bpy)_3{}^{2+}$ and let to stand for 30 min to preconcentration before recording the DPV curves.

Real sample analysis was conducted on human serum from healthy adults using the standard addition method without complex preprocessing. The serum was spiked with a known amount of CA 15-3, followed by 50-times dilution with PBS, and was analyzed by the established methods.

3. Results and Discussions

3.1. Morphology and Permeability of VMSF/ITO

As expected, the VMSF was assembled on the surface of the ITO electrode due to polycondensation of TEOS under the catalysis of hydroxide ions generated from the electroreduction of protons/water and nitrate ions, where CTAB served as the template, and was proved by TEM and SEM images (Figure 1a,b). VMSF presents a 2D structure of intact film with a thickness of 90 nm and is free of any cracks. A too thin VMSF would lead to an inadequate enrichment of $Ru(bpy)_3{}^{2+}$, and would finally fail to achieve a high sensitivity. With the prolongation of electrodeposition time, a much thicker VMSF with more silica aggregates byproducts on its surface would be obtained, which would restrict the diffusion of $Ru(bpy)_3{}^{2+}$ species across the film to the underlying electrode surface. According to the many reported works [23,24,28–30], a procedure of electrodeposition for 10 s at $j = -0.70$ mA/cm^2 was selected here to obtain a VMSF with an appropriate thickness. This not only achieves an effective electrostatic attraction to $Ru(bpy)_3{}^{2+}$, but also guarantees an efficient and rapid response. The film contains numerous nanochannels that are highly ordered and hexagonally packed and whose diameter is 2~3 nm.

Considering that VMSF plays the role of the modification material that is on the electrolyte/electrode interface, it is crucial to determine how VMSF affects the mass transfer. The permeability is investigated by comparing the electrochemical behaviors of $Ru(NH_3)_6{}^{3+}$ and $Fe(CN)_6{}^{3-/4-}$ on ITO, SM@VMSF/ITO and VMSF/ITO. As shown in Figure 1c,d, SM@VMSF/ITO exhibited no peak current but only a non-faradic current towards the $Ru(NH_3)_6{}^{3+}$ probe and a rather large charge transfer resistance (R_{ct}) represented by the equivalent diameter of Nyquist curve obtained using the EIS technique, a powerful tool to study interfacial property [44]. This is due to the hydrophobic SM templates inside the nanochannels that block the mass transfer of charged probes ($Ru(NH_3)_6{}^{3+}$ and $Fe(CN)_6{}^{3-/4-}$) and finally hinder the electron exchange. The nanochannels open after the removal of SM, which facilitates the diffusion of particles, and thus VMSF/ITO shows a pair of well-defined redox peaks to $Ru(NH_3)_6{}^{3+}$ and a much smaller R_{ct}. In addition, VMSF/ITO exhibits a relatively higher peak current towards $Ru(NH_3)_6{}^{3+}$ than that of ITO, which derives from the deprotonation of abundant silanol with a pK_a of ~2 in the nanochannel walls that can create a strong electrostatic attraction to positively charged species [25].

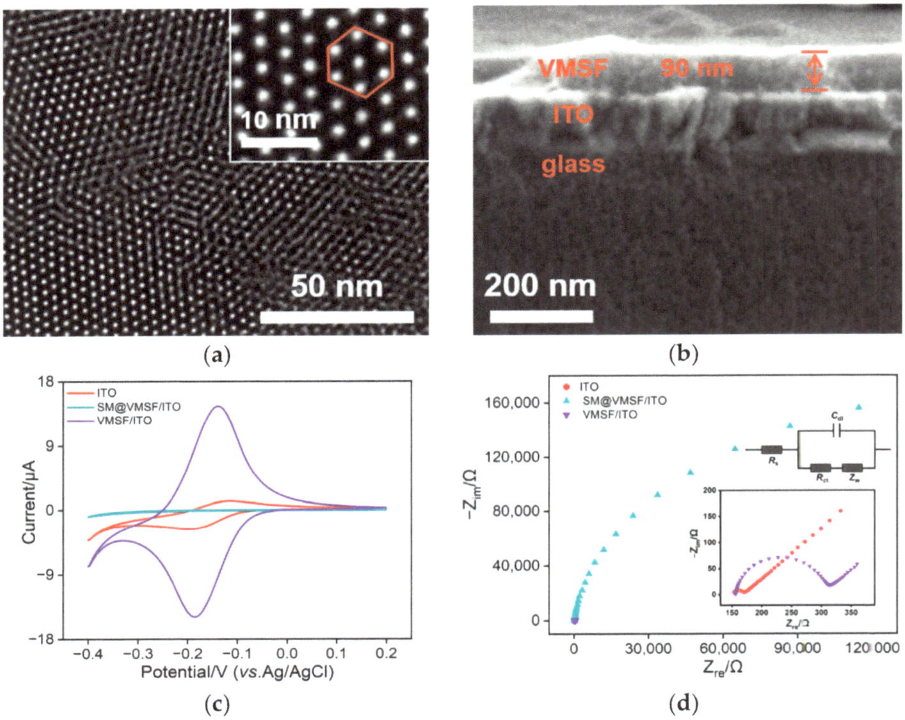

Figure 1. (**a**) Top-view of TEM images and (**b**) SEM image of VMSF. The inset in (**a**) is the corresponding TEM image with high magnification. (**c**) CV plots of ITO, SM@VMSF/ITO, and VMSF/ITO in 0.5 mM $Ru(NH_3)_6{}^{3+}$ and (**d**) EIS plots of ITO, SM@VMSF/ITO, and VMSF/ITO in 0.1 M KCl solution containing 2.5 mM $Fe(CN)_6{}^{3-/4-}$. The scan rate in (**c**) was 0.05 V/s. The insets up and down in (**d**) are the equivalent circuit and magnified EIS plots of ITO and VMSF/ITO, respectively.

The excellent capability of VMSF to enrich cations was further verified by $Ru(bpy)_3{}^{2+}$, a bifunctional probe with outstanding ECL and EC activities. The ECL process was triggered by a continuous CV procedure; the mechanism is as follows:

$$Ru(bpy)_3{}^{2+} - e^- \rightarrow Ru(bpy)_3{}^{3+}$$

$$TPA - e^- \rightarrow TPA^{\cdot +}$$

$$TPA^{\cdot +} - H^+ \rightarrow TPA^{\cdot}$$

$$TPA^{\cdot} + Ru(bpy)_3{}^{3+} \rightarrow Ru(bpy)_3{}^{2+*}$$

$$Ru(bpy)_3{}^{2+*} \rightarrow Ru(bpy)_3{}^{2+} + h\nu\ (\lambda_{max} \approx 620\ nm)$$

During the positive scanning of potential, TPA and $Ru(bpy)_3{}^{2+}$ were electrooxidized to $TPA^{\cdot}$ radical and $Ru(bpy)_3{}^{3+}$. The redox reaction between $TPA^{\cdot}$ radical and $Ru(bpy)_3{}^{3+}$ led to the generation of excited $Ru(bpy)_3{}^{2+*}$ that emitted light ($\lambda_{max} \approx 620$ nm) when turning to the ground state, which was recorded by the photomultiplier. As shown in Figure 2, the ECL and DPV signals produced by 10 μM $Ru(bpy)_3{}^{2+}$ on ITO were very weak while those obtained on VMSF/ITO were remarkable enhanced thanks to the electrostatic preconcentration of the nanochannel layer in the positively charged probe. VMSF/ITO tends to provide an adequate basic signal and exhibits enormous potential in the subsequent immunosensor fabrication.

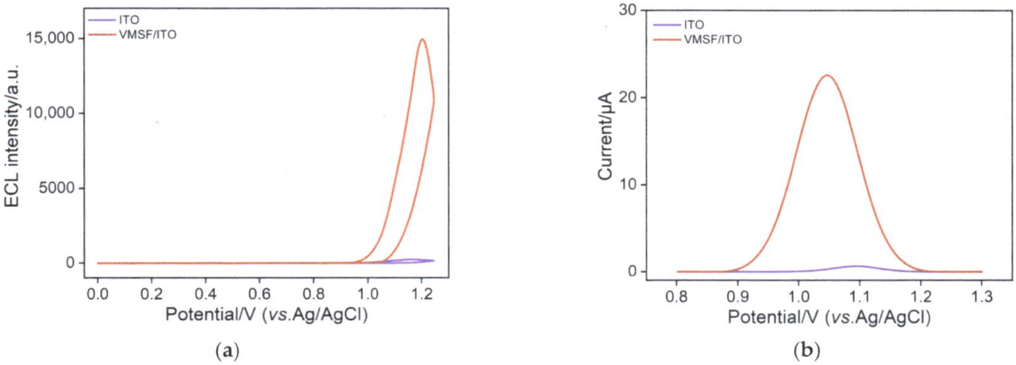

Figure 2. (**a**) ECL curves of ITO and VMSF/ITO in 10 μM $Ru(bpy)_3^{2+}$ and 3 mM TPA; (**b**) DPV curves of ITO and VMSF/ITO in 10 μM $Ru(bpy)_3^{2+}$.

3.2. Characterization of the Immunosensor Construction

The CV and EIS techniques were employed to monitor the interface features during the bioelectrode's stepwise assembly. As shown in Figure 3a, VMSF/ITO showed two well-defined redox peaks, which were decreased after the modification of GPTMS, indicating the presence of epoxy group entities. The peak currents were further reduced with the immobilization of antibodies and the capture of CA 15-3 antigens, suggesting that an insulating protein layer was formed on the electrode surface, creating an obstacle to both mass and electron transfer. Figure 3b shows the EIS results fitted according to the Randles equivalent circuit. As shown by the inset, the Randles equivalent circuit contained a solution resistance (R_s), a double-layer capacitance (C_{dl}), Warburg impedance (Z_w), and an apparent charge transfer resistance (R_{ct}) that is equal to the equivalent diameter of the semicircle in the high-frequency region [45]. The R_{ct} of VMSF/ITO was the smallest (141 Ω) and increased after the crosslinking of GPTMS (329 Ω). The incubation with the antibodies caused an obstruction to the electron transfer of the redox probe, increasing the R_{ct} of anti-CA 15-3/O-VMSF/ITO significantly (493 Ω). The bio-recognition and combination of CA 15-3 further aggravated the blocking effect and CA 15-3/anti-CA 15-3/O-VMSF/ITO had the largest R_{ct} (607 Ω). The results above indicate the successful fabrication of the immunosensor.

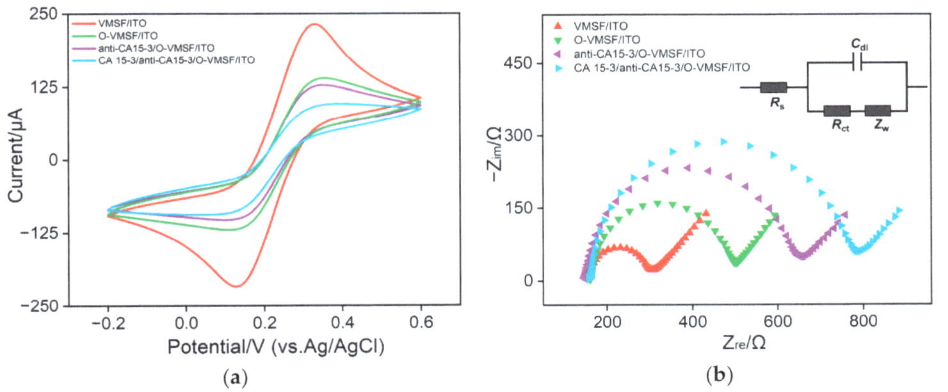

Figure 3. CV (**a**) and EIS (**b**) plots of VMSF/ITO, O-VMSF/ITO, anti-CA 15-3/O-VMSF/ITO, and CA 15-3/anti-CA 15-3/O-VMSF/ITO in a 0.1 M KCl solution containing 2.5 mM $Fe(CN)_6^{3-/4-}$. The scan rate in (**a**) was 0.05 V/s.

3.3. Dual-Mode Tests of CA 15-3

The specific recognition and combination between the anti-CA 15-3 antibody and the CA 15-3 antigen formed an antibody-antigen immunocomplex. In analogy with the aforementioned CV and EIS results, Figure 4a shows decreased ECL responses with the step-by-step assembling of the bioelectrode, confirming the feasibility of the ECL analysis. The inset suggested an outstanding stability of the biosensor under continuous CV scanning with a relative standard deviation (RSD) of less than 2%. The I_{ECL}-time curves recorded on bioelectrodes incubated with a series of different amounts of CA 15-3 are shown in Figure 4b. With the increase in CA 15-3, more immunocomplex was formed, which perturbed the mass transfer of $Ru(bpy)_3^{2+}$ and restrained the electron transfer at the electrode. A linear relationship for CA 15-3 was obtained from 0.1 mU/mL to 100 U/mL using the equation $I_{ECL} = -997.0 \log C_{CA15\text{-}3} + 2537$ ($R^2 = 0.991$). LOD is defined by the International Union of Pure and Applied Chemistry (IUPAC) as the concentration (c_L) that can be detected with a reasonable certainty for a given analytical procedure. It can be derived from the smallest measure (x_L), which is calculated with the equation $x_L = x_{b1} \pm k s_{b1}$, where x_{b1} is the mean of the blank measures, s_{b1} is the standard deviation of the blank measures, and k is a numerical factor commonly recommended as 3. Here, + and − are applied to signal-on and signal-off sensors, respectively [46]. Thus, the LOD was calculated to be 9 μU/mL according to the calibration curve equation. Compared with the other reported methods listed in Table 1, this proposed ECL immunosensor exhibits a relatively wider linear range and an exceedingly lower LOD for CA 15-3 determination without elaborate and complex labor during the construction process. Diverse sophisticated nanomaterials and sandwich strategies are widely used in other related works listed in Table 1. They require a complex preparation process, several steps of incubation, and consume more high-cost biological reagents. Our proposed sensor is simple in preparation, does not need multiple incubations for detection, and is much more economic.

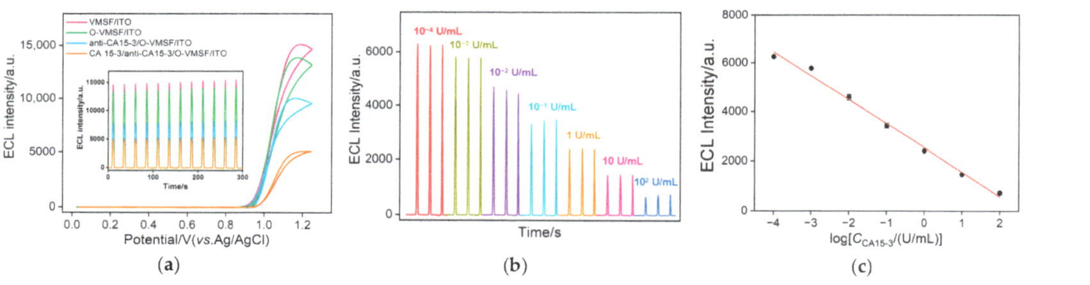

Figure 4. (**a**) IECL-potential curves of VMSF/ITO, O-VMSF/ITO, anti-CA 15-3/O-VMSF/ITO, and CA 15-3/anti-CA 15-3/O-VMSF/ITO in 10 μM Ru(bpy)32+. The inset in (**a**) shows the corresponding IECL-time curves. (**b**) ECL responses of the immunosensor to various concentrations of CA 15-3 from 10^{-4} U/mL, 10^{-3} U/mL, 10^{-2} U/mL, 10^{-1} U/mL, 1 U/mL, 10 U/mL to 100 U/mL. (**c**) The corresponding calibration curve. The error bars represent the standard deviation of three measurements under the same condition.

In addition, considering the excellent electroactivity of $Ru(bpy)_3^{2+}$, CA 15-3 was quantified using the DPV method. Similar to the ECL mode, the feasibility of the EC mode was verified by the gradually reduced DPV response during the bioelectrode fabrication as shown in Figure 5a. Upon the increase in CA 15-3, a group of peak currents which decreased linearly with the logarithm of the antigen concentration from 10 mU/mL to 200 U/mL were obtained (Figure 5b), and the calibration curve was represented in Figure 5c using the equation $I = -0.82 \log C_{CA15\text{-}3} + 4.60$ ($R^2 = 0.992$). Similar to the ECL mode, the LOD was calculated to be 5.4 mU/mL (S/N = 3). Although the EC method was inferior in LOD compared with the ECL mode, a higher detection upper limit was achieved. Although the LOD of our proposal is not as low as 10^{-5} U/mL for the sandwich strategy using

Ab_2/HRP-MBs and Ab_1/AuE, our immunosensor is free of complex labeling operations or several steps of incubation. Dual-mode tests that combine the merits of ECL and EC strategies succeed to broaden the working linear range while achieving an ultrasensitive LOD. The total cost of a single fabricated immunosensor is approximately 0.366 USD, which is much lower in cost compared to ELISA which costs around 1.31 USD for a single sample test. Furthermore, the materials involved here are all commercially available, indicating its potential for industrialization, as well as the possibility of application in microfluidic Point-of-Care devices [47,48]. However, the immunosensor is susceptible to matrix effects, and a dilution is required before testing bio-samples to guarantee the accuracy in clinical application. In addition, an alkali condition may lead to the hydrolysis of VMSF as the proposed immunosensor is applicative in neutral medium to keep its mechanical stability. The current clinical determination of CA 15-3 is mainly based on a chemiluminescence immunoassay. This method not only involves labeling technology, but also requires two antibodies, as well as several steps of incubation and washing, which is high-cost and time-consuming. Our proposed method involves only one step of incubation, which meets the needs of rapid clinical determination of CA 15-3 at low cost.

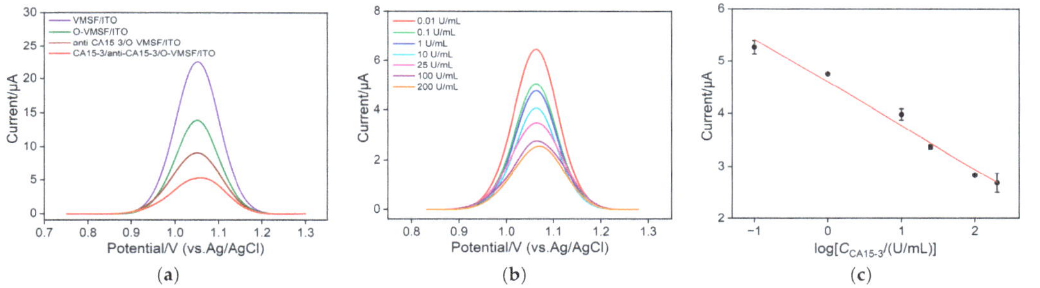

Figure 5. (**a**) DPV curves of VMSF/ITO, O-VMSF/ITO, anti-CA 15-3/O-VMSF/ITO, and CA 15-3/anti-CA 15-3/O-VMSF/ITO in 10 μM $Ru(bpy)_3^{2+}$. (**b**) DPV response of anti-CA 15-3/O-VMSF/ITO in the presence of different concentrations of CA 15-3. (**c**) The corresponding calibration curve. The error bars represent the standard deviation of three measurements under the same condition.

Table 1. Comparison of various methods for the determination of CA 15-3.

Electrode	Mode	Classification	Linear Range (U/mL)	LOD (μU/mL)	Ref.
anti-CA 15-3/NH_2-SiO_2-PTCA/CeO_2/Pt/rGO/GCE	ECL	Label-free	0.012–120	1348	[49]
anti-CA 15-3/$Ru(bpy)_3^{2+}$@UiO-66-NH_2/GCE	ECL	Label-free	5×10^{-4}–500	17.705	[50]
Ab_2/AuNPs/CQDs-PEI-GO and Ab_1/AgNPs-PDA/GCE	ECL	Labeled	5×10^{-3}–500	1.7×10^3	[51]
anti-CA 15-3/CS/PtNi NCs-*L*-Cys-luminol/GCE	ECL	Label-free	5×10^{-4}–500	167	[52]
Ab_2/CdTe@CdS/PAMAM and Ab_2/Fe_3O_4@SiO_2, ITO	ECL	Labeled	10^{-4}–100	10	[53]
Ab_2/HRP-MBs and Ab_1/AuE	EC	Label-free	1.5×10^{-5}–50	15	[54]
anti-CA 15-3/Ag/TiO_2/rGO/GCE	EC	Label-free	0.1–300	7×10^4	[55]
anti-CA 15-3/DAP-AuNPs/P3ABA/2D-$MoSe_2$/GO/SPCE	EC	Label-free	0.14–500	1.4×10^5	[56]
anti-CA 15-3/O-VMSF/ITO	ECL EC	Label-free	10^{-4}–100 10^{-2}–100	9 5.4×10^3	This work

PTCA: 3,4,9,10-perylenetetracarboxylic acid; rGO: reduced graphene oxide; GCE: glassy carbon electrode; Ab: antibody; AuNPs: gold nanoparticles; CQDs: carbon quantum dots; PEI: polyethylenimine; GO: graphene oxide; AgNPs: silver nanoparticles; PDA: polydopamine; PtNi NCs: platinum nickel nanocubes; *L*-Cys: *L*-cysteine; PAMAM: polyamidoamine; HRP: horseradish peroxidase; MBs: magnetic beads; AuE: gold electrode; DAP: deposited redox dye; P3ABA: poly(3-aminobenzylamine); and SPCE: screen-printed carbon electrode.

3.4. Specificity, Selectivity and Stability of the Immunosensor

The proposed immunosensor was expected to be highly specific and selective for CA 15-3 detection due to the antigen-antibody recognition. Therefore, the bioelectrode was tested with different possible clinical cancer biomarkers such as the carcinoembryonic antigen (CEA) and the prostate specific antigen (PSA) (the concentrations of interfering proteins are both 500 ng/mL). As illustrated in Figure 6, the signal fluctuation caused by individual interfering proteins was less than 7%; an evident decrease in signal was observed when incubated with CA 15-3 or the mixture of all proteins. These results indicated the high specificity and outstanding selectivity of the fabricated immunosensor. The stability of the anti-CA 15-3/O-VMSF/ITO was evaluated by comparing the signal after several days of storage at 4 °C with the initial signal on the first day. Anti-CA 15-3/O-VMSF/ITO retained more than 95% of the initial signal after a 15-day storage at 4 °C, suggesting a satisfactory stability.

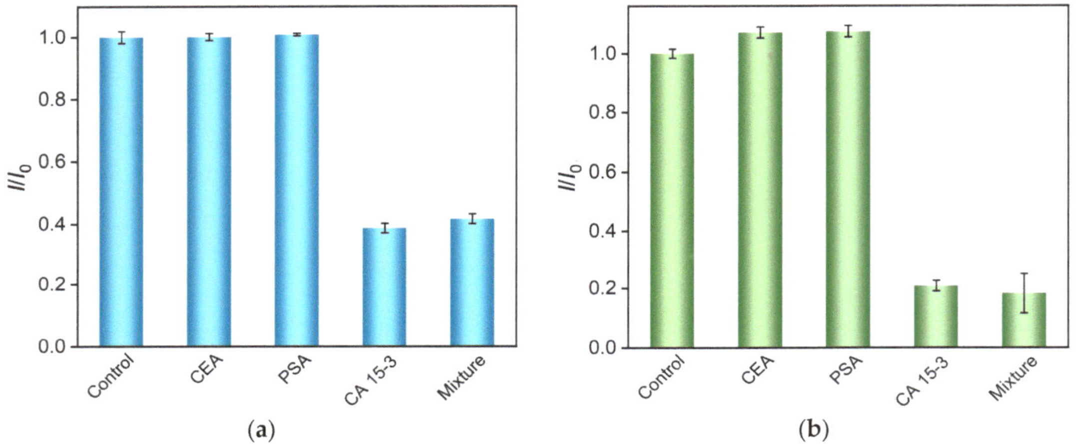

Figure 6. Relative ratio of ECL intensity (**a**) and DPV current (**b**) before (I_0) and after (I) incubation with buffer (Control), CEA (500 ng/mL), PSA (500 ng/mL), CA 15-3 (25 U/mL), or a protein mixture (25 U/mL CA 15-3 + 500 ng/mL other proteins). The error bars represent the standard deviation of three measurements under the same condition.

3.5. Real Sample Analysis

To validate the practical applicability of the established method, the biosensor was used to determine the CA 15-3 spiked in healthy human serum samples. A series of known amount of CA 15-3 was added to the samples and then they were diluted with PBS 50 times. After the incubation with anti-CA 15-3/O-VMSF/ITO, the CA 15-3/anti-CA 15-3/O-VMSF/ITO was obtained for ECL and EC testing. The concentration of CA 15-3 could be calculated by substituting the obtained ECL and EC response into the original linear equation. The results are listed in Table 2. Satisfactory recoveries ranging from 95.4% to 104.0% and an RSD below 2.9% were obtained using the ECL-EC dual-mode tests, which implied the potential for clinical application.

Table 2. Detection of CA 15-3 in human serum.

Mode	Sample *	Added (U/mL)	Found (U/mL)	RSD (%)	Recovery (%)
ECL	1	1.00	0.980	2.1	98.0
		5.00	4.80	1.5	96.0
		10.0	10.1	2.0	101.0
	2	1.00	1.04	2.7	104.0
		5.00	5.12	2.9	102.4
		10.0	9.86	2.0	98.6
EC	1	5.00	4.77	1.0	95.4
		50.0	49.5	1.7	99.0
		100.0	99.6	2.1	99.6
	2	5.00	4.86	1.8	97.2
		50.0	51.5	2.0	103.0
		100.0	102.2	2.7	102.2

* Diluted by PBS for 50 times and CA 15-3 was added before dilution.

4. Conclusions

In summary, an ECL-EC dual-mode immunosensor with a simple fabrication strategy was designed for CA 15-3 determination with an efficient biosensing performance. The VMSF equipped on the electrode surface displays a strong electrostatic attraction to the positively charged bifunctional probes, and provides a large surface area with good biocompatibility for the biorecognition of elements binding with the aid of the crosslinker GPTMS. Integrating the merits of both the sensitivity of electrochemiluminescence and the efficiency of electrochemistry, the immunosensor demonstrated a satisfactory performance with a linear working range of more than 7 orders of magnitude and a LOD of 9 μU/mL. Furthermore, VMSF exhibits good resistance to foulants thanks to the ultrasmall diameter of nanochannels. Real sample analysis in human serum was performed, expecting to provide a new method for clinical CA 15-3 diagnosis.

Author Contributions: Conceptualization, J.L.; investigation, J.H., T.Z. and Y.Z.; data curation, J.H., T.Z. and Y.Z.; writing—original draft preparation, J.H.; writing—review and editing, J.L.; supervision, J.L. All authors have read and agreed to the published version of the manuscript.

Funding: This research was funded by the National Natural Science Foundation of China (21904117) and the Zhejiang Provincial Natural Science Foundation of China (LY20B050007).

Institutional Review Board Statement: Not applicable.

Informed Consent Statement: Not applicable.

Data Availability Statement: The data presented in this study are available on request from the corresponding author.

Conflicts of Interest: The authors declare no conflict of interest.

References

1. Bray, F.; Ferlay, J.; Soerjomataram, I.; Siegel, R.L.; Torre, L.A.; Jemal, A. Global cancer statistics 2018: GLOBOCAN estimates of incidence and mortality worldwide for 36 cancers in 185 countries. *CA-Cancer J. Clin.* **2018**, *68*, 394–424. [CrossRef] [PubMed]
2. DeSantis, C.; Ma, J.; Bryan, L.; Jemal, A. Breast cancer statistics, 2013. *CA-Cancer J. Clin.* **2014**, *64*, 52–62. [CrossRef] [PubMed]
3. Jing, L.; Xie, C.; Li, Q.; Yang, M.; Li, S.; Li, H.; Xia, F. Electrochemical biosensors for the analysis of breast cancer biomarkers: From design to application. *Anal. Chem.* **2022**, *94*, 269–296. [CrossRef] [PubMed]
4. Rifai, N.; Gillette, M.A.; Carr, S.A. Protein biomarker discovery and validation: The long and uncertain path to clinical utility. *Nat. Biotechnol.* **2006**, *24*, 971–983. [CrossRef]
5. Carneiro, M.; Rodrigues, L.; Moreira, F.; Goreti, F.; Sales, M. Paper-based ELISA for fast CA 15-3 detection in point-of-care. *Microchem. J.* **2022**, *181*, 107756. [CrossRef]
6. Marques, R.; Costa-Rama, E.; Viswanathan, S.; Nouws, H.; Costa-García, A.; Delerue-Matos, C.; González-García, M. Voltammetric immunosensor for the simultaneous analysis of the breast cancer biomarkers CA 15-3 and HER2-ECD. *Sens. Actuators B Chem.* **2018**, *255*, 918–925. [CrossRef]

7. Hasanzadeh, M.; Tagi, S.; Solhi, E.; Mokhtarzadeh, A.; Shadjou, N.; Eftekhari, A.; Mahboob, S. An innovative immunosensor for ultrasensitive detection of breast cancer specific carbohydrate (CA 15-3) in unprocessed human plasma and MCF-7 breast cancer cell lysates using gold nanospear electrochemically assembled onto thiolated graphene quantum dots. *Int. J. Biol. Macromol.* **2018**, *114*, 1008–1017. [CrossRef]
8. Ambrosi, A.; Airo, F.; Merkoçi, A. Enhanced gold nanoparticle based ELISA for a breast cancer biomarker. *Anal. Chem.* **2010**, *82*, 1151–1156. [CrossRef]
9. Huang, X.; Liu, H.; Fang, W.; Lin, Y.; Tan, Y. Sensitive and selective immunofluorescence assay for CA15-3 detection using fluorescein derivative A10254. *Protein Pept. Lett.* **2018**, *25*, 776–782. [CrossRef]
10. Yang, Y.; Zhang, H.; Zhang, M.; Meng, Q.; Cai, L.; Zhang, Q. Elevation of serum CEA and CA15-3 levels during antitumor therapy predicts poor therapeutic response in advanced breast cancer patients. *Oncol. Lett.* **2017**, *14*, 7549–7556. [CrossRef]
11. Zhou, H.; Ma, X.; Sailjoi, A.; Zou, Y.; Lin, X.; Yan, F.; Su, B.; Liu, J. Vertical silica nanochannels supported by nanocarbon composite for simultaneous detection of serotonin and melatonin in biological fluids. *Sens. Actuators B Chem.* **2022**, *353*, 131101. [CrossRef]
12. Xuan, L.; Liao, W.; Wang, M.; Zhou, H.; Ding, Y.; Yan, F.; Liu, J.; Tang, H.; Xi, F. Integration of vertically-ordered mesoporous silica-nanochannel film with electro-activated glassy carbon electrode for improved electroanalysis in complex samples. *Talanta* **2021**, *225*, 122066. [CrossRef] [PubMed]
13. Hu, Y.; Zhu, L.; Mei, X.; Liu, J.; Yao, Z.; Li, Y. Dual-mode sensing platform for electrochemiluminescence and colorimetry detection based on a closed bipolar electrode. *Anal. Chem.* **2021**, *93*, 12367–12373. [CrossRef]
14. Mi, X.; Li, H.; Tan, R.; Tu, Y. Dual-modular aptasensor for detection of cardiac troponin I based on mesoporous silica films by electrochemiluminescence/electrochemical impedance spectroscopy. *Anal. Chem.* **2020**, *92*, 14640–14647. [CrossRef] [PubMed]
15. Kuo, S.; Li, H.; Wu, P.; Chen, C.; Huang, Y.; Chan, Y. Dual colorimetric and fluorescent sensor based on semiconducting polymer dots for ratiometric detection of lead ions in living cells. *Anal. Chem.* **2015**, *87*, 4765–4771. [CrossRef]
16. Xue, J.; Zhao, Q.; Yang, L.; Ma, H.; Wu, D.; Liu, L.; Ren, X.; Ju, H.; Wei, Q. Dual-mode sensing platform guided by intramolecular electrochemiluminescence of a ruthenium complex and cationic N,N-Bis(2-(trimethylammonium iodide)propylene) Perylene-3,4,9,10-tetracarboxydiimide for estradiol assay. *Anal. Chem.* **2021**, *93*, 6088–6093. [CrossRef]
17. Gong, J.; Zhang, T.; Chen, P.; Yan, F.; Liu, J. Bipolar silica nanochannel array for dual-mode electrochemiluminescence and electrochemical immunosensing platform. *Sens. Actuators B Chem.* **2022**, *368*, 132086. [CrossRef]
18. Ma, K.; Zheng, Y.; An, L.; Liu, J. Ultrasensitive immunosensor for prostate-specific antigen based on enhanced electrochemiluminescence by vertically ordered mesoporous silica-nanochannel film. *Front. Chem.* **2022**, *10*, 851178. [CrossRef]
19. Zhao, T.; Elzatahry, A.; Li, X.; Zhao, D. Single-micelle-directed synthesis of mesoporous materials. *Nat. Rev. Mater.* **2019**, *4*, 775–791. [CrossRef]
20. Liu, X.; Chen, Z.; Wang, T.; Jiang, X.; Qu, X.; Duan, W.; Xi, F.; He, Z.; Wu, J. Tissue imprinting on 2D nanoflakes-capped silicon nanowires for lipidomic mass spectrometry imaging and cancer diagnosis. *ACS Nano* **2022**, *16*, 6916–6928. [CrossRef]
21. Cui, Y.; Duan, W.; Jin, Y.; Wo, F.; Xi, F.; Wu, J. Ratiometric fluorescent nanohybrid for noninvasive and visual monitoring of sweat glucose. *ACS Sens.* **2020**, *5*, 2096–2105. [CrossRef]
22. Zhou, L.; Hou, H.; Wei, H.; Yao, L.; Sun, L.; Yu, P.; Su, B.; Mao, L. In vivo monitoring of oxygen in rat brain by carbon fiber microelectrode modified with antifouling nanoporous membrane. *Anal. Chem.* **2019**, *91*, 3645–3651. [CrossRef]
23. Deng, X.; Lin, X.; Zhou, H.; Liu, J.; Tang, H. Equipment of vertically-ordered mesoporous silica film on electrochemically pretreated three-dimensional graphene electrodes for sensitive detection of methidazine in urine. *Nanomaterials* **2023**, *13*, 239. [CrossRef] [PubMed]
24. Yang, L.; Zhang, T.; Zhou, H.; Yan, F.; Liu, Y. Silica nanochannels boosting $Ru(bpy)_3^{2+}$-mediated electrochemical sensor for the detection of guanine in beer and pharmaceutical samples. *Front. Nutrit.* **2022**, *9*, 987442. [CrossRef] [PubMed]
25. Mathieu, E.; Alida, Q.; David, G.; Lionel, N.; Clement, S.; Alain, W. Molecular transport into mesostructured silica thin films: Electrochemical monitoring and comparison between *p6m*, $P6_3/mmc$, and *Pm3n* structures. *Chem. Mater.* **2007**, *19*, 844–856.
26. Nasir, T.; Herzog, G.; Hébrant, M.; Despas, C.; Liu, L.; Walcarius, A. Mesoporous silica thin films for improved electrochemical detection of paraquat. *ACS Sens.* **2018**, *3*, 484–493. [CrossRef]
27. Zhu, X.; Xuan, L.; Gong, J.; Liu, J.; Wang, X.; Xi, F.; Chen, J. Three-dimensional macroscopic graphene supported vertically-ordered mesoporous silica-nanochannel film for direct and ultrasensitive detection of uric acid in serum. *Talanta* **2022**, *238*, 123027. [CrossRef]
28. Zhou, H.; Ding, Y.; Su, R.; Lu, D.; Tang, H.; Xi, F. Silica nanochannel array film supported by ß-cyclodextrin-functionalized graphene modified gold film electrode for sensitive and direct electroanalysis of acetaminophen. *Front. Chem.* **2022**, *9*, 812086. [CrossRef]
29. Zhang, M.; Zou, Y.; Zhou, X.; Yan, F.; Ding, Z. Vertically-ordered mesoporous silica films for electrochemical detection of Hg(II) ion in pharmaceuticals and soil samples. *Front. Chem.* **2022**, *10*, 952936. [CrossRef]
30. Yan, F.; Luo, T.; Jin, Q.; Zhou, H.; Sailjoi, A.; Dong, G.; Liu, J.; Tang, W. Tailoring molecular permeability of vertically-ordered mesoporous silica-nanochannel films on graphene for selectively enhanced determination of dihydroxybenzene isomers in environmental water samples. *J. Hazard. Mater.* **2021**, *410*, 124636–124644. [CrossRef]
31. Zhou, C.; Yan, J.; Chen, B.; Li, P.; Dong, X.; Xi, F.; Liu, J. Synthesis and application of ternary photocatalyst with a gradient band structure from two-dimensional nanosheets as precursors. *RSC Adv.* **2016**, *6*, 108955–108963. [CrossRef]

32. Gong, J.; Zhang, T.; Luo, T.; Luo, X.; Yan, F.; Tang, W.; Liu, J. Bipolar silica nanochannel array confined electrochemiluminescence for ultrasensitive detection of SARS-CoV-2 antibody. *Biosens. Bioelectron.* **2022**, *215*, 114563. [CrossRef] [PubMed]
33. Yan, F.; Chen, J.; Jin, Q.; Zhou, H.; Sailjoi, A.; Liu, J.; Tang, W. Fast one-step fabrication of a vertically-ordered mesoporous silica-nanochannel film on graphene for direct and sensitive detection of doxorubicin in human whole blood. *J. Mater. Chem. C* **2020**, *8*, 7113–7119. [CrossRef]
34. Saadaoui, M.; Fernández, I.; Sánchez, A.; Díez, P.; Campuzano, S.; Raouafi, N.; Pingarrón, J.; Villalonga, R. Mesoporous silica thin film mechanized with a DNAzyme-based molecular switch for electrochemical biosensing. *Electrochem. Commun.* **2015**, *58*, 57–61. [CrossRef]
35. Serrano, M.B.; Despas, C.; Herzog, G.; Walcarius, A. Mesoporous silica thin films for molecular sieving and electrode surface protection against biofouling. *Electrochem. Commun.* **2015**, *52*, 34–36. [CrossRef]
36. Lv, N.; Qiu, X.; Han, Q.; Xi, F.; Wang, Y.; Chen, J. Anti-biofouling electrochemical sensor based on the binary nanocomposite of silica nanochannel array and graphene for doxorubicin detection in human serum and urine samples. *Molecules* **2022**, *27*, 8640. [CrossRef]
37. Zheng, W.; Su, R.; Yu, G.; Liu, L.; Yan, F. Highly sensitive electrochemical detection of paraquat in environmental water samples using a vertically ordered mesoporous silica film and a nanocarbon composite. *Nanomaterials* **2022**, *12*, 3632. [CrossRef]
38. Lu, L.; Zhou, L.; Chen, J.; Yan, F.; Liu, J.; Dong, X.; Xi, F.; Chen, P. Nanochannel-confined graphene quantum dots for ultrasensitive electrochemical analysis of complex samples. *ACS Nano* **2018**, *12*, 12673–12681. [CrossRef] [PubMed]
39. Zou, Y.; Zhou, X.; Xie, L.; Tang, H.; Yan, F. Vertically-ordered mesoporous silica films grown on boron nitride-graphene composite modified electrodes for rapid and sensitive detection of carbendazim in real samples. *Front. Chem.* **2022**, *10*, 939510. [CrossRef] [PubMed]
40. Zhou, Z.; Guo, W.; Xu, L.; Yang, Q.; Su, B. Two orders-of-magnitude enhancement in the electrochemiluminescence of $Ru(bpy)_3^{2+}$ by vertically ordered silica mesochannels. *Anal. Chim. Acta* **2015**, *886*, 48–55. [CrossRef]
41. Yan, F.; Ma, X.; Jin, Q.; Tong, Y.; Tang, H.; Lin, X.; Liu, J. Phenylboronic acid-functionalized vertically ordered mesoporous silica films for selective electrochemical determination of fluoride ion in tap water. *Microchim. Acta* **2020**, *187*, 470–477. [CrossRef]
42. Yan, L.; Xu, S.; Xi, F. Disposal immunosensor for sensitive electrochemical detection of prostate-specific antigen based on amino-rich nanochannels array-modified patterned indium tin oxide electrode. *Nanomaterials* **2022**, *12*, 3810. [CrossRef]
43. Huang, L.; Su, R.; Xi, F. Sensitive detection of noradrenaline in human whole blood based on Au nanoparticles embedded vertically-ordered silica nanochannels modified pre-activated glassy carbon electrodes. *Front. Chem.* **2023**, *11*, 1126213. [CrossRef]
44. Fang, D.; Zhang, S.; Dai, H.; Lin, Y. An ultrasensitive ratiometric electrochemiluminescence immunosensor combining photothermal amplification for ovarian cancer marker detection. *Biosens. Bioelectron.* **2019**, *146*, 111768. [CrossRef]
45. Aydin, E.; Aydin, M.; Sezginturk, M. New impedimetric sandwich immunosensor for ultrasensitive and highly specific detection of spike receptor binding domain protein of SARS-CoV-2. *ACS Biomater. Sci. Eng.* **2021**, *7*, 3874–3885. [CrossRef]
46. Chen, L.; Li, Y.; Miao, L.; Pang, X.; Li, T.; Qian, Y.; Li, H. "Lighting-up" curcumin nanoparticles triggered by pH for developing improved enzyme-linked immunosorbent assay. *Biosens. Bioelectron.* **2021**, *188*, 113308. [CrossRef]
47. Yin, B.; Wan, X.; Yang, M.; Qian, C.; Muhtasim Fuad Sohan, A.S.M. Wave-shaped microfluidic chip assisted point-of-care testing for accurate and rapid diagnosis of infections. *Mil. Med. Res.* **2022**, *9*, 8. [CrossRef]
48. Yin, B.; Qian, C.; Wan, X.; Muhtasim Fuad Sohan, A.S.M.; Lin, X. Tape integrated self-designed microfluidic chip for point-of-care immunoassays simultaneous detection of disease biomarkers with tunable detection range. *Biosens. Bioelectron.* **2022**, *212*, 114429. [CrossRef]
49. Lin, Z.; Zheng, S.; Xie, J.; Zhou, R.; Chen, Y.; Gao, W. A sensitive electrochemiluminescence immunosensor for the detection of CA15-3 based on CeO_2/Pt/rGO as a novel co-reaction accelerator. *Talanta* **2023**, *253*, 123912. [CrossRef]
50. Xiong, X.; Zhang, Y.; Wang, Y.; Sha, H.; Jia, N. One-step electrochemiluminescence immunoassay for breast cancer biomarker CA 15-3 based on $Ru(bpy)_3^{2+}$-coated UiO-66-NH_2 metal-organic framework. *Sens. Actuators B Chem.* **2019**, *297*, 126812. [CrossRef]
51. Qin, D.; Jiang, X.; Mo, G.; Feng, J.; Yu, C.; Deng, B. A novel carbon quantum dots signal amplification strategy coupled with sandwich electrochemiluminescence immunosensor for the detection of CA15-3 in human serum. *ACS Sens.* **2019**, *4*, 504–512. [CrossRef] [PubMed]
52. Wang, Y.; Sha, H.; Ke, H.; Jia, N. An ultrasensitive electrochemiluminescence immunosensor based on platinum nickel nanocubes-*L*-cysteine-luminol nanocomposite. *Talanta* **2018**, *186*, 322–329. [CrossRef] [PubMed]
53. Babamiri, B.; Hallaj, R.; Salimi, A. Ultrasensitive electrochemiluminescence immunoassay for simultaneous determination of CA125 and CA15-3 tumor markers based on PAMAM-sulfanilic acid-$Ru(bpy)_3^{2+}$ and PAMAM-CdTe@CdS nanocomposite. *Biosens. Bioelectron.* **2018**, *99*, 353–360. [CrossRef] [PubMed]
54. Akbari Nakhjavani, S.; Khalilzadeh, B.; Samadi Pakchin, P.; Saber, R.; Ghahremani, M.; Omidi, Y. A highly sensitive and reliable detection of CA15-3 in patient plasma with electrochemical biosensor labeled with magnetic beads. *Biosens. Bioelectron.* **2018**, *122*, 8–15. [CrossRef] [PubMed]

55. Shawky, A.; El-Tohamy, M. Signal amplification strategy of label-free ultrasenstive electrochemical immunosensor based ternary Ag/TiO_2/rGO nanocomposites for detecting breast cancer biomarker CA 15-3. *Mater. Chem. Phys.* **2021**, *272*, 124983. [CrossRef]
56. Pothipor, C.; Bamrungsap, S.; Jakmunee, J.; Ounnunkad, K. A gold nanoparticle-dye/poly(3-aminobenzylamine)/two dimensional $MoSe_2$/graphene oxide electrode towards label-free electrochemical biosensor for simultaneous dual-mode detection of cancer antigen 15-3 and microRNA-21. *Colloids Surf. B Biointerfaces* **2022**, *210*, 112260. [CrossRef] [PubMed]

biosensors

MDPI

Article

An Impedimetric Biosensor for Detection of Volatile Organic Compounds in Food

Alessia Calabrese [1,2,3], Pietro Battistoni [4], Seniz Ceylan [4], Luigi Zeni [3], Alessandro Capo [1,2], Antonio Varriale [1,2], Sabato D'Auria [5,*] and Maria Staiano [1]

1 Institute of Food Science, CNR Italy, 83100 Avellino, Italy
2 URT-ISA, CNR at Department of Biology, University of Naples Federico II, 80126 Napoli, Italy
3 Department of Engineering, University of Campania Luigi Vanvitelli, 81031 Aversa, Italy
4 Allinit s.r.l., 84084 Fisciano, Italy
5 Department of Biology, Agriculture, and Food Science, National Research Council of Italy (CNR-DISBA), 00185 Rome, Italy
* Correspondence: sabato.dauria@cnr.it; Tel.: +39-06-4993-7803

Abstract: The demand for a wide choice of food that is safe and palatable increases every day. Consumers do not accept off-flavors that have atypical odors resulting from internal deterioration or contamination by substances alien to the food. Odor response depends on the volatile organic compounds (VOCs), and their detection can provide information about food quality. Gas chromatography/mass spectrometry is the most powerful method available for the detection of VOC. However, it is laborious, costly, and requires the presence of a trained operator. To develop a faster analytic tool, we designed a non-Faradaic impedimetric biosensor for monitoring the presence of VOCs involved in food spoilage. The biosensor is based on the use of the pig odorant-binding protein (pOBP) as the molecular recognition element. We evaluated the affinity of pOBP for three different volatile organic compounds (1-octen-3-ol, trans-2-hexen-1-ol, and hexanal) related to food spoilage. We developed an electrochemical biosensor conducting impedimetric measurements in liquid and air samples. The impedance changes allowed us to detect each VOC sample at a minimum concentration of 0.1 μM.

Keywords: odorant-binding protein (OBP); volatile organic compounds (VOCs); 1-octen-3-ol; trans-2-hexen-1-ol; hexanal; food safety; biosensors

Citation: Calabrese, A.; Battistoni, P.; Ceylan, S.; Zeni, L.; Capo, A.; Varriale, A.; D'Auria, S.; Staiano, M. An Impedimetric Biosensor for Detection of Volatile Organic Compounds in Food. *Biosensors* **2023**, *13*, 341. https://doi.org/10.3390/bios13030341

Received: 3 February 2023
Revised: 23 February 2023
Accepted: 1 March 2023
Published: 4 March 2023

1. Introduction

Food safety has substantially impacted human health, the economy, and society [1]. One of the main concerns of the food industry is represented by taints and off-flavors, which may affect consumer food acceptability. Food quality assessment requires the identification of compounds responsible for taints or off-flavors, defined by a broad, heterogeneous group of chemicals named volatile organic compounds (VOCs). The decomposition process that leads to the production of VOCs in food matrices originates from many sources, such as microbiological degradation [2–4], contaminations from packaging migration [5,6], contaminated water process, and improper food storage [7], involving the breakdown of carbohydrates [8–10], proteins [11–13], or lipids [14,15] into their constituents. The oxidation of fats and oils is one of the most predominant phenomena occurring during food life. This chemical process produces VOCs, which decisively influence the sensory quality of foods [16]. Secondary reaction products [17–19] (aldehydes, alkanes, alcohols, esters, and epoxides) can continue to react, forming tertiary VOCs such as unsaturated 2-alkenals and 2,4-alkadienals, harmful to human health [20–22]. Among the secondary products, hexanal is the most critical indicator for the progress of lipid oxidation [23,24]. Therefore, detecting this rancidity marker can help to characterize the oxidative status of high-fat content food matrices. Furthermore, technological aspects such as the improper use of temperature in the food processing line, packaging conditions, and microbial activity can promote the increase in the concentration of VOCs such as trans-2-hexen-1-ol, described as an astringent and unpleasant odor [25], and 1-octen-3-ol [26] perceived as a moldy, musty, and metallic odor.

Gas chromatography/mass spectrometry (GC/MS) is the methodology of choice for VOC analysis in food. However, despite its proven robustness, it suffers from some drawbacks, such as long-term analysis, time-consuming sample preparation, the need for the presence of a trained operator, and high costs. Hence, novel analytical methods are increasingly required for monitoring food quality in line with the farm-to-fork strategy to ensure consumer confidence. In this scenario, biosensing technologies aim to offer innovative solutions to face the challenges of the food industry.

An example is given by electrochemical biosensors that combine the advantage of specificity with the simplicity of the operation. Furthermore, due to the progress in electronics, these devices can be miniaturized as lab-on-chip for on-site monitoring.

The identification of the biomolecule that specifically recognizes the analyte is the key element in designing a biosensor. Recently, odorant-binding proteins (OBPs) have attracted increasing interest as sensing elements for odors, given their physiological role as odorant carriers to the olfactory neurons. These proteins have been identified in both vertebrates and insects. However, they are structurally unrelated: the first ones belong to the lipocalin superfamily [27,28] and are folded into a β-barrel structure [29], while the second ones show mostly α-helical domains [30]. Furthermore, they are characterized by a compact structure, small size, and extreme thermal and chemical denaturation stabilities [31,32].

In this work, we describe the design of an impedimetric protein-based biosensor for monitoring the presence of three VOCs (hexanal, trans-2-hexen-1-ol, and 1-octen-3-ol) involved in food spoilage using as molecular recognition element (MRE) pOBP. Through direct docking simulation experiments, the binding affinity of pOBP for the selected VOCs was calculated. A competitive assay, based on fluorescence resonance energy transfer (FRET), was performed for each VOC molecule. pOBP was covalently attached to the self-assembled monolayer (SAM) of an α-lipoic acid (ALA)-modified gold surface. The effect of each VOC molecule on the surface electrical properties was monitored by non-Faradaic electrochemical impedance spectroscopy (EIS). The binding phenomena occurring at the electrode interface were observed as a decrease in the imaginary part of impedance with increasing analyte concentrations. The obtained results show the capability of the biosensor to detect the presence of the three selected VOCs in a micromolar range. Therefore, a preliminary study conducted with air contaminated with the selected VOCs shows the maintenance of pOBP capability to recognize the three VOC molecules and suggests that the biosensor could be used as an on-off tool for VOC detection in food.

2. Materials and Methods

2.1. Materials

Hexanal, trans-2-hexen-1-ol, 1-octen-3-ol, α-lipoic acid, and solvents were purchased from Sigma-Aldrich. Sepharose 4 Fast Flow resin for pOBP purification was acquired from GE Healthcare. DEAE–Sepharose for glutamine-binding protein (GlnBP) purification was obtained from Sigma-Aldrich. Thin-film gold single electrodes (ED-SE1-Au) were purchased from Micrux Technologies (Oviedo, Spain). Materials used for protein electrophoresis were obtained from Bio-Rad. All the used reagents were at analytical or higher grades available. All the solutions were prepared using Milli-Q water.

2.2. Direct Docking Analysis

In order to assess the proper binding conformations of the VOC ligands with pOBP, direct docking simulation experiments [33] were performed by MGLTools (http://mgltools.scripps.edu/downloads/ accessed on 8 December 2021), an open-source software suite. After preparing structure files of the protein and the ligands using AutodockTools (ADT) 1.5.6, Autogrid [34] allowed us to calculate the affinity maps through a 3D grid delimiting the ligand-receptor complex. Lastly, with the aid of Autodock, the potential binding poses were predicted, and binding free energies were calculated. The structure files of the ligands were downloaded from PubChem [35] in SDF format and converted into PDB by Open Babel GUI [36]. The structure file of pOBP [37] was obtained by Protein Data Bank [38] (ID: 1E02). The protein crystal structure was processed by removing water and the co-crystallized ligand and by analyzing only chain A. All hydrogen atoms, charges, and atom types were assigned [39,40]. After calculating the ligand torsdof and adding partial charge to atoms, two different PDBQT files were generated. Next, docking simulations were performed considering a grid box of 58 × 66 × 46 points, with a spacing of 0.375 Å. Finally, the affinity maps were computed by Autogrid and saved in a GLG file. The AutoDock Lamarckian genetic algorithm was chosen to perform 100 docking runs, treating the protein as rigid and the ligand as flexible. As a result, 30,000 poses were calculated, and 2,500,000 was set as the maximum energy evaluations. Ultimately, Autodock 4.2 allowed the collection of the docking simulation results in a DLG file consisting of the three-dimensional coordinates of the generated poses. In addition, the binding free energy and the pose clustering were calculated and compared to a reference co-crystallized ligand 1-aminoanthracene. Through AutodockTools, it was possible to identify the amino acid residues involved in binding, exploring the presence of hydrogen bonds or π–π stacking interactions. The lowest binding energy and the higher number of cluster poses were used to identify the best ligand–protein complexes.

2.3. Expression and Purification of Recombinant pOBP

The expression and purification of pOBP were performed in accordance with Capo et al. (2018) [41]. In brief, the recombinant pOBP-GST gene, subcloned into expression vector pGEX-2TK, was transformed into the *Escherichia coli* BL21 (DE3) strain. A single *E. coli* colony was picked from an LB agar plate and inoculated overnight in 10 mL LB broth with the selective antibiotic (ampicillin). This starter culture was inoculated in 500 mL of fresh LB medium containing ampicillin (50 µg/mL) for approximately 3 h at 37 °C under shaking. Cells were grown until the absorbance at 600 nm reached a value of roughly 1.0. At this point, the expression of pOBP was induced by adding 0.5 mM isopropyl β-D-1-thiogalactopyranoside (IPTG) for 3 h at 37 °C. The bacterial suspension was centrifuged at 4000 rpm for 30 min at 4 °C. The cellular pellet obtained was resuspended in phosphate buffer saline (PBS) (137 mM NaCl, 2.7 mM KCl, 10 mM Na_2HPO_4, 1.8 mM KH_2PO_4, pH 7.3) and was incubated at 37 °C for 30 min before with 0.4% Lysozyme, and after with DNase I (50 µg/mL per mL of solution), 5 mM $MgCl_2$ (1 mg per mL of solution). Finally, the cells were lysed using an ultrasonic homogenizer (Misonix Sonicator XL Ultrasonic Processor). Centrifugation removed the nucleic acid fragments and cell debris at the end of this step. After filtration, the soluble fraction collected was loaded on a Glutathione Sepharose 4 Fast Flow resin incubated overnight at 4 °C under shaking. After the incubation phase, to remove protein contaminants unbound from the resin, several washes step with PBS were carried out. Then the column was incubated with thrombin (1 unit of thrombin per 100 µg of fusion protein) for 16 h at 25 °C. Finally, after collecting pOBP, the GST tag was removed from the resin with elution buffer (50 mM Tris-HCl, 10 mM reduced glutathione pH 8.0). The purity of the collected protein samples was evaluated by Sodium Dodecyl Sulphate-Poly-Acrylamide Gel Electrophoresis (SDS-PAGE). A single band at about 21 kDa confirmed that the protein was purified to homogeneity. The protein concentration of fractions containing pOBP was estimated by measuring UV absorbance at 280 nm using Jasco V-730 UV/Vis spectrophotometer (molar extinction coefficient 12,200 M^{-1} cm^{-1}).

2.4. Expression and Purification of Glutamine-Binding Protein

As a negative control of the pOBP-based biosensor, the electrode surface was functionalized with the recombinant glutamine-binding protein (GlnBP). The expression and purification steps were performed following D'Auria et al. (2005) [42]. In brief, *E. coli* cells HB101 expressing GlnBP were grown overnight at 37 °C in LB in the presence of 100 µg/mL ampicillin and then were disrupted by osmotic shock. Next, the crude periplasmic preparations were equilibrated with Na phosphate buffer (pH 7.5) and applied to a DEAE–Sepharose column. Due to the high basicity of GlnBP, there was no adsorption at this pH value, and the protein was collected in the flowthrough fractions. Finally, the protein concentration of GlnBP fractions was estimated by measuring UV absorbance at 280 nm using Jasco V-730 UV/Vis spectrophotometer (molar extinction coefficient 22,920 M^{-1} cm^{-1}).

2.5. Fluorescence Spectroscopy

Steady-state fluorescence measurements were performed using a Jasco FP-8600 fluorescence spectrometer in a 1.0 cm light path quartz cuvette. Tryptophan (Trp) residues were selectively excited at 295 nm. Emission spectra were acquired in the range of 320–600 nm at 1 nm intervals with a scan speed of 500 nm/min by fixing excitation and emission slit widths at 5 nm. Assays were performed in 500 µL of PBS at pH 7.4 with a volume concentration of ethanol equal to 0.01% [43]. Fluorescence experiments were performed on pOBP samples with optical density lower than 0.1 OD at 295 nm [43,44] to avoid the inner filter effect phenomenon. An ethanolic solution of 1-aminoanthracene (1-AMA) was used as an extrinsic fluorophore after determining its concentration by the Beer–Lambert law using an extinction coefficient equal to 35.45 mM^{-1} cm^{-1} at 280 nm. Titration experiments were performed in triplicate by adding increasing amounts of 1-AMA to pOBP samples. Alcoholic solutions of hexanal, trans-2-hexen-1-ol, and 1-octen-3-ol were tested in competitive binding experiments repeated three times and obtained similar results. Excel 2016 by Microsoft and OriginPro 2021b software were used to analyze the data.

2.6. Instruments for Electrochemical Experiments

The impedimetric measurements were carried out with a miniaturized all-in-one electrochemical workstation (MicruX ECStat). The ED-SE1-Au electrochemical sensors are based on a three-electrodes approach: a working, a reference, and an auxiliary electrode.

2.7. Surface Derivatization and Functionalization

The gold electrodes were functionalized by slightly modifying the derivatization protocol described by Capo et al. (2022) [45]. Before functionalization, gold electrodes were cleaned by applying 12 potential cycles between −1.0 and +1.3 V with a 0.1 V/s scan rate in 0.05 M sulfuric acid. Then, a self-assembled monolayer (SAM) of thiols was prepared by immersing the clean gold substrates into a 40 mM solution of α-lipoic acid (ALA) for 20 h. Next, the terminal carboxylic groups of the organo-sulfur molecules immobilized on gold electrodes were activated by dropping 15 µL of 200 mM EDC and 50 mM NHS in PBS buffer (pH 7.4) on the surfaces. After 10 min of incubation, the EDC/NHS solution was removed. Subsequently, 15 µL of a 1.7 mg/mL pOBP sample (or GlnBP, used as a negative control) were deposited on the electrodes; after 2 h of incubation, the surfaces were washed three times with sterilized water. Lastly, 10 µL of 1 M ethanolamine (pH 8.5) solution was placed on the surfaces for 20 min to block unreacted active sites. A schematic diagram of the biosensor fabrication is presented in Figure 1.

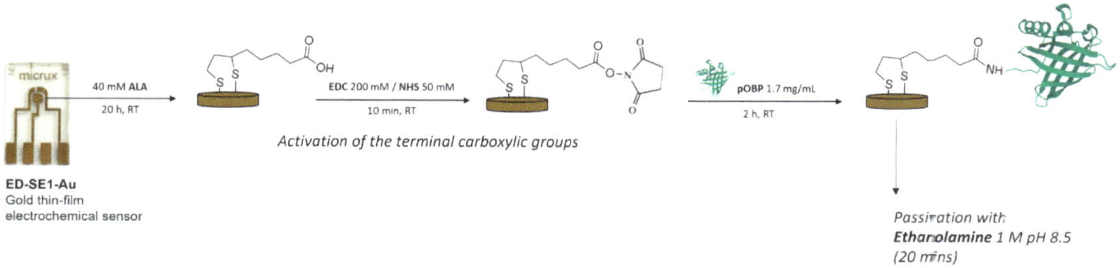

Figure 1. Schematic representation of surface derivatization and functionalization processes.

2.8. Impedance Measurements

Non-Faradaic impedance spectroscopy in the absence of a redox probe was chosen to investigate the biorecognition events at the functionalized electrode surface. Binding experiments were carried out at 25 °C and 44% of humidity by depositing 10 μL of hexanal, trans-2-hexen-1-ol, and 1-octen-3-ol dissolved in PBS, preparing serial dilutions, considering the limit of water solubility, at different concentrations (0.1; 0.5; 1; 5, and 10 μM) on the working area of the sensor for 10 min. After rinsing it thoroughly, EIS measurements were conducted in PBS by superimposing a sinusoidal AC potential (0.1 V) to 0 V DC potential in a frequency range of 0.1 to 100,000 Hz. Impedimetric responses are the means of three replicates.

2.9. Statistical Analysis

Each impedance measurement was performed in triplicate. The mean and standard deviation were calculated from the value of triplicates. The graphs report the mean values gleaned from the blank values, and the error bars represent the calculated standard deviation. The linear calibration curve was obtained by plotting the change in impedance at 0.1 Hz ($\Delta Z = Z_{baseline} - Z_{VOC}$) as a function of the logarithm of the VOC concentration. The limit of detection (LOD = 3.3 σ/S) was determined by considering the slope of the calibration curve, S, and the standard deviation of the response, σ [46]. The graphs were realized in Excel 2016 by Microsoft® and Origin Pro 8.0 software.

3. Results and Discussion

In this work, intending to develop an electrochemical protein biosensor able to detect the presence of VOCs associated with food spoilage (1-octen-3-ol, hexanal, and trans-2-hexen-1-ol), we selected pOBP as MRE of the biosensor. This protein exhibits a typical hydrophobic cavity that binds VOCs (Figure 2a).

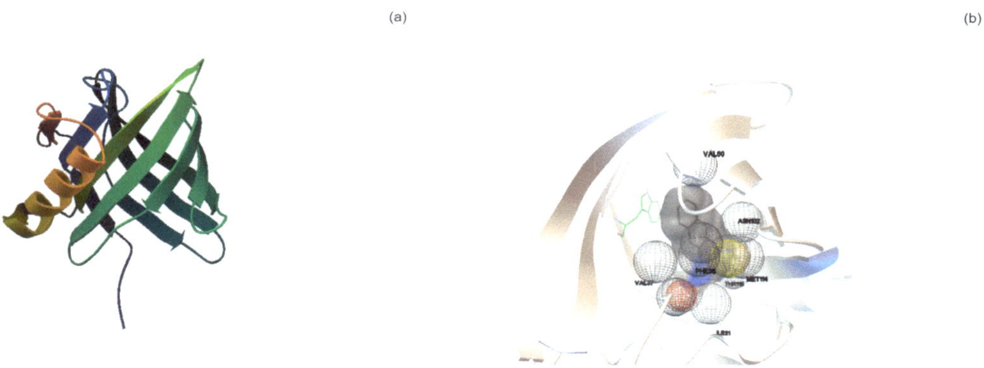

Figure 2. pOBP-1-AMA docking simulation experiments: (**a**) pOBP is a monomer of 157 amino acids containing one disulfide bridge between cysteines at positions 63 and 155 and a single tryptophan residue (Trp) at position 16, accessible to extrinsic fluorophores such as 1-aminoanthracene (1-AMA); (**b**) The image depicts the position of 1-AMA (in gray color) in the β-barrel structure of pOBP. The single tryptophan residue is highlighted in green color.

Direct docking experiments were performed to predict binding affinity between pOBP and the selected VOCs. Conformations of pOBP-VOC complexes (Figure 2b) were ranked based on predicted free energy of binding (ΔG), cluster analysis, and ligand position in the binding site. Thus, firstly, the correct position of each ligand in the binding site was checked. Subsequently, the conformation belonging to the most populous cluster, with the lowest estimated free energy of binding, was selected for each ligand. Moreover, it was possible to obtain information on the interactions and the amino acids involved. Finally, docking results were validated with a co-crystallized ligand, 1-aminoanthracene. Table 1 provides the direct docking results sorted in decreasing order according to the estimated free energy of binding (ΔG). Although the ΔG values are quite similar, it is possible to observe that the portion of the alkyl chain and the presence of un-saturations affect the affinity (1-aminoanthracene > 1-octen-3-ol > trans-2-hexen-1-ol > hexanal).

Table 1. Direct docking simulation results. Ligands are arranged in order of decreasing binding affinity values.

Ligand	ΔG (kcal/mol)	Ki	Amino Acid Residues Involved in Interactions
1-Aminoanthracene	−8.42	674.02 nM	ILE21; PHE35; VAL37; LEU53; VAL80; TYR82; PHE88; ILE100; ASN102; MET114; THR115; GLY116
1-Octen-3-ol	−4.67	379.09 μM	ILE21; PHE35; PHE88; ASN102; THR115; GLY116; LEU118
trans-2-Hexen-1-ol	−3.85	1.50 mM	ILE21; ILE29; PHE35; VAL37; ILE100; SER101; ASN102; MET114; THR115; GLY116
Hexanal	−3.51	2.70 mM	ILE21; PHE35; VAL37; ILE100; SER101; ASN102; MET114; THR115; GLY116

ΔG: Estimated Free Energy of Binding; Ki: Estimated Inhibition Constant; mM millimolar; μM: micromolar; nM: nanomolar.

3.1. Fluorescence Spectroscopy Measurements

Fluorescence spectroscopy measurements were performed to assess the structural conformation and the binding capacity of pOBP.

In proteins, the indole ring of tryptophan is the dominant fluorophore known to emit fluorescence near 340 nm upon excitation at 295 nm. Monitoring the intrinsic protein tryptophan fluorescence offers the possibility to investigate the relationship structure-function. Any changes in the tryptophan microenvironment may affect the fluorescence emission. A red-shifted fluorescence emission to longer wavelengths may suggest the presence of protein unfolding processes with the loss of protein function [44]. The investigation of protein folding and protein–ligand interactions often relies on the use of the fluorescence resonance energy transfer (FRET) technique. In the case of pOBP, the most widely adopted fluorescence reporter for ligand-binding assays is 1-aminoanthracene (1-AMA). In fact, this fluorescence compound is proven to be a strong ligand for pOBP [47]. One of the emissive characteristics of this compound is to dramatically increase the fluorescence emission when it is bound to a protein matrix.

To develop the FRET assay, the capability of pOBP to bind 1-AMA was evaluated. When excited at 295 nm, 1-AMA showed a weak fluorescence emission with a maximum centered at 537 nm. However, when 1-AMA binds to pOBP, we registered two events: (1) the blue-shift of the emission maximum to 481 nm with the increase in fluorescence emission intensity; (2) the decrease in fluorescence emission intensity at 340 nm. These variations are attributable to the resonance energy transfer phenomenon between the single protein tryptophan residue (at position 16 in the protein structure) and 1-AMA intercalated in pOBP binding site.

Figure 3a displays pOBP fluorescence emission intensity increase at 481 nm at increasing 1-AMA concentrations. The analysis of the binding curve indicates the presence of a plateau at 4 mM of 1-AMA, which represents the amount of 1-AMA required to saturate pOBP binding sites (Figure 3b).

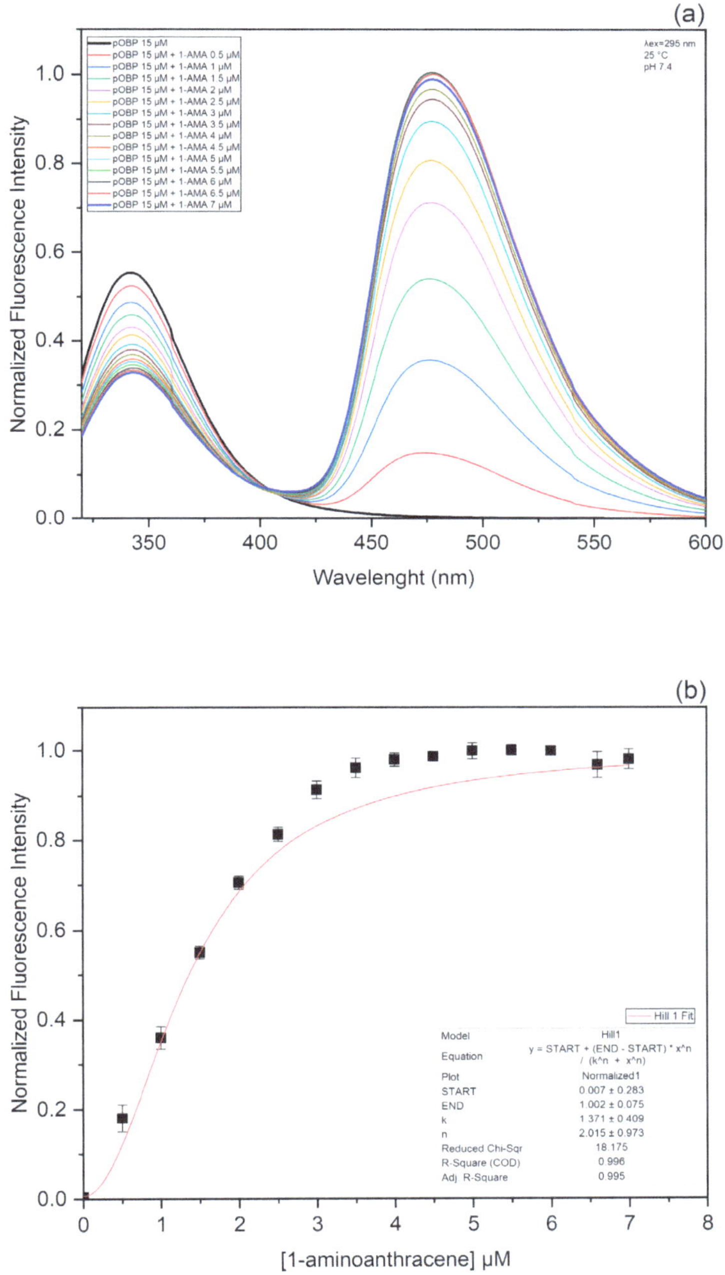

Figure 3. 1-AMA titration curve spectra and titration curves fitting: (**a**) The emission spectra of pOBP upon excitation at 295 nm were acquired in the absence and in the presence of different amounts of 1-AMA; (**b**) fluorescence emission intensity at 481 nm was plotted as a function of 1-AMA concentrations.

Competitive FRET Assay

In order to validate the direct docking results and confirm the protein affinity for the selected VOCs, competitive FRET assays were performed for each VOC molecule. In this assay, the affinity of a ligand was evaluated on its ability to displace the fluorescence probe from the protein complex [48]. More precisely, the competition of 1-octen-3-ol (Figure 4a), trans-2-hexen-1-ol (Figure 4c), and hexanal (Figure 4e) with 1-AMA to the binding site of pOBP were observed by monitoring the displacement of 1-AMA at increasing concentrations of each VOC sample. The titration curves of 1-octen-3-ol, trans-2-hexen-1-ol, and hexanal are presented in Figure 4b,d,f, respectively. The kinetics parameters were calculated by plotting the decrease in the fluorescence intensity at 481 nm as a function of the concentration for each single VOC molecule. The dissociation constant values, calculated through a non-linear fitting function, are 18.5 mM for 1-octen-3-ol, 122.3 mM for trans-2-hexen-1-ol, and 349.7 mM for hexanal.

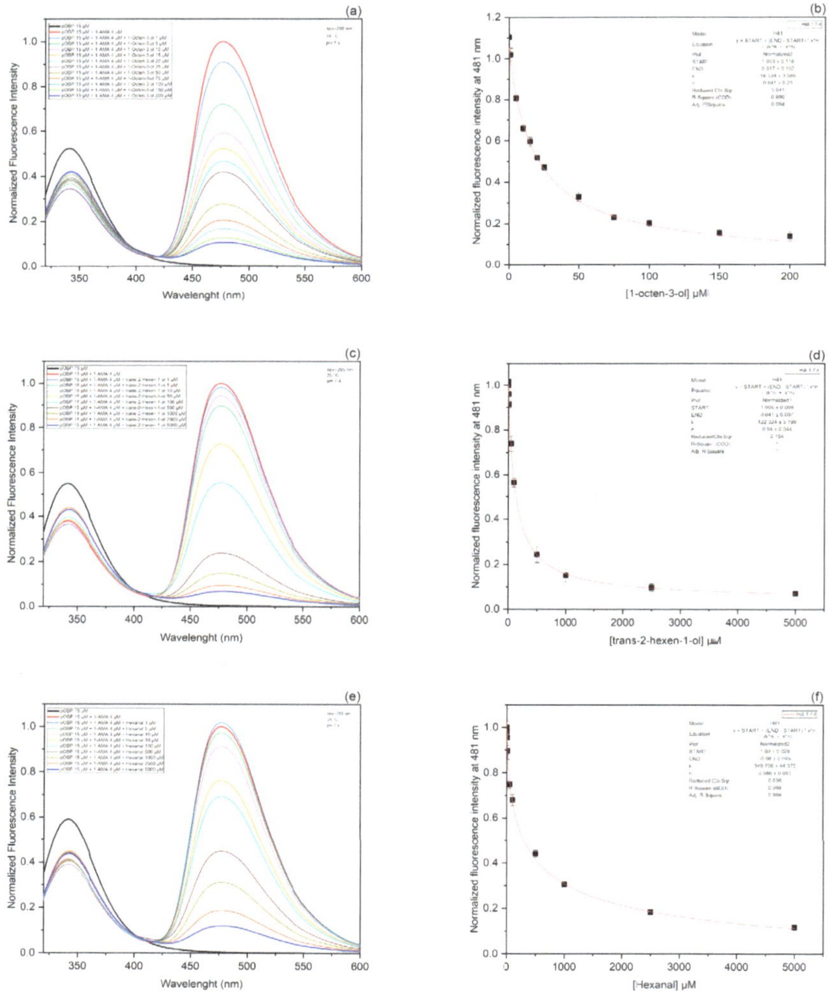

Figure 4. Fluorescence emission spectra and titration curves fitting of 1-octen-3-ol, trans-2-hexen-1-ol, and hexanal. Emission fluorescence spectra of pOBP were acquired in the presence of saturating concentrations of 1-AMA. The addition of increasing amounts of 1-octen-ol (**a**), trans-2-hexen-1-ol (**c**),

and hexanal (**e**) determined the decrease in the peak at 481 nm and the increase in the peak at 340 nm. The decreasing intensity of 1-AMA fluorescence emission at 481 nm was plotted as a function of 1-octen-3-ol (**b**), trans-2-hexen-1-ol (**d**), and hexanal (**f**) concentration. The fitting curves obtained by a non-linear function are highlighted in red color.

3.2. Functionalization and Immobilization of pOBP on Sensor Surface

pOBP was covalently attached to the self-assembled monolayer (SAM) of an α-lipoic acid (ALA)-modified gold surface. The design of the immobilization layout on the gold surface is crucial for biosensor performance. In fact, it significantly influences the responsiveness of the bio-interface. The electrodes were cleaned by applying potential cycles in the presence of sulfuric acid. Afterward, the gold electrodes were treated as described in detail in "Materials and Methods" (Section 2.7). The surface functionalization was monitored using electrochemical impedance spectroscopy (EIS). Figure S1 (see supplementary material section) shows the Nyquist plot of the gold electrode before and after the immobilization procedure. The results show an increase in the value of the impedance, indicating the presence of covalent immobilization of pOBP on the electrode surface.

3.3. Electrochemical Characterization

The electrochemical performance of the biosensor was analyzed by non-Faradaic electrochemical impedance spectroscopy (EIS). This technique allows us to observe binding phenomena at the electrode–electrolyte interface. When a charged electrode surface is placed in an electrolyte, an electrical double layer (EDL) is formed. Its thickness is related to the capacitive variation due to the binding between the target analyte (VOC) and the capture probe (pOBP). A single VOC molecule was dropped from the lowest (0.1 µM) to the highest concentration (10 µM) on the functionalized electrode surface. After incubating for 10 min, the surface was rinsed, and the impedance response was recorded in a PBS buffer.

The same procedure was applied to GlnBP functionalized gold electrodes (the negative control).

The results are presented in the Nyquist diagram, where the imaginary part (Z″ or $-Z_{img}$) is plotted versus the real part (Z′ or Z_{real}). The magnitude impedance is a complex number given by:

$$Z(\omega) = Z_{real}(\omega) - j\, Z_{img}(\omega), \quad (1)$$

where $\omega = 2\pi f$.

A non-Faradaic Nyquist plot exhibits a large incomplete semicircle tending to infinity in the low-frequency region. The lack of a redox probe eliminates the parameters associated with electron transfer, such as charge transfer resistance (R_{ct}) and Warburg impedance. As a result, the partial semicircle, due to the extremely slow electron transfer, is not followed by the typical diffusion tail [49,50]. Moreover, the solution resistance (R_{sol}) in the high-frequency region is attributable to the bulk ionic concentration. Therefore, the impedance of a non-Faradaic sensor is determined by the insulating characteristics of the VOC sample bond to the conductive substrate [51]. The binding phenomena occurring at the electrode interface generate a charge perturbation observable in the Nyquist plot as a decrease in the imaginary part of impedance (−Z″) in the low-frequency regime with increasing VOC concentrations (Figure 5).

The GlnBP-functionalized electrode (negative control) showed no significant impedance variation at 0.1 Hz in the presence of the selected VOC samples in the range of 0.1–10 µM. (Figure S2). These results confirm the specificity of pOBP sensors for the tested VOC molecules.

Since the maximum signal variation was observed at 0.1 Hz, this frequency was selected to represent the further data. Specifically, the binding curve was obtained by plotting the change in impedance at 0.1 Hz ($\Delta Z = Z_{baseline} - Z_{VOC}$) versus the value of VOC concentrations and applying a non-linear fitting function (Figure 5). The limit of detection (LOD) determined as described in "Statistical Analysis" (Section 2.9), was estimated to be 0.49 µM for 1-octen-3-ol, 0.60 µM for hexanal and 0.81 µM for tran-2-hexen-1-ol (Figure 6).

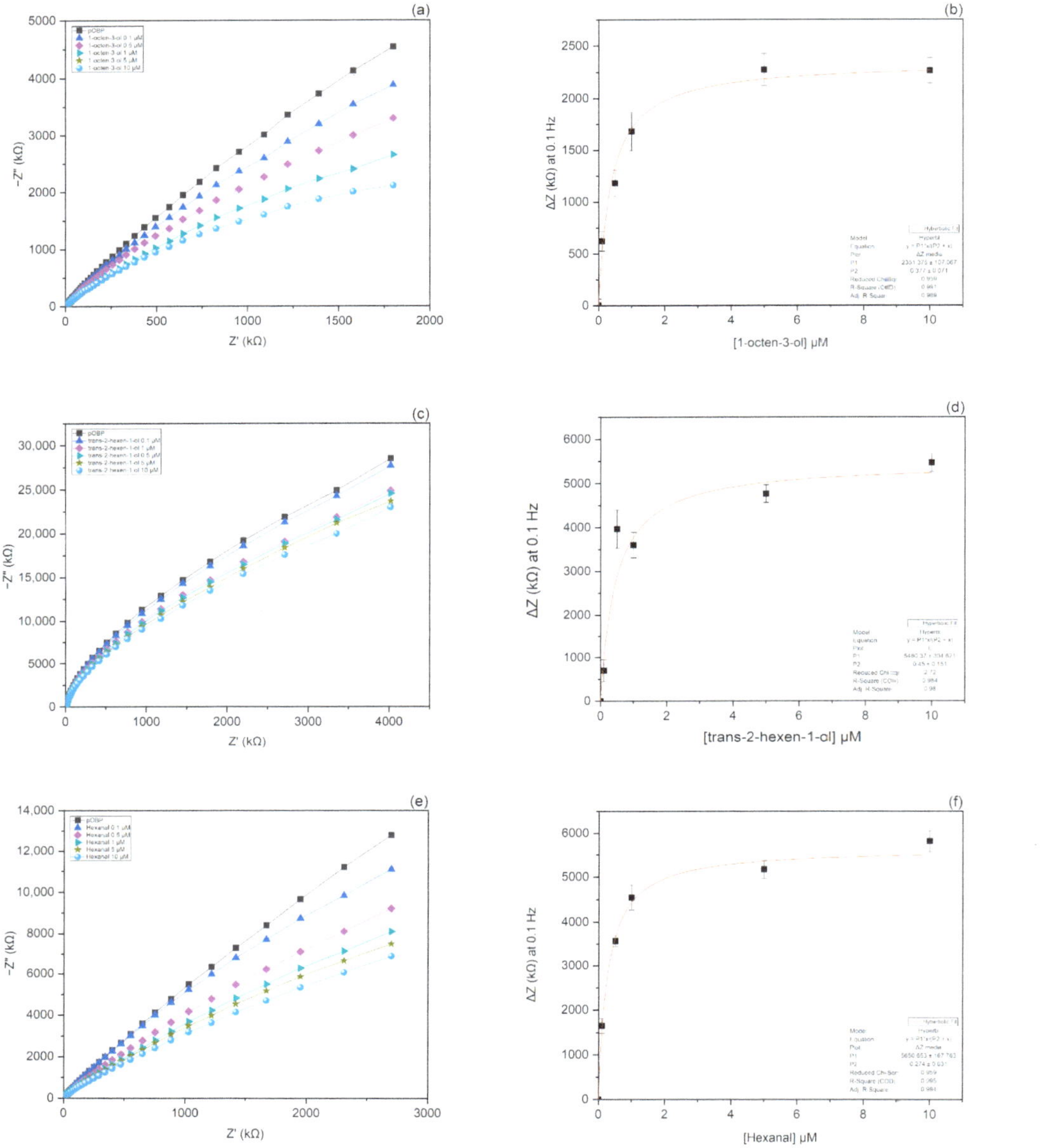

Figure 5. Nyquist plots and binding curves fitting of 1-octen-3-ol, trans-2-hexen-1-ol, and hexanal. Nyquist plots represent the impedance response of the biosensor to increasing concentrations of 1-octen-3-ol (**a**), trans-2-hexen-1-ol (**c**), and hexanal (**e**). As a consequence of the binding events at the interface, the impedance of the electrochemical system increases at increasing concentrations of the VOC samples. Binding curves were obtained by plotting the change in impedance at 0.1 Hz as a function of 1-octen-3-ol (**b**), trans-2-hexen-1-ol (**d**), and hexanal (**f**) concentration. The fitting curves, obtained by a non-linear function, are highlighted in red color.

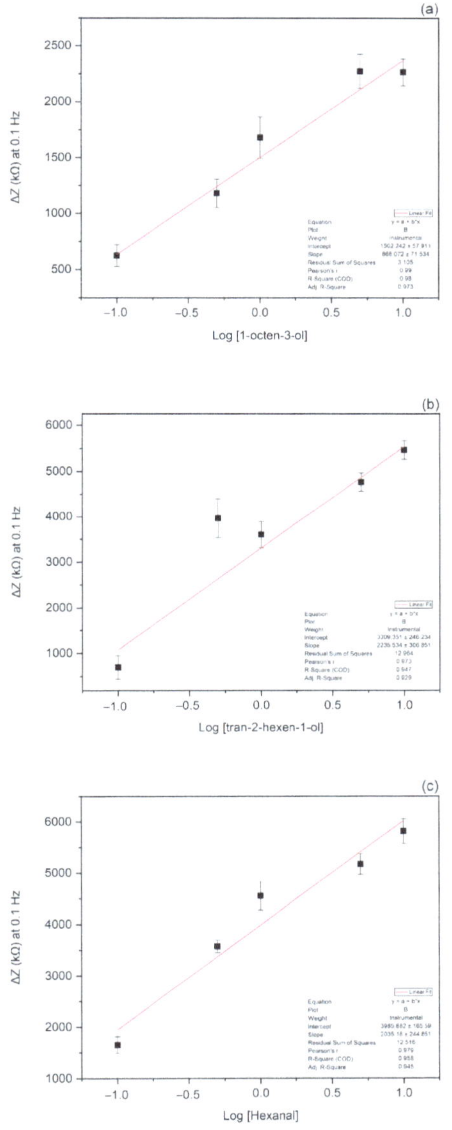

Figure 6. Calibration curves of 1-octen-3-ol, trans-2-hexen-1-ol, and hexanal. The fitting curves were obtained by plotting the logarithmic value of 1-octen-3-ol (**a**), trans-2-hexen-1-ol (**b**), and hexanal (**c**) concentrations versus the change of the impedance value at 0.1 Hz.

The biosensor showed a response time of 15 min, including the incubation time, and a recovery time of 30 min. For evaluating the long-term stability, the functionalized electrodes were stored in PBS for 20 days at 4 °C with a negligible loss of activity of the biosensor.

3.4. EIS Experiments in Gas

pOBP-functionalized gold electrodes were further characterized by electrical impedance spectroscopy (EIS) in air in the absence and in the presence of the selected VOC samples. For this purpose, each electrode was placed inside a suitable gas chamber (Figure S3).

Several measurements were acquired until the steady state was reached. Afterward, the gold electrodes were exposed to the single VOC sample, and impedance spectra were collected after 10 and 20 min.

The Nyquist plots thus obtained (Figure 7) show a variation of the impedance curves in the presence of the volatiles already after 10 min, reaching saturation after 20 min. This result can be ascribed to binding events between pOBP and the VOC samples.

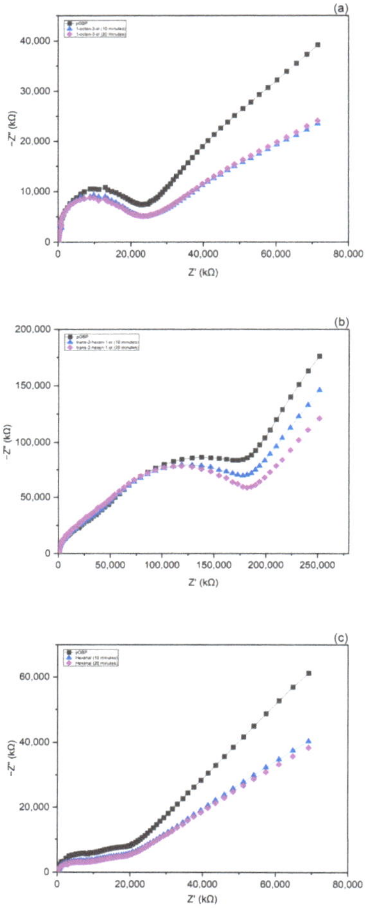

Figure 7. Impedance variations in air as a function of 1-octen-3-ol (**a**), trans-2-hexen-1-ol (**b**), and hexanal (**c**) concentrations.

4. Conclusions

In this work, we designed a non-Faradaic impedimetric biosensor to detect three different VOC samples (1-octen-3-ol, trans-2-hexen-1-ol, and hexanal) involved in food spoilage by immobilizing pOBP onto a SAM-functionalized gold electrode. The changes in impedance values at 0.1 Hz allowed us to detect the VOC samples in the range of 0.1–10 μM. Furthermore, we observed similar behavior when the biosensor was exposed to air contaminated with the VOC samples. These results suggest that pOBP maintains its functionality outside its living ambient, lending itself to being used as a molecular recognition element even in the gas phase. In the future, to enhance biosensor selectivity, we plan to use the collected data to train neural networks, which could help to discriminate between the different VOC samples. Moreover, by gaining more insight into the type of interactions with the help of molecular docking, we intend to synthesize tailored pOBP mutants with improved selectivity.

Supplementary Materials: The following supporting information can be downloaded at: https://www.mdpi.com/article/10.3390/bios13030341/s1, Figure S1: Electrode EIS response before and after the functionalization procedure.; Figure S2: Impedance responses of GlnBP-functionalized electrodes to increasing concentrations of 1-octen-3-ol (a), trans-2-hexen-1-ol (b) and hexanal (c); Figure S3: System used for impedimetric measurements in gas.

Author Contributions: Conceptualization, A.C. (Alessia Calabrese), M.S., A.V. and S.D.; methodology, A.V., P.B. and S.C.; software, A.C. (Alessia Calabrese) and S.C.; validation, A.C. (Alessandro Capo) and L.Z.; formal analysis, A.C. (Alessia Calabrese) and A.C. (Alessandro Capo); investigation, A.C. (Alessandro Capo); resources, M.S.; data curation, A.C. (Alessia Calabrese); writing—original draft preparation, A.V., A.C. (Alessia Calabrese), M.S. and S.D.; writing—review and editing, A.V., M.S. and S.D.; visualization, A.C. (Alessia Calabrese); supervision, M.S., S.D., L.Z. and A.V.; project administration, S.D. and M.S.; funding acquisition, S.D. All authors have read and agreed to the published version of the manuscript.

Funding: This research was funded by CNR and AllinIT s.r.l., as part of the agreement between the National Research Council (CNR) and Confindustria for Industrial Doctorates.

Institutional Review Board Statement: Not applicable.

Informed Consent Statement: Not applicable.

Data Availability Statement: The data presented in this study are available within the article and its Supplementary Materials. Other data that support the findings of this study are available upon request from the corresponding author.

Conflicts of Interest: The authors declare no conflict of interest.

References

1. Garcia, S.N.; Osburn, B.; Russell, M.T.J. One Health for Food Safety, Food Security, and Sustainable Food Production. *Front. Sustain. Food Syst.* **2020**, *4*, 1–9. [CrossRef]
2. Whitfield, F.B. Microbiology of food taints. *Int. J. Food Sci.* **1998**, *33*, 31–51. [CrossRef]
3. Springett, M.B. Formation of off-flavours due to microbiological and enzymic action. In *Food Taints and Off-Flavours*; Saxby, M.J., Ed.; Springer: Boston, MA, USA, 1996; pp. 274–289. [CrossRef]
4. Whitfield, F.B. 5—Microbiologically derived off-flavours. In *Taints and Off-Flavours in Food*, 1st ed.; Baigrie, B., Ed.; Woodhead Publishing: Sawston, UK, 2003; pp. 112–139. [CrossRef]
5. Lord, T. 4—Packaging materials as a source of taints. In *Taints and Off-Flavours in Food*, 1st ed.; Baigrie, B., Ed.; Woodhead Publishing: Sawston, UK, 2003; pp. 64–111. [CrossRef]
6. Koszinowski, J.; Piringer, O. Evaluation of off-odors in food packaging—The role of conjugated unsaturated carbonyl compounds. *J. Plast. Film Sheeting* **1986**, *2*, 40–50. [CrossRef]
7. Valero, E.; Villamiel, M.; Miralles, B.; Sanz, J.; Martínez-Castro, I. Changes in flavour and volatile components during storage of whole and skimmed UHT milk. *Food Chem.* **2001**, *72*, 51–58. [CrossRef]
8. Choe, E.; Min, D.B. Chemistry and reactions of reactive oxygen species in foods. *Crit. Rev. Food Sci. Nutr.* **2006**, *46*, 1–22. [CrossRef] [PubMed]
9. Gottschalk, G. *Bacterial Metabolism*, 2nd ed.; Springer: New York, NY, USA, 1986. [CrossRef]
10. Saier, M.H., Jr. Multiple mechanisms controlling carbon metabolism in bacteria. *Biotechnol. Bioeng.* **1998**, *58*, 170–174. [CrossRef]
11. del Castillo-Lozano, M.L.; Mansour, S.; Tache, R.; Bonnarme, P.; Landaud, S. The effect of cysteine on production of volatile sulphur compounds by cheese-ripening bacteria. *Int. J. Food Microbiol.* **2008**, *122*, 321–327. [CrossRef] [PubMed]
12. Kadota, H.; Ishida, Y. Production of volatile sulfur compounds by microorganisms. *Annu. Rev. Microbiol.* **1972**, *26*, 127–138. [CrossRef]
13. Chin, H.W.; Lindsay, R. Mechanisms of formation volatile sulfur compounds following the action of cysteine sulfoxide lyases. *J. Agric. Food Chem.* **1994**, *42*, 1529–1536. [CrossRef]
14. Forbes, S.L.; Stuart, B.H.; Dent, B.B. The effect of the burial environment on adipocere formation. *Forensic. Sci. Int.* **2005**, *154*, 24–34. [CrossRef] [PubMed]
15. Ubelaker, D.H.; Zarenko, K.M. Adipocere: What is known after over two centuries of research. *Forensic. Sci. Int.* **2011**, *208*, 167–172. [CrossRef]
16. Vass, A.A.; Smith, R.R.; Thompson, C.V.; Burnett, M.N.; Wolf, D.A.; Synstelien, J.A.; Dulgerian, N.; Eckenrode, B.A. Decompositional odor analysis database. *J. Forensic. Sci.* **2004**, *49*, 760–769. [CrossRef]

17. Bottcher, S.; Steinhauser, U.; Drusch, S. Off-flavour masking of secondary lipid oxidation products by pea dextrin. *Food Chem.* **2015**, *169*, 492–498. [CrossRef]
18. Campo, M.M.; Nute, G.R.; Hughes, S.I.; Enser, M.; Wood, J.D.; Richardson, R.I. Flavour perception of oxidation in beef. *Meat Sci.* **2006**, *72*, 303–311. [CrossRef] [PubMed]
19. Brühl, L. Fatty acid alterations in oils and fats during heating and frying. *Eur. J. Lipid Sci. Technol.* **2014**, *116*, 707–715. [CrossRef]
20. Kubow, S. Toxicity of dietary lipid peroxidation products. *Trends Food Sci. Technol.* **1990**, *1*, 67–71. [CrossRef]
21. Perluigi, M.; Coccia, R.; Butterfield, D.A. 4-Hydroxy-2-Nonenal, a Reactive Product of Lipid Peroxidation, and Neurodegenerative Diseases: A Toxic Combination Illuminated by Redox Proteomics Studies. *Antioxid. Redox Signal.* **2011**, *17*, 1590–1609. [CrossRef] [PubMed]
22. Guillen, M.D.; Goicoechea, E. Formation of oxygenated α,β-unsaturated aldehydes and other toxic compounds in sunflower oil oxidation at room temperature in closed receptacles. *Food Chem.* **2008**, *111*, 157–164. [CrossRef]
23. Shahidi, F.; Pegg, R.B. Hexanal as an Indicator of the Flavor Deterioration of Meat and Meat Products. *J Food Lipids* **1994**, *1*, 177–186. [CrossRef]
24. Kaykhaii, M.; Rahmani, M. Headspace liquid phase microextraction for quantitation of hexanal in potato crisps by gas chromatography. *J. Sep. Sci.* **2007**, *30*, 573–578. [CrossRef] [PubMed]
25. Aparicio, R.; Morales, M.T.; Alonso, M.V. Relationship between volatile compounds and sensory attributes of olive oils by the sensory wheel. *J. Am. Oil Chem. Soc.* **1996**, *73*, 1253–1264. [CrossRef]
26. Odeyemi, O.A.; Burke, C.M.; Bolch, C.C.J.; Stanley, R. Seafood spoilage microbiota and associated volatile organic compounds at different storage temperatures and packaging conditions. *Int. J. Food Microbiol.* **2018**, *280*, 87–99. [CrossRef]
27. Flower, D.R.; North, A.C.; Sansom, C.E. The lipocalin protein family: Structural and sequence overview. *Biochim. Biophys. Acta* **2000**, *1482*, 9–24. [CrossRef] [PubMed]
28. Bianchet, M.A.; Bains, G.; Pelosi, P.; Pevsner, J.; Snyder, S.H.; Monaco, H.L.; Amzel, L.M. The three-dimensional structure of bovine odorant binding protein and its mechanism of odor recognition. *Nat. Struct. Biol.* **1996**, *3*, 934–939. [CrossRef]
29. Tegoni, M.; Ramoni, R.; Bignetti, E.; Spinelli, S.; Cambillau, C. Domain swapping creates a third putative combining site in bovine odorant binding protein dimer. *Nat. Struct Biol.* **1996**, *3*, 863–867. [CrossRef]
30. Sandler, B.H.; Nikonova, L.; Leal, W.S.; Clardy, J. Sexual attraction in the silkworm moth: Structure of the pheromone-binding-protein-bombykol complex. *Chem. Biol.* **2000**, *7*, 143–151. [CrossRef] [PubMed]
31. Marabotti, A.; Scirè, A.; Staiano, M.; Crescenzo, R.; Aurilia, V.; Tanfani, F.; D'Auria, S. Wild-type and mutant bovine odorant binding proteins to probe the role of the quaternary structure organization in the protein thermal stability. *J. Proteome Res.* **2008**, *7*, 5221–5229. [CrossRef] [PubMed]
32. Paolini, S.; Tanfani, F.; Fini, C.; Bertoli, E.; Paolo, P. Porcine odorant-binding protein: Structural stability and ligand affinities measured by fourier-transform infrared spectroscopy and fluorescence spectroscopy. *Biochim. Biophys. Acta* **1999**, *1431*, 179–188. [CrossRef]
33. Morris, G.M.; Huey, R.; Lindstrom, W.; Sanner, M.F.; Belew, R.K.; Goodsell, D.S.; Olson, A.J. AutoDock4 and AutoDockTools4: Automated docking with selective receptor flexibility. *J. Comput. Chem.* **2009**, *30*, 2785–2791. [CrossRef]
34. Goodsell, D.S.; Morris, G.M.; Olson, A.J. Automated docking of flexible ligands: Applications of AutoDock. *J. Mol. Recognit. JMR* **1996**, *9*, 1–5. [CrossRef]
35. Kim, S.; Thiessen, P.A.; Bolton, E.E.; Chen, J.; Fu, G.; Gindulyte, A.; Han, L.; He, J.; He, S.; Shoemaker, B.A.; et al. PubChem Substance and Compound databases. *Nucleic Acids Res.* **2016**, *44*, D1202–D1213. [CrossRef] [PubMed]
36. O'Boyle, N.M.; Banck, M.; James, C.A.; Morley, C.; Vandermeersch, T.; Hutchison, G.R. Open Babel: An open chemical toolbox. *J. Cheminform.* **2011**, *3*, 33. [CrossRef] [PubMed]
37. Vincent, F.; Spinelli, S.; Ramoni, R.; Grolli, S.; Pelosi, P.; Cambillau, C.; Tegoni, M. Complexes of porcine odorant binding protein with odorant molecules belonging to different chemical classes. *J. Mol. Biol.* **2000**, *300*, 127–139. [CrossRef] [PubMed]
38. Berman, H.M.; Westbrook, J.; Feng, Z.; Gilliland, G.; Bhat, T.N.; Weissig, H.; Shindyalov, I.N.; Bourne, P.E. The Protein Data Bank. *Nucleic Acids Res.* **2000**, *28*, 235–242. [CrossRef]
39. Gasteiger, J.; Marsili, M. Iterative partial equalization of orbital electronegativity: A rapid access to atomic charges. *Tetrahedron* **1980**, *36*, 3219–3228. [CrossRef]
40. Hosainzadeh, A.; Gharanfoli, M.; Saberi, M.; Chamani, J. Probing the interaction of human serum albumin with bilirubin in the presence of aspirin by multi-spectroscopic, molecular modeling and zeta potential techniques: Insight on binary and ternary systems. *J. Biomol. Struct Dyn.* **2012**, *29*, 1013–1050. [CrossRef]
41. Capo, A.; Pennacchio, A.; Varriale, A.; D'Auria, S.; Staiano, M. The porcine odorant-binding protein as molecular probe for benzene detection. *PLoS ONE* **2018**, *13*, e0202630. [CrossRef]
42. D'Auria, S.; Scire, A.; Varriale, A.; Scognamiglio, V.; Staiano, M.; Ausili, A.; Marabotti, A.; Rossi, M.; Tanfani, F. Binding of glutamine to glutamine-binding protein from Escherichia coli induces changes in protein structure and increases protein stability. *Proteins* **2005**, *58*, 80–87. [CrossRef]
43. Albani, J.R. *Principles and Applications of Fluorescence Spectroscopy*; Blackwell Science: Oxford, UK, 2007. [CrossRef]
44. Lakowicz, J.R. *Principles of Fluorescence Spectroscopy*, 2nd ed.; Plenum Press: New York, NY, USA, 1999. [CrossRef]

45. Capo, A.; Cozzolino, S.; Cavallari, A.; Bruno, U.; Calabrese, A.; Pennacchio, A.; Camarca, A.; Staiano, M.; D'Auria, S.; Varriale, A. The Porcine Odorant-Binding Protein as a Probe for an Impedenziometric-Based Detection of Benzene in the Environment. *Int. J. Mol. Sci.* **2022**, *23*, 4039. [CrossRef]
46. Shrivastava, A.; Gupta, V.B. Methods for the determination of limit of detection and limit of quantitation of the analytical methods. *Chron. Young Sci.* **2011**, *2*, 21–25. [CrossRef]
47. Paolini, S.; Scaloni, A.; Amoresano, A.; Marchese, S.; Napolitano, E.; Pelosi, P. Amino acid sequence, post-translational modifications, binding and labelling of porcine odorant-binding protein. *Chem. Senses* **1998**, *23*, 689–698. [CrossRef] [PubMed]
48. Clegg, R.M.; Murchie, A.I.; Zechel, A.; Lilley, D.M. Observing the helical geometry of double-stranded DNA in solution by fluorescence resonance energy transfer. *Proc. Natl. Acad. Sci. USA* **1993**, *90*, 2994–2998. [CrossRef] [PubMed]
49. Bogomolova, A.; Komarova, E.; Reber, K.; Gerasimov, T.; Yavuz, O.; Bhatt, S.; Aldissi, M. Challenges of electrochemical impedance spectroscopy in protein biosensing. *Anal. Chem.* **2009**, *81*, 3944–3949. [CrossRef] [PubMed]
50. Daniels, J.S.; Pourmand, N. Label-Free Impedance Biosensors: Opportunities and Challenges. *Electroanalysis* **2007**, *19*, 1239–1257. [CrossRef]
51. Barsoukov, E.; Macdonald, J.R. *Impedance Spectroscopy: Theory, Experiment, and Applications*, 2nd ed.; John Wiley & Sons, Inc.: Hoboken, NJ, USA, 2005. [CrossRef]

biosensors

MDPI

Article

Development of Gold-Nanoparticle-Based Lateral Flow Immunoassays for Rapid Detection of TB ESAT-6 and CFP-10

Palesa Pamela Seele [1,*], Busiswa Dyan [1], Amanda Skepu [2], Charlotte Maserumule [1] and Nicole Remaliah Samantha Sibuyi [1,*]

1 Nanotechnology Innovation Centre, Health Platform, Advanced Materials Division, Mintek, Private Bag X3015, Randburg, Johannesburg 2125, South Africa
2 Advanced Chemistry and Life Sciences Division, Next Generation Health Cluster, Council for Scientific and Industrial Research (CSIR), Pretoria 0001, South Africa
* Correspondence: palesas@mintek.co.za (P.P.S.); nicoles@mintek.co.za (N.R.S.S.)

Abstract: The current study reports on the development of a rapid and cost-effective TB-antigen diagnostic test for the detection of *Mycobacterium* biomarkers from non-sputum-based samples. Two gold nanoparticle (AuNP)-based rapid diagnostic tests (RDTs) in the form of lateral flow immunoassays (LFIAs) were developed for detection of immunodominant TB antigens, the 6 kDa early secreted antigen target EsxA (ESAT-6) and the 10 kDa culture filtrate protein EsxB (CFP-10). AuNPs were synthesized using the Turkevich method and characterized by UV-vis spectrophotometer and transmission electron microscope (TEM). The AuNP–detection probe conjugation conditions were determined by comparing the stability of 14 nm AuNPs at different pH conditions, following salt challenge. Thereafter, ESAT-6 and CFP-10 antibodies were conjugated to the AuNPs and used for the colorimetric detection of TB antigens. Selection of the best detection and capture antibody pairs was determined by Dot spotting. The limits of detection (LODs) for the LFIAs were evaluated by dry testing. TEM results showed that the 14 nm AuNPs were mostly spherical and well dispersed. The ESAT-6 LFIA prototype had an LOD of 0.0625 ng/mL versus the CFP-10 with an LOD of 7.69 ng/mL. Compared to other studies in the literature, the LOD was either similar or lower, outperforming them. Moreover, in some of the previous studies, an enrichment/extraction step was required to improve on the LOD. In this study, the LFIAs produced results within 15 min and could be suitable for use at PoCs either in clinics, mobile clinics, hospitals or at home by the end user. However, further studies need to be conducted to validate their use in clinical samples.

Keywords: tuberculosis; rapid diagnostic test; point of care; immunoassay; ESAT-6; CFP-10

Citation: Seele, P.P.; Dyan, B.; Skepu, A.; Maserumule, C.; Sibuyi, N.R.S. Development of Gold-Nanoparticle-Based Lateral Flow Immunoassays for Rapid Detection of TB ESAT-6 and CFP-10. *Biosensors* **2023**, *13*, 354. https://doi.org/10.3390/bios13030354

Received: 2 February 2023
Revised: 24 February 2023
Accepted: 2 March 2023
Published: 6 March 2023

1. Introduction

Tuberculosis (TB) is an infectious disease caused by *Mycobacterium tuberculosis* (*M. tb*), which when detected timeously and with appropriate therapeutic intervention can be curable. The global representation of this crisis is massive, as 3.6 million of TB cases can go undiagnosed or unreported in low–middle income countries. In sub-Saharan Africa, 5 out of 10 active TB cases go undiagnosed, and in South Africa (SA) alone, an estimated 150,000 of cases are not diagnosed, which poses a risk for increasing the rates of transmission, morbidity and mortality [1]. The lack of rapid, accurate diagnostic tools at a point of care (PoC), including the inaccessibility of easily collectible samples, is central to the missed diagnosis and subsequent spread of the disease. In the current study, the immunodominant TB antigens ESAT-6 and CFP-10 were explored as targets for the development of TB-antigen rapid diagnostic test (RDT) prototypes which were based on the lateral flow immunoassays (LFIAs) technology.

To date, microbiological testing such as sputum smear microscopy and sputum culture test remain the most widely used methods of TB diagnosis [2]. The major limitation of the sputum smear microscopy test which exploits staining of acid-fast bacteria (AFB) is that it requires 10,000 bacilli/mL from a sputum sample for a positive TB detection [2], making it less sensitive for paucibacillary cases in HIV positive patients, children who often present with extra-pulmonary TB (EPTB) and very ill patients who cannot expectorate [2–4]. The sputum culture test has a long turnaround time of 2 to 4 weeks. These tests are expensive and need to be carried out by experienced personnel in specialized facilities and with costly instrumentation. Likewise, molecular tests such as the PCR-based GeneXpert MTB/RIF (Cepheid Inc., Sunnyvale, CA, USA) and the TB loop-mediated isothermal amplification (LAMP) (Eiken; Tokyo, Japan) are costly, and not readily accessible to low-income countries nor suitable for PoC testing [5–7]. The blood-based immune response tuberculin skin test (TST) and the IFN-γ release assays (IGRAs) (QuantiFERON-TB Gold, Cellestis Ltd., Australia) have uncircumventable limitations such as their inability to accurately distinguish between active and latent infections [4]. In addition, TST does not differentiate *M. tb* from nontuberculous *Mycobacterium* and the Bacille Calmette–Guérin (BCG) vaccine, making it unsuitable for TB testing in a South African cohort [2].

Lateral flow immunoassays are user friendly, produce results within 10 to 15 min and are suitable for PoC use either at medical facilities or at home by the end user [8,9]. The Alere Determine™ LFIA was introduced in South Africa (SA) in 2018, and used to detect the *Mycobacterial* lipoarabinomannan (LAM) antigen in urine samples within 25 min. This test, however, was found to be more suitable for TB testing in HIV-positive patients. The commercially available LFIA-based TB RDTs have limited endorsement from the World Health Organization (WHO) [10], and in some studies, the reported sensitivity and specificity of the diagnostic kits were inconsistent [11]. The poor specificity may be attributed to cross-reactivity with the BCG vaccine and other *Mycobacterium* species [12].

In contrast, the current TB-antigen RDT explores the immunodominant TB secreted antigens, ESAT-6 and CFP-10, as targets in detecting TB following the research strategy summarized in Scheme 1. In the globally used BCG vaccine, these antigens are not expressed subsequent to the deletion of the region of difference (RD)1 locus in which they are located [13,14]. As a result, cross-reactivity in patients previously vaccinated with BCG and/or those infected with other *Mycobacterium species* will be minimized [15–17]. Additionally, these biomarkers have been previously detected in the serum from HIV-positive and HIV-negative pulmonary TB (PTB) and extra-pulmonary TB (EPTB) patients with similar sensitivities. The antigens were detectable at low concentrations <10 nM in the serum, which was useful for diagnosis and evaluating anti-TB response [14]. To expedite the Stop TB Strategy, TB RDTs are now adopted to detect TB biomarkers in easily accessible samples such as urine [8] and blood samples, and sensing of volatile organic compounds from the breath [9,10]. The presence of ESAT-6 and CFP-10 in urine [15–18] and blood [14,15] samples makes them ideal TB biomarkers, especially for RDTs. However, the methods used to detect antigens such as mass spectrometry (MS) and ELISA are not suitable for POC testing, as the very long sample processing and analysis require highly trained personnel. Therefore, the proposed AuNP-based LFIAs will be optimized for rapid detection of ESAT-6 and CFP-10 from non-sputum samples which will be beneficial for patients who cannot expectorate.

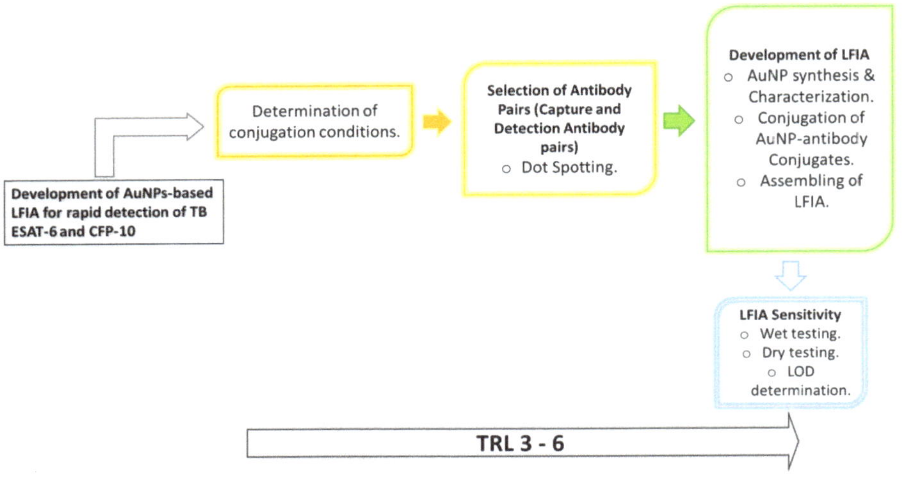

Scheme 1. Simplified representation of the research strategy. The diagram depicts the objectives of the research work.

2. Materials and Methods

2.1. Materials

The ESAT-6 and CFP-10 antibodies were obtained from manufacturers outlined in Table 1. The antibodies were given specific codes in the current study, as indicated in the table. The gold (III) chloride ($HAuCl_4.3H_2O$), trisodium citrate ($Na_3C_6H_5O_7$), and anti-Mouse IgM (μ-chain specific) antibody produced in goat were bought from Sigma-Aldrich (St Louis, Missouri, USA). The anti-Rabbit Affinity purified Goat IgG was acquired from R&D systems (Minneapolis, Minnesota, USA). The nitrocellulose membrane CN95 was purchased from Sartorius (PTY) LTD (Midrand, South Africa), whilst the Millipore conjugate pad G041 and Millipore C083 absorbent pad were bought from Sigma-Aldrich and the backing cards from Diagnostic Consulting Network (Carlsbad, California, USA).

Table 1. The primary antibodies used for CFP-10 and ESAT-6.

Antigen	Antibody Code	Antibody Details (Cat No.)	Supplier Details
CFP-10	Ab1	Mouse Monoclonal *Mycobacterium Tuberculosis* CFP-10 Antibody (KFB16)	Novus Biologicals, LLC (Briarwood Avenue, CO 80112, USA)
	Ab2	Mouse Monoclonal *Mycobacterium Tuberculosis* CFP-10 Antibody (KFB42)	Novus Biologicals, LLC
	Ab3	Mouse Anti-*Mycobacterium tuberculosis* CFP-10 Monoclonal Antibody (DMAB3941)	Creative diagnostics PTY (Ltd) (Shirley, NY, USA)
	Ab4	Mouse Anti-*Mycobacterium tuberculosis* CFP-10 Monoclonal Antibody (DMAB3943)	Creative diagnostics PTY (Ltd)
	Ab5	Rabbit Polyclonal Anti-CFP-10 antibody (AB45073)	Abcam PTY (LTD) (Discovery Drive, Cambridge, UK)
ESAT-6	Ab6	Mouse Anti-*Mycobacterium tuberculosis* ESAT-6 Monoclonal Antibody (DMAB3944)	Creative diagnostics PTY (Ltd)
	Ab7	Mouse Anti-*Mycobacterium tuberculosis* ESAT-6 Monoclonal Antibody (DMAB3945)	Creative diagnostics PTY (Ltd)
	Ab8	Rabbit Polyclonal Anti-ESAT-6 antibody (AB45074)	Abcam PTY (LTD)

2.2. Synthesis of 14 nm AuNPs

The 14 nm AuNPs were synthesized using the Turkevich method (Turkevich et al., 1951), which entails the reduction of $HAuCl_4$ with trisodium citrate as described previously. Trisodium citrate and $HAuCl_4.3H_2O$ solutions were prepared at a concentration of 1% and 0.0289 M, respectively. The solutions were each filtered using 0.2 μm filters and thereafter used in the synthesis of the 14 nm AuNPs [15,16].

2.3. Characterization of the AuNPs Using UV-vis Spectrophotometer and TEM

The optical properties of the 14 nm AuNPs were measured using the UV-vis spectrophotometer (Multiskan^{TM} GO plate reader (Thermo Fischer Scientific, Waltham, Massachusetts, USA)) within a wavelength range of 200 to 800 nm. Transmission electron microscopy (TEM) analysis was carried out on the JEOL JEM-2100F following previous procedures [19,20].

2.4. Conjugation of Antibodies to AuNPs

2.4.1. Stability of AuNP–Antibody Conjugates

Stability of 14 nm AuNP–antibody conjugates was assessed at different pH levels. The pH levels of AuNPs were adjusted to pH 7, 8, 9 and 10 using 0.1 M K_2CO_3. Thereafter, 100 μL of each AuNP solutions was dispensed into a 96-well plate. The antibodies were added to the AuNPs to a final concentration of 10 μg/mL. In negative controls, phosphate-buffered saline (PBS) was added in place of the antibodies. The reactions were incubated for 5 min at room temperature, then 50 μL of 0.1 M NaCl was added and color changes, if any, were recorded.

2.4.2. Conjugation of Antibodies to the AuNPs

To conjugate the antibodies to the AuNPs, 1 mL of AuNP solutions (OD 1 for the 14 nm AuNPs) at the selected optimal pH and 1 μg/mL antibody were added to an Eppendorf tube and incubated for 20 min with agitation. Then, 100 μL of conjugate blocking buffer (10% bovine serum albumin (BSA) prepared in borate buffer, pH 9) was added to the conjugate mixture and further incubated for 30 min. The conjugate mixture was then centrifuged at 15 000 rpm for 20 min at 4 °C. The supernatant was discarded and the AuNPs were re-suspended in diluent buffer and stored at 4 °C until further analysis.

2.5. Selection of Antibody Pairs by Dot Spotting

The detection and capture antibody pairs for ESAT-6 and CFP-10 were tested for compatibility using dot spotting following a sandwich assay. Immobilization of antibodies was carried out by dot spotting 1 μL of 1 mg/mL of antibodies at about 6 mm of spacing between the control and the test line antibody. Thereafter, the membranes were dried for 15 min in an oven at 37 °C. Blocking of the membranes was carried out for 5 min in 0.1% BSA blocking buffer and dried for 45 min at 37 °C in an oven. Strips were cut into 3 mm and analyzed accordingly. Binding of the detection antibodies to the antigen was analyzed by immersing the strips in an Eppendorf tube containing the AuNP–antibody conjugate with buffer (negative test) or AuNP–antibody conjugate with buffer containing antigen (test sample) (refer to Scheme 1). The experiment was allowed to run for 15 min, and signals were then visualized by a red dot.

2.6. Development of the AuNP-Based LFIA and Determination of the Limit of Detection (LOD)

The selected control and test line antibodies were dispensed using XYZ (X = 25 mm; Y = 30 mm and Z = 8 mm) Platform Dispenser HM3030 (Shanghai Kinbio Tech.Co, LTD, Pudong New Strict, Shanghai, China) at a rate of 1 μL/cm on to the nitrocellulose membrane which was laminated onto a backing card. The card was then dried for 30 min at 37 °C in an oven, and the membrane was blocked with 0.1% BSA in phosphate buffer for 5 min with agitation. Drying was carried out for 45 min at 37 °C.

Meanwhile, the conjugate pad was prepared by immersing it in conjugate pad blocking buffer (3% BSA in borate buffer) for 15 min with gentle agitation. Once the conjugate pad had been dried, the AuNP conjugate of choice was sprayed using the KinBio Dispenser instrument (Kinbio Tech.Co, LTD, Pudong New Strict, Shanghai, China) and again dried for 45 min at 37 °C. Finally, the conjugate pad and absorbent pad were laminated on the backing card with the membrane attached and cut into 3 mm strips using the High Speed Cutter ZQ4500 (Kinbio Tech.Co, LTD, Pudong New Strict, Shanghai, China) prior to assembling into cassettes (schematic representation of the assembling process in Scheme 2).

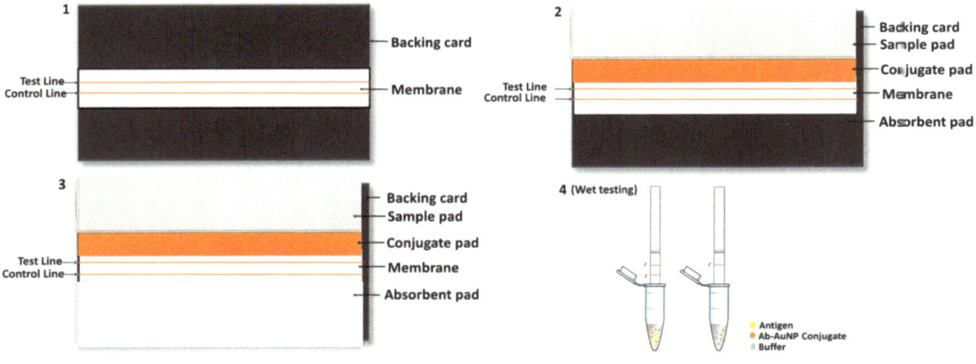

Scheme 2. Schematic diagram showing steps of the assembling and wet testing of LFIA strips. The diagram illustrates the lamination of components of a strip on the backing card (steps 1 to 4). (**1**) The nitrocellulose membrane is the first to be laminated, followed by (**2**) the conjugate and sample pads which have a 2 mm overlap, and finally (**3**) the absorbent pad is embedded on the card. Following the cutting of the strips, wet testing (step 4) was carried out as shown on the diagram. Lines on both control and test lines represent a positive test, while a single line in the control represent a negative test.

To determine the LOD of the prototypes, the recombinant antigens were serially diluted with 1 x PBS at concentrations ranging between 0.0769 to 0.0769 $\times$ 10^6 ng/mL for CFP-10, and 0.0000625 to 62.5 $\times$ 10^2 ng/mL for ESAT-6. The recombinant antigens were provided at different concentrations by suppliers; hence, the starting concentrations dictated the working solution's dilution series. A volume (65 µL) of the prepared samples was loaded onto the assembled test strips and allowed to run for 15 min prior to the analysis of the results.

2.7. Wet and Dry Conjugate Testing of the LFIA

Wet testing: To perform wet conjugate testing, 30 µL of running buffer and 10 µL of AuNP–antibody conjugate were added to an Eppendorf tube. For a test sample, commercial recombinant *M. tb* CFP-10 (227-20144, Biodisc Raybiotech Life, Inc., Peachtree Corners, Georgia, USA) or *M. tb* ESAT-6 (Creative diagnostics PTY (Ltd), Shirley, NY, USA) was also added to separate tubes, after which the sample pad part of the strip was cut and immersed into those tubes. The test was allowed to run for 15 min prior to analysis.

Dry testing: The backing card was assembled with the conjugate pad which contained the AuNP–antibody conjugate. To conduct a negative test, 40 µL of running buffer (without any antigens/proteins) was added to the sample pad on the test strip and allowed to flow for 15 min. The same procedure was followed for a test sample by using a buffer containing the commercial recombinant proteins/antigens. Various concentrations of antigen were tested until a red line/dot was visible. Concentrations of 4 and 2.7 μg/mL for CFP-10 and ESAT-6, respectively, were used as a reference point, as determined in the Dot spotting results. Where necessary and where false negatives were observed higher concentrations were also tested.

3. Results

3.1. Synthesis and Characterization of AuNPs

UV-vis spectroscopy and TEM were used to evaluate the quality of the synthesized AuNPs. The absorption maxima of the 14 nm AuNPs was at 519 nm (Figure 1A), which is a typical wavelength banding pattern for AuNPs owing to the surface plasmon resonance (SPR) [17–20]. The morphology of the AuNPs was mostly spherical, as shown by the TEM micrographs (Figure 1B), with an average core size of 14 nm. Using this method, different sized diameters of the AuNPs, in the range of 14 to 150 nm, can be produced by varying the ratio of the reducing agent to the gold precursor (trisodium citrate: gold chloride) [16,20,21]. AuNPs are able to absorb and scatter light, which enables them to emit different colors depending on their size, shape and degree of aggregation. This SPR phenomenon is known to occur due to the ability of the metal to conduct electrons simultaneously excited by electromagnetic waves striking the metal surface [19].

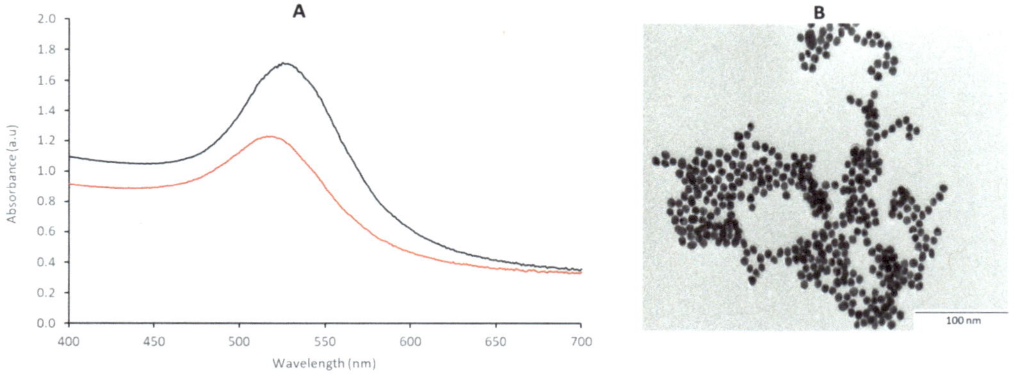

Figure 1. Characterization of 14 nm AuNPs. The nanoparticles were analyzed using (**A**) absorbance spectroscopy and (**B**) TEM. The absorption maxima were at 512 nm.

3.2. Assessment of Conjugation Conditions and Conjugate Stability

The stability of the AuNP–antibody conjugates was visually assessed by color change following the addition of 0.1 M NaCl to the conjugate samples. AuNPs that had no salt nor antibody were used as a reference for unstable conjugates. In cases where color changes were not obvious by visual inspection, a pH that was closer to the protein's storage buffer (prepared by the suppliers) was selected. The 14 nm AuNP–antibody conjugates that were indicative of instability changed color from red to purple or colorless, while the stable conjugates remained red after addition of NaCl. The color changes were as a result of flocculation after adding NaCl to AuNPs; under the tested conditions, this was more evident in the AuNP samples without antibodies (Figure 2). The selected pH for conjugation of Ab1 and Ab2 was pH 8, while Ab3, Ab4 and Ab5 were more stable at pH 9. Similarly, Ab6, Ab7 and Ab8 were stable at pH 9. In the current study, the conjugates were mostly stable at pH 8 and 9 with 14 nm AuNPs, and these were selected as the conjugation conditions for producing the detection probes (antibody–AuNP conjugates).

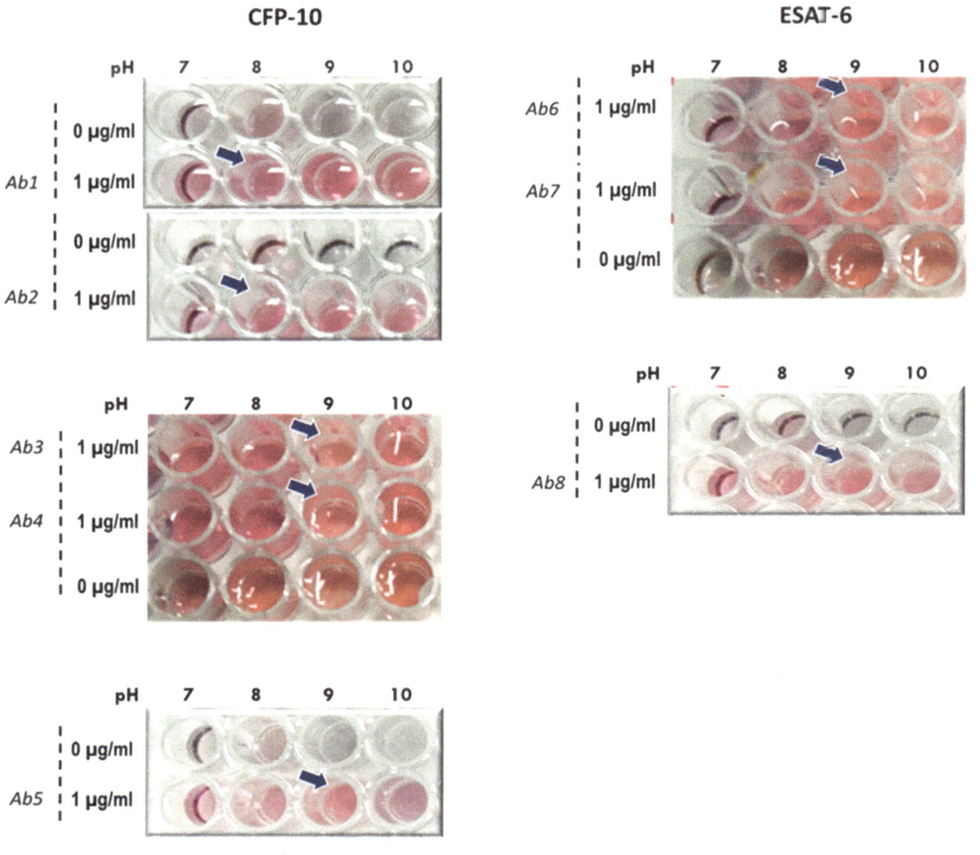

Figure 2. Stability of AuNPs with ESAT-6 and CFP-10 antibodies at different pH values. AuNPs of 14 nm at pH 7, 8, 9 and 10 were dispensed into wells of a 96-well plate. Anti-ESAT-6 and anti-CFP-10 antibodies were added into wells and incubated at room temperature for 5 min prior to addition of NaCl. Color changes, if any, were observed and with reference from AuNPs that had no antibody and NaCl, and optimum conditions for conjugation were chosen. The arrows indicate the selected conditions for conjugation.

3.3. Selection of Antibodies

Pairing and compatibility of the antibodies was evaluated by dot spotting to determine the capture and detection antibodies specific to CFP-10 and ESAT-6. Overall, CFP-10 had eight pairs of compatible capture and detection antibodies, whilst ESAT-6 had only two pairs to choose from (results are summarized in Table 2 and representative strips shown in Figure 3). With respect to CFP-10, the detection AuNP–antibody conjugates that were compatible with the Ab1 capture antibody were Ab1, Ab2 and Ab4 whilst the Ab2 capture antibody was compatible with the Ab1 and to a lesser extent with Ab3 and Ab4 AuNP–antibody conjugates as detection antibodies. Testing the pairs Ab2/Ab3 (capture antibody/detection antibody) and Ab2/Ab4 with double the concentration (from 4 to 8 µg/mL) of the recombinant CFP-10 antigen resulted in very faint dots at the control test and a concentrated dots at the test line.

Table 2. Summary of dot spotting analysis.

Capture Antibody	Control/Test	CFP-10					ESAT-6			Detector Antibody
		Ab1	Ab2	Ab3	Ab4	Ab5	Ab6	Ab7	Ab8	
Ab1	Control	+++	++	++	+++					
	Test	+++	+++	-	+++					
Ab2	Control	++	-	+++	+++					
	Test	+++	-	+++	++					
Ab3	Control	+	+	++	++					
	Test	n/a	n/a	-	+++					
Ab4	Control	+	+	++	+					
	Test	-	-	++	-					
Ab5	Control					+++				
	Test					-				
Ab6	Control						++	+++		
	Test						-	-		
Ab7	Control						+++	+++		
	Test						-	++		
Ab8	Control								+++	
	Test								-	

Key: +++ high intensity; ++ moderate intensity; + low intensity/faint; - no dot visualized.

Consistent with other studies, pairing the same antibody yielded unfavorable results, mostly false negatives, and also a 'ghost line' was depicted by the Ab3/Ab3 pair. 'Ghost lines' are an artefact that is characterized by a red background encircling a white dot where an antibody was immobilized. In contrast, the Ab1/Ab1 and Ab7/Ab7 pairs gave distinct red-colored dots with respect to the negative and test samples, and the Ab4/Ab4 showed a visible but faint dot upon testing with recombinant protein.

The Ab3 and Ab4 capture antibodies paired favorably with the Ab4 and Ab3 as detection antibodies, respectively. Ab5 (anti-CFP-10 antibody) and Ab8 (anti-ESAT-6 antibody) antibodies are rabbit polyclonal antibodies (pAb) that could not be paired with the mouse antibodies due to incompatibility with the secondary antibody at the control line. The Ab5/Ab5 pair gave false negatives whilst the Ab8/Ab8 gave false positives when testing BSA-based buffers. pAb targets multiple epitopes on an antigen, and this wide range makes them less specific when compared to monoclonal antibodies (mAb). Consequently, pAbs are less preferred as capture antibodies, and are rather used as detection antibodies.

The ESAT-6s Ab6 and Ab7 showed compatibility when paired in both orientations. Buffer components also influenced the antibody pairing outcomes. In most cases, BSA-based buffers gave false positives, such as when testing the Ab8/Ab8 pair between BSA concentrations of 0.1 to 1%. Although casein-based buffers were able to eliminate the false positivity, the Ab8 test line could not detect the recombinant protein. 'Ghost lines' were observed on the strips with the Ab3/Ab3 and with Ab8/Ab8 pairs where Tris-based buffers were used for optimization.

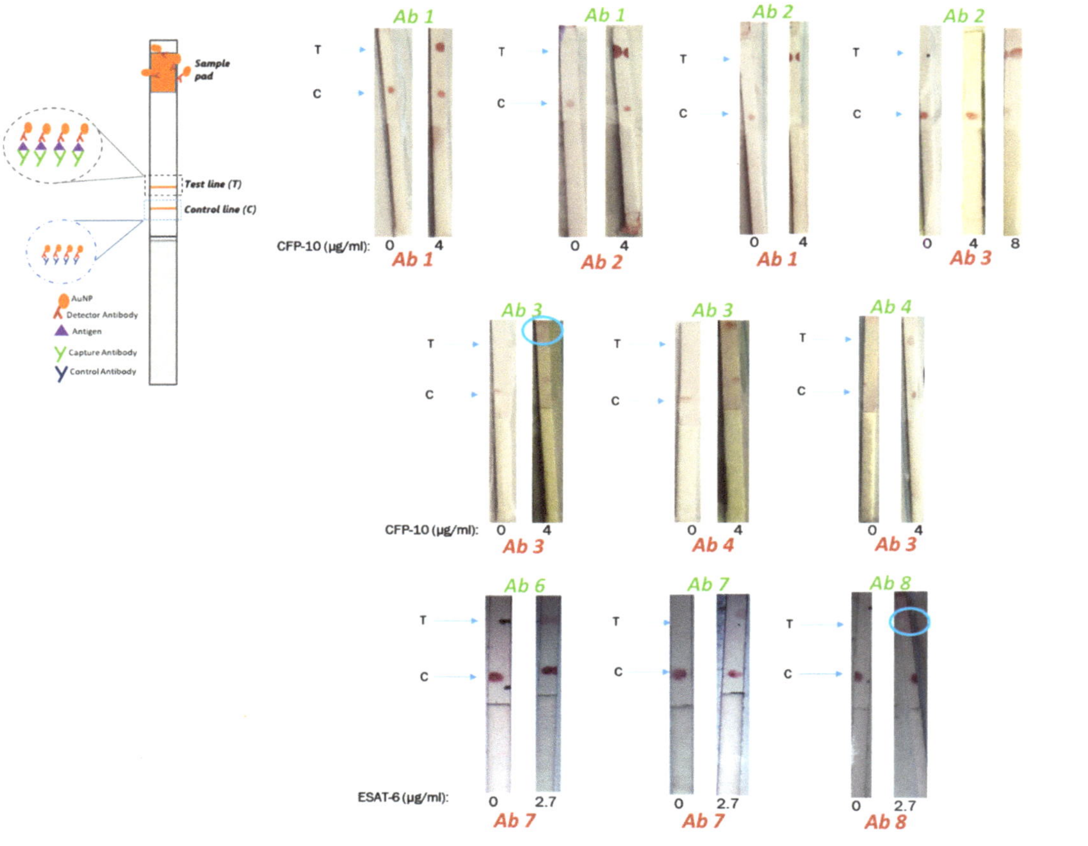

Figure 3. Compatibility of CFP-10 capture and detection antibody pairs. The capture and control antibodies were dotted on the nitrocellulose membrane using 1 μL of 1 mg/mL antibody. The membranes were left to dry for 15 min prior to testing. To test the samples, the negative control contained 30 μL of running buffer and 10 μL of AuNP–antibody conjugate, whilst the test sample also contained recombinant CFP-10 protein (4 μg/mL) or ESAT-6 protein (2.7 μg/mL). Tests were allowed to run for 15 min before analysis. 'Ghost lines' are indicated by blue circles.

3.4. LOD for CFP-10 and ESAT-6

The visual LOD of the LFIA was determined on selected pairs, which showed the most consistent performance in the compatibility pairing studies: the Ab1/Ab1 and Ab1/Ab2 pairs for CFP-10 and Ab6/Ab7 and Ab7/Ab7 pairs for ESAT-6. The LOD for recombinant CFP-10 for the Ab1/Ab1 pair was 0.0769×10^4 ng/mL (Figure 4A), and 0.0769×10^2 ng/mL for the Ab1/Ab2 pair (Figure 4B). The Ab1/Ab2 pair was 100 times more sensitive than the Ab1/Ab1 pair, hence, it was chosen for future validations. ESAT-6 depicted an LOD of 62.5 ng/mL for the Ab6/Ab7 pair and 0.000625×10^2 ng/mL for the Ab7/Ab7 pair (Figure 5A,B, respectively). The latter pair was 1000 times more sensitive to the ESAT-6 antigen and was thus selected for further studies. The CFP-10 and ESAT-6 prototypes showed potential for future clinical evaluations or studies, with LOD values of 7.69 and 0.0625 ng/mL, respectively. The ESAT-6 LFIA LOD was more than 120-fold lower compared to the CFP-10 LFIA.

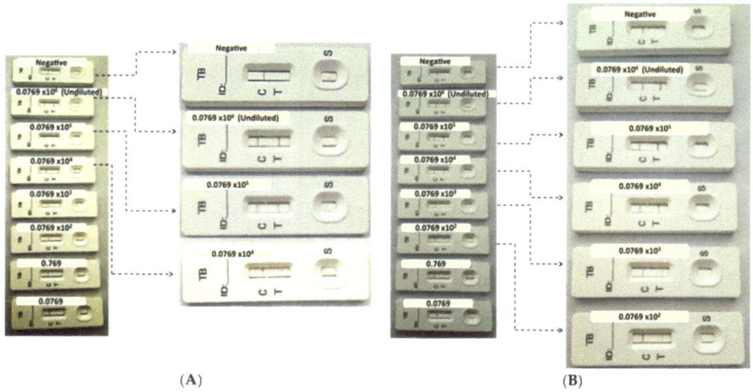

Figure 4. Determination of the LOD of CFP-10 (**A**) Ab1/Ab1 and (**B**) Ab1/Ab2 pairs. Ab1 was dispensed on the test line at a concentration of 0.5 mg/mL, and the anti-Mouse IgM antibody was dispensed at 0.3 mg/mL on the control line. The (**A**) Ab1 and (**B**) Ab2 were conjugated to AuNPs, then sprayed on the conjugate pad which was blocked using 3% BSA in sodium borate buffer. Following assembly of the nitrocellulose membrane, conjugate pad and absorbent pad on the backing cards, 3 mm strips were cut using the Kinbio cutter. Test samples were prepared at a range of 0.0769×10^6 and 0.0769 ng/mL recombinant CFP-10 in running buffer. Samples were then applied at a volume of 65 µL and analyzed after 15 min of incubation. These results are representative of the same test performed in triplicate.

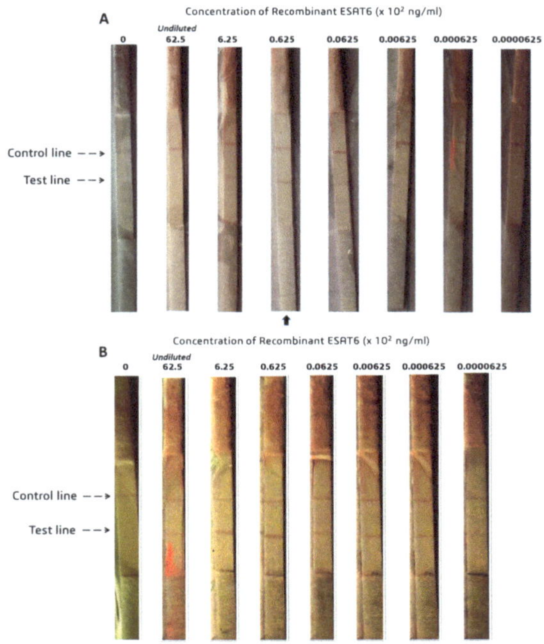

Figure 5. LOD of ESAT-6 (**A**) Ab6/Ab7 and (**B**) Ab7/Ab7 pair. (**A**) The Ab6 and (**B**) Ab7 capture antibodies were dispensed at 1 mg/mL on the nitrocellulose membrane, whilst the Ab7 conjugate was sprayed on the conjugate pad thereafter assembled on the backing cards. Recombinant ESAT-6 was serially diluted in the range indicated in order to compute the LOD. The arrows indicate the strip selected as the LOD. The results are representative of the same test performed in triplicate.

4. Discussion

The attenuation of Mycobacterium bovis to the widely used BCG vaccine resulted in the deletion of the RD1 loci. We postulated that exploiting biomarkers within this region as targets for the TB-antigen RDT may minimize and/or eliminate false positives from previously BCG vaccinated individuals and/or those infected with other Mycobacteria species. The CFP-10 and ESAT-6 are two of the immunodominant antigens that are expressed by the RD1 loci and were explored as targets for the development of AuNP-based LFIA. The stability of antibodies with the 14 nm AuNPs at various pHs was tested, and overall the antibodies were stable at most pHs. Interestingly, Wu and colleagues found their CFP-10 and ESAT-6 antibodies to be suitable for conjugation with AuNPs sized between 18 and 20 nm [11], suggesting that the smaller-sized AuNPs are more stable with these biomarkers.

Initially, Dot spotting was used to determine compatible pairs of detection and capture antibodies for CFP-10 and ESAT-6 from different suppliers. Suitable antibody pairs will then be sourced from the same supplier to maintain consistency. In cases where the test line was visualized but with the faint control line, the antigen could be causing steric hindrance, blocking the binding site for the immobilized secondary antibody of the control line. The kinetics of solid-phase immobilized antibodies have demonstrated that equilibrium between antigen, capture and detection antibodies do not occur simultaneously, suggesting that steric hindrance plays a role in the detection-antigen-capture sandwich formation [22]. However in some instances, antibodies are able to re-organize to allow for further binding, and a high binding affinity may prevent this re-organization, causing saturation of binding sites at a faster rate [22–24]. Thus, in addition to target epitope compatibility, the association (k_{ass}) and dissociation (k_{off}) rate constants of the detection and capture antibodies may have a crucial role in defining suitable binding pairs.

The false negatives for the other pairs suggest that the capture and detection antibodies are competing for the same epitope on the antigen. It is possible that the detection antibody, which is the first to be encountered by the antigen in solution during testing, saturates the same epitope that is recognized and required for binding by the capture antibody. This effect was observed by Cavalera and colleagues, which they referred to as the 'antigen hook effect', and they showed that lowering the concentration of the detection antibody can enhance the sensitivity of LFIA. The 'hook effect' has been previously described and was initially observed for the LFIA-based pregnancy test, which measures the concentration of human chorionic gonadotropin (hCG). High concentrations of hCG led to disappearance of the test line, which yielded false negatives [25,26]. Cavalera et al. also reported that the distance of the test line from the sample pad affects the sensitivity of the LFIA. The further the test line was from the sample pad meant that there was an increase in contact time between the detection of the AuNP conjugate and the antigen which increased the saturation effect, thus augmenting the 'antigen hook effect' [27,28].

False positives were also observed when BSA-based buffers were used, such as when testing the Ab8/Ab8 pair in the 0.1 to 1% BSA concentration. Since BSA is a protein like the antigens, it may present with peptides that act as epitopes which may be causing non-specific binding. This effect became less pronounced at a BSA concentration of 0.1% versus at 1%. Using casein as an alternate buffer eliminated the false positives (results not shown), and this may be because casein is a less structured protein (mostly random coiled or disordered) than BSA (67% helix, 10% turns and 23% extended) [29,30], making it less stable and more prone to denaturation effects by the surfactant than BSA. It is possible that if a higher concentration of surfactant was used with the BSA buffers no false positives would appear. 'Ghost lines' are an artefact characterized by a red background encircling a white dot where an antibody was immobilized. The appearance of 'ghost lines' has been proposed to be as a result of high concentrations of capture antibody on the test line repelling the conjugate. Since only the Tris-based buffers caused the 'ghost line' effect with the Ab8/Ab8 pair, a plausible reason may be that the buffer confers an overall charge to the Ab8–antigen complex that caused it to be repelled by the immobilized capture antibody.

Although CFP-10's Ab1/Ab1 pair showed potential, the Ab1/Ab2 pairing had an LOD that was 100 times more sensitive to the same antigen. Generally, for sandwich-type LFIA, a pair of mAbs targeting different epitopes of the antigen are often used over a single epitope-targeting mAb as both the capture and detection antibody [23,27,28]. The sensitivity of the LFIA is often lower with single epitope-targeting pairs. This does not mean that successful LFIAs have not been developed using the same mAb as both the capture and detection antibody, but some considerations and optimizations have to be performed [27,28]. Wu and colleagues determined the LOD of their CFP-10 test strip to be 2.4 ng/mL and 6.0 ng/mL for ESAT-6 [11], suggesting that the current study's CFP-10 LFIA prototype is only slightly weaker in detecting the antigen. In contrast, the ESAT-6 prototype was 96-fold more sensitive than the one produced by Wu and colleagues, suggesting the prospect of a potentially better performance in clinical sample evaluation.

Other technologies have been developed to improve the sensitivity for detection of ESAT-6, such as the magnetic-bead-coupled AuNP-based immuno-PCR assay (MB GNP-I-PCR assay). Comparing the MB GNP-I-PCR assay to the ELISA, the authors reported an LOD of 10 fg/mL, which was 105-fold higher in sensitivity compared to the 1 ng/mL displayed by ELISA [31]. Simultaneously exploiting different sized AuNPs, 20 and 60 nm, to capture the ESAT-6 antigen may improve the LOD. However, this technique remains unsuitable for PoCs, as it requires additional steps that need skilled personnel and laboratory equipment and is more costly than the lateral flow assays. Previous studies have indicated that LOD values are not the ultimate predictor and/or determinant of sensitivity in clinical sample testing. Despite the reported LODs of 2.4 ng/mL and 6.0 ng/mL for CFP-10 and ESAT-6, respectively, Wu and colleagues found that the positive detection rate was 29.4% for CFP-10 and 41.2% for ESAT-6 in TB-positive plasma samples [11]. Accordingly, test conditions and LOD studies need to be carried out and optimized in the intended clinical sample of choice. Another type of immunoassay previously used for ESAT-6 detection in plasma/serum was electrochemiluminescence (ECL)-based, with a reported LOD of 6 pg/mL and 56% sensitivity [3]. There can be various contributing factors to this phenomenon, including the quantity of the biomarkers or analytes in the samples. For example, the status of expression and/or dysregulation of a potential biomarker is influenced by different factors, including the geography of the sampled population. Genetic variations in the host and of the Mycobacterium can contribute to the differential regulation and expression repertoire of biomarkers in the host [25,28]. Such factors need to be considered in biomarker discovery for TB.

Mehaffy and colleagues used mass spectroscopy (MS) to identify *M. tb* peptides in human serum. Their experimental approach included an exosome enrichment step from the plasma of TB-positive patients originating from four different countries: South Africa, Bangladesh, Peru and Vietnam. By using Multiple Reaction Monitoring MS, the group discovered that their representative peptides for CFP-10 and ESAT-6 were detected in 3% and 15% of the total sample population, respectively [32]. Out of the 10 South Africans, only one was smear positive who was also co-infected with HIV; none of these patients had the CFP-10 peptide in their serum, and the 3 who had the ESAT-6 peptide were also HIV positive. With respect to the South African cohort, these results were inconclusive because of the small sample size tested and due to that 90% of the samples that were collected were smear-negative samples [32].

In an independent study, Nanodisk-MS was used to detect CFP-10 and ESAT-6 in serum. The method involved the enrichment of selected CFP-10 and ESAT-6 peptides using antibody-conjugated Nanodisks prior to MS analysis. The LODs of the recombinant CFP-10 and ESAT-6 in the TB-free serum were 50 pM and 200 pM, respectively [14]. In HIV-negative TB patients' serum, the sensitivity was 100% and 91% in smear-positive and smear-negative patients, respectively. The specificity was 87.1% in healthy controls and 90.6% in nontuberculous Mycobacteria patients. HIV co-infected TB patients exhibit EPTB, therefore, smear-negative and -positive EPTB and PTB patients were tested. In PTB cases the sensitivities were 91.3% and 82.4% in smear-positive and smear-negative patients, respectively, and 92.3% and 75.0% in EPTB patients [14]. The similarity in sensitivities between PTB and EPTB patients was an important outcome of the study, as it illustrated that serum has great potential to be used as an alternative sample for testing TB including EPTB. The key lesson learned from these studies is that a trend exists, wherein ESAT-6 shows a higher LOD compared to CFP-10 but with apparently superior sensitivity in clinical samples. Additionally, the specificity of the immunodominant antigens in healthy controls and more importantly in nontuberculous Mycobacteria patients is essential for developing a device that reduces cross-reactivity with nontuberculous Mycobacteria patients. These studies, including ours, demonstrated that nanotechnology-based immunoassays can greatly improve the simplicity, sensitivity and robustness of diagnostic methods.

5. Conclusions

The immunodominant antigens CFP-10 and ESAT-6 that are expressed by the RD1 loci of *M. tb* are potential biomarkers for developing a TB-antigen RDT that can discriminate previously active TB cases from previously BCG vaccinated and nontuberculous *Mycobacteria* cases. The pairs that were selected for preliminary testing with commercial recombinant protein in the current study showed great promise, with LODs that were either comparable or outperforming ones from those previously reported the in literature. An exhaustive study into biomarker discovery and research is required prior to choosing an appropriate target analyte in developing a diagnostic device for TB. The process is complicated and delayed by a spectrum of factors, such as the disease state, that is, active or latent disease, which can determine the level of expression and/or differential regulation of the targeted biomarker. Other factors such as the TB vaccination status, HIV status, age, and geography of the population tested need to be considered as well. Thus, future studies will explore other targets from the RD1 loci. Additionally, previous studies indicate that the methods used to improve sensitivity required additional steps, such as an enrichment step alongside exploiting immunoassays in combination with nanotechnology. These approaches need to be greatly refined in order to provide rapid results at the point of need/care, hence, our approach still has superiority. Proof-of-concept investigations for the rapid detection of TB in easily accessible clinical samples such as blood and saliva will be carried out both internally and subsequently validated externally to confirm clinical utility. This will greatly improve the diagnostic and therapeutic outcomes in low-income countries.

Author Contributions: Writing—draft preparation, review and editing, P.P.S., C.M., B.D., N.R.S.S. and A.S. Project conceptualization, resources and administration, P.P.S. and A.S. Supervision, N.R.S.S. and A.S. All authors have read and agreed to the published version of the manuscript.

Funding: This project was funded by Mintek (science vote: ADR-42305) and the DSI/National Research Foundation (NRF) Professional Development Program (PDP) (grant 128105).

Institutional Review Board Statement: Not applicable.

Informed Consent Statement: Not applicable.

Data Availability Statement: Not applicable.

Conflicts of Interest: The authors declare no conflict of interest.

References

1. Ajudua, F.I.; Mash, R.J. Implementing Active Surveillance for TB—The Views of Managers in a Resource Limited Setting, South Africa. *PLoS ONE* **2020**, e0239430. [CrossRef]
2. Yong, Y.K.; Tan, H.Y.; Saeidi, A.; Wong, W.F.; Vignesh, R.; Velu, V.; Eri, R.; Larsson, M.; Shankar, E.M. Immune Biomarkers for Diagnosis and Treatment Monitoring of Tuberculosis: Current Developments and Future Prospects. *Front. Microbiol.* **2019**, *10*, 2789. [CrossRef] [PubMed]
3. Broger, T.; Tsionksy, M.; Mathew, A.; Lowary, T.L.; Pinter, A.; Plisova, T.; Bartlett, D.; Barbero, S.; Denkinger, C.M.; Moreau, E.; et al. Sensitive Electrochemiluminescence (ECL) Immunoassays for Detecting Lipoarabinomannan (LAM) and ESAT-6 in Urine and Serum from Tuberculosis Patients. *PLoS ONE* **2019**, *14*, e0215443. [CrossRef]

4. Pai, M.; Denkinger, C.M.; Kik, S.V.; Rangaka, M.X.; Zwerling, A.; Oxlade, O.; Metcalfe, J.Z.; Cattamanchi, A.; Dowdy, D.W.; Dheda, K.; et al. Gamma Interferon Release Assays for Detection of Mycobacterium Tuberculosis Infection. *Clin. Microbiol. Rev.* **2014**, *27*, 3–20. [CrossRef]
5. Martin, D.R.; Sibuyi, N.R.; Dube, P.; Fadaka, A.O.; Cloete, R.; Onani, M.; Madiehe, A.M.; Meyer, M. Aptamer-Based Diagnostic Systems for the Rapid Screening of Tb at the Point-of-Care. *Diagnostics* **2021**, *11*, 1352. [CrossRef]
6. Yadav, R.; Daroch, P.; Gupta, P.; Agarwal, P.; Aggarwal, A.N.; Sethi, S. Diagnostic Accuracy of TB-LAMP Assay in Patients with Pulmonary Tuberculosis–a Case-Control Study in Northern India. *Pulmonology* **2022**, *28*, 449–453. [CrossRef]
7. Ni, J.R.; Yan, P.J.; Liu, S.D.; Hu, Y.; Yang, K.H.; Song, B.; Lei, J.Q. Diagnostic Accuracy of Transthoracic Echocardiography for Pulmonary Hypertension: A Systematic Review and Meta-Analysis. *BMJ Open* **2019**, *9*, e033084. [CrossRef] [PubMed]
8. O'Farrell, B. *Evolution in Lateral Flow–Based Immunoassay Systems*; Wong, R.C., Tse, H.Y., Eds.; Humana Press: New York, NY, USA, 2009; ISBN 9781597452403.
9. Wang, C.; Liu, M.; Song, L.; Yan, D.; Nongyue, H. Point-of-Care Diagnostics for Infectious Diseases: From Methods to Devices. *Nano Today* **2021**, *37*, 19–21. [CrossRef]
10. Walzl, G.; McNerney, R.; du Plessis, N.; Bates, M.; McHugh, T.D.; Chegou, N.N.; Zumla, A. Tuberculosis: Advances and Challenges in Development of New Diagnostics and Biomarkers. *Lancet Infect. Dis.* **2018**, *18*, e199–e210. [CrossRef] [PubMed]
11. Wu, X.; Wang, Y.; Weng, T.; Hu, C.; Wang, F.X.C.; Wu, Z.; Yu, D.; Lu, H.; Yao, H. Preparation of Immunochromatographic Strips for Rapid Detection of Early Secreted Protein ESAT-6 and Culture Filtrate Protein CFP-10 from Mycobacterium Tuberculosis. *Medicine* **2017**, *96*, 4–9. [CrossRef]
12. University Research Co Rapid Test Improves TB Diagnosis in People with HIV in South Africa—URC. Available online: https://www.urc-chs.com/news/rapid-test-improves-tb-diagnosis-in-people-with-hiv-in-south-africa/ (accessed on 9 January 2023).
13. Pym, A.S.; Brodin, P.; Brosch, R.; Huerre, M.; Cole, S.T. Loss of RD1 Contributed to the Attenuation of the Live Tuberculosis Vaccines Mycobacterium Bovis BCG and Mycobacterium Microti. *Mol. Microbiol.* **2002**, *46*, 709–717. [CrossRef]
14. Liu, C.; Zhao, Z.; Fan, J.; Lyon, C.J.; Wu, H.J.; Nedelkov, D.; Zelazny, A.M.; Olivier, K.N.; Cazares, L.H.; Holland, S.M.; et al. Quantification of Circulating Mycobacterium Tuberculosis Antigen Peptides Allows Rapid Diagnosis of Active Disease and Treatment Monitoring. *Proc. Natl. Acad. Sci. USA* **2017**, *114*, 3969–3974. [CrossRef] [PubMed]
15. Mdluli, P.; Tetyana, P.; Sosibo, N.; van der Walt, H.; Mlambo, M.; Skepu, A.; Tshikhudo, R. Gold Nanoparticle Based Tuberculosis Immunochromatographic Assay: The Quantitative ESE Quanti Analysis of the Intensity of Test and Control Lines. *Biosens. Bioelectron.* **2014**, *54*, 1–6. [CrossRef]
16. Turkevich, J.; Stevenson, P.C.; Hillier, J. A Study of the Nucleation and Growth Processes in the Synthesis of Colloidal Gold. *Discuss. Faraday Soc.* **1951**, *11*, 55–75. [CrossRef]
17. Ariffin, N.; Yusof, N.A.; Abdullah, J.; Abd Rahman, S.F.; Ahmad Raston, N.H.; Kusnin, N.; Suraiya, S. Lateral Flow Immunoassay for Naked Eye Detection of Mycobacterium Tuberculosis. *J. Sens.* **2020**, *2020*, 1365983. [CrossRef]
18. Huang, X.; El-Sayed, M.A. Gold Nanoparticles: Optical Properties and Implementations in Cancer Diagnosis and Photothermal Therapy. *J. Adv. Res.* **2010**, *1*, 13–28. [CrossRef]
19. Shafiqa, A.R.; Abdul Aziz, A.; Mehrdel, B. Nanoparticle Optical Properties: Size Dependence of a Single Gold Spherical Nanoparticle. *J. Phys. Conf. Ser.* **2018**, *1083*, 012040. [CrossRef]
20. Herizchi, R.; Abbasi, E.; Milani, M.; Akbarzadeh, A. Current Methods for Synthesis of Gold Nanoparticles. *Artif. Cells, Nanomed. Biotechnol.* **2016**, *44*, 596–602. [CrossRef]
21. Hu, M.; Chen, J.; Li, Z.Y.; Au, L.; Hartland, G.V.; Li, X.; Marquez, M.; Xia, Y. Gold Nanostructures: Engineering Their Plasmonic Properties for Biomedical Applications. *Chem. Soc. Rev.* **2006**, *35*, 1084–1094. [CrossRef] [PubMed]
22. Cavalera, S.; Pezzoni, G.; Grazioli, S.; Brocchi, E.; Baselli, S.; Lelli, D.; Colitti, B.; Serra, T.; Nardo, F.D.; Chiarello, M.; et al. Investigation of the "Antigen Hook Effect" in Lateral Flow Sandwich Immunoassay: The Case of Lumpy Skin Disease Virus Detection. *Biosensors* **2022**, *12*, 739. [CrossRef]
23. Nygren, H.; Stenberg, M. Kinetics of Antibody-Binding to Surface-Immobilized Antigen: Influence of Mass Transport on the Enzyme-Linked Immunosorbent Assay (ELISA). *J. Colloid Interface Sci.* **1985**, *107*, 560–566. [CrossRef]
24. Stenberg, M.; Stiblert, L.; Nygren, H. External Diffusion in Solid-Phase Immunoassays. *J. Theor. Biol.* **1986**, *120*, 129–140. [CrossRef]
25. Priyadarshini, S.; Manas, F.; Prabhu, S. False Negative Urine Pregnancy Test: Hook Effect Revealed. *Cureus* **2022**, *14*, 10–13. [CrossRef]
26. Ross, G.M.S.; Filippini, D.; Nielen, M.W.F.; Salentijn, G.I.J. Unraveling the Hook Effect: A Comprehensive Study of High Antigen Concentration Effects in Sandwich Lateral Flow Immunoassays. *Anal. Chem.* **2020**, *92*, 15587–15595. [CrossRef] [PubMed]
27. Byzova, N.A.; Safenkova, I.V.; Slutskaya, E.S.; Zherdev, A.V.; Dzantiev, B.B. Less Is More: A Comparison of Antibody-Gold Nanoparticle Conjugates of Different Ratios. *Bioconjug. Chem.* **2017**, *28*, 2737–2746. [CrossRef]
28. Cavalera, S.; Russo, A.; Foglia, E.A.; Grazioli, S.; Colitti, B.; Rosati, S.; Nogarol, C.; Di Nardo, F.; Serra, T.; Chiarello, M.; et al. Design of Multiplexing Lateral Flow Immunoassay for Detection and Typing of Foot-and-Mouth Disease Virus Using Pan-Reactive and Serotype-Specific Monoclonal Antibodies: Evidence of a New Hook Effect. *Talanta* **2022**, *240*, 123155. [CrossRef] [PubMed]
29. Lu, R.; Li, W.W.; Katzir, A.; Raichlin, Y.; Yu, H.Q.; Mizaikoff, B. Probing the Secondary Structure of Bovine Serum Albumin during Heat-Induced Denaturation Using Mid-Infrared Fiberoptic Sensors. *Analyst* **2015**, *140*, 765–770. [CrossRef]
30. Swaisgood, H.E. Review and Update of Casein Chemistry. *J. Dairy Sci.* **1993**, *76*, 3054–3061. [CrossRef]

31. Singh, N.; Dahiya, B.; Radhakrishnan, V.S.; Prasad, T.; Mehta, P.K. Detection of Mycobacterium Tuberculosis Purified ESAT-6 (Rv3875) by Magnetic Bead-Coupled Gold Nanoparticle-Based Immuno-PCR Assay. *Int. J. Nanomed.* **2018**, *13*, 8523–8535. [CrossRef] [PubMed]
32. Mehaffy, C.; Dobos, K.M.; Nahid, P.; Kruh-Garcia, N.A. Second Generation Multiple Reaction Monitoring Assays for Enhanced Detection of Ultra-Low Abundance Mycobacterium Tuberculosis Peptides in Human Serum. *Clin. Proteom.* **2017**, *14*, 21. [CrossRef] [PubMed]

biosensors

MDPI

Review

Recent Trends in Biosensing and Diagnostic Methods for Novel Cancer Biomarkers

Jagadeeswara Rao Bommi [1], Shekher Kummari [2], Kavitha Lakavath [2], Reshmi A. Sukumaran [2], Lakshmi R. Panicker [2], Jean Louis Marty [3,*] and Kotagiri Yugender Goud [2,*]

1 School of Medicine, Case Western Reserve University, Cleveland, OH 44106, USA
2 Department of Chemistry, Indian Institute of Technology Palakkad, Palakkad 678 557, Kerala, India
3 Université de Perpignan Via Domitia, 52 Avenue Paul Alduy, 66860 Perpignan, France
* Correspondence: jlmarty@univ-perp.fr (J.L.M.); yugenderkotagiri@iitpkd.ac.in (K.Y.G.)

Abstract: Cancer is one of the major public health issues in the world. It has become the second leading cause of death, with approximately 75% of cancer deaths transpiring in low- or middle-income countries. It causes a heavy global economic cost estimated at more than a trillion dollars per year. The most common cancers are breast, colon, rectum, prostate, and lung cancers. Many of these cancers can be treated effectively and cured if detected at the primary stage. Nowadays, around 50% of cancers are detected at late stages, leading to serious health complications and death. Early diagnosis of cancer diseases substantially increases the efficient treatment and high chances of survival. Biosensors are one of the potential screening methodologies useful in the early screening of cancer biomarkers. This review summarizes the recent findings about novel cancer biomarkers and their advantages over traditional biomarkers, and novel biosensing and diagnostic methods for them; thus, this review may be helpful in the early recognition and monitoring of treatment response of various human cancers.

Keywords: cancer diagnosis; biomarkers; electrochemical biosensors; optical biosensors; aptamers; antibodies; hybrid nanocomposites; recognition elements

Citation: Rao Bommi, J.; Kummari, S.; Lakavath, K.; Sukumaran, R.A.; Panicker, L.R.; Marty, J.L.; Yugender Goud, K. Recent Trends in Biosensing and Diagnostic Methods for Novel Cancer Biomarkers. *Biosensors* **2023**, *13*, 398. https://doi.org/10.3390/bios13030398

Received: 20 February 2023
Revised: 13 March 2023
Accepted: 15 March 2023
Published: 18 March 2023

1. Introduction

1.1. Cancer

One of the major life-threatening health issues in this world is cancer. It has become the second leading cause of death, around 75% of cancer deaths transpiring in low- to middle-income countries. It causes a heavy global economic cost estimated at more than trillion dollars per year [1–3]. In the course of cancer, cells can grow uncontrollably as well as expand to other parts of the organs in body. Cancer can occur from the transformation of normal cells into tumor cells. Tumor cells are classified as benign and malignant. Tumors that stick to their primary location without occupying distant parts of the body are called benign tumors; these are likely to grow slowly and are not problematic. Fibroids in the uterus are an example of benign tumors. Some of the benign tumors can change into malignant tumors (e.g., colon polyps); these can be removed surgically. Tumor cells that can grow uncontrollably and spread from their primary location to distant sites are called malignant tumors; these are cancerous (i.e., invade from other parts). Malignant tumors can rapidly spread to distant parts through blood or lymph stream; this process is called as metastasis. Omnipresent cancers are the primary reason of death in the patients with cancer. Most frequently metastasis can be found in brain, lungs, liver and bone [4].

Cancers can be grouped into different categories. Mostly, the group that the cancer belongs to is determined based on the type of cells or tissues it is producing. The following are some of the following important groups in this category.

Carcinoma: Most common type of cancer. It starts in epithelium, which is the tissue that lines or covers internal organs and passageways in the human body as well as skin. It appears in the form of tumors, which can form on lungs, breasts, skin, pancreas, kidneys, prostate, colon, etc. There are various subtypes, including adeno-carcinoma, SCC (squamous cell carcinoma), ductal carcinoma, TCC (transitional cell carcinoma) and (BCC) basal cell carcinoma.

Sarcoma: This is a type of cancer that arises in connective tissue and/or supportive tissue such as muscle, cartilage, bone, blood vessels or fat.

Leukemia: This type is also known as blood cancer/cancer of WBC (white blood cells). It can appear in tissues which can produce blood cells (e.g., bone marrow).

Myeloma and Lymphoma: This type of cancer occurs in the cells present in the immune system (myeloma: Starts in plasma cells present inside bone marrow; lymphoma: Starts in lymphatic system such as spleen, lymph nodes).

Spinal cord and brain cancers: These cancers occur in spinal cord and brain, and are also known as central nervous system cancers.

Multiple factors can generate cancer. These cancer-causing agents are termed carcinogens (Figure 1). These could be genetic factors (mutated genes pass from parents to children and cause various cancers, e.g., *BRCA1* and *BRCA2*) and external factors (physical carcinogens: Ionizing and UV radiation; tobacco smoke, alcohol, aflatoxin and arsenic are the examples of chemical carcinogens; certain viruses, bacteria, fungus and some type parasites are considered biological carcinogens; life style factors: lack of exercise, smoking, nutrient imbalance) [5] (https://www.who.int/news-room/fact-sheets/detail/cancer) (accessed on 11 November 2022).

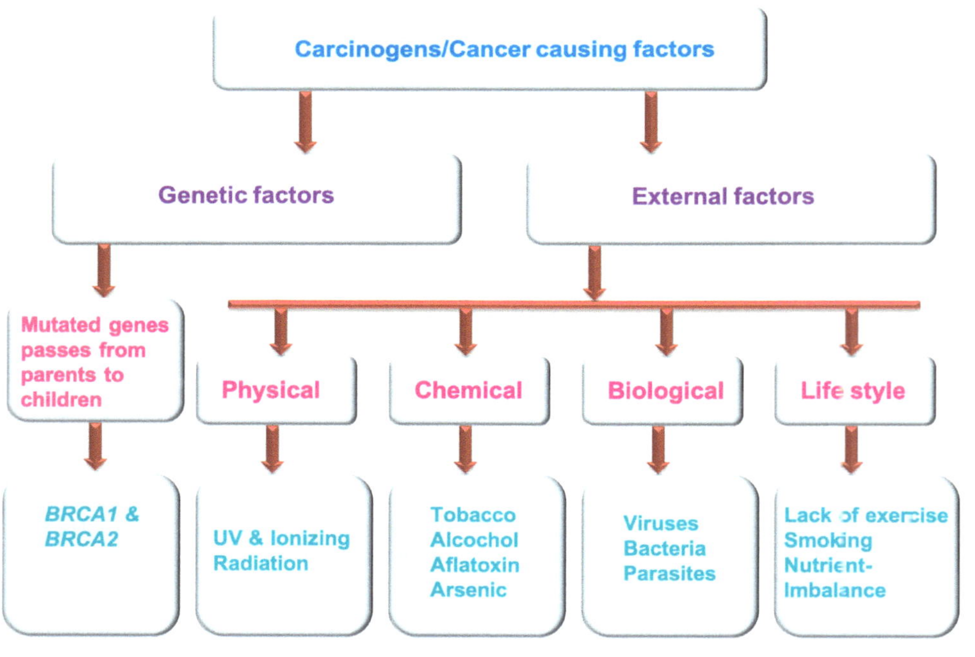

Figure 1. Cancer causing factors/carcinogens.

1.2. *Importance of Cancer Diagnostics*

Early diagnosis of cancer has held great assurance and intuitive interest in the medical community for over 100 years [6,7]. However, delayed identification and imperfect prognosis are major causes for very low survival rate in various patients with cancer [8]. Many cancers can be treated effectively and cured if detected at the primary stage. Patients diagnosed with cancer at earlier stages are not only likely to survive, but, importantly, they also experience better care, minimal treatment morbidity, and enhanced quality of life when compared with patients with late diagnosis [9,10]. Improving earlier detection of cancers is a complex and multifaceted process. Few recent patient behaviors could be helpful for the earlier diagnosis of various cancers. One is attending cancer screening procedures, which aim to identify the cancer in asymptomatic condition (e.g., mammography to detect breast cancer), and the other is promptly introducing the patient with any potential cancer symptoms to primary care providers [11]. The necessity of symptomatic presentation to primary care is emphasized by improving public perception of the early indications of cancer. In recent years, cancer awareness is also increasing in developing and lesser-income countries. Raising public awareness, promoting visits to a healthcare provider, and diagnosing a patient at an earlier stage of the cancer can provide the opportunity to offer better treatment [12–14]. Nowadays, around 50% of cancers are detected at late stages, leading to serious health complications and death [12]. Early diagnosis of cancer significantly increases the efficient treatment and high chances of survival.

1.3. Traditional Screening Techniques

Cancer screening techniques have significantly promoted the decline of morbidity as well as mortality of cancer. Methods to enhance the choice of candidates for cancer screening, to acknowledge the biological foundation of cancer formation and development of novel technologies for the tumor screening would allow for advancement in the tumor screening process over time [15,16]. Screening in healthy as well as high-risk populations provides the chance to detect cancer at an early stage and with an expanded chance for treatment. Nowadays, screening techniques play a crucial role in detecting specific cancer types. However, each screening technique has some limitations, and upgraded screening techniques are very much essential to identify cancer early in healthy populations [17–19]. Identifying tumors at their primary stage often delivers the finest probability for a cure, which is why it is always crucial for the general population to talk with their health care providers about the types of screening that might be required. Research studies explain that early screening process can save many lives by identifying cancer in primary stages. Various medical communities and patient advocacy groups have drafted guidelines for cancer screening. In general, primary health care providers use the following approaches for tumor diagnosis [20–22].

Physical exam: During this exam, a provider can perform a scan to detect any abnormalities in body, such as skin color change, formation of lumps or abnormal growth of a tissue or an organ. This may provide the indication of the tumor.

Laboratory tests: Some lab procedures such as blood and/or urine tests can help to detect any cancer-related anomalies in body. For example, simple blood work called complete blood cell count might disclose any abnormal number or type of WBC (white blood cells) in leukemia patients.

Genetic tests: These are also lab-based tests. In these procedures, cells or tissues are examined to observe any modifications in their genes and/or chromosomes. Any of those differences might be an indication that a person is at risk of encountering a particular problem or condition.

Imaging tests: Imaging tests can examine bones or any interior organs through non-invasive method. Common cancer diagnosing imaging methods are a CT (computerized tomography) scan, X-ray, PET (positron emission tomography) scan, ultrasound, and MRI (magnetic resonance imaging) [23,24].

Biopsy: During this procedure, samples of cells are collected for examining in the laboratory. Collection of a sample can be performed by a variety of methods. The suitable type of biopsy is based on the cancer type and cancer location in body. In many cases, biopsy is the exclusive method for conclusive cancer diagnosis [25–27].

1.3.1. Advantages of Traditional Cancer Screening

1. It can help to detect the cancer before it spreads, when it is easier to treat.
2. It can provide an advantage of early detection, which might lead to lesser recovery time and no intense treatment.
3. It can provide a better chance of survival.
4. It offers flexibility to start early treatment before symptoms appear.
5. It can also reassure a person if the screening result is normal [28–30].

1.3.2. Limitations of Traditional Cancer Screening

1. Sometimes a false-positive test result suggests a cancer-positive status, even though no cancer is present.
2. Sometimes false-negative test results may not detect cancer, even though it is present.
3. Some screening tests might lead to more detection tests and procedures that can be painful.
4. Over-diagnosis causes needless anxiety.
5. Some screenings might cause potential issues (e.g., colon cancer screening may cause tear in colon lining).
6. These screening methods are high cost.
7. Test availability is limited to metro cities only [28–30].

There is a strong necessity to develop rapid and affordable screening methods for cancer diagnosis. Cancer biomarker screening through clinical and point-of-care diagnostic methods is a promising tool for an early diagnosis of cancer.

At present, most of the available literature is specific to biomarkers for certain cancer types such as biomarkers for breast cancer [31], biomarkers for colorectal cancer [32], biomarkers for prostate cancer [33], biomarkers for ovarian cancer [34], biomarkers for cervical cancer [35], etc.

In this review, we focused mainly on more recent cancer-detecting methods such as cancer biomarker detection in various cancer types and ways in which this biosensor-mediated biosensing technology can show advantage over traditional detection methods. We also explored various types of cancer biomarkers, availability of traditional cancer biomarkers, recent developments in finding novel cancer biomarkers and their respective detection methods.

2. Cancer Biomarkers

The conventional diagnostic technologies such as MRI (magnetic resonance imaging), CT (computerized tomography) scan, ultrasound and biopsy were not effective for cancer detection at primary stages; this is because of their dependency on tumorigenic properties or phenotypic characters of a tumor [36–40]. Cancer is a very complex disease, with many epigenetic as well as genetic modifications which might alter the cell signaling process, related to development and resulting in tumorigenic transformation and malignancy [41]. For almost all cancer patients, researchers and clinicians expect tests or methods that might diagnose cancer significantly earlier, provide better prognosis, and that can allow for increased survival rates. Cancer markers have been used over the past few decades in the oncology field. Biomarkers are molecules of biologic emergence found in blood, tissues, various body fluids such as urine, cerebrospinal fluid, or different body tissues that are elevated is the indicative of an abnormal disease or condition with cancer. Human body responsiveness to any therapy can be observed and regulated through biomarkers. These also exist on or in cancer cells. Cancer biomarkers are possibly one of the most potential tools to detect cancer early [42–46].

2.1. Clinical Significance of Cancer Biomarkers

Cancer biomarkers can be utilized for cancer patient evaluation in different clinical levels, as well as disease screening, prognosis, diagnosis, staging, risk assessment, stratification, therapy planning and monitoring (Figure 2). Still, to date, several cancer markers have indicated poor validity and efficacy, especially in the most widespread cancers such as lung and breast cancers. Cancer biomarkers are bio-molecules necessary for remodeling throughout cancer which maintain excessive clinical relevance. These can be enzymes, iso-enzymes, nucleic acids, proteins, metabolites. Biomarkers are classified into three types based on their clinical advantage: prognostic, predictive and diagnostic. Prognostic biomarkers provide details about course of recurrence of the disease; patient response to the treatment is estimated by predictive biomarkers; disease detection can be performed by using diagnostic biomarkers [43,47–49]. The difference in the level of any unique biomarker in a cell or tissue is often used as evidence of tumor expansion. Biomarkers also play potential role in differentiating benign and malignant tumors and one type of malignancy from another type; specific biomarkers are helpful in unique settings, other biomarkers can be involved in multiple settings [48]. Biomarkers might be helpful to estimate a person's chance of developing tumors/cancer. For example, a person having a solid family network (via ancestors) with ovarian cancer might receive a genetic test to decide whether they are acting as a carrier for a specific germ line modification or mutation, such as BRCA1, which could cause potential chance of developing breast and/or ovarian cancer [50,51]. Biomarkers might be helpful to determine malignancy in fit populations. A frequently utilized component for screening is PSA (prostate-specific antigen). It was approved by Food and Drug Administration (FDA) in 1986. Enhanced screening in male population above 50 years old lead to growth in the identification of prostate cancer. These kinds of traditional biomarkers also have limitations. In the previous decade, U.S. Preventive Services Task Force survey noted that an adequate documentation for common diagnosis with PSA [52–54]. Biomarkers were used to monitor prognosis and possibility of cancer reappearance irrespective of therapy/treatment. The clinical and pathological properties of a tumor could be useful for the prediction of cancers. Recently, modern techniques were used to evaluate prognosis of independent tumors; for example, a large number of genetic marks that had been matured in breast cancer might be useful to evaluate the identification for an individual patient depending on tumor assessment [55,56]. During breast cancer (metastatic) condition, circulating tumor cells are indicative of overall survival [57–59]. Biomarkers could be utilized as stimulus changers or prognostic factors for unique type of therapy, as well as to select the effective type of treatment. KRAS is a predictive biomarker for colorectal cancer; mutations occurring to KRAS in somatic cells are related to low response to anti-EGFR-mediated treatment [60,61]. Likewise, HER2 over-expression in the breast cancer as well as gastric cancer anticipates for stimulus to anti-Her2 drugs such as trastuzumab [62–67], and over-expression of estrogen receptor anticipates for stimulus to anti-endocrine therapy or treatment such as tamoxifen in breast cancer [68,69]. Identification of novel cancer biomarkers might help in quick and efficient diagnosis as well as monitoring of cancer progression.

2.2. Identification of Novel Cancer Biomarkers

Possible cancer biomarkers could be recognized through various approaches. An excellent way to recognize novel biomarkers mostly depends on biological nature of tumor and nearby environment of a tumor or metabolism of the drugs or biological products. With most recent studies and new information related to cancers and appearance of latest technology, cancer biomarker detection is performed frequently these days with applying a discovery approach. In this approach, few major areas of research such as gene expression arrays, proteomic technologies (mass spectroscopy, LC-MS/MS, MALDI-MS), and high throughput sequencing can be used to rapidly recognize unique biomarkers or pool of biomarkers which can show difference in the middle of cohorts. Expansion of sophisticated software algorithms for large data analysis has emerged in rapid advancement in the identification of novel cancer biomarkers. Openly available software programs can sort these data and compare the sequences to annotated genome databases to permit quantitative comparative evaluation of proteomes from multiple sources such as tumor area and nearby healthy tissues. Thus, over-expressed or down-regulated proteins in a cancer cell can be identified as putative cancer biomarkers. Huge amount of data produced with these technological methods mean that special attention needs to be directed toward both developing a study plan and conducting a large data analysis. Moreover, it is crucial to reduce the possibility of detecting relationships that are eventually determined to provide false-positive results. The most crucial features of biomarker improvement and identification to consider in depth include mindful study pattern to minimize any kind of bias, extensive evaluation, validation, and accurate communication of the results [48,70–79].

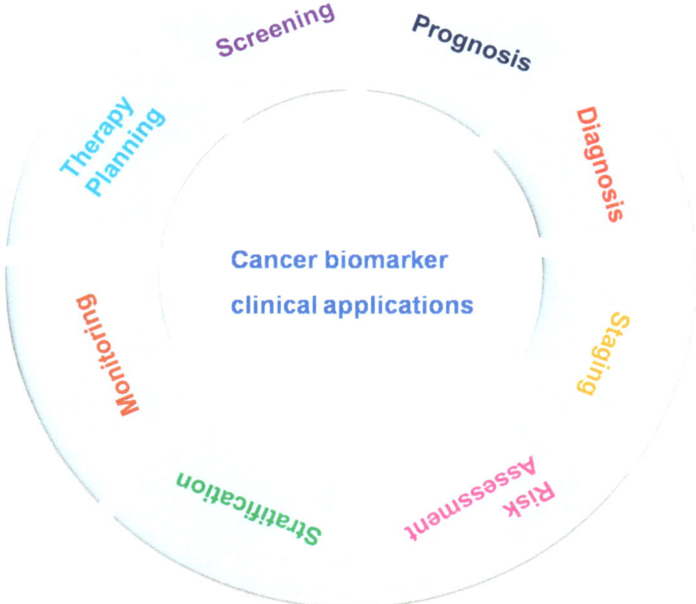

Figure 2. Cancer biomarker clinical applications.

2.3. *Cancer Biomarkers Currently Used in Clinical Settings*

A cancer biomarker is a molecule existing inside and/or generated by tumor cells or surrounding cells in tissues or organs in stimulus to tumor or certain noncancerous (benign) situations, which can provide information about cancer, mostly which stage it is, whether it can respond to treatment, and what type of therapy might be useful. Here, few recent cancer biomarkers used in clinical practice are explained. New cancer biomarkers become available continuously, and they may not be explained below [27,40,44,80–82].

2.3.1. AFP (Alpha-Fetoprotein)

AFP (Alpha-fetoprotein) is one of the leading biomarker. Early fetal life of a baby (mostly yolk sac and liver) produces AFP during pregnancy. AFP can be detected in huge amounts in serum of the patients with specific tumors. According to Yuri Semenovich Tatarinov, AFP was first accepted as an antigen unique for human HCC (hepatocellular carcinoma) [83,84]. The scientific literature has explained that AFPs are classified into subtypes based on their dissimilar affinities to LCA (lens culinaris agglutinin), such as AFP-L1 (LCA unreactive AFP), AFP-L2 (LCA mild active AFP) and AFP-L3 (LCA active AFP). In healthy individual serum, an average level of AFP is less than 20 ng·mL^{-1} [85–87]. AFP is widely established for HCC (hepatocellular carcinoma) diagnosis. Moreover, in congenital tyrosinemia, cirrhosis, hepatitis (alcohol-induced), hepatitis (viral-induced), ataxia-telangiectasia syndrome or in several malignancies such as testicular cancer, liver cancer, gastric cancer and nasopharyngeal cancer elevated AFP levels may also be present. Hence, sensing the AFP values is completely mandatory in clinical settings. High recognition rates of molecular assays have been obtained in quantitative observation of AFP due to their specificity and unalterable affinity of the probes to molecular targets [88,89].

2.3.2. PSA (Prostate Specific Antigen)

PSA (prostate-specific antigen) was one of the first recognized cancer biomarkers, utilized to detect and screen prostate cancer in clinical setting. It has been shown that increased levels of PSA can directly relate to prostate cancer. Human regular PSA level is 4 ng/mL. According to the study by Smith et. al, almost 30% of individuals with PSA values higher than normal (range of 4.1–9.9 ng/mL) were diagnosed with prostate cancer [90]. Along with this, raised PSA values may also indicate benign tumors (non-fatal), prostatitis/prostate inflammation or benign prostatic hyperplasia. Therefore, elevated values of PSA may not consistently suggest malignant tumors. There is a fact that caused reasonable controversy about using regular PSA screening to detect prostate cancer. Small-sized tumors identified by PSA screening may grow very slowly; death caused by small tumor might not be feasible in an individual lifetime. Moreover, it is very expensive to treat such slow-growing tumors. Other frequent issue with PSA screening is false-positives. This limitation of PSA testing can be overcome by modern biosensing technology mediated by biosensors [79,84,91–93]. PSA detection can be performed by various methods; these traditional methods are time-consuming as well as expensive. Yang et al. explained a graphene oxide/ssDNA-based biosensor integrated with dual antibody-modified PLLA NPs to amplify electrochemical signals for the effective and rapid electrochemical capture of PSA in serum samples from prostate cancer patients. The detection limit for PSA was 1 ng/mL, which achieved a wide linear range of 1–100 ng/mL for PSA. This is one of the examples that shows the usefulness of modern biosensing technology mediated by biosensors [94].

2.3.3. RCAS1 (Receptor-Binding Cancer Antigen)

RCAS1 (receptor-binding cancer antigen) overexpression data has been described in many gastric carcinomas; it is related with progression of gastric cancer. Further, RCAS1 is also suggested as a cancer biomarker for poor prediction in breast, esophageal, endometrial, bladder cancers and is associated with tumor weakening in pharyngeal carcinoma and laryngeal cancer. RCAS overexpression is observed in several types of cancers. Thus, it serves as a potential biomarker for cancer detection and prediction [95–98].

2.3.4. CA 15-3 (Cancer Antigen 15-3)

The most predominant cancer marker for breast cancer identification as well as monitoring is cancer antigen 15-3; additional biomarkers that are related to breast cancer are CA 27.29, BRCA1, BRCA2 and (carcinoembryonic antigen) CEA [44,99,100]. This particular marker is frequently used on a clinical level to monitor the therapy for breast cancer in its advanced stages. During breast cancer, CA 15-3 values increased by 10, 20, 30 and 40 % at various stages first, second, third and fourth stage [49]. Tampellini et. al showed the connection between breast cancer and CA 15-3 levels, also mentioning that before treatment, patients with levels of 30U/mL had notably higher survival rate compared to patients with higher values. Raised CA 15-3 values correspond with extensive metastasis [101,102]. Other research data explained that raised CA 15-3 levels post cancer treatment can be a sign of disease recurrence. Nowadays, to determine breast cancer treatment, protocol CA 15-3 values are considered along with risk factors (negative) such as PR/ER condition as well as iHer-2, cancer stage and tumor dimension. Hepatitis, endometriosis, pelvic inflammatory disease, lactation and pregnancy are conditions other than cancer where CA 15-3 levels are increased [103].

2.3.5. Cancer–Testis (CT) Antigens

Cancer–testis antigens are a specific type of cancer biomarkers. They are expressed in various cancers. CT antigen expression is limited to male germ cells of the testis but not shown in normal adult cells. These antigens are also expressed in trophoblast and ovary cells. Therefore, CT antigens have been considered as possible immunogenic targets for cancer vaccines (cancer immunotherapies). CT antigen autoantibodies have been studied as potential cancer biomarkers [104]. NY-ESO-1 (NewYorkesophagealsquamouscellcarcinoma1) is encoded by *CTAG-1B*; this is the class of antigens with high immunogenic nature which induces very robust cellular and humoral immune response in NY-ESO-1-positive cancers. Antibody titer to NY-ESO-1 has been shown to relate with disease progression. One of the best benefits of using CT autoantibodies as tumor biomarkers is the fact that they are easy to obtain and are also more stable proteins present in serum compared to tissues obtained via biopsy. Thus, they can be useful for cancer progression and recurrence [105–109]. Major limitation of CT antigens is the fact that many cancers express CT antigens, and they are rarely tumor-specific. Biosensor technology can obtain a profile for CT antigen, which can enhance the use of these antigens in cancer prognosis and diagnosis [110–112].

2.3.6. CA 125

The rise in CA 125 levels is primarily related to ovarian cancer. It is further correlated with several different cancers such as cervix, lungs, breast, liver, pancreas, uterus, stomach and colon cancers. Enhanced levels of CA 125 are also observed in various non-pathological conditions such as menstruation and pregnancy [8,87,113]. A total of 90% of women with advanced stage ovarian cancer and 40% of humans with intra-abdominal malignant tumors also exhibit high CA 125 levels. Still, approximately 50% of patients diagnosed with primary stage ovarian cancer show normal CA 125 levels [87–89]. Other germ cell origin biomarkers such as alpha-fetoprotein/AFP, human chorionic gonadotrophin/HCG as well as Lactate dehydrogenase/LDH are also connected to ovarian cancer [114,115]. Increase in CA 125 levels can be used to detect the development of benign tumors into malignant tumors. Enhanced CA 125 levels are also used to identify treatment failure as well as disease recurrence (e.g., high CA 125 levels after bilateral salpingo-oophorectomy or total abdominal hysterectomy, which may occur after first line chemotherapy) [116–120]. Altogether, CA 125 is an extremely essential biomarker for detection of cancer, and also for cancer progression monitoring and treatment.

2.3.7. CA 19-9

This antigen was first identified in pancreatic and colon cancer patient's serum in 1981. It is a Lewis antigen of the MUC1 glycoprotein [8,121]. CA 19-9 normal level in serum is less than 37 U/mL. In the recent decade, on a clinical level, CA 19-9 biomarker has become extremely useful for the diagnosis of pancreatic cancer. Normal human serum CA 19-9 levels can play an outstanding role in clinical diagnosis of urothelial and gastric cancers [122,123]. Consequently, there is a necessity to improve highly sensitive methods which can detect CA 19-9 values in patients with cancer.

2.3.8. Nse (Neuron-Specific Enolase)

This neuron-specific enolase is a popular and unique marker for SCLC as well as NSCLC non-small cell lung cancer. It has a crucial role in glycolysis; in 1980s, NSE expression was noted in SCLC cells. From that time, it has been used as potential biomarker for lung cancer, able to detect increased values of NSCLC and acting as crucial predictor for patient survival (it is not dependent on remaining prognostic factors) [124–126]. NSE is also a unique marker for neuro-endocrine cells. Raised NSE values in body fluids might be an indication of tumor proliferation and staging determination in some of brain tumors [127,128]. In the recent decade, the value of NSE for prognosis in cancer patients is debatable. Therefore, it is mandatory to improve sensitive techniques to perceive Nse values in patients with cancer.

2.3.9. Tdt (Terminal Deoxynucleotidyl Transferase)

Tdt is an intracellular marker which has detected in the bone marrow as well as blood (mononucleate) cells in leukemia patients during diagnosis. Overall, TdT values are remarkably raised in several lymphocytic lymphomas. Tdt might be helpful to identify specific leukemia type and supportive sign for the solution of therapy [129–132].

2.3.10. CYFRA21-1

Cytokeratin-19 fragments/CYFRA21-1 have been extensively studied in patients with NSCLC and are widely utilized as predictive, prognostic markers. They have 56% sensitivity as well as 88% specificity when the value is >1.5 ng/mL. Researchers used maximum cut off value for CYFRA21-1 similar to $\geq$3 ng/mL; it had shown increased specificity at 97%. It has potential capability in lung, esophagal cancer prediction. Raised values are certain but barely sensitive. These values are strongly connected with cancer metastasis. Recent reports explained that CYFRA21-1 is used as independent prediction factor for various phases of lung cancer. This might function as definitive distinction between benign and malignant lung cancer along with clinical information [133–137].

Table 1 demonstrates the summary of various cancer biomarkers available in the literature for the detection of cancer diseases. Significant biomarkers are assembled in the table including their respective cancer diseases and their advantages.

3. Importance of Finding Novel Bio-Sensing Methods to Detect Cancer Biomarkers

Biomarkers might have different molecular origins, such as changes to nucleic acids including RNA, DNA (amplification, translocation, point mutation, loss of heterozygosis) and protein (antibodies, tumor suppressors, oncogenes and hormones). Some of these cancer biomarkers are convincing and extremely crucial for early diagnosis of tumors. Some of the biomarkers need to exhibit adequate specificity and sensitivity for translation into clinical use or for monitoring of disease progression. In this area, biosensing technology can potentially play a crucial role to improve early diagnosis of cancer. The traditional PCR (polymerase chain reaction) or ELISA (enzyme-linked immunosorbent assay) techniques for cancer biomarker identification have some technical limitations, including utilization of costly chemicals and expensive machines in every single assay, which could delay the detection. Moreover, these types of techniques are not capable of constant monitoring in patients throughout treatment. Along with this, multiple pathways are interlinked with cancer cells and these cells express more than one biomarker. Therefore, simultaneous identification of different biomarkers for accurate prognosis and diagnosis is indispensable [8,42,48,138–143]. Biosensors offer great potential sensing methodology platforms for the detection of various cancer biomarkers. Specifically electrochemical and optical biosensors based on the affinity, chemical, bio-affinity recognition elements attract great interest. In the following section, we comprehensively discuss the electrochemical and optical sensing strategies for cancer disease diagnosis.

Table 1. Biomarkers used in cancer detection.

Tumor/Cancer Biomarker	Type of Cancer/Infected Location	Application	References
AFP	Liver (HCC)	Identifying recurrence, treatment monitoring, disease diagnosis	[144–147]
PSA	Prostate gland	Screening, identifying recurrence, treatment monitoring, disease diagnosis	[148–150]
CA 15-3	Breast	Treatment monitoring	[103,151,152]

Table 1. *Cont.*

Tumor/Cancer Biomarker	Type of Cancer/Infected Location	Application	References
CT antigens	Prostate, liver, lung,bladder, skin	Diagnosis, prognosis	[105,106]
CA27.29	Breast	Monitoring	[69,153]
RCAS1	Stomach	Detection, prognosis	[96–98]
CA 19-9	Pancreas, colon	Treatment monitoring	[122,154]
CEA (Carcinoembryonic antigen)	Colon, liver	Screening, Identifying recurrence, Treatment monitoring, Disease prognosis	[155,156]
Calcitonin	Thyroid gland	Treatment monitoring, Disease prognosis	[157]
ER & PgR (Estrogen, progesterone receptors)	Breast	Stratification	[158–160]
HER2	Lung, breast	Monitoring therapy	[64,161–163]
CA 125	Ovary	Prognosis, identifying recurrence, treatment monitoring, disease diagnosis	[88,89]
HCG-β	Ovary, testis	Diagnosis, staging, identifying recurrence, treatment monitoring	[164,165]
Tdt	Blood/leukemia	Diagnosis	[129]
NSe	Lung	Prognosis	[125,166]
Thyroglobulin	Thyroid	Treatment monitoring	[167,168]
PCA3	Prostate gland	Prognosis	[169]
NY-eSO-1	Skin/melanoma	Progression monitoring	[170]
EGFR	Lung	Diagnosis and monitoring therapy	[171,172]
KRAS, ALK	Lung	Diagnosis and monitoring therapy	[173,174]
CD30	Blood/Leukemia	Diagnosis and prognosis	[175,176]
NMP 22	Bladder	Screening, treatment monitoring, disease prognosis	[177,178]
CYFRA21-1	Esophagus	Prognosis, Treatment monitoring	[179–181]
BCL2	Blood and breast	Diagnosis, treatment plan	[182–184]
BCR-ABL fusion gene	Bone marrow, blood	Prognosis, treatment determination, monitoring	[185,186]
CD20	Blood	Treatment determination	[187]

Table 1. *Cont.*

Tumor/Cancer Biomarker	Type of Cancer/Infected Location	Application	References
CD22	Blood	Treatment determination, diagnosis	[188]
CD25	Blood	Treatment determination	[189]
FGFR2 & FGFR3	Bladder	Treatment determination, therapy	[190,191]
Fibrin-fibrinogen	Bladder	Treatment determination, monitoring	[192,193]
SMRP	Leukemia	Progression monitoring	[194,195]
ROS1	Lung	Treatment determination	[196]
OVA1	Ovary	Prognosis	[197]
VMA	Brain	Diagnosis	[198]

4. Analytical Diagnostics Methodologies for Cancer Biomarkers Screening

Analytical biosensing methodologies for the detection of various analyte molecules including cancer biomarkers, pharmaceutical drugs, and agricultural toxins are recently rapidly growing. These techniques have several unique advantages such as point-of-care diagnosis, miniaturized portable instrumentation, cost-effectiveness and, moreover, user-friendliness to the end users. In this current section, we discuss some of the recently published works related to the optical- and electrochemical-based biosensors for cancer biomarker detection.

O. Awatef et al. reported a selective, sensitive, and inexpensive aptamer-based SERS biosensor for detection of prostate-specific antigen in human serum. Here, they have been using 1D (1 Dimensional) Silicon nanowires as a transduction material due to their advantage large surface area. These materials prepared from N-doped Si(100) wafers by metal-assisted chemical etching method. AgNPs were deposited on SiNWs through the electroless deposition technique to enhance the optical signal properties of the sensor. AgNPs/SiNWs were further functionalized with self-assembled layer of hexanethiol by incubating the SERS substrate of MCH to avoid the non-specific binding of the analyte. The developed sensor exhibited a good response in the dynamic range from 0.1 to 20 $\mu g \cdot L^{-1}$ with a limit of detection of 0.1 $\mu g \cdot L^{-1}$. Moreover, this sensing platform selectively and sensitively detected PSA in spiked PBS solutions [199].

M. Sachin et al. developed the fabrication of a tailored biofunctionalized interdigitated capacitor electrode (Ti/Pt imprinted) for label-free PSA detection. This sensing platform exhibits rapid detection within 3 s, stability up to many weeks, reusability and reproducibility; it is also requires low volume and easy to operate. Here, interdigitated capacitor (IDC) chip was initially functionalized with Ti/Pt metal by e-beam deposition process. Electrode surface was activated by placing it into piranha solution to formation of hydroxyl groups on the surface. Then, further formation of an amine group on the surface by APTES solution was drop-costed, and glutaraldehyde solution was applied to generate an antibody through cross-linking. Later anti-PSA was fabricated on the electrode surface, and further immobilization of PSA onto the bio surface has performed to verify the interactions through capacitance. APTES and glutaraldehyde increase the positive capacitive response of IDC-based PSA biosensor and are treated as reference value. By using LCR meter, the change in capacitance variables with respect to changing in concentration of target protein has been calculated [200].

Jose Ribeiro et al. developed new biosensing methodology by merging two different techniques, surface plasmon resonance spectroscopy and electrochemical technique for the detection of breast cancer biomarker carbohydrate Antigen 15-3 to monitor disease progression. Two steps are mainly involved in this process: (i) direct SPR monitoring interaction between surface immobilized antibody and CA 15-3 antigen, performed until adsorption reached equilibrium, and (ii) electrochemical measurements at the SPR gold surface and the resulting immunosensor selective detection for the breast cancer biomarker CA 15-3 protein [201].

D. Haihan, et al. performed investigations to explore the construction of paper-based photo electrochemical (PEC) biosensors with 1D self-doping SnO_2 nanotubes for selectively detection of alpha fetoprotein (AFP). (Figure 3) With the template consumption technique, paper-based 1D-domed SnO_2 nanotubes have been created from template ZnO nanorods. Additionally, a method of Sn self-doping was suggested to make it easier to separate photo-induced charge carriers and improve the harvesting of visible light. Additionally, self-doping of Sn can reduce the recombination rate of charge carriers and narrow the band gap of SnO_2 nanotubes, which results in a significant increase in photocurrent intensity under visible light illumination [202].

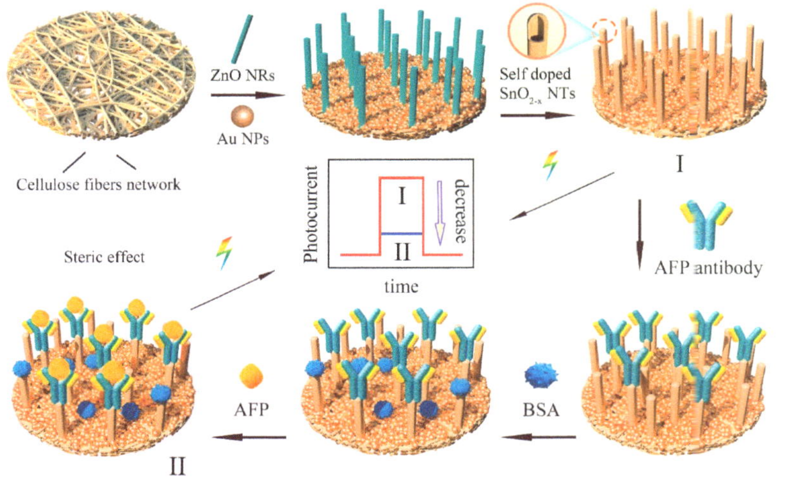

Figure 3. Schematic illustration of the construction of paper-based Sn-doped SnO_{2-x} sample and assay procedures for the specific detection of cancer biomarker alpha fetoprotein. Reproduced with the permission. [202] Copyright 2021, Elsevier.

W. Qiong et al. reported the new class of 2D Ti3C2-MXene nanosheet-based SPR biosensor for detection of carcinoembryonic antigen (CEA) cancer biomarker with high specificity and reproducibility. A novel class of ultrathin Ti3C2-MXene possesses hydrophilic biocompatible surface, which can be used as a biosensing material. Ti3C2 -MXene nanosheets were coated with AuNPs using a chemical reduction method and further modified with SPA to improve detection sensitivity or orient purpose. Later immobilization of anti-monoclonal CEA(Ab1) has performed to capture the analyte CEA. Here, Ti3C2-MXene-based SPR sensing platform exhibited significant performance for detection of CEA in real serum samples [203].

Table 2 demonstrates the summary of various optical screening methods based on different recognition matrices for the screening of cancer biomarkers. Significant articles with the transduction techniques, bio/chemical recognition matrices, dynamic working calibration range and limit of detection are assembled in the table.

Table 2. Optical sensing methodologies for cancer biomarker detection.

No.	Biomarker	Recognition	Method	Linear Range	LOD	Ref.
1	AFP	SiO_2@CQDs/AuNPs/MPBA	ECL	0.001–1000 ng/mL	0.0004 ng/mL	[204]
2	AFP	1D SnO_2 NTs	PEC	10 pg mL^{-1}–200 ng mL^{-1}	3.84 pg mL^{-1}	[202]
3	AFP	aptamer-MCHA	Fluorescence	0.1 ng mL^{-1}–10 $mgmL^{-1}$	0.033 ng. mL $^{-1}$	[205]
4	CA 19-9	luminol-AgNPs@ZIF-67	ECL	0.0001–10 U/mL	31 μU/mL	[206]
5	CA 19-9	Ni NCs-Ab_2	Fluorescence	0.001–48 ng mL^{-1}	0.00013 ng mL^{-1}	[207]
6	CA 125	rGO-based FET-type aptasensor	Fluorescence	1.0×10^{-9}-1.0 U/mL	5.0×10^{-10} U/mL	[208]
7	CA 125	CA 125/MUC16	SPRI	2.2–150 U/ml	-	[209]
8	CEA	Ti_3C_2-MXene/AuNPs/SPA	SPR	2×10^{-16}–2×10^{-8} M	0.07 fM	[203]
9	CEA	HCR and G-quadruplex DNAzyme	Fluorescence	0.25–1.5 nM	0.2 nM	[210]
10	HER2	nanoparticle coated QCM	QCM	10–500 cells/mL	10 cells/mL	[211]
11	HER2	PtAmi	Fluorescence	-	-	[212]
12	HER2	3D DNA walker	Fluorescence	0.5–5 ng mL^{-1}	0.01 ng mL^{-1}	[213]
13	PSA	$NaYF_4:Yb^{3+}$, Er^{3+} UCNPs and $NaYF_4:Yb^{3+}$, Er^{3+}@$NaYF_4:Yb^{3+}$ UCNPs	Fluorescence	0.1 ng/mL–10 ng/mL	0.01 ng/mL	[214]
14	PSA	anti-PSA/MCH/ AgNPs/SiNWs	SERS	0.1–20 μg·L^{-1}	0.1 μg·L^{-1}	[199]
15	PSA	IDC	FET	0.1–10.0 μL/mL	-	[200]
16	CA-15-3	PHMPF	Fluorescence	2.56×10^{-5}–1.28 U mL^{-1}	2.56×10^{-5} U mL^{-1}	[215]
17	CA-15-3	SPR gold substrates	SPR	-	0.0998 U mL^{-1}	[201]
18	NMP 22	orange emitting quantum dot CdTe/CdS	Fluorescence	2–22 pg mL^{-1}	0.05 pg mL^{-1}	[216]
19	NMP 22	NCDs	Fluorescence	1.3–16.3 ng/mL	0.047 ng/mL	[217]

AFP—alpha-fetoprotein; ECL—electrochemiluminescence; PEC—photoelectrochemical; SPRI—Surface Plasmon Resonance Imaging; QCM—quartz crystal microbalance; PtAmi—red-emitting exchanged Pt nanoclusters; UCNPs—upconversion nanoparticles; IDC—interdigitated capacitor; PHMPF—prismatic hollow Metal-polydopamine frameworks; NCDs—-nitrogen-doped carbon dots.

Electrochemical Sensing Methodologies

Electrochemical biosensing methodologies for the detection of various analyte molecules including cancer biomarkers, pharmaceutical drugs [218–221], and agricultural toxins are recently rapidly growing. These techniques have several unique advantages such as point-of-care diagnosis, miniaturized portable instrumentation, cost-effectiveness and, moreover, user-friendliness to the end users. In this current section, we discuss some of the recently published works related to the electrochemical-based biosensors for cancer biomarker detection.

Huiqing Yang's team recently developed electrochemical aptasensor for the specific detection of carcinoembryonic antigen (CEA) biomarker. In this investigation, the authors proposed the future of a new antifouling material MXC-Fe_3O_4-Ru on functional 2D nanomaterial-modified magnetic gold electrode (MGE). The ferrocene-modified carcinoembryonic antigen aptamer sequences were immobilized on the MGE/modified electrode surface with amido bond chemistry. Electrochemical signal of ferrocene decreases and $[Ru(NH3)_6]_3$ signal fixed on the electrode remains unchanged. The ratio of the electrochemical signals of ferrocene and $[Ru(NH3)_6]_3$ is proportional to the CEA concentration. Even in the complex samples, biosensors can reach high accuracy, selectivity and sensitivity for the detection of targets because of excellent antifouling performance and good conductivity [222].

João G. Pacheco et. al. developed an electrochemical sensor based on molecularly imprinted polymer to monitor breast cancer biomarker CA 15-3. In this work, the screen-printed gold electrodes were modified with the MIP recognition matrices. This MIP-based sensor was demonstrated in the electrolytic solution hexacyanoferrate (II/III) as redox probe for measuring the CA protein MIP-binding interactions. Interestingly, the peak current density increases with respect to the CA 15-3 concentration in adynamic range between 5 and 50 U mL^{-1} with the detection limit of 1.5 U mL^{-1}. The prepared MIP sensor is low cost and works efficiently for fast (15 min) analysis [223].

Other efforts have focused on the fabrication of label-free electrochemical aptasensor with the integration of microfluidic paper device for the specific detection of prostate specific antigen (PSA) in clinical samples. Screen-printed gold electrodes were fabricated with wax-printed technology. (Figure 4) Au-SPE surface is modified with the reduced graphene oxide and gold–thionine nanoparticle composites. Then, DNA aptamer was immobilized on top of the modified SPE. Afterward, the fabricated aptasensor was tested for the specific detection of PSA for diagnosing prostate cancer disease [224].

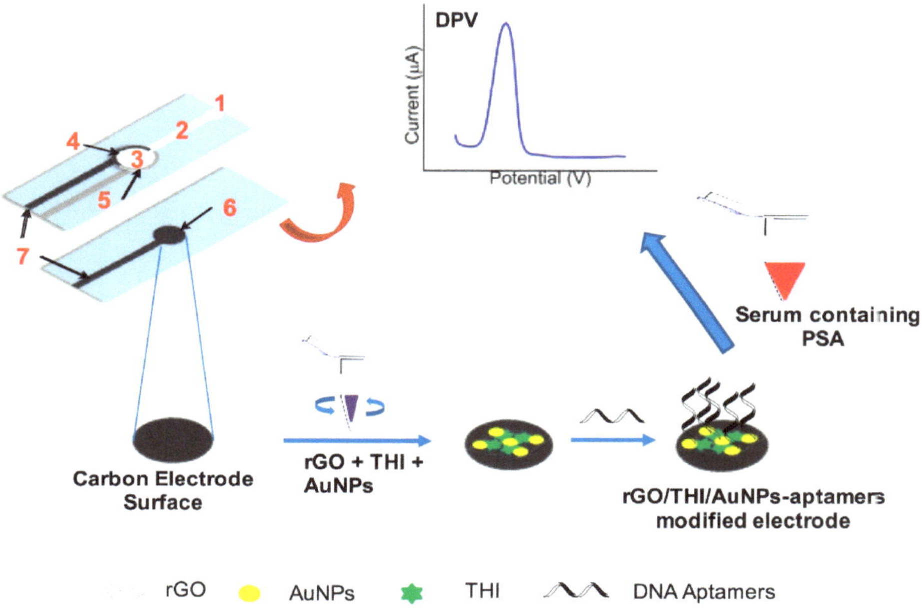

Figure 4. Schematic representation of microfluidic paper-based electrochemical aptasensor for the specific detection of prostate-specific antigen (PSA) in clinical samples. Reproduced with the permission. [224]. Copyright 2018, Elsevier.

In another report, an electrochemical aptasensor was proposed by *Leila Farzin's* research team for the specific detection of CA-125 cancer biomarker. The proposed sensing platform consists of polycrystalline nanofibers coupled with amidoxime-doped silver nanoparticles. The authors reported that hybrid nanomaterial-modified sensor surface helped in better immobilization of aptamer sequences and obtaining sensitive detection limits for CA-125 biomarker detection in ovarian cancer-infected patients [225].

Another interesting system that was established was a sandwich-type electrochemical aptasensor to measure carcinoembryonic antigen and cancer antigen 15-3(CA 15-3). The proposed sensing platform consists of a three-dimensional graphene gel embedded with gold nanoparticles (AuNPs/3DGH). This biosensing transducing layer helps in better immobilization of the redox-labelled aptamer sequences. Affinity interaction between aptamers and respective cancer biomarkers (CEA and CA 15-3) was recorded with square wave voltammetry methods by measuring the change in redox probe electroactive signals. This proposed sandwich aptasesing assay exhibited good limit of detection value in nanomolar linear range. The obtained results are comparable with the ELISA method [226].

In another report, a DNA nano-tweezer-based electrochemical sensor was developed for sensing and specific detection carcinoembryonic antigen biomarker. (Figure 5) DNA nano-tweezer is a DNA nanomachine used to enhance the sensing performance of electrochemical biosensor. Here, three-dimensional DNA nanomachine possesses the more active sites that could help in enhancement of the competence of reaction. This is the first kind of 3D DNA nanoprobe used in electrochemical sensing platform to obtain stable and reproducible results. Sensor exhibited good electroanalytical performance towards the target biomarker (CEA) with a detection limit of 4.88 fg mL^{-1} [227].

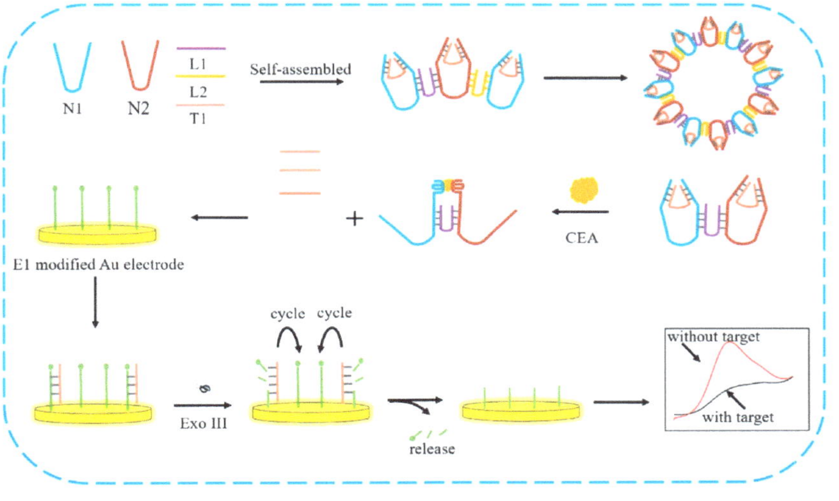

Figure 5. Schematic representation of label-free 3D DNA nanoprobe DNA tweezers-based electrochemical sensor for the detection of carcinoembryonic antigen (CEA) biomarker. Reproduced with the permission [227]. Copyright 2018, Elsevier.

Some other designs have been recently proposed by Navid Taheri et. al. for multiplexed determination of biomarkers alpha-fetoprotein (AFP) and carcinoembryonic antigen (CEA). Here, the sensing assay consisted of an electropolymerized polypyrrole conducting polymer, methyl orange layer and a DMIP layer on FTO surface. The target biomarker and MIP interactions were recorded with the electrochemical impedance spectroscopy by measuring impedance values. The sensor exhibited promising results in the dynamic range of 5–104 and 10–104 pg mL^{-1} and detection limits of 1.6 and 3.3 pg mL^{-1} for CEA and AFP, respectively [228].

M. Samira et al. developed a novel method for ovarian cancer antigen detection by ultrasensitive flexible aptasensor based on functionalized CNT-reduced graphene oxide nanocomposite. (Figure 6) Reduced graphene oxide film was prepared by modified Hummer's method. The CA 125 ssDNA aptamer sequences were immobilized on MWCNTs surface by amide bond formation. Then, fabrication of rGO was performed on the polymethyl methacrylate (PMMA) surface by using the polishing method. Here, gold source and drain electrode deposited on graphene film, and the surface of graphene was modified with MWCNTs/CA 125 aptamer through π-π interaction. Overall, this technique specifically and selectively detects the CA 125 biomarker from serum [208].

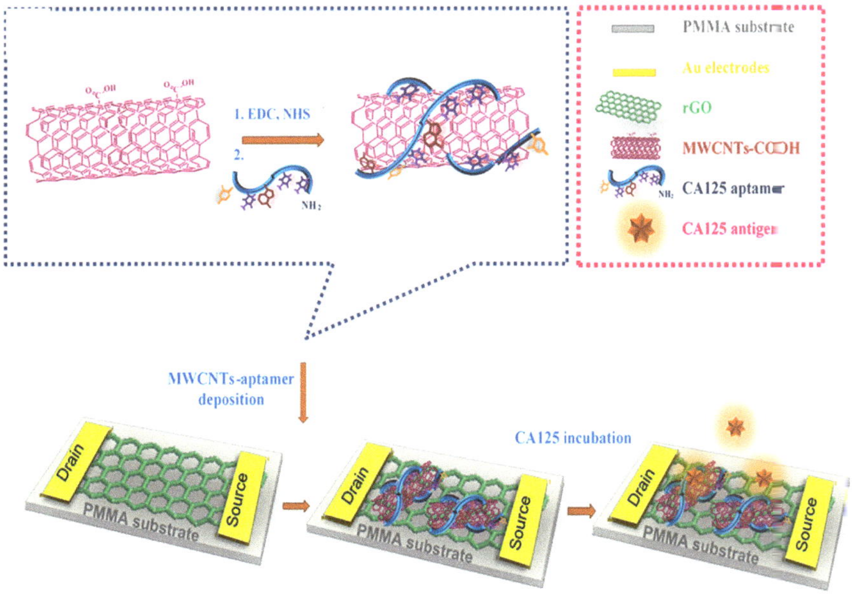

Figure 6. Schematic representation of rGO-FET-based electrochemical aptasensor for the specific detection of CA 125 ssDNA biomarker. Reproduced with the permission [208]. Copyright 2018, Elsevier.

Table 3 summarizes the recently reported literature related to the electrochemical biosensor for the detection of various cancer biomarkers based on various recognition matrices for the detection of cancer biomarkers. The reports are clearly classified and assembled in the table with respect to their recognition matrices, dynamic working ranges and limits of detection.

Table 3. Electrochemical sensing methodologies for cancer biomarker detection.

No.	Biomarker	Recognition	Method	Linear Range	LOD	Ref.
1	AFP	PtNPs/GO-COOH	SWV	3.0–30 ng mL^{-1}	1.22 ng mL^{-1}	[229]
2	AFP	FTO/PPy-MO DMIP	EIS	10–10^4 pg mL^{-1}	[illegible].3 pg mL^{-1}	[228]
3	CA 19-9	Au-SPE/TH	DPV	0.010–10 U/mL	-	[230]
4	CA 19-9	1DMoS$_2$ nanorods/LiNb$_3$O$_8$ and AuNPs@POM	DPV	0.1–10.0 µU mL^{-1}	0.[illegible]30 µU mL^{-1}	[231]
5	CA 125	MIP@AuSPE	SWV	0.01 and 500 U/mL	0.1 U/mL	[232]
6	CA 125	ITO/Ag NPs–PAN-oxime NFs/aptamer/c-DNA–MB	DPV	0.01 to 350 UmL^{-1}	0.0042 UmL^{-1}	[225]
7	CA 125	Tb-MOF-on- Fe-MOF	EIS	1 × 10^2 – 1 × 10^5 cell·mL^{-1}	[illegible] cell·mL^{-1}	[233]
8	CEA	MXC-Fe$_3$O$_4$-Ru	DPV	1 pg/mL–1 µg/mL	[illegible].62 pg/mL	[222]
9	CEA	IEC-BA	DPV	-	4.88 fg mL^{-1}	[227]
10	HER2	SPCE-MWCNT/AuNP	LSV	7.5 and 50 ng/mL	[illegible].16 ng/mL	[234]
11	HER2	polycytosine DNA (dC$_{20}$)	SWV	0.001 – 1 ng/mL	0.5 pg/mL	[235]
12	HER2	GCE/PEDOT/Gel/Ab/HER2	DPV	0.1 ng mL^{-1}–1.0 µg mL^{-1}	[illegible]5 pg mL^{-1}	[236]
13	HER2	MIP-AuSPE	DPV	10–70 ng/mL	1.6 ng/L	[237]
14	PSA	AuNPs/rGO/THI-aptamer	DPV	0.05 to 200 ng mL^{-1}	[illegible]0 pg mL^{-1}	[224]
15	PSA	aptamer PSAG-1	EIS	0.64–62.5 ng/mL	-	[238]

Table 3. *Cont.*

No.	Biomarker	Recognition	Method	Linear Range	LOD	Ref.
16	CA-15-3	AuNPs/3DGH	DPV	5.0×10^{-2}–100.0 U mL^{-1}	-	[226]
17	CA-15-3	MIP/Au-SPE	DPV	5–50 U mL^{-1}	1.5 U mL^{-1}	[223]
18	CA-15-3	CysA/Au NSs/GQDs	SWV	0.16–125 U/mL	0.11 U/ml	[239]
19	NMP 22	Cu-MOFs@SiO2@AgNPs	DPV	0.1 pg·mL^{-1}–1000 ng·mL^{-1}	33.33 fg·mL^{-1}	[240]
20	NMP 22	AuNPs-PtNPs-MOFs	DPV	0.005 ng·mL^{-1} –20 ng·mL^{-1}	1.7 pg·mL^{-1}	[241]
21	NMP 22	Co-MOFs/CuAu NWs	Amperometric	0.1 pg mL^{-1}–1 ng mL^{-1}	33 fg mL^{-1}	[242]

DMIP—dual-template molecularly imprinted polymer; TH—thionine; IEC-BA—ingenious electrochemical aptamer biosensor; PSA—prostate-specific antigen; AuNPs/3DGH—gold nanoparticle three-dimensional graphene hydrogel.

5. Conclusions and Future Perspectives

Overall, this review summarizes the importance of cancer biomarker detection, cancer-causing environments, traditional available screening biomarkers for cancer diagnosis, and novel cancer biomarkers and their advantages over traditional biomarkers. Recent developments in analytical diagnostic strategies including electrochemical and optical transduction methods for cancer biomarkers screening are also addressed. A summary of different cancer biomarkers available in the literature for the detection of cancer diseases are displayed in Table 1, including their respective cancer diseases and their advantages. Summaries of optical and electrochemical screening methods based on various recognition matrices for the detection of cancer biomarkers are assembled in their respective tables.

Cancer biomarker biosensing assay development consists of several critical challenges, including biofluid separation, real sample analysis, sensitivity, multiplex detection and integration of miniaturized instrumentation [243]. Real sample analysis is the particularly significant challenge; we are aiming to detect the target analyte in presence of several protein molecules. To achieve this, we need to integrate the device with the micro/nanofluidic devices and use different coating layers to protect the assay reading from biofouling studies. A recent publication has explained in detailed manner the way in which the fouling occurs, ways to overcome this problem by using different anti-biofouling coatings, as well as the effect of the fouling on electrocatalytic responses [244]. Finally, the device needs to be integrated with the miniaturized instrumentations, and a specific app/software need to be developed to record recognition element and biomarker interactions. The recent review exclusively discussed the integration of biosensing strategies with the electronics devices and wirelessly operated mobile phones for point-of-care diagnostic applications [245]. These sensor integration challenges could be overcome with the help of interdisciplinary approaches, specifically through collaborations between chemistry, electronics, computer, and nanofabrication experts. This review could be help to the early career researchers who are working in the domains of chemistry, biotechnology, nanotechnology, cell biology and biosensors. Overall, the current review provides new insights to the researchers to develop novel biosensing methodologies for the detection of various cancer biomarkers.

Funding: Funding source—Department of Science and Technology, Science and Engineering Research Board (DST-SERB/RJF/2021/000113, Ramanujan Award), New Delhi, India, and the Indian Institute of Technology Palakkad for financial support and research facilities.

Institutional Review Board Statement: Not applicable.

Informed Consent Statement: Not applicable.

Data Availability Statement: Not applicable.

Acknowledgments: K. Yugender Goud would like to thank the Department of Science and Technology, Science and Engineering Research Board (DST-SERB/RJF/2021/000113, Ramanujan Award), New Delhi, India, and the Indian Institute of Technology Palakkad for financial support and research facilities.

Conflicts of Interest: The authors declare no conflict of interest.

Abbreviations

AFP	Alpha-fetoprotein
PSA	Prostate-specific antigen
RCAS1	Receptor-binding cancer antigen
CA 15-3	Cancer antigen 15-3
CT antigen	Cancer–testis antigen
CA 125	Cancer antigen 125
CA 19-9	Cancer antigen 19-9
Nse	Neuron-specific enolase
Tdt	Terminal deoxynucleotidyl transferase
CYFRA21-1	Cytokeratin-19 fragments
Carcinoma	Epithelial cell cancer
Sarcoma	Connective tissue/bone cancer
Lymphoma	Lymphatic system cancer
Myeloma	Plasma cell cancer
Leukemia	Blood cancer
BRCA1	Breast cancer gene 1
BRCA2	Breast cancer gene 2

References

1. Siegel, R.L.; Miller, K.D.; Fuchs, H.E.; Jemal, A. Cancer Statistics, 2022. *CA A Cancer J. Clin.* **2022**, *72*, 7–33. [CrossRef]
2. Siegel, R.L.; Miller, K.D.; Jemal, A. Cancer Statistics, 2020. *CA A Cancer J. Clin.* **2020**, *70*, 7–30. [CrossRef]
3. Yabroff, K.R.; Wu, X.C.; Negoita, S.; Stevens, J.; Coyle, L.; Zhao, J.; Mumphrey, B.J.; Jemal, A.; Ward, K.C. Association of the COVID-19 Pandemic with Patterns of Statewide Cancer Services. *J. Natl. Cancer Inst.* **2022**, *114*, 907–909. [CrossRef]
4. Patel, A. Benign vs Malignant Tumors. *JAMA Oncol.* **2020**, *6*, 1488. [CrossRef]
5. WHO Cancer and Carcinogens. 2022. Available online: https://www.who.int/news-room/fact-sheets/detail/cancer) (accessed on 11 November 2022).
6. Croswell, J.M.; Ransohoff, D.F.; Kramer, B.S. Principles of Cancer Screening: Lessons from History and Study Design Issues. *Semin. Oncol.* **2010**, *37*, 202–215. [CrossRef]
7. Hamilton, W.; Walter, F.M.; Rubin, G.; Neal, R.D. Improving early diagnosis of symptomatic cancer. *Nat. Rev. Clin. Oncol.* **2016**, *13*, 740–749. [CrossRef]
8. Patel, J.; Patel, P. Biosensors and Biomarkers: Promising Tools for Cancer Diagnosis. *Int. J. Biosens. Bioelectron.* **2017**, *3*, 00072. [CrossRef]
9. Neal, R.D.; Tharmanathan, P.; France, B.; Din, N.U.; Cotton, S.; Fallon-Ferguson, J.; Hamilton, W.; Hendry, A.; Hendry, M.; Lewis, R.; et al. Is Increased Time to Diagnosis and Treatment in Symptomatic Cancer Associated with Poorer Outcomes? Systematic Review. *Br. J. Cancer* **2015**, *112*, S92–S107. [CrossRef]
10. Summerfield, C.; Smith, L.; Todd, O.; Renzi, C.; Lyratzopoulos, G.; Neal, R.D.; Jones, D. The Effect of Older Age and Frailty on the Time to Diagnosis of Cancer: A Connected Bradford Electronic Health Records Study. *Cancers* **2022**, *14*, 5666. [CrossRef] [PubMed]
11. Elliss-Brookes, L.; McPhail, S.; Ives, A.; Greenslade, M.; Shelton, J.; Hiom, S.; Richards, M. Routes to Diagnosis for Cancer—Determining the Patient Journey Using Multiple Routine Data Sets. *Br. J. Cancer* **2012**, *107*, 1220–1226. [CrossRef] [PubMed]
12. Whitaker, K. Earlier Diagnosis: The Importance of Cancer Symptoms. *Lancet Oncol.* **2020**, *21*, 6–8. [CrossRef] [PubMed]
13. Kennedy, M.P.T.; Cheyne, L.; Darby, M.; Plant, P.; Milton, R.; Robson, J.M.; Gill, A.; Malhotra, P.; Ashford-Turner, V.; Rodger, K.; et al. Lung Cancer Stage-Shift Following a Symptom Awareness Campaign. *Thorax* **2018**, *73*, 1128–1136. [CrossRef] [PubMed]
14. Power, E.; Wardle, J. Change in Public Awareness of Symptoms and Perceived Barriers to Seeing a Doctor Following Be Clear on Cancer Campaigns in England. *Br. J. Cancer* **2015**, *112*, S22–S26. [CrossRef] [PubMed]
15. Loud, J.T.; Murphy, J. Cancer Screening and Early Detection in the 21 St Century. *Semin. Oncol. Nurs.* **2017**, *33*, 121–128. [CrossRef]
16. Hewitt, K.; Son, J.; Glencer, A.; Borowsky, A.D.; Cooperberg, M.R.; Esserman, L.J. The Evolution of Our Understanding of the Biology of Cancer Is the Key to Avoiding Overdiagnosis and Overtreatment. *Cancer Epidemiol. Biomark. Prev.* **2020**, *29*, 2463–2474. [CrossRef]

17. Loomans-Kropp, H.A.; Umar, A. Cancer Prevention and Screening: The next Step in the Era of Precision Medicine. *Npj Precis. Oncol.* **2019**, *3*, 3. [CrossRef] [PubMed]
18. Shieh, Y.; Eklund, M.; Madlensky, L.; Sawyer, S.D.; Thompson, C.K.; Fiscalini, A.S.; Ziv, E.; Van't Veer, L.J.; Esserman, L.J.; Tice, J.A. Breast Cancer Screening in the Precision Medicine Era: Risk-Based Screening in a Population-Based Trial. *J. Natl. Cancer Inst.* **2017**, *109*, djw290. [CrossRef]
19. Esserman, L.J. The WISDOM Study: Breaking the Deadlock in the Breast Cancer Screening Debate. *Npj Breast Cancer* **2017**, *3*, 34. [CrossRef]
20. Henley, S.J.; King, J.B.; German, R.R.; Richardson, L.C.; Plescia, M.; Centers for Disease Control and Prevention (CDC). Surveillance of Screening-Detected Cancers. *MMWR Surveill. Summ.* **2010**, *59*, 1–25.
21. Negoita, S.; Feuer, E.J.; Mariotto, A.; Cronin, K.A.; Petkov, V.I.; Hussey, S.K.; Benard, V.; Henley, S.J.; Anderson, R.N.; Fedewa, S.; et al. Annual Report to the Nation on the Status of Cancer, Part II: Recent Changes in Prostate Cancer Trends and Disease Characteristics. *Cancer* **2018**, *124*, 2801–2814. [CrossRef]
22. Elmore, L.W.; Greer, S.F.; Daniels, E.C.; Saxe, C.C.; Melner, M.H.; Krawiec, G.M.; Cance, W.G.; Phelps, W.C. Blueprint for Cancer Research: Critical Gaps and Opportunities. *CA A Cancer J. Clin.* **2021**, *71*, 107–139. [CrossRef]
23. Schiffman, J.D.; Fisher, P.G.; Gibbs, P. Early Detection of Cancer: Past, Present, and Future. *Am. Soc. Clin. Oncol. Educ. Book* **2015**, *35*, 57–65. [CrossRef] [PubMed]
24. Saini, S.; Stark, D.D.; Brady, T.J.; Wittenberg, J.; Ferrucci, J.T. Dynamic Spin-Echo MRI of Liver Cancer Using Gadolinium-DTPA: Animal Investigation. *AJR Am. J. Roentgenol.* **1986**, *147*, 357–362. [CrossRef] [PubMed]
25. Lyman, D.F.; Bell, A.; Black, A.; Dingerdissen, H.; Cauley, E.; Gogate, N.; Liu, D.; Joseph, A.; Kahsay, R.; Crichton, D.J.; et al. Modeling and Integration of N-Glycan Biomarkers in a Comprehensive Biomarker Data Model. *Glycobiology* **2022**, *32*, 855–870. [CrossRef]
26. Alimirzaie, S.; Bagherzadeh, M.; Akbari, M.R. Liquid Biopsy in Breast Cancer: A Comprehensive Review. *Clin. Genet.* **2019**, *95*, 643–660. [CrossRef] [PubMed]
27. Nikanjam, M.; Kato, S.; Kurzrock, R. Liquid Biopsy: Current Technology and Clinical Applications. *J. Hematol. Oncol.* **2022**, *15*, 1–14. [CrossRef]
28. Chen, X.; Gole, J.; Gore, A.; He, Q.; Lu, M.; Min, J.; Yuan, Z.; Yang, X.; Jiang, Y.; Zhang, T.; et al. Non-Invasive Early Detection of Cancer Four Years before Conventional Diagnosis Using a Blood Test. *Nat. Commun.* **2020**, *11*, 3475. [CrossRef] [PubMed]
29. Hall, I.J.; Tangka, F.K.L.; Sabatino, S.A.; Thompson, T.D.; Graubard, B.I.; Breen, N. Patterns and Trends in Cancer Screening in the United States. *Prev. Chronic Dis.* **2018**, *15*, 170465. [CrossRef]
30. Pulumati, A.; Pulumati, A.; Dwarakanath, B.S.; Verma, A.; Papineni, R.V.L. Technological advancements in cancer diagnostics: Improvements and limitations. *Cancer Rep.* **2023**, *6*, e1764. [CrossRef]
31. Gamble, P.; Jaroensri, R.; Wang, H.; Tan, F.; Moran, M.; Brown, T.; Flament-Auvigne, I.; Rakha, E.A.; Toss, M.; Dabbs, D.J.; et al. Determining Breast Cancer Biomarker Status and Associated Morphological Features Using Deep Learning. *Commun. Med.* **2021**, *1*, 14. [CrossRef]
32. Zygulska, A.L.; Pierzchalski, P. Novel Diagnostic Biomarkers in Colorectal Cancer. *Int. J. Mol. Sci.* **2022**, *23*, 852. [CrossRef] [PubMed]
33. Porzycki, P.; Ciszkowicz, E. Modern Biomarkers in Prostate Cancer Diagnosis. *Cent. Eur. J. Urol.* **2020**, *73*, 300–306. [CrossRef]
34. Atallah, G.A.; Aziz, N.H.A.; Teik, C.K.; Shafiee, M.N.; Kampan, N.C. New Predictive Biomarkers for Ovarian Cancer. *Diagnostics* **2021**, *11*, 465. [CrossRef] [PubMed]
35. Molina, M.A.; Diatricch, L.C.; Quintana, M.C.; Melchers, W.J.G.; Andralojc, K.M. Cervical Cancer Risk Profiling: Molecular Biomarkers Predicting the Outcome of HrHPV Infection. *Expert Rev. Mol. Diagn.* **2020**, *20*, 1099–1120. [CrossRef]
36. Hussain, T.; Nguyen, Q.T. Molecular Imaging for Cancer Diagnosis and Surgery. *Adv. Drug Deliv. Rev.* **2014**, *66*, 90. [CrossRef]
37. Coleman, R.; Finkelstein, D.M.; Barrios, C.; Martin, M.; Iwata, H.; Hegg, R.; Glaspy, J.; Periañez, A.M.; Tonkin, K.; Deleu, I.; et al. Adjuvant Denosumab in Early Breast Cancer (D-CARE): An International, Multicentre, Randomised, Controlled, Phase 3 Trial. *Lancet Oncol.* **2020**, *21*, 60–72. [CrossRef]
38. Taouli, B.; Beer, A.J.; Chenevert, T.; Collins, D.; Lehman, C.; Matos, C.; Padhani, A.R.; Rosenkrantz, A.B.; Shukla-Dave, A.; Sigmund, E.; et al. Diffusion-Weighted Imaging Outside the Brain: Consensus Statement From an ISMRM-Sponsored Workshop. *J. Magn. Reason. Imaging* **2016**, *44*, 521–540. [CrossRef]
39. Padhani, A.R.; Liu, G.; Mu-Koh, D.; Chenevert, T.L.; Thoeny, H.C.; Takahara, T.; Dzik-Jurasz, A.; Ross, B.D.; Van Cauteren, M.; Collins, D.; et al. Diffusion-Weighted Magnetic Resonance Imaging as a Cancer Biomarker: Consensus and Recommendations. *Neoplasia* **2009**, *1*, 102–125. [CrossRef] [PubMed]
40. Altintas, Z.; Tothill, I. Biomarkers and Biosensors for the Early Diagnosis of Lung Cancer. *Sens. Actuators B Chem.* **2013**, *188*, 988–998. [CrossRef]
41. del Sol, A.; Balling, R.; Hood, L.; Galas, D. Diseases as Network Perturbations. *Curr. Opin. Biotechnol.* **2010**, *21*, 566–571. [CrossRef]
42. Sarhadi, V.K.; Armengol, G. Molecular Biomarkers in Cancer. *Biomolecules* **2022**, *12*, 1021. [CrossRef]
43. Jayanthi, V.S.A.; Das, A.B.; Saxena, U. Recent Advances in Biosensor Development for the Detection of Cancer Biomarkers. *Biosens. Bioelectron.* **2017**, *91*, 15–23. [CrossRef] [PubMed]
44. Tothill, I.E. Biosensors for Cancer Markers Diagnosis. *Semin. Cell Dev. Biol.* **2009**, *20*, 55–62. [CrossRef]

45. Fong, Z.V.; Winter, J.M. Biomarkers in Pancreatic Cancer: Diagnostic, Prognostic, and Predictive. *Cancer J.* **2012**, *18*, 530–538. [CrossRef] [PubMed]
46. Basil, C.F.; Zhao, Y.; Zavaglia, K.; Jin, P.; Panelli, M.C.; Voiculescu, S.; Mandruzzato, S.; Lee, H.M.; Seliger, B.; Freedman, R.S.; et al. Common Cancer Biomarkers. *Cancer Res.* **2006**, *66*, 2953–2961. [CrossRef]
47. Chowdhury, S.G.; Ray, R.; Karmakar, P. Exosomal miRNAs—a diagnostic biomarker acting as a guiding light in the diagnosis of prostate cancer. *Funct. Integr. Genom.* **2023**, *23*, 23. [CrossRef]
48. Henry, N.L.; Hayes, D.F. Cancer Biomarkers. *Mol. Oncol.* **2012**, *6*, 140–146. [CrossRef]
49. Somerfield, M.R. Clinical Practice Guidelines for the Use of Tumor Markers in Breast and Colorectal Cancer. Adopted on May 17, 1996 by the American Society of Clinical Oncology. *J. Clin. Oncol.* **1996**, *14*, 2843–2877. [CrossRef]
50. Meindl, A. Comprehensive Analysis of 989 Patients with Breast or Ovarian Cancer Provides BRCA1 and BRCA2 Mutation Profiles and Frequencies for the German Population. *Int. J. Cancer* **2002**, *97*, 472–480. [CrossRef]
51. Easton, D.F.; Ford, D.; Bishop, D.T.; Haites, N.; Milner, B.; Allan, L.; Easton, D.F.; Ponder, B.A.J.; Peto, J.; Smith, S.; et al. Breast and Ovarian Cancer Incidence in BRCA1-Mutation Carriers. Breast Cancer Linkage Consortium. *Am. J. Hum. Genet.* **1995**, *56*, 265.
52. Grossman, D.C.; Curry, S.J.; Owens, D.K.; Bibbins-Domingo, K.; Caughey, A.B.; Davidson, K.W.; Doubeni, C.A.; Ebell, M.; Epling, J.W.; Kemper, A.R.; et al. Screening for Prostate Cancer USPreventive Services task force recommendation Statement. *JAMA-J. Am. Med. Assoc.* **2018**, *319*, 1901–1913. [CrossRef]
53. Lin, K.; Lipsitz, R.; Miller, T.; Janakiraman, S. Benefits and Harms of Prostate-Specific Antigen Screening for Prostate Cancer: An Evidence Update for the U.S. Preventive Services Task Force. *Ann. Intern. Med.* **2008**, *149*, 192–199. [CrossRef]
54. Calonge, N.; Petitti, D.B.; Dewitt, T.G.; Dietrich, A.J.; Gregory, K.D.; Harris, R.; Isham, G.J.; Lefevre, M.L.; Leipzig, R.; Loveland-Cherry, C.; et al. Screening for Prostate Cancer: U.S. Preventive Services Task Force Recommendation Statement. *Ann. Intern. Med.* **2008**, *149*, 185–191. [CrossRef]
55. Van't Veer, L.J.; Paik, S.; Hayes, D.F. Gene Expression Profiling of Breast Cancer: A New Tumor Marker. *J. Clin. Oncol. Off. J. Am. Soc. Clin. Oncol.* **2005**, *23*, 1631–1635. [CrossRef] [PubMed]
56. Paik, S.; Shak, S.; Tang, G.; Kim, C.; Baker, J.; Cronin, M.; Baehner, F.L.; Walker, M.G.; Watson, D.; Park, T.; et al. A Multigene Assay to Predict Recurrence of Tamoxifen-Treated, Node-Negative Breast Cancer. *N. Engl. J. Med.* **2004**, *351*, 2817–2826. [CrossRef] [PubMed]
57. Cohen, J. Detection and Localization of Surgically Resectable Cancers with a Multi-Analyte Blood Test. *Science* **2018**, *359*, 926–930. [CrossRef]
58. Cristofanilli, M.; Budd, G.T.; Ellis, M.J.; Stopeck, A.; Matera, J.; Miller, M.C.; Reuben, J.M.; Doyle, G.V.; Allard, W.J.; Terstappen, L.W.M.M.; et al. Circulating Tumor Cells, Disease Progression, and Survival in Metastatic Breast Cancer. *N. Engl. J. Med.* **2004**, *351*, 781–791. [CrossRef]
59. Zhao, Q.; Li, B.; Gao, Q.; Luo, Y.; Ming, L. Prognostic Value of Epithelial-Mesenchymal Transition Circulating Tumor Cells in Female Breast Cancer: A Meta-Analysis. *Front. Oncol.* **2022**, *12*, 1024783. [CrossRef] [PubMed]
60. Abe, S.; Kawai, K.; Nozawa, H.; Sasaki, K.; Murono, K.; Emoto, S.; Ozawa, T.; Yokoyama, Y.; Nagai, Y.; Anzai, H.; et al. Clinical Impact of Primary Tumor Sidedness and Sex on Unresectable Post-Recurrence Survival in Resected Pathological Stage II-III Colorectal Cancers: A Nationwide Multicenter Retrospective Study. *BMC Cancer* **2022**, *22*, 1–10. [CrossRef]
61. Allegra, C.J.; Jessup, J.M.; Somerfield, M.R.; Hamilton, S.R.; Hammond, E.H.; Hayes, D.F.; McAllister, P.K.; Morton, R.F.; Schilsky, R.L. American Society of Clinical Oncology Provisional Clinical Opinion: Testing for KRAS Gene Mutations in Patients with Metastatic Colorectal Carcinoma to Predict Response to Anti-Epidermal Growth Factor Receptor Monoclonal Antibody Therapy. *J. Clin. Oncol. Off. J. Am. Soc. Clin. Oncol.* **2009**, *27*, 2091–2096. [CrossRef]
62. Thuss-Patience, P.C.; Shah, M.A.; Ohtsu, A.; Van Cutsem, E.; Ajani, J.A.; Castro, H.; Mansoor, W.; Chung, H.C.; Bodoky, G.; Shitara, K.; et al. Trastuzumab Emtansine versus Taxane Use for Previously Treated HER2-Positive Locally Advanced or Metastatic Gastric or Gastro-Oesophageal Junction Adenocarcinoma (GATSBY): An International Randomised, Open-Label, Adaptive, Phase 2/3 Study. *Lancet Oncol.* **2017**, *18*, 640–653. [CrossRef] [PubMed]
63. Bang, Y.J.; Van Cutsem, E.; Feyereislova, A.; Chung, H.C.; Shen, L.; Sawaki, A.; Lordick, F.; Ohtsu, A.; Omuro, Y.; Satoh, T.; et al. Trastuzumab in Combination with Chemotherapy versus Chemotherapy Alone for Treatment of HER2-Positive Advanced Gastric or Gastro-Oesophageal Junction Cancer (ToGA): A Phase 3, Open-Label, Randomised Controlled Trial. *Lancet* **2010**, *376*, 687–697. [CrossRef] [PubMed]
64. Sawaki, M.; Taira, N.; Uemura, Y.; Saito, T.; Baba, S.; Kobayashi, K.; Kawashima, H.; Tsuneizumi, M.; Sagawa, N.; Bando, H.; et al. Adjuvant Trastuzumab without Chemotherapy for Treating Early HER2-Positive Breast Cancer in Older Patients: A Propensity Score-Adjusted Analysis of a Prospective Cohort Study. *Breast* **2022**, *66*, 245–254. [CrossRef] [PubMed]
65. Piccart-Gebhart, M.J.; Procter, M.; Leyland-Jones, B.; Goldhirsch, A.; Untch, M.; Smith, I.; Gianni, L.; Baselga, J.; Bell, R.; Jackisch, C.; et al. Trastuzumab after Adjuvant Chemotherapy in HER2-Positive Breast Cancer. *N. Engl. J. Med.* **2005**, *353*, 1659–1672. [CrossRef]
66. Perez, E.A.; Ballman, K.V.; Tenner, K.S.; Thompson, E.A.; Badve, S.S.; Bailey, H.; Baehner, F.L. Association of Stromal Tumor-Infiltrating Lymphocytes with Recurrence-Free Survival in the N9831 Adjuvant Trial in Patients with Early-Stage HER2-Positive Breast Cancer. *JAMA Oncol.* **2016**, *2*, 56–64. [CrossRef]

67. Romond, E.H.; Perez, E.A.; Bryant, J.; Suman, V.J.; Geyer, C.E.; Davidson, N.E.; Tan-Chiu, E.; Martino, S.; Paik, S.; Kaufman, P.A.; et al. Trastuzumab plus Adjuvant Chemotherapy for Operable HER2-Positive Breast Cancer. *N. Engl. J. Med.* **2005**, *353*, 1673–1684. [CrossRef]
68. Liu, J.; Gan, M.; Lin, Z.; Deng, Q.; Deng, J.; Zeng, B.; Shi, Y.; Ming, J. Clinical Features and Prognosis Analysis of Hormone Receptor-Positive, HER2-Negative Breast Cancer with Differential Expression Levels of Estrogen and Progesterone Receptors: A 10-Year Retrospective Study. *Breast J.* **2022**, *2022*, 5469163. [CrossRef]
69. Abe, O.; Abe, R.; Enomoto, K.; Kikuchi, K.; Koyama, H.; Masuda, H.; Nomura, Y.; Ohashi, Y.; Sakai, K.; Sugimachi, K.; et al. Relevance of Breast Cancer Hormone Receptors and Other Factors to the Efficacy of Adjuvant Tamoxifen: Patient-Level Meta-Analysis of Randomised Trials. *Lancet* **2011**, *378*, 771–784. [CrossRef]
70. Teutsch, S.M.; Bradley, L.A.; Palomaki, G.E.; Haddow, J.E.; Piper, M.; Calonge, N.; Dotson, W.D.; Douglas, M.P.; Berg, A.O. The Evaluation of Genomic Applications in Practice and Prevention (EGAPP) Initiative: Methods of the EGAPP Working Group. *Genet. Med. Off. J. Am. Coll. Med. Genet.* **2009**, *11*, 3–14. [CrossRef]
71. Hanash, S.M.; Pitteri, S.J.; Faca, V.M. Mining the Plasma Proteome for Cancer Biomarkers. *Nature* **2008**, *452*, 571–579. [CrossRef]
72. Mighton, C.; Shickh, S.; Aguda, V.; Krishnapillai, S.; Adi-Wauran, E.; Bombard, Y. From the Patient to the Population: Use of Genomics for Population Screening. *Front. Genet.* **2022**, *13*, 893832. [CrossRef]
73. Kohsaka, S.; Tada, Y.; Ando, M.; Nakaguro, M.; Shirai, Y.; Ueno, T.; Kojima, S.; Hirai, H.; Saigusa, N.; Kano, S.; et al. Identification of Novel Prognostic and Predictive Biomarkers in Salivary Duct Carcinoma via Comprehensive Molecular Profiling. *Npj Precis. Oncol.* **2022**, *6*, 82. [CrossRef]
74. Wang, N.; Zhang, H.; Li, D.; Jiang, C.; Zhao, H.; Teng, Y. Identification of Novel Biomarkers in Breast Cancer via Integrated Bioinformatics Analysis and Experimental Validation. *Bioengineered* **2021**, *12*, 12431. [CrossRef]
75. Wu, M.; Li, Q.; Wang, H. Identification of Novel Biomarkers Associated with the Prognosis and Potential Pathogenesis of Breast Cancer via Integrated Bioinformatics Analysis. *Technol. Cancer Res. Treat.* **2021**, *20*, 1533033821992081. [CrossRef]
76. Li, C.-J.; Lin, L.-T.; Chu, P.-Y.; Chiang, A.-J.; Tsai, H.-W.; Chiu, Y.-H.; Huang, M.-S.; Wen, Z.-H.; Tsui, K.-H.; Li, C.-J.; et al. Identification of Novel Biomarkers and Candidate Drug in Ovarian Cancer. *J. Pers. Med.* **2021**, *11*, 316. [CrossRef]
77. Islam, S.U.; Ahmed, M.B.; Ahsan, H.; Lee, Y.S. Role of Biomarkers in Personalized Medicine. In *Cancer Biomarkers in Diagnosis and Therapeutics*; Springer: Singapore, 2022; pp. 249–275. [CrossRef]
78. Janjua, K.A.; Shahzad, R.; Shehzad, A. Development of Novel Cancer Biomarkers for Diagnosis and Prognosis. In *Cancer Biomarkers in Diagnosis and Therapeutics*; Springer: Singapore, 2022; pp. 277–343. [CrossRef]
79. Bohunicky, B.; Mousa, S.A. Biosensors: The New Wave in Cancer Diagnosis. *Nanotechnol. Sci. Appl.* **2011**, *4*, 1–10. [CrossRef]
80. Prabhakar, B.; Shende, P.; Augustine, S. Current Trends and Emerging Diagnostic Techniques for Lung Cancer. *Biomed. Pharmacother.* **2018**, *106*, 1586–1599. [CrossRef]
81. Chen, M.; Wu, D.; Tu, S.; Yang, C.; Chen, D.J.; Xu, Y. A Novel Biosensor for the Ultrasensitive Detection of the LncRNA Biomarker MALAT1 in Non-Small Cell Lung Cancer. *Sci. Rep.* **2021**, *11*, 3666. [CrossRef] [PubMed]
82. Reddy, K.K.; Bandal, H.; Satyanarayana, M.; Goud, K.Y.; Gobi, K.V.; Jayaramudu, T.; Amalraj, J.; Kim, H. Recent Trends in Electrochemical Sensors for Vital Biomedical Markers Using Hybrid Nanostructured Materials. *Adv. Sci.* **2020**, *7*, 1902980. [CrossRef]
83. Ferraro, S.; Bussetti, M.; Bassani, N.; Rossi, R.S.; Incarbone, G.P.; Bianchi, F.; Maggioni, M.; Runza, L.; Ceriotti, F.; Panteghini, M. Definition of Outcome-Based Prostate-Specific Antigen (PSA) Thresholds for Advanced Prostate Cancer Risk Prediction. *Cancers* **2021**, *13*, 3381. [CrossRef]
84. Thompson, I.M.; Ankerst, D.P. Prostate-Specific Antigen in the Early Detection of Prostate Cancer. *CMAJ Can. Med. Assoc. J.* **2007**, *176*, 1853–1858. [CrossRef]
85. Wu, M.; Liu, H.; Liu, Z.; Liu, C.; Zhang, A.; Li, N. Analysis of Serum Alpha-Fetoprotein (AFP) and AFP-L3 Levels by Protein Microarray. *J. Int. Med. Res.* **2018**, *46*, 4297–4305. [CrossRef] [PubMed]
86. Yuan, Q.; Song, J.; Yang, W.; Wang, H.; Huo, Q.; Yang, J.; Yu, X.; Liu, Y.; Xu, C.; Bao, H. The Effect of CA125 on Metastasis of Ovarian Cancer: Old Marker New Function. *Oncotarget* **2017**, *8*, 50015–50022. [CrossRef] [PubMed]
87. Meyer, T.; Rustin, G.J. Role of Tumour Markers in Monitoring Epithelial Ovarian Cancer. *Br. J. Cancer* **2000**, *82*, 1535–1538. [CrossRef]
88. Chianca, M.; Panichella, G.; Fabiani, I.; Giannoni, A.; L'Abbate, S.; Aimo, A.; Del Franco, A.; Vergaro, G.; Grigoratos, C.; Castiglione, V.; et al. Bidirectional Relationship Between Cancer and Heart Failure: Insights on Circulating Biomarkers. *Front. Cardiovasc. Med.* **2022**, *9*, 936654. [CrossRef]
89. Tuxen, M.K.; Sölétormos, G.; Dombernowsky, P. Tumor Markers in the Management of Patients with Ovarian Cancer. *Cancer Treat. Rev.* **1995**, *21*, 215–245. [CrossRef] [PubMed]
90. Smith, D.S.; Humphrey, P.A.; Catalona, W.J. The Early Detection of Prostate Carcinoma with Prostate Specific Antigen: The Washington University Experience. *Cancer* **1997**, *80*, 1852–1856. [CrossRef]
91. Bill-Axelson, A.; Holmberg, L.; Ruutu, M.; Häggman, M.; Andersson, S.-O.; Bratell, S.; Spångberg, A.; Busch, C.; Nordling, S.; Garmo, H.; et al. Radical Prostatectomy versus Watchful Waiting in Early Prostate Cancer. *N. Engl. J. Med.* **2005**, *352*, 1977–1984. [CrossRef]

92. Jaffee, E.M.; Van Dang, C.; Agus, D.B.; Alexander, B.M.; Anderson, K.C.; Ashworth, A.; Barker, A.D.; Bastani, R.; Bhatia, S.; Bluestone, J.A.; et al. Future Cancer Research Priorities in the USA: A Lancet Oncology Commission. *Lancet Oncol.* **2017**, *18*, e653–e706. [CrossRef]
93. National Cancer Institute. *Prostate-Specific Antigen (PSA) Test—NCI*; National Cancer Institute: Bethesda, MD, USA, 2022.
94. Kim, J.H.; Suh, Y.J.; Park, D.; Yim, H.; Kim, H.; Kim, H.J.; Yoon, D.S.; Hwang, K.S. Technological Advances in Electrochemical Biosensors for the Detection of Disease Biomarkers. *Biomed. Eng. Lett.* **2021**, *11*, 309–334. [CrossRef]
95. Dutsch-Wicherek, M.; Tomaszewska, R.; Lazar, A.; Wicherek, L.; Skladzien, J. The Association between RCAS1 Expression in Laryngeal and Pharyngeal Cancer and Its Healthy Stroma with Cancer Relapse. *BMC Cancer* **2009**, *9*, 35. [CrossRef]
96. Kazmierczak, W.; Lazar, A.; Tomaszewska, R.; Popiela, T.J.; Koper, K.; Wicherek, L.; Dutsch-Wicherek, M. Analysis of the Intensity of Immune Cell Infiltration and Immunoreactivity of RCAS1 in Diffuse Large B-Cell Lymphoma of the Palatine Tonsil and Its Microenvironment. *Cell Tissue Res.* **2015**, *361*, 823–831. [CrossRef]
97. Kubokawa, M.; Nakashima, M.; Yao, T.; Ito, K.I.; Harada, N.; Nawata, H.; Watanabe, T. Aberrant Intracellular Localization of RCAS1 Is Associated with Tumor Progression of Gastric Cancer. *Int. J. Oncol.* **2001**, *19*, 695–700. [CrossRef] [PubMed]
98. Naito, Y.; Okabe, Y.; Nagayama, M.; Nishinakagawa, T.; Taira, T.; Kawahara, A.; Hattori, S.; MacHida, K.; Ishida, Y.; Kaji, R.; et al. Accuracy of Differential Diagnosis for Pancreatic Cancer Is Improved in the Combination of RCAS1 and CEA Measurements and Cytology in Pancreatic Juice. *Med. Mol. Morphol.* **2011**, *44*, 86–92. [CrossRef] [PubMed]
99. Duffy, M.J. Serum Tumor Markers in Breast Cancer: Are They of Clinical Value? *Clin. Chem.* **2006**, *52*, 345–351. [CrossRef] [PubMed]
100. Hasan, D. Diagnostic Impact of CEA and CA 15-3 on Chemotherapy Monitoring of Breast Cancer Patients. *J. Circ. Biomark.* **2022**, *11*, 57–63. [CrossRef]
101. Tampellini, M.; Berruti, A.; Gerbino, A.; Buniva, T.; Torta, M.; Gorzegno, G.; Faggiuolo, R.; Cannone, R.; Farris, A.; Destefanis, M.; et al. Relationship between CA 15-3 Serum Levels and Disease Extent in Predicting Overall Survival of Breast Cancer Patients with Newly Diagnosed Metastatic Disease. *Br. J. Cancer* **1997**, *75*, 698–702. [CrossRef]
102. Lian, M.; Zhang, C.; Zhang, D.; Chen, P.; Yang, H.; Yang, Y.; Chen, S.; Hong, G. The Association of Five Preoperative Serum Tumor Markers and Pathological Features in Patients with Breast Cancer. *J. Clin. Lab. Anal.* **2019**, *33*, e22875. [CrossRef]
103. Bast, J.; Ravdin, P.; Hayes, D.F.; Bates, S.; Fritsche, H.J.; Jessup, J.M.; Kemeny, N.; Locker, G.Y.; Mennel, R.G.; Somerfield, M.R. 2000 Update of Recommendations for the Use of Tumor Markers in Breast and Colorectal Cancer: Clinical Practice Guidelines of the American Society of Clinical Oncology. *J. Clin. Oncol. Off. J. Am. Soc. Clin. Oncol.* **2001**, *19*, 1865–1878. [CrossRef]
104. Lin, M.J.; Svensson-Arvelund, J.; Lubitz, G.S.; Marabelle, A.; Melero, I.; Brown, B.D.; Brody, J.D. Cancer Vaccines: The next Immunotherapy Frontier. *Nat. Cancer* **2022**, *3*, 911–926. [CrossRef]
105. Chow, S.; Berek, J.S.; Dorigo, O. Development of Therapeutic Vaccines for Ovarian Cancer. *Vaccines* **2020**, *8*, 657. [CrossRef] [PubMed]
106. Jäger, E.; Chen, Y.T.; Drijfhout, J.W.; Karbach, J.; Ringhoffer, M.; Jäger, D.; Arand, M.; Wada, H.; Noguchi, Y.; Stockert, E.; et al. Simultaneous Humoral and Cellular Immune Response against Cancer-Testis Antigen NY-ESO-1: Definition of Human Histocompatibility Leukocyte Antigen (HLA)-A2-Binding Peptide Epitopes. *J. Exp. Med.* **1998**, *187*, 265–270. [CrossRef] [PubMed]
107. Tsuji, K.; Hamada, T.; Uenaka, A.; Wada, H.; Sato, E.; Isobe, M.; Asagoe, K.; Yamasaki, O.; Shiku, H.; Ritter, G.; et al. Induction of Immune Response against NY-ESO-1 by CHP-NY-ESO-1 Vaccination and Immune Regulation in a Melanoma Patient. *Cancer Immunol. Immunother.* **2008**, *57*, 1429–1437. [CrossRef]
108. Jäger, E.; Stockert, E.; Zidianakis, Z.; Chen, Y.T.; Karbach, J.; Jäger, D.; Arand, M.; Ritter, G.; Old, L.J.; Knuth, A. Humoral Immune Responses of Cancer Patients against "Cancer-Testis" Antigen NY-ESO-1: Correlation with Clinical Events. *Int. J. Cancer* **1999**, *84*, 506–510. [CrossRef]
109. Karbach, J.; Neumann, A.; Atmaca, A.; Wahle, C.; Brand, K.; Von Boehmer, L.; Knuth, A.; Bender, A.; Ritter, G.; Old, L.J.; et al. Efficient in Vivo Priming by Vaccination with Recombinant NY-ESO-1 Protein and CpG in Antigen Naive Prostate Cancer Patients. *Clin. Cancer Res. Off. J. Am. Assoc. Cancer Res.* **2011**, *17*, 861–870. [CrossRef]
110. Stockert, E.; Jäger, E.; Chen, Y.T.; Scanlan, M.J.; Gout, I.; Karbach, J.; Arand, M.; Knuth, A.; Old, L.J. A Survey of the Humoral Immune Response of Cancer Patients to a Panel of Human Tumor Antigens. *J. Exp. Med.* **1998**, *187*, 1349–1354. [CrossRef] [PubMed]
111. Imai, N.; Ikeda, H.; Shiku, H. [Targeting Cancer Antigen (MAGE-A4, NY-ESO -1) for Immunotherapy]. *Nihon Rinsho. Jpn. J. Clin. Med.* **2012**, *70*, 2125–2129.
112. Raza, A.; Merhi, M.; Inchakalody, V.P.; Krishnankutty, R.; Relecom, A.; Uddin, S.; Dermime, S. Unleashing the Immune Response to NY-ESO-1 Cancer Testis Antigen as a Potential Target for Cancer Immunotherapy. *J. Transl. Med. J. Transl. Med.* **2020**, *18*, 1–11. [CrossRef]
113. Babaier, A.; Mal, H.; Alselwi, W.; Ghatage, P. Low-Grade Serous Carcinoma of the Ovary: The Current Status. *Diagnostics* **2022**, *12*, 458. [CrossRef]
114. Neesham, D. Ovarian Cancer Screening. *Aust. Fam. Physician* **2007**, *36*, 126–128.
115. Menon, U.; Gentry-Maharaj, A.; Hallett, R.; Ryan, A.; Burnell, M.; Sharma, A.; Lewis, S.; Davies, S.; Philpott, S.; Lopes, A.; et al. Sensitivity and Specificity of Multimodal and Ultrasound Screening for Ovarian Cancer, and Stage Distribution of Detected

Cancers: Results of the Prevalence Screen of the UK Collaborative Trial of Ovarian Cancer Screening (UKCTOCS). *Lancet Oncol.* **2009**, *10*, 327–340. [CrossRef]
116. Wong, N.K.; Easton, R.L.; Panico, M.; Sutton-Smith, M.; Morrison, J.C.; Lattanzio, F.A.; Morris, H.R.; Clark, G.F.; Dell, A.; Patankar, M.S. Characterization of the Oligosaccharides Associated with the Human Ovarian Tumor Marker CA125. *J. Biol. Chem.* **2003**, *278*, 28619–28634. [CrossRef]
117. White, B.; Patterson, M.; Karnwal, S.; Brooks, C.L. Crystal Structure of a Human MUC16 SEA Domain Reveals Insight into the Nature of the CA125 Tumor Marker. *Proteins Struct. Funct. Bioinform.* **2022**, *90*, 1210–1218. [CrossRef]
118. Wang, Y.-S.; Ren, S.-F.; Jiang, W.; Lu, J.-Q.; Zhang, X.-Y.; Li, X.-P.; Cao, R.; Xu, C.-J. CA125-Tn ELISA Assay Improves Specificity of Pre-Operative Diagnosis of Ovarian Cancer among Patients with Elevated Serum CA125 Levels. *Ann. Transl. Med.* **2021**, *9*, 788. [CrossRef] [PubMed]
119. Seelenmeyer, C.; Wegehingel, S.; Lechner, J.; Nickel, W. The Cancer Antigen CA125 Represents a Novel Counter Receptor for Galectin-1. *J. Cell Sci.* **2003**, *116*, 1305–1318. [CrossRef] [PubMed]
120. Liang, L.; Fang, J.; Han, X.; Zhai, X.; Song, Y.; Lu, Y.; Zhang, Q.; Ma, R. Prognostic Value of CEA, CA19-9, CA125, CA724, and CA242 in Serum and Ascites in Pseudomyxoma Peritonei. *Front. Oncol.* **2021**, *11*, 594763. [CrossRef] [PubMed]
121. Li, B.; Li, Y.; Li, C.; Yang, J.; Liu, D.; Wang, H.; Xu, R.; Zhang, Y.; Wei, Q. An ultrasensitive split-type electrochemical immunosensor based on controlled-release strategy for detection of CA19-9. *Biosens. Bioelectron.* **2023**, *227*, 115180. [CrossRef]
122. Poruk, K.E.; Firpo, M.A.; Adler, D.G.; Mulvihill, S.J. Screening for Pancreatic Cancer: Why, How, and Who? *Ann. Surg.* **2013**, *257*, 17–26. [CrossRef]
123. Badheeb, M.; Abdelrahim, A.; Esmail, A.; Umoru, G.; Abboud, K.; Al-Najjar, E.; Rasheed, G.; Alkhulaifawi, M.; Abudayyeh, A.; Abdelrahim, M. Pancreatic Tumorigenesis: Precursors, Genetic Risk Factors and Screening. *Curr. Oncol.* **2022**, *29*, 8693–8719. [CrossRef] [PubMed]
124. Hatzakis, K.D.; Froudarakis, M.E.; Bouros, D.; Tzanakis, N.; Karkavitsas, N.; Siafakas, N.M. Prognostic Value of Serum Tumor Markers in Patients with Lung Cancer. *Respiration* **2002**, *69*, 25–29. [CrossRef]
125. Ferrigno, D.; Buccheri, G.; Giordano, C. Neuron-Specific Enolase Is an Effective Tumour Marker in Non-Small Cell Lung Cancer (NSCLC). *Lung Cancer* **2003**, *41*, 311–320. [CrossRef] [PubMed]
126. Gazdar, A.F.; Carney, D.N.; Nau, M.M.; Minna, J.D. Characterization of Variant Subclasses of Cell Lines Derived from Small Cell Lung Cancer Having Distinctive Biochemical, Morphological, and Growth Properties. *Cancer Res.* **1985**, *45*, 2924–2930. [PubMed]
127. Isgrò, M.A.; Bottoni, P.; Scatena, R. Neuron-Specifi c Enolase as a Biomarker: Biochemical and Clinical Aspects. *Adv. Exp. Med. Biol.* **2015**, *867*, 125–143. [CrossRef]
128. Tian, Z.; Liang, C.; Zhang, Z.; Wen, H.; Feng, H.; Ma, Q.; Liu, D.; Qiang, G. Prognostic Value of Neuron-Specific Enolase for Small Cell Lung Cancer: A Systematic Review and Meta-Analysis. *World J. Surg. Oncol.* **2020**, *18*, 1–8. [CrossRef] [PubMed]
129. Campagnari, F.; Bombardieri, E.; de Braud, F.; Baldini, L.; Maiolo, A.T. Terminal Deoxynucleotidyl Transferase, Tdt, as a Marker for Leukemia and Lymphoma Cells. *Int. J. Biol. Markers* **1987**, *2*, 31–42. [CrossRef] [PubMed]
130. Peng, Y.; Shang, H.; Zheng, Z.; Li, H.; Chen, W.; Xu, J. Self-activation of symbiotic multi-DNA machines transducing exponentially amplified fluorescence for ultrasensitively signaling of terminal deoxynucleotidyl transferase activity. *Sens. Actuators B. Chem* **2023**, *380*, 133400. [CrossRef]
131. Sur, M.; AlArdati, H.; Ross, C.; Alowami, S. TdT Expression in Merkel Cell Carcinoma: Potential Diagnostic Pitfall with Blastic Hematological Malignancies and Expanded Immunohistochemical Analysis. *Mod. Pathol.* **2007**, *20*, 1113–1120. [CrossRef]
132. Mathewson, R.C.; Kjeldsberg, C.R.; Perkins, S.L. Detection of Terminal Deoxynucleotidyl Transferase (TdT) in Nonhematopoietic Small Round Cell Tumors of Children. *Pediatr. Pathol. Lab. Med. J. Soc. Pediatr. Pathol. Affil. Int. Paediatr. Pathol. Assoc.* **1997**, *17*, 835–844.
133. Chapman, M.H.; Sandanayake, N.S.; Andreola, F.; Dhar, D.K.; Webster, G.J.; Dooley, J.S.; Pereira, S.P. Circulating CYFRA 21-1 Is a Specific Diagnostic and Prognostic Biomarker in Biliary Tract Cancer. *J. Clin. Exp. Hepatol.* **2011**, *1*, 6–12. [CrossRef]
134. Zhang, L.; Liu, D.; Li, L.; Pu, D.; Zhou, P.; Jing, Y.; Yu, H.; Wang, Y.; Zhu, Y.; He, Y.; et al. The Important Role of Circulating CYFRA21-1 in Metastasis Diagnosis and Prognostic Value Compared with Carcinoembryonic Antigen and Neuron-Specific Enolase in Lung Cancer Patients. *BMC Cancer* **2017**, *17*, 96. [CrossRef]
135. Dal Bello, M.G.; Filiberti, R.A.; Alama, A.; Orengo, A.M.; Mussap, M.; Coco, S.; Vanni, I.; Boccardo, S.; Rijavec, E.; Genova, C.; et al. The Role of CEA, CYFRA21-1 and NSE in Monitoring Tumor Response to Nivolumab in Advanced Non-Small Cell Lung Cancer (NSCLC) Patients. *J. Transl. Med.* **2019**, *17*, 74. [CrossRef]
136. Rapellino, M.; Niklinski, J.; Pecchio, F.; Furman, M.; Baldi, S.; Chyczewski, L.; Ruffini, E.; Chyczewska, E. CYFRA 21-1 as a Tumour Marker for Bronchogenic Carcinoma. *Eur. Respir. J.* **1995**, *8*, 407–410. [CrossRef]
137. Dall'Olio, F.G.; Abbati, F.; Facchinetti, F.; Massucci, M.; Melotti, B.; Squadrilli, A.; Buti, S.; Formica, F.; Tiseo, M.; Ardizzoni, A. CEA and CYFRA 21-1 as Prognostic Biomarker and as a Tool for Treatment Monitoring in Advanced NSCLC Treated with Immune Checkpoint Inhibitors. *Ther. Adv. Med. Oncol.* **2020**, *12*, 1758835920952994. [CrossRef]
138. Hayakawa, T.; Naruse, S.; Kitagawa, M.; Ishiguro, H.; Kondo, T.; Kurimoto, K.; Fukushima, M.; Takayama, T.; Horiguchi, Y.; Kuno, N.; et al. A Prospective Multicenter Trial Evaluating Diagnostic Validity of Multivariate Analysis and Individual Serum Marker in Differential Diagnosis of Pancreatic Cancer from Benign Pancreatic Diseases. *Int. J. Pancreatol. Off. J. Int. Assoc. Pancreatol.* **1999**, *25*, 23–29. [CrossRef]

139. Chen, G.; Yu, M.; Cao, J.; Zhao, H.; Dai, Y.; Cong, Y.; Qiao, G. Identification of Candidate Biomarkers Correlated with Poor Prognosis of Breast Cancer Based on Bioinformatics Analysis. *Bioengineered* **2021**, *12*, 5149–5161. [CrossRef]
140. Kumar, S.; Mohan, A.; Guleria, R. Biomarkers in Cancer Screening, Research and Detection: Present and Future: A Review. *Biomarkers* **2008**, *11*, 385–405. [CrossRef]
141. Chatterjee, S.K.; Zetter, B.R. Cancer Biomarkers: Knowing the Present and Predicting the Future. *Future Oncol.* **2005**, *1*, 37–50. [CrossRef]
142. Hayes, D.F. Biomarker Validation and Testing. *Mol. Oncol.* **2015**, *9*, 960–966. [CrossRef]
143. Hayes, D.F.; Bast, R.C.; Desch, C.E.; Fritsche, H.; Kemeny, N.E.; Jessup, J.M.; Locker, G.Y.; Macdonald, J.S.; Mennel, R.G.; Norton, L.; et al. Tumor Marker Utility Grading System: A Framework to Evaluate Clinical Utility of Tumor Markers. *J. Natl. Cancer Inst.* **1996**, *88*, 1456–1466. [CrossRef]
144. Hamanishi, J.; Mandai, M.; Ikeda, T.; Minami, M.; Kawaguchi, A.; Murayama, T.; Kanai, M.; Mori, Y.; Matsumoto, S.; Chikuma, S.; et al. Safety and Antitumor Activity of Anti-PD-1 Antibody, Nivolumab, in Patients with Platinum-Resistant Ovarian Cancer. *J. Clin. Oncol.* **2015**, *33*, 4015–4022. [CrossRef]
145. Johnson, P.J.; Williams, R.; Johnson, P.J. Serum Alpha-Fetoprotein Estimations and Doubling Time in Hepatocellular Carcinoma: Influence of Therapy and Possible Value in Early Detection. *J. Natl. Cancer Inst.* **1980**, *64*, 1329–1332. [CrossRef] [PubMed]
146. Ye, K.; Fan, Q.; Yuan, M.; Wang, D.; Xiao, L.; Long, G.; Chen, R.; Fang, T.; Li, Z.; Zhou, L. Prognostic Value of Postoperative Circulating Tumor DNA in Patients with Early- and Intermediate-Stage Hepatocellular Carcinoma. *Front. Oncol.* **2022**, *12*, 834992. [CrossRef] [PubMed]
147. Chen, L.; Ho, D.W.Y.; Lee, N.P.Y.; Sun, S.; Lam, B.; Wong, K.F.; Yi, X.; Lau, G.K.; Ng, E.W.Y.; Poon, T.C.W.; et al. Enhanced Detection of Early Hepatocellular Carcinoma by Serum SELDI-TOF Proteomic Signature Combined with Alpha-Fetoprotein Marker. *Ann. Surg. Oncol.* **2010**, *17*, 2518–2525. [CrossRef]
148. Vickers, A.J.; Brewster, S.F. PSA Velocity and Doubling Time in Diagnosis and Prognosis of Prostate Cancer. *Br. J. Med. Surg. Urol.* **2012**, *5*, 162–168. [CrossRef]
149. Bjartell, A.S. Next-Generation Prostate-Specific Antigen Test: Ready to Use? *Eur. Urol.* **2013**, *64*, 700–702. [CrossRef]
150. Dunn, M.W. Prostate Cancer Screening. *Semin. Oncol. Nurs.* **2017**, *33*, 156–164. [CrossRef] [PubMed]
151. Molina, R.; Barak, V.; Van Dalen, A.; Duffy, M.J.; Einarsson, R.; Gion, M.; Goike, H.; Lamerz, R.; Nap, M.; Sölétormos, G.; et al. Tumor Markers in Breast Cancer- European Group on Tumor Markers Recommendations. *Tumour Biol. J. Int. Soc. Oncodev. Biol. Med.* **2005**, *26*, 281–293. [CrossRef] [PubMed]
152. Leja, M.; Linē, A. Early Detection of Gastric Cancer beyond Endoscopy—New Methods. *Best Pract. Res. Clin. Gastroenterol.* **2021**, *50–51*, 101731. [CrossRef]
153. Jensen, J.L.; Maclean, G.D.; Suresh, M.R.; Almeida, A.; Jette, D.; Lloyd, S.; Bodnar, D.; Krantz, M.; Longenecker, B.M. Possible Utility of Serum Determinations of CA 125 and CA 27.29 in Breast Cancer Management. *Int. J. Biol. Markers* **1991**, *6*, 1–6. [CrossRef]
154. Ludwig, J.A.; Weinstein, J.N. Biomarkers in Cancer Staging, Prognosis and Treatment Selection. *Nat. Rev. Cancer* **2005**, *5*, 845–856. [CrossRef]
155. von Kleist, S. The Clinical Value of the Tumor Markers CA 19/9 and Carcinoembryonic Antigen (CEA) in Colorectal Carcinomas: A Critical Comparison. *Int. J. Biol. Markers* **1986**, *1*, 3–8. [CrossRef] [PubMed]
156. Koizumi, F.; Odagiri, H.; Fujimoto, H.; Kawamura, T.; Ishimori, A. [Clinical evaluation of four tumor markers (CEA, TPA, CA50 and CA72-4) in colorectal cancer]. *Rinsho Byori. Jpn. J. Clin. Pathol.* **1992**, *40*, 523–528.
157. Bae, Y.J.; Schaab, M.; Kratzsch, J. Calcitonin as Biomarker for the Medullary Thyroid Carcinoma. *Recent Results Cancer Res. Fortschr. Der Krebsforschung. Prog. Dans Les Rech. Sur Le Cancer* **2015**, *204*, 117–137. [CrossRef]
158. Paik, S.; Tang, G.; Shak, S.; Kim, C.; Baker, J.; Kim, W.; Cronin, M.; Baehner, F.L.; Watson, D.; Bryant, J.; et al. Gene Expression and Benefit of Chemotherapy in Women with Node-Negative, Estrogen Receptor-Positive Breast Cancer. *J. Clin. Oncol. Off. J. Am. Soc. Clin. Oncol.* **2006**, *24*, 3726–3734. [CrossRef] [PubMed]
159. Porras, L.; Gorse, F.; Thiombane, N.K.; Gaboury, L.; Mader, S. CAXII Is a Surrogate Marker for Luminal Breast Tumors Regulated by ER and GATA3. *Cancers* **2022**, *14*, 5453. [CrossRef]
160. McGuire, W.L.; Horwitz, K.B.; Pearson, O.H.; Segaloff, A. Current Status of Estrogen and Progesterone Receptors in Breast Cancer. *Cancer* **1977**, *39*, 2934–2947. [CrossRef]
161. Varadan, V.; Kamalakaran, S.; Gilmore, H.; Banerjee, N.; Janevski, A.; Miskimen, K.L.S.; Williams, N.; Basavanhalli, A.; Madabhushi, A.; Lezon-Geyda, K.; et al. Brief-Exposure to Preoperative Bevacizumab Reveals a TGF-β Signature Predictive of Response in HER2-Negative Breast Cancers. *Int. J. Cancer* **2016**, *138*, 747–757. [CrossRef]
162. Alexandrov, L.B.; Nik-Zainal, S.; Wedge, D.C.; Aparicio, S.A.J.R.; Behjati, S.; Biankin, A.V.; Bignell, G.R.; Bolli, N.; Borg, A.; Børresen-Dale, A.L.; et al. Signatures of Mutational Processes in Human Cancer. *Nature* **2013**, *500*, 415–421. [CrossRef]
163. Bhatt, A.N.; Mathur, R.; Farooque, A.; Verma, A.; Dwarakanath, B.S. Cancer Biomarkers—Current Perspectives. *Indian J. Med. Res.* **2010**, *132*, 129–149.
164. Perrier D'Hauterive, S.; Berndt, S.; Tsampalas, M.; Charlet-Renard, C.; Dubois, M.; Bourgain, C.; Hazout, A.; Foidart, J.M.; Geenen, V. Dialogue between Blastocyst HCG and Endometrial LH/HCG Receptor: Which Role in Implantation? *Gynecol. Obstet. Investig.* **2007**, *64*, 156–160. [CrossRef]

165. Kurtzman, J.T.; Wilson, H.; Rao, C.V. A Proposed Role for HCG in Clinical Obstetrics. *Semin. Reprod. Med.* **2001**, *19*, 63–68. [CrossRef] [PubMed]
166. Adewole, I.F.; Newlands, E.S. Neuron-Specific Enolase (NSE) as a Tumour Marker and Comparative Evaluation with Carcinoembryonic Antigen (CEA) in Small-Cell Lung Cancer. *Med. Oncol. Tumor Pharmacother.* **1987**, *4*, 11–15. [CrossRef] [PubMed]
167. Darmawan, B.; Sari, M.; Susilo, S.; Kartamihardja, A.H.S. Preradioactive Iodine Thyroglobulin Levels as Predictors of Metastasis in Well-Differentiated Thyroid Carcinoma Patients. *World J. Nucl. Med.* **2022**, *21*, 296–301. [CrossRef]
168. Mazzaferri, E.L.; Robbins, R.J.; Spencer, C.A.; Braverman, L.E.; Pacini, F.; Wartofsky, L.; Haugen, B.R.; Sherman, S.I.; Cooper, D.S.; Braunstein, G.D.; et al. A Consensus Report of the Role of Serum Thyroglobulin as a Monitoring Method for Low-Risk Patients with Papillary Thyroid Carcinoma. *J. Clin. Endocrinol. Metab.* **2003**, *88*, 1433–1441. [CrossRef] [PubMed]
169. Van Gils, M.P.M.Q.; Cornel, E.B.; Hessels, D.; Peelen, W.P.; Witjes, J.A.; Mulders, P.F.A.; Rittenhouse, H.G.; Schalken, J.A. Molecular PCA3 Diagnostics on Prostatic Fluid. *Prostate* **2007**, *67*, 881–887. [CrossRef] [PubMed]
170. Thomas, R.; Al-Khadairi, G.; Roelands, J.; Hendrickx, W.; Dermime, S.; Bedognetti, D.; Decock, J. NY-ESO-1 Based Immunotherapy of Cancer: Current Perspectives. *Front. Immunol.* **2018**, *9*, 947. [CrossRef]
171. Ellison, G.; Zhu, G.; Moulis, A.; Dearden, S.; Speake, G.; McCormack, R. EGFR Mutation Testing in Lung Cancer: A Review of Available Methods and Their Use for Analysis of Tumour Tissue and Cytology Samples. *J. Clin. Pathol.* **2013**, *66*, 79–89. [CrossRef] [PubMed]
172. Da Cunha Santos, G.; Shepherd, F.A.; Tsao, M.S. EGFR Mutations and Lung Cancer. *Annu. Rev. Pathol.* **2011**, *6*, 49–69. [CrossRef]
173. Lee, B.; Lee, T.; Lee, S.H.; Choi, Y.L.; Han, J. Clinicopathologic Characteristics of EGFR, KRAS, and ALK Alterations in 6595 Lung Cancers. *Oncotarget* **2016**, *7*, 23874. [CrossRef]
174. Warth, A.; Penzel, R.; Lindenmaier, H.; Brandt, R.; Stenzinger, A.; Herpel, E.; Goeppert, B.; Thomas, M.; Herth, F.J.F.; Dienemann, H.; et al. EGFR, KRAS, BRAF and ALK Gene Alterations in Lung Adenocarcinomas: Patient Outcome, Interplay with Morphology and Immunophenotype. *Eur. Respir. J.* **2014**, *43*, 872–883. [CrossRef]
175. Pierce, J.M.R.; Mehta, A. Diagnostic, Prognostic and Therapeutic Role of CD30 in Lymphoma. *Expert Rev. Hematol.* **2017**, *10*, 29–37. [CrossRef] [PubMed]
176. van der Weyden, C.A.; Pileri, S.A.; Feldman, A.L.; Whisstock, J.; Prince, H.M. Understanding CD30 Biology and Therapeutic Targeting: A Historical Perspective Providing Insight into Future Directions. *Blood Cancer J.* **2017**, *7*, e603. [CrossRef] [PubMed]
177. Kumar, A.; Kumar, R.; Gupta, N.P. Comparison of NMP22 BladderChek Test and Urine Cytology for the Detection of Recurrent Bladder Cancer. *Jpn. J. Clin. Oncol.* **2006**, *36*, 172–175. [CrossRef] [PubMed]
178. Kibar, Y.; Goktas, S.; Kilic, S.; Yaman, H.; Onguru, O.; Peker, A.F. Prognostic Value of Cytology, Nuclear Matrix Protein 22 (NMP22) Test, and Urinary Bladder Cancer II (UBC II) Test in Early Recurrent Transitional Cell Carcinoma of the Bladder. *Ann. Clin. Lab. Sci.* **2006**, *36*, 31–38. [PubMed]
179. Mei, X.; Zhu, X.; Zuo, L.; Wu, H.; Guo, M.; Liu, C. Predictive Significance of CYFRA21-1, Squamous Cell Carcinoma Antigen and Carcinoembryonic Antigen for Lymph Node Metastasis in Patients with Esophageal Squamous Cancer. *Int. J. Biol. Markers* **2019**, *34*, 200–204. [CrossRef]
180. Nakamura, T.; Ide, H.; Eguchi, R.; Hayashi, K.; Takasaki, K.; Watanabe, S. CYFRA 21–1 as a Tumor Marker for Squamous Cell Carcinoma of the Esophagus. *Dis. Esophagus* **1998**, *11*, 35–39. [CrossRef]
181. Brockmann, J.G.; St Nottberg, H.; Glodny, B.; Heinecke, A.; Senninger, N.J. CYFRA 21-1 Serum Analysis in Patients with Esophageal Cancer. *Clin. Cancer Res. Off. J. Am. Assoc. Cancer Res.* **2000**, *6*, 4249–4252.
182. McCarthy, B.A.; Boyle, E.; Xue, P.W.; Guzowski, D.; Paul, S.; Catera, R.; Trott, J.; Yan, X.J.; Croce, C.M.; Damle, R.; et al. Surface Expression of Bcl-2 in Chronic Lymphocytic Leukemia and Other B-Cell and Lymphomas without a Breakpoint t(14;18). *Mol. Med.* **2008**, *14*, 618–627. [CrossRef]
183. Dawson, S.J.; Makretsov, N.; Blows, F.M.; Driver, K.E.; Provenzano, E.; Le Quesne, J.; Baglietto, L.; Severi, G.; Giles, G.G.; McLean, C.A.; et al. BCL2 in Breast Cancer: A Favourable Prognostic Marker across Molecular Subtypes and Independent of Adjuvant Therapy Received. *Br. J. Cancer* **2010**, *103*, 668–675. [CrossRef]
184. Kulsoom, B.; Shamsi, T.S.; Afsar, N.A.; Memon, Z.; Ahmed, N.; Hasnain, S.N. Bax, Bcl-2, and Bax/Bcl-2 as Prognostic Markers in Acute Myeloid Leukemia: Are We Ready for Bcl-2-Directed Therapy? *Cancer Manag. Res.* **2018**, *10*, 403–416. [CrossRef]
185. Gaafar, T.M.; Raafat, I.I.; Aly, A.A.; Abd El-Ghaffar Mohamed, N.; Farid, R.J.; Saad, N.E.; El-Hawary, R.; Mostafaa, N.; Ahmed, M.M. Detection of BCR/ABL Translocation in Bone Marrow Derived Mesenchymal Stem Cells in Egyptian CML Patients. *Open Access Maced. J. Med. Sci.* **2015**, *3*, 231–236. [CrossRef] [PubMed]
186. Wu, S.Q.; Voelkerding, K.V.; Sabatini, L.; Chen, X.R.; Huang, J.; Meisner, L.F. Extensive Amplification of Bcr/Abl Fusion Genes Clustered on Three Marker Chromosomes in Human Leukemic Cell Line K-562. *Leukemia* **1995**, *9*, 858–862. [PubMed]
187. Boyd, S.D.; Natkunam, Y.; Allen, J.R.; Warnke, R.A. Selective Immunophenotyping for Diagnosis of B-Cell Neoplasms: Immunohistochemistry and Flow Cytometry Strategies and Results. *Appl. Immunohistochem. Mol. Morphol.* **2013**, *21*, 116–131. [CrossRef] [PubMed]
188. Matsushita, K.; Margulies, I.; Onda, M.; Nagata, S.; Stetler-Stevenson, M.; Kreitman, R.J. Soluble CD22 as a Tumor Marker for Hairy Cell Leukemia. *Blood* **2008**, *112*, 2272–2277. [CrossRef] [PubMed]
189. Li, J.; Ran, Q.; Xu, B.; Luo, X.; Song, S.; Xu, D.; Zhang, X. Role of CD25 Expression on Prognosis of Acute Myeloid Leukemia: A Literature Review and Meta-Analysis. *PLOS ONE* **2020**, *15*, e0236124. [CrossRef] [PubMed]

190. Casadei, C.; Dizman, N.; Schepisi, G.; Cursano, M.C.; Basso, U.; Santini, D.; Pal, S.K.; De Giorgi, U. Targeted Therapies for Advanced Bladder Cancer: New Strategies with FGFR Inhibitors. *Ther. Adv. Med. Oncol.* **2019**, *11*, 1758835919890285. [CrossRef]
191. Knowles, M.A. Role of FGFR3 in Urothelial Cell Carcinoma: Biomarker and Potential Therapeutic Target. *World J. Urol.* **2007**, *25*, 581–593. [CrossRef]
192. Topsakal, M.; Karadeniz, T.; Anaç, M.; Dönmezer, S.; Besisik, A. Assessment of Fibrin-Fibrinogen Degradation Products (Accu-Dx) Test in Bladder Cancer Patients. *Eur. Urol.* **2001**, *39*, 287–291. [CrossRef]
193. Glas, A.S.; Roos, D.; Deutekom, M.; Zwinderman, A.H.; Bossuyt, P.M.M.; Kurth, K.H. Tumor Markers in the Diagnosis of Primary Bladder Cancer. A Systematic Review. *J. Urol.* **2003**, *169*, 1975–1982. [CrossRef]
194. Robinson, B.W.S.; Creaney, J.; Lake, R.; Nowak, A.; Musk, A.W.; De Klerk, N.; Winzell, P.; Hellstrom, K.E.; Hellstrom, I. Soluble Mesothelin-Related Protein—A Blood Test for Mesothelioma. *Lung Cancer* **2005**, *49*, S109–S111. [CrossRef]
195. Kaeding, A.J.; Barwe, S.P.; Gopalakrishnapillai, A.; Ries, R.E.; Alonzo, T.A.; Gerbing, R.B.; Correnti, C.; Loken, M.R.; Broderson, L.E.; Pardo, L.; et al. Mesothelin Is a Novel Cell Surface Disease Marker and Potential Therapeutic Target in Acute Myeloid Leukemia. *Blood Adv.* **2021**, *5*, 2350–2361. [CrossRef] [PubMed]
196. Bubendorf, L.; Büttner, R.; Al-Dayel, F.; Dietel, M.; Elmberger, G.; Kerr, K.; López-Ríos, F.; Marchetti, A.; Öz, B.; Pauwels, P.; et al. Testing for ROS1 in Non-Small Cell Lung Cancer: A Review with Recommendations. *Virchows Arch. Int. J. Pathol.* **2016**, *469*, 489–503. [CrossRef] [PubMed]
197. Nolen, B.M.; Lokshin, A.E. Biomarker Testing for Ovarian Cancer: Clinical Utility of Multiplex Assays. *Mol. Diagn. Ther.* **2013**, *17*, 139–146. [CrossRef]
198. Jelski, W.; Mroczko, B. Molecular and Circulating Biomarkers of Brain Tumors. *Int. J. Mol. Sci.* **2021**, *22*, 7039. [CrossRef] [PubMed]
199. Ouhibi, A.; Raouafi, A.; Lorrain, N.; Guendouz, M.; Raouafi, N.; Moadhen, A. Functionalized SERS Substrate Based on Silicon Nanowires for Rapid Detection of Prostate Specific Antigen. *Sens. Actuators B Chem.* **2021**, *330*, 129352. [CrossRef]
200. Mishra, S.; Kim, E.-S.; Sharma, P.K.; Wang, Z.-J.; Yang, S.-H.; Kaushik, A.K.; Wang, C.; Li, Y.; Kim, N.-Y. Tailored Biofunctionalized Biosensor for the Label-Free Sensing of Prostate-Specific Antigen. *ACS Appl. Bio Mater.* **2020**, *3*, 7821–7830. [CrossRef]
201. Ribeiro, J.A.; Sales, M.G.F.; Pereira, C.M. Electrochemistry-Assisted Surface Plasmon Resonance Biosensor for Detection of CA 15–3. *Anal. Chem.* **2021**, *93*, 7815–7824. [CrossRef]
202. Yu, H.; Tan, X.; Sun, S.; Zhang, L.; Gao, C.; Ge, S. Engineering Paper-Based Visible Light-Responsive Sn-Self Doped Domed SnO2 Nanotubes for Ultrasensitive Photoelectrochemical Sensor. *Biosens. Bioelectron.* **2021**, *185*, 113250. [CrossRef]
203. Wu, Q.; Li, N.; Wang, Y.; Liu, Y.; Xu, Y.; Wei, S.; Wu, J.; Jia, G.; Fang, X.; Chen, F.; et al. A 2D Transition Metal Carbide MXene-Based SPR Biosensor for Ultrasensitive Carcinoembryonic Antigen Detection. *Biosens. Bioelectron.* **2019**, *144*, 111697. [CrossRef]
204. Mo, G.; He, X.; Zhou, C.; Ya, D.; Feng, J.; Yu, C.; Deng, B. A Novel ECL Sensor Based on a Boronate Affinity Molecular Imprinting Technique and Functionalized SiO2@CQDs/AuNPs/MPBA Nanocomposites for Sensitive Determination of Alpha-Fetoprotein. *Biosens. Bioelectron.* **2019**, *126*, 558–564. [CrossRef]
205. Li, S.; Liu, X.; Liu, S.; Guo, M.; Liu, C.; Pei, M. Fluorescence Sensing Strategy Based on Aptamer Recognition and Mismatched Catalytic Hairpin Assembly for Highly Sensitive Detection of Alpha-Fetoprotein. *Anal. Chim. Acta* **2021**, *1141*, 21–27. [CrossRef] [PubMed]
206. Mo, G.; He, X.; Qin, D.; Meng, S.; Wu, Y.; Deng, B. Spatially-Resolved Dual-Potential Sandwich Electrochemiluminescence Immunosensor for the Simultaneous Determination of Carbohydrate Antigen 19–9 and Carbohydrate Antigen 24-2. *Biosens. Bioelectron.* **2021**, *178*, 113024. [CrossRef] [PubMed]
207. Bahari, D.; Babamiri, B.; Salimi, A. An Eco-Friendly MIP-Solid Surface Fluorescence Immunosensor for Detection of CA 19-9 Tumor Marker Using Ni Nanocluster as an Emitter Labels. *J. Iran. Chem. Soc.* **2020**, *17*, 2283–2291. [CrossRef]
208. Mansouri Majd, S.; Salimi, A. Ultrasensitive Flexible FET-Type Aptasensor for CA 125 Cancer Marker Detection Based on Carboxylated Multiwalled Carbon Nanotubes Immobilized onto Reduced Graphene Oxide Film. *Anal. Chim. Acta* **2018**, *1000*, 273–282. [CrossRef]
209. Szymańska, B.; Lukaszewski, Z.; Hermanowicz-Szamatowicz, K.; Gorodkiewicz, E. A Biosensor for Determination of the Circulating Biomarker CA125/MUC16 by Surface Plasmon Resonance Imaging. *Talanta* **2020**, *206*, 120187. [CrossRef]
210. Bai, Y.; Zhang, H.; Zhao, L.; Wang, Y.; Chen, X.; Zhai, H.; Tian, M.; Zhao, R.; Wang, T.; Xu, H.; et al. A Novel Aptasensor Based on HCR and G-Quadruplex DNAzyme for Fluorescence Detection of Carcinoembryonic Antigen. *Talanta* **2021**, *221*, 121451. [CrossRef]
211. Yılmaz, M.; Bakhshpour, M.; Göktürk, I.; Pişkin, A.K.; Denizli, A. Quartz Crystal Microbalance (QCM) Based Biosensor Functionalized by HER2/Neu Antibody for Breast Cancer Cell Detection. *Chemosensors* **2021**, *9*, 80. [CrossRef]
212. Tanaka, S.; Wadati, H.; Sato, K.; Yasuda, H.; Niioka, H. Red-Fluorescent Pt Nanoclusters for Detecting and Imaging HER2 in Breast Cancer Cells. *ACS Omega* **2020**, *5*, 23718–23723. [CrossRef]
213. Xie, Y.; Si, S.; Deng, C.; Wu, H. A Sensitive "Turn-on" Fluorescent Biosensor for the Detection of Biomarkers Based on Three-Dimensional DNA Walker. *Talanta* **2023**, *253*, 124071. [CrossRef]
214. Hu, S.; Xu, H.; Zhou, B.; Xu, S.; Shen, B.; Dong, B.; Yin, Z.; Xu, S.; Sun, L.; Lv, J.; et al. Double Stopband Bilayer Photonic Crystal Based Upconversion Fluorescence PSA Sensor. *Sens. Actuators B Chem.* **2021**, *326*, 128816. [CrossRef]

215. Wu, Y.; Chen, X.; Wang, X.; Yang, M.; Xu, F.; Hou, C.; Huo, D. A Fluorescent Biosensor Based on Prismatic Hollow Metal-Polydopamine Frameworks and 6-Carboxyfluorescein (FAM)-Labeled Protein Aptamer for CA15-3 Detection. *Sens. Actuators B Chem.* **2021**, *329*, 129249. [CrossRef]
216. Othman, H.O.; Salehnia, F.; Fakhri, N.; Hassan, R.; Hosseini, M.; Faizullah, A.; Ganjali, M.R.; Kazem Aghamir, S.M. A Highly Sensitive Fluorescent Immunosensor for Sensitive Detection of Nuclear Matrix Protein 22 as Biomarker for Early Stage Diagnosis of Bladder Cancer. *RSC Adv.* **2020**, *10*, 28865–28871. [CrossRef] [PubMed]
217. Othman, H.O.; Salehnia, F.; Hosseini, M.; Hassan, R.; Faizullah, A.; Ganjali, M.R. Fluorescence Immunoassay Based on Nitrogen Doped Carbon Dots for the Detection of Human Nuclear Matrix Protein NMP22 as Biomarker for Early Stage Diagnosis of Bladder Cancer. *Microchem. J.* **2020**, *157*, 104966. [CrossRef]
218. Satyanarayana, M.; Yugender Goud, K.; Koteshwara Reddy, K.; Vengatajalabathy Gobi, K. Conducting Polymer-Layered Carbon Nanotube as Sensor Interface for Electrochemical Detection of Dacarbazine In-Vitro. *Electrocatalysis* **2017**, *8*, 214–223. [CrossRef]
219. Goud, K.Y.; Satyanarayana, M.; Hayat, A.; Gobi, K.V.; Marty, J.L. Nanomaterial-Based Electrochemical Sensors in Pharmaceutical Applications. In *Nanoparticles in Pharmacotherapy*; Elsevier: Amsterdam, The Netherlands, 2019; pp. 195–216. ISBN 9780128165041.
220. Kummari, S.; Sunil Kumar, V.; Vengatajalabathy Gobi, K. Facile Electrochemically Reduced Graphene Oxide-Multi-walled Carbon Nanotube Nanocomposite as Sensitive Probe for In-vitro Determination of Nitrofurantoin in Biological Fluids. *Electroanalysis* **2020**, *32*, 2452–2462. [CrossRef]
221. Shekher, K.; Sampath, K.; Vandini, S.; Satyanarayana, M.; Vengatajalabathy Gobi, K. Gold Nanoparticle Assimilated Polymer Layer on Carbon Nanotube Matrices for Sensitive Detection of Serotonin in Presence of Dopamine In-Vitro. *Inorg. Chim. Acta* **2023**, *549*, 121399. [CrossRef]
222. Yang, H.; Xu, Y.; Hou, Q.; Xu, Q.; Ding, C. Magnetic Antifouling Material Based Ratiometric Electrochemical Biosensor for the Accurate Detection of CEA in Clinical Serum. *Biosens. Bioelectron.* **2022**, *208*, 114216. [CrossRef]
223. Pacheco, J.G.; Silva, M.S.V.; Freitas, M.; Nouws, H.P.A.; Delerue-Matos, C. Molecularly Imprinted Electrochemical Sensor for the Point-of-Care Detection of a Breast Cancer Biomarker (CA 15-3). *Sens. Actuators B Chem.* **2018**, *256*, 905–912. [CrossRef]
224. Wei, B.; Mao, K.; Liu, N.; Zhang, M.; Yang, Z. Graphene Nanocomposites Modified Electrochemical Aptamer Sensor for Rapid and Highly Sensitive Detection of Prostate Specific Antigen. *Biosens. Bioelectron.* **2018**, *121*, 41–46. [CrossRef]
225. Farzin, L.; Sadjadi, S.; Shamsipur, M.; Sheibani, S.; Mousazadeh, M. hasan Employing AgNPs Doped Amidoxime-Modified Polyacrylonitrile (PAN-Oxime) Nanofibers for Target Induced Strand Displacement-Based Electrochemical Aptasensing of CA125 in Ovarian Cancer Patients. *Mater. Sci. Eng. C* **2019**, *97*, 679–687. [CrossRef]
226. Shekari, Z.; Zare, H.R.; Falahati, A. Dual Assaying of Breast Cancer Biomarkers by Using a Sandwich–Type Electrochemical Aptasensor Based on a Gold Nanoparticles–3D Graphene Hydrogel Nanocomposite and Redox Probes Labeled Aptamers. *Sens. Actuators B Chem.* **2021**, *332*, 129515. [CrossRef]
227. Ji, Y.; Guo, J.; Ye, B.; Peng, G.; Zhang, C.; Zou, L. An Ultrasensitive Carcinoembryonic Antigen Electrochemical Aptasensor Based on 3D DNA Nanoprobe and Exo III. *Biosens. Bioelectron.* **2022**, *196*, 113741. [CrossRef] [PubMed]
228. Taheri, N.; Khoshsafar, H.; Ghanei, M.; Ghazvini, A.; Bagheri, H. Dual-Template Rectangular Nanotube Molecularly Imprinted Polypyrrole for Label-Free Impedimetric Sensing of AFP and CEA as Lung Cancer Biomarkers. *Talanta* **2022**, *239*, 123146. [CrossRef]
229. Upan, J.; Youngvises, N.; Tuantranont, A.; Karuwan, C.; Banet, P.; Aubert, P.H.; Jakmunee, J. A Simple Label-Free Electrochemical Sensor for Sensitive Detection of Alpha-Fetoprotein Based on Specific Aptamer Immobilized Platinum Nanoparticles/Carboxylated-Graphene Oxide. *Sci. Rep.* **2021**, *11*, 13969. [CrossRef]
230. Castaño-Guerrero, Y.; Romaguera-Barcelay, Y.; Moreira, F.T.C.; Brito, W.R.; Fortunato, E.; Sales, M.G.F. Poly(Thionine)-Modified Screen-Printed Electrodes for CA 19-9 Detection and Its Properties in Raman Spectroscopy. *Chemosensors* **2022**, *10*, 92. [CrossRef]
231. Yola, M.L.; Atar, N. Carbohydrate Antigen 19-9 Electrochemical Immunosensor Based on 1D-MoS2 Nanorods/LiNb3O8 and Polyoxometalate-Incorporated Gold Nanoparticles. *Microchem. J.* **2021**, *170*, 106643. [CrossRef]
232. Rebelo, T.S.C.R.; Costa, R.; Brandão, A.T.S.C.; Silva, A.F.; Sales, M.G.F.; Pereira, C.M. Molecularly Imprinted Polymer SPE Sensor for Analysis of CA-125 on Serum. *Anal. Chim. Acta* **2019**, *1082*, 126–135. [CrossRef]
233. Wang, M.; Hu, M.; Li, Z.; He, L.; Song, Y.; Jia, Q.; Zhang, Z.; Du, M. Construction of Tb-MOF-on-Fe-MOF Conjugate as a Novel Platform for Ultrasensitive Detection of Carbohydrate Antigen 125 and Living Cancer Cells. *Biosens. Bioelectron.* **2019**, *142*, 111536. [CrossRef]
234. Freitas, M.; Nouws, H.P.A.; Delerue-Matos, C. Electrochemical Sensing Platforms for HER2-ECD Breast Cancer Biomarker Detection. *Electroanalysis* **2019**, *31*, 121–128. [CrossRef]
235. Li, X.; Shen, C.; Yang, M.; Rasooly, A. Polycytosine DNA Electric-Current-Generated Immunosensor for Electrochemical Detection of Human Epidermal Growth Factor Receptor 2 (HER2). *Anal. Chem.* **2018**, *90*, 4764–4769. [CrossRef]
236. Wang, W.; Han, R.; Chen, M.; Luo, X. Antifouling Peptide Hydrogel Based Electrochemical Biosensors for Highly Sensitive Detection of Cancer Biomarker HER2 in Human Serum. *Anal. Chem.* **2021**, *93*, 7355–7361. [CrossRef] [PubMed]
237. Pacheco, J.G.; Rebelo, P.; Freitas, M.; Nouws, H.P.A.; Delerue-Matos, C. Breast Cancer Biomarker (HER2-ECD) Detection Using a Molecularly Imprinted Electrochemical Sensor. *Sens. Actuators B Chem.* **2018**, *273*, 1008–1014. [CrossRef]
238. Díaz-Fernández, A.; Miranda-Castro, R.; De-los-Santos-Álvarez, N.; Lobo-Castañón, M.J.; Estrela, P. Impedimetric Aptamer-Based Glycan PSA Score for Discrimination of Prostate Cancer from Other Prostate Diseases. *Biosens. Bioelectron.* **2021**, *175*, 112872. [CrossRef] [PubMed]

239. Hasanzadeh, M.; Tagi, S.; Solhi, E.; Mokhtarzadeh, A.; Shadjou, N.; Eftekhari, A.; Mahboob, S. An Innovative Immunosensor for Ultrasensitive Detection of Breast Cancer Specific Carbohydrate (CA 15-3) in Unprocessed Human Plasma and MCF-7 Breast Cancer Cell Lysates Using Gold Nanospear Electrochemically Assembled onto Thiolated Graphene Quantum Dots. *Int. J. Biol. Macromol.* **2018**, *114*, 1008–1017. [CrossRef] [PubMed]
240. Rong, S.; Zou, L.; Zhu, Y.; Zhang, Z.; Liu, H.; Zhang, Y.; Zhang, H.; Gao, H.; Guan, H.; Dong, J.; et al. 2D/3D Material Amplification Strategy for Disposable Label-Free Electrochemical Immunosensor Based on RGO-TEPA@Cu-MOFs@SiO2@AgNPs Composites for NMP22 Detection. *Microchem. J.* **2021**, *168*, 106410. [CrossRef]
241. Zhao, S.; Zhang, Y.; Ding, S.; Fan, J.; Luo, Z.; Liu, K.; Shi, Q.; Liu, W.; Zang, G. A Highly Sensitive Label-Free Electrochemical Immunosensor Based on AuNPs-PtNPs-MOFs for Nuclear Matrix Protein 22 Analysis in Urine Sample. *J. Electroanal. Chem.* **2019**, *834*, 33–42. [CrossRef]
242. Li, S.; Yue, S.; Yu, C.; Chen, Y.; Yuan, D.; Yu, Q. A Label-Free Immunosensor for the Detection of Nuclear Matrix Protein-22 Based on a Chrysanthemum-like Co-MOFs/CuAu NWs Nanocomposite. *Analyst* **2019**, *144*, 649–655. [CrossRef]
243. Goud, K.Y.; Reddy, K.K.; Khorshed, A.; Kumar, V.S.; Mishra, R.K.; Oraby, M.; Ibrahim, A.H.; Kim, H.; Gobi, K.V. Electrochemical Diagnostics of Infectious Viral Diseases: Trends and Challenges. *Biosens. Bioelectron.* **2021**, *180*, 113112. [CrossRef]
244. Ruiz-Valdepeñas Montiel, V.; Sempionatto, J.R.; Esteban-Fernández de Ávila, B.; Whitworth, A.; Campuzano, S.; Pingarrón, J.M.; Wang, J. Delayed Sensor Activation Based on Transient Coatings: Biofouling Protection in Complex Biofluids. *J. Am. Chem. Soc.* **2018**, *140*, 14050–14053. [CrossRef]
245. Ghafar-Zadeh, E. Wireless Integrated Biosensors for Point-of-Care Diagnostic Applications. *Sensors* **2015**, *15*, 3236–3261. [CrossRef]

biosensors

MDPI

Communication

Interactions of Amino Group Functionalized Tetraphenylvinyl and DNA: A Label-Free "On-Off-On" Fluorescent Aptamer Sensor toward Ampicillin

Weifu Geng [1,†], Yan Feng [2,†], Yu Chen [2], Xin Zhang [1], Haoyi Zhang [2], Fanfan Yang [2] and Xiuzhong Wang [1,2,*]

1 College of Plant Health and Medicine, Qingdao Agricultural University, Qingdao 266109, China; gwf17854221211@163.com (W.G.)
2 College of Chemistry and Pharmaceutical Sciences, Qingdao Agricultural University, Qingdao 266109, China
* Correspondence: xzwang@qau.edu.cn; Tel.: +86-532-58957442
† These authors contributed equally to this work.

Abstract: As a type of aggregation-induced emission (AIE) fluorescent probe, tetraphenylvinyl (TPE) or its derivatives are widely used in chemical imaging, biosensing and medical diagnosis. However, most studies have focused on molecular modification and functionalization of AIE to enhance the fluorescence emission intensity. There are few studies on the interaction between aggregation-induced emission luminogens (AIEgens) and nucleic acids, which was investigated in this paper. Experimental results showed the formation of a complex of AIE/DNA, leading to the quenching of the fluorescence of AIE molecules. Fluorescent test experiments with different temperatures proved that the quenching type was static quenching. The quenching constants, binding constants and thermodynamic parameters demonstrated that electrostatic and hydrophobic interactions promoted the binding process. Then, a label-free "on-off-on" fluorescent aptamer sensor for the detection of ampicillin (AMP) was constructed based on the interaction between the AIE probe and the aptamer of AMP. Linear range of the sensor is 0.2–10 nM with a limit of detection 0.06 nM. This fluorescent sensor was applied to detect AMP in real samples.

Keywords: aggregation-induced emission; DNA; label-free; biosensor; antibiotic

Citation: Geng, W.; Feng, Y.; Chen, Y.; Zhang, X.; Zhang, H.; Yang, F.; Wang, X. Interactions of Amino Group Functionalized Tetraphenylvinyl and DNA: A Label-Free "On-Off-On" Fluorescent Aptamer Sensor toward Ampicillin. *Biosensors* **2023**, *13*, 504. https://doi.org/10.3390/bios13050504

Received: 20 March 2023
Revised: 22 April 2023
Accepted: 26 April 2023
Published: 27 April 2023

1. Introduction

Twenty years ago, Tang's research group found a new class of organic molecules which had almost no emission in dilute solutions, but strong fluorescence emission in an aggregate or solid state, which they termed aggregation-induced emission luminogens (AIEgens) [1]. AIE molecules could avoid concentration quenching or aggregation-caused quenching (ACQ) of the conventional fluorophores [2]. Therefore, AIEgens molecules have been used to construct chemicals and biosensors [3], bioimaging probes, and have been widely applied in various fields over the past twenty years [4]. As typical AIE dyes, tetraphenylethene (TPE) and its derivatives have been widely used to construct fluorescent sensors and biological probes [5,6]. Li and coworkers [7] developed a fluorescent aptasensor for the detection of exosome tumor-associated proteins, combining aptamers, tertiary amine-containing tetraphenylethene, and graphene oxide (GO) as recognition units, fluorescent probe, and the quencher, respectively. Numerous AIE molecules could bind to the aptamer and form aggregates rapidly, leading to an amplified fluorescence intensity. Zhu et al. [8] constructed a label-free and turn-on AIE-based fluorescence aptamer sensor for the detection of mycotoxin. Hu [9] prepared AIE-Red and AIE-Green fluorescence probes, using the same TPE cores functionalized with the positively charged morpholine groups or vinyl pyridinium. With only one excitation laser, this strategy can facilitate the multicolor monitoring of mitochondria and lysosomes, respectively. Xu et al. [10] reported a TPE-based fluorescent probe by decorating carboxylate coordination groups on TPE for the highly selective and sensitive detection of nitrofuran in aqueous solution. Because TPE molecules are insoluble in water, most of the reports focused on functional modification of the TPE core skeleton to enhance its practicability which led to the complex molecular structures of TPE derivatives, with complex synthesis steps and a high cost [11–13]. Therefore, it is necessary to develop new sensitive and accurate biosensors with simple molecule probes without complex functionalization in the TPE-based fluorescence sensing systems.

Aptamer is a single-stranded DNA (ssDNA), RNA, or modified nucleic acid isolated through SELEX technology, and can specifically bind to corresponding targets, for example, small organic molecules, viruses or proteins with high affinity and selectivity [14,15]. Aptamer-based sensors (Aptasensor) have been widely reported and developed in recent years [16–19]. AIE molecules, generally having a positive charge, would tend to bind to negatively charged aptamers, and the fluorescence emission varies with the change of the molecular aggregation state [8,20]. Zhang et al. [21] developed a turn-on fluorescent sensor for the detection of chloramphenicol in which an AIE 9,10-distyrylanthracene derivative with short alkyl chains and GO function as the fluorescent probe and the fluorescence quencher, respectively. However, the interaction between AIEgens and DNA strands has been rarely reported. To illustrate the binding mechanism of AIEgens to DNA, the interactions between the amino group functionalized TPE (TPE-Am) and single-stranded DNA were studied in this paper. The binding types, constants and types of force were investigated. Experimental results showed that the binding modes between AIE molecule and DNA were intercalation and groove binding. Fluorescent test experiments under different temperatures proved that the quenching type was a static quenching. The quenching constants, binding constants and thermodynamic parameters demonstrated that electrostatic and hydrophobic interactions promoted the binding process.

Antibiotics are generally used to treat infections caused by pathogenic bacteria or fungi in healthcare. However, the overuse of antibiotics may lead to food and environmental pollution [22–24]. Ampicillin (AMP) is widely used to manage and treat certain bacterial infections. Therefore, it is very important to detect AMP residues in natural environments and foodstuffs for human health protection. Up to now, various detection methods such as chromatography, microbiological assay, electrochemical and photochemical sensors have been developed to detect antibiotics [25–30]. Chromatography has great advantages on selectivity; however, it requires professional instruments and cumbersome pretreatment. Sensitivity and specificity of microbiological assay for the detection of antibiotics is not enough. In spite of high detection sensitivity, electrochemical sensors are prone to contamination and have poor reproducibility. Compared with other detection methods, photochemical sensors, especially fluorescent sensors for the detection of antibiotics, have received more and more attention in recent years. However, most of the detection methods need fluorescent labeling which are time-consuming and costly. Therefore, it is urgent to develop a simple, fast and inexpensive detection method for AMP.

Based on the aforementioned developments, in this paper we developed a simple, label-free fluorescent aptasensor, with high specificity and sensitivity, to detect AMP using the AIE probe based on interaction between AIEgens and the aptamer of AMP.

2. Materials and Methods

2.1. Reagents and Materials

The amino group functionalized TPE (TPE-Am) was donated by professor Lei Han (Qingdao Agricultural University). Ampicillin (AMP) was bought from Macklin (Shanghai, China). Other antibiotic and pesticides, including penicillin (PEN), chloramphenicol (CHL), roxithromycin (ROX), tetracycline (TET), imidacloprid (IMI), and methidathion (MET) were bought from Aladdin Chemical Co., Ltd. (Shanghai, China). Dimethyl sulfoxide was bought from Kangde (Laiyang, China). Na_2HPO_4, KH_2PO_4, NaCl and KCl were purchased from Tianjin Basf Chemical Co., Ltd. (Tianjin, China). The DNA sequences 5′-GCGGGCGGTTGTATAGCGG-3′ (aptamer for AMP) was obtained by Sangon Biotech (Shanghai, China). All of the DNA was dispersed in 10 mM phosphate-buffered solution

(PBS) and stored at −18 °C before use. All chemicals and solvents used in the experiment were of analytical grade. Ultrapure water used in the experiments was obtained from a Milli-Q water purification system (Millipore Corp., Burlington, MA, USA).

2.2. Measurement and Apparatus

Fluorescence tests were performed on a Gangdong technology F-380 fluorescence spectrometer (Tianjin, China) equipped with 1.0 cm micro quartz cells. The excitation wavelength was set at 340 nm and the fluorescence spectra were collected from 400 to 625 nm. A fluorescence intensity of 490 nm was used to evaluate the performance of the sensing system. The pH of solutions was adjusted by a PB-10 digital pH meter (Sartorius, Shanghai, China). The ultraviolet-visible absorption spectra were collected at room temperature on a U3900 spectrophotometer (Hitachi, Tokyo, Japan) equipped with 1.0 cm micro quartz cells.

2.3. Principles of Fluorescence Quenching

The fluorescence quenching process is usually described according to the Stern–Volmer Equation (1) [31]:

$$F_0/F = 1 + k_q\tau_0[Q] = 1 + K_{sv}[Q] \quad (1)$$

where F and F_0 represents the steady state fluorescence emission intensities with and without quencher, respectively, k_q is rate constant of the dynamic fluorescence quenching process, τ_0 is the fluorescence lifetime without quencher in the system, $[Q]$ is the concentration of the quencher, and K_{SV} is the quenching constant. Hence, K_{SV} can be determined according to Equation (1) using a linear regression plot of F_0/F against $[Q]$.

2.4. Fluorescence Detection Procedures for AMP

A volume of 10 μL of AMP aptamer (10 μM) and 2 μL TPE-Am (100 μM) were incubated for 30 min in 99% PBS (PBS:DMSO, *v:v*) at 25 °C. Then, a different concentration of AMP was introduced into the above solution and the final solution was fixed to 200 μL. After 45 min incubation at 25 °C, the fluorescence intensity of the reaction solution was measured. As control experiments, penicillin (PEN), chloramphenicol (CHL), roxithromycin (ROX), tetracycline (TET), imidacloprid (IMI) and methidathion (MET) were instead incubated with AMP, respectively.

For AMP content detection in real samples, water samples were collected from the Hongzi river located in the Chengyang District (Qingdao, China) and were pretreated according to the literature [32].

3. Results and Discussion

3.1. Fluorescence Spectroscopy of TPE-Am

The excitation and emission spectra of TPE-Am in 99% PBS (vs. DMSO) were firstly tested. Figure 1A shows that the maximum excitation and emission wavelengths of the AIE molecule are 340 nm and 490 nm, respectively, which is consistent with the literature [33]. Next, aggregation-induced emission characteristics of TPE-Am was investigated. According to the molecular structure of TPE-Am (Figure S1, in the Supplementary Materials), as is well known, it is an insoluble and commercially available molecule. We investigated AIE characteristics of TPE-Am by investigating photoluminescence behaviors in the DMSO and PBS/DMSO mixtures. As shown in Figure 1B, TPE-Am (1 μM) almost has no emission in diluted DMSO solution, but the fluorescence intensity of TPE-Am increases with increasing PBS fraction (f_{PBS}) from 0 to 99% (vs. DMSO). A photograph of TPE-Am in the DMSO and PBS/DMSO mixtures with different PBS fractions (f_{PBS}) is shown in Figure 1C. These phenomena indicate that TPE-Am enhanced emission in PBS/DMSO mixtures is derived from its aggregation in poor solvents.

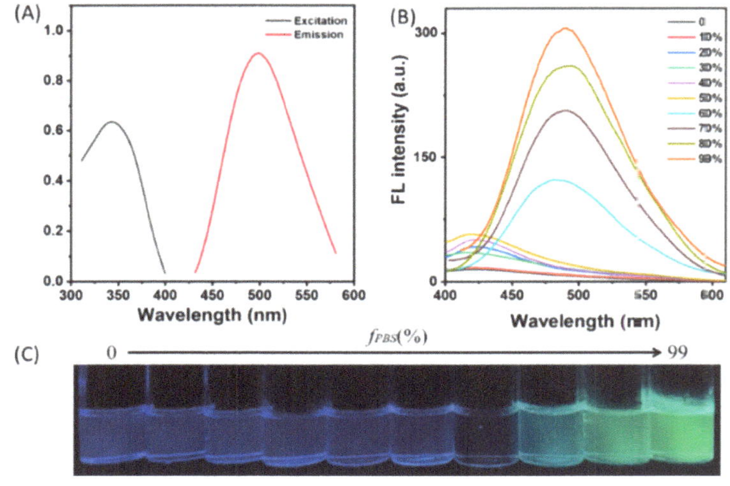

Figure 1. Fluorescence spectroscopy of TPE-Am. (**A**) Normalized excitation and emission profile of TPE-Am in PBS/DMSO mixture (f_{PBS} = 99%); (**B**) the AIE characteristics of TPE-Am in DMSO and PBS/DMSO mixture; (**C**) photograph of TPE-Am in DMSO and PBS/DMSO mixtures with different PBS fractions (f_{PBS}).

3.2. Interaction of TPE-Am and DNA

Binding modes between TPE-Am and DNA were examined by fluorescence spectroscopy. The fluorescent emission spectra of TPE-Am and TPE-Am/DNA strands (the aptamer of AMP) are illustrated in Figure 2A. The fluorescence intensity significantly decreases (curve b) in the presence of 500 nM DNA strands compared with TPE-Am alone (curve a).

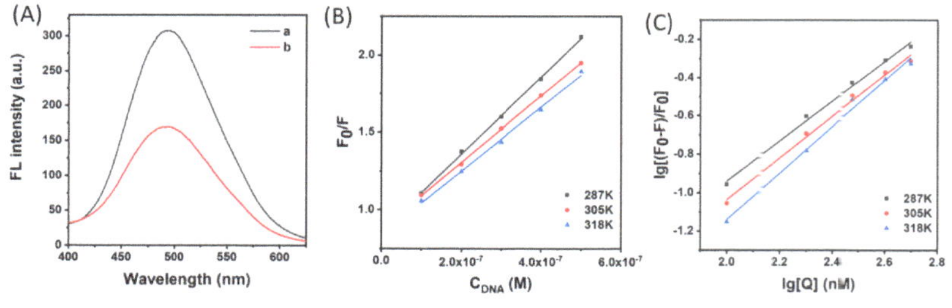

Figure 2. Interaction of TPE-Am and DNA strands (**A**). FL intensity of the system with different conditions: (a) TPE-Am and (b) TPE-Am + DNA. (**B**) The fluorescence quenching tests at different temperature. (**C**) Stern–Volmer plots for the TPE-Am/DNA system with different temperatures.

3.2.1. Quenching Mechanism of Fluorescence

Dynamic or static quenching mechanisms could be judged from constants K_q and K_{sv} at different temperatures. Because of its dependance on diffusion effects, K_q and K_{sv} were larger with increasing temperature. However, non-fluorescing complexes were produced during static quenching process, leading to decreasing static quenching constants with increasing temperature [34,35].

Next, the fluorescence quenching tests were carried out at a temperature of 287, 305 and 318 K between TPE-Am and the aptamer of AMP, respectively, which are shown in Figure 2B. The K_{SV} and K_q values were calculated according to Equation (1) and are exhibited in Table 1. The experimental results demonstrated a classical static quenching process for TPE-Am binding to DNA strands, as the values of K_q were much larger than 2×10^{10} $L{\cdot}mol^{-1}s^{-1}$ which was the maximum scattering collision quenching constant. The

K_{sv} values decreased with increasing temperature; furthermore, the dynamic quenching constant was generally less than 100 L·mol^{-1}.

Table 1. Stern–Volmer quenching constants of TPE-Am binding to DNA.

T (K)	K_{sv} (L·mol^{-1})	K_q (L·mol^{-1}·s^{-1})	R [1]
287	2.488×10^6	2.488×10^{14}	0.9986
305	2.149×10^6	2.149×10^{14}	0.9995
318	2.065×10^6	2.065×10^{14}	0.9949

[1] R is the correlation coefficient.

3.2.2. Binding Constants

For the static quenching process, the binding constant (K_b) and the number of TPE-Am-bound sites (n) per DNA strand could be determined from Equation (2).

$$\lg\frac{F_0 - F}{F} = \lg K_b + n\lg[Q] \quad (2)$$

Parameters F_0, F and [Q] express the same meanings as in Equation (1). K_b and n could be obtained from the intercept and slope by plotting lg $[(F_0 - F)/F]$ against lg [Q] (intercept = lgK_b, slope = n) (Figure 2C), and the values of K_b and n are listed in Table 2. The value of K_b decreased with increasing temperature according to Table 2 which was in accordance with K_{sv}, showing moderate binding between TPE-Am and DNA, and that TPE-Am/DNA complex might be formed [36]. The values of n are equal to approximately 1, indicating that there was one class of binding sites for TPE-Am in DNA.

Table 2. The binding parameters and relevant thermodynamic parameters for the TPE-Am—DNA system.

T (K)	K_b (/10^5 L/mol)	n	R	ΔH (kJ/mol)	ΔS (J/mol)	ΔG (kJ/mol)
287	9.59	1.03	0.9929			−33.07
305	6.26	1.08	0.9902	−29.46	12.59	−33.3
318	2.75	1.21	0.9902			−33.46

3.2.3. Thermodynamic Constants

It is well known that the four major interaction forces between small organic molecules and biomacromolecules are electrostatic interactions, hydrogen bonds, hydrophobic force, and van der Waals force, respectively. The enthalpy change (ΔH) of the interaction usually remains stable if the temperature does not change much. For small molecules binding to DNA, the interaction types and modes can be estimated by thermodynamic parameters, including free energy change (ΔG), enthalpy change (ΔH), and entropy change (ΔS), according to Equations (3) and (4):

$$\ln K = -\frac{\Delta H}{RT} + \frac{\Delta S}{R} \quad (3)$$

$$\Delta G = -RT\ln K = \Delta H - T\Delta S \quad (4)$$

where K is the binding constant (K_b) at 287, 305 and 318 K, respectively, and R is the gas constant which is 8.314 J mol^{-1}·K^{-1}. The corresponding values were obtained from Equations (3) and (4) and listed in Table 2. The electrostatic and hydrophobic interactions were the main interaction force between TPE-Am and DNA, judging from the negative ΔH and positive ΔS value [37].

3.3. Fluorescence Assay for AMP

A facile, sensitive and label-free "on-off-on" fluorescent aptamer sensor for AMP detection was successfully developed based on the interaction between the TPE-Am probe and aptamer of AMP which is illustrated in Scheme 1. There was strong fluorescence in 99% PBS (PBS:DMSO, *v:v*) because of the hydrophobicity of TPE-Am. In the presence of the aptamer of AMP (Apt), the weakened fluorescence intensities of the system was the result of the hydrophobic interaction between the TPE-Am probe and DNA (from on to off). However, the fluorescence recovered once the target AMP molecules were introduced into the solution (from off to on). As is well known, the force between the target and the aptamer is much larger than the hydrogen bonds and hydrophobic interactions. Fluorescence recovery came from the release of the TPE-Am probe because of the formation of an Aptamer-AMP complex. Therefore, the detection of AMP can be realized by monitoring the changes of the fluorescence intensities using this on-off-on fluorescent aptamer sensor.

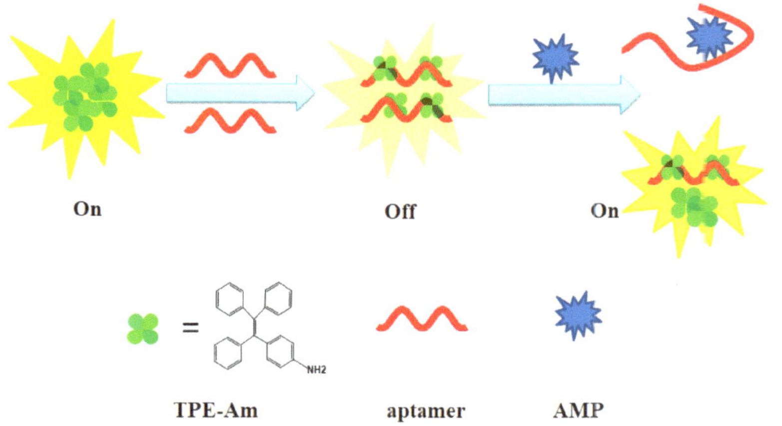

Scheme 1. Sensing mechanism of the proposed aptasensor for detection of AMP.

3.3.1. Characterization of TPE-Am/Aptamer

UV-vis absorption spectra were used to verify the complex of TPE-Am and Aptamer. It could be seen from Figure 3A that both the TPE-Am and the TPE-Am/Aptamer complex had the same absorption bands at the wavelength range of 300–360 nm which is the representative absorption peak of a tetraphenylethylene structure [33]. Compared with the TPE-Am, the absorption band of the TPE-Am/Apt complex were in the wavelength range of 260–280 nm which is the same as that of the typical structural features of DNA.

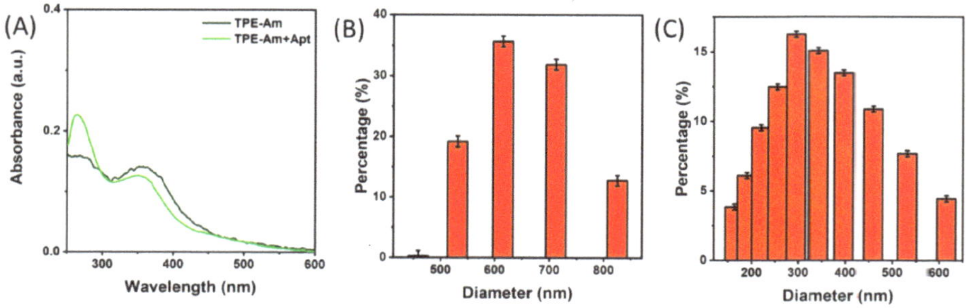

Figure 3. Characterization of TPE-Am/Aptamer. (**A**) UV absorption spectrum of TPE-Am and TPE-Am/Aptamer. (**B**) Size distribution of TPE-Am. (**C**) Size distribution of TPE-Am/Aptamer.

Dynamic light scattering (DLS) was used to test the size distribution of TPE-Am in 99% PBS (PBS:DMSO, *v:v*). The mean particle size changed from 610 nm (Figure 3B) to 300 nm (Figure 3C), and the reason is that the formed TPE-Am aggregates tend to disperse upon the formation of the TPE-Am/Aptamer complex. These results clearly indicated that the TPE-Am probe was successfully grafted on the aptamer of AMP.

3.3.2. Feasibility of the Fluorescence Aptamer Sensor

Figure 4 shows the fluorescent emission spectra of TPE-Am (1.0 μM) under different conditions. The fluorescence intensity was largest with TPE-Am alone in 99% PBS (PBS:DMSO, *v:v*) (curve a). There was little change in the fluorescence intensities when the AMP targets were added (curve b) compared with TPE-Am alone. Fluorescence intensities significantly decreased in the presence of the aptamer of AMP (curve c). Nevertheless, fluorescence obviously recovered when the AMP targets were added into the solution because of the formation of the Aptamer-AMP complex accompanied with the release of the TPE-Am probe (curve d). Therefore, AMP content could be detected by monitoring the changes of the fluorescence intensities using this on-off-on fluorescent aptamer sensor.

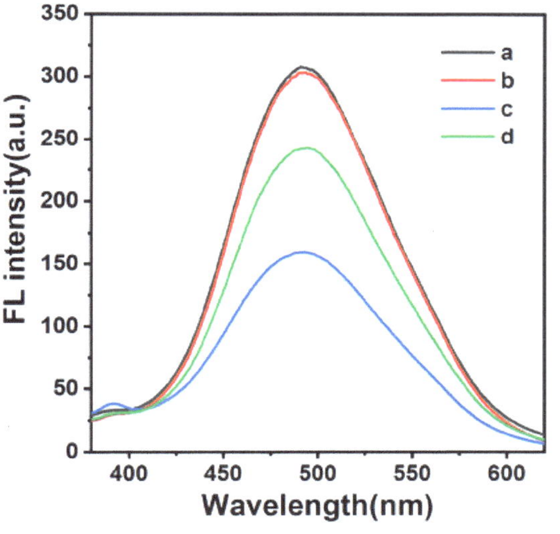

Figure 4. Fluorescence emission under different conditions: (a) TPE-Am, (b) TPE-Am + AMP, (c) TPE-Am + Apt, and (d) TPE-Am + Apt + AMP.

3.3.3. Optimization of the Experimental Conditions

In order to obtain the optimal performance of the proposed fluorescent sensor for detection of AMP, the experimental conditions, including the concentrations, response times, and pH of the buffer solution, were optimized, respectively. As shown in Figures S2–S5, it was observed that the optimal concentration of aptamer, incubation time for TPE-Am/Aptamer, TPE-Am/Aptamer/AMP and pH of the solution were 500 nM, 30 min, 45 min, and 7.4, respectively.

3.3.4. Analytical Performance of Fluorescent Aptasensor

The fluorescent aptasensor was evaluated by monitoring the fluorescence change under different concentrations of AMP. As shown in Figure 5A, the fluorescence intensity increased gradually with increasing AMP concentration. A good linear relationship between the fluorescence intensity and the AMP concentration was found, ranging from 0.2 to 10 nM, which is shown in the inset of Figure 5B. The linear regression equation was $F = 53.46\ lgc + 187.78$ with a correlation coefficient of 0.9959 (F: fluorescence intensity, c: AMP concentration (nM)). The limit of detection was estimated to be 0.06 nM (based on 3σ) which was lower than those of the previously reported assay methods (Table S1).

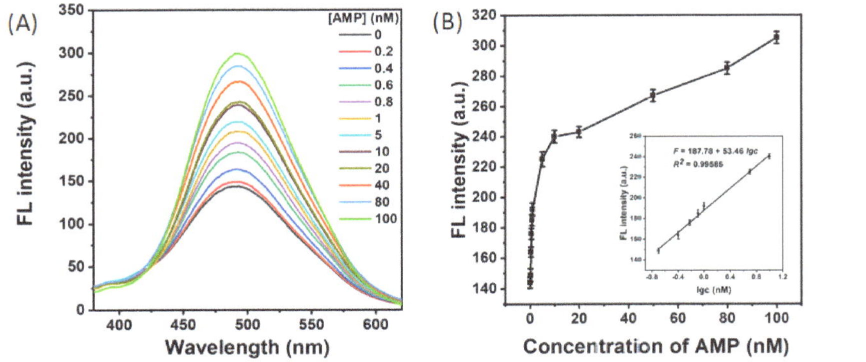

Figure 5. (**A**) Fluorescence responses of the aptasensor to various concentrations of AMP (0, 0.2, 0.4, 0.6, 0.8, 1, 5, 10, 20, 50, 80 and 100 nM). (**B**) The relationship between fluorescence intensity (λem = 493 nm) and the concentration of AMP. Inset: the linear relationship between fluorescence intensity and the logarithm value of AMP concentration.

3.3.5. Selectivity of the Sensor for Detection of AMP

The selectivity of the aptasensor for AMP was investigated. The controlled experiments were carried out by selecting several common interfering substances such as PEN, CHL, ROX, TET, IMI, and MET under the same experimental conditions. As shown in Figure 6, compared with the background signal, the fluorescence intensities were barely changed in the presence of interfering substances; obvious fluorescence intensity change was obtained only in the presence of AMP. These data indicated that this aptasensor exhibited a good performance in discriminating AMP from other interfering antibiotics and pesticides. Therefore, it has good potential to be applied in real complex samples.

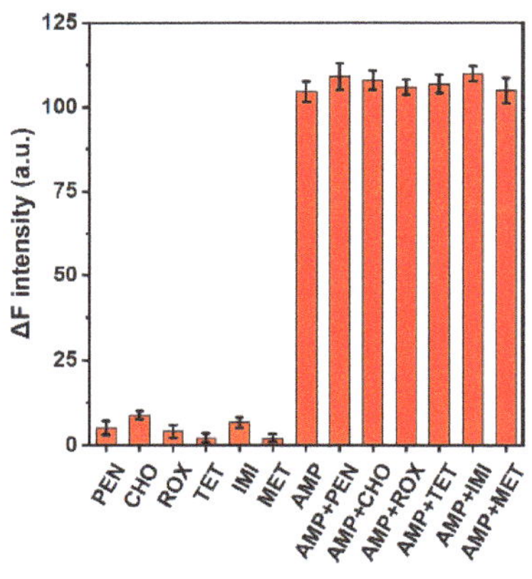

Figure 6. The selectivity of the aptasensor responds to AMP in the presence of PEN, CHO, ROX, TET, IMI, and MET. The concentration of AMP was 10 nM and that of the other disruptors was 100 nM. The error bars represent the standard deviation of three repeated measurements.

3.3.6. AMP Content in Real Samples

To investigate the applicability of the sensing strategy in real samples, it was applied to detect AMP residues in natural water. The experimental results demonstrated that the level of AMP residues in the water samples was below the limit of detection. Next, samples were tested by adding different amounts of AMP (0.2, 0.6 and 1 nM) in the river water samples, with three repeated measurements, and recovery from 99.6% to 102.85% was obtained

(Table S2). These results indicated that the sensing strategy has good anti-interference capabilities and therefore shows strong potential application in food safety fields.

4. Conclusions

In summary, the interaction between the amino group functionalized TPE molecule and nucleic acids was investigated in detail by fluorescence and UV–vis absorption spectroscopy in this paper. Experimental results showed that the amino group functionalized TPE could combine with DNA by intercalation and groove binding force. Binding and thermodynamic constants demonstrated that the fluorescence quenching mechanism of the TPE molecule by DNA is a static quenching procedure and that the electrostatic and hydrophobic interactions were the main driving force between TPE-Am and DNA. Based on the above, a sensitive and label-free "on-off-on" fluorescent aptamer sensor for AMP detection was developed. The sensor has a linear range of 0.2–10 nM with a limit of detection of 0.06 nM. It was used to detect the AMP content in real samples.

Supplementary Materials: The following supporting information can be downloaded at: https://www.mdpi.com/article/10.3390/bios13050504/s1, Figure S1: Molecular structure of TPE-Am; Figure S2: The effect of aptamer concentration; Figure S3: The effect of incubation time for the TPE-Am/Aptamer; Figure S4: The effect of incubation time for the TPE-Am/Aptamer/AMP; Figure S5: The effect of pH to the response of the sensing strategy; Figure S6: Repeatability and stability test of the aptasensor. The concentration of AMP is 10 nM; Table S1: Comparison of various detection methods for ampicillin; Table S2: The recovery and RSD for the detection of AMP in river water samples.

Author Contributions: Conceptualization, W.G. and X.W.; methodology, Y.F.; validation, Y.C. and X.Z.; investigation, H.Z. and F.Y.; writing—original draft preparation, W.G.; writing—review and editing, X.W.; supervision, X.W. All authors have read and agreed to the published version of the manuscript.

Funding: This research was funded by the National Innovation Training Program for College Students (No. 202110435184), the Shandong Provincial Higher Education Research Project on Undergraduate Teaching Reform (No. Z2021220), the Syncretic of Professional and Innovation Education Reform of Qingdao Agricultural University (No. ZCJG202105), and the Youth Innovation Team Project for Talent Introduction and Cultivation in Universities of Shandong Province (096-1622002).

Institutional Review Board Statement: Not applicable.

Informed Consent Statement: Not applicable.

Data Availability Statement: The data that supports the findings of this study are available in the supplementary material of this article.

Conflicts of Interest: The authors declare no conflict of interest. The funders had no role in the design of the study; in the collection, analyses, or interpretation of data; in the writing of the manuscript; or in the decision to publish the results.

References

1. Luo, J.; Xie, Z.; Lam, J.W.Y.; Cheng, L.; Chen, H.; Qiu, C.; Kwok, H.S.; Zhan, X.; Liu, Y.; Zhu, D.; et al. Aggregation-induced emission of 1-methyl-1,2,3,4,5-pentaphenylsilole. *Chem. Commun.* **2001**, 1740–1741. [CrossRef] [PubMed]
2. Li, Y.; Liu, S.; Ni, H.; Zhang, H.; Zhang, H.; Chuah, C.; Ma, C.; Wong, K.S.; Lam, J.W.Y.; Kwok, R.T.K.; et al. ACQ-to-AIE Transformation: Tuning Molecular Packing by Regioisomerization for Two-Photon NIR Bioimaging. *Angew. Chem. Int. Ed.* **2020**, *59*, 12822–12826. [CrossRef] [PubMed]
3. Sun, L.; Wang, X.; Shi, J.; Yang, S.; Xu, L. Kaempferol as an AIE-active natural product probe for selective Al^{3+} detection in Arabidopsis thaliana. *Spectrochim. Acta A Mol. Biomol. Spectrosc.* **2021**, *249*, 119303. [CrossRef] [PubMed]
4. Cai, X.; Liu, B. Aggregation-Induced Emission: Recent Advances in Materials and Biomedical Applications. *Angew. Chem. Int. Ed.* **2020**, *59*, 9868–9886. [CrossRef] [PubMed]
5. Kwok, R.T.K.; Leung, C.W.T.; Lam, J.W.Y.; Tang, B.Z. Biosensing by luminogens with aggregation-induced emission characteristics. *Chem. Soc. Rev.* **2015**, *44*, 4228–4238. [CrossRef] [PubMed]
6. Mei, J.; Leung, N.L.C.; Kwok, R.T.K.; Lam, J.W.Y.; Tang, B.Z. Aggregation-Induced Emission: Together We Shine, United We Soar! *Chem. Rev.* **2015**, *115*, 11718–11940. [CrossRef]

7. Li, B.; Liu, C.; Pan, W.; Shen, J.; Guo, J.; Luo, T.; Feng, J.; Situ, B.; An, T.; Zhang, Y.; et al. Facile fluorescent aptasensor using aggregation-induced emission luminogens for exosomal proteins profiling towards liquid biopsy. *Biosens. Bioelectron.* **2020**, *168*, 112520. [CrossRef]
8. Zhu, Y.; Xia, X.; Deng, S.; Yan, B.; Dong, Y.; Zhang, K.; Deng, R.; He, Q. Label-free fluorescent aptasensing of mycotoxins via aggregation-induced emission dye. *Dyes Pigm.* **2019**, *170*, 107572. [CrossRef]
9. Hu, F.; Cai, X.; Manghnani, P.N.; Kenry; Wu, W.; Liu, B. Multicolor monitoring of cellular organelles by single wavelength excitation to visualize the mitophagy process. *Chem. Sci.* **2018**, *9*, 2756–2761. [CrossRef]
10. Xu, L.; Zhang, S.; Liang, X.; Zhong, S.; Wang, B.; Li, Z.; Cui, X. Novel biocompatible AIEgen from natural resources: Palmatine and its bioimaging application. *Dyes Pigm.* **2021**, *184*, 108860. [CrossRef]
11. Zhu, Z.C.; Xu, L.; Li, H.; Zhou, X.; Qin, J.G.; Yang, C.L. A tetraphenylethene-based zinc complex as a sensitive DNA probe by coordination interaction. *Chem. Commun.* **2014**, *50*, 7060–7062. [CrossRef]
12. Ying, Y.M.; Tao, C.L.; Yu, M.X.; Xiong, Y.; Guo, C.R.; Liu, X.G.; Zhao, Z.J. In situ encapsulation of pyridine-substituted tetraphenylethene cations in metal-organic framework for the detection of antibiotics in aqueous medium. *J. Mater. Chem. C* **2019**, *7*, 8383–8388. [CrossRef]
13. Cai, Y.; Fang, J.; Wang, B.; Zhang, F.; Shao, G.; Liu, Y. A signal-on detection of organophosphorus pesticides by fluorescent probe based on aggregation-induced emission. *Sens. Actuators B Chem.* **2019**, *292*, 156–163. [CrossRef]
14. Wang, X.; Jiang, A.; Hou, T.; Li, H.; Li, F. Enzyme-free and label-free fluorescence aptasensing strategy for highly sensitive detection of protein based on target-triggered hybridization chain reaction amplification. *Biosens. Bioelectron.* **2015**, *70*, 324–329. [CrossRef]
15. Rothlisberger, P.; Hollenstein, M. Aptamer chemistry. *Adv. Drug Del. Rev.* **2018**, *134*, 3–21. [CrossRef]
16. Marrazza, G. Aptamer Sensors. *Biosensors* **2017**, *7*, 5. [CrossRef]
17. Kim, Y.S.; Raston, N.H.A.; Gu, M.B. Aptamer-based nanobiosensors. *Biosens. Bioelectron.* **2016**, *76*, 2–19.
18. Liang, C.X.; Wang, Y.M.; Zhang, T.T.; Nie, H.L.; Han, Y.M.; Bai, J. Aptamer-functionalised metal-organic frameworks as an 'on-off-on' fluorescent sensor for bisphenol S detection. *Talanta* **2023**, *253*, 123942. [CrossRef]
19. Rahbarimehr, E.; Chao, H.P.; Churcher, Z.R.; Slavkovic, S.; Kaiyum, Y.A.; Johnson, P.E.; Dauphin-Ducharme, P. Finding the Lost Dissociation Constant of Electrochemical Aptamer-Based Biosensors. *Anal. Chem.* **2023**, *95*, 2229–2237. [CrossRef]
20. Wang, H.; Ma, K.; Xu, B.; Tian, W.J. Tunable Supramolecular Interactions of Aggregation-Induced Emission Probe and Graphene Oxide with Biomolecules: An Approach toward Ultrasensitive Label-Free and "Turn-On" DNA Sensing. *Small* **2016**, *12*, 6613–6622. [CrossRef]
21. Zhang, S.; Ma, L.; Ma, K.; Xu, B.; Liu, L.; Tian, W. Label-Free Aptamer-Based Biosensor for Specific Detection of Chloramphenicol Using AIE Probe and Graphene Oxide. *ACS Omega* **2018**, *3*, 12886–12892. [CrossRef] [PubMed]
22. Wang, X.; Dong, S.; Gai, P.; Duan, R.; Li, F. Highly sensitive homogeneous electrochemical aptasensor for antibiotic residues detection based on dual recycling amplification strategy. *Biosens. Bioelectron.* **2016**, *82*, 49–54. [CrossRef] [PubMed]
23. Zhang, Z.H.; Li, X.; Liu, H.; Zamyadi, A.; Guo, W.S.; Wen, H.T.; Gao, L.; Nghiem, L.D.; Wang, Q.L. Advancements in detection and removal of antibiotic resistance genes in sludge digestion: A state-of-art review. *Bioresour. Technol.* **2022**, *344*, 126197. [CrossRef] [PubMed]
24. Luan, Y.X.; Wang, N.; Li, C.; Guo, X.J.; Lu, A.X. Advances in the Application of Aptamer Biosensors to the Detection of Aminoglycoside Antibiotics. *Antibiotics* **2020**, *9*, 787. [CrossRef]
25. Zhou, J.J.; Xu, Z.Q. Simultaneous separation of 12 different classes of antibiotics under the condition of complete protonation by capillary electrophoresis-coupled contactless conductivity detection. *Anal. Methods* **2022**, *14*, 174–179. [CrossRef]
26. Li, F.; Li, X.; Zhu, N.; Li, R.; Kang, H.; Zhang, Q. An aptasensor for the detection of ampicillin in milk using a personal glucose meter. *Anal. Methods* **2020**, *12*, 3376–3381. [CrossRef]
27. Simmons, M.D.; Miller, L.M.; Sundstrom, M.O.; Johnson, S. Aptamer-Based Detection of Ampicillin in Urine Samples. *Antibiotics* **2020**, *9*, 655. [CrossRef]
28. Luo, W.; Hansen, E.B., Jr.; Ang, C.Y.W.; Deck, J.; Freeman, J.P.; Thompson, H.C. Simultaneous Determination of Amoxicillin and Ampicillin in Bovine Milk by HPLC with Fluorescence Detection. *J. Agric. Food Chem.* **1997**, *45*, 1264–1268. [CrossRef]
29. Guliy, O.I.; Evstigneeva, S.S.; Bunin, V.D. Bacteria-based electro-optical platform for ampicillin detection in aquatic solutions. *Talanta* **2021**, *225*, 122007. [CrossRef]
30. Mehlhorn, A.; Rahimi, P.; Joseph, Y. Aptamer-Based Biosensors for Antibiotic Detection: A Review. *Biosensors* **2018**, *8*, 54. [CrossRef]
31. Sun, P.; Li, H.-L.; Wang, X.-Z. Investigation of interaction between nitrobenzene and bovine serum albumin by spectroscopic methods. *Chem. Res. Appl.* **2014**, *26*, 804–808.
32. Zeng, Y.; Chang, F.; Liu, Q.; Duan, L.; Li, D.; Zhang, H. Recent Advances and Perspectives on the Sources and Detection of Antibiotics in Aquatic Environments. *J. Anal. Methods Chem.* **2022**, *2022*, 5091181. [CrossRef]
33. Liu, J.; Zhang, H.; Hu, L.; Wang, J.; Lam, J.W.Y.; Blancafort, L.; Tang, B.Z. Through-Space Interaction of Tetraphenylethylene: What, Where, and How. *J. Am. Chem. Soc.* **2022**, *144*, 7901–7910. [CrossRef]
34. Muralisankar, M.; Bhuvanesh, N.S.P.; Sreekanth, A. Synthesis, X-ray crystal structure, DNA/protein binding and DNA cleavage studies of novel copper(II) complexes of N-substituted isatin thiosemicarbazone ligands. *New J. Chem.* **2016**, *40*, 2661–2679. [CrossRef]

35. Yu, X.; Ying, Y.; Lu, S.; Qing, Y.; Liu, H.; Li, X.; Yi, P. The fluorescence spectroscopic study on the interaction between imidazo [2,1-b]thiazole analogues and bovine serum albumin. *Spectrochim. Acta A Mol. Biomol. Spectrosc.* **2011**, *83*, 322–328. [CrossRef]
36. Sun, Y.; Wei, S.; Yin, C.; Liu, L.; Hu, C.; Zhao, Y.; Ye, Y.; Hu, X.; Fan, J. Synthesis and spectroscopic characterization of 4-butoxyethoxy-N-octadecyl-1,8-naphthalimide as a new fluorescent probe for the determination of proteins. *Bioorg. Med. Chem. Lett.* **2011**, *21*, 3798–3804. [CrossRef]
37. Hu, Y.J.; Liu, Y.; Xiao, X.H. Investigation of the interaction between Berberine and human serum albumin. *Biomacromolecules* **2009**, *10*, 517–521. [CrossRef]

biosensors

MDPI

Article

Fiber Optic Localized Surface Plasmon Resonance Sensor Based on Carboxymethylated Dextran Modified Gold Nanoparticles Surface for High Mobility Group Box 1 (HMGB1) Analysis

Chang-Yue Chiang [1,*], Chien-Hsing Chen [2] and Chin-Wei Wu [1,*]

[1] Graduate School of Engineering Science and Technology and Interdisciplinary Program of Engineering, National Yunlin University of Science and Technology, Yunlin 64002, Taiwan
[2] Department of Biomechatronics Engineering, National Pingtung University of Science and Technology, Pingtung 91201, Taiwan; garychc@mail.npust.edu.tw
* Correspondence: chiangcy@yuntech.edu.tw (C.-Y.C.); wucw@yuntech.edu.tw (C.-W.W.); Tel.: +886-5-5342601 (ext. 4014) (C.-Y.C.); +886-5-5342601 (ext. 4015) (C.-W.W.)

Abstract: Rapid, sensitive, and reliable detection of high mobility group box 1 (HMGB1) is essential for medical and diagnostic applications due to its important role as a biomarker of chronic inflammation. Here, we report a facile method for the detection of HMGB1 using carboxymethyl dextran (CM-dextran) as a bridge molecule modified on the surface of gold nanoparticles combined with a fiber optic localized surface plasmon resonance (FOLSPR) biosensor. Under optimal conditions, the results showed that the FOLSPR sensor detected HMGB1 with a wide linear range (10^{-10} to 10^{-6} g/mL), fast response (less than 10 min), and a low detection limit of 43.4 pg/mL (1.7 pM) and high correlation coefficient values (>0.9928). Furthermore, the accurate quantification and reliable validation of kinetic binding events measured by the currently working biosensors are comparable to surface plasmon resonance sensing systems, providing new insights into direct biomarker detection for clinical applications.

Keywords: high mobility group box 1; carboxymethyl-dextran; gold nanoparticle; localized surface plasmon resonance; biosensor; kinetic binding

Citation: Chiang, C.-Y.; Chen, C.-H.; Wu, C.-W. Fiber Optic Localized Surface Plasmon Resonance Sensor Based on Carboxymethylated Dextran Modified Gold Nanoparticles Surface for High Mobility Group Box 1 (HMGB1) Analysis. *Biosensors* **2023**, *13*, 522. https://doi.org/10.3390/bios13050522

Received: 29 March 2023
Revised: 28 April 2023
Accepted: 4 May 2023
Published: 6 May 2023

1. Introduction

In recent years, clinical studies have pointed out chronic inflammation such as cancers, diabetes, cardiovascular diseases, allergies, oral diseases, obesity, strokes, and arthritis, will cause mass mortality and significantly increase treatment costs. It has been proven that these chronic diseases are closely related to high mobility group box 1 (HMGB1) [1]. HMGB1 has a molecular weight of 29 kDa and comprises 216 single-chain amino acid polypeptides connected to an acidic C-terminal tail through a short alkaline hinge [2,3]. HMGB1 is a typical damage-associated molecular pattern (DAMP) and a central mediator of lethal inflammation [4,5]. The activated immunocytes or necrotic cells combine with advanced glycation end products (RAGE) and Toll-like receptors upon release from the cells and initiate immunoreaction (inflammation, repair, and recombination) [6–8]. Finally, the immune reaction process can induce tissue destruction, fibrosis, necrosis, and death. The general methods to detect HMGB1 include enzyme-linked immunosorbent assay (ELISA) [9,10], Western blot [11], liquid chromatography-mass spectrometry (LC-MS) [12], surface plasmon resonance (SPR) [13,14], electrophoretic mobility shift assay (EMSA) [15], and electrochemical immunosensor [16]. ELISA is the most familiar among these techniques. It is regarded as one of the most convenient, accurate, and reliable methods. However, ELISA requires a long processing time, skilled professionals for the operation, and complex sample preparation. Western blot only performs semiquantitative detection and is time-consuming. LC-MS and SPR provide good quantitative sensitivity, but the instruments are expensive, large, and unlikely to be miniaturized. Therefore, a rapid, accurate, and economical HMGB1 detection method is urgently needed.

The localized surface plasmon resonance (LSPR) biosensor technology is a real-time, rapid response, and high-sensitivity technology [17–19]. Gold nanoparticles (AuNPs), as the key components of sensors, have received great attention due to their unique optical and electronic properties and high biocompatibility. They have been widely used to detect biological or chemical molecules, such as protein–protein and protein-nucleic acid binding affinities [18,20,21]. For example, Nath et al. used a label-free optical biosensor to detect streptavidin with a detection limit of 16 nM [22]. Jeon et al. developed a disposable LSPR-based colorimetric sensor for detecting cortisol in the serum [23]. The advantages of LSPR biosensor technology are real time, rapid response, and high sensitivity. However, the interaction between extremely low-concentration samples or low-molecular-weight analytes (e.g., peptides) is still challenging. Due to the refractive index variation induced by molecules being tiny, the difficulty in measurement is increased [24–26]. To improve the sensitivity of LSPR biosensors, the previously reported FOLSPR biosensor platform is a promising approach [27–32].

In previous studies, mixed self-assembled monolayer (SAM) of 11-mercaptoundecanoic acid (MUA)/6-mercapto-1-hexanol (MCH) bridging molecules were used [33,34]. The MUA is a long-chain molecule with a -COOH functional group. It forms a peptide bond (-CO-NH-) with the amino groups on the surface of biological recognition molecules for a chemical covalent bonding reaction. The MCH has a short chain with a -OH functional group. It is a dilution of thiolate, reducing adhesion and non-specific adsorption between protein molecules [30]. However, the challenge is that the film thickness and spatial structure (steric hindrance) after self-assembly functionalization lead to fewer binding sites for antibody connections and worse anti-adhesion effects [35]. As a result, the detection results are unstable, and the sensitivity is reduced. To overcome these limitations, Carboxymethyl-dextran (CM-dextran) as a binder provides a high surface-to-volume ratio and more -COOH functional groups [36,37]. It increases the conjugate binding sites of biomolecules improving the sensor's responses [38,39]. CM-dextran is a branched glucose polysaccharide containing the 1,6-α-d-glucopyranosyl bond [40]. It is biocompatible, highly water soluble, highly anti-adhesive, and not toxic. Moreover, it has been widely used in biosensors [41–45].

This study proposes a novel CM-dextran FOLSPR biosensor related to a functionalized CM-dextran-modified binding peptide technique. Chemical covalent bonding is employed in this technique. The experiment used 1-ethyl-3-(3-dimethylaminopropyl)-carbodiimide hydrochloride (EDC) and N-hydroxysuccinimide (NHS) to perform intermolecular cross-linking of CM-dextran and then to modify the biological recognition molecules. The HMGB1 biological standard was tested in the optimal experimental conditions (including modification time, modification concentration, EDC/NHS concentration, antibody modification concentration, and antibody incubation time). The results show that the HMGB1 sensing chip had a wide linear range, high reproducibility, and excellent limit-of-detection (LOD). In addition, compared with the existing MUA/ MCH mixed self-assembled monolayer, the bridge-based HMGB1 molecular detection doubles the detection sensitivity. Notably, this is the first report on detecting HMGB1 in FOLSPR biosensors using CM-dextran as amine coupling reagents to immobilize recognition molecules. This study suggests that the technique can be widely used in fields like biological analysis and detection in food, environment, and clinical medicine.

2. Materials and Methods

2.1. Materials and Reagents

All chemical reagents are analytical-graded reagents. Hydrogen tetrachloroaurate trihydrate ($HAuCl_4$), Phosphate buffered saline (PBS buffer), trisodium citrate solution ($C_6H_5Na_3O_7$, $\geq$99%), (3-Mercaptopropyl)methyldimethoxysilane (MPDMS, >95%), cystamine dihydrochloride (cystamine), dextran 70 (MW $\approx$ 70,000), sodium periodate ($NaIO_4$),

11-mercaptoundecanoic acid (MUA; ≥95%), 6-mercapto-1-hexanol (MCH; ≥97%), 1-ethyl-3-(3-dimethylaminopropyl)-carbodiimide hydrochloride (EDC), N-hydroxysulfosuccinimide (NHS), ethanolamine (EA), mouse IgG, streptavidin, monoclonal anti-HMGB1 antibody, and high mobility group box 1 (HMGB1) protein were from Sigma-Aldrich (St. Louis, MO, USA). Sulfuric acid (H_2SO_4; ≥98%), hydrogen peroxide (H_2O_2), ethanol, and acetate buffer were purchased from Fluka (Buchs, Switzerland). Ultrapure deionized water (18.2 MΩ·cm^{-1}, Milli-Q pure water purification system, Millipore Ltd., Burlington, MA, USA) was used for preparing solutions. The optical fiber probe was multimode plastic-clad silica optical fiber (model F-MBC, Newport), with core and cladding diameters of 400 and 430 μm, respectively, bought from Instant NanoBiosensors Co., Ltd. (Taipei, Taiwan). Sensing chips (poly (methyl methacrylate) (PMMA) plates) were prepared using a CO_2 laser engraving machine (New Taipei, Taiwan).

2.2. Preparation of AuNP Probe

The synthesis of aqueous spherical AuNPs was prepared by the oxidation-reduction method [34]. A 5.2 mL $HAuCl_4·3H_2O$ solution at a concentration of 2.43 mM was diluted with 14.8 mL distilled (DI) water to make a 20 mL auric salt aqueous solution. This solution was boiled and slowly mixed with 2.4 mL 1% sodium citrate-reducing solution. The solution slowly turned transparent (Au^{2+}) from light yellow (Au^{3+}), then into dark black (Au^{+}), and claret red AuNPs were formed at last. Afterward, the solution was stirred for 10 min to ensure the reduction process, kept still, and cooled to room temperature. The sol-gel method performed immobilization of AuNP on the optical fiber probe. A 2% MPDMS was prepared in toluene and pre-hydrolyzed for 12 h. The fiber surface of the optical fiber probe was cleaned with soapy water, methanol, and DI water using an ultrasonic bath. The surface was then pickled with piranha solution (H_2SO_4 and H_2O_2 solution in a volume ratio of 7:3) allowing the surface of the optical fiber probe to carry more -OH functional groups. Afterward, it was cleaned with DI water. The optical fiber probe was immersed in the pre-hydrolyzed 2% MPDMS/toluene solution for 6 h to perform the hydrolysis-condensation reaction of SiO_2 functional groups. In this order, the optical fiber probe was washed with 1:1 ethanol/toluene solution, ethanol, and DI water and then dried with nitrogen. Afterward, the optical fiber probe was immersed in AuNPs solution for 30 min, and the AuNPs were stably bonded to the surface of the optical fiber probe through S-Au covalent bonding. Finally, the optical fiber probe was washed with DI water to remove the AuNPs not bonded on the optical fiber probe. After nitrogen drying, the UV-Visible/NIR Spectrophotometer (Hitachi UH5700, Tokyo, Japan), transmission electron microscopy (TEM, JEM-2100Plus, Tokyo, Japan), and ultra high-resolution thermal field emission scanning electron microscope (FESEM, JEOL JSM-7610FPlus, Tokyo, Japan) were used to measure the absorption spectra of the synthetic AuNPs and the AuNPs on the surface of the immobilized optical fiber probe. The spectral characteristic peak absorption wavelength position and absorption strength of the spherical AuNPs solution were identified to verify the shape and size. The test was repeated three times for each data point, and the data were represented by an average value and standard deviation (mean ± SD).

2.3. Microfluidic Sensing Chip and FOLSPR Sensing System

Microfluidic sensing chip was composed of a PMMA substrate. It comprises a cover and bottom plates. The dimensions of each cover plate are 25 mm (width) × 50 mm (length) × 2 mm (thickness). The actual surface of AuNP coated zone is 25.1 mm^2 (0.4 mm (the core diameter) × π × 20 mm (length)). The bottom plate involved a microfluidic technique, and the sample micro-channel is 800 μm (depth) × 800 μm (width) to accommodate a 730 μm optical fiber probe with a volume of 35 μL. A CO_2 laser engraving machine processed the cover plate. A micro-channel access hole of about 0.2 cm diameter could be connected to the plastic flow pipe to import and export fluids. The sensing chip was packaged with 3M double-sided adhesive to bond the cover plate and bottom plate. Afterward, DI water was injected into the sensing area for cleaning. Finally, a complete microfluidic sensing chip was obtained, as shown in Figure 1a.

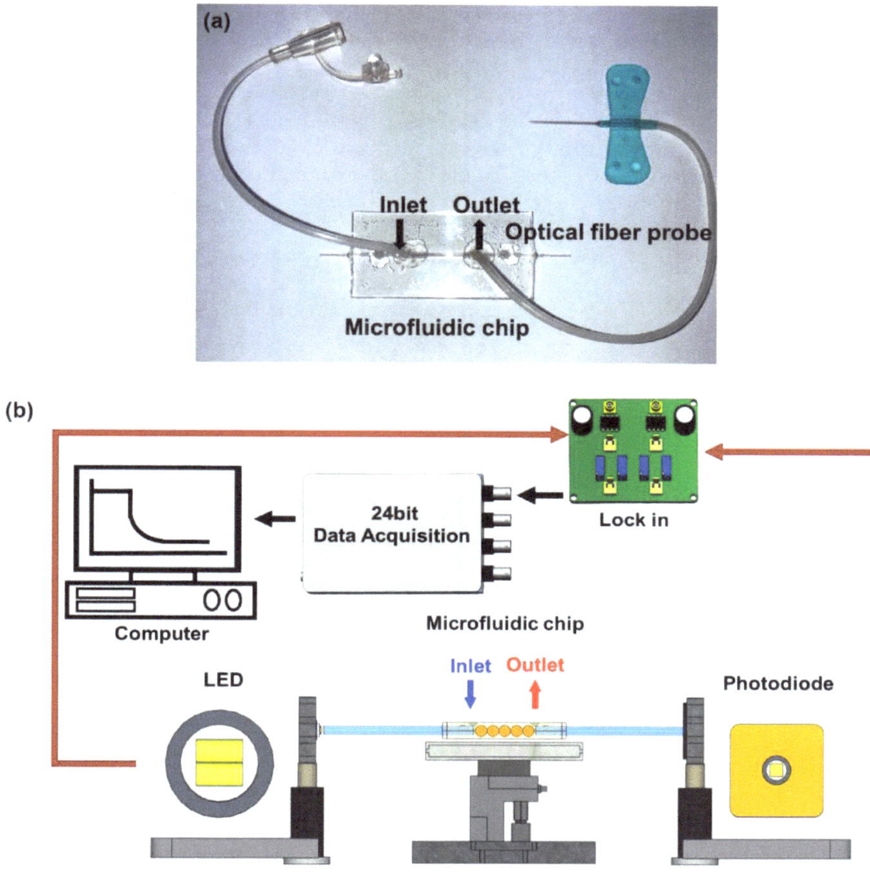

Figure 1. FOLSPR sensor. (**a**) Microfluidic sensing chip, and (**b**) schematic representation of the experimental setup of the sensing system.

Figure 1b is the schematic diagram of a FOLSPR biosensor system. The sensing system used a green LED with a microlens (model IF-E93, Industrial Fiber Optic, Inc., Tempe, AZ, USA) as the light source, with a peak wavelength of 530 nm and a circuit system (including 1 kHz square wave signal and signal amplifier). One end of the light source was coupled to the inside of the optical fiber sensing chip for multiple total internal reflections. This helps generate evanescent waves on the surface of the sensing area. These evanescent waves excite AuNPs to generate the LSPR effect. Meanwhile, the other end was connected to the light exit of photodetection. The experiment used a high-stability photodiode (S1336-18BK, Hamamatsu, Tokyo, Japan) to detect the light intensity. The circuit system was linked with a data acquisition card lock-in module (dynamic signal acquisition module USB-9234 and Lab-VIEW 2019 software, National Instrument, Austin, TX, USA) to transform the light into voltage signals (real-time light intensity). The generated voltage signals were captured and analyzed by a computer to transform the light into voltage signals (real-time light intensity). The generated voltage signals were captured and analyzed by a computer.

2.4. Preparation of CM-Dextran Solution and Functionalization of CM-Dextran Sensing Chips

Carboxymethyl-dextran (CM-dextran) was synthesized using the previously reported protocol with minor modification [46,47]. Briefly, CM-dextran was formed by dissolving 3 g of dextran in 10 mL of NaOH 100 mM, adding 1 M solution of bromoacetic acid, and then shaking at room temperature in the dark for 16 h. The functionalization of the sensing chip was performed by chemical covalent bonding (Figure 2). To form abundant amine (-NH_2) groups on the AuNP surface of the optical fiber, which was modified with cystamine. CM-dextran is then dissolved in phosphate-buffered saline (PBS, pH = 7.4) and mixed with EDC and NHS for activation, allowing cystamine to react with the carboxymethyl groups on CM-dextran. Subsequently, a second activation step using EDC and NHS was used for intermolecular cross-linking of CM-dextran, forming peptide bonds (-CO-NH-) with amino groups of the antibody to complete the antibody fixed. Residual active groups were saturated by injection of ethanolamine. UV/Vis-NIR spectrum and Energy-dispersive X-ray spectroscopy (EDS) were used to identify material analysis in the experimental process.

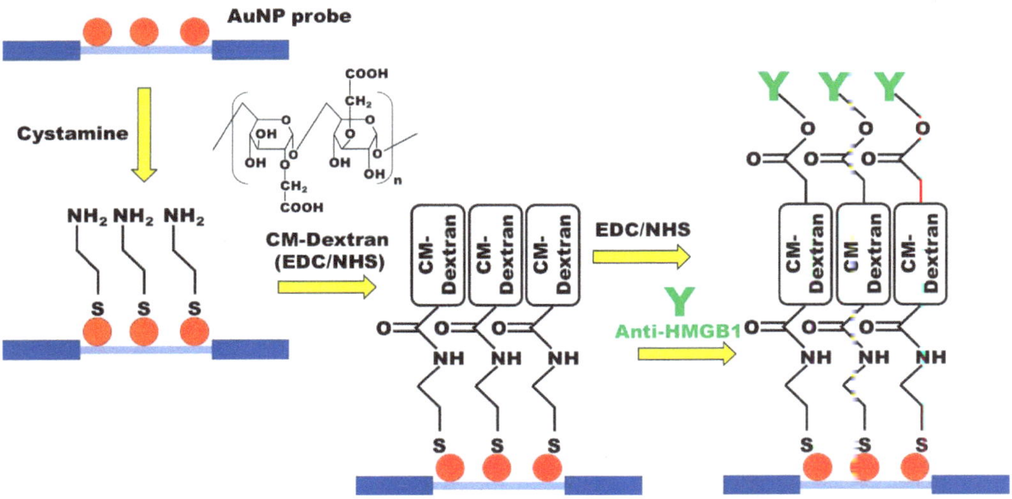

Figure 2. The CM-dextran-based FOLSPR chip experimental conditions.

The optimum conditions for the modification of the sensing chip were discussed in the experiment. First, the modified AuNPs optical fiber was used for sensing chip packaging. Then, 35 μL 0.02 M cystamine solution was injected into the sensing chip for 30 min to form AuNP-cystamine. Afterward, DI water was applied to remove uncombined cystamine. The CM-dextran (1%, 2.5%, 5%, 7.5%, and 10%)/EDC (0.2 M)/NHS (0.05 M) mixed solutions were injected on the surface of AuNP-cystamine, with a CM-dextran immobilization time of 0.5, 1, 1.5, 2, and 2.5 h. The AuNP-cystamine- CM-dextran was formed on the surface. It was removed by using ionized water. The -COOH functional group on the surface of CM-dextran was activated using the mixed liquor of EDC (0.2 M) and NHS (0.05 M) for 20 min. A 50 μL anti-HMGB1 (50, 75, 100, 125, and 150 μg/mL) was applied. The incubation time was 0.5, 1, 1.5, and 2 h for AuNP-cystamine-CM- dextran-anti-HMGB1 functionalization. The uncombined anti-HMGB1 was removed by PBS buffer. Finally, 1 M ethanolamine (pH 8.5) aqueous solution was used for 7.5 min to inactivate the unreacted -COO- and prevent nonspecificity. Inject PBS solution into the sensor chip for verification after step-by-step modification. The gradual modification process of the sensing chip experiment was verified by UV/Vis-NIR spectrum. The FOLSPR system was used for real-time signal measurement.

2.5. Preparation of Functional MUA/MCH Sensing Chip

MUA/MCH sensing chips were functionalized (Figure 3) in accordance with the previous procedure with slight modification [30]. The optical fiber of modified AuNPs was immersed in the ethanol solution with MUA (2 mM) and MCH (2 mM) in a mole ratio of 1:4. It was kept still at room temperature for 12 h to form the mixed SAM. It was then washed with ethanol and dried with N_2. Afterward, the AuNPs-MUA/MCH optical fiber was used for sensing chip packaging. A 35 μL aqueous solution of EDC (0.2 M) and NHS (0.05 M) was injected into the sensing chip to activate the -COOH group for 30 min. Then, the DI water was injected for washing. A 35 μL anti-HMGB1 (100 μg/mL) solution was injected for a 1 h incubation reaction. The PBS buffer was injected into the solution to wash out the uncombined anti-HMGB1. To inactivate the unreacted -COO-, 1 M aqueous ethanolamine solution with pH 8.5 (7.5 min) was used for washing. Finally, the PBS solution was injected to wash the immobilized antibody's surface. The FOLSPR system was used for real-time signal measurement in the gradual modification process.

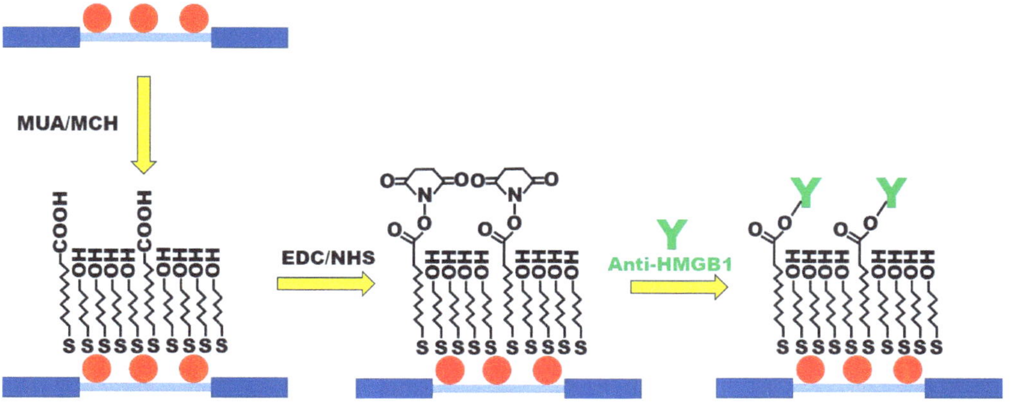

Figure 3. The MUA/MCH-based FOLSPR chip experimental conditions.

2.6. Sample Preparation

The stock HMGB1 standard solution was diluted with PBS and stored in a refrigerator at −20 °C. The sample was prepared and used in the same week to avoid the inactivated HMGB1-inducing errors. The concentration range of the prepared HMGB1 standard solution was 1×10^{-10} to 1×10^{-6} g/mL, which was stored at 4 °C for further use. The PBS buffer of pH 7.4 was injected into the sensing chip as the baseline. The prepared HMGB1 standard solutions (1×10^{-10}, 1×10^{-9}, 1×10^{-8}, 1×10^{-7}, 1×10^{-6} g/mL) were tested to obtain corresponding signal responses. Each was tested by three different sensing chips, and the data were shown by average values and standard deviations (average value ± standard deviation). A linear relationship between signal responses and concentrations was drawn to obtain the calibration curve. Finally, Origin 2021 (OriginLab, Northampton, MA, USA) was used for statistical analyses of data.

3. Results

3.1. Material Analysis

Figure 4a shows the absorption spectrum of synthetic aqueous spherical AuNP solution in a UV-Visible/NIR spectrophotometer. The maximum value of absorption peak occurs at 521 ± 0.6 nm. Figure 4b shows the TEM image of the aqueous spherical AuNP solution. The AuNPs were complete spheres with no aggregation and a mean particle size of 12.2 ± 0.6 nm. Figure 4c shows the extinction spectrum of the fiber surface modified AuNPs measured by the self-mounted fiber-optic spectral system as shown in Figure 4c insert, including the spectrometer (Ocean Optics, QEPro, optical resolution: 1.2 nm), a white light source of the 350 nm ~ 1700 nm (OTO Photonics, LS-HA) and two optical fiber of the core is 400 μm (Ocean Optics, numerical aperture: 0.22) [48]. The absorption peak of AuNPs occurred at 532 ± 0.8 nm. Figure 4d shows AuNPs on the sensing chip fiber surface measured by ultra-high resolution thermal FESEM. It was obvious that the nanoparticles were spherically dispersed on the fiber surface without aggregation, and FESEM calculated the AuNPs particle size. The results show that the mean particle size is 13.4 ± 1.2 nm (200 particles).

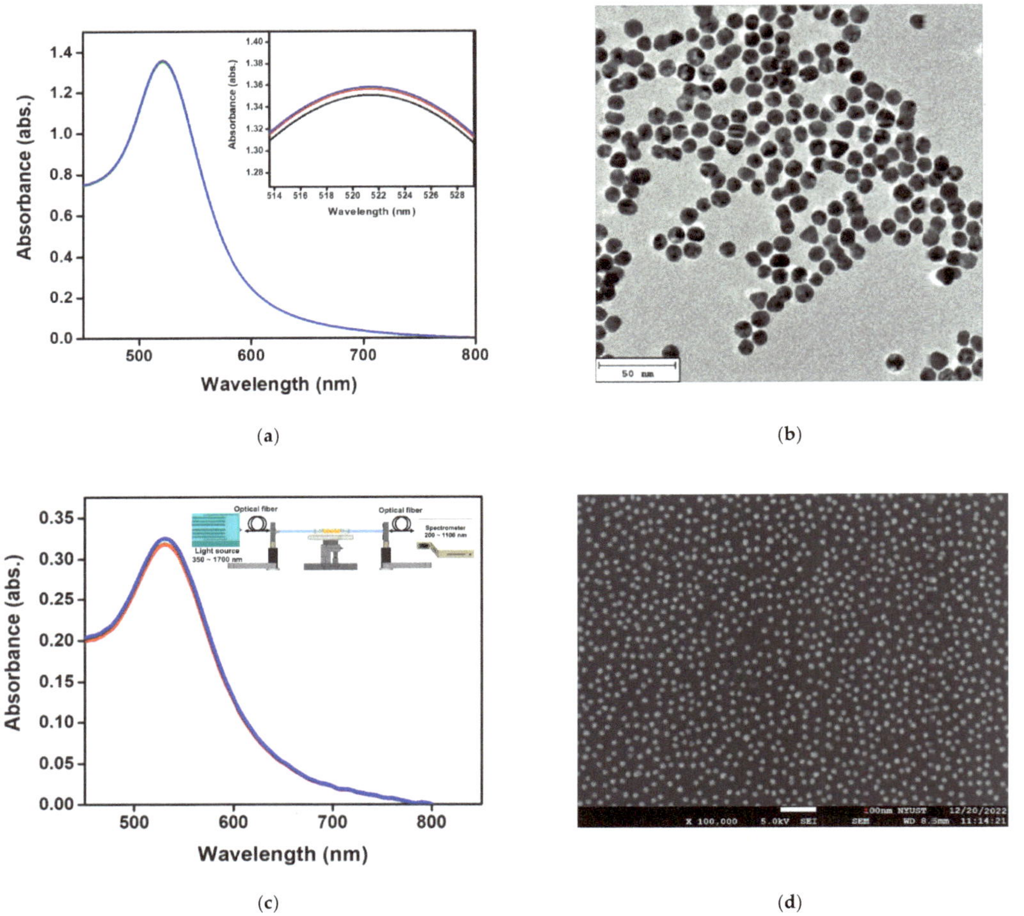

Figure 4. Structural characterizations of materials. (**a**) Absorption spectra of Au nanoparticles (AuNPs) in the aqueous medium in the visible region; (**b**) TEM image of AuNPs; (**c**) a fiber optic spectrometer was used to measure the absorption spectrum of AuNPs on the surface of the probe in the visible region in the aqueous medium; (**d**) FESEM image of AuNPs on the fiber core surface. The test was repeated three times for each data point ($n = 3$), and the data were represented by an average value and standard deviation (average value ± standard deviation).

In this study, we use UV-Vis spectroscopy to monitor the molecular step-by-step modification process of a sensor chip experiment. Figure 5a shows the gradual functionalization process of AuNP, cystamine, CM-dextran/EDC/NHS, EDC/NHS, anti-HMGB1, and ethanolamine in the optical fiber sensing area measured by a fiber optic spectrometer (Measured in the PBS buffer). The figure shows that the characteristic peak and wavelength of AuNPs on the optical fiber surface sensing area were at 531.1 nm at the beginning (absorbance was 0.3125). After gradual modification, it was observed that the last characteristic peak of AuNP-cystamine-CM-dextran/EDC/NHS-EDC/NHS-anti-HMGB1-ethanolamine was at 537.6 nm (absorbance was 0.35886), showing a redshift (displacement was 8 nm). The results show the variation of the local refractive index near the AuNPs surface and the electron loss on the AuNPs surface. It indirectly proves the immobilization of molecules on the surface of AuNPs or the conjugation of recognition molecules analyte. In addition, the absorbance peak increased from 0.3125 to 0.35886. This is consistent with the observation of the absorption increase and spectral redshift after combining chemical and biological molecules proved in prior studies [49–51]. Therefore, these results prove the successful functionalization of the sensing probe modified by AuNP-cystamine-CM-dextran-EDC/NHS-EDC/NHS-anti-HMGB1-ethanolamine. Figure 5b shows the elemental analysis of AuNP-cystamine-CM dextran surface in the optical fiber sensing area measured by Energy-dispersive X-ray spectroscopy (EDS). The C, O, and Au signals from the functionalization of AuNP-cystamine-CM dextran were detected, meaning that the components were CM dextran and Au.

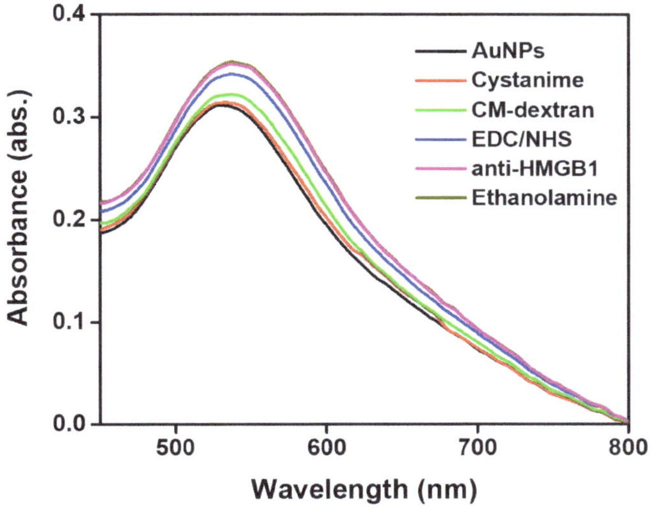

(**a**)

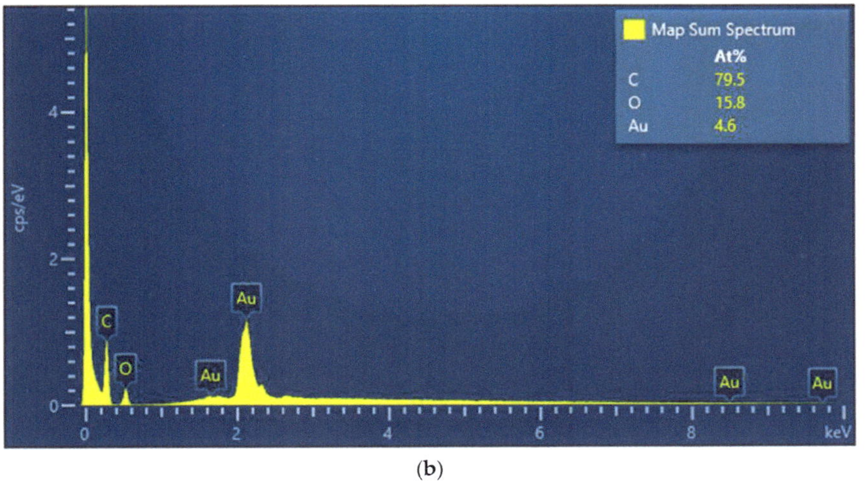

(**b**)

Figure 5. (**a**) Gradual functionalization of AuNP, cystamine, CM-dextran/EDC/NHS, EDC/NHS, anti-HMGB1, and ethanolamine in the probe modification spectrogram (measured in the PBS buffer); (**b**) EDS measurement verification.

3.2. Principle of FOLSPR Sensors

It is well-known that the physical characteristics of LSPR generated on the surface of AuNPs are susceptible to the external refractive index [17,52–54]. Especially during biological detections, the density of recognition molecules and molecular weight will lead to the sensitivity of the biomolecular binding [55,56]. This study is the first to use CM-dextran as bridging molecules to increase the density of recognition molecules and improve the sensitivity of FOLSPR sensing technology. The principle of FOLSPR is that after the light is led in the optical fiber. After multiple total internal reflections in the optical fiber core, the evanescent wave energy is generated on the surface of the cladding-off cylindrical fiber. When the surface AuNPs absorb the evanescent wave energy, the electrons on the surface of AuNPs generate the LSPR effect [57,58]. The intensity of this resonance will vary with the number of modification molecules on the surface of AuNPs, resulting in differences in light intensity. The difference in light intensity will increase the absorbance or change the refractive index of the external environment in the bonding process of the recognition molecules (e.g., capture antibodies, DNA, RNA, primers, or aptamers) and analytes. Therefore, the light intensity change can be recorded instantly as the biosensor analysis tool [48]. In this study, light signal I_0 is the light intensity response in the blank solution under a PBS buffer. I_A is the light intensity response under the analyte solution. The sensor response is defined as $(I_0 - I_A)/I_0 = \Delta I/I_0$ and is implemented by comparing the collected light intensity from a fiber probe immersed in a sample solution (I_A) containing the target protein at defined concentrations to that of immersed in a blank solution (I_0), i.e., the normalized light intensity change. The light's real-time signal response intensity declines as the analyte concentration increases. The quantitative testing analysis was obtained by concentration and signal responses.

In order to understand the optical properties of biomolecules interacting with nanoparticles, this study used discrete dipole approximation (DDA) to simplify that model and simulate nanoparticle surface modification biomolecules' extinction spectrum [59]. It can be used to design a real-time light intensity detection system. The simulation structure and parameters are shown in Figure 6a and Table 1. The simulation wavelength range is 450–800 nm. The simulation structure contained a circular AuNP with a size of 13 nm (reference Figure 4b TEM image). The refractive index ($N_{metal} = n + ik$) had real part n and imaginary part k [60]. Biomolecules were adsorbed on the circular AuNP surface. Biomolecules consist of three spheres with a radius and refractive index of 3.2 nm and 1.45, respectively. Their size is about a 13 nm Y-shaped structure, as shown in Figure 6b. The external environment was a phosphate-buffered saline (PBS buffer) with a refractive index of 1.42. The results of the extinction spectrum of nanoparticle surface modification biomolecules simulated by DDA are shown in Figure 7. The LSPR peak wavelength changes were not apparent before and after the AuNP surface modification of biomolecules. Additionally, the extinction cross-section increased from 108.3 nm^2 to 120.9 nm^2 at the wavelength of 530 nm with a variable rate of about 11.1%. It shows that it is easier to observe the change of light intensity as the indicator of biosensor analysis than the wavelength shift. Based on the experimental results, the before (at 530 nm, Figure 5 black line AuNPs) and after (Figure 5 purple line anti-HMGB1) absorbance was 0.31028 and 0.35317, respectively. The AuNPs surface was modified with biomolecules. With an increased absorbance, the light intensity decreased after the AuNPs surface was modified with biomolecules. The light intensity variation rate was 9.4% ($10^{-0.31028} - 10^{-0.35317}/10^{-0.31028}$), meaning the biomolecules modified the AuNPs surface. This finding matched the trend of simulation results. It is also the reference for choosing the wavelength of 530 nm as the light source for the FOLSPR biosensor system in this study.

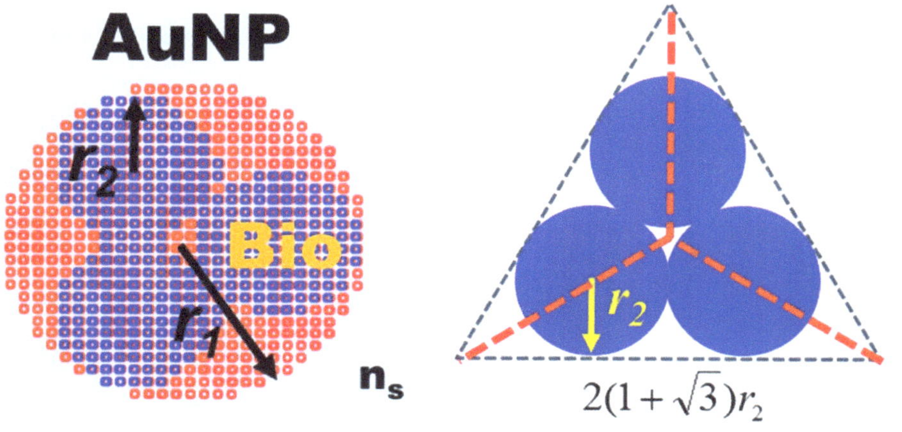

(a) (b)

Figure 6. Structural diagram of AuNP surface modification biomolecules simulated by DDA. (**a**) Structure of AuNP surface modification biomolecules; (**b**) Biomolecular structure.

Table 1. Parameter list of AuNP surface modification biomolecules simulated by DDA.

Gold Sphere Nanoparticle	**Diameter ($2r_1$)**	**13.0 nm**
	Refractive Index (N_{metal})	**Ref [60]**
Biomolecular	Radius (r2)	3.2 nm
	Refractive index (n_B)	1.45
Wavelength	450~800 nm	
Cube size	0.5 nm	
Surrounding medium (n_s)	1.42	

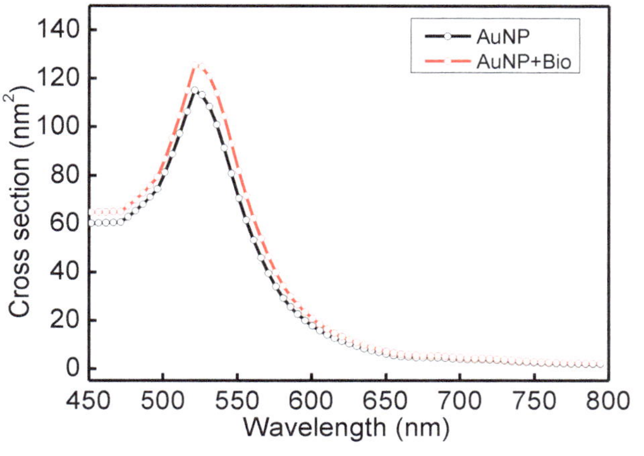

Figure 7. Extinction spectrum of AuNP surface modification biomolecules simulated by DDA.

3.3. Optimization of the Sensor

In this study, the sensing chip formed surface functionalization through chemical bonding reactions. In each experiment, 10^{-8} g/mL HMGB1 was injected for 15 min of molecular binding to obtain the signal response. First, the AuNP-cystamine surface was modified individually with 1%, 2.5%, 5%, 7.5%, and 10% (weight%) CM-dextran solution. As shown in Figure 8a, the increase is apparent when the concentration was 1%, 2.5%, and 5%. When the concentration was 5%, the response of HMGB1 was the largest ($\Delta I/I_0 = 0.419 \pm 0.011$). Signals gradually decreased after 7.5% and 10%. This is because the viscous CM-dextran solution resulted in steric hindrance. Therefore, the concentration of CM-dextran was set at 5% for subsequent experiments. The CM-dextran immobilization time was 0.5, 1, 1.5, 2, and 2.5 h. It was observed that the immobilization gradually reached saturation in 1.5 h (Figure 8b). Therefore, the subsequent work used 1.5 h as the CM-dextran formation time. The AuNP-cystamine-CM-dextran was formed on the surface of the probe, and the unbonded CM-dextran was removed by ionized water. Afterward, the -COOH functional group on the surface of CM-dextran was activated using the mixed liquor of EDC (0.2 M) and NHS (0.05 M) for an amine coupling reaction with activation times of 20 min. At this time, 50 μL anti-HMGB1 (50, 75, 100, 125, and 150 μg/mL) was injected to influence the signal response by antibody modification. It was observed that when the concentration of anti-HMGB1 was 100 μg/mL, the signal response was the highest (Figure 8c). Then, the signal response became smooth. Therefore, the concentration of anti-HMGB1 was set as 100 μg/mL for subsequent experiments. The incubation time of anti-HMGB1 (0.5, 1, 1.5, 2, and 4 h) was closely related to the stability of the chemical covalence of recognition molecules. Different from the traditional monolayer, CM-dextran had multiple binding points. It was observed that the signal response exhibited the maximum signal decline at 1.5 h and then presented a smooth trend (Figure 8d). Therefore, the anti-HMGB1 incubation time was 1.5 h in subsequent experiments.

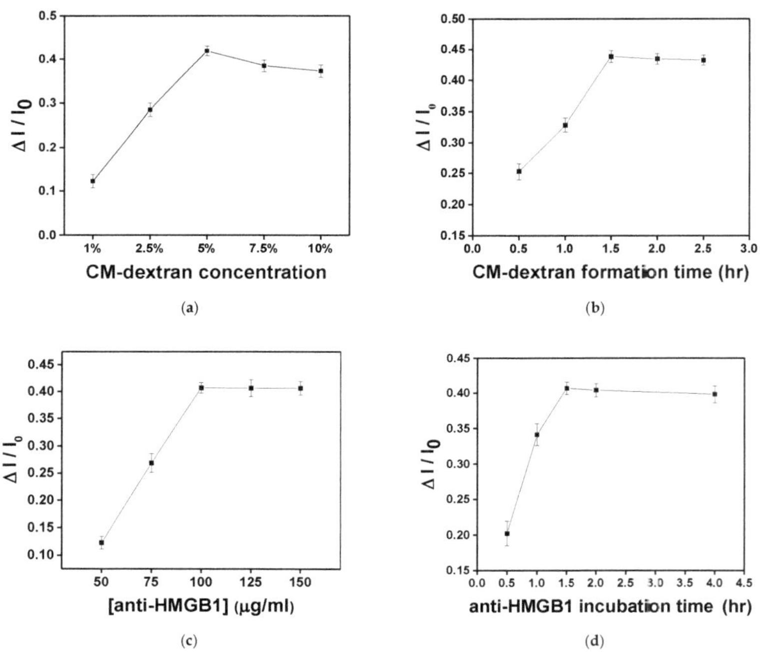

Figure 8. In each experiment, 10^{-8} g/mL HMGB1 was injected for 15 min of molecular binding to obtain the signal response. Each point is the mean of three repeated measurements. (**a**) Effect of CM-dextran concentration ratio (1%, 2.5%, 5%, 7.5%, and 10%) in the starting solution on the sensor response. (**b**) Effect of immersion time (0.5, 1, 1.5, 2, and 2.5 h) given to form CM-dextran on the sensor response. (**c**) Effect of concentration of anti-HMGB1 (50, 75, 100, 125, and 150 μg/mL) used for the bioconjugation process on the sensor response. (**d**) Effect of incubation time (0.5, 1, 1.5, 2, and 4 h) of anti-HMGB1 used for bioconjugation on the sensor response.

3.4. Comparison of CM-Dextran and MUA/MCH Sensing Chip Modification Anti-HMGB1 Responses

The signal responses of anti-HMGB1 modified by different bridging molecules were compared. The experiment used two surface modification methods: (a) fixing by CM-dextran, and (b) fixing by MUA/MCH. Afterward, the -COOH on CM-dextran or MUA was activated by EDC/NHS, which bonded with amines on anti-HMGB1 to modify the surface of the optical fiber probe. A PBS buffer washed out the uncombined anti-HMGB1. To inactivate the unreacted -COO-, 1 M aqueous ethanolamine solution of pH 8.5 was used for washing. The signal response at each stage was evident in the real-time signal graph. It indirectly proves the steps of antibody immobilization on the AuNP surface or the antibody-analyte interactions. The difference in signal responses between the two sensing chips when anti-HMGB1 was applied could be seen from the real-time signal change in Figure 8a,b. The CM-dextran sensing chip developed in this study had relatively abundant surface bonding points. The molecular bonding of the carboxyl group on anti-HMGB1 and CM-dextran was relatively reduced in the overall signal ($\Delta I/I_0$) to 0.072 (Figure 9a). However, the MUA/MCH sensing chip was a mixed monolayer, and the molecules were arranged mainly by the Van der Waals force. The steric hindrance was formed in the arrangement process, leading to fewer bonding points. It was observed that the overall relative signal reduction ($\Delta I/I_0$) is 0.028 when modifying anti-HMGB1 (Figure 9b). This indicates that the relative signalization of the CM-dextran sensing chip was 2.57 times that of the MUA/MCH sensing chip with modifying the recognition molecules.

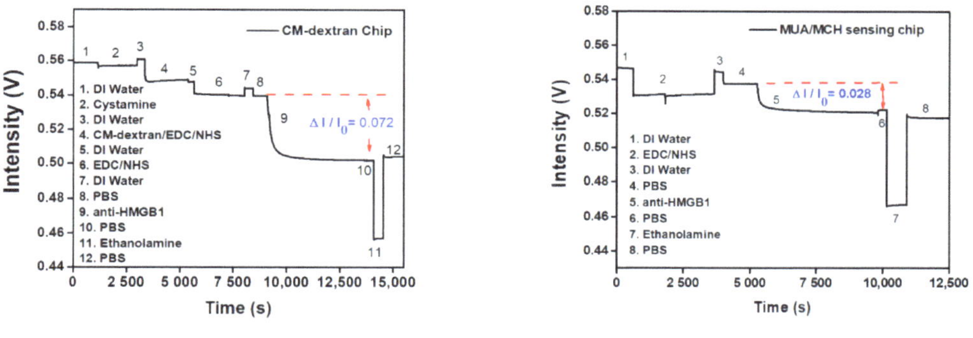

Figure 9. The signal responses of anti-HMGB1 modified by different bridging molecules were compared; (**a**) a CM-dextran sensing chip; (**b**) and an MUA/MCH sensing chip.

3.5. Nonspecific Adsorption Test

The probe experiment has been determined, and the nonspecific adsorption problem will finally be solved. Nonspecific adsorption is a universal problem in biosensor research. The hydroxyl (-OH), methyl (-CH_3), or ethanolamine (pH 8.5) are generally used to inactivate and block excessive NHS ester to avoid or reduce nonspecific adsorption phenomenon. Prior studies have proven that ethanolamine (pH 8.5) can block excess binding sites and reduce the nonspecific adsorption problem. First, the FOLSPR system monitored the recognition molecule-analyte binding reaction in the buffer solution. According to the real-time signal response in Figure 10, there was no signal change after injection of 10^{-7} g/mL IgG and Streptavidin. These results show that nonspecific binding on the CM-dextran sensing chip's surface seemed negligible. The ethanolamine could prevent interference effectively. The signal drop was generated when 10^{-8} g/mL HMGB1 was injected, confirming that the specific bonding interaction between the immobilized anti-HMGB1 on the AuNP surface and the HMGB1 in solution induced the measured signal change.

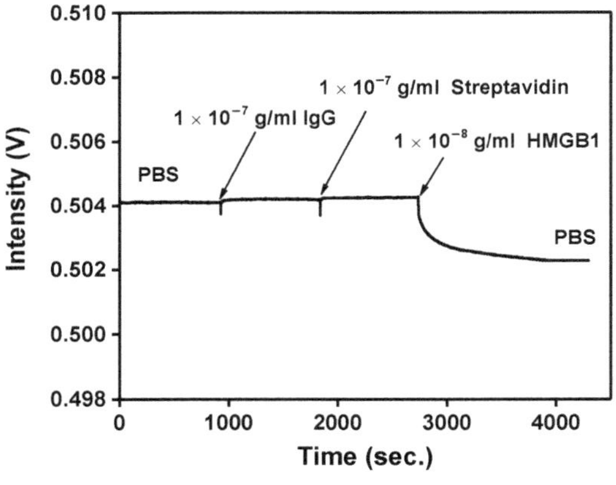

Figure 10. Nonspecific adsorption and specificity tests. The signal response of HMGB1 antibody-functionalized FOPPR sensor in response to IgG (1.0×10^{-10} g/mL), Streptavidin (1.0×10^{-7} g/mL), and HMGB1 (1.0×10^{-8} g/mL) solutions.

3.6. Sensitivity of the Biosensor

The main purpose of this study is to design a rapid, simple, and sensitive real-time detection technology for HMGB1 molecules. Here, the FOLSPR system successfully demonstrated high HMGB1 molecular binding kinetics sensitivity and used CM-dextran as bridging molecules for low-cost detections. The experimental research performed HMGB1 standard solution detection (concentration range was 1×10^{-10} to 1×10^{-6} g/mL) on the FOLSPR sensing chip functionalized by two bridging molecules. First, the PBS solution was injected as the baseline (600 s). Then, the final measured 300 s signals were averaged as I_0, and multi-concentration measurements were performed (each concentration was measured for 600 s). The real-time signal response between the recognition molecules and the analyte is shown in Figure 11a,b. It increased with each analyte concentration (the time to reach 90% of the equilibrium signal level) for about 400 s, and steady signals could be observed within 10 min. The concentration and relative signal quantity were plotted. The performance of the two kinds of sensing chips was linear for all concentrations (Figure 11c). The relative linearity (R^2) was 0.9928 (CM-dextran sensing chip) and 0.986 (MUA/MCH sensing chip), respectively. The standard correction curve was used to estimate the limit of detection (LOD). The CM-dextran sensing chip detection HMGB1 was 43.4 pg/mL (1.7 pM), and the LOD of MUA/MCH sensing chip detection HMGB1 was 3970 pg/mL (158.8 pM). The detected concentration range was enhanced by two orders of magnitude. In the past, Vilma et al. [10] employed ELISA to detect HMGB1 at a LOD of 0.4 ng/mL. Wang et al. [11] used Western blotting to detect HMGB1 at a LOD of 1 ng/mL. Other HMGB1 detection methods include an SPR with a LOD of 900 ng/mL [13] and an electrochemical immunosensor with a LOD of 2 ng/mL [61]. It was apparent that the CM-dextran FOLSPR biosensor had excellent detection sensitivity. This shows that the CM-dextran proposed in this research offered better performance in immune response analysis. In addition, FOPPR sensing methods offer more straightforward procedures, point-of-care detection (which can explore kinetic estimation between capture molecules and analytes), and relatively short analysis times compared to traditional detection methods such as ELISA or Western blotting. Finally, the molecular binding kinetics was estimated to determine the association rate constant (ka) and dissociation rate constant (kd) for anti-HMGB1 on the AuNP surface and HMGB1 in PBS; the values of these constants are $2.66 \pm 0.2 \times 10^{5}$ M^{-1} s^{-1} and $5.71 \pm 0.8 \times 10^{-2}$ s^{-1}, respectively. Afterward, the affinity constant Kf (Kf = ka/kd) was calculated using ka and kd values, i.e., $4.66 \pm 0.59 \times 10^{6}$ M^{-1} (number of samples = 3). It coincides with the constant rate result estimated using a surface plasmon resonance sensor (Biocore T200, GE Healthcare, Anaheim, CA, USA) (ka = $2.86 \pm 0.03 \times 10^{5}$ M^{-1} s^{-1}, kd = 0.119 ± 0.00 s^{-1} and Kf = 2.40 ± 0.03 × 10^{6} M^{-1}) in the literature. The results show that our method has simple operation, quick response, real-time detection, low sample consumption (0.35 mL), good analytical performance, and avoids using enzymes. Hence, it has good prospects in other areas in the future.

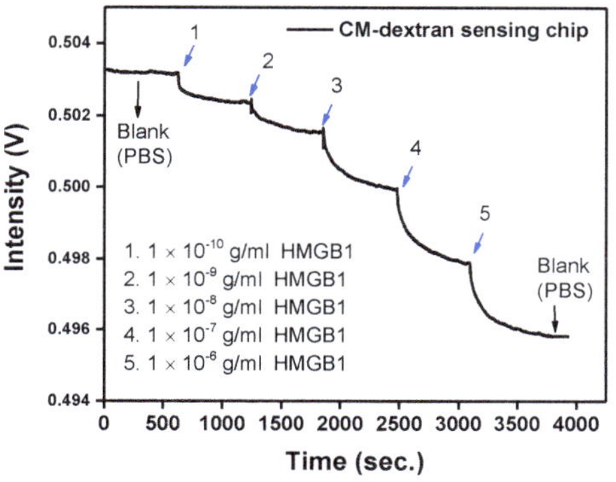

(**a**)

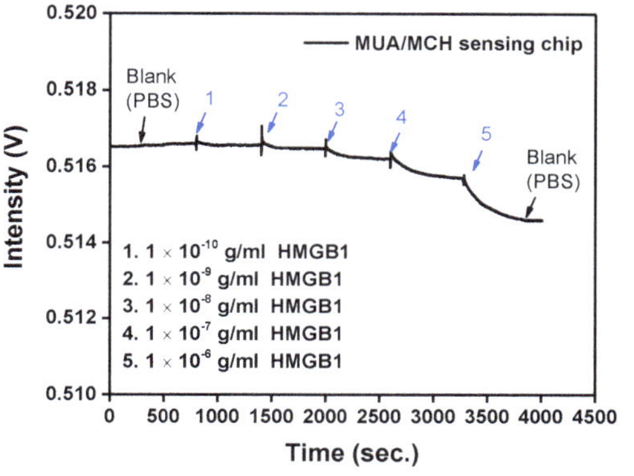

(**b**)

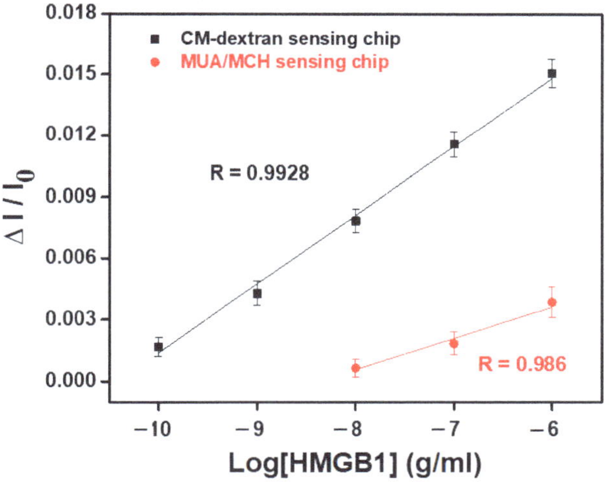

(**c**)

Figure 11. (**a**) The anti-HMGB1 functionalized CM-dextran sensing chip signal with the injection of different HMGB1 concentrations of (1) 1.0 × 10^{-10}, (2) 1.0 × 10^{-9}, (3) 1.0 × 10^{-8}, (4) 1.0 × 10^{-7}, and (5) 1.0 × 10^{-6} g/mL. (**b**) The anti-HMGB1-functionalized MUA/MCH sensing chip with the injection of different HMGB1 concentrations of (1) 1.0 × 10^{-10}, (2) 1.0 × 10^{-9}, (3) 1.0 × 10^{-8}, (4) 1.0 × 10^{-7}, and (5) 1.0 × 10^{-6} g/mL. (**c**) Calibration curve for HMGB1 by CM-dextran and MUA/MCH sensing chip in FOLSPR biosensor. Each point is the mean of three repeated measurements.

3.7. Sensing Chip Repeatability and Stability Tests

To study the repeatability and stability of the CM-dextran sensing chip, the chip was measured three times in the optimum conditions. First, the intra-batch and inter-batch repeatability values of the CM-dextran biosensor chip were checked by repeatedly measuring the HMGB1 at the same concentration. As shown in Figure 12a, the RSD values of relative signal ($\Delta I/I_0$) of 1.0 × 10^{-8} g/mL HMGB1 in the intra-batch and inter-batch tests are 2.51% and 2.63%, respectively. These results show that the assembled CM-dextran sensing chip exhibited acceptable repeatability. Before use, the manufactured chip was stored at 4 °C to study the stability of the proposed CM-dextran sensing chip. In this experiment, 1.0×10^{-8} g/mL HMGB1 was injected to determine the HMGB1 signal response of different storage periods. The experimental results show that after 21 days of storage, the HMGB1 signal response change was still maintained at 86.0% of the original HMGB1 signal response Figure 12b). This result shows that the proposed CM-dextran sensing chip was stable. The results prove that the proposed CM-dextran sensing chip had good stability and repeatability, and the CM-dextran sensing chip could be prepared without an urgent schedule.

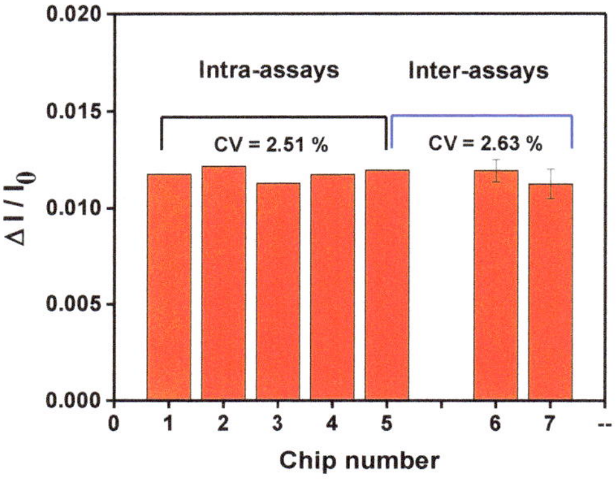

(a)

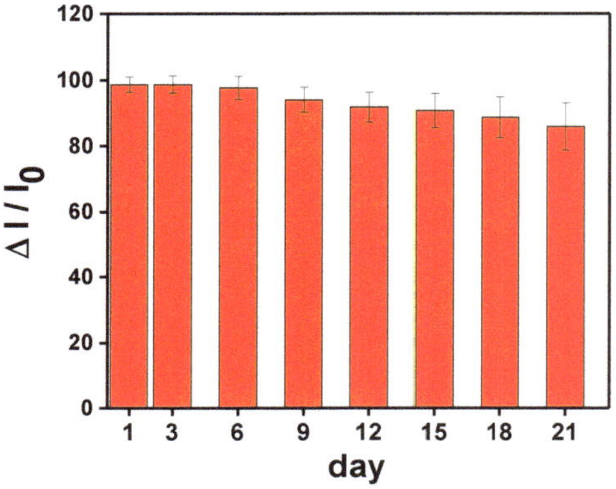

(b)

Figure 12. (**a**)The intra-assay and inter-assay reproducibility values of the CM-dextran biosensor chips were checked by repeated measurements of the same concentration of HMGB1. (**b**) Storage stability of the AuNP-CM-dextran-anti-HMGB1 probe onto sensing chip Each point is the mean of five repeated measurements.

Our results demonstrate the feasibility of constructing a FOLSPR sensor for HMGB1 detection by immobilizing CM-dextran as a bridging molecule on the surface of AuNPs. The FOLSPR sensing technology employs the sensitivity of the noble gold nanoparticle to the refractive index variation in the external environment to detect biological molecules; however, during detections on real samples, it is commonly required to undergo dilution to avoid the change in the refractive index of the external environment. Further, excessive dilution may undesirably degrade the practical detection limits. Since this study mainly verifies the sensing ability of the FOLSPR sensing system combined with CM-dextran, the results show that this method has a lower LOD and a more comprehensive detection range, which can meet the growing number of real sample testing needs. A systematic study is necessary to validate the clinical potential of the FOLSPR for detecting HMGB1 in complex matrices. Therefore, in future clinical applications, we should focus on sample pretreatment and verifying the analytical reliability of actual complex samples, overcome the errors caused by the dilution process, and meet the growing demand for real sample testing.

4. Conclusions

This study first proved the innovative method and feasibility of detecting HMGB1 using CM-dextran as a bridging molecule combined with a FOLSPR biosensor. In optimum conditions, anti-HMGB1 was fixed to the sensor and interacted with HMGB1 at various concentrations to realize target detections. These target detections are highly sensitive, selective, repeatable, and reliable. The linear measurement range of the FOLSPR biosensor is 10^{-10} to 10^{-6} g/mL, R^2 is 0.9928, and LOD is 43.4 pg/mL (1.7 pM). The estimated association rate constant (ka), dissociation rate constant (kd), and affinity constant (kf) of HMGB1 are $2.66 \pm 0.2 \times 10^5$ M^{-1} s^{-1}, $5.71 \pm 0.8 \times 10^{-2}$ s^{-1}, and $4.66 \pm 0.59 \times 10^6$ M^{-1}, respectively. The RSD values of intra-batch and inter-batch detection of the CM dextran sensing chip are smaller than 2.51% and 2.63%, respectively. In addition, after 21 days of storage of the CM-dextran sensing chip proposed in the storage test, the signal response change was still 86.0% of the initial signal response, showing good stability. As far as we know, this is the first kinetic analysis study on FOLSPR biosensors using CM-dextran as bridging molecules to detect the interaction of HMGB1, and the results are equivalent to that of BIAcore T200. Therefore, this study demonstrated a possible universal method, using CM-dextran as bridging molecules to detect clinical, environmental, and food biomarkers, of which the method has a very high application potential.

Author Contributions: Conceptualization, C.-Y.C. and C.-H.C.; Methodology, C.-Y.C. and C.-W.W.; Software, C.-H.C.; Validation, C.-Y.C.; Formal analysis, C.-Y.C., C.-H.C. and C.-W.W.; Investigation, C.-W.W.; Resources, C.-W.W.; Data curation, C.-H.C.; Writing—original draft, C.-Y.C. and C.-H.C.; Writing—review & editing, C.-Y.C. and C.-W.W.; Project administration, C.-Y.C.; Funding acquisition, C.-Y.C. All authors have read and agreed to the published version of the manuscript.

Funding: This work was supported by the National Science and Technology Council of Taiwan through grant numbers: MOST 111-2221-E-020-009-. The joint research program from National Taiwan University of Science and Technology/National Yunlin University of Science and Technology/National Pingtung University of Science and Technology (NPUST-NYUST-NTUST-112-04); Office of Research and Development, National Yunlin University of Science and Technology through grant numbers: 112TAPX-AA (112T05).

Institutional Review Board Statement: Not applicable.

Informed Consent Statement: Not applicable.

Data Availability Statement: Not applicable.

Acknowledgments: The authors would like to thank the Center for Advanced Instrumentation Center (AIC) in National Yunlin University of Science and Technology for the machine equipment support about this study.

Conflicts of Interest: The authors declare no conflict of interest.

References

1. Yang, H.; Antoine, D.J.; Andersson, U.; Tracey, K.J. The many faces of HMGB1: Molecular structure-functional activity in inflammation, apoptosis, and chemotaxis. *J. Leukoc. Biol.* **2013**, *93*, 865–873. [CrossRef] [PubMed]
2. Aki, H.S.; Fujita, M.; Yamashita, S.; Fujimoto, K.; Kumagai, K.; Tsuruta, R.; Kasaoka, S.; Aoki, T.; Nanba, M.; Murata, H.; et al. Elevation of jugular venous superoxide anion radical is associated with early inflammation, oxidative stress, and endothelial injury in forebrain ischemia–reperfusion rats. *Brain Res.* **2009**, *1292*, 180–190. [CrossRef] [PubMed]
3. Agresti, A.; Bianchi, M.E. HMGB proteins and gene expression. *Curr. Opin. Genet. Dev.* **2003**, *13*, 170–178. [CrossRef] [PubMed]
4. Scaffidi, P.; Misteli, T.; Bianchi, M.E. Release of chromatin protein HMGB1 by necrotic cells triggers inflammation. *Nature* **2002**, *418*, 191–195. [CrossRef]
5. Dumitriu, I.E.; Baruah, P.; Valentinis, B.; Voll, R.E.; Herrmann, M.; Nawroth, P.P.; Arnold, B.; Bianchi, M.E.; Manfredi, A.A.; Rovere-Querini, P. Release of High Mobility Group Box 1 by Dendritic Cells Controls T Cell Activation via the Receptor for Advanced Glycation End Products1. *J. Immunol.* **2005**, *174*, 7506–7515. [CrossRef]
6. Yang, H.; Wang, H.; Czura, C.J.; Tracey, K.J. The cytokine activity of HMGB1. *J. Leukoc. Biol.* **2005**, *78*, 1–8. [CrossRef]
7. Venereau, E.; De Leo, F.; Mezzapelle, R.; Careccia, G.; Musco, G.; Bianchi, M.E. HMGB1 as biomarker and drug target. *Pharmacol. Res.* **2016**, *111*, 534–544. [CrossRef]
8. Rocha, D.C.P.; Souza, T.M.A.; Nunes, P.C.G.; Mohana-Borges, R.; Paes, M.V.; Guimarães, G.M.C.; Arcila, J.C.S.; Paiva, I.A.; Azeredo, E.L.d.; Damasco, P.V.; et al. Increased circulating levels of High Mobility Group Box 1 (HMGB1) in acute-phase Chikungunya virus infection: Potential disease biomarker. *J. Clin. Virol.* **2022**, *146*, 105054. [CrossRef]
9. Sobajima, J.; Ozaki, S.; Uesugi, H.; Osakada, F.; Shirakawa, H.; Yoshida, M.; Nakao, K. Prevalence and characterization of perinuclear anti-neutrophil cytoplasmic antibodies (P-ANCA) directed against HMG1 and HMG2 in ulcerative colitis (UC). *Clin. Exp. Immunol.* **1998**, *111*, 402–407. [CrossRef]
10. Urbonaviciute, V.; Fürnrohr, B.G.; Weber, C.; Haslbeck, M.; Wilhelm, S.; Herrmann, M.; Voll, R.E. Factors masking HMGB1 in human serum and plasma. *J. Leukoc. Biol.* **2007**, *81*, 67–74. [CrossRef]

11. Wang, H.; Zhao, L.; Li, J.; Zhu, S.; Yeung, M. Analysis of the Released Nuclear Cytokine HMGB1 in Human Serum. In *Cytokine Bioassays: Methods and Protocols*; Vancurova, I., Ed.; Springer: New York, NY, USA, 2014; pp. 13–25. [CrossRef]
12. Arteaga, A.; Ranathunga, D.T.S.; Qu, J.; Biguetti, C.C.; Nielsen, S.O.; Rodrigues, D.C. Exogenous Protein Delivery of Ionic Liquid-Mediated HMGB1 Coating on Titanium Implants. *Langmuir* **2023**, *39*, 2204–2217. [CrossRef] [PubMed]
13. He, M.; Bianchi, M.E.; Coleman, T.R.; Tracey, K.J.; Al-Abed, Y. Exploring the biological functional mechanism of the HMGB1/TLR4/MD-2 complex by surface plasmon resonance. *Mol. Med.* **2018**, *24*, 21. [CrossRef] [PubMed]
14. Kuroiwa, Y.; Takakusagi, Y.; Kusayanagi, T.; Kuramochi, K.; Imai, T.; Hirayama, T.; Ito, I.; Yoshida, M.; Sakaguchi, K.; Sugawara, F. Identification and Characterization of the Direct Interaction between Methotrexate (MTX) and High-Mobility Group Box 1 (HMGB1) Protein. *PLoS ONE* **2013**, *8*, e63073. [CrossRef]
15. Gaillard, C.; Borde, C.; Gozlan, J.; Maréchal, V.; Strauss, F. A High-Sensitivity Method for Detection and Measurement of HMGB1 Protein Concentration by High-Affinity Binding to DNA Hemicatenanes. *PLoS ONE* **2008**, *3*, e2855. [CrossRef] [PubMed]
16. Madhurantakam, S.; Muthukumar, S.; Prasad, S. Emerging Electrochemical Biosensing Trends for Rapid Diagnosis of COVID-19 Biomarkers as Point-of-Care Platforms: A Critical Review. *ACS Omega* **2022**, *7*, 12467–12473. [CrossRef] [PubMed]
17. Mayer, K.M.; Hafner, J.H. Localized Surface Plasmon Resonance Sensors. *Chem. Rev.* **2011**, *111*, 3828–3857. [CrossRef]
18. Xu, T.; Geng, Z. Strategies to improve performances of LSPR biosensing: Structure, materials, and interface modification. *Biosens. Bioelectron.* **2021**, *174*, 112850. [CrossRef]
19. Behrouzi, K.; Lin, L. Gold nanoparticle based plasmonic sensing for the detection of SARS-CoV-2 nucleocapsid proteins. *Biosens. Bioelectron.* **2022**, *195*, 113669. [CrossRef]
20. Miller, M.M.; Lazarides, A.A. Sensitivity of Metal Nanoparticle Surface Plasmon Resonance to the Dielectric Environment. *J. Phys. Chem. B* **2005**, *109*, 21556–21565. [CrossRef]
21. Meira, D.I.; Proença, M.; Rebelo, R.; Barbosa, A.I.; Rodrigues, M.S.; Borges, J.; Vaz, F.; Reis, R.L.; Correlo, V.M. Chitosan Micro-Membranes with Integrated Gold Nanoparticles as an LSPR-Based Sensing Platform. *Biosensors* **2022**, *12*, 951. [CrossRef]
22. Nath, N.; Chilkoti, A. A Colorimetric Gold Nanoparticle Sensor To Interrogate Biomolecular Interactions in Real Time on a Surface. *Anal. Chem.* **2002**, *74*, 504–509. [CrossRef] [PubMed]
23. Jeon, J.; Uthaman, S.; Lee, J.; Hwang, H.; Kim, G.; Yoo, P.J.; Hammock, B.D.; Kim, C.S.; Park, Y.-S.; Park, I.-K. In-direct localized surface plasmon resonance (LSPR)-based nanosensors for highly sensitive and rapid detection of cortisol. *Sens. Actuators B: Chem.* **2018**, *266*, 710–716. [CrossRef]
24. Chen, S.; Svedendahl, M.; Van Duyne, R.P.; Käll, M. Plasmon-Enhanced Colorimetric ELISA with Single Molecule Sensitivity. *Nano Lett.* **2011**, *11*, 1826–1830. [CrossRef] [PubMed]
25. Liang, Y.; Zhang, H.; Zhu, W.; Agrawal, A.; Lezec, H.; Li, L.; Peng, W.; Zou, Y.; Lu, Y.; Xu, T. Subradiant Dipolar Interactions in Plasmonic Nanoring Resonator Array for Integrated Label-Free Biosensing. *ACS Sens.* **2017**, *2*, 1796–1804. [CrossRef] [PubMed]
26. Liu, J.; Hu, X.; Hu, Y.; Chen, P.; Xu, H.; Hu, W.; Zhao, Y.; Wu, P.; Liu, G.L. Dual AuNPs detecting probe enhanced the NanoSPR effect for the high-throughput detection of the cancer microRNA21 biomarker. *Biosens. Bioelectron.* **2023**, *225*, 115084. [CrossRef]
27. Chang, T.-C.; Wu, C.-C.; Wang, S.-C.; Chau, L.-K.; Hsieh, W.-H. Using A Fiber Optic Particle Plasmon Resonance Biosensor To Determine Kinetic Constants of Antigen–Antibody Binding Reaction. *Anal. Chem.* **2013**, *85*, 245–250. [CrossRef]
28. Cheng, S.-F.; Chau, L.-K. Colloidal Gold-Modified Optical Fiber for Chemical and Biochemical Sensing. *Anal. Chem.* **2003**, *75*, 16–21. [CrossRef]
29. Chang, T.-C.; Sun, A.Y.; Huang, Y.-C.; Wang, C.-H.; Wang, S.-C.; Chau, L.-K. Integration of Power-Free and Self-Contained Microfluidic Chip with Fiber Optic Particle Plasmon Resonance Aptasensor for Rapid Detection of SARS-CoV-2 Nucleocapsid Protein. *Biosensors* **2022**, *12*, 785. [CrossRef]
30. Chiang, C.-Y.; Hsieh, M.-L.; Huang, K.-W.; Chau, L.-K.; Chang, C.-M.; Lyu, S.-R. Fiber-optic particle plasmon resonance sensor for detection of interleukin-1β in synovial fluids. *Biosens. Bioelectron.* **2010**, *26*, 1036–1042. [CrossRef]
31. Li, Y.; Wang, X.; Ning, W.; Yang, E.; Li, Y.; Luo, Z.; Duan, Y. Sandwich method-based sensitivity enhancement of Ω-shaped fiber optic LSPR for time-flexible bacterial detection. *Biosens. Bioelectron.* **2022**, *201*, 113911. [CrossRef]
32. Zhang, H.; Zhou, X.; Li, X.; Gong, P.; Zhang, Y.; Zhao, Y. Recent Advancements of LSPR Fiber-Optic Biosensing: Combination Methods, Structure, and Prospects. *Biosensors* **2023**, *13*, 405. [CrossRef]
33. Huang, Y.-C.; Chiang, C.-Y.; Li, C.-H.; Chang, T.-C.; Chiang, C.-S.; Chau, L.-K.; Huang, K.-W.; Wu, C.-W.; Wang, S.-C.; Lyu, S.-R. Quantification of tumor necrosis factor-α and matrix metalloproteinases-3 in synovial fluid by a fiber-optic particle plasmon resonance sensor. *Analyst* **2013**, *138*, 4599–4606. [CrossRef] [PubMed]
34. Wu, C.-W.; Chiang, C.-Y.; Chen, C.-H.; Chiang, C.-S.; Wang, C.-T.; Chau, L.-K. Self-referencing fiber optic particle plasmon resonance sensing system for real-time biological monitoring. *Talanta* **2016**, *146*, 291–298. [CrossRef] [PubMed]
35. Chen, C.-H.; Chiang, C.-Y.; Wu, C.-W.; Wang, C.-T.; Chau, L.-K. Integrated Graphene Oxide with Noble Metal Nanoparticles to Develop High-Sensitivity Fiber Optic Particle Plasmon Resonance (FOPPR) Biosensor for Biomolecules Determination. *Nanomaterials* **2021**, *11*, 635. [CrossRef]
36. Ambrosetti, E.; Conti, M.; Teixeira, A.I.; Zilio, S.D. Patterned Carboxymethyl-Dextran Functionalized Surfaces Using Organic Mixed Monolayers for Biosensing Applications. *ACS Appl. Bio Mater.* **2022**, *5*, 3310–3319. [CrossRef]
37. Saftics, A.; Türk, B.; Sulyok, A.; Nagy, N.; Gerecsei, T.; Szekacs, I.; Kurunczi, S.; Horvath, R. Biomimetic Dextran-Based Hydrogel Layers for Cell Micropatterning over Large Areas Using the FluidFM BOT Technology. *Langmuir* **2019**, *35*, 2412–2421. [CrossRef] [PubMed]

38. Li, Z.; Narouz, M.R.; Munro, K.; Hao, B.; Crudden, C.M.; Horton, J.H.; Hao, H. Carboxymethylated Dextran-Modified N-Heterocyclic Carbene Self-Assembled Monolayers on Gold for Use in Surface Plasmon Resonance Biosensing. *ACS Appl. Mater. Interfaces* **2017**, *9*, 39223–39234. [CrossRef] [PubMed]
39. Song, S.; Lu, Y.; Li, X.; Cao, S.; Pei, Y.; Aastrup, T.; Pei, Z. Optimization of 3D Surfaces of Dextran with Different Molecule Weights for Real-Time Detection of Biomolecular Interactions by a QCM Biosensor. *Polymers* **2017**, *9*, 409. [CrossRef]
40. Huang, C.-F.; Yao, G.-H.; Liang, R.-P.; Qiu, J.-D. Graphene oxide and dextran capped gold nanoparticles based surface plasmon resonance sensor for sensitive detection of concanavalin A. *Biosens. Bioelectron.* **2013**, *50*, 305–310. [CrossRef]
41. Mannelli, I.; Lecerf, L.; Guerrouache, M.; Goossens, M.; Millot, M.-C.; Canva, M. DNA immobilisation procedures for surface plasmon resonance imaging (SPRI) based microarray systems. *Biosens. Bioelectron.* **2007**, *22*, 803–809. [CrossRef]
42. Polzius, R.; Schneider, T.; Biert, F.F.; Bilitewski, U.; Koschinski, W. Optimization of biosensing using grating couplers: Immobilization on tantalum oxide waveguides. *Biosens. Bioelectron.* **1996**, *11*, 503–514. [CrossRef] [PubMed]
43. Akkoyun, A.; Bilitewski, U. Optimisation of glass surfaces for optical immunosensors. *Biosens. Bioelectron.* **2002**, *17*, 655–664. [CrossRef] [PubMed]
44. Zhang, R.; Tang, M.; Bowyer, A.; Eisenthal, R.; Hubble, J. A novel pH- and ionic-strength-sensitive carboxy methyl dextran hydrogel. *Biomaterials* **2005**, *26*, 4677–4683. [CrossRef] [PubMed]
45. Tabasi, O.; Falamaki, C.; Mahmoudi, M. A Detailed Study on the Fabrication of Surface Plasmon Sensor Chips: Optimization of Dextran Molecular Weight. *Plasmonics* **2019**, *14*, 1145–1159. [CrossRef]
46. Taitt, C.R.; Anderson, G.P.; Lingerfelt, B.M.; Feldstein, M.J.; Ligler, F.S. Nine-Analyte Detection Using an Array-Based Biosensor. *Anal. Chem.* **2002**, *74*, 6114–6120. [CrossRef]
47. Löfås, S.; Johnsson, B. A novel hydrogel matrix on gold surfaces in surface plasmon resonance sensors for fast and efficient covalent immobilization of ligands. *J. Chem. Soc. Chem. Commun.* **1990**, *21*, 1526–1528. [CrossRef]
48. Chiang, C.-Y.; Chen, C.-H.; Wang, C.-T. Fiber Optic Particle Plasmon Resonance-Based Immunoassay Using a Novel Multi-Microchannel Biochip. *Sensors* **2020**, *20*, 3086. [CrossRef]
49. Lin, H.-Y.; Huang, C.-H.; Lu, S.-H.; Kuo, I.T.; Chau, L.-K. Direct detection of orchid viruses using nanorod-based fiber optic particle plasmon resonance immunosensor. *Biosens. Bioelectron.* **2014**, *51*, 371–378. [CrossRef]
50. Mulvaney, P. Surface Plasmon Spectroscopy of Nanosized Metal Particles. *Langmuir* **1996**, *12*, 788–800. [CrossRef]
51. Nath, N.; Chilkoti, A. Label Free Colorimetric Biosensing Using Nanoparticles. *J. Fluoresc.* **2004**, *14*, 377–389. [CrossRef]
52. Stewart, M.E.; Anderton, C.R.; Thompson, L.B.; Maria, J.; Gray, S.K.; Rogers, J.A.; Nuzzo, R.G. Nanostructured Plasmonic Sensors. *Chem. Rev.* **2008**, *108*, 494–521. [CrossRef] [PubMed]
53. Jackman, J.A.; Rahim Ferhan, A.; Cho, N.-J. Nanoplasmonic sensors for biointerfacial science. *Chem. Soc. Rev.* **2017**, *46*, 3615–3660. [CrossRef] [PubMed]
54. Loiseau, A.; Zhang, L.; Hu, D.; Salmain, M.; Mazouzi, Y.; Flack, R.; Liedberg, B.; Boujday, S. Core–Shell Gold/Silver Nanoparticles for Localized Surface Plasmon Resonance-Based Naked-Eye Toxin Biosensing. *ACS Appl. Mater. Interfaces* **2019**, *11*, 46462–46471. [CrossRef] [PubMed]
55. Ruemmele, J.A.; Hall, W.P.; Ruvuna, L.K.; Van Duyne, R.P. A Localized Surface Plasmon Resonance Imaging Instrument for Multiplexed Biosensing. *Anal. Chem.* **2013**, *85*, 4560–4566. [CrossRef] [PubMed]
56. Xie, L.; Yan, X.; Du, Y. An aptamer based wall-less LSPR array chip for label-free and high throughput detection of biomolecules. *Biosens. Bioelectron.* **2014**, *53*, 58–64. [CrossRef]
57. Khijwania, S.K.; Gupta, B.D. Fiber optic evanescent field absorption sensor: Effect of fiber parameters and geometry of the probe. *Opt. Quantum Electron.* **1999**, *31*, 625–636. [CrossRef]
58. Sital, S.; Baliyan, A.; Sharma, E.K.; Gupta, M. Optimization of Multimode Fibers for Surface Plasmon Resonance Based Sensors Under Spectral and Single Wavelength Intensity Interrogation. *Plasmonics* **2022**, *17*, 665–673. [CrossRef]
59. Draine, B.T.; Flatau, P.J. Discrete-dipole approximation for periodic targets: Theory and tests. *J. Opt. Soc. Am. A* **2008**, *25*, 2693–2703. [CrossRef]
60. Alterovitz, S.A.; Amirtharaj, P.M.; Apell, P.; Arakawa, E.T.; Ashok, J.; Barth, J.; Bezuidenhout, D.F.; Birch, J.R.; Birken, H.G.; Blessing, C.; et al. List of Contributors. In *Handbook of Optical Constants of Solids*; Palik, E.D., Ed.; Academic Press: Cambridge, MA, USA, 1998; pp. xv–xviii. [CrossRef]
61. Štros, M.; Polanská, E.V.; Hlaváčová, T.; Skládal, P. Progress in Assays of HMGB1 Levels in Human Plasma-The Potential Prognostic Value in COVID-19. *Biomolecules* **2022**, *12*, 544. [CrossRef]

biosensors

MDPI

Article

Point-of-Care Testing of the MTF1 Osteoarthritis Biomarker Using Phenolphthalein-Soaked Swabs

So Yeon Park [1], Dong-Sik Chae [2], Jae Sun Lee [3], Byung-Ki Cho [4] and Nae Yoon Lee [1,*]

[1] Department of BioNano Technology, Gachon University, 1342 Seongnam-daero, Sujeong-gu, Seongnam-si 13120, Republic of Korea; synb99@gmail.com
[2] Department of Orthopedic Surgery, College of Medicine, Catholic Kwandong University, Incheon 21431, Republic of Korea
[3] Chief Researcher, Healthcontents, Co., Ltd., Cheongju 28119, Republic of Korea
[4] Department of Orthopaedic Surgery, College of Medicine, Chungbuk National University, Cheongju 28644, Republic of Korea
* Correspondence: nylee@gachon.ac.kr

Abstract: Osteoarthritis (OA) is the most common joint disease, which accompanies pain and inconvenience in daily life owing to degradation of cartilage and adjacent tissues. In this study, we propose a simple point-of-care testing (POCT) kit for the detection of the MTF1 OA biomarker to achieve on-site clinical diagnosis of OA. The kit contains an FTA card for patient sample treatments, a sample tube for loop-mediated isothermal amplification (LAMP), and a phenolphthalein-soaked swab for naked eye detection. The MTF1 gene was isolated from synovial fluids using an FTA card and amplified using the LAMP method at 65 °C for 35 min. A test part of the phenolphthalein-soaked swab was decolorized in the presence of the MTF1 gene due to the pH change after the LAMP, but the color remained pink in the absence of the MTF1 gene. The control part of the swab served as a reference color in relation to the test part. When real-time LAMP (RT-LAMP), gel electrophoresis, and colorimetric detection of the MTF1 gene were performed, the limit of detection (LOD) was confirmed at 10 fg/μL, and the overall processes were completed in 1 h. The detection of an OA biomarker in the form of POCT was reported for the first time in this study. The introduced method is expected to serve as a POCT platform directly applicable by clinicians for easy and rapid identification of OA.

Keywords: point-of-care testing (POCT); phenolphthalein-soaked swab; metal regulatory transcription factor 1 (MTF1); loop-mediated isothermal amplification (LAMP); real-time LAMP (RT-LAMP)

Citation: Park, S.Y.; Chae, D.-S.; Lee, J.S.; Cho, B.-K.; Lee, N.Y. Point-of-Care Testing of the MTF1 Osteoarthritis Biomarker Using Phenolphthalein-Soaked Swabs. *Biosensors* **2023**, *13*, 535. https://doi.org/10.3390/bios13050535

Received: 31 March 2023
Revised: 6 May 2023
Accepted: 8 May 2023
Published: 10 May 2023

1. Introduction

Osteoarthritis (OA) is the most common joint disease, causing joint pain, stiffness, and disability in daily life [1–3]. According to the latest version of the Global Burden of Disease (GBD) study, 527.81 million people were affected by OA in 2019, and this number is rising, with a 113.25% increase since 1990 [4,5]. Moreover, OA can occur in any joint of the body, with the knees and hips being the most commonly affected [6,7]. Although radiography has traditionally been used to diagnose OA, it cannot detect early-stage OA [8]. In the case of knee OA, radiography can only detect the disease when more than 10% of the cartilage has already been destroyed [9]. Joint replacement surgery is necessary in the end stage of OA, which emphasizes the importance of early diagnosis [2,10]. In the early stages, changes in the subchondral bone composition can be evaluated using magnetic resonance image (MRI) [11]. Furthermore, biomarkers from serum and urine can be used to provide an early diagnosis of OA pathogenesis before using diagnostic imaging methods [12].

Metal regulatory transcription factor 1 (MTF1) has been identified as a biomarker in the pathogenesis of OA. In particular, inflammatory cytokines and mechanical stress upregulate transporter protein zinc transporter 8 (ZIP8), which leads to an increase in Zn^{2+} levels in chondrocytes and the activation of MTF1. MTF1 then expresses matrix-degrading enzymes, such as matrix metalloproteinases (MMPs) and ADAMTSs, which breakdown the cartilage extracellular matrix (ECM). Additionally, higher levels of MTF1 mRNA and protein were detected in damaged cartilage tissues than in undamaged tissues [13]. Therefore, despite efforts to repair damaged cartilage, the cartilage ECM degrades, and OA develops. By detecting MTF1, the expression of MMPs can be predicted based on this mechanism. However, cartilage breakdown cannot be detected before the nearest subchondral bones or muscles are affected because articular cartilage is aneural. Consequently, OA cannot be diagnosed based on symptoms alone in the early stages [14,15]. Therefore, detecting metabolic signal factors during OA pathogenesis is essential for early diagnosis of OA.

The nucleic acid amplification test (NAAT) is a method for recognizing specific nucleic acid sequences in order to test for genetic diseases and pathogenesis of disease based on genetic information. In general, NAAT entails sequential DNA purification, amplification, and detection [16]. Among the amplification processes, the polymerase chain reaction (PCR) is considered the gold standard [17]. Despite its high specificity, PCR requires a thermocycler to adjust temperatures for denaturation, annealing, and extension, making it difficult to adapt to point-of-care testing (POCT) [18,19]. Isothermal amplification techniques are commonly used as an alternative to PCR due to their rapid readout of the results on site with the naked eye, making them suitable for POCT [20]. Among the various isothermal amplification techniques, such as nucleic acid sequence-based amplification (NASBA) [21], loop-mediated isothermal amplification (LAMP), recombinase polymerase amplification (RPA), and rolling circle amplification (RCA), LAMP shows relatively high specificity because it employs four to six primers that recognize six specific regions in target DNA, and the limit of detection (LOD) is as low as a few copies [22–24]. Moreover, LAMP results can be detected using diverse methods, such as colorimetry [25,26], fluorescence [27], and turbidity [28]. Among the various detection techniques, colorimetric detection is appropriate for POCT because the color can be readily distinguished by the naked eye without requiring a specialized apparatus [29].

In this study, we developed a POCT kit that employs commercially available swabs to detect the MTF1 gene for rapid and simple diagnosis of OA. The POCT kit can provide the testing results directly since it is not necessary to transport patient specimens to experts at a central laboratory. Therefore, it is a more cost effective and time-efficient technique [30]. Using the introduced method, DNA was purified from synovial fluids by employing FTA cards, and LAMP was performed. Subsequently, the LAMP results were visually evaluated using a phenolphthalein-soaked swab, as phenolphthalein is pink in basic conditions and colorless in acidic conditions [31,32]. Colorless phenolphthalein was first converted to a pink color by adding NaOH to make it basic, then the presence of DNA amplicons caused the phenolphthalein-soaked swab to lose its color due to the production of acid, that is, hydrogen ions, which were the byproduct of DNA amplification. A phenolphthalein-soaked swab enables simple detection of the MTF1 gene with only one swab by comparing the color of the test part with that of the control part prepared on either side of the swab. The introduced method can serve as a simple and rapid diagnostic platform for on-site recognition of OA, which is directly applicable to the clinical field.

2. Materials and Methods

2.1. Materials and Reagents

Phenolphthalein, sodium hydroxide (NaOH), ammonium sulfate ($(NH_4)_2SO_4$), potassium chloride (KCl), FTA classic cards, and FTA purification reagent were purchased from Sigma-Aldrich (St. Louis, MO, USA). A LAMP kit which consisted of 10× isothermal amplification buffer, 100 mM $MgSO_4$, and Bst 2.0 DNA polymerase were purchased from New England Biolabs (Ipswich, MA, USA). LAMP 2× master mix was purchased from Elpis-Biotech (Daejeon, Republic of Korea). Primers and fluorescence probe were designed using the PrimerExplorer V5 program and synthesized from Cosmogenetech (Seoul, Republic of Korea). The dNTP mix was purchased from BioFact (Daejeon, Republic of Korea), and agarose was purchased from BioShop (Burlington, ON, Canada). Loading dye (Loading STAR) and a 100-bp DNA ladder were purchased from Dyne Bio (Seongnam, Republic of Korea) and Genes Laboratories (Seongnam, Republic of Korea), respectively. A 5 min Cell/Virus DNA extraction kit was obtained from J&T BIO (Cheonan, Republic of Korea). The swabs were obtained at a pharmacy. The MTF1 plasmid (HG15046-G; GenBank accession no. BC014454) was purchased from Sino Biological (Beijing, China), and SLC23A2 (Sodium-dependent Vitamin C transporter-2) (Plasmid#132025; GenBank accession no. AY380556) was purchased from Addgene (Watertown, MA, USA). A GenePro LAMP cycler (NIR-100G) for performing a real-time LAMP (RT-LAMP) was purchased from NanoBioLife (Seoul, Republic of Korea).

2.2. LAMP Reaction under Low Concentration of Buffer

When new strands of DNA are synthesized during the LAMP reaction, hydrogen ions are released as byproducts. The pH of a normal LAMP can be kept constant by using a reaction buffer. However, the pH can be decreased at low buffer concentrations. To perform pH-dependent colorimetric detection after the LAMP reaction, a low concentration of buffer containing 250 mM $(NH_4)_2SO_4$ and 1.25 M KCl was prepared [31]. Primers consisting of outer primers (F3 and B3), inner primers (FIP and BIP), and loop primers (LF and LB) for amplifying MTF1 and SLC23A2 were designed using the PrimerExplorer V5 program. The primer sequences are shown in Table 1. For each LAMP reaction (total 25 μL), 1.2 mM dNTP (3.0 μL), 10 mM $(NH_4)_2SO_4$ (1.0 μL), 50 mM KCl (1.0 μL), 6 mM $MgSO_4$ (1.5 μL), 8 units/mL of Bst 2.0 DNA polymerase, 0.2 μM outer primers (0.5 μL), 1.6 μM inner primers (0.5 μL), 0.8 μM loop primers (0.5 μL), and water were mixed. The pH of the LAMP mixture was adjusted to 9.5–10 by using 1.7 μL of 0.1 M NaOH. Then, a DNA template (0.5 μL or FTA card containing gDNA from synovial fluid) was added to the positive sample. To ensure the reliability of the kit, a negative control was performed simultaneously, which contained water instead of the DNA template. After mixing the reagents, the mixture was kept at 65 °C for 35 min for the amplification. Finally, the amplification results were analyzed using gel electrophoresis and colorimetric detection. All the experiments were repeated three times to evaluate reproducibility.

Table 1. Primer sequences used for amplifying the MTF1 and SLC23A2 genes.

Target	Primers	Sequences (5′-3′)
MTF1	F3	CAGGACCCTGGCACTTTG
	B3	CTGCAGAGTGAGGGTTGC
	FIP	AAGCCCTCTTCACCCCCTACTAGAGGATGAAGATGACGACGG
	BIP	GTCCCAGGGTTATGTGCAGCACTTGGCATGGGTGTGGAA
	LF	GGCAAGTGTTCTCCGCACTGT
SLC23A2	F3	TGACCATCTTCCTGGTGCT
	B3	CGTACTTTGTGGAATCGGGT
	FIP	GCTTATAGGCTGTCCAGCCCTTCCCAGTACGCCAGAAACG
	BIP	TCCAATCATCCTGGCCATCCTGGGAACACGTCGGTCACTG
	LB	GAGCTGGCTGCTGTGCTTCAT

2.3. Colorimetric Detection Based on pH

For naked-eye detection using phenolphthalein, the LAMP samples were amplified under a low concentration buffer condition because pH of the positive sample can change due to the release of hydrogen ions during amplification. The LAMP samples were then treated with a pink phenolphthalein solution (0.5 μL) containing phenolphthalein and 6 mM NaOH. Figure 1 shows the mechanism of colorimetric detection using this indicator. Because phenolphthalein is initially colorless, no color changes occur when it reacts with a positive sample. As a result, the presence of the target DNA is difficult to discern because the color change is subtle. However, by adding NaOH to phenolphthalein, the solution becomes basic, and the color changes to pink. Therefore, when the target DNA is present, the solution loses its pink color due to the formation of hydrogen ions as byproducts of the

LAMP reaction. In this way, the presence of the MTF1 gene can be readily identified via color changes from pink to colorless within 30 s at room temperature.

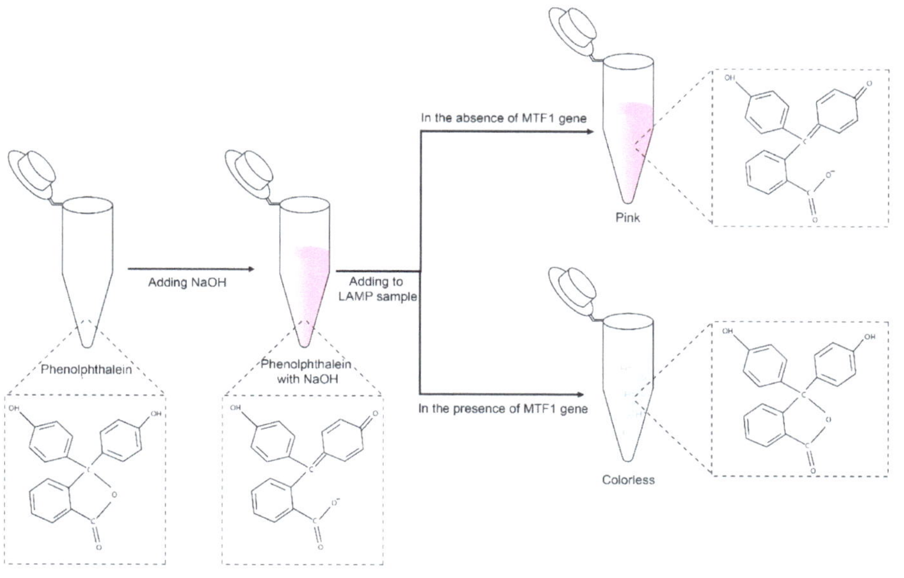

Figure 1. Schematic illustration of the mechanism of colorimetric detection using phenolphthalein.

2.4. Specificity and Sensitivity Tests

For evaluating the specificity of the test, MTF1 gene primer sets were used to amplify both MTF1 and SLC23A2 plasmids. SLC23A2 encodes the sodium-dependent vitamin C transporter (SVCT2), which is a biomarker for L-ascorbate treatment of breast cancer [33]. SLC23A2 primer sets were also used to amplify both SLC23A2 and MTF1 plasmids. For the sensitivity test, MTF1 plasmids were serially diluted 10-fold and amplified using LAMP. Colorimetric detection and gel electrophoresis were conducted to confirm the LOD. The experiments were repeated three times for both specificity and sensitivity tests to assess reproducibility.

2.5. Clinical Samples Analysis

Clinical samples were analyzed using synovial fluids collected from OA patients by clinicians. Genomic DNA containing the MTF1 gene was extracted and purified using a commercial DNA extraction kit and the FTA card. The kit was used to extract DNA from 1 mL of synovial fluid, and 1 µL of the final product was employed for the LAMP reaction. The FTA card was also employed for DNA extraction and purification [34,35]. Specifically, 4 mm of the FTA card was punched, and 10 µL of synovial fluid was applied to it. After thorough drying, the FTA card was washed using FTA card purification reagents (100 µL) and TE buffer (100 µL) to remove impurities. In addition, the FTA card was washed twice for 6 min to thoroughly purify DNA. The FTA card was then directly applied to the LAMP reaction.

2.6. Real-Time LAMP Using POCT Device

To perform RT-LAMP, a fluorescence probe containing 6-FAM and Black Hole Quencher (BHQ) was used. For each reaction (20 µL), 2× LAMP master mix (10 µL), oligo mix (2 µL), a fluorescence probe (0.5 µL), and water were used, and a DNA template (0.5 µL or the FTA card extract) was added to the positive sample. In the negative sample, water was used instead of a DNA template. The results of LAMP reactions were analyzed using

NIR-100G. The samples were kept at 65 °C for 27 min, then for an additional 20 s at 35 °C. All experiments were carried out three times to evaluate the reliability of the tests.

2.7. Preparation of Phenolphthalein-Soaked Swab

To achieve simple detection of the presence of the MTF1 gene, phenolphthalein-soaked swabs were prepared. First, both sides of the swab were dipped into the pink phenolphthalein solution to make both ends of the swab appear pink, then the swabs were dried for 10 min. Next, the test part of the swab was dipped into the sample solution to determine the presence of the target gene. The control part remained pink throughout the experiment. The results were analyzed at room temperature by comparing the color of the test part with that of the control part. Figure 2 shows the overall procedure for the colorimetric detection using a phenolphthalein-soaked swab.

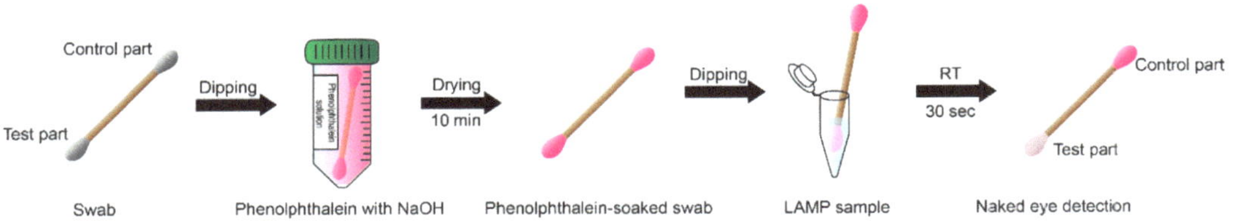

Figure 2. A schematic showing the overall procedure for the preparation of the phenolphthalein-soaked swab and naked-eye detection.

3. Results and Discussion

3.1. The Effect of the Amplification Time

Figure 3 shows the effect of the LAMP reaction time and the performance of the colorimetric detection. To test the effect of amplification time, 1 ng/µL of MTF1 plasmid was used as a DNA template. Based on the gel electrophoresis results, positive signals started to appear when the LAMP reaction was performed over 25 min. However, no distinct color difference was observed between the negative and the positive samples with 25 min of amplification. When the LAMP was performed for 35 min, the positive sample turned colorless, allowing the positive and negative samples to be readily visually distinguished. When the LAMP was performed for 45 min, the color of the positive and negative samples was nearly identical to those when the LAMP was performed for 35 min. Based on these results, 35 min was determined to be the optimum reaction time for LAMP for the colorimetric detection of the MTF1 gene.

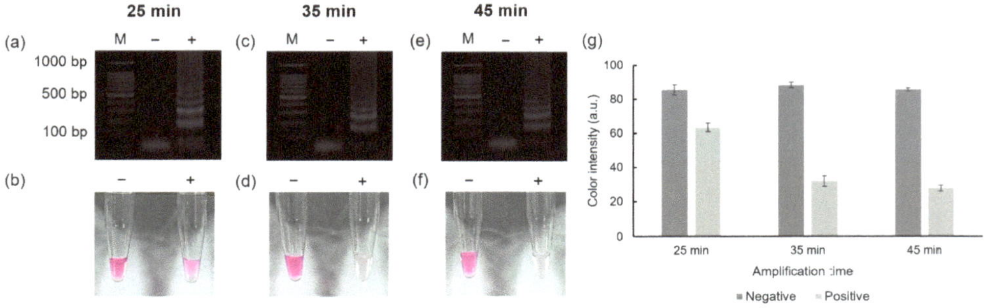

Figure 3. Results showing the agarose gel electrophoresis and color detection. The results of the gel electrophoresis when LAMP reactions were performed for 25, 35, and 45 min are shown in (**a**,**c**,**e**), respectively. The results of the colorimetric detection using phenolphthalein solution are shown in (**b**,**d**,**f**). The color intensity graph obtained by analyzing the colors using ImageJ software (ver 1.53e) is shown in (**g**).

3.2. Specificity Test

Figure 4 shows the specificity of the introduced method by amplifying MTF1 plasmids (1 ng/μL) and SLC23A2 plasmids (1 ng/μL) using each set of primers. Using MTF1 primer sets, MTF1 plasmids were successfully amplified, but the amplification of the SLC23A2 plasmid was not successful. Similarly, SLC23A2 primers amplified only the SLC23A2 plasmid. The results were further evaluated using 0.5 μL of phenolphthalein solution. When MTF1 primer sets were used, only the tubes containing the MTF1 plasmid turned colorless, while the other tubes containing SLC23A2 plasmid remained pink, either partially or entirely. Additionally, when SLC23A2 primer sets were used, only the tubes containing SLC23A2 plasmids turned colorless.

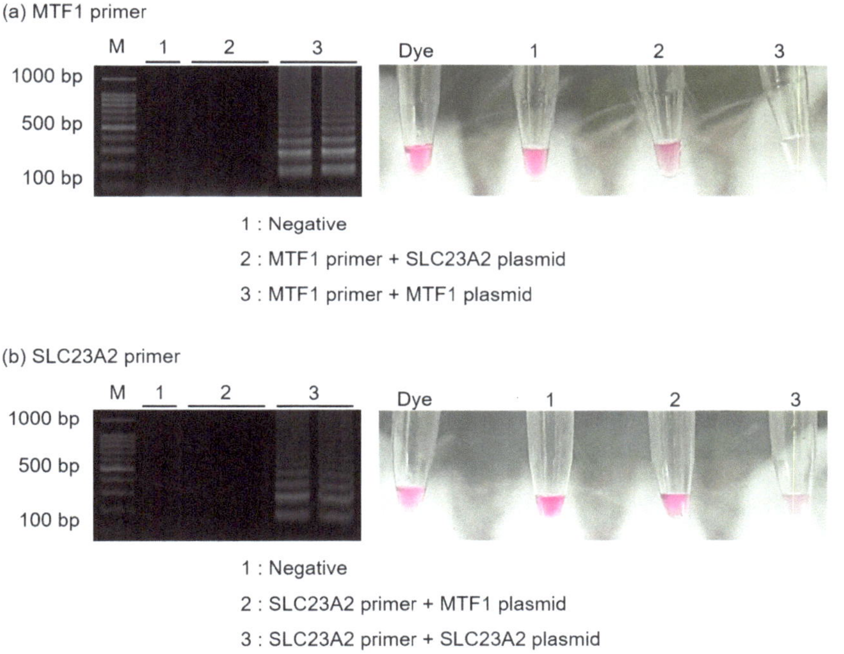

Figure 4. Results showing the agarose gel electrophoresis and color detection when specificity tests were performed. Results showing the amplification of (**a**) MTF1 plasmids and (**b**) SLC23A2 plasmids when both primers were used.

3.3. Sensitivity Tests

Figure 5 shows the sensitivity of the introduced method using MTF1 plasmids. For the sensitivity test, MTF1 plasmids were diluted from a concentration of 1 ng/μL to 0.1 fg/μL through 10-fold serial dilution. As shown in the gel electrophoresis results, ladder-like bands appeared up to 10 fg/μL, indicating that amplification was successful up to 10 fg/μL. No bands appeared below 10 fg/μL, demonstrating that the LOD was approximately 10 fg/μL (Figure 5a). Next, phenolphthalein solution was used for colorimetric detection. Similar to the results of gel electrophoresis, the color changed from pink to colorless for samples containing 1 ng/μL to 10 fg/μL of MTF1 plasmids. Thus, samples containing less than 1 fg/μL of MTF1 plasmids remained pink (Figure 5b). Furthermore, the color intensities of each sample were evaluated using ImageJ software (Figure 5c). As shown in Figure 5, the LODs for both gel electrophoresis and colorimetric detection were confirmed to be 10 fg/μL.

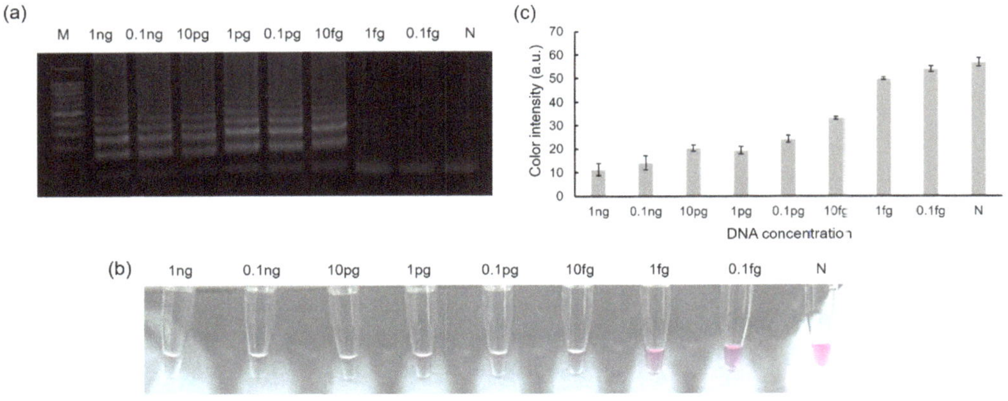

Figure 5. Results showing the sensitivity tests when MTF1 plasmid was amplified. Results of (**a**) gel electrophoresis and (**b**) color detection using phenolphthalein solution obtained for MTF1 plasmid for concentrations ranging from 1 ng/μL to 0.1 fg/μL. (**c**) A color intensity graph obtained based on the color detection results is shown in (**b**).

3.4. Analyses of Clinical Samples

To demonstrate the feasibility of this method for clinical samples, synovial fluid was obtained from OA patients and tested. Figure 6a shows the results of the gel electrophoresis and colorimetric detection when DNA was extracted from 1 mL of synovial fluid using a commercial kit. Negative samples displayed pink, while positive samples containing the MTF1 gene turned colorless as the pH decreased due to gene amplification. In addition, the FTA card was used for DNA extraction after treating it with 10 μL of synovial fluid and washing away impurities. Since the FTA card can capture DNA, it was dipped into tubes containing LAMP reagents (25 μL), and DNA was amplified successfully. As shown in Figure 6b, the DNA was successfully purified when using the FTA card to almost the same extent as when using a commercial kit. These results confirmed that OA diagnosis based on MTF1 gene detection is possible by utilizing synovial fluid. All experiments involved both negative and positive (MTF1 plasmid) control samples for test result reliability.

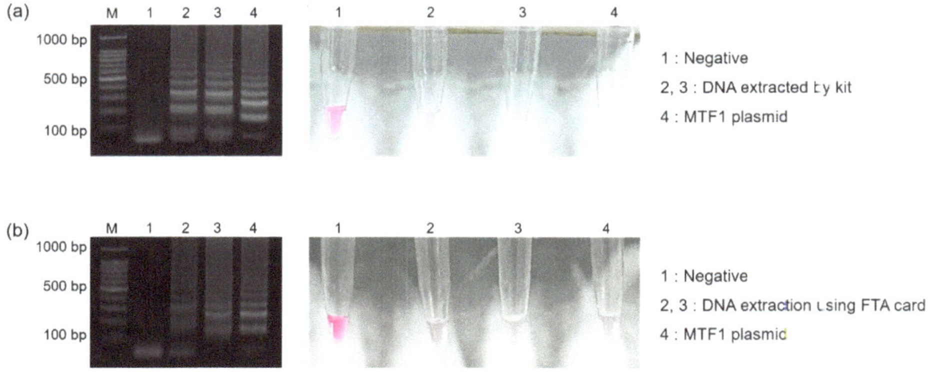

Figure 6. Results showing gel electrophoresis and colorimetric detection when DNA in synovial fluid was extracted and purified using (**a**) a commercial kit and (**b**) the FTA card.

3.5. Real-Time LAMP Using POCT Machine

For the interpretation of RT-LAMP performance, cycle threshold (Ct) values of less than 25 cycles were considered positive, whereas Ct values higher than 25 cycles was considered negative. Figure 7 demonstrates the feasibility of detecting genetic biomarkers using a POCT machine (NIR-100G). Clinical specimens from OA patients were tested using the POCT machine. In Figure 7a, two patient samples were tested and successfully amplified, with Ct values of 15.6 and 18.1, respectively. Fluorescence emissions were also observed under UV illumination, as shown in Figure 7b. Moreover, Figure 7c shows the LOD results when 10-fold serially diluted MTF1 plasmids were used as templates. Ct values were 8.5, 10.3, 11.4, 13.9, 14.9, and 20.6 when DNA concentrations of the MTF1 plasmid were 1 ng/μL, 0.1 ng/μL, 10 pg/μL, 1 pg/μL, 0.1 pg/μL, and 10 fg/μL, respectively. DNA concentrations of 1 fg/μL or lower did not display a peak on the graph and showed no fluorescence emission. Furthermore, Figure 7d shows that when the DNA concentration was greater than 10 fg/μL, the samples emitted green fluorescence, whereas when the DNA concentration was less than 10 fg/μL, no fluorescence was emitted. Based on these results, the LOD was estimated to be approximately 10 fg/μL.

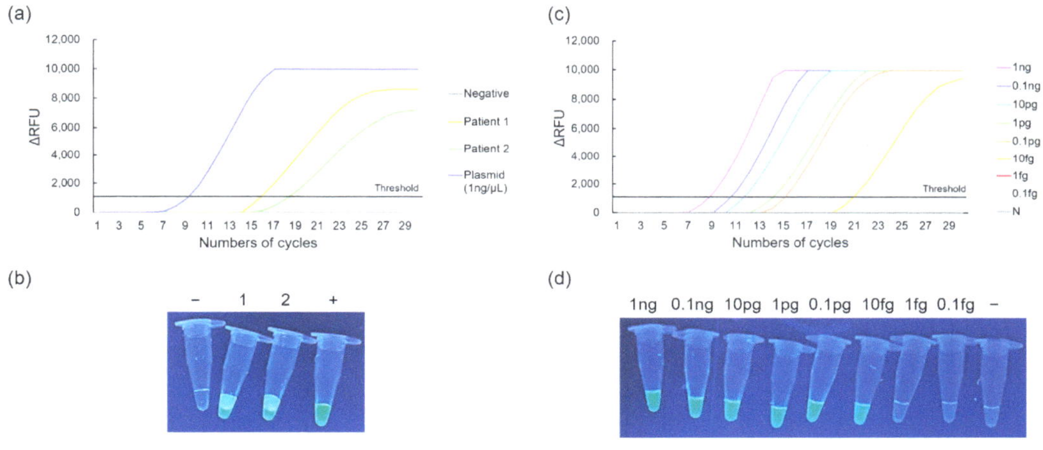

Figure 7. Results of RT-LAMP using NIR-100G. The results of real-time peak analyses and fluorescence measurements amplifying synovial fluid of OA patients are shown in (**a**,**b**). The results of limit of detection using a 10-fold serially diluted MTF1 plasmid are shown in (**c**,**d**).

3.6. Phenolphthalein-Soaked Swab for Naked Eye Detection of MTF1

To enable naked-eye detection using phenolphthalein-soaked swabs, a low concentration buffer was custom-made and used for LAMP reactions. Figure 8 shows the process of MTF1 detection using a phenolphthalein-soaked swab. Figure 8a shows the results of agarose gel electrophoresis. After 35 min of amplification, the test part of the phenolphthalein-soaked swab was dipped into the LAMP sample for 30 s (Figure 8b). Instantly, the color of the test part turned to a lighter pink inside a positive sample containing the MTF1 plasmid with a DNA concentration of 1 ng/uL. However, the test part of the swab remained pink inside a negative sample that did not contain the MTF1 plasmid (Figure 8c). Figure 9a,b shows the results of real sample analyses using synovial fluid from an OA patient. Negative and positive controls (1 ng/μL of MTF1 plasmid) were also carried out to evaluate the reliability of the test. The OA patient sample was amplified successfully, and the test part of the swab was decolorized, while the negative control remained pink. Figure 9c,d shows the results of LOD when the MTF1 plasmid was used as a template. The MTF1 plasmid was diluted to 0.1 fg/μL using 10-fold serial dilution. MTF1 plasmid samples with concentrations ranging from 1 ng/μL to 10 fg/μL were amplified, and the tests part of the swab turned colorless. Therefore, the LOD of the introduced POCT kit can be considered as 10 fg/μL. A specificity test was also conducted, as shown in Figure S1. Furthermore, a stability test was performed to confirm the duration of the color display, and color changes were observed at 1, 2, 5, 10, 20, and 30 min after the reaction (Figure S2). These results indicated that the colors of the negative and positive samples were clearly distinguished for up to 10 min; however, after 10 min had elapsed, the swab color of the negative sample started to decolor, and the color difference between the negative and positive samples was negligible. This is likely because, based on previous study, phenolphthalein is originally colorless. Therefore, color changes occurred inside a negative sample as well, whereas colorlessness remained inside a positive sample [32].

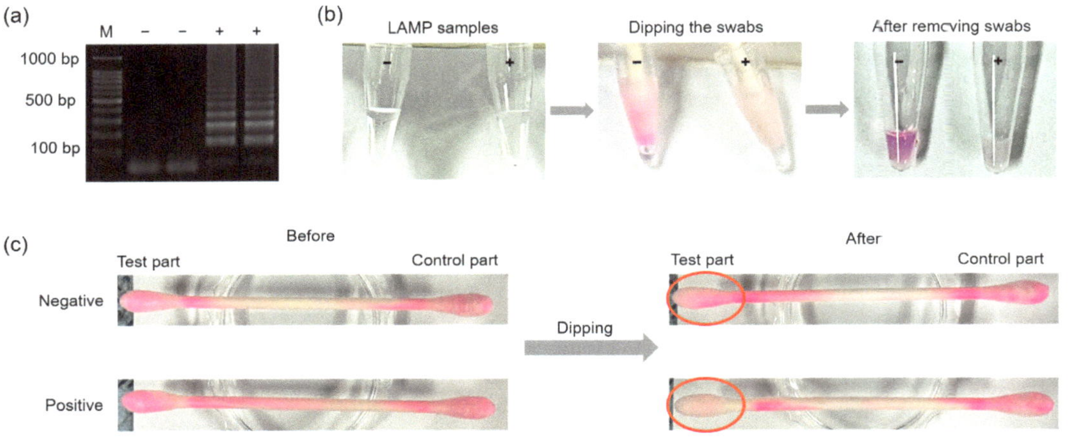

Figure 8. Results showing (**a**) agarose gel electrophoresis, (**b**) the experimental procedures of the swab tests, and (**c**) swab images obtained before and after applying LAMP samples containing MTF1 plasmid with a DNA concentration of 1 ng/μL.

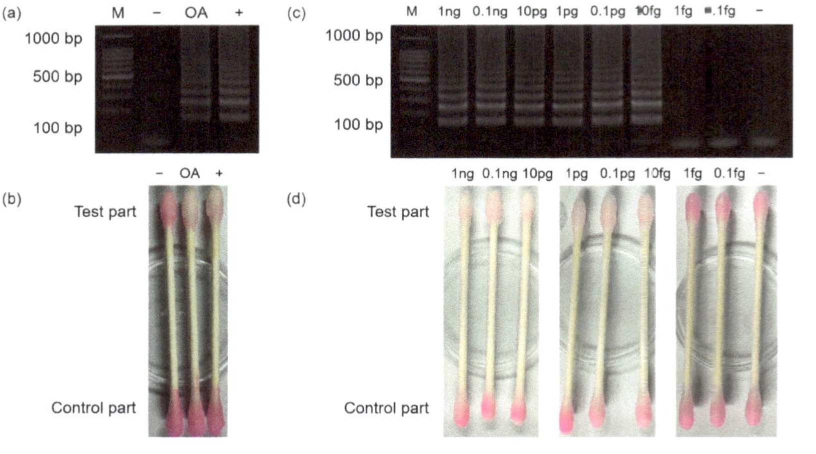

Figure 9. Results showing (**a**) gel electrophoresis and (**b**) phenolphthalein-soaked swabs when synovial fluid from an OA patient was used. Results from the sensitivity test performed using phenolphthalein-soaked swabs are shown in (**c**,**d**). The LOD of the test was confirmed as 10 fg/μL.

4. Conclusions

In this study, we developed a POCT kit that employs phenolphthalein-soaked swab for colorimetric detection integrated with the LAMP method for MTF1 OA biomarker detection. By converting the color of the phenolphthalein to pink prior to the reaction, the color change became more apparent for positive samples. This allows for clear color discrimination between negative and positive samples in one step. By applying patient specimen to the FTA card and amplifying them with the LAMP method, the MTF1 gene was successfully detected using gel electrophoresis and a real-time LAMP machine (NIR-100G). Naked-eye detection was also made possible by pH changes when target DNA was present. In this way, the results can be read out quickly within an hour without the need for sophisticated instruments. Overall, the introduced kit can pave the way for rapid and simple detection of biomarkers in the POCT field. It can also extend to the detection of several disease biomarkers and make the kit directly available to general users in the future.

Supplementary Materials: The following supporting information can be downloaded at: https://www.mdpi.com/article/10.3390/bios13050535/s1. Figure S1. Specificity test of phenolphthalein swab targeting MTF1 gene. Gel electrophoresis results amplified with MTF1 primer sets are shown in (a). Naked eye detection results obtained using a phenolphthalein-soaked swab are shown in (b). Figure S2. Stability test of the phenolphthalein-soaked swab. The gel electrophoresis results are shown in (a), (b) shows the colorimetric detection using phenolphthalein-soaked swab, and (c) shows the color intensity graph of the phenolphthalein-soaked swabs analyzed using ImageJ software. A *t*-test was conducted, and *p*-values were obtained to determine significant differences in color intensity between the negative and positive samples each time. * $p \leq 0.05$, ** $p \leq 0.01$, and ns: $p > 0.05$.

Author Contributions: Conceptualization, S.Y.P. and N.Y.L.; methodology, S.Y.P. and N.Y.L.; software, S.Y.P. and D.-S.C.; validation, S.Y.P. and J.S.L.; formal analysis, S.Y.P. and B.-K.C.; investigation, S.Y.P., D.-S.C., J.S.L. and B.-K.C.; resources, N.Y.L.; data curation, N.Y.L.; writing—original draft preparation, S.Y.P.; writing—review and editing, S.Y.P. and N.Y.L.; supervision, N.Y.L.; project administration, N.Y.L.; funding acquisition, D.-S.C., J.S.L. and N.Y.L. All authors have read and agreed to the published version of the manuscript.

Funding: This work was supported by the National Research Foundation of Korea (NRF) grant funded by the Korean government (MSIT) (RS-2023-00208684) and by the Gachon University research fund of 2021 (GCU-202103270001).

Institutional Review Board Statement: This study was approved by the Institutional Review Board (IRB) of the Chungbuk National University Hospital (CBNUH 2023-01-036-001).

Informed Consent Statement: Informed consent was obtained from all subjects involved in the study.

Data Availability Statement: The data presented in this study are available from the corresponding author upon request.

Conflicts of Interest: The authors declare no conflict of interest.

References

1. Quicke, J.G.; Conaghan, P.G.; Corp, N.; Peat, G. Osteoarthritis year in review 2021: Epidemiology & therapy. *Osteoarthr. Cartil.* **2022**, *30*, 196–206. [CrossRef]
2. Safiri, S.; Kolahi, A.A.; Smith, E.; Hill, C.; Bettampadi, D.; Mansournia, M.A.; Hoy, D.; Ashrafi-Asgarabad, A.; Sepidarkish, M.; Almasi-Hashiani, A.; et al. Global, regional and national burden of osteoarthritis 1990–2017: A systematic analysis of the Global Burden of Disease Study 2017. *Ann. Rheum. Dis.* **2020**, *79*, 819–828. [CrossRef] [PubMed]
3. Palazzo, C.; Nguyen, C.; Lefevre-Colau, M.M.; Rannou, F.; Poiraudeau, S. Risk factors and burden of osteoarthritis. *Ann. Phys. Rehabil. Med.* **2016**, *59*, 134–138. [CrossRef] [PubMed]
4. Long, H.; Liu, Q.; Yin, H.; Wang, K.; Diao, N.; Zhang, Y.; Lin, J.; Guo, A. Prevalence trends of site-specific osteoarthritis from 1990 to 2019: Findings from the Global Burden of Disease Study 2019. *Arthritis Rheumatol.* **2022**, *74*, 1172–1183. [CrossRef] [PubMed]
5. Vos, T.; Lim, S.S.; Abbafati, C.; Abbas, K.M.; Abbasi, M.; Abbasifard, M.; Bhutta, Z.A. Global burden of 369 diseases and injuries in 204 countries and territories, 1990–2019: A systematic analysis for the Global Burden of Disease Study 2019. *Lancet* **2020**, *396*, 1204–1222. [CrossRef]
6. Zhang, Y.; Jordan, J.M. Epidemiology of osteoarthritis. *Clin. Geriatr. Med.* **2010**, *26*, 355–369. [CrossRef]
7. Katz, J.N.; Arant, K.R.; Loeser, R.F. Diagnosis and treatment of hip and knee osteoarthritis: A review. *JAMA* **2021**, *325*, 568–578. [CrossRef]
8. Favero, M.; Ramonda, R.; Goldring, M.B.; Goldring, S.R.; Punzi, L. Early knee osteoarthritis. *RMD Open* **2015**, *1*, e000062. [CrossRef]
9. Jones, G.; Ding, C.; Scott, F.; Glisson, M.; Cicuttini, F. Early radiographic osteoarthritis is associated with substantial changes in cartilage volume and tibial bone surface area in both males and females. *Osteoarthr. Cartil.* **2004**, *12*, 169–174. [CrossRef]
10. Cicuttini, F.M.; Wluka, A.E. Is OA a mechanical or systemic disease? *Nat. Rev. Rheumatol.* **2014**, *10*, 515–516. [CrossRef]

11. Braun, H.J.; Gold, G.E. Diagnosis of osteoarthritis: Imaging. *Bone* **2012**, *51*, 278–288. [CrossRef] [PubMed]
12. Castañeda, S.; Roman-Blas, J.A.; Largo, R.; Herrero-Beaumont, G. Osteoarthritis: A progressive disease with changing phenotypes. *Rheumatology* **2014**, *53*, 1–3. [CrossRef] [PubMed]
13. Kim, J.H.; Jeon, J.; Shin, M.; Won, Y.; Lee, M.; Kwak, J.S.; Lee, G.; Rhee, J.; Ryu, J.H.; Chun, C.H.; et al. Regulation of the catabolic cascade in osteoarthritis by the zinc-ZIP8-MTF1 axis. *Cell* **2014**, *156*, 730–743. [CrossRef] [PubMed]
14. Dijkgraaf, L.C.; de Bont, L.G.; Boering, G.; Liem, R.S. Normal cartilage structure, biochemistry, and metabolism: A review of the literature. *J. Oral Maxillofac. Surg.* **1995**, *53*, 924–929. [CrossRef] [PubMed]
15. Bijlsma, J.W.; Berenbaum, F.; Lafeber, F.P. Osteoarthritis: An update with relevance for clinical practice. *Lancet* **2011**, *377*, 2115–2126. [CrossRef] [PubMed]
16. Craw, P.; Balachandran, W. Isothermal nucleic acid amplification technologies for point-of-care diagnostics: A critical review. *Lab Chip* **2012**, *12*, 2469–2486. [CrossRef]
17. Zhu, H.; Zhang, H.; Xu, Y.; Laššáková, S.; Korabečná, M.; Neužil, P. PCR past, present and future. *Biotechniques* **2020**, *69*, 317–325. [CrossRef]
18. Fakruddin, M.; Mannan, K.S.B.; Chowdhury, A.; Mazumdar, R.M.; Hossain, M.N.; Islam, S.; Chowdhury, M.A. Nucleic acid amplification: Alternative methods of polymerase chain reaction. *J. Pharm. Bioallied Sci.* **2013**, *5*, 245. [CrossRef]
19. Nguyen, D.V.; Nguyen, V.H.; Seo, T.S. Quantification of colorimetric loop-mediated isothermal amplification process. *Biochip J.* **2019**, *13*, 158–164. [CrossRef]
20. Giuffrida, M.C.; Spoto, G. Integration of isothermal amplification methods in microfluidic devices: Recent advances. *Biosens. Bioelectron.* **2017**, *90*, 174–186. [CrossRef] [PubMed]
21. Law, J.W.F.; Ab Mutalib, N.S.; Chan, K.G.; Lee, L.H. Rapid methods for the detection of foodborne bacterial pathogens: Principles, applications, advantages and limitations. *Front. Microbiol.* **2015**, *5*, 770. [CrossRef] [PubMed]
22. Zhao, Y.; Chen, F.; Li, Q.; Wang, L.; Fan, C. Isothermal amplification of nucleic acids. *Chem. Rev.* **2015**, *115*, 12491–12545. [CrossRef] [PubMed]
23. Hassan, Y.; Than, L.T.L. Loop-mediated isothermal amplification (LAMP): Comparative advances over conventional PCR and other molecular techniques. *Annu. Res. Rev. Biol.* **2020**, *35*, 33–44. [CrossRef]
24. Notomi, T.; Okayama, H.; Masubuchi, H.; Yonekawa, T.; Watanabe, K.; Amino, N.; Hase, T. Loop-mediated isothermal amplification of DNA. *Nucleic Acids Res.* **2000**, *28*, e63. [CrossRef] [PubMed]
25. Goto, M.; Honda, E.; Ogura, A.; Nomoto, A.; Hanaki, K.I. Colorimetric detection of loop-mediated isothermal amplification reaction by using hydroxy naphthol blue. *Biotechniques* **2009**, *46*, 167–172. [CrossRef]
26. Kim, D.; Lee, J.; Park, S.; Park, J.; Seo, M.J.; Rhee, W.J.; Kim, E. Target-responsive template structure switching-mediated exponential rolling circle amplification for the direct and sensitive detection of microRNA. *BioChip J.* **2022**, *16*, 422–432. [CrossRef]
27. Tomita, N.; Mori, Y.; Kanda, H.; Notomi, T. Loop-mediated isothermal amplification (LAMP) of gene sequences and simple visual detection of products. *Nat. Protoc.* **2008**, *3*, 877–882. [CrossRef]
28. Mori, Y.; Nagamine, K.; Tomita, N.; Notomi, T. Detection of loop-mediated isothermal amplification reaction by turbidity derived from magnesium pyrophosphate formation. *Biochem. Biophys. Res. Commun.* **2001**, *289*, 150–154. [CrossRef]
29. Moehling, T.J.; Choi, G.; Dugan, L.C.; Salit, M.; Meagher, R.J. LAMP diagnostics at the point-of-care: Emerging trends and perspectives for the developer community. *Expert Rev. Mol. Diagn.* **2021**, *21*, 43–61. [CrossRef]
30. Drain, P.K.; Hyle, E.P.; Noubary, F.; Freedberg, K.A.; Wilson, D.; Bishai, W.R.; Rodriguez, W.; Bassett, I.V. Diagnostic point-of-care tests in resource-limited settings. *Lancet Infect. Dis.* **2014**, *14*, 239–249. [CrossRef]
31. Trinh, T.N.D.; Lee, N.Y. Spinning and fully integrated microdevice for rapid screening of vancomycin-resistant Enterococcus. *ACS Sens.* **2021**, *6*, 2902–2910. [CrossRef] [PubMed]
32. Tanner, N.A.; Zhang, Y.; Evans Jr, T.C. Visual detection of isothermal nucleic acid amplification using pH-sensitive dyes. *Biotechniques* **2015**, *58*, 59–68. [CrossRef]
33. Hong, S.W.; Lee, S.H.; Moon, J.H.; Hwang, J.J.; Kim, D.E.; Ko, E.; Kim, H.S.; Cho, I.J.; Kang, J.S.; Kim, D.J.; et al. SVCT-2 in breast cancer acts as an indicator for L-ascorbate treatment. *Oncogene* **2013**, *32*, 1508–1517. [CrossRef] [PubMed]
34. Trinh, K.T.L.; Trinh, T.N.D.; Lee, N.Y. Fully integrated and slidable paper-embedded plastic microdevice for point-of-care testing of multiple foodborne pathogens. *Biosens. Bioelectron.* **2019**, *135*, 120–128. [CrossRef] [PubMed]
35. Tozzo, P.; Mazzobel, E.; Marcante, B.; Delicati, A.; Caenazzo, L. Touch DNA Sampling Methods: Efficacy Evaluation and Systematic Review. *Int. J. Mol. Sci.* **2022**, *23*, 15541. [CrossRef] [PubMed]

biosensors

MDPI

Article

Direct Electron Transfer of Glucose Oxidase on Pre-Anodized Paper/Carbon Electrodes Modified through Zero-Length Cross-Linkers for Glucose Biosensors

Gilberto Henao-Pabon [1,†], Ning Gao [2,†], K. Sudhakara Prasad [3,4,†] and XiuJun Li [1,3,5,*]

1 Biomedical Engineering, University of Texas at El Paso, 500 W University Ave, El Paso, TX 79968, USA
2 Independent Researcher, 206 Via Morella, Encinitas, CA 92024, USA
3 Department of Chemistry & Biochemistry, University of Texas at El Paso, 500 W University Ave, El Paso, TX 79968, USA
4 Yenepoya Research Centre, Yenepoya University, Mangalore 575018, Karnataka, India
5 Forensic Science & Environmental Science and Engineering, 500 W University Ave, El Paso, TX 79968, USA
* Correspondence: xli4@utep.edu
† These authors contributed equally to this work.

Abstract: A disposable paper-based glucose biosensor with direct electron transfer (DET) of glucose oxidase (GOX) was developed through simple covalent immobilization of GOX on a carbon electrode surface using zero-length cross-linkers. This glucose biosensor exhibited a high electron transfer rate (ks, 3.363 s^{-1}) as well as good affinity (km, 0.03 mM) for GOX while keeping innate enzymatic activities. Furthermore, the DET-based glucose detection was accomplished by employing both square wave voltammetry and chronoamperometric techniques, and it achieved a glucose detection range from 5.4 mg/dL to 900 mg/dL, which is wider than most commercially available glucometers. This low-cost DET glucose biosensor showed remarkable selectivity, and the use of the negative operating potential avoided interference from other common electroactive compounds. It has great potential to monitor different stages of diabetes from hypoglycemic to hyperglycemic states, especially for self-monitoring of blood glucose.

Keywords: microfluidic paper-based analytical device; direct electron transfer; glucose biosensors; electrochemical detection; glucose oxidase immobilization

Citation: Henao-Pabon, G.; Gao, N.; Prasad, K.S.; Li, X. Direct Electron Transfer of Glucose Oxidase on Pre-Anodized Paper/Carbon Electrodes Modified through Zero-Length Cross-Linkers for Glucose Biosensors. *Biosensors* **2023**, *13*, 566. https://doi.org/10.3390/bios13050566

Received: 3 March 2023
Revised: 29 April 2023
Accepted: 8 May 2023
Published: 22 May 2023

1. Introduction

Diabetes is generally defined as a casual plasma glucose concentration that is higher than 200 mg/dL, and the incidence and prevalence of diabetes have been increasing worldwide [1]. In 2015, 9.4% of the total population of the USA, including adults and children, had diabetes (30.3 million people) [2,3]. Of those, 7.2% were diagnosed with diabetes (23.1 million people), and the rest were estimated to be undiagnosed (2.2% of the US population or 7.2 million people) [3]. As of 2015 in the US, about 5% of people diagnosed with diabetes are type 1 diabetics, and the rest are type 2 diabetics [4–6]. Other than diabetics, 33.9% or 84.1 million people are classified as pre-diabetes, when their biomarker glycosylated hemoglobin ($HgbA_{1c}$) is >5.5% but <6.4% [3,7]. Pre-diabetes is defined as a condition in which individuals have fasting plasma glucose (FPG) from 100 mg/dL to 125 mg/dL (5.6–6.9 mmol/L) or HbA_{1c} from 5.7% to 6.5% (39–47 mmol/mol), or impaired glucose tolerance with plasma glucose levels measured by an oral glucose tolerance test (OGTT) that are higher than normal but not high enough to be classified as diabetes [5,8,9]. Moreover, gestational diabetes causes pregnant women to exhibit high blood glucose levels and affects 3–10% of pregnancies [9,10]. The guidelines of the American Diabetes Association (ADA) and the International Diabetes Federation (IDF) recommend that self-monitoring of blood glucose (SMBG) protocols be individualized to meet the specific needs of each patient (T1D and T2D patients) and be performed frequently during pregnancy in diabetic patients [2,11,12]. Hence, the rapid determination of glucose levels in body fluids at the point of care is a key parameter for the diagnosis and management of diabetes [9,13]. Specifically, when a patient's glucose level is higher than 600 mg/DL, it will cause diabetic hyperosmolar syndrome, which is a life-threatening condition, and the patient needs to be quickly diagnosed and hospitalized right away.

GOX-based glucose biosensors have attracted much attention due to their simplicity in fabrication, reagent-less nature, portability, and low operational costs. However, the main limitation is that the commercially available GOX, GDH (glucose dehydrogenase), FAD (flavin-adenine-dinucleotide), and GDH (glucose dehydrogenase)-PQQ (pyrroloquinoline quinone)-based glucometers only cover the range from 20 to 600 mg/dL and cannot address all diabetes scenarios. Only two of them meet the ISO (International Organization for Standardization) 2013 criteria for the international standards for blood glucose monitoring systems [14,15]. In general, such bottlenecks arise in part due to limitations in direct electron transfer (DET) and the inaccessibility of active FAD sites (the active redox center of GOX) because FAD/$FADH_2$ is deeply embedded within a protein shell at approximately a 13 Å depth, which limits the electron transfer rate between the active site of GOX and the electron surface [16–18]. Previously, attempts at increasing the DET from GOX using functionalized carbon nanotube composites with graphene were reported. However, due to structural deformability and non-ohmic contacts, this approach suffered from low electrical conductivity or a low upper detection limit [19–23].

Herein, we report a new simple method for covalent immobilization of GOX on a pre-anodized paper-carbon electrode (PA-PPE) with zero-length cross-linkers for DET of glucose oxidase without utilizing any mediators and demonstrate its application in glucose biosensors. By utilizing EDC (1-ethyl-3-3-dimethylaminopropyl-carbodiimide hydrochloride) and NHS chemistry (1-Hydroxy-2,5-pyrrolidinedione or N-hydroxysuccinimide) [24,25], GOX was covalently linked to PA-PPE. This improved electrode based on the covalent binding of glucose oxidase and its cofactor FAD to a carbon electrode using short-distance co-linkers significantly improved the electron transfer rate (ks, 3.363 s^{-1}) and enabled the rapid quantitation of glucose concentrations from 5.4 mg/dL to 900 mg/dL. This allowed the new DET glucose biosensor to cover the range of hypoglycemic and hyperglycemic states while maintaining the low cost and portability required for a point-of-care testing device.

2. Experimental Procedure

2.1. Materials and Reagents

Glucose oxidase from Aspergillus Niger (EC 1.1.3.4 CAS 9001-37-0), Glucose (D-(+)-glucose (>99.5%, CAS 50-99-7)), 1-ethyl-3(3-dimethylaminopropyl) carbodiimide hydrochloride (EDC) (CAS 25952-53-8), N-hydroxysulfosuccinimide (Sulfo-NHS) (CAS 106627-54-7), $Ru(NH_3)\ Cl_2$ (hexamine-ruthenium III chloride) (98%, CAS 14282-91-8), phosphate-buffered saline (PBS) 10X concentrate (CAS 7558-79-4) with a pH of 7.2–7.6 at 25 °C, ascorbic acid (AA) (99%, CAS 50-81-7), salicylic acid (SA) powder (CAS 69-72-7), and lactic acid (LA) (CAS 50-21-5) were all purchased from Sigma Aldrich, St. Louis, MO, USA.

A highly concentrated graphene oxide (HC-GO) composition (≥79% carbon and ≤20% oxygen with sheet resistance 50 Ω/sq. at 25 mm thickness) was purchased from Graphene Laboratories (Calverton, NY, USA). Ag/AgCl ink (surface resistivity < 75 mΩ/square/mil and viscosity 5570–14,600 CPS) and Carbon ink 79% C-220 carbon resistive ink were obtained from Conductive Compounds. Potassium ferricyanide [$K_3Fe(CN)_6$] (CAS 13746-66-2) was purchased from Fisher Scientific. SU-8 negative photoresist formulation 50 was purchased from Micro-Chem (Newton, MA, USA). A pH 7.4 phosphate buffer solution (PBS) was used in all the studies. Deionized water (>18 MΩcm) was obtained from a Millipore purification system (Millipore, Burlington, MA, USA).

2.2. Electrochemical Measurements

Voltammetry measurements, cyclic voltammetry (CV), and square wave voltammetry (SWV) were carried out with a CHI 730E (CHI, Austin, TX, USA) electrochemical workstation at room temperature (25 °C). Three working electrodes, paper-carbon electrodes (PPE), graphene oxide-PPE (GO-PPE), pre-anodized-PPE (PA-PPE), and modified PA-PPE-GOX-glucose electrodes were studied with cyclic voltammetry at different scan rates ranging from 30 mV/s to 500 mV/s.

The PA-PPE-GOX biosensor was studied in different electrochemical systems, such as 5 mM potassium ferricyanide, 5 mM hexamine ruthenium III chloride, and 0.01 M PBS. The electrochemical oxidation of the biosensor was evaluated to demonstrate the presence of immobilized GOX and, therefore, the analysis of glucose.

2.3. Fabrication and Modification of the Paper-Based Carbon Electrode for GOX Immobilization

As shown in Figure 1A, the PPE was fabricated by pattering electrode designs onto a low-tack paper, which was subsequently pasted onto a piece of SU8-treated chromatography paper [26–30]. Afterwards, a carbon-based working electrode (WE area, 0.03141 cm^2), a counter electrode (CE), and a silver pseudo reference electrode (RE) were stencil printed with conductive carbon and Ag/AgCl ink.

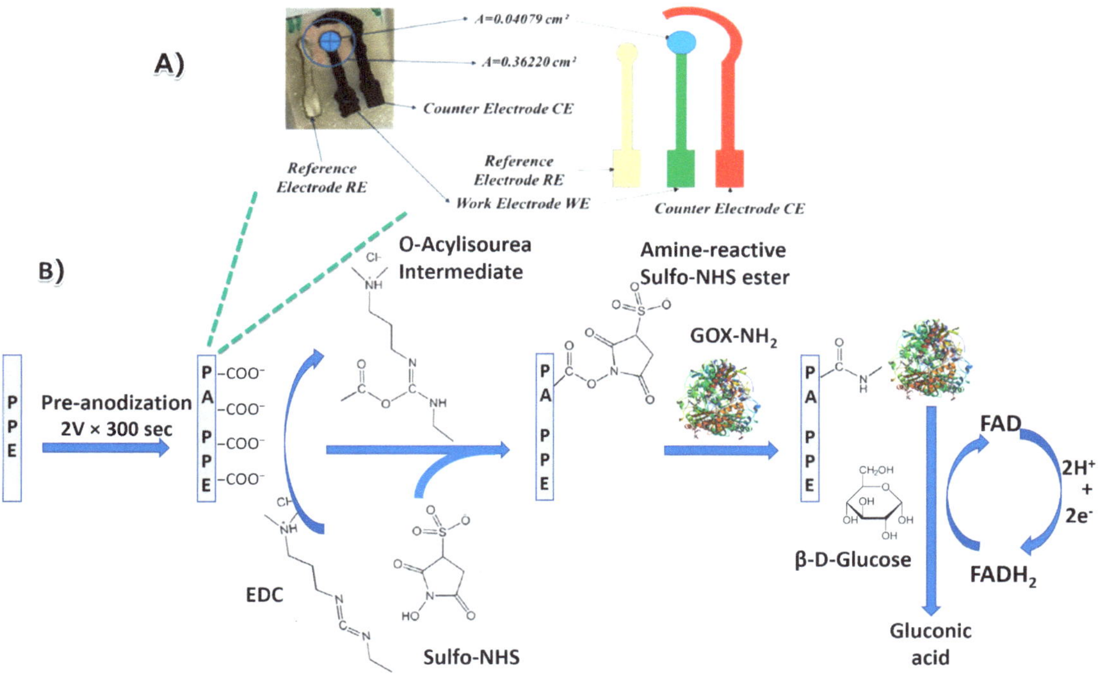

Figure 1. Photograph and schematic (**A**) of the paper-based glucose biosensor via direct electron transfer. The circuitry consisted of a WE, CE, and RE. (**B**) Schematic of the electrode modification process for GOX immobilization on paper-based carbon electrodes (PA-PPE-GOX) for the direct electron transfer glucose biosensor.

Figure 1B depicts the process of GOX immobilization on the electrode surface for DET glucose biosensors. The PPE electrode was prepared and subjected to pre-anodization in pH 7.4 PBS by applying a potential at 2.0 V (vs Ag/AgCl) over the electrode for 300 s. With pre-anodization, the ratio between O_{1s} and C_{1s} (O_{1s}/C_{1s} ratio) changed and created more carbonyl-group functionalities [31,32]. The pre-anodization not only creates more edge plane sites, making the modified carbon electrode more electroactive but also provides low susceptibility to electrode fouling [23,33,34]. Next, 5 mL of an EDC/NHS (0.35 M/0.1 M) solution was dropped over the WE at the PA-PPE and was incubated for 30 min at room temperature. Afterwards, the device was washed with 0.01 M PBS to remove the excess. Then, GOX was coupled with the WE through EDC and NHS cross-linkers. The electrodes (PA-PPE-GOX) were then dried at room temperature for 2 h and subsequently rinsed with 0.01 M PBS to remove the unbound GOX. The prepared electrodes (PA-PPE-GOX) and modified electrodes with GO (PA-PPE-GO-GOX) were further used for the voltammetry measurements and evaluated for the electrochemical oxidation of glucose through direct electron transfer without utilizing any mediators.

2.4. Characterization of the Modified Electrodes

The Fourier transform infrared (FT-IR) spectra of different electrodes were recorded to verify the electrode modification using a Spectrum 100 FT-IR Spectrometer (Perkin Elmer, Waltham, MA, USA). The optical module contains a Class II/2 Helium-Neon (HeNe) laser for continuous radiation at a wavelength of 633 nm.

X-ray photoelectron spectroscopy (XPS) is a powerful technique that provides useful information on the molecular compositions and chemical bonding of the glucose biosensor [35]. XPS analysis was conducted on a PHI 5600 spectrometer with a hemispherical energy analyzer using an aluminum (Kα) source of 1487 eV at 100 Watts. The pressure in the analysis chamber during XPS analysis was maintained in the low range of 10^{-9} Torr. The analyzed area was around 1 mm^2. All the spectra were recorded at a 54° take-off angle, with a 1.0 eV step. The spectra were further corrected using a carbon signal (C1s) at 284.5 eV and analyzed using Casa-XPS software version 2.3.18. The Shirley method was used to extract the background necessary for the curve fitting.

3. Results and Discussion

3.1. Surface Characterization

3.1.1. Analysis of GOX Immobilization via FT-IR Spectroscopy

Using FT-IR spectroscopy, we observed the presence of the amide I band (an overlapping spectrum of α-helices, β-sheets, turns, and random coils, which form the basic structure of the protein) [36] (Figure 2) at 1650 cm^{-1} and 1715 cm^{-1} caused by C=O stretching vibrations, which identified the C=O stretching vibrations of the peptide linkages in the GOX backbone at 1600 cm^{-1} to 1700 cm^{-1} and 1650 cm^{-1} to 1750 cm^{-1} [29,37]. Furthermore, the amide band II peak at 1462 cm^{-1} was observed due to the combination of N-H in-plane bending and C-N stretching linkages for the peptide groups [37–39]. There was another small band, amide band III, at 1239 cm^{-1}, which correlates well with the Delfino data, 2013 [40]. In addition, we found two wide bands in the region 3400 to 2900 cm^{-1}, representing the amide A and bond linkages to -CH_3 stretching at 3400 cm^{-1} and 2958 cm^{-1}, respectively [25]. Therefore, the FT-IR spectra indicate the successful immobilization of GOX on the PA-PPE surface.

3.1.2. Specific Surface Area

The specific surface area of the sample was calculated by the Brunauer–Emmett–Teller (BET) [41] surface adsorption method through the accelerated surface area and porosimetry system (ASAP) 2020 Sorptometer from Micromeritics by measuring the N_2 adsorption on the surface of the sample. The sample was degassed at 150 °C for 70 min at 10 mm Hg to remove any remaining solvent and ensure complete dryness and emptiness of the pores. Hence, the amount of adsorbed N_2 at a bath temperature of 76 K was obtained. The BET surface area report showed 5.07 m^2 per gram of material. The geometric surface area for WE is 0.03141 cm^2. The Brunauer–Emmett–Teller (BET) surface area is 12.675 cm^2/g, with a pore size of 92.607 Å and pore volume of 0.016736 cm^3/g.

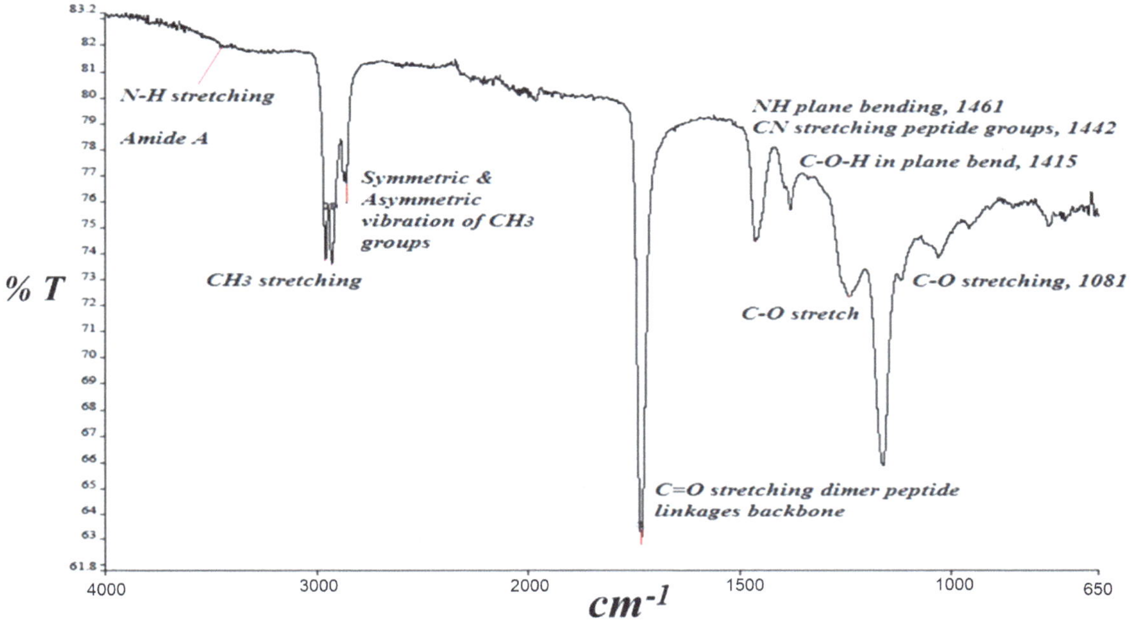

Figure 2. FT-IR spectra for PA-PPE-GOX.

3.1.3. Analysis of GOX Immobilization via XPS

XPS studies were conducted on the PA-PPE-GOX sensor to understand GOX immobilization [36]. The peaks of N and O from the XPS survey scan in Figure 3A indicate the immobilization of GOX on the carbon electrode. The XPS high-resolution scan data in Figure 3 illustrate that the data for C was fitted by a mix of pure C(1s), C-C, and C-O-C bonds in the following proportions. The peak of C(1s) shifted from 284 eV and was focused around 285.8 eV, corresponding to the overlap of the C-C, C-H, and C=C bonds [42,43]. A C(1s) centered peak was observed around 288 eV with two binding energies, 288.3 (N-C-O bond) and 286.3 (C-O bond), which can be assigned to overlap groups such as [R-CH2-NH-(C*O)-] and [(R-CH2*-NH-(CO)-], respectively [42], and presents the overlap of C=N, C≡N and C-O bonds [43]. These findings are consistent with previous XPS characterizations [44–46]. For instance, Li et al. reported two significant changes in the peaks obtained by the binding energy for immobilized GOX on the film surface. The first change corresponded to the reduction in the intensity of the carboxyl group near 288.7 eV, and the second change corresponded to an increase in intensity at the peak of the peptide union at 287.8 eV [44]. Later, Dementjev et al. found a peak with a union energy at 287.9 eV [46] that was identified as C(1s), indicating the interaction between C and N, specifically for carbon atoms (four bonds) that have a single or double bond with nitrogen atoms (C-4N).

In the binding energy of the N(1s) spectrum (Figure 3D), we found a centered peak at 400 eV, which corresponded to the pure N(1s) peak at 399 eV [42]. The N(1s) spectra core-level found the following binding energies at different peaks and was attributed to their respective unions, such as 398.3 eV (overlap N^1 sp^3 as N-C and C-NH_2 bonds) [43,46,47], 399.1 eV (overlap N≡C and C=N-C bonds) [43], 399.9 eV (N=C bond) [43], 400.6 eV (N-C sp^2 bonding) [47], and 400.8 eV [NO (N4)] [45]. These data confirm the presence of organic molecules, due to the configurations of the N(1s) and C(1s) spectra, in the surface of the biosensor and support the immobilization of GOX. The centered peaks overlapping binding energy for 398.5 eV, 398.8 eV, 399.1 eV, and 400.7 eV were attributed to the (C-NH_2 bond) [46], (N-C sp3 bonding) [47], (C=N-C bond) [46], and overlapping N-C sp2 bonding and NH_2 groups [43], respectively. All of them contributed to identifying the different bonds between the N(1s) and C(1s) as a part of the configurations of the carbon nitride compounds. The XPS data we obtained, especially regarding the changes in the spectra of C(1s) and N(1s), related to the chemical group characteristics of the proteins, indicated that GOX was successfully immobilized in our PA-PPE-GOX biosensor.

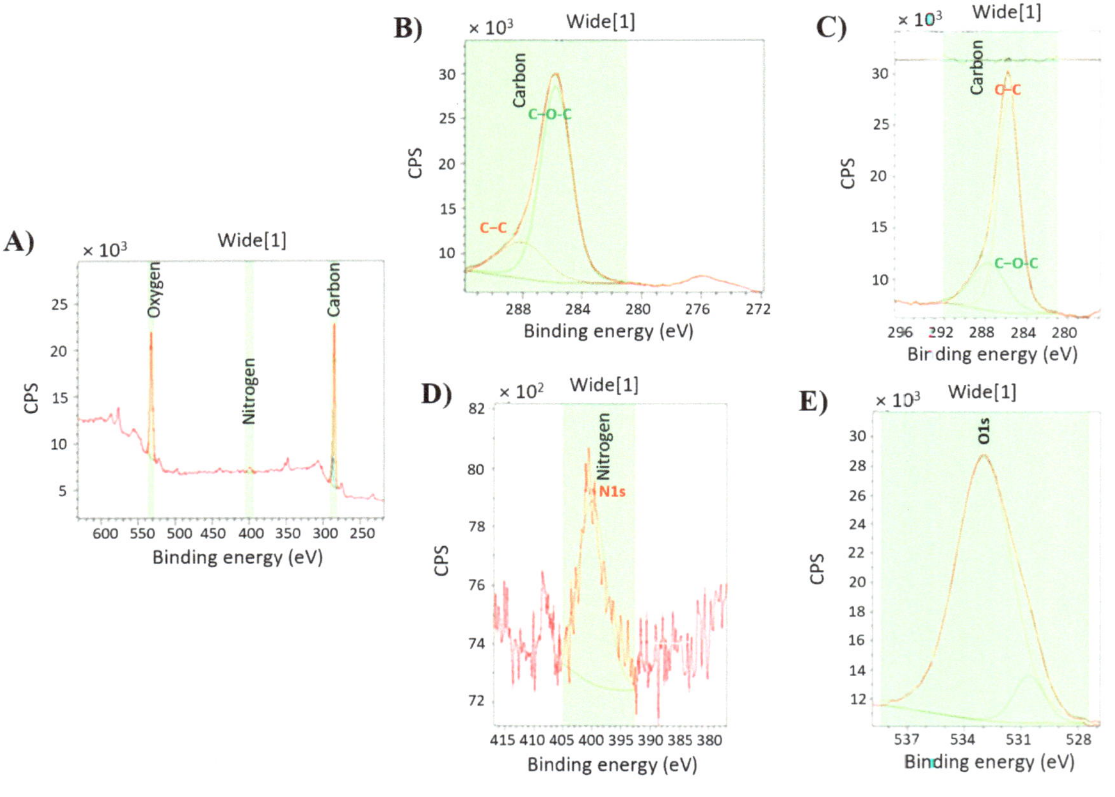

Figure 3. XPS for GOX (**A**) found C(1s) (**B**,**C**), N(1s) (**D**), and O(1s) (**E**) spectra.

3.2. Electrochemical Characterization

After immobilizing GOX onto the PA-PPE, the electrochemical characterization was performed in 0.01 M PBS under N_2. The CV profile for PA-PPE-GOX exhibited a pair of well-defined and nearly symmetric redox peaks with a formal potential of −0.44 V (see Figure 4A) close to the standard electrode potential of GOX [48], which was similar to previous reports [23]. In addition, we compared the CV responses for PPE-PA-GOX at the scan rate of 0.050 V/s between 5 mM potassium ferricyanide, an anionic probe (left), and 5 mM ruthenium-hexamine chloride 5 mM, which exhibited a cationic probe graphics (Figure 4B, right). Two different probes toward the same electrode explained the electron transfer involved and the interaction between the electrodes (negative charge) and the electrostatic interaction between the electrode and probe molecules (specific potential for oxidation and reduction).

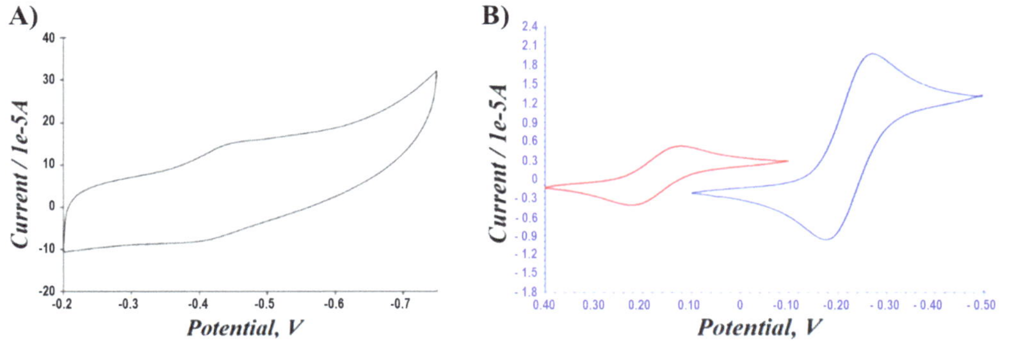

Figure 4. (**A**) CV response of PPE-PA-GOX in 0.01 M PBS at a scan rate of 50 mV/s. (**B**) CV response comparison for PPE-PA-GOX between 5 mM potassium ferricyanide (anionic probe-left side) and 5 mM ruthenium-hexamine chloride (cationic probe-right side).

The difference in the potential between the peaks of the reduction (E_{pc}) and oxidation (E_{pa}) curves was $\Delta E_p = (E_{pa} - E_{pc}) = 0.0599$ V (a value that is close to the theoretical value of 0.059 V for the ferrocene redox pair) for all the scan rates in our experiments. The peak current ratio (i_{pa}/i_{pc}) was equal to 1.01, and the formal electrode potential for a redox process was Eo′ = 0.43145. The electrochemical response of GOX immobilized onto the PA-PPE is attributed to the direct electron transfer of GOX for the conversion of FAD/$FADH_2$ [23,49,50]. The direct electron transfer reaction of the GOX/FAD redox reaction involves two electrons coupled with two protons, as shown by the following chemical reaction:

$$GOX/FAD + 2e^- + 2H^+ = GOX/FADH_2$$

We also investigated the electrochemical responses at different scan rates. The scan rate effect on the PA-PPE-GOX electrodes is shown in Figure 5. When the scan rates (v) increased from 50 to 200 mV/s (Figure 5A), the anodic and cathodic peak currents changed appreciably and increased with respect to the change in the scan rates and exhibited a good linear relationship between both anodic and cathodic current and scan rates (Figure 5B), indicating a surface-diffusion controlled redox electrode process. The relationship between the log peak current versus the log scan rate was linear with a slope of 0.561714 (a = transfer coefficient).

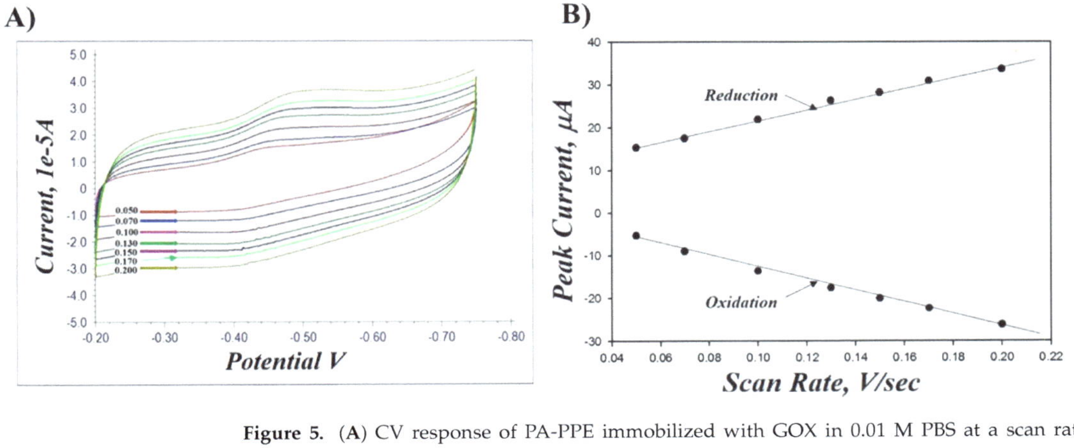

Figure 5. (**A**) CV response of PA-PPE immobilized with GOX in 0.01 M PBS at a scan rate of 50–200 mV/s. (**B**) Plots of anodic (i_{pa}) and cathodic (i_{pc}) peak current versus scan rates.

The electron transfer rate constant, k_s, on the PA-PPE-GOX electrode was calculated with the Laviron equation [51]:

$$Log\ k_s = \alpha\ log\ (1 - \alpha) + (1 - \alpha)\ log\alpha - log\ (RT/nF\ \nu) - \alpha\ (1 - \alpha)\ nF\Delta Ep/2.3RT$$

where the calculated charge transfer coefficient (α) is 0.5617 with two electrons transfer (n = 2), R is the universal gas constant (8.314 J $mol^{-1}K^{-1}$), T is the room temperature in Kelvin degrees (298 °K), F is the Faraday constant (9.64853 × 10^4 °C), and ν is the scan rate.

The electron transfer rate constant, k_s, was calculated to be 3.363 s^{-1}, which was higher than the k_s reported for carbon nanostructured materials, such as a K_s of 1.69 s^{-1} from Janegitz et al. [24], a K_s of 2.83 s^{-1} from Kang et al. [48], a K_s of 3.273 s^{-1} from Hua et al. [52], and a K_s of 1.713 s^{-1} and 1.12 s^{-1} from Razmi et al. [53], and gold nanoparticle incorporated matrices (a Ks of 1.713 s^{-1} from Wu et al. [54] and a Ks of 2.2 s^{-1} from Zhang et al. [55]). From this, it is evident that the carbonyl functionalities and edge plane-like sites formed during the pre-anodization process at PPE play an important role in improving the electron transfer communication between the redox centers of GOX and the electrode.

The surface average concentration of the electroactive GOX (Γ) value was calculated from the equation $\Gamma = Q/nFA$, where (Q) is the charge involved in the reaction, (n) is the number of electrons transferred, (F) is the Faraday constant, and (A) is the area of the WE. According to the equation, Γ of our system was calculated to be Γ = 8.30013 × 10^{-9} mol/cm^2, which was higher than other published values, such as Γ = 2.56 × 10^{-10} mol/cm^2 [55], Γ = 5.1 × 10^{-11} mol/cm^2 [37], Γ = 1.12 × 10^{-9} mol/cm^2 [49], Γ = 1.8 × 10^{-9} mol/cm^2 [53], and Γ = 4.65 ± 0.76 × 10^{-10} mol/cm^2 [56] implying that the higher heterogeneous direct electron transfer rate constant is directly influenced by the multilayer coverage of GOX.

3.3. Square Wave Voltammetry (SWV) Analysis

Additionally, the electrocatalytic activity of the PA-PPE-GOX toward glucose was studied by conducting SWV experiments with different concentrations of glucose ranging from 0.2 mM to 0.84 mM in 0.1 M PBS solution under N_2. The successive addition of glucose resulted in a gradual decrease in the reduction current (Figure 5A), which was linearly proportional to the increased concentration of glucose. This trend can be explained by the fact that the addition of glucose triggers the enzyme-catalyzed reaction between GOX and glucose by the formation of $FADH_2$ from FAD, resulting in a subsequent decrease in the cathodic peak current [24,57]. This reaction causes a decrease in the amount of oxidized GOX on the PA-PPE electrode and reduces the electrode reduction current. To demonstrate the increased sensitivity of the newly developed electrode, PA-PPE-GOX, toward glucose detection, we compared its response with the current response developed for GOX immobilized on graphene oxide through the EDC/NHS cross-coupling method, GO-PPE-GOX. As can be seen from the calibration plots of PA-PPE-GOX and GO-PPE-GOX (Figure 6B), the pre-anodized electrode exhibited a 2.612 times higher current response than the GO-modified electrodes. The improved electrocatalytic activity at PA-PPE-GOX again points toward better electron transfer communication between GOX and the electrodes at the electrochemically activated or pre-anodized electrodes than the GO-modified electrodes.

3.4. Chronoamperometry Analysis (CA)

The electrocatalytic performance of PA-PPE-GOX was further characterized by conducting chronoamperometry experiments for the reduction of different concentrations of glucose in 0.01 M PBS (pH 7.4), ranging from 30 µM to 50 mM. Figure 7 depicts the CA responses of glucose reduction at different concentrations of glucose at −0.55 V. The negative operating potential for the detection of glucose eliminates the interference from other common electroactive species, such as ascorbic acid, uric acid, and dopamine, that can be present in real blood samples, which is another advantage of our direct electron transfer glucose biosensor. As the concentration of glucose increased, there was a shift from the linearity between the concentration and current (Figure 8A), exhibiting typical Michaelis-Menten kinetics [58,59]. The Lineweaver–Burk plot [57], a double reciprocal graphical representation of the enzyme kinetics, shows a linear curve graphic based on the following equation:

$$1/v = (1/v_{max}) + \{(K_m/v_{max})\ (1/[S]),$$

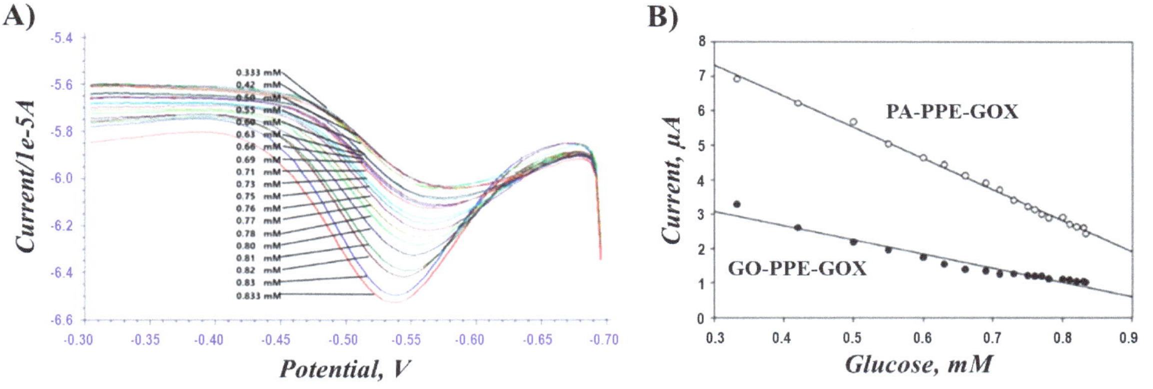

Figure 6. Square wave voltammogram (SWV) for different concentrations of glucose at PA-PPE-GOX (**A**) and the corresponding calibration curve for PA-PPE-GOX and GO-PPE-GOX (**B**).

where v is the initial reaction rate at a given substrate concentration, v_{max} is a constant reaction rate, K_m is the Michaelis–Menten constant, and $[S]$ is the substrate concentration. Vmax = 0.021714, m = 1.38343, and b = −46.05323, which corresponds to the linear equation $y = b + mx$, with a direct correlation between the current *1/v* (*1/mA*) and the glucose concentration $1/[S]$ (*1/mM*) [59].

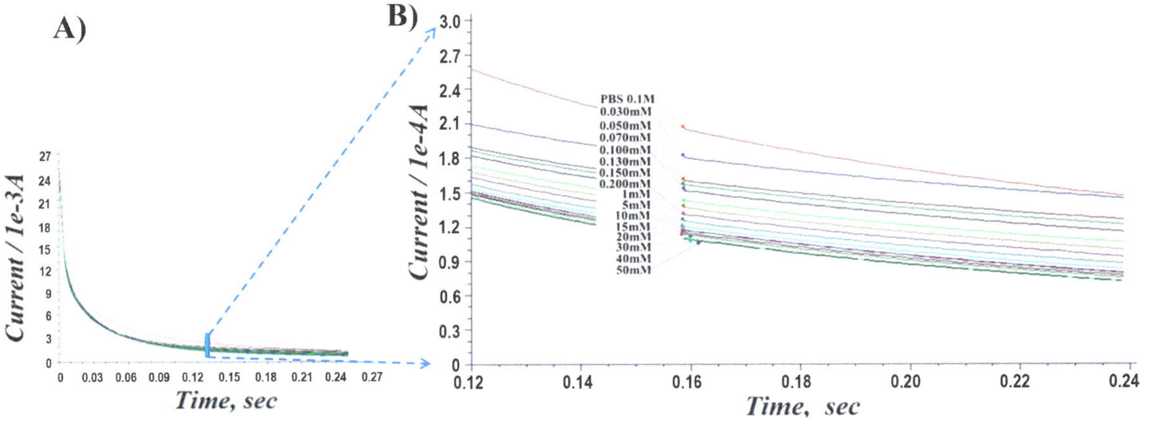

Figure 7. Chronoamperometric responses for PA-PPE-GOX toward different concentrations of glucose (30 μM to 50 mM). Section of (**A**) is zoomed in as (**B**).

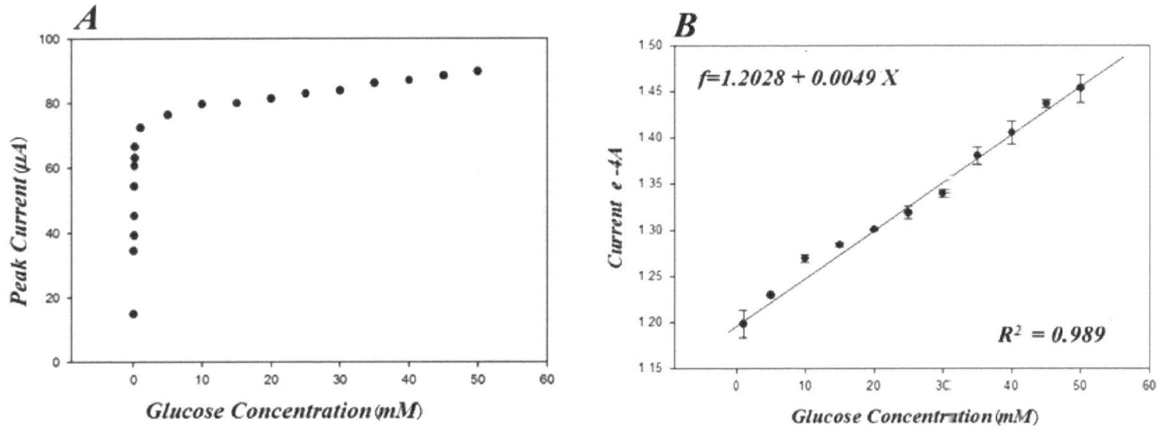

Figure 8. Peak current responses versus different glucose concentrations by the chronoamperometry method (**A**) and their corresponding linearity range from 1 mM through 50 mM (**B**).

The calculated K_m was found to be 0.03 mM, which is lower than that of the GOX immobilized on graphene quantum dots (K_m of 0.76 mM) [53], the reduced graphene oxide (K_m of 0.215 mM) [60], the poly (p-phenylenediamine)-based nanocomposite (K_m of 0.42 mM) [56], and the pre-anodized screen-printed carbon electrodes (K_m of 1.07 mM) [23]. A lower K_m means a higher enzymatic activity of the immobilized GOX at PA-PPE. Thus, the results suggest that the presented PA-PPE-GOX had a high affinity toward glucose, with a limit of detection as low as 5.4 mg/dL (0.3 mM). Figure 8B shows a linearity range for glucose of 0.3 mM through 50 mM (i.e., 5.4–900 mg/dL) with an R^2 value of 0.989, which includes the normal range of blood glucose in humans as well the abnormal range values in hypoglycemia or hyperglycemia. Most commercial glucometers can only achieve quantitative measurement of glucose in a limited range of operation.

3.5. Analytical Validation

There are substances in the blood that are considered to be interfering species and can interfere with glucose readings in biosensors. We conducted specificity tests using the amperometric *i-t* technique to validate the specificity of the biosensor PA-PPE-GOX on glucose, as well as ascorbic acid (AA), salicylic acid (SA), and lactic acid (LA) as interfering species. The concentrations of the interfering species were adjusted to a similar concentration as glucose (0.1 M). The amperometric readings are shown in Figure 9, and the currents of the interfering species were very low. We calculated the current ratios between glucose and the interfering species, which were 1.65%, 5.35%, and [illegible].99% for SA, LA, and AA, respectively. The current ratios between glucose and the interfering species allowed us to infer that our sensor shows great specificity for glucose. The current values of the interfering species represent a higher concentration (0.1 M) than the concentrations that these species can obtain in blood. Therefore, we observed a minimum current response of the interfering species with respect to glucose, indicating high selectivity of the biosensor toward glucose.

The validation for the electrode PA-PPE-GOX towards glucose was carried out through analytical recovery experiments, which were performed with different samples of whole blood, and the percentage of recovery was calculated from the data. The recovery values ranged from 90% to 101%, with an average of 95%, which are in the acceptable range in bioanalytical chemistry. In addition, three human whole blood samples were measured using our method, and the results are shown in Figure S1 and Table S1 (in the Supplementary Materials) which were consistent with the measured values from a commercial glucometer. To review the stability, different electrodes were tested for 3 months. The

PA-PPE-GOX-glucose showed an error percentage of 3.76%, showing that the electrodes have good shelf life and stability.

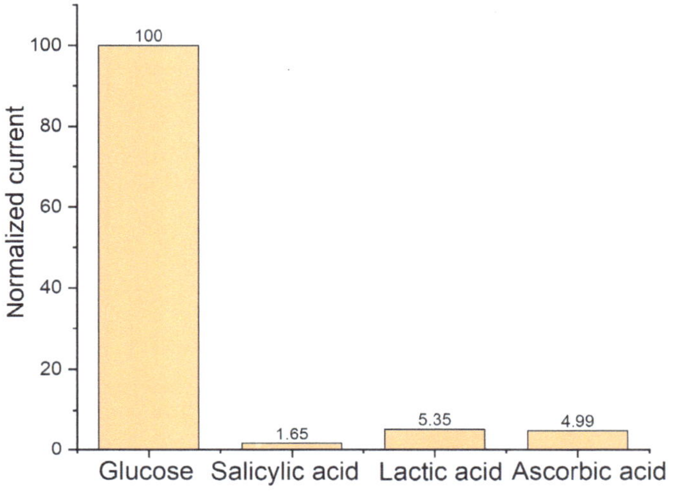

Figure 9. Specificity investigation by testing different interfering species. *E*, 0.5 V.

4. Conclusions

Herein, we developed a simple method for direct electron transfer of glucose oxidase on a disposable pre-anodized paper-carbon electrode modified through zero-length cross-linkers and demonstrated its application for glucose biosensors. FT-IR, XPS, and electrochemical methods were used to characterize the immobilization of GOX. The DET-based glucose biosensor has the following significant features:

(1) The direct electron transfer constant, K_s 3.363 s^{-1}, is much higher than that demonstrated in previous studies of carbon nanostructured materials [24,48,52,53] and gold nanoparticle [54,55] incorporated matrices and comparable with SPCE-Nafion® film, with a rapid detection time of less than 10 s.

(2) The surface average concentration of electroactive GOX (Γ) value was calculated to be 8.30013×10^{-9} mol/cm^2, which was higher than other published values [48,53,55,56]. This indicates that the higher heterogeneous direct electron transfer rate constant is directly influenced by the multilayer coverage of GOX.

(3) The DET-based glucose detection at PA-PPE-GOX showed higher electrocatalytic activity (2.61-fold) than a graphene oxide composite modified electrode.

(4) PA-PPE-GOX has a higher affinity for glucose, K_m 0.03 mM, with a wide range to detect glucose from 5.4 mg/dL to 900 mg/dL, which involves the glucose human range to measure diabetes variability in hypoglycemic and hyperglycemic states. In contrast, most commercially available glucometers can only achieve quantitative measurement of glucose in a limited range of operation. Our PA-PPE-GOX-glucose biosensor meets both recommendations addressed by the FDA and SMBG to minimize the hypoglycemia state and maximize euglycemia [61].

(5) Furthermore, our electrode (PPE-PA-GOX-glucose) showed remarkable stability and selectivity. The use of the negative operating potential eliminates the interference from ascorbic acid, uric acid, and dopamine.

This DET method can be used to immobilize other proteins and enzymes to achieve a higher electrocatalytic activity in various bioassays [62–66]. Our next step in our future work is to conduct clinical validation, especially for self-monitoring of blood glucose (SMBG).

Supplementary Materials: The following supporting information can be downloaded at: https://www.mdpi.com/article/10.3390/bios13050566/s1, Figure S1: Measurement of human whole blood samples using the paper-based glucose biosensor via direct electron transfer; Table S1: Glucose measurement results of human whole blood samples (mM).

Author Contributions: Conceptualization, K.S.P. and X.L.; Methodology, G.H.-P., N.G. and X.L.; Formal analysis, G.H.-P., N.G. and X.L.; Resources, X.L.; Writing—review & editing, N.G., K.S.P. and X.L.; Supervision, X.L.; Project administration, X.L.; Funding acquisition X.L. All authors have read and agreed to the published version of the manuscript.

Funding: This research was funded by the U.S. NIH/NIAID grant number [R41AI162477], the U.S. NSF grant number (IIP2122712, IIP2052347, CHE2216473, and DMR1827745), and CPRIT grant number (RP210165).

Institutional Review Board Statement: Not applicable due to the use of de-identified commercial samples.

Informed Consent Statement: Not applicable.

Data Availability Statement: The data are available on request.

Acknowledgments: We would like to acknowledge financial support from NIH/NIAID (R41AI162477), the U.S. NSF (IIP2122712, IIP2052347, CHE2216473, and DMR1827745), the Cancer Prevention and Research Institute of Texas (CPRIT; RP210165), and the DOT (CARTEEH). We are also grateful for the financial support for our prior research from the National Institute of Allergy and Infectious Diseases of the NIH (R21AI107415), the NIH/NIGMS (SC2GM105584), the NIH/NIMHD RCMI Pilot grant (5G12MD007593-22), the NIH BUILDing Scholar Summer Sabbatical Award, the NSF (IIP1953841), the Philadelphia Foundation, the Medical Center of the Americas Foundation (MCA), the University of Texas (UT) System for the STARS award, and the UTEP for IDR, URI, and MRAP awards.

Conflicts of Interest: A US Patent (#: 10,640,801) by X.L. and G.H. has been granted.

References

1. Centers for Disease Control and Prevention. National Diabetes Fact Sheet. 2011. Available online: http://www.cdc.gov/diabetes/pubs/pdf/ndfs_2011.pdf (accessed on 2 December 2019).
2. American Diabetes Association. Fast Facts: Data and Statistics about Diabetes. Revision 9/2015. Available online: http://professional.diabetes.org/admin/UserFiles (accessed on 2 December 2019).
3. National Diabetes Statistics Report. *Estimates of Diabetes and its Burden in the United States*; National Diabetes Statistics Report; National Center for Chronic Disease Prevention and Health Promotion, CDC, Division of Diabete Translation: Atlanta, GA, USA, 2017.
4. Vijan, S. Type 2 Diabetes. *Ann. Intern. Med.* **2019**, *171*, ITC65–ITC80. [CrossRef] [PubMed]
5. WHO/IDF. *Definition and Diagnosis of DM and Intermediate Hyperglycemia*; WHO: Geneva, Switzerland, 2006; p. 21.
6. American Diabetes Association. Expert committee on the diagnosis and classification of diabetes mellitus. Follow-up report on the diagnosis of diabetes mellitus. *Diabetes Care* **2008**, *31* (Suppl. S1), S55–S60.
7. Centers for Disease Control and Prevention; Department of Health and Human Services. *National Diabetes Statistics Report, 2014 National Diabetes Data Fact Sheet*; CDC: Atlanta, GA, USA, 2014.
8. Centers for Disease Control and Prevention. *Diabetes Report Card 2012: National State Profile of Diabetes and Its Complications*; CDC: Atlanta, GA, USA, 2012; pp. 1–14.
9. American Diabetes Association. Classification and Diagnosis of Diabetes: Standards of Medical Care in Diabetes in 2019. *Diabetes Care* **2019**, *42* (Suppl. 1), S13–S28. [CrossRef] [PubMed]
10. Moore, T.R. Diabetes Mellitus and Pregnancy. Medscape. Available online: https://emedicine.medscape.com/article/127547-overview (accessed on 6 April 2022).
11. Agarwal, M.M. Gestational diabetes mellitus: An update on the current international diagnostic criteria. *World J. Diabetes* **2015**, *6*, 782–791. [CrossRef] [PubMed]
12. Czupryniak, L.; Barkai, L.; Bolgarska, S.; Bronisz, A.; Brož, J.; Cypryk, K.; Honka, M.; Janez, A.; Krnic, M.; Lalic, N.; et al. Self-Monitoring of Blood Glucose in Diabetes: From Evidence to Clinical Reality in Central and Eastern Europe—Recommendations from the International Central-Eastern European Expert Group. *Diabetes Technol. Ther.* **2014**, *16*, 460–475. [CrossRef] [PubMed]
13. Yang, Y.; Ji, W.; Yin, Y.; Wang, N.; Wu, W.; Zhang, W.; Pei, S.; Liu, T.; Tao, C.; Zheng, B.; et al. Catalytic Modification of Porous Two-Dimensional Ni-MOFs on Portable Electrochemical Paper-Based Sensors for Glucose and Hydrogen Peroxide Detection. *Biosensors* **2023**, *13*, 508. [CrossRef]

14. Freckmann, G.; Schmid, C.; Baumstark, A.; Pleus, S.; Link, M.; Haug, C. System Accuracy Evaluation of 43 Blood Glucose Monitoring Systems for Self-Monitoring of Blood Glucose according to DIN EN ISO 15197. *J. Diabetes Sci. Technol.* **2012**, *6*, 1060–1075. [CrossRef]
15. Ekhlaspour, L.; Mondesir, D.; Lautsch, N.; Balliro, C.; Hillard, M.; Magyar, K.; Radocchia, L.G.; Esmaeili, A.; Sinha, M.; Russell, S.J. Comparative Accuracy of 17 Point-of-Care Glucose Meters. *J. Diabetes Sci. Technol.* **2017**, *11*, 558–566. [CrossRef]
16. Wilson, R.; Turner, A.P.F. Glucose oxide: An ideal enzyme. *Biosens. Bioelectron.* **1992**, *7*, 165–185. [CrossRef]
17. Ivnitski, D.; Artyushkova, K.; Rincón, R.A.; Atanassov, P.; Luckarift, H.R.; Johnson, G.R. Entrapment of Enzymes and Carbon Nanotubes in Biologically Synthesized Silica: Glucose Oxidase-Catalyzed Direct Electron Transfer. *Small* **2008**, *4*, 357–364. [CrossRef]
18. Guo, C.X.; Li, C.M. Direct electron transfer of glucose oxidase and biosensing of glucose on hollow sphere-nanostructured conducting polymer/metal oxide composite. *Phys. Chem. Chem. Phys.* **2010**, *12*, 12153–12159. [CrossRef]
19. Chen, J.; Zheng, X.; Miao, F.; Zhang, J.; Cui, X.; Zheng, W. Engineering graphene/carbon nanotube hybrid for direct electron transfer of glucose oxidase and glucose biosensor. *J. Appl. Electrochem.* **2012**, *42*, 875–881. [CrossRef]
20. Mani, V.; Devadas, B.; Chen, S.-M. Direct electrochemistry of glucose oxidase at electrochemically reduced graphene oxide-multiwalled carbon nanotubes hybrid material modified electrode for glucose biosensor. *Biosens. Bioelectron.* **2013**, *41*, 309–315. [CrossRef]
21. Palanisamy, S.; Cheemalapati, S.; Chen, S.-M. Amperometric glucose biosensor based on glucose oxidase dispersed in multiwalled carbon nanotubes/graphene oxide hybrid biocomposite. *Mater. Sci. Eng. C* **2014**, *34*, 207–213. [CrossRef]
22. Terse-Thakoor, T.; Komori, K.; Ramnani, P.; Lee, I.; Mulchandani, A. Electrochemically Functionalized Seamless Three-Dimensional Graphene-Carbon Nanotube Hybrid for Direct Electron Transfer of Glucose Oxidase and Bioelectrocatalysis. *Langmuir* **2015**, *31*, 13054–13061. [CrossRef]
23. Yang, T.H.; Hung, C.L.; Ke, J.H.; Zen, J.M. An electrochemically pre-anodized scree-printed carbon electrode for achieving direct electron transfer to glucose oxidase. *Electrochem. Commun.* **2008**, *10*, 1094–1097. [CrossRef]
24. Prasad, K.S.; Cao, X.; Gao, N.; Jin, Q.; Sanjay, S.T.; Henao-Pabon, G.; Li, X. A low-cost nanomaterial-based electrochemical immunosensor on paper for high-sensitivity early detection of pancreatic cancer. *Sens. Actuators B Chem.* **2020**, *305*, 127516. [CrossRef]
25. Fu, G.; Sanjay, S.T.; Dou, M.; Li, X. Nanoparticle-mediated photothermal effect enables a new method for quantitative biochemical analysis using a thermometer. *Nanoscale* **2016**, *8*, 5422–5427. [CrossRef]
26. Tavakoli, H.; Hirth, E.; Luo, M.; Sharma Timilsina, S.; Dou, M.; Dominguez, D.C.; Li, X. A microfluidic fully paper-based analytical device integrated with loop-mediated isothermal amplification and nano-biosensors for rapid, sensitive, and specific quantitative detection of infectious diseases. *Lab Chip* **2022**, *22*, 4693–4704. [CrossRef]
27. Zuo, P.; Li, X.; Dominguez, D.C.; Ye, B.-C. A PDMS/paper/glass hybrid microfluidic biochip integrated with aptamer-functionalized graphene oxide nano-biosensors for one-step multiplexed pathogen detection. *Lab Chip* **2013**, *13*, 3921–3928. [CrossRef]
28. Dou, M.; Dominguez, D.C.; Li, X.; Sanchez, J.; Scott, G. A Versatile PDMS/Paper Hybrid Microfluidic Platform for Sensitive Infectious Disease Diagnosis. *Anal. Chem.* **2014**, *86*, 7978–7986. [CrossRef] [PubMed]
29. Prasad, K.S.; Abugalyon, Y.; Li, C.; Xu, F.; Li, X. A new method to amplify colorimetric signals of paper-based nanobiosensors for simple and sensitive pancreatic cancer biomarker detection. *Analyst* **2020**, *145*, 5113–5117. [CrossRef] [PubMed]
30. Zhou, W.; Feng, M.; Valadez, A.; Li, X. One-Step Surface Modification to Graft DNA Codes on Paper: The Method, Mechanism, and Its Application. *Anal. Chem.* **2020**, *92*, 7045–7053. [CrossRef] [PubMed]
31. Evans, J.F.; Kuwana, T. Introduction of functional groups onto carbon electrodes via treatment with radio-frequency plasmas. *Anal. Chem.* **1979**, *51*, 358–365. [CrossRef]
32. Notsu, H.; Yagi, I.; Tatsuma, T.; Tryk, D.A.; Fujishima, A. Introduction of oxygen-containing functional groups onto diamond electrode surfaces by oxygen plasma and anodic polarization. *Electrochem. Solid-State Lett.* **1999**, *2*, 522–524. [CrossRef]
33. Prasad, K.; Chen, J.-C.; Ay, C.; Zen, J.-M. Mediatorless catalytic oxidation of NADH at a disposable electrochemical sensor. *Sens. Actuators B Chem.* **2007**, *123*, 715–719. [CrossRef]
34. Prasad, K.S.; Muthuraman, G.; Zen, J.-M. The role of oxygen functionalities and edge plane sites on screen-printed carbon electrodes for simultaneous determination of dopamine, uric acid and ascorbic acid. *Electrochem. Commun.* **2008**, *10*, 559–563. [CrossRef]
35. Jin, Q.; Ma, L.; Zhou, W.; Shen, Y.; Fernandez-Delgado, O.; Li, X. Smart paper transformer: New insight for enhanced catalytic efficiency and reusability of noble metal nanocatalysts. *Chem. Sci.* **2020**, *11*, 2915–2925. [CrossRef]
36. Zhao, H.-Z.; Sun, J.-J.; Song, J.; Yang, Q.-Z. Direct electron transfer and conformational change of glucose oxidase on carbon nanotube-based electrodes. *Carbon* **2010**, *48*, 1508–1514. [CrossRef]
37. Liang, W.; Zhuobin, Y. Direct Electrochemistry of Glucose Oxidase at a Gold Electrode Modified with Single-Wall Carbon Nanotubes. *Sensors* **2003**, *3*, 544–554. [CrossRef]
38. Hui, J.; Cui, J.; Xu, G.; Adeloju, S.B.; Wu, Y. Direct electrochemistry of glucose oxidase based on Nafion-Graphene-GOD modified gold electrode and application to glucose detection. *Mater. Lett.* **2013**, *108*, 88–91. [CrossRef]

39. Malesevic, A.; Vitchev, R.; Schouteden, K.; Volodin, A.; Zhang, L.; Van Tendeloo, G.; Vanhulsel, A.; Van Haesendonck, C. Synthesis of few-layer graphene via microwave plasma-enhanced chemical vapour deposition. *Nanotechnology* **2008**, *19*, 305604. [CrossRef]
40. Delfino, I.; Portaccio, M.; Ventura, B.D.; Mita, D.G.; Lepore, M. Enzyme distribution and secondary structure of sold-gel immobilized glucose oxidase by micro-attenuated total reflection FT-IR spectroscopy. *Mater. Sci. Eng. C* **2013**, *33*, 304–310. [CrossRef]
41. Hu, K.; Ma, L.; Wang, Z.; Fernandez-Delgado, O.; Garay, Y.E.; Lopez, J.A.; Li, X. Facile Synthesis and Integration of Poly(vinyl alcohol) Sponge-Supported Metal Nanocatalysts on a Microfluidic Chip Enable a New Continuous Flow Multireactor Nanocatalysis Platform for High Efficiency and Reusability Catalysis. *ACS Sustain. Chem. Eng.* **2022**, *10*, 10579–10589. [CrossRef]
42. Libertino, S.; Scandurra, A.; Aiello, V.; Giannazzo, F.; Sinatra, F.; Renis, M.; Fichera, M. Layer uniformity in glucose oxidase immobilization on SiO_2 surfaces. *Appl. Surf. Sci.* **2007**, *253*, 9116–9123. [CrossRef]
43. Nie, Z.; Nijhuis, C.A.; Gong, J.; Chen, X.; Kumachev, A.; Martinez, A.W.; Narovlyansky, M.; Whitesides, G.M. Electrochemical sensing in paper-based microfluidic devices. *Lab Chip* **2010**, *10*, 477–483. [CrossRef]
44. Li, Z.; Kang, E.; Neoh, K.; Tan, K. Covalent immobilization of glucose oxidase on the surface of polyaniline films graft copolymerized with acrylic acid. *Biomaterials* **1998**, *19*, 45–53. [CrossRef]
45. Bhattacharyya, S.; Cardinaud, C.; Turban, G. Spectroscopic determination of the structure of amorphous nitrogenated carbon films. *J. Appl. Phys.* **1998**, *83*, 4491–4500. [CrossRef]
46. Dementjev, A.; de Graaf, A.; van de Sanden, M.; Maslakov, K.; Naumkin, A.; Serov, A. X-Ray photoelectron spectroscopy reference data for identification of the C3N4 phase in carbon–nitrogen films. *Diam. Relat. Mater.* **2000**, *9*, 1904–1907. [CrossRef]
47. Boyd, K.J.; Marton, D.; Todorov, S.S.; Al-Bayati, A.H.; Kulik, J.; Zuhr, R.A.; Rabalais, J.W. Formation of C–N thin films by ion beam deposition. *J. Vac. Sci. Technol.* **1995**, *13*, 2110–2122. [CrossRef]
48. Kang, X.; Wang, J.; Wu, H.; Aksay, I.A.; Liu, J.; Lin, Y. Glucose Oxidase–graphene–chitosan modified electrode for direct electrochemistry and glucose sensing. *Biosens. Bioelectron.* **2009**, *25*, 901–905. [CrossRef] [PubMed]
49. Cai, C.; Chen, J. Direct electron transfer of glucose oxidase promoted by carbon nanotubes. *Anal. Biochem.* **2004**, *332*, 75–83. [CrossRef] [PubMed]
50. Wang, J.; Li, M.; Shi, Z.; Li, N.; Gu, Z. Direct Electrochemistry of Cytochrome *c* at a Glassy Carbon Electrode Modified with Single-Wall Carbon Nanotubes. *Anal. Chem.* **2002**, *74*, 1993–1997. [CrossRef] [PubMed]
51. Laviron, E. General Expression of the linear potential sweep voltammogram in the case diffusion less electrochemical systems. *J. Electroanal. Chem.* **1979**, *101*, 19–28. [CrossRef]
52. Hua, L.; Wu, X.; Wang, R. Glucose sensor based on an electrochemical reduced graphene oxide-poly(l-lysine) composite film modified GC electrode. *Analyst* **2012**, *137*, 5716–5719. [CrossRef]
53. Razmi, H.; Mohammad-Rezaei, R. Graphene quatum dots as a new substrate for immobilization and direct electrochemistry of glucose oxidase: Application to sensitive glucose determination. *Biosens. Bioelectron.* **2013**, *41*, 498–504. [CrossRef]
54. Wu, Y.; Hu, S. Direct electrochemistry of glucose oxidase in a colloid Au–dihexadecylphosphate composite film and its application to develop a glucose biosensor. *Bioelectrochemistry* **2007**, *70*, 335–3411. [CrossRef]
55. Zhang, H.; Meng, Z.; Wang, Q.; Zheng, J. A novel glucose biosensor based on direct electrochemistry of glucose oxidase incorporated in bio mediated gold nanoparticles-carbon nanotubes composite film. *Sens. Actuators B Chem.* **2011**, *158*, 23–27. [CrossRef]
56. Baghayeri, M. Glucose sensing by a glassy carbon electrode modified with glucose oxidase and a magnetic polymeric nanocomposite. *RSC Adv.* **2015**, *5*, 18267–18274. [CrossRef]
57. Liu, H.; Tao, C.-A.; Hu, Z.; Zhang, S.; Wang, J.; Zhan, Y. An electrochemical glucose biosensor based on graphene composites: Use of dopamine as reducing monomer and as site for covalent immobilization of enzyme. *RSC Adv.* **2014**, *4*, 43624–43629. [CrossRef]
58. Marcos, J.; Ríos, A.; Valcárcel, M. Automatic determination of Michaelis—Menten constants by the variable flow-rate technique. *Anal. Chim. Acta* **1993**, *283*, 429–438. [CrossRef]
59. Shirale, D.J.; Gade, V.K.; Gaikwad, P.D.; Savale, P.A.; Kharat, H.J.; Kakde, K.P.; Pathan, A.J.; Shirsat, M.D. Studies of immobilized glucose oxidase on galvanostatically synthesized poly(N-methylpyrrole) film with PVS-$NaNO_3$ composite dopant. *Int. J. Electrochem. Sci.* **2006**, *1*, 62–70.
60. Sehat, A.A.; Khodadadi, A.A.; Shemirani, F.; Mortazavi, Y. Fast immobilization of glucose oxidase on graphene oxide for highly sensitive glucose biosensor fabrication. *Int. J. Electrochem. Sci.* **2014**, *10*, 272–286.
61. Federal Drug Administration, USA. Executive Summary; Clinical Chemistry and Clinical Toxicology Devices Panel. 21 July 2016. 351p. Available online: https://www.fda.gov/media/98967/download (accessed on 5 May 2023).
62. Zhou, W.; Dou, M.; Timilsina, S.S.; Xu, F.; Li, X. Recent innovations in cost-effective polymer and paper hybrid microfluidic devices. *Lab Chip* **2021**, *21*, 2658–2683. [CrossRef]
63. Tavakoli, H.; Mohammadi, S.; Li, X.; Fu, G.; Li, X. Microfluidic platforms integrated with nano-sensors for point-of-care bioanalysis. *TrAC Trends Anal. Chem.* **2022**, *157*, 116806. [CrossRef]
64. Tavakoli, H.; Zhou, W.; Ma, L.; Perez, S.; Ibarra, A.; Xu, F.; Zhan, S.; Li, X. Recent advances in microfluidic platforms for single-cell analysis in cancer biology, diagnosis and therapy. *TrAC Trends Anal. Chem.* **2019**, *117*, 13–26. [CrossRef]

65. Sanjay, S.T.; Fu, G.; Dou, M.; Xu, F.; Liu, R.; Qi, H.; Li, X. Biomarker detection for disease diagnosis using cost-effective microfluidic platforms. *Analyst* **2015**, *140*, 7062–7081. [CrossRef]
66. Dou, M.; Sanjay, S.T.; Benhabib, M.; Xu, F.; Li, X. Low-cost bioanalysis on paper-based and its hybrid microfluidic platforms. *Talanta* **2015**, *145*, 43–54. [CrossRef]

biosensors

MDPI

Article

Vertically-Ordered Mesoporous Silica Film Based Electrochemical Aptasensor for Highly Sensitive Detection of Alpha-Fetoprotein in Human Serum

Tongtong Zhang [1,†], Luoxiang Yang [2,†], Fei Yan [2,*] and Kai Wang [1,*]

1 Key Laboratory of Integrated Oncology and Intelligent Medicine of Zhejiang Province, Department of Hepatobiliary and Pancreatic Surgery, Affiliated Hangzhou First People's Hospital, Zhejiang University School of Medicine, Hangzhou 310006, China; tongtongzhang@zju.edu.cn

2 Key Laboratory of Surface & Interface Science of Polymer Materials of Zhejiang Province, Department of Chemistry, Zhejiang Sci-Tech University, Hangzhou 310018, China; 20202010-156@mails.zstu.edu.cn

* Correspondence: yanfei@zstu.edu.cn (F.Y.); kaiw3@zju.edu.cn (K.W.)

† These authors contributed equally to this work.

Abstract: Convenient and rapid detection of alpha fetoprotein (AFP) is vital for early diagnosis of hepatocellular carcinoma. In this work, low-cost (0.22 USD for single sensor) and stable (during 6 days) electrochemical aptasensor was developed for highly sensitive and direct detection of AFP in human serum with the assist of vertically-ordered mesoporous silica films (VMSF). VMSF has silanol groups on the surface and regularly ordered nanopores, which could provide binding sites for further functionalization of recognition aptamer and also confer the sensor with excellent anti-biofouling capacity. The sensing mechanism relies on the target AFP-controlled diffusion of $Fe(CN)_6^{3-/4-}$ redox electrochemical probe through the nanochannels of VMSF. The resulting reduced electrochemical responses are related to the AFP concentration, allowing the linear determination of AFP with a wide dynamic linear range and a low limit of detection. Accuracy and potential of the developed aptasensor were also demonstrated in human serum by standard addition method.

Keywords: vertically-ordered mesoporous silica films; alpha fetoprotein; electrochemistry; aptasensor

Citation: Zhang, T.; Yang, L.; Yan, F.; Wang, K. Vertically-Ordered Mesoporous Silica Film Based Electrochemical Aptasensor for Highly Sensitive Detection of Alpha-Fetoprotein in Human Serum. *Biosensors* **2023**, *13*, 628. https://doi.org/10.3390/bios13060628

Received: 29 March 2023
Revised: 24 May 2023
Accepted: 26 May 2023
Published: 6 June 2023

1. Introduction

As one of the most common cancers in the world, hepatocellular carcinoma (HCC) will cause the liver cirrhosis and even death [1]. Compared with the imaging and histology commonly used in clinic, detection of alpha fetoprotein (AFP) concentration in serum is more simple and also has closely associated with HCC [2,3]. The content of AFP in healthy human is lower than 25 ng/mL [4], however, high levels of AFP (~500 ng/mL) can be observed in the serum of HCC patients [5]. Therefore, developing convenient and highly sensitive methods for monitoring the AFP concentration in human serum is very important for early diagnosis and prognostic evaluation of HCC.

Currently, varieties of analytical approaches have been designed to detect AFP, such as enzyme-linked immunosorption assay (ELISA) [6], electrochemistry [7], resonance light scattering [8], photoelectrochemistry [9], surface enhanced Raman scattering [10], fluorescence [11] and mass spectrometry [12]. Among them, electrochemical sensors have aroused the increasing attention due to their time-saving, low-cost, high sensitivity and easy operations [13–15], showing great clinical application potential in the early diagnosis of HCC [16]. To realize the specific determination, biological recognition elements including antibody, enzyme, and aptamer have been employed to be immobilized on the electrode surface [17–21]. Aptamer artificially screened and synthesized in vitro by systematic evolution of ligand by exponential enrichment (SELEX) displays several advantages with respect to the small size, low-cost, high stability and strong affinity towards preselected targets, which has been widely combined with functional nanomaterials to construct highly specific and sensitive electrochemical aptasensors [22]. In addition, biological macromolecules existed in human serum will cause severe bio-fouling issue on the electrodes, significantly impairing the accuracy and stability of electrodes and limiting their practical application in direct analysis of human serum. Therefore, direct and anti-fouling electrochemical determination of AFP in human serum without tedious pretreatment process still remains as a challenge.

Nanostructured mesoporous silica nanoparticles emerged as therapeutic nanocarriers have been widely utilized in the field of smart chemotheropy [23,24]. In comparison, vertically-ordered mesoporous silica films (VMSF) with perpendicularly oriented nanochannels have an apparent advantage of favored and fast mass transfer from the bulk solution to the electrode surface through nanochannels. During past decade, VMSF have shown great achievements in the construction of anti-fouling and anti-interference electrochemical sensors, due to their unique characteristics of uniform and ultrasmall pore size (usually 2~3 nm), high pore density (~40,000 μm^{-2}), and mechanical stability [25–29]. Apart from direct analysis of small redox targets [30–34], researchers have also employed the recognition elements (e.g., aptamer and antibody) to develop novel VMSF-based electrochemical/electrochemiluminescent sensors for specific determination of ions [35], disease-related biomarkers [36], and cancer cell [37]. These sensors are often realized by targets-controlled diffusion of electrochemical/electrochemiluminescent probes through the nanochannels of VMSF [38–42]. In comparison with homogeneous sensors, immobilization of targets-specific recognition elements on the surface of VMSF spares the reagents and reduces the cost [43,44], which however remains challenge that the mass transfer through the nanochannels to the underlying electrode will be influenced by steric effect. On the basis of above issue, aptamer with small size (one tenth of antibody) becomes the prospective recognition element for developing highly specific and sensitive VMSF-based sensors.

In this work, we present a cost-effective and highly sensitive electrochemical aptasensor for specific determination of AFP by grafting AFP-specific aptamer on the VMSF surface. VMSF attached to the indium tin oxide electrode (ITO) is prepared by Stöber-solution growth method and rich of silanol groups, allowing the further functionalization of aptamer to produce the sensing interface with high specificity and affinity. When generating the complex between aptamer and AFP, the ingress of electrochemical redox ($Fe(CN)_6^{3-/4-}$) to the nanochannels of VMSF was hindered, which could result in the decreased electrochemical responses and realize the quantitative detection of AFP. Moreover, considering the high selectivity of aptamer and anti-fouling characteristic of VMSF, accuracy and potential of proposed electrochemical aptasensor in human serum samples are examined.

2. Materials and Methods

2.1. Materials and Reagents

All antigens including alpha-fetoprotein (AFP), carcinoembryonic antigen (CEA), carbohydrate antigen 125 (CA125) and carbohydrate antigen 199 (CA199) were all obtained from KEY-BIO Biotech Co., Ltd. (Beijing, China). S-100 antigen was purchaed from Proteintech (North America). Amino-functionalized AFP aptamer was synthesized by Sangon Biotech Co., Ltd. (Shanghai, China), with the sequence of 5′-NH_2-GTG ACG CTC CTA ACG CTG ACT CAG GTG CAG TTC TCG ACT CGG TCT TGA TGT GGG TCC TGT CCG TCC GAA CCA ATC-3′. Sodium dihydrogen phosphate dihydrate ($NaH_2PO_4 \bullet 2H_2O$) was purchased from Mackin (Shanghai, China). Other reagents including tetraethyl orthosilicate (TEOS), hexadecyl trimethyl ammonium bromide (CTAB), (3-glycidoxypropyl)methyldiethoxysilane (GPTMS), bovine albumin (BSA) sodium phosphate dibasic dodecahydrate ($Na_2HPO_4 \bullet 12H_2O$) were all bought from Aladdin (Shanghai, China). And phosphate buffer (PBS) was prepared with $NaH_2PO_4 \bullet 2H_2O$ and $Na_2HPO_4 \bullet 12H_2O$. All reagents used were analytical reagent without further treatment. Human blood serum was provided by Affiliated Hangzhou First People's Hospital (Hangzhou, China). The ultrapure water (18.2 MΩ cm) was prepared from Mill-Q system.

2.2. Experiments and Instrumentations

The thickness and the pore structure of VMSF were characterized by scanning electron microscope (SEM, SU8010, Hitachi, Tokyo, Japan) and transmission electron microscope (TEM, HT7700, Hitachi, Japan), respectively. The accelerating voltage used for SEM and TEM measurements were 3 kV and 100 kV. Electrochemical characterization contains cyclic voltammetry (CV), differential pulse voltammetry (DPV), and electrochemical impedance spectroscopy (EIS), which was conducted on the electrochemical workstation (PGSTAT302N, Metrohm, Herisau, Switzerland). The scan rate of CV was 50 mV/s, and the step and modulation amplitude of DPV were 0.005 V and 0.025 V respectively. The error bars in the measurements were calculated from the standard deviation of three experiments.

2.3. Modification of VMSF on ITO Electrode

Prior to the growth of VMSF, bare ITO electrodes need to be pretreated by placing bare ITO electrodes into 1 M NaOH solution overnight, and then into acetone, ethanol and water under ultrasonication for 30 min, respectively. After rinsed by amounts of ultrapure water and dried by N_2 atmosphere, the freshly cleaned bare ITO electrodes were obtained and used to accomplish the growth of VMSF using Stöber-solution growth approach [45,46]. The detailed procedure was as follows: CTAB (160 mg) was dissolved in 100 mL 70% ethanol aqueous solution and ammonia (2.5%, 100 μL) and TEOS (80 μL) were added in turn to achieve silica-based precursor solution. Bare ITO electrodes (2.5 cm × 5 cm) were placed into the above precursor solution with sealing treatment and incubated in a water bath at 60 °C for 24 h. After the reaction finished, the resulting electrode confined surfactant micelles (SM) into the nanochannels, termed as SM@VMSF/ITO electrodes, were washed with ultrapure water and further aged at 100 °C for 10 h. SM could be extracted from silica nanochannels by using 0.1 M HCl-ethanol solution to obtain VMSF/ITO electrodes.

2.4. Construction of Electrochemical Aptasensor

VMSF/ITO based electrochemical aptasensor for AFP detection was constructed as the procedure shown in Figure 1a, which involved in the surface modification of GPTMS and AFP-specific aptamer. Firstly, the SM@VMSF/ITO electrode was placed into ethanol solution containing 0.052% (v/v) GPTMS for 1 h, allowing the modification of GPTMS on the outer surface of SM@VMSF/ITO based on the silanization reaction and producing epoxy groups for further functionalization. Subsequently, in order to remove the micelles from the nanochannels of VMSF, the resulting electrode was stirred in HCl-ethanol (0.1 M) solution for 5 min to obtain GPTMS modified VMSF/ITO, named as O-VMSF/ITO electrode. Secondly, the AFP aptamer (50 μL, 0.1 μM) was dropped onto the surface of O-VMSF/ITO at 4 °C for 2.5 h. After the residual aptamer was washed with PBS (0.01 M, pH = 7.4), BSA (0.5%) was used to block the nonspecific active sites at 4 °C for 0.5 h. Finally, the constructed electrochemical aptasensing interface was called Apt/O-VMSF/ITO.

2.5. Electrochemical Determination of AFP

$K_3[Fe(CN)_6]/K_4[Fe(CN)_6]$ (2.5 mM) in KCl (0.1 M) solution was used as the electrochemical probe for the electrochemical determination of AFP. Different concentration of AFP (50 μL) was dropped onto the electrochemical aptasensing interface (Apt/O-VMSF/ITO) and incubated at 4 °C for 1 h. After rinsing with PBS (0.01 M, pH = 7.4) to remove the residual AFP, AFP was specifically recognized and bounded to the Apt/O-VMSF/ITO electrode surface, resulting in the decreased electrochemical signals of $Fe(CN)_6^{3-/4-}$. The variation of electrochemical signals of $Fe(CN)_6^{3-/4-}$ was recorded using DPV method, realizing the determination of AFP. To further evaluate the reliability of constructed electrochemical aptasensor in clinical detection, standard addition method was chosen to detect AFP in human serum samples.

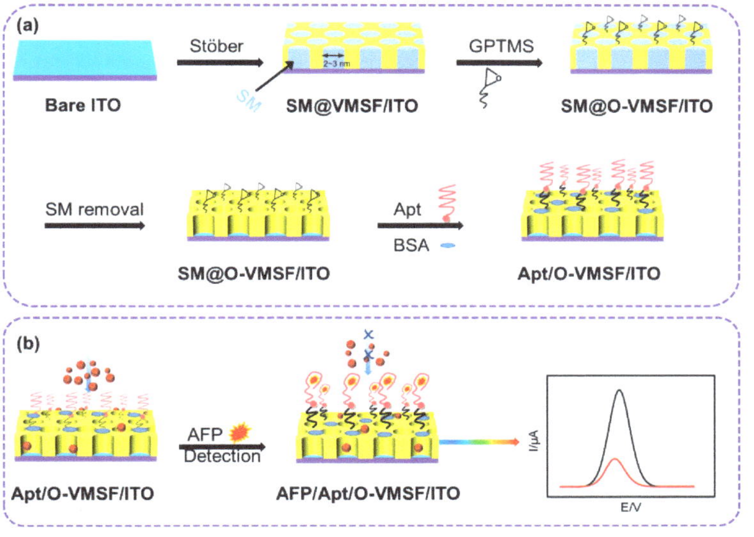

Figure 1. Schematic illustration for the fabrication of electrochemical aptasensing interface (**a**) and AFP detection (**b**).

3. Results and Discussion

3.1. Construction of Electrochemical Aptasensor

Figure 1a shows the construction process of the electrochemical aptasensor, which involves in the surface modification of GPTMS and AFP-specific aptamer. As seen, SM@VMSF/ITO electrode directly prepared by Stöber-solution growth method is first modified with GPTMS, resulting in the epoxy groups onto the outer surface of VMSF (SM@O-VMSF/ITO). Then, SM is excluded from the nanochannels of VMSF to obtain O-VMSF/ITO electrode with open channels. AFP-specific aptamer is immobilized on the O-VMSF/ITO electrode surface through silanization reaction between amino groups of aptamer and epoxy groups of GPTMS. After blocking the nonspecific active sites by blocking agent (BSA), electrochemical aptasensing interface is prepared, designated as BSA/Apt/O-VMSF/ITO. As shown in Figure 1b, the access of $Fe(CN)_6^{3-/4-}$ to the underlying ITO electrode through silica nanochannels is observed. And formed AFP-aptamer complex could inhibit the entrance of $Fe(CN)_6^{3-/4-}$ into the silica nanochannels, leading to the reduced electrochemical responses and ultimately allowing the determination of AFP.

3.2. Morphological and Electrochemical Characterizations of VMSF

SEM and TEM were used to characterize the morphology of VMSF, as shown in Figure 2a,b. The cross-sectional SEM image of VMSF/ITO electrode shows that VMSF with homogeneous thickness (~90 nm) is on the top of ITO coated glass (Figure 2a). The top-view TEM exhibits that regularly oriented and uniform silica nanopores are observed as bright dots and their diameter measured by Image J software are about 2~3 nm (Figure 2b), which is smaller than that of various biological fouling agents in human serum and could be acted as efficient anti-fouling layer. It can be found from the cross-sectional view TEM image that vertical silica nanochannels is parallel to each other and the lengh of silica nanochannels is 90 nm (Figure S1), which is in a accordance to the result from Figure 2a. CV and EIS were then used to examine the integrity of VMSF on the ITO electrode. As shown in Figure 2c, there is almost no redox peak current of the $Fe(CN)_6^{3-/4-}$ probe before SM removal, which is due to the hydrophobic SM filled in the channel inhibits the mass transfer of the hydrophilic and charged $Fe(CN)_6^{3-/4-}$. After SM removal, VMSF possesses open nanochannels and $Fe(CN)_6^{3-/4-}$ probe could access to the underlying ITO electrode through the nanochannels, producing a pair of redox peaks. The semicircle in the high-frequency region refers to the electron transfer-limited process and its equivalent diameter reflects the apparent charge-transfer resistance (R_{ct}). As presented in Figure 2d, R_{ct} of VMSF/ITO electrode is much smaller than that of SM@VMSF/ITO electrode, indicating the integrity and permeability of the VMSF modified on the electrode.

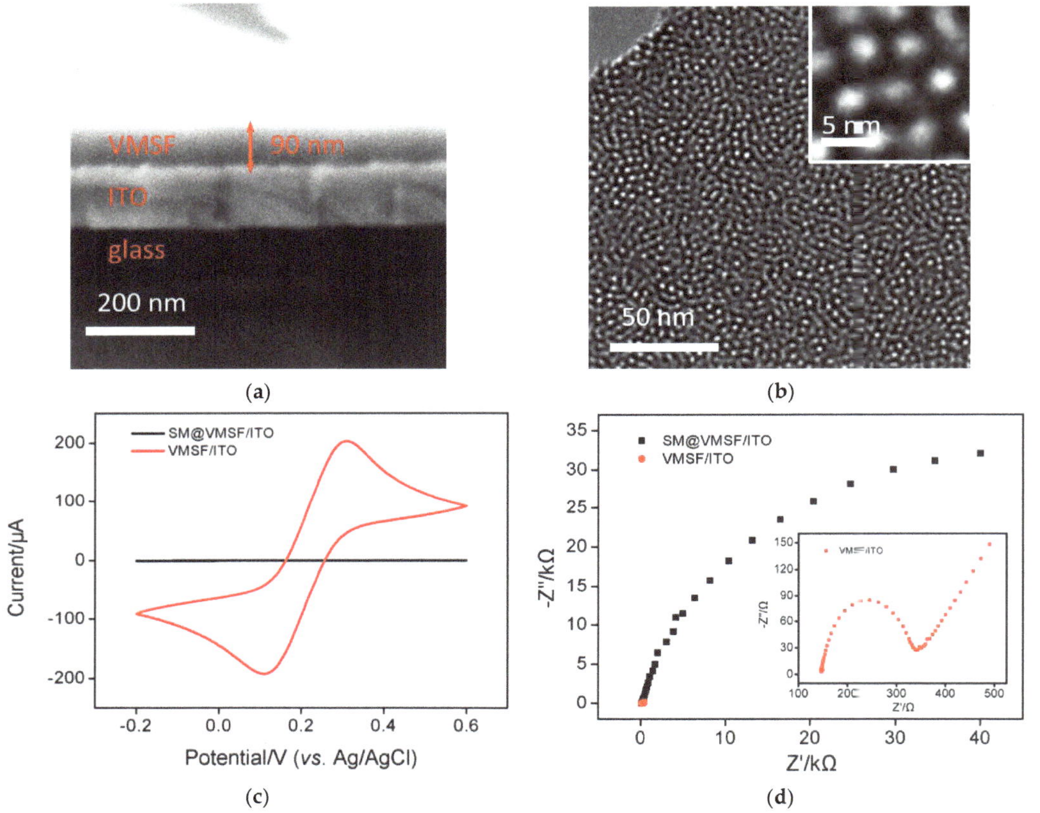

Figure 2. Cross-sectional SEM (**a**) and top-view TEM (**b**) images of VMSF. (**c**) CV curves obtained at SM@VMSF/ITO and VMSF/ITO electrodes in 0.1 M KCl solution containing 2.5 mM $K_3Fe(CN)_6/K_4Fe(CN)_6$. (**d**) Nyquist plots of SM@VMSF/ITO and VMSF/ITO electrodes obtained in 0.1 M KCl solution containing 2.5 mM $K_3Fe(CN)_6/K_4Fe(CN)_6$.

3.3. Electrochemical Characterization of the Construction Process of Electrochemical Aptasensor

Electrochemical probe, $Fe(CN)_6^{3-/4-}$, was employed to investigate the construction process of our proposed electrochemical aptasensor using CV and EIS methods and the results were shown in Figure 3. As seen in Figure 3a, modification of VMSF with GPTMS, aptamer and further BSA leads to the progressively decreased redox responses, which is ascribed to the steric hindrance impedes the diffusion of $Fe(CN)_6^{3-/4-}$ to the underlying ITO electrode through silica nanochannels. And the redox responses of $Fe(CN)_6^{3-/4-}$ further decrease after incubation with AFP, indicating the successful preparation of electrochemical aptasensing interface. Similar phenomenon could be observed from EIS data displayed in Figure 3b, modification of VMSF with GPTMS, aptamer, BSA and further binding of AFP results in a progressive increase of R_{ct}, suggesting the feasibility of our developed electrochemical aptasensor for AFP determination.

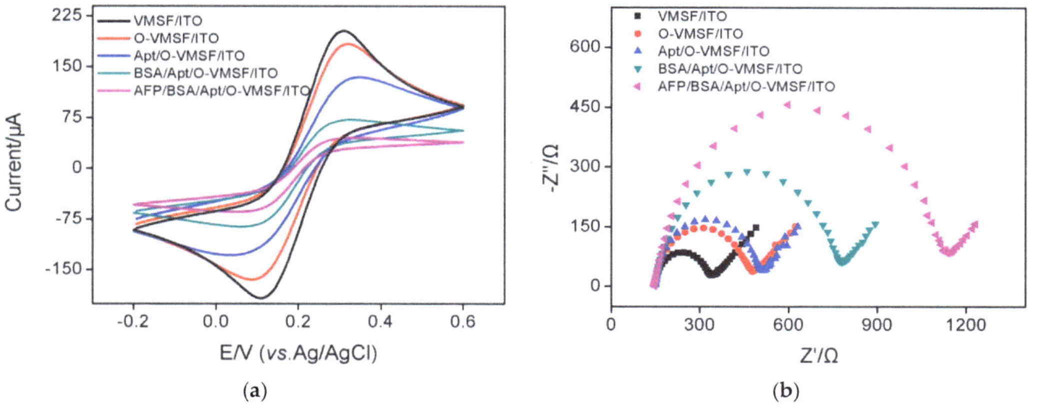

Figure 3. CV curves (**a**) and EIS plots (**b**) of VMSF/ITO, O-VMSF/ITO, Apt/O-VMSF/ITO, BSA/Apt/O-VMSF/ITO and AFP/BSA/Apt/O-VMSF/ITO electrodes obtained in 0.1 M KCl containing 2.5 mM $K_3Fe(CN)_6/K_4Fe(CN)_6$. The concentration of AFP is 1 ng/mL.

3.4. Optimized Conditions for the Construction of Electrochemical Aptasensor

In order to improve the sensitivity of AFP detection, effect of coupling time of aptamer onto the surface of the O-VMSF/ITO electrode was first studied. As shown in Figure 4a, in the range of 0–3 h, the anodic peak current of $Fe(CN)_6^{3-/4-}$ decreases obviously with increasing the coupling time and reaches equilibrium at 2.5 h, proving the outer surface of VMSF has been successfully modified with AFP-specific aptamer. Therefore, 2.5 h is selected as the optimal coupling time for aptamer modification.

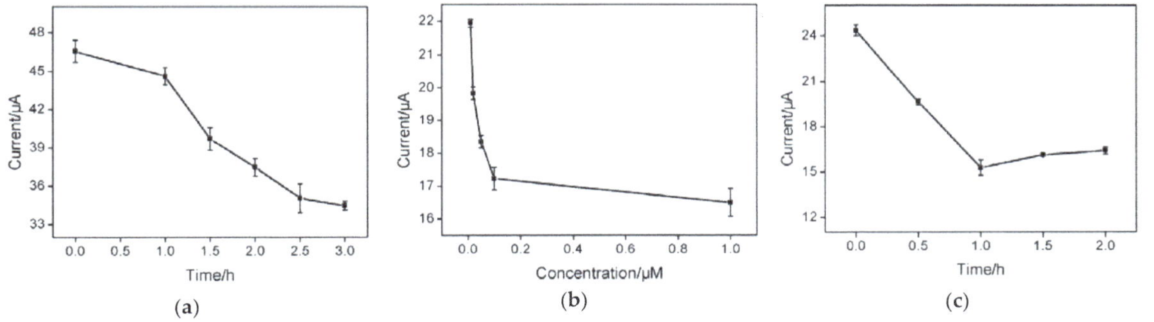

Figure 4. Effect of coupling time for aptamer immobilization (**a**), aptamer concentration (**b**) and incubation time for AFP (**c**) on current signal value.

As AFP-aptamer complex was formed in the recognition process by the developed BSA/Apt/O-VMSF/ITO sensor, aptamer concentration and the incubation time between aptasensor and AFP were investigated. AFP-specific aptamers with five different concentrations were used to prepare BSA/Apt/O-VMSF/ITO aptasensor and their responses on the AFP (1.0 ng/mL) were explored. When the aptamer concentration exceeded 0.1 μM, the anodic peak current remained almost unchanged, indicating that the aptamer was fully immobilized on the sensor surface. Therefore, the aptamer concentration of 0.1 μM is selected for subsequent work (Figure 4b). As depicted in Figure 4c, the anodic peak current decreases with an increase of incubation time and reaches stable at 1.0 h. Thus, 1 h is set as the optimal interaction time between aptamer and AFP.

3.5. Determination of AFP

Under the optimal detection conditions, different concentrations of AFP (1 pg/mL to 1 μg/mL) were incubated with the developed BSA/Apt/O-VMSF/ITO sensor and the electrochemical responses of $Fe(CN)_6^{3-/4-}$ were recorded by DPV. As shown in Figure 5a, anodic peak current decreases as the AFP concentration increases, which arises from the steric hindrances formed by AFP-aptamer complex on the electrode surface. And the calibration curve for AFP detection shown in Figure 5b displays a good linear range between the anodic peak current and the logarithm of AFP concentration ($\log C_{AFP}$), with a regression equation of I (μA) = −2.10 $\log C_{AFP}$ (pg/mL) + 21.9 (R^2 = 0.992) and a limit of detection (LOD) of 0.31 pg/mL (signal to noise ratio is 3). Such a low LOD arises from the fast mass transport inside the vertical silica nanochannels and high affinity between aptamer and AFP. Table 1 presents a comparison of our developed BSA/Apt/O-VMSF/ITO sensor with other electroanalytical techniques for AFP determination. By contrast, our proposed electrochemical aptasensor possesses a wider dynamic linear range and a lower LOD.

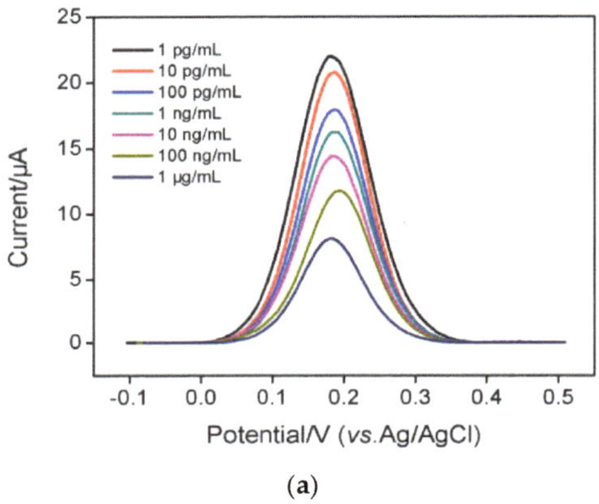

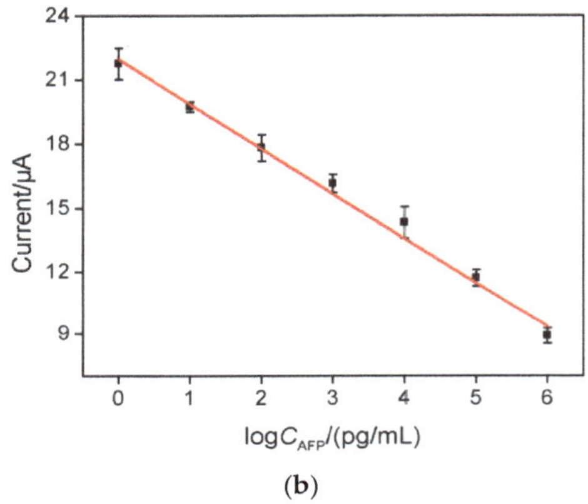

Figure 5. (**a**) DPV curves of the developed electrochemical aptasensor in the presence of different concentrations of AFP (1 pg/mL–1 μg/mL). (**b**) Corresponding calibration curve.

Table 1. Comparison with the existing electroanalytical techniques for the detection of AFP.

Electrode	Method	Liner Rang (ng/mL)	LOD (pg/mL)	Incubation Time (h)	Ref.
AFP/PBNP [1]-Apt [2]/GO/AuE	DPV	0.01–300	6.3	1.5	[47]
Apt/PtNPs/RGO [3]-CS [4]-Fc [5]/AuNPs/SPCE [6]	DPV	1–10^4	3.013 × 10^4	0.5	[48]
Apt/PtNPs/ GO-COOH [7]/SPGE [8]	SWV [15]	3–30	1.22 × 10^3	0.75	[49]
AFP/Apt/Peptide/ PANI [9]/GCE	DPV	10^{-6}–1	5.9 × 10^{-4}	1.0	[50]
ConA [10]-AgNPs/AFP/miR16/ESP [11]/AuE [12]	DPV	0.05–10	8.76	2.0	[51]
Ab_1 [13]/2D [14] $MoSe_2$/2D WSe_2/SPCE	DPV	0.001–5	0.75	0.75	[52]
AFP/Apt/O-VMSF/ITO	DPV	0.001–1000	0.31	1.0	This work

[1] prussian blue nanoparticles; [2] aptamer; [3] reduced graphene oxide; [4] chitosan; [5] ferrocene; [6] screen-printed carbon electrode; [7] carboxylated-graphene oxide; [8] screen-printed graphene-carbon paste electrode; [9] polyaniline; [10] concanavalin A; [11] electrochemical sensing probe; [12] gold electrode; [13] antibody; [14] two dimensional; [15] square wave voltammetry.

3.6. Selectivity, Reproducibility and Stability of the Developed Electrochemical Aptasensor

Selectivity, reproducibility and stability play important roles in evaluation of the developed electrochemical aptasensor. Some common antigens including CEA, CA199, S-100, CA125 and their mixture were incubated with the BSA/Apt/O-VMSF/ITO sensor and their results were shown in Figure 6a. As revealed, except for AFP, no significant responses to other proteins are observed at the proposed aptasensor, implying the satisfactory specificity of the constructed electrochemical aptasensor. To evaluate the reproducibility of the developed BSA/Apt/O-VMSF/ITO sensor, five different modified electrodes were prepared and used to detect 1.0 ng/mL AFP. As presented in Figure 6b, the relative standard deviation (RSD) of anodic peak current calculated from these modified electrodes was 1.8%, indicating the great reproducibility of the electrochemical aptasensor. Figure 6c studies the stability of the developed electrochemical aptasensor within 6 days. It could be found that after six-day storage, our BSA/Apt/O-VMSF/ITO sensor remains almost unchanged responses to 1.0 ng/mL AFP with an RSD of 1.5%, showing the attractive stability of the as-prepared electrochemical aptasensor.

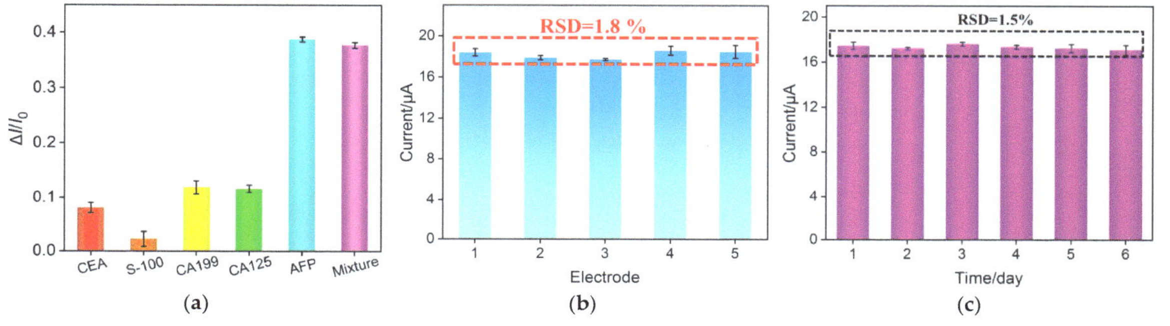

Figure 6. (**a**) Electrochemical responses of BSA/Apt/O-VMSF/ITO sensor to CEA, S-100, CA199, CA125, AFP or their mixture. I_0 and I denote the anodic peak current before (I_0) and after (I) incubation with AFP or other antigens and $\Delta I = I - I_0$. The concentration of AFP and other proteins are 0.1 ng/mL and 1 ng/mL, respectively. (**b**) Anodic peak currents obtained at five different electrodes. (**c**) Stability of the developed electrochemical aptasensor after storage for different days. The concentration of AFP in (**b**,**c**) is 1.0 ng/mL.

3.7. Real Sample Analysis

To access the practical utility of our proposed electrochemical aptasensor, the healthy human serum was spiked with a series of known concentrations of AFP standard solution and tested by our BSA/Apt/O-VMSF/ITO sensor. As shown in Table 2, the initial concentration of AFP in healthy human serum is measured as 1.0 ng/mL, which is close to the result measured by ELISA method (1.7 ng/mL). Recoveries ranging from 98.8% to 101% and RSD values ranging from 1.7% to 2.9% are obtained, verifying that the developed electrochemical aptasensor can be served as a reliable tool for AFP detection in clinical diagnosis of HCC.

Table 2. Determination of AFP in diluted human serum samples.

Sample	Spiked (ng/mL)	Found (ng/mL)	RSD (%) (n = 3)	Recovery (%)
Human serum *	0.100	0.100	2.9	100
	1.00	1.01	1.7	101
	10.0	9.88	1.8	98.8

* Human serum was diluted by 100-fold using 0.01 M PBS (pH 7.4) solution.

4. Conclusions

In summary, we reported the low-cost and simple electrochemical aptasensor for highly sensitive and selective detection of AFP in human serum based on an excellent anti-fouling layer of VMSF and a specific recognition layer of AFP-specific aptamer. Owing to the silanol groups of VMSF, AFP-specific aptamer with small size can not only covalently graft to the surface of VMSF and endow the sensing interface with selective recognition ability, but also remain the effective permeability of VMSF. With the help of electrochemical probe ($Fe(CN)_6^{3-/4-}$), electrochemical response variations have closely relation with AFP concentration, enabling the linear determination of AFP with a wide dynamic linear range and a low limit of detection. Furthermore, the inherent anti-fouling property of VMSF confers the proposed electrochemical aptasensor with promising feasibility in direct and selective analysis of AFP in human serum, which is potential to design a miniaturized and portable sensor integrated with smart phone. And the proposed strategy provides a convenient method for early diagnosis and prognostic evaluation of hepatocellular carcinoma.

Supplementary Materials: The following supporting information can be downloaded at: https://www.mdpi.com/article/10.3390/bios13060628/s1, Figure S1: TEM characterization of VMSF.

Author Contributions: T.Z.: methodology, formal analysis, data curation, writing—original draft preparation, funding acquisition. L.Y.: methodology, formal analysis, validation, writing—original draft preparation. F.Y.: conceptualization, supervision, writing—review and editing, funding acquisition. K.W.: supervision, project administration, writing—review and editing, funding acquisition. All authors have read and agreed to the published version of the manuscript.

Funding: This work was supported by the National Natural Science Foundation of China (3220100216), the Zhejiang Provincial Natural Science Foundation of China (LY21B050003), the Fundamental Research Funds of Zhejiang Sci-Tech University (22062310-Y) and Open Fund of Key Laboratory of Integrated Oncology and Intelligent Medicine of Zhejiang Province.

Institutional Review Board Statement: Not applicable.

Informed Consent Statement: Not applicable.

Data Availability Statement: The data presented in this study are available on request from the corresponding author.

Conflicts of Interest: The authors declare no conflict of interest.

References

1. Bhardwaj, N.; Perera, M.; Silva, M. Current treatment approaches to HCC with a special consideration to transplantation. *J. Transplant.* **2016**, *2016*, 7926264. [CrossRef]
2. Benson, A.B.; D'Angelica, M.I.; Abrams, T.A.; Are, C.; Bloomston, P.M.; Chang, D.T.; Clary, B.M.; Covey, A.M.; Ensminger, W.D.; Iyer, R. Hepatobiliary cancers, version 2.2014. *J. Natl. Compr. Cancer Netw.* **2014**, *12*, 1152–1182. [CrossRef]
3. Li, J.; Gao, T.; Gu, S.; Zhi, J.; Yang, J.; Li, G. An electrochemical biosensor for the assay of alpha-fetoprotein-L3 with practical applications. *Biosens. Bioelectron.* **2017**, *87*, 352–357. [CrossRef] [PubMed]
4. Preechakasedkit, P.; Siangproh, W.; Khongchareonporn, N.; Ngamrojanavanich, N.; Chailapakul, O. Development of an automated wax-printed paper-based lateral flow device for alpha-fetoprotein enzyme-linked immunosorbent assay. *Biosens. Bioelectron.* **2018**, *102*, 27–32. [CrossRef] [PubMed]
5. Dai, H.; Yang, C.; Tong, Y.; Xu, G.; Ma, X.; Lin, Y.; Chen, G. Label-free electrochemiluminescent immunosensor for α-fetoprotein: Performance of nafion–carbon nanodots nanocomposite films as antibody carriers. *Chem. Commun.* **2012**, *48*, 3055–3057. [CrossRef]
6. Wu, Y.; Guo, W.; Peng, W.; Zhao, Q.; Piao, J.; Zhang, B.; Wu, X.; Wang, H.; Gong, X.; Chang, J. Enhanced fluorescence ELISA bsed on HAT triggering fluorescence "turn-on" with enzyme–antibody dual labeled AuNP probes for ultrasensitive detection of AFP and HBsAg. *ACS Appl. Mater. Interfaces* **2017**, *9*, 9369–9377. [CrossRef] [PubMed]
7. Guo, J.; Wang, J.; Wang, Z.; Li, S.; Wang, J. The development of high sensitive alpha-fetoprotein immune-electrochemical detection method using an excellent conductivity 3D-CuFC-C nanocrystals synthesized by solution-grown at room temperature. *Biosens. Bioelectron.* **2022**, *218*, 114766. [CrossRef]
8. Chen, F.; Zhang, F.; Liu, Y.; Cai, C. Simply and sensitively simultaneous detection hepatocellular carcinoma markers AFP and miRNA-122 by a label-free resonance light scattering sensor. *Talanta* **2018**, *186*, 473–480. [CrossRef]
9. Li, X.; Pan, X.; Lu, J.; Zhou, Y.; Gong, J. Dual-modal visual/photoelectrochemical all-in-one bioassay for rapid detection of AFP using 3D printed microreactor device. *Biosens. Bioelectron.* **2020**, *158*, 112158. [CrossRef]
10. Ma, H.; Sun, X.; Chen, L.; Cheng, W.; Han, X.X.; Zhao, B.; He, C. Multiplex immunochips for high-accuracy detection of AFP-L3% Based on surface-enhanced raman scattering: Implications for early liver cancer diagnosis. *Anal. Chem.* **2017**, *89*, 8877–8883. [CrossRef]
11. Chen, P.; Jiang, P.; Lin, Q.; Zeng, X.; Liu, T.; Li, M.; Yuan, Y.; Song, S.; Zhang, J.; Huang, J.; et al. Simultaneous homogeneous fluorescence detection of AFP and GPC3 in hepatocellular carcinoma clinical samples assisted by enzyme-free catalytic hairpin assembly. *ACS Appl. Mater. Interfaces* **2022**, *14*, 28697–28705. [CrossRef] [PubMed]

12. Li, Y.; Huang, Z.; Li, Z.; Li, C.; Liu, R.; Lv, Y. Mass Spectrometric multiplex detection of microRNA and protein biomarkers for liver cancer. *Anal. Chem.* **2022**, *94*, 17248–17254. [CrossRef] [PubMed]
13. Gong, J.; Tang, H.; Wang, M.; Lin, X.; Wang, K.; Liu, J. Novel three-dimensional graphene nanomesh prepared by facile electro-etching for improved electroanalytical performance for small biomolecules. *Mater. Des.* **2022**, *215*, 110506. [CrossRef]
14. Lv, N.; Qiu, X.; Han, Q.; Xi, F.; Wang, Y.; Chen, J. Anti-biofouling electrochemical sensor based on the binary nanocomposite of silica nanochannel array and graphene for doxorubicin detection in human serum and urine samples. *Molecules* **2022**, *27*, 8640. [CrossRef] [PubMed]
15. Liu, Q.; Zhong, H.; Chen, M.; Zhao, C.; Liu, Y.; Xi, F.; Luo, T. Functional nanostructure-loaded three-dimensional graphene foam as a non-enzymatic electrochemical sensor for reagentless glucose detection. *RSC Adv.* **2020**, *10*, 33739–33746. [CrossRef] [PubMed]
16. Liu, S.; Ma, Y.; Cui, M.; Luo, X. Enhanced electrochemical biosensing of alpha-fetoprotein based on three-dimensional macroporous conducting polymer polyaniline. *Sens. Actuators B Chem.* **2018**, *255*, 2568–2574. [CrossRef]
17. Yousefi, M.; Dehghani, S.; Nosrati, R.; Zare, H.; Evazalipour, M.; Mosafer, J.; Tehrani, B.S.; Pasdar, A.; Mokhtarzadeh, A.; Ramezani, M. Aptasensors as a new sensing technology developed for the detection of MUC1 mucin: A review. *Biosens. Bioelectron.* **2019**, *130*, 1–19. [CrossRef]
18. Hasanzadeh, M.; Nahar, A.S.; Hassanpour, S.; Shadjou, N.; Mokhtarzadeh, A.; Mohammadi, J. Proline dehydrogenase-entrapped mesoporous magnetic silica nanomaterial for electrochemical biosensing of L-proline in biological fluids. *Enzym. Microb. Technol.* **2017**, *105*, 64–76. [CrossRef]
19. Yan, L.; Zhang, C.; Xi, F. Disposable amperometric label-free immunosensor on chitosan–graphene-modified patterned ITO electrodes for prostate specific antigen. *Molecules* **2022**, *27*, 5895. [CrossRef]
20. Chang, Q.; Huang, J.; He, L.; Xi, F. Simple immunosensor for ultrasensitive electrochemical determination of biomarker of the bone metabolism in human serum. *Front. Chem.* **2022**, *10*, 940795. [CrossRef]
21. Lin, J.; Li, K.; Wang, M.; Chen, X.; Liu, J.; Tang, H. Reagentless and sensitive determination of carcinoembryonic antigen based on a stable Prussian blue modified electrode. *RSC Adv.* **2020**, *10*, 38316–38322. [CrossRef]
22. Zhang, X.; Wang, F.; Zhi, H.; Zhao, J.; Wan, P.; Feng, L. Electrochemical "signal on/off" paper-based aptasensor for ochratoxin a detection based on MXene-Au and Pt@ NiCo-LDH-catalyzed signal amplification. *Sens. Actuators B Chem.* **2022**, *368*, 132161. [CrossRef]
23. Mazzotta, E.; De Santo, M.; Lombardo, D.; Leggio, A.; Pasqua, L. Mesoporous silicas in materials engineering: Nanodevices for bionanotechnologies. *Mater. Today Bio* **2022**, *17*, 100472. [CrossRef]
24. De Santo, M.; Giovinazzo, A.; Fava, M.; Mazzotta, E.; De Napoli, I.; Greco, M.; Comandé, A.; Nigro, A.; Argurio, P.; Perrotta, I.; et al. Engineered mesoporous silica-based nanoparticles as smart chemotherapy nanodevice for bortezomib administration. *Mater. Chem. Front.* **2023**, *7*, 216–229. [CrossRef]
25. Zhou, H.; Dong, G.; Sailjoi, A.; Liu, J. Facile pretreatment of three-dimensional graphene through electrochemical polarization for improved electrocatalytic performance and simultaneous electrochemical detection of catechol and hydroquinone. *Nanomaterials* **2022**, *12*, 65. [CrossRef] [PubMed]
26. Zheng, W.; Su, R.; Lin, X.; Liu, J. Nanochannel array modified three-dimensional graphene electrode for sensitive electrochemical detection of 2,4,6-trichlorophenol and prochloraz. *Front. Chem.* **2022**, *10*, 954802. [CrossRef] [PubMed]
27. Liang, R.; Jiang, J.; Zheng, Y.; Sailjoi, A.; Chen, J.; Liu, J.; Li, H. Vertically oriented mesoporous silica film modified fluorine-doped tin oxide electrode for enhanced electrochemiluminescence detection of lidocaine in serum. *RSC Adv.* **2021**, *11*, 34669–34675. [CrossRef]
28. Wei, X.; Luo, X.; Xu, S.; Xi, F.; Zhao, T. A flexible electrochemiluminescence sensor equipped with vertically ordered mesoporous silica nanochannel film for sensitive detection of clindamycin. *Front. Chem.* **2022**, *10*, 872582. [CrossRef]
29. Walcarius, A. Electroinduced surfactant self-assembly driven to vertical growth of oriented mesoporous films. *Acc. Chem. Res.* **2021**, *54*, 3563–3575. [CrossRef]
30. Deng, X.; Lin, X.; Zhou, H.; Liu, J.; Tang, H. Equipment of vertically-ordered mesoporous silica film on electrochemically pretreated three-dimensional graphene electrodes for sensitive detection of methidazine in urine. *Nanomaterials* **2023**, *13*, 239. [CrossRef]
31. Huang, L.; Su, R.; Xi, F. Sensitive detection of noradrenaline in human whole blood based on Au nanoparticles embedded vertically-ordered silica nanochannels modified pre-activated glassy carbon electrodes. *Front. Chem.* **2023**, *11*, 1126213. [CrossRef]
32. Zhu, X.; Xuan, L.; Gong, J.; Liu, J.; Wang, X.; Xi, F.; Chen, J. Three-dimensional macroscopic graphene supported vertically-ordered mesoporous silica-nanochannel film for direct and ultrasensitive detection of uric acid in serum. *Talanta* **2022**, *238*, 123027. [CrossRef] [PubMed]
33. Su, R.; Tang, H.; Xi, F. Sensitive electrochemical detection of p-nitrophenol by pre-activated glassy carbon electrode integrated with silica nanochannel array film. *Front. Chem.* **2022**, *10*, 954748. [CrossRef] [PubMed]
34. Zhou, H.; Ding, Y.; Su, R.; Lu, D.; Tang, H.; Xi, F. Silica nanochannel array film supported by ß-cyclodextrin-functionalized graphene modified gold film electrode for sensitive and direct electroanalysis of acetaminophen. *Front. Chem.* **2022**, *9*, 812086. [CrossRef]

35. Liu, J.; He, D.; Liu, Q.; He, X.; Wang, K.; Yang, X.; Shangguan, J.; Tang, J.; Mao, Y. Vertically ordered mesoporous silica films-assisted label-free and universal electrochemiluminescence aptasensor platform. *Anal. Chem.* **2016**, *88*, 11707–11713. [CrossRef] [PubMed]
36. Huang, J.; Zhang, T.; Zheng, Y.; Liu, J. Dual-mode sensing platform for cancer antigen 15-3 determination based on a silica nanochannel array using electrochemiluminescence and electrochemistry. *Biosensors* **2023**, *13*, 317. [CrossRef]
37. Wu, M.; Sun, X.; Zhu, M.; Chen, H.; Xu, J. Mesoporous silica film-assisted amplified electrochemiluminescence for cancer cell detection. *Chem. Commun.* **2015**, *51*, 14072–14075. [CrossRef]
38. Chen, D.; Luo, X.; Xi, F. Probe-integrated electrochemical immunosensor based on electrostatic nanocage array for reagentless and sensitive detection of tumor biomarker. *Front. Chem.* **2023**, *11*, 1121450. [CrossRef]
39. Yan, L.; Xu, S.; Xi, F. Disposal immunosensor for sensitive electrochemical detection of prostate-specific antigen based on amino-rich nanochannels array-modified patterned indium tin oxide electrode. *Nanomaterials* **2022**, *12*, 3810. [CrossRef]
40. Ma, K.; Zheng, Y.; An, L.; Liu, J. Ultrasensitive immunosensor for prostate-specific antigen based on enhanced electrochemiluminescence by vertically ordered mesoporous silica-nanochannel film. *Front. Chem.* **2022**, *10*, 851178. [CrossRef]
41. Ma, N.; Luo, X.; Wu, W.; Liu, J. Fabrication of a disposable electrochemical immunosensor based on nanochannel array modified electrodes and gated electrochemical signals for sensitive determination of C-reactive protein. *Nanomaterials* **2022**, *12*, 3981. [CrossRef] [PubMed]
42. Luo, X.; Zhang, T.; Tang, H.; Liu, J. Novel electrochemical and electrochemiluminescence dual-modality sensing platform for sensitive determination of antimicrobial peptides based on probe encapsulated liposome and nanochannel array electrode. *Front. Nutr.* **2022**, *9*, 962736. [CrossRef]
43. Gong, J.; Zhang, T.; Luo, T.; Luo, X.; Yan, F.; Tang, W.; Liu, J. Bipolar silica nanochannel array confined electrochemiluminescence for ultrasensitive detection of SARS-CoV-2 antibody. *Biosens. Bioelectron.* **2022**, *215*, 114563. [CrossRef] [PubMed]
44. Gong, J.; Zhang, T.; Chen, P.; Yan, F.; Liu, J. Bipolar silica nanochannel array for dual-mode electrochemiluminescence and electrochemical immunosensing platform. *Sens. Actuators B Chem.* **2022**, *368*, 132086. [CrossRef]
45. Teng, Z.; Zheng, G.; Dou, Y.; Li, W.; Mou, C.Y.; Zhang, X.; Asiri, A.M.; Zhao, D. Highly ordered mesoporous silica films with perpendicular mesochannels by a simple Stöber-solution growth approach. *Angew. Chem. Int. Ed.* **2012**, *51*, 2173–2177. [CrossRef]
46. Zhou, X.; Han, Q.; Zhou, J.; Liu, C.; Liu, J. Reagentless electrochemical detection of tumor biomarker based on stable confinement of electrochemical probe in bipolar silica nanochannel film. *Nanomaterials* **2023**, *13*, 1645. [CrossRef]
47. Zhang, B.; Ding, H.; Chen, Q.; Wang, T.; Zhang, K. Prussian blue nanoparticle-labeled aptasensing platform on graphene oxide for voltammetric detection of alpha-fetoprotein in hepatocellular carcinoma with target recycling. *Analyst* **2019**, *144*, 4858–4864. [CrossRef]
48. Li, W.; Chen, M.; Liang, J.; Lu, C.; Zhang, M.; Hu, F.; Zhou, Z.; Li, G. Electrochemical aptasensor for analyzing alpha-fetoprotein using RGO-CS-Fc nanocomposites integrated with gold-platinum nanoparticles. *Anal. Methods* **2020**, *12*, 4956–4966. [CrossRef]
49. Upan, J.; Youngvises, N.; Tuantranont, A.; Karuwan, C.; Banet, P.; Aubert, P.H.; Jakmunee, J. A simple label-free electrochemical sensor for sensitive detection of alpha-fetoprotein based on specific aptamer immobilized platinum nanoparticles/carboxylated-graphene oxide. *Sci. Rep.* **2021**, *11*, 13969. [CrossRef]
50. Liu, N.; Fan, X.; Hou, H.; Gao, F.; Luo, X. Electrochemical sensing interfaces based on hierarchically architectured zwitterionic peptides for ultralow fouling detection of alpha fetoprotein in serum. *Anal. Chim. Acta* **2021**, *1146*, 17–23. [CrossRef]
51. Gao, T.; Zhi, J.; Mu, C.; Gu, S.; Xiao, J.; Yang, J.; Wang, Z.; Xiang, Y. One-step detection for two serological biomarker species to improve the diagnostic accuracy of hepatocellular carcinoma. *Talanta* **2018**, *178*, 89–93. [CrossRef] [PubMed]
52. Chanarsa, S.; Jakmunee, J.; Ounnunkad, K. A sandwich-like configuration with a signal amplification strategy using a methylene blue/aptamer complex on a heterojunction 2D $MoSe_2$/2D WSe_2 electrode: Toward a portable and sensitive electrochemical alpha-fetoprotein immunoassay. *Front. Cell. Infect. Microbiol.* **2022**, *12*, 916357. [CrossRef] [PubMed]

biosensors

Article

A Novel Salivary Sensor with Integrated Au Electrodes and Conductivity Meters for Screening of Diabetes

Chen-Wei Lin [1,†], Yuan-Hsiung Tsai [2,3,†], Yun-Shing Peng [3,4], Jen-Tsung Yang [3,5], Yen-Pei Lu [6], Mei-Yen Chen [7] and Chun-Wu Tung [8,9,10,*]

[1] Department of Medical Education, Chang Gung Memorial Hospital, Chiayi 61363, Taiwan; toddgod7@cgmh.org.tw
[2] Department of Diagnostic Radiology, Chang Gung Memorial Hospital, Chiayi 61363, Taiwan
[3] College of Medicine, Chang Gung University, Taoyuan 33302, Taiwan
[4] Division of Endocrinology and Metabolism, Department of Internal Medicine, Chang Gung Memorial Hospital, Chiayi 61363, Taiwan
[5] Department of Neurosurgery, Chang Gung Memorial Hospital, Chiayi 61363, Taiwan
[6] National Applied Research Laboratories, Taiwan Instrument Research Institute, Hsinchu 30261, Taiwan
[7] Department of Nursing, Chang Gung University of Science and Technology, Chiayi 61363, Taiwan
[8] Department of Nephrology, Chang Gung Memorial Hospital, Chiayi 61363, Taiwan
[9] Chang Gung Medical Education Research Centre, Taoyuan 33302, Taiwan
[10] Department of Biochemical Science and Technology, National Chiayi University, Chiayi 61363, Taiwan
* Correspondence: p122219@cgmh.org.tw; Tel.: +886-(0)5-362-1000
† These authors have contributed equally to this work.

Abstract: The rise in diabetes cases is a growing concern due to the aging of populations. This not only places a strain on healthcare systems but also creates serious public health problems. Traditional blood tests are currently used to check blood sugar levels, but they are invasive and can discourage patients from regularly monitoring their levels. We recently developed nano-sensing probes that integrate Au microelectrodes and conductivity meters, requiring only 50 μL of saliva for measurement. The usage of the co-planar design of coating-free Au electrodes makes the measurement more stable, precise, and easier. This study found a positive correlation between the participant's fasting blood sugar levels and salivary conductivity. We observed a diabetes prevalence of 11.6% among 395 adults under 65 years in this study, using the glycated hemoglobin > 6.5% definition. This study found significantly higher salivary conductivity in the diabetes group, and also a clear trend of increasing diabetes as conductivity levels rose. The prediction model, using salivary conductivity, age, and body mass index, performed well in diagnosing diabetes, with a ROC curve area of 0.75. The study participants were further divided into low and high groups based on salivary conductivity using the Youden index with a cutoff value of 5.987 ms/cm. Individuals with higher salivary conductivity had a 3.82 times greater risk of diabetes than those with lower levels, as determined by the odds ratio calculation. In conclusion, this portable sensing device for salivary conductivity has the potential to be a screening tool for detecting diabetes.

Keywords: diabetes mellitus; salivary conductivity; sensor; non-invasive; blood sugar self-monitoring

Citation: Lin, C.-W.; Tsai, Y.-H.; Peng, Y.-S.; Yang, J.-T.; Lu, Y.-P.; Chen, M.-Y.; Tung, C.-W. A Novel Salivary Sensor with Integrated Au Electrodes and Conductivity Meters for Screening of Diabetes. *Biosensors* **2023**, *13*, 702. https://doi.org/10.3390/bios13070702

Received: 15 May 2023
Revised: 26 June 2023
Accepted: 29 June 2023
Published: 2 July 2023

1. Introduction

Diabetes is a major chronic disease that not only poses significant public health problems but also results in significant economic losses [1–4]. Currently, it is estimated to affect approximately 10.5% of the world's population, and its prevalence is on the rise [5]. Metabolic syndrome is a serious complication of diabetes, but the prolonged state of abnormal blood sugar is also associated with several other comorbidities [6]. Diabetes patients are also more vulnerable to other illnesses, placing them at higher risk of prolonged hospitalization, greater mortality rate, and reduced quality of life [7]. The current standard for diagnosing diabetes and monitoring blood glucose levels involves a blood test, which is invasive and may cause short-term complications, such as pain or local infections [8–10]. Therefore, a rapid, non-invasive, and user-friendly method for measuring blood glucose levels is necessary to enhance patient engagement in monitoring their health.

Saliva is an excellent bodily fluid specimen for health evaluation because it can be obtained non-invasively and contains a diverse array of water, electrolytes, cytokines, antibodies, and metabolites [11–15]. Saliva is a valuable biomarker for monitoring different diseases, and its collection is simple and inexpensive [16–18]. While some studies have shown a correlation between salivary and blood glucose levels, measuring it requires sophisticated instruments [19–21]. Our research team is dedicated to creating a portable and easy-to-use saliva-analyzing device to investigate its potential as a diagnostic biomarker for diabetes.

Salivary electrolyte concentrations, such as sodium, potassium, calcium, chloride, and bicarbonate, significantly increase with the advancement of diabetes [22–25]. It is believed that changes in the permeability of the salivary gland and blood vessel basement membranes caused by prolonged hyperglycemia can lead to significant increases in the salivary electrolytes. Additionally, patients with diabetes often experience xerostomia due to low saliva secretion rates, which can alter the electrical conductivity of saliva [24,25]. Our team has developed a novel sensing device with miniaturized probes for measuring salivary conductivity [26]. The portable device is easy-to-use and does not require specialized personnel to operate. Our initial data suggests a positive correlation between higher salivary conductivity and elevated fasting glucose levels, as well as an increased risk of developing chronic kidney disease [27–29]. However, further research is necessary to validate salivary conductivity as a reliable biomarker for diabetes. Therefore, this pilot study aims to confirm the association between salivary conductivity and diabetes in healthy adults and explore the potential of salivary conductivity as a screening tool for diabetes.

2. Materials and Methods

2.1. The Sensing Device

A portable device has been developed for determining the conductivity of saliva. It comprises two components, a conductivity meter, and a printed circuit board (PCB) equipped with electroless nickel immersion gold electrodes to prevent the oxide layer over the probe (shown in Figure 1A–C). The micro-fabricated gold electrodes are 2×2 mm^2 in size and connected to the PCB using nickel immersion gold wire. The compact size of the electrodes means that only 50 μL of the test sample is needed to cover the electrodes adequately (as illustrated in Figure 1B). The conductivity meter, measuring $10 \times 5.5 \times 2.2$ cm^3, includes a temperature sensor from Aosong (Guangzhou, China), a micro-control unit (MCU system) from STMicroelectronics (Geneva, Switzerland), an analog-to-digital converter (ADC) from Analog Devices (Norwood, MA, USA), and an organic light-emitting diode (OLED) from Zhongjingyuan (Henan, China). The conductivity meter can be activated using the ADC with 1 App and 1 kHz sine waves, and the conductivity parameters can be computed through a discrete Fourier transform. The temperature compensation through the MCU system allows the measurement signal to be obtained at 25 °C, and the result can be displayed on the OLED within 10 s (Figure 1D). Our device is highly reusable, as supported by previous research [29].

2.2. Saliva Collection and Analysis

The collection and analysis of saliva samples follow established protocols from prior research [27–29]. In brief, the process involves four key steps (Figure 1E). Firstly, participants are instructed to rinse their mouths with tap water for 30 s and empty any excess saliva. Secondly, they hold a saliva sample collection swab under their tongue for 10 s to collect saliva. Thirdly, the swab is inserted into the sample well of the sensing probe for analysis. Lastly, the conductivity is measured and displayed on the conductivity meter. The entire process, including preparation, takes approximately 3 min to complete.

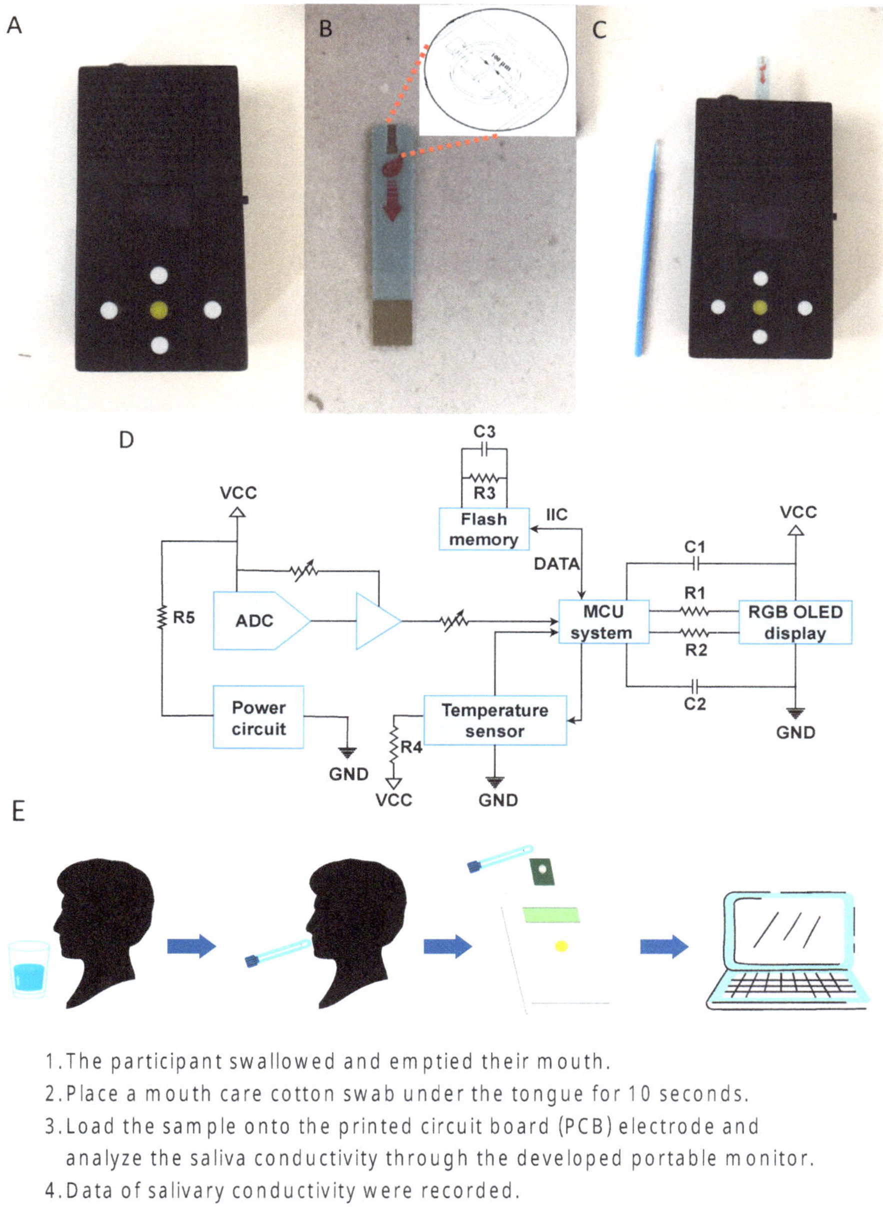

Figure 1. The portable device and the principles for measuring salivary conductivity. (**A**) This portable sensing system has dimensions of $10 \times 5.5 \times 2.2$ cm^3. (**B**) A PCB equipped with co-planar electroless nickel immersion gold electrodes that are free of coating and located in the sample well. (**C**) Connection of the two components and the saliva sample collection swab. (**D**) Electronic schematics of the fabricated conductivity meter. (**E**) Steps for saliva collection and measurement of conductivity. Abbreviations: ADC, analog-to-digital converter; (**C**) capacitance; GND, ground; IIC, inter-integrated circuit; MCU, micro control unit; OLED, organic light-emitting diode; PCB, printed-circuit-board; R, resistance; RGB, red, green, blue color model; VCC, volt current condenser.

2.3. Study Participants

This pilot study, with a cross-sectional design, enrolled adults aged between 18 and 65 years. A total of 421 subjects were recruited during their annual health check-up at the Yunlin Branch of the Chang Gung Memorial Hospital, a regional teaching hospital located in central Taiwan, in August 2021. After excluding participants with poor renal function, 408 eligible subjects were enrolled, but 13 of them were later excluded due to difficulties in cooperating with the saliva collection process, such as rinsing the mouth or placing cotton swabs under the tongue. Ultimately, this study was completed by 395 adult participants (Figure 2). This study adhered to the guidelines of the Declaration of Helsinki and was approved by the Medical Ethics Committee of Chang Gung Memorial Hospital (institutional review board numbers: 202002186B0). All participants provided informed consent before participating in this study.

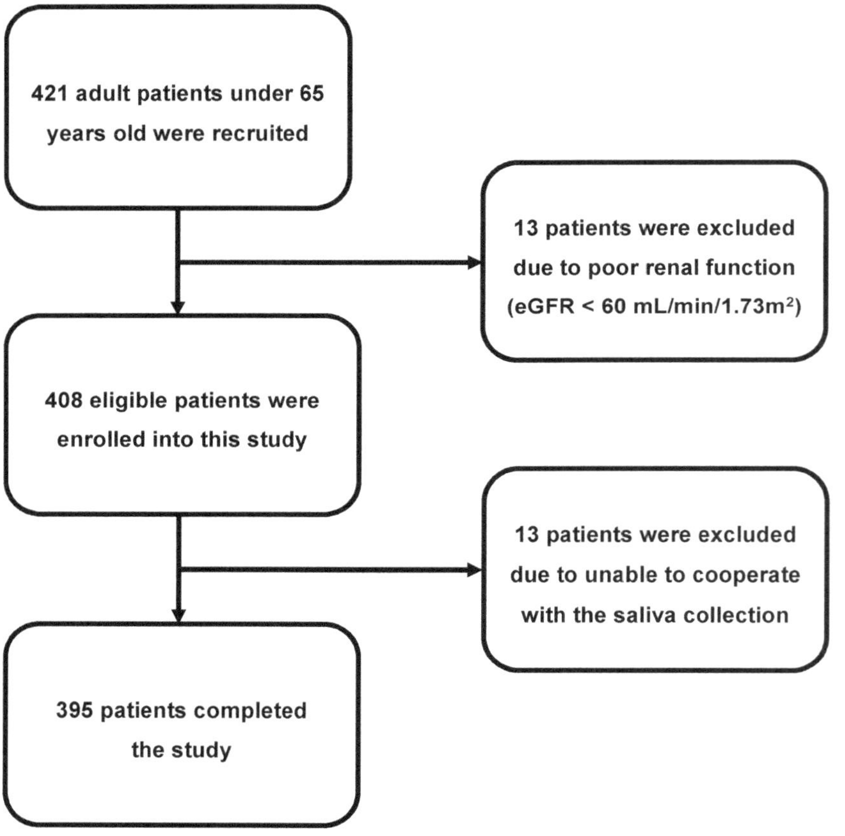

Figure 2. Flow chart of patient enrollment. The diagram illustrates the enrollment status of the participants.

2.4. Clinical Study Design

Before their health examination, participants were asked to fill out a comprehensive questionnaire provided by trained nurses. The questionnaire covered basic health information, including medical histories, such as diabetes, chronic kidney disease, ischemic heart disease, stroke, dyslipidemia, hypertension, gout, and cancer. Body height, weight, and corresponding body mass index (BMI) were measured, and blood pressure was taken in a quiet environment. Participants were instructed to fast overnight before the examination.

Both blood and saliva samples were collected and analyzed. The blood samples were processed using an automatic chemistry analyzer (Beckman DXC880i, Brea, CA, USA) following established laboratory procedures. The saliva samples were collected twice to calculate the average of the salivary conductivity. If the results of the two tests were inconsistent, the collection steps would be repeated to make sure the saliva samples were collected in the right way. Based on our previous work, we have shown that our device can yield great reusability, with the mean absolute percentage error (MAPE) being 0.88% after measuring the same saliva sample 20 times [29].

2.5. Definition of Diabetes

Diabetes is defined as having a glycated hemoglobin A1c (HbA1c) level higher than 6.5%, according to the recommendation of the American Diabetes Association.

2.6. Statistical Analysis

In this study, continuous variables are expressed as means with standard deviations, while categorical variables are presented as the number and percentage of observations. For comparing two groups, the independent Student's *t*-test was used for continuous variables with a normal distribution, and the Mann–Whitney U test was used for continuous variables that were not normally distributed. The normality of numerical variables was assessed using the Kolmogorov–Smirnov method. The results of the normality tests and the distribution plots of the salivary conductivity among different groups were shown in the Supplementary materials (Tables S1 and S2, Figure S1). Pearson's chi-square test was employed for comparing multiple groups of categorical variables. To evaluate the diagnostic accuracy of salivary conductivity in diabetes, receiver operating characteristic (ROC) curve analysis was used, and multivariable logistic regression analysis with backward selection was applied to improve its diagnostic performance. The diagnostic power was determined by calculating the area under the ROC curve (AUROC). The null hypothesis was rejected at a 95% confidence interval, and all statistical analyses were two-sided. The statistical analysis was performed using the Statistical Program for Social Sciences (SPSS) version 26 (IBM Corporation, Armonk, NY, USA) and Python version 3.10.

3. Results

3.1. Demographic Characteristics of the Study Group

The results of this study showed that the mean age of the participants was 51.78 years with a standard deviation of 11.31 years, and the age ranged from 15 to 65 years old. Of the total participants, 126 (31.9%) were male, while the rest were female. The survey revealed that 46 participants (11.6%) had diabetes, 95 (24.1%) had hypertension, 47 (11.9%) had dyslipidemia, and 11 (2.8%) had gout. The mean fasting glucose and hemoglobin A1c levels were 108.40 mg/dL and 5.88%, respectively. The mean value of salivary conductivity was 5.58 ms/cm. The remaining anthropometric parameters, blood pressure, and biochemical data are presented in Table 1.

3.2. Comparison of DM versus Non-DM Study Group

The participants were divided into the DM (Diabetes mellitus) and non-DM groups based on their HbA1c levels. DM group was defined by an HbA1c level greater than 6.5%, while the non-DM group was the rest of the participants. In the DM group (n = 46), the mean salivary conductivity value was 6.29 $\pm$ 1.58 ms/cm, which was higher than that in the non-DM group (5.48 $\pm$ 1.59 ms/cm, n = 349). Additionally, the DM group had significantly higher age and BMI, and a higher prevalence of comorbidities such as hypertension and dyslipidemia than the non-DM group. Furthermore, the DM group had significantly higher serum osmolarity and fasting glucose levels compared to the non-DM group. Hemoglobin A1c levels were also significantly higher in the DM group (7.80 $\pm$ 1.51%) than in the non-DM group (5.63 $\pm$ 0.35) (Table 1). To minimize any potential bias that may have resulted from the unequal distribution of confounding variables, we utilized subgroup analysis to compare the male and female groups (Tables S3 and S4) and propensity score matching analysis (Table S5). These results were consistent with the initial analysis that gender or other demographic data would not be the confounding factors. It suggests that the salivary conductivity value could serve as a potential biomarker for diabetes mellitus.

Table 1. Baseline characteristics of participants stratified by Diabetes (n = 395).

	All (N = 395)	**DM (N = 46)**	**Non-DM (N = 349)**	***p*-Value**
Salivary conductivity, ms/cm	5.58 ± 1.61	6.29 ± 1.58	5.48 ± 1.59	<0.01
Demographics				
Age, years	51.78 ± 11.31	56.96 ± 6.78	51.10 ± 11.51	<0.01
Gender (male), *n* (%)	126 (31.9)	15 (32.7)	111 (31.8)	0.91
Body weight, kg	64.43 ± 12.53	70.41 ± 12.90	63.65 ± 12.29	<0.01
Body height, cm	160.65 ± 7.84	160.66 ± 7.94	160.65 ± 7.33	0.91
Body mass index, kg/m^2	24.90 ± 4.15	27.19 ± 4.01	24.60 ± 4.08	<0.01
Systolic blood pressure, mmHg	128.36 ± 21.15	138.87 ± 24.84	126.97 ± 20.26	<0.01
Diastolic blood pressure, mmHg	78.72 ± 12.61	82.24 ± 12.61	78.26 ± 12.55	0.04
Comorbid conditions, *n* (%) @				
Known history of DM	46 (11.6)	26 (56.5)	20 (5.7)	<0.01
Hypertension	95 (24.1)	19 (41.3)	76 (21.8)	<0.01
Dyslipidemia	47 (11.9)	10 (21.7)	37 (10.6)	0.03
Gout	11 (2.8)	1 (2.2)	10 (2.9)	1.00
Laboratory parameters				
BUN, mg/dL	13.97 ± 4.11	14.85 ± 4.65	13.86 ± 4.03	0.11
Creatinine, mg/dL	0.74 ± 0.16	0.73 ± 0.16	0.74 ± 0.16	0.44
eGFR, mL/min/1.73 m^2	101.54 ± 21.18	101.97 ± 22.20	101.49 ± 21.08	0.86
Serum osmolality, mOsm/kgH_2O	291.41 ± 6.63	294.91 ± 5.94	290.95 ± 6.58	<0.01
Fasting glucose, mg/dL	108.40 ± 35.69	170.15 ± 56.35	100.26 ± 21.47	<0.01
Hemoglobin A1c, %	5.88 ± 0.93	7.80 ± 1.51	5.63 ± 0.35	<0.01

Values are expressed as mean ± standard deviation or number (percentage). @ The information on comorbid conditions was obtained by questionnaires. Abbreviations: BUN, blood urea nitrogen; eGFR, estimated glomerular filtration rate; DM, diabetes mellitus.

3.3. Relationship between Salivary Conductivity and Diabetes Mellitus

This study then evaluated the association between salivary conductivity and diabetes by categorizing the participants into four groups according to their salivary conductivity levels (Figure 3). As the salivary conductivity levels increased, the incidence of diabetes also increased, with diabetes prevalence being 4.5%, 6.9%, 8.3%, and 20.1% for the four groups, respectively (with a p-value for the trend < 0.01).

3.4. The Use of Salivary Conductivity to Diagnose Participants with Diabetes Mellitus

We further assessed the diagnostic ability of salivary conductivity in detecting diabetes by performing a ROC curve analysis. Using only salivary conductivity as the predictor, the AUROC was 0.654 (95% CI: 0.563–0.744). To enhance the predictive performance of the model, we incorporated age and BMI as additional predictors since they can be measured non-invasively. The AUROC increased to 0.698 (95% CI: 0.610–0.787) when salivary conductivity and age were combined as predictors. When BMI was also included, the AUROC further increased to 0.749 (95% CI: 0.664–0.833) (Figure 4).

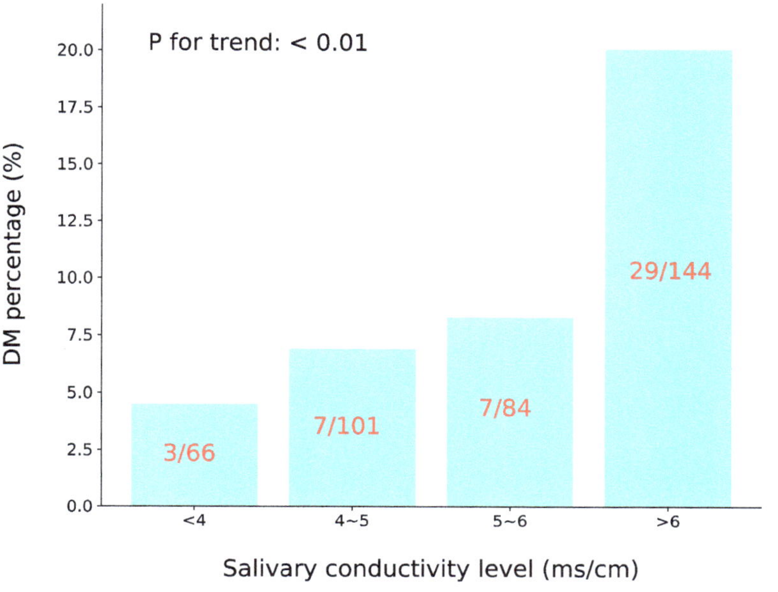

Figure 3. A bar chart showing the percentage of participants with DM at different levels of salivary conductivity. The number of subjects with DM and each group is also displayed in the graph (*p* for trend < 0.01). DM, diabetes mellitus.

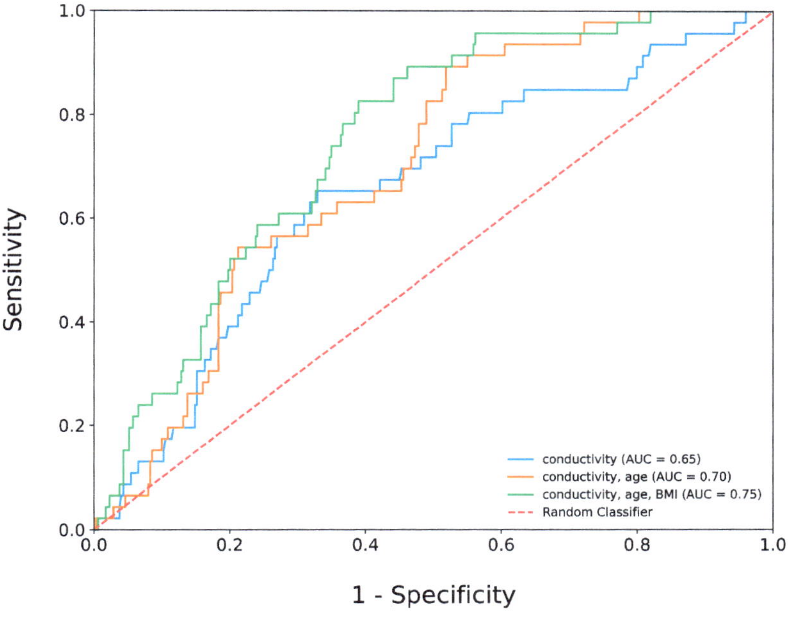

Figure 4. Three receiver operating characteristic curves. The AUROC is equal to 0.65 when using salivary conductivity as the only predicting factor. When combing salivary conductivity, age, and BMI as the predicting factors, the AUROC can be increased to 0.75. AUROC, the area under the receiver operating characteristic curve.

3.5. Comparison of Low versus High Salivary Conductivity Study Group

The study participants were further divided into two groups based on their salivary conductivity levels, using a cutoff value (5.987 ms/cm) determined by the Youden index of the ROC curve. The low group had a mean salivary conductivity level of 4.57 ± 0.81 ms/cm, while the high group had a mean of 7.33 ± 1.06 ms/cm. Table 2 shows a comparison of different variables between the two groups. The results showed that participants in the low salivary conductivity group were younger and had lower body weight, BMI, systolic and diastolic blood pressure, and a lower likelihood of having diabetes, hypertension, and dyslipidemia, in comparison to those in the high salivary conductivity group. Additionally, they had lower levels of BUN, creatinine, fasting glucose, and HbA1c and higher levels of eGFR than those in the high group.

Table 2. Population characteristics of low and high salivary conductivity groups.

	Low Salivary Conductivity Group * (N = 251)	High Salivary Conductivity Group (N = 144)	*p*-Value
Salivary conductivity, ms/cm	4.57 ± 0.81	7.33 ± 1.06	<0.01 #
Demographics			
Age, years	50.57 ± 11.35	53.90 ± 10.95	<0.01 #
Gender (male), *n* (%)	79 (31.5)	47 (32.6)	0.81
Body weight, kg	62.47 ± 11.32	67.85 ± 13.79	<0.01 #
Body height, cm	160.37 ± 7.44	161.15 ± 8.49	0.61
Body mass index, kg/m^2	24.23 ± 3.67	26.07 ± 4.65	<0.01 #
Systolic blood pressure, mmHg	125.81 ± 20.06	132.80 ± 22.32	<0.01 #
Diastolic blood pressure, mmHg	77.76 ± 12.46	80.40 ± 12.73	0.05 #
Comorbid conditions, *n* (%)			
Known history of DM	17 (6.8)	29 (20.1)	<0.01 #
Hypertension	51 (20.3)	44 (30.6)	0.03 #
Dyslipidemia	23 (9.2)	24 (16.7)	0.03 #
Gout	7 (2.8)	4 (2.8)	1.00
Laboratory parameters			
BUN, mg/dL	13.71 ± 4.20	14.44 ± 3.93	0.02 #
Creatinine, mg/dL	0.73 ± 0.16	0.76 ± 0.16	0.05 #
eGFR, $mL/min/1.73\ m^2$	103.20 ± 20.48	98.65 ± 22.14	0.01 #
Serum osmolality, $mOsm/kgH_2O$	291.09 ± 6.83	291.97 ± 6.23	0.12
Fasting glucose, mg/dL	105.02 ± 33.83	114.30 ± 38.12	<0.01 #
Hemoglobin A1c, %	5.77 ± 0.88	6.08 ± 1.00	<0.01 #

Values are expressed as mean ± standard deviation or number (percentage). * Study populations were stratified into low and high groups according to the cutoff value of salivary conductivity (5.987 ms/cm). # indicates *p* value < 0.05. Abbreviations: BUN, blood urea nitrogen; eGFR, estimated glomerular filtration rate; DM, diabetes mellitus.

3.6. Associations between Salivary Conductivity and the Risk of Diabetes

The odds of having diabetes were determined for participants with high and low salivary conductivity. The crude odds ratio was 3.82 (95% CI: 1.44–5.56), indicating that subjects with higher salivary conductivity were 3.82 times more likely to have diabetes compared to those with lower conductivity. The adjusted odds ratio, calculated using two different multivariate models, was 3.35 and 2.69, respectively (Table 3). Table 4 displays the crude and adjusted odds ratios for other clinical parameters.

Table 3. The crude and adjusted odds ratios for the association between salivary conductivity and the risk of diabetes.

Model	Odds Ratio	(95% CI)	*p*-Value
Crude	3.82	1.44–5.56	<0.01
Model 1 *	3.35	1.74–6.46	<0.01
Model 2 #	2.69	1.36–5.32	<0.01

Abbreviations: CI, confidence interval. * Model 1 was adjusted for age and gender. # Model 2 was adjusted for age, gender, body mass index, systolic blood pressure, diastolic blood pressure, and estimated glomerular filtration rate.

Table 4. Crude and adjusted odds ratios for the association between the clinical parameters and the risk of diabetes.

Parameters	Unadjusted		Adjusted			
	Crude		Model 1 *		Model 2 #	
	Odds Ratio	*p*-Value	Odds Ratio	*p*-Value	Odds Ratio	*p*-Value
Age	1.06	<0.01	1.05	<0.01	1.05	0.02
Gender	1.04	0.91	1.02	0.95	0.98	0.97
BMI	1.15	<0.01			1.11	<0.01
SBP	1.03	<0.01			1.02	0.05
DBP	1.02	0.05			0.99	0.46
eGFR	1.00	0.88			1.01	0.22
Salivary conductivity	3.82	<0.01	3.35	<0.01	2.65	<0.01

Abbreviations: BMI, body mass index; DBP, diastolic blood pressure; eGFR, estimated glomerular filtration rate; SBP, systolic blood pressure. * Model 1 was adjusted for age, gender, and salivary conductivity. # Model 2 was adjusted for age, gender, body mass index, systolic blood pressure, diastolic blood pressure, estimated glomerular filtration rate, and salivary conductivity.

4. Discussion

The presented results indicate a positive correlation between salivary conductivity and HbA1c and fasting glucose, implying that salivary conductivity may serve as a potential biomarker for detecting diabetes and monitoring blood sugar levels. Blood tests for glucose or HbA1c are typically regarded as the gold standard for diabetes diagnosis. However, such tests are invasive and may involve procedures such as venipuncture and fingertip pricking, which patients often find burdensome and are, therefore, less willing to perform regularly. In our previous research, a strong positive correlation between salivary conductivity and fasting glucose or HbA1c was found through Pearson's correlation test [29]. Additionally, as salivary conductivity levels increased, so did the prevalence of diabetes. Thus, salivary conductivity may offer a promising alternative to traditional methods of monitoring blood sugar levels.

In our study, we found that the prevalence rate of diabetes, as defined by the American Diabetes Association, was 11.6%. This is consistent with the national prevalence rate of 10.10% reported in a large-scale cohort study conducted in Taiwan in 2014, suggesting that our study participants can be considered representative of the general population [30]. Our previous study also showed a positive correlation between salivary conductivity and age or creatinine, and a negative correlation with eGFR [27,28]. These findings align with those of the current study. Participants with high salivary conductivity were older, had higher fasting sugar and HbA1c levels, and were at a higher risk of developing chronic kidney disease. It is known that the properties and secretion of saliva change with age, leading to an increase in electrolytes, as well as a direct link between salivary conductivity and age [28,31–34]. Furthermore, serum glucose levels have been shown to directly reflect salivary glucose concentration, which supports the use of saliva as a predictor of blood sugar levels [35]. Salivary conductivity was also found to be positively correlated with body weight, BMI, and blood pressure. Given that diabetes patients are often overweight, this may partly explain why they have higher salivary conductivity levels [23].

We have developed a sensor that is utilized for measuring salivary conductivity, which is determined by the electrical admittance between the electrodes. This measurement predominantly reflects the concentration of electrolytes. To enhance selectivity, we have implemented specific design features in our device. Firstly, considering that the signal follows the path of least resistance, electroimpedance spectroscopy primarily occurs at the periphery of the microelectrodes. As a result, interference effects caused by larger particles like food debris and nasal secretion, which typically settle on the top surface of the co-planar electrode, are minimized, leading to improved selectivity. Secondly, to evaluate the stability of the saliva sample, we conducted additional experiments since interfering particles may accumulate over time on the surface of the salivary solution. The findings indicated that the saliva sample exhibited good stability over a prolonged duration. Thirdly, a mere 50 μL saliva sample is required for conducting the test. Consequently, we can assume that the sample temperature quickly reaches equilibrium with the surrounding ambient temperature.

The ROC curve analysis of our study indicated that salivary conductivity, in combination with age and BMI, can be used as a good predictive model with an AUROC of 0.75. Although body weight and BMI can both serve as predictive factors, we selected BMI as it provides more information about an individual's body shape. Our model's sensitivity and specificity were not as good as those of traditional blood glucose meters, but it still represents a useful tool for self-monitoring sugar levels. Pain is a significant obstacle to self-monitoring blood glucose among patients with type 2 diabetes [10], and a device that can indirectly measure blood sugar levels non-invasively, even if less accurately, may be more useful than a device that is more accurate but also invasive if patients need to check their blood sugar multiple times per day.

According to a review article written by Tang et al., they have divided non-invasive blood glucose monitoring technology into three categories: optics, microwave, and electrochemistry [21]. The optics method can be subdivided into five categories, including near-infrared spectroscopy, optical polarimetry, Raman spectroscopy, fluorescence method, and optical coherence tomography. The electrochemistry method can be subdivided into reverse iontophoresis technology and non-invasive biofluid-based glucose monitoring. The advantages of optics and microwave methods are that they are more non-invasive and provide the possibility to continuous monitoring blood sugar levels, while less accuracy and poor correlation to actual blood glucose are the disadvantages. The electrochemistry method can predict more accurate blood sugar levels but delay in measurement results, the need for calibration, and biofluid collection are the defects. A meta-analysis by Lindner et al. has proposed that the diagnostic accuracy of non-invasive glucose monitoring devices is still not sufficiently accurate for blood sugar monitoring due to low sensitivity [36]. Although our prediction model result is similar to the previous work, we still provided a different and potential idea for monitoring blood sugar non-invasively. In addition, the same method has been proposed for detecting chronic kidney disease in our previous article [29]. It means that saliva could indeed reflect the physiological change in our body.

Localized surface plasmon resonance (LSPR) technology is a promising label-free biosensing technique. It utilizes the sensitivity of the plasmon frequency to changes in the local index of refraction at the nanoparticle surface. It has been used to detect several biomolecules, such as creatinine, troponin I, and aflatoxin [37–39]. Using LSPR technology to detect blood sugar is under investigation but still lacks robust evidence. Therefore, combining our salivary conductivity meter with the LSPR technology could be a possible method to increase accuracy when monitoring blood glucose levels in the future.

There are several limitations to this study. Firstly, it is a cross-sectional study, and we did not collect longitudinal data on each participant's blood glucose level, so we cannot definitively confirm the association between salivary conductivity and blood sugar levels. Secondly, we did not distinguish between type 1 and type 2 diabetes, although the prevalence of type 1 diabetes among mostly adult participants is likely to be very low. Thirdly, diabetes was defined as HbA1c higher than 6.5%, which is a narrower definition than the gold standard and may underestimate the prevalence of diabetes. However, our study results, both from the questionnaire survey and blood test, align with the national prevalence rate in Taiwan, suggesting that our study participants are still representative. Fourth, the study participants were collected during an annual health examination, which might attract individuals who are more health conscious. This could limit the generalizability of the findings to the broader population. Fifth, the saliva was collected when the patients were mildly dehydrated in the morning. We did not compare the saliva collected from different instances in which the flow rate or composition of the saliva may alter. Finally, we did not analyze the components of saliva, such as electrolytes, in the diabetes versus the normal group, so we were unable to clearly compare the underlying reason for their difference in salivary conductivity and justify our mechanism hypothesis. Further studies should focus on the limitations mentioned above to enhance the robustness and generalizability of the findings, strengthen the relationship between salivary conductivity and diabetes, and prove that salivary conductivity can be a reliable biomarker to monitor blood glucose levels.

5. Conclusions

Regular monitoring of blood glucose and prompt diagnosis of diabetes is crucial in reducing the associated health risks. The findings of this study demonstrate the relationship between salivary conductivity and blood sugar levels, with higher salivary conductivity associated with higher fasting glucose or HbA1c levels. Our research also shows that using salivary conductivity as a biomarker can be a potential method for screening diabetes, with a prediction model having adequate diagnostic accuracy and sensitivity. In conclusion, using a non-invasive, easy-to-use, and portable device to measure salivary conductivity has the possibility to serve as an alternative method for monitoring blood glucose levels, and could be a valuable tool for diabetes screening in the future.

Supplementary Materials: The following supporting information can be downloaded at: https://www.mdpi.com/article/10.3390/bios13070702/s1, Figure S1: Distribution plots of salivary conductivity among different groups; Table S1: Normality tests of the sample data in Table 1 using Kolmogorov-Smirnov method; Table S2: Normality tests of the sample data in Table 2 using Kolmogorov-Smirnov method; Table S3: Baseline characteristics of male participants stratified by Diabetes (n = 126); Table S4: Baseline characteristics of female participants stratified by Diabetes (n = 269) Table S5 Comparison of demographics and laboratory parameters between diabetes and non-diabetes after propensity-score matching (n = 230).

Author Contributions: Conceptualization, C.-W.L., Y.-S.P., J.-T.Y. and C.-W.T.; Data curation, Y.-P.L.; Formal analysis, Y.-S.P. and C.-W.T.; Funding acquisition, Y.-H.T., J.-T.Y. and M.-Y.C.; Methodology, C.-W.L., Y.-P.L. and C.-W.T.; Project administration, Y.-H.T. and M.-Y.C.; Resources, Y.-H.T., J.-T.Y. and M.-Y.C.; Software, C.-W.L. and Y.-P.L.; Supervision, Y.-H.T., Y.-S.P., M.-Y.C. and C.-W.T.; Validation, Y.-S.P.; Writing—original draft, C.-W.L. and Y.-H.T.; Writing—review and editing, J.-T.Y. and C.-W.T. All authors will be informed about each step of manuscript processing including submission, revision, revision reminder, etc. via emails from our system or assigned Assistant Editor. All authors have read and agreed to the published version of the manuscript.

Funding: This research was supported by grants from the Chang Gung Medical Research Project to Yuan-Hsiung Tsai and Jen-Tsung Yang (CORPG6K0153, CORPG6K0172, CORPG6K0193), and from the Taiwan Formosa Plastic Group (FCRPF6M0011).

Institutional Review Board Statement: This study was conducted following the ethical principles of the Declaration of Helsinki, and it was approved by the Ethics Committee of Chang Gung Memorial Hospital (institutional review board number: 202002186B0).

Informed Consent Statement: Written informed consent was obtained from all subjects involved in this study.

Data Availability Statement: The data presented in this study are available on request from the corresponding author. The data are not publicly available due to privacy.

Conflicts of Interest: The authors declare no conflict of interest. The funders had no role in the design of this study; in the collection, analyses, or interpretation of data; in the writing of the manuscript, or in the decision to publish the results.

References

1. Aekplakorn, W.; Chariyalertsak, S.; Kessomboon, P.; Assanangkornchai, S.; Taneepanichskul, S.; Putwatana, P. Prevalence of diabetes and relationship with socioeconomic status in the Thai population: National Health Examination Survey, 2004–2014. *J. Diabetes Res.* **2018**, *2018*, 1654530. [CrossRef]
2. Khan, A.; Uddin, S.; Srinivasan, U. Comorbidity network for chronic disease: A novel approach to understand type 2 diabetes progression. *Int. J. Med. Inform.* **2018**, *115*, 1–9. [CrossRef] [PubMed]
3. Orr, C.J.; Keyserling, T.C.; Ammerman, A.S.; Berkowitz, S.A. Diet quality trends among adults with diabetes by socioeconomic status in the US: 1999–2014. *BMC Endocr. Disord.* **2019**, *19*, 54. [CrossRef] [PubMed]
4. Volaco, A.; Cavalcanti, A.M.; Précoma, D.B. Socioeconomic status: The missing link between obesity and diabetes mellitus? *Curr. Diabetes Rev.* **2018**, *14*, 321–326. [CrossRef]
5. Sun, H.; Saeedi, P.; Karuranga, S.; Pinkepank, M.; Ogurtsova, K.; Duncan, B.B.; Stein, C.; Basit, A.; Chan, J.C.; Mbanya, J.C. IDF Diabetes Atlas: Global, regional and country-level diabetes prevalence estimates for 2021 and projections for 2045. *Diabetes Res. Clin. Pract.* **2022**, *183*, 109119. [CrossRef]
6. Jung, C.-H.; Son, J.W.; Kang, S.; Kim, W.J.; Kim, H.-S.; Kim, H.S.; Seo, M.; Shin, H.-J.; Lee, S.-S.; Jeong, S.J. Diabetes fact sheets in Korea, 2020: An appraisal of current status. *Diabetes Metab. J.* **2021**, *45*, 1–10. [CrossRef]
7. Ida, S.; Kaneko, R.; Imataka, K.; Murata, K. Relationship between frailty and mortality, hospitalization, and cardiovascular diseases in diabetes: A systematic review and meta-analysis. *Cardiovasc. Diabetol.* **2019**, *18*, 81. [CrossRef]
8. Association, A.D. Introduction: Standards of medical care in diabetes—2022. *Diabetes Care* **2022**, *45*, S1–S2. [CrossRef]
9. Celec, P.; Tóthová, Ľ.; Šebeková, K.; Podracká, Ľ.; Boor, P. Salivary markers of kidney function—Potentials and limitations. *Clin. Chim. Acta* **2016**, *453*, 28–37. [CrossRef]
10. Polonsky, W.H.; Fisher, L.; Hessler, D.; Edelman, S.V. What is so tough about self-monitoring of blood glucose? Perceived obstacles among patients with Type 2 diabetes. *Diabet. Med.* **2014**, *31*, 40–46. [CrossRef]
11. Chojnowska, S.; Baran, T.; Wilińska, I.; Sienicka, P.; Cabaj-Wiater, I.; Knaś, M. Human saliva as a diagnostic material. *Adv. Med. Sci.* **2018**, *63*, 185–191. [CrossRef] [PubMed]
12. Czumbel, L.M.; Kiss, S.; Farkas, N.; Mandel, I.; Hegyi, A.; Nagy, Á.; Lohinai, Z.; Szakács, Z.; Hegyi, P.; Steward, M.C. Saliva as a candidate for COVID-19 diagnostic testing: A meta-analysis. *Front. Med.* **2020**, *7*, 465. [CrossRef]
13. Gleerup, H.S.; Hasselbalch, S.G.; Simonsen, A.H. Biomarkers for Alzheimer's disease in saliva: A systematic review. *Dis. Mrk.* **2019**, *2019*, 4761054. [CrossRef] [PubMed]
14. Hegde, M.N.; Attavar, S.H.; Shetty, N.; Hegde, N.D.; Hegde, N.N. Saliva as a biomarker for dental caries: A systematic review. *J. Conserv. Dent.* **2019**, *22*, 2. [PubMed]
15. Wang, J.; Zhao, Y.; Ren, J.; Xu, Y. Pepsin in saliva as a diagnostic biomarker in laryngopharyngeal reflux: A meta-analysis. *Eur. Arch. Oto-Rhino-Laryngol.* **2018**, *275*, 671–678. [CrossRef]
16. Kułak-Bejda, A.; Waszkiewicz, N.; Bejda, G.; Zalewska, A.; Maciejczyk, M. Diagnostic value of salivary markers in neuropsychiatric disorders. *Dis. Mrk.* **2019**, *2019*, 4360612. [CrossRef]
17. Maciejczyk, M.; Bielas, M.; Zalewska, A.; Gerreth, K. Salivary biomarkers of oxidative stress and inflammation in stroke patients: From basic research to clinical practice. *Oxid. Med. Cell. Longev.* **2021**, *2021*, 5545330. [CrossRef]
18. Matczuk, J.; Zalewska, A.; Łukaszuk, B.; Knaś, M.; Maciejczyk, M.; Garbowska, M.; Ziembicka, D.M.; Waszkiel, D.; Chabowski, A.; Żendzian-Piotrowska, M.; et al. Insulin Resistance and Obesity Affect Lipid Profile in the Salivary Glands. *J. Diabetes Res.* **2016**, *2016*, 8163474. [CrossRef]
19. Mishra, N.; Trivedi, A.; Gajdhar, S.K.; Bhagwat, H.; Khutwad, G.K.; Mall, P.E.; Kulkarni, D. Correlation of blood glucose levels, salivary glucose levels and oral colony forming units of Candida albicans in type 2 diabetes mellitus patients. *J. Contemp. Dent. Pr.* **2019**, *20*, 494–498.
20. Mohammadnejad, P.; Asl, S.S.; Aminzadeh, S.; Haghbeen, K. A new sensitive spectrophotometric method for determination of saliva and blood glucose. *Spectrochim. Acta Part A Mol. Biomol. Spectrosc.* **2020**, *229*, 117897. [CrossRef]
21. Tang, L.; Chang, S.J.; Chen, C.-J.; Liu, J.-T. Non-invasive blood glucose monitoring technology: A review. *Sensors* **2020**, *20*, 6925. [CrossRef] [PubMed]
22. Karthigadevi, G.; Duraisamy, R.; Vishnu Priya, V.; Santhosh Kumar, M. Analysis of salivary electrolytes in diabetic and non-diabetic patients. *Drug Invent. Today* **2018**, *10*, 1801–1805.
23. Kolahdouz-Mohammadi, R.; Soltani, S.; Clayton, Z.S.; Salehi-Abargouei, A. Sodium status is associated with type 2 diabetes mellitus: A systematic review and meta-analysis of observational studies. *Eur. J. Nutr.* **2021**, *60*, 3543–3565. [CrossRef] [PubMed]
24. Mrag, M.; Kassab, A.; Omezzine, A.; Belkacem, C.R.; Ben, F.I.F.; Douki, N.; Laouani, K.C.; Bouslema, A.; Ben, A.F. Saliva diagnostic utility in patients with type 2 diabetes: Future standard method. *J. Med. Biochem.* **2020**, *39*, 140. [CrossRef]
25. Onyemaechi, E.V.; Peggy, O.E.; Elias, D.T.M.; Ogheneyoma, O.O.; Chukwuebuka, N.B.; Ojugbeli, E.T.; Osuagwu, U.L.; Kingsley, N.E. Comparative Analysis of Fasting Blood Glucose and Salivary Electrolytes Concentrations in Type II Diabetic Individuals: A Randomized Controlled Hospital Based Study. *Toxicol. Rep.* **2022**, *9*, 1268–1272.

26. Lu, Y.-P.; Huang, J.-W.; Lee, I.; Weng, R.-C.; Lin, M.-Y.; Yang, J.-T.; Lin, C.-T. A portable system to monitor saliva conductivity for dehydration diagnosis and kidney healthcare. *Sci. Rep.* **2019**, *9*, 14771. [CrossRef]
27. Chen, C.-H.; Lu, Y.-P.; Lee, A.-T.; Tung, C.-W.; Tsai, Y.-H.; Tsay, H.-P.; Lin, C.-T.; Yang, J.-T. A Portable Biodevice to Monitor Salivary Conductivity for the Rapid Assessment of Fluid Status. *J. Pers. Med.* **2021**, *11*, 577. [CrossRef]
28. Lee, A.-T.; Lu, Y.-P.; Chen, C.-H.; Chang, C.-H.; Tsai, Y.-H.; Tung, C.-W.; Yang, J.-T. The Association of Salivary Conductivity with Cardiomegaly in Hemodialysis Patients. *Appl. Sci.* **2021**, *11*, 7405. [CrossRef]
29. Lin, C.-W.; Tsai, Y.-H.; Lu, Y.-P.; Yang, J.-T.; Chen, M.-Y.; Huang, T.-J.; Weng, R.-C.; Tung, C.-W. Application of a Novel Biosensor for Salivary Conductivity in Detecting Chronic Kidney Disease. *Biosensors* **2022**, *12*, 178. [CrossRef]
30. Sheen, Y.-J.; Hsu, C.-C.; Jiang, Y.-D.; Huang, C.-N.; Liu, J.-S.; Sheu, W.H.-H. Trends in prevalence and incidence of diabetes mellitus from 2005 to 2014 in Taiwan. *J. Formos. Med. Assoc.* **2019**, *118*, S66–S73. [CrossRef]
31. Ashton, N.J.; Ide, M.; Zetterberg, H.; Blennow, K. Salivary biomarkers for Alzheimer's disease and related disorders. *Neurol. Ther.* **2019**, *8*, 83–94. [CrossRef] [PubMed]
32. Labat, C.; Thul, S.; Pirault, J.; Temmar, M.; Thornton, S.N.; Benetos, A.; Bäck, M. Differential associations for salivary sodium, potassium, calcium, and phosphate levels with carotid intima media thickness, heart rate, and arterial stiffness. *Dis. Mrk.* **2018**, *2018*, 3152146. [CrossRef]
33. Tomás, I.; Marinho, J.; Limeres, J.; Santos, M.; Araújo, L.; Diz, P. Changes in salivary composition in patients with renal failure. *Arch. Oral Biol.* **2008**, *53*, 528–532. [CrossRef]
34. Xu, F.; Laguna, L.; Sarkar, A. Aging-related changes in quantity and quality of saliva: Where do we stand in our understanding? *J. Texture Stud.* **2019**, *50*, 27–35. [CrossRef] [PubMed]
35. Jung, D.G.; Jung, D.; Kong, S.H. A lab-on-a-chip-based non-invasive optical sensor for measuring glucose in saliva. *Sensors* **2017**, *17*, 2607. [CrossRef] [PubMed]
36. Lindner, N.; Kuwabara, A.; Holt, T. Non-invasive and minimally invasive glucose monitoring devices: A systematic review and meta-analysis on diagnostic accuracy of hypoglycaemia detection. *Syst. Rev.* **2021**, *10*, 145. [CrossRef]
37. Lerdsri, J.; Chananchana, W.; Upan, J.; Sridara, T.; Jakmunee, J. Label-free colorimetric aptasensor for rapid detection of aflatoxin B1 by utilizing cationic perylene probe and localized surface plasmon resonance of gold nanoparticles. *Sens. Actuators B Chem.* **2020**, *320*, 128356. [CrossRef]
38. Menon, P.S.; Said, F.A.; Mei, G.S.; Berhanuddin, D.D.; Umar, A.A.; Shaari, S.; Majlis, B.Y. Urea and creatinine detection on nano-laminated gold thin film using Kretschmann-based surface plasmon resonance biosensor. *PLoS ONE* **2018**, *13*, e0201228. [CrossRef]
39. Sinha, R.K. Wavelength modulation based surface plasmon resonance sensor for detection of cardiac marker proteins troponin I and troponin T. *Sens. Actuators A Phys.* **2021**, *332*, 113104. [CrossRef]

biosensors

MDPI

Article

Evaluation of the Chewing Pattern through an Electromyographic Device

Alessia Riente [1,2], Alessio Abeltino [1,2], Cassandra Serantoni [1,2], Giada Bianchetti [1,2], Marco De Spirito [1,2], Stefano Capezzone [3], Rosita Esposito [4] and Giuseppe Maulucci [1,2,*]

1 Metabolic Intelligence Lab, Department of Neuroscience, Università Cattolica del Sacro Cuore, Largo Francesco Vito, 1, 00168 Rome, Italy; alessia.riente@unicatt.it (A.R.); alessio.abeltino@unicatt.it (A.A.); cassandra.serantoni@unicatt.it (C.S.); giada.bianchetti@unicatt.it (G.B.); marco.despirito@unicatt.it (M.D.S.)
2 Fondazione Policlinico Universitario "A. Gemelli" IRCCS, 00168 Rome, Italy
3 Gruppo Fastal Blu Sistemi, Via Nomentana 263, 00161 Rome, Italy; stefano@capezzone.it
4 Digital Innovation Hub Roma, Chirale S.r.l., Via Ignazio Persico 32-46, 00154 Rome, Italy; rosita.esposito@chirale.it
* Correspondence: giuseppe.maulucci@unicatt.it

Abstract: Chewing is essential in regulating metabolism and initiating digestion. Various methods have been used to examine chewing, including analyzing chewing sounds and using piezoelectric sensors to detect muscle contractions. However, these methods struggle to distinguish chewing from other movements. Electromyography (EMG) has proven to be an accurate solution, although it requires sensors attached to the skin. Existing EMG devices focus on detecting the act of chewing or classifying foods and do not provide self-awareness of chewing habits. We developed a non-invasive device that evaluates a personalized chewing style by analyzing various aspects, like chewing time, cycle time, work rate, number of chews and work. It was tested in a case study comparing the chewing pattern of smokers and non-smokers, as smoking can alter chewing habits. Previous studies have shown that smokers exhibit reduced chewing speed, but other aspects of chewing were overlooked. The goal of this study is to present the device and provide additional insights into the effects of smoking on chewing patterns by considering multiple chewing features. Statistical analysis revealed significant differences, as non-smokers had more chews and higher work values, indicating more efficient chewing. The device provides valuable insights into personalized chewing profiles and could modify unhealthy chewing habits.

Keywords: mastication; chewing profile; EMG device; smoking; chewing features; statistical analysis

Citation: Riente, A.; Abeltino, A.; Serantoni, C.; Bianchetti, G.; De Spirito, M.; Capezzone, S.; Esposito, R.; Maulucci, G. Evaluation of the Chewing Pattern through an Electromyographic Device. *Biosensors* **2023**, *13*, 749. https://doi.org/10.3390/bios13070749

Received: 7 June 2023
Revised: 11 July 2023
Accepted: 18 July 2023
Published: 20 July 2023

1. Introduction

Chewing is a fundamental regulator of the entire metabolic process [1] (cap 22, pp. 716–743). When food is introduced into the oral cavity, the mechanical activity induced by the masticatory muscles and the concomitant salivary secretion (salivary amylase and small quantities of lipase) dissolve the chemical substances present in the food, starting the chemical digestion of some macronutrients. This phase of digestion is important as it triggers an anticipatory response of subsequent sections of the digestive system known as the cephalic phase [1] (cap 22, pp. 716–743), conditioning and regulating the whole digestive process and nutrient absorption.

In the literature, there are several works whose purpose was the analysis of chewing. Various aspects of this activity can be evaluated [2], for example, the force exerted by the masticatory muscles [2–4]. Frequently, dental implants with strain gauges are used [4]. However, these devices could alter oral sensation and cannot be used for long-term monitoring. In some solutions, the sounds emitted during the act of chewing are analyzed using acoustic sensors and microphones [5–7]. The collected data of these devices suffer from noisy components due to the surrounding environment. Another possible approach consists of measuring the contraction of the masseter and temporal muscles (the main muscles of chewing). For this purpose, piezoelectric sensors [8–10] are used. Piezoelectric sensors produce a voltage value when subjected to physical stress [8]. They are, therefore, suitable for detecting muscular activities, such as chewing and even swallowing. A wearable piezoelectric strain sensor in the form of glasses positioned on the temporalis muscle was engineered in [9] to detect chewing features. Similarly, in [10], a method was presented for the automatic quantification of chewing episodes captured by a piezoelectric sensor system. Also, in [8] a device was realized using a necklace with a piezoelectric sensor positioned at the throat level to recognize chewing activity. Although these examples use piezoelectric sensors for chewing analysis, piezoelectric sensors are not able to reliably distinguish movements of masticatory activity from head and neck movements unrelated to this act [11]. Some devices exploit photoplethysmography (PPG) technology [12,13] to detect changes in reflected light levels due to altered venous blood characteristics [14]. The lower jaw moves to open and close the mouth, causing the ear canal to expand as the mandibular condyle slides back and forth. A PPG signal is not completely free of noise; indeed, sudden changes in ambient lighting can produce significant artefacts and cause signal saturation [14].

Among the options mentioned, surface electromyography (EMG) [15–18] is the most useful and accurate solution for practical purposes. However, standard EMG practice has not been established, and some problems remain [14]. One of the disadvantages of using EMG is the requirement for the direct attachment of sensors to the skin [19]. In [15], the authors exploited EMG technology to analyze the variations in jaw movements when the subject varied the masticatory region, while the authors of [16] presented a 3D glass with bilateral EMG electrodes placed on the temporal muscles. The purpose of this device is to automatically monitor the subject's diet by recognizing the chewing act and classifying the data relating to the ingested food, obtaining good performance. In [20], an EMG system capable of identifying masticatory events from other activities and their duration was presented. The invention presents an interface with various sections in which different characteristics of the act are analyzed. For example, there is a section capable of determining the maximum value reached by the recorded signal.

The devices described above, which exploit EMG technology, analyze mastication to ascertain the presence of the act or to classify foods based on the masticatory signal. However, no electromyographic device describes chewing performance in its entirety with the aim of defining a precise chewing profile of the user, making him self-aware of his chewing habits to eventually modify them. To obtain more insights into the chewing process, we implemented 'Chewing', a device representing a solution to evaluate the chewing style, as it can characterize the masticatory activity in a non-invasive way. 'Chewing' analyzes mastication considering different aspects of the process: the chewing time, cycle time, work rate, number of chews and masticatory work. In this way, it is possible to analyze the masticatory process itself, considering different aspects of it, and define personalized chewing profiles. We tested our device by performing a case study in which the chewing profile of smokers and non-smokers was analyzed and identified. Smoking can indeed alter a person's chewing pattern. The constant exposure to tobacco chemicals leads to damage to oral tissue, lesions of the lips and mucous membranes, loss of teeth and a weakening of the perception of taste and smell [21–23], leading to a change in the chewing pattern. In [22], the presence of oral alterations in smokers was verified, and the impact of the alterations on masticatory function was evaluated compared to subjects who have never smoked. However, the variables considered in the study were halitosis, malocclusion, chewing speed and others. It was noted that the chewing speed was significantly reduced in smokers. Similarly, in [24], the authors identified differences in the chewing patterns between smokers and non-smokers considering aspects such as the chewing speed, atypical muscle contractions, and orbicularis and mental contraction muscles during swallowing. Also, in this case, considering the masticatory speed, it has been verified that smokers have a slower pattern. However, in these previous works, only the masticatory speed was evaluated, and other chewing features were neglected. Our aim is, therefore, to add additional insights to this important case study.

2. Materials and Methods

2.1. Development of a Device to Assess Chewing Behavior

The masticatory behavior is evaluated using the Chewing device, which uses electromyographic technology. Electromyography (EMG) is a technique that allows the measurement of the electrical activity produced by muscles during contraction [25]. Since the masseter and temporal muscles are the most important muscles in mastication, the first one was chosen for monitoring as it is more easily accessible. The signal is picked up using surface electrodes, which are commonly used with this technique. These electrodes are made of conductive materials, such as silver or silver chloride, which are characterized by high electrical conductivity and can adhere to the skin securely.

This device consists of an Arduino nano BLE 33 microprocessor (Microcontroller: nRF52840, Operating Voltage: 3.3 V, DC Current per I/O Pin: 15 mA, Length: 45 mm, Width: 18 mm, Weight: 5 g, Digital Input/Output Pins: 14), shown in Figure 1a(5), connected to a PC via a cable in Figure 1a(6), two Arduino muscle v3 modules (Voltage range: $\pm$3.5–$\pm$18 V, Gain settings: 0.01–100 kΩ, Output signal voltage: 0–+Vs) in Figure 1a(1), a 9 volt battery in Figure 1a(3) and a resistive divider in Figure 1a(4). The signal is obtained using the six electrodes (Figure 1a(2)), which are positioned on both masseter muscles of the subject (Figure 1b(1,2)). Specifically, one is positioned on the central part of the right muscle (red), one at the end of the right muscle (green) and one (yellow) on a bone not involved in movement (right cheekbone). The other three are positioned on the left side in the same way. It is possible to see the circuit diagram and the correct placement of the electrodes in Figure 1a,b, respectively.

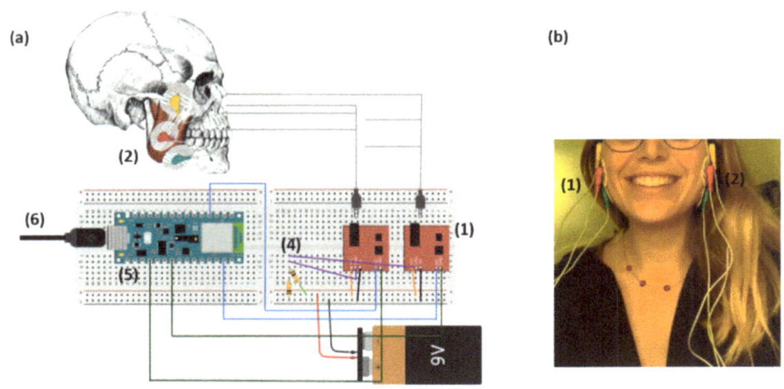

Figure 1. Chewing device. (**a**) Scheme of the device and all the parts that it includes: (5) Arduino nano 33 BLE microprocessor connected to a PC via cable (6), two Arduino muscle v3 modules (1) connected to the microprocessor through a resistive divider (4) and a 9 volt battery (3). The signal is taken through the electrodes (2) connected to the Arduino muscle v3 modules. (**b**) Placement of EMG electrodes on both the masseters of a subject (1–2): the red ones on the central part of the muscles, the green ones at the end of the masseters and the yellow ones on the cheekbones.

As demonstrated in [25], the data collected using EMG technology, from which the masticatory features are extracted, are repeatable and reproducible. Regarding the noise of the signal, the muscle v3 modules are used because they contain a circuit with diodes to rectify the signal, an active low-pass filter to eliminate noise and another active amplifier to obtain an accurate signal (Muscle_Sensor_v3_users_manual, https://www.pololu.com/file/0J745/Muscle_Sensor_v3_users_manual.pdf, accessed on 6 June 2023).

2.2. Sample Preparation and Characterization

In this study, Conad bread is used as a food sample and is pre-cut into cubes of 1 cm^3. The hardness (Young's modulus) of the bread is equal to 0.87 N/m and is calculated using the "univert mechanical tester" tool. Table 1 shows the grams of macronutrients in this food sample.

Table 1. Macronutrients of the food used in the test (bread).

Food	Salt (100 g)	Fats (100 g)	Carbohydrates (100 g)	Proteins (100 g)	Sugars (100 g)	Fiber (100 g)
Bread	0	6.5	58	10	4	2.5

While seated comfortably, the subjects are asked to eat the sample. They are asked not to speak and not to move during the recording, and, finally, they are asked to indicate with a gesture when they have completed chewing. In addition, they are asked to freely consume the food following their typical chewing style. All the subjects provide voluntary informed consent to participate in the study and are informed about the ingredients present in the food administered to avoid allergic reactions.

First, the electrodes are placed as described in Section 2.1. Then, the chewing signals in tension $v_{dx}\ (t)$ and $v_{sx}\ (t)$ are acquired, which represent the electromyographic signal of the right and left masseter, respectively. Finally, these data are processed through Python software. In the algorithm, raw signals from the right $v_{dx}\ (t)$ and left $v_{sx}\ (t)$ masseter are analyzed separately and then averaged with the masticatory features defined later.

Before analysis, the signals $v_{dx}\ (t)$ and $v_{sx}\ (t)$ are rectified, amplified and filtered by the circuit modules. To eliminate any bias, the mean value of the first five samples acquired while the subject is not chewing is respectively subtracted from each signal.

Figure 2 shows an example of signal recordings $v_{dx}(t)$ and $v_{sx}(t)$ of a subject under test while eating the bread sample (respectively (a) and (b)). Each peak represents the masseter contraction.

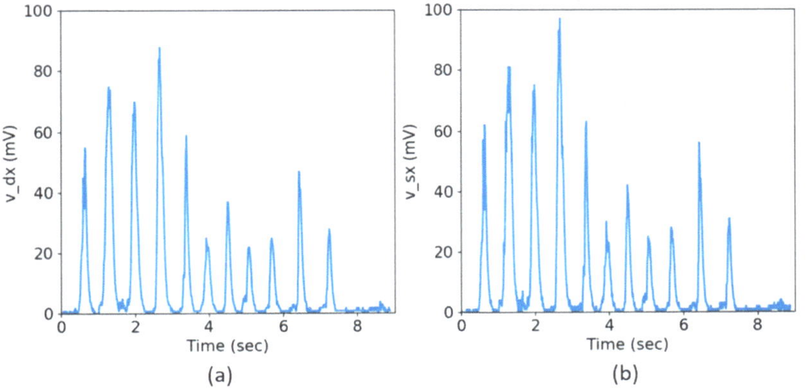

Figure 2. Chewing registration of the signals $v_{dx}(t)$ and $v_{sx}(t)$. The signals shown in the figures have been rectified, amplified and filtered by the circuit modules, and any bias has been eliminated via software. (**a**) Right masseter activity $v_{dx}(t)$. (**b**) Left masseter activity $v_{sx}(t)$.

The sampling time of the signal acquired is $\Delta t = 10$ milliseconds. For $v_{dx}\ (t)$, we define the time interval of a $\Delta t_{chew\_dx} = t_e - t_s$ where t_e is the ending time and t_s is the starting time. We define the threshold σ^*_{dx}, which is the standard deviation of the first five samples of $v_{dx}\ (t)$ before the subject starts to chew. So, the time t_s is determined as the value in which two consecutive $v_{dx}(t_s)$ and $v_{dx}(t_s + \Delta t)$ exceed σ^*_{dx}. The time t_e is determined as the value by which $v_{dx}(t_e) > \sigma^*_{dx}$ and $v_{dx}(t_e + \Delta t) < \sigma^*_{dx}$. This condition is verified by performing a cycle for every t of the signal $v_{dx}\ (t)$. The same is performed for $v_{sx}\ (t)$.

From each time series, the chewing behavior is described in terms of five features: 'Chewing time', 'Number of chews', 'Cycle time', 'Work' and 'Work rate' [25]:

1. Number of chews (n_{chew} adimensional): the number of chews made by the subject (). The number of detected chewing cycles Δt_{chew_dx} is called n_{chew_dx}. The whole process is repeated to calculate n_{chew_sx}. The average of n_{chew_dx} and n_{chew_sx}, gives an estimate of the number of chews n_{chew}:

$$n_{chew} = \frac{(n_{chew\_dx} + n_{chew\_sx})}{2} \tag{1}$$

2. Cycle Time (t_{cyc}, second): the time spent on a single bite in seconds. t_{cyc_dx} is calculated as the ratio of the sum of all the time intervals of the chews Δt_{chew_dx} and the number of chews n_{chew_dx}. The whole process is repeated to calculate t_{cyc_sx}. The average of t_{cyc_dx} and t_{cyc_sx}, gives an estimate of the cycle time t_{cyc}. This parameter is a good estimate of the chewing rate (in seconds). In the following, the full formula used to calculate t_{cyc} is reported.

$$t_{cyc} = \frac{\left(t_{cyc\_dx} + t_{cyc\_sx}\right)}{2} = \frac{\left(\frac{\sum \Delta t_{chew\_dx}}{n_{chew\_dx}}\right) + \left(\frac{\sum \Delta t_{chew\_sx}}{n_{chew\_sx}}\right)}{2} \tag{2}$$

3. Chewing Time (t_{chew}, second): the effective time in which the subject has chewed in seconds (), as expressed by the product between the number of chews and the cycle time, calculated according to the following equation:

$$t_{chew} = \frac{(t_{chew\_dx} + t_{chew\_sx})}{2} = \frac{\left(n_{chew\_dx} \cdot t_{cyc\_dx}\right) + \left(n_{chew\_sx} \cdot t_{cyc\_sx}\right)}{2} \tag{3}$$

4. Work (w, volts * second): the estimated area under the masticatory signal. Right work w_{dx} is the sum of the products between the mean voltage ($\overline{v_{dx}}$) and Δt_{chew_dx}. Dually, it is calculated as w_{sx}. The average between w_{dx} and w_{sx} is the work w.

$$w = \frac{(w_{dx} + w_{sx})}{2} = \frac{\left(\sum\left(\overline{v_{dx}} \cdot \Delta t_{chew\_dx}\right)\right) + \left(\sum\left(\overline{v_{sx}} \cdot \Delta t_{chew\_sx}\right)\right)}{2} \tag{4}$$

5. Work rate (wr, volt): indicates the power exerted by the masticatory muscles (in volts), which is expressed as the ratio between the work and chewing time. This feature is calculated as the ratio between the work and chewing time:

$$wr = \frac{(wr_{dx} + wr_{sx})}{2} = \frac{\left(\left(\frac{w_{dx}}{t_{chew\_dx}}\right)\right) + \left(\left(\frac{w_{sx}}{t_{chew\_sx}}\right)\right)}{2} \tag{5}$$

6. Asymmetry index i_{as}: is related to the number of chews of the masticatory, assessing whether it is balanced or not, and is calculated as follows:

$$i_{as} = mean(n_{chew\_dx}) - mean(n_{chew\_sx}) \tag{6}$$

A chewing flag is associated with the index, distinguishing between:

- balanced if $-1 < i_{as} < 1$,
- slightly unbalanced to the right if $0 < i_{as} < 5$ or to left if $-5 < i_{as} < 0$
- unbalanced to the right $i_{as} > 5$ or to the left if $i_{as} < -5$

2.3. Statistics

In this work, a case study to assess the chewing profiles of smokers and non-smokers was conducted to test the chewing pattern evaluation. To quantify the differences between the two groups through statistical analysis, the *t*-test and the Mann–Whitney test were used to clarify the relationship between the chewing behaviors and characteristics of the subjects involved in the work. Also, the FDR correction was performed to minimize the risk of obtaining false positive results when many hypothesis tests were performed simultaneously. The significance level was set at $p = 0.05$. The statistical power of the study was also determined using the "smp.NormalIndPower" function from the statsmodels library (stats-models.stats.power.NormalIndPower.solve_power). This function takes three inputs: the effect size, which is calculated as twice the U test statistic divided by the product of the sample size, as the features follow a normal distribution; the number of samples (22); and the alpha level (0.05). The resulting statistical power is equal to 0.8 (0.7 for the number of chews and 0.9 for work), indicating a high likelihood of detecting true effects in the study. The test included 25 subjects, but 3 subjects were excluded from the analysis because they presented more than one masticatory feature equal to zero (this indicates that the subject did not correctly chew the test food but directly swallowed it). The other subjects were divided into two categories: 0 are smokers (14) and 1 are non-smokers (8).

3. Results

3.1. Study Population

We involved 25 subjects (16 men and 8 women) with an average age of 44.04 ± 16.92 years (range 17–80 years) and an average body mass index (BMI) equal to 25.05 ± 2.83 kg/m^2. The subjects were divided into two groups: smokers (0) and non-smokers (1), as shown in Figure 3. Each category was divided into four sub-categories based on age (17–24; 25–40, 41–60, >60), and the number of female (women) and male (men) subjects in the various sub-categories is indicated.

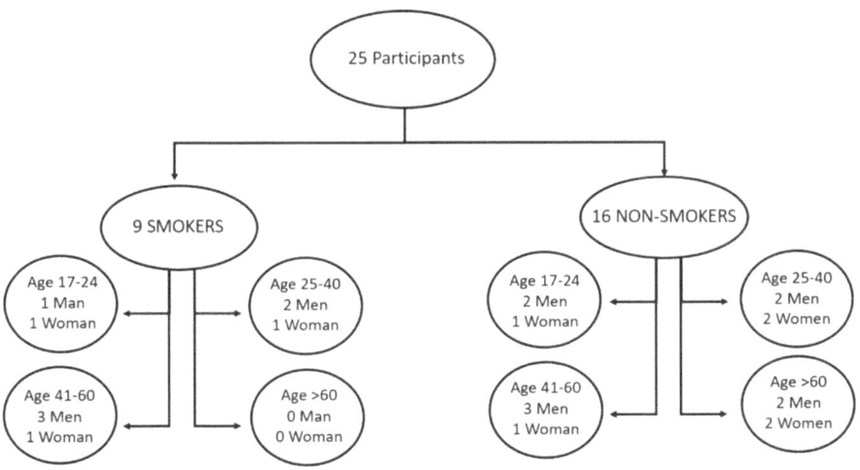

Figure 3. Characteristics of the group of people undergoing the test. The 25 subjects were divided into two macro-categories: smokers (9) and non-smokers (16). Each macro-category was divided into four sub-categories based on the age of the subjects (17–24, 25–40, 41–60, >60), of which the number of female (women) and male (men) subjects is indicated.

3.2. Chewing Profiles of Smokers and Non-Smokers

Figure 4 shows two representative chewing profiles of a smoker and a non-smoker who chewed the tested food (bread). It can be seen from these two profiles that the n_{chew}

and t_{chew} of the non-smoker are higher than those of the smoker (greater number of peaks and time of chewing).

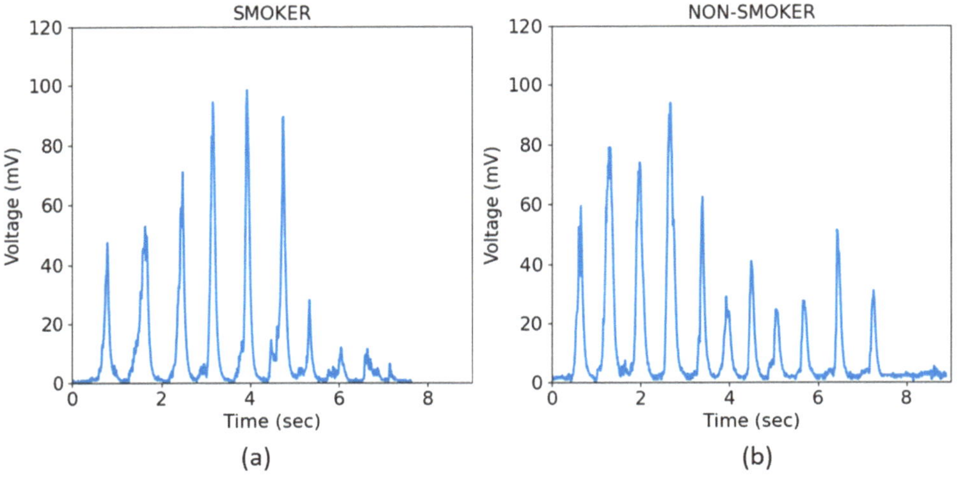

Figure 4. Examples of the chewing profiles of a smoker and a non-smoker. (**a**) Chewing profile for the sample of bread of a smoker; (**b**) Chewing profile for the sample of bread of a non-smoker. Chewing time and number of bites appear to be greater in the pattern of (**b**).

A complete statistical analysis of the chewing characteristics detected by our device has produced the results shown in Table 2.

Table 2. Results of the statistical analysis.

Features	Mean Smokers	Standard Dev. Smokers	Mean Non-Smokers	Standard Dev. Non-Smokers	Statistical Test	*p*-Value [1]
Age	43.88	14.4	46.79	20.73	−0.33	0.32
Sex	0.62	0.48	0.64	0.48	0.14	0.49
BMI	25.64	3.02	24.38	2.02	1.11	0.16
t_{chew}	3.15	1.66	4.39	1.39	−1.78	**0.07** °
n_{chew}	6.31	2.6	9.82	3.13	−2.56	**0.02** *
t_{cyc}	0.5	0.25	0.57	0.38	53	0.34
w	0.06	0.03	0.11	0.04	−3.12	**0.01** *
wr	0.02	0.01	0.03	0.01	−0.99	0.17
i_{as}	2	2.96	−0.57	3.66	37	**0.14**

[1] *t*-test and Mann–Whitney test based on Shapiro's test for data normality. Statistically significant differences are reported in bold. (*) stands for *p*-value < 0.05; (°) stands for *p*-value ≤ 0.1

The non-smokers bite more often (n_{chew} = 6.31 for smokers and 9.8 for non-smokers, $p = 0.02$), the work is higher (w = 0.06 V*sec for smokers and 0.11 V*sec for non-smokers, $p < 0.01$) and the chewing time is higher (t_{chew} = 3.15 s for smokers and 4.39 s for non-smokers, $p < 0.1$) than the smokers. However, these results indicate that what is significant between the two groups is not the chewing time but rather the number of chews and the mechanical work exerted to break up the food.

Also, the statistical differences reported in Table 2 are represented as boxplots in Figure 5, where the smokers are represented in blue and the non-smokers are represented in red.

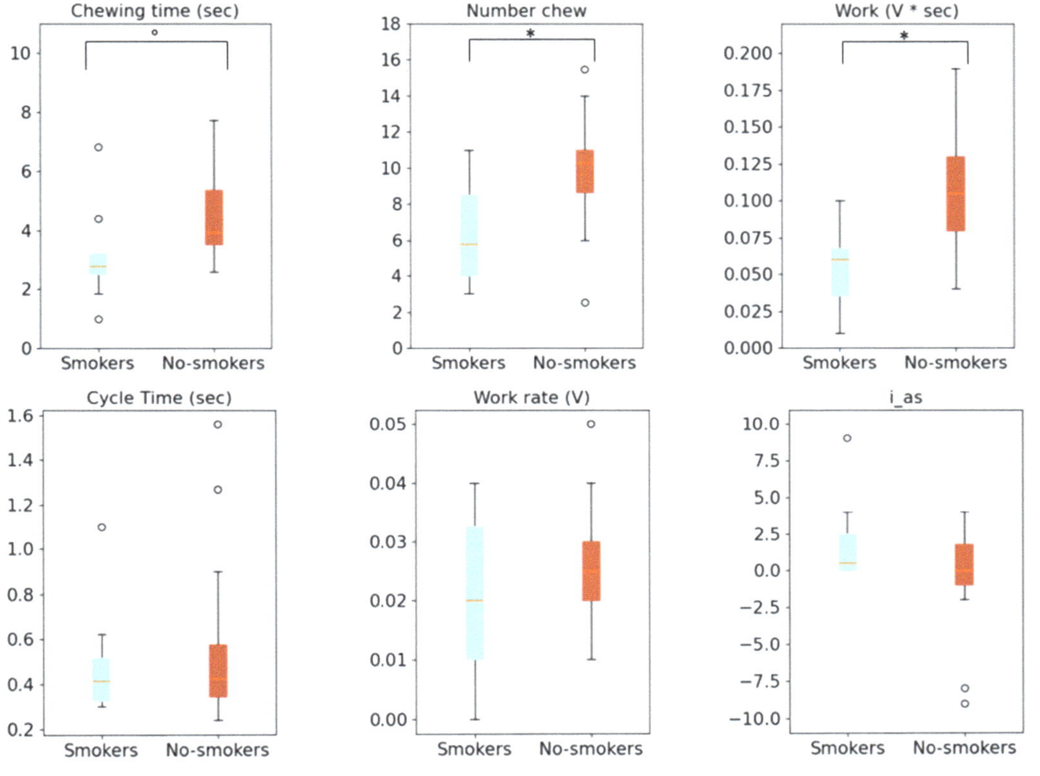

Figure 5. Boxplots of the chewing features of smokers (in blue) and non-smokers (in red): in the first row of the figure, there are the boxplots of chewing time, number of chews and work features; in the second row, there are boxplots of cycle time, work rate and asymmetry index. The asterisks represent a significance level ≤ 0.05. The dot represents a significance level ≤ 0.1.

3.3. Graphical Clustering of the Chewing Profiles of Smokers and Non-Smokers

Furthermore, the device allows one to perform a clusterization in this space of the masticatory features. In Figure 6, the smokers are shown in blue, and the non-smokers are shown in red. A visual inspection of the graphs shows that the non-smokers are placed on the right side of the space, having a higher number of chews and higher work values than the smokers. Furthermore, it is possible to observe the presence of two crosses, which represent the geometric centers of the two distributions: the smokers (blue) and non-smokers (red). Specifically, the geometric center of the smokers has a number of chews of 5.71 and work of 0.05 V*sec, while the geometric center of the non-smokers has a number of chews of 8.97 and work of 0.10 V*sec.

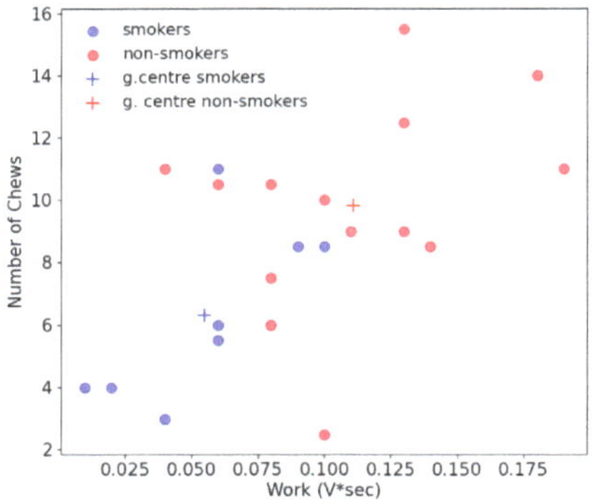

Figure 6. Representation of the data distribution in 2D space realized with n_{chew} and w. The crosses in blue and red represent the geometric center of the distribution of smokers and non-smokers, respectively.

4. Discussion

We have presented an innovative device, "Chewing", which allows one to carry out a chewing analysis in a non-invasive and objective way. Compared to the other devices used for this scope, "Chewing" exploits EMG technology [18]. Once the data has been obtained, it is able to analyze them automatically, extracting the features of interest and producing a complete masticatory profile based exclusively on the performance itself. In the previous works described [22,24], the only masticatory variable considered was the speed, and, furthermore, no instrument was used to measure it. These problems have been overcome using "Chewing", mainly for two reasons: (1) different masticatory features are defined (chewing time, number of chews, masticatory work, masticatory power, time for a single bite) and (2) it uses a gold standard technology for the evaluation. Although the use of the electrodes is uncomfortable because there are several cables connecting them to the device, this problem could be overcome by using wireless data transmission technologies (such as NFC or others).

Furthermore, our device has been tested on a use case: we considered smokers and non-smokers. The results obtained from this analysis showed how the chewing pattern of non-smokers is characterized by a higher number of chews ($p = 0.02$) and work ($p = 0.01$). These results suggest that smokers have an inefficient chewing pattern. Indeed, smokers take fewer bites than non-smokers. In relation to this aspect, in [24], it was shown that the chewing pattern of non-smokers allows for the ingestion of larger and less saliva-moistened fragments, resulting in greater effort during chewing and swallowing, which may be accompanied by compensatory movements of the head and face muscles. Similarly, here, we observed that the work performed by the smokers was less than that of the non-smokers, again indicating that they worked the bolus very little before swallowing.

In addition, smoking impacts olfactory and gustatory sensory perception. It can structurally and functionally alter the ability to perceive different stimuli in a subject. Previous studies have already demonstrated that a smoker's ability to recognize taste is lower than that of non-smokers, with the need to increase the concentration of the tested stimulus to be recognized correctly. This results in variations in an individual's acts of chewing and digestion. Olfactory and gustatory stimulation allows the preparation of the oral and gastrointestinal motor apparatus for the reception of food. This stimulation induces an increase in salivary secretion and gastric juice, favors the correct positioning of the oropharyngeal structures for swallowing and generates nervous and muscular excitability for the passage of food into the stomach [22]. It has also been explained how this difficulty occurs due to changes in the shape, number and vascularity of the taste buds that affect sensing ability and the perception of taste by inducing an increase in the sensory recognition threshold [22]. Continuous exposure to smoke can result in structural and functional changes in the neuroepithelium, leading to a decrease in the production of sensory cells and impaired odor recognition. This reduction in olfactory abilities, combined with a decreased gustatory capacity in smokers, may explain their tendency to chew for shorter durations and take fewer bites compared to non-smokers. The diminished perception of taste due to reduced sample quantities in this case study further hampers their ability to adequately sense flavors. The major limitation of this study is the number of subjects. As a future development, the authors aim to expand the participant pool to conduct additional analyses and differentiate chewing patterns among various subject categories. Increasing the number of subjects involved will allow a more comprehensive understanding of chewing behavior in different populations. Another limitation is the small quantities of food samples and the use of only one test food in the protocol. Indeed, this analysis can be considered the first exploratory analysis of the differences in terms of chewing patterns between smokers and non-smokers to be expanded using different types of foods and larger quantities.

As mentioned above, the areas of application of the tool can be different. First, it is a device useful to realize a specific chewing pattern. This result can be useful for the user himself to know and possibly improve his chewing style but also for carrying out specific studies for research purposes. Also, it can be used as a support for the diagnosis of dental problems, for example, to ascertain the presence of temporomandibular joint pain syndrome in the patient. Furthermore, in speech therapy, it can be a useful device because it could realize a further evaluation of the functionality of oro-facial muscles. In this way, the speech therapist can obtain more complete information about the user, and this allows him to set precise objectives of the therapeutic project based on the objective starting data to be monitored during treatment. The device could also be integrated into environments analyzing metabolic digital twins, such as Personalized Metabolic Avatar (PMA) [26,27] allowing one to estimate the subject's energy balance and predict the weight in a personalized way.

5. Conclusions

In conclusion, this work demonstrates that non-smokers exhibit a more extensive and efficient chewing pattern compared to smokers, as evidenced by the higher values regarding the number of chews and work. The developed device provides valuable insights into personalized chewing profiles and can potentially contribute to modifying unhealthy chewing habits. In addition, it is non-invasive, requiring only the application of electrodes on masticatory muscles. The system presented in this paper is an alternative tool to characterize the user's chewing and make him aware of his eating habits.

Author Contributions: Conceptualization, G.M. and A.R.; methodology, G.M., A.R., A.A., G.B., C.S., M.D.S., S.C. and R.E.; software, A.R. and G.M.; validation, G.M., A.R., A.A., G.B., C.S., M.D.S., S.C. and R.E.; formal analysis, G.M. and A.R.; investigation, G.M., A.R., A.A., G.B., C.S., M.D.S., S.C. and R.E.; resources, G.M. and A.R.; data curation, A.R. and G.M.; writing—original draft preparation, G.M. and A.R.; writing—review and editing, A.R. and G.M.; visualization, A.R. and G.M.; supervision, G.M.; project administration, G.M.; funding acquisition, G.M. All the authors have read and agreed to the published version of the manuscript. All authors have read and agreed to the published version of the manuscript.

Funding: This project was supported in part by a research grant awarded to G.M. from Regione Lazio PO FSE 2014-2020 "QUaD2: Una piattaforma e-Health potenziata da algoritmi di machine learning QUantistico per la prevenzione di complicazioni macrovascolari e microvascolari nel Diabete di tipo 2", co-funded by Blu Sistemi s.r.l, and by a research grant awarded to G.M. by Università Cattolica del Sacro Cuore-Linea D1 2021.

Institutional Review Board Statement: This study was conducted in accordance with the Declaration of Helsinki and approved by the Ethics Committee of Università Cattolica del Sacro Cuore (Protocol Code diab_mf).

Informed Consent Statement: Informed consent was obtained from all the subjects involved in the study. Written informed consent was obtained from the participants to publish this paper.

Data Availability Statement: The data presented in this study are available upon request from the corresponding author.

Conflicts of Interest: The authors declare no conflict of interest.

References

1. Silverthon, D.U. *Fisiologia: Un Approccio Integrato*, 2nd ed.; Casa Editrice Ambrosiana: Rozzano, Italy, 2005.
2. Hossain, D.; Imtiaz, M.H.; Sazonov, E. Comparison of Wearable Sensors for Estimation of Chewing Strength. *IEEE Sens. J.* **2020**, *20*, 5379–5388. [CrossRef] [PubMed]
3. Chung, J.; Oh, W.; Baek, D.; Ryu, S.; Lee, W.G.; Bang, H. Design and Evaluation of Smart Glasses for Food Intake and Physical Activity Classification. *J. Vis. Exp.* **2018**, *2018*, e56633. [CrossRef]
4. Selamat, N.A.; Ali, S.H.M. A Novel Approach of Chewing Detection based on Temporalis Muscle Movement using Proximity Sensor for Diet Monitoring. In Proceedings of the 2020 IEEE EMBS Conference on Biomedical Engineering and Sciences, IECBES 2020, Langkawi Island, Malaysia, 1–3 March 2021; Institute of Electrical and Electronics Engineers Inc.: New York, NY, USA, 2021; pp. 12–17. [CrossRef]
5. Amft, O. A Wearable Earpad Sensor for Chewing Monitoring. In Proceedings of the Sensors, 2010 IEEE, Waikoloa, HI, USA, 1–4 November 2010; IEEE: New York, NY, USA, 2010.
6. Shuzo, M.; Komori, S.; Takashima, T.; Lopez, G.; Tatsuta, S.; Yanagimoto, S.; Warisawa, S.; Delaunay, J.-J.; Yamada, I. Wearable Eating Habit Sensing System Using Internal Body Sound. *J. Adv. Mech. Des. Syst. Manuf.* **2010**, *4*, 158–166. [CrossRef]
7. Bi, Y.; Lv, M.; Song, C.; Xu, W.; Guan, N.; Yi, W. AutoDietary: A Wearable Acoustic Sensor System for Food Intake Recognition in Daily Life. *IEEE Sens. J.* **2016**, *16*, 806–816. [CrossRef]
8. Kalantarian, H.; Alshurafa, N.; Le, T.; Sarrafzadeh, M. Monitoring eating habits using a piezoelectric sensor-based necklace *Comput. Biol. Med.* **2015**, *58*, 46–55. [CrossRef]
9. Farooq, M.; Sazonov, E. Segmentation and Characterization of Chewing Bouts by Monitoring Temporalis Muscle Using Smart Glasses with Piezoelectric Sensor. *IEEE J. Biomed. Health Inform.* **2017**, *21*, 1495–1503. [CrossRef]
10. Farooq, M.; Sazonov, E. Automatic Measurement of Chew Count and Chewing Rate during Food Intake. *Electronics* **2016**, *5*, 62. [CrossRef]
11. Farooq, M.; Sazonov, E. A Novel Wearable Device for Food Intake and Physical Activity Recognition. *Sensors* **2016**, *16*, 1067. [CrossRef]
12. Papapanagiotou, V.; Diou, C.; Zhou, L.; Boer, J.v.D.; Mars, M.; Delopoulos, A. A novel approach for chewing detection based on a wearable PPG sensor. *IEEE Eng. Med. Biol. Soc.* **2016**, *2016*, 6485–6488.
13. Papapanagiotou, V.; Diou, C.; Zhou, L.; Boer, J.V.D.; Mars, M.; Delopoulos, A. A Novel Chewing Detection System Based on PPG, Audio, and Accelerometry. *IEEE J. Biomed. Health Inform.* **2017**, *21*, 607–618. [CrossRef]
14. Wei, Y.; Minhad, K.N.; Selamat, N.A.; Ali, S.H.M.; Bhuiyan, M.A.S.; Ooi, K.J.A.; Samdin, S.B. A review of chewing detection for automated dietary monitoring. *J. Chin. Inst. Eng.* **2022**, *45*, 331–341. [CrossRef]
15. Hashii, K.; Tomida, M.; Yamashita, S. Influence of changing the chewing region on mandibular movement. *Aust. Dent. J.* **2009**, *54*, 38–44. [CrossRef] [PubMed]
16. Zhang, R.; Amft, O. Monitoring Chewing and Eating in Free-Living Using Smart Eyeglasses. *IEEE J. Biomed. Health Inform.* **2018**, *22*, 23–32. [CrossRef] [PubMed]
17. Zhang, R.; Actlab, O.A. Bite Glasses-Measuring Chewing Using EMG and Bone Vibration in Smart Eyeglasses. Available online: http://www.bitalino.com (accessed on 6 June 2023).
18. Nicolas, E.; Veyrune, J.L.; Lassauzay, C.; Peyron, M.A.; Hennequin, M. Validation of video versus electromyography for chewing evaluation of the elderly wearing a complete denture. *J. Oral Rehabil.* **2007**, *34*, 566–571. [CrossRef]
19. Adachi, S.; Morikawa, K. Interface System Utilizing Musticatory Electromyogram. U.S. Patent US20100160808A1, 24 June 2010.
20. Santos, K.W.; Echeveste, S.S.; Vidor, C.G.M. Influence of gustatory and olfactory perception in the oral. *CoDAS* **2014**, *26*, 68–75. [CrossRef]
21. Santos, K.; Maahs, M.A.P.; Vidor, D.C.G.M.; Rech, R.S. Masticatory Changes as a Result of Oral Disorders in Smokers. *Int. Arch. Otorhinolaryngol.* **2014**, *18*, 369–375. [CrossRef]
22. Reibel, J. Tobacco and oral diseases: Update on the evidence, with recommendations. In *Medical Principles and Practice*; Karger Publisher: Berlin, Germany, 2003; pp. 22–32. [CrossRef]
23. Da Silva, G.R.; Rech, R.S.; Vidor, D.C.G.M.; dos Santos, K.W. Influence of Masticatory Behavior on Muscle Compensations During the Oral Phase of Swallowing of Smokers. *Int. Arch. Otorhinolaryngol.* **2019**, *23*, e317–e321. [CrossRef]
24. Brown, W.E.; Shearn, M.; Macfie, H.J.H. Method to investigate differences in chewing behaviour in humans: II. Use of electromyography during chewing to assess chewing behaviour. *Science* **1993**, *25*, 17–31. [CrossRef]
25. Abeltino, A.; Bianchetti, G.; Serantoni, C.; Ardito, C.F.; Malta, D.; De Spirito, M.; Maulucci, G. Personalized Metabolic Avatar: A Data Driven Model of Metabolism for Weight Variation Forecasting and Diet Plan Evaluation. *Nutrients* **2022**, *14*, 3520. [CrossRef]

26. Bianchetti, G.; Abeltino, A.; Serantoni, C.; Ardito, F.; Malta, D.; De Spirito, M.; Maulucci, G. Personalized Self-Monitoring of Energy Balance through Integration in a Web-Application of Dietary, Anthropometric, and Physical Activity Data. *J. Pers. Med.* **2022**, *12*, 568. [CrossRef]
27. Abeltino, A.; Bianchetti, G.; Serantoni, C.; Riente, A.; De Spirito, M.; Maulucci, G. Putting the Personalized Metabolic Avatar into Production: A Comparison between Deep-Learning and Statistical Models for Weight Prediction. *Nutrients* **2023**, *15*, 1199. [CrossRef] [PubMed]

MDPI AG
Grosspeteranlage 5
4052 Basel
Switzerland
Tel.: +41 61 683 77 34

Biosensors Editorial Office
E-mail: biosensors@mdpi.com
www.mdpi.com/journal/biosensors

www.ingramcontent.com/pod-product-compliance
Lightning Source LLC
Chambersburg PA
CBHW042014260926
51850CB00020B/3407
* 9 7 8 3 7 2 5 8 2 6 0 1 8 *